STARS AND GALAXIES

21ST CENTURY ASTRONOMY

FOURTH EDITION

STARS AND GALAXIES

21ST CENTURY ASTRONOMY

FOURTH EDITION

LAURA KAY • Barnard College

STACY PALEN • Weber State University

BRAD SMITH • Santa Fe, New Mexico

GEORGE BLUMENTHAL • University of California–Santa Cruz

W. W. NORTON & COMPANY
NEW YORK • LONDON

W. W. Norton & Company has been independent since its founding in 1923, when William Warder Norton and Mary D. Herter Norton first published lectures delivered at the People's Institute, the adult education division of New York City's Cooper Union. The firm soon expanded its program beyond the Institute, publishing books by celebrated academics from America and abroad. By midcentury, the two major pillars of Norton's publishing program—trade books and college texts—were firmly established. In the 1950s, the Norton family transferred control of the company to its employees, and today—with a staff of 400 and a comparable number of trade, college, and professional titles published each year—W. W. Norton & Company stands as the largest and oldest publishing house owned wholly by its employees.

Printed in Canada
Fourth Edition

Editor: Erik Fahlgren
Marketing Manager: Stacy Loyal
Editorial Assistant: Renee Cotton
Managing Editor, College: Marian Johnson
Associate Managing Editor, College: Kim Yi
Copy Editor: Stephanie Hiebert
Developmental Editor: Erin Mulligan
Science Media Editor: Rob Bellinger
Assistant Editor, Supplements: Jennifer Barnhardt
Production Manager: Eric Pier-Hocking
Art Director and Designer: Rubina Yeh
Photo Researchers: Stephanie Romeo, Rona Tuccillo
Permissions Manager: Megan Jackson, Bethany Salminen
Compositor: Precision Graphics
Manufacturing: Transcontinental

ISBN: 978-0-393-92057-4 (pbk.)
W. W. Norton & Company, Inc., 500 Fifth Avenue, New York, N.Y. 10110
www.wwnorton.com
W. W. Norton & Company Ltd., Castle House, 75/76 Wells Street, London W1T 3QT
1 2 3 4 5 6 7 8 9 0

Laura Kay thanks her partner, M.P.M. She dedicates this book to her late uncle, Lee Jacobi, for an early introduction to physics, and to her late colleagues at Barnard College, Tally Kampen and Sally Chapman.

Stacy Palen thanks the wonderful colleagues in her department and the crowd at Bellwether Farm, all of whom made room for this project in their lives, even though it wasn't their project.

Brad Smith dedicates this book to his patient and understanding wife, Diane McGregor.

George Blumenthal gratefully thanks his wife, Kelly Weisberg, and his children, Aaron and Sarah Blumenthal, for their support during this project. He also wants to thank Professor Robert Greenler for stimulating his interest in all things related to physics.

Brief Contents

CHAPTERS 8–12 ARE NOT INCLUDED IN THIS EDITION.

Part I Introduction to Astronomy

Chapter 1 Why Learn Astronomy? 3

Chapter 2 Patterns in the Sky—Motions of Earth 25

Chapter 3 Motion of Astronomical Bodies 63

Chapter 4 Gravity and Orbits 89

Chapter 5 Light 117

Chapter 6 The Tools of the Astronomer 151

Part II The Solar System

Chapter 7 The Birth and Evolution of Planetary Systems 185

Part III Stars and Stellar Evolution

Chapter 13 Taking the Measure of Stars 395

Chapter 14 Our Star—The Sun 427

Chapter 15 Star Formation and the Interstellar Medium 459

Chapter 16 Evolution of Low-Mass Stars 491

Chapter 17 Evolution of High-Mass Stars 521

Chapter 18 Relativity and Black Holes 551

Part IV Galaxies, the Universe, and Cosmology

Chapter 19 The Expanding Universe 581

Chapter 20 Galaxies 611

Chapter 21 The Milky Way—A Normal Spiral Galaxy 641

Chapter 22 Modern Cosmology 667

Chapter 23 Large-Scale Structure in the Universe 695

Chapter 24 Life 723

Contents

CHAPTERS 8–12 ARE NOT INCLUDED IN THIS EDITION.

Preface xxv

About the Authors xxxiv

PART I Introduction to Astronomy

Chapter 1 Why Learn Astronomy? 3

1.1 Getting a Feel for the Neighborhood 4

1.2 Astronomy Involves Exploration and Discovery 7

1.3 Science Is a Way of Viewing the World 9
Process of Science: The Scientific Method 10

1.4 Patterns Make Our Lives and Science Possible 13

Math Tools 1.1 Mathematical Tools 15

Math Tools 1.2 Reading a Graph 16

1.5 Thinking like an Astronomer: What Is a Planet? 17

1.6 Origins: An Introduction 17

Summary 18

Unanswered Questions 18

Questions and Problems 18

Exploration: Logical Fallacies 23

Chapter 2 Patterns in the Sky—Motions of Earth 25

2.1 A View from Long Ago 26

2.2 Earth Spins on Its Axis 27

Connections 2.1 Relative Motions and Frame of Reference 28

Math Tools 2.1 How to Estimate the Size of Earth 37

2.3 Revolution about the Sun Leads to Changes during the Year 37
Process of Science: Theories Must Fit All the Known Facts 42

2.4 The Motions and Phases of the Moon 45

2.5 Cultures and Calendars 49

2.6 Eclipses: Passing through a Shadow 50

2.7 Origins: The Obliquity of Earth 55

Summary 56

Unanswered Questions 57

Questions and Problems 57

Exploration: The Phases of the Moon 61

Chapter 3 Motion of Astronomical Bodies 63

3.1 The Motions of Planets in the Sky 64

3.2 Earth Moves 65

Connections 3.1 How Copernicus Scaled the Solar System 66

Math Tools 3.1 Sidereal and Synodic Periods 68

3.3 An Empirical Beginning: Kepler's Laws 69

Process of Science: Theories Are Falsified 71

Math Tools 3.2 Kepler's Third Law 74

3.4 Galileo: The First Modern Scientist 75

3.5 Newton's Laws of Motion 76

Math Tools 3.3 Proportionality 79

Math Tools 3.4 Using Newton's Laws 81

3.6 Origins: Planets and Orbits 82

Summary 82

Unanswered Questions 83

Questions and Problems 83

Exploration: Kepler's Laws 87

Chapter 4 Gravity and Orbits 89

4.1 Gravity Is a Force between Any Two Objects Due to Their Masses 90

Process of Science: Universality 91

Math Tools 4.1 Playing with Newton's Laws of Motion and Gravitation 94

Connections 4.1 Gravity Differs from Place to Place within an Object 96

4.2 Orbits Are One Body "Falling around" Another 97

Math Tools 4.2 Circular Velocity and Orbital Periods 100

Math Tools 4.3 Calculating Escape Velocities 102

4.3 Tidal Forces on Earth 103

Math Tools 4.4 Tidal Forces 105

4.4 Tidal Effects on Solid Bodies 108

Connections 4.2 Gravity Affects the Orbits of Three Bodies 110

4.5 Origins: Tidal Forces and Life 110

Summary 111

Unanswered Questions 112

Questions and Problems 112

Exploration: Newton's Laws 115

Chapter 5 Light 117

5.1 The Speed of Light 118

5.2 Light Is an Electromagnetic Wave 119

Process of Science: Agreement Between Fields 121

Math Tools 5.1 Working with Electromagnetic Radiation 125

5.3 The Quantum View of Matter 126

5.4 The Doppler Effect—Motion Toward or Away from Us 134

Math Tools 5.2 Making Use of the Doppler Effect 136

Connections 5.1 Equilibrium Means Balance 136

5.5 Light and Temperature 137

Math Tools 5.3 Working with the Stefan-Boltzmann and Wien's Laws 141

5.6 Light and Distance 141

5.7 Origins: Temperatures of Planets 143

Math Tools 5.4 Using Radiation Laws to Calculate Equilibrium Temperatures of Planets 144

Summary 144

Unanswered Questions 146

Questions and Problems 146

Exploration: Light as a Wave, Light as a Photon 149

Chapter 6 The Tools of the Astronomer 151

6.1 The Optical Telescope 152

Math Tools 6.1 Telescope Aperture and Magnification 153

Connections 6.1 When Light Doesn't Go Straight 156

Math Tools 6.2 Diffraction Limit 161

6.2 Optical Detectors and Instruments 162

Connections 6.2 Interference and Diffraction 166

6.3 Radio and Infrared Telescopes 166

6.4 Getting above Earth's Atmosphere: Orbiting Observatories 170

6.5 Getting Up Close with Planetary Spacecraft 171

6.6 Other Astronomical Tools 174
Process of Science: Technology and Science Are Symbiotic 177

6.7 Origins: Microwave Telescopes That Detect Radiation from the Big Bang 178

Summary 179

Unanswered Questions 179

Questions and Problems 180

Exploration: Geometric Optics and Lenses 183

PART II The Solar System

Chapter 7 The Birth and Evolution of Planetary Systems 185

7.1 Stars Form and Planets Are Born 186

7.2 The Solar System Began with a Disk 188
Process of Science: Converging Lines of Inquiry 189

Math Tools 7.1 Angular Momentum 191

7.3 The Inner Disk Is Hot; the Outer Disk Is Cold 195

Connections 7.1 Conservation of Energy 196

7.4 A Tale of Eight Planets 198

7.5 Planetary Systems Are Common 200

Math Tools 7.2 Estimating the Size of the Orbit of a Planet 204

Math Tools 7.3 Estimating the Size of an Extrasolar Planet 206

7.6 Origins: Kepler's Search for Earth-Sized Planets 206

Summary 208

Unanswered Questions 208

Questions and Problems 208

Exploration: Formation of the Solar System 213

Chapter 8 The Terrestrial Planets and Earth's Moon 215

8.1 Four Main Processes Shape the Inner Planets 216
8.2 Impacts Help Shape the Evolution of the Planets 219
Connections 8.1 Determining the Ages of Rocks 222
Math Tools 8.1 Computing the Ages of Rocks 223
8.3 The Interiors of the Terrestrial Planets 224
Process of Science: Certainty Is Sometimes Out of Reach 229
Math Tools 8.2 How Planets Cool Off 230
8.4 Tectonism, Volcanism, and Erosion 233
8.5 The Geological Evidence for Water 244
8.6 Origins: The Death of the Dinosaurs 248

Summary 250
Unanswered Questions 250
Questions and Problems 250
Exploration: Exponential Behavior 255

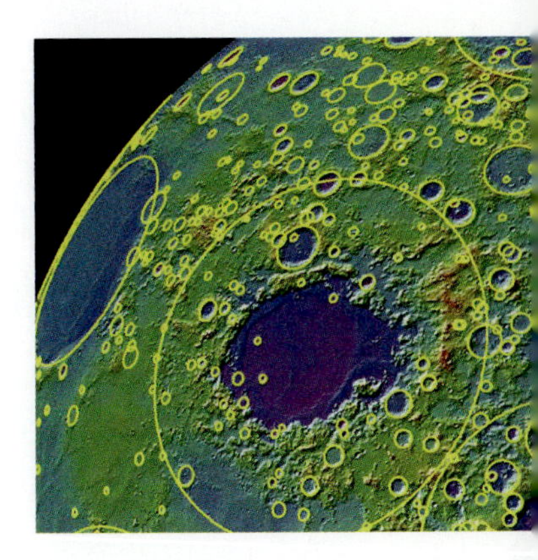

Chapter 9 Atmospheres of the Terrestrial Planets 257

9.1 The Gain and Loss of Atmospheres 258
Connections 9.1 What Is a Gas? 260
Math Tools 9.1 Atmosphere Retention 262
9.2 The Evolution of Secondary Atmospheres 263
9.3 Earth's Atmosphere 266
Connections 9.2 When Convection Runs Amok 274
9.4 Atmospheres of the Other Terrestrial Planets 274
9.5 Planetary Climate Change 278
Process of Science: Thinking About Complexity 281
9.6 Origins: Our Special Planet (or Why Are We Here?) 283

Summary 284
Unanswered Questions 285
Questions and Problems 285
Exploration: Climate Change 289

Chapter 10 Worlds of Gas and Liquid—The Giant Planets 291

10.1 The Giant Planets—Distant Worlds, Different Worlds 292
Process of Science: Theories and Laws 294

10.2 How Giant Planets Differ from Terrestrial Planets 295

10.3 A View of the Clouds 299

10.4 Weather on the Giant Planets 304

Math Tools 10.1 Measuring Wind Speeds on Distant Planets 306

Math Tools 10.2 Internal Thermal Energy Heats the Giant Planets 307

10.5 The Interiors of the Giant Planets Are Hot and Dense 309

10.6 The Giant Planets Are Magnetic Powerhouses 311

Connections 10.1 Synchrotron Radiation 314

10.7 Origins: Giant Planet Migration and the Inner Solar System 316

Summary 317

Unanswered Questions 318

Questions and Problems 318

Exploration: Estimating Rotation Periods of the Giant Planets 321

Chapter 11 Planetary Adornments—Moons and Rings 323

11.1 Moons in the Solar System 324

Math Tools 11.1 Moons and Kepler's Law 327

11.2 The Geological Activity of Moons 327

Math Tools 11.2 Tidal Forces on the Moons 332

11.3 The Discovery of Rings around the Giant Planets 339

Connections 11.1 The Backlighting Phenomenon 344

11.4 The Composition of Ring Material 346

11.5 Gravity and Ring Systems 346

Math Tools 11.3 Feeding the Rings 347
Process of Science: Following Up on the Unexpected 350

11.6 Origins: Extreme Environments and an Organic Deep Freeze 352

Summary 353

Unanswered Questions 353

Questions and Problems 354

Exploration: Measuring Features on Io 357

Chapter 12 Dwarf Planets and Small Solar System Bodies 359

12.1 Leftover Material: From the Small to the Tiniest 360

12.2 Dwarf Planets: Pluto and Others 360
Process of Science: Objectivity 363

Math Tools 12.1 Eccentric Orbits 364

12.3 Asteroids—Pieces of the Past 364

Connections 12.1 Gaps in the Asteroid Belt 365

12.4 Comets: Clumps of Ice 371

12.5 Solar System Debris 380

Math Tools 12.2 Impact Energy 381

12.6 Origins: Comets, Asteroids, and Life 387

Summary 388

Unanswered Questions 388

Questions and Problems 389

Exploration: Asteroid Discovery 393

PART III Stars and Stellar Evolution

Chapter 13 Taking the Measure of Stars 395

13.1 Measuring the Distance, Brightness, and Luminosity of Stars 396

Math Tools 13.1 Parallax and Distance 399

Connections 13.1 The Magnitude System 401

13.2 The Temperature, Size, and Composition of Stars 402

Math Tools 13.2 Estimating Sizes of Stars 407

13.3 Measuring Stellar Masses 408

Math Tools 13.3 Measuring the Masses of an Eclipsing Binary Pair 411

13.4 The H-R Diagram Is the Key to Understanding Stars 412
Process of Science: Science Is Collaborative 414

13.5 Origins: Habitable Zones 419

Summary 421

Unanswered Questions 421

Questions and Problems 421

Exploration: The H-R Diagram 425

Chapter 14 **Our Star—The Sun 427**

14.1 The Structure of the Sun 428

14.2 The Sun Is Powered by Nuclear Fusion 429

Math Tools 14.1 The Source of the Sun's Energy 431

Connections 14.1 The Proton-Proton Chain 432

14.3 The Interior of the Sun 437

Process of Science: Learning from Failure 439

Connections 14.2 Neutrino Astronomy 440

14.4 The Atmosphere of the Sun 441

Math Tools 14.2 Sunspots and Temperature 446

14.5 Origins: The Solar Wind and Life 451

Summary 452

Unanswered Questions 452

Questions and Problems 453

Exploration: The Proton-Proton Chain 457

Chapter 15 **Star Formation and the Interstellar Medium 459**

15.1 The Interstellar Medium 460

Process of Science: Unknown Unknowns 464

Math Tools 15.1 Dust Glows in the Infrared 465

15.2 Molecular Clouds Are the Cradles of Star Formation 470

15.3 The Protostar Becomes a Star 473

Connections 15.1 Brown Dwarfs 478

Math Tools 15.2 Luminosity, Temperature, and Radius of Protostars 479

15.4 Not All Stars Are Created Equal 480

15.5 Origins: Star Formation, Planets, and Life 483

Summary 485

Unanswered Questions 485

Questions and Problems 485

Exploration: The Stellar Thermostat 489

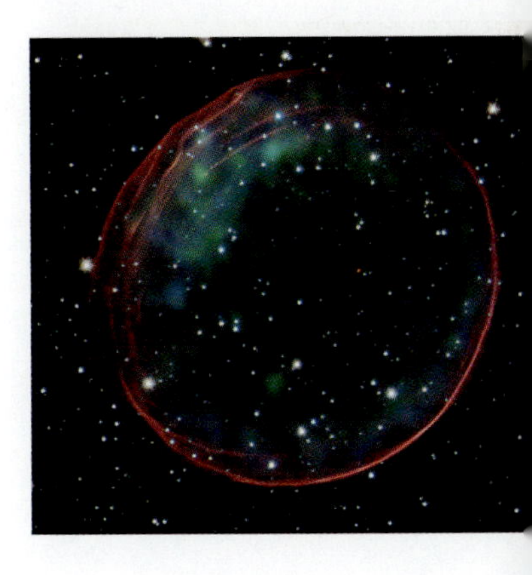

Chapter 16 Evolution of Low-Mass Stars 491

16.1 The Life of a Main-Sequence Star 492
Math Tools 16.1 Estimating Main-Sequence Lifetimes 494
16.2 A Star Runs Out of Hydrogen and Leaves the Main Sequence 494
16.3 Helium Begins to Burn in the Degenerate Core 498
16.4 The Low-Mass Star Enters the Last Stages of Its Evolution 502
Math Tools 16.2 Escaping the Surface of an Evolved Star 504
Connections 16.1 What Happens to the Planets? 508
16.5 Binary Star Evolution 508
Process of Science: Science Is Not Finished 513
16.6 Origins: Stellar Lifetimes and Biological Evolution 514

Summary 515
Unanswered Questions 515
Questions and Problems 515
Exploration: Low-Mass Stellar Evolution 519

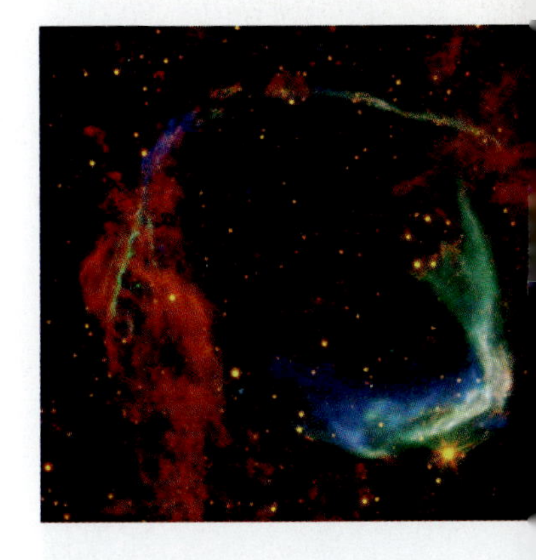

Chapter 17 Evolution of High-Mass Stars 521

17.1 High-Mass Stars Follow Their Own Path 522
17.2 High-Mass Stars Go Out with a Bang 527
Math Tools 17.1 Binding Energy of Atomic Nuclei 528
17.3 The Spectacle and Legacy of Supernovae 531
Connections 17.1 Variations by Stellar Mass 523
Math Tools 17.2 Gravity on a Neutron Star 536
Process of Science: Occam's Razor 539
17.4 Star Clusters Are Snapshots of Stellar Evolution 540
17.5 Origins: Seeding the Universe with New Chemical Elements 545

Summary 546
Unanswered Questions 546
Questions and Problems 546
Exploration: The CNO Cycle 549

Chapter 18 Relativity and Black Holes 551

18.1 Beyond Newtonian Physics 552
Connections 18.1 Aberration of Starlight 552
18.2 Special Relativity 553
Math Tools 18.1 The Boxcar Experiment 557
Math Tools 18.2 The Twin Paradox 560
18.3 Gravity Is a Distortion of Spacetime 561
Connections 18.2 When One Physical Law Supplants Another 565
Process of Science: New Science Includes the Old 566
18.4 Black Holes 570
Math Tools 18.3 Masses in X-Ray Binaries 572
18.5 Origins: Gamma-Ray Bursts 573

Summary 575
Unanswered Questions 575
Questions and Problems 575
Exploration: Black Holes 579

PART IV Galaxies, the Universe, and Cosmology

Chapter 19 The Expanding Universe 581

19.1 Twentieth Century Astronomers Discovered the Universe of Galaxies 582
19.2 The Cosmological Principle 584
19.3 The Universe is Expanding 584
Math Tools 19.1 Redshift—Calculating the Recessional Velocity and Distance of Galaxies 587
Process of Science: Authority Is Irrelevant 588
Math Tools 19.2 Finding the Distance from a Type Ia Supernova 592
19.4 The Universe Began in the Big Bang 593
Math Tools 19.3 Expansion and the Age of the Universe 594
Connections 19.1 When Redshift Exceeds 1 599
19.5 Astronomers Observe Radiation Left Over from the Big Bang 600
19.6 Origins: Big Bang Nucleosynthesis 603

Summary 605

Unanswered Questions 605

Questions and Problems 605

Exploration: Hubble's Law for Balloons 609

Chapter 20 **Galaxies 611**

20.1 Galaxies Come in Many Types 612
Process of Science: Wrong Ideas Are Sometimes Useful 615

20.2 In Spiral Galaxies, Stars Form in the Spiral Arms 619

20.3 Galaxies Are Mostly Dark Matter 622

20.4 Most Galaxies Have a Supermassive Black Hole at the Center 626

Math Tools 20.1 Supermassive Black Holes 629

Math Tools 20.2 Feeding an AGN 631

Connections 20.1 Unified Model of AGN 632

20.5 Origins: Habitability in Galaxies 634

Summary 635

Unanswered Questions 635

Questions and Problems 636

Exploration: Galaxy Classification 639

Chapter 21 **The Milky Way—A Normal Spiral Galaxy 641**

21.1 Measuring the Shape and Size of the Milky Way 642
Process of Science: Unknown Unknowns 644

Connections 21.1 Nightfall 645

21.2 Dark Matter in the Milky Way 646

Math Tools 21.1 The Mass of the Milky Way inside the Sun's Orbit 648

21.3 Stars in the Milky Way 648

21.4 The Milky Way Hosts a Supermassive Black Hole 654

Math Tools 21.2 The Mass of the Milky Way's Central Black Hole 656

21.5 The Milky Way Offers Clues about How Galaxies Form 657

Connections 21.2 Will Andromeda and the Milky Way Collide? 659

21.6 Origins: The Galactic Habitable Zone 660

Summary 661

Unanswered Questions 661

Questions and Problems 661

Exploration: The Center of the Milky Way 665

Chapter 22 **Modern Cosmology 667**

22.1 The Universe Has a Destiny and a Shape 668

Math Tools 22.1 Critical Density 669

22.2 The Accelerating Universe 670

Process of Science: Never Throw Anything Away 672

22.3 Inflation 675

22.4 The Earliest Moments 679

Math Tools 22.2 Pair Production in the Early Universe 681

Connections 22.1 Superstring Theory 684

22.5 Multiple Multiverses 686

22.6 Origins: Our Own Universe Must Support Life 688

Summary 689

Unanswered Questions 689

Questions and Problems 690

Exploration: Understanding Orders of Infinity 693

Chapter 23 **Large-Scale Structure in the Universe 695**

23.1 Galaxies Form Groups, Clusters, and Larger Structures 696

Math Tools 23.1 Mass of a Cluster of Galaxies 699

23.2 The Origin of Structure 700

Process of Science: Nature Does What Nature Does 704

23.3 First Light 705

Math Tools 23.2 Observing High-Redshift Objects 707

23.4 Galaxy Evolution 708

Connections 23.1 Parallels between Galaxy and Star Formation 711

23.5 The Deep Future 714

23.6 Origins: We Are the 4 or 5 Percent 716

Summary 717

Unanswered Questions 717

Questions and Problems 718

Exploration: The Story of a Proton 721

Chapter 24 Life 723

24.1 Life's Beginnings on Earth 724

Connections 24.1 Forever in a Day 729

Math Tools 24.1 Exponential Growth 730

24.2 The Chemistry of Life 731

Connections 24.2 Life, the Universe, and Everything 732

Process of Science: All of Science Is Interconnected 733

24.3 Life beyond Earth 734

24.4 The Search for Signs of Intelligent Life 738

Math Tools 24.2 Putting Numbers into the Drake Equation 739

24.5 Origins: The Fate of Life on Earth 742

Summary 743

Unanswered Questions 743

Questions and Problems 744

Exploration: Fermi Problems and the Drake Equation 747

APPENDIX 1 Mathematical Tools A-1

APPENDIX 2 Physical Constants and Units A-7

APPENDIX 3 Periodic Table of the Elements A-9

APPENDIX 4 Properties of Planets, Dwarf Planets, and Moons A-10

APPENDIX 5 Nearest and Brightest Stars A-13

APPENDIX 6 Observing the Sky A-16

APPENDIX 7 Uniform Circular Motion and Circular Orbits A-25

APPENDIX 8 IAU 2006 Resolutions: Definition of a Planet in the Solar System, and Pluto A-27

Glossary G-1

Credits C-1

Index I-1

Math Tools

CHAPTERS 8–12 ARE NOT INCLUDED IN THIS EDITION.

1.1 Mathematical Tools 15
1.2 Reading a Graph 16
2.1 How to Estimate the Size of Earth 37
3.1 Sidereal and Synodic Periods 68
3.2 Kepler's Third Law 74
3.3 Proportionality 79
3.4 Using Newton's Laws 81
4.1 Playing with Newton's Laws of Motion and Gravitation 94
4.2 Circular Velocity and Orbital Periods 100
4.3 Calculating Escape Velocities 102
4.4 Tidal Forces 105
5.1 Working with Electromagnetic Radiation 125
5.2 Making Use of the Doppler Effect 136
5.3 Working with the Stefan-Boltzmann and Wien's Laws 141
5.4 Using Radiation Laws to Calculate Equilibrium Temperatures of Planets 144
6.1 Telescope Aperture and Magnification 153
6.2 Diffraction Limit 161
7.1 Angular Momentum 191
7.2 Estimating the Size of the Orbit of a Planet 204
7.3 Estimating the Size of an Extrasolar Planet 206
8.1 Computing the Ages of Rocks 223
8.2 How Planets Cool Off 230
9.1 Atmosphere Retention 262
10.1 Measuring Wind Speeds on Distant Planets 306
10.2 Internal Thermal Energy Heats the Giant Planets 307
11.1 Moons and Kepler's Law 327
11.2 Tidal Forces on the Moons 332
11.3 Feeding the Rings 347
12.1 Eccentric Orbits 364
12.2 Impact Energy 381
13.1 Parallax and Distance 399
13.2 Estimating Sizes of Stars 407
13.3 Measuring the Masses of an Eclipsing Binary Pair 411
14.1 The Source of the Sun's Energy 431
14.2 Sunspots and Temperature 446
15.1 Dust Glows in the Infrared 465
15.2 Luminosity, Surface Temperature, and Radius of Protostars 479
16.1 Estimating Main-Sequence Lifetimes 494
16.2 Escaping the Surface of an Evolved Star 504
17.1 Binding Energy of Atomic Nuclei 528
17.2 Gravity on a Neutron Star 536
18.1 The Boxcar Experiment 557
18.2 The Twin Paradox 560
18.3 Masses in X-Ray Binaries 572
19.1 Redshift-Calculating the Recession Velocity and Distance of Galaxies 587
19.2 Finding the Distance from a Type Ia Supernova 592
19.3 Expansion and the Age of the Universe 594
20.1 Supermassive Black Holes 629
20.2 Feeding an AGN 631
21.1 The Mass of the Milky Way inside the Sun's Orbit 648
21.2 The Mass of the Milky Way's Central Black Hole 656
22.1 Critical Density 669
22.2 Pair Production in the Early Universe 681
23.1 Mass of a Cluster of Galaxies 699
23.2 Observing High-Redshift Objects 707
24.1 Exponential Growth 730
24.2 Putting Numbers into The Drake Equation 739

Connections

CHAPTERS 8–12 ARE NOT INCLUDED IN THIS EDITION.

2.1 Relative Motions and Frame of Reference 28
3.1 How Copernicus Scaled the Solar System 66
4.1 Gravity Differs from Place to Place within an Object 96
4.2 Gravity Affects the Orbits of Three Bodies 110
5.1 Equilibrium Means Balance 136
6.1 When Light Doesn't Go Straight 156
6.2 Interference and Diffraction 166
7.1 Conservation of Energy 196
8.1 Determining the Ages of Rocks 222
9.1 What Is a Gas? 260
9.2 When Convection Runs Amok 274
10.1 Synchrotron Radiation 314
11.1 The Backlighting Phenomenon 344
12.1 Gaps in the Main Asteroid Belt 365
13.1 The Magnitude System 401
14.1 The Proton-Proton Chain 432
14.2 Neutrino Astronomy 440
15.1 Brown Dwarfs 478
16.1 What Happens to the Planets? 508
17.1 Variations by Stellar Mass 533
18.1 Abberation of Starlight 552
18.2 When One Physical Law Supplants Another 565
19.1 When Redshift Exceeds 1 599
20.1 Unified Model of AGN 632
21.1 Nightfall 645
21.2 Will Andromeda and the Milky Way Collide? 659
22.1 Superstring Theory 684
23.1 Parallels between Galaxy and Star Formation 711
24.1 Forever in a Day 729
24.2 Life, the Universe, and Everything 732

AstroTours

CHAPTERS 8–12 ARE NOT INCLUDED IN THIS EDITION.

Earth Spins and Revolves 30
The View from the Poles 32
The Celestial Sphere and the Ecliptic 35
The Moon's Orbit: Eclipses and Phases 46
Kepler's Laws 72
Velocity, Acceleration, Inertia 76
Newton's Laws and Universal Gravitation 97
Elliptical Orbits 99
Tides and the Moon 104
Light as a Wave, Light as a Photon 126
Atomic Energy Levels and the Bohr Model 127
Atomic Energy Levels and Light Emission and Absorption 132
The Doppler Effect 135
Geometric Optics and Lenses 154
Solar System Formation 187
Traffic Circle 193
Processes That Shape the Planets 217
Continental Drift 233
Hot Spot Creating a Chain of Islands 240
Atmospheres: Formation and Escape 259
Greenhouse Effect 265
Cometary Orbits 376
Stellar Spectrum 403
The H-R Diagram 415
The Solar Core 432
Star Formation 471
Hubble's Law 586
Big Bang Nucleosynthesis 603
Dark Matter 623
Active Galactic Nuclei 629

AstroTour animations are available from the free StudySpace student website, and they are also integrated into assignable SmartWork exercises. Offline versions of the animations for classroom presentation are available from the Instructor's Resource disc.

Nebraska Simulations

CHAPTERS 8–12 ARE NOT INCLUDED IN THIS EDITION.

Look Back Time Simulator 5
Celestial-Equatorial (RA/DEC) Demonstrator 30
Longitude/Latitude Demonstrator 30
Celestial and Horizon Systems Comparison 31
Rotating Sky Explorer 31
Meridional Altitude Simulator 33
Declination Ranges Simulator 33
Big Dipper Clock 36
Big Dipper 3D 36
Rotating Sky Explorer 36
Coordinate Systems Comparison 36
Ecliptic (Zodiac) Simulator 38
Seasons and Ecliptic Simulator 39
Sun's Rays Simulator 40
Paths of the Sun 40
Sun Motions Overview 40
Daylight Hours Explorer 40
Union Seasons Demonstrator 40
Daylight Simulator 40
Lunar Phase Vocabulary 49
Basketball Phases Simulator 49
Three Views Simulator 49
Lunar Phases Simulator (WAAP) 49
Moon Phases and the Horizon Diagram 49
Moon Phases with Bisectors 49
Phase Positions Demonstrator 49
Lunar Phase Quizzer 49
Synodic Lag 49
Moon Inclinations 54
Eclipse Shadow Simulator 54
Eclipse Table 55
Obliquity Simulator 56
Ptolemaic Orbit of Mars 65
Planetary Configurations Simulator 67
Retrograde Motion 69
Eccentricity Demonstrator 72
Planetary Orbit Simulator 74
Phases of Venus 75
Ptolemaic Phases of Venus 75
Gravity Algebra 94
Law of Gravity Calculator 95
Earth Orbit Plot 99
Tidal Bulge Simulation 106
EM Spectrum Module 125
Three Views Spectrum Demonstrator 132
Hydrogen Atom Simulator 132
Doppler Shift Demonstrator 135
Blackbody Curves 141
Telescope Simulator 154
Snell's Law Demonstrator 156
CCD Simulator 141
Zodiac Simulator 164
EM Spectrum Module 166
Influence of Planets on the Sun 202
Radial Velocity Graph 202
Exoplanet Transit Simulator 202
Radial Velocity Graph 204
Radial Velocity Simulator 204
Gas Retention Simulator 262
Driving through Snow 383
Parallax Calculator 398
Stellar Luminosity Calculator 400

Center of Mass Simulator 408
Eclipsing Binary Simulator 409
Hertzsprung-Russell Diagram Explorer 413
Spectroscopic Parallax Simulator 416
Proton-Proton Animation 433
CNO Cycle Animation 523
H-R Explorer 525
H-R Diagram Star Cluster Fitting Explorer 544
Galactic Redshift Simulator 585
NAAP Lab: Spectroscopic Parallax Simulator 590
NAAP Lab: Supernova Light Curve Fitting Explorer 592
Traffic Density Analogy 622
Milky Way Rotational Velocity 648
Milky Way Habitability Explorer 660

Preface

This introductory astronomy course could be the only science course a student will take in college. As we wrote this book, we considered basic questions such as these: What will students remember 5 years from now about astronomy and about how science works? How should this course change the students who take it? And how can we, as scientists and educators, facilitate those changes within our students?

We believe an introductory astronomy course should help students learn to ask questions, make observations, identify patterns and relationships that go beyond the specifics of a particular object or setting, and apply those patterns and relationships broadly. We want students to challenge and test what they learn. This is what it means to think like a scientist.

We wrote this textbook and built its supporting ancillary package with one goal in mind: to help students understand the world through the eyes of a scientist.

In order to meet that goal, we have tried to tell a story in each chapter and have worked hard to link chapters using a few common threads. The process of science is one of those threads. Helping a student understand a concept as a scientist means guiding that student through the concept, making heavy use of examples and analogies, and tying the concept back to everyday phenomena and experiences that the student can relate to.

The process of science is in the fabric of the text and incorporates the recurring themes of the physics of matter, energy, radiation, and motion. Why did Newton choose the form that he did for his universal law of gravitation? What are the fundamental differences between Kepler's empirical "laws" and Newton's theoretical derivation of the same relationships? And if Einstein was "right," why wasn't Newton "wrong"? In the Fourth Edition, we have emphasized the process of science in several additional ways: new Process of Science Figures, Unanswered Questions boxes, and expanded Origins sections.

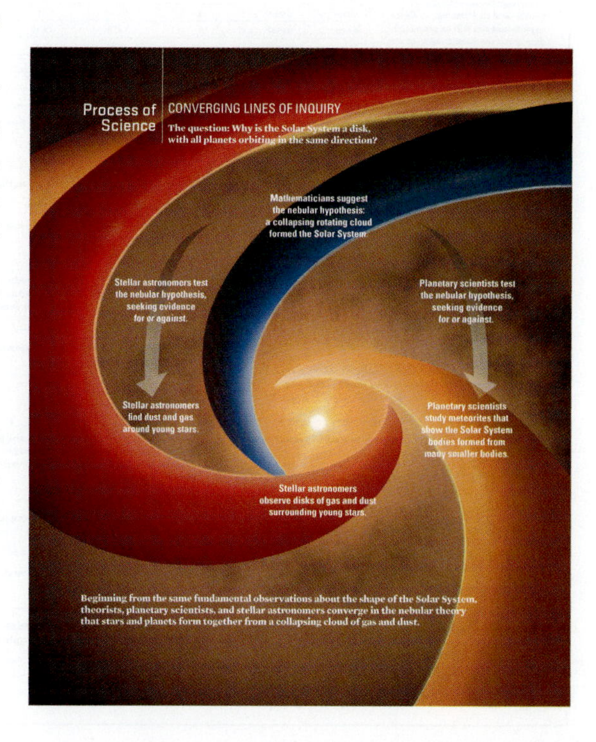

- In each chapter we have chosen one discovery and provided a visual representation of the process used to make that discovery in one of the new **Process of Science** figures. Because science is not a tidy process, we try to illustrate that discoveries are sometimes made by disparate groups, sometimes by accident, but always because people are trying to answer a question and show why or how we think something is the way it is. One example is Chapter 7, where we show how three groups of scientists were all working on the question "Why is the Solar System a disk?" and came to the same conclusion independently.

- At the end of every chapter, an **Unanswered Questions** box poses questions like "How typical is the Solar System?" and "How common are Earth-like planets?" to show that we don't have all the answers and that science is an ongoing process.

Unanswered Questions 

- How typical is the Solar System? Only within the past few years have astronomers found other systems containing four or more planets, and so far the distributions of large and small planets in these multiplanet systems have looked different from those of the Solar System. Computer simulations of planetary system formation suggest that a system with an orbital stability and a planetary distribution like those of the Solar System may develop only rarely. Improved supercomputers can run more complex simulations, which can be compared with the observations.
- How common are Earth-like planets, and how Earth-like must a planet be before scientists declare it to be "another Earth"? An editorial in the science journal *Nature* cautioned that scientists should define "Earth-like" in advance—before multiple discoveries of planets "similar" to Earth are announced (and a media frenzy ensues). Must a planet be of similar size and mass (and thus similar density), be located in the habitable zone, and have spectroscopic evidence of liquid water before we call it "Earth 2.0"?

A second major thread, **Origins**, shows how astronomers relate the topic of each chapter to the study of the origin of the universe or the origin of life. Since no life outside of Earth has been detected, these sections often illustrate how astrobiologists and other scientists approach the study of a scientific question, using the process of science rather than providing actual answers or results.

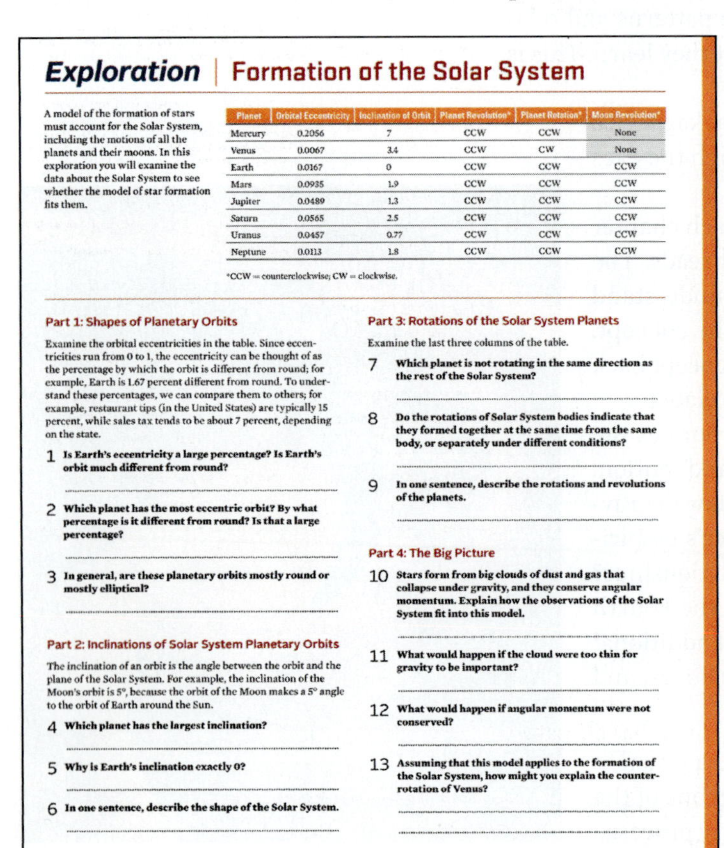

Exploration | Formation of the Solar System

A model of the formation of stars must account for the Solar System, including the motions of all the planets and their moons. In this exploration you will examine the data about the Solar System to see whether the model of star formation fits them.

Planet	Orbital Eccentricity	Inclination of Orbit	Planet Revolution*	Planet Rotation*	Moon Revolution*
Mercury	0.2056	7	CCW	CCW	None
Venus	0.0067	3.4	CCW	CW	None
Earth	0.0167	0	CCW	CCW	CCW
Mars	0.0935	1.9	CCW	CCW	CCW
Jupiter	0.0489	1.3	CCW	CCW	CCW
Saturn	0.0565	2.5	CCW	CCW	CCW
Uranus	0.0457	0.77	CCW	CCW	CCW
Neptune	0.0113	1.8	CCW	CCW	CCW

*CCW = counterclockwise; CW = clockwise.

Part 1: Shapes of Planetary Orbits

Examine the orbital eccentricities in the table. Since eccentricities run from 0 to 1, the eccentricity can be thought of as the percentage by which the orbit is different from round; for example, Earth is 1.67 percent different from round. To understand these percentages, we can compare them to others; for example, restaurant tips (in the United States) are typically 15 percent, while sales tax tends to be about 7 percent, depending on the state.

1 **Is Earth's eccentricity a large percentage? Is Earth's orbit much different from round?**

2 **Which planet has the most eccentric orbit? By what percentage is it different from round? Is that a large percentage?**

3 **In general, are these planetary orbits mostly round or mostly elliptical?**

Part 2: Inclinations of Solar System Planetary Orbits

The inclination of an orbit is the angle between the orbit and the plane of the Solar System. For example, the inclination of the Moon's orbit is 5°, because the orbit of the Moon makes a 5° angle to the orbit of Earth around the Sun.

4 **Which planet has the largest inclination?**

5 **Why is Earth's inclination exactly 0?**

6 **In one sentence, describe the shape of the Solar System.**

Part 3: Rotations of the Solar System Planets

Examine the last three columns of the table.

7 **Which planet is not rotating in the same direction as the rest of the Solar System?**

8 **Do the rotations of Solar System bodies indicate that they formed together at the same time from the same body, or separately under different conditions?**

9 **In one sentence, describe the rotations and revolutions of the planets.**

Part 4: The Big Picture

10 **Stars form from big clouds of dust and gas that collapse under gravity, and they conserve angular momentum. Explain how the observations of the Solar System fit into this model.**

11 **What would happen if the cloud were too thin for gravity to be important?**

12 **What would happen if angular momentum were not conserved?**

13 **Assuming that this model applies to the formation of the Solar System, how might you explain the counter-rotation of Venus?**

In addition to helping students think like a scientist, we have provided a few opportunities for them to actually do science. We have added **Using the Web** questions at the end of each chapter. Some of these send students to websites of space missions, observatories, experiments, or archives to access recent observations, results, or press releases. Other websites are for "citizen science" projects (for example, Zooniverse), in which students can contribute to the analysis of new data. These web problems can be used for homework, lab exercises, recitations, "news" exercises, or "writing across the curriculum" projects. (Updated Web addresses are posted on StudySpace as needed).

Explorations, also new to the Fourth Edition, are either pencil-and-paper activities or media-based activities that ask students to use Nebraska Simulations or Norton's AstroTours to work through a series of guided questions and apply the concepts they learned in the chapter.

To assess student understanding, versions of the end-of-chapter Explorations, as well as Process of Science Guided Inquiry assignments, based on the Process of Science figures, are available in Norton's online homework and tutorial system, **SmartWork**.

Although mathematics is the language of science, we understand that the amount of math used differs from school to school and instructor to instructor. In order to make the text more accessible to a wider variety of students, the math has been moved out of the main text into **Math Tools** boxes. Each box provides a succinct quantitative explanation of the concept being discussed and can be skipped without losing any qualitative understanding.

We have made some organizational changes to the Fourth Edition. Discussion of basic physics is now contained in Part I to accommodate courses that use the *Solar System* or *Stars and Galaxies* volumes. A "just-in-time" approach to introducing the physics is still possible by bringing in material from Chapters 2–6 as needed. For example, the sections on tidal forces in Chapter 4 can be taught along with the moons of the Solar System in Part II, or with mass transfer in binary stars in Part III, or with galaxy interactions in Part IV. Spectral lines in Chapter 5 can be taught with planetary atmospheres in Part II or with stellar spectral types in Part III, and so on.

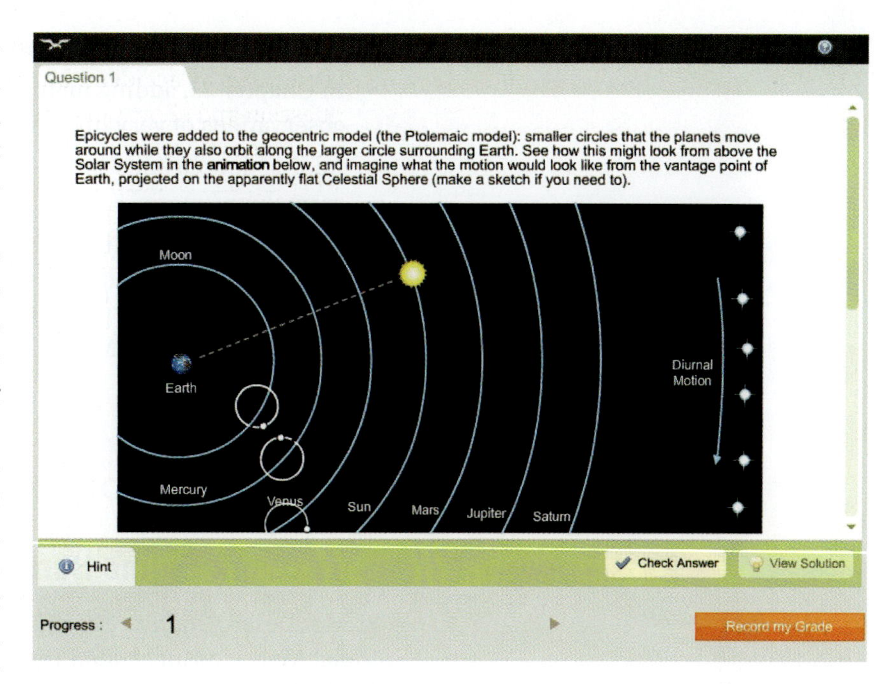

We start Parts II, III, and IV with the big picture before diving into the smaller details that make up that picture. We cover the development of planetary systems in general before discussing our own Solar System , and the basic properties of stars before the Sun. Part IV begins with the historical discovery of extragalactic objects and Hubble's law, which led to the Big Bang theory. At this point in the school year, we find that student interest is greatly renewed by the introduction of Hubble Deep Field images and the concept of the expanding universe. The next chapter continues with the basics of galaxies, including active galactic nuclei. Then, when the Milky Way is discussed in the following chapter, students have the background for understanding the exciting observational data about the Milky Way's central black hole.

In this edition we made pedagogical upgrades, as well as numerous updates and revisions throughout the book to reflect contemporary research and scientific thought. Some of those changes include:

- Updating each chapter's Learning Goals and correlating them with the end-of-chapter Summary, to help students review what is most important in each chapter.
- Expanding discussions of Copernicus, Tycho Brahe, and Galileo in Chapter 3, "Motions of Astronomical Bodies." The chapter now ends with Newton's laws of motion.
- Revising Chapter 4, "Gravity and Orbits," to include all of gravity, including tides.
- Thoroughly updating Part II with the latest information about the Solar System. We added material in Chapter 9 to cover climate change on the terrestrial planets, and how planetary science aids in the study of global climate change on Earth.
- Adding a new chapter, "Relativity and Black Holes" (Chapter 18), which separates out this material and expands some examples in Math Tools.

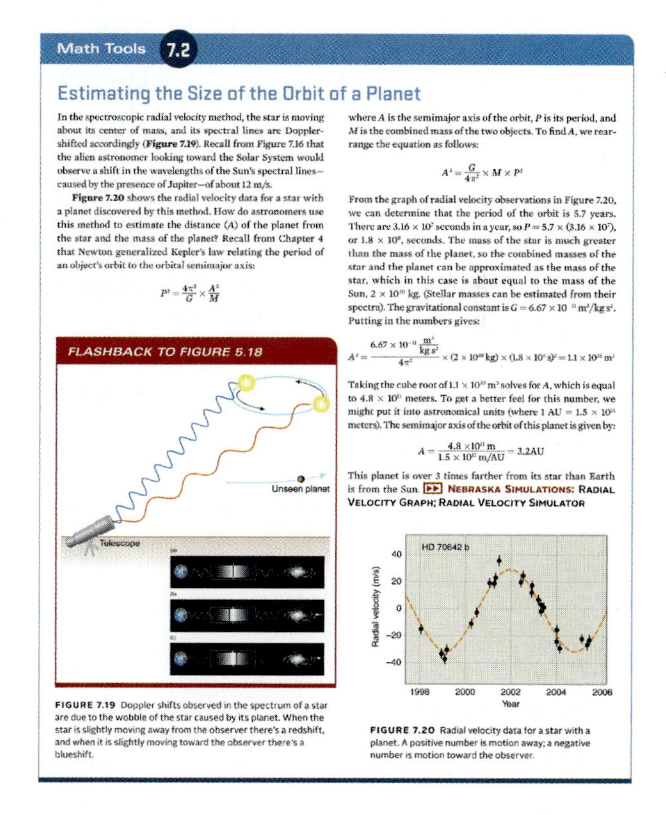

Math Tools 7.2

Estimating the Size of the Orbit of a Planet

In the spectroscopic radial velocity method, the star is moving about its center of mass, and its spectral lines are Doppler-shifted accordingly (**Figure 7.19**). Recall from Figure 7.16 that the alien astronomer looking toward the Solar System would observe a shift in the wavelengths of the Sun's spectral lines—caused by the presence of Jupiter—of about 12 m/s.

Figure 7.20 shows the radial velocity data for a star with a planet discovered by this method. How do astronomers use this method to estimate the distance (A) of the planet from the star and the mass of the planet? Recall from Chapter 4 that Newton generalized Kepler's law relating the period of an object's orbit to the orbital semimajor axis:

$$P^2 = \frac{4\pi^2}{G} \times \frac{A^3}{M}$$

where A is the semimajor axis of the orbit, P is its period, and M is the combined mass of the two objects. To find A, we rearrange the equation as follows:

$$A^3 = \frac{G}{4\pi^2} \times M \times P^2$$

From the graph of radial velocity observations in Figure 7.20, we can determine that the period of the orbit is 5.7 years. There are 3.16×10^7 seconds in a year, so $P = 5.7 \times (3.16 \times 10^7)$, or 1.8×10^8, seconds. The mass of the star is much greater than the mass of the planet, so the combined masses of the star and the planet can be approximated as the mass of the star, which in this case is about equal to the mass of the Sun, 2×10^{30} kg. (Stellar masses can be estimated from their spectra). The gravitational constant is $G = 6.67 \times 10^{-11}$ m^3/kg s^2. Putting in the numbers gives:

$$A^3 = \frac{6.67 \times 10^{-11} \frac{\text{m}^3}{\text{kg s}^2}}{4\pi^2} \times (2 \times 10^{30}\ \text{kg}) \times (1.8 \times 10^8\ \text{s})^2 = 1.1 \times 10^{35}\ \text{m}^3$$

Taking the cube root of 1.1×10^{35} m^3 solves for A, which is equal to 4.8×10^{11} meters. To get a better feel for this number, we might put it into astronomical units (where 1 AU $= 1.5 \times 10^{11}$ meters). The semimajor axis of the orbit of this planet is given by:

$$A = \frac{4.8 \times 10^{11}\ \text{m}}{1.5 \times 10^{11}\ \text{m/AU}} = 3.2\text{AU}$$

This planet is over 3 times farther from its star than Earth is from the Sun. ▶▶ NEBRASKA SIMULATIONS: RADIAL VELOCITY GRAPH; RADIAL VELOCITY SIMULATOR

FLASHBACK TO FIGURE 5.18

FIGURE 7.19 Doppler shifts observed in the spectrum of a star are due to the wobble of the star caused by its planet. When the star is slightly moving away from the observer there's a redshift, and when it is slightly moving toward the observer there's a blueshift.

FIGURE 7.20 Radial velocity data for a star with a planet. A positive number is motion away; a negative number is motion toward the observer.

- Revising Chapter 22 to include the latest ideas about the accelerating universe.
- In Chapter 23, adding material on the first stars, first galaxies, and the recent observations of very high-redshift objects.
- Adding new Origins sections that reflect current thinking in astrobiology or cosmology.
- Significantly upgrading and expanding the types of problems at the end of each chapter, including new True/False and Multiple Choice questions, Applying the Concepts problems that use graphs from the chapter, and problems associated with the Math Tools boxes.

Learning Resources for Students

SmartWork Online Homework: smartwork.wwnorton.com

Steven Desch, *Guilford Technical Community College*
Violet Mager, *Susquehanna University*
David A. Wood, *San Antonio College*
Todd Young, *Wayne State College, Nebraska*

Over 1,500 questions support the Fourth Edition of *21st Century Astronomy*—all with answer-specific feedback, hints, and ebook links. Questions include Summary Self-Tests, Process of Science Guided Inquiry assignments (based on the concept discussed in the Process of Science figure in each chapter), and versions of the Explorations (based on AstroTours and the Nebraska Simulations). Interactive, image-based questions based on both book art and NASA images help instructors to assess students' conceptual understanding.

StudySpace: wwnorton.com/studyspace

W. W. Norton's free and open student website has the following features:

- Study plans and outlines for each chapter.
- Twenty-eight AstroTour animations, which now include audio. These animations, some of which are interactive, use art from the text to help students visualize important physical and astronomical concepts.
- University of Nebraska Simulations (sometimes called applets; or NAAPs, for Nebraska Astronomy Applet Programs), organized to match the goals of the text. Nebraska Simulations enable students to manipulate variables and see how physical systems work.
- Quiz+ diagnostic multiple-choice quizzes, which provide students with feedback on any incorrect answers, also include links to the ebook, AstroTours, and Nebraska Simulations.
- Vocabulary flashcards.
- "Astronomy in the News" feed.
- Updated website addresses for the end-of-chapter problems.

Starry Night Planetarium Software (College Version) and Workbook

Steven Desch, *Guilford Technical Community College*
Donald Terndrup, *Ohio State University*

Starry Night is a realistic, user-friendly planetarium simulation program designed to allow students in urban areas to perform observational activities on a computer. Norton's unique accompanying workbook offers observation assignments that guide students' virtual explorations and help them apply what they've learned from the text reading assignments. The workbook is fully integrated with *21st Century Astronomy*, Fourth Edition.

For Instructors

Instructor's Manual

Ana M. Larson, *University of Washington*
Gregory D. Mack, *Ohio Wesleyan University*
Ben Sugerman, *Goucher College*

Revised and expanded for the Fourth Edition, this is now the most complete and innovative Instructor's Manual available for introductory astronomy. This impressive resource contains suggested classroom demonstrations, class-tested classroom activities with handouts, and additional Explorations to help facilitate collaborative learning and conceptual understanding. It also contains brief chapter overviews and discussion points, notes on the AstroTour animations contained on the Norton Resource Disc and StudySpace, and worked solutions to all end-of-chapter problems.

Interactive Instructor's Guide

This online, searchable database places all of Norton's astronomy resources at instructors' fingertips. Included are the contents of the Instructor's Manual, the lecture PowerPoint slides with lecture notes, all art and tables in JPEG and PowerPoint formats, the AstroTour animations, and the Nebraska Simulations. With its search tools and export capability, the Interactive Instructor's Guide will help instructors search for exactly the resources they need by topic and resource type, and will alert subscribing instructors as new resources are made available.

Test Bank

Carol Hood, *California State University–San Bernardino*
Michael Hood, *Mt. San Antonio College*
Michael Lopresto, *Henry Ford Community College*
Tammy Smecker-Hane, *University of California–Irvine*
Donald Terndrup, *Ohio State University*

The Test Bank has been developed using the Norton Assessment Guidelines and provides a high-quality bank of over 2,000 items. Each chapter of the Test

Bank consists of three question types classified according to Norton's taxonomy of knowledge types:

1. Factual questions test students' basic understanding of facts and concepts.
2. Applied questions require students to apply knowledge in the solution of a problem.
3. Conceptual questions require students to engage in qualitative reasoning and to explain why things are as they are.

Questions are further classified by section and difficulty, making it easy to construct tests and quizzes that are meaningful and diagnostic. Each chapter contains short-answer, multiple-choice, and true/false questions.

PowerPoint Lecture Slides

Gregory D. Mack, *Ohio Wesleyan University*

These ready-made lecture slides integrate selected art from the text, "clicker" questions, and links to the AstroTour animations. Designed with accompanying lecture outlines, these lecture slides are fully editable and are available in Microsoft PowerPoint format.

Norton Instructor's Resource Site

This Web resource contains the following teaching aids to download:

- Test Bank, available in ExamView, Word RTF, and PDF formats.
- Instructor's Manual in PDF format.
- Lecture PowerPoint slides with lecture notes.
- All art and tables in JPEG and PowerPoint formats.
- Twenty-eight AstroTour animations. These animations, some of which are interactive, use art from the text to help students visualize important physical and astronomical concepts.
- Nebraska Simulations. These interactive simulations enable students to manipulate variables and see how physical systems work.
- Coursepacks, available in BlackBoard, Angel, Desire2Learn, and Moodle formats.

Coursepacks

Norton's Coursepacks, available for use in various Learning Management Systems (LMSs), feature all Quiz+ and Test Bank questions, along with links to the AstroTours and applets. Coursepacks are available in BlackBoard, Angel, Desire2Learn, and Moodle formats.

Instructor's Resource Folder

This two-disc set contains the Instructor's Resource DVD—which contains the same files as the Instructor's Resource website—and the Test Bank on CD-ROM in ExamView format.

Acknowledgments

W. W. Norton & Company, our partner in this endeavor, is made up of talented, professional, and thoughtful individuals who are passionate about science publishing. We would like to thank our editor, Erik Fahlgren, for his direction and patience; Stacy Loyal, marketing manager, for her enthusiasm and sales savvy; Rob Bellinger and Jennifer Barnhardt for developing a creative teaching and learning package that extends and enhances what we do in the text; Kim Yi for her attention to detail and flexibility; developmental editor Erin Mulligan for her suggestions on improving the text and the figures; Renee Cotton for moving things along at every step; copy editor Stephanie Hiebert for her frequent perceptive and helpful advice; and Stephanie Romeo for managing the photos that are on nearly every page of this book.

We would also like to thank Ron Proctor of the Ott Planetarium for the artwork for the Process of Science figures, and Daniel Boice, Bram Boroson, David Branning, Kate Dellenbusch, Christopher Gay, Greg Gowens, Charles Hawkins, Steven Kawaler, Charles Kerton, Matthew Lister, Amanda Maxham, Stanimir Metchev, Jason Smolinski, Donald Terndrup, Frank Timmes, Nilakshi Veerabathina, and James Webb for their care and attention to detail during the accuracy-checking stage of this project.

We gratefully acknowledge the contributions of the authors who worked on previous editions of *21st Century Astronomy*: Dave Burstein, Ron Greeley, Jeff Hester, Howard Voss, and Gary Wegner, with special thanks to Dave for starting the project and to Jeff for leading the original authors through the first edition.

Laura Kay
Stacy Palen
Brad Smith
George Blumenthal

Production of a book like this is a far larger enterprise than any of us imagined when we jumped on board. We would like to thank the instructors who reviewed portions of the manuscript along the way and helped us build a stronger book:

Reviewers for the Fourth Edition

David Bennum, University of Nevada–Reno
William Blass, University of Tennessee–Knoxville
Daniel Boice, University of Texas at San Antonio
Bram Boroson, Clayton State University
David Branning, Trinity College
Julie Bray-Ali, Mt. San Antonio College
Suzanne Bushnell, McNeese State University
Paul Butterworth, George Washington University
Michael Carini, West Kentucky University
Gerald Cecil, University of North Carolina at Chapel Hill
Supriya Chakrabarti, Boston University
Damian Christian, California State University–Northridge
Micol Christopher, Mt. San Antonio College
David Cinabro, Wayne State University
Debashis Dasgupta, University of Wisconsin–Milwaukee
Kate Dellenbusch, Bowling Green State University
Matthias Dietrich, Ohio State University
Gregory Dolise, Harrisburg Area Community College

Yuri Efremenko, University of Tennessee–Knoxville
David Ennis, Ohio State University
Jason Ferguson, Wichita State University
John Finley, Purdue University
Todd Gary, O'More College of Design
Christopher Gay, Santa Fe College
Parviz Ghavamian, Towson University
Martha Gilmore, Wesleyan University
Greg Gowens, University of West Georgia
Javier Hasbun, University of West Georgia
Charles Hawkins, Northern Kentucky University
Sebastian Heinz, University of Wisconsin–Madison
Olenka Hubickyj-Cabot, San Jose State University
James Imamura, University of Oregon
Douglas Ingram, Texas Christian University
Steven Kawaler, Iowa State University
Charles Kerton, Iowa State University
Monika Kress, San Jose State University
Jessica Lair, Eastern Kentucky University
Alex Lazarian, University of Wisconsin–Madison
Matthew Lister, Purdue University
Jack MacConnell, Case Western Reserve University
Kevin Mackay, University of South Florida
Dale Mais, Indiana University–South Bend
Michael Marks, Bristol Community College
Stephan Martin, Bristol Community College
Justin Mason, Ivy Tech Community College
Amanda Maxham, University of Nevada–Las Vegas
Ben McGimsey, Georgia State University
Charles McGruder, West Kentucky University
Janet E. McLarty-Schroeder, Cerritos College
Stanimir Metchev, Stony Brook University
Kent Montgomery, Texas A&M University–Commerce
Ylva Pihlström, University of New Mexico
Dora Preminger, California State University–Northridge
Judit Györgyey Ries, University of Texas
Allen Rogel, Bowling Green State University
Kenneth Rumstay, Valdosta State University–Main Campus
Samir Salim, Indiana University–Bloomington
Eric Schlegel, University of Texas at San Antonio
Paul Schmidtke, Arizona State University
Ohad Shemmer, University of North Texas
Allyn Smith, Austin Peay State University
Jason Smolinski, State University of New York at Oneonta
Roger Stanley, San Antonio College
Angelle Tanner, Mississippi State University
Christopher Taylor, California State University–Sacramento
Donald Terndrup, Ohio State University
Todd Thompson, Ohio State University
Glenn Tiede, Bowling Green State University
Frances Timmes, Arizona State University
Walter Van Hamme, Florida International University
Nilakshi Veerabathina, University of Texas at Arlington
Ezekiel Walker, University of North Texas
James Webb, Florida International University

Reviewers of Previous Editions

Scott Atkins, University of South Dakota
Timothy Barker, Wheaton College
Peter A. Becker, George Mason University
Timothy C. Beers, Michigan State University
Edwin Bergin, University of Pittsburgh
Steve Bloom, Hampden Sydney College
Jack Brockway, Radford University
Juan E. Cabanela, Minnesota State University–Moorhead
Amy Campbell, Louisiana State University
Robert Cicerone, Bridgewater State College
Judith Cohen, California Institute of Technology
Eric M. Collins, California State University–Northridge
John Cowan, University of Oklahoma–Norman
Robert Dick, Carleton University
Tom English, Guilford Technical Community College
Matthew Francis, Lambuth University
Kevin Gannon, College of Saint Rose
Bill Gutsch, St. Peter's College
Karl Haisch, Utah Valley University
Barry Hillard, Baldwin Wallace College
Paul Hintzen, California State University–Long Beach
Paul Hodge, University of Washington
William A. Hollerman, University of Louisiana at Lafayette
Hal Hollingsworth, Florida International University
Olencka Hubickyj-Cabot, San Jose State University
Kevin M. Huffenberger, University of Miami
Adam Johnston, Weber State University
Bill Keel, University of Alabama
Kevin Lee, University of Nebraska–Lincoln
M. A. K. Lodhi, Texas Tech University
Leslie Looney, University of Illinois at Urbana–Champaign
Norm Markworth, Stephen F. Austin State University
Kevin Marshall, Bucknell University
Chris McCarthy, San Francisco State University
Chris Mihos, Case Western University
Milan Mijic, California State University–Los Angeles
J. Scott Miller, University of Louisville
Scott Miller, Sam Houston State University
Andrew Morrison, Illinois Wesleyan University
Edward M. Murphy, University of Virginia
Kentaro Nagamine, University of Nevada–Las Vegas
Jascha Polet, California State Polytechnic University
Daniel Proga, University of Nevada–Las Vegas
Laurie Reed, Saginaw Valley State University
Masao Sako, University of Pennsylvania
Ata Sarajedini, University of Florida
Ann Schmiedekamp, Pennsylvania State University
Jonathan Secaur, Kent State University
Caroline Simpson, Florida International University
Paul P. Sipiera, William Rainey Harper College
Ian Skilling, University of Pittsburgh
Tammy Smecker-Hane, University of California–Irvine
Ben Sugerman, Goucher College
Neal Sumerlin, Lynchburg College
Trina Van Ausdal, Salt Lake Community College
Karen Vanlandingham, West Chester University
Paul Voytas, Wittenberg University
Paul Wiita, Georgia State University
Richard Williamon, Emory University
David Wittman, University of California–Davis

About the Authors

Laura Kay is Ann Whitney Olin professor and chair of the Department of Physics and Astronomy at Barnard College, where she has taught since 1991. She received a BS degree in physics from Stanford University, and MS and PhD degrees in astronomy and astrophysics from the University of California–Santa Cruz. She studies active galactic nuclei, using ground-based and space telescopes. She teaches courses in astronomy, astrobiology, women and science, and polar exploration.

Stacy Palen received her PhD from the University of Iowa in 1998. Her astronomical research has focused on multiwavelength studies of dying Sun-like stars. Currently, she is an associate professor at Weber State University in Ogden, Utah. In addition to the usual professorial duties, she runs the Ott Planetarium, an astronomy outreach center that produces planetarium content for all ages.

Brad Smith is a retired professor of planetary science. He has served as an associate professor of astronomy at New Mexico State University, as a professor of planetary sciences and astronomy at the University of Arizona, and as a research astronomer at the University of Hawaii. Through his interest in Solar System astronomy, he participated as a team member or imaging team leader on several U.S. and international space missions, including Mars *Mariners 6, 7,* and *9*; *Viking*; *Voyagers 1* and *2*; and the Soviet *Vega* and *Phobos* missions. He later turned his interest to extrasolar planetary systems, investigating circumstellar debris disks as a member of the Hubble Space Telescope NICMOS experiment team. Smith has four times been awarded the NASA Medal for Exceptional Scientific Achievement. He is a member of the IAU Working Group for Planetary System Nomenclature and is chair of the Task Group for Mars Nomenclature.

George Blumenthal is chancellor at the University of California–Santa Cruz, where he has been a professor of astronomy and astrophysics since 1972. Chancellor Blumenthal received his BS degree from the University of Wisconsin–Milwaukee and his PhD in physics from the University of California–San Diego. As a theoretical astrophysicist, Chancellor Blumenthal's research encompasses several broad areas, including the nature of the dark matter that constitutes most of the mass in the universe, the origin of galaxies and other large structures in the universe, the earliest moments in the universe, astrophysical radiation processes, and the structure of active galactic nuclei such as quasars. Besides teaching and conducting research, Chancellor Blumenthal has served as the chair of the UC–Santa Cruz Astronomy and Astrophysics Department, has chaired the Academic Senate for both the UC–Santa Cruz campus and the entire University of California system, has been both the chair and vice chair of the California Association for Research in Astronomy (which runs Keck Observatory), and has served as the faculty representative to the UC Board of Regents.

STARS AND GALAXIES

21ST CENTURY ASTRONOMY

FOURTH EDITION

The Virgo cluster of galaxies at a distance of about 50 million light-years..

01 Why Learn Astronomy?

The most beautiful thing we can experience is the mysterious.
It is the source of all true art and all science.
He to whom this emotion is a stranger,
who can no longer pause to wonder and stand rapt in awe,
is as good as dead: his eyes are closed.

Albert Einstein (1879–1955)

LEARNING GOALS

We'll begin this chapter by sketching out a rough map of the universe and our place within it. Then we'll present some of the tools that you will need to take along as you look at the wonders of the universe through the eyes of a scientist. By the conclusion of this chapter, you should be able to:

- Identify our planet Earth's place in the universe.
- Explain the process of science.
- Describe the scientific approach to understanding our world and the universe.

1.1 Getting a Feel for the Neighborhood

The title of this book—*21st Century Astronomy*—emphasizes that this is the most fascinating time in history to be studying this most ancient of sciences. Loosely translated, the word **astronomy** means "patterns among the stars." But modern astronomy—the astronomy we will talk about in this book—has progressed beyond merely looking at the sky and cataloging what is visible there. Our intent is to provide reliable answers to many of the questions that you might have asked yourself as a child when you looked at the sky. What are the Sun and Moon made of? How far away are they? What are stars? How do they shine? Do they have anything to do with me?

The origin and fate of the universe, and the nature of space and time, have become the subjects of rigorous scientific investigation. Humans have long speculated about our beginnings, or *origins*. Who or what is responsible for our existence? How did the Sun, stars, and Earth form? The topic of scientific origins is a recurring theme in this book. The answers that scientists are finding to these questions are changing not only our view of the cosmos, but our view of ourselves.

Glimpsing Our Place in the Universe

Most people have a permanent address—building number, street, city, state, country. It is where the mail carrier delivers our postal mail. But let's expand our view for a moment. We also live somewhere within an enormously vast universe. What, then, is our "cosmic address"? It might look something like this: planet, star, galaxy, galaxy group, galaxy cluster.

We all reside on a planet called Earth, which is orbiting under the influence of gravity about a star called the Sun. The **Sun** is an ordinary, middle-aged star, more massive and luminous than some stars but less massive and luminous than others. The Sun is extraordinary only because of its importance to us within our own **Solar System**. Our Solar System consists of eight planets: Mercury, Venus, Earth, Mars, Jupiter, Saturn, Uranus, and Neptune. It also contains many smaller bodies, such as dwarf planets, asteroids, and comets.

The Sun is located about halfway out from the center of a flattened collection of stars, gas, and dust referred to as the **Milky Way Galaxy**. Our Sun is just one among approximately 200–400 billion stars scattered throughout the galaxy, and many of these stars are themselves surrounded by planets, suggesting that other planetary systems may be common.

The Milky Way is a member of a collection of a few dozen galaxies called the **Local Group**. Looking farther outward, the Local Group is part of a vastly larger collection of thousands of galaxies—a **supercluster**—called the Virgo Supercluster.

We can now define our cosmic address—Earth, Solar System, Milky Way Galaxy, Local Group, Virgo Supercluster—as illustrated in **Figure 1.1**. Yet even this address is not complete, because the vast structure we just described is only the *local universe*. Astronomers use the term **light-year** to refer to the *distance* that light travels within one year, about 9.5 trillion kilometers (km) or 6 trillion miles. The part of the universe that we can see extends far beyond the local universe—13.7 billion light-years—and within this volume we estimate that there are *hundreds of billions* of galaxies, roughly as many galaxies as there are stars in the Milky Way. In addition, scientists have concluded that our universe contains much more than the observed planets, stars, and galaxies. Up to 95 percent of the universe is made up of matter that does not emit light (called *dark matter*) and a form of energy that permeates all of space (*dark energy*)—neither of which is well understood.

What is Earth's cosmic address?

The Scale of the Universe

One of the first conceptual hurdles that we face as we begin to think about the universe is its sheer size. If a hill is big, then a mountain is very big. If a mountain is very big, then Earth is enormous. But where do we go from there? We quickly run out of superlatives as the scale begins to dwarf our human experience. One technique that can help us develop a sense for the size of things in the universe is to discuss time as well as distance. If you are driving down the highway at 60 kilometers per hour (km/h), a kilometer is how far you travel in a minute. Sixty kilometers is how far you travel in an hour. Six hundred kilometers is how far you travel in 10 hours. So to get a feeling for the difference in size between 600 km and 1 km, you can think about the difference between 10 hours and a single minute.

The travel time of light helps us understand the scale of the universe.

We can think this same way about astronomy, but the speed of a car on the highway is far too slow to be useful. Instead we use the greatest speed in the universe—the speed of light. Light travels at 300,000 kilometers per second (km/s), circling Earth (a distance of 40,000 km) in

FIGURE 1.1 Our cosmic address is: Earth, Solar System, Milky Way Galaxy, Local Group, Virgo Supercluster. We live on Earth, a planet orbiting the Sun in our Solar System, which is a star in the Milky Way Galaxy. The Milky Way is a large galaxy within the Local Group of galaxies, which in turn is located in the Virgo Supercluster.

just under 1⁄7 of a second—about the time it takes you to snap your fingers. So we say that the circumference of Earth is 1⁄7 a light-second. Fix that comparison in your mind. The size of Earth is like a snap of your fingers.

Now follow along in **Figure 1.2** as we move outward into the universe. The first thing we encounter is the Moon, 384,000 km away, or a bit over 1¼ second when we're moving at the speed of light. If the size of Earth is a snap of your fingers, the distance to the Moon is about the time it takes to turn a page in this book. Continuing further, we find that at this speed the Sun is 8⅓ minutes away, or the duration of a hurried lunch at the student union. Crossing from one side of the orbit of Neptune, the outermost planet in our Solar System, to the other takes about 8.3 hours. Think about that for a minute. Comparing the size of Neptune's orbit to the circumference of Earth is like comparing the time of a good night's sleep to a single snap of your fingers.

In crossing Neptune's orbit, however, we have only just begun to consider the scale of the universe. Many steps remain. It takes a bit more than 4 years—the time between leap years—to cover the distance from Earth to the nearest star (other than the Sun). At this point, our analogy of using the travel time of light can no longer bring astronomical distance to a human scale. Light takes about 100,000 years to travel across our galaxy (the Milky Way). To reach the nearest large galaxies takes a few million years. To reach the limits of the currently observable universe takes 13.7 billion years—the age of the universe, or about 3 times the age of Earth. ▶▶ **NEBRASKA SIMULATION: LOOK BACK TIME SIMULATOR**

The Origin and Evolution of the Chemical Elements

While seeking knowledge about the universe and how it works, modern astronomy and physics have repeatedly come face-to-face with a number of age-old questions long thought to be solely within the domain of religion or philosophy. The nature of the chemical evolution of the universe is such a case. Theory and observation indicate that the universe was created in a "Big Bang" some 13.7 billion years ago. As a result of both observation and theoretical work, scientists now know that the only chemical elements found in substantial amounts in the early universe were the lightest elements: hydrogen and helium, plus tiny amounts of lithium, beryllium, and boron. Yet we live on a planet with a central core consisting mostly of very heavy elements such as iron and nickel, surrounded by outer layers made up of rocks containing large amounts of silicon and various other elements, all heavier than the original elements. Our bodies are built of carbon, nitrogen, oxygen, calcium, phosphorus, and a host of other chemical elements—again all heavier than hydrogen and

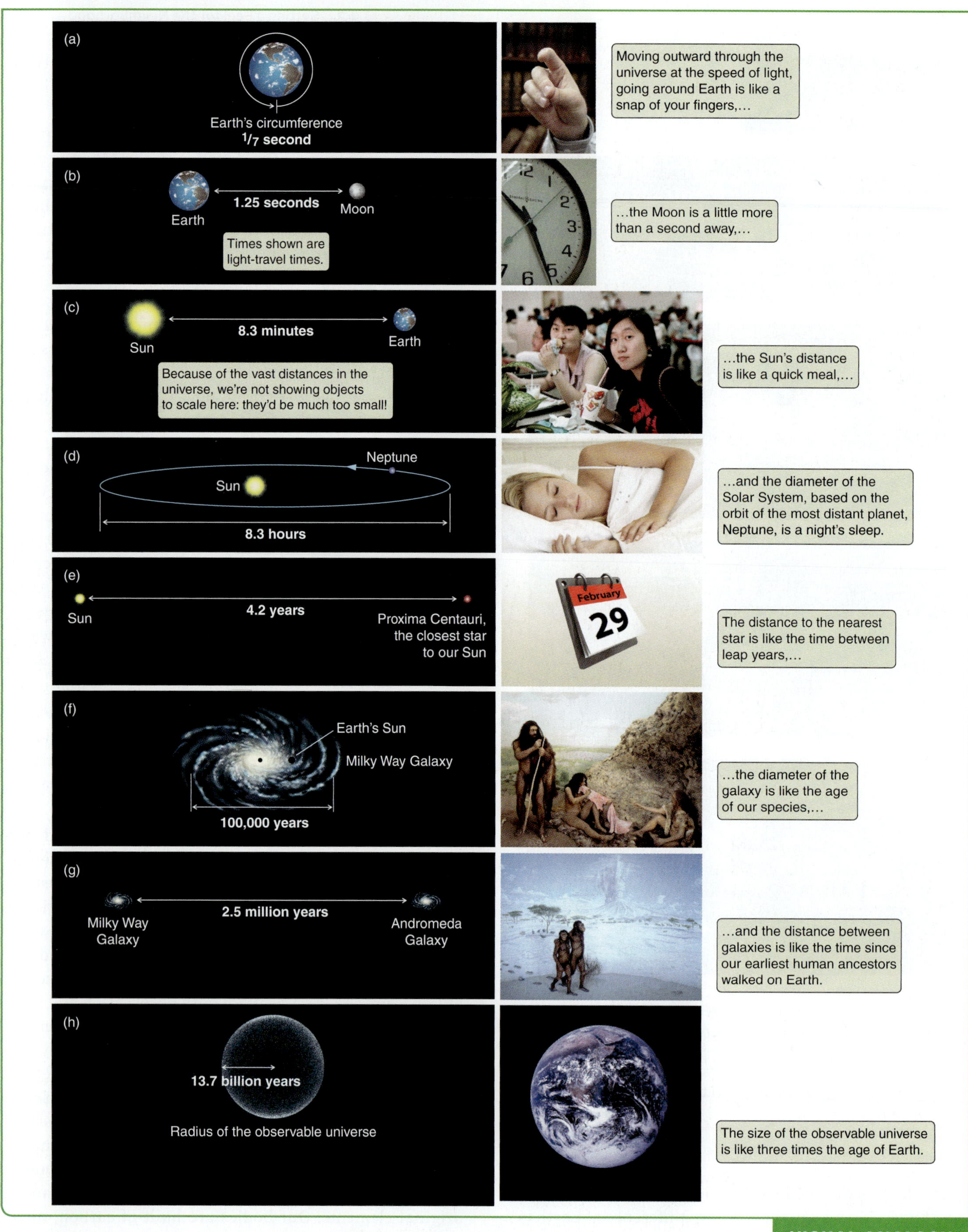

FIGURE 1.2 Thinking about the time it takes for light to travel between objects helps us comprehend the vast distances in the universe. (Figures such as this one, with "Visual Analogy" tags, are images that make analogies between astronomical phenomena and everyday objects more concrete.)

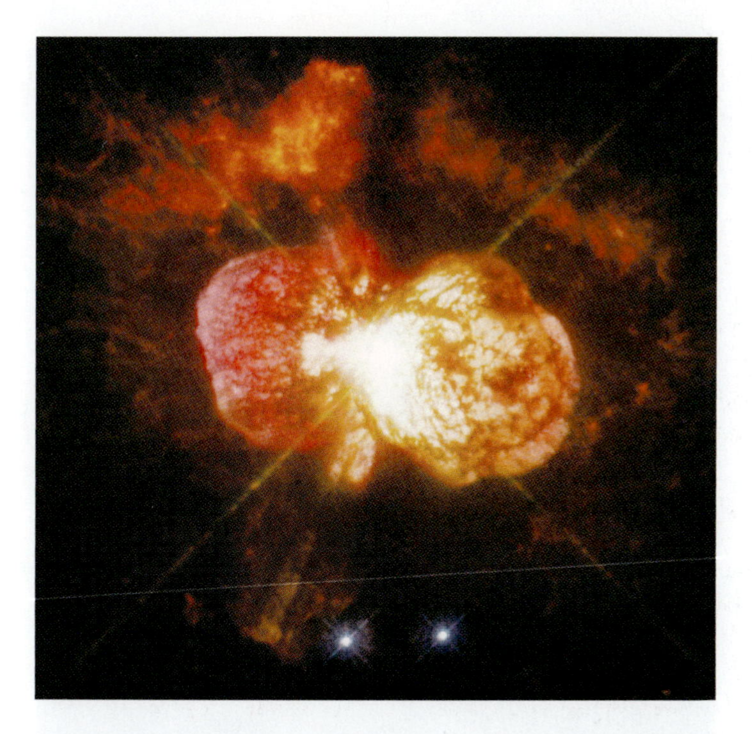

FIGURE 1.3 You and everything around you are composed of atoms forged in the interior of stars that lived and died before the Sun and Earth were formed. The supermassive star Eta Carinae, shown here, is currently ejecting a cloud of chemically enriched material just as earlier generations of stars once did to enrich our Solar System.

helium. If these heavier elements that make up Earth and our bodies were not present in the early universe, where did they come from?

The answer to this question lies within the stars (**Figure 1.3**). Nuclear fusion reactions occurring deep within the interiors of stars combine atoms of light elements such as hydrogen to form more massive atoms. When a star exhausts its nuclear fuel and nears the end of its life, it often loses much of its mass—including some of the new atoms formed in its interior—by blasting it back into interstellar space. We will talk later about the life and death of stars. For now it is enough to note that our Sun and Solar System are recycled—formed from a cloud of interstellar gas and dust that had been "seeded" by earlier generations of stars. This chemical legacy supplies the building blocks for the interesting chemical processes that go on around us—chemical processes such as life. The atoms that make up much of what we see were formed in the hearts of stars. The singer-songwriter Joni Mitchell wrote, "We are stardust," and this is not just poetry. Literally, we are made of the stuff of stars.

We are stardust.

1.2 Astronomy Involves Exploration and Discovery

As you look at the universe through the eyes of astronomers, you can also learn something of how science works. It is beyond the scope of this book to provide a detailed justification for all that we will say. However, we will try to offer some explanation of where key ideas come from and why scientists think these ideas are valid. We will be honest when we are on uncertain, speculative ground, and we will admit it when the truth is that we really do not know. This book is not a compendium of revealed truth or a font of accepted wisdom. Rather, it is an introduction to a body of knowledge and understanding that was painstakingly built (and sometimes torn down and rebuilt) brick by brick.

Science is vitally important to our civilization. Electricity, cars, computers—all of these technologies are derived from science. Another manifestation of science is the technology that has enabled us to explore well beyond our planet. Since the 1957 launch of Sputnik, the first human-made **satellite** (an object in orbit about a more massive body), we have lived in an age of space exploration. Nearly six decades later, satellites are used for weather observation, communication, and global positioning (GPS); humans have walked on the Moon (**Figure 1.4**); and unmanned probes have visited planets. Spacecraft have flown past asteroids, comets, and even the Sun. Human inventions have landed on Mars, Venus, Titan (Saturn's largest moon), and asteroids, and have plunged into the atmosphere of Jupiter. Most of what we know of the Solar System has resulted from these past six decades of exploration.

Satellite observatories in orbit around Earth have also given us many new perspectives on the universe. Space astronomy continues to show us vistas hidden from the gaze of ground-based telescopes by the protective but obscuring blanket

Space exploration has expanded our view of the universe.

FIGURE 1.4 *Apollo 15* astronaut James B. Irwin stands by the lunar rover during an excursion to explore and collect samples from the Moon.

(a)

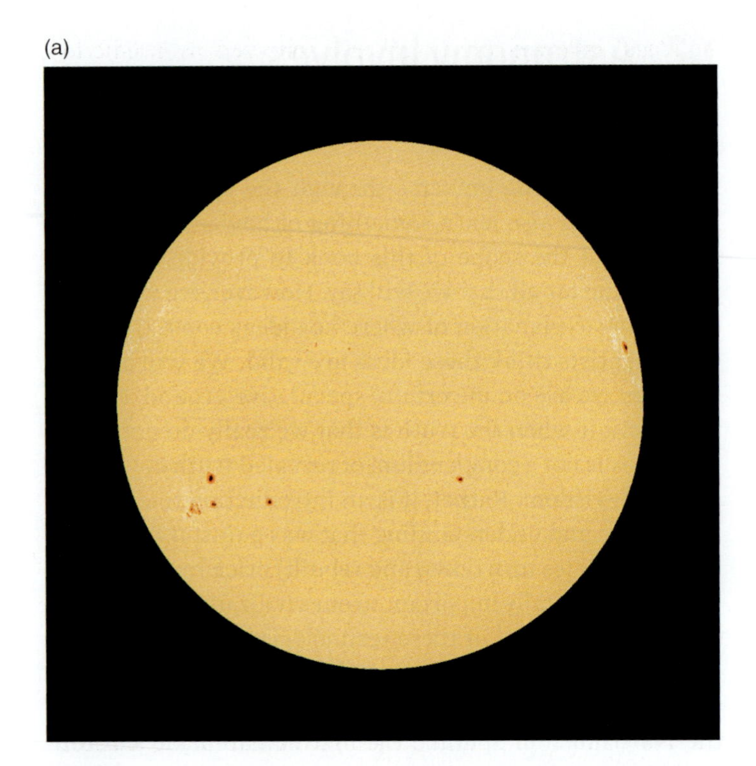

(b)

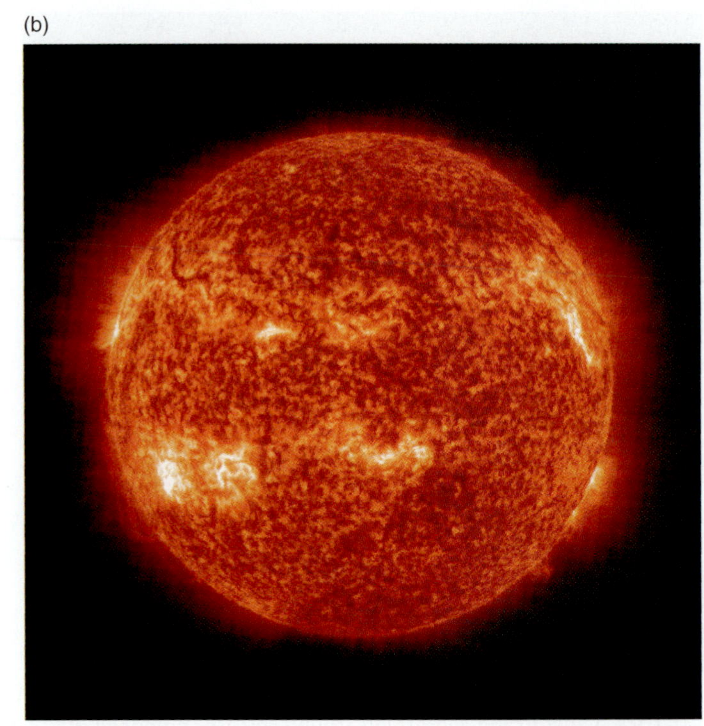

FIGURE 1.5 Visible-light (a) and X-ray (b) telescopic images of the Sun.

of our atmosphere. Satellites capable of detecting the full spectrum of radiation—from the highest-energy gamma rays and X-rays, through ultraviolet and infrared radiation, to the lowest-energy microwaves—have brought surprising discovery after surprising discovery. Since the beginning of the 21st century, large astronomical observatories have been constructed on the ground as well. The objects in the sky are now seen by gamma-ray, X-ray, infrared, and radio telescopes (**Figure 1.5**), extending our observations into light that has shorter or longer wavelengths than we can see with our eyes.

A great deal of frontline astronomy is now carried out in large physics facilities like the particle collider shown in **Figure 1.6**. Today astronomers work along with their colleagues in related fields, such as physics, chemistry, geology, and planetary science, to sharpen their understanding of the physical laws that govern the behavior of matter and energy and to use this understanding to make sense of our observations of the cosmos. Astronomy has also benefited enormously from the computer revolution. The 21st century astronomer spends far more time staring at a computer screen than peering through the eyepiece of a telescope. Astronomers use computers to collect and analyze data from telescopes, calculate physical models of astronomical objects, and prepare and disseminate the results of their work.

FIGURE 1.6 The Large Hadron Collider (which is buried along the path indicated by the red circle) is a particle accelerator near Geneva, Switzerland, that provides clues about the physical environment during the birth of the universe. Laboratory astrophysics, in which astronomers model important physical processes under controlled conditions as they do at this facility, has become an important part of astronomy.

1.3 Science Is a Way of Viewing the World

The Scientific Method and Scientific Principles

What is the scientific method? Consider a scientist coming up with an idea that might explain a particular observation or phenomenon. She presents the idea to her colleagues as a hypothesis. Her colleagues then look for testable predictions capable of *disproving* her hypothesis. *This is an important property of the scientific method*: a scientific hypothesis must be *falsifiable*—in other words, *disprovable*. (Note that a falsifiable hypothesis—one capable of being shown false—may not be testable using current technology, but scientists must at least be able to outline an experiment or observation that could prove the idea wrong.) If continuing tests fail to disprove a hypothesis, the scientific community will come to accept it as a theory and, after enough confirmation, eventually treat it as a law of nature. Scientific theories are accepted only as long as their predictions are borne out. A classic example is Einstein's theory of relativity, which we cover in some depth in Chapter 18. The theory of relativity has withstood a century of scientific efforts to disprove its predictions.

The scientific method includes trying to *falsify* ideas.

Science is sometimes misunderstood because of the ways that scientists use everyday words. An example is the word *theory*. In everyday language, *theory* may mean a conjecture or a guess: "Do you have a theory about who might have done it?" "My theory is that a third party could win the next election." In everyday parlance a theory is something worthy of little serious regard. "After all," people say, "it's only a theory."

In stark contrast, a *scientific* **theory** is a carefully constructed proposition that takes into account all the relevant data and all our understanding of how the world works. A theory makes testable predictions about the outcome of future observations and experiments. It is a well-developed idea that is ready to be tested by what is observed in nature. A well-corroborated theory is a theory that has survived many such tests. Far from being simple speculation, scientific theories represent and summarize bodies of knowledge and understanding that provide fundamental insights into the world around us. A successful and well-corroborated theory is the pinnacle of human knowledge about the world.

In science, a **hypothesis** is an idea that leads to testable predictions. The scientific method consists of observation or ideas, followed by hypothesis, followed by prediction, followed by further observation or experiments to test the prediction, and ending with a tested theory (see the **Process of Science Figure** on the next page). A hypothesis may be the forerunner of a scientific theory, or it may be based on an existing theory, or both. Scientists build **theoretical models** that are used to connect theories with the behavior of complex systems. Ultimately, the basis for deciding among competing theories is the success of their predictions. Some theories become so well tested and are of such fundamental importance that people refer to them as **physical laws**.

A scientific **principle** is a general idea or sense about how the universe is that guides the construction of new theories. **Occam's razor**, for example, is a guiding principle in science stating that when we are faced with two hypotheses that explain a particular phenomenon equally well, we should adopt the simpler of the two, unless the more complicated answer better matches the results of observations or experiment. Another principle comes from the late astronomer Carl Sagan (1934–1996) and is often phrased as "Extraordinary Claims Require Extraordinary Evidence," meaning that when making a new and truly extraordinary claim that has not been tested, confirmed, or proven, extraordinary evidence is required.

At the heart of modern astronomy is the adoption of an additional principle: the **cosmological principle**. The cosmological principle states that on a large scale, the universe looks the same everywhere. That is, when people look out around in every direction, what they see is representative of what the universe is generally like. In other words, there is nothing special about our particular location. By extension, the cosmological principle asserts that matter and energy obey the same physical laws throughout space and time as they do today on Earth. This assumption is important because it means that the same physical laws that we observe and apply in terrestrial laboratories can be used to understand what goes on in the centers of stars or in the hearts of distant galaxies. Each new success that comes from applying the cosmological principle to observations of the universe around us adds to our confidence in the validity of this cornerstone of our worldview. We will discuss the cosmological principle in more detail in Chapter 19.

There is nothing special about our place in the universe.

Science as a Way of Knowing

The path to scientific knowledge is solidly based on the **scientific method**. This concept is so important to an understanding of how science works that we should emphasize it once again. The scientific method consists of observation

Process of Science

THE SCIENTIFIC METHOD

The scientific method is a representation of the path by which an idea or observation is tested for validity.

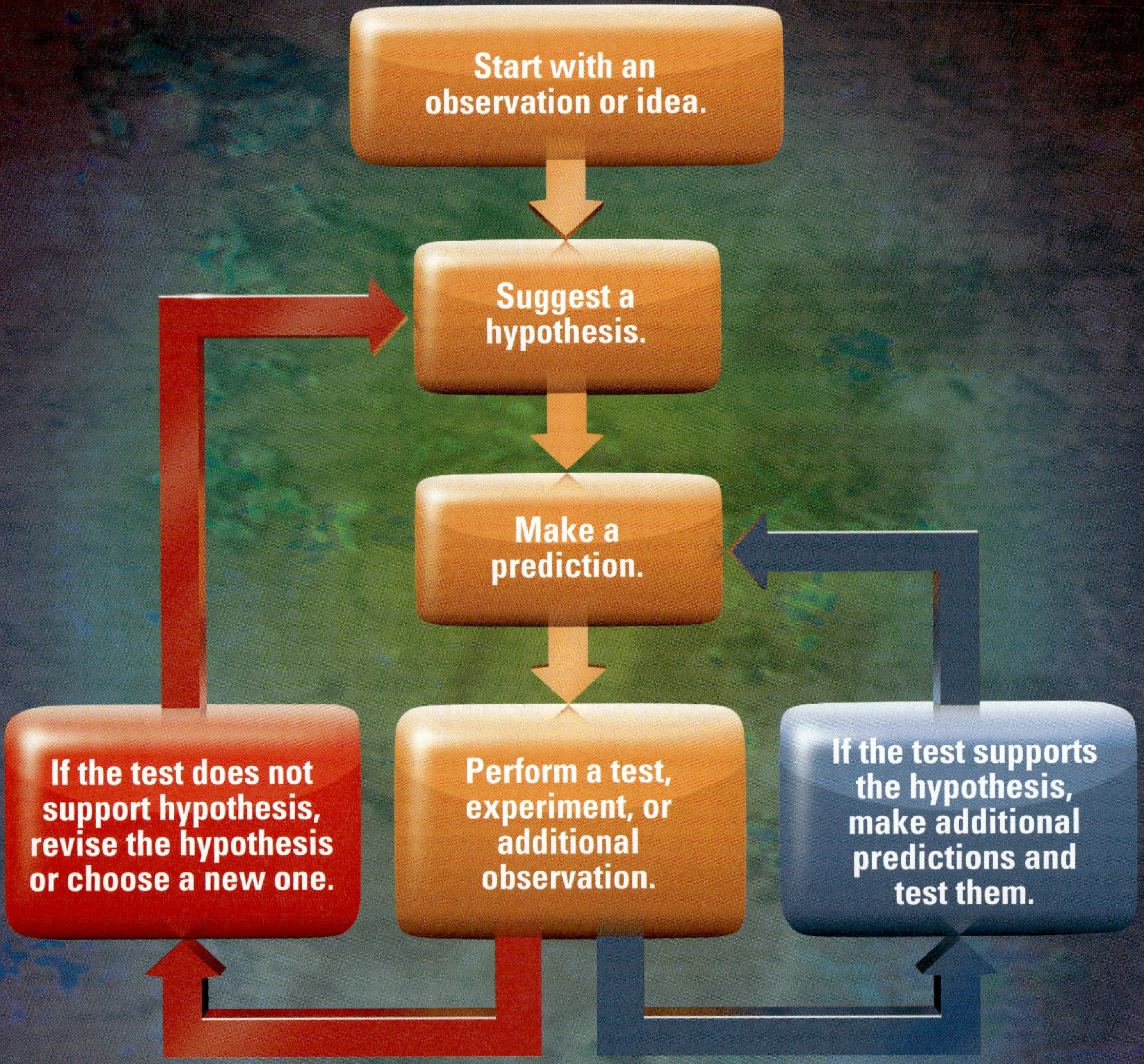

Start with an observation or idea.

Suggest a hypothesis.

Make a prediction.

Perform a test, experiment, or additional observation.

If the test does not support hypothesis, revise the hypothesis or choose a new one.

If the test supports the hypothesis, make additional predictions and test them.

An idea or observation leads to a falsifiable hypothesis that is either accepted as a tested theory or rejected, on the basis of observational or experimental tests of its predictions. The blue loop goes on indefinitely as scientists continue to test the theory.

or idea, followed by hypothesis, followed by prediction, followed by further observation or experiments to test the prediction, and ending as a tested theory, as illustrated in the Process of Science figure. For all practical purposes, the scientific method defines what we mean when we use the verb *to know* scientifically. It is sometimes said that the scientific method is how scientists prove things to be true, but actually it is a way of confirming things to be *false*. Scientific theories are accepted only as long as they are able to be tested and are not shown to be false. In this sense, *all scientific knowledge is provisional.*

All scientific knowledge is provisional.

We should point out here that astronomy differs from other sciences in that astronomers often *cannot* conduct controlled experiments, with varying hypotheses, as is done in physics, chemistry, biology, and other sciences. Instead, astronomers make multiple observations using different methods, and create mathematical and physical models based on established science.

The scientific method provides the rules for asking nature whether an idea is false, but it offers no insight into where the idea came from in the first place or how an experiment was designed. If you were to listen to a group of scientists discussing their work, you might be surprised to hear them using words like *insight*, *intuition*, and *creativity*. Scientists speak of a beautiful theory in the same way that an artist speaks of a beautiful painting or a musician speaks of a beautiful performance. Yet science is not the same as art or music in one important respect: whereas art and music are judged by a human jury alone, in science it is nature (through application of the scientific method) that provides the final decisions about which theories can be kept and which theories must be discarded. Nature is completely unconcerned about what we *want* to be true. In the history of science, many a beautiful and beloved theory has been abandoned. At the same time, however, there is an aesthetic to science that is as human and as profound as any found in the arts (**Figure 1.7**).

Scientific Revolutions

Scientists spend most of their time working within an established framework of understanding, continually extending and refining that framework and testing its boundaries. Occasionally, however, major shifts occur in the framework of an entire scientific field. Much has been written about how science progresses; *The Structure of Scientific Revolutions* (1962) by Thomas Kuhn is one of the earlier influential books. In this work, Kuhn emphasizes the constant tension between the scientist's human need to construct a system within which to interpret the world and the occasional need to drastically overhaul that system of ideas.

A scientific revolution is not a trivial thing. A new theory or way of viewing the world must be able to explain everything that the previous theory could, while extending this understanding to new territory into which the earlier theory could not go. By the middle of the 19th century, many physicists thought that their fundamental understanding of physical law was more or less complete. For over a century, the classical physics developed by Sir Isaac Newton to explain motion, forces, and gravity (discussed in detail in Chapters 3 and 4) had withstood the scrutiny of scientists the world over. It seemed that little remained but cleanup work—filling in the details. Yet during the late 19th and early 20th centuries, physics was rocked by a series of scientific revolutions that shook the very foundations of an understanding of the nature of reality.

If one face can be said to represent these scientific revolutions—and modern science—it is that of Albert Einstein (1879–1955). Einstein's special and general theories of relativity replaced the 200-year-old edifice of Newtonian mechanics—not by proving Newton wrong, but by showing

FIGURE 1.7 The scientific worldview is as aesthetically pleasing as that of art, music, or literature. Unlike the arts, however, science relies on nature alone to determine which scientific theories have lasting value.

FIGURE 1.8 Albert Einstein is perhaps the most famous scientist of the 20th century, and he was *Time* magazine's selection for Person of the Century. Einstein helped to usher in two different scientific revolutions, one of which he himself was never able to accept.

that Newton's mechanics was a special case of a far more general and powerful set of physical laws. Einstein's new ideas unified the concepts of mass and energy and destroyed the conventional notion of space and time as separate. Einstein (**Figure 1.8**) actually helped to start two scientific revolutions. He saw the first of these—relativity—through and embraced the world that it opened. Yet Einstein was unable to accept the implications of the second revolution he helped start—quantum mechanics—and he went to his grave unwilling to embrace the view of the world it offered. Together, these revolutions led to the birth of what has come to be known as **modern physics**. Although modern physics *contains* Newtonian physics, the understanding of the universe offered by modern physics is far more sublime and powerful than the earlier understanding that it subsumed.

As you read this book, you will encounter many discoveries and ideas that forced scientists to abandon earlier notions that ultimately failed the test of observation and experiment. The rigorous standards of scientific knowledge respect no authorities. No theory, no matter how central or how strongly held, is immune from the rules.

For those of us who grew up in a world transformed by science, the scientific worldview might seem anything but revolutionary. Throughout much of history, however, knowledge was sought in the pronouncements of "authority" by theologians, philosophers, royalty, and politicians rather than through observation of nature. This authoritarian view slowed the advance of knowledge throughout Europe for the millennium prior to the European Renaissance, and it was largely the Chinese and Arab cultures that kept the spark of inquiry alive during this time. The greatest scientific revolution of all was the one that overthrew "authority" and replaced it with rational inquiry and the scientific method.

From the perspective of this great scientific revolution, it is clear why science is more than simply a body of facts. Perhaps more than anything, science is a way of thinking about the world. It is a way of relating to nature. It is a search for the relationships that make our world what it is. It is the idea that nature is not capricious, but instead operates by consistent, explicable, inviolate rules. It is a collection of ideas about how the universe works, coupled with an acceptance of the fact that what is known today may be superseded tomorrow. The scientist's assumptions are that there is an order in the universe and that the human mind is capable of grasping the essence of the rules underlying that order. The scientist's creed is that nature, through observation and experiment, is the final arbiter of the only thing worthy of the term *objective truth*. Science is an exquisite blend of aesthetics and practicality. And, in the final analysis, science has found such a central place in our civilization because *science works*.

Challenges to Science

Science did not achieve its prominence without criticism and controversy. Philosophers and theologians of the 16th century did not easily give up the long-held belief that Earth is the center of the universe. In the early 20th century, astronomers were arguing among themselves over the size of the universe. For centuries, astronomers had thought that the Milky Way was all that there was—a vast collection of stars and "fuzzy" objects known as *nebulae*. But better observations from newer and larger telescopes showed that some of these fuzzy objects were actually huge distant groups of stars—galaxies ("island universes")—similar to but far beyond the confines of our own Milky Way (**Figure 1.9**). Astronomers concluded that our Milky Way does not, in fact, constitute the entire universe—that it is only an infinitesimal part of a greatly larger universe.

Scientific theories must be consistent with all that we know of how nature works, and turning a clever idea into a real theory with testable predictions is a matter of careful thought and effort. One of the most remarkable aspects of scientific knowledge is its *independence* from culture. Scientists are people, and politics and culture enter into the day-to-day practice of science. But in the end, *scientific theories must be judged not by cultural norms or by*

Nature is the arbiter of science.

FIGURE 1.9 Until the early 20th century, astronomers believed that all "fuzzy" celestial objects were located within our Milky Way Galaxy. The distant spiral galaxy M51 is shown here in a modern photograph taken by the Hubble Space Telescope.

political opinion, but by whether their predictions are borne out by observation and experiment. Science is not just one of many possible worldviews; science is the most successful worldview in the history of our species because the foundations of the scientific worldview have withstood centuries of fine minds trying to prove them false.

It is significant that no serious critic of science has ever offered a viable alternative for obtaining reliable knowledge of the workings of nature. Furthermore, no other category of human knowledge is subject to standards as rigorous and unforgiving as those of science. For this reason, scientific knowledge is reliable in a way that no other form of knowledge can claim. Whether you want to design a building that will not fall over, consider the most recent medical treatment for a disease, or calculate the orbit of a spacecraft on its way to the Moon, you had better consult a scientist rather than a psychic—regardless of your cultural and religious background or political affiliation.

Political, religious, economic, and cultural considerations can sometimes influence which scientific research projects are funded. This choice of funding prioritizes and channels the directions in which scientific knowledge advances. For example, decisions regarding funding for research on stem cells, environmental regulation, human space exploration, and the Search for Extraterrestrial Intelligence (SETI) have been affected by political infighting.

Finally, disreputable scientists can purposefully try to influence results by inventing or ignoring data, often when claiming a "major breakthrough" or challenging a well-established scientific principle. Fortunately, attempts by others to repeat the experiment will eventually expose such scientific misconduct.

1.4 Patterns Make Our Lives and Science Possible

We experience patterns in our everyday life, and we find comfort in them. Imagine what life would be like if sometimes when you let go of an object it fell up instead of down. What if, unpredictably, one day the Sun rose at noon and set at 1:00 P.M., the next day it rose at 6:00 A.M. and set at 10:00 P.M., and the next day the Sun did not rise at all? In fact, objects do fall toward the ground. The Sun rises, sets, and then rises again at predictable times. Spring turns into summer, summer turns into autumn, autumn turns into winter, and winter turns into spring. The rhythms of nature produce patterns in our lives, and we count on these patterns for our very survival. If nature did not behave according to regular patterns, then our lives—indeed, life itself—would not be possible.

The goal of science is to identify and characterize these patterns and to use them to understand the world around us. Some of the most regular and easily identified patterns in nature are the patterns we see in the sky. What in the sky will look different or the same a week from now? A month from now? A year from now? Most of us lead an indoor and metropolitan existence, removed from an everyday awareness of the patterns in the sky. Away from the lights and glare of our cities, however, the patterns and rhythms of the sky can be seen today as they were in ancient times. Patterns in the sky mark the changing of the seasons (**Figure 1.10**) and share the rhythms of our lives. It is no surprise that astronomy, which is the expression of our human need to understand these patterns, is the oldest of all sciences. With large telescopes, we see patterns of symmetry in some astronomical objects (**Figure 1.11**), which can give astronomers clues to the physical processes that created them.

One important tool that astronomers use to analyze these patterns is mathematics. There are many branches of mathematics, most of which deal with more than just numbers. Arithmetic is about counting things. Algebra is about manipulating symbols and the relationships between things. Geometry is about shapes. What do all these branches have in common? Why do we consider all of them to be part of a single discipline called mathematics? The common thread is that they all deal with patterns.

Mathematics is the language of patterns.

If patterns are the heart of science and mathematics is the language of patterns, it should come as no surprise that *mathematics is the language of science.* Trying to study science while avoiding mathematics is the practical equivalent of trying to study Shakespeare while avoiding the written or spoken word. It quite simply cannot be done, or at least cannot be done meaningfully. Some people prefer the elegance of a simple

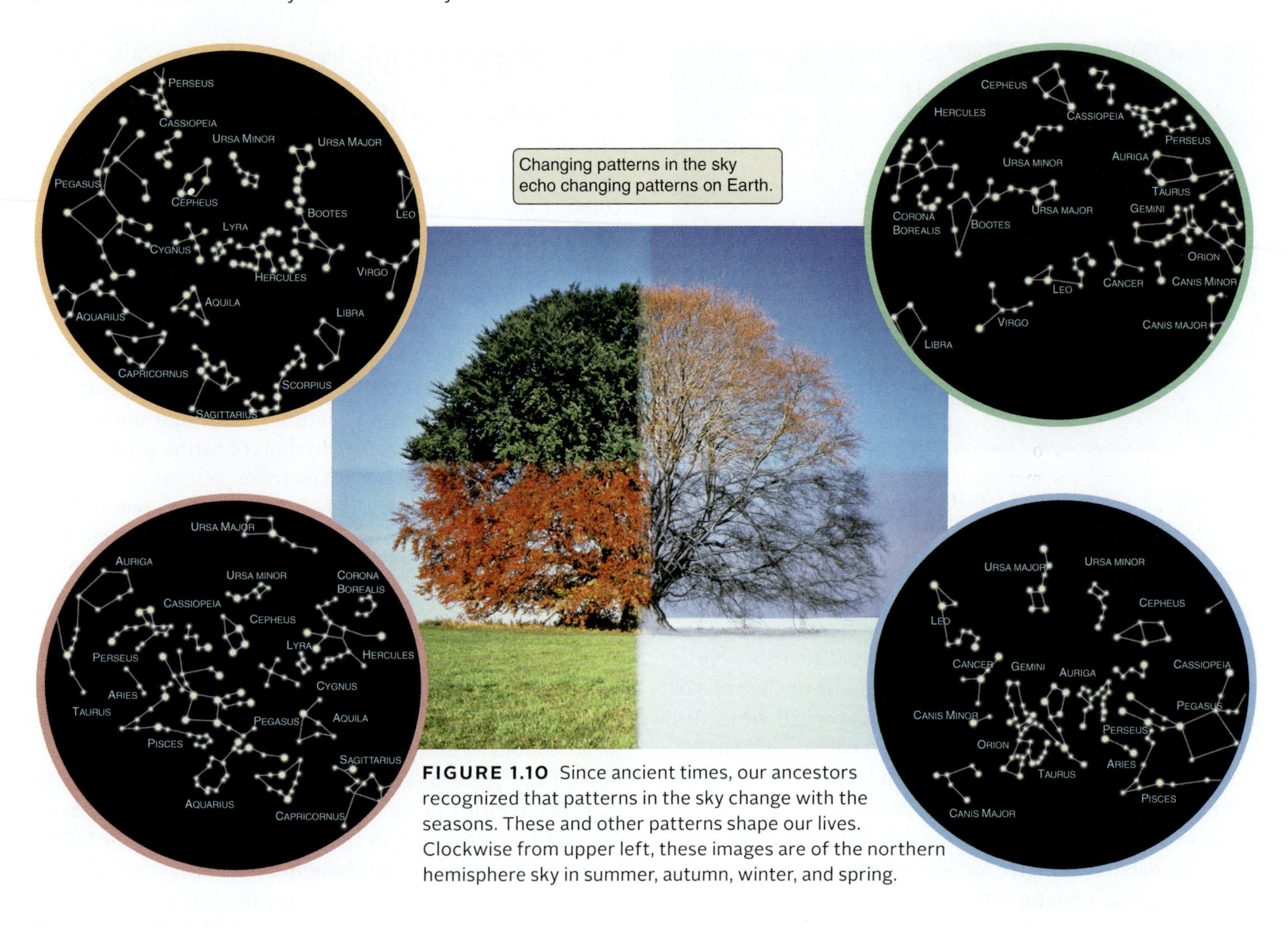

FIGURE 1.10 Since ancient times, our ancestors recognized that patterns in the sky change with the seasons. These and other patterns shape our lives. Clockwise from upper left, these images are of the northern hemisphere sky in summer, autumn, winter, and spring.

mathematical equation, whereas others find mathematics to be a major obstacle to an appreciation of the beauty and elegance of the world as seen through the eyes of a scientist.

While we cannot and will not avoid math, when we use mathematics in this book, we will explain in everyday language what the equations mean and try to show you how equations express concepts that you can connect to the world. We will also limit the mathematics to a few basic tools.

These basic mathematical tools, as described in **Math Tools 1.1**, **Math Tools 1.2**, and Appendixes 1 and 2, enable scientists to communicate complex information. *Basic* does

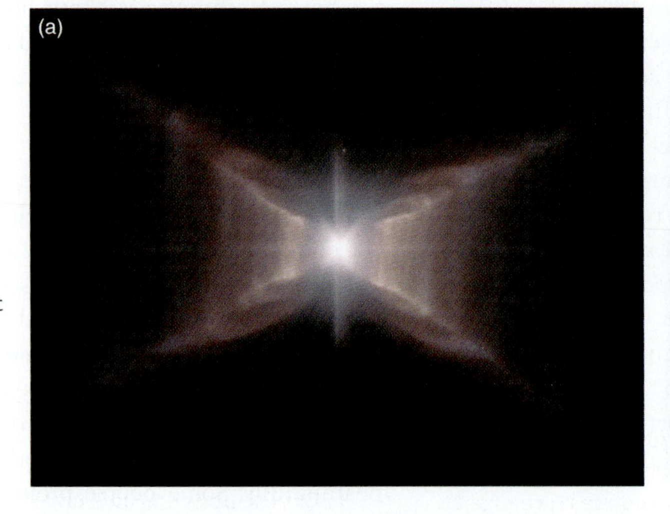

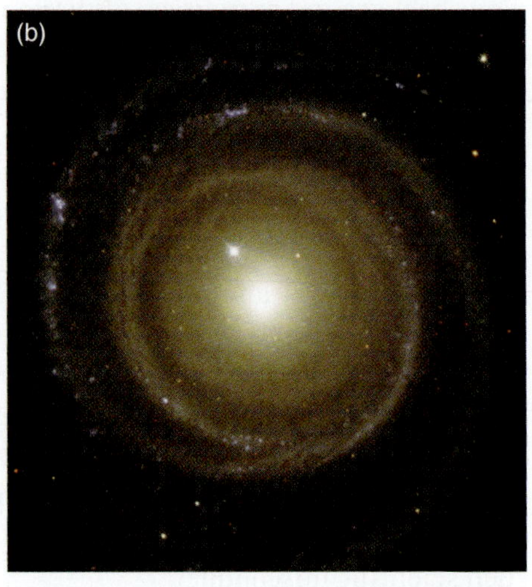

FIGURE 1.11 The patterns in these objects tell us about the physical processes that created them. (a) The Red Rectangle, a protoplanetary nebula around the star HD 44179. (b) The spiral galaxy NGC 4622.

not necessarily mean easy, but it does mean that we will use the most accessible tools that will make our exploration of astronomy as comfortable and informative as possible. It is likely that you know what it means to square a number, or to take its square root, or to raise it to the third power. The mathematics in this book is on a par with what it takes to balance a checkbook, build a bookshelf that stands up straight, check your gas mileage, estimate how long it will take to drive to another city, figure your taxes, or buy enough food to feed an extra guest or two at dinner.

Math Tools 1.1

Mathematical Tools

Mathematics provides scientists many of the tools that they need to understand the patterns they see and to communicate that understanding to others. As the authors of this text, we have worked to keep the math in this book to a minimum. Even so, there are a few tools that we will need in our study of astronomy:

Scientific notation. Scientific notation is how we handle numbers of vastly different sizes. Rather than writing out 7,540,000,000,000,000,000,000 in standard notation, we can express the same number in simpler form as 7.54×10^{21}. Rather than writing out 0.000000000005, we write 5×10^{-12}. For example, the distance to the Sun is 149,600,000 km, but astronomers usually express it as 1.496×10^8 km. (For a more detailed explanation of the use of scientific notation and significant figures, see Appendix 1.)

Ratios. Ratios are a useful way to compare things. A star may be "10 times as massive as the Sun" or "10,000 times as luminous as the Sun." These expressions are ratios.

Geometry. To describe and understand objects in astronomy and physics, we use concepts such as distance, shape, area, and volume. Apparent separations between objects in the sky are expressed as *angles*. Earth's orbit is an *ellipse* with the Sun at one *focus*. The planets in the Solar System lie close to a *plane*. Geometry provides the tools for working with these concepts.

Algebra. In algebra we use and manipulate symbols that represent numbers or quantities. We express relationships that are valid not just for a single case, but for many cases. Algebra lets us conveniently express ideas such as "the distance that you travel is equal to the speed at which you are moving multiplied by the length of time you go at that speed." Written as an algebraic expression, this idea is

$$d = s \times t$$

where d is distance, s is speed, and t is time.

Proportionality. Often, understanding a concept amounts to understanding the *sense* of the relationships that it predicts or describes. "If you have twice as far to go, it will take you twice as long to get here." "If you have half as much money, you will be able to buy only half as much gas." These are examples of proportionality.

A proportion relevant in astronomy is the relationship among speed, time, and distance. If you are traveling at a constant speed, then time is proportional to distance. We write

$$t \propto d$$

where $\propto$ means "is proportional to."

Proportionalities often involve quantities raised to a power. For example, a circle of radius R has an area A equal to πR^2, so we say that the area is proportional to the square of the radius, and we write

$$A \propto R^2$$

This means that if you make the radius of a circle 3 times as large, its area will grow by a factor of 3^2, or 9.

Let's consider a simple astronomical example. The surface area of a sphere is proportional to the square of its radius, or $4\pi R^2$. The radius of the Moon is only about one-quarter that of Earth. How does the surface area of the Moon compare with that of Earth? First, we write the equation for the surface area of a sphere:

$$\text{Surface area} = 4\pi R^2$$

Then we take a ratio of this equation for Earth and the Moon in order to compare:

$$\frac{\text{Surface area}_{\text{Moon}}}{\text{Surface area}_{\text{Earth}}} = \frac{4\pi R^2_{\text{Moon}}}{4\pi R^2_{\text{Earth}}}$$

We can cancel out the constants, 4 and π, from the top and bottom, and rewrite as follows:

$$\frac{\text{Surface area}_{\text{Moon}}}{\text{Surface area}_{\text{Earth}}} = \frac{R^2_{\text{Moon}}}{R^2_{\text{Earth}}} = \left(\frac{R_{\text{Moon}}}{R_{\text{Earth}}}\right)^2$$

We use the given number that the radius of the Moon R_{Moon} is 1/4 of the radius of Earth R_{Earth}; then

$$\frac{R_{\text{Moon}}}{R_{\text{Earth}}} = \frac{1}{4}$$

and then,

$$\frac{\text{Surface area}_{\text{Moon}}}{\text{Surface area}_{\text{Earth}}} = \left(\frac{1}{4}\right)^2 = \frac{1}{16}$$

Thus the Moon has 1/16 of the surface area of Earth.

Units. In this book we use the metric system of units. The United States remains one of very few countries in the world still using the English system of units. Conversions to English units can be found in Appendix 2.

Math Tools 1.2

Reading a Graph

Scientists often convey complex information about relationships through graphs. Graphs typically have two axes: a horizontal axis and a vertical axis. The independent variable (the one you might have control over, in some experiments) is shown on the horizontal axis, while the dependent variable (the one you're studying, in some experiments) is on the vertical axis.

Suppose we plot the distance a car travels over a period of time, as shown in **Figure 1.12a**. In a linear graph, each tick mark on the axis represents the same-sized step. Each step on the horizontal axis of the graph in Figure 1.12a means 1 minute has passed. Each step on the vertical axis means the car has traveled 1 km. Data are plotted on the graph, with one dot for each observation—the distance the car had traveled after 20 minutes, for example.

Drawing a line through these data indicates the trend of the data. Finding the slope of the line gives

$$\text{Slope} = \frac{\text{Change in vertical axis}}{\text{Change in horizontal axis}}$$

$$\text{Slope} = \frac{(15-10)\text{ kilometers}}{(15-10)\text{ minutes}} = 1\text{ km/min}$$

The car is traveling at 1 km per minute, or 60 km/h. The slope of a line often contains extra information that is useful. (*Note:* When finding the slope of the line, you must use the values on the graph itself; do not just count the boxes on the graph paper!)

Many observations of natural processes do not make a straight line on a graph. Of these "nonlinear" processes, one of the most important is the exponential. Think about getting sick. When you get up in the morning at 7:00 A.M. you feel fine, and then at 9:00 A.M. you feel a little tired. By 11:00 A.M., you have a bit of a sore throat or a sniffle and think, "I wonder if I'm getting sick"; and by 1:00 P.M. you are very sick, with a runny nose and congestion and fever and chills. This is a nonlinear process, because the virus that has infected you reproduces nonlinearly.

For the sake of this discussion, suppose the virus produces one copy of itself each time it invades a cell. (In fact, viruses produce between 1,000 and 10,000 copies each time they invade a cell, so the exponential is actually much steeper.) One virus infects a cell and multiplies, so now there are two viruses (the original and a copy). These viruses invade two new cells, and each one produces a copy. Now there are four viruses. After the next cell invasion, there are eight. Then 16, 32, 64, 128, 256, 512, 1,024, 2,048, and so on. This behavior is plotted in **Figure 1.12b**. As a rule, each time you see a graph with a shape like this, you should think "Uh-oh," because the viruses (or whatever else is being observed) are growing at such a rate that they are about to cause serious trouble!

It can be difficult to see what's happening in the early stages of an exponential curve, because the later numbers are so much larger than the earlier ones. For this reason we sometimes plot this type of data *logarithmically*, by putting the logarithm (the power of 10) of the data on the vertical axis (**Figure 1.12c**). Now each step on the axis represents 10 times as many viruses as the previous step. Even though we draw all the steps the same size on the page, they represent different-sized steps in the data—the number of viruses, for example. We often use this technique in astronomy because it has a second, related advantage: very large variations in the data can be easily fit on the same graph.

Each time you see a graph, you should first understand the axes—what data are plotted on this graph? Then you should check whether the axes are linear or logarithmic. Finally, you can look at the actual data or lines in the graph to understand how the system behaves.

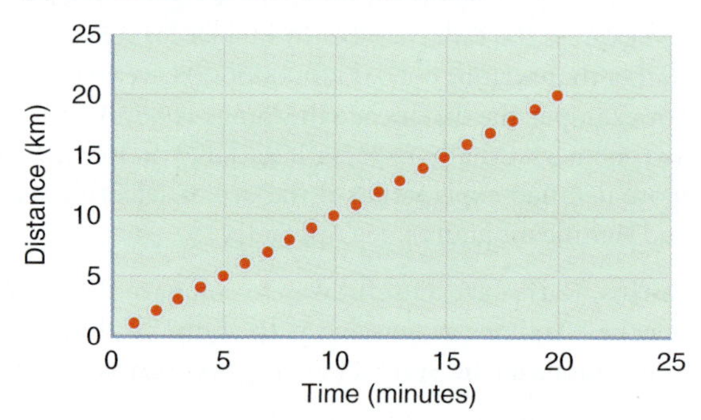

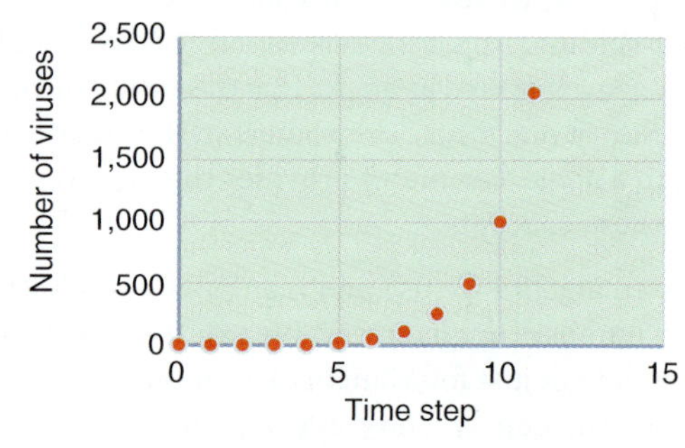

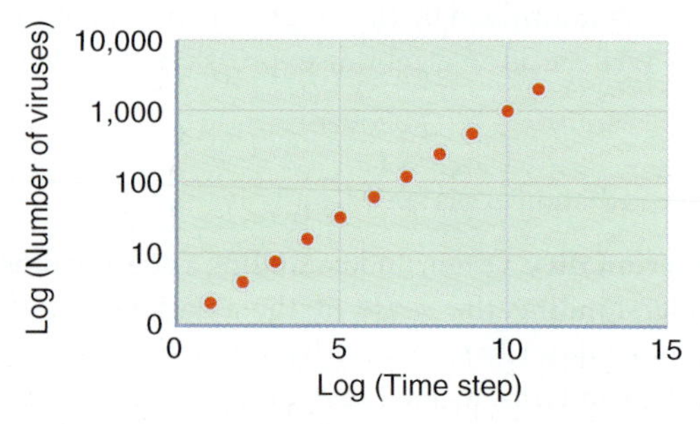

FIGURE 1.12 Graphs like these show relationships between quantities—in (a), time and distance; in (b) and (c), time and number of viruses.

1.5 Thinking like an Astronomer: What Is a Planet?

As the first of many examples of how astronomers think, consider Pluto, which was reclassified from **planet** to **dwarf planet** in 2006. What exactly is a planet? Recall from Section 1.1 that we currently describe our Solar System as having eight planets: Mercury, Venus, Earth, Mars, Jupiter, Saturn, Uranus, and Neptune. The word *planet* comes from the Greek and means "wanderer," because planets were originally observed to change position against the fixed patterns of stars. The only planets the ancient Greeks could observe in those pre-telescopic times were Mercury, Venus, Mars, Jupiter, and Saturn. Other ancient cultures in addition to the Greeks recognized these objects as different from the stars because of their changing positions in the sky.

The more distant and faint planet Uranus had been noticed visually before its official discovery, but it moved only slowly among the stars and thus was not recognized as a planet until the astronomer William Herschel observed it with his telescope in 1781. Neptune is not visible to the naked eye. An astronomer discovered it with his telescope in 1846 after mathematical predictions told him where to look. After the discovery of Uranus and before Neptune, several smaller objects were found circling the Sun between the orbits of Mars and Jupiter—a region now called the **asteroid belt**. Astronomers discovered the first object, called Ceres, in 1801 and then Pallas in 1802. Herschel noted that these objects were small and appeared "starlike" and named them **asteroids**. Other astronomers disagreed, and included them in their list of planets. Next to be discovered was Juno, in 1804, and Vesta in 1807, both small and in similar orbits. Astronomy texts of the early 19th century listed 11 planets! Another asteroid was found in 1845, followed by three in 1847, and yet another seven by 1851. The U.S. Naval Observatory used the term *asteroids* to label these small objects until 1868, then *small planets* until 1892, then *minor planets* in 1900, and then back to *asteroids* in 1929.

When first discovered in 1930, Pluto was thought to be about the same size and mass as Earth, and it was designated the "ninth planet." As astronomers' ability to measure Pluto's properties improved over time, however, they realized it had to be much smaller than Earth. By 1948 its mass was estimated to be no more than 10 percent of Earth's mass; and by 1976, only 1 percent. Following the discovery of its moon Charon in 1978, astronomers were eventually able to determine that Pluto is only about a fifth as large as Earth and has a mass that is only 0.2 percent that of Earth. Pluto is larger than the asteroids but smaller than the other planets by far. Some astronomers began to question whether Pluto should be considered a true planet. A debate ensued within the astronomical community. In 2005 that dispute heated up with the discovery of a distant planetary body about the same size as Pluto. The newly discovered object, eventually named Eris (for the Greek goddess of discord), was thought to have a mass even greater than Pluto's.

Astronomers were forced to decide whether Eris should be considered a tenth planet or whether instead Pluto should be redefined as a new and distinctly different type of planetary body. In August 2006 the International Astronomical Union (IAU) officially redefined what astronomers mean when they call something a planet (see Appendix 8). Under this new definition, Pluto became a "dwarf planet," along with Ceres and Eris and several other small bodies discussed in Chapter 12. To astronomers, Pluto is still Pluto; only its label has changed. It still holds the same high degree of scientific interest that it held before the IAU reclassification. (Many cartoons about this change appeared in the media; see, for example, **Figure 1.13**). Because astronomers reclassified Pluto as a dwarf planet in light of new evidence regarding an additional kind of Solar System body, this reclassification is an example of the scientific method in action.

1.6 Origins: An Introduction

How and when did the universe begin? What combination of events led to humans' existence as sentient beings living on a small rocky planet orbiting a typical middle-aged star? Are there others like us scattered throughout the galaxy?

Recall that earlier in this chapter we mentioned the theme of origins. Throughout this book you will see that this theme involves much more than how humans came to be on Earth. We'll look into the origin of life on Earth, but we will also examine the possibilities of life elsewhere in the Solar

FIGURE 1.13 Many cartoons similar to this one appeared after Pluto was reclassified as a dwarf planet.

System and beyond—a subject called **astrobiology**. Our origins theme will include the discovery of planets around other stars and how they compare with the planets of our own Solar System.

For example, you learned in Section 1.1 that the early universe was made up almost entirely of two elements—hydrogen and helium—with just a touch of lithium, beryllium, and boron thrown in. Further along you will see that hydrogen and helium were the big winners in a minutes-old universe that had barely cooled down from the Big Bang, and how they were later transformed into heavier elements such as carbon, nitrogen, oxygen, sulfur, and phosphorus—the very atoms that make up the molecules of life on Earth. You will find that some elements were created in the cores of stars like the Sun, whereas others had their origin in the violence of the explosions of more massive stars. When all this is put together, you will see that terrestrial life, Earth itself, the other planets in the Solar System, and those beyond have a common origin: All are made of recycled stardust.

The study of origins provides examples of the process of science as well. Many of the physical processes in chemistry, geology, planetary science, physics, and astronomy that are seen on Earth or in the Solar System are observed across the galaxy and throughout the universe. But as of this writing, the only biology we know about is that which exists on planet Earth. Thus, at this point in human history, much of what scientists can say about the origin of life on Earth and the possibility of life elsewhere is reasoned extrapolation and educated speculation. In the origins sections we'll address some of these hypotheses, and try to be clear about which are speculative and what is known. ■

Summary

1.1 We reside on a planet orbiting a star at the center of a solar system in a vast galaxy that is one of many galaxies in the universe. We are a product of the universe; the very atoms we're made of were formed in stars that died long before the Sun and Earth were formed.

1.2 Astronomers use telescopes, spacecraft, particle colliders, and computers to study the universe and understand the physical laws that govern the behavior of matter and energy.

1.3 The scientific method consists of observation, followed by hypothesis, prediction, further observation or experiments to test the prediction, and ultimately a tested theory. The scientific method is a way of trying to *falsify*, not prove, ideas. *All* scientific knowledge is provisional. Like art, literature, and music, science is a creative human activity; it is also a remarkably powerful, successful, and aesthetically beautiful way of viewing the world.

1.4 Mathematics provides many of the tools that we need to understand the patterns we see and to communicate that understanding to others.

1.5 Pluto's reclassification as a dwarf planet is an example of the scientific method in action.

1.6 All life on Earth and the planets around us are recycled stardust; our atoms were processed inside stars to become the elements that form our bodies today.

Unanswered Questions

- What makes up the universe? We have listed planets, stars, and galaxies as making up the cosmos, but recall from Section 1.1 that astronomers now have evidence that 95 percent of the universe is composed of dark matter and dark energy, and we do not understand what those are. Scientists are using the largest telescopes and particle colliders on Earth, as well as telescopes and experiments in space, to explore what makes up dark matter and what constitutes dark energy.
- Is the type of structure we call life as common in the universe as the planets, stars, and galaxies that we observe? At the time of this writing there is no scientific evidence that life exists on any other planet. Our universe is enormously large and has existed for a great length of time. What if life is too far away, or existed too long ago, for us to ever "meet"?

Questions and Problems

Summary Self-Test

1. Rank the following in order of increasing size.
   a. Local Group
   b. Milky Way
   c. Solar System
   d. universe
   e. Sun
   f. Earth
   g. Virgo Supercluster

2. A light-year is a unit of distance most comparable to the distance from Earth to the
   a. Moon.
   b. Sun.
   c. outer Solar System.
   d. nearest star.
   e. nearest galaxy.

3. The following astronomical events led to the formation of life on Earth—including you. Place them in order of their occurrence over astronomical time.
   a. Stars died and distributed heavy elements into the space between the stars.
   b. Hydrogen and helium were made in the Big Bang.
   c. Enriched dust and gas gathered into clouds in interstellar space.
   d. Stars were born, and they processed light elements into heavier ones.
   e. The Sun and planets formed from a cloud of interstellar dust and gas.

4. Suppose you set your keys on the table and left the room. When you came back, they were on the ground. Which of the following is most likely to be true, given the application of Occam's razor?
   a. Invisible flying elephants carried them to the floor.
   b. They just walked off by themselves.
   c. Someone bumped the table and knocked them off.
   d. A violent wind blew them off the table.

5. Suppose the cosmological principle did *not* apply. Then we could say that
   a. planets orbit other stars because of gravity.
   b. other stars shine by the same process as the Sun.
   c. other galaxies have the same chemical elements in them.
   d. light travels in a straight line until it hits something.
   e. none of the above

6. The scientific method is a way of trying to ________, not prove, ideas; thus, all scientific knowledge is provisional.

7. *Understanding* in science means that
   a. we have accumulated lots of facts.
   b. we are able to connect facts through an underlying idea.
   c. we are able to predict events on the basis of accumulated facts.
   d. we are able to predict events on the basis of an underlying idea.

8. The fact that scientific revolutions take place means that
   a. all the science we know is wrong.
   b. the science we know now is more correct than it was in the past.
   c. scientists start out lying, and then get caught.
   d. you can never really know anything about the universe; it's all relative.
   e. the universe keeps changing the rules.

9. If we compare Earth's place in the universe with a very distant place, we can say that
   a. the laws of physics are different in each place.
   b. some laws of physics are different in each place.
   c. all the laws of physics are the same in each place.
   d. some laws of physics are the same in each place, but we don't know about others.

10. ________ is the language of patterns and of science.

True/False and Multiple Choice

11. **T/F:** A scientific theory can be tested by observations and proven to be true.

12. **T/F:** A pattern in nature can reveal an underlying physical law.

13. **T/F:** A theory, in science, is a guess about what might be true.

14. **T/F:** Once a theory is proven in science, scientists stop testing it.

15. **T/F:** Scientific disputes are solved by debate.

16. The Sun is part of
    a. the Solar System.
    b. the Milky Way Galaxy.
    c. the universe.
    d. all of the above

17. A light-year is a measure of
    a. distance.
    b. time.
    c. speed.
    d. mass.

18. Which of the following was *not* made in the Big Bang?
    a. hydrogen
    b. lithium
    c. beryllium
    d. carbon

19. Occam's razor states that
    a. the universe is expanding in all directions.
    b. the laws of nature are the same everywhere in the universe.
    c. if two hypotheses fit the facts equally well, the simpler one is the more likely to apply.
    d. patterns in nature are really manifestations of random occurrences.

20. The circumference of Earth is $\frac{1}{7}$ of a light-second. Therefore,
    a. if you were traveling at the speed of light, you would travel around Earth 7 times in 1 second.
    b. light travels a distance equal to Earth's circumference in $\frac{1}{7}$ second.
    c. neither a nor b
    d. both a and b

21. "We are stardust" means that
    a. Earth exists because of the collision of two stars.
    b. the atoms in our bodies have passed through (and in many cases formed in) stars.
    c. Earth is formed of material that used to be in the Sun.
    d. Earth and the other planets will eventually form a star.

22. According to the graphs in Figures 1.12b and c, by how much did the number of viruses increase in four time steps?
    a. It doubled.
    b. It tripled.
    c. It quadrupled.
    d. It went up more than 10 times.

23. Any explanation of a phenomenon that includes a supernatural influence is not scientific because
    a. it does not have a hypothesis.
    b. it is wrong.
    c. people who believe in the supernatural are not credible.
    d. science is the study of the natural world.

24. "All scientific knowledge is provisional." In this context, *provisional* means
    a. "wrong."
    b. "relative."
    c. "tentative."
    d. "incomplete."

25. When we observe a star that is 10 light-years away, we are seeing that star
    a. as it is today.
    b. as it was 10 days ago.
    c. as it was 10 years ago.
    d. as it was 20 years ago.

Thinking about the Concepts

26. Suppose you lived on imaginary planet Zorg orbiting Alpha Centauri, a nearby star. How would you write your cosmic address?

27. Imagine yourself living on a planet orbiting a star in a very distant galaxy. What does the cosmological principle tell you about the physical laws at this distant location?

28. If the Sun suddenly exploded, how soon after the explosion would we know about it?

29. If a star exploded in the Andromeda Galaxy, how long would it take that information to reach Earth?

30. It is said that we are made of stardust. Explain why this statement is true.

31. Some people have proposed the theory that Earth was visited by extraterrestrials (aliens) in the remote past. Can you think of any tests that could support or refute that theory? Is it falsifiable? Would you regard this proposal as science or pseudoscience?

32. What does the word *falsifiable* mean? Provide an example of an idea that is not falsifiable. Provide an example of an idea that is falsifiable.

33. Explain how the word *theory* is used differently in the context of science than in common everyday language.

34. What is the difference between a *hypothesis* and a *theory* in science?

35. Suppose the tabloid newspaper at your local supermarket claimed that, compared to average children, the children born under a full Moon become better students.
    a. Is this theory falsifiable?
    b. If so, how could it be tested?

36. A textbook published in 1945 stated that light takes 800,000 years to reach Earth from the Andromeda Galaxy. In this book we assert that it takes 2,500,000 years. What does this difference tell you about a scientific "fact" and how our knowledge evolves with time?

37. Astrology makes testable predictions. For example, it predicts that the horoscope for your star sign on any day should fit you better than do horoscopes for other star signs. Read the daily horoscopes for all of the astrological signs in a newspaper or online. How many of them might fit the day you had yesterday? Repeat the experiment every day for a week and keep a record of which horoscopes fit your day each day. Was your horoscope sign consistently the best description of your experiences?

38. A scientist on television states that it is a known fact that life does not exist beyond Earth. Would you consider this scientist reputable? Explain your answer.

39. Some astrologers use elaborate mathematical formulas and procedures to predict the future. Does this show that astrology is a science? Why or why not?

40. Why was Pluto reclassified as a dwarf planet?

Applying the Concepts

41. Convert the following numbers to scientific notation:
    a. 7,000,000,000
    b. 0.00346
    c. 1,238

42. Convert the following numbers to standard notation:
    a. 5.34×10^8
    b. 4.1×10^3
    c. 6.24×10^{-5}

43. If a car is traveling at 35 km/h, how far does it travel in
    a. one hour?
    b. half an hour?
    c. one minute?

44. The surface area of a sphere is proportional to the square of its radius. How many times larger is the surface area if the radius is
    a. doubled?
    b. tripled?
    c. halved (divided by 2)?
    d. divided by 3?

45. The average distance from Earth to the Moon is 384,000 km. How many days would it take you, traveling at 800 km/h—the typical speed of jet aircraft—to reach the Moon?

46. (a) If it takes about 8 minutes for light to travel from the Sun to Earth, and Pluto is 40 times farther from Earth than the Sun is, how long does it take light to reach Earth from Pluto? (b) Radio waves travel at the speed of light. What does this fact imply about the problems you would have if you tried to conduct a two-way conversation between Earth and a spacecraft orbiting Pluto?

47. Study the Process of Science figure on page 10. Make a similar flowchart for your behavior while sitting at the computer for an hour. It might include, for example: checking Facebook, looking at blogs, working on homework, checking email, looking things up on Wikipedia. Many of these things are done multiple times. Make sure you show this repetition on the flowchart.

48. The surface area of a sphere is proportional to the square of its radius. The radius of the Moon is only about one-quarter that of Earth. How does the surface area of the Moon compare with that of Earth?

49. A remote Web page may sometimes reach your computer by going through a satellite orbiting approximately 3.6×10^4 km above Earth's surface. What is the minimum delay, in seconds, that the Web page takes to show up on your computer?

50. Imagine that you have become a biologist, studying rats in Indonesia. Most of the time, Indonesian rats maintain a constant population. Every half century, however, these rats suddenly begin to multiply exponentially! Then the population crashes back to the constant level. Sketch a graph that shows the rat population over two of these episodes.

51. The number of pages in a science journal published every year increased exponentially from 625 pages in 1940 to 14,000 pages in 1980. Sketch two graphs of this increase: one with linear axes, and one with logarithmic axes.

52. Some theorize that a tray of hot water will freeze more quickly than a tray of cold water when both are placed in a freezer.
    a. Does this theory make sense to you?
    b. Is the theory falsifiable?
    c. Do the experiment yourself. Note the results. Was your intuition borne out?

53. A pizzeria offers a 9-inch-diameter pizza for $12 and an 18-inch-diameter pizza for $24. Are both offerings equally economical? If not, which is the better deal? Explain your reasoning.

54. The circumference of a circle is given by $C = 2\pi r$, where r is the radius of the circle.
    a. Calculate the approximate circumference of Earth's orbit around the Sun, assuming that the orbit is a circle with a radius of 1.5×10^8 km.
    b. Noting that there are 8,766 hours in a year, how fast, in kilometers per hour, does Earth move in its orbit?
    c. How far along in its orbit does Earth move in one day?

55. Gasoline is sold by the gallon in the United States and by the liter nearly everywhere else in the world. There are approximately 3.8 liters in a gallon. If the price of gas is $4 per gallon, how much does it cost per liter?

Using the Web

56. Go to the Astronomy Picture of the Day (APOD) app or website (http://apod.nasa.gov/apod) and click on "Archive" to look at the recent pictures and videos. Submissions to this website come from all around the world. Pick one and read the explanation. Was the image or video taken from Earth or from space? Is it a combination of several images? Does it show Earth, our Solar System, objects in our Milky Way Galaxy, more distant galaxies, or something else? Is the explanation understandable to someone who has not studied astronomy? Do you think this website promotes a general interest in astronomy?

57. Go to the Hayden Planetarium website (http://www.haydenplanetarium.org/universe) and view the short video *The Known Universe*, which takes the viewer on a journey from the Himalayan mountains to the most distant galaxies. Do you think the video is effective for showing the size and scale of the universe?

58. A similar film produced in 1996 in IMAX, *Cosmic Voyage*, can be found online at http://topdocumentaryfilms.com/cosmic-voyage. Watch the "powers of ten" zoom out to the cosmos, starting at the 7-minute mark, for about 5 minutes. Do the "powers of ten" circles add to your understanding of the size and scale of the universe? (The original film *Powers of Ten*, a 1968 documentary, can be viewed online at http://www.powersof10.com/film, but notably it extends only a hundredth as far as the newer ones.)

59. Throughout this book we will examine how discoveries in astronomy and space are covered in the media. Go to your favorite news website (or to one assigned by your instructor) and find a recent article about astronomy or space. Does this website have a separate section for science? Is the article you selected based on a press release, on interviews with scientists, or on an article in a scientific journal? Use Google News or the equivalent to see how widespread the coverage of this story is. Have many newspapers carried it? Has it been picked up internationally? Has it been discussed in blogs? Do you think this story was interesting enough to be covered?

60. Go to a blog about astronomy or space. There are good collections at http://www.scienceblogs.com and http://blogs.discovermagazine.com. Is the blogger a scientist, a science writer, a student, or an enthusiastic amateur astronomer? What is the current topic of interest? Is it controversial? Are readers making many comments? Is this blog something you would want to read again?

STUDYSPACE is a free and open website that provides a Study Plan for each chapter of *21st Century Astronomy*. Study Plans include animations, reading outlines, vocabulary flashcards, and multiple-choice quizzes, plus links to premium content in SmartWork and the ebook. Visit **wwnorton.com/studyspace**.

SMARTWORK Norton's online homework system, includes algorithmically generated versions of these questions, plus additional conceptual exercises. If your instructor assigns questions in SmartWork, log in at **smartwork.wwnorton.com.**

Exploration | Logical Fallacies

Logic is fundamental to the study of science and to scientific thinking. A logical fallacy is an error in reasoning, which good scientific thinking avoids. For example, "because Einstein said so" is not an adequate argument. No matter how famous the scientist is (even if he is Einstein), he or she must still supply a logical argument and evidence to support a claim. Anyone who claims that something must be true because Einstein said it has committed the logical fallacy known as an "appeal to authority." There are many types of logical fallacies, but a few of them crop up often enough in discussions about science that you should be aware of them.

Ad hominem. In an *ad hominem* fallacy, you attack the person who is making the argument, instead of the argument itself. Here is an extreme example of an *ad hominem* argument: "A famous scientist says Earth is warming. But I think the scientist is an idiot. So Earth can't be warming."

Appeal to belief. This fallacy has the general pattern "Most people believe X is true; therefore X is true." For example, "Most people believe Earth orbits the Sun. Therefore, Earth orbits the Sun." Note that even if the conclusion is correct, you may still have committed a logical fallacy in your argument.

Biased sample. If a sample drawn from a pool has a bias, then conclusions about the sample cannot be applied to a larger pool. For example, imagine you poll students at your university and find that 30 percent of them visit the library one or more times per week. Then you conclude that 30 percent of Americans visit the library one or more times per week. You have committed the "biased sample" fallacy, because university students are not a representative sample of the American public.

Begging the question. In this fallacy (also known as circular reasoning), you assume the claim is true and then use this assumption as evidence to prove the claim is true. For example, "I am trustworthy; therefore I must be telling the truth." No real evidence is presented for the conclusion.

Post hoc ergo propter hoc. *Post hoc ergo propter hoc* is Latin for "after this, therefore because of this." Just because one thing follows another doesn't mean that one caused the other. For example, "There was an eclipse and then the king died. Therefore, the eclipse killed the king." This fallacy is often connected to related inverse reasoning: "If we can prevent an eclipse, the king won't die."

Slippery slope. In this fallacy you claim that a chain reaction of events will take place, inevitably leading to a conclusion that no one could want. For example, "If I don't get A's in all of my classes, I will not ever be able to get into graduate school, and then I won't ever be able to get a good job, and then I will be living in a van down by the river until I'm old and toothless and starve to death." None of these steps actually follows inevitably from the one before.

Following are some examples of logical fallacies. Identify the type of fallacy represented. Each of the fallacies we just discussed is represented once.

1 **You get a chain email threatening terrible consequences if you break the chain. You move it to your spam box. Later that day you get in a car accident. The following morning, you retrieve the chain email and send it along.**

....

....

....

2 **If I get question 1 on the assignment wrong, then I'll get question 2 wrong as well, and before you know it, I will never catch up in the class.**

....

....

....

3 **All my friends love the band Degenerate Electrons. Therefore, all people my age love this band.**

....

....

....

4 **Eighty percent of Americans believe in the tooth fairy. Therefore, the tooth fairy exists.**

....

....

....

5 **My professor says that the universe is expanding. But my professor is a geek, and I don't like geeks. So the universe can't be expanding.**

....

....

....

6 **When applying for jobs, you use a friend as a reference. Prospective employers ask you how they know your friend is trustworthy, and you say, "I can vouch for her."**

....

....

....

The Moon over the Parthenon
in Athens, Greece.

02 Patterns in the Sky—Motions of Earth

. . . marking the conclave of all the night's stars,

those potentates blazing in the heavens

that bring winter and summer to mortal men

the constellations, when they wane, when they rise.

Aeschylus (525–456 BCE)

LEARNING GOALS

The stars first found a special place in legend and mythology as the realm of gods and goddesses. From these legends and myths, people created constellations—pictures in the sky formed by bright stars—and noticed how the constellations in the sky changed over the course of a year. Now we can explain how changes in the sky are the unavoidable consequences of the motions of Earth, the Sun, and the Moon. By the conclusion of this chapter, you should be able to:

- Explain how the stars appear to move through the sky, and how the motion of the stars differs when viewed from different latitudes on Earth.
- Explain why there are different seasons throughout the year.
- Summarize how the motion of the Moon in its orbit about Earth, together with the motion of Earth and the Moon around the Sun, shapes the phases of the Moon and the spectacle of eclipses.
- Sketch the alignment of Earth, Moon, and Sun during different phases of the Moon.

2.1 A View from Long Ago

Ancient humans may not have known that they were "stardust," but they had some sense that there was a connection between their lives on Earth and the sky above. Before our modern technological civilization, people's lives were more attuned to the ebb and flow of nature, and the patterns in the sky are a part of that ebb and flow. The repeating patterns of the Sun, Moon, and stars echo the rhythms that have long defined the lives of humans. By watching the patterns in the sky, people found that they could predict when the seasons would change and the rains would come and the crops would grow. Ancient astronomers with knowledge of the sky, priests and priestesses, natural philosophers and explorers—all had additional knowledge of the world, and knowledge of the world was power. Some of these early observations and ideas about these patterns live on today in the names of stars and constellations, in calendars based on the Moon and Sun that are still used by many cultures, in the astronomical names of the seven days of the week, and in belief in **astrology**.

How have patterns in the sky been important to our species?

From Mesopotamia to Africa, from Europe to Asia, from the Americas to the British Isles, the archaeological record holds evidence of early humans who projected ideas from their own cultures onto what they saw in the sky. There have been as many different sets of **constellations** (groups of stars that form recognizable patterns), and stories to go with them, as there have been cultural traditions in human history (**Figure 2.1**). But how were these winged horses, dragons, and other imaginary images formed from patterns of stars? If you look at the sky, no clear pictures of these images emerge. Instead there is only the random pattern of stars—about 5,000 of them visible to the naked eye—spread out across the sky. Constellations are creations of the human imagination. They are the ideas and pictures that humans impose on the lights in the sky in an effort to connect their lives on Earth with the workings of the "heavens."

Modern constellations visible from the Northern Hemisphere draw heavily from the list of constellations compiled 2,000 years ago by the Alexandrian astronomer Ptolemy. Modern constellation names in the southern sky are drawn from the lists put together by European explorers visiting the Southern Hemisphere during the 17th and 18th centuries. Today astronomers use an officially sanctioned set of constellations as a kind of road map of the sky. The entire sky is broken up into 88 different constellations, much as continental landmasses are divided into countries by invisible lines. Every star in the sky lies within the borders of a single constellation, and the names of constellations are used in naming the stars that lie within their boundaries. For example, Sirius, the brightest star in the sky, lies within the boundaries of the constellation Canis Major (meaning the "big dog"). The official name of Sirius is Alpha Canis Majoris (see Appendix 6), indicating that it is the brightest star in that constellation and earning it its nickname: the "Dog Star." Appendix 5 contains a list of the nearest and brightest stars. Appendix 6 provides sky maps showing the constellations.

Ancient cultures built structures that were sometimes used to study astronomical positions and events. One of the most famous of these is Stonehenge, a massive prehistoric stone artifact in the English countryside built in stages during the years 3000–1600 BCE (**Figure 2.2a**). Historians are still debating the cultural function of the structure, but it does seem to have an astronomical purpose, such as predicting the eclipses, solstices, and equinoxes discussed later in this chapter. Archaeology with astronomical alignments is found in the ziggurats in Mesopotamia, the pyramids in Egypt and Mexico, the structure housing the Black Stone in Mecca,

(a)

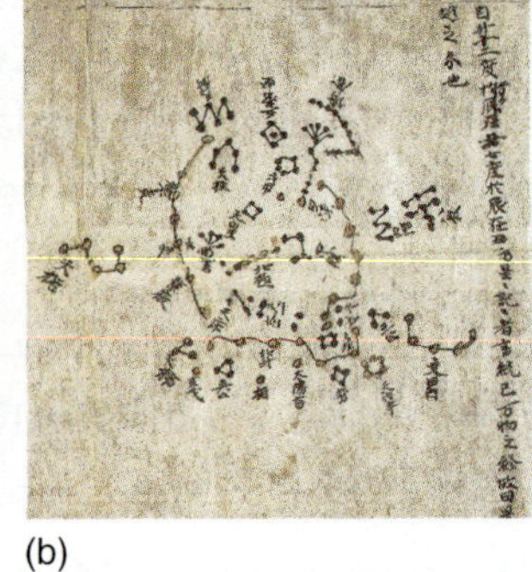
(b)

(c)

FIGURE 2.1 The region of the sky near what we now call the Big Dipper (Ursa Major, or the "Great Bear") as viewed by three different civilizations: (a) Egyptian, 1275 BCE; (b) Chinese, 840 CE; (c) European, 1540 CE. Constellations are groupings of stars whose images, such as a dipper or bear, exist only in the human mind.

(a)

(b)

(c)

FIGURE 2.2 (a) Stonehenge is an ancient construction in the English countryside. One of its suspected uses 4,000 years ago was to keep track of celestial events. (b) Chichén Itzá observatory in Mexico. (c) Beijing Ancient Observatory in China.

the Forbidden City of China, the temple of Angkor Wat in Cambodia, the Pantheon of Rome and Parthenon of Athens, and structures in the American Southwest. Pre-telescopic astronomical observatories to study the sky for timekeeping, navigation, and so forth are found across the globe, including Mount Qasioun near Damascus (830 CE), the Mayan El Caracol at Chichén Itzá in Mexico (906 CE—**Figure 2.2b**), the Maragheh observatory in Iran (1259), the Mongol observatory (1279) and the Beijing Ancient Observatory (1442—**Figure 2.2c**) in China, Ulugh Beg's observatory in Samarkand (1420), and Jantar Mantar in India (1727). Many of these are national historical or UNESCO World Heritage sites.

Ancient cultures built structures to study the skies.

2.2 Earth Spins on Its Axis

When the earliest humans first noticed the sky, it is likely that the daily motion of the Sun is what drew their attention. As civilizations became more aware of the often complex motions of the Sun, Moon, planets, and stars, early astronomers developed models to explain what they saw. The ancient Greeks, for example, devised **geocentric** models of the universe in which heavenly bodies were embedded in a celestial sphere that revolved around a stationary Earth. The most successful model was Ptolemy's, which survived for nearly 1,500 years, until finally overthrown by Copernicus in the early Renaissance (see Chapter 3).

Despite the stories you may have learned in grade school, Christopher Columbus did not discover that the world is not flat. Long before his journey to the New World, anyone who had read Aristotle or other Greek philosophers (as had Columbus) knew that Earth is a ball. Far more difficult to accept was the idea that the changes occurring in the sky from day to day and month to month are the result of the motions of Earth rather than the motion of the Sun and stars around Earth. The most apparent among these motions is Earth's rotation on its axis, which determines the very rhythm of life on Earth—the passage of day and night.

One reason the ancients did not believe that Earth rotates is that they could not perceive the spinning motion of Earth. In fact, as a result of Earth's rotation, the surface of Earth is moving along at a respectable speed—1,674 kilometers per hour (km/h) at the equator (which we calculate by dividing the circumference of Earth by the period of its rotation). Even so, we do not feel that motion any more than we would "feel" the speed of a car with a perfectly smooth ride cruising down a straight highway. (Some measurable effects of Earth's rotation are discussed in **Connections 2.1**.)

Connections 2.1

Relative Motions and Frame of Reference

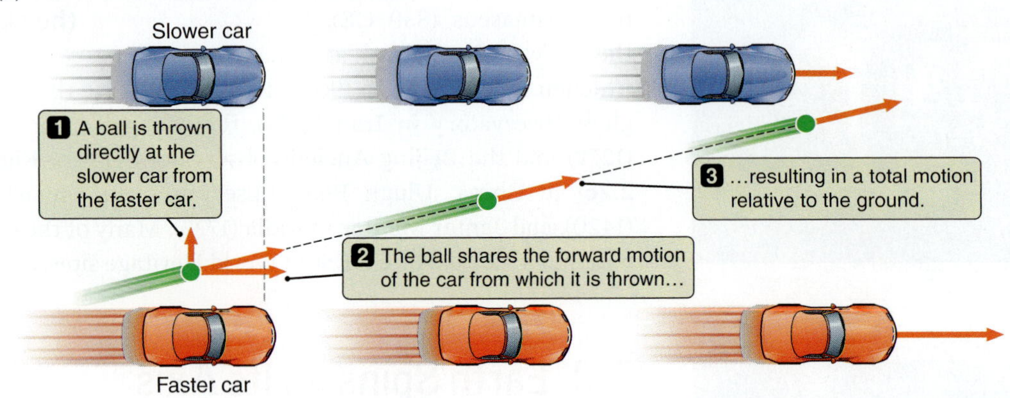

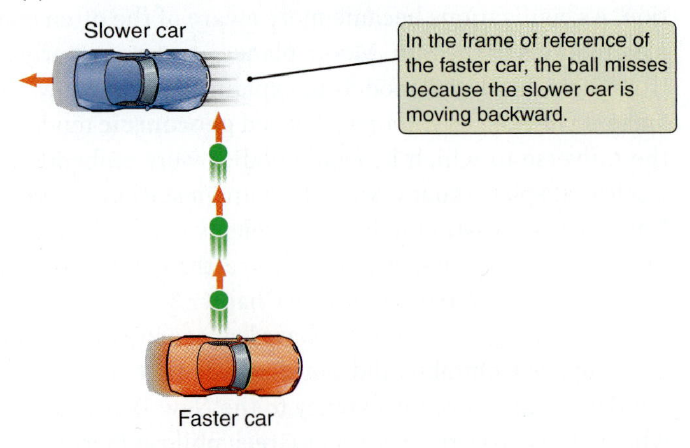

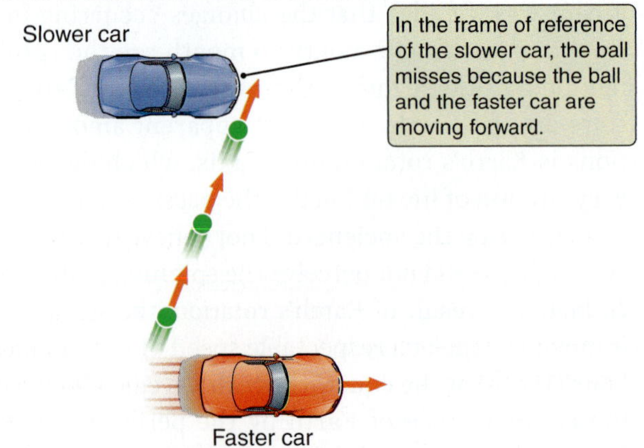

FIGURE 2.3 The motion of an object depends on the frame of reference of the observer.

Earth's rotation actually influences things as diverse as the motion of weather patterns on Earth and how an artillery gunner must aim at a distant target. Any object sitting on the surface of Earth follows a circle each day as Earth rotates on its axis. This circle is larger for objects near Earth's equator and smaller for objects closer to one of Earth's poles. But because Earth is a solid body, all objects must complete their circular motion in exactly one day. Because an object closer to the equator has farther to go each day than does an object nearer a pole, the object nearer the equator must be moving *faster* than the object at a higher latitude (**Figure 2.4a**). If an object starts out at one latitude and then moves to another, its apparent motion over the surface of Earth is influenced by this difference in speed.

Imagine that you are riding in a car traveling down a straight section of highway at a constant speed. The windows are blacked over so that you cannot see the scenery go by. If you are not looking out the window or feeling road vibrations, there is no experiment you can easily do to tell the difference between riding in a car down a straight section of highway at constant speed and sitting in the car while it is parked in your driveway. Because everything in the car is moving together, the **relative motions** between objects in the car are all that count.

The idea that only relative motions count occurs again and again in astronomy and physics. There are numerous examples in this chapter alone. For example, even though Earth is spinning on its axis and flying through space in its orbit about the Sun, the resulting relative motions between objects that are near each other on Earth are very small. This is why we do not notice Earth's motion. It's also helpful to understand the concept of a **frame of reference**. Briefly, a frame of reference is a coordinate system within which an observer measures positions and motions.

Now imagine that two cars are driving down the road at *different* speeds, as shown in **Figure 2.3**. Ignoring for the moment any real-world complications, like wind resistance, if you were to throw a ball from the faster-moving car directly at the slower-moving car as the two cars passed, you would miss. The ball shares the forward motion of the faster car, so the ball outruns the forward motion of the slower car. From your perspective in the faster car (**Figure 2.3b**), the slower car lagged behind the ball. From the slower car's perspective (**Figure 2.3c**), your car and the ball sped on ahead.

Now do the same experiment again, but instead of two cars, think about two locations at different latitudes. Suppose you fire a cannon directly north from a point in the Northern Hemisphere, as shown in **Figure 2.4b**. Because the cannon is located nearer to the equator than its target is, the cannon itself is moving toward the east faster than its target. Even though the cannonball is fired toward the north, it shares the eastward velocity of the cannon itself. This means that the cannonball is *also* moving toward the east faster than its target. Recall how the ball thrown from the faster car outpaced the slower-moving car in Figure 2.3. Similarly, as the cannonball flies north, it moves toward the east faster than the ground underneath it does. To an observer on the ground, the cannonball appears to curve toward the east as it outruns the eastward motion of the ground it is crossing. The farther north the cannonball flies, the greater is the difference between its eastward velocity and the eastward velocity of the ground. Thus, the cannonball follows a path that appears to curve more and more to the east the farther north it goes. If you are located in the Northern Hemisphere and fire a cannonball *south* toward the equator (**Figure 2.4c**), the opposite effect will occur. Now the cannon is moving toward the east more slowly than its target. As the cannonball flies toward the south, its eastward motion lags behind that of the ground underneath it, and the cannonball appears to curve toward the west.

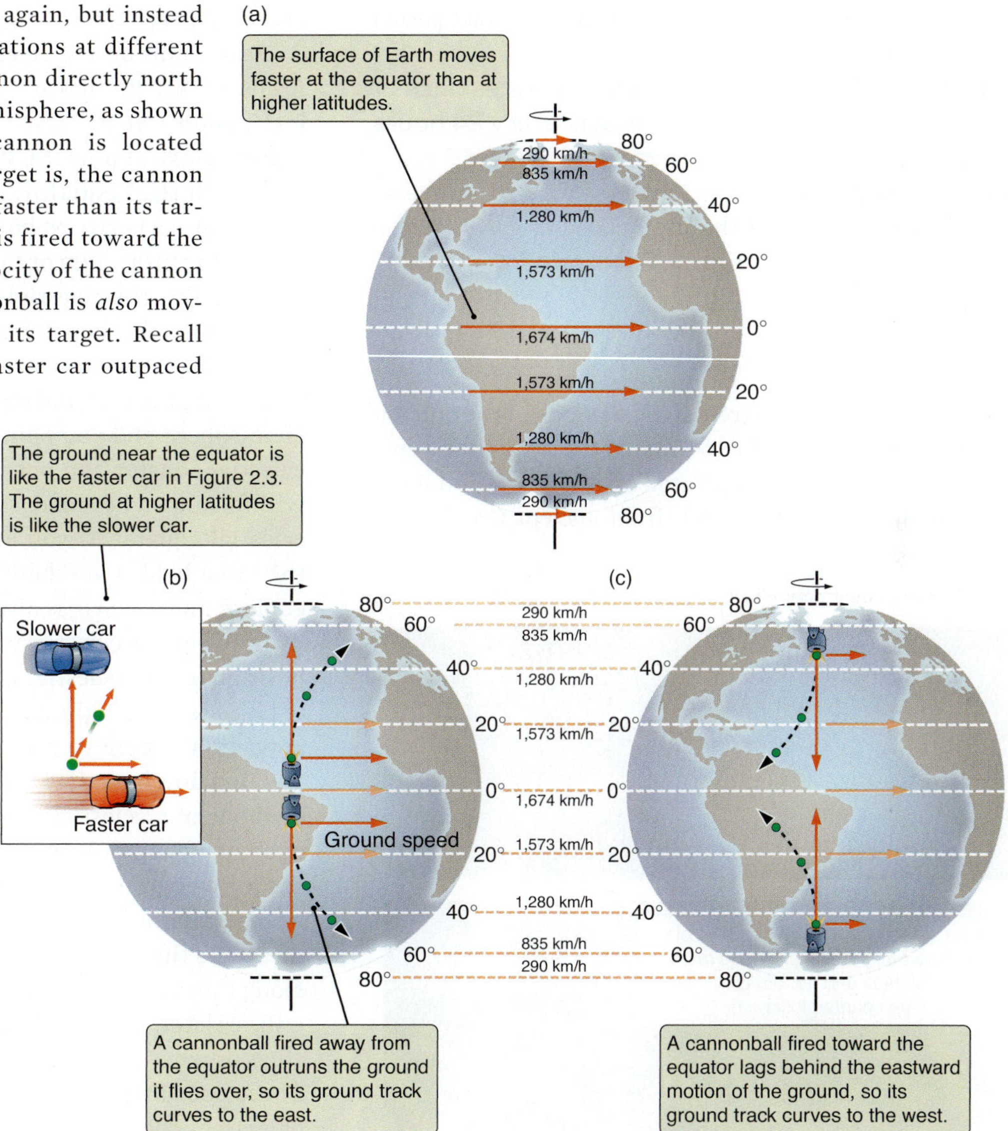

FIGURE 2.4 The Coriolis effect causes objects be deflected as they move across the surface of Earth.

This effect of Earth's rotation is called the **Coriolis effect**. In the Northern Hemisphere the Coriolis effect causes a cannonball fired north to drift to the east as seen from the surface of Earth. In other words, the cannonball appears to curve to the right. A cannonball fired south appears to curve to the west, which also gives it the appearance of curving to the right. In the Northern Hemisphere the Coriolis effect seems to deflect things to the *right*. If you think through this example for the Southern Hemisphere, you will see that south of the equator the Coriolis effect seems to deflect things to the *left*. In between, at the equator itself, the Coriolis effect vanishes.

We can see this Coriolis effect in action in weather systems. On Earth the effect is enough to deflect a fly ball hit north or south into deep left field in a stadium in the northern United States by about a half a centimeter. At some time or other the Coriolis effect has probably determined the outcome of a ball game.

Nor do we directly sense the *direction* of Earth's spin, although it is clearly revealed by the hourly motion of the Sun, Moon, and stars across the sky. As viewed from above Earth's **North Pole**, Earth rotates in a counterclockwise direction (**Figure 2.5**), completing one rotation in a 24-hour period. As the rotating Earth carries us from west to east, objects in the sky *appear* to move in the other direction, from east to west. As seen from Earth's surface, the path each celestial body takes across the sky is called its *apparent daily motion*. ▶II ASTROTOUR: EARTH SPINS AND REVOLVES

Earth rotates once on its axis every 24 hours.

The Celestial Sphere Is a Useful Fiction

To help visualize the apparent daily motions of the Sun and stars, it is sometimes useful to think of the sky as if it were a huge sphere with the stars painted on its surface and Earth at its center. (As we have said, from ancient Greek times to the Renaissance, most people believed this to be a true representation of the heavens.) Astronomers refer to this imaginary sphere as the **celestial sphere** (**Figure 2.6a**). The celestial sphere is a useful concept because it is easy to draw and visualize, but never forget that it is, in fact, imaginary.

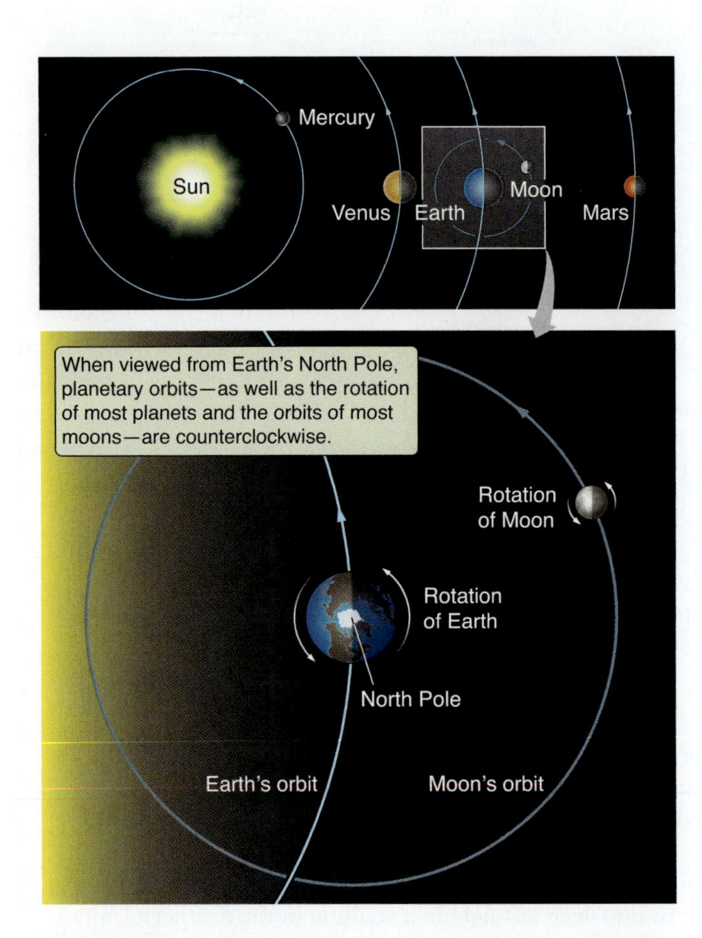

FIGURE 2.5 The rotation of Earth and the Moon, the revolution of Earth and the planets about the Sun, and the orbit of the Moon about Earth are counterclockwise as viewed from above Earth's North Pole. (Not drawn to scale.)

Each point on the celestial sphere indicates a *direction* in space. Directly above Earth's North Pole is the **north celestial pole (NCP)**. Directly above Earth's South Pole is the **south celestial pole (SCP)**. Directly above Earth's equator is the **celestial equator**, an imaginary circle that divides the sky into a northern half and a southern half. Just as the NCP is the projection of the direction of Earth's North Pole into the sky, the celestial equator is the projection of the plane of Earth's equator into the sky. And just as Earth's North Pole is 90° away from Earth's equator, the NCP is 90° away from the celestial equator. If you are in the Northern Hemisphere and you point one arm toward a point on the celestial equator and one arm toward the NCP, your arms will form a right angle, so the NCP is 90° away from the celestial equator. If you are in the Southern Hemisphere, the same holds true there: the angle between the celestial equator and the SCP is always 90° as well.

Between the celestial poles and the equator, objects have positions on the celestial sphere with coordinates analogous to latitude and longitude on Earth. **Latitude** is a measure of a location's north or south distance from Earth's equator. **Declination** is similar to latitude and tells you the angular distance of a celestial body north or south of the celestial equator (from 0° to ±90°). **Longitude** measures how far east or west you are from the Royal Observatory in Greenwich, England. **Right ascension** is similar to longitude and measures the angular distance of a celestial body eastward along the celestial equator from the vernal equinox. (As we will see later, the vernal equinox is the point on the celestial equator where the Sun's path crosses from south to north.) These coordinates are used at telescopes to quickly locate objects in the sky. More detailed descriptions of latitude and longitude, and of celestial coordinates used with the celestial sphere, are in Appendix 6. ▶▶ NEBRASKA SIMULATIONS: CELESTIAL-EQUATORIAL (RA/DEC) DEMONSTRATOR; LONGITUDE/LATITUDE DEMONSTRATOR

From any location on Earth, you can also divide the sky into an east half and a west half with an imaginary north–south line called the **meridian**. This line runs from due north through a point directly overhead, called the **zenith**, to a point due south. It then continues around the far side of the celestial sphere, through the **nadir** (the point directly below you), and back to the starting point due north (**Figure 2.6b**). The

The ecliptic is the path the Sun takes along the celestial sphere.

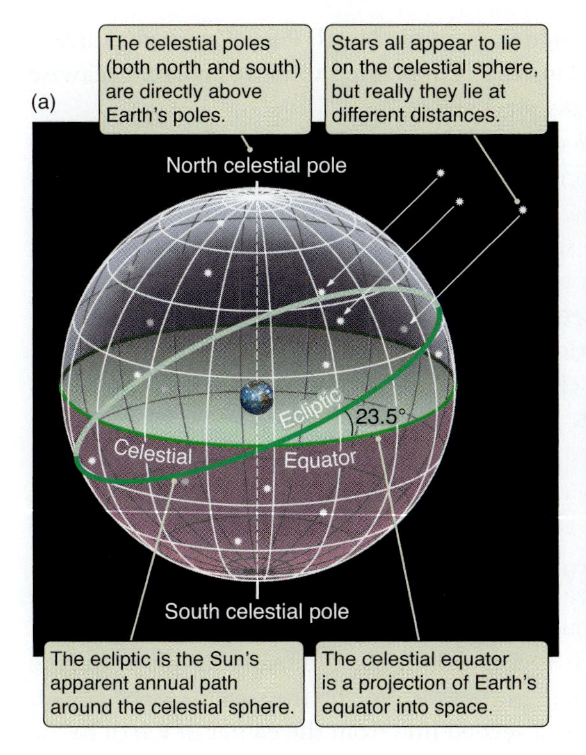

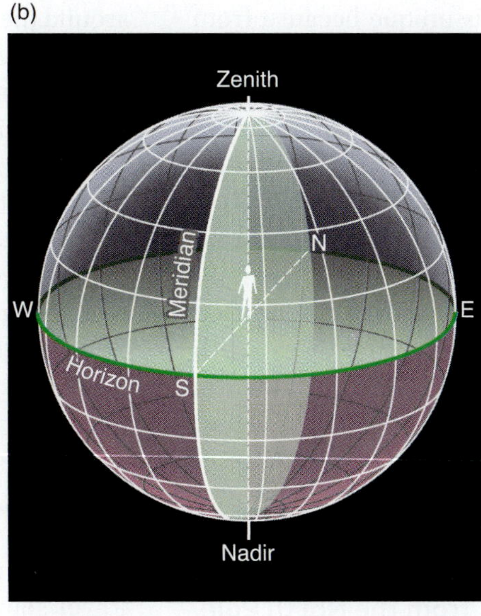

FIGURE 2.6 (a) The celestial sphere is a useful fiction for thinking about the appearance and apparent motion of the stars in the sky. (b) At any location on Earth, the sky is divided into an east half and a west half by an imaginary meridian projected onto the celestial sphere.

path that the Sun takes along the celestial sphere is called the **ecliptic**. This imaginary circle is inclined 23.5° to the celestial equator (see Section 2.3).

To see how the concept of the celestial sphere can be used, let's consider the Sun at noon and at midnight. From your perspective on Earth, and as modeled in the celestial sphere, the Sun appears to move across the sky and reach its highest point in the sky at noon. Astronomers define true *local noon* as the time when the Sun crosses the meridian at their location. True *local midnight* occurs when the Sun again crosses the meridian at its lowest point below the northern horizon. From your perspective on Earth, the celestial sphere appears to rotate, carrying the Sun across the sky to its highest point at noon and around through the meridian again at midnight (see Figure 2.6). What is really happening? The Sun remains in the same place in space through the entire 24-hour period. Earth rotates, so your spot faces a different direction at every moment. At noon, your spot on Earth has rotated to face most directly toward the Sun. Half a day later, at midnight, your spot on Earth has rotated to face most directly away from the Sun.

We will continue to use the celestial sphere model throughout the chapter to show apparent motions of the Sun and stars from different points on Earth. ▶▶ **NEBRASKA SIMULATIONS:** CELESTIAL AND HORIZON SYSTEMS COMPARISON; ROTATING SKY EXPLORER

The View from the Poles

The apparent daily motions of the stars and the Sun that you see depend on where on the surface you live. For example, the apparent daily motions of celestial objects in Alaska are quite different from the apparent daily motions seen from Hawaii. To understand how location affects the perception of apparent daily motions of celestial bodies, let's look at the daily motions of the stars when viewed from one of the "ends of the Earth," the North Pole. (In science we often start by working out the "easy" or "limiting" cases—the view of the stars from the poles, for example—and then use these to guide our thinking about what happens in more complicated situations.)

Imagine that you are standing on the North Pole watching the sky as shown in **Figure 2.7a**. (Ignore the Sun for the moment and suppose that you can always see stars in the sky.) You are standing where Earth's axis of rotation intersects its surface, which is much the same as standing at the center of a rotating carousel. As Earth rotates, the spot directly above you seems to remain fixed while everything else in the sky appears to revolve in a counterclockwise direction around this spot (**Figure 2.7b**). (If you are having trouble visualizing this, find a globe and, as you spin it, imagine standing at the pole of the globe.) Objects close to the pole appear to follow small circles, while objects nearest to the horizon follow the largest circles. **AstroTour: The View from the Poles**

The view from the North Pole is unique because from there you always see the *same* half of the sky (**Figure 2.7c**). Nothing rises or sets as Earth turns beneath you. Regardless of where you are on Earth's surface, you can never see more than half of the sky at any one time. The other half of the sky is blocked from view by Earth. The boundary between the part of the sky you can see and the part that is blocked by Earth is the **horizon**. From most locations on Earth (unlike from the North Pole), the half of the sky that you can see above the horizon changes constantly as Earth rotates. (If you are not at the North Pole, the direction in space in which your zenith points right now is different from what it was 12 hours ago, or even 12 seconds ago. In contrast, Earth's North Pole points in the *same* direction, hour after hour and day after day.) For this reason, if you look off toward the horizon, you will see that the objects visible there follow circular paths that keep them always the same distance above the horizon.

The same half of the sky is always visible from the North Pole.

The view from Earth's **South Pole** is much the same, but with two major differences. First, the South Pole is on the opposite side of Earth from the North Pole, so the half of the sky you see overhead at the South Pole is precisely the half that is hidden from the North Pole. The second difference is that instead of appearing to move counterclockwise around the sky, stars appear to move *clockwise* around the south celestial pole. (To see this, sit in a swivel chair and spin it around from right to left. As you look at the ceiling, things appear to move in a counterclockwise direction; but as you look at the floor, they appear to be moving clockwise.) **AstroTour: The View from the Poles**

From the South Pole, the other half of the sky is visible, and stars circle in a clockwise direction.

Latitude Determines the Part of the Sky That You See

What do you see in the sky as you leave the North Pole and travel south to lower latitudes? Imagine a line from the center of Earth to your location on the surface of the planet. Now imagine a second line from the center of Earth to the point on the equator closest to you (refer to **Figure 2.8** for help imagining these lines). The angle between these two lines is your latitude. At the North Pole, for example, these two imaginary lines form a 90° angle. At the equator, they form a 0° angle. So the latitude of the North Pole is 90° north, and the latitude of the equator is 0°. The South

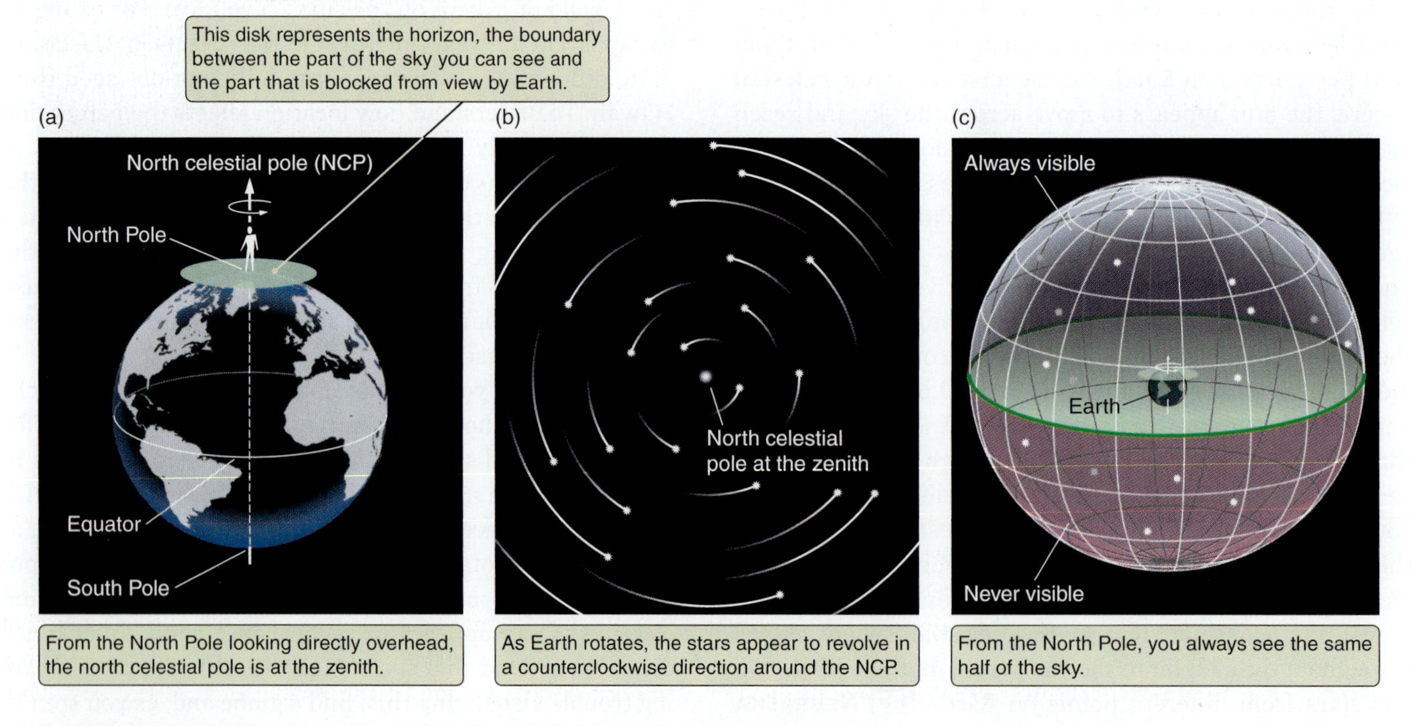

FIGURE 2.7 As viewed from Earth's North Pole (a), stars move throughout the night on counterclockwise, circular paths about the zenith (b). (c) The same half of the sky is always visible from the North Pole.

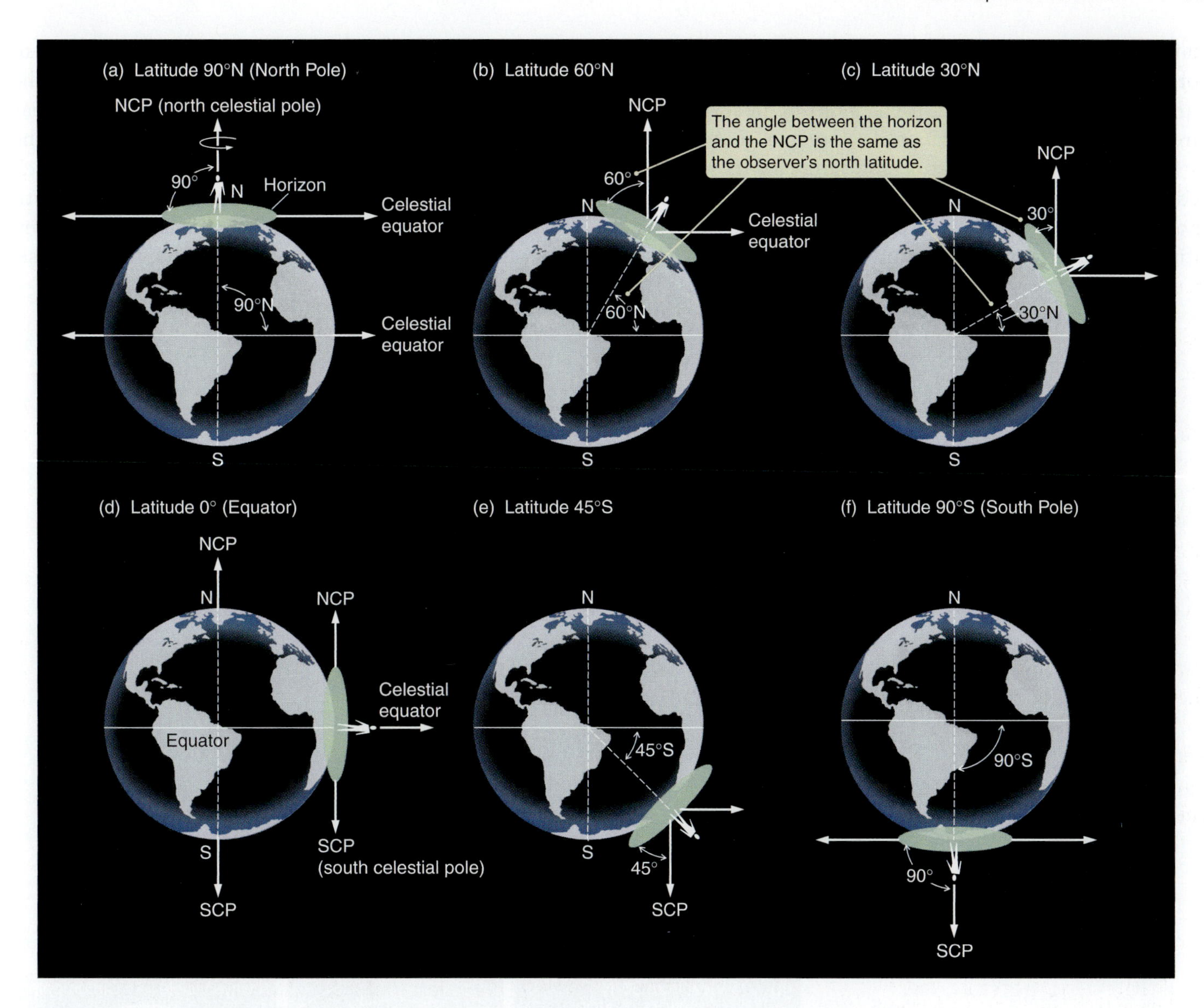

FIGURE 2.8 Your perspective on the sky depends on your location on Earth. Here you can see how the locations of the celestial poles and the celestial equator depend on the observer's latitude.

Pole is at latitude 90° south. ▶▶ **NEBRASKA SIMULATIONS: MERIDIONAL ALTITUDE SIMULATOR; DECLINATION RANGES SIMULATOR**

Your latitude determines the part of the sky that you can see throughout the year. If you follow the curve of Earth south from the North Pole, the horizon tilts and your zenith moves away from the north celestial pole. At a latitude of 60° north (as shown in **Figure 2.8b**), your horizon is tilted 60° from the NCP. The north latitude and the height of the NCP above the northern horizon are equal everywhere. In **Figure 2.8d**, you have reached Earth's equator, at a latitude of 0°. Notice that the NCP is now sitting on the northern horizon. At the same time, you get your first look at the south celestial pole, which is sitting opposite the NCP on the southern horizon. As you continue into the Southern Hemisphere, the SCP is now visible above the southern horizon, while the NCP is hidden from view by the northern horizon. At a latitude of 45° south (**Figure 2.8e**), the SCP lies 45° above the southern horizon. At the South Pole (latitude 90° south—**Figure 2.8f**), the SCP is at the zenith, 90° above the horizon.

One way to cement your understanding of the view of the sky at different latitudes is to draw pictures like those in Figure 2.8. If you can draw a picture like this for any latitude—filling in the values for each of the angles in the drawing and imagining what the sky looks like from that location—then you will be well on your way to developing a working knowledge of the appearance of the sky. That knowledge will prove useful later, when we discuss a variety of phenomena, such as the changing of the seasons. When practicing your sketches, however, take care not to make the common mistake illustrated in **Figure 2.9**. The north celestial pole is not a location in space, hovering over Earth's North Pole. Instead, it is a *direction* in space—the direction parallel to Earth's axis of rotation.

Now that we have shown you how the horizon is oriented at different latitudes, let's look at how the apparent motions of the stars about the celestial poles differ from latitude to latitude. **Figure 2.10a** shows your view if you are at latitude 30° north. As Earth rotates, the part of the sky visible to you is constantly changing. From this perspective it is the horizon that seems to remain fixed, while the stars appear to move past overhead. If you focus your attention on the north celestial pole, from this perspective you still see much the same thing you saw from Earth's North Pole. The north celestial pole remains fixed in the sky, and all of the stars appear to move throughout the night in counterclockwise,

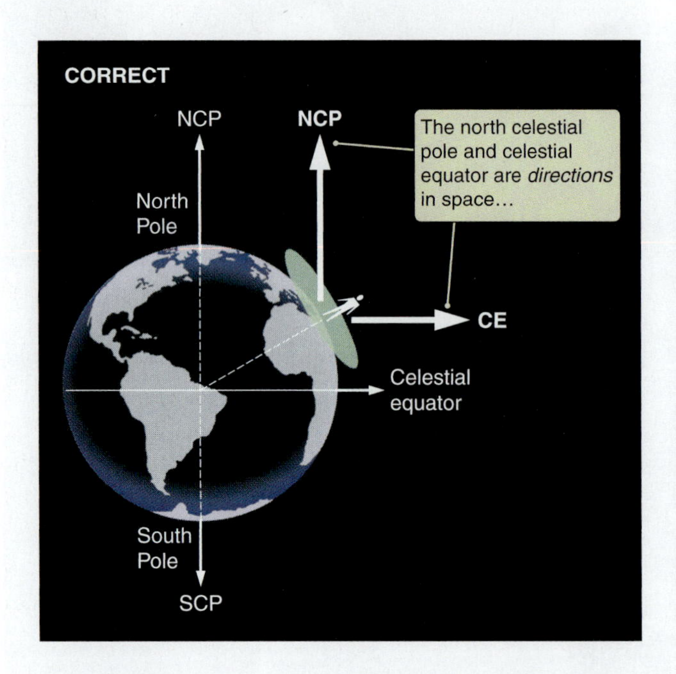

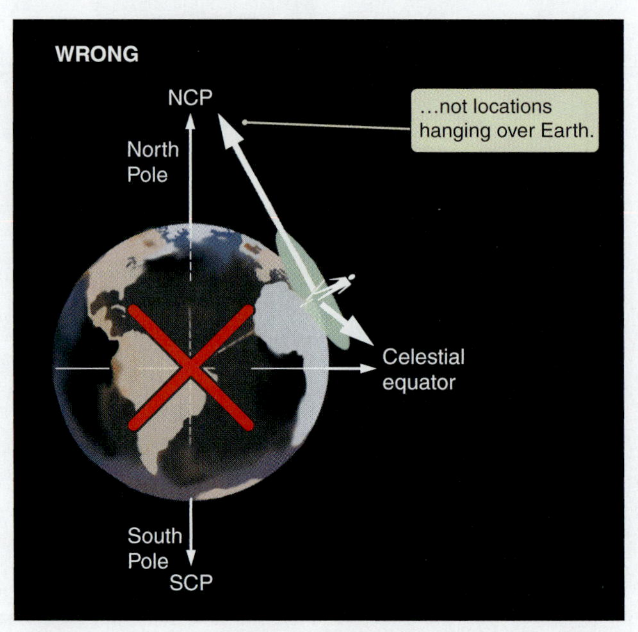

FIGURE 2.9 The celestial poles and the celestial equator are directions in space, not fixed locations hanging above Earth. (The red *X* means "avoid this misconception.")

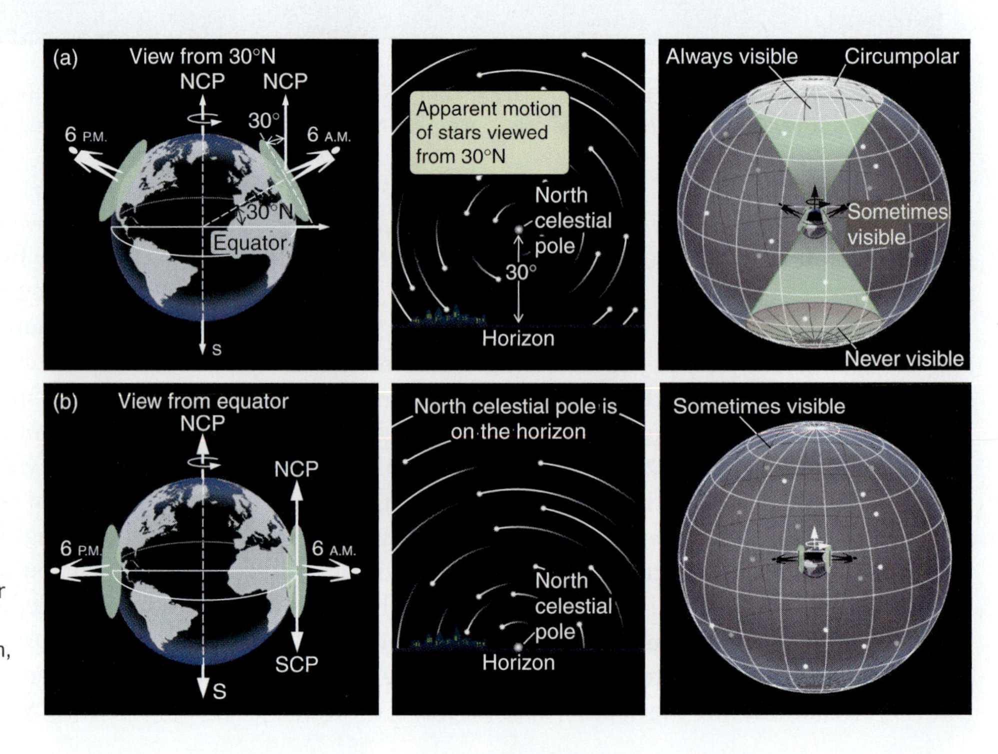

FIGURE 2.10 (a) As viewed from latitude 30° north, the north celestial pole is 30° above the northern horizon. Stars appear to move on counterclockwise paths around this point. At this latitude some parts of the sky are always visible, while others are never visible. (b) From the equator, the north and south celestial poles are seen on the horizon, and the entire sky is visible over a period of 24 hours.

circular paths around that point. But because the north celestial pole is no longer directly overhead as it was at the North Pole, the apparent circular paths of the stars are now tipped relative to the horizon. (More correctly, your horizon is now tipped relative to the apparent circular paths of the stars.) ▶II **ASTROTOUR: THE CELESTIAL SPHERE AND THE ECLIPTIC**

Stars located close enough to the north celestial pole are above the horizon 24 hours a day (even if you can't see them in the daytime) as they complete their apparent paths around the pole (see Figure 2.10a and **Figure 2.11**). This always-visible region of our sky is referred to as being **circumpolar**, which means "around the pole." There is also a part of the sky that can *never* be seen from this latitude. This is the part of the sky near the south celestial pole that never rises above your horizon. And between this region and the always-visible circumpolar region lies a portion of the sky that can be seen for *part but not all* of each day. Stars in this intermediate region appear to rise above and set below Earth's shifting horizon as Earth turns. The only place on Earth where you can see the entire sky over the course of 24 hours is the equator. From the equator (**Figure 2.10b**), the north and south celestial poles sit on the northern and southern horizons, respectively, and the whole of the heavens passes through the sky each 24-hour day.

Circumpolar stars are always above the horizon.

Look at the location of the celestial equator in **Figure 2.12**. The points where the celestial equator intersects the horizon are always due east and due west. (The only exception is at the poles, where the celestial equator is coincident with the horizon.) An object on the celestial equator rises due east and sets due west. Objects that are north of the celestial equator rise north of east and set north of west. Objects that are south of the celestial equator rise south of east and set south of west.

Figure 2.12 also shows that regardless of where you are on Earth (again with the exception of the poles), half of the celestial equator is always visible above the horizon. Because half of the celestial equator is always visible, it follows that you can see any object that lies in the direction of the celestial equator half of the time. An object that is in the direction of the celestial equator rises due east, is above the horizon for exactly 12 hours, and sets due west. This is not true for objects that are not on the celestial equator. A look at **Figure 2.12b** shows that from the Northern Hemisphere, you can see more than half of the apparent circular path of any star that is north of the celestial equator. And if you can see more than half of a star's path, then the star is above the horizon for more than half of the time.

As seen from the Northern Hemisphere, stars north of the celestial equator remain above the horizon for more than 12 hours each day. The farther north the star is, the longer it stays up. The circumpolar stars near the north celestial pole that we mentioned already are the extreme example of this phenomenon; they are up 24 hours a day. In contrast, objects south of the celestial equator are above the horizon for less than 12 hours a day, and the farther south you look, the less time a star is visible. Stars that are located close to the south celestial pole never rise above our horizon.

If you were an observer in the Southern Hemisphere (**Figure 2.12d**), the reverse would be true: objects on the celestial equator would still be up for 12 hours a day, but now objects south of the celestial equator would be up more than 12 hours, and objects north of the celestial equator would be up less than 12 hours.

Since ancient times, travelers, including sailors at sea, have used the stars for navigation. We can find the north

FIGURE 2.11 Time exposures of the sky showing the apparent motions of stars through the night. Note the difference in the circumpolar portion of the sky as seen from the two different latitudes.

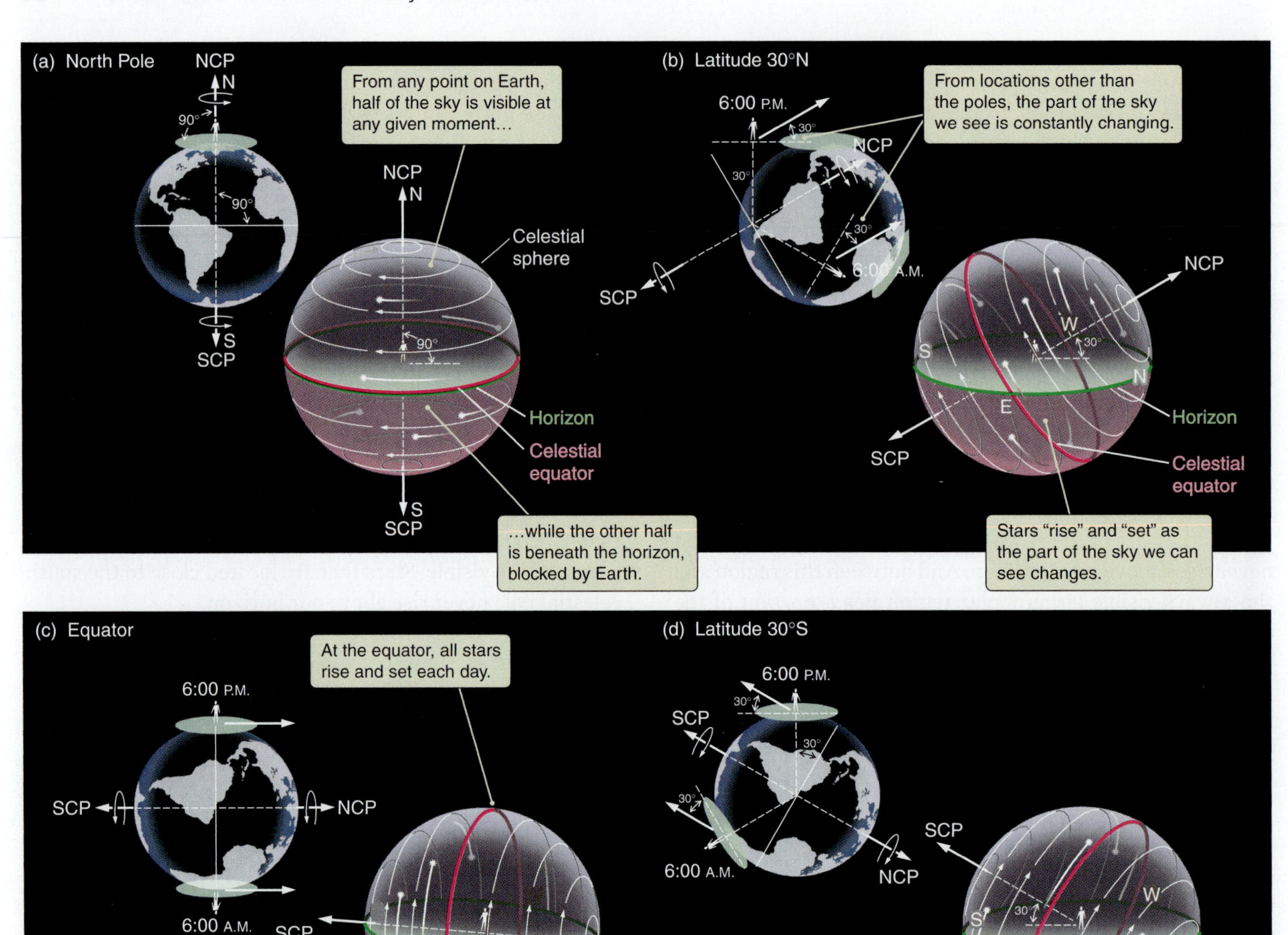

FIGURE 2.12 The celestial sphere is shown here as viewed by observers at four different latitudes. At all locations other than the poles, stars rise and set as the part of the celestial sphere that an observer sees change during the day.

and south celestial poles by recognizing the stars that surround them. In the Northern Hemisphere, a moderately bright star happens by chance to be located within 0.7° of the north celestial pole. This star is called Polaris, or more commonly, the "North Star." If you can find Polaris in the sky and measure the angle between the north celestial pole and the horizon, then you know your latitude. For example, if you are in Phoenix, Arizona (latitude 33.5° north), you will find the north celestial pole 33.5° above your northern horizon. On the other hand, in Fairbanks, Alaska (latitude 64.6° north), the north celestial pole sits much higher overhead, 64.6° above the horizon in the north. The location of the north celestial pole in the sky can also be used to measure the size of Earth, as described in **Math Tools 2.1**. Determining your longitude by astronomical methods is much more complicated because of Earth's rotation. **NEBRASKA SIMULATIONS: BIG DIPPER CLOCK; BIG DIPPER 3D; ROTATING SKY EXPLORER; COORDINATE SYSTEMS COMPARISON**

Math Tools 2.1

How to Estimate the Size of Earth

We can use the location of the north celestial pole in the sky to estimate the size of Earth. Suppose we start out in Phoenix, Arizona, and we observe the north celestial pole to be 33.5° above the horizon. If we head north, by the time we reach the Grand Canyon, about 290 km from Phoenix, we notice that the north celestial pole has risen to about 36° above the horizon. This difference between 33.5° and 36° (2.5°) is 1/144 of the way around a circle. (A circle is 360°, and 2.5°/360° = 1/144.)

This means that we must have traveled 1/144 of the way around the circumference, C, of Earth in traveling the 290 km between Phoenix and the Grand Canyon. In other words,

$$\frac{1}{144} \times C = 290 \text{ km}$$

Rearranging the expression, the circumference of Earth is given by

$$C = 144 \times 290 \text{ km} \approx 42{,}000 \text{ km}$$

The actual circumference of Earth is just a shade over 40,000 km, so our simple calculation was not too bad. The circumference of a circle is equal to 2π multiplied by its radius. So, the radius of Earth is given by

$$\text{Radius} = \frac{C}{2\pi} = \frac{40{,}000 \text{ km}}{2\pi} = 6{,}400 \text{ km}$$

It was in much this same way that the Greek astronomer Eratosthenes (276–194 BCE) made the first accurate measurements of the size of Earth, in about 230 BCE (**Figure 2.13**). Eratosthenes used the distance between his home city of Alexandria and the city of Syene (currently Aswân, in Egypt), which was 5,000 "stadia." He noticed that on the first day of summer in Syene, the sunlight reflected directly off the water in a deep well, so the Sun must have been nearly at the zenith. By measuring the shadow of the Sun from an upright stick in Alexandria, he saw that the Sun was about 7.2° south of the zenith on the same date. Assuming Earth was spherical and Syene was directly south of Alexandria, he determined the distance between the two cities to be 7.2° divided by 360, or 1/50 of the circumference of Earth.

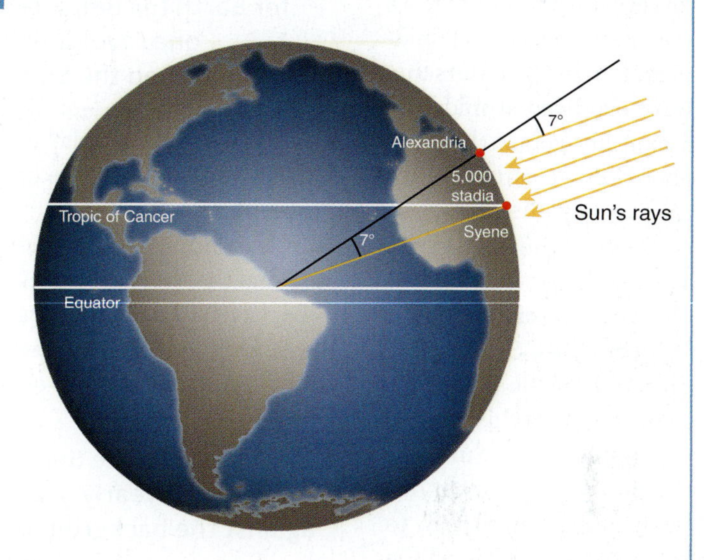

FIGURE 2.13 Eratosthenes estimated the size of Earth using observations and basic calculations.

It was difficult to estimate distances accurately in those days, and although historians know Eratosthenes concluded that the cities were 5,000 stadia apart, they are still not at all sure of the value of his stadion unit. The standard stadion was 185 meters, and then Eratosthenes would have worked the math in a similar way:

$$\frac{1}{50} \times C = 5{,}000 \text{ stadia} \times 185 \text{ meters/stadion}$$
$$= 925{,}000 \text{ meters} = 925 \text{ km}$$

Eratosthenes would have found the circumference of Earth to be:

$$C = 50 \times 925 \text{ km} = 46{,}250 \text{ km}$$

only about 16 percent higher than the modern value. (Some have argued that if the Egyptian stadion, about 157.5 meters had been used, then Eratosthenes' number would have been only a few percentage points off from the modern value.)

2.3 Revolution about the Sun Leads to Changes during the Year

Earth's average distance from the Sun is 1.50×10^8 km. This distance is called an **astronomical unit** (**AU**) and is used for measuring distances in the Solar System. Earth revolves around the Sun in the same direction that Earth spins about its axis—counterclockwise as viewed from above Earth's North Pole. A **year** is the time it takes for Earth to complete one revolution around the Sun. The motion of Earth around the Sun is responsible for many of the patterns of change we see in the sky and on Earth, including changes in the stars we see at night. As Earth moves around the Sun, the stars we see overhead at midnight change. Six months from now, Earth will be on the other side of the Sun, and the stars that we see overhead at midnight will be in nearly the opposite direction from the stars we see near overhead at midnight tonight. The stars that were overhead at midnight 6 months ago are the same stars that are overhead today at noon, but we cannot see them today because of the glare of the Sun. AstroTour: Earth Spins and Revolves

A year is the time it takes for Earth to complete one revolution around the Sun.

If you could note the position of the Sun relative to the stars each day for a year, you would find that the Sun traces out a **great circle** against the background of the stars (**Figure 2.14**). On September 1 the Sun appears to be in the direction of the constellation of Leo. Six months later, on March 1, Earth is on the other side of the Sun, and the Sun appears from our perspective on Earth to be in the direction of the constellation of Aquarius. The apparent path that the Sun follows against the background of the stars is called the ecliptic. The 12 constellations that lie along the ecliptic through which the Sun appears to move are the constellations of the **zodiac**. Ancient astrologers, who believed the positions of the Sun, Moon, and planets in the sky affected individual people's lives on Earth, assigned special mystical significance to these stars. Actually, the constellations of the zodiac are simply random patterns of distant stars that happen by chance to lie near the plane of Earth's orbit about the Sun. AstroTour: The Celestial Sphere and the Ecliptic; Nebraska Simulation: Ecliptic (Zodiac) Simulator

The ecliptic is the Sun's apparent yearly path against the background of stars.

Seasons Are Due to the Tilt of Earth's Axis

What causes the seasons? Let's make a hypothesis: Earth is closer to the Sun in the summer and farther away in the winter, and this change in distance is the cause for the seasons. Can this hypothesis be falsified? If the distance from Earth to the Sun caused the seasons, we could predict that all of Earth should experience summer at the same time of year. But the United States experiences summer in June,

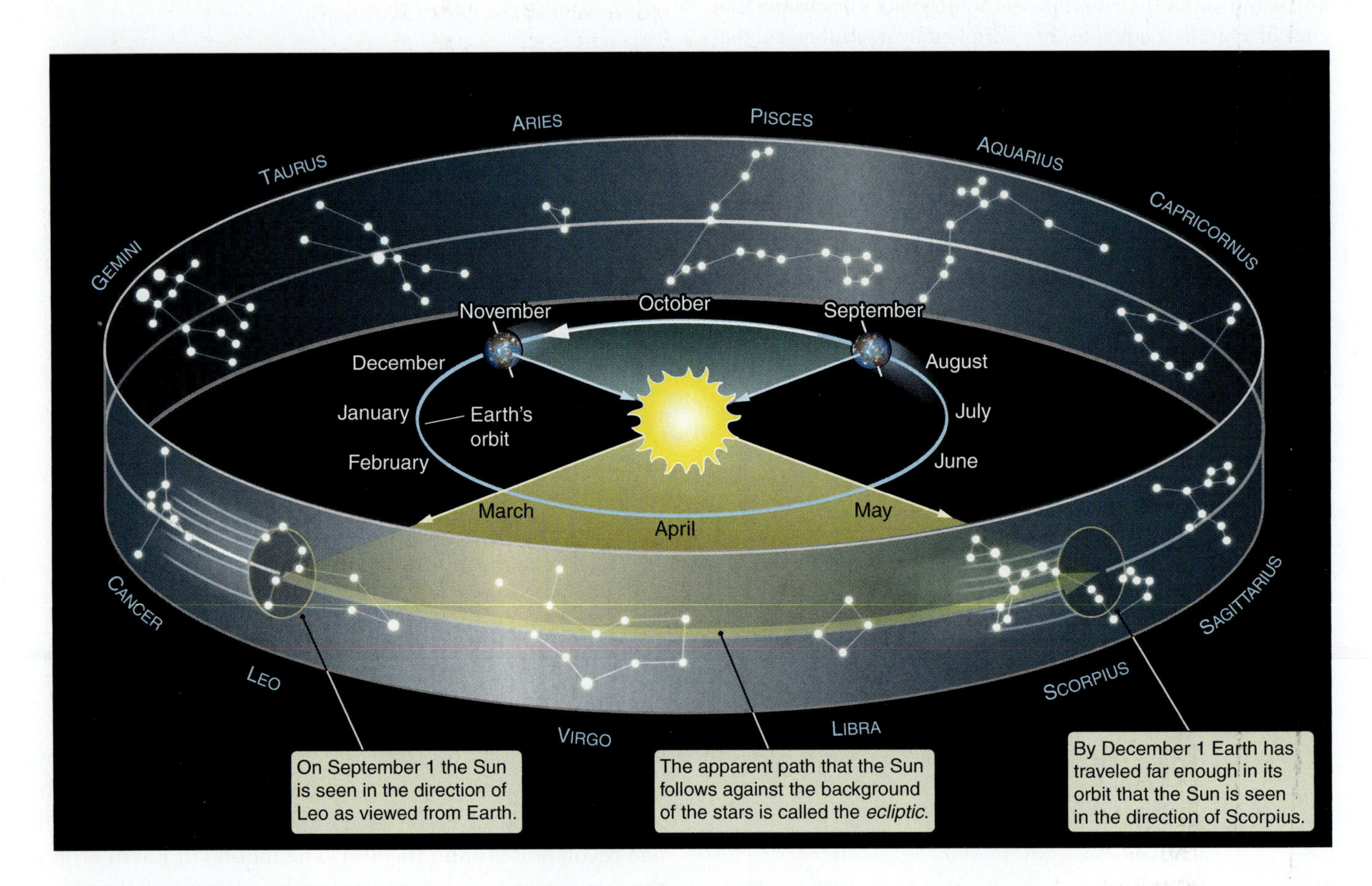

FIGURE 2.14 As Earth orbits about the Sun, the Sun's apparent position against the background of stars changes. The imaginary circle traced by the annual path of the Sun is called the ecliptic. Constellations along the ecliptic form the zodiac.

while Australia experiences summer in December. We have just falsified this hypothesis, and we need to look for another one that explains *all* of the available facts.

We will find a new hypothesis by investigating how the combination of Earth's axial tilt and its annual path around the Sun create seasons. If Earth's spin axis were exactly perpendicular to the plane of Earth's orbit (the **ecliptic plane**), then the Sun would always appear to lie on the celestial equator. Because the position of the celestial equator is determined by our latitude, the Sun would follow the same path through the sky every day, rising due east each morning and setting due west each evening. If the Sun were always on the celestial equator, it would be above the horizon for exactly half the time, and days and nights would always be exactly 12 hours long. In short, if Earth's axis were exactly perpendicular to the plane of Earth's orbit, each day would be just like the last, and there would be no seasons.

However, Earth's axis of rotation is *not* exactly perpendicular to the plane of the ecliptic. Instead it is tilted by 23.5° from the perpendicular. As Earth moves around the Sun, its axis points in almost exactly the same direction throughout the year and from one year to the next. (We say *almost* exactly because, as you will soon learn, Earth's axis wobbles.) As a result, sometimes Earth's North Pole is tilted more toward the Sun, and other times it is pointed more away from the Sun. When Earth's North Pole is tilted toward the Sun, an observer on Earth sees the Sun as lying north of the celestial equator. Six months later, when Earth's North Pole is tilted away from the Sun, an observer in the same place sees the Sun as lying south of the celestial equator. If we look at the circle of the Sun's apparent path through the stars—the ecliptic—we see that it is tilted by 23.5° with respect to the celestial equator. **ASTROTOUR: EARTH SPINS AND REVOLVES; NEBRASKA SIMULATION: SEASONS AND ECLIPTIC SIMULATOR**

To understand the effect that this tilt has on Earth, begin by looking at **Figure 2.15a**. This figure shows the situation on June 20, the first day of summer in the Northern Hemisphere, when Earth is on the side of the Sun where the tilt of the North Pole is toward the Sun. Note first that the Sun is

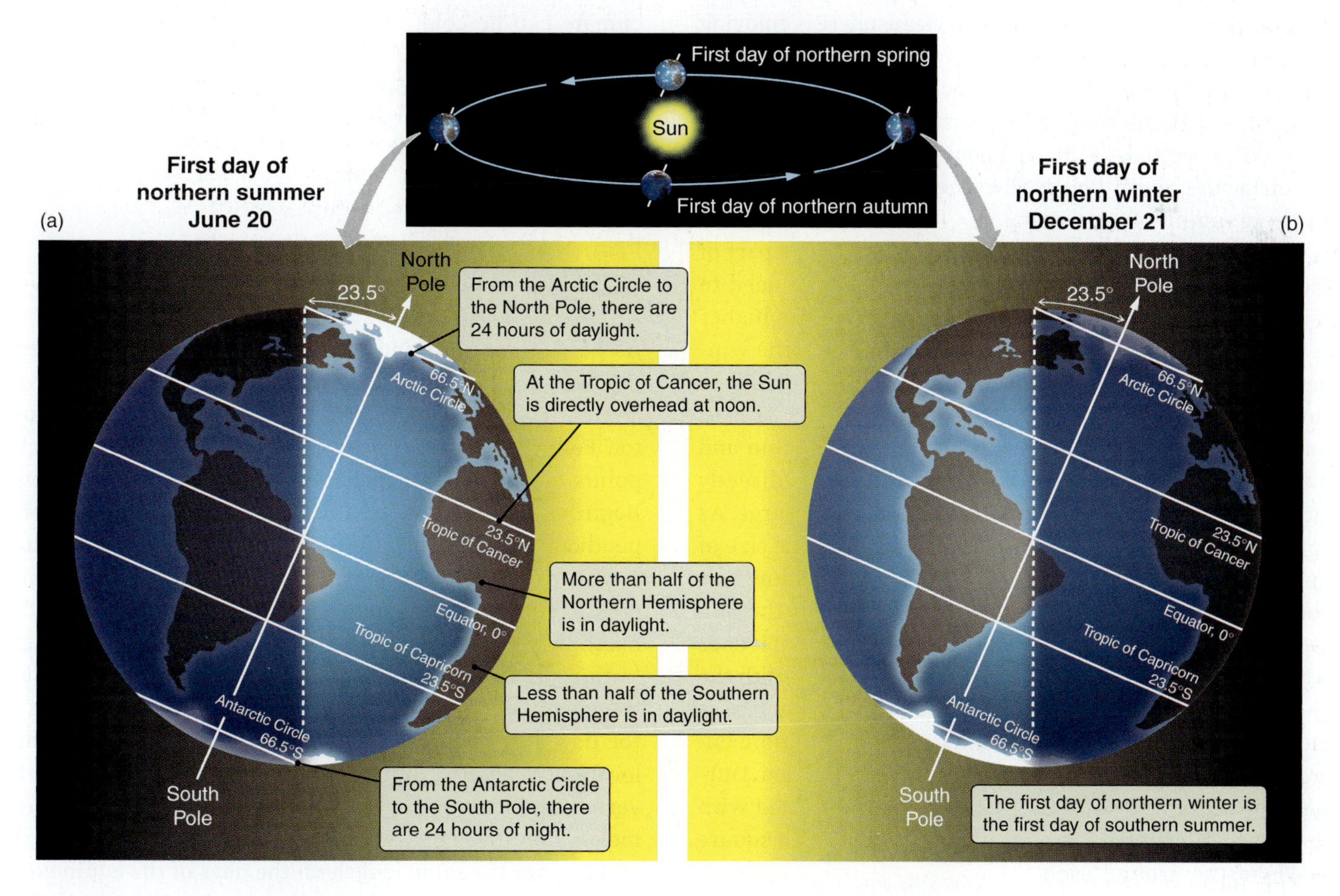

FIGURE 2.15 (a) On the first day of the northern summer (around June 20, the summer solstice), the northern end of Earth's axis is tilted most nearly toward the Sun, while the Southern Hemisphere is tipped away. (b) Six months later, on the first day of the northern winter (around December 21, the winter solstice), the situation is reversed. Seasons are opposite in the Northern and Southern hemispheres.

north of the celestial equator. Recall from earlier in the chapter that, from the perspective of an observer in the Northern Hemisphere, an object north of the celestial equator can be seen above the horizon for more than half the time. This is true for the Sun, as well as for any other celestial object. Saying that the Sun is above the horizon for more than half the time is just another way of saying that the days are longer than 12 hours. In Figure 2.15a, note that when Earth's North Pole is tilted toward the Sun, over half of Earth's Northern Hemisphere is illuminated by sunlight. These are the long days of the northern summer. Six months later the situation is very different. Around December 21 (**Figure 2.15b**), the first day of winter in the Northern Hemisphere, Earth has moved in its orbit to the opposite side of the Sun. Then the tilt of the North Pole is away from the Sun, and the Sun appears in the sky south of the celestial equator. Someone in the Northern Hemisphere will see the Sun for less than 12 hours each day. Less than half of the Northern Hemisphere is illuminated by the Sun. It is winter in the north.

Things in the Southern Hemisphere are exactly reversed from what is going on in the north. Look again at Figure 2.15. Around June 20, while the Northern Hemisphere is enjoying long days and short nights of summer, Earth's South Pole is tilted in the direction away from the Sun. Less than half of the Southern Hemisphere is illuminated by the Sun, and the winter days are shorter than 12 hours. Similarly, on December 21 Earth's South Pole is tilted toward the Sun, and the southern summer days are long.

The differing length of days through the year is part of the explanation for the changing seasons, but we need to consider another important effect: the Sun appears higher in the sky during the summer than it does during the winter, and sunlight strikes the ground *more directly* during the summer than during the winter. To understand why this is important, hold a piece of cardboard toward the Sun and look at the size of its shadow. If the cardboard is directly face-on to the Sun, then the cardboard's shadow is large. As you turn the cardboard more edge-on, however, the size of its shadow shrinks. The size of the cardboard's shadow tells you that the cardboard catches less energy from the Sun each second when it is tilted relative to the Sun than it does when it is face-on to the Sun. This is what happens with the changing seasons. During the summer, Earth's surface is more nearly face-on to the incoming sunlight, so more energy falls on each square meter of ground each second. During the winter, the surface of Earth is more inclined with respect to the sunlight, so less energy falls on each square meter of the ground each second. That's the primary reason why it is hotter in the summer and colder in the winter.

The angle of sunlight to the ground changes with the seasons.

Figure 2.16 shows the direction of incoming sunlight striking Earth at latitude 40° north, which stretches across Middle America from northern California to New Jersey. At noon on the first day of summer, the Sun is high in the sky—73.5° above the horizon and only 16.5° away from the zenith. Sunlight strikes the ground almost face-on (**Figure 2.16a**). In contrast, at this same latitude at noon on the first day of winter, the Sun is only 26.5° above the horizon, or 63.5° from the zenith. Sunlight strikes the ground at a rather shallow angle (**Figure 2.16b**). As a result of these differences, more than twice as much solar energy falls on each square meter of ground per second at noon on June 20 as falls there at noon on December 21 (see the **Process of Science Figure**).

We do not have to wait for the seasons to change to see the effect that the height of the Sun in the sky has on terrestrial climate. We need only compare the climates found at different latitudes on Earth. Near the equator, the Sun passes high overhead every day, regardless of the season. As a result, the climate is warm throughout the year. At high latitudes, however, the Sun is *never* high in the sky, and the climate can be cold and harsh even during the summer. ▶▶ **NEBRASKA SIMULATIONS:** SUN'S RAYS SIMULATOR; PATHS OF THE SUN; SUN MOTIONS OVERVIEW; DAYLIGHT HOURS EXPLORER; UNION SEASONS DEMONSTRATOR; DAYLIGHT SIMULATOR

Four Special Days Mark the Passage of the Seasons

As Earth travels around the Sun over the course of a year, the Sun appears to trace a path along the ecliptic, a great circle that is tilted 23.5° with respect to the celestial equator. Follow along in **Figure 2.17** and note the four special points on this path that mark the passage of the seasons. Begin with the point in March when Earth's axis is perpendicular to the direction of the Sun (point 1 in **Figure 2.17a**). Here the Sun's apparent motion along the ecliptic crosses the celestial equator moving from the south to the north (point 1 in **Figure 2.17b**). That direction on the celestial sphere, located in the constellation Pisces, is called the **vernal equinox**. The term *vernal equinox* also refers to the day—around March 20—when the Sun appears at this location. When the Sun lies on the celestial equator on the vernal equinox, days are 12 hours long. (The term **equinox** means literally "equal night"; everywhere on Earth, night and day are the same length on the days of the equinoxes.) In the Northern Hemisphere, the vernal equinox is the first day of spring.

As Earth continues its journey around the Sun, the Sun climbs higher into the northern sky. The northern end of

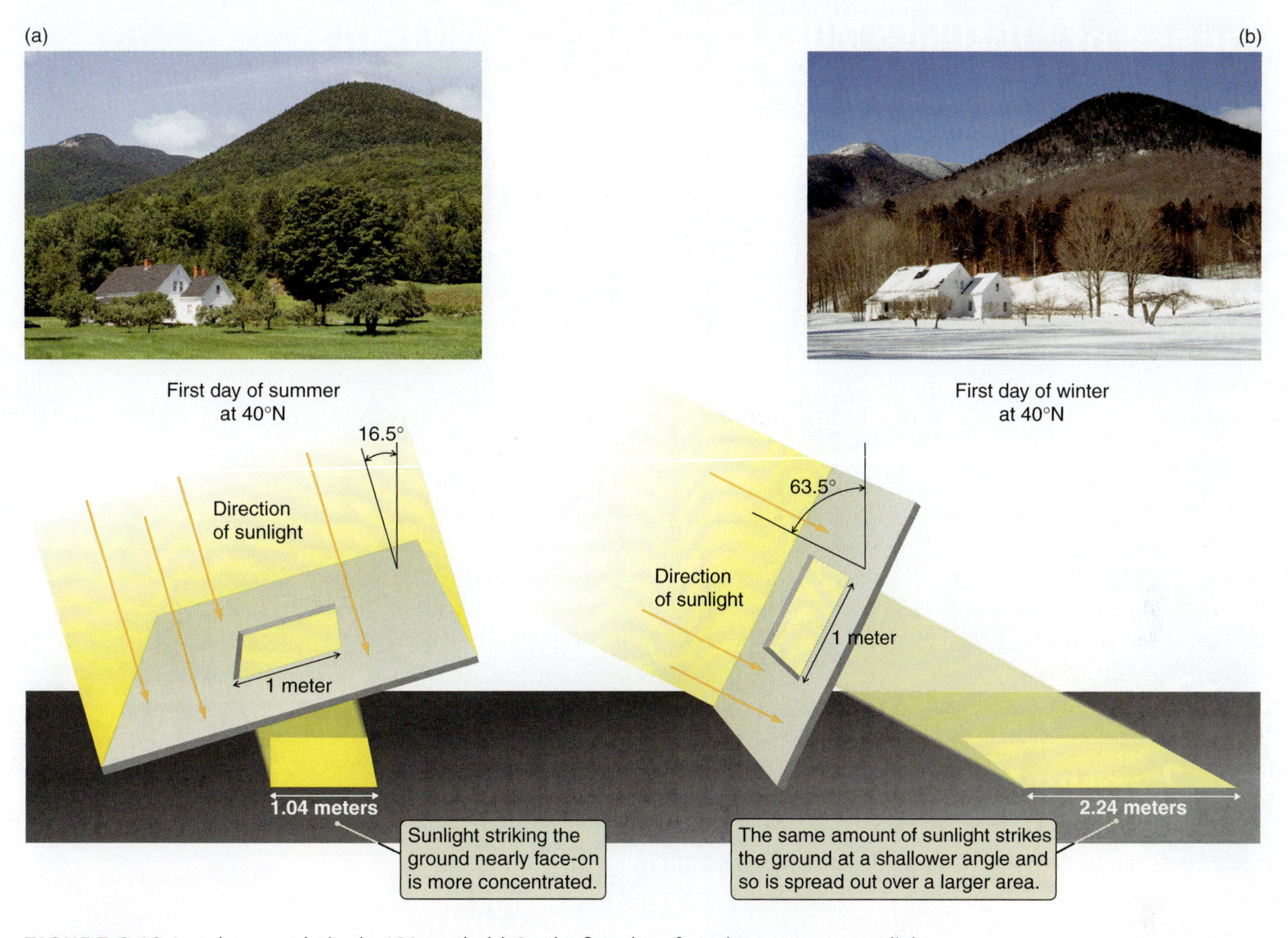

FIGURE 2.16 Local noon at latitude 40° north. (a) On the first day of northern summer, sunlight strikes the ground almost face-on. (b) On the first day of northern winter, sunlight strikes the ground more obliquely, and less than half as much sunlight falls on each square meter of ground each second.

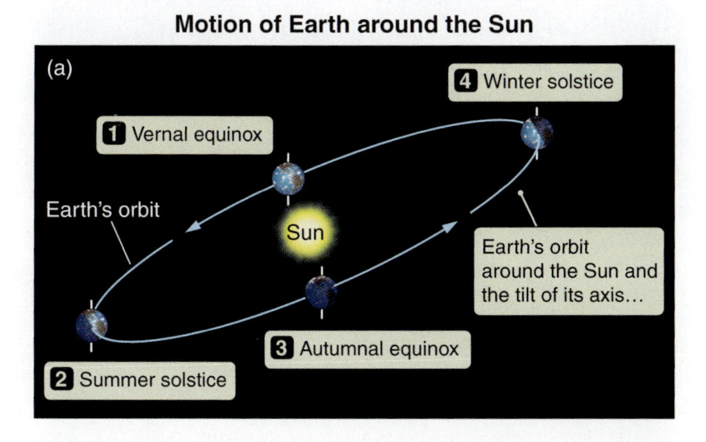

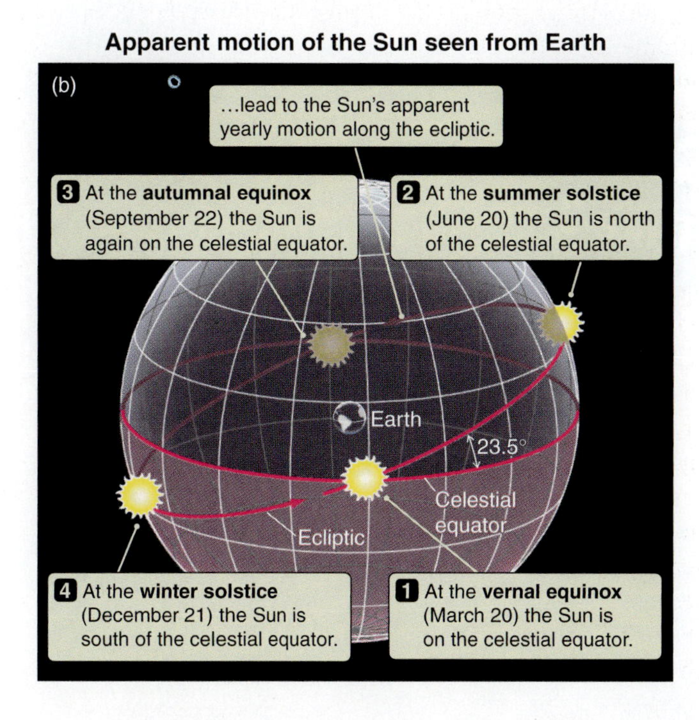

FIGURE 2.17 The motion of Earth about the Sun as seen from the frame of reference of (a) the Sun and (b) Earth.

Process of Science

THEORIES MUST FIT ALL THE KNOWN FACTS

The reason for the seasons is a common misconception that demonstrates how new information can change our minds.

The Hypothesis
We have seasons because Earth is closer to the Sun in summer and farther away in winter.

The Test
If this is true, both the Northern and Southern hemispheres would have summer in July.
The Northern and Southern hemisphere experience opposite seasons.

The Conclusion
The hypothesis is falsified.

The Hypothesis
We have seasons because the tilt of Earth's axis causes one hemisphere to be significantly closer to the Sun than the other.

The Test
If this is true, the distances must be very different to cause such a large effect.
Because Earth is spherical, the difference in distance between the hemispheres is insignificant.

The Conclusion
The hypothesis is falsified.

The Hypothesis
We have seasons because Earth's tilt changes the distribution of energy—one hemisphere receives more light than the other.

The Test
If this is true, the amount of light striking the ground in the summer should be more than in the winter.

The Conclusion
Seasons are caused primarily by a change in illumination due to Earth's tilt. During winter, less energy falls on each square meter of ground per second.

Understanding why a phenomenon occurs often requires conceptual change.

FIGURE 2.18 Most cultural traditions in the Northern Hemisphere include a major celebration in late December, around the time when days begin to grow longer.

Earth's axis tilts toward the Sun about 3 months after the vernal equinox. When this happens, the Sun reaches its northernmost point in the sky, located in the constellation Taurus, near its border with Gemini. This day, which is the longest day of the year and marks the beginning of summer in the Northern Hemisphere, occurs around June 20. This is the **summer solstice** (point 2 in Figures 2.17a and b). (**Solstice** means literally "sun standing still.") Note that on this same day, the southern end of Earth's axis is tipped directly away from the Sun. This is the shortest day of the year in the Southern Hemisphere, marking the beginning of the southern winter.

Three months later, around September 22, the Sun is again crossing the celestial equator. This point on the Sun's apparent path, located in the constellation Virgo, is the **autumnal equinox** (point 3 in Figures 2.17a and b). The term refers both to the Sun's location on the celestial sphere and the date when this happens. Because the Sun is again on the celestial equator, days and nights are again exactly 12 hours long. It is the first day of autumn in the Northern Hemisphere and the first day of spring in the Southern Hemisphere.

Around December 21, the Sun reaches its southernmost point in the sky as its apparent path takes it through the constellation Sagittarius. This day is the northern **winter solstice** (point 4 in Figures 2.17a and b). Earth's North Pole is tipped most directly away from the Sun on the winter solstice. In the Northern Hemisphere this is the shortest day of the year—the first day of winter. As the Sun passes the winter solstice and moves on toward the vernal equinox, the northern days begin growing longer again. Almost all cultural traditions in the Northern Hemisphere include a major celebration of some sort in late December (**Figure 2.18**). Christmas, for example, is celebrated just a few days after the winter solstice. These winter festivals have many different meanings to their various celebrants, but they all share one thing: they celebrate the return of the source of Earth's light and warmth. The days have stopped growing shorter and are beginning to get longer. Spring will come again.

Interestingly, there is more to what we feel during the different seasons than the amount of energy we are receiving from the Sun. Just as it takes time for a pot of water on a stove to heat up when the burner is turned up and time for the pot to cool off when the burner is turned down, it takes time for Earth to respond to changes in heating from the Sun. The hottest months of northern summer are usually July and August, which come *after* the summer solstice, when the days are growing shorter. Similarly, the coldest months of northern winter are usually January and February, which occur *after* the winter solstice, when the days are growing longer. The climatic seasons on Earth lag behind changes in the amount of heating we receive from the Sun.

Seasonal temperatures lag behind changes in the directness of sunlight.

Our picture of the seasons must be modified somewhat near Earth's poles. At latitudes north of 66.5° north and south of 66.5° south, the Sun is circumpolar for a part of the year surrounding the first day of summer. These lines of latitude are the **Arctic Circle** and the **Antarctic Circle**. When the Sun is circumpolar, it is above the horizon 24 hours a day, earning the polar regions the nickname "land of the midnight Sun" (**Figure 2.19**). The Arctic and Antarctic regions pay for these long days, however, with an equally long period surrounding the first day of winter when the Sun never rises and the nights are 24 hours long. The Sun never rises high in the Arctic or Antarctic sky, which means that sunlight is never very direct. This is why, even with the long days at the height of summer, the Arctic and Antarctic regions remain relatively cool.

The seasons are also different near the equator. Recall from Figure 2.12c that for an observer on the equator, *all* stars are above the horizon 12 hours a day, and the Sun is no exception. On the equator, days and nights are 12 hours long throughout the year. Changes in the directness of sunlight through the year are also different on the equator. Here the Sun passes directly overhead on the first day of spring and the first day of autumn because these are the days when the Sun is on the celestial equator. Sunlight is most direct at the equator on these days. At the summer solstice, the Sun is at its northernmost point along the ecliptic. It is on this day, and on the winter solstice, that the Sun is farthest from the zenith at noon, and therefore sunlight is least direct at the equator. Strictly speaking, the equator experiences only two seasons: summer, when the Sun passes directly overhead; and winter, when the Sun is at its northernmost and southernmost points on the ecliptic. However, summer and winter are not very different. The Sun is up for 12 hours a day year-round, and the Sun is always so close to being overhead at noon that the directness of sunlight changes only slightly throughout the year.

If you live between the latitudes of 23.5° south and 23.5° north—in Rio de Janeiro or Honolulu, for example—the Sun will be directly overhead at noon twice during the year. The band between these two latitudes is called the **Tropics**. The northern limit of this region is the Tropic of Cancer; the southern limit is the Tropic of Capricorn (see Figure 2.15).

We have already pointed out that the Sun appears in the constellation of Taurus when at its northernmost point and in the constellation of Sagittarius at its southernmost point. Why, then, is the Tropic of Cancer not called the Tropic of Taurus? Why is the Tropic of Capricorn not called the Tropic of Sagittarius? The answer is historical. When these latitude limits were named about 2,000 years ago, the Sun's northernmost point *was* in Cancer. Since then, it has slowly drifted from Cancer to Gemini and is now in Taurus. Likewise, the Sun's southernmost point has drifted from Capricornus to Sagittarius.

Earth's Axis Wobbles, and the Seasons Shift through the Year

When the Alexandrian astronomer Ptolemy and his associates were formalizing their knowledge of the positions and motions of objects in the sky 2,000 years ago, the Sun appeared in the constellation of Cancer on the first day of northern summer and in the constellation of Capricornus on the first day of northern winter—hence the names of the Tropics. Which leads us to this question: Why have the constellations in which solstices appear changed? The answer has to do with the fact that *two* motions are associated with Earth and its axis: Earth spins on its axis, but its axis also wobbles like the axis of a spinning top (**Figure 2.20**). The wobble is very slow, taking about 26,000 years to complete one cycle. During this time the north celestial pole makes one trip around a large circle centered on the north ecliptic pole. Polaris is a modern

Earth's axis wobbles like the axis of a spinning top.

FIGURE 2.19 The midnight Sun, seen in latitudes above 66.5° north (or south). In the 360 degree panoramic view, the Sun moves 15° each hour.

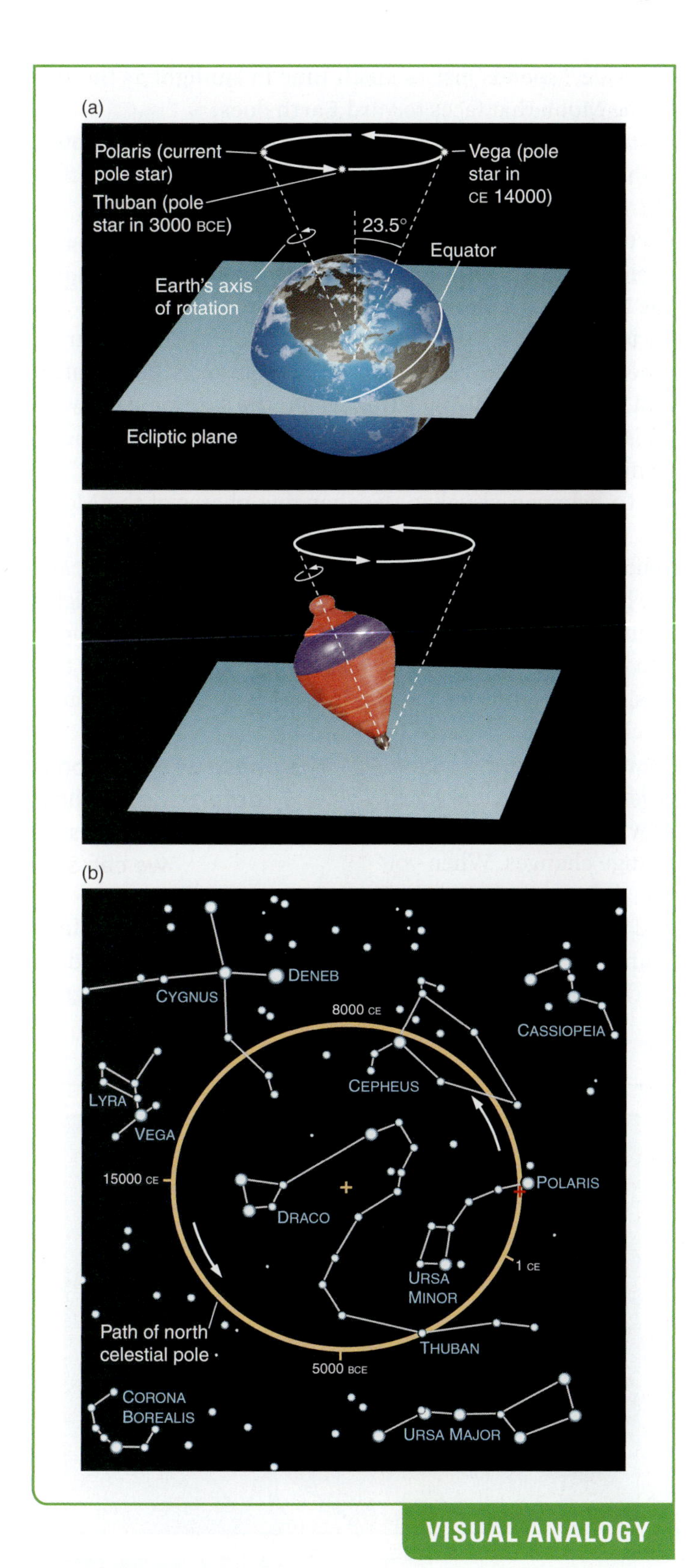

FIGURE 2.20 (a) Earth's axis of rotation changes orientation in the same way that the axis of a spinning top changes orientation. (b) This precession causes the projection of Earth's rotation axis to move in a circle, centered on the north ecliptic pole (orange cross in the center). The red cross marks the projection of Earth's axis on the sky in the early 21st century.

name for the star we see near the north celestial pole. If you could travel several thousand years into the past or future, you would find that the point about which the northern sky appears to rotate is no longer near Polaris.

Recall that the celestial equator is the set of directions in the sky that are perpendicular to Earth's axis. As Earth's axis wobbles, then, the celestial equator must appear to tilt with it. And as the celestial equator wobbles, the locations where it crosses the ecliptic—the equinoxes—change as well. During each 26,000-year wobble of Earth's axis, the locations of the equinoxes make one complete circuit around the celestial equator. Together, these shifts in position are called the **precession of the equinoxes**.

A 26,000-year wobble causes the position of the equinoxes to shift gradually.

2.4 The Motions and Phases of the Moon

The most prominent object in our sky after the Sun is the Moon. Just as Earth orbits about the Sun, the Moon orbits around Earth. (Actually, Earth and the Moon orbit around each other, and together they orbit the Sun, as we will see in Chapter 3.) In some respects, the appearance of the Moon is constantly changing, but we begin our discussion of the motion of the Moon by talking about an aspect of the Moon's appearance that does *not* change.

We Always See the Same Face of the Moon

The Moon constantly changes its lighted shape and position in the sky, but one thing that does not change is the face of the Moon that we see. If we were to go outside next week or next month, or 20 years from now, or 20,000 *centuries* from now, we would still see the same side of the Moon that we see tonight. This fact is responsible for the common misconception that the Moon does not rotate. The Moon does rotate on its axis—exactly once for each revolution that it makes about Earth.

The Moon rotates on its axis once for each orbit around Earth.

Imagine walking around the Washington Monument while keeping your face toward the monument at all times (a reasonable thing to do—you want to get a good look at it). By the time you complete one circle around the monument, your head has turned completely around once. (When you were south of the monument, you were facing north; when you were west of the monument, you were facing east; and so on.) But someone looking at you from the monument would never have seen anything other than your face. The Moon does exactly the same thing, rotating on its axis once per revolution around Earth, always keeping the same face toward Earth (**Figure 2.21**). This phenomenon is referred to as the Moon's **synchronous rotation**. ▶II **AstroTour: The Moon's Orbit: Eclipses and Phases**

The Changing Phases of the Moon

The Moon and its changing aspects have long fascinated humans. We speak of the "man in the Moon," the "harvest Moon," and sometimes a "blue Moon." In mythology, the Moon was the Roman goddess Diana, the Greek goddess Artemis, and the Inuit god Igaluk. The Moon has been the frequent subject of mythology, art, literature, and music. Sometimes the Moon appears as a circular disk in the sky. Other times it is nothing more than a thin sliver or its face appears dark. Popular culture often refers to the side of the Moon away from Earth as the "dark side of the Moon" (this is even the title of one of the most influential rock albums of the 20th century). In fact, however, there is no side of the Moon that is always dark. At any given time, half of the Moon is in sunlight and half is in darkness—just as at any given time, half of Earth is in sunlight and half is in darkness. The side of the Moon that faces away from Earth, the "far side," spends just as much time in sunlight as the side of the Moon that faces toward Earth does.

Unlike the Sun, the Moon has no light source of its own; it shines by reflected sunlight as the planets do. The different **phases** of the Moon occur because as the Moon orbits Earth, the portion of the Moon that is illuminated by the Sun is constantly changing. Sometimes (during a new Moon) the side facing away from us is illuminated, and sometimes (during a full Moon) the side facing toward us is illuminated. The rest of the time, only part of the illuminated portion can be seen from Earth.

The Moon shines by reflected sunlight.

To help you visualize the changing phases of the Moon, use an orange, a desk lamp (with the shade removed), and your head. Your head is Earth, the orange is the Moon, and the lamp is the Sun (**Figure 2.22**). Turn off all the other lights in the room, and push your chair back as far from the lamp as you can. Hold up the orange slightly above your head so that it is illuminated from one side by the lamp. Move the orange clockwise (this would be counterclockwise if viewed from the ceiling), and watch how the appearance of the orange changes. When you are between the orange and the lamp, the face of the orange that is toward you is fully illuminated. The orange appears to be a bright, circular disk. As the orange moves around its circle, you will see a progres-

The phase of the Moon is determined by how much of its bright side we can see.

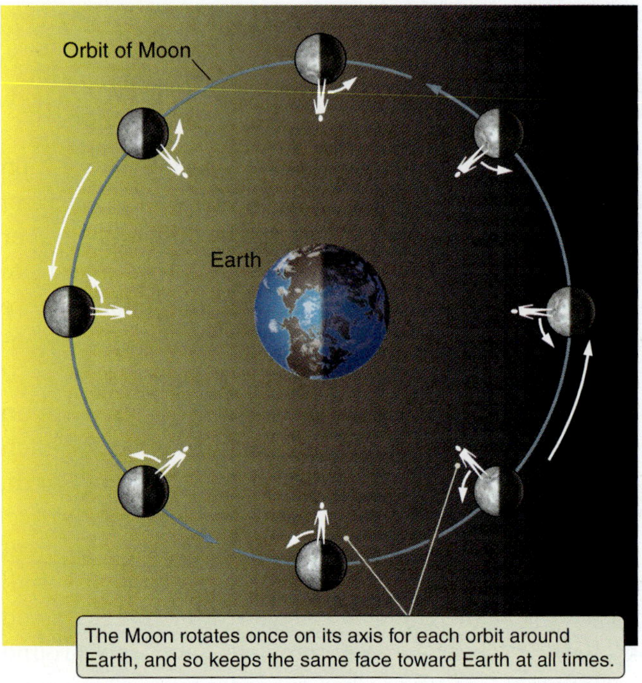

Orbit of Moon

Earth

The Moon rotates once on its axis for each orbit around Earth, and so keeps the same face toward Earth at all times.

FIGURE 2.21 The Moon rotates once on its axis for each orbit around Earth—an effect called synchronous rotation. The Sun is on the left.

FIGURE 2.22 You can experiment with illumination effects by using an orange as the Moon, a lamp with no shade as the Sun, and your own head as Earth. As you move the orange around your head, viewing it in different relative locations, you will see that the illuminated part of the orange mimics the phases of the Moon.

sion of lighted shapes, depending on how much of the bright side and how much of the dark side of the orange you can see. This progression of shapes exactly mimics the changing phases of the Moon.

Figure 2.23 shows the changing phases of the Moon. When the Moon is between Earth and the Sun, the sunlit side of the Moon faces away from us, and we see only its night side. This phase is called a **new Moon**. (Study Figure 2.23 to understand that a new Moon can be "seen" only from the illuminated side of Earth.) The new Moon appears close to the Sun in the sky, so it rises in the east at sunrise, crosses the meridian near noon, and sets in the west near sunset. A new Moon is never visible in the nighttime sky. **AstroTour: The Moon's Orbit: Eclipses and Phases**

As the Moon continues on its orbit around Earth, a small part of its illuminated hemisphere becomes visible. This shape is called a **crescent**, from the Latin *crescere*, meaning "to grow." Because the Moon appears to be "filling out" from night to night at this time, the full name for this phase of the Moon is a *waxing crescent Moon*. (**Waxing** here means "growing in size and brilliance.") From our perspective, the Moon has also moved away from the Sun in the sky. Because the Moon travels around Earth in the same direction in which Earth rotates, we now see the Moon located to the east of the Sun. A waxing crescent Moon is visible in the western sky in the evening, near the setting Sun but remaining above the horizon after the Sun sets. The "horns" of the crescent always point directly away from the Sun.

As the Moon moves farther along in its orbit, more and more of its illuminated side becomes visible each night, so the crescent continues to fill out. At the same time, the angular separation in the sky between the Moon and the Sun grows. After about a week the Moon has moved a quarter of the way around Earth. We now see half the Moon as illuminated and half the Moon as dark—a phase that we

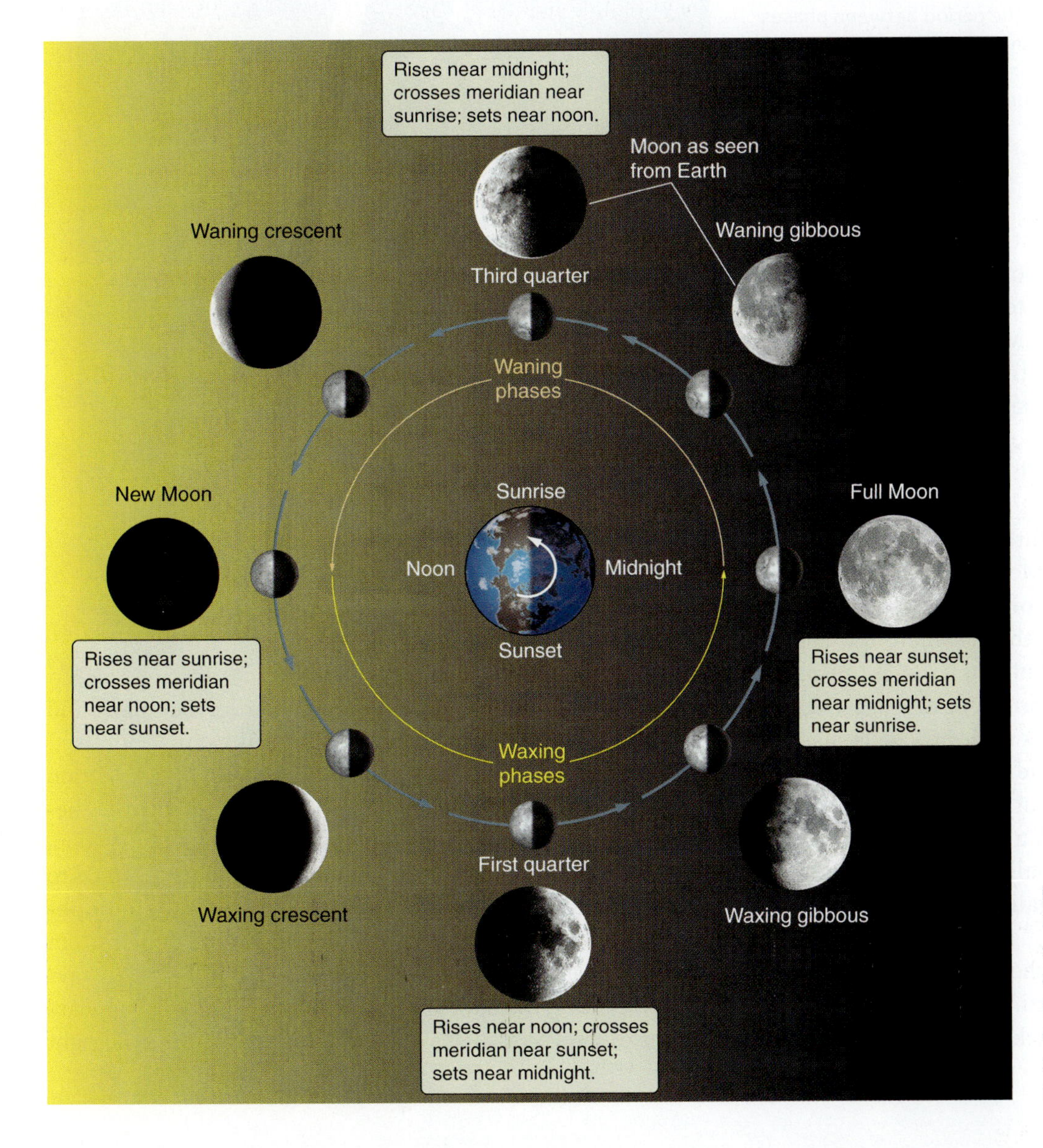

FIGURE 2.23 The inner circle of images (connected by blue arrows) shows the Moon as it orbits Earth, as seen by an observer far above Earth's North Pole. The outer ring of images shows the corresponding phases of the Moon as seen from Earth. The Sun is on the left.

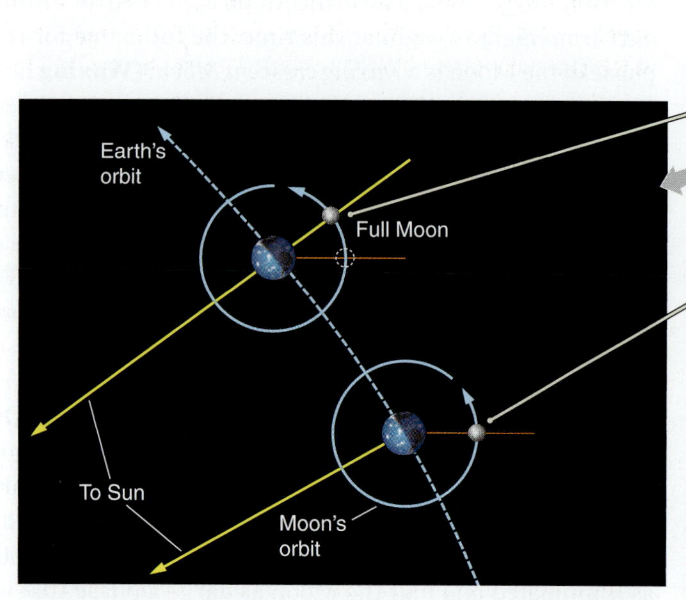

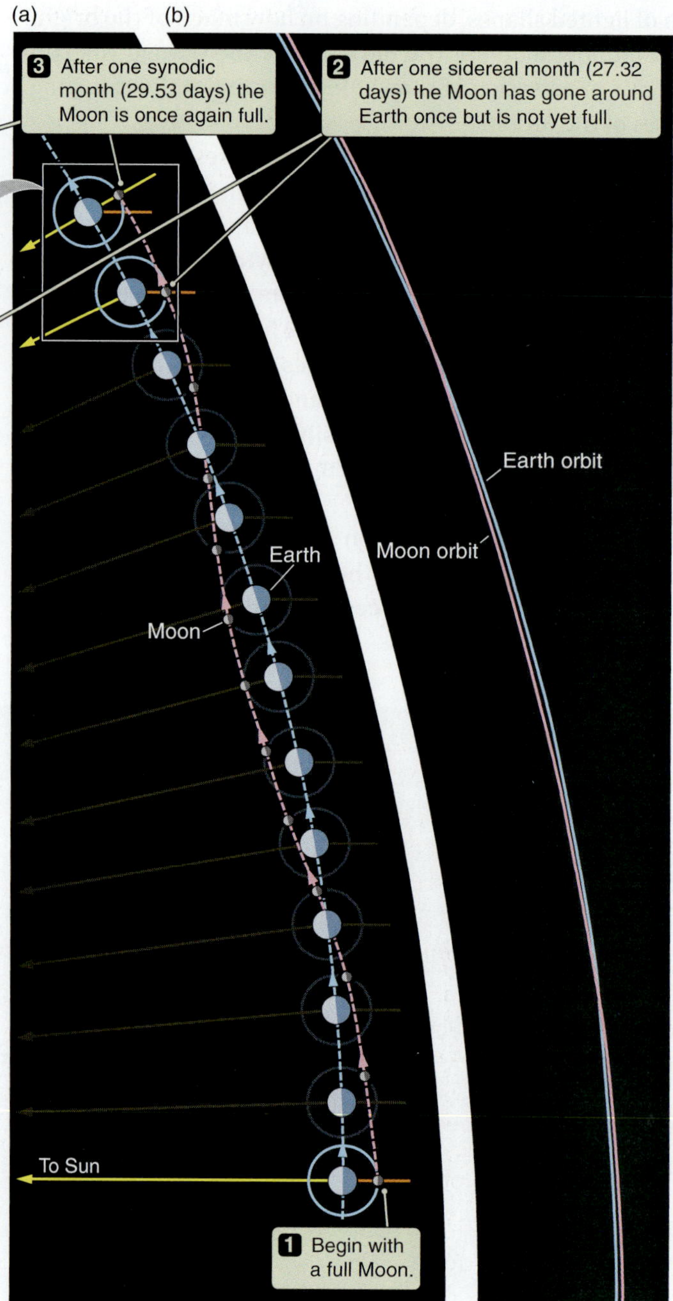

FIGURE 2.24 (a) The Moon completes one sidereal orbit in 27.32 days, but the synodic period (the period between phases seen from Earth) from one full Moon to the next is 29.53 days. (The horizontal orange line to the right of the Moon indicates a fixed direction in space.) (b) The orbits of Earth and the Moon are shown here to scale.

call **first quarter Moon**. A look at Figure 2.23 shows that the first quarter Moon rises at noon, crosses the meridian at sunset, and sets at midnight. Note that *first quarter* refers not to how much of the face of the Moon that we see illuminated, but rather to the fact that the Moon has completed the first quarter of its cycle from new Moon to new Moon.

As the Moon moves beyond first quarter, we are able to see more than half of its bright side. This phase is called a *waxing gibbous Moon*, from the Latin *gibbus*, meaning "hump." The **gibbous** Moon continues nightly to "grow" until finally Earth is between the Sun and the Moon and we see the entire bright side of the Moon—a **full Moon**. The Sun and the Moon now appear opposite each other in the sky when viewed from Earth. The full Moon rises as the Sun sets, crosses the meridian at midnight, and sets in the morning as the Sun rises.

The second half of the Moon's orbit proceeds just like the first half, but in reverse. The Moon continues in its orbit, again appearing gibbous but now becoming smaller each night. This phase is called a *waning gibbous Moon*. (**Waning** means "becoming smaller.") A **third quarter Moon** occurs when we once again see half of the sunlit part of the Moon and half of the night side. A third quarter Moon rises at midnight, crosses the meridian near sunrise, and sets at noon. The Moon continues on its path, visible now as a *waning crescent Moon* in the morning sky, until the Moon again appears as nothing but a dark circle rising and setting with the Sun, and the cycle begins again.

It takes the Moon 27.32 days to complete one revolution about Earth; this is called its **sidereal period**. However, because of the changing relationships among Earth, the Moon, and the Sun due to Earth's orbital motion, it takes 29.53 days to go from one full Moon to the next; this is called its **synodic period** (**Figure 2.24**). You can always tell a waxing Moon from a waning Moon because the side that is illuminated is always the side facing the Sun. When the Moon is waxing, it appears in the evening sky, so its western side is illuminated (the right side as viewed from the Northern Hemisphere). Conversely, when the Moon is waning, the eastern side (the left side as viewed from the Northern Hemisphere) appears bright.

Do not try to memorize all possible combinations of where the Moon is in the sky at what phase and at what time of day. Instead, work on *understanding* the motion and phases of the Moon, and then use your understanding to figure out the specifics of any given case. To study the phases of the Moon, draw a picture like Figure 2.23, and use it to follow the Moon around its orbit. From your drawing, figure out what phase you would see and where it would appear in the sky at a given time of day. You might also try the simulations described in "Exploration: The Phases of the Moon" at the end of the chapter. ▶▶ **NEBRASKA SIMULATIONS: LUNAR PHASE VOCABULARY; BASKETBALL PHASES SIMULATOR; THREE VIEWS SIMULATOR; LUNAR PHASES SIMULATOR (WAAP); MOON PHASES AND THE HORIZON DIAGRAM; MOON PHASES WITH BISECTORS; PHASE POSITIONS DEMONSTRATOR; LUNAR PHASE QUIZZER**

2.5 Cultures and Calendars

Historians tell us that the development of agriculture was crucial for the rise of human civilization, and keeping track of the seasons and best times of the year to plant and harvest was critical to successful farming. Records going back to the dawn of humanity suggest that people kept track of time by following the patterns in the sky, especially those of the Sun and the Moon. Some anthropologists have speculated that notches on fragments of bone found in southern France represent a 33,000-year-old lunar calendar. As civilizations developed around the globe, different cultures tried to solve the "problem" of the calendar. Specifically, the number of days does not fit neatly into months (29.5 days) or years (365.24 days), and the number of lunar cycles (full Moon to full Moon) does not fit neatly into the solar cycle (the time it takes for the Sun to leave and return to its highest possible point in the sky at noon on the summer solstice). Twelve months of 29.5 days each comes to a total of 354 days, 11.24 days short of a solar year.

Some of the oldest known calendars come from the Babylonians, the Egyptians, and the Chinese. The ancient Egyptians kept track of their history with a rather interesting calendar. They used 12 months of 30 days each—which added up to 360 days—and then added five "festival days" to the end of the year. Without leap years, this practice led to a drift of the seasons, so an extra month was added when necessary. When we consider how we celebrate the days between the modern December holidays and the New Year, an end-of-year calendar break for festivals seems like a good solution!

The Babylonians started the 24-hour day and 7-day week—7 for the Sun, the Moon, and the 5 planets visible with the naked eye. They created the first lunisolar calendar, in which a month began with the first sighting of the lunar crescent, and a 13th month was added when needed to catch up to the solar year. As the Babylonians developed mathematics, they discovered the Metonic cycle, in which 235 lunar months equals 19 solar years equals 6,940 days. In the Metonic scheme, a regular cycle consists of 19 years in which 12 years have 12 months and 7 years have 13 months, and then the cycle repeats after 19 years. The Hebrew calendar adopted the Metonic cycle and uses it to this day.

The ancient Chinese occasionally added a 13th month to their calendar, which dates back several thousand years. By about 500 BCE the Chinese were using a year of 365.25 days and a system similar to the Metonic cycle. Other cultures used lunar and stellar calendars, paying attention to the position of a bright star like Sirius or to certain prominent groups of stars in their sky, such as the Pleiades or the Big Dipper, to mark out a year.

The Islamic calendar is a purely lunar calendar, with no 13th lunar month added in. The 12 months of 29 or 30 days each add up to 354 days—11.24 days short of the solar year. For this reason, the Islamic New Year and all other holidays occur 11 days earlier in each solar year. In the Islamic calendar, a holiday may fall in the winter in some years, and then a few years later it will have moved back to autumn.

The civil calendar we use today is known as the **Gregorian calendar** and is based on the **tropical year**, which is 365.242 solar days long. A **solar day** is the 24-hour period of Earth's rotation that brings the Sun back to the same local meridian. (This is in contrast to the **sidereal day**, which is the time it takes for Earth to make one rotation and face the same part of the sky, the vernal equinox. The sidereal day is about 23 hours 56 minutes, and differs from the solar day because of Earth's motion around the Sun). The tropical year measures the time from one vernal equinox to the next—from the start of spring to the start of spring. Notice that the tropical year is not an integral number of days long, but is approximately ¼ day longer than 365. The Gregorian calendar includes a system of **leap years** decreed by Julius Caesar in 45 BCE—years in which a 29th day is added to the month of February—to make up for the extra fraction of a day. Leap years prevent the seasons from slowly sliding through the year (becoming increasingly out of sync with the months), so we don't end up experiencing winter in December one year and in August other years. ▶▶ **NEBRASKA SIMULATION: SYNODIC LAG**

The Gregorian calendar is named for Pope Gregory XIII, who was concerned that the Easter holiday (which by definition falls on the first Sunday after the first full Moon following March 21) was drifting from the Spring Equinox. Julius

Caesar's rule of one leap year every 4 years resulted in an average year of 365.25 days, but the actual year is 365.242 days. This difference of 0.008 day is about 11.5 minutes per year, or 3 days every 400 years, and by Gregory's time it had caused the spring equinox to drift by about 10 days. To address this problem, in 1582 Pope Gregory decreed that 10 days would be deleted from the calendar to move the vernal equinox back to March 21. And to make this work out better in the future, he decreed that only century years divisible by 400 are leap years, thereby deleting 3 leap years (and the 3 days) every 400 years. Catholic countries followed this system immediately, but Protestant countries did not adopt it until the 1700s. Eastern Orthodox countries, including Russia, did not switch from the Julian to the Gregorian calendar until the 1900s. The Gregorian calendar is now the internationally accepted civil calendar. One slight further revision—making years divisible by 4,000 into common 365-day years—was eventually made, and now the modern Gregorian calendar slips by about only 1 day in 20,000 years.

Despite international adoption of the Gregorian calendar, billions of people still celebrate holidays and festivals according to a lunisolar or lunar calendar. Chinese New Year, Passover, Easter, Ramadan, Rosh Hashanah, and Diwali, among others, have dates that change from one year to the next because they are based on lunar months from these different calendars. The astronomy of people from long ago is still in use today.

2.6 Eclipses: Passing through a Shadow

For ancient peoples attuned to the patterns of the sky, it must have been terrifying to look up during an eclipse and see the Sun being eaten away as if by a giant dragon, or the full Moon turning the color of blood. An **eclipse** is the total or partial obscuration of one celestial body, or the light from that body, by another celestial body. Archaeological evidence suggests that ancient peoples put great effort into trying to figure out the pattern of eclipses and thereby bring them into the orderly scheme of the heavens. Stonehenge, pictured in Figure 2.2a, may have enabled its builders to predict when eclipses might occur. Ancient Chinese, Babylonian, and Greek astronomers had figured out that eclipses occur in cycles, and they were able to use their knowledge of calendars to make predictions about when and where eclipses would occur.

Varieties of Eclipses

There are three types of solar eclipses: total, partial, and annular.

An eclipse in which Earth moves through the shadow of the Moon is a **solar eclipse**. Three different types of solar eclipses are possible: *total*, *partial*, and *annular*. To learn why, consider the structure of the shadow of the Sun cast by a round object such as the Moon, as shown in **Figure 2.25**. An observer at point A would be unable to see any part of the surface of the Sun. This darkest, inner part of the shadow is called the **umbra**. If a point on Earth passes through the Moon's umbra, the Sun's light is totally blocked by the Moon. This is a **total solar eclipse** (**Figure 2.26**). At points B and C in Figure 2.25, an observer can see one side of the disk of the Sun but not the other. This outer region, which is only partially in shadow, is the **penumbra**. If a point on the surface of Earth passes through the Moon's penumbra, viewers at that point will observe a **partial solar eclipse**, in which the disk of the Moon blocks the light from a portion of the Sun's disk.

In the third type of eclipse, called an **annular solar eclipse**, the Sun appears as a bright ring surrounding the

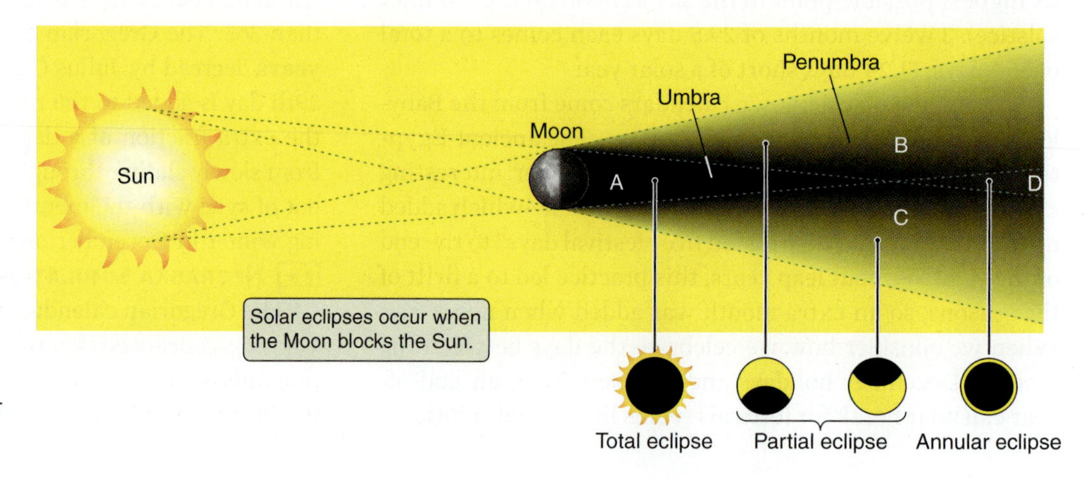

FIGURE 2.25 Different parts of the Sun are blocked at different places within the Moon's shadow. An observer on Earth in the umbra (A) sees a total solar eclipse, observers in the penumbra (B and C) see a partially eclipsed Sun, and observers at point D see an annular solar eclipse.

FIGURE 2.26 The full spectacle of a total eclipse of the Sun.

FIGURE 2.27 An annular eclipse, in which the Moon does not quite cover the Sun. Note that Earth's atmosphere distorts the shape of the Sun when close to the horizon.

dark disk of the Moon (**Figure 2.27**). An observer at point D in Figure 2.25 is far enough from the Moon that the Moon's apparent size in the sky is smaller than the Sun's. The actual size of the Moon and Sun do not change, but one eclipse can be total and another annular. Two things make this possible. One is a fluke of nature: The diameter of the Sun is about 400 times the diameter of the Moon, and the Sun is about 400 times farther away from Earth than the Moon is. As a result, the Moon and Sun have almost exactly the same apparent size in the sky. The second factor is that the Moon's orbit is not a perfect circle. When the Moon and Earth are a bit closer together than on average, the Moon appears larger in the sky than the Sun. An eclipse occurring at that time will be total. When the Moon and Earth are farther apart than on average, the Moon appears smaller than the Sun, so eclipses occurring during this time will be annular. Sequential pictures of total and annular solar eclipses are shown in **Figure 2.28**. Among all solar eclipses, one-third are total, one-third are annular, and one-third are seen only as a partial eclipse.

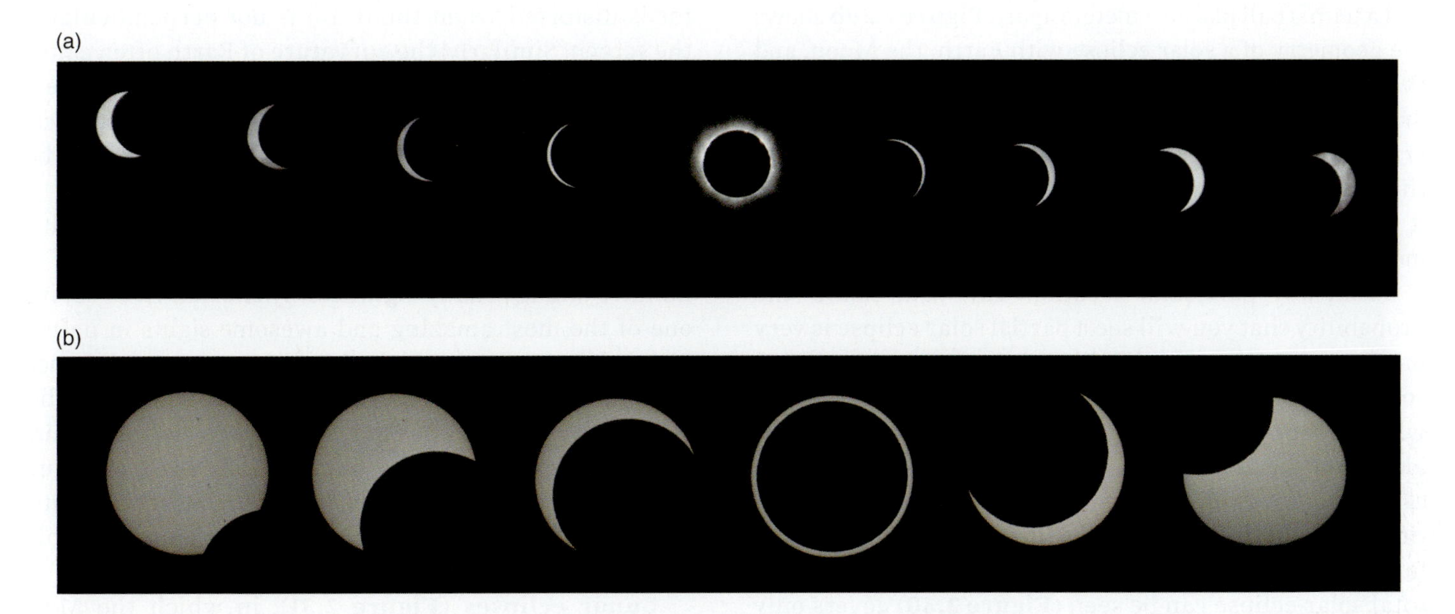

FIGURE 2.28 Time sequences of images of the Sun taken during a total solar eclipse (a) and during (b) the annular solar eclipse of May 20, 2012. The Sun set during the ending phases.

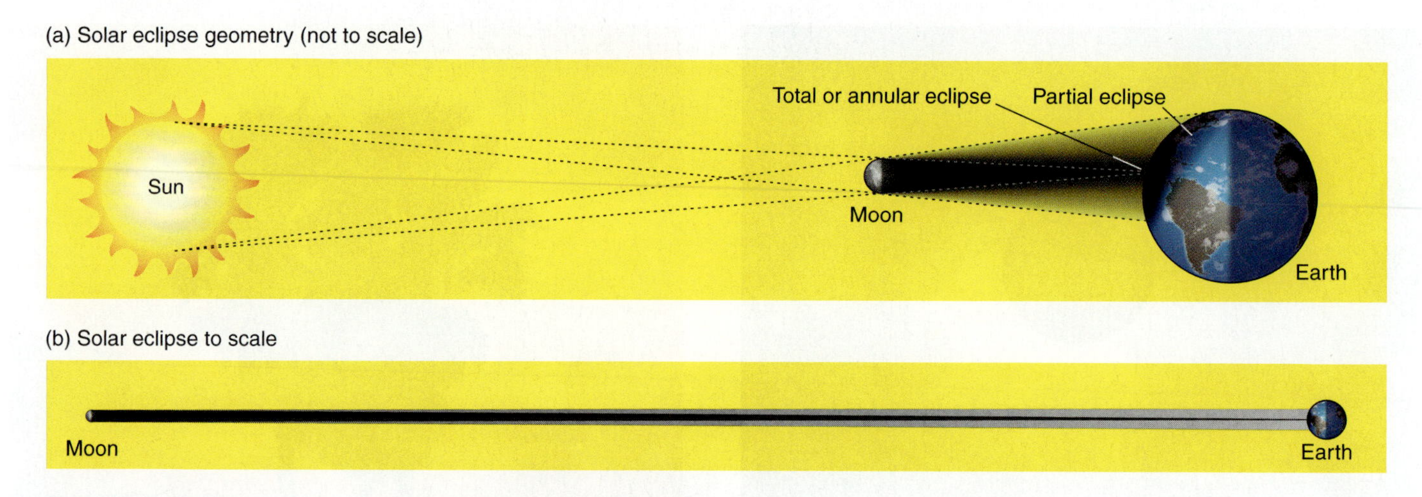

FIGURE 2.29 A solar eclipse occurs when the shadow of the Moon falls on the surface of Earth. Note that (b) is drawn to proper scale.

Figure 2.29 illustrates the geometry of a solar eclipse, with the Moon's shadow falling on the surface of Earth. Note that figures like this (or like Figures 2.21 and 2.23) are seldom drawn to scale. Instead, they show Earth and the Moon much closer together than they are in reality. The reason for distorting figures in this way is simple: there is not enough room on the page to draw them to scale and keep the smaller details visible. The 384,400-km distance between Earth and the Moon is over 60 times the radius of Earth and over 220 times the radius of the Moon. The relative sizes and distances between Earth and the Moon are roughly equivalent to the difference between a basketball and a tennis ball placed 7 meters apart. **Figure 2.29b** shows the geometry of a solar eclipse with Earth, the Moon, and the separation between them drawn to scale. Compare this drawing to **Figure 2.29a** and you will understand why artistic license is normally taken in drawings of Earth and the Moon. If the Sun were drawn to scale in Figure 2.29a, it would be 6/10 of a meter across and located almost 64 meters off the left side of the page.

From any particular location—say, your home—the probability that you will see a partial solar eclipse is very much greater than the likelihood of your seeing a total solar eclipse. The reason is that the Moon's penumbra where it touches Earth has a diameter about twice the diameter of the Moon itself, or almost 7,000 km. This part of the shadow is large enough to cover a substantial fraction of Earth, so partial solar eclipses are often seen from much of the planet. In contrast, the path along which a total solar eclipse can be seen (**Figure 2.30**) covers only a tiny fraction of Earth's surface. Earth is so close to the tip of the Moon's umbra that even when the distance between Earth and the Moon is at a minimum, the umbra is only 270 km wide at the surface of Earth. As the Moon moves along in its orbit, this tiny shadow sweeps across the face of Earth at speeds of a few thousand kilometers per hour. The Moon moves in its orbit around Earth at a speed of about 3,400 km/h, and its shadow sweeps across the disk of Earth at about the same rate. Earth is also rotating on its axis with a velocity of 1,670 km/h at the equator (and less than that at other latitudes). The situation is further complicated by the fact that the Moon's shadow falls on the curved surface of Earth. You may have noticed that the image displayed by an LCD projector is distorted when the beam is not perpendicular to the screen. Similarly, the curvature of Earth often causes the region shaded by the Moon during a solar eclipse to be elongated by differing amounts. The curvature can even cause an eclipse that started out as annular to become total.

When all of these effects are considered, the result is that a total solar eclipse can never last longer than 7½ minutes and is usually significantly shorter. Even so, it is one of the most amazing and awesome sights in nature. People all over the world flock to the most remote corners of Earth to witness the fleeting spectacle of the bright disk of the Sun blotted out of the daytime sky, leaving behind the eerie glow of the Sun's outer atmosphere (see Figure 2.27). The first total solar eclipse in the continental United States since 1979 will take place in 2017. Viewing a solar eclipse should be on your lifetime to-do list!

Lunar eclipses (**Figure 2.31**), in which the Moon moves through the shadow of Earth, are very different in character from solar eclipses. The geometry of a lunar

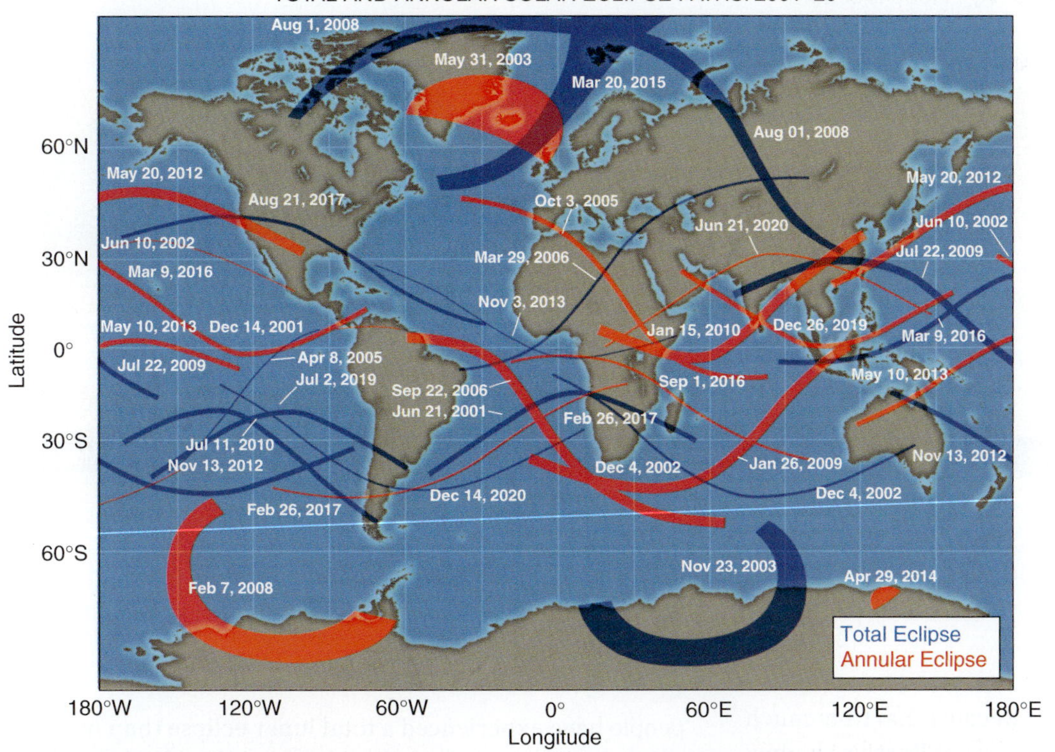

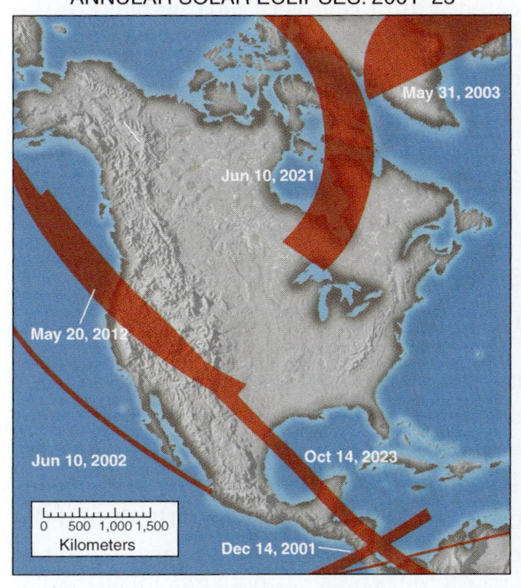

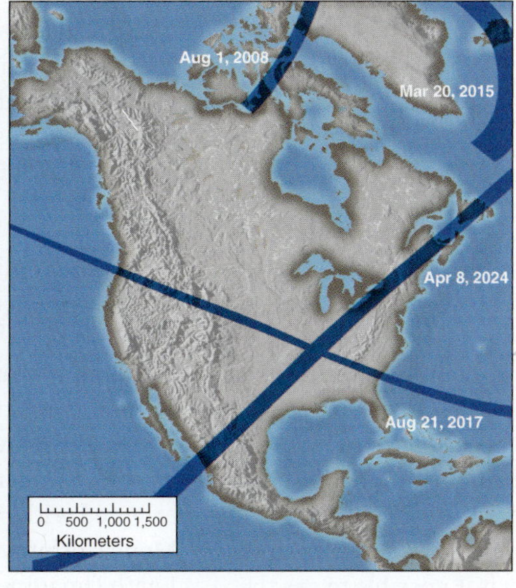

FIGURE 2.30 The paths of total and annular solar eclipses predicted for the early 21st century. Solar eclipses occurring in Earth's polar regions cover more territory because the Moon's shadow hits the ground obliquely.

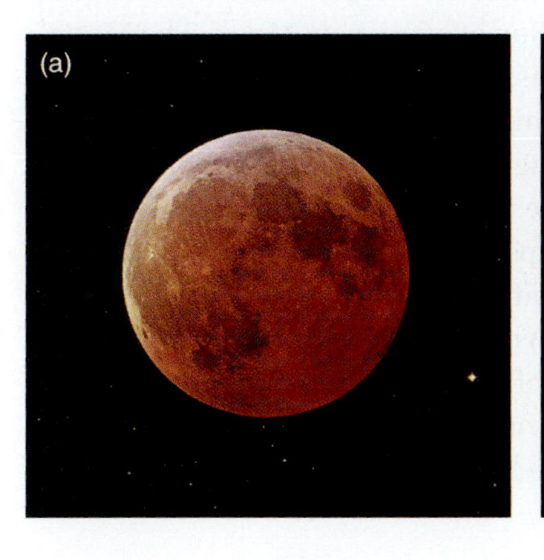

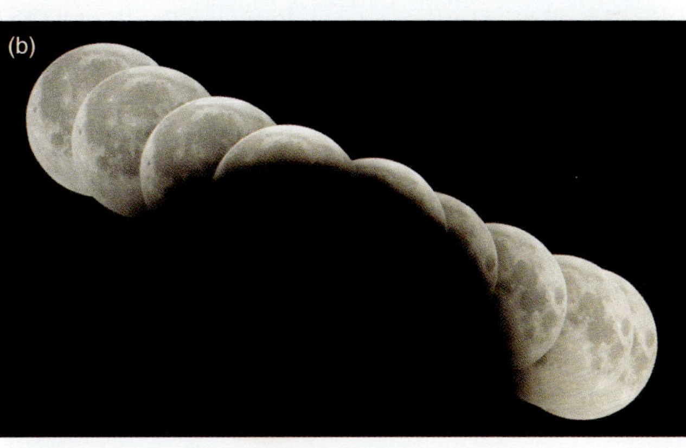

FIGURE 2.31 (a) A total lunar eclipse. (b) The progression of a partial lunar eclipse. Note the size of Earth's shadow compared to the size of the Moon.

FIGURE 2.32 A lunar eclipse occurs when the Moon passes through Earth's shadow. Note that (b) is drawn to proper scale

eclipse is shown in **Figure 2.32**. Because Earth is much larger than the Moon, the dark umbra of Earth's shadow is over 2½ times the diameter of the Moon. A **total lunar eclipse** is a much more leisurely affair than a total solar eclipse, with the Moon spending as long as 1 hour 40 minutes in the umbra of Earth's shadow. A **penumbral lunar eclipse** occurs when the Moon passes through the penumbra of Earth's shadow. A penumbral eclipse can be unspectacular: its appearance from Earth is nothing more than a fading in the brightness of the full Moon. Although the penumbra of Earth is 16,000 km across at the distance of the Moon—over 4 times the diameter of the Moon—a penumbral eclipse is noticeable only when the Moon passes within about 1,000 km of the umbra.

Total lunar eclipses last much longer but are less dramatic than total solar eclipses.

Total eclipses of the Moon are relatively common, and you may have seen at least one. How is it possible to see the Moon when it is completely immersed in Earth's shadow? Take another look at Figure 2.32, and imagine standing on the Moon during a total lunar eclipse. What would you see? One thing you would not see is the Sun, because the Sun would be behind Earth's disk. What you *would* see is Earth's dark disk surrounded by a thin, bright, reddish ring. That ring is sunlight coming from behind Earth and being scattered by dust and other small atmospheric particles. From Earth that bright reddish ring, Earth's atmosphere, lights up the Moon's surface during a total lunar eclipse.

Once you understand the geometry of solar and lunar eclipses, you will also understand why so many more people have experienced a total lunar eclipse than have experienced a total solar eclipse. To see a total solar eclipse, you must be located within that very narrow band of the Moon's shadow as it moves across Earth's surface. On the other hand, when the Moon is immersed in Earth's shadow, anyone located in the hemisphere of Earth that is facing the Moon can see it. ▶▶ **NEBRASKA SIMULATIONS: MOON INCLINATIONS; ECLIPSE SHADOW SIMULATOR**

Eclipse Seasons Occur Roughly Twice Every 11 Months

You know from experience that you don't see a lunar eclipse every time the Moon is full, nor do you observe a solar eclipse every time the Moon is new. These facts tell us something about how the Moon's orbit around Earth is oriented with respect to Earth's orbit around the Sun. If the Moon's orbit were in exactly the same plane as the orbit of Earth (imagine Earth, the Moon, and the Sun all sitting on the same flat tabletop), then the Moon would pass directly between Earth and the Sun at every new Moon. The Moon's shadow would pass across the face of Earth, and we would see a solar eclipse. Similarly, Earth would pass directly between the Sun and the Moon every synodic month, and each full Moon would be marked by a lunar eclipse.

Solar and lunar eclipses do not happen every month, because the Moon's orbit does not lie in exactly the same plane as the orbit of Earth. Study **Figure 2.33** to see how this works. The plane of the Moon's orbit about Earth is inclined by about 5.2° with respect to the plane of Earth's orbit about the Sun. The line along which the orbital plane of the Sun and the orbital plane of the Moon intersect is called the **line of nodes**. For part of the year, the line of nodes points in the general direction of the Sun. During these times, called **eclipse seasons**, a new Moon passes between the Sun and Earth, casting its shadow on Earth's surface and causing a solar eclipse. Similarly, a full Moon occurring during an eclipse season passes through Earth's shadow, and a lunar eclipse results. An eclipse season lasts only 38 days. That's how long the Sun is close enough to the line of nodes for eclipses to occur. Most of the time the line of nodes points farther away from the Sun, and Earth, Moon, and Sun cannot line up closely enough for an eclipse to occur. A solar eclipse cannot take place, because the shadow of a new Moon passes "above" or "below" Earth. Similarly, no lunar eclipse can occur, because a full Moon passes "above" or "below" the shadow of Earth.

If the plane of the Moon's orbit always had the same orientation, then eclipse seasons would occur twice a year, as suggested by Figure 2.33. In actuality, eclipse seasons occur about every 5 months and 20 days. The roughly 10-day difference is due to the fact that the plane of the Moon's orbit slowly wobbles, much like the wobble of a spinning plate balanced on the end of a circus performer's stick. As it does so, the line of nodes changes direction. This wobble rotates in the direction opposite the direction of the Moon's motion in its orbit. (That is, the line of nodes moves clockwise as viewed from above Earth's orbital plane.) It takes the Moon's orbit 18.6 years to complete one wobble of 360°, so we say that the line of nodes *regresses* at a rate of 360° every 18.6 years, or 19.4° per year. This amounts to about a 20-day regression each year. If January 1 marks the middle of an eclipse season, the next eclipse season will be centered around June 20, and the one after that around December 10.

▶▶ **NEBRASKA SIMULATION: ECLIPSE TABLE**

2.7 Origins: The Obliquity of Earth

The various motions of Earth give rise to the most basic of patterns faced by life on Earth. Earth's rotation is responsible for the coming of night and day. Earth's axial tilt and its passage around the Sun bring the changing of the seasons. Life evolving on Earth had to adapt to these patterns.

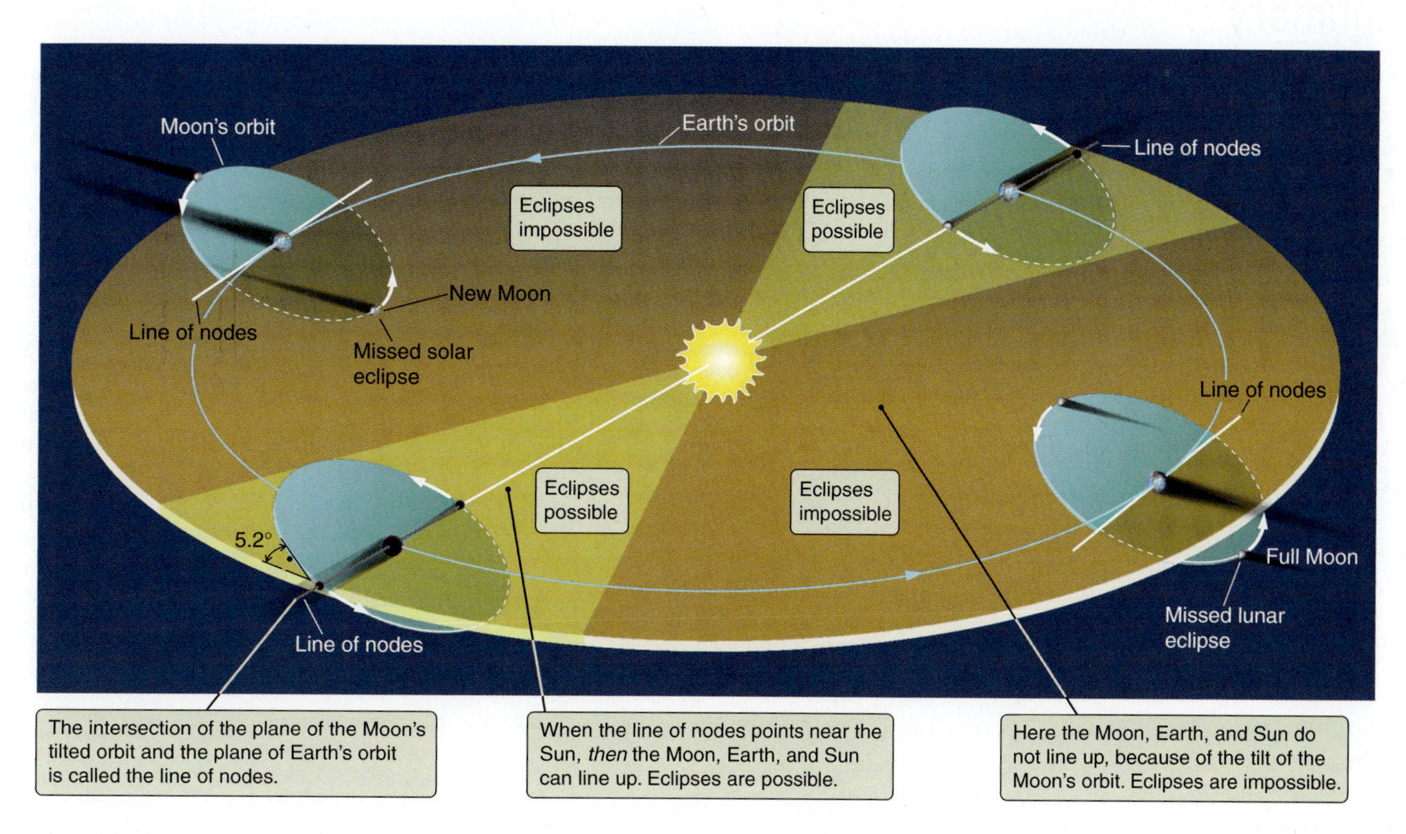

FIGURE 2.33 Eclipses are possible only when the Sun, Moon, and Earth lie along (or very close to) an imaginary line known as the line of nodes. When the Sun does not lie along the line of nodes, Earth passes under or over the shadow of a new Moon, and a full Moon passes under or over the shadow of Earth.

Astronomers use the term **obliquity** to refer to the angle between a planet's equatorial and orbital planes. We have discussed how seasonal variations on Earth come from the obliquity, or "tilt," of Earth's axis. Changes in the direction in which sunlight falls on Earth cause dramatic differences in climate from the equator to the poles. If Earth's tilt were larger than 23.5°, the seasonal variation would be even stronger. If the tilt were smaller, the seasonal variation would be weaker. How might this obliquity have affected the development and evolution of life on our planet? The range of climate on Earth based on distance from the equator likely has contributed to the broad diversity of life. Earth's biodiversity includes life that adapted to the long, cold polar nights, and life in equatorial latitudes that adapted to much higher temperatures. Earth's life adapted to seasonal patterns in rain and drought, leading to acquired seasonal patterns of migration and reproduction.

The tilt of Earth's axis was estimated in ancient times by measuring the length of the shadow from a vertical pole on the day of the solstice. Measurements of the tilt go back 3,000 years from Chinese, Greek, and Arabic records. We now know that Earth's tilt actually varies from 22.1° to 24.5° over a 41,000-year cycle. The Moon is responsible for maintaining the tilt within this small range over the past half-billion years—about the time since animal life greatly diversified on Earth. Currently the tilt is about midway between the two extremes and getting smaller. It will reach its minimum value of 22.1° in about 10,000 years. Scientists are studying how this variation in tilt might correlate with periods of climate change on Earth, especially the times of ice ages. ▶▶ **NEBRASKA SIMULATION: OBLIQUITY SIMULATOR**

Throughout this book we will encounter many of the motions discussed in this chapter, but from a rather different perspective. So far, we have concentrated on describing the motion of Earth about the Sun and of the Moon about Earth. From here we will take the step that separates post-Renaissance science from all that came before, by taking up the question of *why* these motions are as they are. The search for understanding will lead us to a discovery that changed perceptions of the universe and our place in it: that the same natural forces that dictate the path of a well-hit softball also govern the clockwork motions of Earth, the Moon, and planets. ■

Summary

2.1 Constellations are patterns of stars that reappear in the same place in the sky at the same time of night each year. You can determine your latitude from the altitude of the pole star.

2.2 A day is the time it takes for Earth to complete one rotation on its axis. This rotation causes the apparent daily motion of the Sun, Moon, and stars. Our location on Earth and Earth's location in its orbit about the Sun determine which stars we see at night.

2.3 A year is the time it takes for Earth to complete one revolution around the Sun. The ecliptic is the path that the Sun appears to take through the stars. A frame of reference is a coordinate system within which an observer measures positions and motions. The tilt of Earth's axis determines the seasons, by changing the angle at which sunlight strikes the surface in different locations. The changing angle of sunlight and the differing length of the day is the reason temperatures change with the seasons in most regions of Earth. The changing seasons are marked by equinoxes and solstices.

2.4 The relative locations of the Sun, Earth, and Moon determine the phases of the Moon. A month is the time it takes for the Moon to complete one revolution about Earth. The phase of the Moon is determined by how much of its bright side is visible from Earth

2.5 Different cultures created various calendars based on the day (sunrise to sunrise), the month (full Moon to full Moon) and the year (for example, fall equinox to fall equinox, or winter solstice to winter solstice).

2.6 Special alignments of the Sun, Earth, and Moon result in solar and lunar eclipses. A solar eclipse occurs when the new Moon is in the plane of Earth and the Sun. A lunar eclipse occurs when the full Moon is in the plane of Earth and the Sun and the shadow of Earth falls on the Moon.

2.7 The tilt of Earth's axis determines its seasons, causing the variation in climate at different latitudes and over different times of the year. This tilt varies slightly over tens of thousands of years. Life on Earth adapted to these seasonal variations.

Unanswered Questions

- How long will Earth continue to have total solar eclipses? These occur because the Moon and the Sun are coincidentally the same size in our sky, but will that always be the case? The observed size of an object in the sky depends on its actual diameter and its distance from us. And these can change. The Moon is slowly moving away from Earth by about 4 meters per century. Over time the Moon will appear smaller in the sky, and it won't be able to cover the full disk of the Sun. While we can measure the current rate of the Moon's movement away from Earth, we are less certain of how this rate may change with time. A lesser and more uncertain effect comes from the Sun—which will continue to brighten slowly, as it has throughout its history. With this brightening, the actual diameter of the Sun may slightly increase, and it will appear larger in our sky. A more distant Moon and a larger Sun will eventually result in an end to total eclipses on Earth.

Questions and Problems

Summary Self-Test

1. The Sun, Moon, and stars
   a. appear to move each day because Earth rotates.
   b. change their relative positions over time.
   c. rise north or south of east and set north or south of west, depending on their location on the celestial sphere.
   d. all of the above
2. The stars we see at night depend on
   a. our location on Earth.
   b. Earth's location in its orbit.
   c. the time of the observation.
   d. all of the above
3. On the summer solstice in July, the Sun will be directly above ________ and all locations north of ________ will experience daylight all day.
   a. The Tropic of Cancer; the Antarctic Circle
   b. The Tropic of Capricorn; the Arctic Circle
   c. The Tropic of Cancer; the Arctic Circle
   d. The Tropic of Capricorn; the Antarctic Circle
4. If you look due East, you will see the point at which
   a. the Sun rises.
   b. the celestial equator intersects the horizon.
   c. all the stars rise.
   d. the Sun sets.
5. The seasons are caused by ________.
6. You see the Moon rising just as the Sun is setting. What phase is the Moon in?
   a. full
   b. new
   c. first quarter
   d. third quarter
   e. none of the above
7. You see the first quarter Moon on the meridian. Where is the Sun?
   a. on the western horizon
   b. on the eastern horizon
   c. below the horizon
   d. on the meridian
   e. none of the above
8. You do not see eclipses every month because
   a. you are not very observant.
   b. the Sun, Earth, and Moon line up only about twice a year.
   c. the Sun, Earth, and Moon line up only about once a year.
   d. eclipses happen randomly and are unpredictable.
9. If the Moon were in its same orbital plane but twice as far from Earth, which of the following would happen?
   a. The phases of the Moon would remain unchanged.
   b. Total eclipses of the Sun would not be possible.
   c. Total eclipses of the Moon would not be possible.
10. A frame of reference is a coordinate system within which an observer measures ________ and ________.

True/False and Multiple Choice

11. **T/F:** The celestial sphere is not an actual object in the sky.
12. **T/F:** Eclipses happen somewhere on Earth every month.
13. **T/F:** The phases of the Moon are caused by the relative positions of Earth, the Moon, and the Sun.
14. **T/F:** If a star rises north of east, it will set south of west.
15. **T/F:** From the North Pole, all stars in the night sky are circumpolar stars.
16. Constellations are groups of stars that
    a. are close to each other in space.
    b. are bound to each other by gravity.
    c. are close to each other in Earth's sky.
    d. all have the same composition.
17. Just from the name, you can tell that the star Alpha Andromedae is the __________ star in the constellation Andromeda.
    a. brightest
    b. hottest
    c. closest
    d. biggest
18. Day and night are caused by
    a. the tilt of Earth on its axis.
    b. the rotation of Earth on its axis.
    c. the revolution of Earth about the Sun.
    d. the revolution of the Sun about Earth.
19. Polaris, the North Star, is unique because
    a. it is the brightest star in the night sky.
    b. it is the only star in the sky that doesn't move throughout the night.
    c. it is always located at the zenith, for any observer.
    d. it has a longer path above the horizon than any other star has.
20. There is an angle between the ecliptic and the celestial equator because
    a. Earth's axis is tilted with respect to its orbit.
    b. Earth's orbit is tilted with respect to the orbits of other planets.
    c. the Sun follows a rising and falling path through space.
    d. the Sun's orbit is tilted with respect to Earth's.
21. The tilt of Earth's axis causes the seasons because
    a. one hemisphere of Earth is closer to the Sun in summer.
    b. the days are longer in summer.
    c. the rays of light strike the ground more directly in summer.
    d. both a and b
    e. both b and c
22. On the vernal and autumnal equinoxes,
    a. every place on Earth has 12 hours of daylight and 12 hours of darkness.
    b. the Sun rises due east and sets due west.
    c. the Sun is located on the celestial equator.
    d. all of the above
    e. none of the above
23. We always see the same side of the Moon because
    a. the Moon does not rotate on its axis.
    b. the Moon rotates on its axis once for each revolution around Earth.
    c. when the other side of the Moon is facing Earth, it is unlit.
    d. when the other side of the Moon is facing Earth, it is on the opposite side of Earth.
    e. none of the above
24. You see the Moon on the meridian at sunrise. The phase of the Moon is
    a. waxing gibbous.
    b. full.
    c. new.
    d. first quarter.
    e. third quarter.
25. A lunar eclipse occurs when __________ shadow falls on __________.
    a. Earth's; the Moon
    b. the Moon's; Earth
    c. the Sun's; the Moon
    d. the Sun's; Earth

Thinking about the Concepts

26. Earth has a North Pole, a South Pole, and an equator. What are their equivalents on the celestial sphere?
27. Seafaring sailors such as Columbus used Polaris for navigation as they sailed from Europe to the New World. When Magellan sailed the South Seas, he could not use Polaris for navigation. Explain why.
28. If you were standing at Earth's North Pole, where would you see the north celestial pole relative to your zenith?
29. Is there a location on Earth where, over the course of a year, you see the entire sky? If so, where is it? If not, why not?
30. Certain constellations are associated with certain times of year. For example, we see the zodiacal constellation Gemini in the Northern Hemisphere's winter (Southern Hemisphere's summer) and the zodiacal constellation Sagittarius in the Northern Hemisphere's summer. Why do we not see Sagittarius in the Northern Hemisphere's winter or Gemini in the Northern Hemisphere's summer?
31. Imagine that you are flying along in a jetliner.
    a. Define your frame of reference.
    b. What relative motions take place within your frame of reference?
32. The tilt of Jupiter's rotational axis is 3°. Explain how Earth's seasons would be different if Earth's axis had this tilt.
33. Why is the winter solstice *not* the coldest time of year?
34. Earth spins on its axis but wobbles like a top.
    a. How long does it take to complete one spin?
    b. How long does it take to complete one wobble?

35. What is the approximate time of day when you see the full Moon near the meridian? At what time is the first quarter (waxing) Moon on the eastern horizon? Use a sketch to help explain your answers.

36. Assume that the Moon's orbit is circular. Imagine that you are standing on the side of the Moon that faces Earth.
    a. How would Earth appear to move in the sky as the Moon made one revolution around Earth?
    b. How would the "phases of Earth" appear to you, compared to the phases of the Moon as seen from Earth?

37. Solar and lunar eclipses are recurring phenomena.
    a. From any given location, why are you more likely to witness a partial eclipse of the Sun than a total eclipse?
    b. Why do we not see a lunar eclipse each time the Moon is full or witness a solar eclipse each time the Moon is new?

38. Why does the fully eclipsed Moon appear reddish?

39. Stonehenge was erected roughly 4,000 years ago. Referring to the zodiacal constellations shown in Figure 2.14, identify the constellation in which these ancient builders saw the vernal equinox.

40. Which, if either, occur more often: (a) total eclipses of the Moon seen from Earth, or (b) total eclipses of the Sun seen from the Moon? Explain your answer.

Applying the Concepts

41. Earth is spinning along at 1,674 km/h at the equator. Use this fact, along with the length of the day, to calculate Earth's equatorial diameter.

42. Assume that rain is falling at a speed of 5 meters per second (m/s) and you are driving in the rain at the same speed, a leisurely 5 m/s (18 km/h). Estimate the angle from the vertical at which the rain appears to be falling.

43. The Moon's orbit is tilted by about 5° relative to Earth's orbit around the Sun. What is the highest altitude in the sky that the Moon can reach, as seen in Philadelphia (latitude 40° north)?

44. Imagine that you are standing on the South Pole at the time of the southern summer solstice.
    a. How far above the horizon will the Sun be at noon?
    b. How far above (or below) the horizon will the Sun be at midnight?

45. Determine the latitude where you live. Draw and label a diagram showing that your latitude is the same as (a) the altitude of the north celestial pole and (b) the angle (along the meridian) between the celestial equator and your local zenith. What is the altitude of the Sun at noon as seen from your home at the times of the winter solstice and the summer solstice?

46. The southernmost star in a group of stars known as the Southern Cross lies approximately 65° south of the celestial equator. What is the farthest-north latitude for which the entire Southern Cross is visible? Can it be seen in any U.S. states? If so, which ones?

47. Using a protractor, you estimate an angle of 40° between your zenith and Polaris. What is your latitude? Are you in the continental United States or Canada?

48. Suppose the tilt of Earth's equator relative to its orbit were 10° instead of 23.5°. At what latitudes would the Arctic and Antarctic circles and the Tropics of Cancer and Capricorn be located?

49. Suppose you would like to witness the midnight Sun (when the Sun appears just above the northern horizon at midnight), but you don't want to travel any farther north than necessary.
    a. How far north (that is, to which latitude) would you have to go?
    b. At what time of year would you make this trip?

50. The vernal equinox is now in the zodiacal constellation of Pisces. Wobbling of Earth's axis will eventually cause the vernal equinox to move into Aquarius. How long, on average, does the vernal equinox spend in each of the 12 zodiacal constellations?

51. Referring to Figure 2.20a , estimate when Vega, the fifth-brightest star in our sky (excluding the Sun), will once again be the northern pole star.

52. The apparent diameter of the Moon is approximately 1/2°. About how long does it take the Moon to move a distance equal to its own diameter across the sky?

53. The Moon has a radius of 1,737 km, with an average distance of 3.780×10^5 km from Earth's surface. The Sun has a radius of 696,000 km, with an average distance of 1.496×10^8 km from Earth. Show why the apparent sizes of the Moon and Sun in our sky are approximately the same.

54. Earth has an average radius of 6,371 km. If you were standing on the Moon, how much larger would Earth appear in the lunar sky than the Moon appears in our sky?

55. In what way would the length of the eclipse season change if the plane of the Moon's orbit were inclined less than its current 5.2° to the plane of Earth's orbit? Explain your answer.

Using the Web

56. Go to the U.S. Naval Observatory website (USNO "Data Services," at http://aa.usno.navy.mil/data). Look up the times for sunrise and sunset for your location for the current week. (You can change the dates one at a time, or bring up a table for the entire month.) How are the times changing from one day to the next? Are the days getting longer or shorter? Bring up the "Duration of Days/Darkness Table for One Year" page for your location. When do the shortest and the longest days occur? Look up a location in the opposite hemisphere (Northern or Southern). When are the days shortest and longest?

57. Go to the "Earth and Moon Viewer" website (http://fourmilab.ch/earthview). Under "Viewing the Earth," click on "latitude, longitude and altitude" and enter your approximate latitude and longitude, and 40,000 for altitude; then select "View Earth." Are you in daytime or nighttime? Now play with the locations; keep the same latitude but change to the opposite hemisphere (Northern or Southern). Is it still night or day? Go back to your latitude, and this time enter 180° minus your longitude, and change from west to east, or from east to west, so that you are looking at the opposite side of Earth. Is it night or day there? What do you see at the North Pole (latitude 90° north) and the South Pole (latitude 90° south)? At the bottom of your screen you can play with the time. Move back 12 hours. What do you observe at your location and at the poles?

58. Go to the U.S. Naval Observatory website (USNO "Data Services," at http://aa.usno.navy.mil/data). Look up the Moon data for the current day. When will it rise and set? What is the phase? How will it change over the next 4 weeks. Enter one day at a time or look at the yearlong tables for moonrise and moonset and for the dates of primary phases. What time of day does a quarter Moon rise? When (in what phases) can you see the Moon in the daytime?

59. Using the times of moonrise and moonset that you located in question 58, make a plan to observe the Moon directly at least once a day for a week. Take a picture of the Moon (or make a sketch) every day. How is the brightness of the Moon changing? If it's daytime, how far is the Moon from the Sun in the sky? If it's nighttime, are the stars that are near the Moon in the sky the same every night?

60. Go to the "NASA Eclipse" website (http://eclipse.gsfc.nasa.gov/eclipse.html). When is the next lunar eclipse? Will it be visible at your location if the skies are clear? Is it a total or partial eclipse? How about the next solar eclipse? Will it be visible at your location? Compare the fraction of Earth that the solar eclipse will affect with the fraction for the lunar eclipse. Why are lunar eclipses visible in so many more locations?

StudySpace is a free and open website that provides a Study Plan for each chapter of *21st Century Astronomy*. Study Plans include animations, reading outlines, vocabulary flashcards, and multiple-choice quizzes, plus links to premium content in SmartWork and the ebook. Visit **wwnorton.com/studyspace**.

SmartWork Norton's online homework system, includes algorithmically generated versions of these questions, plus additional conceptual exercises. If your instructor assigns questions in SmartWork, log in at **smartwork.wwnorton.com.**

Exploration | The Phases of the Moon

Visit StudySpace, and open the Lunar Phase Simulator applet in Chapter 2.

Study the diagrams shown in the simulator. The largest window shows a view of the Earth-Moon system as seen from above Earth's North Pole. The Sun is far off the screen to the left. An observer stands on Earth. The small window at upper right shows the appearance of the Moon, as seen from the Northern Hemisphere. The small window at lower right shows the observer's location, with the Sun and Moon pictured as flat disks in the sky.

1 **Given the relative positions of the observer and the Sun, approximately what time is it for this observer: 6:00 A.M., noon, 6:00 P.M., or midnight?**

....

2 **Where is the Moon in the observer's sky: on the eastern horizon, on the western horizon, below the horizon, or crossing the meridian?**

....

3 **What is the phase of the Moon?**

....

4 **Imagine yourself on the Moon in the image shown in the larger window. If you looked toward Earth, what phase of Earth would you see?**

....

Now select "start animation." Allow the animation to run until the Moon is 20 percent illuminated (as shown in the upper small window on the right).

5 **In which direction does the Moon orbit Earth: clockwise or counterclockwise?**

....

6 **For observers in the Northern Hemisphere, which side of the Moon is illuminated first after a new Moon: right or left?**

....

7 **If you observe a crescent Moon with the horns of the crescent pointing right, is the Moon waxing or waning?**

....

Grab the Moon with your mouse, and drag it to first quarter. Drag the observer so that her local time is approximately midnight.

8 **Where is the first quarter Moon in the observer's sky: on the eastern horizon, on the western horizon, below the horizon, or crossing the meridian?**

....

These three things are related: the time, the phase of the Moon, and the Moon's location in the sky.

9 **Arrange the observer and the Moon so that the Moon is full and crossing the meridian. What time is it for the observer: 6:00 A.M., noon, 6:00 P.M., or midnight?**

....

10 **Arrange the observer and the Moon so that the Moon is in third quarter and the time for the observer is approximately noon (the Sun is on the meridian). Where is the Moon in the observer's sky: on the eastern horizon, on the western horizon, below the horizon, or crossing the meridian?**

....

11 **Arrange the observer and the Moon so that it is approximately 6:00 P.M. (sunset) for the observer, and the Moon is just rising on the eastern horizon. What is the phase of the Moon?**

....

There are many other combinations of the time, phase of the Moon, and Moon's location to play with. Challenge yourself to be able to set up any two of the three and find the third. When you can do this without the simulator, just by making the picture in your head, you will really understand the phases of the Moon.

NASA astronaut in space.

03 Motion of Astronomical Bodies

Eppur si muove ("And yet it moves").

Galileo Galilei (1564–1642)

LEARNING GOALS

In this chapter we embark on the story of the birth of modern science, which begins when astronomers and mathematicians discovered regular patterns in the motions of the planets. By the conclusion of this chapter, you should be able to:

- Sketch and contrast the geocentric and heliocentric models of the Solar System.
- Summarize the laws developed by Kepler that describe the motion of objects in the Solar System.
- Describe the evidence that the Earth and planets orbit the Sun.
- Explain the physical laws discovered by Isaac Newton and Galileo Galilei that govern the motion of all objects.

3.1 The Motions of Planets in the Sky

What is Earth's relationship to the rest of the objects in its celestial neighborhood? Ancient astronomers did not have the benefit of telescopes to see objects close-up, nor did they have the option of looking up or down at Earth from a spacecraft. All they could do was observe the sky and try to understand what they saw from the ground. What they saw was that the Sun, Moon, and stars rose in the east and set in the west, and they appeared to be moving around Earth. Ancient peoples had long been aware of five planets ("wandering stars") that moved in a generally eastward direction from one night to the next among the "fixed stars." But they did not know that Earth was similar to these planets. A successful theory of how Earth moved and how it fits in with its neighbors in the Solar System was the first step to understanding Earth's place in the universe.

How might the motions of Earth affect its conditions for life?

Some astronomers and philosophers in ancient times hypothesized that the Sun might be the center of the Solar System, but they did not have the tools to test the hypothesis or the mathematical insight to formulate a more complete and testable model. Instead, in large part because we can't feel Earth's motion through space, early astronomers developed a **geocentric** (Earth-centered) model of the Solar System to explain what they observed in the world around them. When people looked up at the sky, the Sun, Moon, and stars appeared to be moving around Earth. As you saw in the previous chapter, hard evidence of the motions of Earth was remarkably difficult to come by. So for nearly 1,500 years, most educated people believed that the Sun, the Moon, and the known planets (Mercury, Venus, Mars, Jupiter, and Saturn) all moved in circles around a stationary Earth.

However, ancient astronomers knew that the planets would occasionally exhibit **retrograde motion**, in which they stop their eastward motion, move westward across the sky for a while, and then return to their normal eastward travel. This odd behavior of the five "naked eye" planets—Mercury, Venus, Mars, Jupiter, and Saturn—created a puzzling problem for the geocentric model as it was summarized in 150 CE by the Alexandrian astronomer Ptolemy (Claudius Ptolemaeus, 90–168 CE). A few ancient Greek astronomers had considered the possibility that Earth revolved around the Sun—including Aristarchus of Samos (310–230 BCE), who also calculated the

(a)

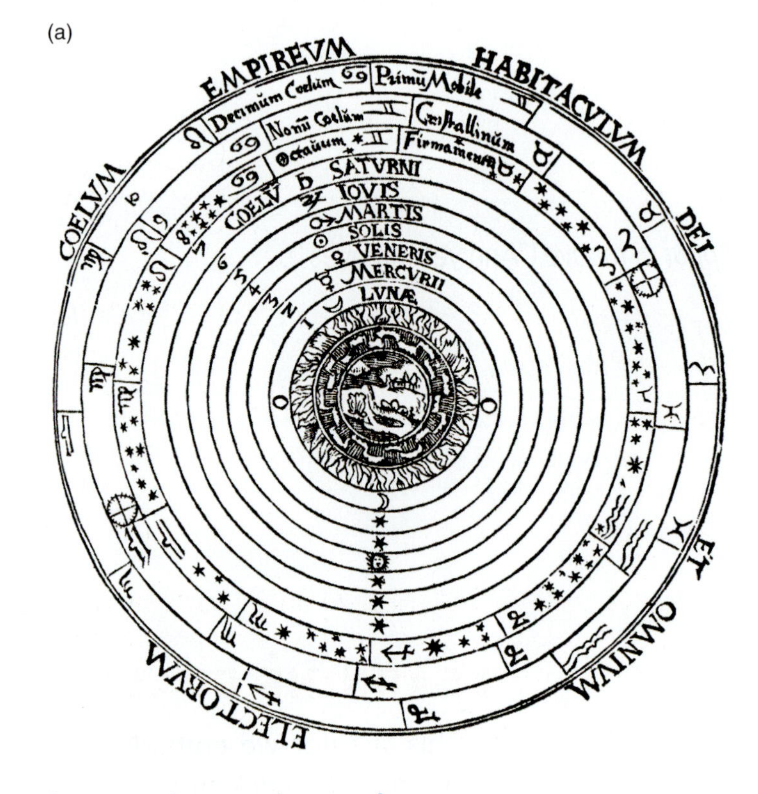

(b)

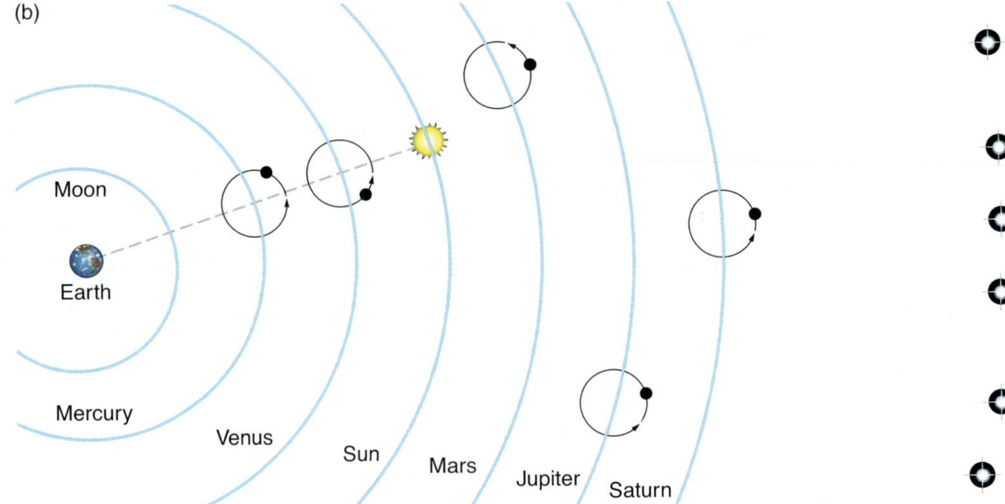

FIGURE 3.1 (a)The Ptolemaic view of the Heavens. (b) Additional loops called epicycles were added to each planet's circle around Earth to explain its retrograde motion in the geocentric model of the Solar System.

relative distances of the Sun and the Moon. But most other astronomers of the time were skeptical, because they thought they should feel Earth's motion. Therefore they preferred the geocentric model, in which the Sun, Moon, and planets all moved in perfect circles around a stationary Earth, with the "fixed stars" being located somewhere way out there beyond the planets (**Figure 3.1a**). Although a stationary Earth may have "felt right" to astronomers in those early times, the geocentric model in its simplest form failed to explain retrograde motion of the planets.

To account for this retrograde motion, Ptolemy had to resort to an embellishment called an *epicycle*, a small circle superimposed on each planet's larger circle (**Figure 3.1b**). While traveling along its larger circle, a planet would at the same time be moving along its smaller circle. When its motion along the smaller circle was in a direction opposite to that of the forward motion of the larger circle, its forward motion would be reversed. **NEBRASKA SIMULATION: PTOLEMAIC ORBIT OF MARS**

There are indications that others throughout history questioned the geocentric model, but they did not have actual observations or a mathematical model that disproved it. For nearly 1,500 years, Ptolemy's model of the heavens was the accepted paradigm in the Western world.

Early astronomers pondered Earth's place in the universe.

3.2 Earth Moves

Nicolaus Copernicus (1473–1543—**Figure 3.2**) is famous for placing the Sun rather than Earth at the center of the Solar System. He was not the first person to consider the idea that Earth orbited the Sun, but he was the first to develop a mathematical model that made predictions that later astronomers would be able to test. This work was the beginning of what was later called the Copernican Revolution. Through the work of 16th- and 17th-century scientists such as Tycho Brahe, Galileo Galilei, Johannes Kepler, and Sir Isaac Newton, the **heliocentric** (Sun-centered) theory of the Solar System has become one of the best-corroborated theories in all of science.

Copernicus's heliocentric cosmological model knocked humankind from the center of the universe.

Copernicus was multilingual and highly educated; he studied philosophy, canon (Catholic) law, medicine, economics, mathematics, and astronomy in his native Poland and in Italy. After his schooling he returned to Poland, where he worked for his uncle the prince-bishop (and subsequently his uncle's successors) as an economist, a diplomat, and a physician. Copernicus conducted astronomical observations from a small tower, and sometime around 1514 he started writing about heliocentricity. He completed his manuscript around 1532, but he knew his ideas would be controversial because philosophical and religious views of the time held that humanity and thus Earth must be the center of the universe. Late in his life, Copernicus was finally persuaded to publish his ideas, and his great work *De revolutionibus orbium coelestium* ("On the Revolutions of the Heavenly Spheres") did not appear until 1543, the year of his death. This work pointed the way toward the modern cosmological principle introduced in Chapter 1 (the idea that on a large scale the universe looks the same everywhere).

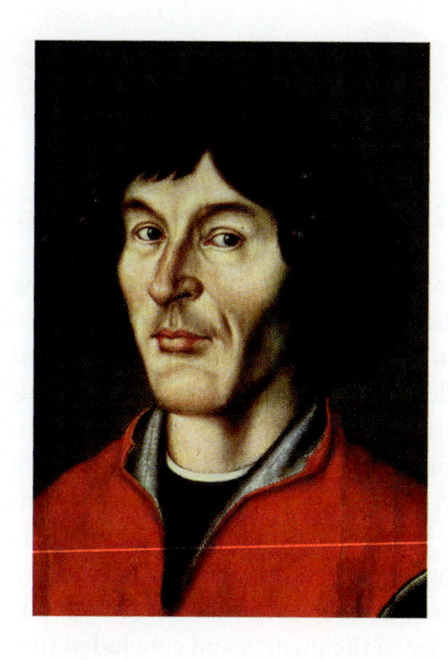

FIGURE 3.2 Nicolaus Copernicus rejected the ancient Greek belief in an Earth-centered universe and replaced it with one centered on the Sun.

Copernicus's heliocentric model explained the observed motions of Earth, the Moon, and the planets, including retrograde motion, much more simply than the geocentric model did. Combining geometry with his observations of the positions of the planets in the sky (their altitudes, and the times they rose and set), Copernicus determined the time when each planet, Earth, and the Sun were in physical alignment, forming a straight line or a right triangle in space (see **Connections 3.1**). This information helped him estimate the planet–Sun distances in terms of the Earth–Sun distance. These relative distances estimated by Copernicus are remarkably close to those obtained by modern methods. The time between these alignments enabled him to estimate how long it took each planet to orbit the Sun. This model made testable predictions of the location of each planet on a given night, which were at least as accurate (although not necessarily more so than) as those of the geocentric model. Copies of *De Revolutionibus* and Copernicus's ideas slowly spread through Europe and excited other scientists, beginning a scientific revolution.

Copernicus's heliocentric model provided a simpler explanation of retrograde motion (see Figure 3.3). In his model, the outermost planets known at that time—Mars, Jupiter, and Saturn—appear to interrupt their eastward **prograde motion**

Connections 3.1

How Copernicus Scaled the Solar System

One of Copernicus's major accomplishments was computing the relative distances of the planets from the Sun, compared to the distance of Earth from the Sun. He didn't know the actual Earth–Sun distance in miles or kilometers, but the relative values he obtained for the other planets are remarkably correct. He found these values by making simple observations without a telescope. Copernicus devised a model with the Sun at the center and the planets in orbit around the Sun (**Figure 3.3**). Copernicus's model assumed that the planets traveled around the Sun in circular orbits with constant speeds. From his observations he deduced the correct order of the planets and concluded that the ones closer to the Sun traveled faster than the ones farther from the Sun. He also realized there were two categories of planets: those closer to the Sun than Earth, or *inferior* planets; and those farther from the Sun, or *superior* planets.

Periodically a planet, Earth, and the Sun line up as shown in **Figure 3.4a**. When a superior planet is opposite the Sun in the sky, we call the configuration *opposition*, and like a full Moon, when a superior planet is in opposition it rises when the Sun sets and sets when the Sun rises. When a superior planet is in line with the Sun but behind it, we call the configuration *conjunction*. A superior planet in conjunction will rise and set in the sky with the Sun. Note that when a superior planet is in conjunction, it is the farthest away from Earth that it gets and thus is at its faintest (you won't see the planet at all exactly at conjunction, because it's behind the Sun). When the planet is in opposition, it is the closest it gets to us and thus is at its brightest; therefore, opposition is the best time to observe the planet in the sky. Opposition is also the time when the planet exhibits retrograde motion. The other mathematically useful point in the orbit of a superior planet is called a *quadrature*—one of two points at which Earth, the Sun, and a superior planet form a right triangle in space.

FIGURE 3.3 The Copernican heliocentric view of the Solar System (II–VII) and the fixed stars (I).

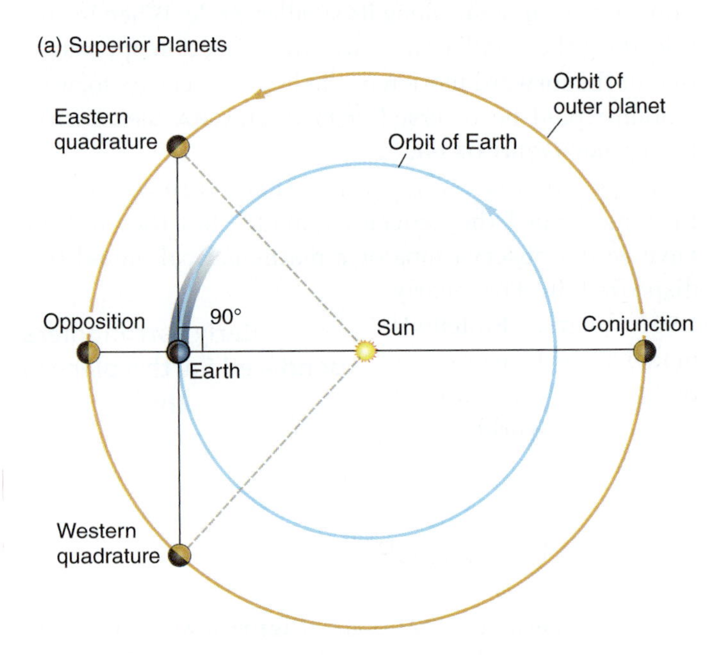

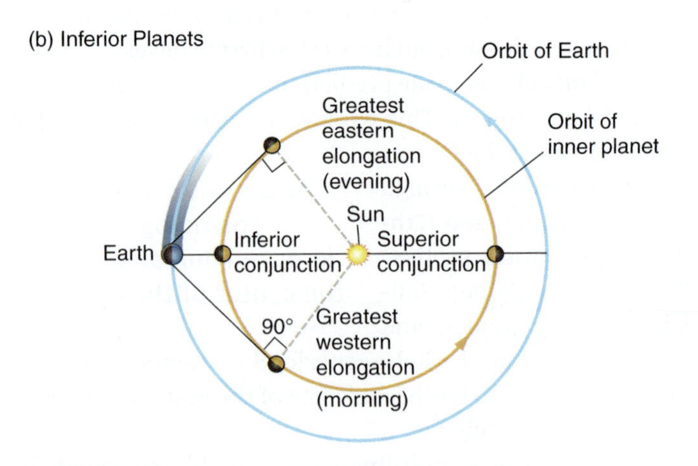

FIGURE 3.4 Planetary configurations for superior (outer) planets (a) and inferior (inner) planets (b).

For an inferior planet, the configurations are slightly different (**Figure 3.4b**). If the planet is between Earth and the Sun, we call the configuration *inferior conjunction*; if it is on the other side of the Sun from Earth, we call the configuration *superior conjunction*; and when it is at a right angle with Earth and the Sun—and thus the farthest it gets from the Sun in the sky—we call the configuration *greatest elongation*. Think about standing on Earth and looking toward the inner planets—they would never get very far from the Sun in the sky, so you would see Mercury and Venus only within a few hours of sunrise or sunset.

Copernicus realized there were two types of orbital periods. A *sidereal* period is how long it takes the planet to make one orbit around the Sun with respect to the stars, and return to the same point in space. A *synodic* period is how long it takes the planet to return to the same configuration with the Sun and Earth, such as inferior conjunction to inferior conjunction, or opposition to opposition. The synodic period is what can be observed from Earth. Superior planets move around the Sun more slowly than Earth does, so Earth will complete one orbit around the Sun and then catch up to a superior planet to make the configuration. For superior planets, the sidereal period is longer than the synodic period (**Figure 3.5a**). An inferior planet moves around the Sun faster than Earth does, so it completes one sidereal period and then must continue in its orbit to catch up to Earth. Thus, for inferior planets the synodic period is longer than the sidereal period (**Figure 3.5b**). **▶▶ NEBRASKA SIMULATION: PLANETARY CONFIGURATIONS SIMULATOR**

These two periods can be related mathematically in simple equations, as described in **Math Tools 3.1**. Copernicus determined the sidereal periods from the observed synodic periods of the planets in this way. He let the Earth–Sun distance equal 1 (which we defined in Chapter 2 as one astronomical unit, or AU). For the inferior planets, he had a right triangle at the point of greatest elongation, and then he could use right-triangle trigonometry to solve for the planet–Sun distance. For the superior planets, he measured the fraction of the circular orbit that the planet completed in the time between opposition and quadrature, and again used right-triangle trigonometry to solve for the planet–Sun distance. Copernicus's values are compared with modern values in **Table 3.1**. Note how remarkably similar they are. Copernicus still did not know the actual value of the Earth–Sun distance, but he was able to show the relative distances of the planets from the Sun, and thereby scale the Solar System for the first time.

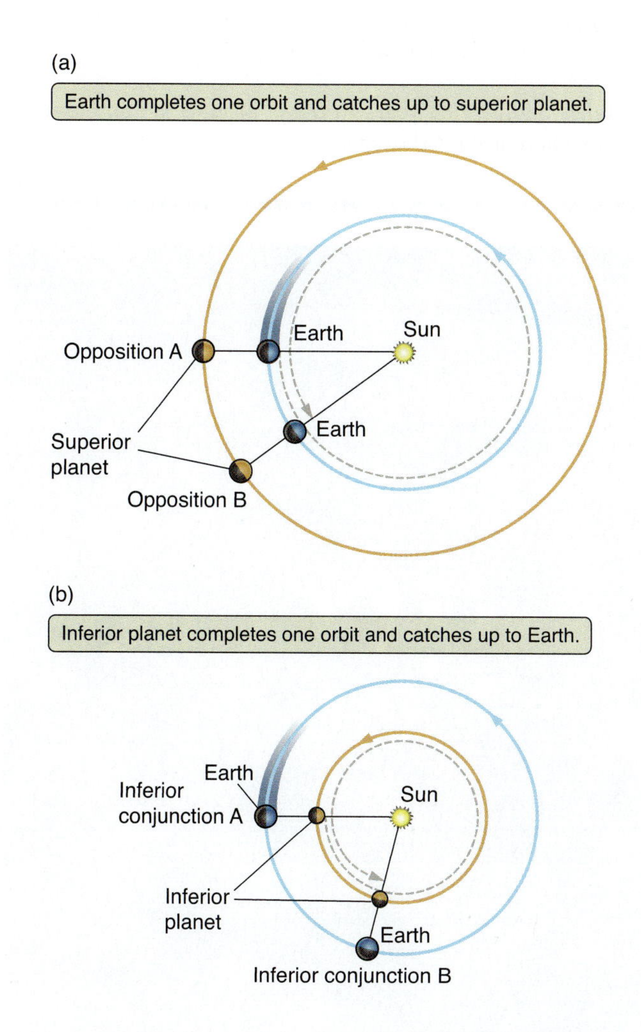

FIGURE 3.5 The synodic periods of planets. (a) Earth completes one orbit first, and then catches up to the superior planet. (b) Inferior planets complete a full orbit first and then catch up to Earth.

TABLE 3.1 | Copernicus's Scale of the Solar System

Planet	Copernicus's Value (AU)	Modern Value (AU)
Mercury	0.38	0.39
Venus	0.72	0.72
Earth	1.00	1.00
Mars	1.52	1.52
Jupiter	5.22	5.20
Saturn	9.17	9.58

Math Tools 3.1

Sidereal and Synodic Periods

How long does it take a planet to orbit around the Sun? Copernicus showed that this sidereal period could be calculated from observations of the synodic period. Let P be the sidereal period and S the synodic period of a planet. E is the sidereal period of Earth, which equals 1 year, or 365.25 days. An inferior planet orbits the Sun in less time than Earth does. So, with P, E, and S all in the same units of days or years,

$$\frac{1}{P} = \frac{1}{E} + \frac{1}{S}$$

for an inferior planet. Similarly, Earth orbits the Sun in less time than a superior planet does, so the planet has traveled only part of its orbit around the Sun after 1 Earth year. Therefore,

$$\frac{1}{P} = \frac{1}{E} - \frac{1}{S}$$

for a superior planet.

Synodic periods can be observed from Earth; for example, the time that passes between one opposition of Saturn and the next is a synodic period. Noting the time that Saturn rises in the east and measuring the time that passes between its days of maximum brightness shows that Saturn's synodic period is 378 days, or $378/365.25 = 1.035$ years. Then, using $S = 1.035$ and $E = 1$ to compute its sidereal period, P, gives this result:

$$\frac{1}{P} = \frac{1}{1\text{ yr}} - \frac{1}{1.035\text{ yr}}$$
$$= 1 - 0.966 = 0.034\text{ yr}^{-1}$$

and

$$P = \frac{1}{0.034} = 29.4\text{ years}$$

So, it takes Saturn 29.4 years to travel around the Sun and return to where it started in space.

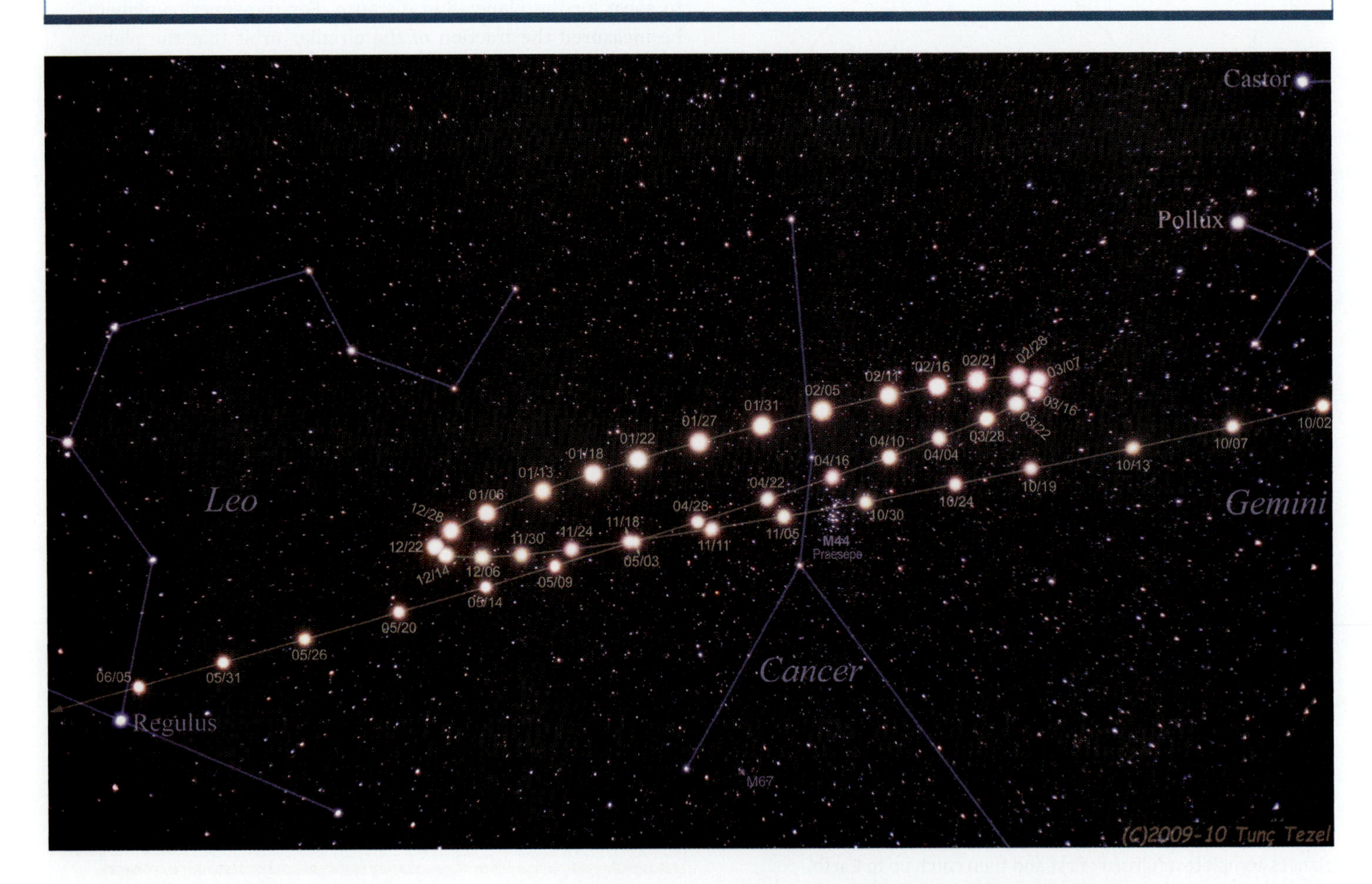

FIGURE 3.6 The apparent retrograde motion of Mars as Earth overtook Mars in its orbit in 2009–10. Each bright dot shows the location of Mars on the date indicated.

and move backward in a westward direction when Earth overtakes them in their orbits. Likewise, Mercury and Venus appear to reverse direction in their inner orbits when overtaking Earth. With the exception of the Sun and the Moon, all Solar System objects exhibit retrograde motion, although the magnitude of the effect diminishes with increasing distance from Earth. **Figure 3.6** shows a time-lapse sequence of Mars going through its retrograde "loop." Copernicus still conceived of the planets as moving in circular orbits (instead of the elliptical orbits you will learn about in Section 3.3), and as a result he needed to use some epicycles to match the observations. But this heliocentric model was, overall, simpler and more elegant than the geocentric model.

Retrograde motion is only apparent, not real. We have all experienced how relative motions can fool us. If you are in a car or train and you pass a slower-moving car or train, it can seem as if the other vehicle is moving backward. Without an external frame of reference, it can be hard to tell which vehicle is moving and in what way. Copernicus provided that frame of reference for the Sun and its planets.

▶▶ NEBRASKA SIMULATION: RETROGRADE MOTION

3.3 An Empirical Beginning: Kepler's Laws

In Chapter 1 we described science as a worldview in which physical laws govern nature and mathematical descriptions of these physical laws are used to explain natural phenomena. But how do scientists go about discovering these physical laws? When facing phenomena as complex and puzzling as the motions of the planets in the heavens, scientists carefully observe the phenomenon under study, systematically recording as much information as they can as accurately as possible. As observations start piling up, scientists look for patterns in those observations and start formulating rules that describe those patterns.

The quest to accurately describe patterns in nature is called **empirical science**. Empirical science often involves a great deal of creativity and pure (but educated) guesswork. Copernicus's theory that Earth and the planets move in circular orbits about the Sun is an example of empirical science. Copernicus did not understand *why* the planets move about the Sun, but he did realize that his heliocentric picture provided a better description of the observed motions of planets than did a model with Earth at its center. Copernicus's work was revolutionary because he was able to see beyond the geocentric prejudice of his time and to think the unthinkable—that perhaps Earth is "merely" one planet among many. Copernicus's work paved the way for other great empiricists.

Tycho Brahe

Tycho Brahe (1546–1601—**Figure 3.7a**) was a Danish astronomer of noble birth who entered university at age 13 to study philosophy and law but became interested in astronomy after seeing a partial solar eclipse in 1560. A few years later, Tycho (conventionally referred to by his first name) observed

(a)

(b)

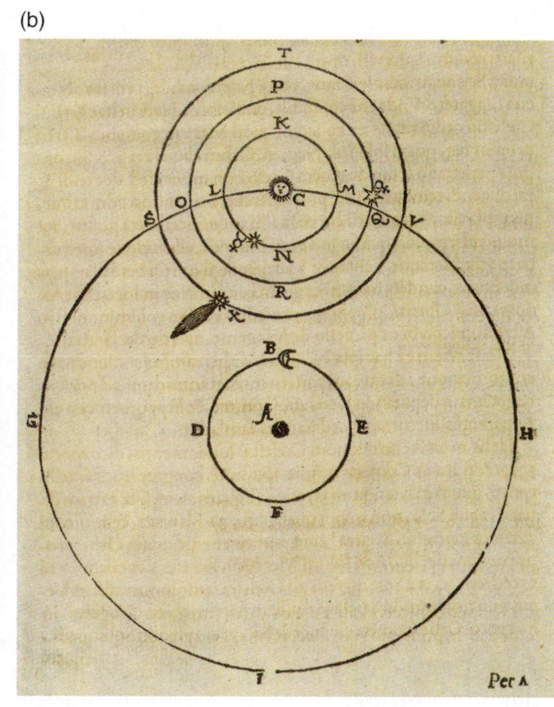

FIGURE 3.7 (a) Tycho Brahe, known commonly as Tycho, was one of the greatest astronomical observers before the invention of the telescope. (b) Tycho's geo-heliocentric model, with the planets orbiting the Sun and the Sun orbiting Earth.

Jupiter and Saturn near each other in the sky, but not in the exact positions predicted by the astronomical tables based on Ptolemy's model. Tycho gave up studying law and devoted himself to making better astronomical tables of the planets.

The Danish king granted Tycho the island of Ven (known at that time as Hven), located between Sweden and Denmark, to build an observatory. Tycho designed and built new instruments, operated a printing press, and taught students and others how to conduct observations. With the assistance of his sister Sophie, Tycho carefully measured the precise positions of planets in the sky over several decades, developing the most comprehensive set of planetary data available at that time. He created his own geo-heliocentric model, in which the planets orbited the Sun but then the Sun along with the planets orbited Earth (**Figure 3.7b**). This model gained limited acceptance among people who preferred to keep Earth at the center for philosophical or religious reasons. Tycho lost his financial support when the king died, and in 1600 he relocated to Prague.

Tycho Brahe spent decades making accurate observations of planets.

In 1600, Tycho hired a more mathematically inclined astronomer, Johannes Kepler (1571–1630—**Figure 3.8**), as his assistant. Kepler had studied the ideas of Copernicus, and he acquired Tycho's data records when Tycho died the next year. Kepler was responsible for the next major step toward understanding the motions of the planets. Working first with Tycho's observations of Mars, Kepler was able to deduce three empirical rules that elegantly and accurately describe the motions of the planets. These three rules are now generally referred to as **Kepler's laws**.

FIGURE 3.8 Johannes Kepler explained the motions of the planets with three empirically based laws.

Kepler's First Law: Planets Move on Elliptical Orbits with the Sun at One Focus

When Kepler used Copernicus's model to calculate where in the sky a planet should be at a particular time, he expected Tycho's data to confirm the circular shape of orbits, but instead he found disagreement between his predictions and the observations. Kepler was not the first to notice such discrepancies. Rather than discarding Copernicus's ideas, however, Kepler played with Copernicus's heliocentric model until it matched Tycho's observations.

Kepler discovered that if he replaced Copernicus's circular orbits with *elliptical* orbits, his predictions fit Tycho's observations almost perfectly. You might think of an ellipse as an oval shape, but to make sense of Kepler's discovery we need to be more precise about what an ellipse is. A concrete way to define an **ellipse** is to call it the shape that results when you attach the two ends of a piece of string to a piece of paper, stretch the string tight with the tip of a pencil, and then draw around those two points while keeping the string taut (**Figure 3.9**). Each of the points at which the string is attached is a **focus** (plural: *foci*) of the ellipse. The closer together the two foci are, the more nearly circular the ellipse is. In fact, a circle is just an ellipse with its two foci at the same place. (To visualize this, think about the shape you would draw if the two ends of the string were attached at the same spot. In this case, each half of the string would become a radius of the circle.) As the two foci are moved farther apart, however, the ellipse becomes more and more elongated. Kepler determined that *the orbit of each planet is an ellipse with the Sun located at one focus and nothing but empty space at the other focus*. This result is now known as **Kepler's first law** of planetary motion (see the **Process of Science Figure**).

Planetary orbits are ellipses.

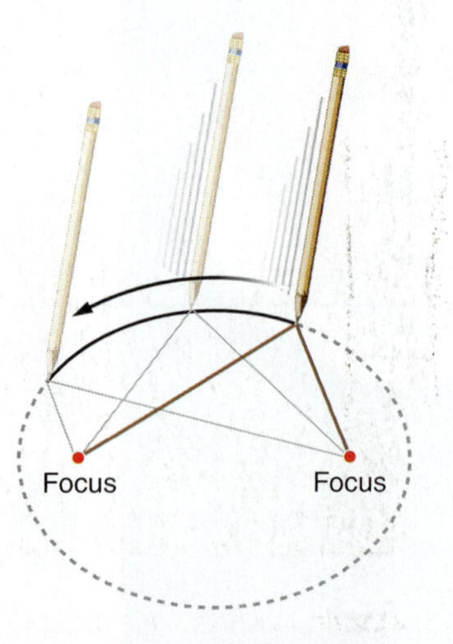

FIGURE 3.9 We can draw an ellipse by attaching a length of string to a piece of paper at two points (called foci) and then pulling the string around as shown.

Process of Science

THEORIES ARE FALSIFIABLE

Early Astronomers studied the motions of the planets but did not understand why they behave as they did.

The Big Idea:
Copernicus proposes that planets move in circular heliocentric orbits, with epicycles.

The Observation:
Tycho observes and collects lots of data about planet positions.

The Prediction:
Kepler uses Copernicus's model to predict where planets should be.

The Test:
Kepler compares predictions with data—they disagree. Copernicus's idea is falsified!

The New Big Idea:
Planet orbits are not circular. They are elliptical.

Model gains acceptance with a physical understanding of Newton's laws.

In order for a theory to be "scientific," it must be possible, in principle, to falsify it. Disproving an old theory always leads to deeper understanding.

Figure 3.10 illustrates Kepler's first law and shows how the features of an ellipse elegantly match observed planetary motions. The dashed lines in **Figure 3.10a** represent the two main axes of the ellipse. Half of the length of the long axis of the ellipse is its **semimajor axis**, often denoted by the letter A. The semimajor axis of an orbital ellipse turns out to be a handy way to describe the orbit because, apart from being half the longer dimension of the ellipse, it is also the average distance between one focus and the ellipse itself. The average distance between the Sun and Earth, for example, equals the semimajor axis of Earth's orbit. The same is true for the orbits of all the other planets. **ASTROTOUR: KEPLER'S LAWS**

In the case of a circular orbit, the semimajor axis is the radius of the circle. Some ellipses, however, are very elongated. **Eccentricity** describes the shape of an ellipse. The eccentricity of an ellipse is the separation between the two foci divided by the length of the long axis. A circle has an eccentricity of 0 because the two foci coincide at the center of the circle. The more elongated the ellipse becomes, the closer its eccentricity gets to 1 (**Figure 3.10b**). Most planets have nearly circular orbits with eccentricities close to 0. The eccentricity of Earth's orbit, for example, is 0.017, which means that the distance between the Sun and Earth departs from its average value by only 1.7 percent. You may recall from Chapter 2 that the times when Earth is closest to or farthest from the Sun have almost nothing to do with our seasons; a look at **Figure 3.11a** may help you understand why. It is difficult to distinguish the difference between the orbit of Earth and a circle centered on the Sun. By contrast, one of the many characteristics that distinguish the dwarf planet Pluto from the planets is its highly eccentric orbit (**Figure 3.11b**). With an eccentricity of 0.249, the distance between the Sun and Pluto varies by 24.9 percent from its average value. Pluto's orbit is noticeably elongated, and its center is noticeably displaced from the Sun. **NEBRASKA SIMULATION: ECCENTRICITY DEMONSTRATOR**

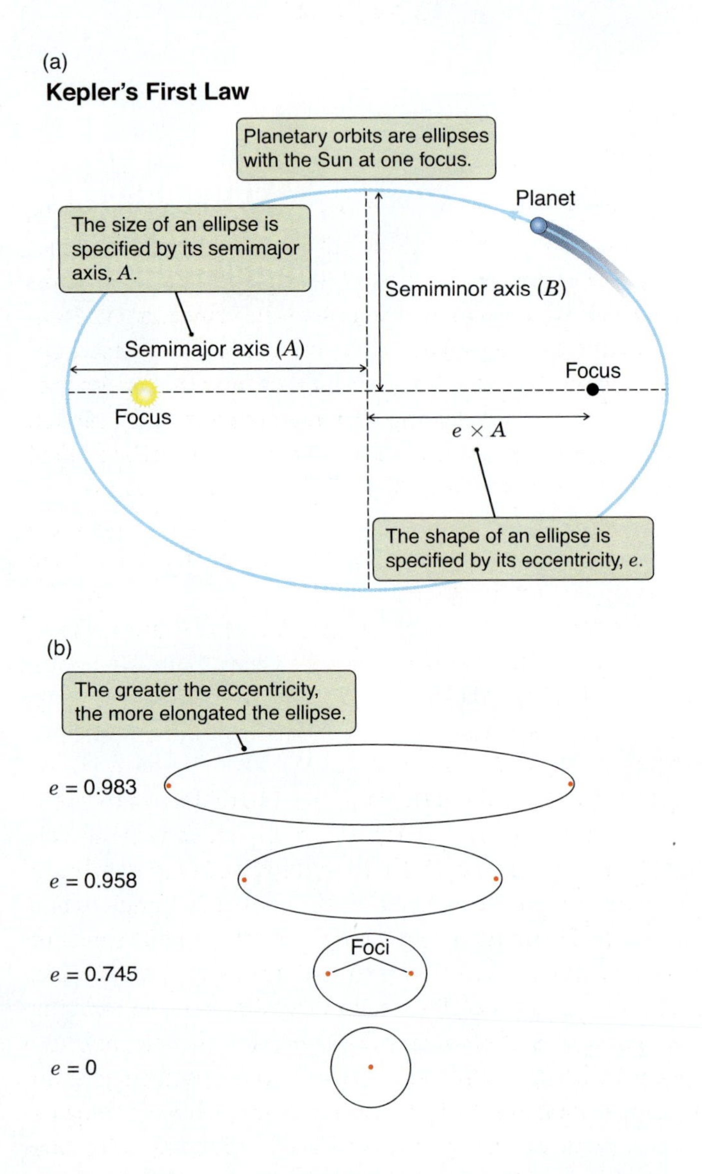

FIGURE 3.10 (a) Planets move on elliptical orbits with the Sun at one focus. (b) Ellipses range from circles to elongated eccentric shapes. e = eccentricity.

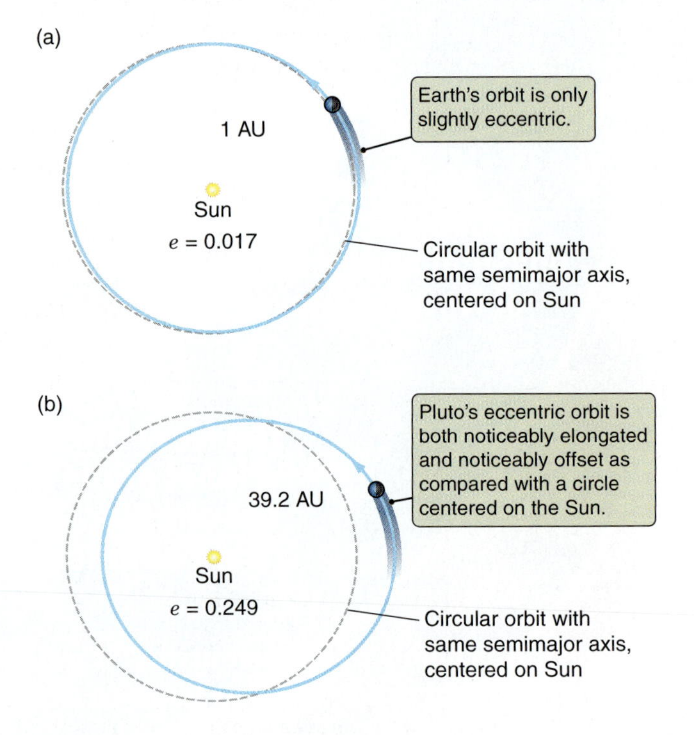

FIGURE 3.11 The shapes of the orbits of Earth (a) and Pluto (b) compared with circles centered on the Sun. e = eccentricity.

Kepler's Second Law: Planets Sweep Out Equal Areas in Equal Times

The next empirical rule that Kepler found has to do with how fast planets move at different places on their orbits. A planet moves most rapidly when it is closest to the Sun and is at its slowest when it is farthest from the Sun. The average speed of Earth in its orbit about the Sun is 29.8 kilometers per second (km/s). When Earth is closest to the Sun, it travels at 30.3 km/s. When it is farthest from the Sun, it travels at 29.3 km/s.

Planets move fastest when they are closest to the Sun.

Kepler found an elegant way to describe the changing speed of a planet in its orbit about the Sun. **Figure 3.12** shows a planet at six different points in its orbit (t_1 to t_6). Imagine a straight line connecting the Sun with this planet. We can think of this line as "sweeping out" an area as it moves with the planet from one point to another. Area A (in red) is swept out between times t_1 and t_2, area B (in blue) is swept out between times t_3 and t_4, and area C (in green) is swept out between times t_5 and t_6. When the planet is closest to the Sun (area A), it is moving rapidly but the distance between the planet and the Sun is small. Kepler realized that changes in the distance between the Sun and a planet and changes in the speed of a planet work together to produce a surprising result: the area swept out by a planet in the same amount of time is always the same, regardless of the location of the planet in its orbit. This means that if the three time intervals in the figure are equal (that is, $t_1 \rightarrow t_2 = t_3 \rightarrow t_4 = t_5 \rightarrow t_6$), then the three areas A, B, and C will be equal as well.

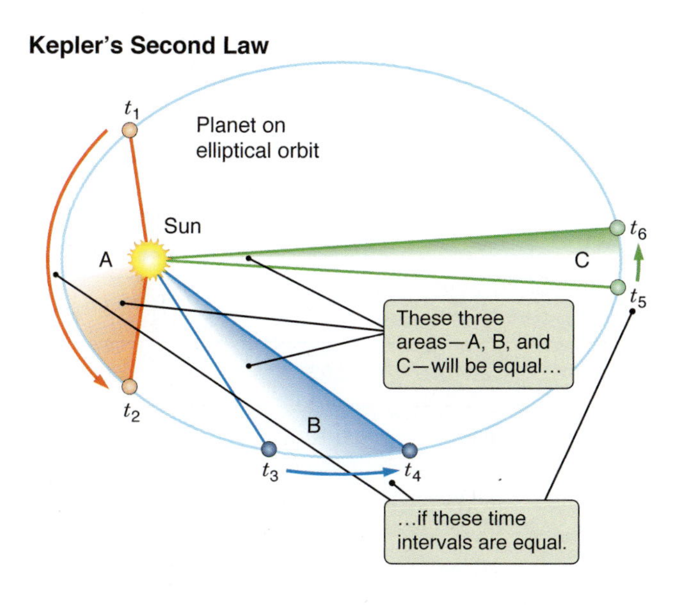

FIGURE 3.12 An imaginary line between a planet and the Sun sweeps out an area as the planet orbits. Kepler's second law states that if the three intervals of time shown are equal, then the three areas A, B, and C will be the same.

This is **Kepler's second law**, which is also referred to as Kepler's **law of equal areas**. It states that *the imaginary line connecting a planet to the Sun sweeps out equal areas in equal times, regardless of where the planet is in its orbit.* This law applies to only one planet at a time. The area swept out by Earth in a given time is always the same. Likewise, the area swept out by Mars in a given time is always the same. But the area swept out by Earth and the area swept out by Mars in a given time are *not* the same.

A planet sweeps out equal areas in equal times.

AstroTour: Kepler's Laws

Kepler's Third Law: Planetary Orbits Reveal a Harmony of the Worlds

Kepler's first law describes the shapes of planetary orbits, and Kepler's second law describes how the speed of a planet changes as it travels along its orbit. But neither law indicates how long it takes a planet to complete one orbit about the Sun (referred to as the **period** of the orbit, and determined observationally). Nor do these laws tell us how this orbital period depends on the distance between the Sun and a planet. Kepler looked for patterns, and discovered a mathematical relationship between orbital period and semimajor axis.

Planets that are closer to the Sun do not have as far to go to complete one orbit as do planets that are farther from the Sun. Jupiter, for example, has an average distance of 5.2 AU from the Sun—5.2 times as far from the Sun as Earth is. That means Jupiter must travel 5.2 times farther in its orbit about the Sun than Earth does in its orbit. You might guess, then, that if the two planets were traveling at the same speed, Jupiter would complete one orbit in 5.2 years. But Jupiter takes almost 12 years to complete one orbit. Clearly, Jupiter not only has farther to go in its orbit but must be moving more slowly than Earth as well. This trend holds true for all the planets. Farther out from the Sun, the circumferences of planetary orbits become greater, and the speeds at which the planets travel decrease. Mercury, at an average distance of 0.387 AU from the Sun, whizzes along its short orbit at an average speed of 47.9 km/s, completing one revolution in only 88 days. At a distance of 30.0 AU from the Sun, Neptune lumbers along at an average speed of 5.40 km/s, taking 164.8 years to make it once around the Sun.

Outer planets have farther to go and move more slowly in their orbits around the Sun.

Kepler discovered a mathematical relationship between the sidereal period of a planet's orbit and its distance from the Sun. **Kepler's third law** states that *the square of the sidereal period of a planet's orbit, measured in years, is equal to the cube of the semimajor axis of the planet's orbit, measured in astronomical units*. This relationship is explored in **Math Tools 3.2**.

The square of a planet's orbital period is proportional to the cube of the orbit's semimajor axis.

To judge for yourself how well Kepler's third law works, look at **Table 3.2**. Here you can see the periods and semimajor axes of the orbits of the eight classical planets and three of the dwarf planets, along with the values of the ratio P^2 divided by A^3. These data are also plotted in **Figure 3.13**. This relationship was so beautiful to Kepler that he referred to it as his **harmonic law** or, more poetically, as the "Harmony of the Worlds." ASTROTOUR: KEPLER'S LAWS NEBRASKA SIMULATION: PLANETARY ORBIT SIMULATOR

In recognition of his contribution to our understanding of planetary orbits and movements, a NASA telescope designed to look for planets orbiting other stars is named Kepler.

Math Tools 3.2

Kepler's Third Law

Kepler's third law states that the square of the period of a planet's orbit, P_{years}, measured in years, is equal to the cube of the semimajor axis of the planet's orbit, A_{AU}, measured in astronomical units. Translated into math, the law says

$$(P_{\text{years}})^2 = (A_{\text{AU}})^3$$

Here, astronomers use nonstandard units as a matter of convenience. Years are handy units for measuring the periods of orbits, and astronomical units are handy for measuring the sizes of orbits. When we use years and astronomical units as our units, we get the relationship just shown. It is important to realize that *the choice of units in no way changes the physical relationship* being studied. For example, if we instead choose seconds and meters as our units, we use the fact that 1 year $= 3.16 \times 10^7$ seconds and 1 AU $= 1.496 \times 10^{11}$ meters to make this relationship read:

$$(3.17 \times 10^{-8}\ \text{yr/s} \times P_{\text{seconds}})^2 = (6.68 \times 10^{-12}\ \text{AU/m} \times A_{\text{meters}})^3$$

which simplifies to:

$$(P_{\text{seconds}})^2 = 3 \times 10^{-19}\ (A_{\text{meters}})^3$$

Suppose that you want to know the average radius of Neptune's orbit, in astronomical units. First you need to find out how long Neptune's period is in Earth years, which you can determine by observing the synodic period and then computing its sidereal period from that. Neptune's sidereal period is 165 years. Plugging this number into Kepler's third law gives this result:

$$(P_{\text{years}})^2 = (165)^2 = 27{,}225 = (A_{\text{AU}})^3$$

To solve this equation, you must first square the left side to get 27,225 and then take its cube root (see Appendix 1 for calculator hints). Solving the equation gives this result:

$$A_{\text{AU}} = \sqrt[3]{27{,}225} = 30.1$$

So the average distance between Neptune and the Sun is 30.1 AU.

TABLE 3.2 | Kepler's Third Law: $P^2 = A^3$

The Orbital Properties of the Classical and Dwarf Planets

Planet	Period P (years)	Semimajor Axis A (AU)	$\frac{P^2}{A^3}$
Mercury	0.241	0.387	$\frac{0.241^2}{0.387^3} = 1.00$
Venus	0.615	0.723	$\frac{0.615^2}{0.723^3} = 1.00$
Earth	1.000	1.000	$\frac{1.000^2}{1.000^3} = 1.00$
Mars	1.881	1.524	$\frac{1.881^2}{1.524^3} = 1.00$
Ceres	4.599	2.765	$\frac{4.559^2}{2.765^3} = 1.00$
Jupiter	11.86	5.204	$\frac{11.86^2}{5.204^3} = 1.00$
Saturn	29.46	9.582	$\frac{29.46^2}{9.582^3} = 0.99$*
Uranus	84.01	19.201	$\frac{84.01^2}{19.201^3} = 1.00$
Neptune	164.79	30.047	$\frac{164.79^2}{30.047^3} = 1.00$
Pluto	247.68	39.236	$\frac{247.68^2}{39.236^3} = 1.02$*
Eris	557.00	67.696	$\frac{557.00^2}{67.696^3} = 1.00$

*Slight perturbations from the gravity of other planets are the reason that these ratios are not exactly 1.00.

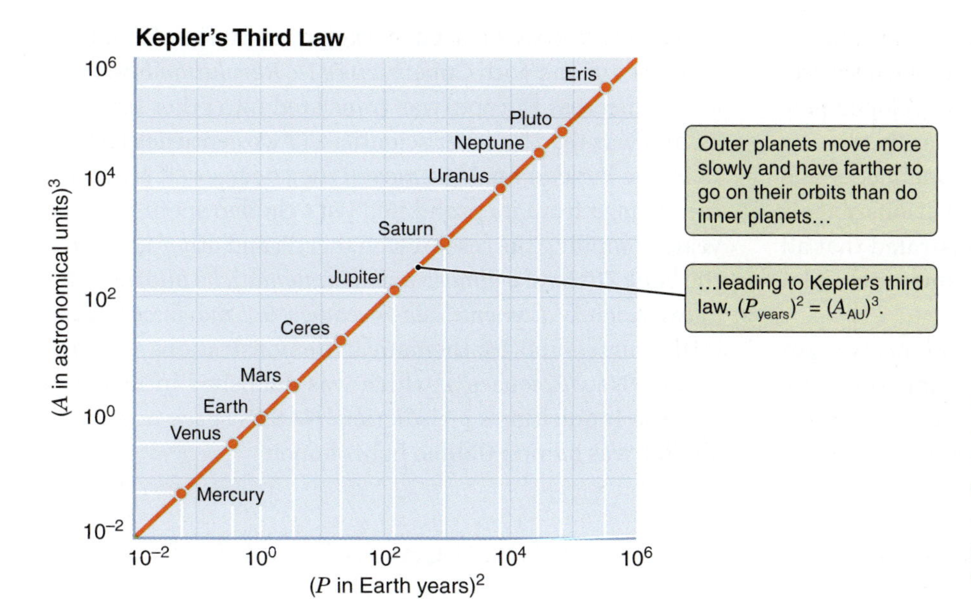

FIGURE 3.13 A plot of A^3 versus P^2 for the eight planets and three dwarf planets in our Solar System shows that they obey Kepler's third law. (Note that by plotting powers of 10 on each axis, we are able to fit both large and small values on the same plot. We will do this frequently.)

3.4 Galileo: The First Modern Scientist

Galileo Galilei (1564–1642—**Figure 3.14**) is one of the heroes of astronomy—the first to use a telescope to conduct and report on significant discoveries about astronomical objects. Much has been written about the considerable danger in which Galileo (as he is commonly known) found himself as a result of these discoveries. Galileo's telescopes were relatively small, yet sufficient for him to observe spots on the Sun, the uneven surface and craters of the Moon, and the large number of stars in the band of light in the sky called the Milky Way.

Other sets of observations, however, are what made Galileo famous. When he turned his telescope to the planet Jupiter, he observed several "stars" in a line near Jupiter, and over time he saw that there were actually four of these stars, and that their positions changed from night to night. Galileo correctly realized that these were moons in orbit around Jupiter, and thus he provided the first observational evidence that some objects in the sky do not orbit Earth. (The four moons that Galileo observed are the largest of the many moons of Jupiter and are still referred to as the Galilean moons.) Galileo also observed that the planet Venus went through phases, and that the phases were correlated with the size of the image of Venus in his telescope (**Figure 3.15**). The phases of Venus are not easily explained in a geocentric model, but they make sense in the heliocentric model. These observations in particular convinced Galileo that Copernicus was correct in placing the Sun at the center of our Solar System. ▶▶ **NEBRASKA SIMULATIONS: PHASES OF VENUS; PTOLEMAIC PHASES OF VENUS**

Galileo's work on the motion of objects was at least as fundamental a contribution as his astronomical observations. He conducted actual experiments with falling and rolling objects—a process that differed from the way natural

FIGURE 3.14 Galileo Galilei, known commonly as Galileo, was among the first to make telescopic astronomical observations, laying the physical framework for Newton's laws.

FIGURE 3.15 Modern photographs of the phases of Venus show that when we see Venus more illuminated, it also appears smaller, implying that Venus is farther away at that time.

philosophers had approached the question, *thinking* about objects in motion but not actually experimenting with them. As with his telescopes, Galileo improved or developed new technology to enable him to conduct his experiments. For example, by carefully rolling balls down an inclined plane (and by thinking about or actually dropping various objects from the Leaning Tower of Pisa), he demonstrated that all objects falling to Earth accelerate at the same rate, independent of their mass.

Galileo's observations and experiments with many types of moving objects, such as carts and balls, led him to disagree with the Greek philosophers about when and why objects continue to move or come to rest. He agreed with them that an object at rest remains at rest unless something causes it to move. But in disagreement, Galileo asserted that, left on its own, an object in motion will remain in motion. Specifically, Galileo said that *an object in motion will continue moving along a straight line with a constant speed until an unbalanced force acts on it to change its state of motion.* When we speak of an **unbalanced force**—we mean the *net* force acting on an object. (The net, or "resultant," force is a single force that represents a combination of all of the individual forces acting on an object.) Galileo referred to this resistance to change in an object's state of motion from an unbalanced force as **inertia**. ▶II ASTROTOUR: VELOCITY, ACCELERATION, INERTIA

Galileo found that an object left in motion remains in motion.

Galileo developed the idea of inertia relatively early in his career, but much of his later life was consumed by conflict with the Catholic Church over his support of the Copernican system. In 1632, Galileo published his best-selling book, *Dialogo sopra i due massimi sistemi del mondo* ("Dialogue Concerning the Two Chief World Systems"). In the *Dialogo*, the champion of the Copernican heliocentric view of the universe is a brilliant philosopher named Salviati. The defender of an Earth-centered universe, Simplicio, uses arguments made by the classical Greek philosophers and the pope, and he sounds silly and ignorant.

Galileo was a religious man who had two daughters in a convent, and he actually thought he had the tacit approval of the Vatican for his book. But when he placed a number of the pope's geocentric arguments in the unflattering mouth of Simplicio, he found that the Vatican's tolerance had limits. The perceived attack on the pope attracted the attention of the Church, and Galileo was put on trial for heresy, sentenced to prison, and eventually placed under house arrest instead. To escape a harsher sentence from the Inquisition, Galileo was forced to publicly recant his belief in the Copernican theory that Earth moves around the Sun. In one of the great apocryphal stories of the history of astronomy, it is said that as he left the courtroom following his sentencing, Galileo stamped his foot on the ground and muttered, "And yet it moves!"

The *Dialogo* was placed on the pope's Index of Prohibited Works, along with Copernicus's *De Revolutionibus*, but it traveled across Europe, was translated into other languages, and was read by other scientists. (Two centuries later, in 1835, the Vatican finally removed the uncensored version of the *Dialogo* from its prohibited list.) Galileo spent his final years compiling his research on inertia and other ideas into the book *Discorsi e dimostrazioni matematiche intorno à due nuove scienze attenenti alla mecanica & í movimenti locali* ("Discourses and Mathematical Demonstrations Relating to Two New Sciences"), which was published in 1638 outside the Inquisition's jurisdiction. NASA's space mission to Jupiter was named Galileo in his honor.

3.5 Newton's Laws of Motion

Empirical rules like those that Kepler spent his life developing are only the first step in understanding a phenomenon. Empirical rules *describe*, but they do not explain. Kepler described the orbits of planets as ellipses, but he did not explain why they should be so. The next step in the scientific process is to try to understand empirical rules in terms of more general physical principles or laws. Beginning with basic physical principles and armed with the tools of mathematics, a scientist works to *derive* the empirically determined rules. Or sometimes a scientist starts with physical laws and predicts relationships, which are then verified (or falsified) by experiment and observation. If these predictions are indeed borne out by experiment and observation, then the scientist just may have determined something fundamental about how the universe works.

One of the earliest great advances in theoretical science was also arguably one of the greatest intellectual accomplishments in human history. In many ways, the work of Sir Isaac Newton (1642–1727—**Figure 3.16**) on the nature of motion set the standard for what is now referred to as *scientific theory* and *physical law*. Newton was a student of mathematics at Cambridge University when it closed down because of the Great Plague and students were sent home to the safer countryside. Over the next 2 years he continued to study on his own, and at the age of 23 or so, he invented calculus. (The German mathematician Gottfried Leibniz independently developed calculus around the same time.) After the plague abated, Newton returned to Trinity College of Cambridge University, becoming a professor of mathematics at age 27.

Building on the work of Kepler, Galileo, and others, Newton proposed three physical laws that he asserted govern the motions of all objects in the sky and on Earth. **Newton's laws** themselves are beautifully elegant, describing the relationships among ev-

Newton's laws of motion are the basis of the physics of motion.

FIGURE 3.16 Sir Isaac Newton formulated three laws of motion.

eryday concepts such as force, velocity, acceleration, and mass. Newton's laws of motion are essential to our understanding of the motions of the planets and all other celestial bodies. These laws enabled him to observe the motion of a shot fired from a cannon and apply what he learned from those observations to the motions of the planets on their orbits around the Sun.

Newton's First Law: Objects at Rest Stay at Rest; Objects in Motion Stay in Motion

Newton's first law of motion states that *an object will remain at rest or will continue moving along a straight line at a constant speed until an unbalanced force acts on it.* It is a tribute to Galileo that his law of inertia became the cornerstone of physics as Newton's first law. You can understand inertia in terms of what we discussed about relative motion and frames of reference in Connections 2.1. Recall that within a frame of reference, only the *relative* motions between objects have any meaning. There is no perceptible difference between an object at rest and an object in uniform motion. The object at rest beside you on the front seat of your car as you drive down the highway is moving at 100 kilometers per hour (km/h) according to a bystander along the side of the road, but it is moving at 200 km/h according to the viewpoint of someone in a car in oncoming traffic. All of these perspectives are equally valid, so motion itself can be described only in relation to a particular frame of reference.

The connection between inertia and the relative nature of motion is so fundamental that a frame of reference moving in a straight line at a constant speed is referred to as an **inertial frame of reference**. The realization that the laws of physics are the same in any inertial frame of reference is one of the deepest insights ever made into the nature of the universe. Thinking along these lines, you can see that an object moving in a straight line at a constant speed remains in motion. As illustrated in **Figure 3.17a**, in the inertial frame of reference of a cup of coffee, the cup is at rest relative to itself even if the car is speeding down the road.

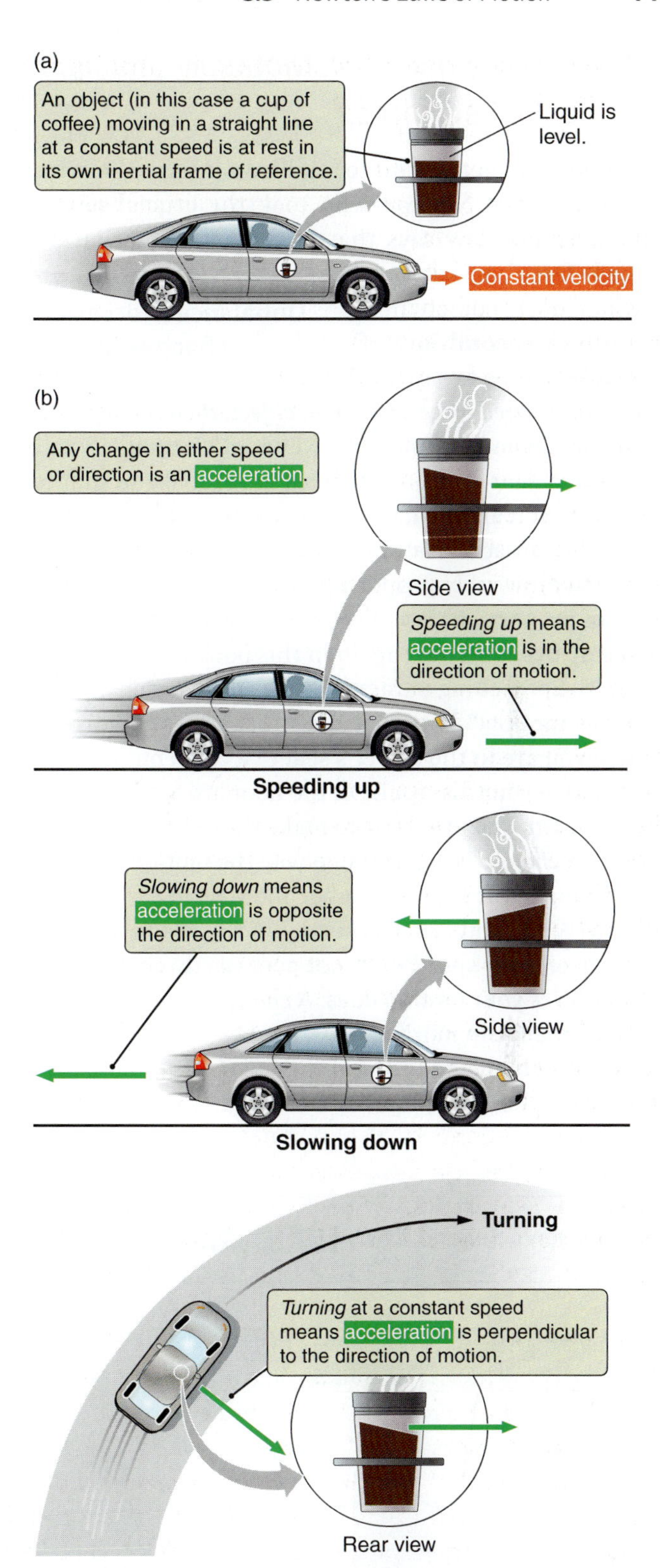

FIGURE 3.17 (a) An object moving in a straight line at a constant speed is at rest in its own inertial frame of reference. (b) Any change in the velocity of an object is an acceleration. When you are driving, for example, any time your speed changes or you follow a curve in the road, you are experiencing an acceleration. (Throughout the text, velocity arrows will be shown as red, and acceleration arrows will be shown as green.)

Newton's Second Law: Motion Is Changed by Unbalanced Forces

Newton often gets credit for Galileo's insight about inertia because it was Newton who took the crucial next step. Newton's first law says that in the absence of an unbalanced force, an object's motion does not change. **Newton's second law of motion** goes on to say that *if an unbalanced force acts on an object, then the object's motion does change*. Even more, Newton's second law tells us *how* the object's motion changes in response to that force, stating that if an unbalanced force acts on a body, that body will have an acceleration proportional to the unbalanced force and inversely proportional to the object's mass. (The idea of proportionality, discussed in **Math Tools 3.3**, will come up over and over again in this book.)

Unbalanced forces cause changes in motion.

In the preceding paragraphs we spoke of "changes in an object's motion," but what does that phrase really mean? When you are in the driver's seat of a car, a number of controls are at your disposal. On the floor are a gas pedal and a brake pedal. You use these to make the car speed up or slow down. A *change in speed* is one way the motion of an object can change. But also remember the steering wheel in your hands. When you are moving down the road and you turn the wheel, your speed does not necessarily change, but the direction of your motion does. A *change in direction* is also a kind of change in motion.

Together, the **speed** and direction of an object's motion are called the object's **velocity**. The rate at which the velocity of an object changes is called **acceleration**. For example, if you go from 0 to 100 km/h in 4 seconds in a car, you feel a strong push from the seat back as it shoves your body forward, causing you to accelerate along with the car. But if you take 2 minutes to go from 0 to 100 km/h, the acceleration is so slight that you hardly notice it. Because the gas pedal on a car is often called the accelerator, people sometimes think *acceleration* means that an object is speeding up. But it is important to stress that, in physics, *any* change in speed or direction is an acceleration. **Figure 3.17b** illustrates the point. Slamming on your brakes and going from 100 to 0 km/h in 4 seconds is just as much acceleration as going from 0 to 100 km/h in 4 seconds. Similarly, the acceleration you experience as you go through a fast, tight turn at a constant speed is every bit as real as the acceleration you feel when you slam your foot on the gas pedal or the brake pedal. Speeding up, slowing down, turning left, turning right—if you are not moving in a straight line at a constant speed, you are experiencing an acceleration.

Acceleration measures how quickly a change in motion takes place.

According to Newton's second law of motion, changes in motion—accelerations—are caused by unbalanced forces. The acceleration that an object experiences depends on two things (**Figure 3.18**). First, it depends on the strength of the unbalanced force acting on the object to change its motion. When all the forces acting on an object balance each other—making the total force on the object zero—the object is not accelerating. If the forces acting on the object do *not* add up to zero, then there is an unbalanced force and the object accelerates (**Figure 3.18a**).

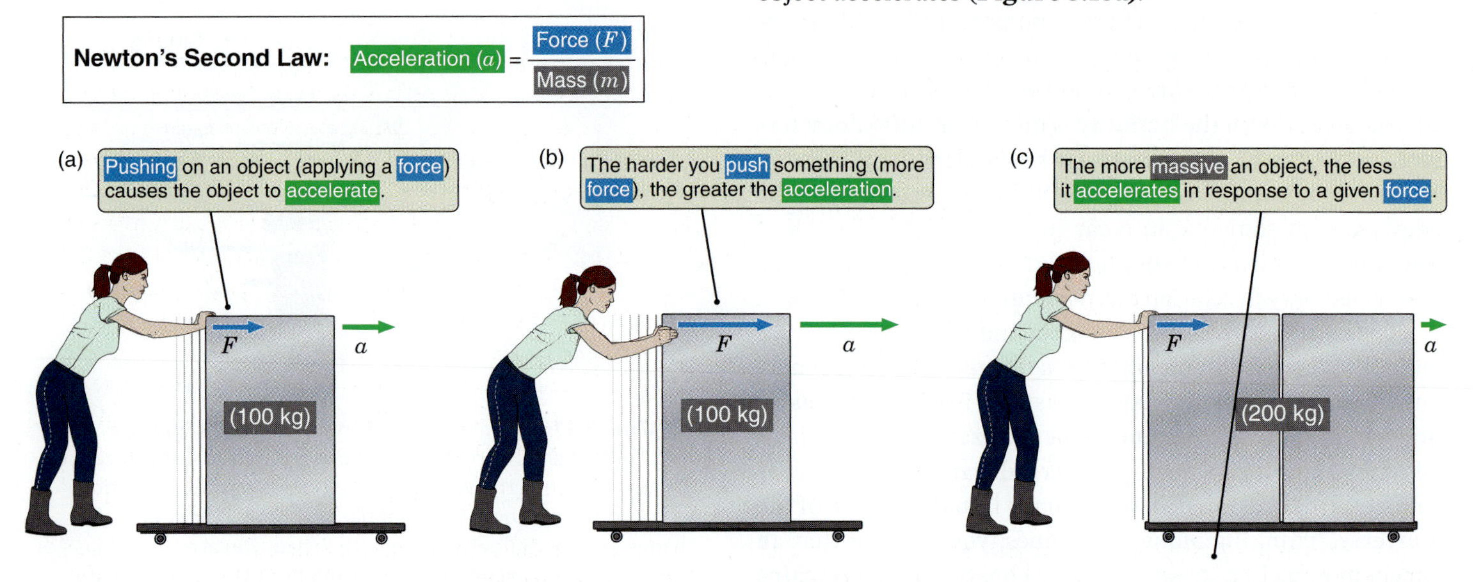

FIGURE 3.18 Newton's second law of motion says that the acceleration experienced by an object is determined by the force acting on the object, divided by the object's mass. (Throughout the text, force arrows will be shown as blue.)

Math Tools 3.3

Proportionality

Often in this text we will say that one quantity is *proportional* to another. Proportionality is a way of getting the gist of how something works—understanding the relationships between things—without having to actually calculate the details of one case after another.

Proportionality

If two quantities are **proportional** to each other, then making one of them larger means making the other quantity larger by the same factor. In other words, the ratio between the two remains constant. For example, think about the weight of a bag of apples and how much the bag costs. Double the weight of the bag of apples and you double the cost. Increase the weight of the bag of apples by a factor of 5, and the cost goes up by a factor of 5 as well. The cost of a bag of apples is *proportional* to the weight of the bag of apples. We write this relationship as:

$$\text{Cost} \propto \text{Weight}$$

where the symbol $\propto$ means "is proportional to." This expression captures the essence of the relationship between the cost and the weight of apples. It tells us that the more apples we buy, the more we will pay.

Mass and acceleration provide another example of proportionality. Mass is measured in units of kilograms (kg). An object with a mass of 2 kg is twice as hard to accelerate as an object with a mass of 1 kg. An object with a mass of 9 kg is 3 times as hard to accelerate as an object with a mass of 3 kg.

Inverse Proportionality

If two quantities are inversely proportional to each other, then making one of them smaller means that the other quantity becomes larger by the same factor. For example, think about a trip to your grandmother's house. If your driving speed is cut in half, the time it takes you to get there is doubled. The time of travel is *inversely proportional* to the average travel speed. We write this relationship as:

$$\text{Time for a trip} \propto \frac{1}{\text{Average speed driven}}$$

This expression tells us that if we drive half as fast, it will take us twice as long to get there. And if we drive twice as fast, our time will be cut in half.

Constants of Proportionality

Sometimes it is enough to know that two quantities are proportional to each other, but sometimes it is not. What if you need to know how much one of those bags of apples will actually cost? We know that the full relationship between the cost and weight of a bag of apples is that the cost is equal to the price per pound of apples, multiplied by the weight of the bag. We write:

$$\text{Cost} = \text{Price per pound} \times \text{Weight}$$

Compare this expression with the previous one. When we say that two quantities are proportional, what we mean is that one quantity equals a particular number *multiplied by* the other quantity. The number by which one quantity is multiplied to get the other number is called the **constant of proportionality**. In our example, the constant of proportionality is just the price per pound of apples.

The fact that the cost of a bag of apples is proportional to the weight of the apples is a statement about the relationship between things. More apples cost more, not less, than fewer apples. The constant of proportionality—here, the price per pound of apples—carries information about factors such as growing, transporting, and selling apples. Very often, physical laws work in this same fashion. Proportionalities tell us about relationships—how two things vary with each other. They let us get a feel for the "how" in how something works.

For example, Galileo wrote the equations for a falling object and found that the velocity of a falling object at any given second is proportional to the time it has been falling:

$$v = gt$$

and the distance traveled by the falling object is proportional to the square of the time it has been falling:

$$d = \frac{1}{2}gt^2$$

where d is the distance fallen, t is the amount of time falling, v is the velocity, and $(\frac{1}{2})g$ is the constant of proportionality.

Galileo did not have a value for g. That came later, when it was shown to be 9.8 meters per second per second. (This is the same as saying "9.8 meters per second squared," which is written as 9.8 m/s^2 or 9.8 m s^{-2}.) So if something did fall off of the Leaning Tower of Pisa, at the end of the first second it would be falling at a rate of 9.8 meters per second (m/s), at the end of the next second it would be falling at 19.6 m/s, at the end of the third second it would be falling at 29.4 m/s, and so on. The distance fallen increases as the square of the time, so at the end of the first second the object falls 4.9 meters, at the end of the next second it falls 19.6 meters:

$$d = \frac{1}{2} \times 9.8\ \text{m/s}^2 \times (2\ \text{s})^2 = 19.6\ \text{meters}$$

at the end of the third second it falls 4.9 meters $\times$ (3 seconds)^2 = 44.1 meters

The stronger the unbalanced force, the greater the acceleration. In fact, the acceleration of an object is *proportional* to the unbalanced force applied (**Figure 3.18b**). Push on something twice as hard and it experiences twice as much acceleration. Push on something 3 times as hard and its acceleration is 3 times as great. The resulting change in motion occurs in the direction in which the unbalanced force is imposed. Push something forward and it moves ahead. Push it to the left and it veers in that direction.

The second thing affecting the acceleration that an object experiences is the degree to which the object resists changes in motion—that is, its inertia. Some objects—for example, an empty box from a refrigerator delivery—are easily shoved around by humans. However, the actual refrigerator, even though it is about the same size, is not so easy to move around. For our purposes here, an object's mass is interchangeable with its inertia. An object has inertia because it has mass. The greater the mass, the greater the inertia, and the *less* acceleration will occur in response to the same unbalanced force (**Figure 3.18c**).

If we introduce a convenient bit of shorthand—a for acceleration, F for force, and m for mass—we get:

$$a = \frac{F}{m}$$

Newton's second law is often written as $F = ma$, giving force as units of mass multiplied by units of acceleration, or kilograms times meters per second squared (kg m/s²). These units of force are aptly named **newtons**, abbreviated **N**.

The equation $F = ma$ is the succinct mathematical statement of Newton's second law of motion. This elegant expression may speak to you clearly and directly. If not, when you see this equation, remind yourself that Newton's second law is nothing more than the embodiment of three commonsense ideas: (1) When you push on an object, that object accelerates in the direction you are pushing. (2) The harder you push on an object, the more it accelerates. (3) The more massive the object is, the harder it is to change its state of motion.

This relationship is directly proportional, so if you have one object with twice the mass of a second object, it will take twice the force to accelerate the first object at the same rate as the second.

Newton's Third Law: Whatever Is Pushed, Pushes Back

Imagine that you are standing on a skateboard and pushing yourself along with your foot. Each shove of your foot against the ground sends you faster along your way. But why? Your muscles flex, and your foot exerts a force on the ground. (Earth does not noticeably accelerate, because its great mass gives it great inertia.) Yet all this does not explain why *you* experience an acceleration. The fact that you accelerate means that as you push on the ground, the ground must be pushing back on you.

Part of Newton's genius was his ability to see sublime patterns in such everyday events. Newton realized that *every* time one object exerts a force on another, the second object exerts a matching force on the first. That second force is exactly as strong as the first force but is in exactly the opposite direction. You push backward on Earth, and Earth pushes you forward. A canoe paddle pushes backward through the water, and the water pushes forward on the paddle, sending the canoe along its way. A rocket engine pushes hot gases out of its nozzle, and those hot gases push back on the rocket, propelling it into space.

All of these examples illustrate **Newton's third law of motion**, which says that *forces always come in pairs, and those pairs are always equal in strength but opposite in direction*. The forces in these action-reaction pairs always act on two different objects. Your weight pushes down on the floor, and the floor pushes back up on your feet with the same amount of force. For every force there is *always* an equal force in the opposite direction. This is one of the few times we can say "always" and really mean it. **Figure 3.19** gives a few examples. There is a great game hiding in Newton's third law. It is called "find the force." Look around you at all the forces at work in the world, and for each force find its pair. It will *always* be there.

For every force there is an equal force in the opposite direction.

To see how Newton's three laws of motion work together, think about the situation illustrated in Figure 3.20 and described in **Math Tools 3.4**.

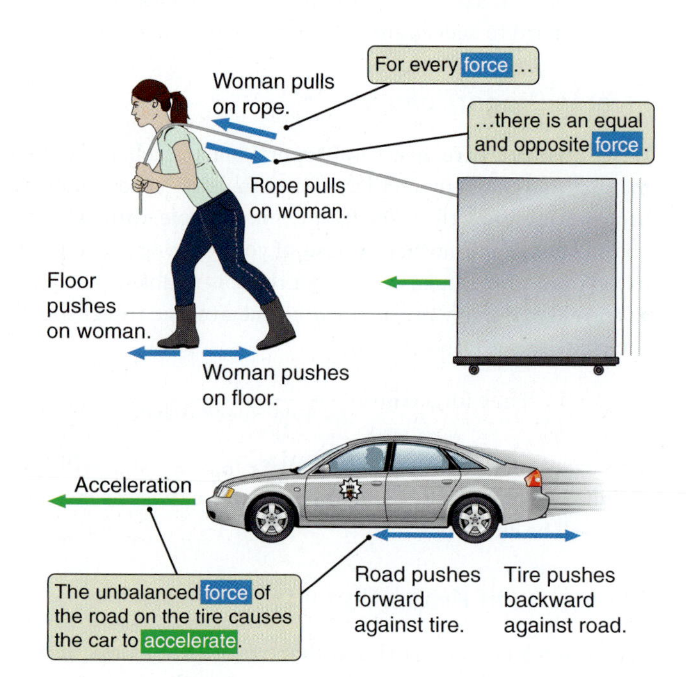

FIGURE 3.19 Newton's third law states that for every force there is always an equal and opposite force. These opposing forces always act on the two different objects in the same pair.

Math Tools 3.4

Using Newton's Laws

Your acceleration is determined by how much your velocity changes, divided by how long it takes for that change to happen:

$$\text{Acceleration} = \frac{\text{How much velocity changes}}{\text{How long the change takes to happen}}$$

For example, if an object's speed goes from 5 to 15 m/s, then the change in velocity is 10 m/s. If that change happens over the course of 2 seconds, then the acceleration is 10 m/s divided by 2 seconds, which equals 5 m/s². ▶II **AstroTour: Velocity, Acceleration, Inertia**

If we want to know, then, how an object's motion is changing, we need to know two things: what unbalanced force is acting on the object, and what is the resistance of the object to that force? We can put the idea into equation form as follows:

$$\left(\begin{array}{c}\text{The}\\ \text{acceleration}\\ \text{experienced}\\ \text{by an object}\end{array}\right) = \frac{\begin{array}{c}\text{The force acting to change}\\ \text{the object's motion}\end{array}}{\begin{array}{c}\text{The object's resistance}\\ \text{to that change}\end{array}} = \frac{\text{Force}}{\text{Mass}}$$

As a simple example, suppose you are holding two blocks. The block in your right hand has twice the mass of the block in your left hand. When you drop the blocks, they both fall with the same acceleration, as shown by Galileo, and they hit your two feet at the same time. Which will hit with more force: the block falling onto your right foot or the one falling onto your left foot? The block in your right hand, with twice the mass, will hit your right foot with twice the force that the other block hits your left foot.

Now consider the following example (**Figure 3.20**). An astronaut is adrift in space, motionless with respect to the nearby space shuttle. With no tether to pull on, how can the astronaut get back to the ship? The answer: throw something. Suppose the 100-kg astronaut throws a 1-kg wrench directly away from the shuttle at a speed of 10 m/s. Newton's second law says that in order to cause the motion of the wrench to change, the astronaut has to apply a force to it in the direction away from the shuttle. Newton's third law says that the wrench must therefore push back on the astronaut with as much force but in the opposite direction. The force of the wrench on the astronaut causes the astronaut to begin drifting toward the shuttle. How fast will the astronaut move? Turn to Newton's second law again. A force that causes the 1-kg wrench to accelerate to 10 m/s will have much less effect on the 100-kg astronaut. Because acceleration equals force divided by mass, the 100-kg astronaut will experience only 1/100 as much acceleration as the 1-kg wrench. The astronaut will drift toward the shuttle, but only at the leisurely rate of $1/100 \times 10$ m/s, or 0.1 m/s.

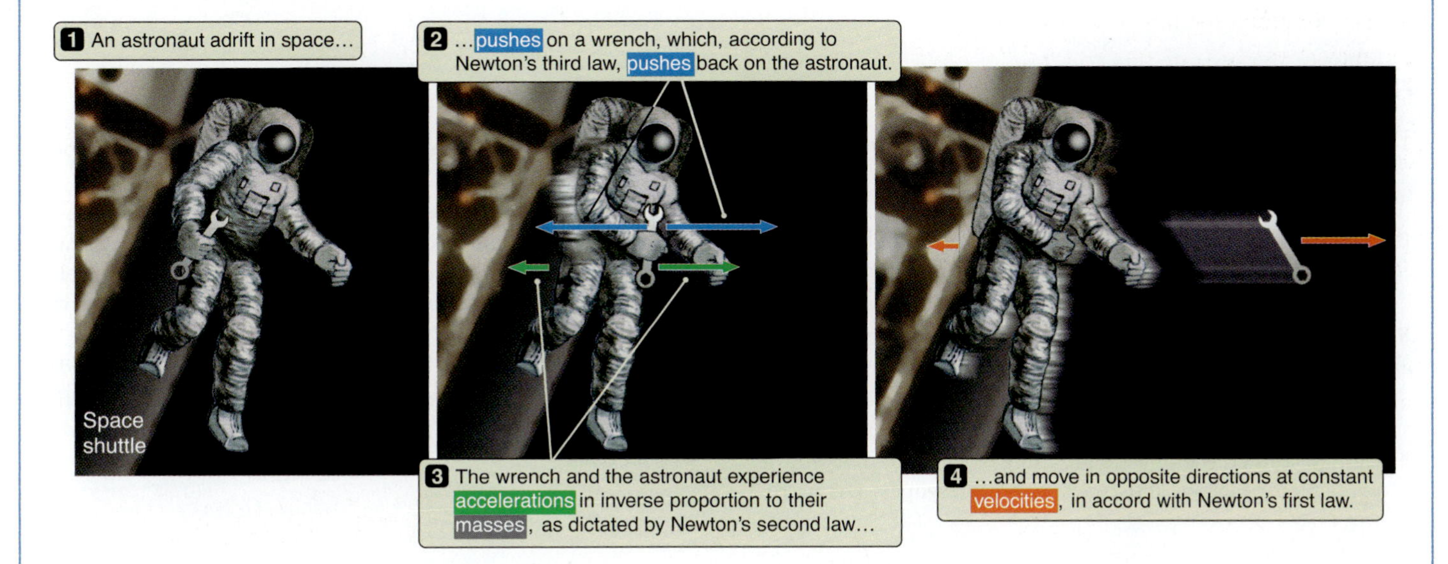

FIGURE 3.20 According to Newton's laws, if an astronaut adrift in space throws a wrench, the two will move in opposite directions at speeds that are inversely proportional to their masses. (Acceleration and velocity arrows are not drawn to scale.)

3.6 Origins: Planets and Orbits

In this chapter you learned how the planets in our Solar System, including Earth, orbit the Sun. Can something as basic as the orbit of a planet about its star affect its chances of developing life? Astrobiologists think that this is an important factor. Hundreds of extrasolar planets have been detected orbiting stars other than our own, and their orbits can be calculated and understood by applying the same three laws of Kepler that we have discussed here in terms of our own Solar System.

Consider the average distance of the planet from its star. You might intuitively guess that a planet close to its star will be hotter than a planet far from its star. If Earth were significantly closer to the Sun, it would be hotter throughout the year—perhaps so hot that water would evaporate and not exist as a liquid. If Earth were significantly farther from the Sun, it would be colder all year long and perhaps all water would freeze. We know that liquid water was a crucial element for the formation of life on Earth. So some astronomers prioritize looking for planets that are situated at a distance from their star such that liquid water can exist (this distance will vary depending on the temperature of the star).

We might also think about the eccentricity of a planet's orbit: recall from Figure 3.11a that Earth's orbit differs from a circle by less than 2 percent. Thus, Earth's distance from the Sun does not vary much throughout the year; and as we saw in Chapter 2, seasonal variation on Earth is caused by the tilt of Earth's axis, not by the slight changes in its distance from the Sun. If we look at our neighboring planet Mars, which has about the same axial tilt as Earth, we see a greater seasonal variation because of its more eccentric orbit. The distance between Mars and the Sun varies from 1.38 AU (207 million km) to 1.67 AU (249 million km)—an eccentricity of 9 percent. As a result, the seasons on Mars are not equal. They are shorter when Mars is closer to the Sun and moving faster (currently, these are the conditions when the northern hemisphere of Mars is in winter and the southern hemisphere is in summer) and longer when Mars is farther from the Sun and moving slower (with the northern hemisphere in summer and the southern hemisphere in winter). The inequality of the seasons on Mars has an effect on the overall stability of its temperature and climate. When we look at planets orbiting other stars, we see that many have orbit eccentricities even higher than that of Mars, and therefore large variations in temperature.

Earth's semimajor axis is at the right distance from the Sun to have temperatures that permit water to be liquid, and its orbital eccentricity is low enough that the temperature remains roughly constant. These orbital characteristics may have contributed to making the conditions on Earth suitable for the development of life.

Now that we have developed a fundamental understanding of motion, in Chapter 4 we will turn to the concept of gravity. ■

Summary

- **3.1** Retrograde motion was difficult to understand in a geocentric model of the Solar System, in which all objects orbit about the Earth.
- **3.2** Copernicus created the first mathematical model of the Solar System with the Sun at the center, called a heliocentric model.
- **3.3** Using Tycho's observational data, Kepler developed empirical solutions to describe the motions of the planets (in short: planets in elliptical orbits about the Sun, moving fastest when closest and slowest when farthest, and $P^2 = A^3$).
- **3.4** Galileo studied the physics of falling objects, discovered the principle of inertia, and constructed telescopes. His astronomical observations were difficult to explain with a geocentric model.
- **3.5** Newton's three laws (in shorthand: law of inertia, $F = ma$, and "every action has an equal and opposite reaction") govern the motion of all objects. Unbalanced forces cause accelerations—that is, changes in motion. Proportionality describes patterns and relationships in nature. Mass is the property of matter that gives it inertia. Inertia resists changes in motion.
- **3.6** Orbital semimajor axis, eccentricity, and stability may affect the likelihood that a planet will have conditions that can foster life.

Unanswered Questions

- Would the history of scientific discoveries in physics and astronomy be different if the Catholic Church had not prosecuted Galileo? Galileo wrote the *Dialogo* after being ordered by the Catholic Church in 1616 not to "hold or defend" the idea that Earth moves and the Sun is still. And he wrote his equally famous *Discorsi e Dimostrazioni Matematiche* (often shortened in English to "Two New Sciences") while under house arrest after his trial. However undeterred Galileo appeared to be, the effects of the decrees, prohibitions, and prosecutions might have dissuaded other scientists in Catholic countries from pursuing this type of work. Indeed, after Galileo the center of the scientific revolution moved north to Protestant Europe.
- What percentage of planets are in unstable orbits? In younger planetary systems, planets might migrate in their orbits because of the presence of other massive planets nearby. We will see in Chapter 7 that Uranus and Neptune probably migrated in this way. Some planets have been discovered moving through the galaxy without any obvious orbit about a star—and therefore are not stable orbits.

Questions and Problems

Summary Self-Test

1. Most ancient Greek astronomers were skeptical about the idea that Earth revolved around the Sun because
   a. of their religious beliefs.
   b. it feels as though Earth is stationary.
   c. Earth is so important.
   d. the heavens were thought to be perfect.

2. When Earth catches up to a slower-moving outer planet and passes it in its orbit in the same way that a faster runner overtakes a slower runner in an outside lane, the planet
   a. exhibits retrograde motion.
   b. slows down because it feels Earth's gravitational pull.
   c. decreases in brightness as it passes through Earth's shadow.
   d. moves into a more elliptical orbit.

3. Copernicus's model of the Solar System was superior to Ptolemy's because
   a. it had a mathematical basis that could be used to predict the positions of planets.
   b. it was more accurate.
   c. it did not require epicycles.
   d. it was simpler.

4. A planet with an eccentricity of 0.5 has
   a. nearly equal semimajor and semiminor axes.
   b. a longer semiminor axis than semimajor axis.
   c. a longer semimajor axis than semiminor axis.
   d. the Sun at the center of its orbit.

5. Suppose a planet is discovered orbiting a star in a highly elliptical orbit. While the planet is close to the star it moves ________, but while it is far away it moves ________.
   a. faster; slower
   b. slower; faster
   c. retrograde; prograde
   d. prograde; retrograde

6. For Earth, $P^2/A^3 = 1.0$ (in appropriate units). Suppose a new dwarf planet is discovered that is 14 times as far from the Sun as Earth is. For this planet,
   a. $P^2/A^3 = 1.0$.
   b. $P^2/A^3 > 1.0$.
   c. $P^2/A^3 < 1.0$.
   d. you can't know the value of P^2/A^3 without more information.

7. Place the following in order from largest to smallest semimajor axis.
   a. a planet with a period of 84 Earth days
   b. a planet with a period of 1 Earth year
   c. a planet with a period of 2 Earth years
   d. a planet with a period of 0.5 Earth year

8. Suppose you watch a car sliding on ice down a hill. As it slides, it slowly drifts to your right across the road, speeding up in that direction until it bumps into the curb. In this situation, which of the following is more likely, and why?
   a. The car was already traveling diagonally across the street before it hit the patch of ice.
   b. The road surface is rounded (or "crowned"), causing the car to be pushed across the street.

9. Imagine you are walking along a forest path. Which of the following is *not* an action-reaction pair in this situation?
   a. the gravitational force between you and Earth; the gravitational force between Earth and you
   b. your shoe pushing back on Earth; Earth pushing forward on your shoe
   c. your foot pushing back on the inside of your shoe; your shoe pushing forward on your foot
   d. you pushing down on Earth; Earth pushing you forward

10. An unbalanced force must be acting when an object
    a. accelerates.
    b. changes direction but not speed.
    c. changes speed but not direction.
    d. changes speed and direction.
    e. all of the above

True/False and Multiple Choice

11. **T/F:** Copernicus's determinations of the distances between the planets and the Sun were quite accurate.
12. **T/F:** Retrograde motion is the counterclockwise motion of Solar System objects, as seen from Earth.
13. **T/F:** Kepler obtained accurate data on the positions of the planets in the sky over time, which Galileo used to prove that planets revolve around the Sun.
14. **T/F:** Planets with circular orbits travel at the same speed at all points in their orbits; planets with elliptical orbits change their speeds at different points in their orbits.
15. **T/F:** If an object moves at constant speed, a net force is acting on it.
16. A ________ model of the Solar System puts Earth at the center, while a ________ model of the Solar System puts the Sun at the center.
    a. geocentric; heliocentric
    b. heliocentric; geocentric
    c. heliocentric; Copernican
    d. geocentric; Ptolemaic
17. Kepler's first law replaced Copernicus's perfect circles with ellipses, thus shattering the idea that
    a. Tycho's data were accurate.
    b. the Sun is at a focus.
    c. the heavens were perfect, with perfectly round objects and perfectly round orbits.
    d. Earth goes around the Sun.
18. Kepler's second law says that
    a. planetary orbits are ellipses with the Sun at one focus.
    b. the square of a planet's orbital period equals the cube of its semimajor axis.
    c. for every action there is an equal and opposite reaction.
    d. unbalanced forces cause changes in motion.
    e. planets move fastest when they are closest to the Sun.
19. Suppose you read in the newspaper that a new planet has been found. Its average speed in orbit is 33 km/s. When it is closest to its star it moves at 31 km/s, and when it is farthest from its star it moves at 35 km/s. This story is in error because
    a. the average speed is far too fast.
    b. Kepler's third law says the planet has to sweep out equal areas in equal times, so the speed of the planet cannot change.
    c. planets stay at a constant distance from their stars; they don't move closer or farther away.
    d. Kepler's second law says the planet must move fastest when it's closest, not when it is farthest away.
    e. using these numbers, the square of the orbital period will not be equal to the cube of the semimajor axis.
20. Galileo observed that Jupiter has moons. From this information, you may conclude that
    a. Jupiter is the center of the Solar System.
    b. Jupiter orbits the Sun.
    c. Jupiter orbits Earth.
    d. some things do not orbit Earth.
21. Galileo observed that Venus had phases that correlated with its size in his telescope. From this information, you may conclude that Venus
    a. is the center of the Solar System.
    b. orbits the Sun.
    c. orbits Earth.
    d. orbits the Moon.
22. A planet with a large orbit ________ than a planet with a smaller orbit.
    a. is colder
    b. moves faster
    c. has a longer period
    d. all of the above
23. Planets with high eccentricity may be unlikely candidates for life because
    a. the speed varies too much.
    b. the period varies too much.
    c. the temperature varies too much.
    d. the orbit varies too much.
24. Imagine you are pulling a small child on a sled by means of a rope. Which of the following are action-reaction pairs in this situation?
    a. You pull forward on the rope; the rope pulls backward on you.
    b. The sled pushes down on the ground; the ground pushes up on the sled.
    c. The sled pushes forward on the child; the child pushes backward on the sled.
    d. The rope pulls forward on the sled; the sled pulls backward on the rope.
25. Suppose you read about a new car that can go from 0 to 100 km/h in only 2.0 seconds. What is this car's acceleration?
    a. about 50 km/h
    b. about 14 m/s^2
    c. about 50 km/s^2
    d. about 200 km
    e. about 0.056 km/h^2

Thinking about the Concepts

26. Copernicus and Kepler engaged in what is called empirical science. What do we mean by *empirical*?
27. Study Figure 3.6. During normal motion, does Mars move toward the east or west? Which direction does it travel when moving retrograde? For how many days did Mars move retrograde?
28. Each ellipse has two foci. The orbits of the planets have the Sun at one focus. What is at the other focus?
29. Ellipses contain two axes: major and minor. Half the major axis is called the semimajor axis. What is especially important about the semimajor axis of a planetary orbit?

30. What is the eccentricity of a circular orbit?

31. The speed of a planet in its orbit varies in its journey around the Sun.
    a. At what point in its orbit is the planet moving the fastest?
    b. At what point is it moving the slowest?

32. The distance that Neptune has to travel in its orbit around the Sun is approximately 30 times greater than the distance that Earth must travel. Yet it takes nearly 165 years for Neptune to complete one trip around the Sun. Explain why.

33. Galileo came up with the concept of inertia. What do we mean by *inertia*? How is it related to mass?

34. If Kepler had lived on Mars, would he have deduced the same empirical laws for the motion of the planets?

35. What is the difference between speed and acceleration?

36. You push your book across the desk, and it slides off the edge and falls to the floor. Identify the forces that act on the book during the three parts of this scenario: (a) while you are pushing the book, (b) while the book is sliding, and (c) while the book is falling.

37. Imagine you throw a ball straight up in the air. At the top of its flight, it stops and then falls again. What is the acceleration of the ball at the top of its flight?

38. Consider the ball in question 37. Neglecting air resistance, compare the velocity of the ball just after it leaves your hand to the velocity of the ball just before you catch it. (Assume that your hand is at the same height above the ground at both times.)

39. When involved in an automobile collision, a person not wearing a seat belt will move through the car and often strike the windshield directly. Which of Newton's laws explains why the person continues forward, even though the car stopped?

40. When riding in a car, we can sense changes in speed or direction through the forces that the car applies on us. Do we wear seat belts in cars and airplanes to protect us from speed or from acceleration? Explain your answer.

Applying the Concepts

41. Study the graph in Figure 3.13. Is this graph linear or logarithmic? From the data on the graph, find the approximate semimajor axis and period of Saturn. Show your work.

42. Study Figure 3.15, which shows that the apparent size of Venus changes as it goes through phases. Approximately how many times larger is Venus in the sky at the tiniest crescent than at the gibbous phase shown? Therefore, approximately how many times closer is Venus to us at the phase of that tiniest crescent than at the gibbous phase?

43. Assume a new dwarf planet is discovered. It is located at 18.3 AU. Its synodic period is 370 days.
    a. Use Math Tools 3.1 to find the sidereal period.
    b. Use Kepler's law to find the sidereal period.
    c. Compare your results for (a) and (b).

44. Suppose a new dwarf planet is discovered orbiting the Sun with a semimajor axis of 50 AU. What would be the orbital period of this new dwarf planet?

45. Suppose you discover a planet around a Sun-like star. From careful observation over several decades, you find that its period is 12 Earth years. Find the semimajor axis.

46. Suppose you read in a tabloid newspaper that "experts have discovered a new planet with a distance from the Sun of 2 AU and a period of 3 years." Use Kepler's third law to argue that this is impossible.

47. Show, as Galileo did, that Kepler's third law applies to the four moons of Jupiter that he discovered, by calculating P^2 divided by A^3 for each moon. (Data on the moons can be found in Appendix 4.)

48. A sports car accelerates from a stop to 100 km/h in 4 seconds.
    a. What is its acceleration?
    b. If it went from 100 km/h to a stop in 5 seconds, what would be its acceleration?
    c. Suppose the car has a mass of 1,200 kg. How strong is the force on the car?
    d. What supplies the "push" that accelerates the car?

49. A train pulls out of a station accelerating at 0.1 m/s^2. What is its speed, in kilometers per hour, 2 1/2 minutes after leaving the station?

50. A Fermi problem involves making reasonable estimates to arrive at a reasonable answer, even if you do not know something in detail. Estimate the mass of a train, and then estimate the force that accelerates the train described in question 49.

51. Flybynite Airlines takes 3 hours to fly from Baltimore to Denver at a speed of 800 km/h. To save fuel, management orders the airline's pilots to reduce their speed to 600 km/h. How long will it now take passengers on this route to reach their destination?

52. You are driving down a straight road at a speed of 90 km/h, and you see another car approaching you at a speed of 110 km/h along the road.
    a. Relative to your own frame of reference, how fast is the other car approaching you?
    b. Relative to the other driver's frame of reference, how fast are you approaching the other driver's car?

53. You are riding along on your bicycle at 20 km/h and eating an apple. You pass a bystander.
    a. How fast is the apple moving in your frame of reference?
    b. How fast is the apple moving in the bystander's frame of reference?
    c. Whose perspective is more valid?

54. During the latter half of the 19th century, a few astronomers thought there might be a planet circling the Sun inside Mercury's orbit. They even gave it a name: Vulcan. We now know that Vulcan does not exist. If a planet with an orbit one-fourth the size of Mercury's actually existed, what would be its orbital period relative to that of Mercury?

55. Suppose you are pushing a small refrigerator of mass 50 kg on wheels. You push with a force of 100 N.
    a. What is the refrigerator's acceleration?
    b. Assume the refrigerator starts at rest. How long will the refrigerator accelerate at this rate before it gets away from you (that is, before it is moving faster than you can run—on the order of 10 m/s)?

Using the Web

56. Go to the Web page "This Week's Sky at a Glance" (http://skyandtelescope.com/observing/ataglance) on *Sky & Telescope* magazine's website. Which planets are visible in your sky this week? Why are Mercury and Venus visible in the morning before sunrise or in the evening just after sunset? Before telescopes, how did people know the planets were different from the stars?

57. Look up the dates for the next opposition of Mars, Jupiter, or Saturn. One source is the NASA "Sky Events Calendar" at http://eclipse.gsfc.nasa.gov/SKYCAL/SKYCAL.html. Check only the "Planet Events" box in "Section 2: Sky Events"; and in Section 3, generate a calendar or table for the year. As noted in Connections 3.1, opposition means that the planet will be opposite the Sun in the sky, so it will rise at sunset and set at sunrise. It is also during opposition that the planet is closest to Earth and you can observe retrograde motion. If you are coming up on an opposition, take pictures of the planet over the next few weeks. Can you see its position move in retrograde fashion with respect to the background stars?

58. Refer to the website from question 57 to find the current observational positions of all the planets.
    a. Which ones are in or near to conjunction, opposition, or greatest elongation?
    b. Which are visible in the morning sky? In the evening sky?
    c. Sketch the Solar System with Earth, Sun, and planets as it looks from "above." Check your result using NASA's "Solar System Simulator" (http://space.jpl.nasa.gov): Set it for Solar System as seen from above, and look at the field of view of 2°, 5°, and 30° to see the inner and then the outer planets. Does the simulator agree with your sketch?

59. Go to the Museo Galileo website (http://brunelleschi.imss.fi.it/telescopiogalileo/index.html), and view the exhibit on Galileo's telescope. What did his telescope look like? What other instruments did he use? From the museum page you can link to short videos (in English) on his science and his trial (http://catalogue.museogalileo.it). Why is Galileo considered the first modern scientist? Why is his middle finger on display in the museum?

60. Go to the online "Extrasolar Planets Encyclopedia" (http://exoplanet.eu/catalog-all.php), and find a planet with an orbital period similar to that of Earth. What is the semimajor axis of its orbit? If it is very different from 1 AU, then the mass of the star is different from that of the Sun. Click on the star name in the first column to see the star's mass. What is the orbital eccentricity? Now select a star with multiple planets. Verify that Kepler's third law applies by showing that the value of P^2/A^3 is about the same for each of the planets of this star. How eccentric are the orbits of the multiple planets?

StudySpace is a free and open website that provides a Study Plan for each chapter of *21st Century Astronomy*. Study Plans include animations, reading outlines, vocabulary flashcards, and multiple-choice quizzes, plus links to premium content in SmartWork and the ebook. Visit **wwnorton.com/studyspace**.

SmartWork Norton's online homework system, includes algorithmically generated versions of these questions, plus additional conceptual exercises. If your instructor assigns questions in SmartWork, log in at **smartwork.wwnorton.com.**

Exploration | Kepler's Laws

In this Exploration we will examine how Kepler's laws apply to the orbit of Mercury. Visit StudySpace and open the Planetary Orbit Simulator applet. This simulator animates the orbits of the planets, enabling you to control the simulation speed, as well as a number of other parameters. Here we focus on exploring the orbit of Mercury, but you may wish to spend some time examining the orbits of other planets as well.

Kepler's First Law

To begin exploring the simulation, in the "Orbit Settings" panel, use the drop-down menu next to "set parameters for" to select "Mercury" and then click "OK." Click the "Kepler's 1st Law" tab at the bottom of the control panel. Use the radio buttons to select "show empty focus" and "show center."

1 **How would you describe the shape of Mercury's orbit?**

....

....

....

Deselect "show empty focus" and "show center," and select "show semiminor axis" and "show semimajor axis." Under "Visualization Options," select "show grid."

2 **Use the grid markings to estimate the ratio of the semiminor axis to the semimajor axis.**

....

....

....

3 **Calculate the eccentricity of Mercury's orbit from this ratio using $e = [1 - (\text{Ratio})^2]^{1/2}$.**

....

....

....

Kepler's Second Law

Click on "reset" near the top of the control panel, set parameters for Mercury and click OK. Then click on the "Kepler's 2nd Law" tab at the bottom of the control panel. Slide the "adjust size" slider to the right, until the fractional sweep size is $1/8$.

Click on "start sweeping." The planet moves around its orbit, and the simulation fills in area until $1/8$ of the ellipse is filled. Click on "start sweeping" again as the planet arrives at the rightmost point in its orbit (that is, at the point in its orbit farthest from the Sun). You may need to slow the animation rate using the slider under "Animation Controls." Click on "show grid" under the visualization options. (If the moving planet annoys you, you can pause the animation.) One easy way to estimate an area is to count the number of squares.

4 **Count the number of squares in the yellow area and in the red area. You will need to decide what to do with fractional squares. Are the areas the same? Should they be?**

....

....

Kepler's Third Law

Click on "reset" near the top of the control panel, set parameters for Mercury, and then click on the "Kepler's 3rd Law" tab at the bottom of the control panel. Select "show solar system orbits" in the "Visualization Options" panel. Study the graph. Use the eccentricity slider to change the eccentricity of the simulated planet. Make the eccentricity first smaller and then larger.

5 **Did anything in the graph change?**

....

....

6 **What do your observations of the graph tell you about the dependence of the period on the eccentricity?**

....

....

Set parameters back to those for Mercury. Now use the semimajor axis slider to change the semimajor axis of the simulated planet.

7 **What happens to the period when you make the semimajor axis smaller?**

....

....

8 **What happens when you make it larger?**

....

....

9 **What do these results tell you about the dependence of the period on the semimajor axis?**

....

....

The International Space Station in orbit about Earth.

04 Gravity and Orbits

The Newtonian principle of gravitation is now more firmly established, on the basis of reason, than it would be were the government to step in, and to make it an article of necessary faith. Reason and experiment have been indulged, and error has fled before them.

Thomas Jefferson (1743–1826)

LEARNING GOALS

In this chapter we continue the story of the birth of modern science and explore the physical laws that astronomers and physicists discovered to explain the regular patterns in the motions of the planets. By the conclusion of this chapter, you should be able to:

- Synthesize the concepts of motion and gravitation to explain planetary orbits.
- Describe how and why objects must achieve a certain speed to go into orbit.
- Explain how tidal forces are caused by gravity.
- Illustrate how the relationship between the Moon, Earth, and Sun causes tides on Earth.

4.1 Gravity Is a Force between Any Two Objects Due to Their Masses

Having explored both Kepler's empirical description of the motions of planets about the Sun and Newton's laws of motion in Chapter 3, we turn now to Newton's universal law of gravitation. Where did Newton get his ideas about gravity, which united empirical and theoretical science? What guided him in his development of those ideas, and how did he turn them into a theory with testable predictions? By answering these questions, rather than simply stating Newton's law of gravitation, we will gain some insight into how science is done.

Why does Earth have tides?

Drop a ball and the ball falls toward the ground, picking up speed as it falls. It accelerates toward Earth. Newton's second law says that where there is acceleration, there is force. But where is the force that causes the ball to accelerate? Many forces that we see in everyday life involve direct contact between objects. The cue ball slams into the eight ball, knocking it into the pocket. The shoe of the child pushing along a scooter presses directly against the surface of the pavement. When there is physical contact between two objects, the source of the forces between them is easy to see. But the ball falling toward Earth is an example of a different kind of force, one that acts at a distance across the intervening void of space. The ball falling toward Earth is accelerating in response to the force of **gravity**, one of the fundamental forces of nature.

Gravity is "force at a distance."

Where Do Theories Come From? Newton's Law of Gravitation

Recall from Chapter 3 the insight and observation of Galileo, who discovered experimentally that all freely falling objects accelerate toward Earth at the same rate, regardless of their mass. Drop a marble and a cannonball at the same time and from the same height, and they will hit the ground together. (If proof was needed that this free-fall behavior is not solely a property of Earth's gravity, it was provided by astronaut David Scott on the lunar surface—**Figure 4.1**; see the **Process of Science Figure** on the next page.) The gravitational acceleration near the surface of Earth, also measured experimentally by Galileo, is usually written as g (lowercase) and has a value of 9.8 meters per second squared (m/s^2) on average. (The value varies with location because Earth's poles are closer to Earth's center than is its equator and because Earth rotates; g ranges from 9.78 m/s^2 at the equator to 9.83 m/s^2 at the poles.) As you saw in Math Tools 3.3, whether you drop a marble or a cannonball, after 1 second it will be falling at a speed of 9.8 m/s, after 2 seconds at 19.6 m/s, and after 3 seconds at 29.4 m/s. (These numbers assume that we can neglect air resistance on Earth, which is reasonably negligible for relatively dense objects moving at relatively slow speeds. On the Moon, there is no air or air resistance.)

All objects on Earth fall with the same acceleration, g.

After working out the laws governing the motion of objects, Newton saw something deeper in Galileo's findings. Newton realized that if all objects fall with the same acceleration, then the gravitational *force* on an object must be determined by the object's *mass*. To see why, look back at Newton's second law (acceleration equals force divided by mass, or $a = F/m$). Gravitational acceleration can be the same for all objects only if the value of the force divided by the mass is the same for all objects. Since force and mass are proportional (see Math Tools 3.3), greater mass *must* be accompanied by a stronger gravitational force. In other words, the gravitational force on an object on Earth is, according to Newton's second law, the object's mass multiplied by the acceleration due to gravity, or $F_{grav} = mg$. So an object twice as massive has double the gravitational force

FIGURE 4.1 Astronaut Alan Bean's portrait of fellow astronaut David Scott standing on the Moon and dropping a hammer and a falcon feather together. The two objects reached the lunar surface simultaneously. (The astronauts' lunar module was nicknamed "Falcon.")

Process of Science

UNIVERSALITY

The laws of physics are the same everywhere and at all times. The principle governs our understanding of the natural world.

Galileo determined that all objects have the same gravitational acceleration.

Newton's law of universal gravitation extended this observation to the Solar System.

Apollo 15 commander David Scott tested the law with a feather and a hammer on the Moon. With no air resistance, the feather and the hammer fell at the same rate.

The same physical laws apply to falling objects, to planets orbiting the Sun, to stars orbiting within the galaxy, and to galaxies orbiting each other.

acting on it. An object 3 times as massive has triple the gravitational force acting on it.

In precise terms, the gravitational force acting on an object is commonly referred to as the object's **weight**. On the surface of Earth, weight is mass multiplied by Earth's gravitational constant, *g*. Because of our everyday use of language, it is easy to see why people confuse mass and weight. We often say that an object with a mass of 2 kilograms (kg) "weighs" 2 kg, but it is more correct scientifically to express a weight in terms of newtons (N):

$$F_{\text{weight}} = m \times g,$$

where F_{weight} is an object's weight in newtons, m is the object's mass in kilograms, and g is Earth's gravitational acceleration in meters per second squared.

Your mass is a property of you, but your weight depends on where you are.

On Earth, an object with a mass of 2 kg has a *weight* of 2 kg × 9.8 m/s^2, or 19.6 N. On the Moon, where the gravitational acceleration is 1.6 m/s^2, the 2-kg mass would have a weight of 2 kg × 1.6 m/s^2, or 3.2 N (**Figure 4.2**). Although your mass remains the same everywhere, your weight depends on where you are. For example, on the Moon your weight is about one-sixth of your weight on Earth.

Newton's next great insight came from applying his third law of motion to gravity. Recall that Newton's third law states that for every force there is an equal and opposite force. Therefore, if Earth exerts a force of 19.6 N on a 2-kg mass sitting on its surface, then that 2-kg mass must exert a force of 19.6 N on Earth as well. Drop a 7-kg bowling ball and it falls toward Earth, but at the same time Earth falls toward the 7-kg bowling ball. The reason we do not notice the motion of Earth is that Earth is very massive. It has a lot of resistance to changes in its motion. In the time it takes a 7-kg bowling ball to fall to the ground from a height of 1 kilometer (km), Earth has "fallen" toward the bowling ball by only a tiny fraction of the size of an atom.

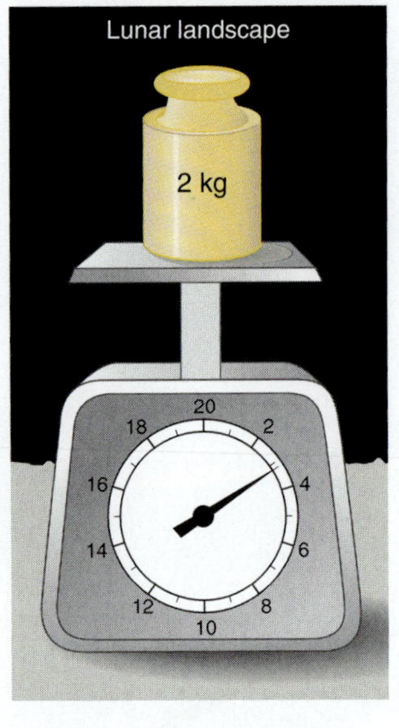

FIGURE 4.2 On the Moon, a mass of 2 kg has ⅙ the weight (displayed in newtons) that it has on Earth.

Newton reasoned that this relationship should work with either object. If doubling the mass of an object doubles the gravitational force between the object and Earth, then doubling the mass of Earth ought to do the same thing. In short, the gravitational force between Earth and an object must be equal to the product of the two masses multiplied by something:

$$\text{Gravitational force} = \text{Something} \times \text{Mass of Earth} \times \text{Mass of object}$$

If the mass of the object were 3 times greater, then the force of gravity would be 3 times greater. Likewise, if the mass of Earth were 3 times what it is, the force of gravity would have to be 3 times greater as well. If *both* the mass of Earth and the mass of the object were 3 times greater, the gravitational force would increase by a factor of 3 × 3, or 9, times. Because objects fall toward the center of Earth, we know that this force is an attractive force acting along a line between the two masses.

As the story goes, when Newton saw an apple fall from a tree to the ground, he reasoned that if gravity is a force that depends on mass, then there should be a gravitational force between *any* two masses, including between a falling apple and Earth. Suppose we have two masses—call them mass 1 and mass 2, or m_1 and m_2 for short. The gravitational force between them is something multiplied by the product of the masses:

The force of gravity is proportional to the product of two masses.

$$\text{Gravitational force between two objects} = \text{Something} \times m_1 \times m_2.$$

We have gotten this far just by combining Galileo's observations of falling objects with (1) Newton's laws of motion and (2) Newton's belief that Earth is a mass just like any other mass. But what about that "something" in the previous expression? Today we have instruments sensitive enough that we can put two masses close to each other in a laboratory, measure the force between them, and determine the value of that something directly. But Newton had no such instruments. He had to look elsewhere to continue his exploration of gravity.

Kepler had already thought about this question. He reasoned that because the Sun is the focal point for planetary orbits, the Sun must be responsible for exerting an influence over the motions of the planets. Kepler speculated that whatever this influence is, it must grow weaker with distance from the Sun. (After all, it must surely require a stronger influence to keep Mercury whipping around in its tight, fast orbit than it does to keep the outer planets lumbering along their paths around the Sun.) Kepler's speculation went even further. Although he did not know about forces or inertia or gravity as the cause of celestial motion, he did know quite a lot about geometry, and geometry alone suggested how this solar "influence" might change for planets progressively farther from the Sun.

Imagine you have a certain amount of paint to spread over the surface of a sphere. If the sphere is small, you will get a thick coat of paint. But if the sphere is larger, the paint has to spread farther and you get a thinner coat. Geometry tells us that the surface area of a sphere depends on the square of the sphere's radius. Double the radius of a sphere, and the sphere's surface area becomes 4 times what it was. If you paint this new, larger sphere, the paint must cover 4 times as much area and the thickness of the paint will be only a fourth of what it was on the smaller sphere. Triple the radius of the sphere and the sphere's surface will be 9 times as large, and the thickness of the coat of paint will be only one-ninth as thick.

Kepler thought the influence that the Sun exerts over the planets might be like the paint in this example. As the influence of the Sun extended farther and farther into space, it would have to spread out to cover the surface of a larger and larger imaginary sphere centered on the Sun. If so, then, like the thickness of the paint, the influence of the Sun should be proportional to 1 divided by the square of the distance between the Sun and a planet. We call this relationship an **inverse square law**.

Gravity is governed by an inverse square law.

Kepler had an interesting idea, but not a scientific hypothesis with testable predictions. He lacked an explanation for how the Sun influences the planets, as well as the mathematical tools to calculate how an object would move under such an influence. Newton had both. If gravity is a force between *any* two objects, then there should be a gravitational force between the Sun and each of the planets. If this gravitational force were the same as Kepler's "influence," then the "something" in Newton's expression for gravity might be a term that diminishes according to the square of the distance between two objects. In essence, gravity might behave according to an inverse square law.

Newton's expression for gravity came to look like this:

$$\text{Gravitational force between two objects} = \text{Something} \times \frac{m_1 \times m_2}{(\text{Distance between objects})^2}$$

There is still a "something" left in this expression, and that something is a constant of proportionality (see Math Tools 3.3). Newton hypothesized that the constant was a measure of the intrinsic strength of gravity between objects and that it would turn out to be the same for all objects. He named it the **universal gravitational constant**, written as G (uppercase). Today the value of G is accepted as 6.67×10^{-11} N m^2/kg^2 (or its equivalent, m^3/kg s^2).

Putting the Pieces Together: A Universal Law of Gravitation

Newton had good reasons every step of the way in his thinking about gravity—reasons directly tied to observations of how things in the world behave. Through logic and reason he arrived at what has come to be known as Newton's **universal law of gravitation**. This law, illustrated in **Figure 4.3**, states that gravity is a force between any two objects having mass, and that it has the following properties:

1. It is an attractive force acting along a straight line between the two objects.

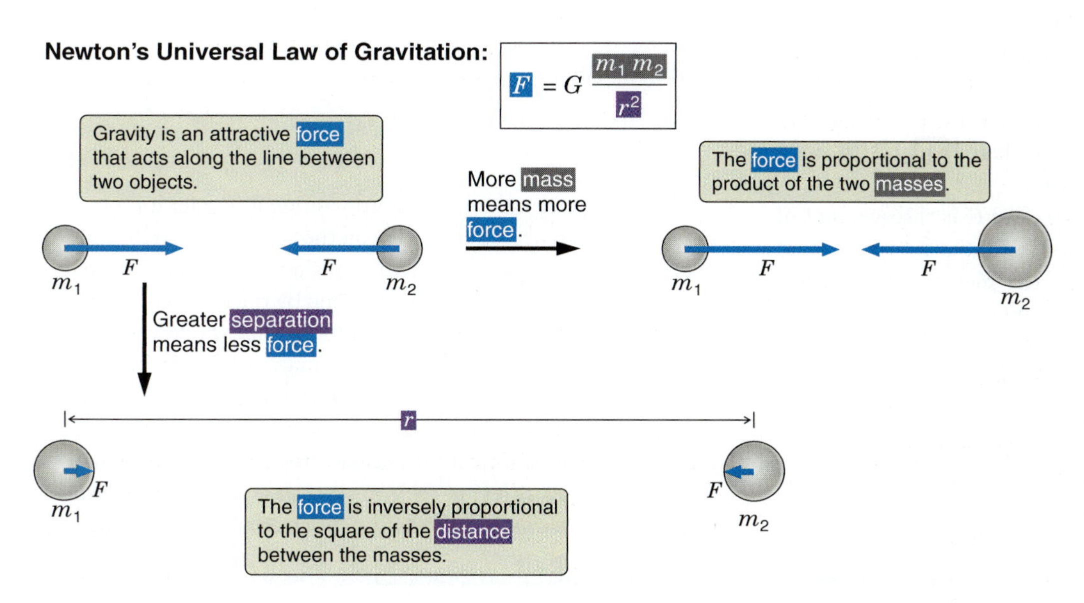

FIGURE 4.3 Gravity is an attractive force between two objects. The force of gravity depends on the masses of the objects and the distance between them.

2. It is proportional to the mass of one object multiplied by the mass of the other object: $F_{grav} \propto m_1 \times m_2$.
3. It is inversely proportional to the square of the distance r between the centers of the two objects: $F_{grav} \propto 1/r^2$.

Written as a mathematical formula, the universal law of gravitation states:

$$F_{grav} = G \times \frac{m_1 \times m_2}{r^2}$$

where F_{grav} is the force of gravity between two objects, m_1 and m_2 are the masses of objects 1 and 2, r is the distance between the centers of mass of the two objects, and G is the universal gravitational constant.

Any time you run across a statement like this, get into the habit of pulling it apart to be sure it makes sense to you. You can break this statement down as follows: First, gravity is an attractive force between two masses that acts along the straight line between the two masses. Regardless of where

Math Tools 

Playing with Newton's Laws of Motion and Gravitation

If you are an artist, you play with colors and patterns of light and dark as you create new works. If you are a musician, you might play with tones and rhythms as you compose. Writers play with combinations of words. All of these uses of the term *play* mean the same thing. Play is very serious business because it is by playing that you explore the world. And in exactly the same sense, scientists often play with the equations describing natural laws as they seek new insights into the world around them. Let's play a bit with Newton's laws of gravitation and motion and see what interesting things turn up.

Gravity obeys what is called an inverse square law. This means that the force of gravity is *inversely proportional* to the square of the distance between two objects, or:

$$F_{grav} \propto \frac{1}{r^2}$$

If two objects are moved so that they are twice as far apart as they were originally, the force of gravity between them becomes only ¼ of what it was. If two objects are moved 3 times as far apart, the force of gravity drops to ⅑ of its original value. For example, double the distance between the Sun and a planet, and the force of gravity declines by a factor of $2 \times 2 = 4$, to ¼ of its original strength. Triple the distance and this influence declines by a factor of $3^2 = 9$, becoming ⅑ of its initial strength (**Figure 4.4**). Gravity is only one of several phenomena in nature that obey an inverse square law. ▶▶ **NEBRASKA SIMULATION: GRAVITY ALGEBRA**

The universal gravitational constant G is the constant of proportionality that characterizes the intrinsic strength of gravitational interactions and enables calculation of the numerical value of this force. G has a value of 6.67×10^{-11} N m^2/kg^2, indicating that gravity is a *very* weak force. The gravitational force between two heavy (7-kg) bowling balls sitting 0.3 meter (about a foot) apart is only:

$$F_{grav} = 6.67 \times 10^{-11} \frac{\text{N m}^2}{\text{kg}^2} \times \frac{7.0 \text{ kg} \times 7.0 \text{ kg}}{(0.3 \text{ m})^2}$$

$$= 3.6 \times 10^{-8} \text{ N}$$

or 0.000000036 N. This value is about equal to the weight on Earth of a small eyebrow hair. Gravity is such an important force in everyday life only because Earth is so very massive.

There are two different ways to think about the gravitational force that Earth exerts on an object with mass m. The first is to look at gravitational force from the perspective of Newton's second law of motion: gravitational force equals mass multiplied by gravitational acceleration, or:

$$F_{grav} = m \times g$$

The other way to think about the force is from the perspective of the universal law of gravitation, which says:

$$F_{grav} = G \times \frac{M_\oplus \times m}{R_\oplus^2}$$

(The symbol ⊕ signifies Earth. Here, $M_\oplus$ is the mass of Earth and $R_\oplus$ is the radius of Earth.) The two expressions describing this force must be equal to each other. $F_{grav} = F_{grav}$, so:

$$m \times g = G \times \frac{M_\oplus \times m}{R_\oplus^2}$$

The mass m is on both sides of the equation, so we can cancel it out. The equation then becomes:

$$g = G \times \frac{M_\oplus}{R_\oplus^2}$$

We set out to calculate the gravitational acceleration experienced by an object of mass m on the surface of Earth. The expression we arrived at says that this acceleration (g) is determined by the mass of Earth ($M_\oplus$) and by the radius of Earth ($R_\oplus$). But the mass of the object itself (m) appears nowhere in this expression. So, according to this equation, changing m has no effect on the gravitational acceleration experienced by an object on Earth. In other words, our play with Newton's laws has shown that all objects experience the same gravitational acceleration, regardless of their mass. *This is just what Galileo found in his experiments with falling objects.* Earlier you learned you stand on the surface of Earth, Earth's gravity pulls you *toward the center of Earth*. Second, the force of gravity depends on the product of the two masses. If you make m_1 twice as large, then the gravitational force between m_1 and m_2 becomes twice as large. Again, this relationship should make sense. Doubling the mass of an object also doubles its weight. A subtle—but in retrospect very important—point lurks in this statement. The mass that appears in the universal law of gravitation is the same mass that appears in Newton's laws of motion. *The same property of an object that gives it inertia is the property of the object that makes it interact gravitationally.*

The third part of Newton's universal law of gravitation tells us that the force of gravity is inversely proportional to the *square* of the distance between two objects. Gravity is only one of several laws we will encounter in which the strength of a particular effect diminishes in proportion to the square of the distance (see **Math Tools 4.1**).

that Galileo's work shaped Newton's thinking about gravity. Now you also know that Galileo's discoveries about gravity are contained within Newton's laws of motion and gravitation.

What else can be revealed? Rearranging the previous equation a bit, so that the mass of Earth is on the left and everything else is on the right, gives this equation:

$$M_\oplus = \frac{g \times R_\oplus^2}{G}$$

Everything on the right side of this equation is known. The value for g (the acceleration due to gravity on the surface of Earth) was measured a few hundred years ago; and you saw in Chapter 2 that in about 230 BCE, Eratosthenes measured the radius of Earth. The universal gravitational constant G can be measured in the laboratory. For example, we can determine the value of G by measuring the slight gravitational forces between two large metal spheres. ▶▶ **NEBRASKA SIMULATION: LAW OF GRAVITY CALCULATOR** With everything on the right side now known, we can calculate the mass of Earth:

$$M_\oplus = \frac{\left(9.80\ \frac{\text{m}}{\text{s}^2}\right) \times (6.37 \times 10^6\ \text{m})^2}{6.67 \times 10^{-11}\ \frac{\text{m}^3}{\text{kg s}^2}}$$
$$= 5.96 \times 10^{24}\ \text{kg}$$

This is how Earth's mass was determined. After all, no one can just pick up a planet and set it on a bathroom scale. Newton actually guessed at the mass of Earth by assuming it had about the same density as typical rocks have. Then he used this mass and the previous equation to get a rough idea of the value of G.

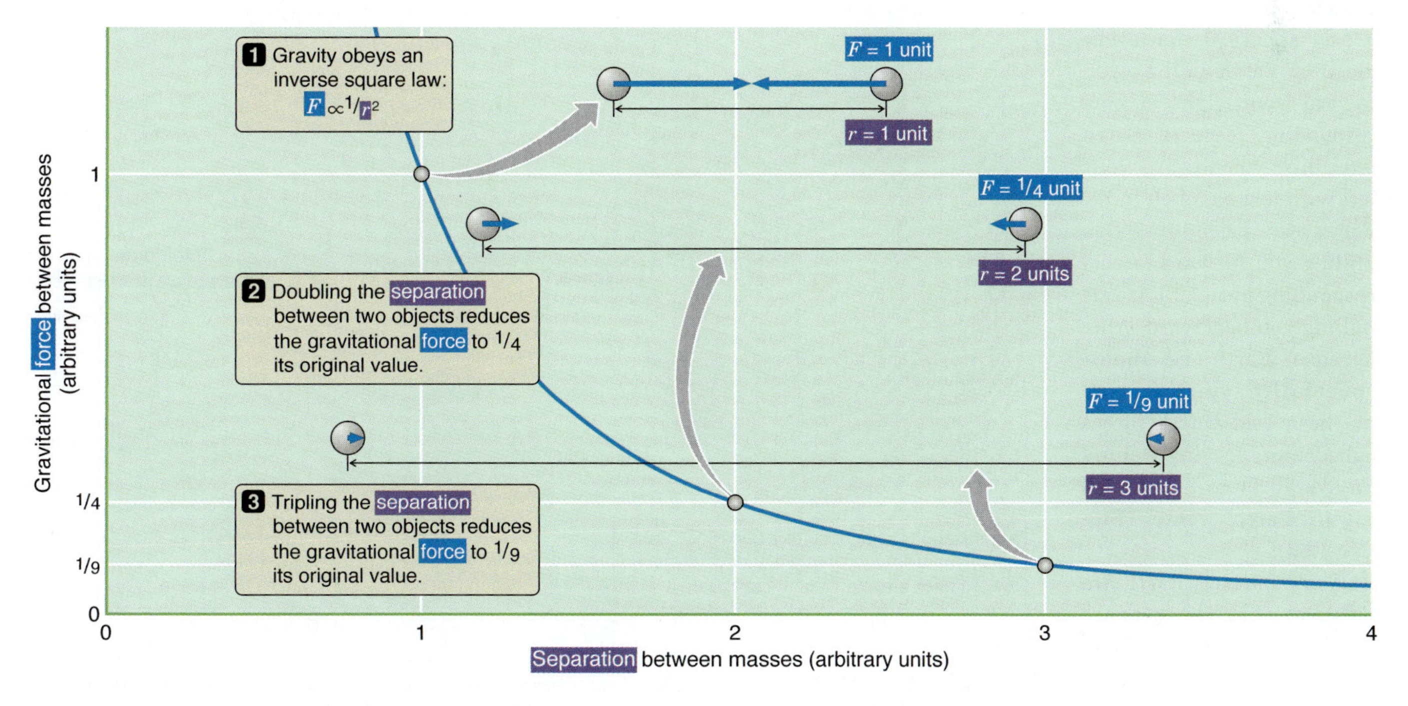

FIGURE 4.4 As two objects move apart, the gravitational force between them decreases by the inverse square of the distance between them.

We will provide examples of gravity in most chapters in this book. In **Connections 4.1** we discuss how an object interacts gravitationally with itself. Gravity holds the planets and stars together and keeps the thin blanket of air we breathe close to Earth's surface. The planets, including Earth, orbit around the Sun, and gravity holds them in orbit. Gravity caused a vast interstellar cloud of gas and dust to collapse 4.5 billion years ago to form the Sun, Earth, and the rest of the Solar System. Gravity binds colossal groups of stars into galaxies. Gravity shapes space and time, and it will affect the ultimate fate of the universe. We will return often to the concept of gravity and find it central to an understanding of the universe.

Connections 4.1

Gravity Differs from Place to Place within an Object

How does an object interact gravitationally with itself? You can think of Earth, for example, as a collection of small masses, each of which feels a gravitational attraction toward every other small part of Earth. The mutual gravitational attraction that occurs among all parts of the same object is called **self-gravity**.

As you sit reading this book, you are exerting a gravitational attraction on every other fragment of Earth, and every other fragment of Earth is exerting a gravitational attraction on you. Your gravitational interaction is strongest with the parts of Earth closest to you. The parts of Earth that are on the other side of our planet are much farther from you, so their pull on you is correspondingly less.

The net effect of all these forces is to pull you (or any other object) toward the center of Earth. If you drop a hammer, it falls directly toward the ground. Because Earth is nearly spherical, for every piece of Earth pulling you toward your right, a corresponding piece of Earth is pulling you toward your left with just as much force. For every piece of Earth pulling you forward, a corresponding piece of Earth is pulling you backward. And, because Earth is almost **spherically symmetric**, all of these "sideways" forces cancel out, leaving behind an overall force that points toward Earth's center (**Figure 4.5a**).

Some parts of Earth are closer to you and others are farther away, but there is an average distance between you and each of the small fragments of Earth that is pulling on you. This average distance turns out to be the distance between you and the center of Earth. So, *the overall pull that you feel is the same as if all the mass of Earth were concentrated at a single point located at the very center of the planet* (**Figure 4.5b**).

This relationship is true for any spherically symmetric object. As far as the rest of the universe is concerned, the gravity from such an object behaves as if all the mass of that object were concentrated at a point at its center. This relationship will be important in many applications. For example, when you estimate your weight on another planet, you are calculating the force of gravity between you and the planet. The "distance" in the gravitational equation will be the distance between you and the center of the planet, which is just the radius of the planet.

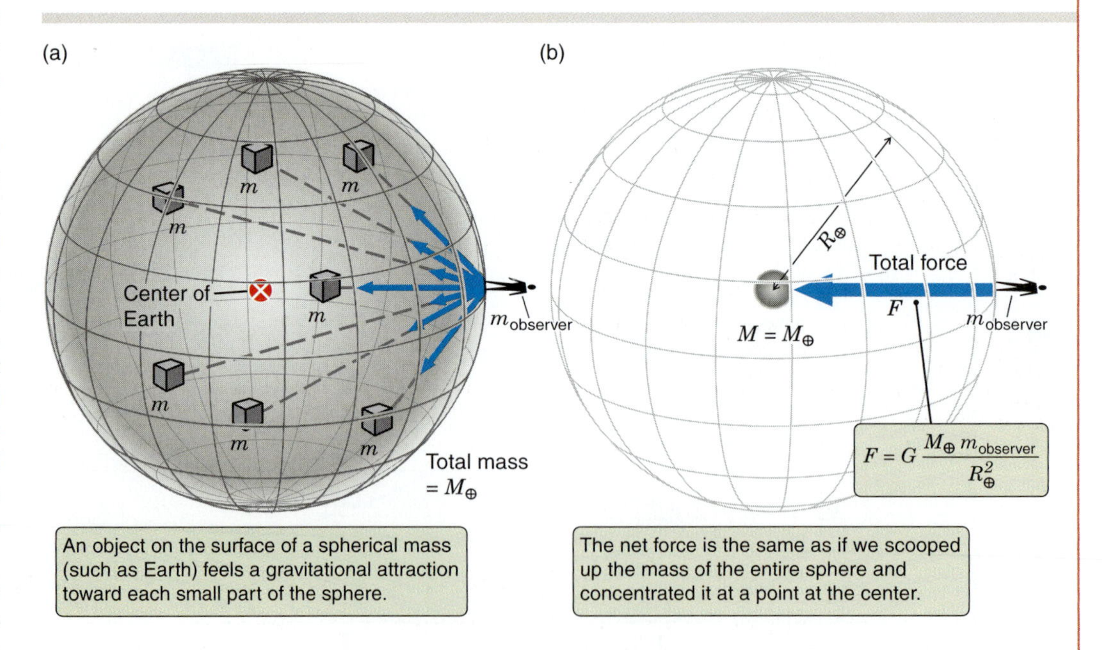

FIGURE 4.5 The net gravitational force due to a spherical mass is the same as the gravitational force from the same mass concentrated at a point at the center of the sphere.

4.2 Orbits Are One Body "Falling around" Another

Kepler may have speculated about the dependence of the solar "influence" that holds the planets in their orbits, and Newton may have speculated that this influence is gravity, but physical law is *not* a matter of speculation! Newton could not measure the gravitational force between two objects in the laboratory directly, so how did he test his universal law of gravitation?

Newton used his laws of motion and his proposed law of gravity to *calculate* the paths that planets should follow as they move around the Sun. His calculations predicted that planetary orbits should be ellipses with the Sun at one focus, that planets should move faster when closer to the Sun, and that the square of the period of a planet's orbit should vary as the cube of the semimajor axis of that elliptical orbit. In short, Newton's universal law of gravitation *predicted* that planets should orbit the Sun in just the way that Kepler's empirical laws described. This was the moment when it all came together. By explaining Kepler's laws, Newton found important corroboration for his law of gravitation. And in the process he moved the cosmological principle out of the realm of interesting ideas and into the realm of testable scientific theories. To see how this happened, we need to look below the surface of how scientists go about connecting their theoretical ideas with events in the real world.

Newton's laws tell us how an object's motion changes in response to forces and how objects interact with each other through gravity. To go from statements about how an object's motion is changing to more practical statements about where an object *is*, we have to carefully "add up" the object's motion over time. Newton used the calculus he co-invented as a student to do this, but we will aim just for a conceptual understanding. In Newton's time, the closest thing to making a heavy object fly was shooting cannonballs out of a cannon, so he used cannonballs in his thought experiments to help his understanding of planetary motions.

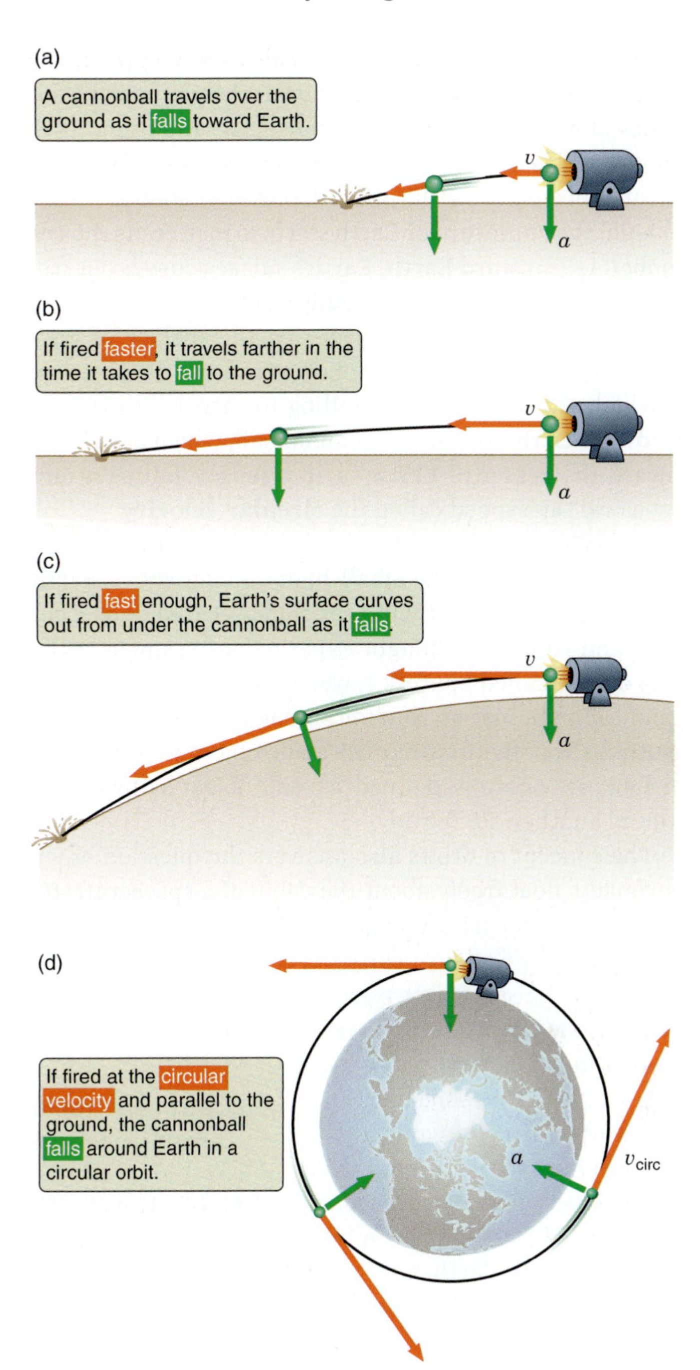

FIGURE 4.6 Newton realized that a cannonball fired at the right speed would fall around Earth in a circle. Velocity (v) is indicated by a red arrow, and acceleration (a) by a green arrow.

Newton Fires a Shot around the World

Drop a cannonball and it falls directly to the ground, just as any other mass does. If instead, however, we fire the cannonball out of a cannon that is level with the ground, as shown in **Figure 4.6a**, it behaves differently. The ball still falls to the ground in the same time as before, but while it is falling it is also traveling over the ground, following a curved path that carries it some horizontal distance before it finally lands. The faster the cannonball is fired from the cannon (**Figure 4.6b**), the farther it will go before hitting the ground. **AstroTour: Newton's Laws and Universal Gravitation**

In the real world this experiment reaches a natural limit. To travel through air the cannonball must push the air out of its way—an effect normally referred to as *air resistance*—which slows it down. But because this is only a thought experiment, we can ignore such real-world complications. Instead imagine that, having inertia, the cannonball continues along its course until it runs into something. If the cannonball is fired faster and faster, it goes farther and farther before hitting the ground. If the cannonball flies far enough, the curvature of Earth starts to matter. As the cannonball falls toward Earth, Earth's surface curves out from under it (**Figure 4.6c**). Eventually a point is reached where the cannonball is flying so fast that the surface of Earth curves away from the cannonball at exactly the same rate at which the cannonball is falling toward Earth (**Figure 4.6d**). When this occurs, the cannonball, which always falls toward the center of Earth, is, in a sense, "falling around the world" at a speed called the **circular velocity**.

In 1957 the Soviet Union used a rocket to lift an object about the size of a basketball high enough above Earth's upper atmosphere that wind resistance ceased to be a concern, and Newton's thought experiment became a reality. This object, called Sputnik 1, was moving so fast that it fell around Earth, just as Newton's imagined cannonball did. Sputnik 1 was the first human-made object to orbit Earth. In fact, an **orbit** is defined as one object falling freely around another.

The concept of orbits also answers the question of why astronauts float freely about the cabin of a spacecraft. It is not because they have escaped Earth's gravity. Instead the answer lies in Galileo's early observation that all objects fall in just the same way, regardless of their mass. The astronauts and the spacecraft are both in orbit around Earth, moving in the same direction, at the same speed, and experiencing the same gravitational acceleration, so *they fall around Earth together*. **Figure 4.7** demonstrates this point. The astronaut is orbiting Earth just as the spacecraft is orbiting Earth. On the surface of Earth your body tries to fall toward the center of Earth, but the ground gets in the way. You experience your weight when you are standing on Earth because the ground pushes on you to counteract the force of gravity, which pulls you downward. In the spacecraft, however, nothing interrupts the astronaut's fall, because the spacecraft is falling around Earth in just the same orbit. The astronaut is not truly weightless; instead the astronaut is in **free fall**.

Orbiting is "falling around a world."

When one object is falling around another, much more massive object, we say that the less massive object is a satellite of the more massive object. Planets are satellites of the Sun, and moons are natural satellites of planets. Newton's imaginary cannonball is a **satellite**. Sputnik 1, the first artificial satellite (*sputnik* means "satellite" in Russian), was the early forerunner of spacecraft and astronauts, which are independent satellites of Earth that conveniently happen to share the same orbit.

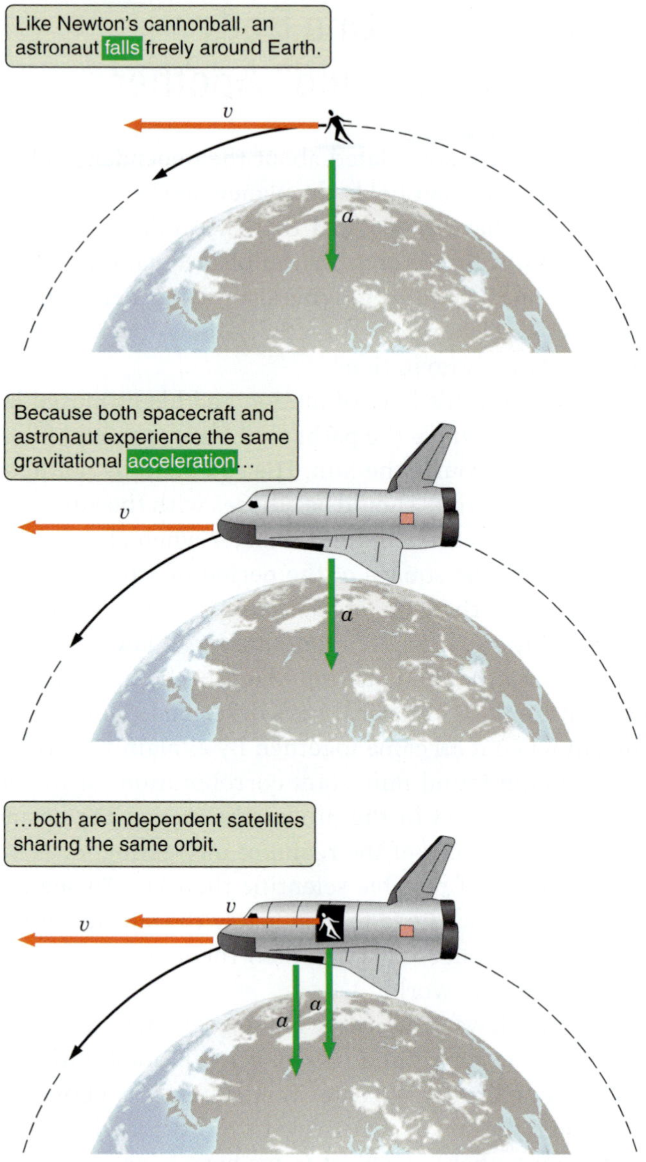

FIGURE 4.7 A "weightless" astronaut has not escaped Earth's gravity. Rather, an astronaut and a spacecraft share the same orbit as they fall around Earth together.

What Velocity Is Needed to Reach Orbit?

How fast must Newton's cannonball be fired for it to fall around the world? The cannonball moves along a circular path at constant speed. This type of motion, referred to as **uniform circular motion**, is discussed in more depth in Appendix 7. Another example of uniform circular motion is a ball whirling around your head on a string (**Figure 4.8a**).

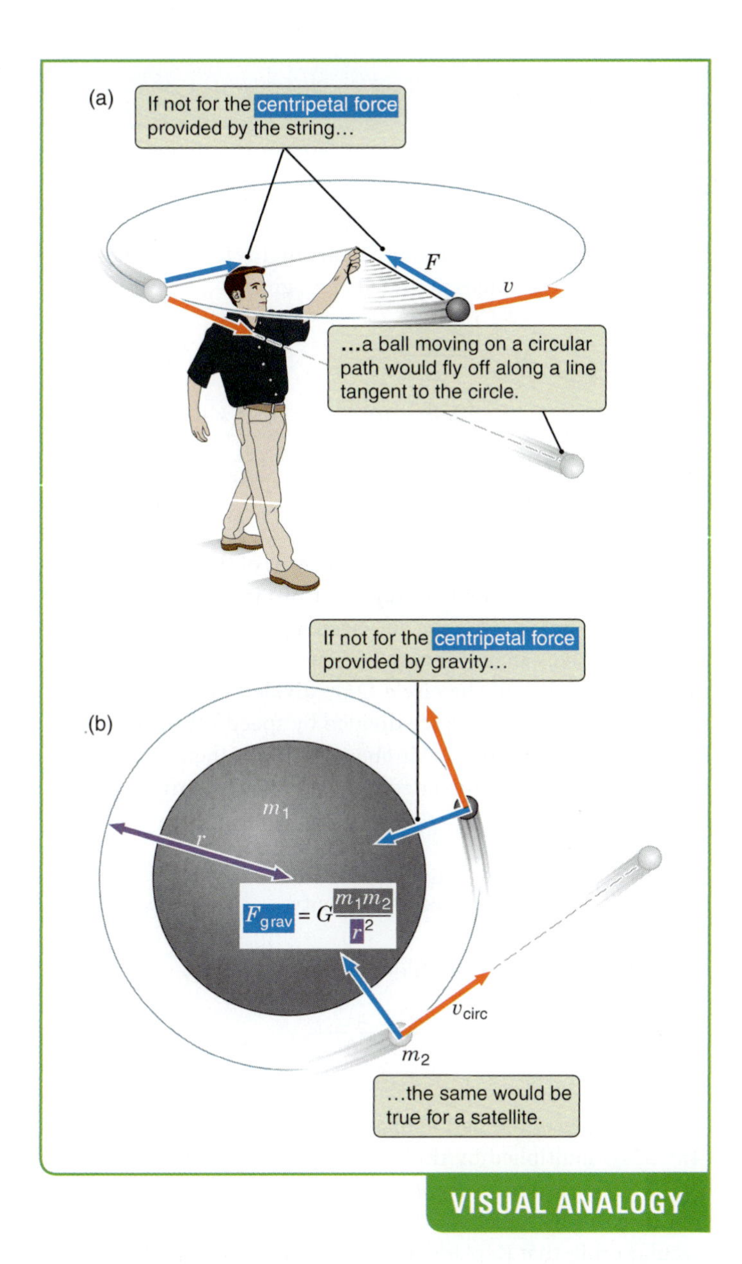

FIGURE 4.8 (a) A string provides the centripetal force that keeps a ball moving in a circle. (We are ignoring the smaller force of gravity that also acts on the ball.) (b) Similarly, gravity provides the centripetal force that holds a satellite in a circular orbit.

If you were to let go of the string, the ball would fly off in a straight line in whatever direction it was traveling at the time, just as Newton's first law says. It is the string that keeps this from happening. The string exerts a steady force on the ball, causing it constantly to change the direction of its motion, always bending its flight toward the center of the circle. This central force is called a **centripetal force**. Using a more massive ball, speeding up its motion, making the circle smaller so that the turn is tighter—all of these actions increase the force needed to keep the ball from being carried off in a straight line by its inertia.

Centripetal forces maintain circular motion.

In the case of Newton's cannonball (or a satellite), there is no string to hold the ball in its circular motion. Instead, the centripetal force is provided by gravity (**Figure 4.8b**). For Newton's thought experiment to work, the force of gravity must be just right to keep the cannonball moving on its circular path. Because this force has a specific strength, it follows that the satellite must be moving at a particular speed around the circle, which is its circular velocity (v_{circ}). If the satellite were moving at any other velocity, it would not be moving in a circular orbit. Remember the cannonball; if it were moving too slowly, it would drop below the circular path and hit the ground. Similarly, if the cannonball were moving too fast, its motion would carry it above the circular orbit. Only a cannonball moving at just the right velocity—the circular velocity—will fall around Earth on a circular path (see Figure 4.6d). Circular velocity is explored further in **Math Tools 4.2**.

Gravity holds a satellite in its orbit by centripetal force.

NEBRASKA SIMULATION: EARTH ORBIT PLOT

Most Orbits Are Not Perfect Circles

Some Earth satellites—like Newton's orbiting cannonball—move along a circular path at constant speed. Satellites traveling at the circular velocity remain the same distance from Earth at all times, neither speeding up nor slowing down in orbit. But what if the satellite were in the same place in its orbit and moving in the same direction, but traveling *faster* than the circular velocity? The pull of Earth is as strong as ever, but because the satellite has greater speed, its path is not bent by Earth's gravity sharply enough to hold it in a circle. So the satellite begins to climb above a circular orbit.

ASTROTOUR: ELLIPTICAL ORBITS

As the distance between Earth and the satellite begins to increase, the satellite slows down. Think about a ball thrown into the air (**Figure 4.9a**). As the ball climbs higher, the pull of Earth's gravity opposes its motion, slowing the ball down. The ball climbs more and more slowly until its vertical motion stops for an instant and then is reversed; the ball then begins to fall back toward Earth, picking up speed along the way. A satellite does exactly the same thing. As the satellite climbs above a circular orbit and begins to move away from Earth, Earth's gravity opposes the satellite's outward motion, slowing the satellite down. The farther the satellite pulls away from Earth, the more slowly the satellite moves—just as happened with the ball thrown into the air. And just like the ball, the satellite reaches a

Math Tools 4.2

Circular Velocity and Orbital Periods

The centripetal force needed to keep an object moving in a circle at a steady speed equals the force provided by gravity. In Appendix 7 we derive the following expression for the circular velocity—the velocity at which an object in a circular orbit *must* be moving:

$$v_{\text{circ}} = \sqrt{\frac{GM}{r}}$$

where M is the mass of the orbited object and r is the radius of the circular orbit. Remember that a cannonball moving at just the right velocity—the circular velocity—will fall around Earth on a circular path. The circular velocity at Earth's surface is about 8 km/s.

We can put some values into the equation for circular velocity to show how fast Newton's cannonball would really have to travel to stay in its circular orbit. Recall that the average radius of Earth is 6.37×10^6 meters, the mass of Earth is 5.97×10^{24} kg, and the gravitational constant is 6.67×10^{-11} m^3/kg s^2. Inserting these values into the expression for v_{circ}, we get:

$$v_{\text{circ}} = \sqrt{\frac{(6.67 \times 10^{-11}\ \text{m}^3/\text{kg s}^2) \times (5.97 \times 10^{24}\ \text{kg})}{6.37 \times 10^6\ \text{m}}} = 7{,}900\ \text{m/s} = 7.9\ \text{km/s}$$

Newton's cannonball would have to be traveling about 8 km/s—over 28,000 kilometers per hour (km/h)—to stay in its circular orbit. That's well beyond the reach of a typical cannon, but these speeds are routinely attained by rockets.

Suppose the goal is to put a satellite into orbit about the Moon just above the lunar surface. How fast will it have to fly? The radius of the Moon is 1.737×10^6 meters, and its mass is 7.35×10^{22} kg. These values give the following circular velocity:

$$v_{\text{circ}} = \sqrt{\frac{(6.67 \times 10^{-11}\ \text{m}^3/\text{kg s}^2) \times (7.35 \times 10^{22}\ \text{kg})}{1.737 \times 10^6\ \text{m}}} = 1{,}680\ \text{m/s} = 1.680\ \text{km/s}$$

This same idea can be applied to the motion of Earth around the Sun. Earth travels at a speed of 2.98×10^4 m/s (or 29.8 km/s) in its orbit about the Sun; and the radius of Earth's orbit, r, is 1.50×10^{11} meters. So we know everything about the nearly circular orbit except for the mass of the Sun. Squaring both sides of the equation for v_{circ} gives:

$$v^2_{\text{circ}} = \frac{GM_\odot}{r}$$

(The symbol ⊙ signifies the Sun.) Rearranging to solve for $M_\odot$, and putting in the values provided here, gives:

$$\begin{aligned} M_\odot &= \frac{v^2_{\text{circ}} \times r}{G} \\ &= \frac{\left(2.98 \times 10^4\ \frac{\text{m}}{\text{s}}\right)^2 \times (1.50 \times 10^{11}\ \text{m})}{6.67 \times 10^{-11}\ \frac{\text{m}^3}{\text{kg s}^2}} \\ &= 2.00 \times 10^{30}\ \text{kg} \end{aligned}$$

This is how astronomers estimate the mass of the Sun.

Let's go one step further. Kepler's third law addresses the time for a planet to complete one circular orbit about the Sun. The time it takes an object to make one trip around a circle is the circumference of the circle ($2\pi r$) divided by the object's speed. (Time equals distance divided by speed.) If the object is a planet in a circular orbit about the Sun, then its speed must be equal to the calculated circular velocity. Bringing all this together, we get:

$$\text{Period} = \frac{\text{Circumference of orbit}}{\text{circular velocity}}\ \frac{2\pi r}{\sqrt{\frac{GM_\odot}{r}}}$$

Squaring both sides of the equation gives:

$$P^2 = \frac{4\pi^2 r^2}{\frac{GM_\odot}{r}} = \frac{4\pi^2}{GM_\odot} \times r^3$$

The square of the period of an orbit is equal to a constant ($4\pi^2/GM_\odot$) multiplied by the cube of the radius of the orbit. *This is Kepler's third law applied to circular orbits.* This derivation is how Newton showed, at least in the special case of a circular orbit, that Kepler's third law—his beautiful "Harmony of the Worlds"—is a direct consequence of the way objects move under the force of gravity. A more complete treatment of the problem (for which Newton invented calculus) shows that Newton's laws of motion and gravitation predict *all* of Kepler's empirical laws of planetary motion—for elliptical as well as circular orbits.

maximum height on its curving path and then begins falling back toward Earth. As the satellite falls back toward Earth, Earth's gravity pulls it along, causing it to pick up more and more speed as it gets closer and closer to Earth. The satellite's orbit has changed from circular to elliptical.

What is true for a satellite orbiting Earth in such an elliptical orbit is also true for any object in an elliptical orbit, including a planet orbiting the Sun. Recall from Chapter 3 that Kepler's law of equal areas says that a planet moves fastest when it is closest to the Sun and slowest when it is farthest

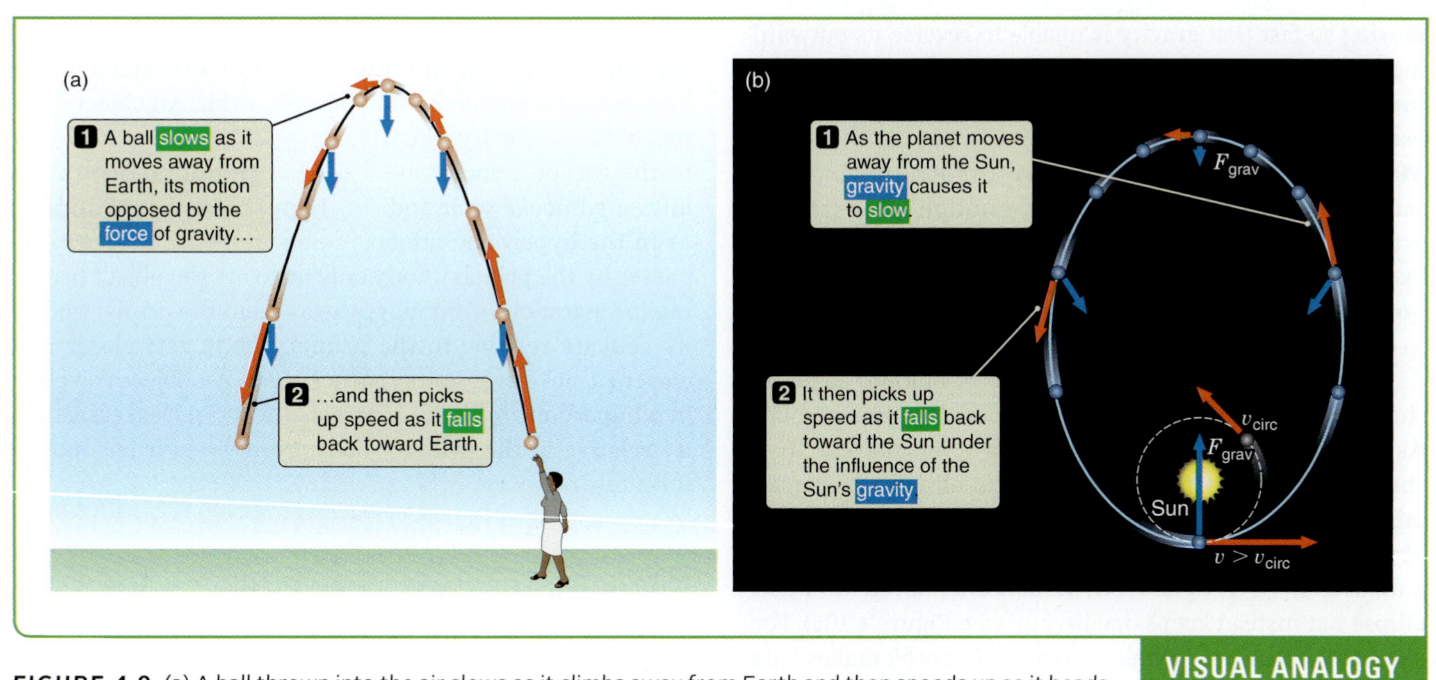

FIGURE 4.9 (a) A ball thrown into the air slows as it climbs away from Earth and then speeds up as it heads back toward Earth. (b) A planet on an elliptical orbit around the Sun does the same thing. (Although no planet has an orbit as eccentric as the one shown here, the orbits of comets can be far more eccentric.)

from the Sun. Now you know why. Planets lose speed as they pull away from the Sun and then gain that speed back as they fall inward toward the Sun (**Figure 4.9b**).

Newton's laws do more than explain Kepler's laws; they predict different types of orbits beyond Kepler's empirical experience. **Figure 4.10a** shows a series of satellite orbits, each with the same point of closest approach to Earth but with different velocities at that point, as indicated in **Figure 4.10b**. The greater the speed a satellite has at its closest approach to Earth, the farther the satellite is able to pull away from Earth, and the more eccentric its orbit becomes. As long as it remains elliptical, no matter how eccentric, the orbit is called a **bound orbit** because the satellite is gravitationally bound to the object it is orbiting.

Bound orbits are circular or elliptical.

In this sequence of faster and faster satellites there comes a point of no return—a point at which the satellite is

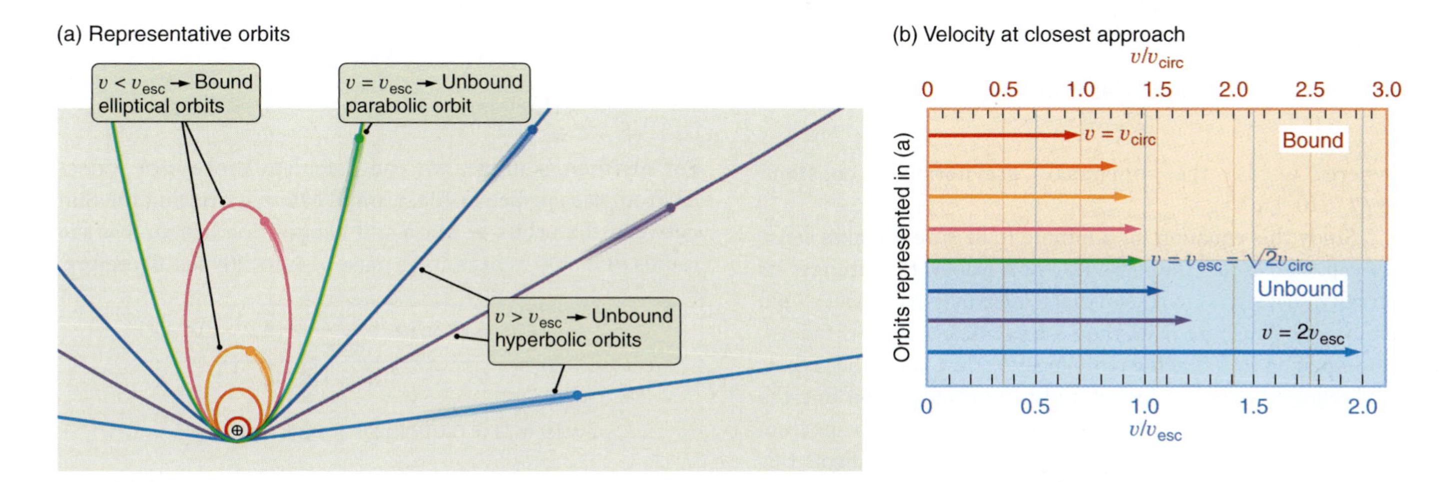

FIGURE 4.10 (a) A range of different orbits that share the same point of closest approach but differ in velocity at that point. (b) Closest-approach velocities for the orbits in (a). An object's velocity at closest approach determines the orbit shape and whether the orbit is bound or unbound. v_{circ} = circular velocity; v_{esc} = escape velocity.

moving so fast that gravity is unable to reverse its outward motion, so the object travels away from Earth, never to return. The lowest speed at which this happens is called the **escape velocity**, v_{esc} (see **Math Tools 4.3**). Once a satellite's velocity at closest approach equals or exceeds v_{esc}, it is in an **unbound orbit**. The object is no longer gravitationally bound to the body that it was orbiting.

A satellite moving fast enough will escape a planet's gravity.

When Newton solved his equations of motion, he found that unbound orbits are shaped as hyperbolas or parabolas. As Figure 4.10 shows, an object with a velocity *less* than the escape velocity (v_{esc}) will be on an elliptically shaped orbit. Elliptical orbits "close" on themselves; that is, an object traveling in an elliptical orbit will follow the same path over and over again. A hyperbola does not close like an ellipse but instead keeps opening up (see Figure 4.10a). For example, a comet traveling on a *hyperbolic orbit* makes only a single pass around the Sun and then is back off into deep space, never to return. The third type of orbit is the borderline case in which the orbiting object moves at exactly the escape velocity. If it had any less velocity it would be traveling in a bound elliptical orbit; any more, and it would be moving on an unbound hyperbolic orbit. An object moving with a velocity equal to the escape velocity follows a *parabolic orbit* and, as in the hyperbolic orbit, passes by the primary body only once. As the object traveling in a parabolic orbit moves away from the primary body, its velocity relative to the primary body gets closer and closer to, but never quite reaches, zero. An object traveling in a hyperbolic orbit, by contrast, always has excess velocity relative to the primary body, even when it has moved infinitely far away.

Unbound orbits are hyperbolas or parabolas.

Newton's Theory Is a Powerful Tool for Measuring Mass

Newton showed that the same physical laws that describe the flight of a cannonball on Earth—or the legendary apple falling from a tree—also describe the motions of the planets

Math Tools 4.3

Calculating Escape Velocities

The concept of escape velocity occurs frequently in astronomy. Sending a spacecraft to another planet requires launching it with a velocity greater than Earth's escape velocity. Working through the calculation shows that the escape velocity is a factor of $\sqrt{2}$, or approximately 1.414, times the circular velocity. This relation can be expressed as:

$$v_{esc} = \sqrt{\frac{2GM}{R}} = \sqrt{2}\, v_{circ}$$

where G is the universal gravitational constant, $6.67 \times 10^{-11}\ \mathrm{m^3/kg\ s^2}$.

Study this equation for a minute to be sure it makes sense to you. The larger the mass (M) of a planet, the stronger its gravity, so it stands to reason that a more massive planet would be harder to escape from than a less massive planet. Indeed, the equation says that the more massive the planet, the greater the required escape velocity. It also stands to reason that the closer you are to the planet, the harder it will be to escape from its gravitational attraction. Again, the equation confirms this intuition. As the distance R becomes larger (that is, as we get farther from the center of the planet), v_{esc} becomes smaller (in other words, it is easier to escape from the planet's gravitational pull). Using this equation, we can now calculate the escape velocity from Earth's surface, noting that Earth has an average radius ($R_\oplus$) of 6.37×10^6 meters and a mass ($M_\oplus$) of 5.97×10^{24} kg:

$$v_{esc} = \sqrt{\frac{2 \times \left(6.67 \times 10^{-11}\ \frac{\mathrm{m^3}}{\mathrm{kg\ s^2}}\right) \times (5.97 \times 10^{24}\ \mathrm{kg})}{6.37 \times 10^6\ \mathrm{m}}}$$

$$= 11{,}180\ \mathrm{m/s} = 11.18\ \mathrm{km/s} = 40{,}250\ \mathrm{km/h}$$

For another example, we can calculate the escape velocity from the surface of Ida, a small asteroid orbiting the Sun between the orbits of Mars and Jupiter. Ida has an average radius of 15,700 meters and a mass of 4.2×10^{16} kg. Therefore:

$$v_{esc} = \sqrt{\frac{2 \times \left(6.67 \times 10^{-11}\ \frac{\mathrm{m^3}}{\mathrm{kg\ s^2}}\right) \times (4.2 \times 10^{16}\ \mathrm{kg})}{1.57 \times 10^4\ \mathrm{m}}}$$

$$= 19\ \mathrm{m/s} = 0.019\ \mathrm{km/s} = 68\ \mathrm{km/h}$$

This means that a professionally thrown baseball traveling at about 130 km/h would easily escape from Ida's surface and fly off into interplanetary space.

through the heavens. Newton's calculations opened up an entirely new way of investigating the universe. Copernicus may have dislodged Earth from the center of the universe and started us on the way toward the cosmological principle. It was Newton who moved the cosmological principle out of the realm of philosophy and into the realm of testable scientific theory by extending his theory of gravity on Earth to the gravitational interaction between astronomical objects.

Besides being more philosophically satisfying than simple empiricism, Newton's method is far more powerful. For example, Newton's laws can be used to measure the mass of Earth and the Sun. This calculation could never be done with Kepler's empirical rules alone. If we can measure the size and period of an orbit—any orbit—then we can use Newton's law of universal gravitation to calculate the mass of the object being orbited. To do so, we rearrange Newton's form of Kepler's third law to read:

$$M = \frac{4\pi^2}{G} \times \frac{A^3}{P^2}$$

Everything on the right side of this equation is either a constant (4, π, and G) or a quantity that we can measure (the semimajor axis A and period P of an orbit). The left side of the equation is the mass of the object at the focus of the ellipse.

Note that a couple of corners were cut to get to this point. For one thing, this relationship was arrived at by thinking about circular orbits and then simply asserting that the relationship holds for elliptical orbits as well. It was also assumed that a low-mass object such as a cannonball is orbiting a more massive object such as Earth. Earth's gravity has a strong influence on the cannonball but, as you have seen, the cannonball's gravity has little effect on Earth. Therefore, Earth remains motionless while the cannonball follows its elliptical orbit. Similarly, it is a good approximation to say that the Sun remains motionless as the planets orbit about it.

This picture changes, though, when two objects are closer to having the same mass. In this case, both objects experience significant accelerations in response to their mutual gravitational attraction. The mass M in this case refers to the sum of the masses of the two objects, which are both orbiting about a common **center of mass** located between them. We now must think of the two objects as falling around each other, with each mass moving on its own elliptical orbit around the two objects' mutual center of mass. This means that if we can measure the size and period of an orbit—any orbit—then we can use this equation to calculate the mass of the orbiting objects.

Objects having similar mass orbit each other.

This relationship is true not only for the masses of Earth and the Sun, but also for the masses of other planets, distant stars, our galaxy and distant galaxies, and vast clusters of galaxies. In fact, it turns out that *almost all knowledge about the masses of astronomical objects comes directly from the application of this one equation.*

Newton found that his laws of motion and gravitation predicted elliptical orbits that agree exactly with Kepler's empirical laws. *This is how Newton tested his theory that the planets obey the same laws of motion as cannonballs and how he confirmed that his law of gravitation is correct.* Had Newton's predictions not been borne out by observation, he would have had to throw them out and start again.

Kepler's empirical rules for planetary motion pointed the way for Newton and provided the crucial observational test for Newton's laws of motion and gravitation. At the same time, Newton's laws of motion and gravitation provided a powerful new understanding of why planets and satellites move as they do. Theory and empirical observation work together, and understanding of the universe moves forward.

Newton's laws provided a physical explanation for Kepler's empirical results.

4.3 Tidal Forces on Earth

The rise and the fall of the oceans are called Earth's **tides**. Coastal dwellers long ago noted that the tides are most pronounced when the Sun and the Moon are either together in the sky or 180° apart (that is, during a new Moon or a full Moon) and are more subdued when the Sun and the Moon are separated in the sky by 90° (first quarter Moon or third quarter Moon). Today it is known that tides are the result of the gravitational pull of the Moon and the Sun on Earth. Tides are the result of *differences* between how hard the Moon and Sun pull on one part of Earth in comparison to their pull on other parts of Earth.

Tides are due to differences in the gravitational pull from external objects.

In Connections 4.1 you saw that each small part of an object feels a gravitational attraction toward every other small part of the object, and this self-gravity differs from place to place. In addition, each small part of an object feels a gravitational attraction toward every other mass in the universe, and these *external* forces differ from place to place within the object as well.

The most notable local example of an external force is the effect of the Moon's gravity on Earth. Overall, the Moon's gravity pulls on Earth as if the mass of Earth were concentrated at the planet's center. When astronomers calculate the orbits of Earth and the Moon assuming that the mass of each is concentrated at the center point of each body, their calculations perfectly match the observed orbits. Yet there is more going on. The side of Earth that faces the Moon is closer to the Moon than is the rest of Earth, so it feels a stronger-than-average gravitational attraction toward the Moon. In contrast, the side of Earth facing away from the Moon is farther than average from the Moon, so it feels a weaker-than-average attraction toward the Moon. These effects have been evaluated numerically, and it turns out that the pull on the near side of Earth is about 7 percent larger than the pull on the far side of Earth.

Imagine three rocks being pulled by gravity toward the Moon. A rock closer to the Moon feels a stronger force than a rock farther from the Moon. Now suppose the three rocks are connected by springs (**Figure 4.11a**). As the rocks are pulled toward the Moon, the differences in the gravitational forces they feel will stretch *both* of the springs. Now imagine that the rocks are at different places on Earth (**Figure 4.11c**). The strength of the force imposed by the Moon's gravity varies at different points on Earth. On the side of Earth away from the Moon the force is smaller (as indicated by the shorter arrow), so that part gets left behind. These differences in the Moon's gravitational attraction on different parts of Earth try to stretch Earth out along a line pointing toward the Moon. ASTROTOUR: TIDES AND THE MOON

The Moon's gravity pulls harder on the side of Earth facing the Moon.

Tidal forces are caused by the change in the strength of gravity across an object (**Figure 4.12**). For example, the Moon's pull on Earth is strongest on the part of Earth closest to the Moon and weakest on the part farthest from the Moon. This difference stretches Earth. The Moon is not pushing the far side of Earth away; rather, it simply is not pulling on the far side of Earth as hard as it is pulling on the planet as a whole. The far side of Earth is "left behind" as the rest of the planet is pulled more strongly toward the Moon. Figure 4.12 shows that there is also a net force squeezing inward on Earth in the direction perpendicular to the line between Earth and the Moon. Together, the stretching of tides along the line between Earth and the Moon and the squeezing of the tides perpendicular to this line distort the shape of Earth like a rubber ball caught in the middle of a tug-of-war.

Tidal forces stretch out Earth in one direction and squeeze it in the other.

Earlier you learned that the strength of the gravitational force between two bodies is proportional to their masses and inversely proportional to the square of the distance between. The strength of tidal forces caused by one body acting on another, however, is more complicated. Tidal influence is also proportional to the mass of the body that is raising the tides, but it is inversely proportional to the cube of the distance between them (see **Math Tools 4.4**).

The gravitational pull of the Moon causes Earth to stretch along a line that points approximately in the direction of the Moon. In the idealized case—in which the surface of Earth is perfectly smooth and covered with a uniform ocean, and Earth does not rotate—the **lunar tides** (tides on Earth due to the gravitational pull of the Moon) would pull our oceans into an elongated **tidal bulge** like that shown

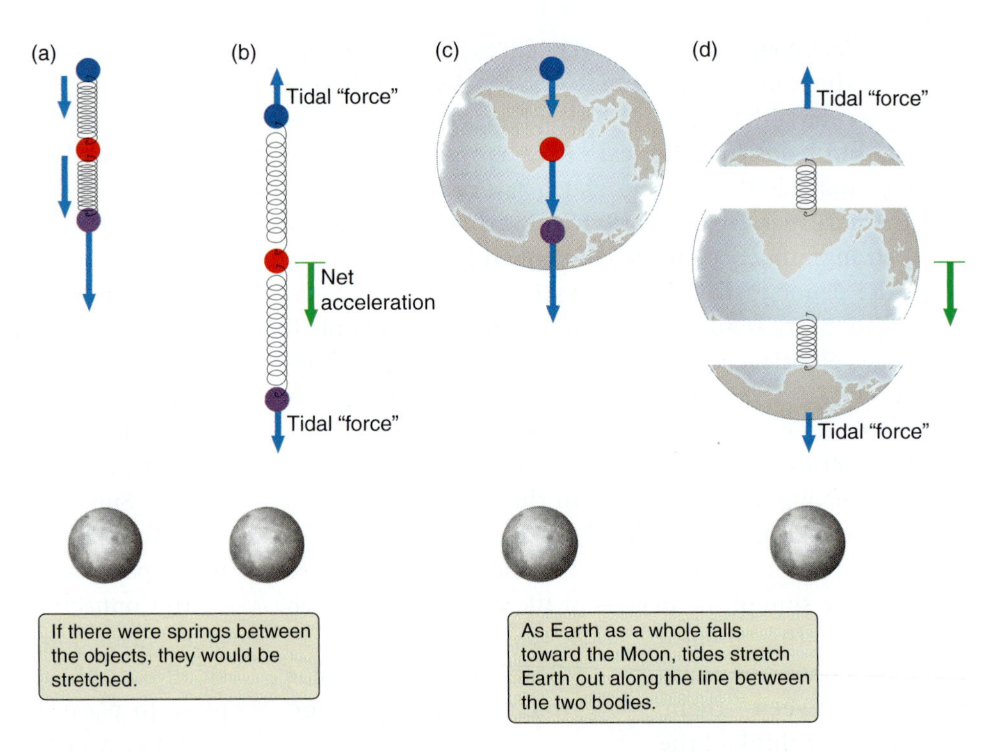

FIGURE 4.11 (a) Imagine three objects connected by springs. (b) The springs are stretched as if there were forces pulling outward on each end of the chain. (c) Similarly, three objects on Earth experience different gravitational attractions toward the Moon. (d) And the difference in the Moon's gravitational attraction between the near and far sides of Earth is the cause of Earth's tides.

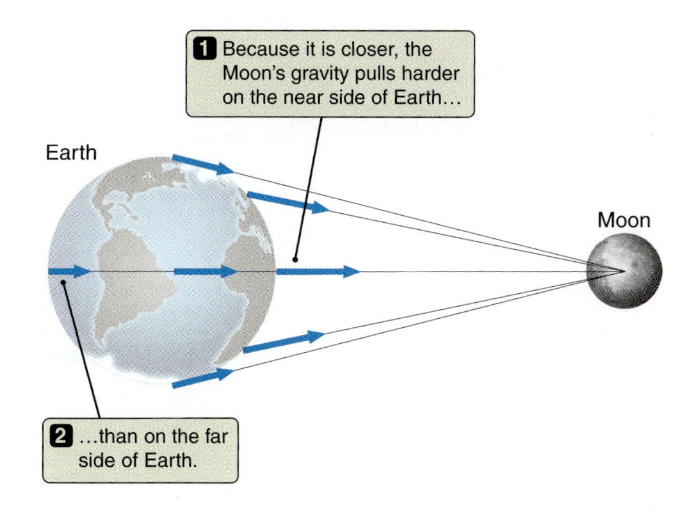

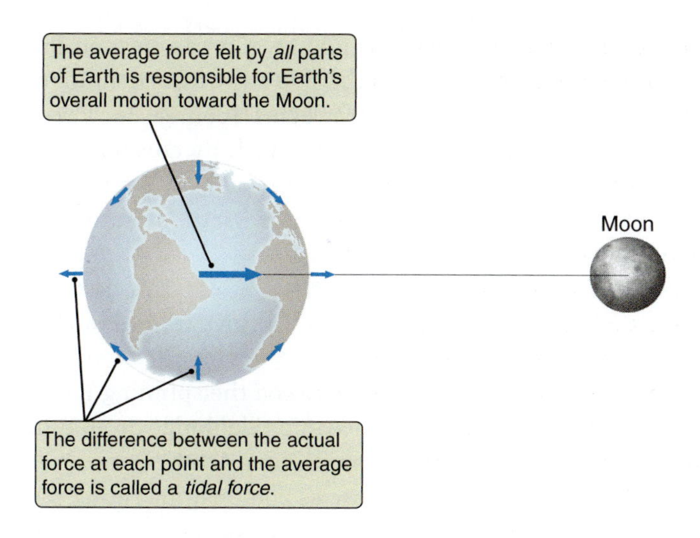

FIGURE 4.12 Tidal forces stretch Earth along the line between Earth and the Moon but compress Earth perpendicular to this line.

Math Tools 4.4

Tidal Forces

The equation for the tidal force comes from the difference of the gravitational force on one side of a body compared with the force on the other side. Consider, for example, the tidal force acting on Earth by the Moon:

$$F_{\text{tidal}}(\text{Moon}) = \frac{2GM_{\oplus}M_{\text{Moon}}R_{\oplus}}{d^3_{\text{Earth-Moon}}}$$

where $R_{\oplus}$ is Earth's radius and $d_{\text{Earth-Moon}}$ is the distance between Earth and the Moon. Billions of years ago, the Moon was much closer to Earth than it is today. Had the Moon's distance been half of what it is now, oceanic tides would have been 8 (2^3) times as large. Billions of years from now, as the Moon moves farther away from Earth, the oceanic tides will be weaker.

As an example, let's compare the tidal force acting on Earth by the Moon with the tidal force acting on Earth by the Sun. F_{tidal} from the Moon is given in the preceding equation; F_{tidal} from the Sun is given by:

$$F_{\text{tidal}}(\text{Sun}) = \frac{2GM_{\oplus}M_{\odot}R_{\oplus}}{d^3_{\text{Earth-Sun}}}$$

We know the Moon is much closer to Earth than the Sun is, but the Sun is much more massive than the Moon. To see how these factors play out, we can take a ratio of the tidal forces:

$$\frac{F_{\text{tidal}}(\text{Moon})}{F_{\text{tidal}}(\text{Sun})} = \frac{2GM_{\oplus}M_{\text{Moon}}R_{\oplus}/d^3_{\text{Earth-Moon}}}{2GM_{\oplus}M_{\odot}R_{\oplus}/d^3_{\text{Earth-Sun}}}$$

Canceling out the constant G and the terms common in both equations ($M_{\oplus}$ and $R_{\oplus}$) gives:

$$\frac{F_{\text{tidal}}(\text{Moon})}{F_{\text{tidal}}(\text{Sun})} = \frac{M_{\text{Moon}}}{M_{\odot}} \times \frac{d^3_{\text{Earth-Sun}}}{d^3_{\text{Earth-Moon}}} = \frac{M_{\text{Moon}}}{M_{\odot}} \times \left(\frac{d_{\text{Earth-Sun}}}{d_{\text{Earth-Moon}}}\right)^3$$

Using the values $M_{\text{Moon}} = 7.35 \times 10^{22}$ kg, $M_{\odot} = 2 \times 10^{30}$ kg, $d_{\text{Earth-Moon}} = 384{,}400$ km, and $d_{\text{Earth-Sun}} = 149{,}598{,}000$ km gives:

$$\frac{F_{\text{tidal}}(\text{Moon})}{F_{\text{tidal}}(\text{Sun})} = \frac{7.35 \times 10^{22}\ \text{kg}}{2 \times 10^{30}\ \text{kg}} \times \left(\frac{149{,}598{,}000\ \text{km}}{384{,}400\ \text{km}}\right)^3 = 2.2$$

So the tidal force from the Moon is 2.2 times stronger than the tidal force from the Sun; or, to put it another way, solar tides on Earth are only about half as strong as lunar tides. This is why the Moon is said to cause the tides. But the Sun is an important factor too, and that's why the tides change depending on the alignment of the Moon and the Sun with Earth.

in **Figure 4.13a**. The water would be at its deepest on the side toward the Moon and on the side away from the Moon, and at its shallowest midway between. Of course, our Earth is *not* a perfectly smooth, nonrotating body covered with perfectly uniform oceans, and many effects complicate this picture. One complicating effect is Earth's rotation. As a point on Earth rotates through the ocean's tidal bulges, that point experiences the ebb and flow of the tides. In addition, friction between the spinning Earth and its tidal bulge drags the oceanic tidal bulge around in the direction of Earth's rotation (**Figure 4.13b**).

Earth's oceans have a tidal bulge.

Follow along in **Figure 4.13c** as you imagine riding the planet through the course of a day. You begin as the rotating Earth carries you through the tidal bulge on the Moonward side of the planet. Because the tidal bulge is not exactly aligned with the Moon, the Moon is not exactly overhead but is instead high in the western sky. When you are at the high point in the tidal bulge, the ocean around you is deeper than average—a condition referred to as *high tide*. About $6\frac{1}{4}$ hours later, probably somewhat after the Moon has settled beneath the western horizon, the rotation of Earth carries you through a point where the ocean is shallowest. It is *low tide*. If you wait another $6\frac{1}{4}$ hours, it is again high tide. You are now passing through the region where ocean water is "pulled" (relative to Earth as a whole) into the tidal bulge on the side of Earth that is away from the Moon. The Moon, which is responsible for the tides you see, is itself at that time hidden from view on the far side of Earth. About $6\frac{1}{4}$ hours later, probably sometime after the Moon has risen above the eastern horizon, it is low tide. About 25 hours after you started this journey—the amount of time the Moon takes to return to the same point in the sky from which it started—you again pass through the tidal bulge on the near side of the planet. This is the age-old pattern by which mariners have lived their lives for millennia: the twice-daily coming and going of high tide, shifting through the day in lockstep with the passing of the Moon.

Tides rise and fall twice each day.

AstroTour: Tides and the Moon

This description of daily tides is often simpler than what people living near the ocean actually observe. The shapes of Earth's shorelines and ocean basins complicate the simple picture of tides. In addition, there are oceanwide oscillations similar to water sloshing around in a basin. As they respond to the tidal forces from the Moon, Earth's oceans flow around the various landmasses that break up the water covering our planet. Some places, like the Mediterranean Sea and the Baltic Sea, are protected from tides by their relatively small sizes and the narrow passages connecting these bodies of water with the larger ocean. In other places, the shape of the land funnels the tidal surge from a large region of ocean into a relatively small area, concentrating its effect, as at the Bay of Fundy (**Figure 4.14**).

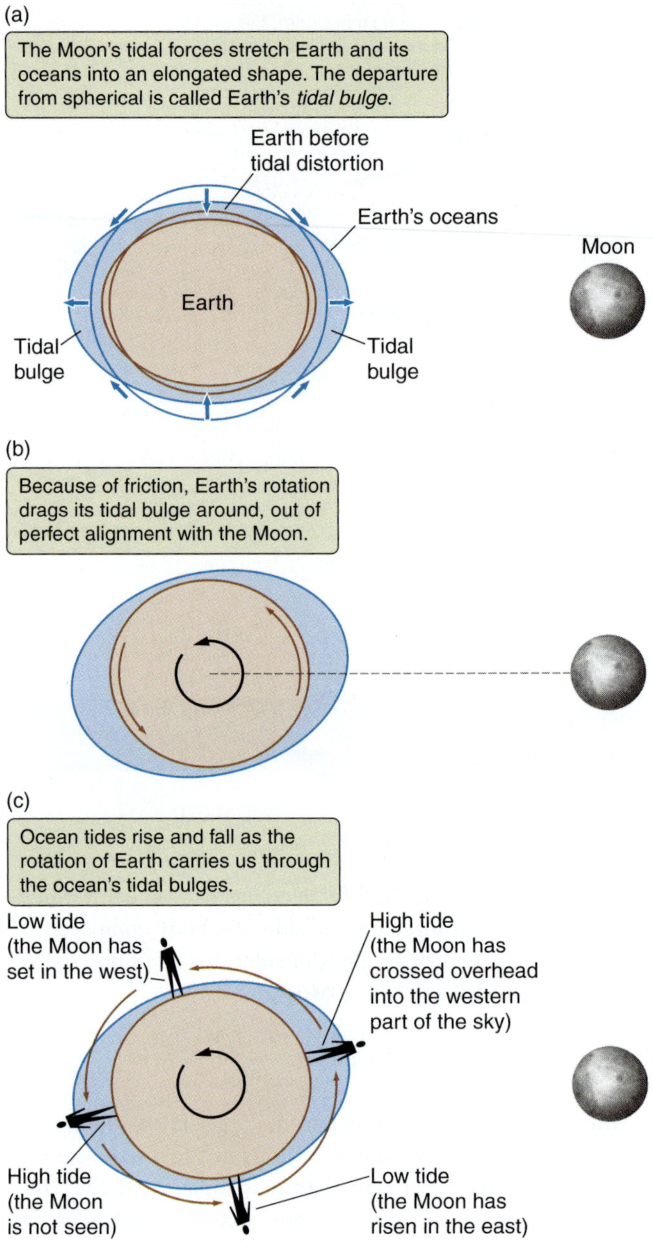

The Sun also influences Earth's tides. The side of Earth closer to the Sun is pulled toward the Sun more strongly

(a) The Moon's tidal forces stretch Earth and its oceans into an elongated shape. The departure from spherical is called Earth's *tidal bulge*.

Earth before tidal distortion; Earth's oceans; Moon; Earth; Tidal bulge; Tidal bulge

(b) Because of friction, Earth's rotation drags its tidal bulge around, out of perfect alignment with the Moon.

(c) Ocean tides rise and fall as the rotation of Earth carries us through the ocean's tidal bulges.

Low tide (the Moon has set in the west); High tide (the Moon has crossed overhead into the western part of the sky); High tide (the Moon is not seen); Low tide (the Moon has risen in the east)

FIGURE 4.13 (a) Tidal forces pull Earth and its oceans into a tidal bulge. (b) Earth's rotation pulls its tidal bulge slightly out of alignment with the Moon. (c) As Earth's rotation carries you through these bulges, you experience the well-known ocean tides. The magnitude of the tides has been exaggerated in these diagrams for clarity. In these figures the observer is looking down from above Earth's North Pole. Sizes and distances are not to scale.

Nebraska Simulation: Tidal Bulge Simulation

FIGURE 4.14 The world's most extreme tides are found in the Bay of Fundy, located between Nova Scotia and New Brunswick, Canada. This bay, along with the Gulf of Maine, forms a great basin in which water naturally rocks back and forth with a period of about 13 hours. This amount of time is close to the 12½-hour period of the rising and falling of the tides. The characteristics of the basin amplify the tides, sending the water sloshing back and forth like the water in a huge bathtub. The difference in water depth between low tide (a) and high tide (b) is typically about 14.5 meters and may reach as much as 16.6 meters.

than is the side of Earth away from the Sun, just as the side of Earth closest to the Moon is pulled more strongly toward the Moon. Although the absolute strength of the Sun's pull on Earth is nearly 200 times greater than the strength of the Moon's pull on Earth, the Sun's gravitational attraction does not change by much from one side of Earth to the other, because the Sun is much farther away than the Moon. (Recall that tidal effects are inversely proportional to the *cube* of the distance to the tide-raising body.) As a result, **solar tides**—tides on Earth due to differences in the gravitational pull of the Sun—are only about half as strong as lunar tides (see Math Tools 4.4).

Solar and lunar tides interact. When the Moon and the Sun are lined up with Earth, as occurs at either new Moon or full Moon, the tides on Earth due to the Sun are in the same direction as the tides due to the Moon. At these times the solar tides reinforce the lunar tides, so the tides are about 50 percent stronger than average. The especially strong tides near the time of the new Moon or full Moon are referred to as **spring tides** (**Figure 4.15a**). Conversely, around the first quarter Moon and the third quarter Moon, the stretching due to the solar tide is at right angles to the stretching due to the lunar tide. The solar tides pull water into the dip in the lunar tides and away from the tidal bulge due to the Moon. The effect is that low tides are higher and high tides are lower. Such tides, when lunar tides

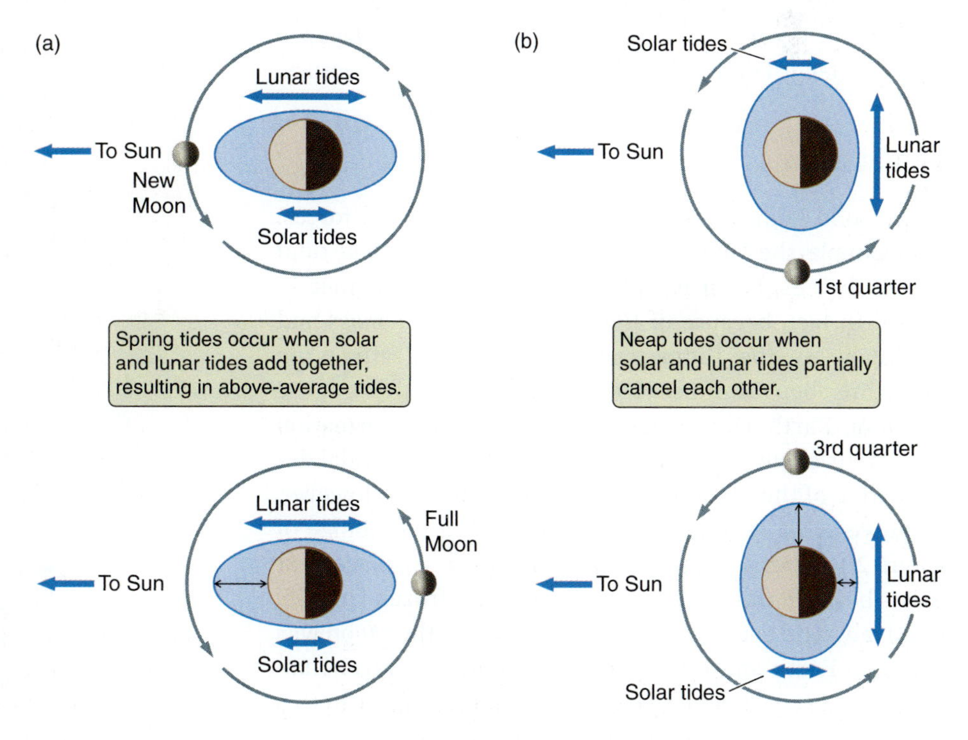

FIGURE 4.15 Earth experiences solar tides that are about half as strong as tides due to the Moon. The interactions of solar and lunar tides result in either spring tides when they are added together (a) or neap tides when they partially cancel each other (b).

are diminished by solar tides, are called **neap tides** (**Figure 4.15b**). Neap tides are only about half as strong as average tides and only a third as strong as spring tides.

Solar tides may reinforce or diminish lunar tides.

The daily ebb and flow of the tides allows beachgoers to experience the wonders of tide pools, the natural aquariums teeming with marine life. As the tides roll in and out, they recharge the tide pools with fresh seawater while sometimes exchanging the pool's inhabitants.

4.4 Tidal Effects on Solid Bodies

So far, the discussion in this chapter has focused on the sometimes dramatic movements of the liquid of Earth's oceans in response to the tidal forces from the Moon and Sun. But these tidal forces also affect the *solid* body of Earth. As Earth rotates through its tidal bulge, the solid body of the planet is constantly being deformed by tidal forces. It takes energy to deform the shape of a solid object. (If you want a practical demonstration of this fact, hold a rubber ball in your hand, and squeeze and release it a few dozen times.) The energy of the tidal forces is converted into thermal energy by *friction* in Earth's interior. This friction opposes and takes energy from the rotation of Earth, causing Earth to gradually slow. Earth's internal friction adds to the slowing caused by friction between Earth and its oceans as the planet rotates through the oceans' tidal bulge. As a result, Earth's days are currently lengthening by about 0.0015 second every century.

Tidal Locking

Other solid bodies besides Earth experience tidal forces. For example, the Moon has no bodies of liquid to make tidal forces obvious, but it is distorted in the same manner as Earth. In fact, because of Earth's much greater mass and the Moon's smaller radius, the tidal effects of Earth on the Moon are about 20 times as great as the tidal effects of the Moon on Earth. Given that the average tidal deformation of Earth is about 30 centimeters (cm), the average tidal deformation of the Moon should be about 6 meters. However, what we actually observe on the Moon is a tidal bulge of about 20 meters. This unexpectedly large displacement exists because the Moon's tidal bulge was "frozen" into its relatively rigid crust at an earlier time, when the Moon was closer to Earth and tidal forces were much stronger than they are today. Planetary scientists sometimes refer to this deformation as the Moon's *fossil tidal bulge*.

The synchronous rotation of the Moon discussed in Chapter 2 is a result of the **tidal locking** (in which an object's rotation period exactly equals its orbital period) of the Moon to Earth. Early in its history, the period of the Moon's rotation was almost certainly different from its orbital period. As the Moon rotated through its extreme tidal bulge, however, friction within the Moon's crust was tremendous, rapidly slowing the Moon's rotation. After a fairly short time, the period of the Moon's rotation equaled the period of its orbit. When its orbital and rotation periods became equalized, the Moon no longer rotated with respect to its tidal bulge. Instead, the Moon and its tidal bulge rotated *together*, in lockstep with the Moon's orbit about Earth. This scenario continues today as the tidally distorted Moon orbits Earth, always keeping the same face and the long axis of its tidal bulge toward Earth (**Figure 4.16**).

Tidal forces lock the Moon's rotation to its orbit around Earth.

Tidal forces affect not only the rotations of the Moon and Earth, but also their orbits. Because of its tidal bulge, Earth is not a perfectly spherical body. Therefore, the material in Earth's tidal bulge on the side nearer the Moon pulls on the Moon more strongly than does material in the tidal bulge on the back side of Earth. Because the tidal bulge on the Moonward side of Earth "leads" the Moon somewhat, as shown in **Figure 4.17**, the gravitational attraction of the bulge causes the Moon to accelerate slightly along the direction of its orbit about Earth. It is as if the rotation of Earth were dragging the Moon along with it, and in a sense this is exactly what is happening. The angular momentum lost by Earth as its rotation

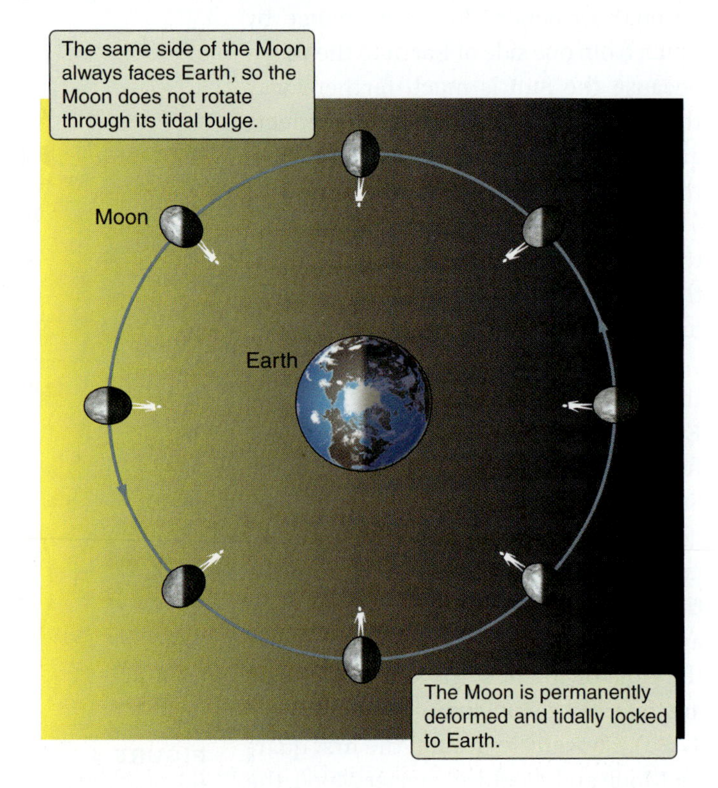

FIGURE 4.16 Tidal forces due to Earth's gravity lock the Moon's rotation to its orbital period.

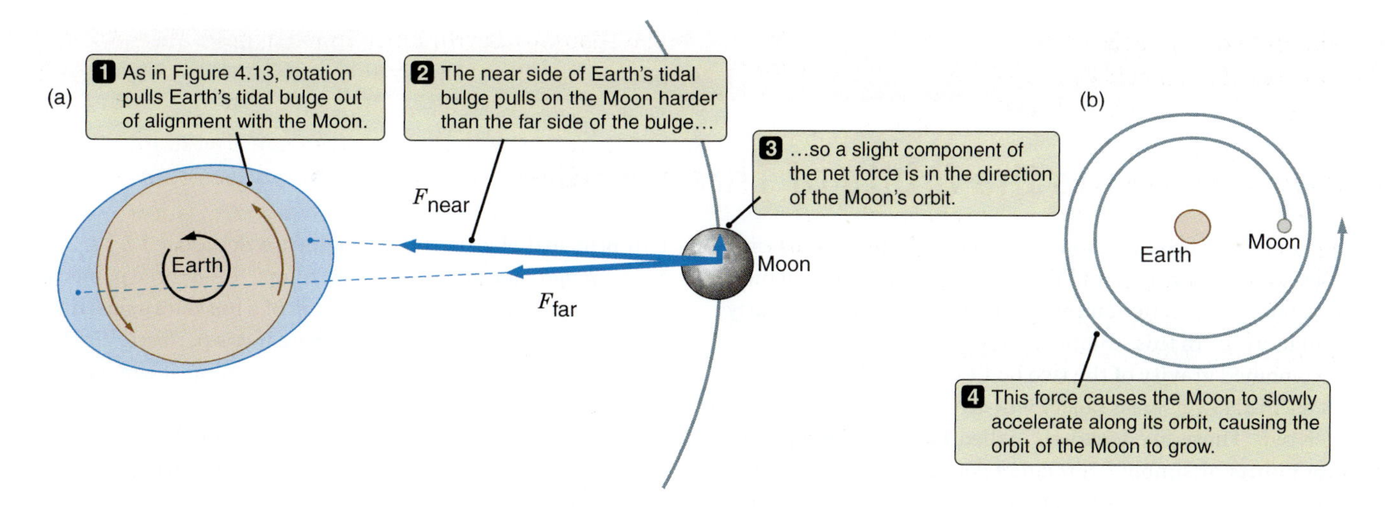

FIGURE 4.17 Interaction between Earth's tidal bulge and the Moon causes the Moon to accelerate in its orbit and the Moon's orbit to grow.

slows is exactly equal to the angular momentum gained by the Moon as it accelerates in its orbit. The acceleration of the Moon in the direction of its orbital motion causes the orbit of the Moon to grow larger. At present, the Moon is drifting away from Earth at a rate of 3.83 cm per year.

As the Moon grows more distant, the length of the lunar month increases by about 0.014 second each century. If this increase in the radius of the Moon's orbit were to continue long enough (about 50 billion years), Earth would become tidally locked to the Moon, just as the Moon is now tidally locked to Earth. At that point the period of rotation of Earth, the period of rotation of the Moon, and the orbital period of the Moon would all be exactly the same—about 47 of our present days—and the Moon would be about 43 percent farther from Earth than it is today. However, this situation will never actually occur—at least not before the Sun itself has burned out.

The Moon's orbit is growing larger, and Earth's rotation is slowing.

The effects of tidal forces can be seen throughout the Solar System. Most of the moons in the Solar System are tidally locked to their parent planets, and in the case of Pluto and its largest moon, Charon, each is tidally locked to the other.

Tidal locking is only one example of **spin-orbit resonance** (see **Connections 4.2**). Other types of spin-orbit resonance are also possible. For example, Mercury is in a very elliptical orbit about the Sun. As with the Moon, tidal forces have coupled Mercury's rotation to its orbit. Yet unlike the Moon with its synchronous rotation, Mercury rotates in a 3:2 spin-orbit resonance, spinning on its axis three times for every two trips around the Sun. The period of Mercury's orbit—87.97 Earth days—is exactly 1½ times the 58.64 days that it takes Mercury to spin once on its axis. Thus, each time Mercury comes to **perihelion**—the point in its orbit that is closest to the Sun—first one hemisphere and then the other faces the Sun.

Tides on Many Scales

We normally think of the effects of tidal forces as small compared to the force of gravity holding an object together, yet this is not always the case. Under certain conditions, tidal effects can be extremely destructive. Consider for a moment the fate of a small moon, asteroid, or comet that wanders too close to a massive planet such as Jupiter or Saturn. All objects in the Solar System larger than about a kilometer in diameter are held together by their self-gravity. However, the self-gravity of a small object such as an asteroid, a comet, or a small moon is feeble. In contrast, the tidal forces close to a massive object such as Jupiter can be fierce. If the tidal forces trying to tear an object apart become greater than the self-gravity trying to hold the object together, the object will break into pieces.

The **Roche limit** is the distance at which a planet's tidal forces exceed the self-gravity of a smaller object—such as a moon, asteroid, or comet—causing the object to break apart. For a smaller object having the same density as the planet and having no internal mechanical strength, the Roche limit is about 2.45 times the planet's radius. Such an object bound together solely by its own gravity can remain intact when it is outside a planet's Roche limit, but not when it is inside the limit. (Very small objects generally have mechanical strength that is much stronger than their own self-gravity. This is why the International Space Station and other Earth satellites are not torn apart, even though they orbit well within Earth's Roche limit.)

Inside the Roche limit, tidal forces exceed self-gravity.

Connections 4.2

Gravity Affects the Orbits of Three Bodies

An example of orbital resonance occurs when two massive bodies (such as a planet and the Sun, or a moon and a planet) move about their common center of mass in circular or nearly circular orbits. In this situation, there are five locations where the combined gravity of the two bodies adds up in such a way that a third, lower-mass object will orbit in lockstep with the other two. These five locations, called **Lagrangian equilibrium points** or simply *Lagrangian points*, are named for the Italian-born French mathematician Joseph-Louis Lagrange (1736–1813), who first called attention to them.

The exact locations of the Lagrangian points depend on the ratio of the masses of the two primary bodies. The Lagrangian points for the Sun-Earth system are shown in **Figure 4.18**. Three of the Lagrangian points lie along the line between the two principal objects. These points are designated L_1, L_2, and L_3. Although they are equilibrium points, these three points are unstable in the same way that the top of a hill is unstable. A ball placed on the top of a hill will sit there if it is perfectly perched, but give it the slightest bump and it goes rolling down one side. Similarly, an object displaced slightly from L_1, L_2, or L_3 will move away from that point. Thus, although L_1, L_2, and L_3 are equilibrium points, they do not capture and hold objects in their vicinity.

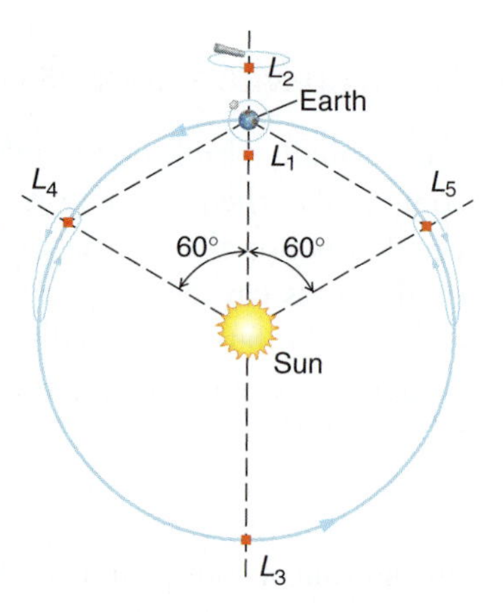

FIGURE 4.18 The pattern of Lagrangian equilibrium points of a two-body system, in this case composed of Earth and the Sun. A third, lower-mass object will orbit in lockstep with two more massive bodies at these five Lagrangian points. The entire pattern of Lagrangian points rotates solidly like a spinning disk.

Even so, L_1, L_2, and L_3 can be useful places to locate scientific instruments. For example, as of this writing, the *SOHO* (Solar and Heliospheric Observatory) spacecraft sits near the L_1 point of the Sun-Earth system, where tiny nudges from its onboard jets are enough to keep it orbiting directly between Earth and the Sun, giving a constant, unobscured view of the side of the Sun facing Earth. Several spacecraft currently sit near the L_2 point—including the Wilkinson Microwave Anisotropy Probe (*WMAP*), the Planck observatory, and the Herschel Space Observatory—and it will also be the location of the upcoming missions Gaia and the James Webb Space Telescope.

L_4 and L_5 are located 60° in front of and 60° behind the less massive of the two main bodies in its orbit about the center of mass of the system. These are *stable* equilibrium points, like the stable equilibrium that exists at the bottom of a bowl. If you bump a marble sitting at the bottom of a bowl, it will roll around a bit but it will stay in the bowl. Objects near L_4 or L_5 follow elongated, tadpole-shaped paths relative to these gravitational "bowls."

A 1970s group that advocated human colonization of space called itself the L5 Society because the L_4 and L_5 Lagrangian points of the Earth-Moon system would make stable locations for large orbital space colonies.

We have concentrated on the role that tidal forces play on Earth and the Moon. But we find tidal forces throughout the Solar System and the universe. Any time two objects of significant size or two collections of objects interact gravitationally, the gravitational forces will differ from one place to another within the objects, giving rise to tidal effects.

In later chapters you'll see tidal forces between the planets in the Solar System and their moons and rings. Tidal disruption of small bodies is thought to be the source of the particles that make up the rings of the giant planets. Comparison of the major ring systems around the giant planets shows that most rings lie within their respective planets' Roche limits.

You will learn that stars are often members of binary pairs in which two stars orbit each other. Tidal interactions can cause material from one star to overflow and be pulled onto the other star. Tidal effects can strip stars from clusters consisting of thousands of stars. **Galaxies**—vast collections of hundreds of billions of stars that all orbit each other under the influence of gravity—can pass close enough together to strongly interact gravitationally. When this happens, as in **Figure 4.19**, tidal effects can grossly distort both galaxies taking part in the interaction. Tidal forces even play a role in shaping huge collections of galaxies—the largest known structures in the universe.

4.5 Origins: Tidal Forces and Life

In this chapter we noted that Earth's rotation is slowing down as the Moon slowly moves away into a larger orbit.

FIGURE 4.19 Tidal interactions distort the appearance of two galaxies. The tidal "tails" seen here are characteristic of tidal interactions between galaxies.

Elsewhere in the Solar System, the giant planets Jupiter and Saturn are far from the Sun, and thus very cold. They each have many moons, the closest of which would experience strong tidal forces from their planet. Several of these moons are thought to have a liquid ocean underneath an icy surface. Tidal forces from Jupiter or Saturn provide the heat to keep the water in a liquid state. Astrobiologists think that these subsurface liquid oceans are perhaps the most probable location for life elsewhere in the Solar System.

On Earth, the tidal forces from the Sun are about half as strong as those from the Moon. A planet with a closer orbit would experience much stronger tidal forces from its star. In Chapter 7 you will see that many of the planets detected outside of the Solar System have orbits very close to their stars. These planets would experience strong tidal forces, and they might be tidally locked so that they have synchronous rotation as the Moon does around Earth, with one side of the planet always facing the star and one side facing away. How might life on Earth have evolved differently if half of the planet was in perpetual night and half in perpetual day?

In the distant past—when Earth was young—the Moon was closer and Earth rotated faster, so tides would have been stronger and the interval between high tides would have been shorter. It is not known precisely how much faster Earth rotated billions of years ago. But the stronger and more frequent tides would have provided additional energy to the oceans of the young Earth.

Scientists are debating whether life on Earth originated deep in the ocean, on the surface of the ocean, or on land (see Chapter 24). The tides shaped the regions in the margins between land and ocean, such as tide pools and coastal flats. Some think that these border regions, which alternated between wet and dry with the tides, could have been places where concentrations of biochemicals periodically became high enough for more complex reactions to take place. These complex reactions were important to the earliest life. Later, these border regions may have been important as advanced life moved from the sea to the land.

Newton's laws have enabled robotic spacecraft to be launched to the planets within the Solar System. But even the nearest stars are thousands of times more distant than the most far-flung of these robotic planetary explorers. Scientists' knowledge of the rest of the universe depends on signals reaching Earth from across space. The most common of these signals is electromagnetic radiation, which includes light. The ability to interpret these signals depends on what is known about light, which we discuss in Chapter 5. ■

Summary

4.1 Gravity, as one of the fundamental forces of nature, binds the universe together. Gravity is a force between any two objects due to their masses. The force of gravity is proportional to the mass of each object, and inversely proportional to the square of the distance between them.

4.2 Planets orbit the Sun in bound, elliptical orbits. Objects "fall around" the Sun and Earth on elliptical, parabolic, or hyperbolic paths. Orbits are ultimately given their shape by the gravitational attraction of the objects involved, which in turn is a reflection of the masses of these objects.

4.3 Tides on Earth are the result of differences between how hard the Moon and Sun pull on one part of Earth in comparison with their pull on other parts of Earth. Both the Moon and the Sun create tides on Earth. As Earth rotates, tides rise and fall twice each day.

4.4 Tidal forces lock the Moon's rotation to its orbit around Earth.

4.5 Tidal forces provide energy to Earth's oceans. Tide pools on Earth may have been a site of early biochemical reactions.

Unanswered Questions

- Why is the gravitational force only attractive? Other forces in nature, as we'll see later, can be attractive or repulsive. Are there conditions under which gravity is repulsive, perhaps when acting at extremely large distances? Today there is evidence that gravity attracts any form of energy. Stunning evidence suggests that there is also energy associated with empty space, and that this energy leads to a repulsive force. Astronomers and physicists are actively working to understand how gravity, mass, and this energy of empty space work to shape the overall fate of the universe.
- What range of gravities will support human life? Humans have evolved to live on Earth's surface, but what happens when humans go elsewhere? What are the limits for our hearts, lungs, eyes, and bones? At the higher end of human tolerance, fighter pilots have been trained to experience about 10 times the normal surface gravity on Earth for very short periods of time (too long and they black out.) Astronauts who spend several months in near-weightless conditions experience medical problems such as bone loss. On the Moon or Mars, humans will weigh much less than on Earth. Numerous science fiction tales have been written about what happens to children born on a space station or on another planet or moon with low surface gravity: would their hearts and bodies ever be able adjust to the higher surface gravity of Earth, or must they stay in space forever?

Questions and Problems

Summary Self-Test

1. If you drop two balls of the same radius, they will accelerate toward the ground. If one ball is more massive than another, its acceleration
   a. will be larger than that of the less massive ball.
   b. will be smaller than that of the less massive ball.
   c. will be the same as that of the less massive ball.
   d. will depend on other factors.
2. The force with which you pull upward on Earth is
   a. zero.
   b. equal to your weight.
   c. equal to your mass.
   d. equal to Earth's weight.
3. In Newton's universal law of gravity, the force is
   a. proportional to both masses.
   b. proportional to the radius.
   c. proportional to the radius squared.
   d. inversely proportional to the orbiting mass.
4. Suppose you were transported to a planet with twice the mass of Earth but the same radius that Earth has. Your weight would ________ by a factor of ________.
   a. increase; 2
   b. increase; 4
   c. decrease; 2
   d. decrease; 4
5. Rank the following objects, in order of their circular velocities, from smallest to largest.
   a. a 5-kg object orbiting Earth halfway to the Moon
   b. a 10-kg object orbiting Earth just above Earth's surface
   c. a 15-kg object orbiting Earth at the same distance as the Moon
   d. a 20-kg object orbiting Earth one-quarter of the way to the Moon
6. Rank the following types of orbits in terms of the maximum speed of the orbiting object, from smallest to largest.
   a. elliptical, with semimajor axis *R*
   b. hyperbolic
   c. circular, with radius *R*
   d. parabolic
7. Which of the following objects would escape from Earth's gravity?
   a. a tennis ball traveling at 41,000 km/h, straight up
   b. a bear traveling at 41,000 km/h, straight up
   c. a car traveling at 41,000 km/h, straight up
   d. an airplane traveling at 41,000 km/h, straight up
8. Once you know the semimajor axis and the period of an orbit, you can determine
   a. the mass of the orbiting object.
   b. the speed of the orbiting object.
   c. the mass of the object being orbited.
   d. both b and c
9. Spring tides occur
   a. in March, April, or May.
   b. when the Moon's phase is new or full.
   c. when the Moon's phase is first quarter or third quarter.
   d. in either spring or fall.
10. If an object crosses the Roche limit, it
    a. can no longer be seen.
    b. begins to accelerate very quickly.
    c. slows down.
    d. may be torn apart.

True/False and Multiple Choice

11. **T/F:** When an object falls to Earth, Earth also falls up to the object.
12. **T/F:** If a particle existed all alone in the universe, it could exert a gravitational force.
13. **T/F:** A satellite orbiting faster than the circular velocity will spiral in.
14. **T/F:** The acceleration due to gravity is the same throughout the interior of Earth.

15. **T/F:** Earth has two tides each day because one is caused by the Moon and the other is caused by the Sun.

16. Venus has about 80 percent of Earth's mass and about 95 percent of Earth's radius. Your weight on Venus will be
   a. 20 percent more than on Earth.
   b. 20 percent less than on Earth.
   c. 10 percent more than on Earth.
   d. 10 percent less than on Earth.

17. The escape velocity is approximately ________ as large as the circular velocity.
   a. 0.4 times
   b. 1.4 times
   c. 14 times
   d. 140 times

18. If the Moon had twice the mass that it does, how would the strength of lunar tides change?
   a. The highs would be higher, and the lows would be lower.
   b. Both the highs and the lows would be higher.
   c. The highs would be lower, and the lows would be higher.
   d. Nothing would change.

19. If Earth had half of its current radius, how would the strength of lunar tides change?
   a. The highs would be higher, and the lows would be lower.
   b. Both the highs and the lows would be higher.
   c. The highs would be lower, and the lows would be higher.
   d. Nothing would change.

20. If the Moon were 2 times closer to Earth than it is now, the gravitational force between Earth and the Moon would be
   a. 2 times stronger.
   b. 4 times stronger.
   c. 8 times stronger.
   d. 16 times stronger.

21. If the Moon were 2 times closer to Earth than it is now, the tides would be
   a. 2 times stronger.
   b. 4 times stronger.
   c. 8 times stronger.
   d. 16 times stronger.

22. If two objects are tidally locked to each other,
   a. the tides always stay on the same place on each object.
   b. the objects always remain in the same place in each other's sky.
   c. the objects are falling together.
   d. both a and b

23. The strongest tides occur during ________ of the Moon.
   a. only the full phase
   b. only the new phase
   c. the full and new phases
   d. the first quarter and third quarter phases

24. Self-gravity is
   a. the gravitational pull of a person.
   b. the force that holds objects like people and lamps together.
   c. the gravitational interaction of all the parts of a body.
   d. the force that holds objects on Earth.

25. An object in a(n) ________ orbit in the Solar System will remain in its orbit forever. An object in a(n) ________ orbit will escape from the Solar System.
   a. unbound; bound
   b. circular; elliptical
   c. bound; unbound
   d. elliptical; circular

Thinking about the Concepts

26. Both Kepler's laws and Newton's laws tell us something about the motion of the planets, but there is a fundamental difference between them. What is that difference?

27. An astronaut standing on Earth could easily lift a wrench having a mass of 1 kg, but not a scientific instrument with a mass of 100 kg. In the International Space Station, she is quite capable of manipulating both, although the scientific instrument responds much more slowly than the wrench. Explain why the scientific instrument responds more slowly.

28. Explain the difference between weight and mass.

29. Weight on Earth is proportional to mass. On the Moon, too, weight is proportional to mass, but the constant of proportionality is different on the Moon than it is on Earth. Why?

30. Two comets are leaving the vicinity of the Sun, one traveling in an elliptical orbit and the other in a hyperbolic orbit. What can you say about the future of these two comets? Would you expect either of them to return eventually?

31. What is the advantage of launching satellites from spaceports located near the equator? Would you expect satellites to be launched to the east or to the west? Why?

32. We speak of Earth and the other planets all orbiting about the Sun. Under what circumstances do we have to consider bodies orbiting about a "common center of mass"?

33. If you could live in a house at the center of Earth, you could float from room to room as though you were living in the International Space Station. Explain why this statement is true.

34. What determines the strength of gravity at various radii between Earth's center and its surface?

35. The best time to dig for clams along the seashore is when the ocean tide is at its lowest. What phases of the Moon and times of day would be best for clam digging?

36. The Moon is on the meridian at your seaside home, but your tide calendar does not show that it is high tide. What might explain this apparent discrepancy?

37. We may have an intuitive feeling for why lunar tides raise sea level on the side of Earth facing the Moon, but why is sea level also raised on the side facing away from the Moon?

38. Tides raise and lower the level of Earth's oceans. Can they do the same for Earth's landmasses? Explain your answer.

39. Lunar tides raise the ocean surface less than 1 meter. How can tides as large as 5–10 meters occur?

40. Most commercial satellites are well inside the Roche limit as they orbit Earth. Why are they not torn apart?

Applying the Concepts

41. A typical house cat has a mass of roughly 5 kg. What is the weight of a typical house cat in newtons?

42. Mars has about one-tenth the mass of Earth, and about half of Earth's radius. What is the value of gravitational acceleration

on the surface of Mars compared to that on Earth? Estimate your mass and weight on Mars compared with your mass and weight on Earth. Do Hollywood movies showing people on Mars accurately portray this change in weight?

43. Earth speeds along at 29.8 km/s in its orbit. Neptune's nearly circular orbit has a radius of 4.5×10^9 km, and the planet takes 164.8 years to make one trip around the Sun. Calculate the speed at which Neptune plods along in its orbit.

44. An Earth-like planet is orbiting the bright star Vega at a distance of 1 astronomical unit (AU). The mass of Vega is twice that of the Sun.
    a. How fast does the planet travel in its orbit?
    b. What is the period of the planet?

45. Venus's circular velocity is 35.03 km/s, and its orbital radius is 1.082×10^8 km. Calculate the mass of the Sun.

46. At the surface of Earth, the escape velocity is 11.2 km/s. What would be the escape velocity at the surface of a very small asteroid having a radius 10^{-4} that of Earth's and a mass 10^{-12} that of Earth's?

47. How long does it take Newton's cannonball, moving at 7.9 km/s just above Earth's surface, to complete one orbit around Earth?

48. When a spacecraft is sent to Mars, it is first launched into an Earth orbit with circular velocity.
    a. Describe the shape of this orbit.
    b. What minimum velocity must we give the spacecraft to send it on its way to Mars?

49. Earth's mean radius and mass are 6,370 km and 5.97×10^{24} kg, respectively. Show that the acceleration of gravity at the surface of Earth is 9.81 m/s^2.

50. Using 6,370 km for Earth's radius, compare the gravitational force acting on a NASA rocket when it's sitting on its launchpad with the gravitational force acting on it when it is orbiting 350 km above Earth's surface.

51. The International Space Station travels on a nearly circular orbit 350 km above Earth's surface. What is its orbital speed?

52. As described in Math Tools 4.4, tidal force is proportional to the masses of the two objects and is inversely proportional to the cube of the distance between them. Some astrologers claim that your destiny is determined by the "influence" of the planets that are rising above the horizon at the moment of your birth. Compare the tidal force of Jupiter (mass 1.9×10^{27} kg; distance 7.8×10^{11} meters) with that of the doctor in attendance at your birth (mass 80 kg, distance 1 meter).

53. The asteroid Ida (mass 4.2×10^{16} kg) is attended by a tiny asteroidal moon, Dactyl, which orbits Ida at an average distance of 90 km. Neglecting the mass of the tiny moon, what is Dactyl's orbital period in hours?

54. The two stars in a binary star system (two stars orbiting about a common center of mass) are separated by 10^9 km and revolve around their common center of mass in 10 years.
    a. What is their combined mass, in kilograms?
    b. The mass of the Sun is 1.99×10^{30} kg. What is the combined mass of the two stars, in solar masses?

55. Suppose you go skydiving.
    a. Just as you fall out of the airplane, what is your gravitational acceleration?
    b. Would this acceleration be bigger, smaller, or the same if you were strapped to a flight instructor, and so had twice the mass?
    c. Just as you fall out of the airplane, what is the gravitational force on you? (Assume your mass is 70 kg.)
    d. Would the gravitational force be bigger, smaller, or the same if you were strapped to a flight instructor, and so had twice the mass?

Using the Web

56. Go to NASA's "Apollo 15 Hammer-Feather Drop" Web page (http://nssdc.gsfc.nasa.gov/planetary/lunar/apollo_15_feather_drop.html) and watch the video from Apollo 15 of astronaut David Scott dropping the hammer and falcon feather on the Moon. (You might find a better version on YouTube.) What did this experiment show? What would happen if you tried this on Earth with a feather and a hammer? Would it work? What would you see? Suppose instead you dropped the hammer and a big nail. How would they fall? How fast do things fall on the Moon compared to on Earth?

57. Go to the Exploratorium's "Your Weight on Other Worlds" page (http://exploratorium.edu/ronh/weight), which will calculate your weight on other planets and moons in our Solar System. On which objects would your weight be higher than it is on Earth? What difficulties would human bodies have in a higher-gravity environment? For example, would it be easy to get up out of bed and walk? What are the possible short-term and long-term effects of lower gravity on the human body? Can you think of some types of life on Earth that might adapt well to a different gravity?

58. Go to a website that will show you the times for high and low tides in your area—for example, http://saltwatertides.com. Pick a location and bring up the tide table for today and the next 2 weeks. Why are there two high tides and two low tides every day? What is the difference in the height of the water between high and low tides?

59. Go to a website with the phases of the Moon and the times of moonrise and moonset for your location—for example, http://aa.usno.navy.mil/data/docs/RS_OneYear.php. Does the time of the high tide lead or follow the position of the Moon?

STUDYSPACE is a free and open website that provides a Study Plan for each chapter of *21st Century Astronomy*. Study Plans include animations, reading outlines, vocabulary flashcards, and multiple-choice quizzes, plus links to premium content in SmartWork and the ebook. Visit **wwnorton.com/studyspace**.

SMARTWORK Norton's online homework system, includes algorithmically generated versions of these questions, plus additional conceptual exercises. If your instructor assigns questions in SmartWork, log in at **smartwork.wwnorton.com.**

Exploration | Newton's Laws

In the last Exploration we used the Planetary Orbit Simulator to explore Kepler's laws for Mercury. Now that we know how Newton's laws explain why Kepler's laws describe orbits, we will revisit the simulator to explore the Newtonian features of Mercury's orbit. Visit StudySpace and open the Planetary Orbit Simulator applet.

Acceleration

To begin exploring the simulation, set parameters for "Mercury" in the "Orbit Settings" panel and then click "OK." Click the "Newtonian Features" tab at the bottom of the control panel. Select "show solar system orbits" and "show grid" under "Visualization Options." Change the animation rate to 0.01, and click on the "start animation" button.

Examine the graph at the bottom of the panel.

1 **Where is Mercury in its orbit when the acceleration is smallest?**

....

....

2 **Where is Mercury in its orbit when the acceleration is largest?**

....

....

3 **What are the values of the largest and smallest accelerations?**

....

....

In the "Newtonian Features" graph, mark the boxes for vector and line that correspond to the acceleration. Specifying these parameters will insert an arrow that shows the direction of the acceleration and a line that extends the arrow.

4 **To what Solar System object does the arrow point?**

....

....

5 **In what direction is the force on the planet?**

....

....

Velocity

Examine the graph at the bottom of the panel again.

6 **Where is Mercury in its orbit when the velocity is smallest?**

....

....

7 **Where is Mercury in its orbit when the velocity is largest?**

....

8 **What are the values of the largest and smallest velocities?**

....

Add the velocity vector and line to the simulation by clicking on the boxes in the graph window. Study the resulting arrows carefully.

9 **Are the velocity and the acceleration always perpendicular (is the angle between them always 90°)?**

....

10 **If the orbit were a perfect circle, what would be the angle between the velocity and the acceleration?**

....

....

Hypothetical Planet

In the "Orbit Settings" panel, change the semimajor axis to 0.8 AU.

11 **How does this imaginary planet's orbital period now compare to Mercury's?**

....

....

....

Now change the semimajor axis to 0.1 AU.

12 **How does this planet's orbital period now compare to Mercury's?**

....

....

....

13 **Summarize your observations of the relationship between the speed of an orbiting object and the semimajor axis.**

....

....

....

The visible part of the electromagnetic spectrum is laid out in all its glory in the colors of this rainbow.

05 Light

Light brings us the news of the Universe.

William Henry Bragg (1862–1942)

LEARNING GOALS

Unlike the physicist or the chemist, who has control over the conditions in a laboratory, the astronomer must try to glean the secrets of the universe from the light and other particles that reach Earth from distant objects. We now turn our attention to light, a most informative messenger. By the conclusion of this chapter, you should be able to:

- Explain how light acts sometimes like a wave and sometimes like a particle.
- List the particles that compose atoms and summarize how the energy levels of an atom determine the wavelength of the light that the atom emits and absorbs.
- Describe how to measure the composition, properties, and motion of distant objects using the unique spectral lines of different types of atoms.
- Analyze how temperature measures the thermal energy of an object and determines the amount and spectrum of light that an object emits.
- Differentiate luminosity from brightness, and illustrate how distance affects each.
- Recognize how astronomical observations of different kinds of light provide different kinds of information about the universe.

5.1 The Speed of Light

Our knowledge of the universe beyond Earth comes overwhelmingly from **light** given off or reflected by astronomical objects. For any given object, light carries information about that object's temperature, composition, speed, and even the nature of the material that the light passed through on its way to Earth. Yet light plays a far larger role in astronomy than just being a messenger. Light is one of the primary means by which **energy** is transported throughout the universe. Objects as diverse as stars, planets, and vast clouds of gas and dust filling interstellar space heat up as they absorb light and cool off as they emit light. Light carries energy generated in the heart of a star outward through the star and off into space. Light transports energy from the Sun outward through the Solar System, heating the planets; and light carries energy away from each planet, allowing it to cool. The balance between these two processes establishes each planet's temperature and therefore the planet's possible suitability for life.

How does the light from a star affect the temperature of its planets?

Historically it was difficult for scientists to understand light. Galileo attempted to measure the speed of light by sending an assistant to a hilltop far away. In his experiment, Galileo uncovered a lantern, and the instant his assistant saw the light from the lantern, he uncovered his own lantern. The amount of time from the moment Galileo uncovered his lantern to the moment he saw his assistant's light would be the light's round-trip travel time—plus, of course, the assistant's reaction time. Galileo could not measure any delay, and concluded that the speed of light must be very great, possibly even infinite. In fact, you will learn in this book that although the speed is finite, nothing can travel faster than light. Light provides the ultimate speed limit in the universe.

Because light travels so rapidly, measuring its speed requires having either very good clocks or access to very large unobstructed distances over which to measure its flight. Galileo had neither at his disposal, but by the end of the 17th century astronomers had both. The great distances were the distances between the planets, and Kepler and Newton provided the good "clock." Recall from Chapter 4 that according to Newton's derivation of Kepler's laws, orbital periods are constant, with each orbit taking exactly as much time as the orbit before. This property applies to moons orbiting planets just as it applies to planets orbiting the Sun.

In the 1670s, Danish astronomer Ole Rømer (1644–1710) was studying the moons of Jupiter, measuring the times when each moon disappeared behind the planet. Much to his amazement, Rømer found that rather than maintaining a regular schedule, the observed times of these events would slowly drift in comparison with predictions. Sometimes the moons disappeared behind Jupiter too soon; other times they went behind Jupiter later than expected. Rømer realized that the difference depended on where Earth was in its orbit. If he began tracking the moons when Earth was closest to Jupiter, then by the time Earth was farthest from Jupiter the moons were a bit over 16 ½ minutes "late." But if he waited until Earth was once again closest to Jupiter, the moons once again passed behind Jupiter at the predicted times.

Often in science, a difference between theoretical predictions and experimental results points the way to new knowledge, and Rømer's work was no exception. Rømer correctly surmised that rather than a failure of Kepler's laws, he was seeing the first clear evidence that light travels at a finite speed. The moons appeared "late" when Earth was farther from Jupiter because of the time needed for light to travel the extra distance between the two planets (**Figure 5.1**). Over the course of Earth's yearly trip around the Sun, the distance between Earth and Jupiter changes by 2 astronomical units (AU), which is about 3×10^{11} meters. The speed of light equals this distance divided by Rømer's 16.7-minute delay, or about 3×10^8 meters per second (m/s). The value that Rømer actually announced in 1676 was a bit on the low side—2.25×10^8 m/s—because the length of 1 AU was not well known at that time. But Rømer's result was more than adequate to make the point that the speed of light is very large.

The International Space Station moves around Earth at a speed of about 28,000 kilometers per hour (km/h), almost 8,000 m/s, taking 91 minutes to complete one orbit. Light travels almost 40,000 times faster. Light can circle Earth in only ⅐ of a second, which explains why Galileo's attempts to measure the speed of light failed.

A good deal of work has been done to improve on Rømer's original result. Modern measurements of the speed of light made with the benefit of high-speed electronics and lasers give a value of 2.99792458×10^8 m/s in a vacuum (a region of space devoid of matter). As of 1983, the length of a meter is now defined as the distance traveled by light in a vacuum in 1/299,792,458 of a second.

The speed of light in a vacuum—about 300,000 km/s (given above in meters per second)—is one of nature's fundamental constants, usually written as c (lowercase). Keep in mind, however, that this is the speed in a vacuum. The speed of light

The speed of light is 300,000 km/s in a vacuum.

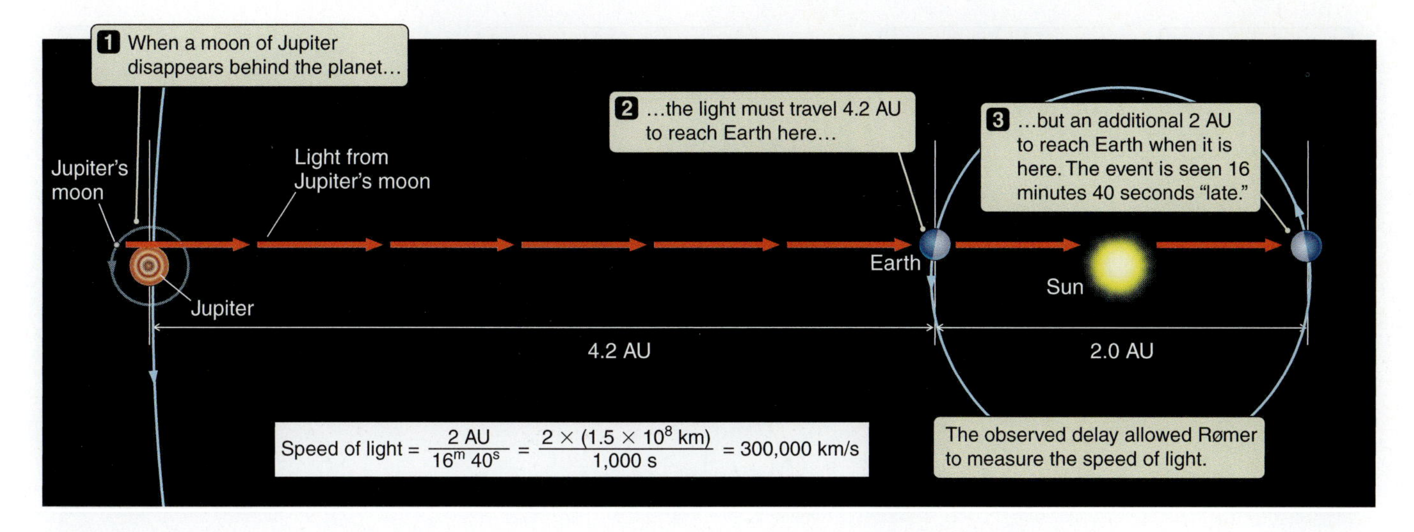

FIGURE 5.1 Danish astronomer Ole Rømer realized that apparent differences between the predicted and observed orbital motions of Jupiter's moons depend on the distance between Earth and Jupiter. He used these observations to measure the speed of light. (The superscript letters in the expression "16^m40^s" stand for minutes and seconds of time, respectively.)

through any medium, such as air or glass, is always less than *c*.

Recall that in Chapter 1 we spoke of distances expressed not in kilometers or miles, but in units of time. For example it takes light 1¼ seconds to travel between Earth and the Moon. In other words, we can say that the Moon is 1¼ light-*seconds* from Earth. The Sun is 8⅓ light-*minutes* away, and the next-nearest star is 4⅓ light-*years* distant. The travel time of light is a convenient way of expressing cosmic distances, and the basic unit is the **light-year**. A light-year is defined as the distance traveled by light in 1 year, or about 9.5 trillion km. Although it's sometimes misused as a measure of time, *the light-year is a measure of distance*. (Astronomers frequently use another unit to describe stellar and galactic distances: the **parsec**, abbreviated **pc**. One parsec is equal to 3.26 light-years.)

A light-year is the distance that light travels in 1 year.

5.2 Light Is an Electromagnetic Wave

Since the earliest investigations of light, there has been disagreement over the question of whether light is composed of particles, as Newton believed, or is instead a wave. (A **wave** is a disturbance that travels from one point to another.) In 1873 the Scottish physicist James Clerk Maxwell (1831–1879) seemingly put this controversy to rest by showing that light is a form of **electromagnetic radiation**. One of Maxwell's many accomplishments was his introduction of the concept that electricity and magnetism are actually two components of the same physical phenomenon. An **electric force** is the push or pull between electrically charged particles such as protons and electrons, arising from their electric charges. Opposite charges attract, and like charges repel. A **magnetic force**, on the other hand, is a force between electrically charged particles arising from their motion.

A wave is a disturbance that travels away from a source.

To describe electric and magnetic forces, Maxwell introduced the concepts of the **electric field** and the **magnetic field**. A charged particle creates an electric field that points away from the charge if the charge is positive (**Figure 5.2a**), or toward it if the charge is negative. Because the electric field points directly away from a positively charged particle (or directly toward a negatively charged particle), the force that a second charge feels is either directly toward or directly away from the first charged particle. Experiments show that if one charged particle (q_1) is moved, there is no immediate change in the force felt by the second charged particle (q_2) (**Figure 5.2b**). Only later does the second particle feel the change in location of the first (**Figure 5.2c**).

The situation is similar to what happens if you are holding one end of a long piece of rope and a friend is holding the other end. When you yank your end of the rope up and down, your friend does not feel the result immediately. Instead your yank starts a pulse—a wave—that travels along the rope. Your friend notices the yank only when the wave arrives at his end. Similarly, when a charged particle moves, information

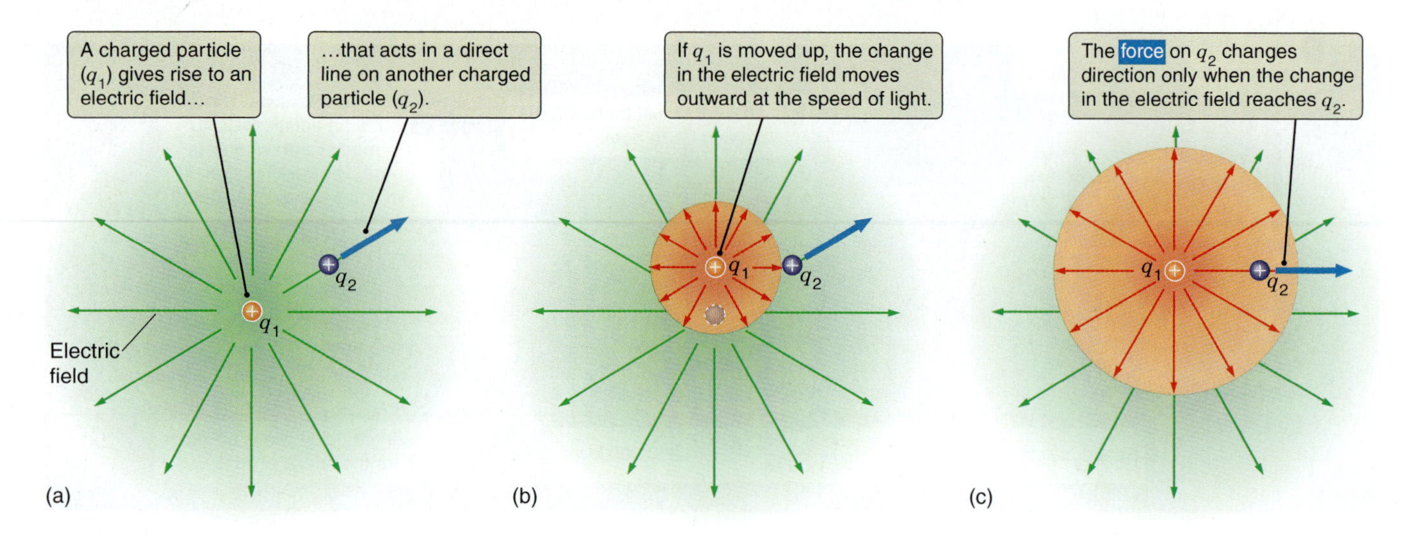

FIGURE 5.2 Maxwell's theory of electromagnetic radiation describes how electrically charged particles move and interact. (a) A positively charged particle, q_1, has an electric field (shown in orange) that acts outward along the field lines. (b) When q_1 accelerates, changes in the electric field move outward at the speed of light. For simplicity, the charge is shown moving instantly from one place to another, but in reality this could not happen. (c) Particle q_2 does not respond until it feels the change in q_1.

about the change travels outward through space as a wave in the electric field. Other charged particles are not affected by the first particle's movement until the wave reaches them.

Maxwell summarized the behavior of electric and magnetic fields in four elegant equations. Among other things, these equations say that a changing electric field causes a magnetic field, and that a changing magnetic field causes an electric field. A change in the motion of a charged particle causes a changing electric field, which causes a changing magnetic field, which causes a changing electric field, and so on. Once the process starts, a self-sustaining procession of oscillating electric and magnetic fields moves out in all directions through space. In other words, an accelerating charged particle gives rise to an **electromagnetic wave** (**Figure 5.3**). These electromagnetic waves, and the accelerating charges that generate them, are the sources of electromagnetic radiation. By contrast, an electric charge that is moving at a constant velocity is stationary in its inertial frame of reference, so it does not produce electromagnetic waves.

Changing electric and magnetic fields produce a self-sustaining electromagnetic wave.

In addition to predicting that electromagnetic waves should exist, Maxwell's equations predict how rapidly the disturbance in the electric and magnetic fields should move—that is, the speed at which an electromagnetic wave should travel. When Maxwell carried out this calculation, he discovered that electromagnetic waves should travel at 3×10^8 m/s—which is the measured speed of light (c) (see the **Process of Science Figure**). This agreement means that Maxwell had shown that light is an electromagnetic wave.

Maxwell's wave description of light also gives us an idea of how light originates and how it interacts with matter. Imagine a sink full of still water. A drop falls from the faucet into the sink, causing a disturbance, or wave, that

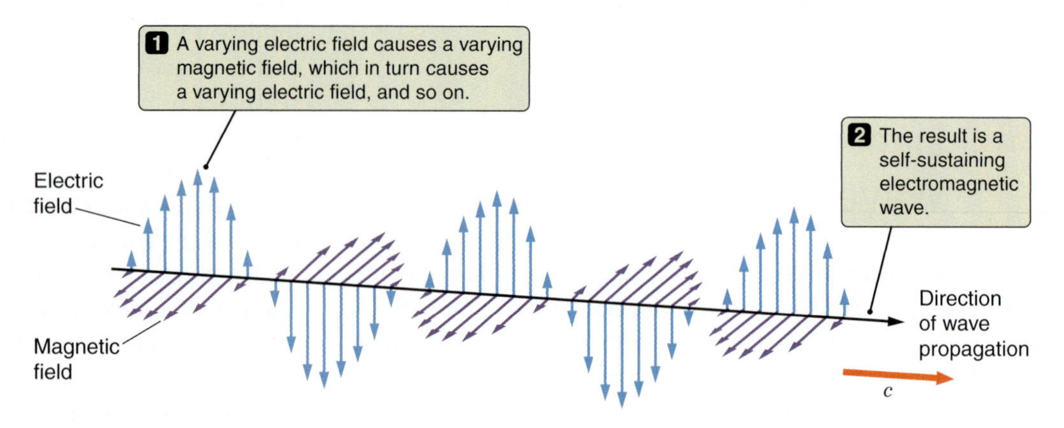

FIGURE 5.3 An electromagnetic wave consists of oscillating electric and magnetic fields that are perpendicular both to each other and to the direction in which the wave travels.

Process of Science

AGREEMENT BETWEEN FIELDS

Scientists working on very different problems in different fields all find the same result: light has a speed and can be measured.

Ole Rømer studies eclipses of Jupiter's moons.

Rømer calculates the speed of light from eclipse delays of Jupiter's moons.

No one believes it... ...Rømer dies.

James Bradley studies the apparent motion of stars, known as the "aberration of starlight."

Twenty years after Rømer's death, Bradley's motion studies show that the speed of light is finite.

James Clerk Maxwell studies electricity and magnetism.

Einstein builds on Maxwell's theory in 1905 to claim that the speed of light is the same for all observers.

Astronomers and physicists converge on an understanding that photons travel at the speed of light– a fundamental constant of the universe that is the same for all observers.

moves outward as a ripple on the surface of the water (**Figure 5.4a**). In much the same way, an oscillating (and therefore accelerating) electric charge causes a disturbance that moves outward through space as an electromagnetic wave (**Figure 5.4b**). However, whereas the ripples in the sink result from *mechanical* distortions of the water's surface and require a medium (in this case water) to travel through, Maxwell showed that light waves are a fundamentally different type of wave and do not require a medium. Light waves result not from mechanical distortion of a medium, but from periodic changes in the strength of the electric and magnetic fields.

Accelerating charges cause electromagnetic waves.

Now imagine that a soap bubble is floating in the sink (**Figure 5.5a**). The bubble remains stationary until the ripple from the dripping faucet reaches it. As the ripple passes by, the rising and falling action of the water causes the bubble to rise and fall. Similarly, the oscillating electric field of an electromagnetic wave causes an oscillating force on any charged particle that the wave encounters, and this force causes the particle to move about as well (**Figure 5.5b**). It takes energy to produce an electromagnetic wave, and that energy is carried through space by the wave. Matter far from the source of the wave can absorb this energy. In this way, some of the energy lost by the particles generating the electromagnetic wave is transferred to other charged particles. The *emission* and *absorption* of

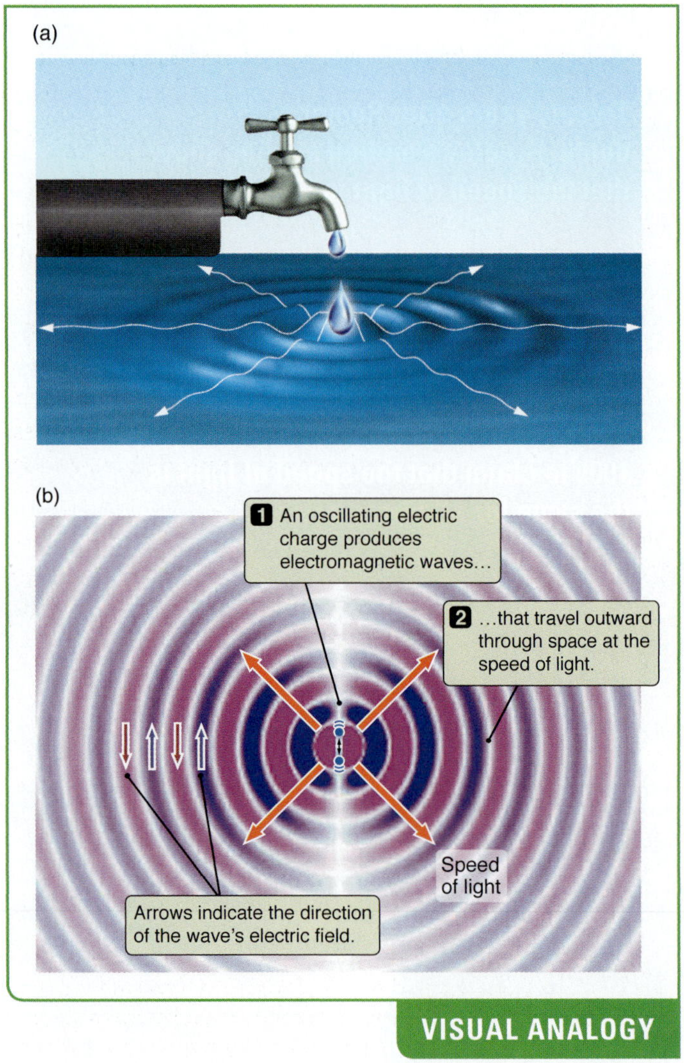

(a)

(b)

1 An oscillating electric charge produces electromagnetic waves...

2 ...that travel outward through space at the speed of light.

Speed of light

Arrows indicate the direction of the wave's electric field.

VISUAL ANALOGY

FIGURE 5.4 (a) A drop falling into water generates waves that move outward across the water's surface. (b) In similar fashion, an oscillating (accelerated) electric charge generates electromagnetic waves that move away at the speed of light.

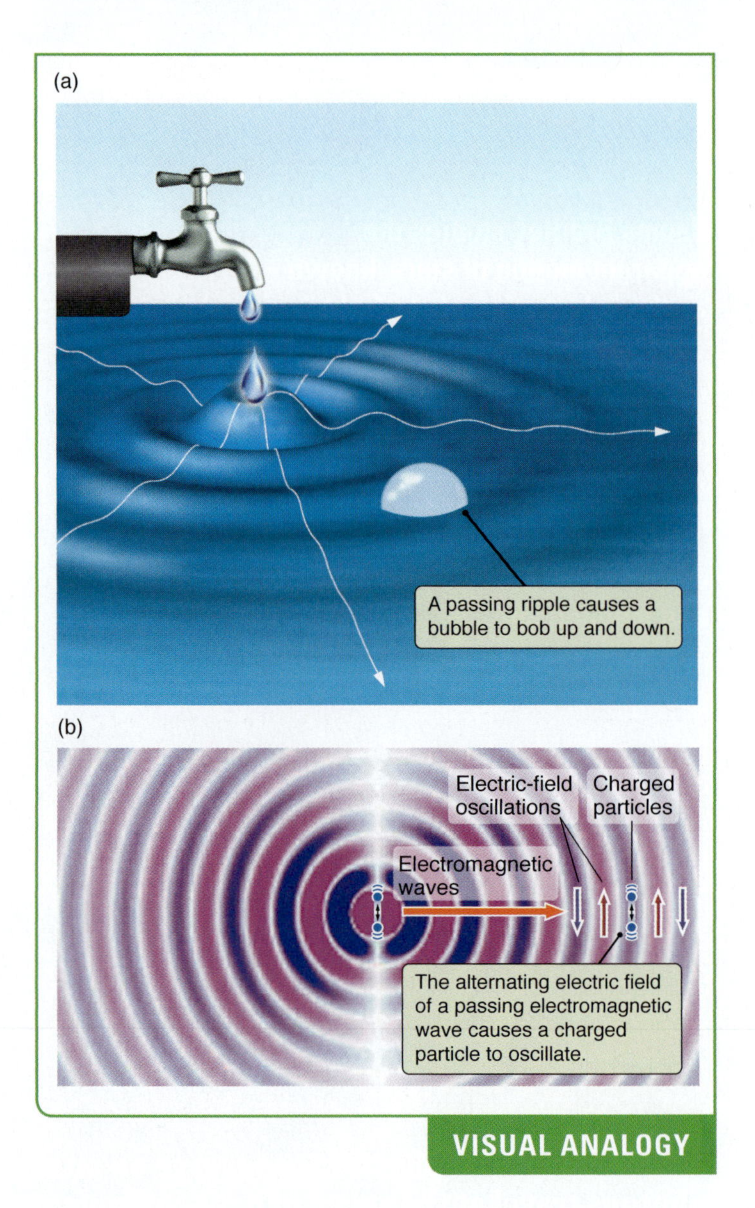

FIGURE 5.5 (a) When waves moving across the surface of water reach a bubble, they cause the bubble to bob up and down. (b) Similarly, a passing electromagnetic wave causes an electric charge to oscillate in response to the wave.

light by matter are the result of the interaction of electric and magnetic fields with electrically charged particles.

Waves Are Characterized by Amplitude, Speed, Frequency, and Wavelength

In this book you will learn about waves of different kinds, including electromagnetic waves crossing the vast expanse of the universe and earthquakes traveling through Earth. Waves are generally characterized by four quantities: *amplitude, speed, frequency,* and *wavelength* (**Figure 5.6**). The **amplitude** of a wave is the height of the wave above the equilibrium position. In the case of light waves, the amplitude is an indication of the intensity or brightness of the radiation. All types of waves travel at a particular speed, and light waves, as we have already shown, travel at 300,000 km/s in a vacuum.

The number of wave crests passing a point in space each second is called the wave's **frequency**, f. The unit of frequency is cycles per second, which is called **hertz** (abbreviated **Hz**) after the 19th century physicist Heinrich Hertz (1857–1894), who was the first to experimentally confirm Maxwell's predictions about electromagnetic radiation. The time taken for one complete cycle is called the **period**, P, which is measured in seconds. The distance that a wave travels during one complete oscillation is its **wavelength**. Wavelength is the distance from one wave crest to the next, or the distance from one wave trough to the next. Wavelength is usually denoted by the Greek letter lambda, λ.

There is a clear relationship between the frequency of a wave and its period. If the period of a wave is ½ second—that is, if it takes ½ second for one wave to pass by, crest to crest—then two waves will go by in 1 second. So a wave with a period of ½ second per cycle has a frequency of 2 cycles per second (or 2 Hz). Similarly, if a wave has a period of 1/100 second per cycle, then 100 waves will pass by each second, and the wave has a frequency of 100 Hz. More generally, the frequency of a wave is just 1 divided by its period:

$$\text{Frequency} = \frac{1}{\text{Period}} \quad \text{or} \quad f = \frac{1}{P}$$

There is also a relationship between the period of a wave and its wavelength. The period of a wave is the time between the arrival of one wave crest and the next. During this time the wave travels one wavelength. Recall that distance traveled equals speed multiplied by time taken. Changing "distance traveled" to one wavelength and changing "time taken" to one period shows that the wavelength of a wave equals the speed at which the wave is traveling multiplied by the period of the wave:

$$\text{Wavelength} = \text{Speed} \times \text{Period}$$

As noted earlier, the letter c represents the speed of light, so:

$$\lambda = c \times P$$

Using the relationship between period and frequency just given, you can also write the equation like this:

$$\text{Wavelength} = \frac{\text{Speed}}{\text{Frequency}} \quad \text{or} \quad \lambda = \frac{c}{f}$$

Accordingly, if you know the speed of a wave and the value of one of its three other properties—its wavelength, period, or frequency—you can determine the values of the remaining two properties (see Math Tools 5.1 later in the chapter).

This relationship indicates that the longer the length of a wave, the longer the time between wave crests, and the lower the frequency of the wave. A shorter wavelength means less distance between wave crests, which means a shorter wait until the next wave comes along. Therefore, a shorter wavelength means a higher frequency. A tremendous amount of information can be carried by waves—for

Long wavelengths mean low frequency; short wavelengths mean high frequency.

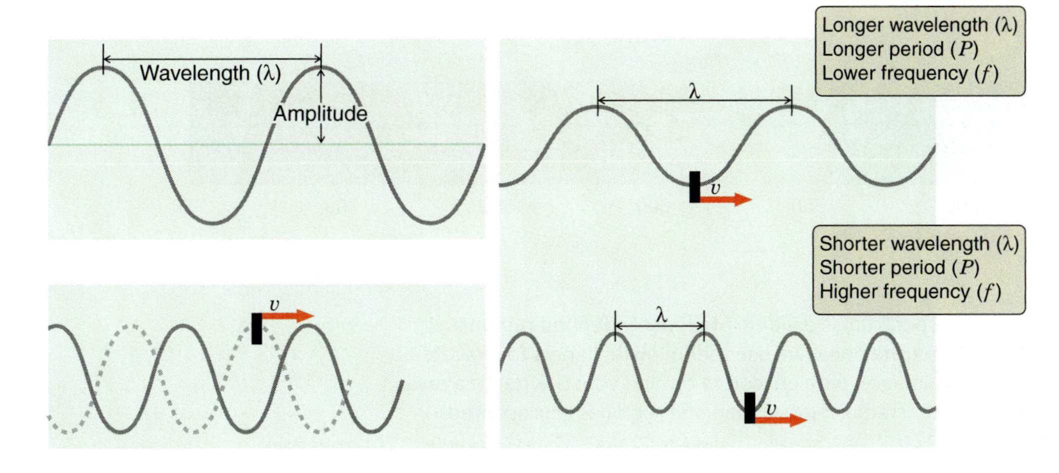

FIGURE 5.6 A wave is characterized by the distance from one peak to the next (wavelength, λ), the frequency of the peaks (f), the maximum deviations from the medium's undisturbed state (amplitude), and the speed (v) at which the wave pattern travels from one place to another. In an electromagnetic wave, the amplitude is the maximum strength of the electric field, and the speed of light is written as c.

example, complex and beautiful music. As you continue your study of the universe, time and time again you will find that the information you receive, whether about the interior of Earth or about a distant star or galaxy, rides in on a wave.

A Wide Range of Wavelengths Make Up the Electromagnetic Spectrum

You have almost certainly seen a rainbow like the one in the chapter-opening figure, spread out across the sky. This sorting of light by colors is really a sorting by wavelength. Light spread out according to wavelength is called a **spectrum**. The colors in the visible spectrum, in order of decreasing wavelength, are red, orange, yellow, green, blue, and violet.

The spectrum of visible light is seen as the colors of the rainbow.

At the long-wavelength (and therefore low-frequency) end of the visible spectrum is red light. At the other end is violet light. A commonly used unit for the wavelength of visible light is the **nanometer**, abbreviated **nm**. A nanometer is one-billionth (10^{-9}) of a meter. Human eyes can see light between violet (about 380 nm) and red (750 nm). Stretched out between the two, in a rainbow, is the rest of the visible spectrum.

When we say *visible light*, we mean "the light that the light-sensitive cells in our eyes respond to." But this is not the whole range of possible wavelengths for electromagnetic radiation. Radiation can have wavelengths that are much shorter or much longer than your eyes can perceive. The whole range of different wavelengths of light is collectively referred to as the **electromagnetic spectrum**.

Visible light is only one small segment of the electromagnetic spectrum.

The electromagnetic spectrum is illustrated in **Figure 5.7**. Below the shortest visible wavelength is **ultraviolet (UV) radiation**, with wavelengths between 40 and 380 nm. The prefix *ultra-* means "extreme," so ultraviolet light is light that is more extremely violet than violet. Ultraviolet light is fundamentally no different from visible light, any more than high C on a piano is fundamentally different from middle C.

Wavelengths shorter than 40 nm, or 4×10^{-8} meter, are called **X-ray** radiation. This distinction arose for historical reasons. When X-rays were discovered in the late 19th century, they were given the name *X* by their discoverer, physicist Wilhelm Conrad Roentgen (1845–1923), to indicate that they were "a new kind of ray." Electromagnetic radiation with the very shortest wavelengths (less than about 10^{-10} meter) is called **gamma ray** radiation.

Wavelengths longer than about 750 nm and shorter than 500 micrometers (μm) are called **infrared (IR) radiation**, for light that is "redder than red." The prefix *infra-* means "below," and infrared light has a frequency lower than

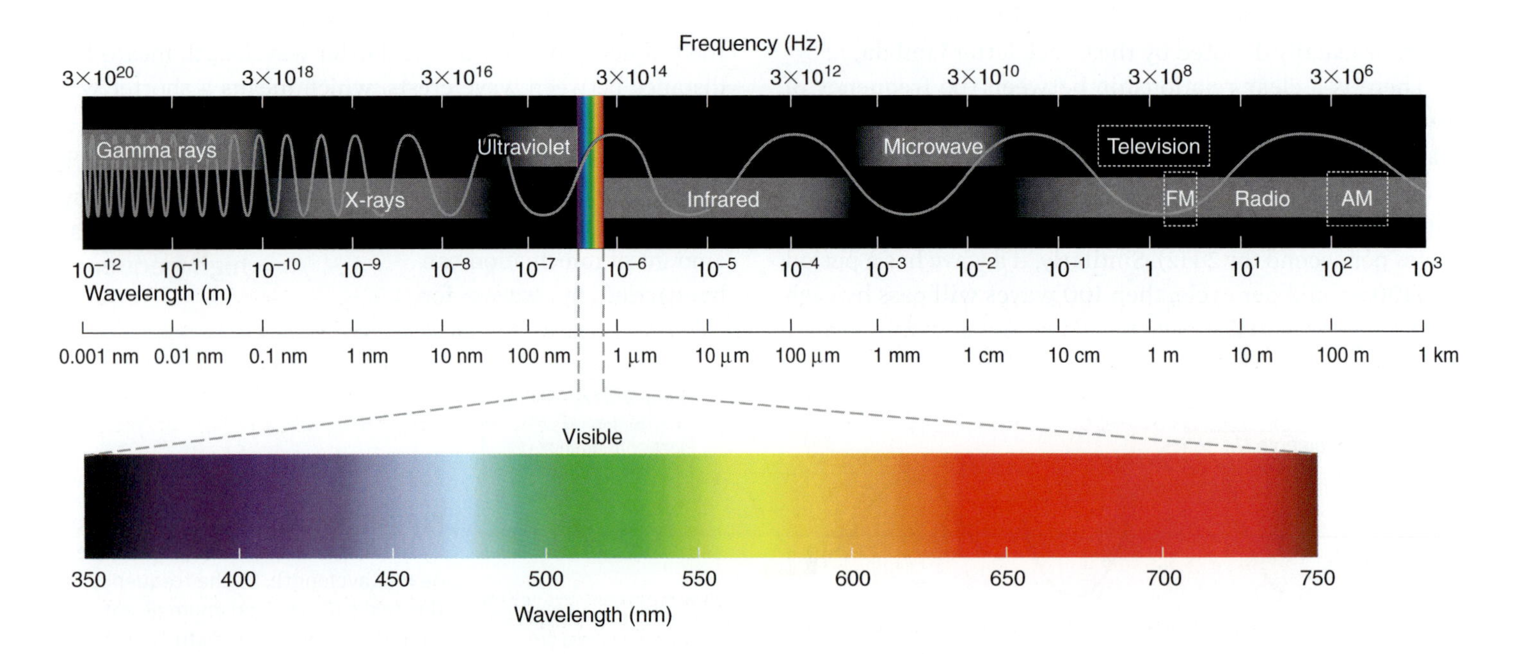

FIGURE 5.7 By convention, the electromagnetic spectrum is divided into loosely defined regions ranging from gamma rays to radio waves. Throughout the book, we use the following labels to indicate the form of radiation used to produce astronomical images, with an icon to remind you: G = gamma rays; X = X-rays; U = ultraviolet; V = visible; I = infrared; R = radio. If more than one region is represented, multiple labels are highlighted.

(below) that of red light. The **micrometer**, or **micron**, is one-millionth (10^{-6}) of a meter (abbreviated **μm**, where μ is the Greek letter mu). Light with even longer wavelengths is called **microwave radiation**. The longest-wavelength (and therefore lowest-frequency) electromagnetic radiation, which has wavelengths ranging from longer than a few centimeters up to arbitrarily long wavelengths, is **radio waves** (see **Math Tools 5.1**). In Chapter 6 we will discuss the various kinds of telescopes used by astronomers to capture and analyze the wide range of electromagnetic radiation. ▶▶ NEBRASKA SIMULATION: EM SPECTRUM MODULE

Light Is a Wave, but It Is Also a Particle

By showing that light is an electromagnetic wave, Maxwell's work seemed to put to rest the issue of whether light consists of waves or particles. However, although the electromagnetic wave theory of light has successfully described many phenomena, there are also phenomena that it does not describe well. Many of the difficulties with the wave model of light have to do with the way in which light interacts with small particles such as atoms and molecules. Scientists working in the late 19th and early 20th centuries discovered that many of the puzzling aspects of light could be better understood if light energy came in *discrete packages*.

In 1905, Albert Einstein published a paper in which he argued that light consists of particles. He based his argument on the **photoelectric effect**, the emission of electrons from surfaces that occurs when the surfaces are illuminated by electromagnetic radiation greater than a certain frequency. Einstein showed that the rate at which electrons are ejected depends only on the amount of light of the incident radiation, and that the electron velocity depends only on the frequency of the incident radiation. (It was this work on the photoelectric effect that earned Einstein a Nobel Prize in 1921.) The work of Einstein and other scientists modified our understanding of light to show that, although it can sometimes be explained as acting like a wave, light can also be explained as acting like a particle. In this model, light is described as being made up of particles called **photons** (*phot-* means "light," as in *photograph*; and *-on* signifies a particle). Photons have no mass, always travel at the speed of light, and carry energy.

Recognition of the particle theory of light, however, did not mean that the wave theory had been discarded. The particle description of light is tied to the wave description of light by a relationship between the energy of a photon and the frequency or wavelength of the wave. Specifically, they are linked with the following equation:

$$E = hf \quad \text{or} \quad E = \frac{hc}{\lambda}$$

The h in this equation is called **Planck's constant** (named after physicist Max Planck, 1858–1947) and has the value 6.63×10^{-34} joule-second. (The **joule**, abbreviated **J**, is a unit of energy.)

The energy of a photon is proportional to its frequency.

Math Tools 5.1

Working with Electromagnetic Radiation

When you tune to a radio station at, say, 770 AM, you are receiving an electromagnetic signal that travels at the speed of light and is broadcast at a frequency of 770 kilohertz (kHz), or 7.7×10^5 Hz. We can use the relationship between wavelength and frequency to calculate the wavelength of the AM signal:

$$\lambda = \frac{c}{f} = \left(\frac{3 \times 10^8 \text{ m/s}}{7.7 \times 10^5\text{/s}}\right) = 390 \text{ m}$$

This AM wavelength is about 4 times the length of a football field.

We can compare this wavelength to that of a typical FM broadcast signal: 99.5 FM, or 99.5 megahertz (MHz), or 9.95×10^7 Hz:

$$\lambda = \frac{c}{f} = \left(\frac{3 \times 10^8 \text{ m/s}}{9.95 \times 10^7\text{/s}}\right) = 3 \text{ m}$$

Therefore, this FM wavelength is 3 meters (about 10 feet). FM wavelengths are much shorter than AM wavelengths.

The human eye is most sensitive to light in green and yellow wavelengths, about 500–590 nm. If we examine green light with a wavelength of 530 nm, we can compute its frequency:

$$f = \frac{c}{\lambda} = \left(\frac{3.00 \times 10^8 \text{ m/s}}{530 \times 10^{-9}\text{/s}}\right) = \frac{5.66 \times 10^{14}}{\text{s}} = 5.66 \times 10^{14} \text{ Hz}$$

This frequency corresponds to 566 *trillion* wave crests passing by each second.

According to the particle description of light, the electromagnetic spectrum is a spectrum of photon energies. The higher the frequency of the electromagnetic wave, the greater the energy carried by each photon. Photons of shorter wavelength (higher frequency) carry more energy than do photons of longer wavelength (lower frequency). For example, photons of blue light carry more energy than do photons of longer-wavelength red light. Ultraviolet photons carry more energy than do photons of visible light, and X-ray photons carry more energy than do ultraviolet photons. The lowest-energy photons are radio wave photons.

The *total* amount of energy that a beam of the light carries is called its **intensity**. A beam of red light can be just as intense as a beam of blue light—that is, it can carry just as much energy—but because the energy of a red photon is less than the energy of a blue photon, maintaining that same intensity requires more red photons than blue photons. This relationship is a lot like money: $10 is $10, but it takes a lot more pennies (low-energy photons) than quarters (high-energy photons) to make up $10 (**Figure 5.8**).

When the energy of light is described as broken into discrete packets called photons, the light energy is said to be **quantized**. The word *quantized,* which has the same root as the word *quantity,* means that something is subdivided into discrete units. A photon is referred to as a **quantum of light**. The branch of physics that deals with the quantization of energy and of other properties of matter is called **quantum mechanics**.

Photons are the quantum mechanical description of light.

Although the predictions of quantum mechanics have been confirmed over and over again by experiment, its fundamental assumptions seem counterintuitive. The wave-particle description of light conflicts with everyday, commonsense ideas about the world. It is hard to imagine a single thing sharing the properties of a wave on the ocean *and* a beach ball. The wave model of light is clearly the correct description to use in many instances (think of Maxwell's work). At the same time, the particle description of light is also clearly the correct description to use in other cases (as scientists like Planck and Einstein demonstrated). But how can the same thing—light—be described as both a wave and a particle?

The trouble with thinking of light as both a wave and a particle hints at the puzzling and philosophically troublesome world of quantum mechanics. In Section 5.3 you will learn that light is not the only thing that shares wave and particle properties. In fact, all matter shares wave and particle properties.

Like light, all matter can behave as waves and particles.

AstroTour: Light as a Wave, Light as a Photon

5.3 The Quantum View of Matter

To a physicist, **matter** is anything that occupies space and has mass. Virtually all of the matter we have direct experience with is composed of **atoms**. Atoms are tiny—so tiny that a single teaspoon of water contains about 10^{23} of them. (There are as many atoms in a single teaspoon of water as there are stars in the observable universe.) The interaction of light with matter is really the interaction of light with atoms and their components.

Virtually all ordinary matter is composed of atoms.

Atoms are the fundamental building blocks of matter, consisting of a central massive **nucleus** of **protons** (which are positively charged) and **neutrons** (which have no

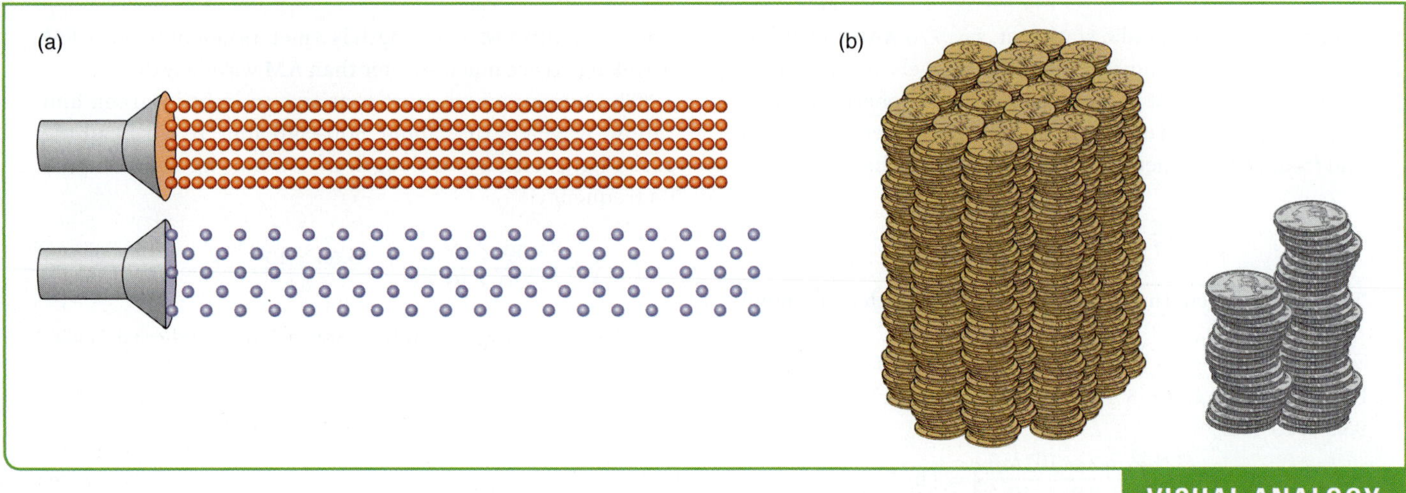

FIGURE 5.8 (a) Red light carries less energy than blue light, so it takes more red photons than blue photons to make a beam of a particular intensity. (b) Similarly, pennies are worth less than quarters, so it takes more pennies than quarters to add up to $10.

charge). A cloud of negatively charged **electrons** surrounds the nucleus. Atoms with the same number of protons are all of the same type, known as an **element**. For example, an atom with two protons is the element helium (**Figure 5.9a**). An atom with six protons is the element carbon, one with eight protons is the element oxygen, and so forth. **Molecules** are groups of atoms bound together by shared electrons.

For an atom to be electrically neutral, it must have the same number of electrons as protons. Electrons have much less mass than protons or neutrons have, so almost all the mass of an atom is found in its nucleus. This description led to a model of an atom as a tiny "solar system," with the massive nucleus sitting in the center and the smaller electrons orbiting about it, much as planets orbit about the Sun (**Figure 5.9b**). It is called the **Bohr model** after the Danish physicist Niels Bohr (1885–1962), who proposed it in 1913. **AstroTour: Atomic Energy Levels and the Bohr Model**

The Bohr model, however, has a flaw. In this view, an electron whizzing about in an atom is constantly undergoing an acceleration: the *direction* of its motion is constantly changing. But the wave description of electromagnetic radiation says that any electrically charged particle that is accelerating must also be giving off electromagnetic radiation. This electromagnetic radiation, in the form of photons, should be carrying away the orbital energy of the electron. (Imagine that electron as the wiggling electric charge in Figure 5.4b.) Calculating how much energy should be radiated away from the electron shows that only a tiny fraction of a second should be needed for the electrons in an atom to lose all their energy and fall into the atom's nucleus. But this does not happen. Atoms are stable; they exist for very long periods of time, and electrons never "fall into" the nuclei of atoms. So Bohr's model is not a complete description of an atom.

Just as waves of light have particle-like properties, particles of matter also have wavelike properties. With this realization, the miniature solar system model of the atom was modified so that the positively charged nucleus is surrounded by electron "clouds" or "waves" (**Figure 5.9c**). In the quantum model, it is not possible to know precisely where the electron is in its orbit. This uncertainty about the electron's location is expressed by the **Heisenberg uncertainty principle**, named for physicist Werner Heisenberg (1901–1976). The wave characteristics of particles make it impossible to simultaneously pin down both their exact location and their exact **momentum** (lowercase p), defined as the product of mass and velocity ($p = m \times v$). There will always be some uncertainty about either the particle's position or its momentum. This is why a featureless cloud is used to represent electrons in orbit around an atomic nucleus.

The uncertainty principle says an observer cannot know both the precise momentum and position of a particle.

The one absolute certainty is that there is *uncertainty* at the root of everything physical. You have learned that all electromagnetic radiation behaves as both waves and particles. And things like electrons and protons, which are often visualized as "solid" particles, also have wave characteristics. There is a nice symmetry here. Waves have particle-like characteristics and particles have wavelike characteristics. This conclusion has huge implications in both science and technology. For instance, the wave-particle property is the principle by which electron microscopes work, as you will see in the next chapter.

Remember that velocity includes both speed and direction. In view of this fact, there is a more quantitative way to express the uncertainty principle. The product of the

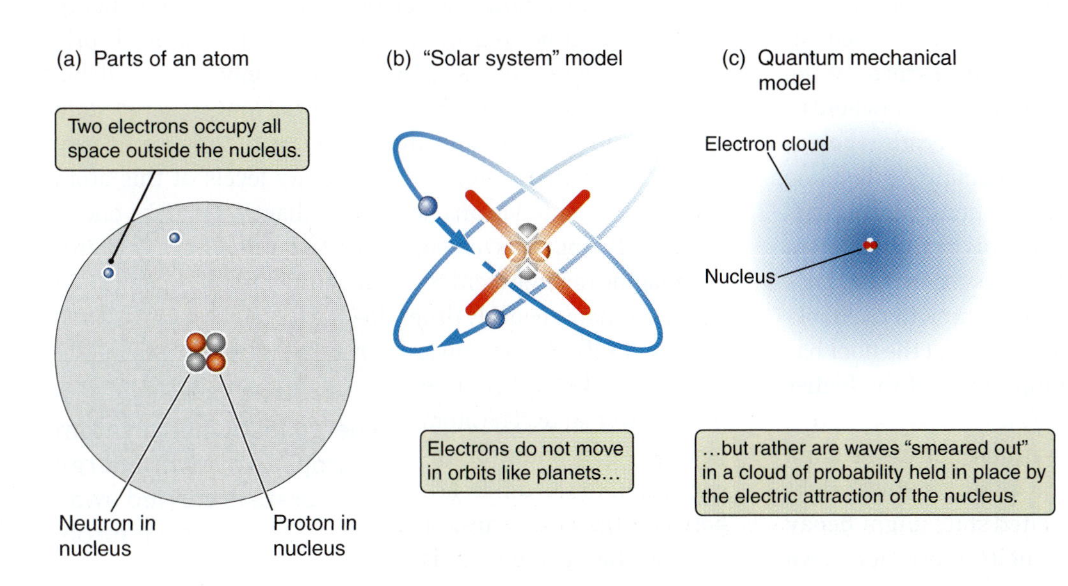

FIGURE 5.9 (a) An atom (in this case helium) is made up of a nucleus consisting of positively charged protons and electrically neutral neutrons and surrounded by less massive negatively charged electrons. (b) Atoms are often drawn as miniature "solar systems," but this model is incorrect, as the red *X* indicates. (c) Electrons are actually smeared out around the nucleus in quantum mechanical clouds of probability.

uncertainty in a particle's position (Δx) and the uncertainty in its momentum (Δp) is always equal to or greater than a particular constant, which is on the order of Planck's constant, h. This relationship is reflected in a simple equation: $\Delta x \times \Delta p \approx h$. In other words, the more you know about where something is (Δx approaching zero), the less you can know about how fast it is moving and in what direction (Δp approaching infinity). Conversely, the better you know the momentum of something, the less you know about its location. This is not a matter of making inferior measurements. It is simply not possible to do better, no matter how precisely you measure (**Figure 5.10**).

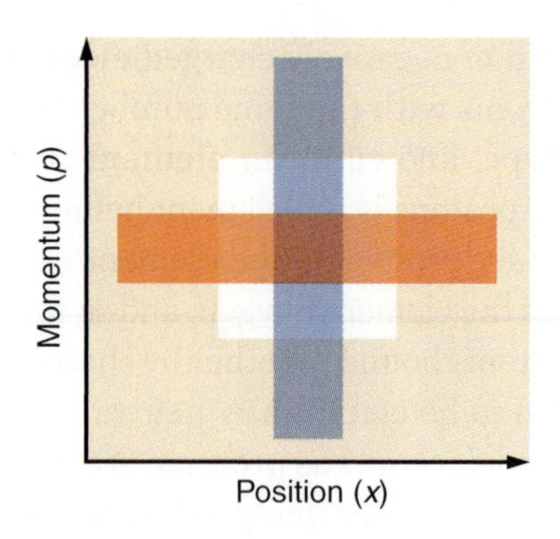

FIGURE 5.10 Heisenberg's uncertainty principle says that if momentum is measured with great precision, then the measurement of position will be less precise. The product of the two quantities will always be ≈ h. Suppose the white square represents the product of position times momentum. Confining the position to the width of the blue bar means momentum can range anywhere along the height. Confining momentum to the height of the red bar means position can range anywhere along the width.

Atoms Have Discrete Energy Levels

Another aspect of the wave-particle nature of electrons is that electron waves in an atom can assume only certain specific forms. (Analogously, the strings on a guitar can vibrate only at certain discrete frequencies, giving rise to the distinct notes we hear.) The form that the electron waves take depends on the possible energy states of atoms. We can imagine the energy states of atoms as being like a bookcase with a series of shelves (**Figure 5.11a**). The energy of an atom might correspond to the energy of one state or to the energy of the next state, but the energy of the atom is never found between the two states, just as a book can be on only one shelf at a time and cannot be partly on one shelf and partly on another. A given atom may have a tremendous number of different energy states available to it, but these states are *discrete*. An atom has the energy of one of these allowed states, or it has the energy of the next allowed state, but *it cannot have an energy somewhere in between*.

Atoms can have only certain discrete energies.

The lowest possible energy state of an atom—the "floor"—is the **ground state** of the atom (**Figure 5.11b**). Allowed energy levels above the ground state are **excited states** of the atom. When the atom is in its ground state, it has nowhere to go. An electron cannot "fall" into the nucleus because there is no allowed state with less energy for it to occupy. At the same time, it cannot move up to a higher energy state without getting extra energy from somewhere. For this reason an atom will remain in its ground state forever, unless something happens to knock it into an excited state. Returning to the bookcase analogy, a book sitting on the bottom shelf at the floor has nowhere left to fall, and it cannot jump to one of the higher shelves of its own accord.

An atom in an excited state is very different from an atom in the ground state. Just as a book on an upper shelf might fall to a lower shelf, an atom in an excited state might **decay** to a lower state by getting rid of some of its extra energy. An important difference between the atom and the book on the shelf, however, is that whereas a snapshot might catch the book falling between the two shelves, the atom will never be caught between two energy states. When the transition from one state to another occurs, the difference in energy between the two states is carried off all at once. A common way for an atom to do this is to emit a photon. The photon emitted by the atom carries away exactly the amount of energy lost by that atom as it goes from the higher energy state to the lower energy state. In a similar fashion, atoms moving from a lower energy state to a higher energy state can absorb only certain specific energies.

An Atom's Energy Levels Determine the Wavelengths It Can Emit and Absorb

To better understand the relationship between the energy levels of an atom and the radiation it can emit or absorb, imagine a hypothetical atom that has only two available energy states. Let's call the energy of the lower energy state (the ground state) E_1 and the energy of the higher energy state (the excited state) E_2. The energy levels of this atom can be represented in an energy level diagram like the one in Figure 5.11b, but this one has only two levels (**Figure 5.12a**).

Imagine that the atom begins in the higher energy state (E_2) and then spontaneously drops down to the lower state (E_1). In **Figure 5.12b**, the downward arrow indicates that the atom went from the higher state to the lower state. The atom just lost an amount of energy equal to the difference between the two states, or $E_2 - E_1$. Because energy is

Energy lost when an atom drops to a lower energy state is carried away as a photon.

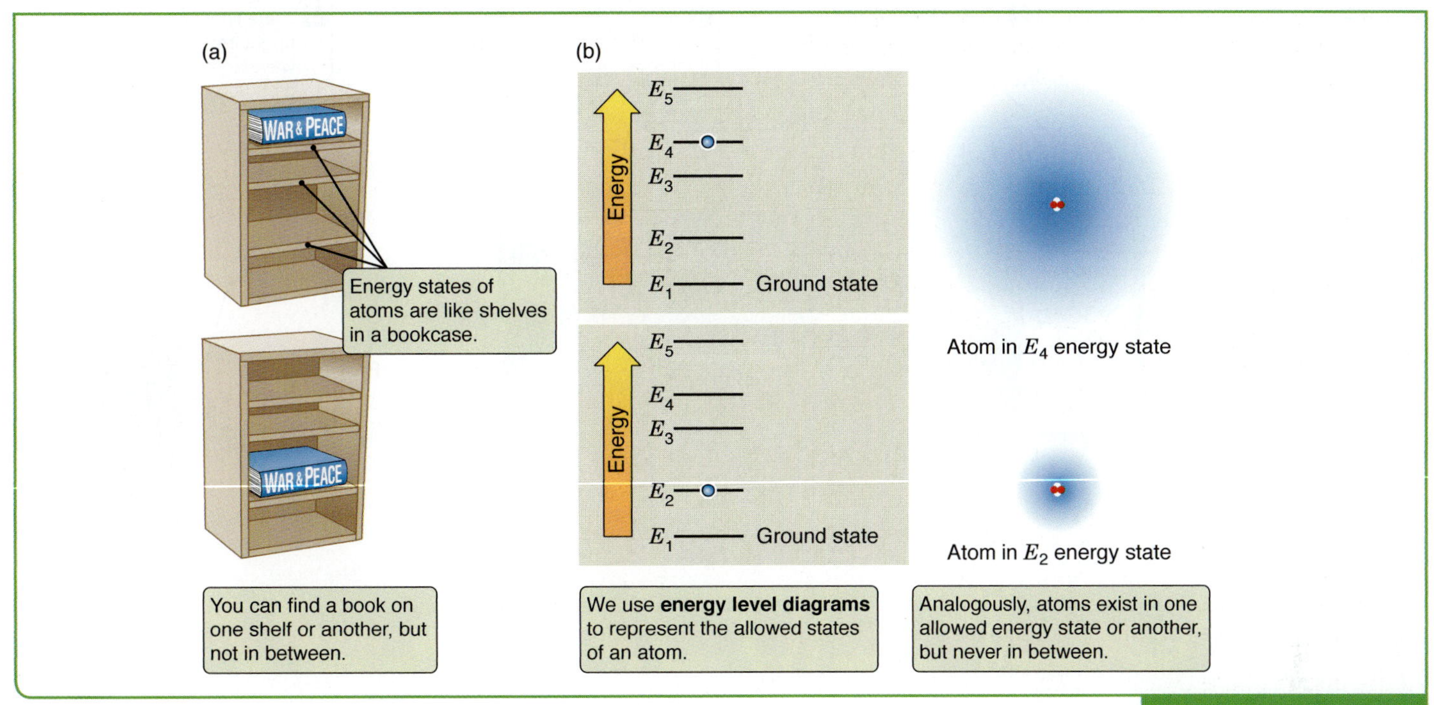

FIGURE 5.11 (a) Energy states of an atom are analogous to shelves in a bookcase. You can move a book from one shelf to another, but books can never be placed between shelves. (b) Atoms exist in one allowed energy state or another but never in between. There is no level below the ground state.

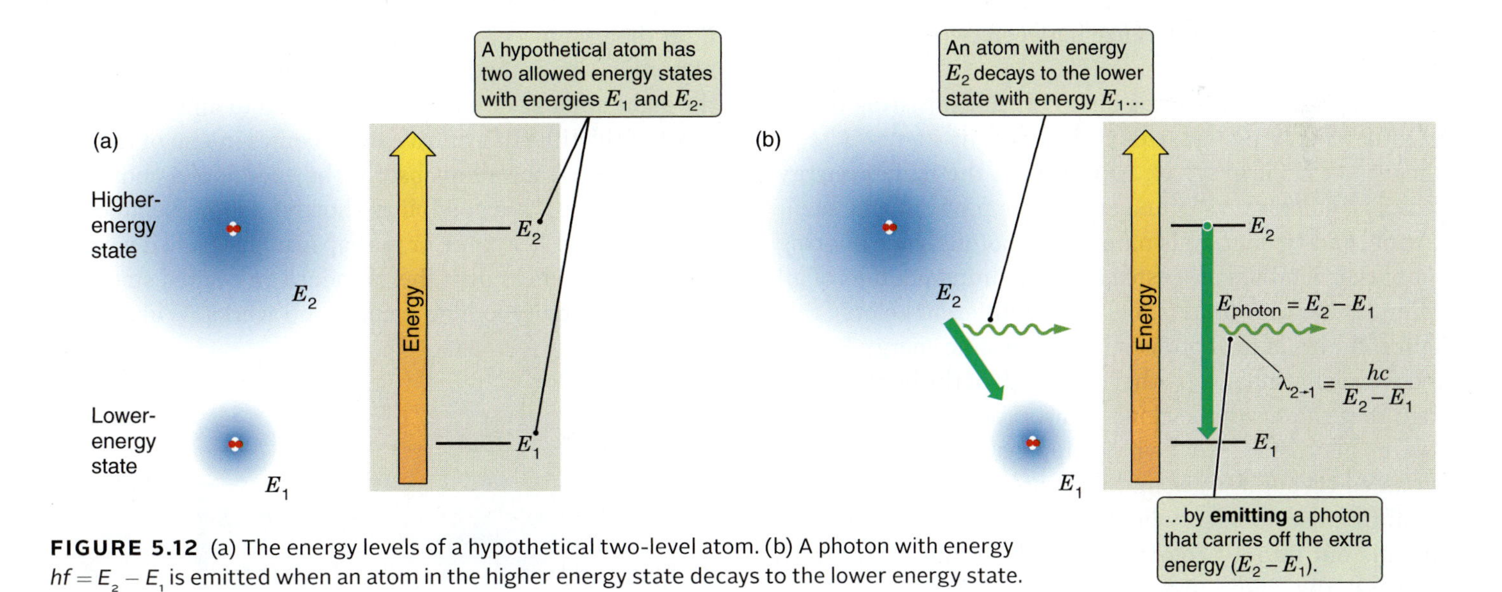

FIGURE 5.12 (a) The energy levels of a hypothetical two-level atom. (b) A photon with energy $hf = E_2 - E_1$ is emitted when an atom in the higher energy state decays to the lower energy state.

never truly lost or created, the energy lost by the atom has to show up somewhere. In this case the energy shows up in the form of a photon that is emitted by the atom. The energy of the photon emitted exactly matches the energy lost by the atom; that is, $E_{\text{photon}} = E_2 - E_1$.

Recall that the energy of a photon is related to the frequency or wavelength of electromagnetic radiation. Using this relationship, we can say that the frequency of the photon emitted by a transition from E_2 to E_1, which we will denote as $f_{2\to1}$, is the energy difference divided by Planck's constant (h):

$$f_{2\to1} = \frac{E_{\text{photon}}}{h} = \frac{E_2 - E_1}{h}$$

Similarly, the wavelength of the photon is $\lambda = c/f$, or:

$$\lambda_{2\to1} = \frac{c}{f_{2\to1}} = \frac{hc}{E_2 - E_1}$$

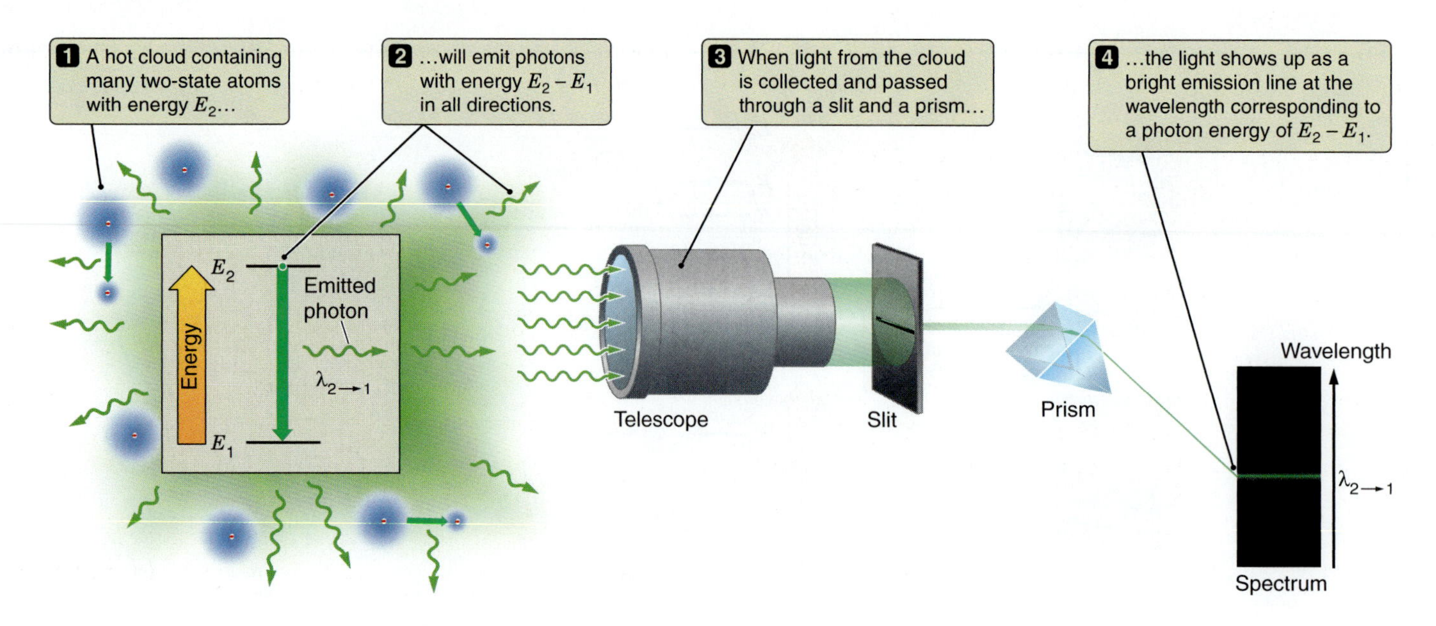

FIGURE 5.13 A cloud of gas containing atoms with two energy states, E_1 and E_2, emits photons with an energy $E = hf = E_2 - E_1$, which appear in the spectrum (right) as a single *emission line*.

The energy level structure of the atom in this way determines the wavelengths of the photons emitted by an atom (which in turn dictate the color of the light that the atom gives off). An atom can emit photons with energies corresponding only to the difference between two of its allowed energy states.

The light coming from a cloud of gas consisting of the hypothetical two-state atoms depicted in Figure 5.12 is illustrated in **Figure 5.13**. Any atom in the higher energy state (E_2) quickly decays and emits a photon in a random direction, and an enormous number of photons come pouring out of the cloud of gas. Instead of containing photons of all different energies (that is, light of all different colors), this light contains only photons with the specific energy $E_2 - E_1$ and wavelength $\lambda_{2\to1}$. In other words, all of the light coming from these atoms in the cloud is the same color.

You may have seen what happens to sunlight when it passes through a prism. Sunlight contains photons of all different colors, so when it passes through a prism it spreads out into all colors of a rainbow. But if you were to pass the light from your hypothetical cloud of gas through a slit and a prism, as in Figure 5.13, the results would be very different. This time there would be no rainbow. Instead, all of the light from the cloud of gas would show up on the screen as a single bright line. The process just described—the production of a photon when an atom decays to a lower energy state—is referred to as **emission**. The bright, single-colored feature in the spectrum of the cloud of gas is an **emission line**.

The spectrum of a cloud of glowing gas contains emission lines.

Until now in this discussion, we have ignored an important question: How did the atom get to be in the excited state E_2 in the first place? An atom sitting in its ground state will remain in the ground state unless it is somehow given just the right amount of energy to kick it up to an excited state. Most of the time this extra energy comes in one of two possible forms: (1) the atom absorbs the energy of a photon (we will talk about this possibility shortly); or (2) the atom collides with another atom, or perhaps an unattached electron, and the collision knocks the atom into an excited state. This second possibility is how a neon sign works. When a neon

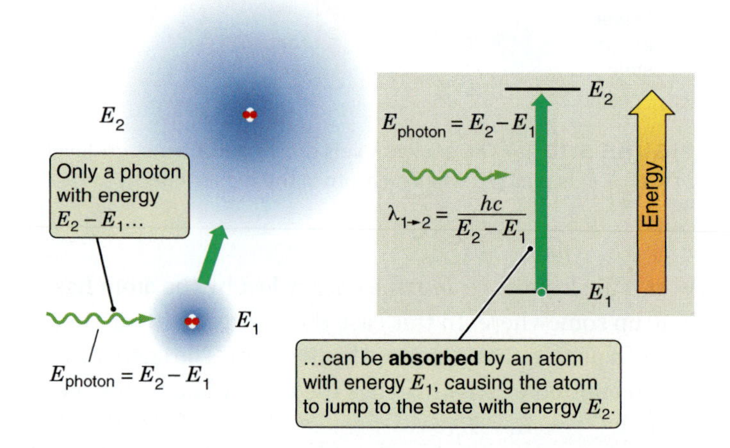

FIGURE 5.14 An atom in a lower energy state may absorb a photon of energy $hf = E_2 - E_1$, leaving the atom in a higher energy state.

sign is turned on, an alternating electric field inside the glass tube pushes electrons in the gas back and forth through the neon gas inside the tube. Some of these electrons crash into atoms of the gas, knocking them into excited states. The atoms then drop back down to their ground states by emitting photons, causing the gas inside the tube to glow.

So far, we have focused on the emission of photons by atoms in an excited state, but what about the opposite process? An atom in a low energy state can absorb the energy of a passing photon and jump up to a higher energy state as shown in **Figure 5.14**, but not just any photon can be absorbed by the atom. Once again, the energy required to go from E_1 to E_2 is the difference in energy between the two states, or $E_2 - E_1$. For a photon to cause an atom to jump from E_1 to E_2, it must provide exactly this much energy. The relationship $E_{\text{photon}} = hf$, or $f = E_{\text{photon}}/h$, shows that the only photons capable of exciting atoms from E_1 to E_2 are photons whose frequency and wavelength are, respectively:

$$f_{1\to2} = \frac{E_{\text{photon}}}{h} = \frac{E_2 - E_1}{h}$$

and

$$\lambda_{1\to2} = \frac{c}{f_{1\to2}} = \frac{hc}{E_2 - E_1}$$

These photons have exactly the same energy—the same color of light—emitted by the atoms when they decay from E_2 to E_1. This is not a coincidence. The energy difference between the two levels is the same whether the atom is emitting a photon or absorbing one, so the energy of the photon involved will be the same in either case.

When you shine light from a lightbulb directly though a glass prism, a rainbow of colors comes out (**Figure 5.15a**).

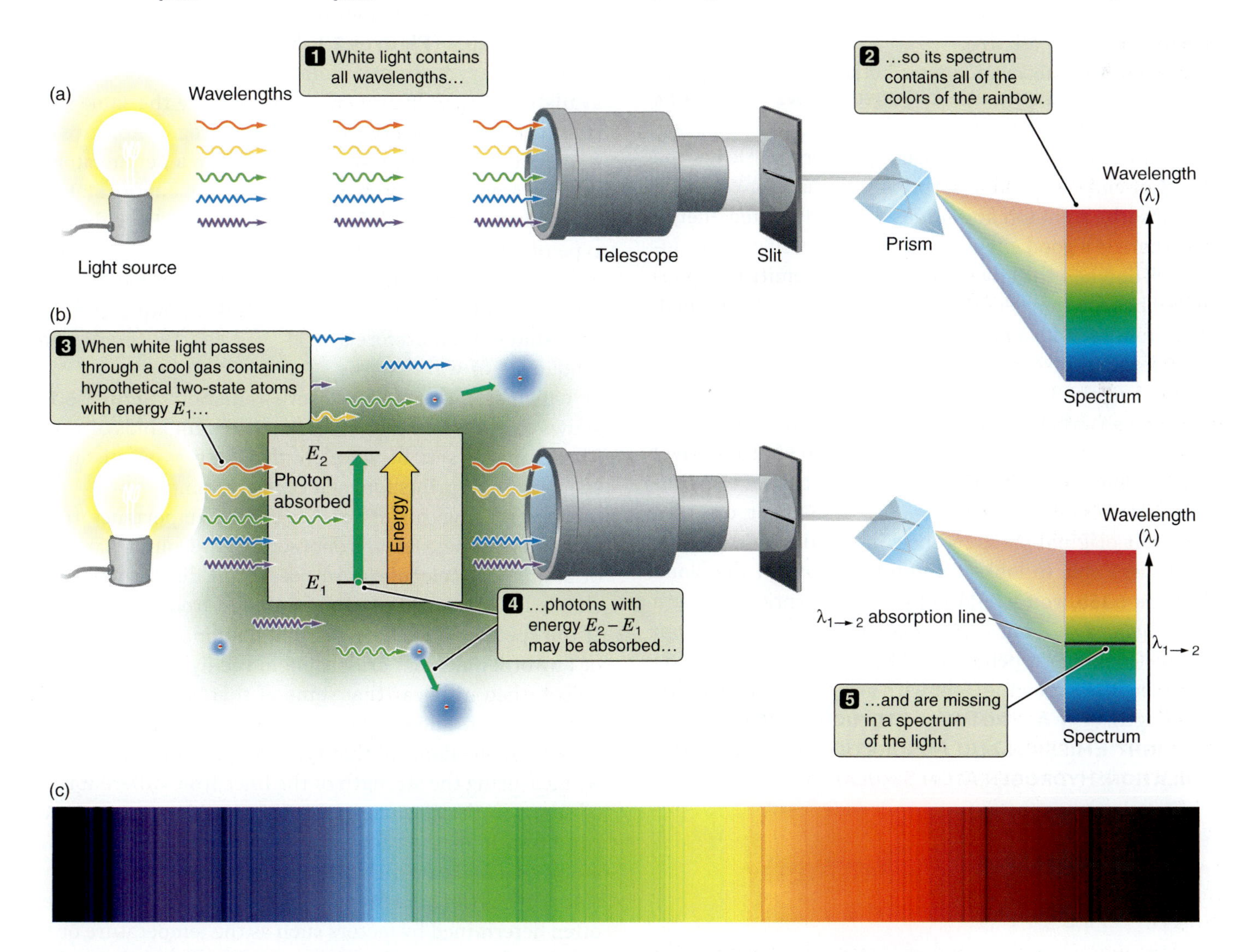

FIGURE 5.15 (a) When passed through a prism, white light produces a spectrum containing all colors. (b) When light of all colors passes through a cloud of hypothetical two-state atoms, photons with energy $hf = E_2 - E_1$ may be absorbed, leading to the dark absorption line in the spectrum. (c) Absorption lines in the spectrum of a star.

What does the spectrum of light look like when viewed through a cloud composed of our hypothetical gas of two-state atoms? If you shine photons of all different wavelengths (that is, light of all different colors) through the gas from one side, almost all of these photons will pass through the cloud of gas unscathed. There is only one exception. Rather than passing through the gas, some of the photons with just the right energy ($E_2 - E_1$) instead are absorbed by atoms. As a result, these photons are missing in the light passing through the prism. Where the color corresponding to each of these missing photons should be, there is a sharp, dark line instead (**Figure 5.15b**). The process by which atoms capture the energy of passing photons is referred to as **absorption**, and the dark feature seen in the spectrum is called an **absorption line**. **Figure 5.15c** shows such absorption lines in the spectrum of a star. ▶▶ NEBRASKA SIMULATION: THREE VIEWS SPECTRUM DEMONSTRATOR

Viewed through a cloud of gas, the spectrum of a lightbulb contains absorption lines.

One final point is worth making before we leave the subject of emission and absorption of radiation. When an atom absorbs a photon and jumps up to an excited energy state, there is a good chance that the atom will quickly decay back down to the lower energy state by emitting a photon that has the same energy as the photon it just absorbed. If the atom reemits a photon just like the one it absorbed, why does the absorption matter? The photon that was taken out of the passing light was replaced, but whereas all of the absorbed photons were originally traveling in the same direction, the photons that are reemitted travel off in *random directions*. In other words, some of the photons with energies equal to $E_2 - E_1$ are, in effect, diverted from their original paths by their interaction with atoms. If you were able to look at a lightbulb *through* the cloud in Figure 5.15b, you would notice an absorption line at a wavelength of $\lambda_{1\rightarrow2}$; but if you could look at the cloud from the side (looking perpendicular to the original beam), you would see it as a glowing light with an emission line at this wavelength. ▶II ASTROTOUR: ATOMIC ENERGY LEVELS AND LIGHT EMISSION AND ABSORPTION ▶▶ NEBRASKA SIMULATION: HYDROGEN ATOM SIMULATOR

Emission and Absorption Lines Are the Spectral Fingerprints of Atoms

In the previous subsection we considered a hypothetical atom with only two allowed energy states. Real atoms have many more than just two possible energy states that they might occupy; therefore, any given type of atom will be capable of emitting and absorbing photons at many different wavelengths. An atom with three energy states, for example, might jump from state 3 to state 2, or from state 3 to state 1, or from state 2 to state 1. The emission lines from a gas made up of these atoms would have wavelengths of $hc/(E_3 - E_2)$, $hc/(E_3 - E_1)$, and $hc/(E_2 - E_1)$, respectively.

The allowed energy states of an atom are determined by the complex quantum mechanical interactions among the electrons and the nucleus. Every hydrogen atom consists of a nucleus containing one proton, plus a single electron in a cloud surrounding the nucleus. Therefore, every hydrogen atom has the same energy states available to it. It follows that all hydrogen atoms are capable of emitting and absorbing photons that have the same wavelengths. **Figure 5.16a** shows the energy level diagram of hydrogen, along with the spectrum of emission lines for hydrogen in the visible part of the spectrum (**Figures 5.16b and c**).

Since every hydrogen atom has the same energy states available to it, all hydrogen atoms produce the same spectral lines. But the energy states of a hydrogen atom are different from the energy states available to a helium atom, a lithium atom, or a boron atom, just as the energy states of these kinds of atoms differ from each other. Each different type of atom (that is, each chemical element) has a unique set of available energy states and therefore a unique set of wavelengths at which it can emit or absorb radiation. **Figure 5.16d** shows the emission spectra of four different kinds of atoms. These unique sets of wavelengths serve as unmistakable spectral fingerprints for each type of atom.

Each chemical element has a unique spectral fingerprint.

Spectral fingerprints are of crucial importance to astronomers. They let astronomers figure out what types of atoms (or molecules) are present in distant objects by simply looking at the spectrum of light from those objects. If the spectral lines of hydrogen, helium, carbon, oxygen, or any other element are visible in the light from a distant object, then it's clear that some of that element is present in that object. The strength of a line is determined in part by how many atoms of that type are present in the source. By measuring the strength of the lines from different types of atoms in the spectrum of a distant object, astronomers can often infer the relative amounts of different types of atoms that make up the object. In additional, the fraction of atoms of a given kind that are in a particular energy state is often determined by factors such as the temperature or the density of the gas. By looking at the relative strength of different lines from the same kind of atom, it is often possible to determine the temperature, density, and pressure of the material as well.

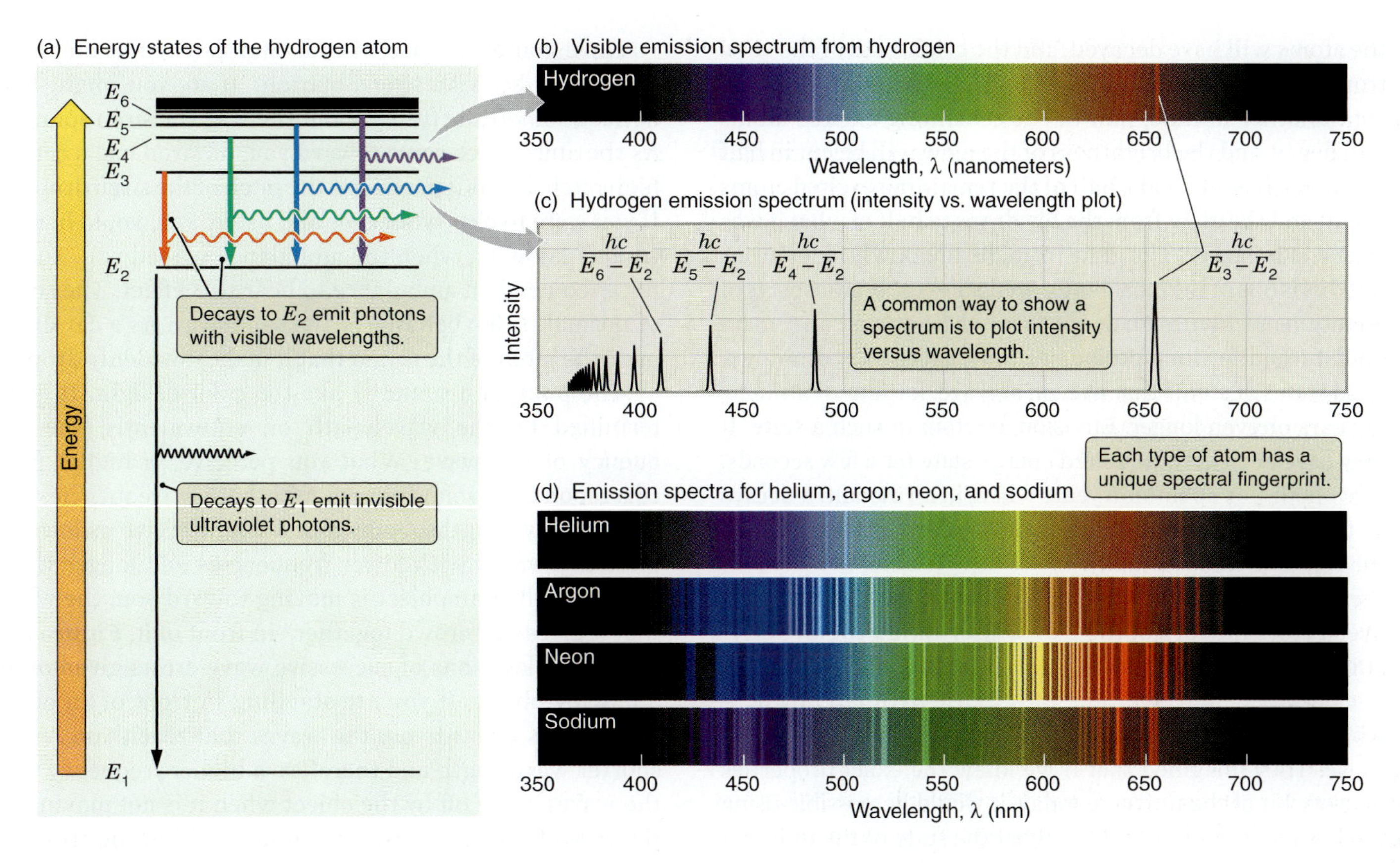

FIGURE 5.16 (a) The energy states of the hydrogen atom. Decays to level E_2 emit photons in the visible part of the spectrum. (b) This spectrum is what you might see if you looked at the light from a hydrogen lamp projected through a prism onto a screen. (c) This graph of the brightness (intensity) of spectral lines versus their wavelength illustrates how spectra are traditionally plotted. (d) Emission spectra from several other gases: helium, argon, neon, and sodium.

How Are Atoms Excited, and Why Do They Decay?

Earlier in this discussion we sidestepped an aspect of the emission process that has troubled physicists and philosophers alike since the earliest days of quantum mechanics. To appreciate this question, return to the analogy between the emission of a photon and a book falling off a shelf. If you place a book on a level shelf and do not disturb it, the book will sit there forever. Once the book is resting on an upper shelf, something must *cause* the book to fall off the shelf. So what about the atom? Once an atom is in an excited state, what causes it to jump down to a lower energy state and emit a photon? Sometimes an atom in a higher energy state can be "tickled" into emitting a photon—a process called *stimulated emission*—but under most circumstances the answer is that nothing causes the atom to jump to the lower energy state. Instead the atom decays *spontaneously*. And while scientists can say approximately how long the atom is likely to remain in the excited state, the rules of quantum mechanics say (and experimentation shows) that exactly when a given atom will decay cannot be known until after the decay has happened. An atom decays at a random time that is not influenced by anything in the universe and cannot be known ahead of time.

You have seen an example of this rather amazing phenomenon if you are familiar with toys that glow in the dark. Photons in sunlight or from a lightbulb are absorbed by certain phosphorescent atoms in the toy, knocking those atoms into excited energy states. Unlike the excited energy states of many atoms that tend to decay in a small fraction of a second, the excited states of the atoms in the toy live for many seconds. Suppose, for example, that on average these atoms tend to remain in their excited state for 1 minute before decaying and emitting a photon. In other words, if we wait 60 seconds, there is a 50-50 chance that any particular atom in the toy will have decayed and a 50-50 chance that the atom will remain in its excited state. There are trillions upon trillions of such atoms in the toy. Although it is impossible to say exactly which atoms will decay after a minute, we can say with certainty that about half of them will decay within 60 seconds. If we wait 1 minute, half of the atoms will have decayed, and the brightness of the glow from the toy will have dropped to half of what it was. If we wait another minute, half of the remaining excited atoms will decay, and the brightness of the glow will be cut in half again. Each 60 seconds, half of the remaining excited atoms decay, and the glow from the toy drops to half of what it was 60 seconds earlier. The glow from the toy slowly fades away.

This is one of the most philosophically troubling aspects of quantum mechanics. In deep space, where atoms can remain undisturbed for long periods of time, there are certain excited states of atoms that live, on average, for tens of millions of years or even longer. Envision an atom in such a state. It may have been in that excited energy state for a few seconds, a few hours, or 50 million years when, in an instant, it decays to the lower energy state *without anything causing it to do so*. Newton and virtually every other physicist who lived before the turn of the 20th century envisioned a clockwork universe in which every effect had a cause. They imagined that if we knew the exact properties of every bit of the universe today, it would be possible using the laws of physics to predict what the state of the universe would be tomorrow. Then, quantum mechanics came along and turned this view on its head. Instead of dealing with strict cause-and-effect relationships, physicists found themselves calculating the *probabilities* that certain events would take place and facing fundamental limitations on what can ever be known about the state of the universe.

Quantum mechanics undermines the orderly, causal universe of Newtonian physics.

Although Einstein helped start the scientific revolution of quantum mechanics, in the end he could never shake his firm belief in Newton's clockwork, causal universe. "God does not play dice with the universe!" he insisted. As more of the predictions of quantum mechanics were borne out by experiment, most physicists came to accept the implications of the strange new theory. Einstein, however, went to his grave looking unsuccessfully for a way to save his notion of order in the universe.

5.4 The Doppler Effect—Motion Toward or Away from Us

To an astronomer, light is a tightly packed bundle of information that, when spread into its component wavelengths, can reveal a wealth of information about the physical state of material located tremendous distances away. Light can, for example, be used to measure one of the most straightforward questions about a distant astronomical object: is it moving away from us or toward us, and at what speed?

Have you ever stood outside and listened as an ambulance sped by with sirens blaring? If so, you might have noticed something funny about the way the siren sounded. As the ambulance came toward you, its siren had a certain high pitch; but as it passed by, the pitch of the siren dropped. If you were to close your eyes and listen, you would have no trouble knowing when the ambulance passed you. You do not even need an ambulance to hear this effect. The sound of normal traffic behaves in the same way. As a car drives past, the pitch of the sound that it makes suddenly drops.

The pitch of a sound is like the color of light. It is determined by the wavelength or, equivalently, the frequency of the wave. What you perceive as higher pitch corresponds to sound waves with higher frequencies and shorter wavelengths. Sounds that you perceive as lower in pitch are waves with lower frequencies and longer wavelengths. When an object is moving toward you, the waves that it gives off "crowd together" in front of it. **Figure 5.17** shows the locations of successive wave crests given off by a moving object. If you are standing in front of an object as it moves toward you, the waves that reach you have a shorter wavelength and therefore a higher frequency than the waves given off by the object when it is not moving. In the case of sound waves, the sound reaching you from the object has a higher pitch than the sounds that are given off by the object when it is stationary. Conversely, if an object is moving away from you, the waves reaching you from the object are spread out. In the case of sound, this means that the pitch of the sound drops, in accordance with your experience with the ambulance. This phenomenon

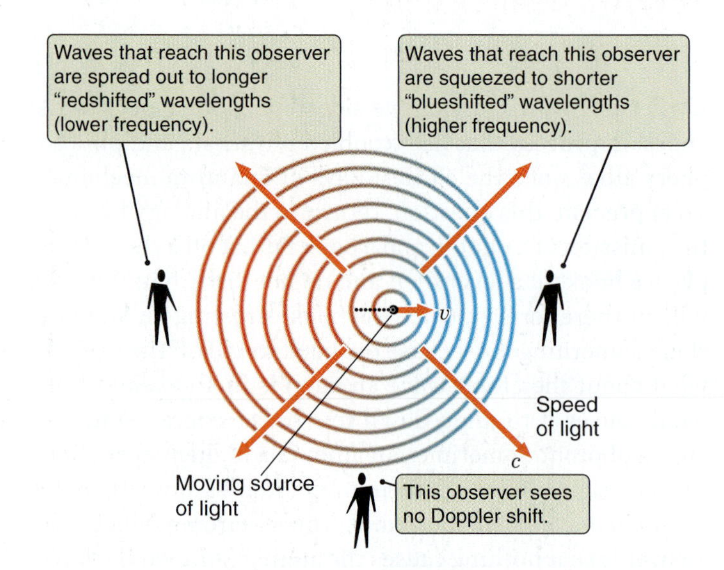

FIGURE 5.17 Motion of a light or sound source relative to an observer may cause waves to be spread out (*redshifted*, or lower in pitch) or squeezed together (*blueshifted*, or higher in pitch). A change in the wavelength of light or the frequency of sound is called a *Doppler shift*.

is referred to as the **Doppler effect**, named after physicist Christian Doppler (1803–1853). ASTROTOUR: THE DOPPLER EFFECT NEBRASKA SIMULATION: DOPPLER SHIFT DEMONSTRATOR

The Doppler effect occurs with light as well as sound. The wavelength of the light as measured from the source's frame of reference is its **rest wavelength** (λ_{rest}), as shown in **Figure 5.18a**. If an object is moving toward you, the wavelength of the light reaching you from the object is shorter than its rest wavelength. In other words, the light that you see is shifted to shorted wavelengths—or bluer—than if the source were not moving toward you. The light from an object moving toward you is thus described as **blueshifted** (**Figure 5.18b**). In contrast, light from a source that is moving away from you is shifted to longer wavelengths. The light that you see is redder than if the source were not moving away from you, so the light is described as **redshifted** (**Figure 5.18c**). The amount by which the wavelength of light is shifted by the Doppler effect is called the **Doppler shift** of the light.

Light from approaching objects is blueshifted. Light from receding objects is redshifted.

The variable v_r stands for **radial velocity**, which is the rate at which the distance between you and the object is changing. As long as the speed of an object is much less than the speed of light, the observed wavelength of the Doppler-shifted light, λ_{obs}, is given by the following equation:

$$\lambda_{obs} = \left(1 + \frac{v_r}{c}\right)\lambda_{rest}$$

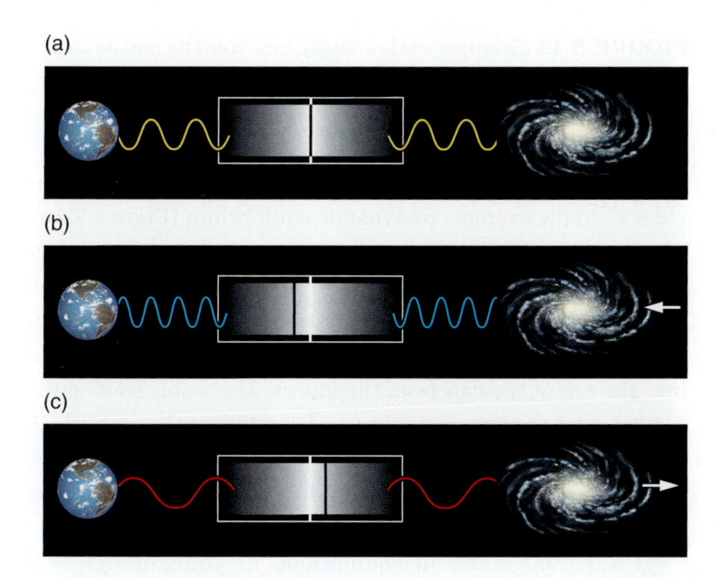

FIGURE 5.18 From their rest wavelength (a), spectral lines of astronomical objects are blueshifted if they are moving toward the observer (b) and redshifted if they are moving away from the observer (c).

In this expression, λ_{rest} is the rest wavelength of the light. The velocity v_r is the velocity at which the object is moving relative to you. To be more precise, v_r is the rate at which the distance between you and the object is changing. If v_r is positive, the object is getting farther away from you. If v_r is negative, the object is getting closer.

Take a moment to be sure this expression makes sense to you. If an object is moving away from you, the wavelength of the light coming to you from that object is greater than it would be if the object were at rest (not moving relative to you). If the object is moving away, v_r is greater than 0. If v_r is greater than 0, then $1 + v_r/c$ is greater than 1, and therefore $\lambda_{obs} = (1 + v_r/c)\lambda_{rest}$ is larger than λ_{rest}, so you see a redshift. What about the opposite case? If an object is moving toward you, v_r is less than 0, so $(1 + v_r/c)$ is less than 1. In this case, λ_{obs} is shorter than λ_{rest}. Because the observed wavelength is shorter than the rest wavelength, you see a blueshift.

The Doppler shift provides information only about whether an object is moving toward you or away from you. That is what the subscript *r* in v_r signifies. At the moment the ambulance is passing you, it is getting neither closer nor farther away, so the pitch you hear is the same as the pitch heard by the crew riding on the truck. You can see this directly by looking at Figure 5.17 or by referring to the equation giving the wavelength for the Doppler-shifted source. If an object is moving perpendicular to your line of sight, then $v_r = 0$ and $\lambda_{obs} = \lambda_{rest}$. The observed wavelength equals the rest wavelength. The light is neither redshifted nor blueshifted.

Radial velocity is measured from Doppler shifts of emission or absorption lines.

Doppler shifts become especially useful when you are looking at an object that has emission or absorption lines in its spectrum (see **Math Tools 5.2**). These spectral lines enable astronomers to determine how rapidly the object is moving toward or away from Earth. To determine the speed, astronomers first identify the spectral line as being from a certain chemical element, which has a unique rest wavelength (λ_{rest}) measured in a lab on Earth. They then measure the wavelength (λ_{obs}) of the spectrum of the distant object. The difference between the rest wavelength and the observed wavelength determines the object's radial velocity. Turning the preceding expression around a bit gives this modified equation:

$$v_r = \frac{\lambda_{obs} - \lambda_{rest}}{\lambda_{rest}} \times c$$

If you know λ_{rest}, just measure λ_{obs}, plug both values into this expression, and you will know what v_r is.

Math Tools 5.2

Making Use of the Doppler Effect

A prominent spectral line of hydrogen atoms has a rest wavelength, λ_{rest}, of 656.3 nm (see Figure 5.16b). Suppose that, using a telescope, you measure the wavelength of this line in the spectrum of a distant object and find that, instead of seeing the line at 656.3 nm, you see the line at a wavelength, λ_{obs}, of 659.0 nm. You can then infer that the object is moving at a velocity of:

$$v_r = \frac{\lambda_{obs} - \lambda_{rest}}{\lambda_{rest}} \times c = \frac{659.0 \text{ nm} - 656.3 \text{ nm}}{656.3 \text{ nm}} \times (3 \times 10^8 \text{ m/s})$$

$$= 1.2 \times 10^6 \text{ m/s}$$

In this way you determine that the object is moving away from you with a radial velocity of 1.2×10^6 m/s, or 1,200 km/s.

Consider the case of Earth's nearest stellar neighbor, Alpha Centauri, which is moving toward us at a radial velocity of −21.6 km/s (-2.16×10^4 m/s.) What is the observed wavelength, λ_{obs}, of a magnesium line in Alpha Centauri's spectrum that has a rest wavelength, λ_{rest}, of 517.27 nm? Using the first equation of Section 5.4 (see page 135) gives the answer:

$$\lambda_{obs} = \left(1 + \frac{v_r}{c}\right)\lambda_{rest} = \left(1 + \frac{-2.16 \times 10^4 \text{ m/s}}{3 \times 10^8 \text{ m/s}}\right) \times 517.27 \text{ nm} = 517.23 \text{ nm}$$

Although the observed Doppler blueshift ($\lambda_{obs} - \lambda_{rest}$ = 517.23 − 517.27) in this case is only −0.04 nm, it is easily measured with modern instrumentation.

Connections 5.1

Equilibrium Means Balance

Equilibrium is the term used to refer to systems that are in balance. Imagine two well-matched teams struggling in a tug-of-war contest. Each team pulls steadfastly on the rope. However, the force of one team's pull is only enough to match, not overcome, the force exerted by the other team. Muscles flex, but the scene does not change. A picture taken now and another taken 5 minutes from now would not differ in any significant way. In this book you will frequently encounter this kind of **static equilibrium**, in which opposing forces just balance each other. The equilibrium between the downward force of gravity and the pressure that opposes it will play a central role in the stories of planetary interiors, planetary atmospheres, and the interiors of stars. Static equilibrium can be stable, unstable, or neutral (**Figure 5.19**). Consider a book standing on its edge, unsupported on either side. If you nudged the book, it would fall over rather than settling back into its original position. This is an example of an **unstable equilibrium** (see Figure 5.19b). When an unstable equilibrium is disturbed, it moves farther away from equilibrium rather than back toward it.

Equilibrium can also be dynamic, in which the system is constantly changing. In **dynamic equilibrium**, one source of change is exactly balanced by another source of change, so the configuration of the system remains the same. Placing a can

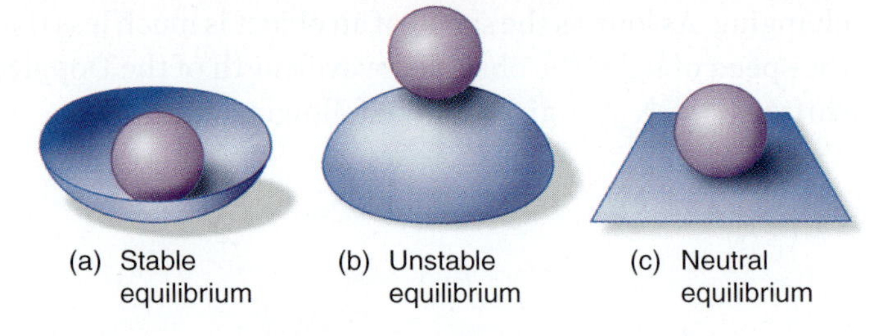

FIGURE 5.19 Examples of stable (a), unstable (b), and neutral (c) equilibrium. Imagine what would happen to the ball if, in each case, you gave it a small nudge.

with a hole cut in the bottom under an open water faucet provides a simple example of dynamic equilibrium (**Figure 5.20**). The depth of the water in the can determines how fast water pours out through the hole in the bottom of the can. When the water reaches just the right depth (**Figure 5.20a**), water pours out of the bottom of the can at exactly the same rate it pours into the top of the can from the faucet. The water leaving the can balances the water entering, and equilibrium is established. If you took a picture now and another picture in a few minutes, little of the water in the can would be the same. Even so, the pictures would be indistinguishable.

If a system is not in equilibrium, its configuration will change. If the level of the water in the can is too low (**Figure 5.20b**), water will not flow out of the bottom of the can fast

5.5 Light and Temperature

Spectral lines also enable astronomers to determine a distant object's temperature. This will be important when we talk about the temperatures of planets and stars. Two things determine the temperature of any object: what is trying to heat up the object and what is trying to cool it down. If an object's temperature is constant, then these two must be in balance with each other.

Your body, for instance, is heated by the release of chemical energy from inside. It is also sometimes heated by energy from your surroundings. If you are standing in sunshine on a hot day, the hot air around you and the sunlight falling on you both are working to heat you. In response to this heating, your body must have some way to cool itself off. When you perspire, water seeps from the pores in your skin and evaporates. It takes energy to evaporate water, and much of this energy comes from your body. Thus, as the perspiration evaporates, it carries away your body's **thermal energy**, cooling your body down. For your body temperature to remain stable, the heating must be balanced by the cooling. Such balance is referred to as **thermal equilibrium**. If your body is out of thermal equilibrium because there is more heating than cooling, then your body temperature climbs. If your body is out of thermal equilibrium because there is more cooling than heating, then your body temperature drops. The basic properties of equilibrium are described in **Connections 5.1**.

If an object's temperature is constant, heating and cooling must be in balance.

Planets also have a thermal equilibrium, and electromagnetic radiation plays a crucial role in maintaining that **equilibrium**. The energy from sunlight heats the surface of a planet, driving its temperature up. This is one side of the equilibrium. The other side is also

A planet is heated by sunlight and cooled by the radiation of energy back into space.

enough to balance the water flowing in, and the water level will begin to rise. A picture taken now and another taken a short time later would not look the same. Conversely, if the water level in the can is too high (**Figure 5.20c**), water will flow out of the can faster than it flows into the can, and the water level will begin to fall. Once again, if the system is not in equilibrium, its configuration will change.

Water passing through a can is an example of a **stable equilibrium** (see Figure 5.19a). When a stable equilibrium is disturbed (forced away from its equilibrium configuration), it will tend to return to its equilibrium state. If the water level is too high or too low, it will move back toward its equilibrium level.

The equilibrium discussed in this chapter—between sunlight falling on a planet and thermal energy radiated away into space—is a stable equilibrium. If you take the example illustrated in Figure 5.20 and imagine sunlight being absorbed instead of water flowing in, energy being radiated by the planet instead of water flowing out, and the planet's temperature as equivalent to the water level in the can, then the stable equilibrium that sets the level of the water in the can becomes the stable equilibrium that sets the temperature of a planet.

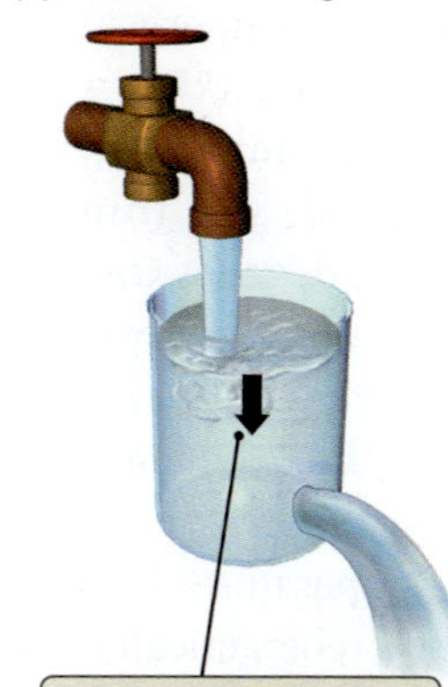

FIGURE 5.20 Water flowing into and out of a can determines the water level in the can. This is an example of dynamic equilibrium.

controlled by light energy: the planet radiates energy back into space, thereby cooling itself. Your eyes are not sensitive to the infrared wavelengths of light that planets radiate, but that light is emitted nonetheless. Overall, a planet must radiate away just as much energy into space as it absorbs from the Sun. If there were more heating than cooling, the temperature of a planet would climb. If there were more cooling than heating, the temperature would fall. For a planet to remain at the same average temperature over time, the energy it radiates into space must exactly balance the energy it absorbs from the Sun. Thermal equilibrium must be maintained. Equilibrium is an important concept in science. There are many kinds of equilibrium besides thermal equilibrium, some of which we will encounter later in the book.

To turn this idea of thermal equilibrium into a real prediction for the temperatures of the planets, you will need to know more about light and temperature and the relationship between the two. A good start is to improve your understanding of what is meant by *temperature*.

Temperature Is a Measure of How Energetically Particles Are Moving

In everyday life, *hot* and *cold* are defined in terms of our subjective experiences. Something is hot when it *feels* hot or cold when it *feels* cold. When we talk about measuring **temperature**, we speak of degrees on a thermometer, but the way we define a degree is arbitrary. If you grew up in the United States, for example, you probably think of temperatures in degrees **Fahrenheit** (**°F**), whereas if you grew up almost anywhere else in the world, you think of temperatures in degrees **Celsius** (**°C**). Both of these are reasonable scales for measuring temperatures. But what does the thermometer actually measure?

Temperature is a measurement of how energetically the atoms that make up an object are moving about. The air around us is composed of vast numbers of atoms and molecules. Those molecules are moving about every which way. Some move slowly; some move more rapidly. Similarly, the atoms that make up the chair that you are sitting in or the floor that you are walking on are constantly in motion. We can characterize these motions by talking about the average **kinetic energy** (E_K), the energy of motion:

$$E_K = \frac{1}{2}mv^2$$

where m is the mass of a particle and v is its velocity.

The more energetically the atoms or molecules in something are bouncing about, the higher is its temperature. In fact, the random motions of atoms and molecules are often referred to as their **thermal motions**, to emphasize the connection between these motions and temperature.

If something is hotter than you are, thermal energy flows from that object into you. At the atomic level, that means the object's atoms are bouncing more energetically than are the atoms in your body, so if you touch the object, its atoms collide with your atoms, causing the atoms in your body to move faster. Your body gets hotter as thermal energy flows from the object to you. (At the same time, these collisions rob the particles in the object of some of their energy. Their motions slow down, and the hotter object becomes cooler.) When physicists talk about *heating*, they mean processes that increase the average thermal energy of an object's particles; and when they talk about *cooling*, they mean processes that decrease the average thermal energy of those particles (**Figure 5.21**).

The change in thermal energy associated with a change of one unit, or degree, is arbitrary on any temperature scale. On the Fahrenheit scale there are 180 degrees between the freezing point (32°F) and the boiling point (212°F) of wa-

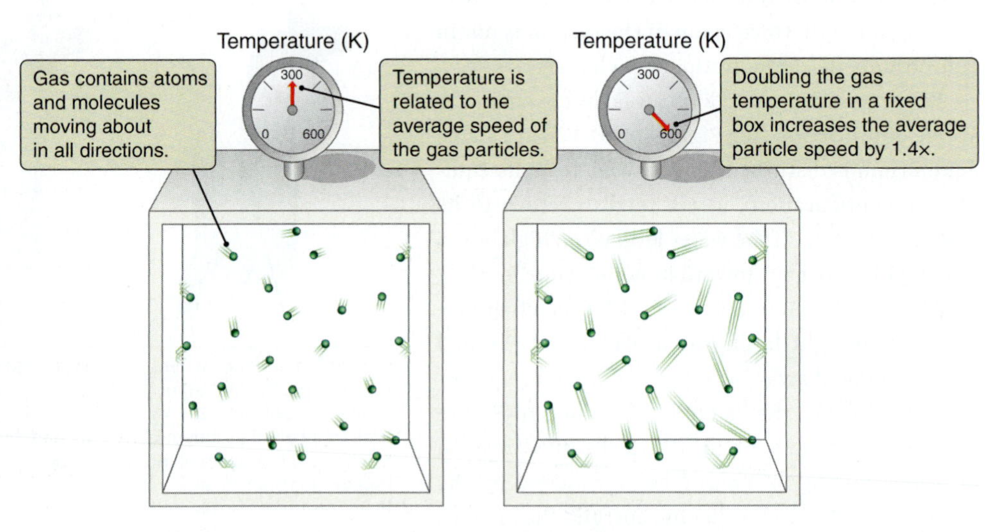

FIGURE 5.21 Hotter gas temperatures correspond to faster motions of atoms. K stands for *kelvin*, the unit in the Kelvin scale of temperature, which is commonly used in scientific investigations.

ter at sea level. On the Celsius scale there are 100 degrees between those two temperatures. For these two scales, the temperature corresponding to zero degrees is also arbitrary. On the Celsius scale, 0°C was chosen to be the temperature at which water freezes; on the Fahrenheit scale, 0°F corresponds to −17.78°C. However, there is a lowest possible physical temperature below which no object can fall. As the motions of the atoms in an object slow down, the temperature drops lower and lower. When the motions of the particles finally stop, things have gotten as cold as they can get. This lowest possible temperature, where thermal motions have come to a standstill, is called **absolute zero**. Absolute zero corresponds to −273.15°C, or −459.57°F.

The preferred temperature scale for most scientists is the **Kelvin scale**. For convenience, the size of one unit on the Kelvin scale, called a **kelvin** (abbreviated **K**) is the same as the Celsius degree. What makes the Kelvin scale special is that 0 K is set equal to that absolute lowest temperature where thermal motions stop—absolute zero. There are no negative temperatures on the Kelvin scale. *When temperatures are measured in kelvins, you know that the average thermal energy of particles is proportional to the measured temperature.* The average thermal energy of the atoms in an object with a temperature of 200 K is twice the average thermal energy of the atoms in an object with a temperature of 100 K.

Absolute zero, the temperature at which thermal motions stop, is zero on the Kelvin scale.

Hotter Means More Luminous and Bluer

So far, this discussion has focused on the way discrete atoms emit and absorb radiation, leading to a useful understanding of emission lines and absorption lines and what these lines reveal about the physical state and motion of distant objects. But not all objects have spectra that are dominated by discrete spectral lines. For instance, if you pass the light from an incandescent lightbulb through a prism, as in Figure 5.15a, instead of discrete bright and dark bands you will see light spread out smoothly from the blue end of the spectrum to the red. Similarly, if you look closely at the spectrum of the Sun, you will see absorption lines, but mostly you will see light smoothly spread out across all colors of the spectrum—red through violet. What is the origin of such **continuous radiation**, and what clues does this kind of radiation carry about the objects that emit it?

We can think of a dense material as being composed of a collection of charged particles that are being jostled as their thermal motions cause them to run into their neighbors. The hotter the material is, the more violently its particles are being jostled. Recall that anytime a charged particle is subjected to an acceleration, it radiates. So the jostling of particles due to their thermal motions causes them to give off a continuous spectrum of electromagnetic radiation. This is why any material that is sufficiently dense for its atoms to be jostled by their neighbors emits light *simply because of its temperature*. Radiation of this sort is called **thermal radiation**.

Objects like incandescent lightbulbs emit continuous radiation at all wavelengths.

You can think about how the radiation from an object changes as the object heats up or cools down. To a physicist or an astronomer, **luminosity** is the amount of light *leaving* a source—that is, the total amount of light emitted (energy per second, measured in watts, W). The hotter the object, the more energetically the charged particles within it move, and the more energy they emit in the form of electromagnetic radiation. So as an object gets hotter, the light that it emits becomes more intense. Here is our first point about thermal radiation: *hotter means more luminous*.

Now let's move to the question of what color light an object emits. As the object gets hotter, the thermal motions of its particles become more energetic. These more energetic motions are capable of producing more energetic photons. So as an object gets hotter, the average energy of the photons that it emits increases. In other words, the average wavelength of the emitted photons gets shorter, and the light from the object gets bluer. Here is our second point about thermal radiation: *hotter means bluer*.

Making an object hotter also makes its thermal radiation bluer and more luminous.

You might have used a Bunsen burner in a chemistry class. As you heated up a piece of metal, you saw the metal glow—first a dull red, then orange, then yellow. The hotter the metal became, the more the highly energetic blue photons became mixed with the less energetic red photons, and the whiter and more intense the light became. The color of the light shifted from red toward blue, confirming our second point: hotter means bluer.

Around the year 1900, Max Planck was thinking about objects called **blackbodies** that emit electromagnetic radiation only because of their temperature, not their composition. Blackbodies

A blackbody emits thermal radiation that has a Planck spectrum.

emit just as much thermal radiation as they absorb from their surroundings. Planck graphed the intensity of the emitted radiation across all wavelengths and obtained the characteristic curves that we now call **Planck spectra** or **blackbody spectra**. **Figure 5.22** shows Planck spectra for objects at several different temperatures.

In the real world, light sources are not perfect blackbody radiators, but they can come close to having blackbody spectra. Most of the light emitted by charged particles within the filament of an incandescent lightbulb is absorbed by other charged particles within the filament. This is the same assumption that Planck made when calculating the shape of a blackbody spectrum, so the spectrum of radiation from the filament of an incandescent lightbulb is very close to a Planck spectrum. More importantly for astronomy, the light from stars such as the Sun and the thermal radiation from a planet also often come close to having blackbody spectra.

The Stefan-Boltzmann Law Says That Hotter Means Much More Luminous

As the temperature of an object increases, Planck's theory says that the object gives off more radiation at every wavelength, so the luminosity of the object should increase.

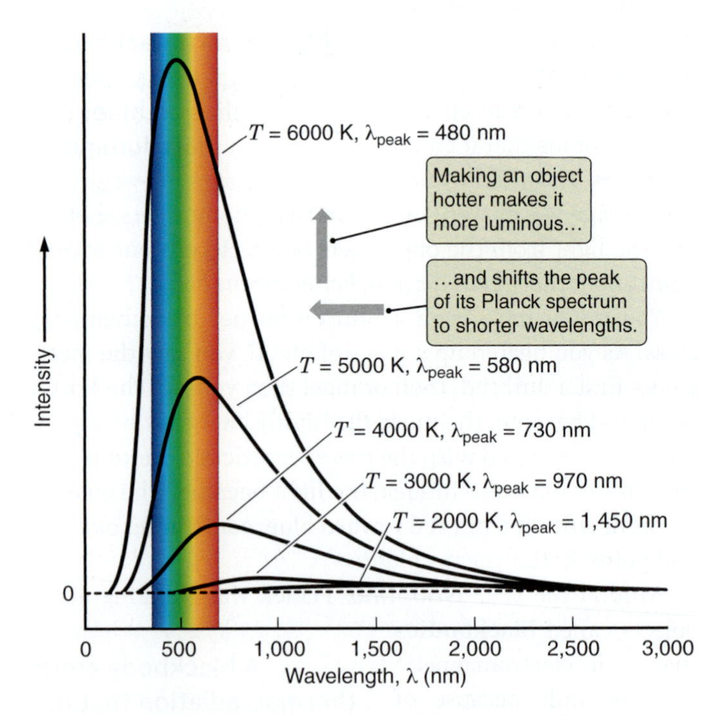

FIGURE 5.22 Planck spectra emitted by sources with temperatures of 2000 K, 3000 K, 4000 K, 5000 K, and 6000 K. At higher temperatures the peak of the spectrum shifts toward shorter wavelengths, and the amount of energy radiated per second from each square meter of the source increases.

Adding up all of the energy in a Planck spectrum shows that the increase in luminosity is proportional to the fourth power of the temperature: Luminosity $\propto T^4$. This result is known as the **Stefan-Boltzmann law** because it was discovered in the laboratory by physicist Josef Stefan (1835–1893) and derived by his student Ludwig Boltzmann (1844–1906) before Planck's theory came along to explain it.

The luminosity of a blackbody is proportional to T^4.

The amount of energy radiated by each square meter of the surface of an object each second is called the **flux**, abbreviated $\mathcal{F}$. The flux is proportional to the luminosity, but easier to measure. (Think about trying to catch all the photons emitted by Earth in all directions, versus just the ones from the particular square meter under your feet. You can find the luminosity by multiplying the flux by the total surface area.) The Stefan-Boltzmann law says that the flux is given by the following equation:

$$\mathcal{F} = \sigma T^4$$

The constant σ (the Greek letter sigma), which is called the **Stefan-Boltzmann constant**, equals 5.67×10^{-8} W/(m^{2}K^4), where 1 W = 1 joule per second (J/s).

The Stefan-Boltzmann law says that an object rapidly becomes more luminous as its temperature increases (see **Math Tools 5.3**). If the temperature of an object doubles, the amount of energy being radiated each second increases by a factor of 2^4, or 16. If the temperature of an object goes up by a factor of 3, then the energy being radiated by the object each second goes up by a factor of 3^4, or 81. A lightbulb with a filament temperature of 3000 K radiates 16 times as much light as it would if the filament temperature were 1500 K. Even modest changes in temperature can result in large changes in the amount of luminosity radiated by an object.

Slight changes in temperature mean large changes in luminosity.

Wien's Law Says That Hotter Means Bluer

Look again at Figure 5.22. Notice where the peak of each curve lines up along the horizontal axis. As the temperature, T, increases, the peak of the spectrum shifts toward shorter wavelengths. Photon energy and wavelength are inversely related, so as the wavelength becomes shorter, the average photon energy becomes greater. The object becomes bluer. The shift in the location of the

The peak wavelength of a blackbody is inversely proportional to its temperature.

Planck spectrum's peak wavelength with increasing temperature is given by the following equation:

$$\lambda_{peak} = \frac{2{,}900{,}000 \text{ nm K}}{T}$$

This result is referred to as **Wien's law**, named for German physicist Wilhelm Wien (1864–1928). In this equation, λ_{peak} is the wavelength where the Planck spectrum is at its peak, where the electromagnetic radiation from an object is greatest. Wien's law says that the peak wavelength in the spectrum is inversely proportional to the temperature of the object. If you double the temperature, the peak wavelength becomes half of what it was. If you increase the temperature by a factor of 3, the peak wavelength becomes a third of what it was (see Math Tools 5.3). We will return to these laws later in the chapter when we use them to estimate the temperatures of the planets. ▶▶ **NEBRASKA SIMULATION: BLACKBODY CURVES**

5.6 Light and Distance

We have consistently spoken of the luminosity of objects, but in everyday language we probably would say that one object is "brighter" than another. This is a case where everyday language is too inexact for science. Recall that *luminosity* refers to the amount of light *leaving* a source. By contrast, the **brightness** of electromagnetic radiation is the amount of light that is *arriving* at a particular location, such as the page of the book you are reading or the pupil of your eye. The concept of brightness is certainly related to the concept of luminosity. For example, replacing a lightbulb having a luminosity of 50 W with a 100-W bulb succeeds in making a room twice as bright because it doubles the light reaching any point in the room. But brightness also depends on the distance from a source of electromagnetic radiation. If you needed more light to read this book, you could replace the bulb in your lamp

Math Tools 5.3

Working with the Stefan-Boltzmann and Wien's Laws

The Stefan-Boltzmann law can be used to estimate the flux and luminosity of Earth. Earth's average temperature is 288 K, so the flux from its surface is:

$$\begin{aligned} \mathcal{F} &= \sigma T^4 \\ &= (5.67 \times 10^{-8} \text{ W/m}^2\text{K}^4) \times (288 \text{ K})^4 \\ &= 390 \text{ W/m}^2 \end{aligned}$$

The luminosity is the flux multiplied by the surface area (A) of Earth. Surface area is given by $4\pi R^2$, and the radius of Earth is 6,378,000 meters, or 6.378×10^6 meters. So the luminosity is:

$$\begin{aligned} L &= \mathcal{F} \times A \\ &= \mathcal{F} \times 4\pi R^2 \\ &= (390 \text{ W/m}^2) \times [4\pi (6.378 \times 10^6 \text{ m})^2] \\ &\approx 2 \times 10^{17} \text{ W} \end{aligned}$$

This means that Earth emits the equivalent of 2,000,000,000,000,000 (2 million billion) 100-W lightbulbs. This is an enormous amount of energy, but not anywhere close to the amount emitted by the Sun.

Wien's law also proves useful to astronomers as they study the universe. If they measure the spectrum of an object emitting thermal radiation and find where the peak in the spectrum is, they can use Wien's law to calculate the temperature of the object. For example, astronomers cannot drop a thermometer onto the Sun to *directly* measure its surface temperature, but they can observe the spectrum of the light coming from the Sun and estimate its surface temperature. The peak in the Sun's spectrum occurs at a wavelength of about 500 nm. Wien's law can be written as:

$$T = \frac{2{,}900{,}000 \text{ nm K}}{\lambda_{peak}}$$

Plugging the observed peak of the spectrum of the Sun ($\lambda_{peak} = 500$ nm) into this equation gives:

$$T = \frac{2{,}900{,}000 \text{ nm K}}{500 \text{ nm}} = 5800 \text{ K}$$

This is how you can know the surface temperature of the Sun.

Suppose you want to calculate the peak wavelength at which Earth radiates. Plugging Earth's average temperature of 288 K into Wien's law gives:

$$\lambda_{peak} = \frac{2{,}900{,}000 \text{ nm K}}{288 \text{ K}} = 10{,}100 \text{ nm} = 10.1 \text{ }\mu\text{m}$$

Earth's radiation peaks in the infrared region of the spectrum.

with a more luminous one, but it might be easier just to move the book closer to the light. Conversely, if a light was too bright for you, you could move away from it. Our everyday experience says that as we move away from a light, its brightness decreases.

The particle description of light provides a convenient way to think about the brightness of radiation and how brightness depends on distance. Suppose you had a piece of cardboard that measured 1 meter by 1 meter. To make the light falling on the cardboard twice as bright would mean doubling the number of photons that hit the cardboard each second. Tripling the brightness of the light would mean increasing the number of photons hitting the cardboard each second by a factor of 3, and so on. Brightness depends on the number of photons falling on each square meter of a surface each second.

Brightness measures how much light falls per square meter per second.

Now imagine a lightbulb sitting at the center of a spherical shell (**Figure 5.23**). Photons from the bulb travel in all directions and land on the inside of the shell. To find the number of photons landing on each square meter of the shell during each second (that is, to determine the brightness of the light), take the *total* number of photons given off by the lightbulb each second and divide by the number of square meters over which those photons have to be spread. The surface area of a sphere is given by the formula $A = 4\pi R^2$, where R is the distance between the bulb and the surface of the sphere (that is, R = the radius of the sphere). Combining all these factors reveals that:

$$\text{Number of photons striking one square meter each second} = \frac{\text{Total number of photons emitted per second}}{\text{Number of square meters the photons are spread over}} = \frac{\text{Total number of photons emitted each second}}{4\pi R^2}$$

Now think about what happens if you change the size of the spherical shell while keeping the total number of photons given off by the lightbulb each second the same. As the shell becomes larger, the photons from the lightbulb must spread out to cover a larger surface area. Each square meter of the shell receives fewer photons each second, so the brightness of the light decreases. If the shell's surface is moved twice as far from the light, the area over which the light must spread increases by a factor of $2^2 = 2 \times 2 = 4$. The photons from the bulb spread out over 4 times as much area, so the number of photons falling on each square meter each second becomes ¼ of what it was. If the surface of the sphere is 3 times as far from the light, the area over which the light must spread increases by a factor of $3^2 = 3 \times 3 = 9$, and the number of photons per second falling on each square meter becomes 1/9 of what it was originally. You encountered just this kind of inverse square relationship when we talked about gravity in Chapter 4. The brightness of the light from an object is inversely proportional to the square of the distance from the object. *Twice as far means one-fourth as bright.*

Like gravity, light obeys an inverse square law.

This idea of photons streaming and spreading onto a surface from a light explains why brightness follows an inverse square law. In practice, however, it is usually more convenient to talk about the *energy* coming to a surface each second, rather than the number of photons arriving.

The luminosity of an object is the total number of photons given off by the object multiplied by the energy of each photon. Instead of talking about how the number of photons must spread out to cover the surface of a sphere (brightness), we now talk about how the energy carried by the photons must spread out to cover the surface of a sphere. When speaking of brightness in this way, we mean the amount of energy falling on a square meter in a second. If L is the luminosity of the bulb, then the brightness of the light at a distance r from the bulb is given by this equation:

$$\text{Brightness} = \frac{\text{Energy radiated per second}}{\text{Area over which energy is spread}}$$

$$= L/4\pi R^2$$

Usually, the only information that astronomers have to work with is the light from a distant object. We will return to this concept of luminosity and brightness to determine the distances to stars.

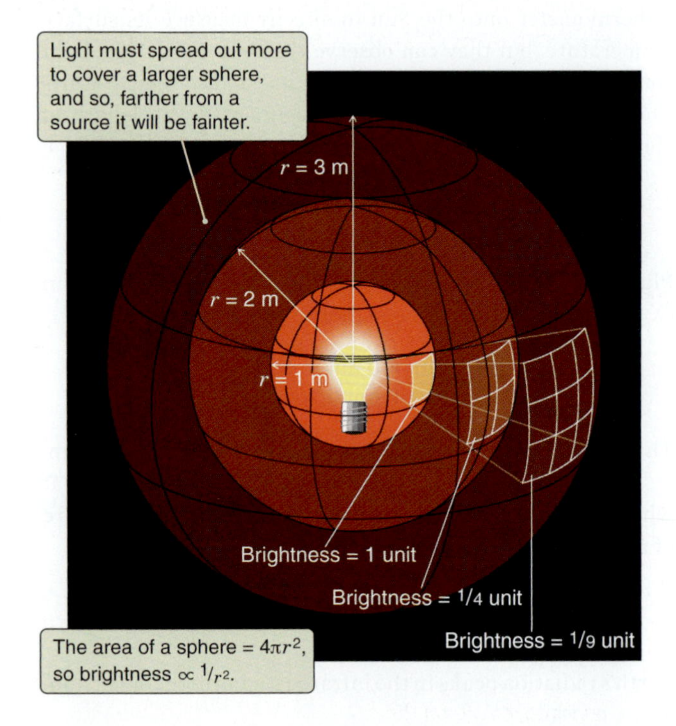

FIGURE 5.23 Light obeys an inverse square law as it spreads away from a source. Twice as far means one-fourth as bright.

5.7 Origins: Temperatures of Planets

In the previous chapters we discussed in general how a planet's axial tilt and its orbital shape can affect its temperature and thus perhaps its prospects for life. Now let's get more specific about the temperatures of planets, using what you learned in this chapter about thermal radiation. For a planet in equilibrium, T is the temperature at which the energy radiated by a planet exactly balances the energy absorbed by the planet. If the planet is hotter than this equilibrium temperature, it will radiate energy away faster than it absorbs sunlight, and its temperature will fall. If the planet is cooler than this temperature, it will radiate away less energy than is falling on it in the form of sunlight, and its temperature will rise. Only at the equilibrium temperature do the two balance.

As the distance from the Sun increases the temperature of the planet decreases, and the temperature should be inversely proportional to the square root of the distance (**Math Tools 5.4**). **Figure 5.24** plots the actual and predicted temperatures of the planets (and the dwarf planet Pluto). Each vertical orange bar shows the range of temperatures found on the surface of the planet (or, in the case of the giant planets, at the top of the planet's clouds). The black dots show the predictions made using the equation in Math Tools 5.4. Overall, the predictions are not too far off, indicating that our basic understanding of *why* planets have the temperatures they have is probably fairly close to the mark. The data for Mercury, Mars, and Pluto agree particularly well.

Balancing cooling and heating sets an equilibrium temperature.

In some cases, however, the predictions are wrong. For Earth and the giant planets, the actual measured temperatures are a bit higher than the predicted temperatures, and for Venus the actual surface temperature is much higher than the prediction. Note that the predicted values are based on the assumption that the temperature of the planet is the same everywhere. This is clearly not true, since planets are likely to be hotter on the day side than on the night side. The predictions also assume that a planet's only source of energy is the sunlight falling on it, and that **albedo** (the fraction of sunlight reflected) is constant for each planet. There is also the assumption that the planets absorb and radiate energy into space as blackbodies, and, as noted earlier, real objects are not perfect blackbody radiators.

The discrepancies between the calculated and the measured temperatures of some of the planets indicate that for these planets, some or all of these assumptions must be incorrect. The question of *why* these planets are hotter than the prediction will lead to a number of new and interesting insights into how these planets work. For example, astronomers will need to consider whether the planet has its own source of energy besides sunlight, and whether the planet has an atmosphere. Understanding the temperatures of planets makes it possible to think about why life may have evolved here on Earth, instead of on a different planet in the Solar System.

In the next chapter you will learn about the many and varied tools that astronomers use to detect the light signals coming from the Solar System and beyond, and to decipher the meaning of those signals. ■

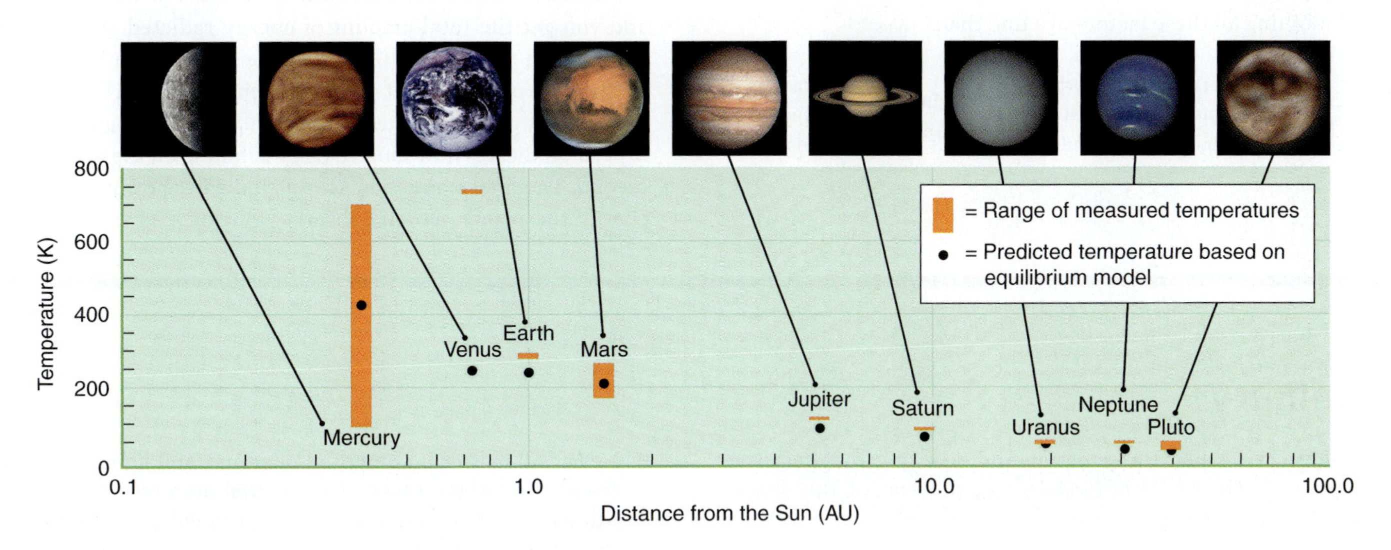

FIGURE 5.24 Predicted temperatures for the planets and the dwarf planet Pluto, based on the equilibrium between absorbed sunlight and thermal radiation into space, are compared with ranges of observed surface temperatures. Some predictions are correct; others are not.

Math Tools 5.4

Using Radiation Laws to Calculate Equilibrium Temperatures of Planets

Why does a planet have the temperature that it has? Qualitatively, the temperature of a planet is determined by a balance between the amount of sunlight being absorbed and the amount of energy being radiated back into space. We now have the tools we need to turn this qualitative idea into a real prediction of the temperatures of the planets.

We begin with the amount of sunlight being absorbed. From the Sun's perspective, a planet looks like a circular disk with a radius equal to the radius of the planet, R_{planet}. The area of the planet that is lit by the Sun is:

$$\text{Absorbing area of planet} = \pi R^2_{\text{planet}}$$

The amount of energy striking a planet also depends on the brightness of sunlight at the distance at which the planet orbits. The brightness of sunlight at a distance d from the Sun is equal to the luminosity of the Sun ($L_\odot$, in watts) divided by $4\pi d^2$, as noted in Section 5.6 (d here is the same as r was in Section 5.6; we use d in this case to avoid confusion with the planet's radius, R_{planet}):

$$\text{Brightness of sunlight} = \frac{L_\odot}{4\pi d^2}$$

We must also consider one additional factor. A planet does not absorb all the sunlight that falls on it. The fraction of the sunlight that is reflected from a planet is called the albedo, a, of the planet. The corresponding fraction of the sunlight that is absorbed by the planet is 1 minus the albedo. A planet covered entirely in snow would have a high albedo, close to 1; a planet covered entirely by black rocks would have a low albedo, close to 0:

$$\text{Fraction of sunlight absorbed} = 1 - a$$

Combining all these factors, we find that:

$$\begin{pmatrix}\text{Energy absorbed}\\ \text{by the planet}\\ \text{each second}\end{pmatrix} = \begin{pmatrix}\text{Absorbing}\\ \text{area of}\\ \text{planet}\end{pmatrix} \times \begin{pmatrix}\text{Brightness}\\ \text{of sunlight}\end{pmatrix} \times \begin{pmatrix}\text{Fraction}\\ \text{of sunlight}\\ \text{absorbed}\end{pmatrix}$$

$$= \pi R^2_{\text{planet}} \times \frac{L_\odot}{4\pi d^2} \times (1 - a)$$

Now let's turn to the other piece of the equilibrium: the amount of energy that the planet radiates away into space each second. We can calculate this amount by multiplying the number of square meters of the planet's total surface area by the energy radiated by each square meter each second. The surface area for the planet is given by $4\pi(R_{\text{planet}})^2$. The Stefan-Boltzmann law tells us that the energy radiated by each square meter each second is given by σT^4. So we can say that:

$$\begin{pmatrix}\text{Energy radiated}\\ \text{by planet each}\\ \text{second}\end{pmatrix} = \begin{pmatrix}\text{Surface}\\ \text{area of}\\ \text{planet}\end{pmatrix} \times \begin{pmatrix}\text{Energy radiated}\\ \text{per square meter}\\ \text{per second}\end{pmatrix}$$

$$= 4\pi R^2 \times \sigma T^4$$

Remember that if the planet's temperature is to remain stable—if it is to keep from heating up or cooling down—it must be radiating away just as much energy into space as it is absorbing in the form of sunlight (**Figure 5.25**). Therefore, we can equate the two preceding expressions by setting the energy radiated equal to the energy absorbed:

$$\begin{pmatrix}\text{Energy radiated by}\\ \text{planet each second}\end{pmatrix} = \begin{pmatrix}\text{Energy absorbed by}\\ \text{the planet each second}\end{pmatrix}$$

or

$$4\pi R^2_{\text{planet}} \sigma T^4 = \pi R^2_{\text{planet}} \frac{L_\odot}{4\pi d^2} (1 - a)$$

On the left side of the equation, $4\pi R^2_{\text{planet}}$ indicates how many square meters of the planet's surface are radiating energy back into space, while σT^4 indicates how much energy each one of those square meters radiates each second. Put them together and you get the total amount of energy radiated away by the planet each second. On the right side of the equation, πR^2_{planet} is the area of the planet as seen from the Sun. That amount multiplied by the brightness of the sunlight reaching the planet, $L_\odot/4\pi d^2$, indicates how much energy is falling on the planet each second. The final expression, $(1 - a)$, indicates how much of that energy the planet actually absorbs. Putting everything on the

Summary

5.1 Light carries both information and energy throughout the universe. The speed of light in a vacuum is 300,000 km/s; nothing can travel faster.

5.2 From gamma rays to visible light to radio waves, all radiation is an electromagnetic wave. Light is simultaneously an electromagnetic wave and a stream of particles called photons.

5.3 Nearly all matter is composed of atoms, and light can reveal the identity of the types of atoms that are present in matter. Atoms absorb and emit radiation at unique wavelengths like spectral fingerprints.

5.4 Because of the Doppler effect, light from receding objects is redshifted and light from approaching objects is blueshifted.

right side of the equation together, we get the amount of energy absorbed by the planet each second. The equal sign says that the energy radiated away needs to balance the sunlight absorbed. Canceling out R^2_{planet} on both sides, and rearranging this equation to put T on one side and everything else on the other gives:

$$T^4 = \frac{L_\odot(1-a)}{16\sigma\pi d^2}$$

If we take the fourth root of each side, we get:

$$T = \left(\frac{L_\odot(1-a)}{16\sigma\pi d^2}\right)^{1/4}$$

Putting in the appropriate numbers for the luminosity of the Sun, $L_\odot$, and the constants π and σ yields this simpler equation:

$$T = 279\ \text{K} \times \left(\frac{1-a}{d^2_{\text{AU}}}\right)^{1/4}$$

where d_{AU} is the distance of the planet from the Sun in astronomical units.

To use this equation, we would need to know a planet's distance from the Sun and its average albedo. For a blackbody ($a = 0$) at 1 AU from the Sun, the temperature is 279 K. For Earth, with an albedo of 0.3 and a distance from the Sun of 1 AU, the temperature is:

$$T = 279\ \text{K} \times \left(\frac{1-0.3}{1^2}\right)^{1/4} = 255\ \text{K}$$

Earth is cooler than a blackbody at 1 AU from the Sun because its average albedo is greater than zero.

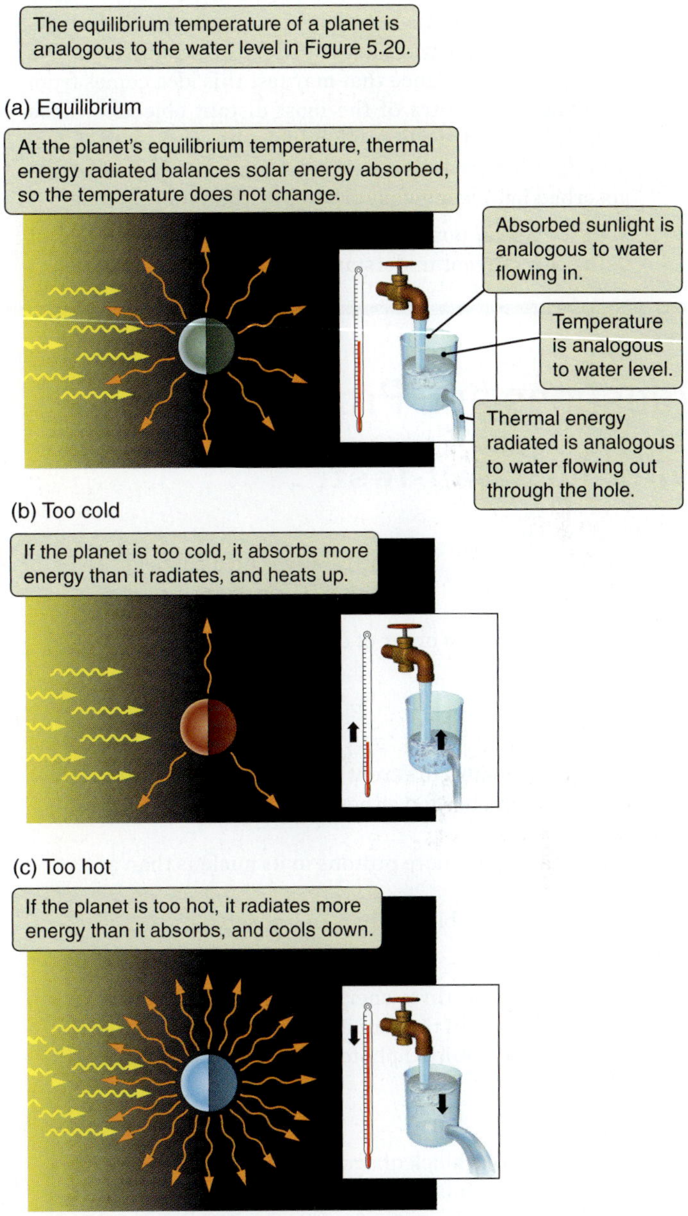

FIGURE 5.25 Planets are heated by absorbing sunlight (and sometimes internal heat sources) and cooled by emitting thermal radiation into space. If there are no other sources of heating or means of cooling, then the equilibrium between these two processes determines the temperature of the planet.

The wavelength shifts of the spectral lines indicate whether an astronomical object is moving toward or away from Earth.

5.5 Temperature is a measure of how energetically particles are moving in an object. A light source that emits electromagnetic radiation only because of its temperature, not its composition, is called a blackbody.

5.6 The brightness of an object depends on its light output—its luminosity—divided by its distance squared.

5.7 A planet's temperature depends on its albedo and its distance from its star.

Unanswered Questions

- Has the speed of light always been 300,000 km/s? Some theoretical physicists have questioned whether light traveled much faster earlier in the history of our universe. The observational evidence that may test this idea comes from studying the spectra of the most distant objects—whose light has been traveling for billions of years—and determining whether billions of years ago chemical elements absorbed light somewhat differently than they do today.
- Will it ever be possible to travel faster than the speed of light? Our current understanding of the science says no. A staple of science fiction films and stories is spaceships that go into "warp speed" or "hyperdrive"—moving faster than light—to traverse the huge distances of space (and visit a different planetary system every week). If this premise is simply fictional and the speed of light is a true universal limit, then travel between the stars will take many years. Since all electromagnetic radiation travels at the speed of light, even an electromagnetic signal sent to another planetary system would take many years to get there. Interstellar visits (and interstellar conversations) will be quite difficult.

Questions and Problems

Summary Self-Test

1. Light acts like
   a. a wave.
   b. a particle.
   c. both a wave and a particle.
   d. neither a wave nor a particle.

2. Rank the following in order of decreasing wavelength.
   a. gamma rays
   b. visible light
   c. infrared light
   d. ultraviolet light
   e. radio waves

3. Why is an iron atom a different element from a sodium atom?
   a. A sodium atom has fewer neutrons in its nucleus than an iron atom has.
   b. An iron atom has more protons in its nucleus than a sodium atom has.
   c. A sodium atom is bigger than an iron atom.
   d. all of the above

4. Suppose an atom has three energy levels, specified in arbitrary units as 10, 7, and 5. In these units, which of the following energies might an emitted photon have? (Select all that apply.)
   a. 3
   b. 2
   c. 5
   d. 4

5. Suppose you have a block of a material in which half the atoms decay in 2 minutes. After 6 minutes, what fraction of the original material remains?
   a. ½
   b. ¼
   c. ⅙
   d. ⅛
   e. 1/64

6. When a boat moves through the water, the waves in front of the boat bunch up, while the waves behind the boat spread out. This is an example of
   a. the Bohr model.
   b. the Heisenberg uncertainty principle.
   c. emission and absorption.
   d. the Doppler effect.

7. As a blackbody becomes hotter, it also becomes ________ and ________.
   a. more luminous; redder
   b. more luminous; bluer
   c. less luminous; redder
   d. less luminous; bluer

8. Which of the following factors does *not* directly influence the temperature of a planet?
   a. the luminosity of the Sun
   b. the distance from the planet to the Sun
   c. the albedo of the planet
   d. the size of the planet

9. Two stars are of equal luminosity. Star A is 3 times as far from you as star B. Star A appears ________ star B.
   a. 9 times brighter than
   b. 3 times brighter than
   c. the same brightness as
   d. ⅓ as bright as
   e. 1/9 as bright as

10. When less energy is radiated from a terrestrial planet, its ________ increases until a new ________ is achieved.
    a. temperature; equilibrium
    b. size; temperature
    c. equilibrium; size
    d. temperature; size

True/False and Multiple Choice

11. **T/F:** The frequency of a photon is related to the energy of the photon.

12. **T/F:** Blue light has more energy than red light.

13. **T/F:** An atom can emit or absorb a photon of any wavelength.

14. **T/F:** The emission spectrum of a helium atom is identical to that of a carbon atom.

15. **T/F:** Cooler objects radiate more of their total light at shorter wavelengths than do hotter objects.

16. Suppose you have two monochromatic light beams. Beam 1 has half the wavelength of beam 2. How do their frequencies compare?
    a. Beam 1 has 4 times the frequency of beam 2.
    b. Beam 1 has 2 times the frequency of beam 2.
    c. They are the same.
    d. Beam 1 has ½ the frequency of beam 2.
    e. Beam 1 has ¼ the frequency of beam 2.

17. How does the speed of light in a medium compare to the speed in a vacuum?
   a. It's the same, since the speed of light is a constant.
   b. The speed in the medium is always faster than the speed in a vacuum.
   c. The speed in the medium is always slower than the speed in a vacuum.
   d. The speed in the medium may be faster or slower, depending on the medium.

18. Momentum is the product of
   a. mass and position.
   b. position and energy.
   c. mass and energy.
   d. mass and velocity.
   e. mass and acceleration.

19. At what wavelength does your body radiate the most energy? (Assume that your temperature is approximately that of Earth, which is 300 K.)
   a. 10^{-5} meter
   b. 10^{-3} meter
   c. 10^{-2} meter
   d. 10 meters
   e. 1,000 meters

20. When an electron moves from a higher energy level in an atom to a lower energy level,
   a. the atom is ionized.
   b. a continuous spectrum is emitted.
   c. a photon is emitted.
   d. a photon is absorbed.

21. Study Figure 5.16. In this figure the red light comes from the transition from E_3 to E_2. These photons will have the ________ wavelengths because they have the ________ energy.
   a. shortest; least
   b. shortest; most
   c. longest; least
   d. longest; most

22. Star A and star B appear equally bright in the sky. Star A is twice as far away from Earth as star B. How do the luminosities of stars A and B compare?
   a. Star A is 4 times as luminous as star B.
   b. Star A is 2 times as luminous as star B.
   c. Star B is 2 times as luminous as star A.
   d. Star B is 4 times as luminous as star A.

23. What is the surface temperature of a star that has a peak wavelength of 290 nm?
   a. 1,000 K
   b. 2,000 K
   c. 5,000 K
   d. 10,000 K
   e. 100,000 K

24. If a planet is in thermal equilibrium,
   a. no energy is leaving the planet.
   b. no energy is arriving on the planet.
   c. the amount of energy leaving equals the amount of energy arriving.
   d. the temperature is very low.

25. The temperature of an object has a very specific meaning as it relates to the object's atoms. A high temperature means that the atoms
   a. are very large.
   b. are moving very fast.
   c. are all moving together.
   d. are very massive.
   e. have a lot of energy.

Thinking about the Concepts

26. We know that the speed of light in a vacuum is 3×10^8 m/s. Is it possible for light to travel at a lower speed? Explain your answer.

27. Is light a wave or a particle or both? Explain your answer.

28. In what way can a charged particle create a magnetic field?

29. If photons of blue light have more energy than photons of red light, how can a beam of red light carry as much energy as a beam of blue light?

30. What is a continuous spectrum?

31. Patterns of emission or absorption lines in spectra can uniquely identify individual atomic elements. Explain how positive identification of atomic elements can be used as one way of testing the validity of the cosmological principle.

32. Why is it impossible to know the exact position and the exact velocity of an electron simultaneously?

33. An atom in an excited state can drop to a lower energy state by emitting a photon. Is it possible to predict exactly how long the atom will remain in the higher energy state? Explain your answer.

34. Spectra of astronomical objects show both bright and dark lines. Describe what these lines indicate about the atoms responsible for the spectral lines.

35. Astronomers describe certain celestial objects as being *redshifted* or *blueshifted*. What do these terms indicate about the objects?

36. An object somewhere near you is emitting a pure tone at middle C on the octave scale (262 Hz). You, having perfect pitch, hear the tone as A above middle C (440 Hz). Describe the motion of this object relative to where you are standing.

37. During a popular art exhibition, the museum staff finds it necessary to protect the artwork by limiting the total number of viewers in the museum at any particular time. New viewers are admitted at the same rate that others leave. Is this an example of static equilibrium or of dynamic equilibrium? Explain.

38. A favorite object for amateur astronomers is the double star Albireo, with one of its components a golden yellow and the other a bright blue. What do these colors tell you about the relative temperatures of the two stars?

39. Study Figure 5.24. For which planet is the range of measured temperatures farthest from the predicted value? What accounts for this difference?

40. The stars you see in the night sky cover a large range of brightness. What does that range tell you about the distances of the various stars? Explain your answer.

Applying the Concepts

41. You are tuned to 790 on AM radio. This station is broadcasting at a frequency of 790 kHz (7.90×10^5 Hz). What is the wavelength of the radio signal? You switch to 98.3 on FM radio. This station is broadcasting at a frequency of 98.3 MHz (9.83×10^7 Hz). What is the wavelength of this radio signal?

42. Your microwave oven cooks by vibrating water molecules at a frequency of 2.45 gigahertz (GHz), or 2.45×10^9 Hz. What is the wavelength, in centimeters, of the microwave's electromagnetic radiation?

43. You observe a spectral line of hydrogen at a wavelength of 502.3 nm in a distant galaxy. The rest wavelength of this line is 486.1 nm. What is the radial velocity of this galaxy?

44. Assume that an object emitting a pure tone of 440 Hz is on a vehicle approaching you at a speed of 25 m/s. If the speed of sound at this particular atmospheric temperature and pressure is 340 m/s, what will be the frequency of the sound that you hear? (Hint: Keep in mind that frequency is inversely proportional to wavelength.)

45. If half of the phosphorescent atoms in a glow-in-the-dark toy give up a photon every 30 minutes, how bright (relative to its original brightness) will the toy be after 2 hours?

46. How bright would the Sun appear from Neptune, 30 AU from the Sun, compared to its brightness as seen from Earth? The spacecraft *Voyager 1* is now about 124 AU from the Sun and heading out of the Solar System. Compare the brightness of the Sun seen by *Voyager 1* with that seen from Earth.

47. On a dark night you notice that a distant lightbulb happens to have the same brightness as a firefly that is 5 meters away from you. If the lightbulb is a million times more luminous than the firefly, how far away is the lightbulb?

48. Two stars appear to have the same brightness, but one star is 3 times more distant than the other. How much more luminous is the more distant star?

49. A panel with an area of 1 square meter (m^2) is heated to a temperature of 500 K. How many watts is it radiating into its surroundings?

50. The Sun has a radius of 6.96×10^5 km and a blackbody temperature of 5780 K. Calculate the Sun's luminosity. (Hint: The area of a sphere is $4\pi R^2$.)

51. Some of the hottest stars known have a blackbody temperature of 100,000 K. What is the peak wavelength of their radiation? Sketch the blackbody curves from Figure 5.22, adding an approximated curve for these hottest stars.

52. Your body, at a temperature of about 37°C (98.6°F), emits radiation in the infrared region of the spectrum.
    a. What is the peak wavelength, in micrometers, of your emitted radiation?
    b. Assuming an exposed body surface area of 0.25 m^2, how many watts of power do you radiate?

53. A planet with no atmosphere at 1 AU from the Sun would have an average blackbody surface temperature of 279 K if it absorbed all the Sun's electromagnetic energy falling on it (albedo = 0).
    a. What would be the average temperature on this planet if its albedo were 0.1, typical of a rock-covered surface?
    b. What would be the average temperature if its albedo were 0.9, typical of a snow-covered surface?

54. Earth's average albedo is approximately 0.3.
    a. Calculate Earth's average temperature in degrees Kelvin (and then Celsius and Fahrenheit).
    b. Does this temperature meet your expectations? Explain why or why not.
    c. How would this temperature change if Earth's albedo were lower? Higher?

55. The orbit of Eris, a dwarf planet, carries it out to a maximum distance of 97.7 AU from the Sun. Assuming an albedo of 0.8, what is the average temperature of Eris when it is farthest from the Sun?

Using the Web

56. Go to NASA's "Astronomy Picture of the Day" (APOD) Web page (http://apod.nasa.gov/apod/ap101027.html), and study the picture of the Andromeda Galaxy in visible light and in ultraviolet light. Which light represents a hotter temperature? What differences do you see in the two images? Go to the APOD archive (http://apod.nasa.gov/cgi-bin/apod/apod_search), and enter "false color" in the search box. Examine a few images that come up in the search. What does *false color* mean in this context? What wavelength(s) were the pictures exposed in? What is the color coding; that is, what wavelength does each color in the image represent?

57. Crime scene investigators may use different types of light to examine a crime scene. Search on "forensic lighting" in your browser. What wavelengths of light are used to search for blood and saliva? For fingerprints? Why is it useful for an investigator to have access to different kinds of light? Search on "forensic spectroscopy" and select a recent report. How is spectroscopy being used in crime scene investigations?

58. Using Google Images or an equivalent website, search on "night vision imaging" and "thermal imaging." How do night-vision goggles and thermal-imaging devices work differently from regular binoculars or cameras?

59. The Transportation Security Administration (TSA) uses several types of imaging devices to screen passengers in airports. Search on "TSA imaging" in your browser. What wavelengths of light are being used in these devices? What concerns do passengers have about some of these imaging devices?

60. Go to the NASA Earth Observations (NEO) website (http://neo.sci.gsfc.nasa.gov/Search.html?group=72), and look at the current map of Earth's albedo (click on "albedo" in the menu for "energy" or "land" if it didn't come up). Compare this map with those of 2, 4, 6, 8, and 10 months ago. Which parts of Earth have the lowest and highest albedos? In which parts do the albedos seem to change the most with the time of the year? Would you expect ice, snow, oceans, clouds, forests, and deserts to add or subtract in each case from the total Earth albedo? Which parts of Earth are not showing up on this map?

STUDYSPACE is a free and open website that provides a Study Plan for each chapter of *21st Century Astronomy*. Study Plans include animations, reading outlines, vocabulary flashcards, and multiple-choice quizzes, plus links to premium content in SmartWork and the ebook. Visit **wwnorton.com/studyspace**.

SMARTWORK Norton's online homework system, includes algorithmically generated versions of these questions, plus additional conceptual exercises. If your instructor assigns questions in SmartWork, log in at **smartwork.wwnorton.com.**

Exploration | Light as a Wave, Light as a Photon

Visit StudySpace, and open the "Light as a Wave, Light as a Photon" AstroTour in Chapter 5. Watch the first section and then click through, using the "Play" button, until you reach "Section 2 of 3."

Here we will explore the following questions: How many properties does a wave have? Are any of these properties related to each other?

Work your way to the experimental section, where you can adjust the properties of the wave. Watch for a moment to see how fast the frequency counter increases.

1 **Increase the wavelength by pressing the arrow key. What happens to the rate of the frequency counter?**

2 **Reset the simulation and then decrease the wavelength. What happens to the rate of the frequency counter?**

3 **How are the wavelength and frequency related to each other?**

4 **Imagine that you increase the frequency instead of the wavelength. How should the wavelength change when you increase the frequency?**

5 **Reset the simulation, and increase the frequency. Did the wavelength change in the way you expected?**

6 **Reset the simulation, and increase the amplitude. What happens to the wavelength and the frequency counter?**

7 **Decrease the amplitude. What happens to the wavelength and the frequency counter?**

8 **Is the amplitude related to the wavelength or frequency?**

9 **Why can't you change the speed of this wave?**

The twin 10-meter Keck reflectors on Mauna Kea, Hawaii, have a multimirror, compact design.

06 The Tools of the Astronomer

All truths are easy to understand once they are discovered; the point is to discover them.

Galileo Galilei (1564–1642)

LEARNING GOALS

In the previous chapter you learned that electromagnetic radiation carries information of the physical and chemical properties of distant planets, stars, and galaxies. These data must first be collected and processed before they can be analyzed and converted to useful knowledge. Here you will learn about the tools that astronomers use to capture and scrutinize that information. By the conclusion of this chapter, you should be able to:

- Compare and contrast the two main types of optical telescopes, and summarize how these telescopes utilize the behavior of light.
- Describe how telescopes of various types collect radiation over the range of the electromagnetic spectrum.
- Explain the advantages and disadvantages of ground-based telescopes and telescopes in orbit.
- Summarize why spacecraft are sent to study the planets and moons of our Solar System.
- Recount how particle accelerators, neutrino and gravitational-wave detectors, and high-speed computers contribute to the body of information we have about the universe.
- Discuss the types of observations that led to the discovery of leftover radiation from the Big Bang.

6.1 The Optical Telescope

Have you ever experienced a close-up view of the Moon's surface? Perhaps your family or a neighbor had a backyard **telescope** like the one shown in **Figure 6.1**. Or the Moon might have been the featured attraction during a visit to your local planetarium. With that first view came recognition of what the Moon really is—a nearby planetary world covered with craters and vast, lava-flooded basins. In a dark sky with a small telescope you can view the Orion Nebula, a giant assemblage of gas and dust 1,500 light-years away, or the larger Andromeda Galaxy 2 million light-years away. With the largest telescopes, astronomers can detect light that has been traveling across space for billions of years—even electromagnetic radiation from soon after the Big Bang, the origin of the universe itself. The telescope is the astronomer's most important instrument. Yet it is only within the past century and a half that its capabilities have been fully utilized.

How are telescopes used to investigate how the universe began?

As long ago as the late 13th century, craftsmen in Venice were making small lentil-shaped disks of glass that could be mounted in frames and worn over the eyes to improve vision. Known as *lenses* (from *lens*, which is Latin for "lentil"), these glass disks were convex on both sides. In retrospect, it's remarkable that more than 300 years would pass before these lenses would be employed for something other than spectacles.

Refractors and Reflectors

Hans Lippershey (1570–1619) was a German-born spectacle maker living in the Netherlands. Legend has it that in 1608, children playing with his lenses put two of them together and saw a distant object magnified. Lippershey mounted the lenses together in a tube to produce a *kijker* ("looker"), as he called it. News of Lippershey's invention spread rapidly, soon reaching the Italian instrument maker Galileo Galilei. Galileo at once saw the potential of the "looker" for studying the heavens, and he constructed one of his own (**Figure 6.2**). Recall from Chapter 3 that by 1610, Galileo had become the first to see the phases of Venus and the moons of Jupiter, and among the first to see craters on the Moon. He was also the first to realize that the Milky Way is made up of countless numbers of individual stars. As the story goes, Galileo was demonstrating his instrument to guests when one of them christened it the *telescope*, from the Greek meaning "farseeing," and the name stuck. With its ability to see far beyond the range of the human eye, the **refracting telescope**—one that uses lenses—quickly revolutionized the science of astronomy.

FIGURE 6.1 Amateur astronomers with a modern reflecting telescope. There are several hundred thousand amateur astronomers in the United States alone. Advances in telescope design and electronics have enabled amateur observers to discover new celestial objects and help with long-term research projects of professional astronomers.

FIGURE 6.2 A replica of Galileo's refracting telescope.

Refraction is the change in direction of light entering a new medium (see Connections 6.1). A refracting telescope has a simple convex lens, called the **objective lens**, whose curved surfaces refract the light from a distant object (**Figure 6.3**). This

A telescope's aperture determines its light-gathering power.

refracted light forms an image on the telescope's **focal plane**, which is perpendicular to the optical axis—the path that light takes through the center of the lens (or mirror in the case of a *reflecting* telescope). The diameter of the objective lens, known as the telescope's **aperture**, determines the light-gathering power of the lens. The light-gathering power of a telescope is proportional to the area of its aperture—that is, to the square of its diameter. The distance between the telescope lens and the images formed is referred to as the **focal length** of the telescope. Longer focal lengths increase the size and separation of objects in the focal plane. Aperture and focal length are the two most important parameters of a telescope. The aperture determines a telescope's light-collecting power, and the focal length establishes the size of the image (**Math Tools 6.1**).

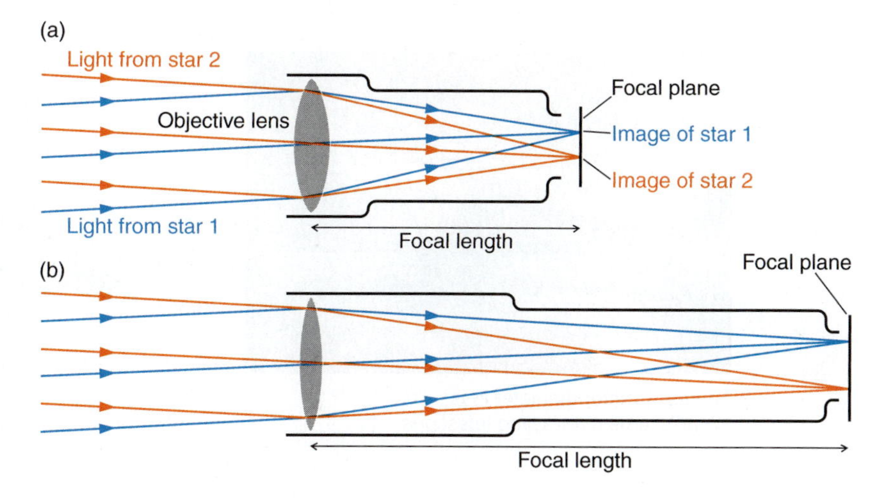

FIGURE 6.3 (a) A refracting telescope uses a lens to collect and focus light from two stars, forming images of the stars in its focal plane. (b) Telescopes with longer focal length produce larger, more widely separated images.

Math Tools 6.1

Telescope Aperture and Magnification

If you were shopping for a telescope, you would likely be told that the most important specifications are aperture and magnification. The aperture is the diameter of the primary lens or mirror, and the light-gathering power is proportional to the square of the aperture size. So a telescope with a larger aperture collects more light than does one with a smaller aperture. As an example, here's how the light-gathering power of a telescope with a 200-millimeter (mm), or 8-inch, diameter compares with the light-gathering power of the pupil of your eye, which is about 6 mm in the dark:

$$\text{Light-gathering power of telescope} \propto (200\text{ mm})^2$$

and:

$$\text{Light-gathering power of eye} \propto (6\text{ mm})^2$$

So, to compare:

$$\frac{\text{Light-gathering power of telescope}}{\text{Light-gathering power of eye}} = \frac{(200\text{ mm})^2}{(6\text{ mm})^2} = \left(\frac{200}{6}\right)^2 = 1{,}100$$

A typical 8-inch telescope has over a thousand times the light-gathering power of your eye.

Comparing this 8-inch telescope to the Keck 10-meter telescope shows why bigger is better. Since 200 mm = 0.2 meter,

$$\frac{\text{Light-gathering power of Keck}}{\text{Light-gathering power of 8-inch telescope}} = \frac{(10\text{ m})^2}{(0.2\text{ m})^2} = \left(\frac{10}{0.2}\right)^2 = 2{,}500$$

Plans are under way to build even larger telescopes, 25–40 meters in diameter.

Most telescopes have a set focal length and come with a collection of eyepieces. The magnification is given by:

$$\text{Magnification} = \frac{\text{Telescope focal length}}{\text{Eyepiece focal length}}$$

Suppose the focal length of the amateur telescope in the preceding example is 2,000 mm. (Sometimes focal length is specified on a telescope as "focal ratio," which equals focal length divided by aperture size; in this example, the focal ratio would be $f = 2{,}000\text{ mm}/200\text{ mm} = 10$.) Combined with the focal length of a standard eyepiece, 25 mm, this telescope will give the following magnification:

$$\text{Magnification} = \frac{2{,}000\text{ mm}}{25\text{ mm}} = 80$$

This telescope and eyepiece combination has a magnifying power of 80, meaning that a crater on the Moon will appear 80 times larger in the telescope's eyepiece than it does when viewed by the naked eye. An eyepiece that has a focal length of 8 mm will have about 3 times more magnifying power, or 250.

But a higher magnification will not necessarily let you see the object better. A faint and fuzzy image will not look clearer when magnified, and you will see later in the chapter that the atmosphere of Earth and the *angular resolution* of the telescope are what limit how clearly you can see an object. Therefore, when discussing any telescope in this book, we will refer only to the size of its aperture.

(a)

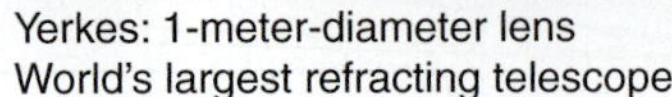

(b)

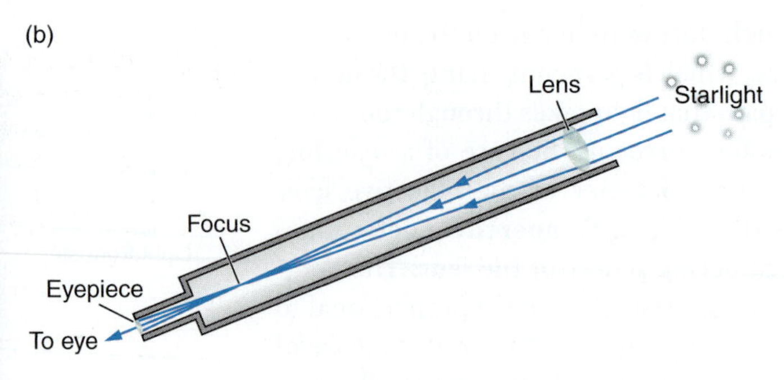

FIGURE 6.4 The Yerkes 1-meter refractor uses a lens to collect light. The parts of a refractor are sketched in (b).

Unfortunately, refracting telescopes have two major problems. The first is that all simple-lens telescopes (those using a pair of single lenses) suffer from *chromatic aberration* (see Connections 6.1), which results in blue halos around bright objects. As light passes through a simple lens, blue light is brought to a shorter focus than are the longer visible wavelengths, causing an out-of-focus blue halo effect. To minimize chromatic aberration, refracting telescopes now use *compound lenses*, which leave only small, residual effects of the blue halo. Refracting telescopes grew in size throughout the 19th century, up to the 1897 completion of the Yerkes 1-meter (40-inch) refractor (**Figure 6.4**), the world's largest operational refracting telescope. Located in Williams Bay, Wisconsin, the Yerkes telescope carries a 450-kilogram (kg) objective lens mounted at the end of a 19.2-meter tube. The second major problem with refractors is that as the objective lenses get larger (to increase the light-gathering power and produce the largest images possible), they get heavier, and the massive piece of glass at the end of a very long tube sags too much under the force of gravity. ▶▶ **NEBRASKA SIMULATION: TELESCOPE SIMULATOR**

To address the issue of chromatic aberration, in 1668 Isaac Newton designed a **reflecting telescope**—using mirrors instead of lenses (**Figure 6.5**). Because the direction of a reflected ray does not depend on the wavelength of light, chromatic aberration is not a problem in reflecting telescopes. To make his first reflecting telescope, Newton cast a 2-inch mirror made of copper and tin and polished it to a special curvature. He then placed this primary mirror at the bottom of a tube with a secondary flat mirror mounted above it at a 45° angle. The second mirror directed the focused light to an eyepiece on the outside of the tube.

There are two types of optical telescopes: refractors and reflectors.

A reflecting telescope forms an image in its focal plane when light is reflected from a specially curved mirror rather than from a convex lens as in the refracting telescope. This mirror is called the **primary mirror** to distinguish it from the additional mirrors usually employed in modern reflecting telescopes. The light path from the primary mirror to the focal plane can be folded by introducing a **secondary mirror**, which enables a significant reduction in the length and weight of the telescope. In a modern Cassegrain telescope like the one pictured in Figure 6.1, the primary mirror has a hole so that light can pass back through it; the eyepiece is on the back, and the tube can be even shorter (**Figure 6.6**).

Connections 6.1 explores the optics of refracted and reflected light (see pages 156–157). ▶II **ASTROTOUR: GEOMETRIC OPTICS AND LENSES**

The spherical mirror surface used by Newton works only for small mirrors. Larger mirrors require a *parabolic* surface (shaped like a parabola, the open curve we discussed in Chapter 4) to produce a sharply focused image, and parabolic surfaces are much more difficult to fabricate. Not until the latter half of the 18th century did large reflect-

(a) (b)

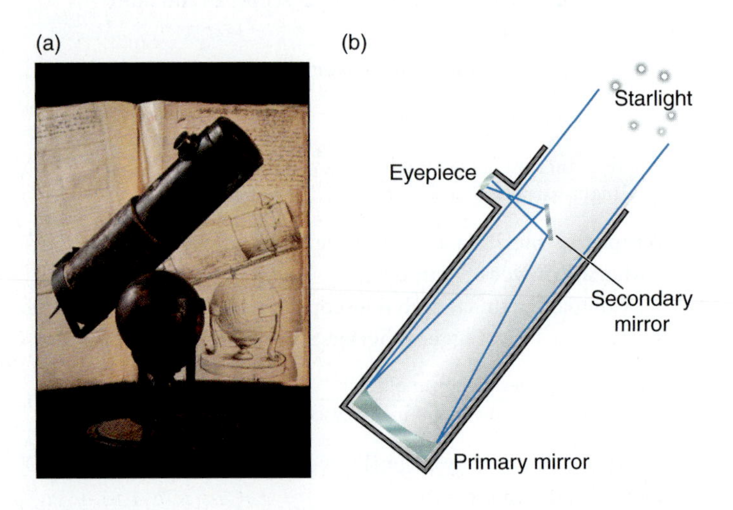

FIGURE 6.5 Newton's reflecting telescope. The parts of a reflector are sketched in (b).

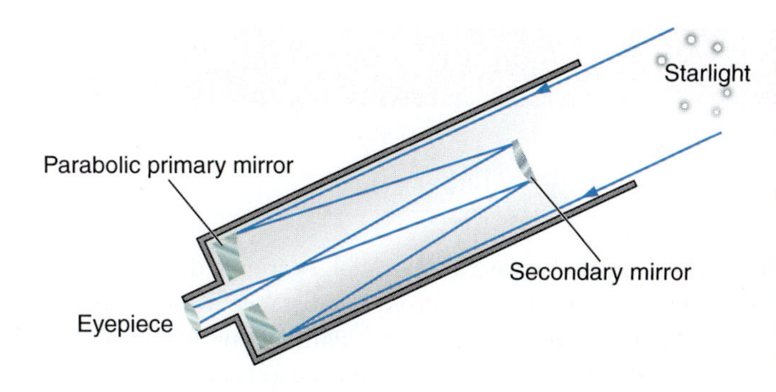

FIGURE 6.6 Reflecting telescopes use mirrors to collect and focus light. Large telescopes typically use a secondary mirror that directs the light back through a hole in the primary mirror to an accessible focal plane behind the primary mirror.

ing telescopes became more common. But then the size of the primary mirrors in reflecting telescopes continued to grow; they became larger every decade. Now primary mirrors can be supported from the back, and they can be made thinner and therefore less massive than the objective lenses found in refracting telescopes. The limitation on the size of reflecting telescopes is the cost of their fabrication and support structure. **Table 6.1**

TABLE 6.1 | The World's Largest Optical Telescopes

Mirror Diameter (meters)	Telescope	Sponsor(s)	Location	Operational Date
39.3	European Extremely Large Telescope (E-ELT)	European Southern Observatory (Europe, Chile, Brazil)	Cerro Amazones, Chile	Planning stage
30.0	Thirty Meter Telescope (TMT)	International collaboration led by Caltech, University of California, and Canada	Mauna Kea, Hawaii	Planning stage
24.5	Giant Magellan Telescope (GMT)	Carnegie Institution, Harvard University, Smithsonian Institution, University of Arizona, University of Texas, Texas A&M, University of Chicago, Australian National University, Astronomy Australia Ltd., Korea Astronomy and Space Science Institute	Cerro Las Campanas, Chile	Under construction
11.0	South African Large Telescope (SALT)	South Africa, USA, UK, Germany, Poland, New Zealand, India	Sutherland, South Africa	2005
10.4	Gran Telescopio CANARIAS (GTC)	Spain, Mexico, University of Florida	Canary Islands	2007
10	Keck I	Caltech, University of California, NASA	Mauna Kea, Hawaii	1993
10	Keck II	Caltech, University of California, NASA	Mauna Kea, Hawaii	1996
9.2	Hobby-Eberly Telescope (HET)	University of Texas, Penn State, Stanford, Germany	Mount Fowlkes, Texas	1999
8.4 × 2	Large Binocular Telescope (LBT)	University of Arizona, Ohio State, Italy, Germany, Arizona State, and others	Mount Graham, Arizona	2008
8.3	Subaru Telescope	Japan	Mauna Kea, Hawaii	1999
8.2 × 4	Very Large Telescope (VLT)	European Southern Observatory	Cerro Paranal, Chile	2000
8.1	Gemini North	USA, UK, Canada, Chile, Brazil, Argentina, Australia	Mauna Kea, Hawaii	1999
8.1	Gemini South	USA, UK, Canada, Chile, Brazil, Argentina, Australia	Cerro Pachón, Chile	2000
6.5	MMT	Smithsonian Institution, University of Arizona	Tucson, Arizona	2000
6.5	Magellan I	Carnegie Institution, University of Arizona, Harvard, University of Michigan, MIT	Cerro Las Campanas, Chile	2000
6.5	Magellan II	Carnegie Institution, University of Arizona, Harvard, University of Michigan, MIT	Cerro Las Campanas, Chile	2002

Connections 6.1

When Light Doesn't Go Straight

Refraction

Recall from Chapter 5 that the speed of light is constant in a vacuum: 300,000 kilometers per second (km/s). But through a medium such as air or glass, the speed of light is always less. The ratio of light's speed in a vacuum, c, to its speed in a medium, v, is the medium's **index of refraction**, n—that is, $n = c/v$. For example, n is approximately 1.5 for typical glass, so the speed of light in glass is about 200,000 km/s.

Because the speed of light changes as it enters a medium, light that enters glass at an angle is bent. The amount of bend depends on the angle and on the index of refraction of the medium—in this case, glass. **Figure 6.7a** shows a schematic diagram of light **wavefronts** (a series of parallel waves) striking a medium at an angle. **Figure 6.7b** shows an actual light ray passing into and out of a medium. The ray is bent each time the index of refraction changes. This change of direction, called **refraction**, is the basis for the refracting telescope (see Figure 6.3). Because the telescope's glass lens is curved, light at the outer edges of the lens is refracted more than light near its center. The difference in refraction concentrates the light rays entering the telescope, bringing them to a sharp focus in the telescope's focal plane and creating an image. ▶▶ NEBRASKA SIMULATION: SNELL'S LAW DEMONSTRATOR

Reflection

Another property of light is **reflection**. When light encounters a different medium—in this case going from air to glass—there will be a certain amount of reflection from the surface of the new medium. In other words, some of the light will reverse its direction of travel. The most common example occurs when light encounters an ordinary flat mirror. In this case, the angle of the incoming light (*angle of incidence*) and the angle of outgoing light (*angle of reflection*) are always equal (**Figure 6.8**). What reflects from the mirror is a good representation of what falls on it, although left and right are interchanged. That's what makes a flat mirror so convenient for inspecting your appearance.

In astronomical telescopes, curved mirrors are very useful. Mirrors with a surface that curves inward toward the incoming light are called **concave**, and this curve can be spherical like a circle or parabolic like a parabola. The same rules of incidence and reflection hold here for each ray, but in this case the reflected rays do not maintain the same angle with respect to each other as they do with a flat mirror. Concave mirrors will reflect the rays so that they converge to form an image (**Figure 6.9**). If the incoming rays are parallel, as from a distant source like a star, the reflected rays cross at a distance from the mirror called the *focal length* of the mirror. Parallel rays from a distant source on the axis come together on the axis at a point called the **focus**. The surface at which *all* parallel rays cross is called the *focal plane*.

Dispersion

Shine white light through a glass prism and you'll get a rainbow-like spectrum (**Figure 6.10**). This effect demonstrates that the refraction, and therefore the speed of light in glass, depends on wavelength. Like most other transparent materials, glass has an index of refraction that increases with decreasing wavelength. This means that shorter wavelengths (those toward the blue) are refracted more strongly than longer wavelengths (those toward the red). This wavelength-dependent difference in refraction, which spreads the white light out into its spectral colors, is called **dispersion**. Although dispersion is helpful in producing prism spectra, it creates a serious problem called **chromatic aberration** (**Figure 6.11a**). Chromatic aberration causes blue light to come to a shorter focus than do the longer visible wavelengths. You can see this effect when you look at a bright object such as a distant streetlight through an inexpensive telescope. (Low-priced telescopes usually have only a simple convex objective lens.) The streetlight will appear to be

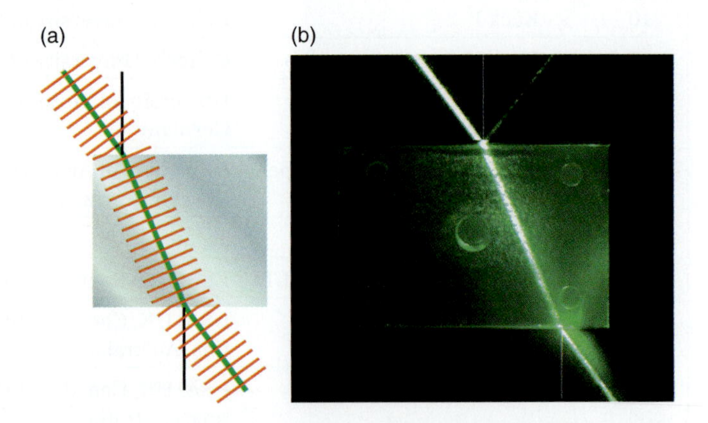

FIGURE 6.7 (a) Light waves are refracted (bent) when entering a medium with a higher index of refraction (in this case, glass). They are refracted again as they reenter the medium with the lower index of refraction (here, air). (b) Light from a green laser beam is refracted as it enters and exits a plastic block.

surrounded by a blue halo, because the blue component of the light is out of focus. Manufacturers of quality cameras and telescopes avoid the use of simple convex lenses in favor of the **compound lens (Figure 6.11b)**. By using two types of glass, a compound lens corrects for chromatic aberration.

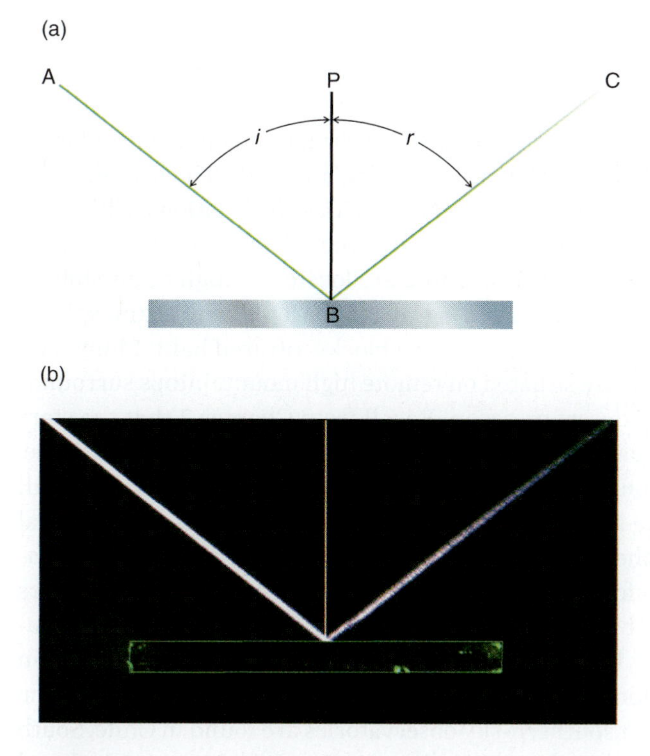

FIGURE 6.8 (a) A ray of incident light shines on a flat surface. The incoming or incident ray AB reflects from the surface, becoming the reflected ray BC. The angle between AB and PB, the perpendicular to the surface, is the angle of incidence (*i*). The angle between BC and PB is the angle of reflection (*r*). In the case of a flat mirror, the angles of incidence and reflection are always equal. (b) Light from a laser beam is reflected from a flat glass surface.

FIGURE 6.10 White light is dispersed into its component colors as it passes through a glass prism.

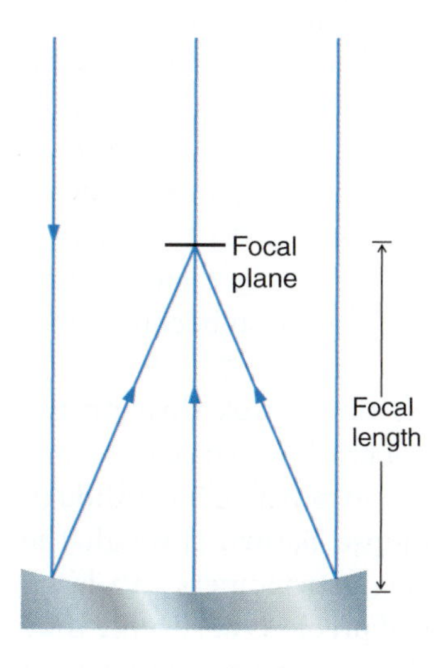

FIGURE 6.9 Parallel rays of light incident on a concave parabolic mirror are brought to a focus in the mirror's focal plane.

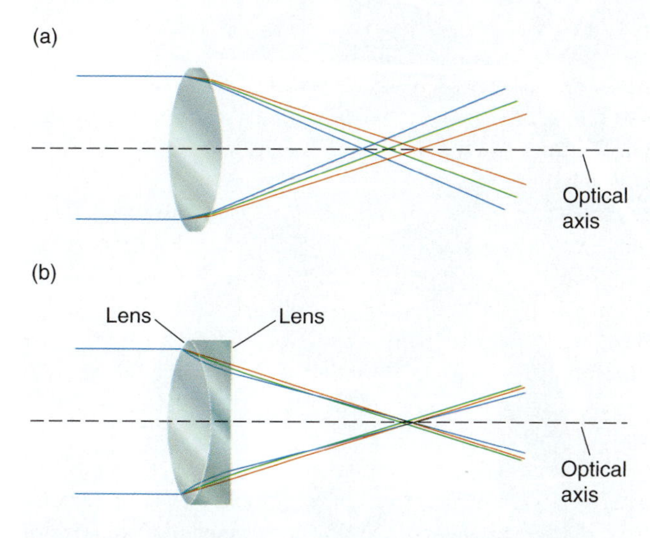

FIGURE 6.11 (a) Light of different wavelengths (different colors) comes to different foci along the optical axis of a simple lens, causing chromatic aberration. (b) A compound lens using two types of glass with different indices of refraction can compensate for much of the chromatic aberration, so different colors of light all come to a focus at the same point.

lists the world's largest optical telescopes. All are reflecting telescopes. The largest single mirrors constructed today are 8 meters in diameter, but telescopes even bigger than this are designed to make use of an array of smaller segments. Each of the 10-meter twin Keck telescopes, shown in both the chapter-opening figure and **Figure 6.12**, is made up of 36 hexagon-shaped segments each 1.8 meter in diameter.

The world's largest telescopes are reflecting telescopes.

FIGURE 6.12 Each of the Keck 10-meter reflectors uses an aligned group of 36 hexagonal mirrors to collect light.

Located on 4,100-meter-high Mauna Kea in Hawaii, the Keck telescopes are among the world's largest reflecting telescopes. Each one has 4 million times the light-gathering power of the human eye.

Observatory Locations

What makes a good location for a telescope? Astronomers look for sites that are high, dry, and dark. The best sites are far away from the lights of cities, in locations with little moisture, humidity, or rain, and where the atmosphere is relatively still. Telescopes are located as high as possible so that they get above a significant part of Earth's atmosphere, which distorts images and blocks infrared light. Many telescopes are situated on remote high mountaintops surrounded by desert or ocean. Recall from Chapter 2 that the stars that can be seen throughout the year depend on latitude, and only at the equator would a telescope have access to all of the stars in the sky. But equatorial latitudes have tropical weather—wet, humid, and stormy—and thus are poor locations for a telescope. So, to cover the entire sky, astronomers have built telescopes in both northern and southern locations. In the United States, large telescopes are located in California, Arizona, New Mexico, Texas, and Hawaii. The largest southern-sky observatories are found in Chile, South Africa, and Australia. The twin Gemini telescopes, designed to be a matched pair, are located in Hawaii in the Northern Hemisphere, and in Chile in the Southern Hemisphere.

Newer and larger telescopes are planned for many of the same locations that are listed in Table 6.1. The 8-meter Large Synoptic Survey Telescope (LSST) is headed for Cerro Pachón in Chile, current site of the Gemini South telescope. The Giant Magellan Telescope (GMT), consisting of seven 8-meter mirrors in a pattern equivalent to a 24.5-meter mirror, is planned to be constructed at Cerro Las Campanas in Chile. The Thirty Meter Telescope (TMT) is being designed for Mauna Kea in Hawaii (**Figure 6.13**), current site of the twin Keck telescopes; and the European Southern Observatory (ESO) is planning to build the 42-meter European Extremely Large Telescope (E-ELT) at Cerro Amazones in Chile. As these telescopes get larger—and more expensive—international collaboration becomes ever more important.

Today's working astronomers rarely look through the eyepiece of a telescope, because operating a large telescope for just a single night can be very expensive. So, although it might be exhilarating to glimpse Saturn through the eyepiece of a really big telescope, astronomers can learn much more and make better use of precious observing time by permanently recording the planet's image at a variety of wavelengths or seeing its light spread out into a revealing

FIGURE 6.13 Artist's rendering of the Thirty Meter Telescope, a reflecting telescope proposed by a collaboration of Californian and Canadian institutions.

spectrum. Some astronomers no longer travel to telescopes at all, instead observing remotely from the base of the mountain or far away at their own institutions.

Professional and amateur astronomers are concerned about the loss of the dark sky. As cities and suburbs grow and expand around the world, the use of outdoor artificial light becomes more widespread. Pictures from space show how bright many areas of Earth are at night. Recent photos of specific regions indicate that many parts of the world already are much brighter than they appear in the photo from 2000 shown in **Figure 6.14**. In the United States, two-thirds of the population resides in an area that is too bright for seeing the Milky Way at night, and it has been estimated that by 2025 there will be almost no dark skies in the continental United States. Air pollution, which also dims the view of the night sky, is increasing in many locations as well. The U.S. National Park Service now advertises evening astronomy programs in natural, unpolluted dark skies as one of the reasons to visit some parks. Several international astronomy associations are working with UNESCO (the United Nations Educational, Scientific and Cultural Organization) to promote the "right to starlight," arguing that for historical, cultural, and scientific reasons, it would be a huge loss if humanity could no longer view the stars. These organizations are encouraging countries to create starlight reserves and starlight parks where people can experience increasingly rare dark skies and a natural nocturnal environment.

Optical and Atmospheric Limitations

Another important characteristic of a telescope is its **resolution**. To astronomers, the term *resolution* is defined as how close two points of light can be to each other before a telescope is no longer able to split the light into two separate images. Unaided, the human eye can resolve objects separated by an angle as small as 1 arcminute (arcmin),[1] or 1/30 the diameter of the full Moon. This angular size may seem small, and in

[1] A more in-depth description of angular units—radians, degrees, arcminutes, and arcseconds—can be found in Appendix 1.

FIGURE 6.14 A view of Earth at night, as seen from space in the year 2000. Many satellite images were combined to produce this composite picture.

our daily lives it is; yet when we look at the sky, thousands of stars and galaxies may hide within the smallest area that the unaided human eye can resolve. Review Figure 6.3a to see the path followed by rays of light from two distant stars as they pass through the lens of a refracting telescope. Figure 6.3b illustrates that increasing the focal length increases the size and separation between the images that a telescope produces. This is one important reason why telescopes provide a much clearer view of the stars than the naked eye can. The focal length of a human eye is typically about 20 mm, whereas telescopes used by professional astronomers often have focal lengths of tens or even hundreds of meters. Such telescopes make images that are far larger than those formed by the human eye, and consequently they contain far more detail.

However, focal length explains only one difference between the resolution of telescopes and the unaided eye. The other difference results from the wave nature of light. As light waves pass through the aperture of a telescope, they spread out from the edges of the lens or mirror (**Figure 6.15**). The distortion of the wavefront as it passes the edge of an opaque object is called **diffraction**. Diffraction "diverts" some of the light from its path, slightly blurring the image made by the telescope. The degree of blurring depends on the wavelength of the light compared with the diameter of the telescope aperture. If the telescope aperture is smaller, the diffracted image grows larger, causing more blurring and limiting how close two images can be and still be resolved. The angular size of the diffraction pattern depends on the ratio between the wavelength of light (λ) and the size of the telescope aperture (D): λ/D. The best resolution that a given telescope can achieve is known as the **diffraction limit** (see **Math Tools 6.2**).

The diffraction limit depends on the ratio of wavelength to telescope aperture.

The diffraction limit tells us that larger telescopes get better resolution. Theoretically, the 10-meter Keck telescopes have a diffraction-limited resolution of 0.0113 arcsec in visible light, which would be good enough for you to read newspaper headlines 60 km away. But for telescopes with apertures larger than about a meter, Earth's atmosphere stands in the way of better resolution. If you have ever looked out across a large asphalt parking lot on a summer day, you have seen the distant horizon shimmer as light from that horizon is constantly bent this way and that by turbulent bubbles of warm air rising off the ground.

The problem is less pronounced when you look overhead, but the twinkling of stars in the night sky is caused by the same phenomenon. As telescopes magnify the angular diameter of a planet, they also magnify the shimmering effects of the atmosphere. The limit on the resolution of a telescope on the surface of Earth caused by this atmospheric distortion is called **astronomical seeing**. One advantage of launching telescopes such as the Hubble Space Telescope into orbit around Earth is that from their vantage point above the atmosphere, there is a much clearer view of the universe, unhampered by the limits of astronomical seeing. Still, ground-based telescopes are not obsolete. Modern technology has come to their rescue with computer-controlled **adaptive optics**, which compensate for much of the atmosphere's distortion.

Earth's atmosphere distorts images.

To better understand how adaptive optics work, we need to look more closely at how Earth's atmosphere smears out an otherwise perfect stellar image. In **Figure 6.15a**, light from a distant star arrives at the top of Earth's atmosphere as a flat, parallel wavefront. If Earth's

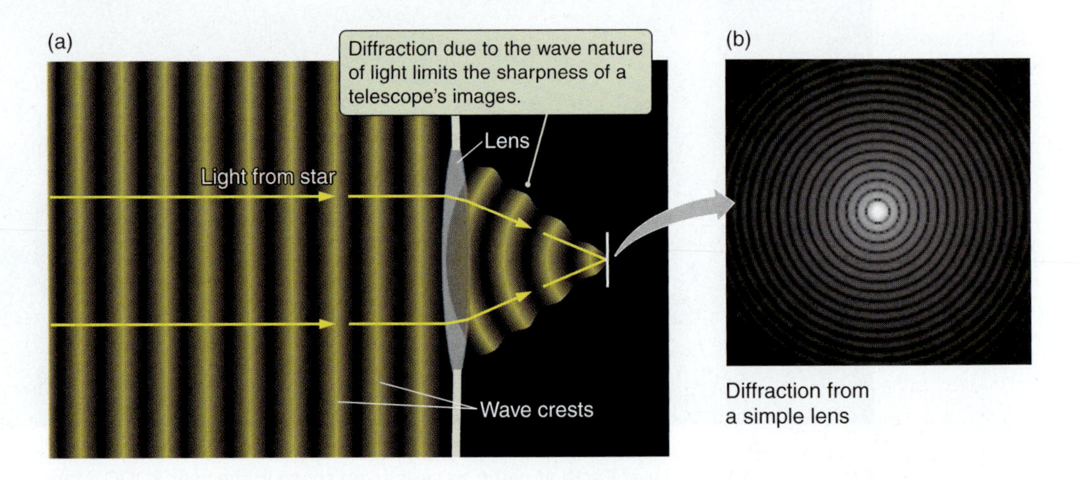

FIGURE 6.15 (a) Light waves from a star are diffracted by the edges of a telescope's lens or mirror. (b) This diffraction causes the stellar image to be blurred, limiting a telescope's ability to resolve objects.

Math Tools 6.2

Diffraction Limit

The ultimate limit on the angular resolution of a telescope, called the *diffraction limit*, is determined by the ratio of the wavelength of light passing through it to the diameter of the aperture:

$$\theta = 2.06 \times 10^5 \left(\frac{\lambda}{D}\right) \text{ arcsec}$$

where θ is the diffraction-limited angular resolution in arcseconds (an arcsecond [arcsec] is 1/1,800 of the size of the Sun in the sky, or about the size of a tennis ball if you could see it from 8 miles away); λ is the wavelength of light; D is the diameter, or aperture, of the telescope; and 2.06×10^5 is the number of arcseconds in a radian. The wavelength and diameter (λ and D) must be expressed in the same units, usually meters. The smaller the ratio of λ to D (λ/D), the better the resolution. For example, the size of the human pupil (see Figure 6.18) can range from about 2 mm in bright light to 8 mm in the dark. A typical pupil size is about 4 mm, or 0.004 meter. Visible (green) light has a wavelength (λ) of 550 nanometers (nm)—that is, 550×10^{-9} meter or 5.5×10^{-7} meter. Putting in the numbers gives:

$$\theta = 2.06 \times 10^5 \left(\frac{5.5 \times 10^{-7}\text{ m}}{0.004\text{ m}}\right) \text{ arcsec} = 28.3 \text{ arcsec}$$

or about 0.5 arcmin. The best resolution that the human eye can achieve is about 1 arcmin; however, a more typical resolution is 2 arcmin. The reason you do not achieve the theoretical resolution of your eyes is that the optical properties of the lens and the physical properties of the retina are not perfect.

How does the resolution of the human eye compare to that of a telescope? Consider the Hubble Space Telescope (HST), which operates in the visible (and near-infrared) part of the spectrum. Its primary mirror has a diameter of 2.4 meters. Substituting this value for D and again using visible (green) light gives:

$$\theta = 2.06 \times 10^5 \left(\frac{5.5 \times 10^{-7}\text{ m}}{2.4\text{ m}}\right) \text{ arcsec} = 0.047 \text{ arcsec}$$

or about 600 times better than the theoretical resolving power of the human eye.

Electron microscopes take advantage of this property of diffraction to achieve very high resolution by using electrons instead of photons to illuminate the target. Recall the dual wave-particle nature of electromagnetic radiation, discussed in Chapter 5. Electrons can behave both as particles and as waves, with wavelengths shorter than 0.1 nm. As a result, electron microscopes have more than five *thousand* times better resolution than conventional microscopes, which use visible light ($\lambda \approx 550$ nm).

atmosphere were perfectly uniform, the wavefront would remain flat as it reached the objective lens or primary mirror of a ground-based telescope. After making its way through the telescope's optical system, the wavefront would produce a tiny diffraction disk in the focal plane, as shown in **Figure 6.15b**. But Earth's atmosphere is not uniform. It is filled with small bubbles of air that have slightly different temperatures than their surroundings. Different temperatures mean different densities, and different densities mean different refractive properties, so each bubble bends light differently.

Adaptive optics can correct telescopic images of atmospheric distortion.

These air bubbles act as weak lenses, and by the time the wavefront reaches the telescope it is far from flat (**Figure 6.16**). Instead of a tiny diffraction disk, the image in the telescope's focal plane is distorted and swollen, degrading the resolution. Adaptive optics flatten out this distortion. First an optical device within the telescope constantly measures the wavefront. Then, before reaching the telescope's focal plane, the light is reflected off yet another mirror, which has a flexible surface. (Astronomers sometimes call this a "rubber" mirror, although it is actually made of glass.) A computer analyzes the wavefront distortion and bends the flexible mirror's surface so that it accurately corrects for the distortion of the wavefront. **Figure 6.17** shows examples of images corrected by adaptive optics. The widespread use of adaptive optics has made the image quality of ground-based telescopes competitive with Hubble images. But image distortion is not the only problem caused by Earth's atmosphere. As we will see in Section 6.3, large regions of the electromagnetic spectrum are partially or completely absorbed by various atmospheric molecules.

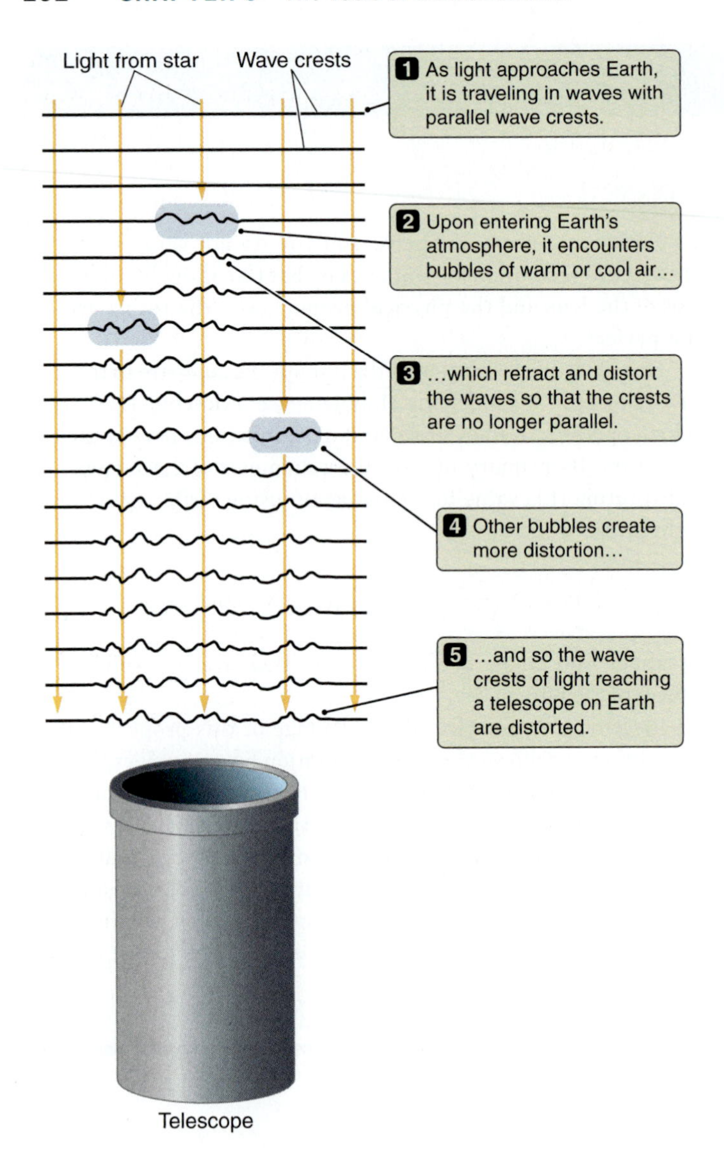

FIGURE 6.16 Bubbles of warmer or cooler air in Earth's atmosphere distort the wavefront of light from a distant object.

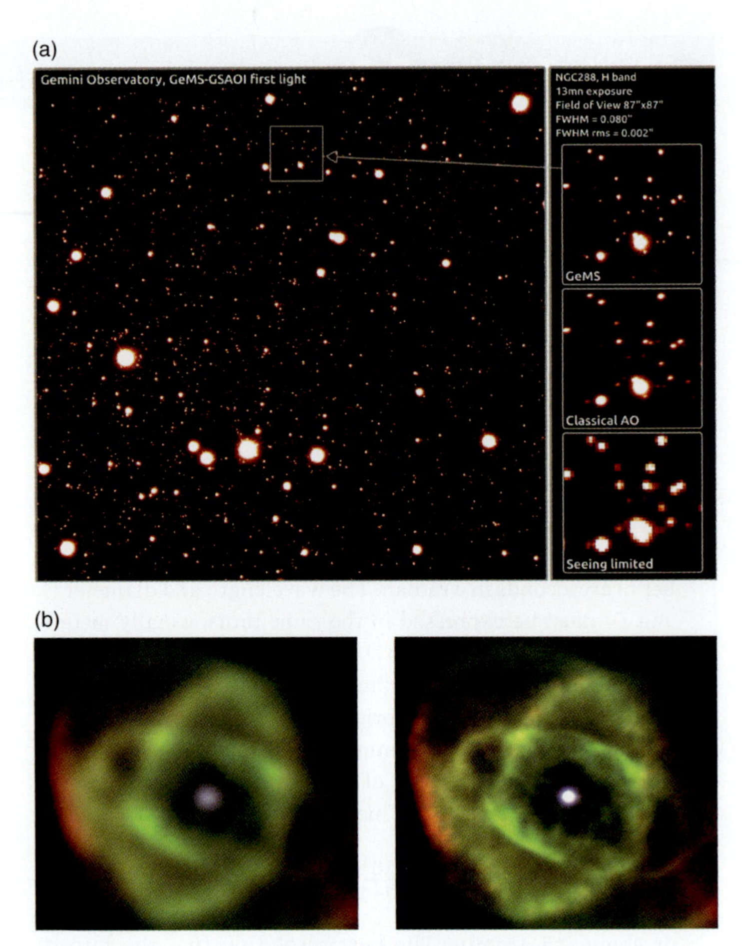

FIGURE 6.17 (a) A new adaptive optics (AO) system at the Gemini Observatory makes images much sharper (top right) than without AO (bottom right) or with standard AO (middle right). (b) Images of the Cat's Eye Nebula from the Palomar Observatory telescope without (left) and with (right) adaptive optics.

6.2 Optical Detectors and Instruments

The sole purpose of an astronomical telescope is to collect light waves from the cosmos and bring them to a focus. We turn now to the various ways in which astronomers capture these light waves and convert them to useful information. Detectors in a telescope's focal plane transform light waves into images that we can see and record.

Humans are sensitive to light with wavelengths ranging from about 380 nm (deep violet) to 750 nm (far red). The retina is the light detector in the human eye (**Figure 6.18**), and the individual receptor cells that respond to light falling on the retina are called rods and cones. Cones are located near the eye's optical axis at the center of your vision. They provide the highest resolution and enable you to recognize color. The size and spacing of cones—not the size of the pupil—determine the 1- to 2-arcmin resolution of the human eye. Rods, which are located away from the eye's optical axis and are responsible for our peripheral vision, provide the highest sensitivity to low light levels, but they have poorer resolution and cannot distinguish color.

As photons from a star enter the aperture of the eye (the pupil), they fall on and excite cones at the center of vision. The cones then send a signal to the brain, which interprets this message as "I see a star." So, what limits the faintest stars we can see with our unaided eyes, assuming a clear, dark night and good eyesight? This limit is determined in part by two factors that are characteristic of all detectors: integration time and quantum efficiency.

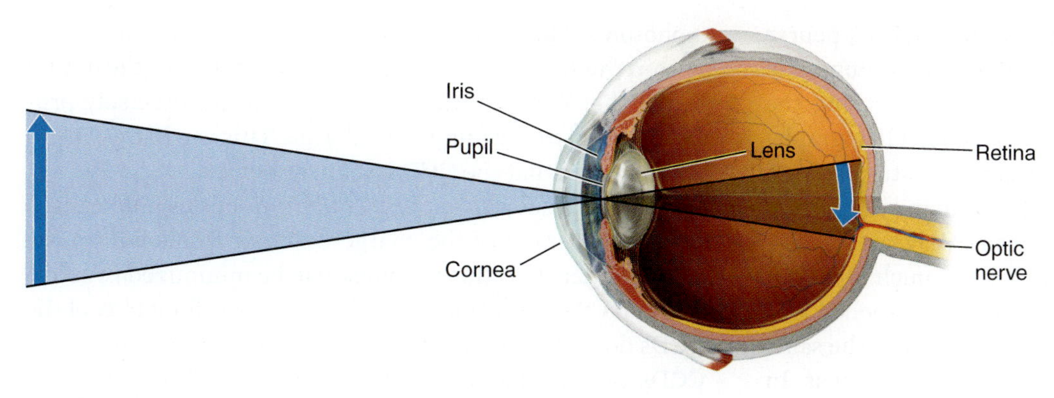

FIGURE 6.18 A schematic view of the human eye.

Integration time is the limited time interval over which the eye can add up photons. The brain "reads out" the information gathered by the eye about every 100 milliseconds (ms). Anything faster appears to happen all at once. If two images on a computer screen appear 30 ms apart, you will see them as a single image because your eyes will sum (or integrate) whatever they see over an interval of 100 ms or less. But if the images occur 200 ms apart, you will see them as separate images. This relatively brief integration time is the biggest factor limiting your nighttime vision. Stars too faint to be seen with the unaided eye are those from which you receive too few photons for your eyes to process in 100 ms.

Quantum efficiency is the likelihood that a particular photon landing on the retina will, in fact, produce a response. For the human eye, 10 photons must strike a cone within 100 ms to activate a single response. So the quantum efficiency of our eyes is about 10 percent: for every 10 events, the eye sends one signal to the brain. Together, integration time and quantum efficiency determine the rate at which photons must arrive at the retina before the brain says, "Aha, I see something."

Photographic Plates

Until 1840, the retina of the human eye was the only astronomical detector. Permanent records of astronomical observations were limited to what an experienced observer could sketch on paper while working at the eyepiece of a telescope. For more than two centuries after the invention of the telescope, astronomers struggled with the problem of **surface brightness**. In a telescope (or binoculars) the stars appear both closer and brighter, but only *point sources* such as stars appear brighter in a telescope. If you look at a distant landscape, or the Moon, or other extended (non-point-source) astronomical objects, they appear bigger in the eyepiece, but their surfaces are no brighter than they appear to the unaided eye. (Recall the discussion in Math Tools 6.1 about aperture and magnification). Even when astronomers built larger telescopes, nebulae and galaxies appeared larger but the details of these faint objects remained elusive. The problem was not with the telescopes but with the limitations of optics and the human eye. Only with the longer exposure times made possible by the invention of photography and the later development of electronic cameras were astronomers finally able to discern the faint but intricate fabric of the cosmos.

In 1840, John W. Draper (1811–1882), a New York chemistry professor, created the earliest known astronomical photograph (**Figure 6.19**). His subject was the Moon. Early photography was slow and very messy, however, and astronomers were reluctant to use it. In the late 1870s, a faster, simpler process was invented and astronomical photography took off. Astronomers could now create permanent images of planets, nebulae, and galaxies. Thousands of photographic plates soon filled the plate vaults of major observatories. Photography had created its own astronomical revolution. The quantum efficiency of most photographic systems used in

Photography opened the door to modern astronomy.

FIGURE 6.19 A photograph of the Moon taken by John W. Draper in 1840.

astronomy was very low—typically 1–3 percent, even poorer than that of the human eye. But unlike the eye, photography can overcome poor quantum efficiency by increasing the integration time to many hours of exposure. Photography made it possible for astronomers to record and study objects that had previously been invisible to the human eye.

Photography was not without problems. Very faint objects often required long exposures that took up much of an observing night, and a 10-hour exposure could be spoiled by a mishap. Photographic plates were sensitive to only the same parts of the electromagnetic spectrum that the eye was. In addition, the response of photography to light was not linear, especially at long exposures, so if you doubled the exposure time, you did not get twice the effect on your image. Finally, each expensive photographic plate was used only once.

Charge-Coupled Devices

Throughout the latter half of the 20th century, astronomers employed various electronic detectors to overcome the sensitivity, spectral range, and nonlinearity problems of photography. Devices called photometers receive photons from objects and convert them to an electronic signal that is proportional to the brightness of the object, just as a camera measures the amount of available light before taking a picture. Photometers enabled measurements of the variable brightness of objects but did not create images. In 1969, scientists at Bell Laboratories were developing "picture phones," telephones containing a small camera and viewing screen that displayed an image of the person at the other end of the conversation. Public opinion of the time declared the picture phones an invasion of personal privacy, and they were not commercially produced then, but the Bell research led to the invention of a detector called a **charge-coupled device**, or **CCD**. Astronomers soon realized that this was the detector they had been waiting for. By the late 1970s, the CCD had become the detector of choice in almost all astronomical imaging applications. CCDs are linear, so doubling the exposure means you record twice as much light. They are therefore good for measuring objects that vary in brightness, as well as faint objects that require long exposures. CCDs have a higher quantum efficiency, up to 70 or 80 percent at some wavelengths. The output from a CCD is a digital signal that can be sent directly from the telescope to image-processing software or stored electronically for later analysis.

The CCD is the astronomer's detector of choice.

A CCD is an ultrathin wafer of silicon, less than the thickness of a human hair, that is divided into a two-dimensional array of picture elements, or **pixels** (**Figure 6.20a**). When a photon strikes a pixel, it liberates a small electric charge within the silicon. As each CCD pixel is read out, the digital signal that flows to the computer is almost precisely proportional to the accumulated charge. This is what we mean when we say that the CCD is a linear device.

CCDs are subject to *thermal noise*—a false signal caused by the movement of the charge-carrying atoms within the silicon wafer. This thermal noise can be minimized in astronomical CCDs by using liquid nitrogen or helium to cool the CCDs down to very low temperatures, below 80 kelvins (K). CCDs can also be saturated, at which point they lose their linearity. The first astronomical CCDs were small arrays containing no more than a few hundred thousand pixels. The larger CCDs used in astronomy today may contain as many as 100 million pixels (**Figure 6.20b**). As CCDs get larger, more computing power is needed to process the images taken with them. Nearly every spectacular astronomical image in ultraviolet, optical, or infrared wavelengths that you find online was produced by a CCD, from telescopes either on the ground or in space. CCDs are found in many nonastronomical devices too, such as digital cameras, digital video cameras, and cell phone cameras (**Figure 6.20c**).

Your cell phone takes color pictures by using a grid of CCD pixels arranged in groups of three. Each pixel in a group is constructed to respond to only a particular range of colors—only to red light, for example. This limitation reduces the resolution of the camera because each spot in the final image requires three pixels of information. Astronomers choose instead to use all the pixels on the camera to measure the number of photons that fall on each pixel, without regard to color. They put filters in front of the camera to allow light of only particular wavelengths to pass through, such as the light of a specific spectral line. Color pictures like those from the Hubble Space Telescope are constructed by taking multiple pictures, coloring each one, and then carefully aligning and overlapping them to produce beautiful and informative images. Sometimes the colors are "true"; that is, they are close to the colors you would see if you were actually looking at the object with your eyes. Other times "false" colors represent different portions of the electromagnetic spectrum and tell you the temperature or composition of different parts of the object. Using changeable filters instead of designated color pixels gives astronomers greater flexibility and greater resolution.

▶▶ NEBRASKA SIMULATION: CCD SIMULATOR

Spectrographs

Spectroscopy is the study of an object's **spectrum** (plural: *spectra*)—its electromagnetic radiation split into its component wavelengths. **Spectrographs** (sometimes called

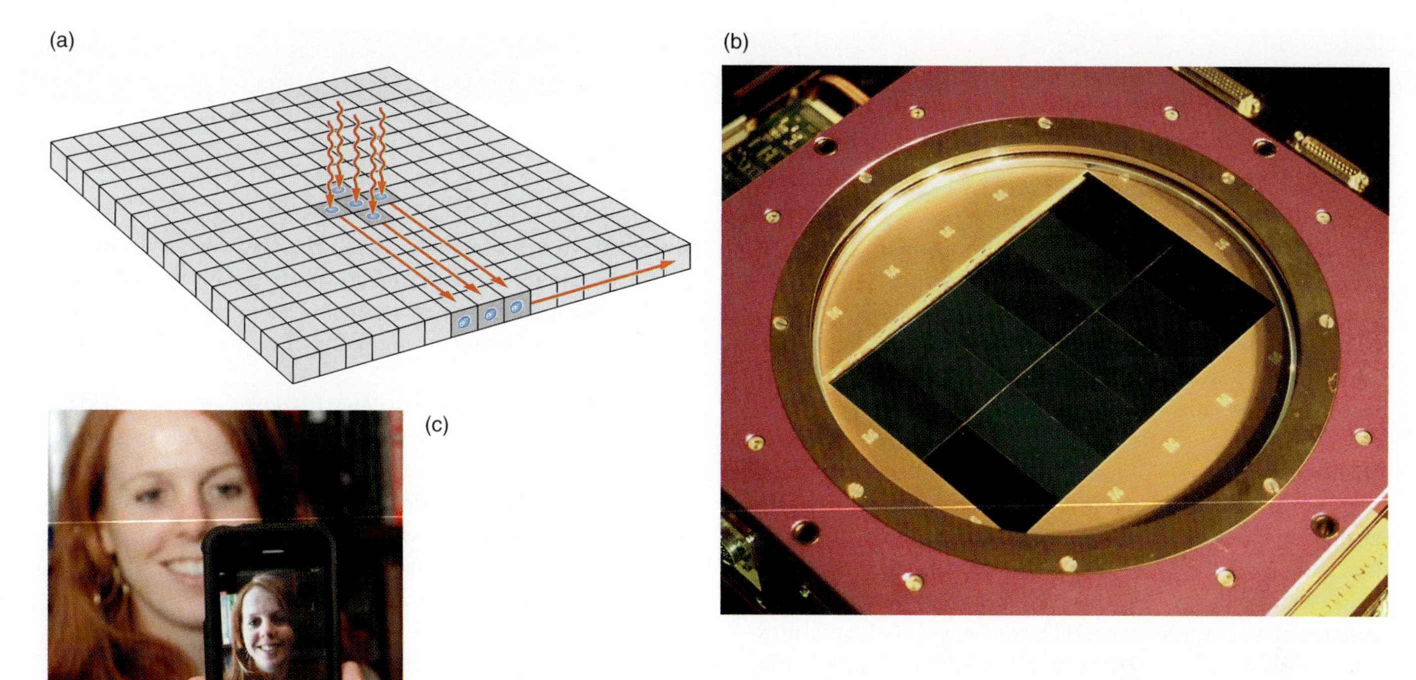

FIGURE 6.20 (a) A simplified diagram of a charge-coupled device (CCD). Photons from a star land on pixels (each gray square represents a single pixel) and produce free electrons within the silicon. The electron charges are electronically moved sequentially to the collecting register at the bottom. Each row is then moved out to the right to an electronic amplifier, which converts the electric charge of each pixel into a digital signal. (b) This large CCD (about 6 inches across) contains 12,288 × 8,192 pixels. (c) CCDs are used in many consumer electronic devices, such as cell phone cameras.

spectrometers) are tools that enable astronomers to probe the chemical and physical properties of distant objects by studying their spectra. Early spectrographs used glass prisms to disperse the incoming light into its component wavelengths (**Figure 6.21**), creating a spectrum like the one shown in Figure 5.16c. These spectra were recorded on photographic plates for precise measurement of the wavelengths of their spectral lines. Glass-prism spectrographs produce spectra with more dispersion at the shorter-wavelength (violet) end of the spectrum than at the longer-wavelength (red) end. Glass does not transmit short-wavelength ultraviolet and long-wavelength infrared light; thus, prism spectrographs are limited to the visible part of the spectrum. Most modern spectrographs use a *diffraction grating* (see **Connections 6.2**) to disperse the light and a CCD to record the spectrum.

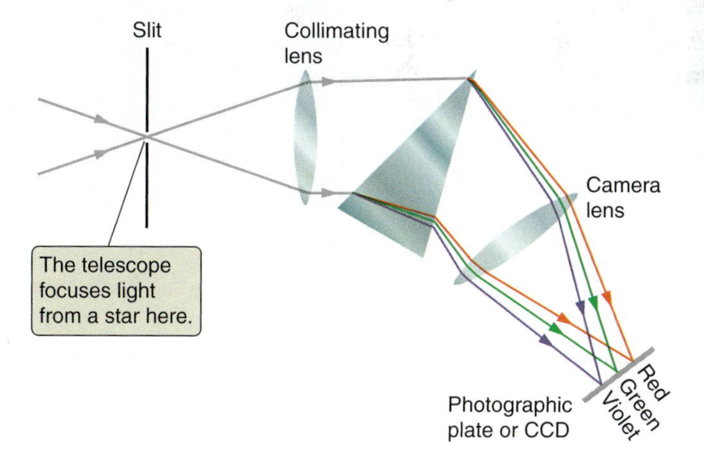

FIGURE 6.21 A diagram of a prism spectrograph.

Spectrographs may be designed for either low or high dispersion. Astronomers use low-dispersion spectrographs to identify the chemical components of faint light sources, or to measure the reflected spectral energy distribution of faint Solar System objects, such as small or distant asteroids. Measurements of temperature and radial velocity usually require very high dispersion. The high-dispersion spectrographs associated with most major telescopes tend to be huge and weigh several metric tons—think SUV size (too heavy to be attached directly to the telescope). A system of mirrors feeds light from the telescope's focal plane into the spectrograph, which is located nearby. Some modern spectrographs are "multi-slit," and use bundles of optical fibers to simultaneously obtain spectra from multiple objects in the field of view of the telescope.

Connections 6.2

Interference and Diffraction

The intersection of two sets of electromagnetic waves can produce patterns of high and low intensity called **interference** patterns. Suppose you have a pair of slits and an opaque screen. **Figure 6.22a** illustrates monochromatic light (light having a single wavelength or a very narrow range of wavelengths) going through the slits. Each slit now becomes a source of wavefronts. Notice the regular pattern on the screen where the wavefronts from the two slits intersect. If the intersection point occurs where the amplitudes of both waves are at their maximum positive or maximum negative value, the two add together and the light will be bright (the intensity of light is actually equal to the *square* of the amplitude). This is **constructive interference**. If instead, one wave is at its maximum positive value and the other is at its maximum negative value, the sum is zero and the result will be darkness. This is **destructive interference**. When the two slits are replaced with a large number of very narrow, very closely spaced parallel slits, the result is called a transmission **grating**. The same effect can be achieved by engraving closely spaced lines on a mirror.

Substituting a multiwavelength source of light yields a similar pattern for every wavelength, as shown in **Figure 6.22b** for a reflection grating. For each wavelength, constructive interference takes place at a different point on the screen. In other words, the grating produces a spectrum. Modern spectrographs use a grating to disperse incoming light into its constituent wavelengths. You can see this effect for yourself. Look at light reflected from a CD or DVD; the closely spaced tracks act as a grating and create a respectable spectrum (**Figure 6.22c**).

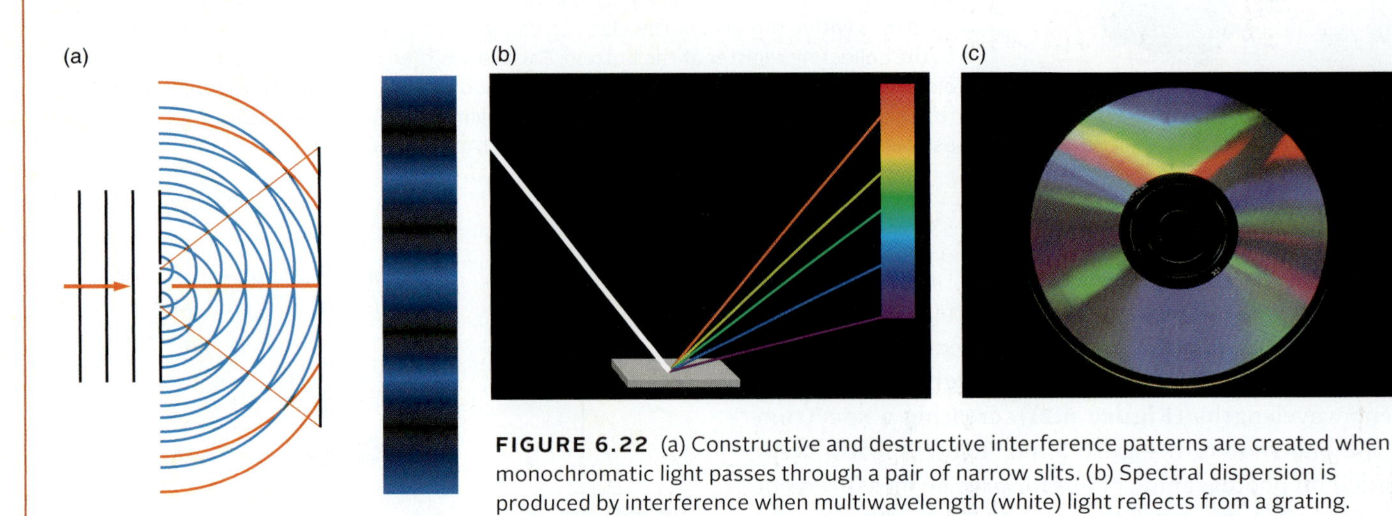

FIGURE 6.22 (a) Constructive and destructive interference patterns are created when monochromatic light passes through a pair of narrow slits. (b) Spectral dispersion is produced by interference when multiwavelength (white) light reflects from a grating. (c) A spectrum is created by the reflection of light from the closely spaced tracks of a CD.

6.3 Radio and Infrared Telescopes

The human eye is sensitive to light in only the visible part of the electromagnetic spectrum. (That's why we call it *visible*.) We evolved on a planet with an atmosphere that is transparent to light in the visible part of the spectrum but is opaque to most other wavelengths, except radio. Gamma-ray, X-ray, ultraviolet, and most of the infrared light arriving at Earth fails to reach the ground because it is partially or completely absorbed by ozone, water vapor, carbon dioxide, and other atmospheric molecules. Light at these wavelengths has to be observed from space. The visible part of the spectrum, which is not blocked by our atmosphere, has a fairly narrow range of wavelengths, or "window," through which we can view the universe. There are a few other **atmospheric windows** in the spectrum as well (**Figure 6.23**). The largest window is in radio wavelengths, including microwaves at the short-wavelength end of the radio window.

Earth's atmosphere blocks much of the electromagnetic spectrum.

NEBRASKA SIMULATION: EM SPECTRUM MODULE

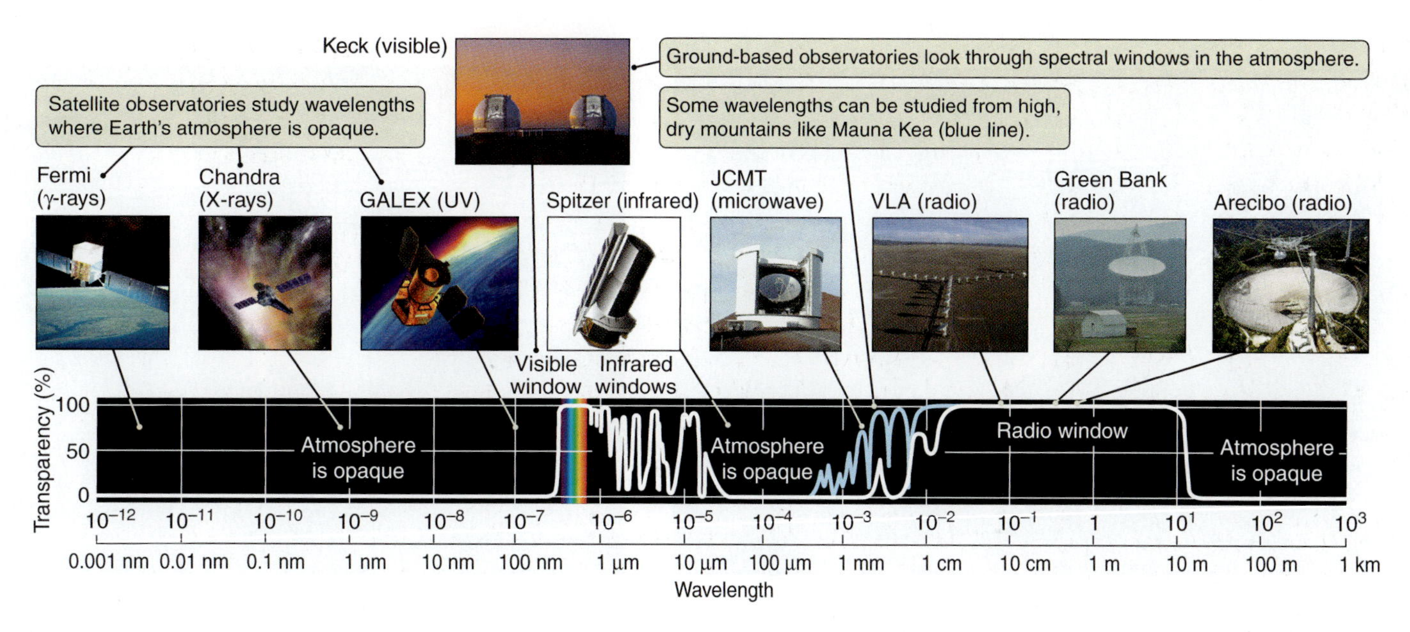

FIGURE 6.23 Earth's atmosphere blocks most electromagnetic radiation. Fermi = Fermi Gamma-ray Space Telescope (orbiting); Chandra = Chandra X-ray Observatory (orbiting); GALEX = Galaxy Evolution Explorer (orbiting); Keck = Keck Observatory (Hawaii); Spitzer = Spitzer Space Telescope (orbiting); JCMT = James Clerk Maxwell Telescope (Hawaii); VLA = Very Large Array (New Mexico); Green Bank = Robert C. Byrd Green Bank Telescope (West Virginia); Arecibo = Arecibo Observatory (Puerto Rico).

Radio Telescopes

Karl Jansky (1905–1950) was a young physicist working for Bell Laboratories in the early 1930s when he was assigned the job of identifying sources of static in transatlantic radiotelephone service. He built a pointable antenna and soon identified the major sources of static as nearby and distant thunderstorms. But one source remained a mystery: a faint, steady hiss rose and fell once every 23 hours and 56 minutes. Recall from Chapter 2 that this length of time is the same period as Earth's rotation with respect to the stars. In 1932, Jansky identified the mysterious source. It was located in the Milky Way in the direction of the galactic center, in the constellation of Sagittarius. Excited by his discovery, Jansky submitted a request to build a large dish antenna, or **radio telescope**, to study these signals in more detail. Bell Labs turned down the request. After all, Jansky had already provided the information they needed. Nevertheless, Jansky's discovery marked the birth of radio astronomy. In his honor, the basic unit for the strength of a radio source is called the **jansky (Jy)**.

In 1937, Grote Reber (1911–2002), an American radio engineer and ham radio operator, decided to build his own radio telescope. It consisted of a parabolic sheet of metal, 9 meters in diameter, with a radio receiver mounted at the focus. With this instrument, Reber conducted the first survey of the sky at radio frequencies, and he published the first radio frequency map of the galaxy in 1944. Reber was largely responsible for the rapid advancement in radio astronomy that blossomed in the post–World War II era.

Radio telescopes are another indispensable tool for astronomers. From our Solar System to the most distant galaxies, the penetrating power of radio waves unlocks secrets that cannot be revealed with shorter-wavelength optical or infrared telescopes. Look at Figure 6.23 and notice the wide radio window in Earth's atmosphere, covering wavelengths ranging all the way from about a centimeter to 10 meters. This ability of radio waves to pass undiminished through our atmosphere is also the property that enables us to peer through the vast amounts of gas and dust found in many galaxies. Most radio telescopes are large, steerable dishes, typically tens of meters in diameter, like the one shown in **Figure 6.24a**. The world's largest single-dish radio telescope is the 305-meter Arecibo dish built into a natural bowl-shaped depression in Puerto Rico (**Figure 6.24b**). (A 500-meter single-dish radio telescope with a similar design is under construction in China.) This huge structure is too big to steer. Arecibo can observe only those sources that pass within 20° of the zenith as Earth's rotation carries them overhead.

Radio telescopes enable astronomers to "see" through obscuring gas and dust.

As large as radio telescopes are, they have relatively poor angular resolution. Recall that a telescope's angular resolution is determined by the ratio λ/D, where λ is the wavelength of electromagnetic radiation and D is the telescope's aperture. A larger ratio means poorer resolution. Radio telescopes have diameters much larger than the apertures of

(a)

(b)

FIGURE 6.24 (a) The Parkes radio telescope in Australia. (b) The Arecibo radio telescope is the world's largest. The steerable receiver suspended above the dish permits limited pointing toward celestial targets as they pass close to the zenith.

most optical telescopes, and that helps. But the wavelengths of radio waves are typically several hundred thousand times greater than the wavelengths of visible light, and that hurts. Radio telescopes are thus hampered by the very long wavelengths they are designed to receive. For example, the resolution of the huge Arecibo dish in Figure 6.24b is typically about 1 arcmin, no better than the unaided human eye. So radio astronomers have had to develop their own bag of tricks, and one of the cleverest is the **interferometer**.

Single-dish radio telescopes have relatively poor resolution.

Recall the discussion about interference in Connections 6.2. Mathematically combining the signals from two radio telescopes makes them act like a telescope with a diameter equal to the separation between them. For example, if two 10-meter telescopes are located 1,000 meters apart, the D in λ/D is 1,000, not 10. Such an arrangement is called an interferometer because it makes use of the wavelike properties of electromagnetic radiation, in which signals from the individual telescopes constructively interfere with one another. Usually several telescopes are employed, in an arrangement called an **interferometric array**. By using very large arrays, radio astronomers can attain and exceed the angular resolution achieved by their optical colleagues.

One of the larger radio interferometric arrays is the Very Large Array (VLA) in New Mexico (**Figure 6.25**). The VLA is made up of 27 individual movable dishes spread out in a Y-shaped configuration up to 36 km across. At a wavelength of 10 centimeters (cm), this array can achieve resolutions of less than 1 arcsec. The Very Long Baseline Array (VLBA) employs 10 radio telescopes spread out over more than 8,000 km from the Virgin Islands in the Caribbean to Hawaii in the Pacific. At a wavelength of 10 cm, this array can attain resolutions of better than 0.003 arcsec. A radio telescope put into near-Earth orbit as part of a Space Very Long Baseline Interferometer (SVLBI) overcomes even this

Interferometric arrays improve resolution.

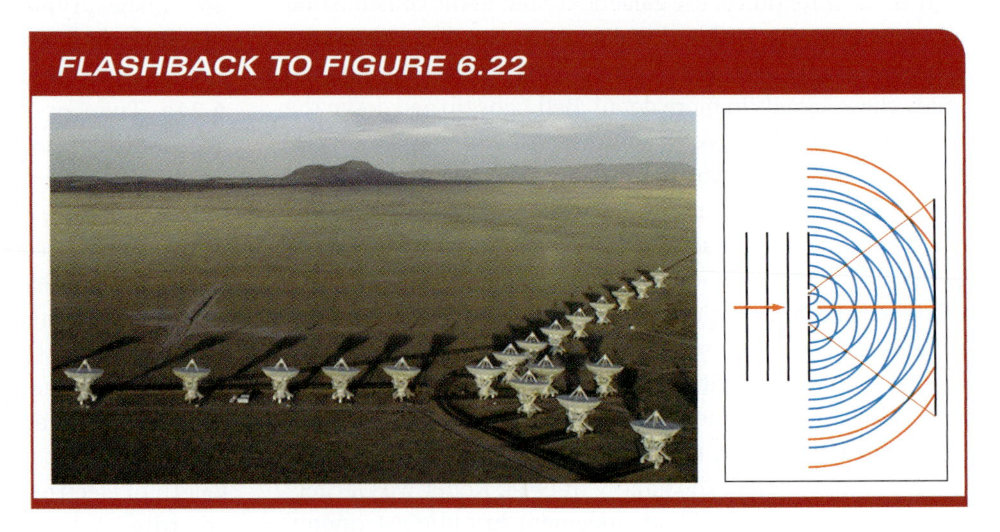

FIGURE 6.25 The Very Large Array (VLA) in New Mexico.

limit. Future SVLBI projects would extend the baseline to as much as 100,000 km, yielding resolutions far exceeding those of any existing optical telescope. The Russian RadioAstron observatory, launched in 2011, works with ground-based radio telescopes in this way.

Newer radio telescopes may use much larger numbers of small dishes. The Allen Telescope Array (ATA) near San Francisco was designed to have 350 six-meter dishes. It opened with 42 dishes in 2007, and efforts are being made to acquire additional funding. The Square Kilometre Array (SKA) is designed to have *thousands* of small radio dishes, which together will act as one dish with a collecting area of 1 square km. Twenty countries are supporting this telescope, which will be located in Australia, New Zealand, and South Africa.

Optical telescopes can also be arrayed to yield resolutions greater than those of single telescopes, although for technical reasons the individual units cannot be spread as far apart as radio telescopes can. The Very Large Telescope Interferometer (VLTI), operated by the European Southern Observatory (ESO) in Chile, combines the four VLT 8-meter telescopes (**Figure 6.26**) with four movable 1.8-meter auxiliary telescopes. It has a baseline of up to 200 meters, yielding angular resolution in the milliarcsecond range.

A newer interferometer, the Atacama Large Millimeter/submillimeter Array (ALMA), opened recently in the Atacama Desert in Chile, at an elevation of 5,000 meters. This project, an international collaboration of astronomers from Europe, North America, East Asia, and Chile, consists of 66 twelve- and seven-meter dishes for observations in the 0.3- to 9.6-mm wavelength range.

FIGURE 6.26 The Very Large Telescope (VLT), operated by ESO in Chile. Movable auxiliary telescopes allow the four large telescopes to operate as an optical interferometer.

Infrared Telescopes

Molecules such as water vapor in Earth's atmosphere block infrared (IR) photons from reaching astronomical telescopes on the ground, so telescopes that observe in the infrared window (0.75–30 microns, μm)—like those pictured in Figure 6.23—are at the highest locations. Mauna Kea, a dormant volcano and home of the Mauna Kea Observatories (MKO), rises 4,200 meters above the Pacific Ocean. At this altitude the MKO telescopes sit above 40 percent of Earth's atmosphere; but more important, 90 percent of Earth's atmospheric water vapor lies below. Still, for the infrared astronomer the remaining 10 percent is troublesome.

One way to solve the water vapor problem is to make use of high-flying aircraft. NASA's Kuiper Airborne Observatory (KAO), a modified C-141 cargo aircraft, carried a 90-cm telescope and was among the first of these flying observatories. It cruised at altitudes of 12–14 km, above 98 percent of Earth's water vapor. NASA retired KAO in 1995 and replaced it in 2010 with the Stratospheric Observatory for Infrared Astronomy (SOFIA), a joint project with the German Aerospace Center (DLR). SOFIA is a modified 747 wide-body airplane that carries a 2.5-meter telescope and works in the far-infrared region of the spectrum, from 1 to 650 μm. It flies in the stratosphere at an altitude of about 12 km, above 99 percent of the water vapor in Earth's lower atmosphere (**Figure 6.27**). Because airplanes are highly mobile, SOFIA can observe in both the Northern and Southern hemispheres.

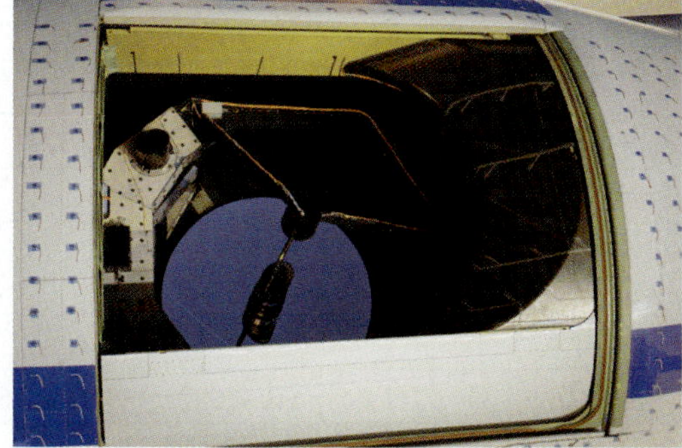

FIGURE 6.27 SOFIA is a 2.5-meter infrared telescope that is mounted in a Boeing 747 aircraft.

6.4 Getting above Earth's Atmosphere: Orbiting Observatories

Airborne observatories are able to overcome atmospheric absorption of infrared light by placing telescopes above most of the water vapor in the atmosphere. But gaining full access to the complete electromagnetic spectrum is yet another matter. This means getting completely above Earth's atmosphere. In the late 1940s, scientists put ultraviolet (UV) instruments in the nose cones of captured German V-2 rockets and launched them from the White Sands Proving Ground in New Mexico to altitudes greater than 100 km. Observations made by these instruments had to be brief because gravity brought the rockets carrying them back down. The next step was to put astronomical instruments into orbit. The first astronomical satellite was the British Ariel 1, launched in 1962 to study solar UV and X-ray radiation. Today there are a multitude of orbiting astronomical telescopes covering the electromagnetic spectrum from gamma rays to microwaves, with more in the planning stage (see **Table 6.2**).

Optical telescopes, such as the 2.4-meter Hubble Space Telescope (HST), operate successfully at modest altitudes in what is called low Earth orbit (LEO), 600 km above Earth's surface. Launched in 1990, HST has been the workhorse for UV, visible, and IR space astronomy for over two decades. LEO is also the region where the International Space Station (ISS) and many scientific satellites orbit. For certain other satellites and space telescopes, 600 km is not nearly high enough. The Chandra X-ray Observatory, NASA's X-ray telescope, cannot tolerate even the tiniest traces of atmosphere; it orbits more than 16,000 km above Earth's surface. And even this is not distant enough for some telescopes. NASA's Spitzer Space Telescope, an infrared telescope, is so sensitive that it needs

Orbiting observatories explore regions of the spectrum inaccessible from the ground.

TABLE 6.2 Selected Present and Future Space Observatories

Telescope	Sponsor(s)	Description	Launch Year
Hubble Space Telescope (HST)	NASA, ESA	Optical, infrared, ultraviolet observations	1990
Chandra X-ray Observatory	NASA	X-ray imaging and spectroscopy	1999
X-ray Multi-Mirror Mission (XMM-Newton)	ESA	X-ray spectroscopy	1999
Galaxy Evolution Explorer (GALEX)	NASA	Ultraviolet observations	2003
Spitzer Space Telescope	NASA	Infrared observations	2004
Swift Gamma-Ray Burst Mission	NASA	Gamma-ray bursts	2004
Convection Rotation and Planetary Transits (COROT) space telescope	CNES (France)	Planet finder	2006
Fermi Gamma-ray Space Telescope	NASA, European partners	Gamma-ray imaging and gamma-ray bursts	2008
Planck telescope	ESA	Cosmic microwave background	2009
Herschel Space Observatory	ESA	Far-infrared and submillimeter observations	2009
Kepler telescope	NASA	Planet finder	2009
Solar Dynamics Observatory (SDO)	NASA	Sun, solar weather	2010
RadioAstron	Russia	Radio space VLBI	2011
Nuclear Spectroscopic Telescopic Array (NuSTAR)	NASA	High-energy X-ray	2012
Gaia	ESA	Optical, digital 3D space camera	2013
James Webb Space Telescope (JWST)	NASA, ESA, CSA (Canada)	Optical and infrared; replacement for HST	2018

to be completely free from Earth's own infrared radiation. The solution was to put it into a *solar* orbit, trailing tens of millions of kilometers behind Earth. The James Webb Space Telescope (JWST), NASA's replacement for HST, will be placed 1.5 million miles away from Earth, at the L_2 Lagrangian point (see Connections 4.2), orbiting the Sun at a fixed distance from the Sun and Earth.

Orbiting telescopes located above the atmosphere are not affected by atmospheric image distortions, weather, or brightening night skies. But space observatories are much more expensive than ground-based observatories, and many of these projects have had cost overruns leading to lengthy delays. In addition, space telescopes can be difficult or impossible to fix. The Hubble Space Telescope required several servicing missions with the space shuttle to install optics to compensate for errors in the mirror and to replace failing parts and install improved instruments, but such missions are not possible for the other observatories in higher orbits. Ground-based telescopes at even the most remote mountaintop locations (except for Antarctica) can receive shipments of replacement parts in a few days; space telescopes cannot. Of course, some wavelengths can be observed from space only. But issues of cost and repair are the reason why ground-based telescopes are much more prevalent.

6.5 Getting Up Close with Planetary Spacecraft

Only in the past half century has the technology existed to explore Earth's local corner of space. Spacecraft have now visited most of the planets and some of their moons, as well as a few comets and asteroids, providing the first close-up views of these distant worlds. The general strategy for exploring the Solar System began with a reconnaissance phase, using spacecraft to fly by or orbit a planet or other body. (A **flyby** is a spacecraft that first approaches and then continues flying past a planet or moon.) As they sped by, instruments aboard these spacecraft briefly probed the physical and chemical properties of their targets and their environments.

Reconnaissance spacecraft employ remote-sensing instrumentation like that used by Earth-orbiting satellites to study our own planet. These instruments include tools such as cameras that take images at different wavelength ranges, radar that can map surfaces hidden beneath obscuring layers of clouds, and spectrographs that analyze the electromagnetic spectrum. Remote sensing enables planetary scientists to map other worlds, measure the heights of mountains, identify geological features and rock types, watch weather patterns develop, measure the composition of atmospheres, and get a general sense of the place. Additional instruments make measurements of the extended atmospheres and space environment through which they travel.

Planetary spacecraft take scientific instruments directly to the planets.

The study of the Solar System from space is an international collaboration involving NASA, the European Space Agency (ESA), the Russian Federal Space Agency (Roscosmos), the Japan Aerospace Exploration Agency (JAXA), the China National Space Administration (CNSA), and the Indian Space Research Organisation (ISRO). Other countries may soon join the endeavor.

Flybys and Orbiters

Recall from Chapter 2 that everyone always sees the same face of the Moon from Earth because the Moon's orbital and rotational periods are equal. With one side of the Moon permanently facing Earth, the other, "far" side (sometimes inappropriately called the "dark side") is hidden from view. On October 18, 1959 the Soviet flyby probe *Luna 3* sent back humanity's first view of the far side of our nearest celestial neighbor. The curious features of the Moon's other side, so different from its Earth-facing half, amazed viewers around the world. No matter how powerful we make our ground-based or Earth-orbiting telescopes, *Luna 3* showed us there is nothing quite like sending a spacecraft for a different view.

Flyby missions have distinct advantages in the reconnaissance phase of exploration. First, they are relatively inexpensive and the easiest missions to design and execute. Second, flyby spacecraft such as *Voyager* (**Figure 6.28a**) are sometimes able to visit several different worlds during their travels (**Figure 6.28b**). The downside of flyby missions is that, because of the physics of orbits, these spacecraft must move by very swiftly. They are limited to just a few hours or at most a few days in which to conduct close-up studies of their targets. Yet flyby spacecraft provide astronomers with their first intimate views of neighboring planets. These images and data can then be used for planning follow-up studies.

More detailed reconnaissance work is done by spacecraft that orbit around planets. These missions are intrinsically more difficult than flyby missions, but **orbiters** can linger, looking in detail at more of the surface of the objects they are orbiting and studying things that change with time, like planetary weather. Spacecraft have orbited the Moon, Mercury, Venus, Mars, Jupiter, Saturn, and even an asteroid. **Figure 6.29** shows the *Messenger* spacecraft, which is in orbit around Mercury.

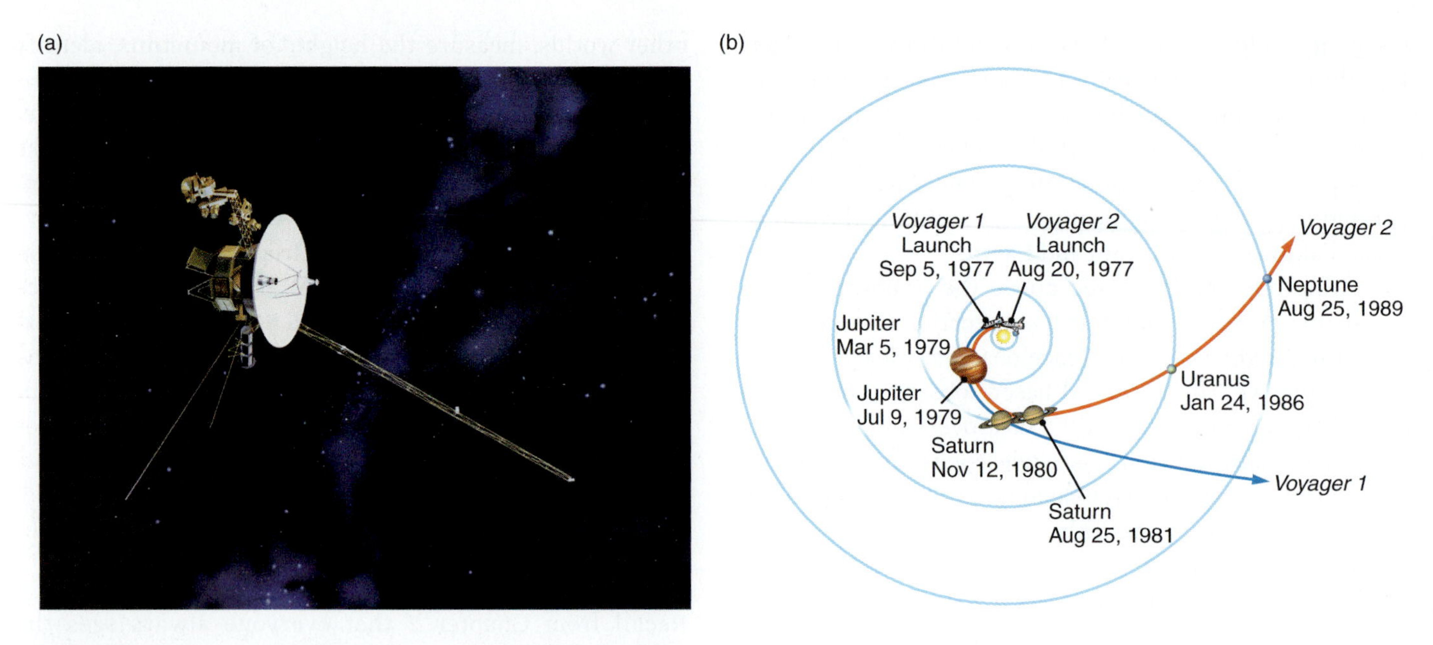

FIGURE 6.28 Explorers of the planets. *Voyager* flew past Jupiter, Saturn, Uranus, and Neptune.

FIGURE 6.29 Artist's depiction of *Messenger*, which is currently in orbit about Mercury; the enhanced color image of Mercury is from its 2008 flyby.

Landers, Rovers, and Atmospheric Probes

Reconnaissance spacecraft provide a wealth of information about a planet, but there is no better way of obtaining "ground truth" than to put our instruments where they can get right to the heart of it—within a planet's atmosphere or on solid ground. Spacecraft have landed on the Moon, Mars, Venus, Saturn's large moon Titan, and the asteroid Eros. These spacecraft have taken pictures of planetary surfaces, measured surface chemistry, and conducted experiments to determine the physical properties of the surface rocks and soils.

One disadvantage of using **landers**—spacecraft that touch down and remain on the surface—is that only a few landings in limited areas are practical, because of the expense, and the results may apply only to the small area around the landing site. Imagine, for example, what a different picture of Earth you might get from a spacecraft that landed in Antarctica, as opposed to a spacecraft that landed in a volcano or the floor of a dry riverbed. Sites to be explored with landed spacecraft must be very carefully chosen on the basis of reconnaissance data. Some of the limitations of landers can be mitigated by putting them on wheels and sending them from place to place, to explore the vicinity of the landing site. Such remote-controlled vehicles, called **rovers**, were used first by the Soviet Union on the Moon four decades ago, and more recently by the United States on Mars. **Figure 6.30** shows an artist's view of the *Curiosity* rover on Mars.

Spacecraft have landed on several moons and planets.

FIGURE 6.30 The robotic rover *Curiosity* on the surface of Mars.

Probes have been sent into the atmospheres of Venus, Jupiter, and Titan. As they descend, **atmospheric probes** continually measure and send back data on temperature, pressure, and wind speed, along with other properties, such as chemical composition. Meteorologists take measurements of Earth's atmosphere from the surface up by sending their instruments aloft in balloons. By contrast, planetary scientists must work from the top down by suspending their instruments from parachutes. Atmospheric probes have survived all the way to the solid surfaces of Venus and Titan, sending back streams of data during their descent. An atmospheric probe sent into Jupiter's atmosphere never reached that planet's surface because, as we will discuss later in the book, Jupiter does not have a solid surface in the same sense that terrestrial planets and moons do. After sending back its data, the Jupiter probe eventually melted and vaporized as it descended into the hotter layers of the planet's atmosphere.

Sample Returns

If you pick up a rock from the side of a mountain road, you might learn a lot from the rock using tools that you could carry in your pocket or your car. But it would be much better to pick up a few samples and carry them back to a laboratory equipped with a full range of state-of-the-art instruments capable of measuring chemical composition, mineral type, age, and other information needed to reconstruct the story of your rock sample's origin and evolution. The same is true of Solar System exploration. One of the most powerful methods for investigating remote objects is to collect samples of the objects and bring them back to Earth for detailed study. So far, only samples of the Moon, a comet, and the **solar wind** (a stream of charged particles from the Sun) have been collected and returned to Earth. Scientists have found meteorites on Earth that are likely pieces of Mars that were blasted loose by bodies that had crashed into Mars. Someday there may be unmanned "sample and return" missions to Mars.

The missions discussed so far in this section have all been conducted with robotic spacecraft. The only spacecraft that took people to another world were the *Apollo* missions to the Moon. This program ran from 1961 to 1972 and included several missions before the actual Moon landings. The *Apollo 8* astronauts brought back the now-famous picture of Earth rising over the surface of the Moon (**Figure 6.31**). Each mission from *Apollo 11* through *Apollo 17* had three astronauts—two to land on the Moon and one to remain in orbit. One mission (*Apollo 13*) did not reach the Moon but returned to Earth safely. Twelve American astronauts walked on the Moon between 1969 and 1972, and they brought back 382 kg of rocks and other material.

The return of extraterrestrial samples to Earth is governed by international treaties and standards to ensure that these samples do not contaminate Earth. For example, before the lunar samples brought back by the *Apollo* missions could be studied, they (and the astronauts) had to be placed in quarantine and tested for alien life-forms. The

FIGURE 6.31 The first view of Earth seen from deep space. In December 1968, *Apollo 8* astronauts photographed Earth rising above the Moon's limb.

same international standards apply to spacecraft landing on planets. The goal of these standards is to avoid *forward contamination*, or transporting life-forms from Earth to another planet. If there is life on other planets, there is concern about introducing potential harm, and from a scientific perspective we do not want to "discover" life that we, in fact, introduced.

With numerous missions under way and others on the horizon, unmanned exploration of the Solar System is an ongoing, dynamic activity. **Table 6.3** summarizes some recent and current missions. Information on the latest discoveries can be found on mission websites, and in science news sources (such as the news feed on the *21st Century Astronomy* website).

6.6 Other Astronomical Tools

High-profile space missions have sent back stunning images and data from across the electromagnetic spectrum, but astronomers use other tools as well, including particle accelerators and colliders, neutrino and gravitational-wave detectors, and high-speed computers.

Ever since the early years of the 20th century, physicists have been peering into the structure of the atom by observing what happens when small particles collide. By the 1930s, physicists had developed the technology to accelerate charged subatomic particles such as protons to very high speeds and then observe what happens when they slam into a target. From such experiments (which are continuing even today), physicists have discovered many kinds of subatomic particles and learned about their physical properties. High-energy particle colliders have proved to be an essential tool for physicists studying the basic building blocks of matter.

Astronomers have realized that to understand the very largest structures seen in the universe—indeed, the universe itself—it is important to understand the physics that took place during the earliest moments in the universe, when everything was unbelievably hot and dense. High-energy particle colliders that physicists use today are designed to approach the energies of the early universe.

Results from particle colliders are used to understand the early universe and the formation of structure.

Two factors determine the effectiveness of particle accelerators: the energy they can achieve and the number of particles they can accelerate. Modern particle colliders such as the Large Hadron Collider near Geneva, Switzerland (**Figure 6.32**) reach very high energies. Particles can also be studied from space. The Alpha Magnetic Spectrometer, installed on the International Space Station in 2011, will search for some of the most exotic forms of matter, such as dark matter, antimatter, and high-energy particles called cosmic rays.

The **neutrino** is an elusive elementary particle that plays a major role in the physics of the interiors of stars. You cannot see beneath the surface of stars like the Sun, but observations of neutrinos can provide important insight into what is happening deep within. However, neutrinos are extremely difficult to detect, and are nearly impossible to catch. In less time than it takes you to read this sentence, a thousand trillion (10^{15}) solar neutrinos from the Sun are passing through your body. Neutrinos are so nonreactive with matter that they can pass right through Earth (and you) as though it (or you) weren't there at all. A neutrino has to interact with a detector to be observed. Neutrino detectors typically record only one out of every 10^{22} (10 billion trillion) neutrinos passing though them, but that's enough to reveal processes deep within the Sun or the violent death of a star 160,000 light-years away.

Experiments designed to look for neutrinos originating outside of Earth are buried deep underground in mines or caverns, or under the ocean or ice to ensure that only neutrinos are detected. For example, the ANTARES experiment uses the Mediterranean Sea as a neutrino telescope. Detectors located 2.5 km under the sea, off the coast of France, observe neutrinos that originated in southern skies and have passed through Earth. In the IceCube neutrino observatory, a neutrino telescope located at the South Pole

FIGURE 6.32 The ATLAS particle detector at CERN's Large Hadron Collider near Geneva, Switzerland. The enormous size of this instrument is evident from the person standing near the bottom center of the picture.

TABLE 6.3 | Selected Recent and Current Solar System Missions

Spacecraft	Sponsoring Nation(s)*	Destination	Launch Year	Type	Status (mid-2012)
Voyager 1 and *2*	USA	Jupiter, Saturn, Uranus (2), Neptune (2)	1977	Flyby	Actively exploring outer edge of Solar System
Galileo	USA	Jupiter	1989	Orbiter/probe	Ended 2003
Ulysses	USA, Europe	Sun	1990	Solar polar orbiter	Ended 2008
Mars Global Surveyor	USA	Mars	1996	Orbiter	Ended 2006
Cassini-Huygens	USA, Europe, Italy	Saturn, Titan	1997	Saturn orbiter, Titan probe/lander	Orbiter active
Stardust	USA	Comets	1999	Sample return/flyby	Ended 2011
Mars Odyssey	USA	Mars	2001	Orbiter	Active
Mars Exploration Rover	USA	Mars	2003	Two landers	One rover active
Hayabusa	Japan	Asteroid	2003	Sample return	Ended 2010
Mars Express	Europe	Mars	2003	Orbiter	Active
Messenger	USA	Mercury (2011)	2004	Orbiter	Active
Venus Express	Europe	Venus	2005	Orbiter	Active
Mars Reconnaissance Orbiter (MRO)	USA	Mars	2005	Orbiter	Active
Deep Impact/EPOXI	USA	Comet Hartley (2010)	2005	Impactor/flyby	Ended 2010
SOHO	USA, Europe	Sun	2005	Orbiter	Active
STEREO	USA	Sun	2006	Two orbiters	Active
New Horizons	USA	Pluto (2015)	2006	Flyby	En route
Chang'e 1	China	Moon	2007	Orbiter	Ended 2009
Kayuga	Japan	Moon	2007	Orbiter	Ended 2009
Artemis	USA	Moon, solar wind	2007		Active
Dawn	USA	Vesta (2011), Ceres (2015)	2007	Orbiter	Active
Chandrayaan	India	Moon	2008	Orbiter/impactor	Ended 2009
Lunar Reconnaissance Orbiter (LRO)	USA	Moon	2009	Orbiter	Active
Lunar Crater Observation and Sensing Satellite (LCROSS)	USA	Moon	2009	Impactor	Ended 2009
Chang'e 2	China	Moon	2010	Orbiter	Ended 2011
Juno	USA	Jupiter (2016)	2011	Orbiter	En route
Gravity Recovery and Interior Laboratory (GRAIL)	USA	Moon	2011	Two orbiters	Active
Mars Science Laboratory	USA	Mars	2011	Lander	Active
Mars Atmosphere and Volatile EvolutioN (MAVEN) mission	USA	Mars	2013	Orbiter	Set to launch in late 2013

*Countries are represented by the following agencies: China = CNSA (China National Space Administration); Europe = ESA (European Space Agency); India = ISRO (Indian Space Research Organisation); Italy = Italian Space Agency; Japan = JAXA (Japan Aerospace Exploration Agency); USA = NASA (National Aeronautics and Space Administration).

FIGURE 6.33 The IceCube neutrino telescope at the South Pole, Antarctica.

FIGURE 6.34 The Laser Interferometer Gravitational-Wave Observatory (LIGO) near Richmond, Washington. A similar telescope is located in Livingston, Louisiana.

in Antarctica, the detectors are 1.5–2.5 km under the ice, and they observe neutrinos that originated in northern skies (**Figure 6.33**).

Another elusive phenomenon is the **gravitational wave**. Gravitational waves are disturbances in a gravitational field, similar to the waves that spread out from the disturbance you create when you toss a pebble onto the quiet surface of a pond. There is strong, although indirect, observational evidence for the existence of gravitational waves, but they are so elusive that they have not yet actually been detected. (See the **Process of Science Figure**.) Several facilities, including the Laser Interferometer Gravitational-Wave Observatory, or LIGO (**Figure 6.34**), have been constructed to detect gravitational waves. Scientists are eager to detect gravitational waves—to confirm their existence and to study the physical phenomena they are likely to reveal, such as the birth and evolution of the universe, stellar evolution, or the very force of gravity itself.

A survey of the astronomer's tools would not be complete without a discussion of the essential role of computers. Data gathering, analysis, and interpretation depend largely on computers—and the more powerful, the better. Consider, for example, analyzing a night's worth of astronomical images recorded by a very large CCD. A single image may contain as many as 100 million pixels, with each pixel displaying roughly 30,000 levels of brightness. That adds up to a several *trillion* pieces of information in each image. And that's only one image. To analyze their data, astronomers typically do calculations on *every single pixel* of an image in order to remove unwanted contributions from Earth's atmosphere or to correct for instrumental effects.

High-speed computers also play an essential role in generating and testing theoretical models of astronomical objects. Even when we completely understand the underlying physical laws that govern the behavior of a particular object, often the object is so complex that it would be impossible to calculate its properties and behavior without the assistance of high-speed computers. For example, as you learned in Chapter 4, you can use Newton's laws to compute the orbits of two stars that are gravitationally bound to one another, because their orbits take the form of simple ellipses. However, it is not so easy to understand the orbits of the several hundred billion stars that make up the Milky Way Galaxy, even though the underlying physical laws are the same.

Computer modeling has worked well for determining the interior properties of stars and planets, including Earth. Although astronomers cannot see beneath their surfaces, they have a surprisingly good understanding of their interiors, as we will describe in later chapters. Astronomers start a model by assigning well-understood physical properties to tiny volumes within a planet or star. The computer assembles an enormous number of these individual elements into an overall representation. When it is all put together, the result is a rather good picture of what the interior of the star or planet is like.

Astronomers also utilize supercomputers to study the evolution of astronomical objects or systems over time. For example, astronomers create models of galaxies, and then run computer simulations to study how those galaxies might change over billions of years. For example, **Figure 6.35** shows a simulation of the collision of two galaxies. The results of the computer simulations are then compared with telescopic observations.

Process of Science

TECHNOLOGY AND SCIENCE ARE SYMBIOTIC

Scientists have been searching for waves that carry gravitational information for nearly 100 years, but the accuracy of their measurements is limited by the available technology.

Weber Bar

Precision-machined bars of metal that should "ring" as a gravitational wave passes by.

Sensitive only to extremely powerful gravitational waves.

No detection.

Take 2

LIGO

New Technology: Lasers

Lasers should interfere as gravitational wave passes by.

Roughly 100 times more sensitive than Weber bar measurements.

No detection (yet).

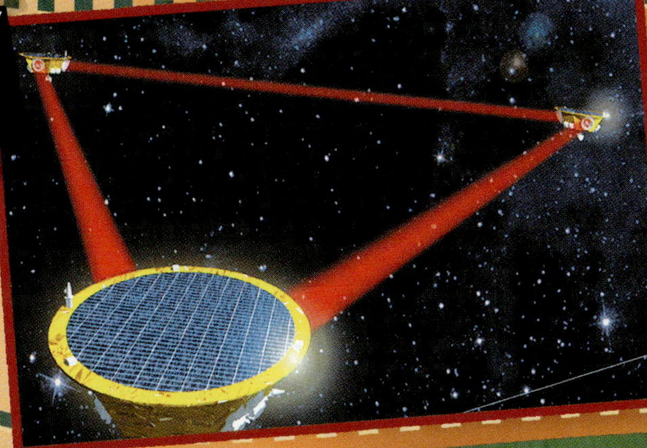

LISA (Future Science Mission)

Lasers will interfere as gravitational waves pass by.

Sensitive to more types of objects than LIGO is.

Technology and science develop together. New technologies enable humans to ask new scientific questions. Asking scientific questions pushes the development of better instrumentation. Deeper scientific understanding leads to new technologies.

FIGURE 6.35 Supercomputer simulations of the collision of two galaxies. Astronomers compare simulations like these with telescopic observations.

6.7 Origins: Microwave Telescopes That Detect Radiation from the Big Bang

In this chapter we explored the tools of the astronomer, from basic optical telescopes to instruments that observe in different wavelengths. Now let's examine in more detail one type of telescope that has aided in the study of the origin of the universe. Recall from Chapter 1 that astronomers think the universe originated with a hot Big Bang. The multiple strands of evidence for this conclusion will be discussed in Chapter 19; here we look at one piece: the observation of faint microwave radiation left over from this early hot universe. Two Bell Labs physicists, Arno Penzias (1933–) and Robert Wilson (1936–), were working on satellite communications when they first accidentally detected this radiation in 1964 with a microwave dish antenna in New Jersey. Today we routinely use cell phones and handheld GPS systems that communicate directly with satellites, but at the time, this capability was at the limit of technology.

Penzias and Wilson needed a very sensitive microwave telescope for the work they were doing for Bell, because any spurious signals coming from the telescope itself might wash out the faint signals bounced off a satellite. To that end, they were working very hard to eliminate all possible sources of interference originating from within their instrument (including keeping the telescope free of bird droppings). They found that no matter how carefully they tried to eliminate sources of extraneous noise, they always still detected a faint signal at microwave wavelengths. Penzias and Wilson shared the 1978 Nobel Prize in Physics for the discovery of this **cosmic microwave background radiation (CMB)** left over from the Big Bang itself.

In the nearly 50 years since, astronomers from around the world have designed increasingly precise instruments to measure this radiation from the ground, from high-altitude balloons, from rockets, and from satellites. The Russian experiment RELIKT-1, launched in 1983, found some limits on the variation of the CMB. The COBE (Cosmic Background Explorer) satellite, launched in 1989, showed that the spectrum of this radiation precisely matched that of a blackbody with a temperature of 2.73 K—exactly what was predicted for the radiation left over from the Big Bang (**Figure 6.36**). (When the results were presented at an astronomy conference in Washington DC in 1990, the large audience stood up and cheered.) The data also show some slight differences in temperature—small fractions of a degree—over the map of the sky. These slight variations tell us about how the universe evolved from one that was dominated by radiation to one that contains structures such as galaxies, stars, planets, and us. John Mather and George Smoot shared the 2006 Nobel Prize in Physics for this work.

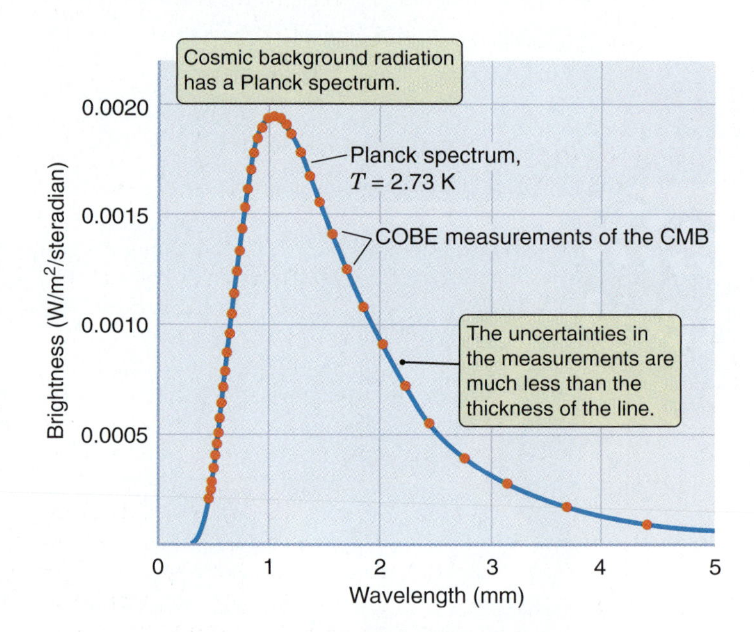

FIGURE 6.36 The spectrum of the cosmic microwave background radiation (CMB) as measured by the COBE satellite (red dots). A steradian is a unit of solid angle. The uncertainty in the measurement at each wavelength is much less than the size of a dot. The line running through the data is a Planck blackbody spectrum with a temperature of 2.73 K.

In 1998 and 2003, a high-altitude balloon experiment called BOOMERANG (short for "balloon observations of millimetric extragalactic radiation and geophysics") flew over Antarctica at an altitude of 42 km to study CMB variations and estimate the overall geometry of the universe. The *WMAP* (Wilkinson Microwave Anisotropy Probe) satellite, launched in 2001, created an even more detailed map of the temperature variations in this radiation, yielding more precise values for the age and shape of the universe, and the presence of dark matter and dark energy. The Atacama Cosmology Telescope (Chile) and the South Pole Telescope (Antarctica) study this radiation to look for evidence of when galaxy clusters formed. The newest microwave observatory in space, the Planck telescope, was launched in 2009 by the European Space Agency. Planck has much greater sensitivity than *WMAP*, and it is expected to be able to study these CMB variations in even more precise detail. These experiments and observations have opened up the current era of precision cosmology, in which astronomers can make detailed models of how the universe was born.

The many technological tools used by astronomers, along with Kepler's laws, Newton's laws, Doppler shifts, Wien's law, the Stefan-Boltzmann law, energy states, spectral lines, and the rest, are the keys to an understanding of planets, stars, galaxies, and ultimately the universe itself. The first place we will bring these tools to bear will be our immediate neighborhood as we consider the nature and origin of the Solar System. ■

Summary

6.1 The telescope is the astronomer's most important tool. Optical telescopes come in two basic types: refractors and reflectors. All large astronomical telescopes are reflectors. Large telescopes collect more light and have greater resolution. The diffraction limit is the limiting resolution of a telescope.

6.2 Photography improved astronomers' ability to record details of faint objects seen in telescopes. Modern CCDs are today's astronomical detector of choice because they are much more linear, have broader spectral response, and can send electronic images directly to the computer.

6.3 Earth's atmosphere blocks many spectral regions and distorts telescopic images. Radio and infrared telescopes are able to see through our atmosphere and through vast clouds of cosmic gas and dust. Radio and optical telescopes can be arrayed to greatly increase angular resolution.

6.4 Putting telescopes in space solves problems created by Earth's atmosphere.

6.5 Most of what is known about the planets and moons comes from observations by spacecraft. Flyby and orbiting missions obtain data from space, and landers and rovers collect data from the ground.

6.6 Astronomers also use particle accelerators, neutrino detectors, and gravitational-wave detectors to study the universe. High-speed computers are essential to the acquisition, analysis, and interpretation of astronomical data.

6.7 Telescopes observing at microwave wavelengths have detected radiation left over from the Big Bang.

Unanswered Questions

- Will telescopes be placed on the Moon? The Moon has no atmosphere to make stars twinkle, cause weather, or block certain wavelengths of light from reaching its surface. The far side of the Moon faces away from the light and radio radiation of Earth, and all parts of the Moon have nights that last for two Earth weeks. One proposal calls for a Lunar Array for Radio Cosmology (LARC), an array of hundreds of radio telescopes that would be deployed on the Moon—after the year 2025—to study the earliest formation of stars and galaxies. Another proposal is for the Lunar Liquid Mirror Telescope (LLMT), with a diameter of 20–100 meters, to be located at one of the Moon's poles. Gravity would settle the rotating liquid into the necessary parabolic shape, and these telescopes are much simpler than are arrays of telescopes with large glass mirrors. The LLMT would observe extremely distant protostars and protogalaxies in infrared wavelengths. Astronomers debate whether telescopes on the Moon would be easier to service and repair than those in space, and whether problems caused by lunar dust would outweigh any advantages.
- Will there be human exploration of the Solar System within your lifetime? Since the *Apollo* program, humans have not returned to the Moon or traveled to other planets or moons in the Solar System. Sending humans to the worlds of the Solar System is much more complicated, risky, and expensive than sending robotic spacecraft. Humans need life support such as air, water, and food; and radiation in space can be dangerous. Furthermore, human explorers would expect to return to Earth, whereas most spacecraft do not come back. Astronomers and space scientists have heated debates about human spaceflight versus robotic exploration. Some argue that true exploration requires that human eyes and brains actually go there; others argue that the costs and risks are too high for the potential additional scientific knowledge. Beyond basic exploration, we also do not know whether humans will ever permanently colonize space.

Questions and Problems

Summary Self-Test

1. Suppose that you are handed a telescope. The tube is roughly as long as your arm, but it is difficult to hold because you can barely get your arms around it. This telescope is most likely a
   a. refractor.
   b. reflector.

2. All large astronomical telescopes are reflectors because
   a. chromatic aberration is minimized.
   b. they are not as heavy as refracting telescopes.
   c. they can be shorter than refracting telescopes.
   d. all of the above

3. You are shopping for telescopes online. You find two in your price range. One of these has an aperture of 20 cm, and one has an aperture of 30 cm. Which should you choose, and why?
   a. The 20 cm, because the light-gathering power will be better.
   b. The 20 cm, because the image size will be larger.
   c. The 30 cm, because the light-gathering power will be better.
   d. The 30 cm, because the image size will be larger.

4. The Kepler Mission telescope observes primarily in visible light. This telescope is located in space because
   a. visible light does not make it through Earth's atmosphere.
   b. it is closer to the targets it is observing.
   c. it is above atmospheric distortion.
   d. it is safe from weather-related disasters.

5. Which of the following can be observed from Earth's surface?
   a. radio waves
   b. gamma radiation
   c. far UV light
   d. X-ray light
   e. visible light

6. Match the following properties of telescopes (lettered) with their corresponding definitions (numbered).
   a. aperture
   b. resolution
   c. focal length
   d. chromatic aberration
   e. diffraction
   f. interferometer
   g. adaptive optics

   (1) two or more telescopes connected to act as one
   (2) distance from lens to focal plane
   (3) diameter
   (4) ability to distinguish close objects
   (5) computer-controlled atmospheric distortion correction
   (6) rainbow-making effect
   (7) smearing effect due to sharp edge

7. The major advantage CCDs have over other imaging techniques is that they
   a. have a higher quantum efficiency.
   b. have a linear response to light.
   c. yield output in digital format.
   d. operate at visible and near-infrared wavelengths.
   e. all of the above

8. Spacecraft are the most effective way to study planets in our Solar System because
   a. planets move too fast across the sky for us to image them well from Earth.
   b. planets cannot be imaged from Earth.
   c. they can collect more information than is available just from images.
   d. space missions are easier than long observing campaigns.

9. Which of the following Solar System objects has yet to be observed by flyby spacecraft?
   a. Uranus
   b. Venus
   c. Pluto
   d. Neptune

10. Which of the following are used by astronomers to understand the universe? (Select all that apply.)
    a. telescopes
    b. particle accelerators
    c. neutrino detectors
    d. supercomputers
    e. gravitational-wave detectors
    f. microscopes
    g. papers

True/False and Multiple Choice

11. **T/F:** Chromatic aberration is a problem that limits the image quality of reflecting telescopes.

12. **T/F:** In the past, astronomers placed telescopes in high-flying aircraft in an effort to rise above the water vapor in Earth's atmosphere.

13. **T/F:** Radio telescopes are often used in interferometric arrays to increase light-gathering power.

14. **T/F:** One of the reasons why the Hubble Space Telescope has better spatial resolution than a 4-meter ground-based telescope is that it has a larger mirror.

15. **T/F:** Robotic spacecraft have visited most of the planets and moons of the Solar System.

16. Refraction is caused by
    a. light bouncing off a surface.
    b. light changing colors as it enters a new medium.
    c. light speeding up as it enters a new medium.
    d. light slowing down as it enters a new medium.

17. The light-gathering power of a 4-meter telescope is ________ than that of a 2-meter telescope.
    a. 8 times larger
    b. 4 times larger
    c. 16 times smaller
    d. 2 times smaller

18. Improved resolution is helpful to astronomers because
    a. they often want to look in detail at small features of an object.
    b. they often want to look at very distant objects.
    c. they often want to look at many objects close together.
    d. all of the above

19. The part of the human eye that acts as the detector is the
    a. retina.
    b. pupil.
    c. lens.
    d. iris.

20. Astronomers put telescopes in space to
    a. get closer to the stars.
    b. avoid atmospheric effects.
    c. look primarily at radio wavelengths.
    d. improve quantum efficiency.

21. The advantage of an interferometer is that
    a. the resolution is dramatically increased.
    b. the focal length is dramatically increased.
    c. the light-gathering power is dramatically increased.
    d. diffraction effects are dramatically decreased.
    e. chromatic aberration is dramatically decreased.

22. The angular resolution of a ground-based telescope is usually determined by
    a. diffraction.
    b. the focal length.
    c. refraction.
    d. atmospheric seeing.

23. A grating is able to spread white light out into a spectrum of colors because of the property of
    a. reflection.
    b. refraction.
    c. dispersion.
    d. interference.

24. Which causes the biggest problem in detecting infrared photons from an astronomical object?
    a. smog
    b. carbon dioxide
    c. water vapor
    d. light pollution

25. Robotic landers are more common than sample-return spacecraft because
    a. there is nothing to be learned by bringing back a bunch of rocks.
    b. carrying enough fuel to get back to Earth makes sample return expensive.
    c. the science instruments on landers are equal to any we have here on Earth.
    d. we are nervous about cross-contamination of life-forms.

Thinking about the Concepts

26. Galileo's telescope used simple lenses. What is the primary disadvantage of using a simple lens in a refracting telescope?

27. The largest astronomical refractor has an aperture of 1 meter. List several reasons why it would be impractical to build a larger refractor with twice this aperture.

28. Your camera may have a zoom lens, ranging between wide angle (short focal length) and telephoto (long focal length). How does the size of an object in the camera's focal plane differ between wide angle and telephoto?

29. Optical telescopes reveal much about the nature of astronomical objects. Why do astronomers also need information provided by gamma-ray, X-ray, infrared, and radio telescopes?

30. For light reflecting from a flat surface, the angles of incidence and reflection are the same. This is also true for light reflecting from the curved surface of a reflecting telescope's primary mirror. Sketch a curved mirror and several of these reflecting rays.

31. Explain constructive and destructive interference.

32. Consider two optically perfect telescopes having different diameters but the same focal length. Is the image of a star larger or smaller in the focal plane of the larger telescope? Explain your answer.

33. Explain why stars twinkle.

34. Explain adaptive optics, and how they improve a telescope's image quality.

35. Explain integration time and quantum efficiency, and how each contributes to the detection of faint astronomical objects.

36. Some people believe that we put astronomical telescopes on high mountaintops or in orbit because doing so gets them closer to the objects they are observing. Explain what is wrong with this popular misconception, and give the actual reason telescopes are located in these places.

37. Humans have sent various kinds of spacecraft—including flybys, orbiters, and landers—to all of the planets in our Solar System. Explain the advantages and disadvantages of each of these types of spacecraft.

38. If there are meteorites that are pieces of Mars on Earth, why is it so important to go to Mars and bring back samples of the martian surface?

39. Humans had a first look at the far side of the Moon as recently as 1959. Why had we not seen it earlier—when Galileo first observed the Moon with his telescope in 1610?

40. Where are neutrino detectors located? Why are neutrinos so difficult to detect?

Applying the Concepts

41. (a) Study Figure 6.3. For a single-lens system like this, is the image of the sky upright or inverted? If you wished to make the image of the sky right side up, how could you do it? (b) Answer the same questions for the double-mirror system shown in Figure 6.6.

42. Study the photograph of light entering and leaving a block of refractive material in Figure 6.7b. Use a protractor to measure the angles of the green light as it enters the block and as it leaves the block. How are these angles related?

43. Many amateur astronomers start out with a 4-inch (aperture) telescope and then graduate to a 16-inch telescope. By what factor does the light-gathering power of the telescope increase with this upgrade? How much fainter are the faintest stars that can be seen in the larger telescope?

44. The resolution of the human eye is about 1.5 arcmin. What would the aperture of a radio telescope (observing at 21 cm) have to be to have this resolution? Even though the atmosphere is transparent at radio wavelengths, humans do not see light in the radio range. Using your calculations and a little logic, explain why.

45. Assume that you have a telescope with an aperture of 1 meter. Compare the telescope's theoretical resolution when you are observing in the near-infrared region of the spectrum ($\lambda = 1{,}000$ nm) with that when you are observing in the violet region of the spectrum ($\lambda = 400$ nm).

46. Assume that the maximum aperture of the human eye, D, is approximately 8 mm and the average wavelength of visible light, λ, is 5.5×10^{-4} mm.
    a. Calculate the diffraction limit of the human eye in visible light.
    b. How does the diffraction limit compare with the actual resolution of 1–2 arcmin (60–120 arcsec)?
    c. To what do you attribute the difference?

47. The diameter of the full Moon in the focal plane of an average amateur's telescope (focal length 1.5 meters) is 13.8 mm. How big would the Moon be in the focal plane of a very large astronomical telescope (focal length 250 meters)?

48. One of the earliest astronomical CCDs had 160,000 pixels, each recording 8 bits (256 levels of brightness). A new generation of astronomical CCDs may contain a billion pixels, each recording 15 bits (32,768 levels of brightness). Compare the number of bits of data that each of these two CCD types produces in a single image.

49. Consider a CCD with a quantum efficiency of 80 percent and a photographic plate with a quantum efficiency of 1 percent. If an exposure time of 1 hour is required to photograph a celestial object with a given telescope, how much observing time would be saved by substituting a CCD for the photographic plate?

50. The VLBA employs an array of radio telescopes ranging across 8,000 km of Earth's surface from the Virgin Islands to Hawaii.
    a. Calculate the angular resolution of the array when radio astronomers are observing interstellar water molecules at a microwave wavelength of 1.35 cm.
    b. How does this resolution compare with the angular resolution of two large optical telescopes separated by 100 meters and operating as an interferometer at a visible wavelength of 550 nm?

51. When operational, the SVLBI may have a baseline of 100,000 km. What will be the angular resolution when studying interstellar molecules emitting at a wavelength of 17 mm from a distant galaxy?

52. The *Mars Reconnaissance Orbiter* (*MRO*) flies at an average altitude of 280 km above the martian surface. If its cameras have an angular resolution of 0.2 arcsec, what is the size of the smallest objects that the *MRO* can detect on the martian surface?

53. The martian rover *Curiosity* can move across the landscape of Mars at speeds of up to 4 cm/s. In contrast, our typical walking speed is about 4 km per hour (km/h).
    a. How long would it take *Curiosity* to cross a soccer field (110 meters)?
    b. How long would it take you to walk the same distance?

54. *Voyager 1* is now about 125 astronomical units (AU) from Earth, continuing to record its environment as it approaches the limits of our Solar System.
    a. How far away is *Voyager 1*, in kilometers?
    b. How long does it take observational data to come back to us from *Voyager 1*?
    c. How does *Voyager 1*'s distance from Earth compare with that of the nearest star (other than the Sun)?

55. The speed, wavelength, and frequency of gravitational waves are related as $c = \lambda \times f$. If we were to observe a gravitational wave from a distant cosmic event with a frequency of 10 hertz (Hz), what would be the wavelength of the gravitational wave?

Using the Web

56. A webcast for the International Year of Astronomy 2009 called "Around the World in 80 Telescopes" can be accessed at http://eso.org/public/events/special-evt/100ha.html. The 80 telescopes are situated all over, including Antarctica and space. Pick two of the telescopes and watch the videos. Do you think these videos are effective for public outreach for the observatory in question or for astronomy in general? For each telescope you choose, answer the following questions: Does the telescope observe in the Northern Hemisphere or the Southern Hemisphere? What wavelengths does the telescope observe? What are some of the key science projects at the telescope?

57. Most major observatories have their own websites. Use the link in question 56 to find a master list of telescopes, and click on a telescope name to link to an observatory website (or run a search on names from Tables 6.1 and 6.2). For the telescope you choose, answer the following questions: (a) What is this telescope's "claim to fame"—is it the largest? at the highest altitude? at the driest location? with the darkest skies? the newest? (b) Does the observatory website have news releases? What is a recent discovery from this telescope?

58. In the chapter we mentioned several radio telescopes under construction. Do a search to find the status of the Allen Telescope Array (ATA), the Square Kilometer Array (SKA), and the Five-hundred-meter Aperture Spherical Telescope (FAST). When is each one scheduled to be completed?

59. What is the current status of the James Webb Space Telescope, JWST (http://jwst.nasa.gov); that is, when is the expected launch date? How will this telescope be different from the Hubble Space Telescope? What are some of the instruments for the JWST and its planned projects? What is the current estimated cost of the JWST?

60. Pick a mission from Table 6.3, go to its website, and see what's new. For the mission you choose, answer the following questions: Is the spacecraft still active? Is it sending images? What new science is coming from this mission?

Exploration | Geometric Optics and Lenses

Visit StudySpace, and open the "Geometric Optics and Lenses" animation in Chapter 6. Read through the animation until you reach the optics simulation, pictured in **Figure 6.37**. The simulator shows a converging lens and a pencil. Rays come from the pencil on the left of the converging lens, pass through the lens, and make an image to the right of the lens. The view that would be seen by an observer at the position of the eye is shown in the circle at upper right. Initially, when the pencil is at position 2.3 and the eye is at position 2.0, the pencil is out of focus and blurry.

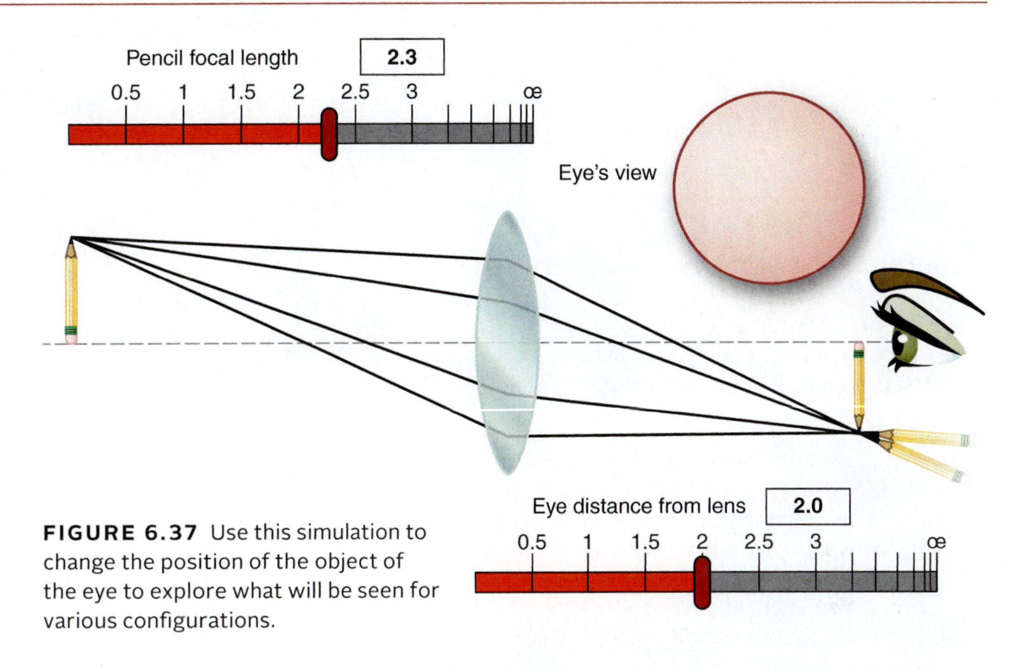

FIGURE 6.37 Use this simulation to change the position of the object of the eye to explore what will be seen for various configurations.

1 **Is the eraser at the top or the bottom of the actual pencil? (This becomes important later.)**

....

Using the red slider in the upper left of the window, try moving the pencil to the right. Pause when the observer's eye sees a recognizable pencil (even if it's still blurry).

2 **Does the eye see the pencil right side up, or upside down?**

....

3 **This is somewhat analogous to the view through a telescope. The objects are very far from the lenses, and the observer sees things upside down in the telescope. If an object in your field of view is at the top of the field and you want it in the center, which way should you move the telescope: up or down?**

....

Now return the pencil to position 2.3. Use the red slider in the lower right of the window to move the eye closer to the lens (to the left).

4 **At what distance does the image of the pencil first become crisp and clear?**

....

5 **Is the pencil right side up or upside down?**

....

6 **In practice at the telescope, we do not move the observer back (away from the eyepiece) to bring the image into focus. Why not?**

....

7 **Instead of moving the observer, we use a focusing knob to move the lens in the eyepiece, which brings the image into focus. Imagine that you are looking through the eyepiece of a telescope and the image is blurry. You turn the focusing knob and things get blurrier! What should you try next?**

....

8 **Now imagine that you get the image focused just right, so it is crisp and sharp. The next person to use the telescope wears glasses and insists that the image is blurry. But when you look through the telescope again, the image is still crisp. Explain why your experiences differ.**

....

Step through the animation to the next picture. Carefully study the two telescopes shown, and the path the light takes through them.

9 **Which telescope has a longer focal length: the top one or the bottom one?**

....

10 **Which telescope produces an image with the red and the blue stars more separated: the top one or the bottom one?**

....

11 **A longer focal length is an advantage in one sense, but it's not the entire story. What are some disadvantages of a telescope with a very long focal length?**

....

From clouds of gas and dust, solar systems are born.

07 The Birth and Evolution of Planetary Systems

There are countless suns and countless earths all rotating around their suns in exactly the same way as the seven planets of our system. We see only the suns because they are the largest bodies and are luminous, but their planets remain invisible to us because they are smaller and non-luminous. The countless worlds in the universe are no worse and no less inhabited than our Earth.

Giordano Bruno (1548–1600), De l'Infinito, Universo e Mondi

LEARNING GOALS

The solar system containing planet Earth—*our* Solar System—is an unmistakable by-product of the birth of the Sun. The discovery of planetary systems surrounding other stars has shown that the Solar System is not unique. The physical processes that shaped its formation also led to the formation of other multiplanet systems. By the conclusion of this chapter, you should be able to:

- Summarize the role that gravity, energy, and angular momentum play in the formation of stars and planets.
- Describe the modern theory of planetary system formation.
- Explain how temperature in the disk that surrounds a forming star affects the composition and location of planets, moons, and other bodies.
- Compare and contrast the processes that resulted in the inner and outer planets that form the Solar System.
- Describe how astronomers find planets around other stars and what those discoveries tell us about our own and other solar systems.

7.1 Stars Form and Planets Are Born

Earth is part of a collection of objects surrounding an ordinary star, the Sun. Astronomers refer to such a system of planets surrounding a star as a **planetary system**, and the planetary system that includes Earth is called the **Solar System** (**Figure 7.1**). Planetary systems are infinitesimally small compared to the universe as a whole. For example, light takes about 4 hours to travel to Earth from Neptune, the outermost planet in the Solar System. Light from the most distant galaxies takes nearly *14 billion years* to reach Earth!

How do planetary systems form?

Giordano Bruno (1548–1600), the Italian friar, philosopher, and astronomer quoted at the beginning of the chapter, believed that Copernicus was correct about Earth and the planets orbiting the Sun. Bruno argued that an infinite universe would have an infinite number of stars and planets, including inhabited planets. Unfortunately, Bruno was charged with heresy for his theological and his astronomical opinions, and he spent 7 years in a Roman prison before being executed.

Over the past century, with the aid of spectroscopy, astronomers have demonstrated that the Sun is a typical star, one of hundreds of billions in its galaxy (the Milky Way), and that the Milky Way is a typical galaxy, one of hundreds of billions in the universe. But only within the last two decades have astronomers obtained evidence that the Solar System is but one of a large number of other planetary systems scattered throughout the galaxy. Before we study the planets and moons of the Solar System, we begin Part II by learning how planetary systems, including the Solar System, are created.

The Solar System is one of many planetary systems in the Milky Way galaxy.

Until the latter part of the 20th century, the origin of the Solar System remained speculative. In the last few decades, stellar astronomers studying the formation of stars, and planetary scientists analyzing clues about the history of the Solar System, have found themselves arriving at the *same* picture of the early Solar System—but from

The Solar System is tiny compared to the universe.

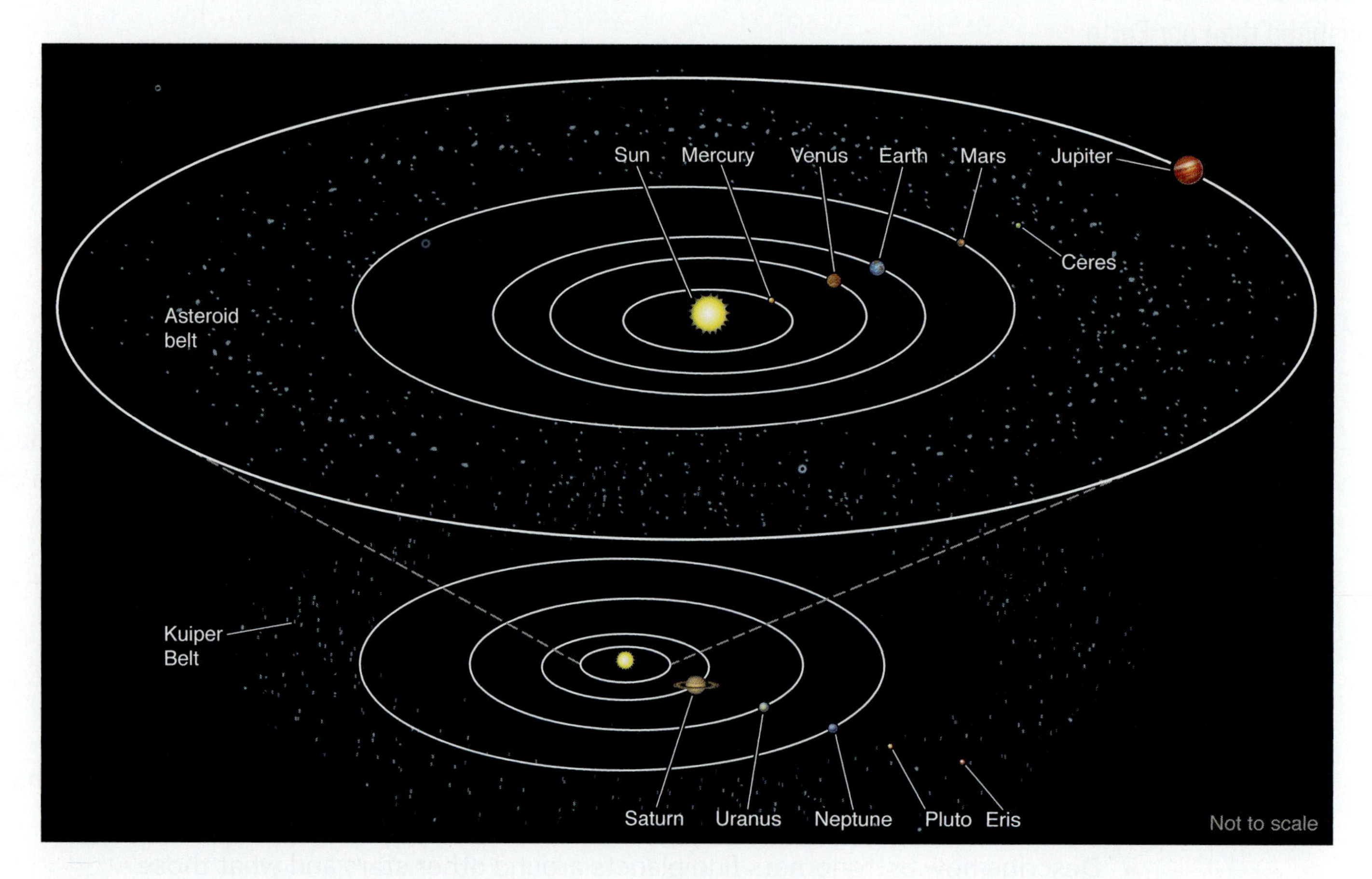

FIGURE 7.1 The Solar System. Besides Pluto, Ceres, and Eris, two other dwarf planets have been identified but are not shown here.

two very different directions. This unified understanding provides the foundation for the way astronomers now think about the Sun and the myriad objects that orbit it.

The first plausible theory for the formation of the Solar System, known as the **nebular hypothesis**, was proposed as early as 1734 by the German philosopher Immanuel Kant (1724–1804) and conceived independently a few years later by the French astronomer Pierre-Simon Laplace (1749–1827). Kant and Laplace argued that a rotating cloud of interstellar gas, or **nebula** (Latin for "cloud"), gradually collapsed and flattened to form a disk with the Sun at its center. Surrounding the Sun were rings of material from which the planets formed. This configuration would explain why the planets orbit the Sun in the same direction. Although the nebular hypothesis remained popular throughout the 19th century, it had serious problems. Even so, the basic principles of the hypothesis are retained today in the modern theory of planetary system formation.

Modern theory of planetary system formation suggests that, when conditions are right, clouds of interstellar gas collapse under the force of their own self-gravity to form stars. Recall from Connections 4.1 that self-gravity is the gravitational attraction between the parts of an object such as a planet or star that pulls all the parts toward the object's center. This inward force is opposed by either structural strength (in the case of rocks that make up terrestrial planets), or the outward force resulting from gas pressure within a star. If the outward force is less than self-gravity, the object contracts; if it is greater, the object expands. In a stable object, the inward and outward forces are balanced.

Young stars are surrounded by rotating disks.

In support of the nebular hypothesis, disks of gas and dust have been observed surrounding young stellar objects (**Figure 7.2**). From this observational evidence, stellar astronomers have shown that, much like a spinning ball of pizza dough spreads out to form a flat crust, the cloud that produces a star—the Sun, for example—collapses first into a rotating disk. Material in the disk eventually suffers one of three fates: it travels inward onto the forming star at its center, it remains in the disk itself to form planets and other objects, or it is thrown back into interstellar space. ▶II **ASTROTOUR: SOLAR SYSTEM FORMATION**

During the same years that astronomers were working to understand star formation, other groups of scientists with very different backgrounds—mainly geochemists and geologists—were piecing together the history of the Solar System. Planetary scientists looking at the current structure of the Solar System inferred some of its early characteristics. The orbits of all the planets lie very close to a single plane, which tells us that the early Solar System must have been flat. The fact that all the planets orbit the Sun in the same direction says that the material from which the planets formed must have been swirling about the Sun in the same direction.

Other clues about what the early Solar System was like are harder to puzzle out. **Meteorites**, for example, include bits and pieces of material that are left over from the Solar System's youth. These fragments of the early Solar System are sometimes captured by Earth's gravity and fall to the ground, where they can be studied. Many meteorites look something like a piece of concrete in which pebbles and sand are mixed with a much finer filler (**Figure 7.3**). This structure surely tells us something about how these pieces of interplanetary debris formed, but what?

Beginning in the 1960s, a flood of information about Earth and other objects in the Solar System poured in from a host of sources, including space probes, ground-based telescopes, laboratory analysis of meteorites, and theoretical calculations. Scientists working with this wealth of

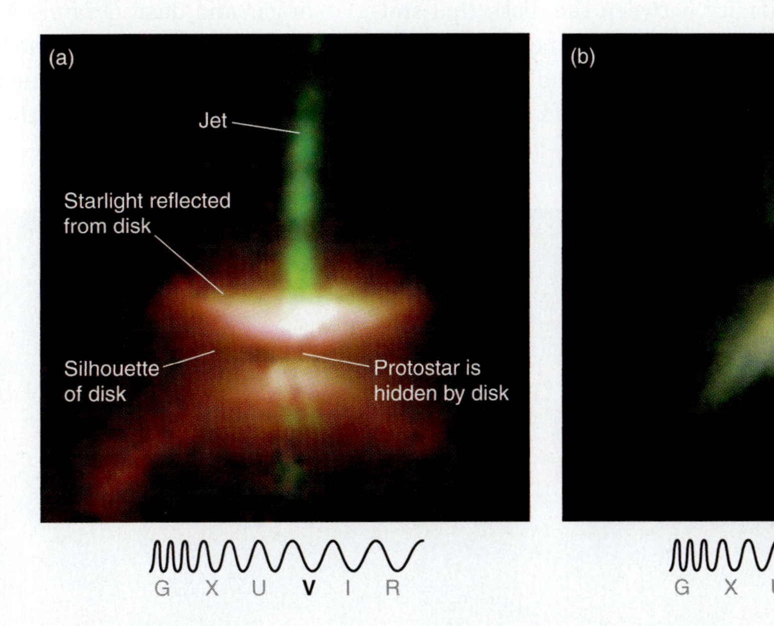

FIGURE 7.2 Hubble Space Telescope images showing disks around newly formed stars. The dark bands are the shadows of the disks seen more or less edge on. Bright regions are dust illuminated by starlight. Some disk material may be expelled in a direction perpendicular to the plane of the disk in the form of violent jets.

FIGURE 7.3 Meteorites are the surviving pieces of young Solar System fragments that land on the surfaces of planets. It is clear from this cross section that this meteorite formed from many smaller components that stuck together.

information began to see a pattern. What they were learning made sense only if they assumed that the larger bodies in the Solar System had grown from the aggregation of smaller bodies. Following this chain of thought back in time, they came to envision an early Solar System in which the young Sun was surrounded by a flattened disk of both gaseous and solid material. This swirling disk of gas and dust provided the raw material from which the objects in the Solar System would later form.

The Solar System formed from a rotating disk of gas and dust.

The remarkable similarity between the disks that stellar astronomers find around young stars and the disk that planetary scientists hypothesize as the cradle of the Solar System is not a coincidence. As astronomers and planetary scientists compared notes, they realized they had arrived at the same picture of the early Solar System from two completely different directions. The rotating disk from which the planets formed was none other than the remains of the disk that had accompanied the formation of the Sun. Earth, along with all the other orbiting bodies that make up the Solar System, evolved from the remnants of an *interstellar cloud* that collapsed to form the local star, the Sun. The connection between the formation of stars and the origin and subsequent evolution of the Solar System has become one of the cornerstones of both astronomy and planetary science—a central theme around which a great deal of astronomical knowledge of the Solar System revolves **(Process of Science Figure).**

7.2 The Solar System Began with a Disk

Consider the newly formed Sun (**Figure 7.4**), roughly 5 billion years ago, adrift in interstellar space. The Sun was not yet a star in the true sense of the word. It was still a **protostar**—a large ball of gas but not yet hot enough in its center to be a star. (The often-used prefix *proto-* means "early form" or "in the process of formation.") As the cloud of interstellar gas collapsed to form the protostar, its gravitational energy was converted into heat energy and radiation.

Surrounding the protostellar Sun was a flat, rotating disk of gas and dust. *Orbiting* is perhaps a better word than *rotating*. Each bit of the material in this thin disk was orbiting around the Sun according to the same laws of motion and gravitation that govern the orbits of the planets today. The

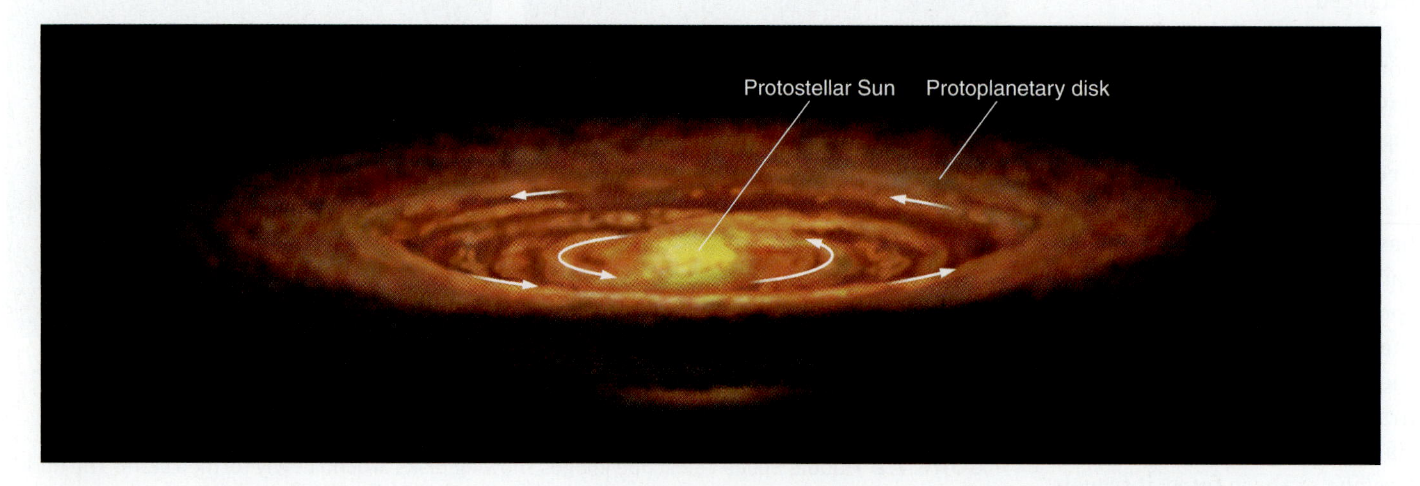

FIGURE 7.4 Think of the young Sun as being surrounded by a flat, rotating disk of gas and dust that was flared at its outer edge.

Process of Science

CONVERGING LINES OF INQUIRY

The question: Why is the Solar System a disk, with all planets orbiting in the same direction?

Mathematicians suggest the nebular hypothesis: a collapsing rotating cloud formed the Solar System.

Stellar astronomers test the nebular hypothesis, seeking evidence for or against.

Planetary scientists test the nebular hypothesis, seeking evidence for or against.

Stellar astronomers find dust and gas around young stars.

Planetary scientists study meteorites that show the Solar System bodies formed from many smaller bodies.

Stellar astronomers observe disks of gas and dust surrounding young stars.

Beginning from the same fundamental observations about the shape of the Solar System, theorists, planetary scientists, and stellar astronomers converge in the nebular theory that stars and planets form together from a collapsing cloud of gas and dust.

disk around the Sun, like the disks that astronomers see today surrounding protostars elsewhere in our galaxy, is called a **protoplanetary disk**. The disk probably contained less than 1 percent of the mass of the star forming at its center, but this amount was more than enough to account for the bodies that make up the Solar System today.

The Collapsing Cloud Rotates

The reason the Solar System formed from a protoplanetary disk and similar disks are seen around newly formed stars lies in **angular momentum**. Consider a figure-skater spinning on the ice (**Figure 7.5**). Like any other rotating object or isolated group of objects, the spinning ice-skater has some amount of angular momentum. The amount of angular momentum that an object possesses depends on three things:

1. **How fast the object is rotating.** The faster an object is rotating, the more angular momentum it has. A top that is spinning rapidly has more angular momentum than the same top does when it is spinning slowly.
2. **The mass of the object.** Imagine two spinning tops. Both tops have the same size, shape, and rate of spin. If one top is made of metal while the other top is made of wood, the more massive metal top has more angular momentum.
3. **How the mass of the object is distributed**—how "spread out" the object is. For an object of a given mass and rate of rotation, the more spread out the object is, the more angular momentum it has. An object that is shaped so that its mass is very spread out has more angular momentum than an object of the same mass that is more compact.

VISUAL ANALOGY

FIGURE 7.5 A figure-skater relies on the principle of conservation of angular momentum to change the speed with which she spins.

The *rotational* or *spin* angular momentum of a single object, such as a spinning top or a rotating planet or interstellar gas cloud, is discussed in **Math Tools 7.1**.

An object's angular momentum remains the same unless an external force acts on the object. This statement is called the law of **conservation of angular momentum**. The combined angular momentum of an isolated group of objects is also conserved. When physicists say that something is "conserved," they mean that the quantity does not change by itself. This idea might remind you of Newton's first law of motion (see Chapter 3), which says that in the absence of an external force, an object continues to move in a straight line at a constant speed. Both Newton's first law and the conservation of angular momentum are examples of **conservation laws**.

Angular momentum is conserved.

Both the ice-skater and the collapsing interstellar cloud are affected by conservation of angular momentum. If one of spin rate, mass, or mass distribution changes, then at least one of the others must change to keep the same angular momentum. A compact object must spin more rapidly to have the same amount of angular momentum as a more extended object with the same mass. For example, an ice-skater can control how rapidly she spins by pulling in or extending her arms or legs. As the skater spins, her angular momentum does not change much. (The slow decrease is due to friction, an external force.) When her arms and leg are fully extended, she spins slowly; but as she pulls her arms and leg in, she spins faster and faster. When her arms are held tightly in front of her and one leg is wrapped around the other, the skater's spin becomes a blur. She finishes by throwing her arms and leg out—an action that abruptly slows her spin. Despite the dramatic effect, her angular

Math Tools 7.1

Angular Momentum

We will discuss angular momentum in its various forms many times in this book. In its simplest form, the angular momentum of a system is given by:

$$L = m \times v \times r$$

where m is the mass, v is the speed at which the mass is moving, and r represents how spread out the mass is.

As an example, we can apply this relationship to the angular momentum, $L_{orbital}$, of Jupiter in its orbit about the Sun. The angular momentum from one body orbiting another is called *orbital* angular momentum. The mass (m) of Jupiter is 1.90×10^{27} kilograms (kg), the speed of Jupiter in orbit (v) is 1.307×10^4 meters per second (m/s), and the radius of Jupiter's orbit (r) is 7.786×10^{11} meters. Putting all this together gives:

$$L_{orbital} = (1.90 \times 10^{27} \text{ kg}) \times (1.31 \times 10^4 \text{ m/s}) \times (7.79 \times 10^{11} \text{ m})$$
$$= 1.94 \times 10^{43} \text{ kg m}^2/\text{s}$$

Calculating the *spin* angular momentum of a spinning object, such as a top, a planet, a star, or an interstellar galactic cloud, is far more complicated. Here we must add up the individual angular momenta of *every tiny mass element* within the object. In the case of a uniform sphere, the spin angular momentum is proportional to the square of its radius and inversely proportional to its rotation period:

$$L_{spin} = \frac{4\pi m R^2}{5P}$$

where R is the radius of the sphere and P is the rotation period of its spin.

Let's compare Jupiter's orbital angular momentum with the Sun's spin angular momentum. This comparison will tell us about the distribution of angular momentum in the Solar System. Appendix 1 provides the Sun's radius (6.96×10^8 meters), mass (1.99×10^{30} kg), and rotation period (24.5 days = 2.12×10^6 seconds). Assuming then that the Sun is a uniform sphere, the spin angular momentum of the Sun is:

$$L_{spin} = \frac{4 \times \pi \times (1.99 \times 10^{30} \text{ kg}) \times (6.96 \times 10^8 \text{ m})^2}{5 \times (2.12 \times 10^6 \text{ s})}$$
$$= 1.14 \times 10^{42} \text{ kg m}^2/\text{s}$$

You can see that Jupiter's orbital angular momentum is about 17 times greater than the spin angular momentum of the Sun. This difference demonstrates that the bulk of the angular momentum of the Solar System was in the disk and now resides in the orbits of its major planets.

Another point is that for a collapsing sphere to conserve spin angular momentum (L_{spin} stays constant), its rotation period (P) must be proportional to the square of its radius (R), and its spin rate ($\propto 1/P$) must therefore be *inversely proportional* to the square of its radius. So when a sphere decreases in radius, its spin rate increases.

momentum remains the same throughout. This impressive athletic spectacle comes courtesy of the law of conservation of angular momentum and from the difference between an extended object and a compact object.

The cloud of interstellar gas that collapsed under the force of its own gravity to form the protostellar Sun was also affected by the conservation of angular momentum. It might seem most natural for the cloud to collapse directly into a ball—and so it would, if the cloud didn't have its own angular momentum. Interstellar clouds are truly vast objects, light-years in size. (Recall that a light-year is the distance traveled by light in 1 year, or about 9.5 trillion kilometers [km], or about 63,000 astronomical units [AU].) As interstellar clouds orbit about the galaxy's center, they are constantly being pushed around by stellar explosions or by collisions with other interstellar clouds. This constant "stirring" guarantees that all interstellar clouds have *some* amount of rotation. As spread out as an interstellar cloud is, even a tiny amount of rotation corresponds to a huge amount of angular momentum. Imagine our ice-skater now with arms that reach from here to the other side of Earth. Even if she were rotating very slowly at first, think how fast she would be spinning by the time she pulled those long arms to her sides.

Interstellar clouds have far more angular momentum than the stars they form.

Just as the ice-skater speeds up when she pulls in her arms, the cloud of interstellar gas rotates faster and faster as it collapses. However, there is a puzzle here. Suppose, for example, that the Sun formed from a typical cloud that was about a light-year across—10^{16} meters—and was rotating so slowly that it took a million years to complete one rotation. If the collapsing spherical cloud conserved spin angular momentum (see Math Tools 7.1), by the time it collapsed to the size of the Sun—1.4×10^9 meters across, or only one 10-millionth the size of the original cloud—it would be spinning 50 trillion times faster, completing a rotation in only 0.6 second. This is over 3 million times faster than the Sun actually spins. At this rate of rotation, the Sun's self-gravity would have to be almost 200 million times stronger to hold the Sun together. It appears that angular momentum was not conserved in this model (in which stars form from collapsing interstellar clouds) or that this description is incomplete.

An Accretion Disk Forms

The key to solving the riddle of angular momentum in a collapsing interstellar cloud lies in realizing that the *direction* of the collapse is important. Imagine that the ice-skater bends her knees, compressing herself downward instead of bringing her arms toward her body. As she does this, she again makes herself less spread out, but her speed does not change, because no part of her body has become any closer to the axis of spin.

Similarly, a collapsing cloud can flatten out without speeding up by collapsing parallel to its axis of rotation (**Figure 7.6**). Instead of collapsing into a ball, the interstellar cloud flattens into a disk. As the cloud collapses, its self-gravity increases, and the inner parts of the flattening cloud begin to fall freely inward, raining down on the growing object at the center. As this happens, the outer portions of the cloud lose the support of the collapsed inner portion, and they start falling inward too. The whole cloud collapses inward, much like a house of cards with the bottom layer knocked out. As this material makes its final inward plunge, it lands on a thin, rotating structure called an **accretion disk**. The accretion disk serves as a way station for material en route to becoming part of the star that is forming at its center.

The formation of accretion disks is common in astronomy. Using what you learned about orbits in Chapter 4 will help you understand what happens during this final stage of the collapse of an interstellar cloud. As the material falls toward the forming star, it travels on curved—almost always elliptical—paths, just as Kepler's laws say it should. These orbits would carry the material around the forming star and back into interstellar space, except for one problem: the path inward toward the forming star is a one-way street. When material nears the center of the cloud, it runs headlong into material that is falling in from the *other* side. Material falling onto the disk from opposite directions comes together at the plane of the accretion disk, which is the plane perpendicular to the cloud's axis of rotation.

The cloud collapses into a disk rather than directly into a star.

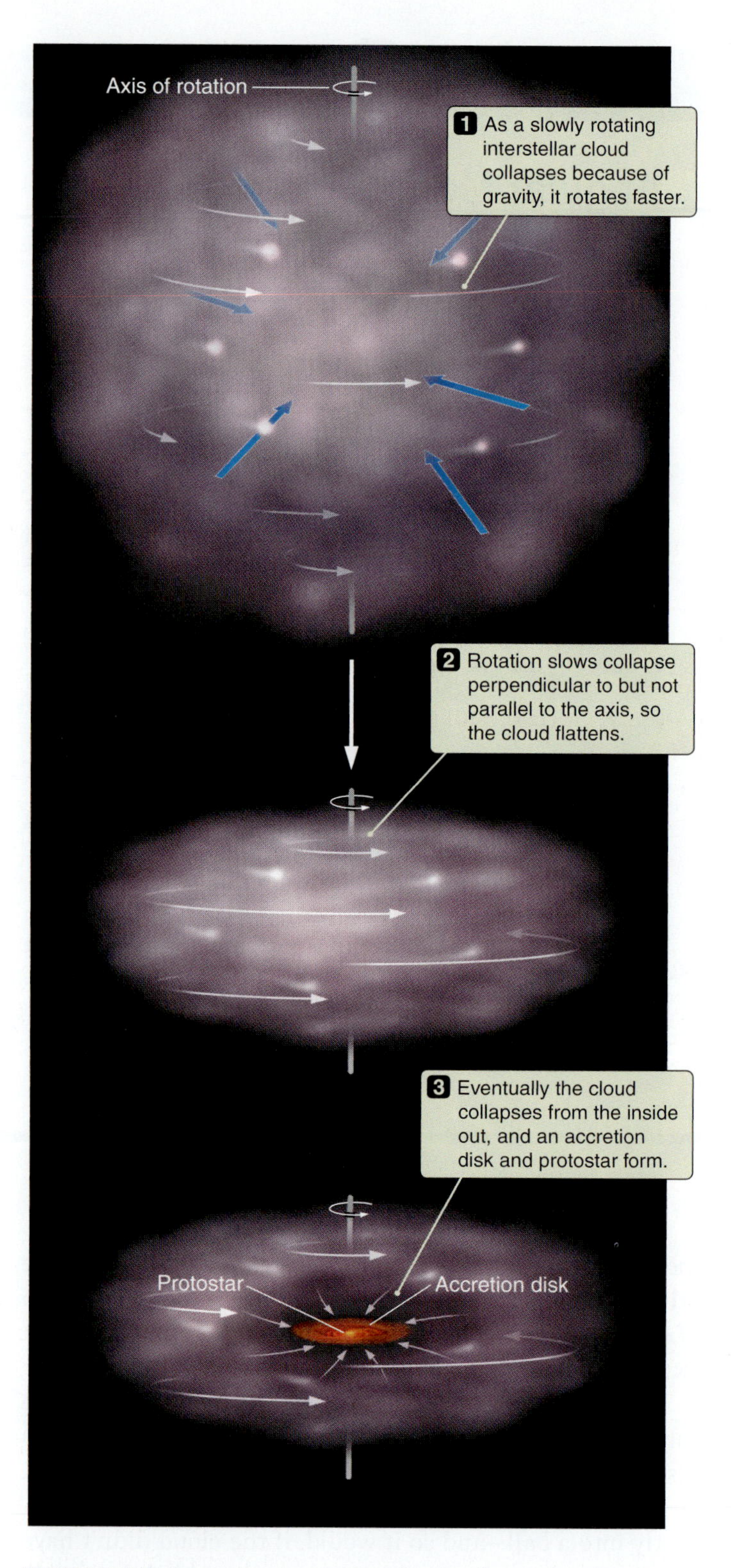

FIGURE 7.6 A rotating interstellar cloud collapses in a direction parallel to its axis of rotation, thus forming an accretion disk.

To understand how interstellar material collects on the accretion disk, a visual analogy might be helpful. Imagine a traffic engineer's nightmare: a huge roundabout, or traffic circle, with multiple entrances but with all exits blocked by incoming traffic (**Figure 7.7**). **AstroTour: Traffic Circle** As traffic flows into the traffic circle, it has nowhere else to go, resulting in a continuous, growing line of traffic driving around and around in an increasingly crowded circle. Eventually, as more and more cars try to pack in, the traffic piles up. This situation is roughly analogous to an accretion disk. Of course, traffic in a roundabout moves on a flat surface, whereas the accretion disk around a protostar forms from material coming in from all directions in three-dimensional space. As material falls onto the disk, its motion perpendicular to the disk stops abruptly, but its mass motion *parallel* to the surface of the disk adds to the disk's total angular momentum. In this way, the angular momentum of the infalling material is transferred to the accretion disk. The rotating accretion disk has a radius of hundreds of astronomical units and is thousands of times greater than the radius of the star that will eventually form at its center. Therefore, most of the angular momentum in the original interstellar cloud ends up in the accretion disk rather than in the central protostar. (See Math Tools 7.1 for an example of the relevant calculation).

Most of the matter that lands on the accretion disk either ends up as part of the star or is ejected back into interstellar space, sometimes in the form of violent jets, as seen in Figure 7.2a. However, a small amount of material is left behind in the disk. It is this leftover disk—the dregs of the process of star formation—to which we next turn our attention. Figure 7.2 shows Hubble Space Telescope (HST) images of edge-on accretion disks around young stars. The dark bands are the shadows of the edge-on disks, the top and the bottom of which are illuminated by light from the forming star. Astronomers cannot go back 5 billion years and watch as the Sun formed from a cloud of interstellar gas, but they can look at objects like the ones in Figure 7.2 to know what they would have seen.

Small Objects Stick Together to Become Large Objects

The chain of events that connects the accretion disk around a young star to a planetary system such as the Solar System begins with random motions of the gas within the protoplanetary disk. These motions push the smaller grains of solid material back and forth past larger grains; and as this happens, the smaller grains stick to the larger grains. The "sticking" process among smaller grains is due to the same static electricity that causes dust and hair to cling

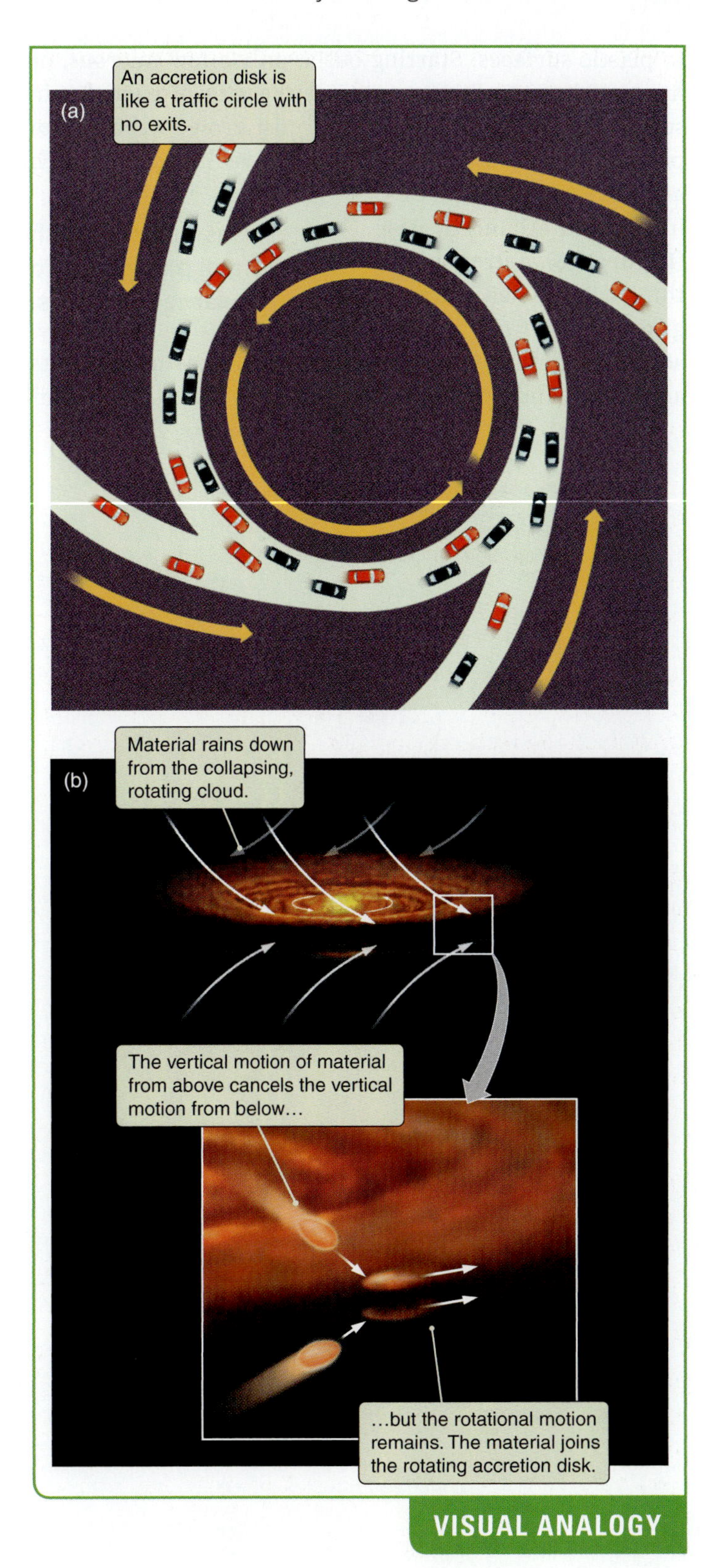

FIGURE 7.7 (a) Traffic piles up in a traffic circle with entrances but no exits. (b) Similarly, gas from a rotating cloud falls inward from opposite sides, piling up onto a rotating disk.

to plastic surfaces. Starting out at only a few microns, or micrometers (μm), across—about the size of particles in smoke—the slightly larger bits of dust grow to the size of pebbles and then to clumps the size of boulders, which are less susceptible to being pushed around by gas (**Figure 7.8**). Astronomers think that when clumps grow to about 100 meters across, their rate of growth slows down. These large objects are so few and far between in the disk that chance collisions become less and less frequent. Even so, the process of growth continues, albeit at a slower pace, as 100-meter clumps eventually join together to produce still larger bodies. Within a protoplanetary disk, the larger dust grains become larger at the expense of the smaller grains.

Gas motions push small particles into larger particles.

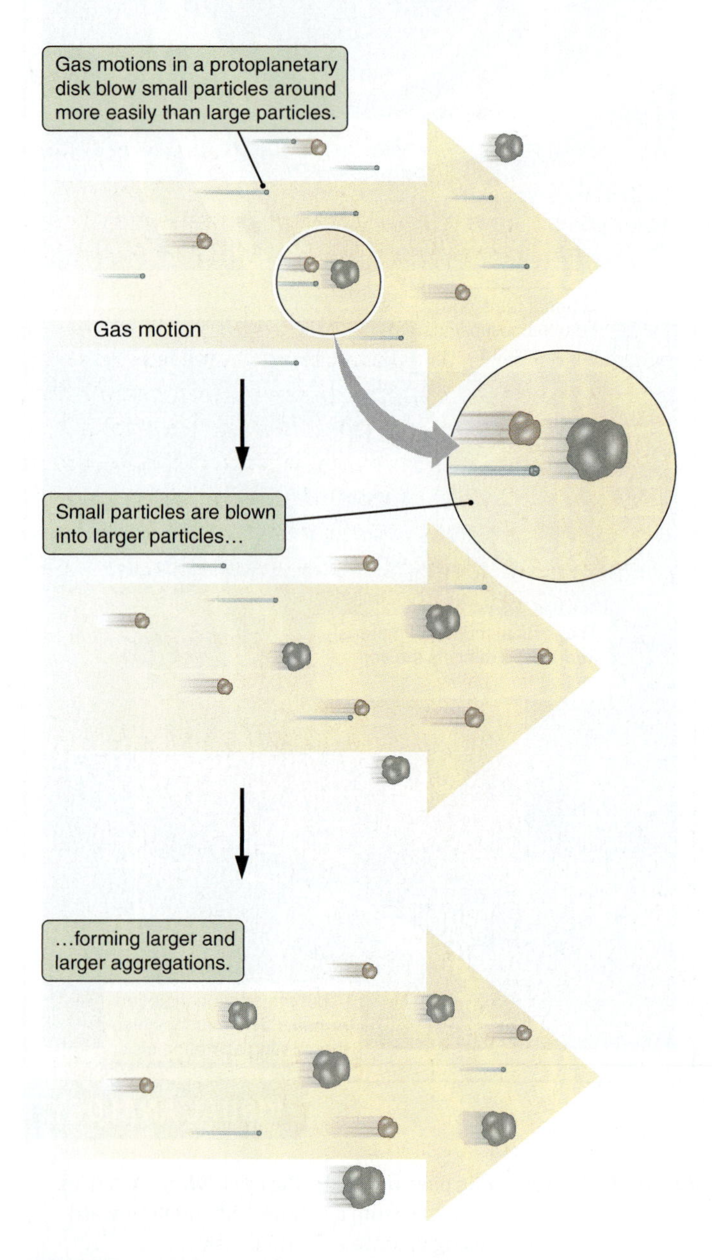

FIGURE 7.8 Motions of gas in a protoplanetary disk blow smaller particles of dust into larger particles, making the larger particles larger still. This process continues, eventually creating objects many meters in size.

In order to stick together, clumps must bump into each other gently—very gently. Otherwise the energy of collision would cause the two colliding bodies to fragment into many smaller pieces instead of forming a single larger one. Collision speeds cannot be much greater than 0.1 meter per second (m/s) for colliding boulders to stick together. If you were to walk that slowly, it would take you 15 minutes to travel the length of a soccer field. In a real accretion disk, collisions more violent than this certainly happen on occasion, breaking these clumps back into smaller pieces and causing many reversals in clump growth.

Up to this point, larger objects have grown mainly by "sweeping up" smaller objects that run into them or that get in their way. As the clumps reach the size of about a kilometer, a different process becomes important. These kilometer-sized objects, now called **planetesimals** (literally "tiny planets"), are massive enough that their gravity begins to exert a significant attraction on nearby bodies (**Figure 7.9**). No longer is growth of the planetesimal fed only by chance collisions with other objects; the planetesimal's gravity can now pull in and capture other smaller planetesimals that lie

Gravity helps planetesimals grow into planets.

FIGURE 7.9 The gravity of a planetesimal is strong enough to attract surrounding material, causing the planetesimal to grow.

outside its direct path. The growth of planetesimals speeds up, with the larger planetesimals quickly consuming most of the remaining bodies in the vicinity of their orbits. The final survivors of this process are now large enough to be called **planets**. As with the major bodies in orbit about the Sun, some of the planets may be small and others quite large.

7.3 The Inner Disk Is Hot; the Outer Disk Is Cold

The accretion disks surrounding young stars form from interstellar material that may have a temperature of only a few kelvins, but the disks themselves reach temperatures of hundreds of kelvins or more. What heats up the disk around a forming star? The answer lies with gravity. Material from the collapsing interstellar cloud falls inward toward the protostar, but because of its angular momentum it misses the protostar and instead falls onto the surface of the disk. When this material reaches the surface of the disk, its infalling motion comes to an abrupt halt, and the velocity that the atoms and molecules in the gas had before hitting the disk is suddenly converted into random *thermal* velocities instead. The cold gas that was falling toward the disk heats up when it lands on the disk.

Imagine dumping a box of marbles from the top of a tall ladder onto a rough, hard floor below (**Figure 7.10a**). The marbles fall, picking up speed as they go. Even though the falling marbles are speeding up, however, they are all speeding up *together*. As far as one marble is concerned, the other marbles are not moving very fast at all. (If you were riding on one of the marbles, the other marbles would not appear to you to be moving very much; it would be the rest of the room that was whizzing by.) The atoms and molecules in the gas falling toward the protostar are like these marbles. They are picking up speed as they fall as a group toward the protostar, but the gas is still cold because the random thermal velocities of atoms and molecules with respect to each other are still low. Now imagine what happens when the marbles hit the rough floor. They bounce every which way. They are still moving rapidly, but they are no longer moving together. A change has taken place from the ordered motion of marbles falling together to the random motions of marbles traveling in all directions. The atoms and molecules in the gas falling toward the central star behave in the same fashion when they hit the accretion disk (**Figure 7.10b**). They are no longer moving as a group, but their random thermal velocities are now very large. The gas is now hot.

Another way to think about why the gas that falls on the disk makes the disk hot is to apply another conservation law. The law of **conservation of energy** states that, unless energy is added to or taken away from a system from the outside, the total amount of energy in the system must remain constant. But the form the energy takes can change.

Imagine you are working against gravity by lifting a heavy object—for example, a brick. It takes energy to lift

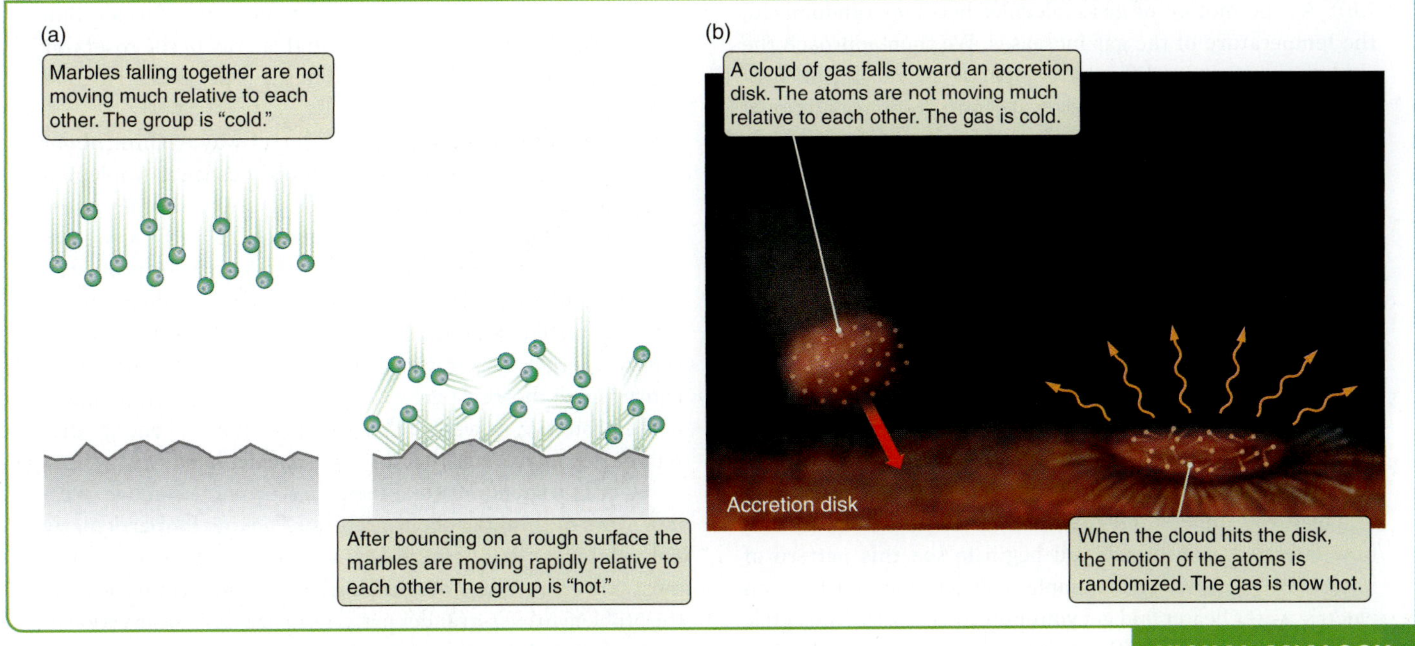

FIGURE 7.10 (a) Marbles dropped as a group fall together until they hit a rough floor, at which point their motions become randomized. (b) Similarly, atoms in a gas fall together until they hit the accretion disk, at which point their motions become randomized, raising the temperature of the gas.

the brick, and conservation of energy says that energy is never lost. Where does that energy go? It is changed into a form called **gravitational potential energy**. The energy is stored in a way that resembles how energy is stored in a battery. Potential energy is energy that has potential—it is waiting to show up in a more obvious form. If you drop the brick it falls, and as it falls it speeds up. The gravitational potential energy that was stored is converted to energy of motion, which, as you may recall from Chapter 5, is called kinetic energy. When the brick hits the floor, it stops suddenly. The brick loses its energy of motion, so what form does this energy take now?

The gravitational energy of infalling material turns into thermal energy.

If the brick cracks, part of the energy goes into breaking the chemical bonds that hold it together. Some of the energy is converted into the sound the brick makes when it hits the floor. Some goes into heating and distorting the floor. But most of the energy is converted into thermal energy. The atoms and molecules that make up the brick are moving about within the brick a bit faster than they were before the brick hit; so the brick and its surroundings, including the floor, grow a tiny bit warmer. Similarly, as gas falls toward the disk surrounding a protostar, gravitational potential energy is converted first to kinetic energy, causing the gas to pick up speed. When the gas hits the disk and stops suddenly, that kinetic energy is turned into thermal energy.

Connections 7.1 discusses why it can be useful to think about the same thing—in this case, energy—in different ways.

In the manner described in the previous subsection, material falling onto the accretion disk around a forming star causes the disk to heat up. The amount of heating depends on *where* the material hits the disk. Material hitting the inner part of the disk (the "inner disk") has fallen farther and picked up greater speed within the gravitational field of the forming star than has material hitting the disk far-

Connections 7.1

Conservation of Energy

In the text we discuss how the gas falling onto a protostellar disk heats the disk. First we present the analogy of marbles falling on a floor, noting that the marbles are jumbled up when they hit, like the atoms in the gas hitting the disk (see Figure 7.10). As the motion of gas molecules becomes randomized, the temperature of the gas increases. We then approach the explanation from a different perspective, focusing on how energy is conserved but changes its form from gravitational potential energy to kinetic energy and finally to thermal energy. The disk heats up as a result of the same physical process, even if these explanations offer two different ways to think about why the disk gets hot.

Both ways of thinking about the process are discussed here because sometimes scientists need to look at the same thing from several different angles before they understand it. In this case, however, most scientists would agree that the second way of thinking about the problem is far more powerful than the first. Properly stated, the heating of protostellar disks is an example of one of the most far-reaching patterns in nature: the conservation of energy.

Once you understand the different forms of energy and how energy is conserved, you begin to see this pattern of nature everywhere. For example, let's assume you have an electric water heater and get your power from a hydroelectric plant. When water falls through the turbines in a hydroelectric generator, it turns the generator and produces electric energy that is carried out over power lines to your home. The energy to power your lights and heat your water comes from the gravitational potential energy of the water in a reservoir near you—just as the energy to heat the newly formed Sun came from the gravitational potential energy of the reservoir of gas from which the Sun formed in the process described in the text.

Conservation of energy is a powerful way to think about the heating of disks around young stars because it enables astronomers to calculate how hot these disks get. If they know the mass of the protostar and the surrounding disk, as well as the distance between the source of the gas and the protostar, then they can calculate the amount of gravitational energy the gas started out with. According to the law of conservation of energy, this gravitational energy must eventually be converted to thermal energy. Then the expected temperature of the protostar, as well as the amount of thermal energy that will be deposited onto the disk, can be calculated.

Scientists spend much of their time trying to come up with new ways of thinking about problems, looking for particularly powerful ways that point to new insights and discoveries. The most powerful means of thinking about a problem usually tie the problem to ever-grander patterns in nature. Conservation of energy is one of the grandest and most useful patterns around.

ther out. Like a rock dropped from a tall building, material striking the inner disk is moving quite rapidly when it hits, so it heats the inner disk to high temperatures. In contrast, material falling onto the outer part of the disk (the "outer disk") is moving much more slowly (like a rock dropped from just a foot or so). So the temperature at the outermost parts of the disk is not much higher than that of the original interstellar cloud. Stated another way, material falling onto the inner disk converts more gravitational potential energy into thermal energy than does material falling onto the outer disk.

The inner disk is hotter than the outer disk.

The energy released as material falls onto the disk is not the only source of thermal energy in the disk. Even before the nuclear reactions that will one day power the new star have ignited, conversion of gravitational energy into thermal energy drives the temperature at the surface of the protostar to several thousand kelvins, and it also drives the luminosity of the huge ball of glowing gas to many times the luminosity of the present-day Sun. For the same reasons that Mercury is hot while Pluto is not (see Section 5.5), the radiation streaming outward from the protostar at the center of the disk drives the temperature in the inner parts of the disk even higher, increasing the difference in temperature between the inner and outer parts of the disk.

Rock, Metal, and Ice

Temperature affects which materials can and cannot exist in a solid form. On a hot summer day, ice melts and water quickly evaporates; on a cold winter night, even the water in your breath freezes before your eyes. Some materials, such as iron, **silicates** (minerals containing silicon and oxygen), and carbon—metals and rocky materials—remain solid even at quite high temperatures. Materials like these, which are capable of withstanding high temperatures without melting or being vaporized, are called **refractory materials**. Other materials, such as water, ammonia, and methane, can remain in a solid form only if their temperature is quite low. These less refractory substances are called **volatile materials** (or "volatiles" for short). Astronomers generally refer to the solid form of any volatile material as an **ice**.

Differences in temperature from place to place within the protoplanetary disk have a significant effect on the makeup of the dust grains in the disk (**Figure 7.11**). In the hottest parts of the disk (closest to the protostar), only the most refractory substances can exist in solid form. In the inner disk, dust grains are composed of refractory materials only. Somewhat farther out in the disk, some hardier volatiles, such as water ice and certain **organic** substances (*organic* refers to a large class of chemical compounds containing the element carbon), can survive in solid form. These add to the materials that make up dust grains. In the coldest, outermost parts of the accretion disk, far from the central protostar, highly volatile components such as methane, ammonia, and carbon monoxide ices and other organic molecules survive only in solid form. The different composition of dust grains within the disk determines the composition of the planetesimals formed from the dust. Planets that form closer to the central star tend to be made up mostly of refractory materials such as rock and metals. Planets that form farther from the central star also contain refractory materials, but in addition they contain large quantities of ices and organic materials.

Volatile ices survive in the outer disk, but only refractory solids survive in the inner disk.

In the Solar System, the inner planets are composed of rocky material surrounding metallic cores of iron and nickel. Objects in the outer Solar System, including moons, giant planets, and comets, are composed largely of ices of various types. In other planetary systems, and possibly the Solar

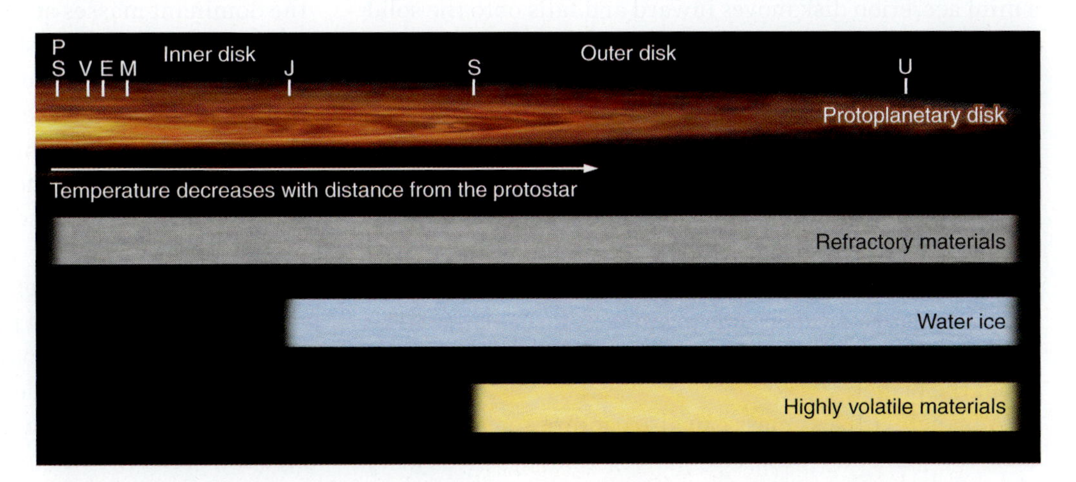

FIGURE 7.11 Differences in temperature within a protoplanetary disk determine the composition of dust grains that then evolve into planetesimals and planets. Refactory materials are found throughout the disk, while water ice is found only outside Jupiter's orbit, and highly volatile materials only outside Saturn's. Shown here are the protostellar Sun (PS) and the orbits of Venus (V), Earth (E), Mars (M), Jupiter (J), Saturn (S), and Uranus (U).

System too, **chaotic** encounters can change this organization of planetary compositions. In a process called **planet migration**, gravitational scattering can force some planets to end up far from the place of their birth. For example, many planetary scientists think that Uranus and Neptune originally formed nearer to the orbits of Jupiter and Saturn, but were then driven outward to their current locations by gravitational encounters with Jupiter and Saturn. A planet can also migrate when it gives up some of its orbital angular momentum to the disk material that surrounds it. Such a loss of angular momentum causes the planet to slowly spiral inward toward the central star. You will see examples when we discuss *hot Jupiters* in Section 7.5.

Solid Planets Gather Atmospheres

Once a solid planet has formed, it may continue growing by capturing gas from the protoplanetary disk. To do so, though, it must act quickly. Young stars and protostars are known to be sources of strong *stellar winds* and intense radiation that can quickly disperse the gaseous remains of the accretion disk. Gaseous planets such as Jupiter probably have only about 10 million years or so to form and to grab whatever gas they can. Tremendous mass is a great advantage in a planet's ability to accumulate and hold on to the hydrogen and helium gases that make up the bulk of the disk. Because of their strong gravitational fields, more massive young planets are thought to create their own mini accretion disks as gas from their surroundings falls toward them. What follows is much like the formation of a star and protoplanetary disk, but on a smaller scale. Just as happens in the accretion disk around the star, gas from a mini accretion disk moves inward and falls onto the solid planet.

The gas that is captured by a planet at the time of its formation is called the planet's **primary atmosphere**. The primary atmosphere of a large planet can become great enough to dominate the mass of the planet, as in the case of giant planets such as Jupiter. Some of the solid material in the mini accretion disk might stay behind to coalesce into larger bodies in much the same way that particles of dust in the protoplanetary disk came together to form planets. The result is a mini "solar system"—a group of moons that orbit about the planet.

A planet with less mass can also capture some gas from the protoplanetary disk, only to lose it later. Here again, more massive planets have the advantage. The gravity of small planets may not be strong enough to prevent less massive atoms and molecules such as hydrogen or helium from escaping back into space. Even if a small planet is able to gather some hydrogen and helium from its surroundings, this temporary primary atmosphere will be short-lived. The atmosphere that remains around a small planet like Earth is a **secondary atmosphere**. A secondary atmosphere forms later in the life of a planet. Carbon dioxide and other gases released from the planet's interior by volcanoes can be one important source of a planet's secondary atmosphere. In addition, volatile-rich comets that formed in the outer parts of the disk continue to fall inward toward the new star long after its planets have formed, and sometimes collide with planets. These comets may provide a significant source of water, organic compounds, and other volatile materials on planets close to the central star.

Less massive planets lose their primary atmospheres and then form secondary atmospheres.

7.4 A Tale of Eight Planets

Nearly 5 billion years ago, the Sun was still a protostar surrounded by a protoplanetary disk of gas and dust. During the next few hundred thousand years, much of the dust in the disk had collected into planetesimals—clumps of rock and metal near the emerging Sun and aggregates of rock, metal, ice, and organic materials in the parts of the disk that were more distant from the Sun. Within the inner few astronomical units of the disk, several rock and metal planetesimals, probably fewer than a half dozen, quickly grew in size to become the dominant masses at their respective distances from the Sun. With their ever-strengthening gravitational fields, they either captured most of the remaining planetesimals or ejected them from the inner part of the disk. **Figure 7.12** shows a model of how this might happen.

Rocky terrestrial planets formed in the inner Solar System.

These dominant planetesimals had now become planet-sized bodies with masses ranging between that of Earth and about 1/20 of that value. They were to become the **terrestrial planets**. Mercury, Venus, Earth, and Mars are the surviving terrestrial planets. Planetary scientists think that one or two others may have formed in the young Solar System but were later destroyed. (As you will see later in this section, one of them may have been responsible for the creation of Earth's Moon.)

For several hundred million years following the formation of the four surviving terrestrial planets, leftover pieces of debris orbiting around the Sun continued to rain down on the surfaces of these planets. Today we can still see the scars of these postformation impacts on the cratered sur-

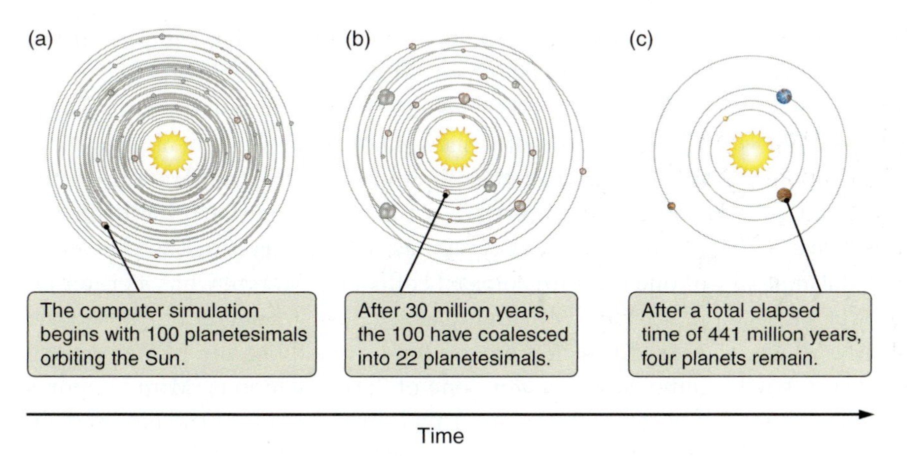

FIGURE 7.12 Computer models simulate how material in the protoplanetary disk became clumped into the planets over time.

faces of all the terrestrial planets (**Figure 7.13**). This rain of debris continues today, albeit at a much lower rate.

Before the proto-Sun emerged as a true star, gas in the inner part of the protoplanetary disk was still plentiful. During this early period the two larger terrestrial planets—Earth and Venus—may have held on to weak primary atmospheres of hydrogen and helium. If so, these thin atmospheres were soon lost to space. For the most part the terrestrial planets were all born devoid of thick atmospheres and remained so until the formation of the secondary atmospheres that now surround Venus, Earth, and Mars. Mercury's proximity to the Sun and small mass must have prevented it from retaining significant secondary atmospheres. It remains nearly airless today.

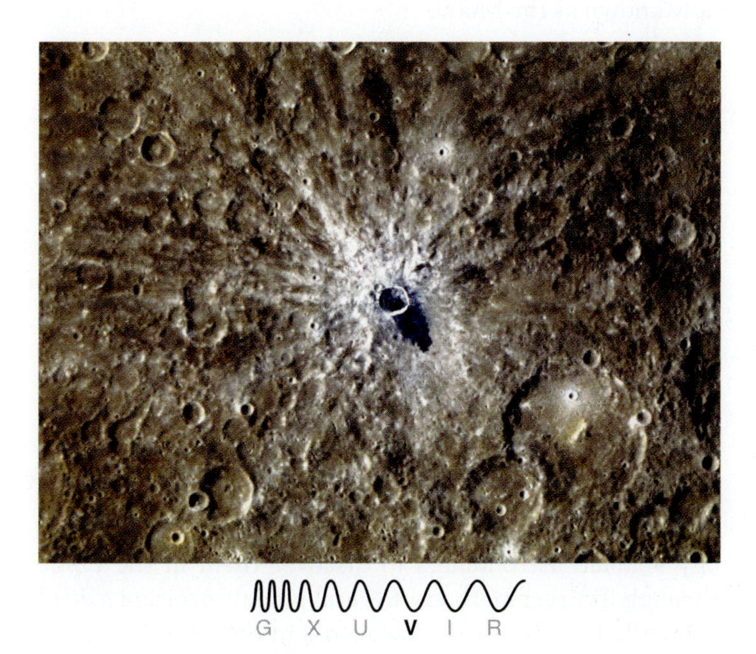

FIGURE 7.13 Large impact craters on Mercury (and on other solid bodies throughout the Solar System) record the final days of the Solar System's youth, when planets and planetesimals grew as smaller planetesimals rained down on their surfaces.

Farther out in the nascent Solar System, 5 AU from the Sun and beyond, planetesimals coalesced to form a number of bodies with masses about 5–10 times that of Earth. Why such large bodies formed in the region beyond the terrestrial planets remains an unanswered question. Located in a much colder part of the accretion disk, these planet-sized objects formed from planetesimals containing volatile ices and organic compounds in addition to rock and metal. In a process astronomers call **core accretion–gas capture**, mini accretion disks formed around these planetary cores, capturing massive amounts of hydrogen and helium and funneling this material onto the planets. Four such massive bodies became the cores of the **giant planets**—Jupiter, Saturn, Uranus, and Neptune.

The giant planets formed cores from planetesimals and then captured gaseous hydrogen and helium.

Jupiter's massive solid core was able to capture and retain the most gas—a quantity roughly 300 times the mass of Earth, or 300 $M_{\oplus}$. (The symbol ⊕ signifies Earth.) The other outer planetary cores captured lesser amounts of hydrogen and helium, perhaps because their cores were less massive or because there was less gas available to them. Saturn ended up with less than 100 $M_{\oplus}$ of gas, and Uranus and Neptune were able to grab only a few Earth masses' worth of gas. Some planetary scientists think that all of the giant planets formed closer to where Jupiter is now, and that their mutual gravitational interactions caused them to migrate to their present orbits.

Some planetary scientists do not think that our protoplanetary disk could have survived long enough to form gas giants such as Jupiter through the general process of core accretion. The core accretion model indicates that it could take up to 10 million years for a Jupiter-like planet to accumulate. Because all the gas in the protoplanetary disk likely dispersed in a little more than half that time, Jupiter's supply of hydrogen and helium would have run out.

This is not just a Solar System predicament. The apparent conflict between the time needed for a Jupiter-type planet to form and the availability of gases within that time period applies as well to other protoplanetary disks and to the formation of their massive planets. To get around this time dilemma, some scientists have proposed a process called *disk instability*, in which the protoplanetary disk suddenly and quickly fragments into massive clumps equivalent to those of the large planets. Although core accretion and disk instability appear to be competing processes, they may not be mutually exclusive. It is possible that both played a role in the formation of our own and other planetary systems.

For the same reasons that a forming protostar is hot—namely, conversion of gravitational energy into thermal energy—the gas surrounding the cores of the giant planets became compressed under the force of gravity and grew hotter. Proto-Jupiter and proto-Saturn probably became so hot that they actually glowed a deep red color (think of the heating element on an electric stove). Their internal temperatures may have reached as high as 50,000 kelvins (K).

Some of the material remaining in the mini accretion disks surrounding the giant planets coalesced into small bodies, which became moons. A **moon** is any natural satellite in orbit about a planet or asteroid. The composition of the moons that formed around the giant planets followed the same trend as the planets that formed around the Sun: the innermost moons formed under the hottest conditions and therefore contained the smallest amounts of volatile material. For example, the closest of Jupiter's many moons may have experienced high temperatures from nearby Jupiter glowing so intensely that it would have evaporated most of the volatile substances in the inner part of its mini accretion disk.

Moons formed from the mini accretion disks around the giant planets.

Not all planetesimals in the protoplanetary disk went on to become planets. Jupiter is our local giant planet. Its gravity kept the region of space between it and Mars so stirred up that most planetesimals there never coalesced into a single planet. (The one exception is Ceres, once considered to be the largest asteroid but now redefined, along with Pluto, as a dwarf planet—see Appendix 8.) The region between Mars and Jupiter, the **asteroid belt**, contains many planetesimals that remain from this early time. In the outermost part of the Solar System as well, planetesimals persist to this day. Born in a "deep freeze," these objects retain most of the highly volatile materials present at the formation of the protoplanetary disk. Far from the crowded inner part of the disk, planetesimals in the outermost parts were too sparsely distributed for large planets to grow. Icy planetesimals such as the dwarf planets Pluto and Eris in the outer Solar System remain today as **comet nuclei**—relatively pristine samples of the material from which our planetary system formed.

Asteroids and comet nuclei are planetesimals that survive to this day.

The early Solar System must have been a remarkably violent and chaotic place. Many Solar System objects show evidence of cataclysmic impacts. The dramatic difference in the terrain of the northern and southern hemispheres on Mars, for example, has been interpreted as the result of one or more colossal collisions. Mercury has a crater on its surface from an impact so devastating that it caused the crust to buckle on the opposite side of the planet. In the outer Solar System, one of Saturn's moons, Mimas, sports a crater roughly one-third the diameter of the moon itself. Uranus suffered one or more collisions that were violent enough to literally knock the planet on its side. Today, as a result, its axis of rotation is tilted at an almost right angle to its orbital plane.

Earth itself did not escape devastation by these cataclysmic events. In addition to the four terrestrial planets that remain, current theory is that there was at least one other large terrestrial object in the early Solar System—with about the same size and mass as Mars. As the newly formed planets were settling into their present-day orbits, this object suffered a grazing collision with Earth and was completely destroyed. The remains of the object, together with material knocked from Earth's outer layers, formed a huge cloud of debris encircling Earth. For a brief period Earth may have displayed a magnificent group of rings like those of Saturn. In time, this debris coalesced into the single body now known as the Moon.

7.5 Planetary Systems Are Common

When astronomers turn their telescopes to young nearby stars, they see disks of the same type from which the Solar System formed (**Figure 7.14**). The physical processes that led to the formation of the Solar System should be commonplace wherever new stars are being born. Compared to stars, however, planets are small and dim objects. They shine by reflection and therefore are millions to billions of times fainter than their host star. Thus, they were difficult to identify until advances in telescope detector technology enabled astronomers to discover them in the 1990s through indirect methods. Several planets orbiting a dead star called a *pulsar* were discovered in 1992. But it was the discovery in 1995 of a Jupiter-sized planet orbiting the Sun-like star 51 Pegasi that initiated the current profusion of **extrasolar planet** detections. Today the number

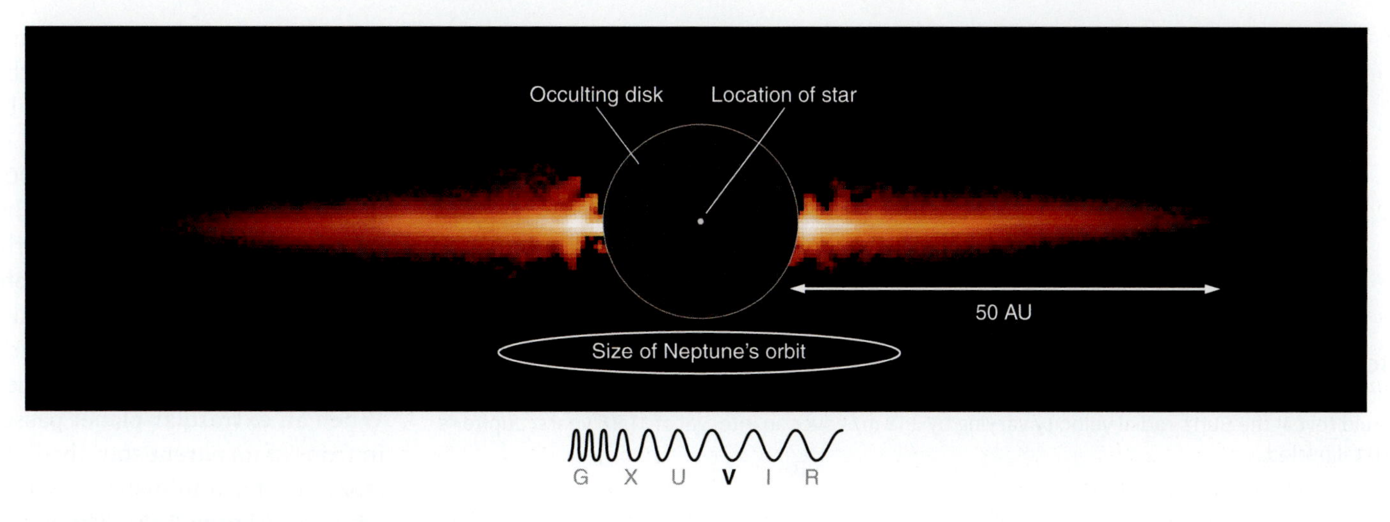

FIGURE 7.14 An edge-on circumstellar dust disk is seen extending outward to 60 AU from the young (12-million-year-old) star AU Microscopii. The star itself, whose brilliance would otherwise overpower the circumstellar disk, is hidden behind an occulting disk (opaque mask) placed in the telescope's focal plane. Its position is represented by the dot.

of known extrasolar planets, sometimes called *exoplanets*, is close to a thousand, with new discoveries occurring almost daily.

The discovery of extrasolar planets, including very large ("supermassive") planets, raises the question of what we mean by the term *planet*. Within the Solar System, planets and dwarf planets are defined as in Appendix 8. The International Astronomical Union (IAU) defines an extrasolar planet as an object that orbits a star and has a mass less than 13 Jupiters (13 M_J). There are intermediate-sized objects called *brown dwarfs*, which are not massive enough to be considered stars, yet are too massive to be called planets. Although the distinction between the most massive planets and the least massive brown dwarfs is somewhat arbitrary, these brown dwarfs are considered by the IAU to be more massive than 13 M_J, but less massive than 0.08 solar mass ($M_\odot$)—about 80 M_J (**Figure 7.15**).

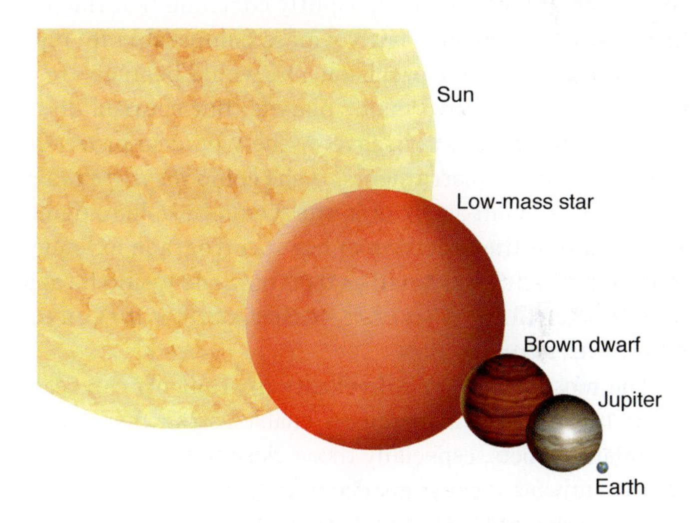

FIGURE 7.15 Compare the sizes of the Sun, a low-mass star, a brown dwarf, Jupiter, and Earth.

The Search for Extrasolar Planets

Currently, over 100 projects are focused on searching for extrasolar planets from the ground and from space. Here we review the most common methods of finding these planets. The first of the successful detection techniques is the *spectroscopic radial velocity method*. Hundreds of extrasolar planets have been detected with this technique. As a planet orbits about a star, the planet's gravity causes the star to wobble back and forth ever so slightly. This motion toward or away from us (radial velocity) creates an observable Doppler shift (see Chapter 5) in the spectrum of the star. After detecting this wobble, astronomers can infer the planet's mass and its distance from the star.

We can see how this works by using the Solar System as an example. To simplify, let's start by assuming that the Solar System consists only of the Sun and Jupiter. (This simplification makes sense because Jupiter's mass is greater than the mass of all the other planets, asteroids, and comets combined.) Imagine an alien astronomer pointing a spectrograph toward the Sun. The Sun and Jupiter orbit

Several techniques are being used to find extrasolar planets.

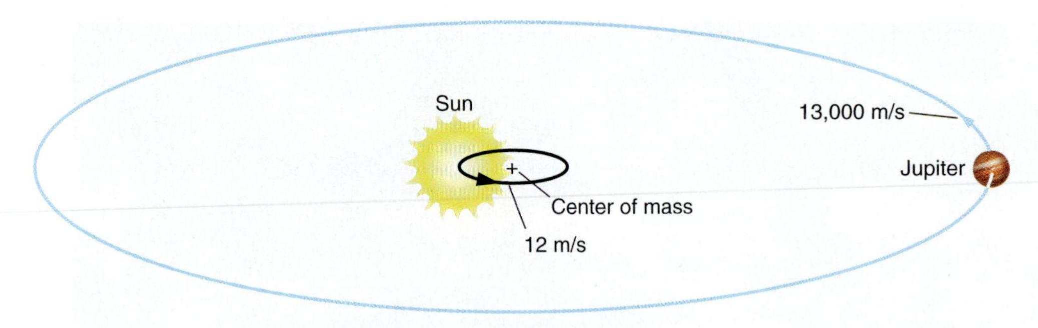

FIGURE 7.16 The Sun and Jupiter orbit around a common center of mass, which lies just outside the Sun's surface. Spectroscopic measurements made by an extrasolar astronomer would reveal the Sun's radial velocity varying by ±12 m/s over an interval of 11.86 years, Jupiter's orbital period.

a common center of mass that lies just outside the surface of the Sun (**Figure 7.16**). The alien astronomer would find that the Sun's radial velocity varies by ±12 m/s, with a period equal to Jupiter's orbital period of 11.86 years. From this information, she would rightly conclude that the Sun has at least one planet with a mass comparable to Jupiter's. Without greater precision, she would be unaware of the other less massive major planets. But spurred on by the excitement of her discovery, she would improve the sensitivity of her instruments. If she could measure radial velocities as small as 2.7 m/s, she would be able to detect Saturn, and if the precision of her spectrograph extended to radial velocities as small as 0.09 m/s, she would be able to detect Earth. ▶▶ NEBRASKA SIMULATION: INFLUENCE OF PLANETS ON THE SUN

The most precise radial velocity instruments can reach about 1 m/s. These instruments enable astronomers to detect giant planets, especially those close to their solar-type stars, but they are not as good at finding planets with masses similar to that of Earth. **Math Tools 7.2** provides additional explanation of the spectroscopic radial velocity method. ▶▶ NEBRASKA SIMULATION: RADIAL VELOCITY GRAPH

Instead of measuring the movement of the star from its radial velocity, might it be possible to directly observe the star's change in position over time as it orbits the common center of mass? This is called the *astrometric* method, but so far it has not led to discoveries of new planets, because the changes in a star's position are very small and there is too much atmospheric distortion from the ground. The upcoming European Space Agency mission Gaia will search for planets using this method (see problem 60 at the end of the chapter).

Another technique for finding extrasolar planets is the *transit method*, in which the effect of a planet passing in front of its parent star is observed. From Earth it is sometimes possible to see the inner planets—Mercury and Venus—transiting the Sun. An alien astronomer could infer the existence of Earth if she were located somewhere in the plane of Earth's orbit (that's the only way she would be able to see Earth pass in front of the Sun) and could detect an 0.009 percent drop in the Sun's brightness during the transit. Similarly, for astronomers on Earth to observe a planet pass in front of a star, Earth must necessarily lie nearly in the orbital plane of that planet. When an extrasolar planet passes in front of its parent star, the light from the star diminishes by a tiny amount (**Figure 7.17**). The radial velocity method yields the mass of a planet and its orbital distance from a star; the transit method provides a measure of the size of a planet. ▶▶ NEBRASKA SIMULATION: EXOPLANET TRANSIT SIMULATOR

Several hundred extrasolar planets have been detected using the transit method from ground-based and space telescopes. Current ground-based technology limits the sensitivity of the transit method to about 0.1 percent of a star's brightness. Amateur astronomers have confirmed the existence of several extrasolar planets by observing transits using charge-coupled device (CCD) cameras mounted on telescopes with apertures as small as 20 centimeters (cm). Two telescopes in space are looking for planets employing the transit method. The French COROT satellite is a small telescope (27 cm) that has discovered two dozen planets and has found hundreds of planet candidates. NASA's 0.95-meter telescope Kepler has discovered many planets and has found thousands more candidates that are being investigated further. Multiplanet systems have been identified with this method; if one planet is found, then observations of the variations in timing of the transit can indicate that there are other planets orbiting the same star (**Figure 7.18**).

Still another technique for exoplanet discovery is *microlensing*, which takes advantage of an effect called *gravitational lensing*, in which the gravitational field of an unseen planet bends light from a distant star in such a way that it causes the star to brighten temporarily while the planet is passing in front of it. Like the radial velocity method, microlensing provides an estimate of the mass of the planet. So far, about a dozen extrasolar planets have been found with this technique. Like the transit method used in space, lensing is also capable of detecting Earth-sized planets.

A fifth method is *direct imaging*. This is a difficult technique because it involves searching for a relatively faint planet in the overpowering glare of a bright star—a challenge

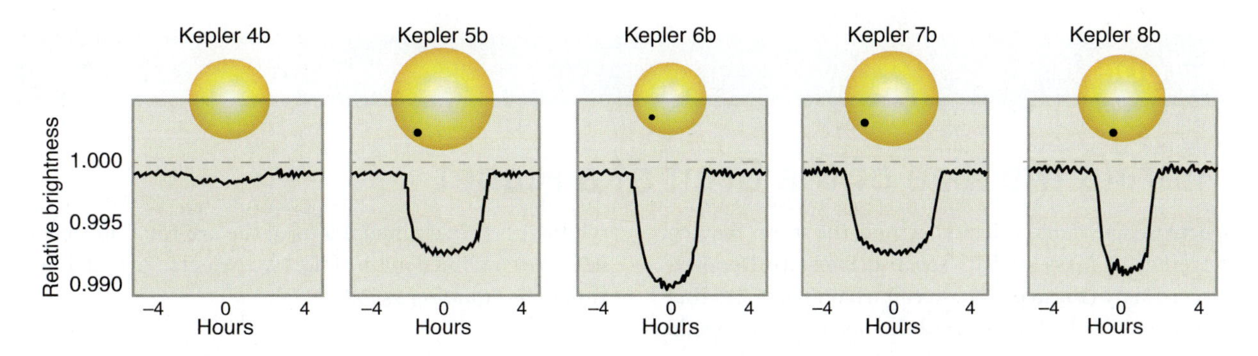

FIGURE 7.17 As a planet passes in front of a star, it blocks some of the light coming from the star's surface, causing the brightness of the star to decrease slightly. These observations are from the Kepler space observatory.

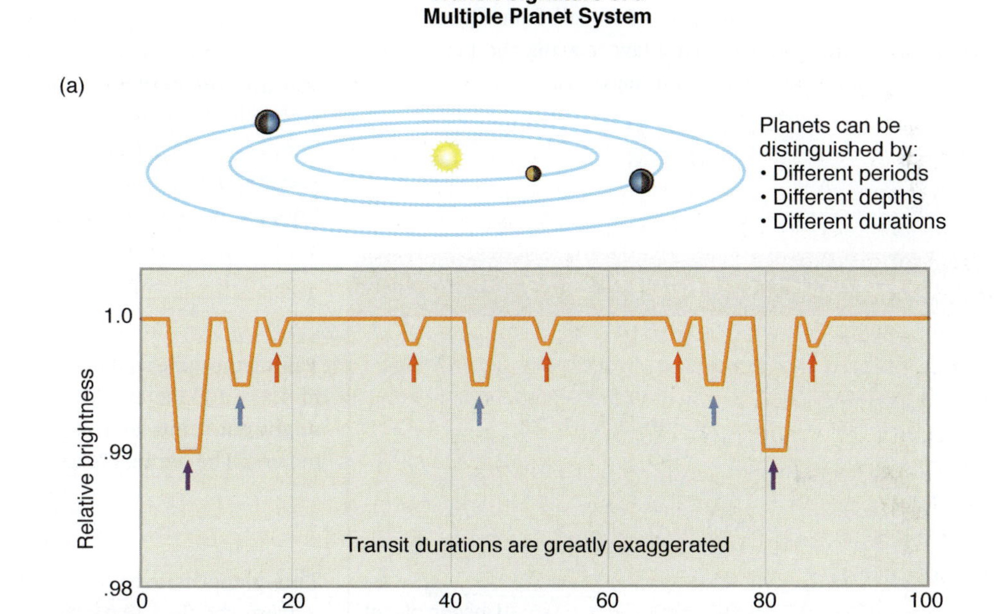

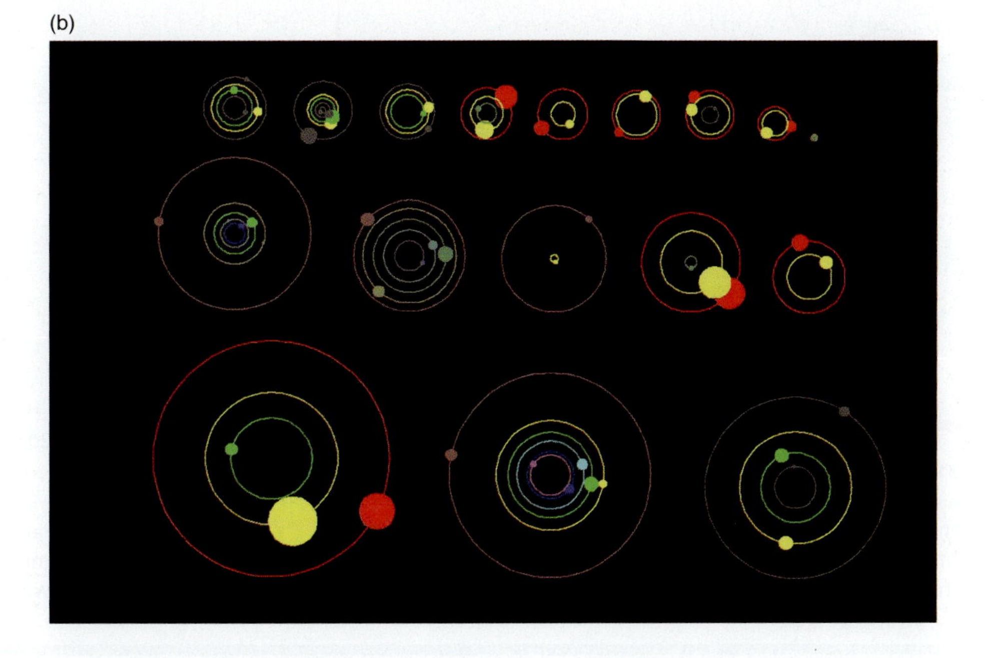

FIGURE 7.18 (a) Multiple planets can be detected by multiple transits with different brightness changes. (b) The Kepler Mission has identified 16 planetary systems that have more than one transiting planet. The relative sizes of the orbits and of the planets are correct, but they are not on the same scale here.

Math Tools 7.2

Estimating the Size of the Orbit of a Planet

In the spectroscopic radial velocity method, the star is moving about its center of mass, and its spectral lines are Doppler-shifted accordingly (**Figure 7.19**). Recall from Figure 7.16 that the alien astronomer looking toward the Solar System would observe a shift in the wavelengths of the Sun's spectral lines—caused by the presence of Jupiter—of about 12 m/s.

Figure 7.20 shows the radial velocity data for a star with a planet discovered by this method. How do astronomers use this method to estimate the distance (A) of the planet from the star and the mass of the planet? Recall from Chapter 4 that Newton generalized Kepler's law relating the period of an object's orbit to the orbital semimajor axis:

$$P^2 = \frac{4\pi^2}{G} \times \frac{A^3}{M}$$

where A is the semimajor axis of the orbit, P is its period, and M is the combined mass of the two objects. To find A, we rearrange the equation as follows:

$$A^3 = \frac{G}{4\pi^2} \times M \times P^2$$

From the graph of radial velocity observations in Figure 7.20, we can determine that the period of the orbit is 5.7 years. There are 3.16×10^7 seconds in a year, so $P = 5.7 \times (3.16 \times 10^7)$, or 1.8×10^8, seconds. The mass of the star is much greater than the mass of the planet, so the combined masses of the star and the planet can be approximated as the mass of the star, which in this case is about equal to the mass of the Sun, 2×10^{30} kg. (Stellar masses can be estimated from their spectra). The gravitational constant is $G = 6.67 \times 10^{-11}\ \text{m}^3/\text{kg s}^2$. Putting in the numbers gives:

$$A^3 = \frac{6.67 \times 10^{-11} \dfrac{\text{m}^3}{\text{kg s}^2}}{4\pi^2} \times (2 \times 10^{30}\ \text{kg}) \times (1.8 \times 10^8\ \text{s})^2 = 1.1 \times 10^{35}\ \text{m}^3$$

Taking the cube root of $1.1 \times 10^{35}\ \text{m}^3$ solves for A, which is equal to 4.8×10^{11} meters. To get a better feel for this number, we might put it into astronomical units (where 1 AU $= 1.5 \times 10^{11}$ meters). The semimajor axis of the orbit of this planet is given by:

$$A = \frac{4.8 \times 10^{11}\ \text{m}}{1.5 \times 10^{11}\ \text{m/AU}} = 3.2\text{AU}$$

This planet is over 3 times farther from its star than Earth is from the Sun. ▶▶ **NEBRASKA SIMULATIONS: RADIAL VELOCITY GRAPH; RADIAL VELOCITY SIMULATOR**

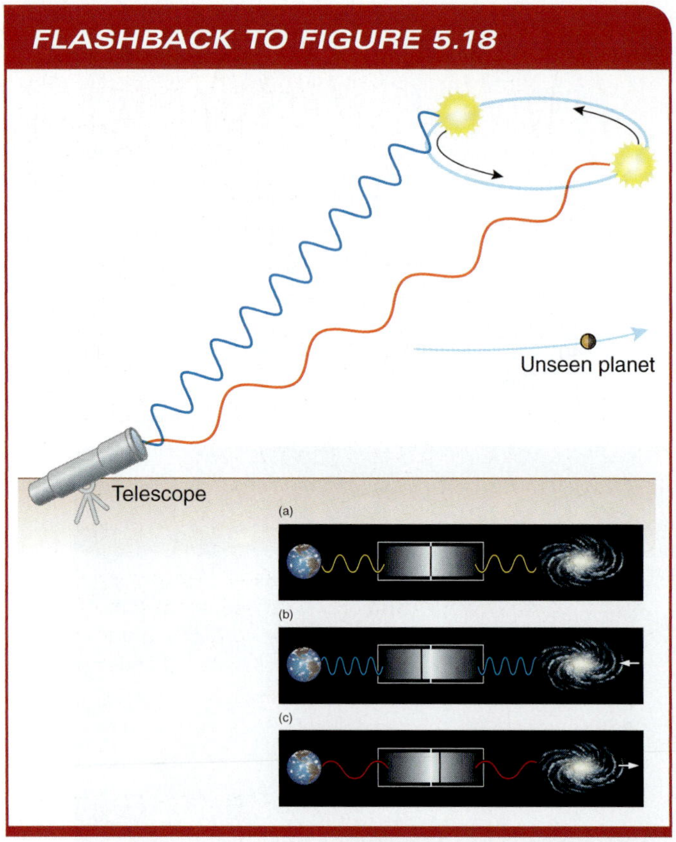

FLASHBACK TO FIGURE 5.18

FIGURE 7.19 Doppler shifts observed in the spectrum of a star are due to the wobble of the star caused by its planet. When the star is slightly moving away from the observer there's a redshift, and when it is slightly moving toward the observer there's a blueshift.

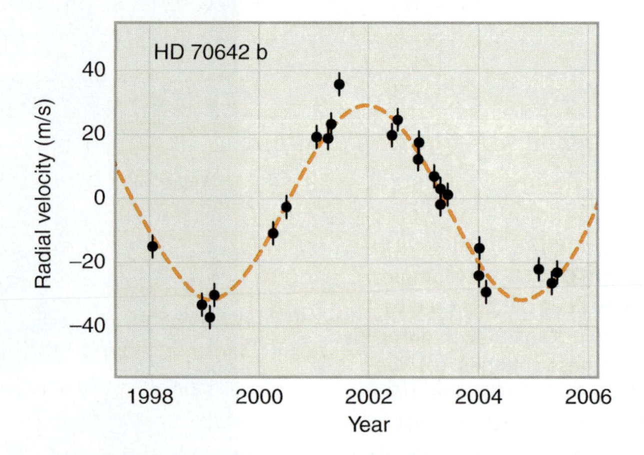

FIGURE 7.20 Radial velocity data for a star with a planet. A positive number is motion away; a negative number is motion toward the observer.

far more difficult than looking for a firefly in the dazzling brilliance of a searchlight. This method is most likely to find large planets far from their stars. As of this writing, astronomers have identified a few dozen planets by direct imaging. These planets are all more massive than Jupiter, and most are very far from their star. Some of the planets (including four around a single star) were discovered by large, ground-based telescopes operating in the infrared region of the spectrum with adaptive optics (**Figures 7.21 and 7.22**).

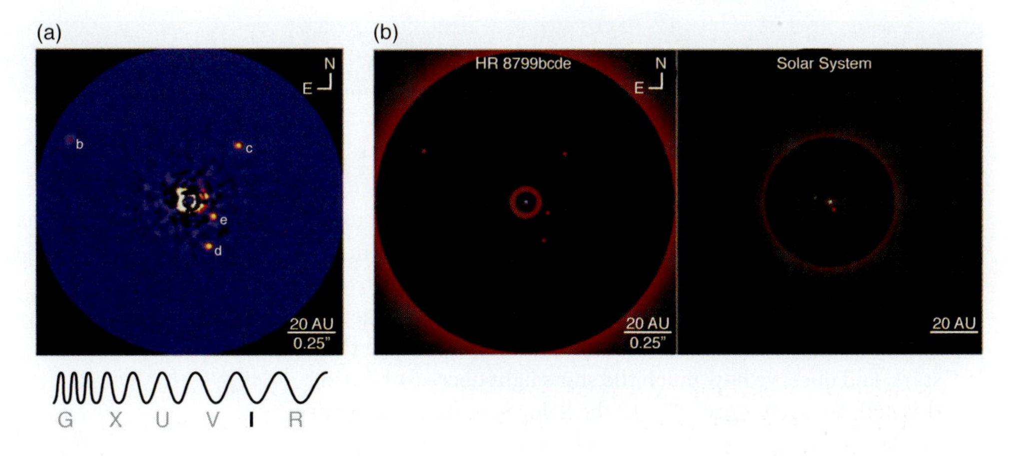

FIGURE 7.21 (a) A direct infrared image shows four planets (labeled "b," "c," "d," and "e"), each with a mass several times that of Jupiter, orbiting the star HR 8799 (which is hidden behind a mask). (b) The scale of this system compared with that of the Solar System.

Hundreds of Extrasolar Planets Have Been Discovered So Far

Searches for extrasolar planets have been remarkably successful. Since the first was identified in 1995, hundreds more have been confirmed, and thousands of candidates are under investigation. As the number of observed systems with single and multiple planets increases, astronomers can compare them with those of the Solar System, and they have found more variation than they expected.

The first discoveries of exoplanets included many **hot Jupiters**, which are Jupiter-type planets orbiting solar-type stars in circular or highly eccentric orbits that get closer to their parent stars than Mercury is to our own Sun. A massive planet orbiting very close to its parent star tugs the star very hard, creating large radial velocity variations in the star. This means that hot Jupiters are relatively easy to find by the spectroscopic radial velocity method. In addition, large planets orbiting close to their parent stars are more likely to move in front of the star periodically and reveal themselves via the transit method. Astronomers realized that these hot Jupiter systems are not representative of most planetary systems; they were just easier to find. Scientists call this bias a *selection effect.*

Astronomers were surprised by the existence of hot Jupiters because, according to the planetary system formation theory available at the time (based only on the Solar System), these giant, volatile-rich planets should not have been able to form so close to their parent stars. The expectation was that Jupiter-type planets should form in the more distant, cooler regions of the protoplanetary disk, where the volatiles that make up much of their composition are able to survive. So astronomers suggested that perhaps hot Jupiters formed much farther away from their parent star and subsequently migrated inward to a closer orbit.

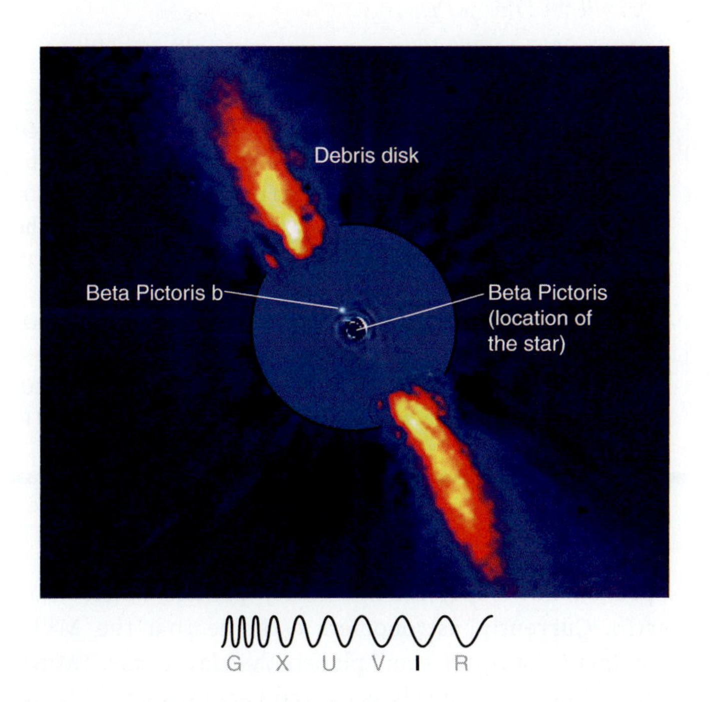

FIGURE 7.22 Beta Pictoris b is seen orbiting within a dusty debris disk that surrounds the bright naked-eye star Beta Pictoris. The planet's estimated mass is 8 times that of Jupiter. The star is hidden behind an opaque mask, and the planet appears through a semitransparent mask used to subdue the brightness of the dusty disk.

The mechanism by which a planet could migrate over such a distance must involve an interaction with gas or planetesimals in which orbital angular momentum is somehow transferred from the planet to its surroundings, allowing it to spiral inward.

Some new planets are even larger than Jupiter, but most of those being discovered now are planets with masses of 2–10 $M_{\oplus}$: mini-Neptunes (gaseous planets smaller than

Math Tools 7.3

Estimating the Size of an Extrasolar Planet

The masses of extrasolar planets can often be estimated using Kepler's laws and the conservation of angular momentum. When planets are detected by the transit method, astronomers can estimate the size (radius) of an extrasolar planet. In this method, astronomers look for planets that eclipse their stars, and observe how much the star's light decreases during this eclipse (see Figure 7.17). In the Solar System when Venus or Mercury transits the Sun, a black circular disk is visible on the face of the circular Sun. During the transit, the amount of light from the transited star is reduced by the area of the circular disk of the planet divided by the area of the circular disk of the star:

$$\text{Percentage reduction in light} = \frac{\text{Area of disk of planet}}{\text{Area of disk of star}} = \frac{\pi R^2_{\text{planet}}}{\pi R^2_{\text{star}}} = \frac{R^2_{\text{planet}}}{R^2_{\text{star}}}$$

Then, to solve for the radius of the planet, astronomers need an estimate of the radius of the star and a measurement of the percentage reduction in light during the transit. The radius of a star is estimated from the surface temperature and the luminosity of the star. The decrease in the light of the star can be measured with a telescope.

Let's consider an example. Kepler-11 is a system of at least six planets that transit a star. The radius of the star, R_{star} is estimated to be 1.1 times the radius of the Sun, or $1.1 \times (7.0 \times 10^8$ meters) $= 7.7 \times 10^8$ meters. The light from planet Kepler-11c is observed to decrease by 0.077 percent, or 0.00077 (**Figure 7.23**). What is Kepler-11c's size?

$$0.00077 = \frac{R^2_{\text{Kepler-11c}}}{R^2_{\text{star}}} = \frac{R^2_{\text{Kepler-11c}}}{(7.7 \times 10^8 \text{ m})^2}$$

$$R^2_{\text{Kepler-11c}} = 4.5 \times 10^{14} \text{ m}^2$$

$$R_{\text{Kepler-11c}} = 2.1 \times 10^7 \text{ m}$$

Dividing Kepler-11c's radius by the radius of Earth (6.4×10^6 meters) shows that the planet Kepler-11c has a radius of 3.3 $R_\oplus$.

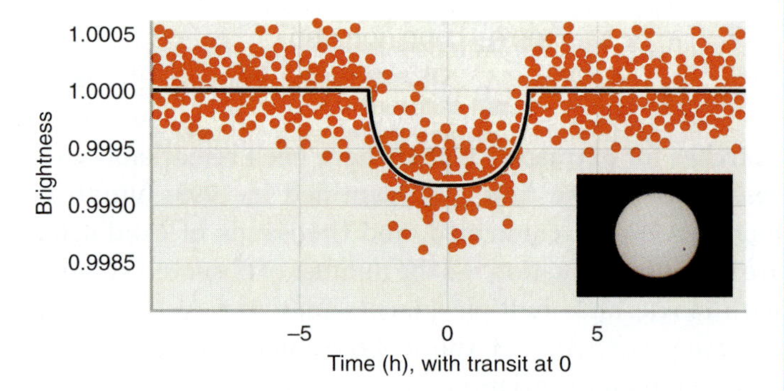

FIGURE 7.23 Light curve for Kepler-11c. Inset: Venus passed in front of the Sun in June of 2012, similar to this transit of Kepler-11c.

Neptune), and super-Earths (rocky planets larger than Earth). Currently, astronomers estimate that the Milky Way Galaxy has more small planets than large ones. (**Math Tools 7.3** demonstrates how the radii are estimated.) Some of the extrasolar planets have highly eccentric orbits compared with those in the Solar System. Planets have been found with orbits that are highly tilted compared with the plane of the rotation of their star, and some planets move in orbits whose direction is opposite that of their star's rotation. Multiple-planet systems have been observed in which the larger mini-Neptunes alternate with smaller Earth-sized planets. The multiple-planet systems that have been found by the transit method reside in flat systems like our own, offering further evidence that the planets formed in a flat accretion disk around a young star. But the current hypothesis to explain the Solar System's inner small rocky planets and outer large gaseous planets may not be applicable in these other planetary systems.

Most planetary systems found to date do *not* resemble our own.

In addition, a dozen planets that have been found by microlensing seem to be wandering freely through the Milky Way. These planets may have been ejected from their solar systems during formation and are no longer in gravitationally bound orbits to their stars. The frequent new discoveries requiring revisions of existing theories make extrasolar planets one of the most exciting topics in astronomy today.

7.6 Origins: Kepler's Search for Earth-Sized Planets

The study of planetary systems, many unlike the Solar System, challenge some aspects of scientists' understanding of planet formation. Yet one message conveyed by these discoveries is clear: the formation of planets frequently,

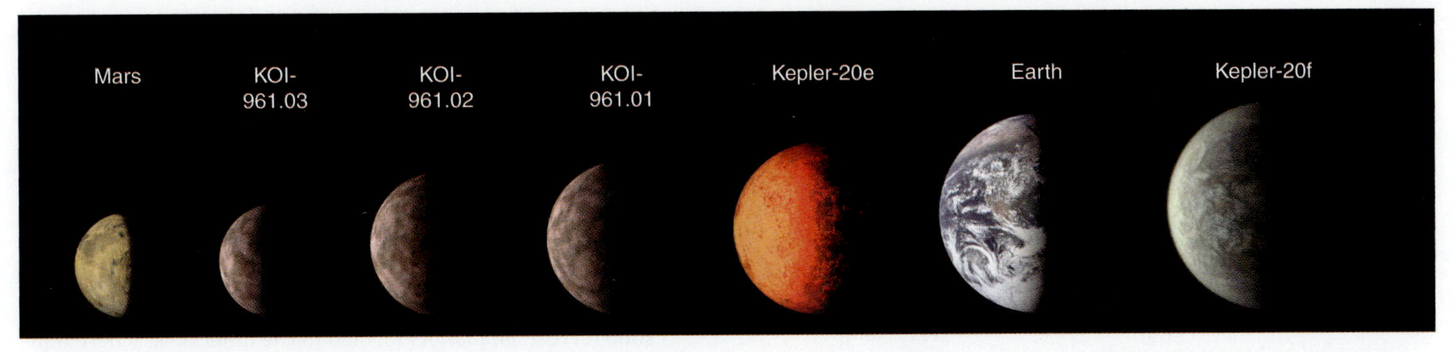

FIGURE 7.24 Artistic impressions of some of the first Earth-sized planets identified with the Kepler telescope, in multiplanet systems Kepler-20 and KOI-961. For comparison, Mars and Earth are shown to scale.

and perhaps always, accompanies the formation of stars. The implications of this conclusion are profound. Planets are a common by-product of star formation. In a galaxy of 200 billion stars, and a universe of hundreds of billions of galaxies, how many planets (or even moons) might exist? And with all of these planets in the universe, how many might have suitable conditions for the particular category of chemical reactions that we refer to as "life"?

The Kepler Mission was developed by NASA to find Earth-sized and larger planets in orbit about a variety of stars. Kepler is a 1-meter telescope with 42 CCD detectors, designed to observe approximately 150,000 stars in 100 square degrees of sky and look for planetary transits. To confirm a planetary detection, the transits need to be observed three times with repeatable changes in brightness, duration of transit times, and computed orbital period. Kepler can detect a dip in the brightness of a star of 0.01 percent—which is sensitive enough to detect an Earth-sized planet. This goal was reached in late 2011, when Kepler identified the first Earth-sized planets. Some of the first small planets that Kepler detected are illustrated in **Figure 7.24**. Stars with transiting planets detected by Kepler are also observed spectroscopically to obtain radial velocity measurements that can lead to an estimate of the planet's mass. Then the planet's density (mass per volume) can be estimated too. From the density, astronomers can get a sense of whether the planet is composed of primarily gas, rock, ice, water, or a mixture of some of these.

On Earth, liquid water was essential for the formation and evolution of life. Since life on Earth is the only example of life for which we have evidence, we do not know whether liquid water is a cosmic requirement, but it is a place to start. The primary scientific goal of the Kepler Mission is to look for rocky planets at the right distance from their star to permit the existence of liquid water. If a planet is too close to its star, water will exist only as a vapor; if it is too far, water will be frozen as ice. This range is called the **habitable zone**, and in the Solar System, Earth is in it. Although announcements of new planets often state whether the planet is in the habitable zone, just being in the zone doesn't guarantee that the planet actually has liquid water—or that the planet is inhabited!

Kepler has identified thousands of planet candidates, some in the habitable zone of their star. One example is Kepler-22b (**Figure 7.25**). The candidates must be confirmed by follow-up observations of more transits or of radial velocities before they are officially announced as planet detections. Amateur astronomers can access the candidate lists online (at the "Exoplanet Transit Database," http://var2.astro.cz/ETD) and conduct their own observations. Anyone with Internet access can go to PlanetHunters.org, examine some Kepler data, and contribute to the search.

The link between the accretion disks that surround young stars and the local collection of planets is the start-

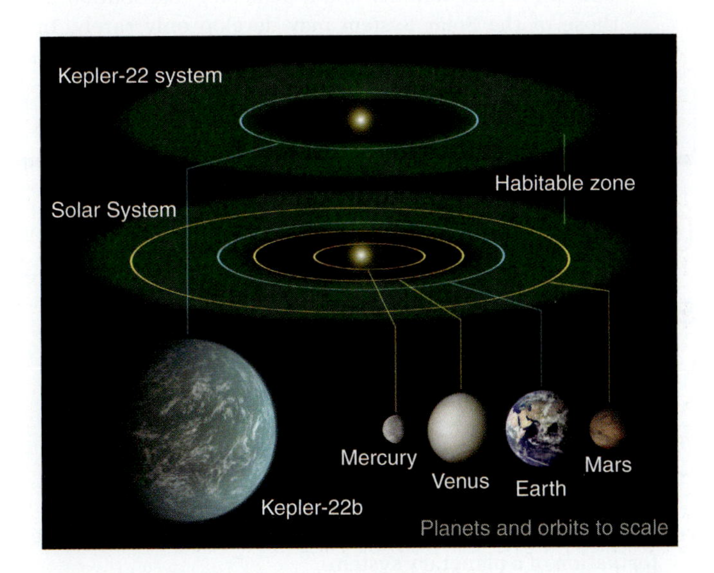

FIGURE 7.25 The Kepler-22 system, located about 600 light-years distant, has a planet in its habitable zone. The Solar System is shown for comparison. The planet is 0.85 AU from its star, which is very similar to the Sun.

ing point for the study of the Solar System. In the next chapters we will look at the planets, moons, asteroids, and comets that orbit the Sun, and you will see that what is known about the Sun and planets makes sense when viewed within the context of the evolving universe as a whole. ■

Summary

7.1 Stars and their planetary systems form from collapsing interstellar clouds of gas and dust, following the laws of gravity and conservation of angular momentum. Planets are a common by-product of star formation, and many stars are surrounded by planetary systems.

7.2 The Solar System formed a little less than 5 billion years ago, nearly 9 billion years after the birth of the universe. Planets grew from a protoplanetary disk of gas and dust that surrounded the forming Sun.

7.3 Dust grains in the disk that surrounds a forming star stick together to form larger and larger solid objects and eventually become planets. Temperatures in the disk affect the composition and location of planets, moons, and other bodies. The gas that is captured by a planet at the time of its formation is the planet's primary atmosphere. Less massive planets lose their primary atmospheres and then form secondary atmospheres.

7.4 In the current model of the formation of the Solar System, solid terrestrial planets formed in the inner disk, where temperatures were high; and giant gaseous planets formed in the outer disk, where temperatures were low. Dwarf planets like Pluto formed in the asteroid belt and in the region beyond the orbit of Neptune. Asteroids and comet nuclei remain today as leftover debris.

7.5 Hundreds of extrasolar planets have been found orbiting other stars within the Milky Way Galaxy. Extrasolar planets are located by the spectroscopic radial velocity method, the transit method, microlensing, and direct imaging.

7.6 The Kepler Mission uses the transit method to search for "Earth-like" planets orbiting other stars.

Unanswered Questions

- How typical is the Solar System? Only within the past few years have astronomers found other systems containing four or more planets, and so far the distributions of large and small planets in these multiplanet systems have looked different from those of the Solar System. Computer simulations of planetary system formation suggest that a system with an orbital stability and a planetary distribution like those of the Solar System may develop only rarely. Improved supercomputers can run more complex simulations, which can be compared with the observations.
- How common are Earth-like planets, and how Earth-like must a planet be before scientists declare it to be "another Earth"? An editorial in the science journal *Nature* cautioned that scientists should define "Earth-like" in advance—before multiple discoveries of planets "similar" to Earth are announced (and a media frenzy ensues). Must a planet be of similar size and mass (and thus similar density), be located in the habitable zone, and have spectroscopic evidence of liquid water before we call it "Earth 2.0"?

Questions and Problems

Summary Self-Test

1. Rank the following in increasing order of size.
   a. protosolar nebula
   b. orbit of Earth around the Sun
   c. universe
   d. orbit of the Moon around Earth
   e. galaxy
   f. Solar System
2. Place the following events in the order they occur during the formation of a planetary system.
   a. Gravity collapses a cloud of interstellar gas.
   b. A rotating disk forms.
   c. Small bodies collide to form larger bodies.
   d. A stellar wind "turns on" and sweeps away gas and dust.
   e. Primary atmospheres form.
   f. Primary atmospheres are lost.
   g. Secondary atmospheres form.
   h. Dust grains stick together by static electricity.
3. If the radius of an object's orbit is halved, what must happen to the speed so that angular momentum is conserved?
   a. It must be halved.
   b. It must stay the same.
   c. It must be doubled.
   d. It must be squared.

4. Unlike the giant planets, the terrestrial planets formed when
   a. the inner Solar System was richer in heavy elements than the outer Solar System.
   b. the inner Solar System was hotter than the outer Solar System.
   c. the outer Solar System took up a bigger volume than the inner Solar System, so there was more material to form planets.
   d. the inner Solar System was moving faster than the outer Solar System.
5. Planetary systems in the Milky Way Galaxy are probably
   a. universal (every star has planets).
   b. common (many stars have planets).
   c. rare (few stars have planets).
   d. exceedingly rare (only one star has planets).
6. Extrasolar planets have been detected by
   a. the spectroscopic radial velocity method.
   b. the transit method.
   c. microlensing.
   d. direct imaging.
   e. all of the above
7. The terrestrial planets and the giant planets have different compositions because
   a. the giant planets are much larger.
   b. the terrestrial planets are closer to the Sun.
   c. the giant planets are made mostly of solids.
   d. the terrestrial planets have few moons.
8. The spectroscopic radial velocity method preferentially detects
   a. large planets close to the central star.
   b. small planets close to the central star.
   c. large planets far from the central star.
   d. small planets far from the central star.
   e. none of the above (The method detects all of these equally well.)
9. The concept of disk instability was developed to solve the problem that
   a. Jupiter-like planets migrate after formation.
   b. there was not enough gas in the Solar System to form Jupiter.
   c. the early solar nebula likely dispersed too soon to form Jupiter.
   d. Jupiter consists mostly of volatiles.
10. When extrasolar planets were first discovered, they were surprising, and astronomers had to
    a. throw out the theories of the formation of stellar systems.
    b. modify the theories of the formation of stellar systems.
    c. modify the data to fit the theories of the formation of stellar systems.
    d. throw out the new data, because they represented outliers that did not fit the theory.

True/False and Multiple Choice

11. **T/F:** A cloud of interstellar gas is held together by gravity.
12. **T/F:** Gravity and angular momentum are both important in the formation of planetary systems.
13. **T/F:** Volatile materials are solid only at low temperatures.
14. **T/F:** The Solar System formed from a giant cloud of dust and gas that collapsed under gravity.
15. **T/F:** Microlensing is similar to the transit method in that both require the planet to pass in front of a bright object.
16. Since angular momentum is conserved, an ice-skater who throws her arms out will
    a. rotate more slowly.
    b. rotate more quickly.
    c. rotate at the same rate.
    d. stop rotating entirely.
17. Clumps grow into planetesimals by
    a. gravitationally pulling in other clumps.
    b. colliding with other clumps.
    c. attracting other clumps with opposite charge.
    d. conserving angular momentum.
18. Two distinct groups of scientists arrived at the same description of how the Solar System formed from two different avenues of investigation. This is an example of the ________ nature of science.
    a. conspiratorial
    b. simplistic
    c. coincidental
    d. self-consistent
19. If the radius of a spherical object is halved, what must happen to the period so that the spin angular momentum is conserved?
    a. It must be divided by 4.
    b. It must be halved.
    c. It must stay the same.
    d. It must double.
    e. It must be multiplied by 4.
20. The amount of angular momentum in an object depends on
    a. its radius.
    b. its mass.
    c. its rotation speed.
    d. all of the above
21. The spectroscopic radial velocity method of planet discovery measures a motion that is most like the motion of
    a. a tetherball around a pole.
    b. Earth around its axis.
    c. a swinger on a swing.
    d. an ice-skater spinning his partner.
22. The planets in the inner part of the Solar System are made primarily of refractory materials; the planets in the outer Solar System are made primarily of volatiles. The reason for the difference is that
    a. refractory materials are heavier than volatiles, so they sank farther into the nebula.
    b. there were no volatiles in the inner part of the accretion disk.
    c. the volatiles were lost soon after the planet formed.
    d. the outer Solar System has gained more volatiles from space since formation.

23. Which of the following planets still has its primary atmosphere?
    a. Mercury
    b. Earth
    c. Mars
    d. Jupiter

24. If scientists want to find out about the composition of the early Solar System, the best objects to study are
    a. the terrestrial planets.
    b. the giant planets.
    c. the Sun.
    d. asteroids and comets.

25. The discovery of "hot Jupiter" planets close to their central star led to the theoretical model of
    a. radial velocities.
    b. disk instabilities.
    c. planet migration.
    d. protoplanetary disks.

Thinking about the Concepts

26. What is the source of the material that now makes up the Sun and the rest of the Solar System?

27. Describe the different ways by which stellar astronomers and planetary scientists each came to the same conclusion about how planetary systems form.

28. What is a protoplanetary disk? There are two reasons why the inner part of the disk is hotter than the outer part. What are they?

29. Physicists describe certain properties, such as angular momentum and energy, as being *conserved*. What does this mean? Do these conservation laws imply that an individual object can never lose or gain angular momentum or energy? Explain your reasoning.

30. How does the law of conservation of angular momentum control a figure-skater's rate of spin?

31. Explain why the law of conservation of angular momentum posed problems for early versions of the nebular hypothesis.

32. What is an accretion disk?

33. Describe the process by which tiny grains of dust grow to become massive planets.

34. Look under your bed for "dust bunnies." If there aren't any, look under your roommate's bed, the refrigerator, or any similar place that might have some. Once you find them, blow one toward another. Watch carefully and describe what happens as they meet. What happens if you repeat this action with additional dust bunnies? Will these dust bunnies ever have enough gravity to begin pulling themselves together? If they were in space instead of on the floor, might that happen? What force prevents their mutual gravity from drawing them together into a "bunny-tesimal" under your bed?

35. Why do we find rocky material everywhere in the Solar System, but large amounts of volatile material only in the outer regions?

36. Why were the four giant planets able to collect massive gaseous atmospheres, whereas the terrestrial planets could not? Explain the source of the atmospheres now surrounding three of the terrestrial planets.

37. What happened to all the leftover Solar System debris after the last of the planets formed?

38. Describe four methods that astronomers use to search for extrasolar planets. What are the limitations of each method; that is, what circumstances are necessary to detect a planet by each method?

39. Why is it so difficult for astronomers to obtain an image of an extrasolar planet?

40. Many of the first exoplanets that astronomers found orbiting other stars are giant planets with Jupiter-like masses, and with orbits located very close to their parent stars. Explain why these characteristics could be a selection effect of the discovery method.

Applying the Concepts

41. Study Figure 7.20. What is the maximum radial velocity of HD 70642 in meters per second? Convert this number to miles per hour (mph). How does this compare to the speed at which Earth orbits the Sun (67,000 mph)?

42. Use information about the planets given in Appendix 4 to answer the following:
    a. What is the total mass of all the planets in the Solar System, expressed in Earth masses ($M_{\oplus}$)?
    b. What fraction of this total planetary mass does Jupiter represent?
    c. What fraction does Earth represent?

43. Compare Earth's orbital angular momentum with its spin angular momentum using the following values: $m = 5.97 \times 10^{24}$ kg, $v = 29.8$ kilometers per second (km/s), $r = 1$ AU, $R = 6{,}378$ km, and $P = 1$ day. Assume Earth to be a uniform body. What fraction does each component (orbital and spin) contribute to Earth's total angular momentum? Refer to Math Tools 7.1 for help.

44. Venus has a radius 0.949 times that of Earth and a mass 0.815 times that of Earth. Its rotation period is 243 days. What is the ratio of Venus's spin angular momentum to that of Earth? Assume that Venus and Earth are uniform spheres.

45. Jupiter has a mass equal to 318 times Earth's mass, an orbital radius of 5.2 AU, and an orbital velocity of 13.1 km/s. Earth's orbital velocity is 29.8 km/s. What is the ratio of Jupiter's orbital angular momentum to that of Earth?

46. In the text we give an example of an interstellar cloud having a diameter of 10^{16} meters and a rotation period of 10^6 years collapsing to a sphere the size of the Sun (1.4×10^9 meters

in diameter). We point out that if all the cloud's angular momentum went into that sphere, the sphere would have a rotation period of only 0.6 second. Do the calculation to confirm this result.

47. The asteroid Vesta has a diameter of 530 km and a mass of 2.7×10^{20} kg.
    a. Calculate the density (mass/volume) of Vesta.
    b. The density of water is 1,000 kg/m^3, and that of rock is about 2,500 kg/m^3. What does this difference tell you about the composition of this primitive body?

48. To an alien astronomer who observes Jupiter passing in front of the Sun, by how much does the Sun's brightness drop during the transit?

49. The best current technology can measure radial velocities of about 1 m/s. Suppose you are observing a spectral line with a wavelength of 575 nanometers (nm). How large a shift in wavelength would a radial velocity of 1 m/s produce?

50. Earth tugs the Sun around as it orbits, but it has a much smaller effect (only 0.09 m/s) than that of any known extrasolar planet. How large a shift in wavelength does this effect cause in the Sun's spectrum at 500 nm?

51. A hot Jupiter nicknamed "Osiris" was found around a solar-mass star, HD 209458. It orbits the star in only 3.525 days.
    a. What is the orbital radius of this extrasolar planet?
    b. Compare this planet's orbit with that of Mercury around the Sun. What environmental conditions must Osiris experience?

52. The extrasolar planet Osiris passes directly in front of its solar-type parent star, HD 209458 (diameter $= 1.7 \times 10^6$ km), every 3.525 days, decreasing the brightness of the star by about 1.7 percent (0.017).
    a. What is the diameter of Osiris?
    b. Compare the diameter of this extrasolar planet with that of Jupiter (mean diameter = 139,800 km).

53. One of the planets orbiting the star Kepler-11 (radius 1.1 solar radii, or $R_\odot$) has a radius of 4.5 Earth radii ($R_\oplus$). By how much does the brightness of Kepler-11 decrease when this planet transits the star?

54. The French COROT satellite has detected a planet with a diameter of 1.7 Earth diameters ($D_\oplus$).
    a. How much larger is the volume of this planet than Earth's?
    b. Assume that the density of the planet is the same as Earth's. How much more massive is this planet than Earth?

55. Consider the planet COROT-11b. It was discovered using the transit method, and astronomers have followed up with radial velocity measurements, so both its size (radius 1.43 Jupiter radii, or R_J) and its mass (2.33 M_J) are known. Using this information, you can find the density, which provides a clue about whether the object is gaseous or rocky.
    a. What is the mass of this planet in kilograms?
    b. What is the planet's radius in meters?
    c. What is the planet's volume?
    d. What is the planet's density? How does this density compare to the density of water (1,000 kg/m^3)? Is the planet likely to be rocky or gaseous?

Using the Web

56. Go to the "Extrasolar Planets Global Searches" Web page (http://exoplanet.eu/searches.php) of the Extrasolar Planets Encyclopaedia, which lists many of the current and future projects looking for planets. Click on one ongoing project under "Ground" and one ongoing project under "Space." What method is used to detect planets in each case? Has the selected project found any planets, and if so, what type are they? Now click on one of the future projects. When will the one you chose be ready to begin? What will be the method of detection?

57. Using the exoplanet catalogs:
    a. Go to the "Catalog" Web page (http://exoplanet.edu/catalog) of the Extrasolar Planets Encyclopaedia and click on "All Planets detected." Look for a star (in the left column) that has multiple planets. Make a graph showing the distances of the planets from their star, and note the masses and sizes of the planets. Put the Solar System planets on the same axis. How does this extrasolar planet system compare with the Solar System?
    b. Go to the "Exoplanets Data Explorer" website (http://exoplanets.org), and click on "Table." This website lists planets that have detailed orbital data published in scientific journals, and it may have a smaller total count than the site in (a). Pick a planet that was discovered this year or last, as specified in the "First Reference" column. What is the planet's minimum mass? What is its semimajor axis and the period of its orbit? Is its orbit circular or more elliptical? Click on the star name in the first column to get more information. Is there a radial velocity curve for this planet? Was it observed in transit, and if so, what is the planet's radius and density? Is it more like Jupiter or more like Earth?

58. Go to the website for the Kepler Mission (http://kepler.nasa.gov).
    a. How many confirmed planets has Kepler discovered? How many planet candidates? What kinds of follow-up observations are being done to verify whether the candidates are planets?
    b. Click on "News" and note the options. "Manager Updates" reports on issues with the spacecraft and telescope hardware; is the telescope working? "NASA Kepler News" includes press releases and conference presentations. "Kepler in the News" has reports about Kepler in the media. Read a recent story in each category. What is being reported? Why is it news? Did the general media pick up this story?

59. Go to http://planethunters.org. PlanetHunters is part of the Zooniverse, a citizen science project that lets individuals participate in a major science project using their own computers. To participate in this or any of the other Zooniverse projects in later chapters, you will need to sign up for an

account. Read through the sections under "About," including the FAQ. What are some of the advantages to having many people look at these Kepler data, instead of just one person or a computer program? Back on the PlanetHunters home page, click on "Tutorial" and watch the "Introduction" and "Tutorial Video." When you're ready to try looking for planets, click on "Classify" and begin. Remember to save a copy of your stars if required for your homework assignment.

60. Go to the website for the European Space Agency (ESA) mission Gaia (http://sci.esa.int/science-e/www/area/index.cfm?fareaid=26). This mission is scheduled for launch in 2013. Is it in space or delayed? Click on the "Extra-solar Planets" link on the left-hand side. What method(s) will Gaia use to look for planets? What are the science goals? If the mission is already in space, click on "News." Have some planets been found?

STUDYSPACE is a free and open website that provides a Study Plan for each chapter of *21st Century Astronomy*. Study Plans include animations, reading outlines, vocabulary flashcards, and multiple-choice quizzes, plus links to premium content in SmartWork and the ebook. Visit **wwnorton.com/studyspace**.

SMARTWORK Norton's online homework system, includes algorithmically generated versions of these questions, plus additional conceptual exercises. If your instructor assigns questions in SmartWork, log in at **smartwork.wwnorton.com.**

Exploration | Formation of the Solar System

A model of the formation of stars must account for the Solar System, including the motions of all the planets and their moons. In this exploration you will examine the data about the Solar System to see whether the model of star formation fits them.

Planet	Orbital Eccentricity	Inclination of Orbit	Planet Revolution*	Planet Rotation*	Moon Revolution*
Mercury	0.2056	7	CCW	CCW	None
Venus	0.0067	3.4	CCW	CW	None
Earth	0.0167	0	CCW	CCW	CCW
Mars	0.0935	1.9	CCW	CCW	CCW
Jupiter	0.0489	1.3	CCW	CCW	CCW
Saturn	0.0565	2.5	CCW	CCW	CCW
Uranus	0.0457	0.77	CCW	CCW	CCW
Neptune	0.0113	1.8	CCW	CCW	CCW

*CCW = counterclockwise; CW = clockwise.

Part 1: Shapes of Planetary Orbits

Examine the orbital eccentricities in the table. Since eccentricities run from 0 to 1, the eccentricity can be thought of as the percentage by which the orbit is different from round; for example, Earth is 1.67 percent different from round. To understand these percentages, we can compare them to others; for example, restaurant tips (in the United States) are typically 15 percent, while sales tax tends to be about 7 percent, depending on the state.

1 **Is Earth's eccentricity a large percentage? Is Earth's orbit much different from round?**

....

2 **Which planet has the most eccentric orbit? By what percentage is it different from round? Is that a large percentage?**

....

3 **In general, are these planetary orbits mostly round or mostly elliptical?**

....

Part 2: Inclinations of Solar System Planetary Orbits

The inclination of an orbit is the angle between the orbit and the plane of the Solar System. For example, the inclination of the Moon's orbit is 5°, because the orbit of the Moon makes a 5° angle to the orbit of Earth around the Sun.

4 **Which planet has the largest inclination?**

....

5 **Why is Earth's inclination exactly 0?**

....

6 **In one sentence, describe the shape of the Solar System.**

....

Part 3: Rotations of the Solar System Planets

Examine the last three columns of the table.

7 **Which planet is not rotating in the same direction as the rest of the Solar System?**

....

8 **Do the rotations of Solar System bodies indicate that they formed together at the same time from the same body, or separately under different conditions?**

....

9 **In one sentence, describe the rotations and revolutions of the planets.**

....

Part 4: The Big Picture

10 **Stars form from big clouds of dust and gas that collapse under gravity, and they conserve angular momentum. Explain how the observations of the Solar System fit into this model.**

....

11 **What would happen if the cloud were too thin for gravity to be important?**

....

12 **What would happen if angular momentum were not conserved?**

....

13 **Assuming that this model applies to the formation of the Solar System, how might you explain the counter-rotation of Venus?**

....

....

The constellation Orion. Betelgeuse is the red star in the upper left. Rigel is the blue star in the lower right.

13 Taking the Measure of Stars

To man, that was in th' evening made,
Stars gave the first delight;
Admiring, in the gloomy shade,
Those little drops of light.

Edmund Waller (1606–1687)

LEARNING GOALS

To all but the most powerful of telescopes, a star is just a point of light in the night sky. Astronomers apply their understanding of light, matter, and motion to what they see and are able to build a remarkably detailed picture of the physical properties of stars. By the conclusion of this chapter, you should be able to:

- Use the brightness of nearby stars and their distances from Earth to determine their luminosity.
- Infer the temperatures and sizes of stars from their colors.
- Determine the composition of stars from their spectra.
- Estimate the masses of stars.
- Classify stars, and organize this information on an H-R diagram.

13.1 Measuring the Distance, Brightness, and Luminosity of Stars

Although humans have been able to directly explore the Solar System, space probes cannot be sent to stars to take close-up pictures or land on their surfaces. Instead, astronomers study the stars by observing their light, by using the current understanding of the laws of physics discussed in earlier chapters, and by finding patterns in subgroups of stars that are extrapolated to other stars. Astronomers use knowledge of geometry, radiation, and orbits to begin to answer basic questions about stars, such as how they are similar to or different from the Sun, and whether they might have planets orbiting around them as the Sun does.

How might the basic properties of a star affect its chance of having a planet with life?

Your two eyes have different views that depend on the distance to the object you are viewing. Hold up your finger in front of you, quite close to your nose. View it with your right eye only and then with your left eye only. Each eye sends a slightly different image to your brain, so your finger appears to move back and forth relative to the background behind it. Now hold up your finger at arm's length, and blink your right eye, then your left. Your finger appears to move much less. This difference in perspective at different distances is the basis of **stereoscopic vision**, one of the ways that humans perceive distances. **Figure 13.1** shows that if each eye is shown a different view, the brain can even be fooled into perceiving distance where none exists.

Your stereoscopic vision enables you to judge the distances of objects as far away as a few hundred meters, but beyond that it is of little use. Your right eye's view of a mountain several kilometers away is indistinguishable from the view seen by your left eye. Comparing the two views tells your brain only that the mountain is too far away for you to judge its distance. The distance over which your stereoscopic vision works is limited by the separation between your eyes, which is only about 6 centimeters (cm). If you could somehow move your eyes farther apart, you could increase the differences between the views your two eyes see. If you could separate your eyes by several meters, their perspectives would be different enough to allow you to judge the distances to objects that were kilometers away.

Of course, you cannot separate your eyes like this, but you can compare pictures taken with a camera from two widely separated locations. The way to achieve the greatest possible separation without leaving Earth is to let Earth's orbital motion carry the camera from one side of the Sun to the other. If you take a picture of the sky tonight and then wait 6 months and take another picture, your point of view between the two pictures will have changed by the diameter of Earth's orbit, or 2 astronomical units (AU). With 2 AU separating your two "eyes," you should have very powerful stereoscopic vision indeed. **Figure 13.2** shows how the view of a field of stars changes as the perspective changes during the year. This change in perspective enables astronomers to measure the distances to nearby stars.

Distances to nearby stars are measured by comparing the view of them from opposite sides of Earth's orbit.

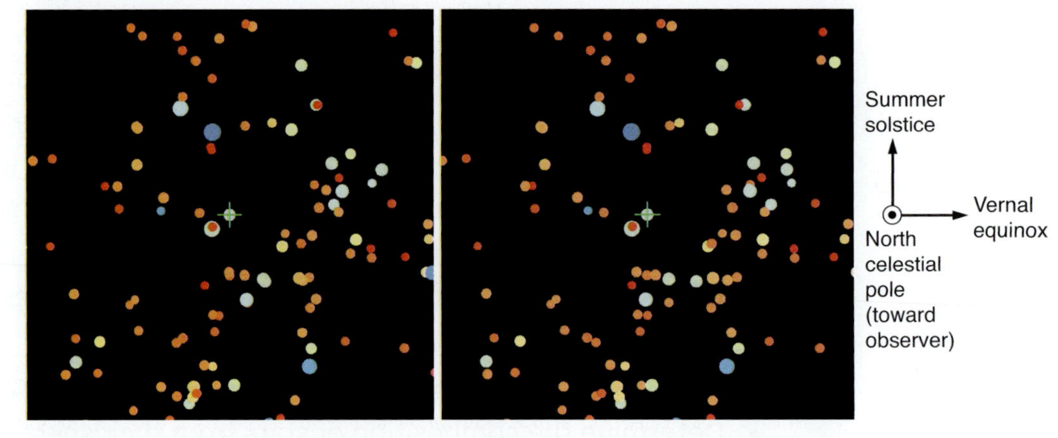

FIGURE 13.1 Your brain uses the slightly different views offered by your two eyes to "see" the distances and three-dimensional character of the world around you. This stereoscopic pair shows the stars in the neighborhood of the Sun as viewed from the direction of the north celestial pole. The field shown is 40 light-years on a side. The Sun is at the center, marked with a green cross. The observer is 400 light-years away and has "eyes" separated by about 30 light-years. To get the stereoscopic view, hold a card between the two images and look at them from about a foot and a half away. Relax and look straight at the page until the images merge into one.

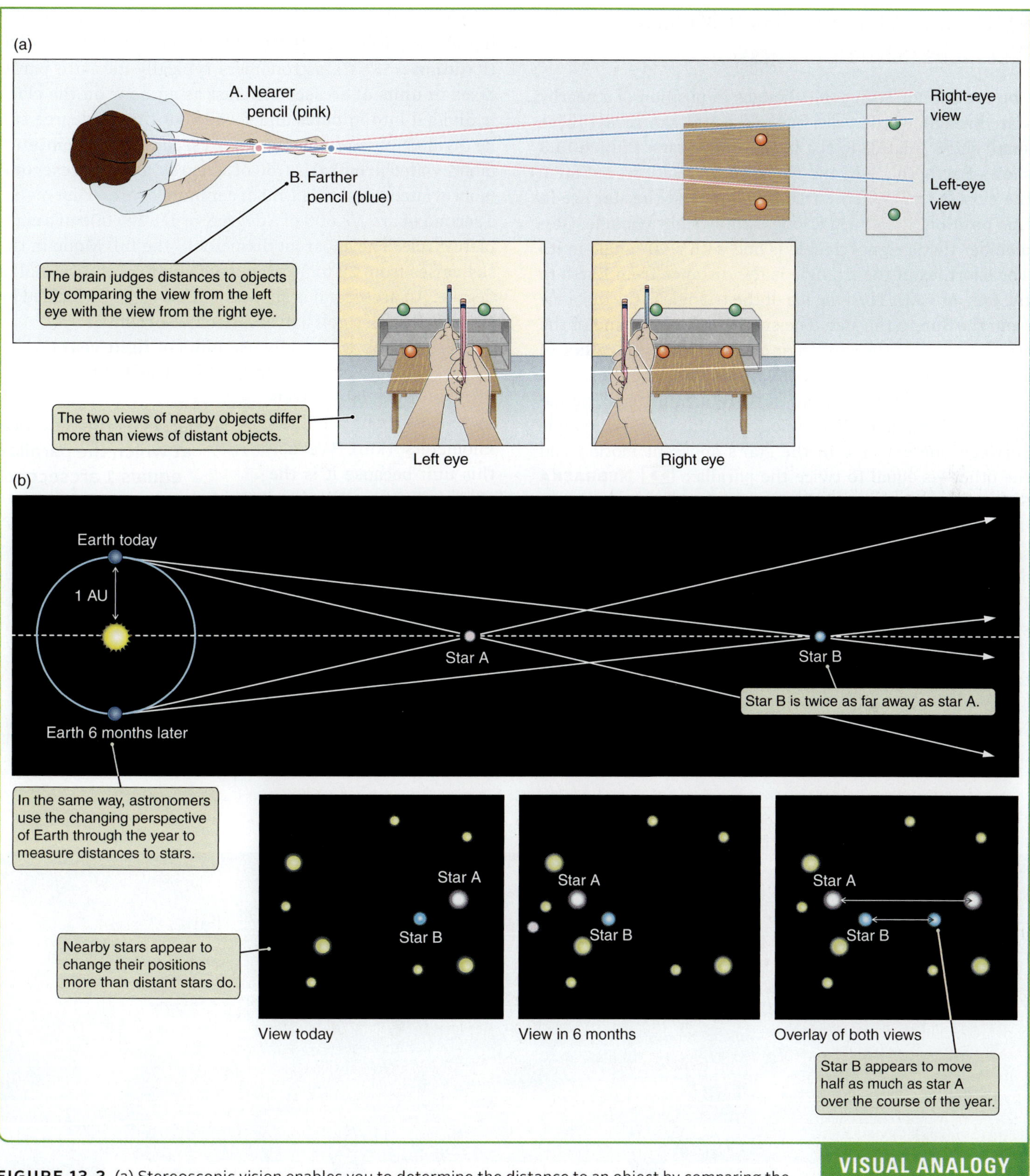

FIGURE 13.2 (a) Stereoscopic vision enables you to determine the distance to an object by comparing the view from each eye. (b) Similarly, comparing views from different places in Earth's orbit enables astronomers to determine the distance to stars. As Earth moves around the Sun, the apparent positions of nearby stars change more than the apparent positions of more distant stars. (The diagram is not to scale.) This is the starting point for measuring the distances to stars.

Astronomers Use Parallax to Measure Distances to Nearby Stars

The eye cannot detect the changes in position of a nearby star throughout the year, but telescopes can reveal these small shifts relative to the background stars. **Figure 13.3** shows Earth, the Sun, and three stars. Look first at Star 1, the closest star. When Earth, the Sun, and the star are in this position, they form a long, skinny right triangle. (Remember that a right triangle is one with a 90° angle in it.) The short leg of the triangle is the distance from Earth to the Sun, or 1 AU. The long leg of the triangle is the distance from the Sun to the star. The small angle at the end of the triangle is called the *parallactic angle*, or simply **parallax**, of the star. Over the course of a year, the star's position in the sky appears to shift back and forth, returning to its original position 1 year later. The amount of this shift—the angle between one extreme in the star's apparent motion and the other—is equal to twice the parallax. ▶▶ NEBRASKA SIMULATION: PARALLAX CALCULATOR

More distant stars make longer and skinnier triangles, and smaller parallaxes. Look again at Figure 13.3. Star 2 is twice as far away as star 1, and its parallax is only half as great. If you were to draw a number of such triangles for different stars, you would find that increasing the distance to the star always reduces the star's parallax. Moving a star 3 times farther away (see star 3 in Figure 13.3) reduces its parallax to ⅓ of its original value. Moving a star 10 times farther away reduces its parallax to 1/10 of its original value. The parallax of a star (p) is inversely proportional to its distance (d).

The parallaxes of real stars are tiny. Rather than talking about parallaxes of 0.0000028° or 4.8×10^{-8} radian (1 **radian** = 57.3°), astronomers typically measure parallaxes in units of arcseconds. Just as an hour on the clock is divided into minutes and seconds of time, a degree can be divided into arcminutes and arcseconds. An **arcminute** (abbreviated **arcmin**) is 1/60 of a degree, and an **arcsecond** (abbreviated **arcsec**) is 1/60 of a minute of arc.[1] That makes a second of arc 1/3,600 of a degree, or 1/1,296,000 of a complete circle. The apparent diameter of the full Moon in the sky varies from 29 to 34 arcmin, averaging just over half a degree. An arcsecond is about equal to the angle formed by the diameter of a golf ball at a distance of 5 miles.

In this book we usually use units of **light-years** to indicate distances to stars and galaxies. One light-year is the distance that light travels in 1 year—about 9 trillion kilometers (km). We use this unit because it is the unit you are most likely to see online or in a popular book about astronomy. When astronomers discuss distances to stars and galaxies, however, the unit they often use is the **parsec** (short for *parallax second* and abbreviated **pc**), which is equal to 3.26 light-years.

A parsec is the distance at which the parallax equals 1 arcsecond.

When astronomers began to apply the parallax technique, they discovered that stars are very distant objects

[1] Arcseconds are often denoted by the symbol ″ and arcminutes by ′; one second of arc is written as 1″, and one minute of arc is written as 1′. In this text, however, we will spell out these units.

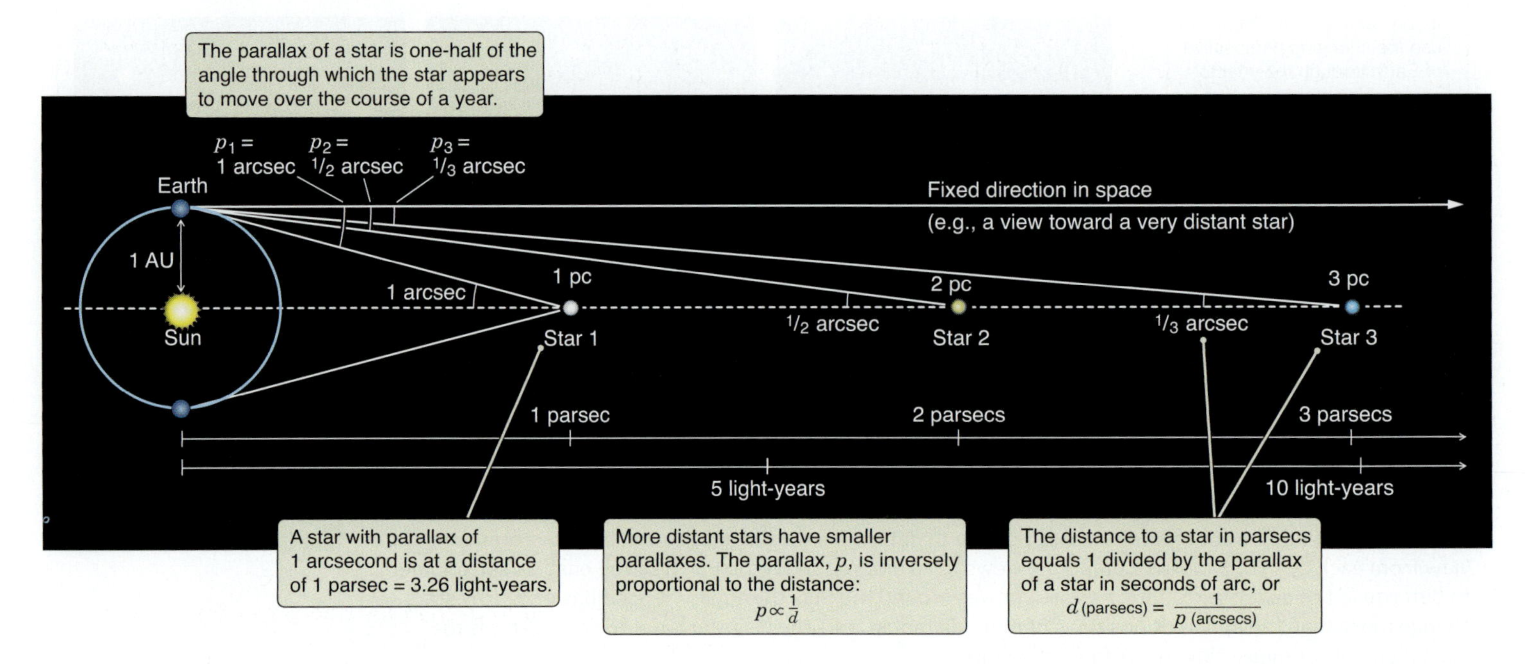

FIGURE 13.3 The parallaxes of three stars at different distances. (The diagram is not to scale.) Parallax is inversely proportional to distance.

(**Math Tools 13.1**). The first successful measurement of the parallax of a star was made by F. W. Bessel, who in 1838 reported a parallax of 0.314 arcsec for the star 61 Cygni. This finding implied that 61 Cygni was 3.2 pc away, or 660,000 times as far away as the Sun. With this one measurement, Bessel greatly increased the known volume of the universe. Today astronomers know of about 70 stars in 50 single-, double-, or triple-star systems within 5 pc (16.3 light-years) of the Sun. A sphere with a radius of 16.3 light-years has a volume of about 18,000 cubic light-years, corresponding to a local density of 50 systems per 18,000 cubic light-years. That is about 0.003 of a star system per cubic light-year. Stated another way, in the neighborhood of the Sun each system of stars has, on average, about 360 cubic light-years of space (a volume about 4.4 light-years in radius) all to itself.

Stars are few and far between in the Sun's neighborhood.

Knowledge of the Sun's stellar neighborhood took a tremendous step forward during the 1990s, when the European Space Agency's (ESA's) Hipparcos satellite measured the positions and parallaxes of 120,000 stars. These measurements, taken from a satellite well above Earth's obscuring atmosphere, are better than the measurements made from telescopes located on the surface of Earth. But even the data from this satellite has its limits. The accuracy of any given Hipparcos parallax measurement is about ±0.001 arcsec. Because of this **observational uncertainty**, measurements of the distances to stars are not perfect. For example, a star with a parallax measured by Hipparcos of 0.004 arcsec really has a parallax of anywhere between 0.003 and 0.005 arcsec. So instead of knowing that the distance to the star is exactly 250 pc (1/0.004 arcsec), astronomers know only that the star is probably between about 200 pc (1/0.005 arcsec) and 333 pc (1/0.003 arcsec). As an analogy, consider your speed while driving down the road. If your digital speedometer says 10 kilometers per hour (km/h), you might actually be traveling 10.4 km/h or 9.6 km/h. The precision of your speedometer is limited to whole numbers, but that doesn't mean you don't know your speed *at all*. You certainly know you are not traveling 100 km/h, for example.

With current technology, astronomers cannot reliably measure stellar distances of more than a few hundred parsecs using parallax.

Math Tools 13.1

Parallax and Distance

The parallax of a star (p) is inversely proportional to its distance (d); when one quanity goes up, the other goes down:

$$p \propto \frac{1}{d} \quad \text{or} \quad d \propto \frac{1}{p}$$

If the angle at the apex of a triangle is 1 arcsec, and the base of the triangle is 1 AU, then the length of the triangle is 206,265 AU (see Appendix 1). This distance, which corresponds to 3.09×10^{16} meters, or 3.26 light-years, is a parsec. The relationship between distance measured in parsecs and parallax measured in arcseconds is illustrated in Figure 13.3. Knowing that a star with a parallax of 1 arcsec is at a distance of 1 pc, we can turn the inverse proportionality between distance and parallax into an equation:

$$\left(\begin{array}{c}\text{Distance measured}\\ \text{in parsecs}\end{array}\right) = \frac{1}{\left(\begin{array}{c}\text{Parallax measured}\\ \text{in arcseconds}\end{array}\right)}$$

or

$$d(\text{pc}) = \frac{1}{p(\text{arcsec})}$$

Suppose that the parallax of a star is measured to be 0.5 arcsec. Then the distance can be found by

$$d(\text{pc}) = \frac{1}{0.5} = 2 \text{ pc}$$

Similarly, a star with a measured parallax of 0.01 arcsec is located at a distance of $1/0.01 = 100$ pc.

The star closest to Earth (other than the Sun) is Proxima Centauri. Located at a distance of 4.22 light-years, Proxima Centauri is a faint member of a system of three stars called Alpha Centauri. What is this star's parallax? First we convert the distance to parsecs:

$$d = 4.22 \text{ light-years} \times \frac{1 \text{ parsec}}{3.26 \text{ light-years}} = 1.29 \text{ parsecs}$$

Then:

$$p(\text{arcsec}) = \frac{1}{1.29 \text{ pc}} = 0.77 \text{ arcsec}$$

Even the closest star to the Sun has a parallax of only about ¾ arcsec.

Distance and Brightness Yield Luminosity

The stars in Earth's sky appear to have different levels of brightness. In Chapter 5 you saw that brightness corresponds to the amount of energy falling on a square meter of area each second in the form of electromagnetic radiation. (When astronomers talk about the brightness of stars, they usually use a comparative system called *magnitude*, discussed in **Connections 13.1** and Appendix 6.) Although the brightness of a star can be measured directly, it does not immediately give much information about the star itself. As illustrated in **Figure 13.4**, a bright star in the night sky may in fact be a dim one that appears bright only because it is nearby. Conversely, a faint star may be a powerful beacon, still visible despite its tremendous distance.

Brightness depends on the observer's perspective; luminosity does not.

To learn about the stars themselves, astronomers need to know the total energy radiated by a star each second—the star's **luminosity**. In Chapter 5 you learned how the brightness, luminosity, and distance of objects are related. Recall that the brightness of an object that has a known luminosity and is located at a distance d is given by the following equation:

$$\text{Brightness} = \frac{\text{Total light emitted per second}}{\text{Area of a sphere of radius } d} = \frac{\text{Luminosity}}{4\pi d^2}$$

You can rearrange this equation, moving the quantities you know how to measure (distance and brightness) to the right-hand side and the quantity you would like to know (luminosity) to the left, to get:

$$\text{Luminosity} = 4\pi d^2 \times \text{Brightness}$$

This equation is used to find how much total light a star must be giving off to appear as bright as it does at its distance from Earth. Two measurable quantities that depend on a particular perspective (distance and brightness) can be used to calculate a quantity that is a property of the star itself—namely, luminosity. ▶▶ **NEBRASKA SIMULATION: STELLAR LUMINOSITY CALCULATOR**

Different stars give off different amounts of light. The Sun provides a convenient yardstick for measuring the properties of stars, including their luminosity. The luminosity of the Sun is

Some stars are 10 billion times more luminous than the least luminous stars.

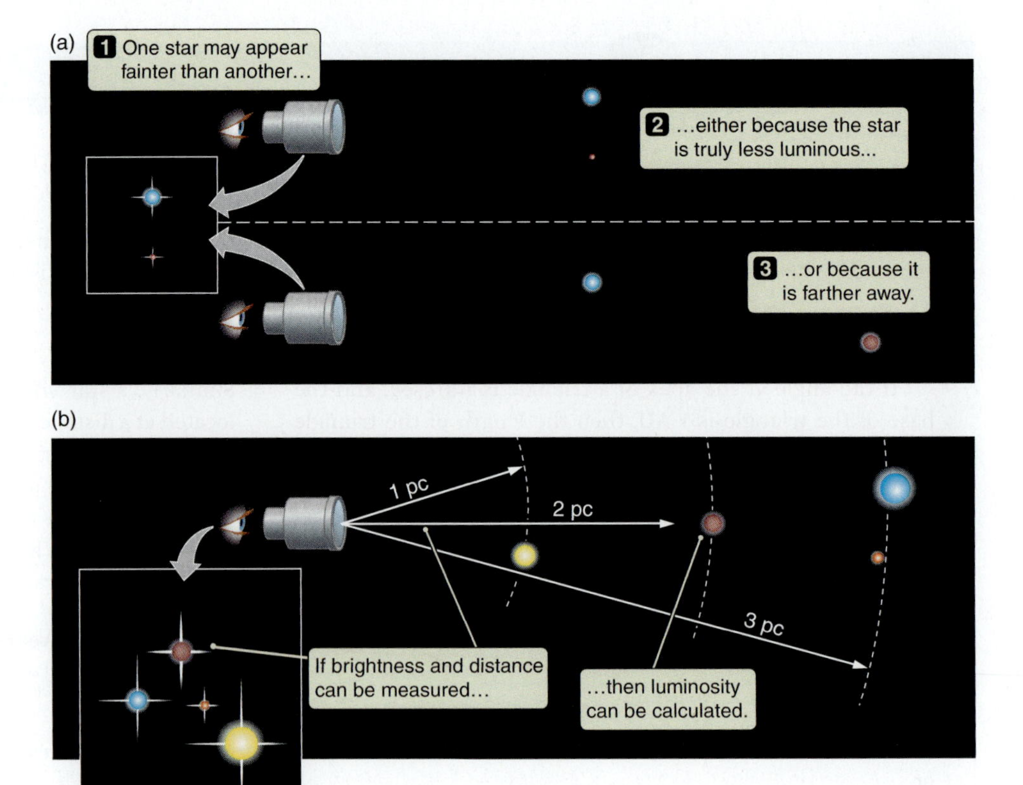

FIGURE 13.4 (a) The brightness of a visible star depends on both its luminosity and its distance. (b) When brightness and distance are measured, luminosity can be calculated.

Connections 13.1

The Magnitude System

The **magnitude** system of brightness for celestial objects can be traced back 2,100 years to the ancient Greek astronomer Hipparchus, who classified the brightest stars he could see as being "of the first magnitude" and the faintest as being "of the sixth magnitude." Later, astronomers defined Hipparchus's 1st magnitude stars as being exactly 100 times brighter than his 6th magnitude stars. With five steps between 1st and 6th magnitudes, each step is equal to the fifth root of 100, or $100^{1/5}$, which is approximately 2.512. This system is logarithmic, but instead of the usual base 10, it is base 2.512, so a 5-magnitude difference in brightness equals $(2.512)^5 = 100$ times difference in brightness.

In the magnitude system, 5th magnitude stars are 2.512 times brighter than 6th magnitude stars, and 4th magnitude stars are $2.512 \times 2.512 = 6.310$ times brighter than 6th magnitude stars. Continuing with this progression, 3rd and 2nd magnitude stars are, respectively, $(2.512)^3$ and $(2.512)^4 = 15.85$ and 39.81 times brighter than 6th magnitude stars. This progression can be represented by a more useful expression. The brightness ratio between any two stars is equal to $(2.512)^N$, where N is the magnitude difference between them. Note that in this system, a larger magnitude refers to a fainter object.

Hipparchus must have had typical eyesight, because an average person under dark skies can see stars only as faint as 6th magnitude. How does this visual limit compare with that of the most powerful telescopes? The Hubble Space Telescope can integrate for long exposures, and detect stars as faint as 30th magnitude. Since $N = 30 - 6 = 24$, HST can detect stars that are $(2.512)^{24} = 4 \times 10^9$, or 4 billion, times fainter than the naked eye can see.

Stars brighter than 1st magnitude have magnitudes that are less than 1, including zero and negative numbers. For example, Sirius, the brightest star in the sky, has a magnitude of -1.46. Venus can be as bright as magnitude -4.4, or about 15 times brighter than Sirius and bright enough to cast a shadow. The magnitude of the full Moon is -12.6, and that of the Sun is -26.7. Thus, the Sun is 14.1 magnitudes—or more than 400,000 times—brighter than the full Moon. (More detailed calculations and a table of magnitudes and brightness differences are located in Appendix 6.)

A star's **apparent magnitude** gives the brightness of the star as it *appears* in Earth's sky. Stars are located at different distances from Earth, so a star's apparent brightness does not provide a clue to its luminosity. But suppose all stars were located at the same distance from Earth. The brightness of each star would then be representative of its luminosity. In principle, astronomers have developed a system that makes exactly that assumption. If the distance to a star is known, they assign a brightness to each star as if it were located at a standard distance—in this case, exactly 10 pc (32.6 light-years). Astronomers refer to this fundamental property of a star as its **absolute magnitude**.

The brightness of astronomical objects generally varies with spectral region (color), and astronomers use special symbols to represent magnitudes in various colors. For example, V and B represent magnitudes in the visual (green) and blue, respectively, regions of the spectrum. The first solar-type star to be found to have an orbiting planet, 51 Pegasi, has a visual magnitude, V, of 5.49. Its blue magnitude, B, is 6.16 (a larger number), indicating that it is less bright in blue light than in visual (green) light. With these numbers, astronomers can assign the star's **color index**, $B - V$, as $6.16 - 5.49 = 0.67$ magnitude (see Appendix 6). (Because magnitudes are logarithmic, the color index is found from their *difference*. When calculated in other ways, the color index is often a *ratio* of two numbers on a linear scale.)

written as $L_\odot$. The most luminous stars exceed 1,000,000 $L_\odot$, or 10^6 $L_\odot$ (a million times the luminosity of the Sun). The least luminous stars have luminosities less than 0.0001 $L_\odot$, or 10^{-4} $L_\odot$ (1/10,000 that of the Sun). The most luminous stars are over 10 billion (10^{10}) times more luminous than the least luminous ordinary stars.

Only a very small fraction of stars are near the upper end of this range of luminosities. The vast majority of stars are at the faint end of this distribution, less luminous than the Sun. **Figure 13.5** shows the relative number of stars compared to their luminosity in solar units. (Distances for the nearest stars are obtained from their parallaxes; other methods—to be discussed later—are used for the more distant stars.)

There are many more low-luminosity stars than high-luminosity stars.

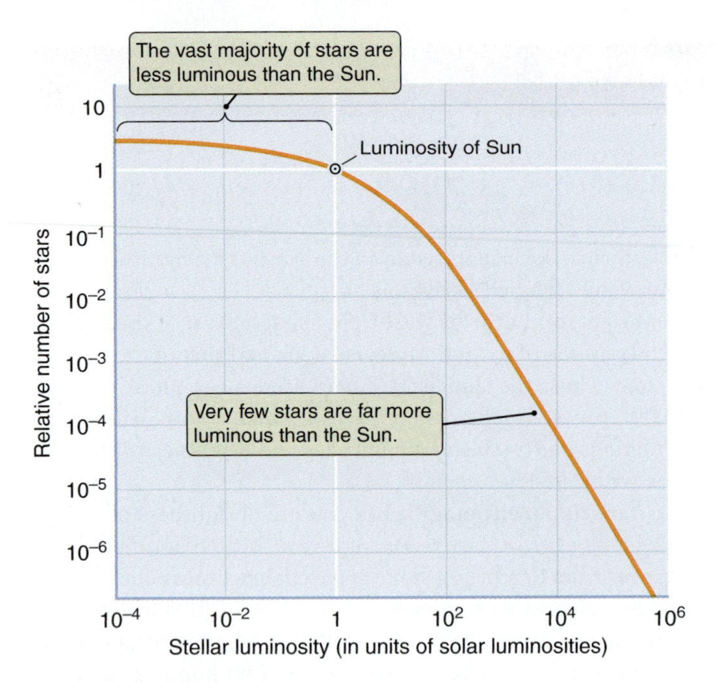

FIGURE 13.5 The distribution of the luminosities of stars. Both axes are plotted logarithmically.

13.2 The Temperature, Size, and Composition of Stars

Two everyday concepts—stereoscopic vision, and the fact that closer objects appear brighter—have provided the tools needed to measure the distance and luminosity of relatively nearby stars. With these two steps, stars have gone from being merely faint points of light in the night sky to being extraordinarily powerful beacons located at distances truly vast compared to direct human experience. The laws of radiation that we described in Chapter 5 reveal still more about stars.

Stars are gaseous, but they are fairly dense—dense enough that the radiation from a star comes close to obeying the same laws as the radiation from objects like the heating element on an electric stove or the filament in a lightbulb. Therefore, an understanding of Planck blackbody radiation—results such as the Stefan-Boltzmann law (hotter at the same size means more luminous) and Wien's law (hotter means bluer)—enables astronomers to measure the temperatures and sizes of the Sun's stellar neighbors.

Wien's Law Revisited: The Color and Surface Temperature of Stars

Wien's law (see Math Tools 5.3) shows that the temperature of an object determines the peak wavelength of its spectrum. The hotter the surface of an object, the bluer the light that it emits. Stars with especially hot surfaces are blue, stars with especially cool surfaces are red, and yellow-white stars such as the Sun are in the middle. Wien's law relates the wavelength at which a spectrum peaks to the temperature, T, of the star's surface. The color of a star reveals the temperature only at the surface, because it is this layer that is giving off the radiation observed. Stellar interiors are far hotter than exteriors.

The color of a star indicates its surface temperature.

In practice, it is usually not necessary to obtain a complete spectrum of a star to determine its temperature. Instead, stellar temperatures are often measured in a way that is similar to the way your eyes see color. Your eyes and brain distinguish color by comparing how bright an object is at one wavelength with how bright it is at another wavelength. The human eye contains nearly 100 million light-sensitive cells, and among them are certain color-sensitive cells called cones (Chapter 6). Cones come in three color-discriminating types. The cone types respond the most to the red, green, and blue regions of the spectrum, respectively. If you look at a very red lightbulb, the red-sensitive cones in your eye report a bright light, and the blue- and green-sensitive cones see less light. If you look at a yellow light, the green- and red-sensitive cones are strongly stimulated, but those with blue sensitivity are not. White light stimulates all three kinds of color-sensitive cells. By combining and comparing the signals from cells that are sensitive to different wavelengths of light, your brain can distinguish among fine shades of color.

Astronomers often measure the colors of stars by comparing the brightness at two different wavelengths. The brightness of a star is usually measured through a **filter**—sometimes just a piece of colored glass—that lets through only a certain range of wavelengths. Two of the most common filters used by astronomers are a blue filter that allows through light with wavelengths around 440 nanometers (nm), and a yellow-green filter that allows light with wavelengths around 550 nm to pass through. The first of these filters is (sensibly enough) called a "blue" filter. By convention, however, the second of these filters is referred to as a "visual" filter rather than a "yellow-green" filter because it allows through roughly the range of wavelengths to which the human eye is most sensitive.

Figure 13.6 shows four Planck spectra with temperatures of 2500–20,000 kelvins (K), adjusted so that they are all the same brightness at 550 nm (the wavelength of the center of the range transmitted by the visual filter). A hot star, with a spectrum like the 20,000-K Planck spectrum shown, gives off more light in the blue part of the spectrum than in the visual part of the spectrum. Dividing the brightness of the star

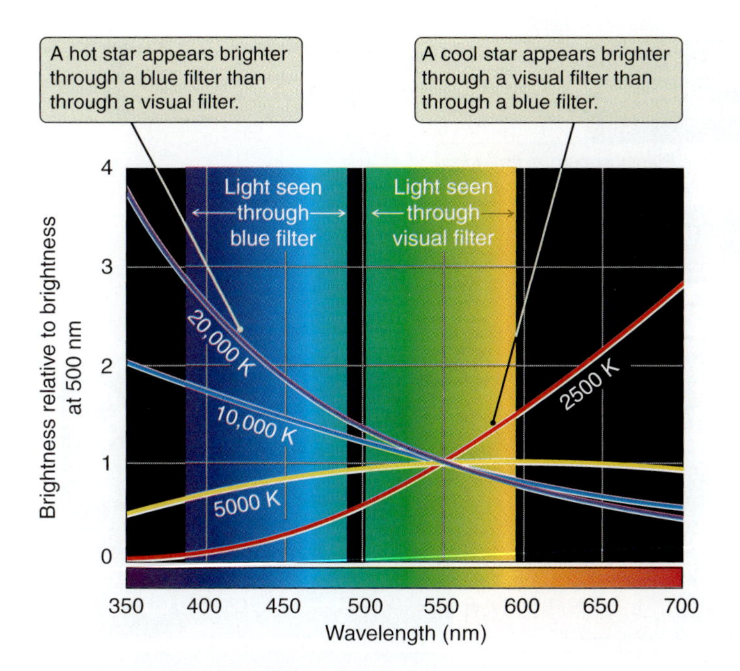

FIGURE 13.6 A star's color depends on its temperature. These four lines represent Planck spectra (blackbody curves) for stars at different temperatures. The curves are adjusted so that they have the same brightness at 550 nm.

as seen through the blue filter by the brightness of the star as seen through the visual filter gives a ratio *greater* than 1. In contrast, a cool star with a spectrum more like the 2500-K Planck spectrum shown is much fainter in the blue part of the spectrum than in the visual part of the spectrum. The ratio of blue light to visual light is *less* than 1 for the cool star. This ratio of brightness between the blue and visual filters is referred to as the *color index* of the star (discussed in Connections 13.1).

The fact that the color of a star depends on the star's temperature is an extremely handy tool for astronomers. It means that a pair of snapshots of a group of stars, each taken through a different filter, provides a measurement of the surface temperature of every star in the camera's field of view. This technique makes it possible to measure the temperatures of hundreds or even thousands of stars at once. This type of analysis shows that just as there are many more low-luminosity stars than high-luminosity stars, there are also many more cool stars than hot stars. In addition, most stars have surface temperatures lower than that of the Sun. **AstroTour: Stellar Spectrum**

Stars Are Classified According to Surface Temperature

So far, we have considered only what can be learned about stars by applying an understanding of thermal radiation. However, the spectra of stars are not smooth, continuous blackbodies. Instead, the spectra of stars contain a wealth of dark absorption lines and occasionally bright emission lines. In Chapter 5 you learned that absorption lines occur when light passes through a cloud of gas: the atoms and molecules of the gas absorb light of certain specific wavelengths, characteristic of the kind of atom or molecule that is doing the absorbing. The atoms and molecules in diffuse hot gas both emit and absorb light of specific wavelengths. Both of these processes are at work in stars.

Absorption lines form as light escapes through a star's atmosphere.

Although the hot "surface" of a star emits radiation with a spectrum very close to a smooth Planck curve, this light then escapes through the outer layers of the star's **atmosphere**. The atoms and molecules in the cooler layers of the star's atmosphere leave their absorption line fingerprints in the escaping light (**Figure 13.7**). Under some circumstances, the atoms and molecules in the star's atmosphere, as well as any gas in the vicinity of the star, can produce emission lines in stellar spectra. Although absorption and emission lines complicate how astronomers use the laws of Planck radiation to interpret light from stars, spectral lines more than make up for this trouble by providing a wealth of information about the state of the gas in a star's atmosphere.

The spectra of stars were first classified during the late 1800s, long before stars, atoms, or radiation were well understood. Stars were classified not by their physical properties, but by the appearance of the dark bands (now known as absorption lines) seen in their spectra. The original ordering of this classification was arbitrarily based on the prominence of particular absorption lines known to be associated with the element hydrogen. Stars with the strongest hydrogen lines were labeled *A stars*, stars with somewhat weaker hydrogen lines were labeled *B stars*, and so on.

Stars are classified by the appearance of their spectra.

This classification scheme was refined and turned into a real sequence of stellar properties in the early 20th century. The new system, originally proposed in 1901, was the work of Annie Jump Cannon (1863–1941), who led an effort at the Harvard College Observatory to systematically examine and classify the spectra of hundreds of thousands of stars. She dropped many of the earlier spectral types, keeping only seven, which she reordered into a sequence that was no longer arbitrary, but instead was based on surface temperatures. Spectra of stars of different types are shown in the horizontal bars in **Figure 13.8**. The hottest stars are at the top of the figure. These have surface

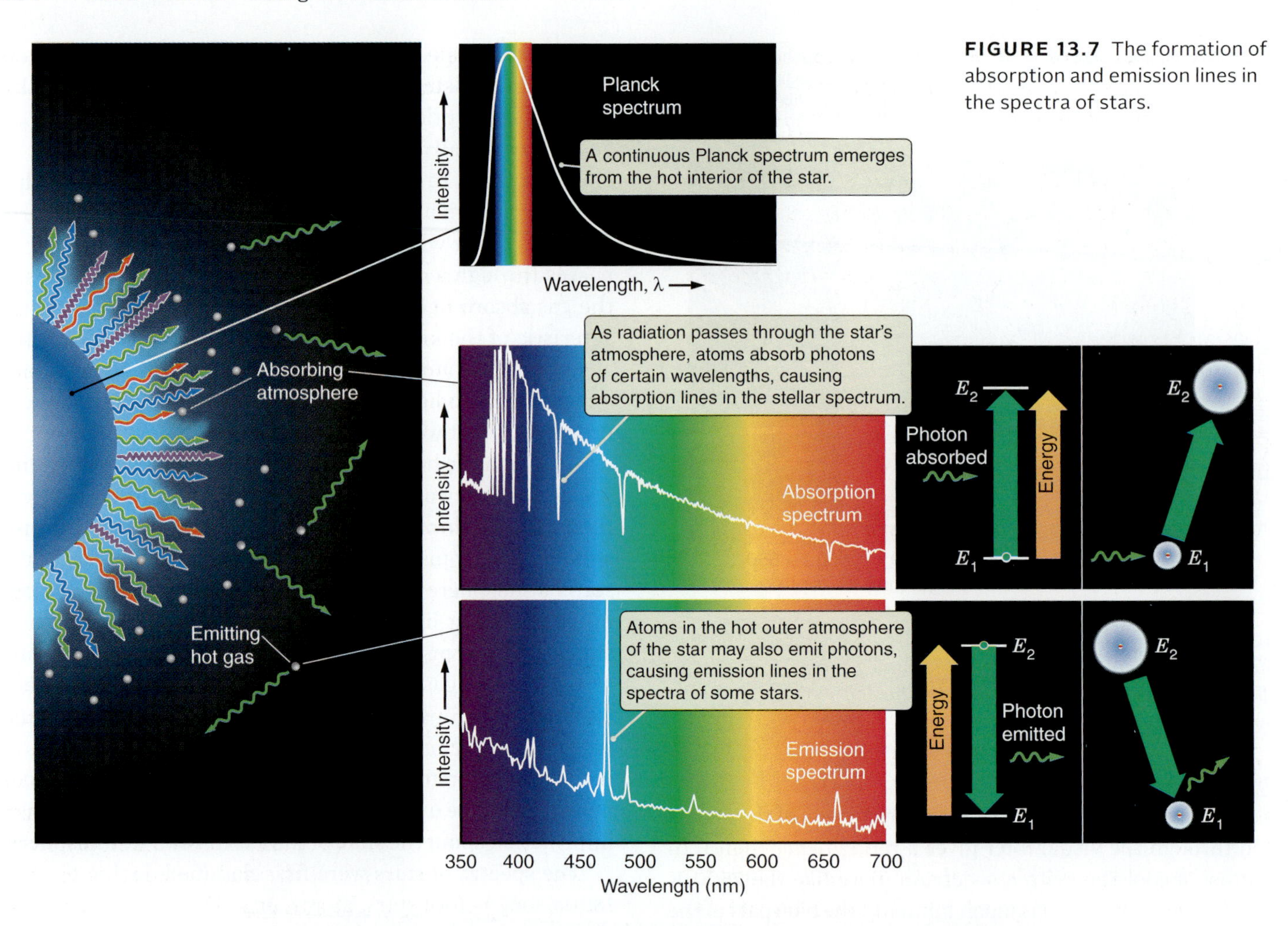

FIGURE 13.7 The formation of absorption and emission lines in the spectra of stars.

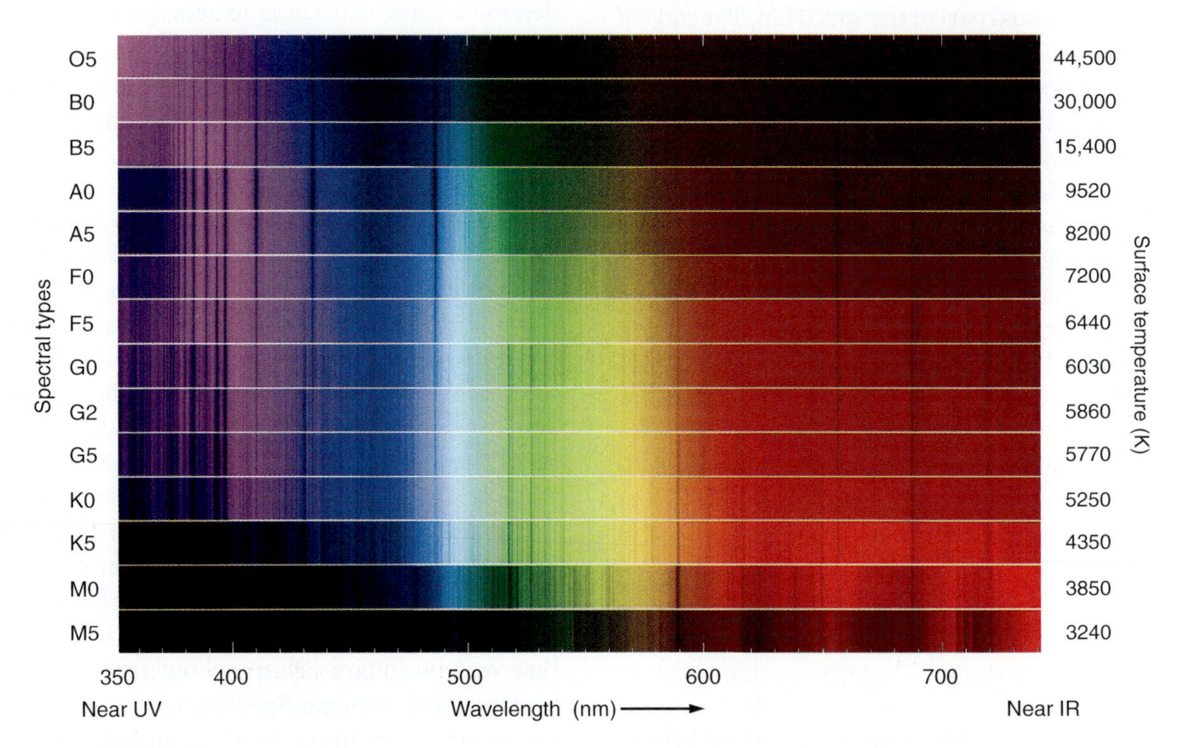

FIGURE 13.8 Spectra of stars with different spectral types, ranging from hot blue O stars to cool red M stars. Hotter stars are brighter at shorter wavelengths. The dark lines are absorption lines.

temperatures of more than 30,000 K and are *O stars*. O stars have relatively featureless spectra, with only weak absorption lines from hydrogen and helium. The coolest stars—*M stars*—have temperatures as low as about 2800 K. M stars show myriad absorption lines from many different types of atoms and molecules. The complete sequence of **spectral types** of stars, from hottest to coolest, is O, B, A, F, G, K, M. This sequence has undergone several modifications over time.

It should be stressed that the boundaries between spectral types are not precise. A hotter-than-average G star is very similar to a cooler-than-average F star. Astronomers divide the main spectral types into a finer sequence of subclasses by adding numbers to the letter designations. For example, the hottest B stars are B0 stars, slightly cooler B stars are B1 stars, and so on. The coolest B stars are B9 stars, which are only slightly hotter than A0 stars. The Sun is a G2 star.

In Figure 13.8, notice that not only are hot stars bluer than cool stars, but the absorption lines in their spectra are quite different as well. The reason is that differences in the temperature of the gas in the atmosphere of a star affect the state of the atoms in that gas, which in turn affects the energy level transitions available to absorb radiation. (See Section 5.3 to review the concept of atomic energy levels). At the hot end of the spectral classification scheme, the temperatures in the atmospheres of O stars are so high that most atoms have had one or more electrons stripped from them by energetic collisions within the gas. Only a few of the transitions available in these **ionized** atoms cause absorption lines in the visible part of the electromagnetic spectrum. Therefore, the visible spectrum of an O star is relatively featureless. At lower temperatures more of the atoms are in energy states that can absorb light in the visible part of the spectrum. For that reason, the visible spectra of cooler stars are far more complex than the spectra of O stars.

Most absorption lines have a temperature at which they are strongest. For example, absorption lines from hydrogen are most prominent at temperatures of about 10,000 K, which is the surface temperature of an A star. At the very lowest stellar temperatures, atoms in the atmosphere of a star begin to react with each other, forming molecules. Molecules such as titanium oxide (TiO) are responsible for much of the absorption in the atmospheres of cool M stars.

Because different spectral lines are formed at different temperatures, astronomers can use these absorption lines to measure a star's temperature. The surface temperatures of stars measured in this way agree extremely well with the surface temperatures of stars measured using Wien's law, again confirming a prediction of the cosmological principle—namely, that the physical laws that apply on Earth apply to stars as well. ▶II **AstroTour: Stellar Spectrum**

Stars Consist Mostly of Hydrogen and Helium

Most of the variation in the lines seen in stellar spectra are due to temperature, but the details of the absorption and emission lines found in starlight carry a wealth of other information as well. By applying the physics of atoms and molecules to the study of stellar absorption lines, astronomers can accurately determine not only surface temperatures of stars, but also pressures, chemical compositions, magnetic-field strengths, and other physical properties of stars. In addition, by making use of the Doppler shift of emission and absorption lines, astronomers can measure rotation rates, atmospheric motions, expansion and contraction, "winds" driven away from stars, and other dynamic properties of stars.

Spectral lines are used to measure many properties of stars, including chemical composition.

As you continue to learn about stars, you will come to appreciate that one of the most interesting and important things about a star is its chemical composition. Recall from Chapter 5 that if a source of thermal radiation is viewed through a cloud of gas, the strength of various absorption lines indicates what kinds of atoms are present in the gas and in what abundance. Although astronomers must be careful in interpreting spectra to properly account for the temperature and density of the gas in the atmosphere of a star, in most respects stars are ready-made laboratories for carrying out just such an experiment.

Most stars have atmospheres that consist predominantly of the least massive elements: hydrogen and helium. Hydrogen typically makes up over 90 percent of the atoms in the atmosphere of a star, with helium accounting for most of what remains. All of the other chemical elements, which are collectively referred to as **heavy elements** or (more properly) as *massive elements*, are present in only trace amounts. **Table 13.1** shows the chemical composition of the atmosphere of the Sun. The Sun's composition is fairly typical for stars in its vicinity, but the percentages of various heavy elements can vary tremendously from star to star. In particular, some stars show even smaller amounts of elements other than hydrogen and helium in their spectra. The existence of such stars, all but devoid of more massive elements, provides important clues about the origin of chemical elements and the chemical evolution of the universe.

TABLE 13.1 | The Chemical Composition of the Sun's Atmosphere*

Element	Percentage by Number†	Percentage by Mass††
Hydrogen	92.5	74.5
Helium	7.4	23.7
Oxygen	0.064	0.82
Carbon	0.039	0.37
Neon	0.012	0.19
Nitrogen	0.008	0.09
Silicon	0.004	0.09
Magnesium	0.003	0.06
Iron	0.003	0.16
Sulfur	0.001	0.04
Total of others	0.001	0.03

*The relative amounts of different chemical elements in the atmosphere of the Sun.
†The percentage of atoms in the Sun accounted for by the listed element.
††The percentage of the Sun's mass consisting of the listed element.

The Stefan-Boltzmann Law Revisited: Finding the Sizes of Stars

The temperature of a star can be found directly, either from Wien's law (**Figure 13.9a**) or from the strength of its spectral lines. The temperature of the surface of a star is related to how much radiation each square meter of the star emits each second. Each square meter of the surface of a hot blue star gives off more radiation per second than does a square meter of the surface of a cool red star. Therefore, a hot star is more luminous than a cool star of the same size (**Figure 13.9b**). A small hot star might even be more luminous than a larger cool star. The relationship between temperature and luminosity of each square meter of a surface (**Figure 13.9c**) can be used to infer the size of the star from the Stefan-Boltzmann law and the relationship between flux and luminosity (see **Math Tools 13.2**).

Astronomers have used the luminosity-temperature-radius relationship to measure the sizes of many thousands of stars. The radius of the Sun, written $R_{\odot}$, is 696,000 km, or about 700,000 km. The smallest stars, called white dwarfs, have radii that are only about 1 percent of the Sun's radius ($R = 0.01\ R_{\odot}$). The largest stars, called red supergiants, can have radii more than 1,000 times that of the Sun. There are many more stars toward the small end of this range, smaller than the Sun, than there are giant stars.

There are many more small stars than large stars.

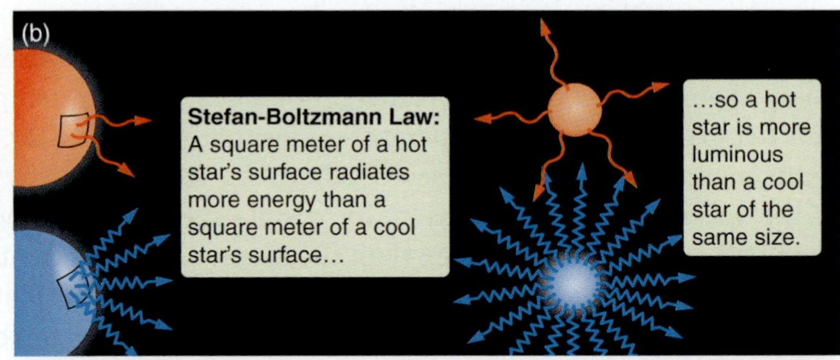

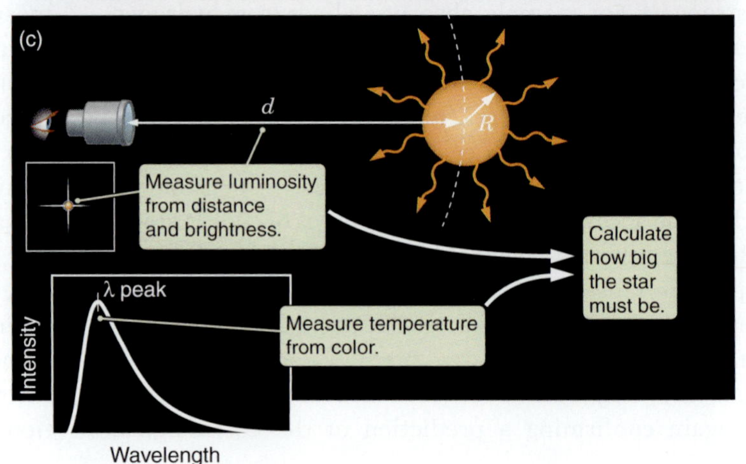

FIGURE 13.9 Astronomers measure the temperature and size of stars by applying their understanding of Planck blackbody radiation.

Math Tools 13.2

Estimating Sizes of Stars

In Chapter 5 you learned that according to the Stefan-Boltzmann law, the amount of energy radiated each second by each square meter of the surface of a star is equal to the constant σ multiplied by the surface temperature of the star raised to the fourth power. Written as an equation, this relationship says:

$$\begin{pmatrix}\text{Energy radiated each} \\ \text{second by 1 m}^2\text{ of surface}\end{pmatrix} = \sigma T^4$$

To find the total amount of light radiated each second by the star, we need to multiply the radiation per second from each square meter by the number of square meters of the star's surface:

$$\begin{pmatrix}\text{Energy radiated} \\ \text{each second}\end{pmatrix} = \begin{pmatrix}\text{Energy radiated each} \\ \text{second by 1 m}^2\text{ of surface}\end{pmatrix} \times \begin{pmatrix}\text{Surface} \\ \text{area}\end{pmatrix}$$

The left-hand item in this equation—the total energy emitted by the star per second (in units of joules per second, = watts)—is the star's luminosity, L. The middle item—the energy radiated by each square meter of the star in a second (in units of joules per square meter per second, or $J/m^2/s$)—can be replaced with the σT^4 factor from the Stefan-Boltzmann law. The remaining item—the number of square meters covering the surface of the star—is the surface area of a sphere, $A_{\text{sphere}} = 4\pi R^2$ (in units of square meters, or m^2), where R is the radius of the star.

If we replace the words in the equation with the appropriate mathematical expressions for the Stefan-Boltzmann law and the area of a sphere, our equation for the luminosity of a star looks like this:

$$\text{Luminosity}\left(\frac{\text{J}}{\text{s}}\right) = \sigma T^4\left(\frac{\text{J}}{\text{m}^2\text{s}}\right) \times 4\pi R^2(\text{m}^2)$$

Combining gives:

$$L = 4\pi R^2 \sigma T^4 \text{ J/s (W)}$$

Because the constants (4, π, and σ) do not change, the luminosity of a star is proportional only to R^2T^4. Make a star 3 times as large, and its surface area becomes $3^2 = 9$ times as large. There is 9 times as much area to radiate, so there is 9 times as much radiation. Make a star twice as hot, and each square meter of the star's surface radiates $2^4 = 16$ times as much energy. Larger, hotter stars are more luminous than smaller, cooler stars.

Now turn this question around and ask, How large must a star of a given temperature be to have a total luminosity of L? The star's luminosity (L) and temperature (T) are quantities that we can measure, and the star's radius (R) is what we want to know. We can rearrange the previous equation, moving the things that we know how to measure (temperature and luminosity) to the right-hand side of the equation, and the thing that we would like to know (the radius of the star) to the left side. After a couple of steps of algebra, we find:

$$R = \sqrt{\frac{L}{4\pi\sigma} \times \frac{1}{T^2}}$$

Again, the right-hand side of the equation contains only things that we know or can measure. The constants 4, π, and σ are always the same. We can find L, the luminosity of the star, by combining measurements of the star's brightness and parallax (although only for those nearer stars with known parallax). T is the surface temperature of the star, which can be measured from its color. From the relationship of these measurements we now know something new: the size of the star. We refer to this last equation as the **luminosity-temperature-radius relationship** for stars.

Often we compare two stars and the constants all cancel out, leaving L, T, and R:

$$\frac{L_{\text{star 1}}}{L_{\text{star 2}}} = \frac{R^2_{\text{star 1}}}{R^2_{\text{star 2}}} \times \frac{T^4_{\text{star 1}}}{T^4_{\text{star 2}}} \quad \text{or} \quad \frac{R_{\text{star 1}}}{R_{\text{star 2}}} = \sqrt{\frac{L_{\text{star 1}}}{L_{\text{star 2}}}} \times \frac{T^2_{\text{star 2}}}{T^2_{\text{star 1}}}$$

Suppose we compare the Sun to the second brightest star in the constellation of Orion, a red star called Betelgeuse (see the chapter-opening photo). From its spectrum, we know that Betelgeuse's surface temperature T is about 3500 K. Its distance is about 200 parsecs, and from that and its brightness, its luminosity is estimated to be 140,000 times that of the Sun. What can we say about the size of Betelgeuse? Using the preceding equation, we can determine the following:

$$\frac{R_{\text{Betelgeuse}}}{R_{\text{Sun}}} = \sqrt{\frac{L_{\text{Betelgeuse}}}{L_{\text{Sun}}}} \times \frac{T^2_{\text{Sun}}}{T^2_{\text{Betelgeuse}}}$$

$$= \sqrt{\frac{140{,}000}{1}} \times \frac{5{,}800^2}{3{,}500^2} = 374 \times 2.7 = 1{,}030$$

Betelgeuse has a radius over 1,000 times larger than that of the Sun. As you will see in Section 13.4, such stars are called *supergiants*.

13.3 Measuring Stellar Masses

Determining the mass of an object can be a tricky business. Massive objects can be large or small, faint or luminous. The only thing that *always* goes with mass is gravity. Mass is responsible for gravity, and gravity in turn affects the way masses move. When astronomers are trying to determine the masses of astronomical objects, they almost always wind up looking for the effects of gravity.

To measure mass, astronomers look for the effects of gravity.

In Chapter 4 you learned that Kepler's laws of planetary motion are the result of gravity and that the properties of the orbit of a planet about the Sun can be used to measure the mass of the Sun. Similarly, astronomers can study two stars that orbit each other to determine their masses. About half of the higher-mass stars in the sky are actually multiple systems consisting of several stars moving about under the influence of their mutual gravity. Most of these are **binary stars** in which two stars orbit each other in elliptical orbits as predicted by Kepler's laws. However, most low-mass stars are single, and low-mass stars far outnumber higher-mass stars, so *most stars are single*.

Binary Stars Orbit a Common Center of Mass

Because the Sun is so much more massive than the planets, it is not necessary to consider the effect of the planets' gravity on the Sun's motion. If two stars have similar masses, however, this simple picture does not suffice. In this case, each object's motion is noticeably affected by the gravitational force from the other, and the two objects *orbit each other*. (As you learned in Chapter 7, much of astronomers' knowledge of planets beyond the Solar System comes from the wobbles they cause in the motions of stars.)

Think about what happens if two unequal masses—m_1 and m_2—are adrift in space, initially with no motion of one mass relative to the other. As soon as the two masses are free to move, gravity begins to pull them together. The force of mass 1 on mass 2 equals the force of mass 2 on mass 1, and each mass begins to fall toward the other. But even though the forces on each mass are equal, the accelerations experienced by the two masses are not. Acceleration equals force divided by mass, so the *less massive object experiences the greater acceleration*.

If m_1 is 3 times as massive as m_2, the acceleration of m_2 will be 3 times as great as the acceleration of m_1. At any given point in time, m_2 will be moving toward m_1 3 times as fast as m_1 is moving toward m_2. When the two objects collide, m_2 will have fallen 3 times as far as m_1. The point where the two objects meet, called the **center of mass** of the two objects, will be 3 times as far from the original position of the less massive m_2 as from the original position of the more massive m_1. If the two objects were sitting on a seesaw in a gravitational field, the support of the seesaw would have to be directly under the center of mass for the objects to balance (**Figure 13.10**). NEBRASKA SIMULATION: CENTER OF MASS SIMULATOR

Suppose now that the two masses have some motion perpendicular to the line between them; instead of simply falling into each other, they will orbit around each other. When Newton applied his laws of motion to the problem of orbits, he found that two objects move in elliptical orbits with their common center of mass at one focus of both of the ellipses (**Figure 13.11**). The center of mass, which lies along the line between the two objects, remains stationary. The two objects are always on exactly opposite sides of the center of mass, and the elliptical orbit of the more massive object is just a smaller version of the elliptical orbit of the less massive object.

Each star in a binary system follows an elliptical orbit around the system's center of mass.

Because the orbit of the less massive star is larger than the more massive star's orbit, the less massive star has farther to go than the more massive one. But it must cover that distance in the same amount of time, so the less massive star must move *faster* than the more massive star. As a result, the ratio of the velocities (v) of the stars in a binary system is inversely proportional to the ratio of their masses:

$$\frac{v_1}{v_2} = \frac{m_2}{m_1}$$

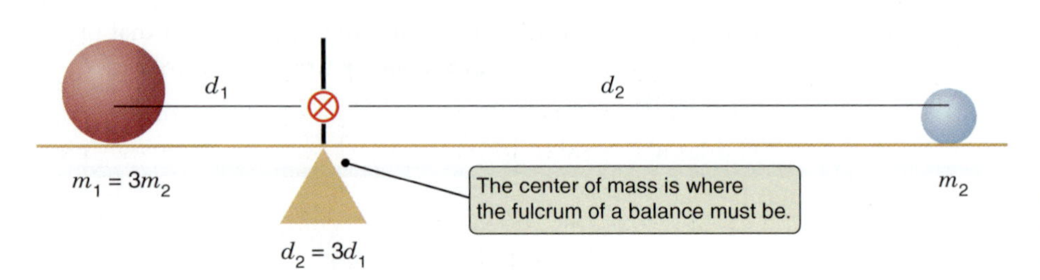

FIGURE 13.10 The center of mass of two objects is the "balance" point on a line joining the centers of two masses.

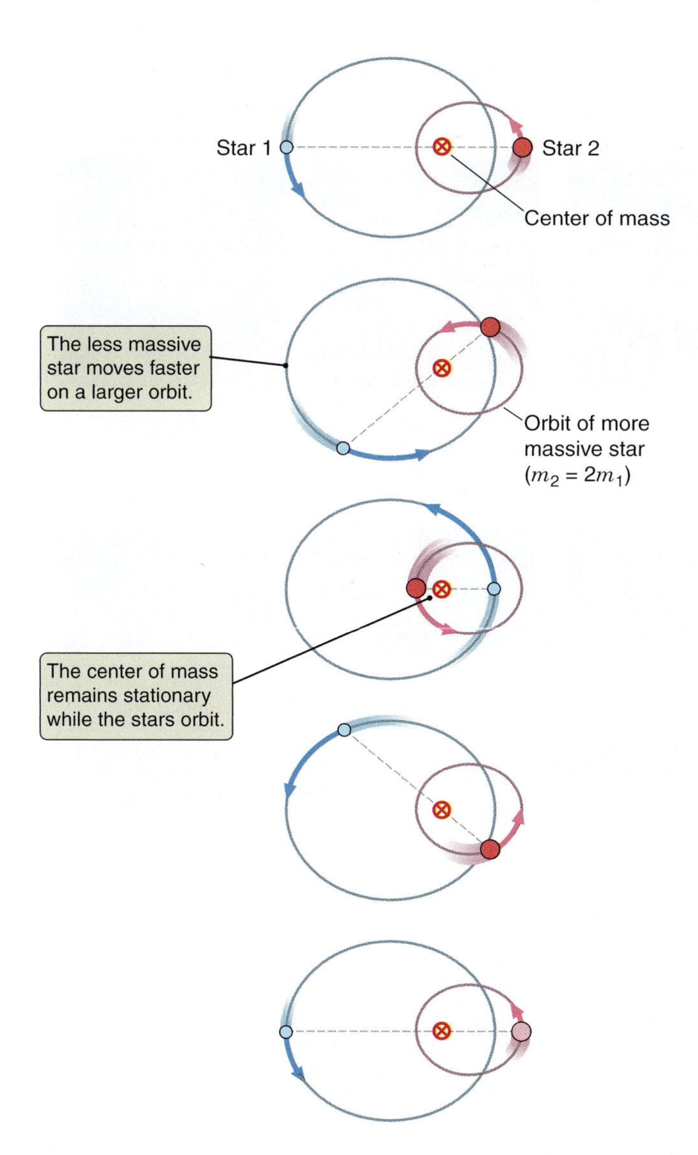

FIGURE 13.11 In a binary star system, the two stars orbit on elliptical paths about their common center of mass. In this case, the red star has twice the mass of the blue one. The eccentricity of the orbits is 0.5. There are equal time steps between the frames.

This relationship between the velocities and the masses of the stars in a binary system is a crucial part of how binary stars are used to measure stellar masses. **Figure 13.12** shows a binary system observed from above. Imagine the system as it would be seen edge on. As one star approached the observer, the other star would be moving away, and vice versa. At a particular instant in the spectra of the two stars, the absorption lines from star 1 might be blueshifted relative to the overall center-of-mass velocity, while the absorption lines from star 2 were redshifted. Half an orbital period later, the situation would be reversed: lines from star 2 would be blueshifted, and lines from star 1 would be Doppler-shifted to the red. Because the two stars are always exactly opposite one another about their common center of mass, they are always moving in opposite directions. Astronomers measure the ratio of the masses of the two stars by comparing the size of the Doppler shift in the spectrum of star 1 with the size of the Doppler shift in the spectrum of star 2.

Kepler's Third Law Gives the Total Mass of a Binary System

Recall Newton's derivation of Kepler's laws (Chapter 4): from the period of the binary system and the average separation between the two stars, Kepler's third law gives the total mass. If either the sizes of the orbits of the two stars or their velocities can be measured independently, then the ratios of the masses of the two stars can be determined. Astronomers use these two numbers (the total mass and the ratio of masses) to determine the mass of each star separately. In other words, if star 1 is 2 times as massive as star 2, and star 1 and star 2 together are 3 times as massive as the Sun, then enough is known to calculate separate values for the masses of star 1 and star 2.

There are several ways to measure the needed orbital properties. If a binary system is a **visual binary** (**Figure 13.13**)—that is, if telescopic images show the two stars separately—then astronomers can observe over time as the stars orbit each other. From these observations they can directly measure the shapes and periods of the orbits of the two stars. These can be combined with Doppler measurements of the radial (line-of-sight) velocities of the stars to solve for the ratio of the two masses.

In most binary systems, however, the two stars are so far away from Earth and so close together that they appear as one star. The identification of these stars as binary systems is more indirect, and comes from observing periodic variations in the light from the star, or from observing periodic changes in the spectrum of the star. If a binary system is an **eclipsing binary**, a repeating dip in brightness occurs as one star passes in front of the other. If the stars are of different brightnesses, there will be a repeating pattern of a smaller dip in brightness when the brighter star eclipses the fainter one, and then a larger dip in brightness when the fainter star eclipses the brighter one (**Figure 13.14**). The pattern of these dips also gives an estimate of the relative sizes (radii) of the two stars. This procedure for identifying binary systems is similar to the transit method for finding extrasolar planets discussed in Chapter 7, and it works only when the system is viewed nearly edge on. The Kepler space telescope has been observing and discovering thousands of eclipsing binaries in addition to finding new extrasolar planets. ▶▶ **NEBRASKA SIMULATION: ECLIPSING BINARY SIMULATOR**

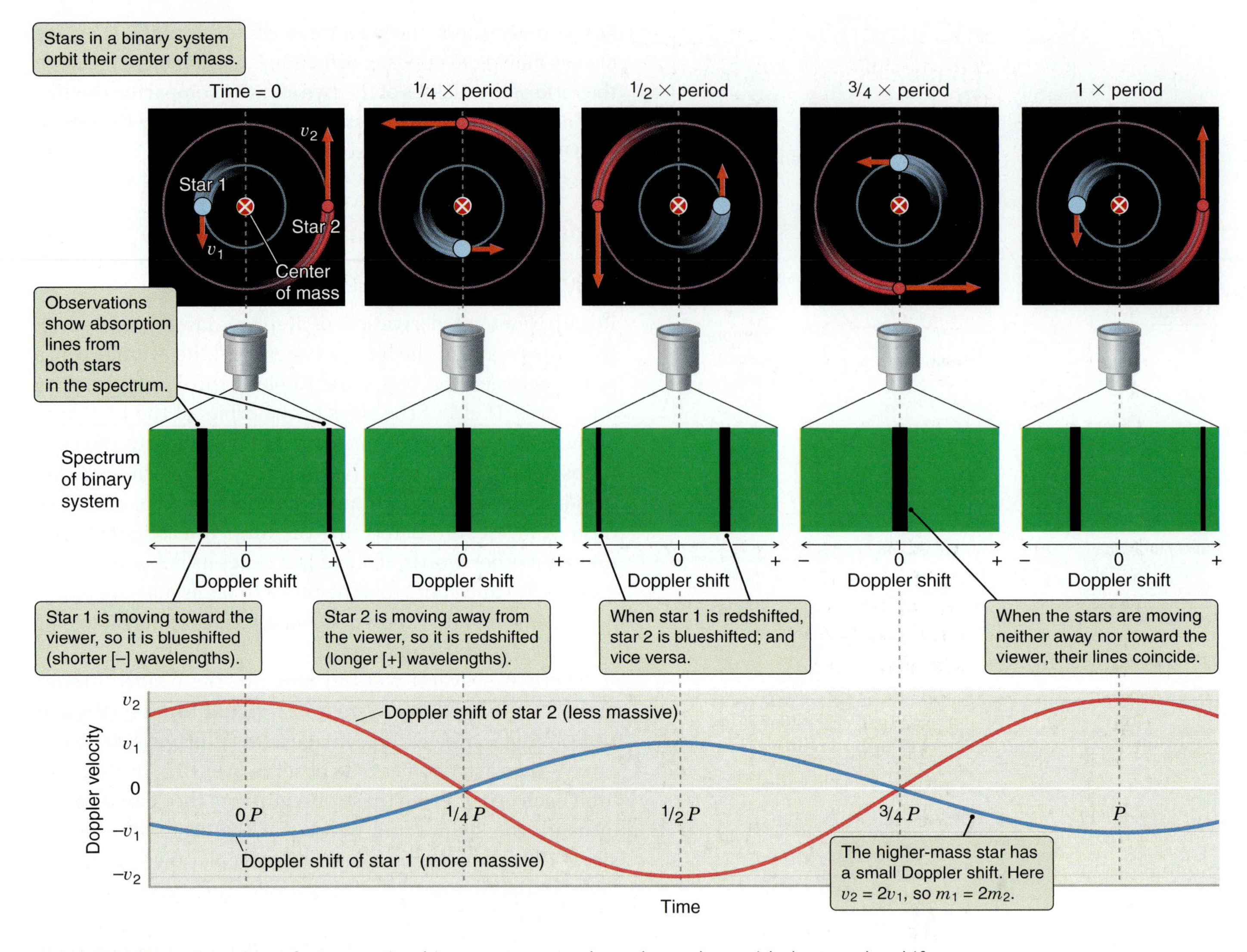

FIGURE 13.12 The orbits of two stars in a binary system are shown here, along with the Doppler shifts of the absorption lines in the spectrum of each star. Star 1 is twice as massive as star 2 and so has half the Doppler shift.

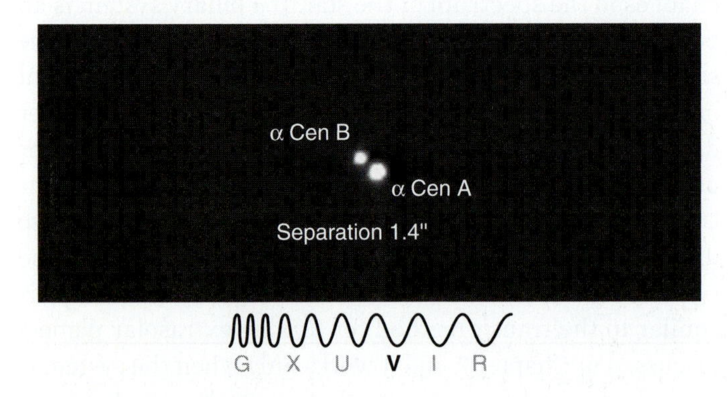

FIGURE 13.13 The two stars of a visual binary (Alpha Centauri A and Alpha Centauri B) are resolved. This is the nearest star system to the Solar System.

If a binary system is a **spectroscopic binary**, the spectral lines of the two stars show periodic changes as they are Doppler-shifted away from each other, first in one direction and then in the other, as shown in Figure 13.12. The period of the orbit is determined from the time it takes for a set of spectral lines to go from approaching to receding and back again. The orbital velocities of the stars and the period of the orbit give the size of the orbit because distance equals velocity multiplied by time. Consequently, astronomers can estimate the combined masses of the two stars. To calculate the individual masses, an estimate of the tilt of the orbit is needed. Thus spectroscopic binary masses are more approximate than those in eclipsing binary systems.

A binary system can fall into more than one of these three categories, regardless of how it was originally

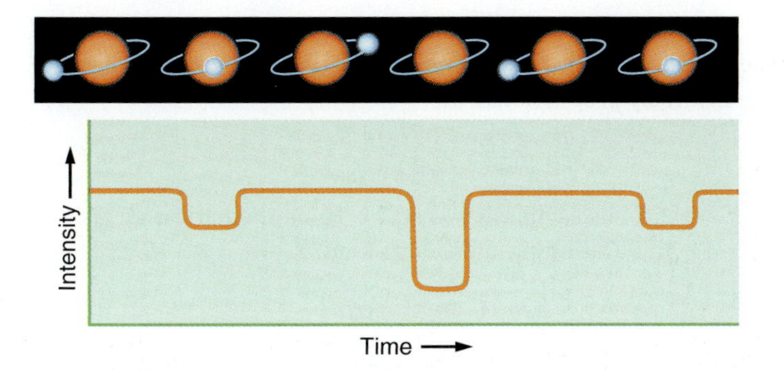

FIGURE 13.14 The shape of the light curve of an eclipsing binary can reveal information about the relative size and surface brightness of the two stars. Even though the blue star here is smaller, it is significantly more luminous because its temperature is higher.

discovered. If a spectroscopic binary system is also a visual or eclipsing binary, then the orbit and masses of the stars can be completely solved (**Math Tools 13.3**). Historically, most stellar masses were measured for stars in eclipsing binary systems, rather than for those in visual or spectroscopic binaries. But new observational capabilities have increased the number of visual binaries, by greatly improving the ability to see the stars in a binary directly. Accurate measurements of masses have been obtained for several hundred binary stars, about half of which are eclipsing binaries. These measurements show that some stars are less massive than the Sun, whereas others are more massive.

Math Tools 13.3

Measuring the Masses of an Eclipsing Binary Pair

Newton showed that if two objects with masses m_1 and m_2 are in orbit about each other, then the period of the orbit, P, is related to the average distance between the two masses, the semimajor axis A, by the equation

$$P^2 = \frac{4\pi^2 A^3}{G(m_1 + m_2)}$$

Rearranging this equation a bit turns it into an expression for the sum of the masses of the two objects:

$$m_1 + m_2 = \frac{4\pi^2}{G} \times \frac{A^3}{P^2}$$

We can ignore the value of $4\pi^2/G$ by applying what we know about Earth's orbit around the Sun, and then expressing m, A, and P as ratios. If $A = 1$ AU and $P = 1$ year, $M_\odot =$ mass of the Sun and $M_\oplus =$ mass of the Earth, then we know that $m_1 + m_2 = M_\odot + M_\oplus \approx M_\odot$ (because $M_\odot$ is much larger than $M_\oplus$). So if we express the masses m_1 and m_2, A, and P in that equation in terms of Solar System units—such as $m_1 = m_1/(1\ M_\odot)$, $A_{AU} = A/(1\ \text{AU})$, and $P_{years} = P/(1\ \text{year})$—then this equation simplifies to:

$$\frac{m_1}{M_\odot} + \frac{m_2}{M_\odot} = \frac{(A_{AU})^3}{(P_{years})^2}$$

Therefore, if we know both m_1/m_2 and $m_1 + m_2$, we can solve for the separate values of m_1 and m_2.

Suppose you are an astronomer studying a binary star system. After observing the star for several years, you accumulate the following information about the system:

1. The star is an eclipsing binary.
2. The period of the orbit is 2.63 years. (You learned this by observation.)
3. Star 1 has a Doppler velocity that varies between +20.4 and −20.4 km/s.
4. Star 2 has a Doppler velocity that varies between +6.8 and −6.8 km/s.
5. The stars are in circular orbits. You know this because the Doppler velocities about the star are symmetric; the approach and recession speeds of the star are equal.

These data are summarized in **Figure 13.15**. You begin your analysis by noting that the star is an eclipsing binary, which tells you that the orbit of the star is edge-on to your line of sight. The Doppler velocities tell you the total orbital velocity of each star, and you determine the size of the orbits using the relationship Distance = Speed × Time. In one orbital period, star 1 travels around a circle—a distance of

$$d = (20.4\ \text{km/s}) \times (2.63\ \text{yr}) = 53.7\ \text{km} \times \text{yr/s}$$

You multiply by the number of seconds in a year:

$$d = 53.7 \frac{\text{km} \times \text{yr}}{\text{s}} \times \frac{3.16 \times 10^7\ \text{s}}{\text{yr}} = 1.70 \times 10^9\ \text{km}$$

This distance is the circumference of the star's orbit, or 2π times the radius of the star's orbit, A_1. Thus, star 1 is following an orbit with a radius of

$$A_1 = \frac{d}{2\pi} = \frac{1.70 \times 10^9\ \text{km}}{2\pi} = 2.7 \times 10^8\ \text{km}$$

To convert this to astronomical units, you use the relation $1\ \text{AU} = 1.50 \times 10^8$ km:

$$A_1 = 2.7 \times 10^8\ \text{km} \times \frac{1\ \text{AU}}{1.50 \times 10^8\ \text{km}} = 1.8\ \text{AU}$$

Continued on next page

A similar analysis of star 2 shows that its orbit has a radius of $A_2 = 0.6$ AU.

Next you apply Newton's version of Kepler's third law, which says that the masses of the stars in solar masses, m_1 and m_2, are related to the average distance between them in AU, A_{AU}, and the period of their orbit in years, P_{years}. Since the stars are always on opposite sides of the center of mass, $A_{AU} = 1.8$ AU + 0.6 AU = 2.4 AU. Because you know A and the period P (measured as 2.63 years), you can calculate the total mass of the two stars:

$$\frac{m_1}{M_\odot} + \frac{m_2}{M_\odot} = \frac{(A_{AU})^3}{(P_{years})^2} = \frac{(2.4)^3}{(2.63)^2} = 2.0$$

So you have learned that the combined mass of the two stars is twice the mass of the Sun.

To sort out the individual masses of the stars, you use the measured velocities and the fact that the mass and velocity are inversely proportional:

$$\frac{m_2}{m_1} = \frac{v_1}{v_2} = \frac{20.4 \text{ km/s}}{6.8 \text{ km/s}} = 3.0$$

Star 2 is 3 times as massive as star 1. In mathematical terms, $m_2 = 3 \times m_1$. Substituting into the equation $m_1 + m_2 = 2.0\, M_\odot$ gives:

$$m_1 + 3m_1 = 2.0\, M_\odot$$

Or $4m_1 = 2.0\, M_\odot$, so $m_1 = 0.5\, M_\odot$. Since $m_2 = 3 \times m_1$, then $m_2 = 1.5\, M_\odot$.

Star 1 has a mass of $0.5\, M_\odot$, and star 2 has a mass of $1.5\, M_\odot$. You have just found the masses of two distant stars.

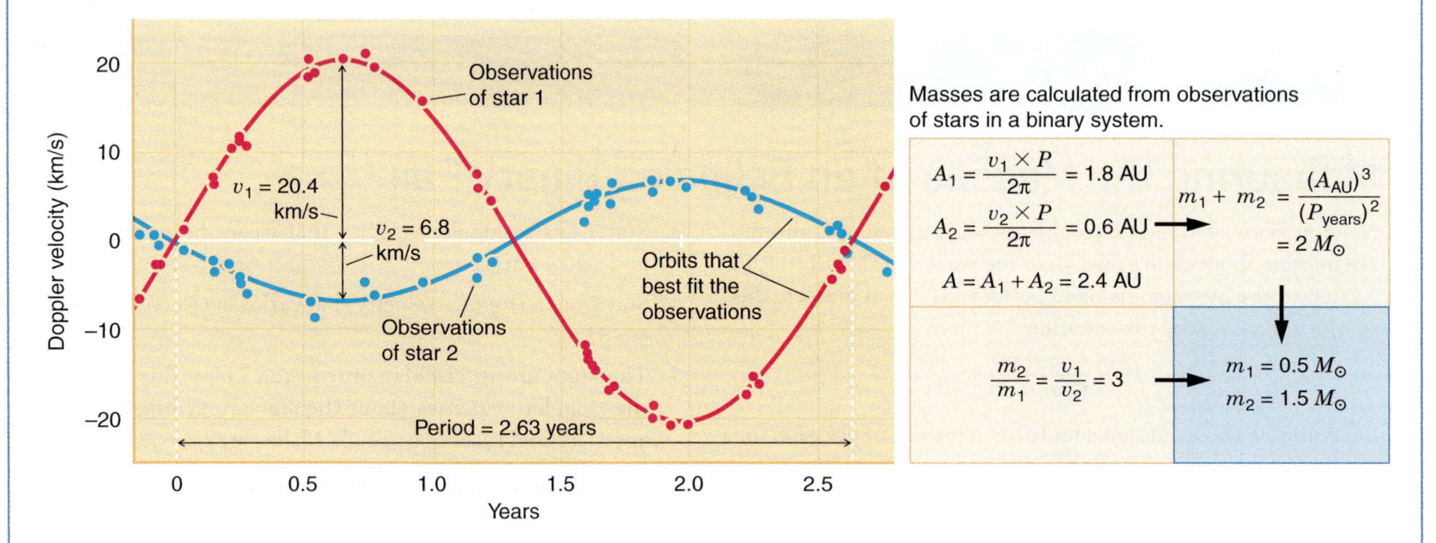

FIGURE 13.15 Doppler velocities of the stars in an eclipsing binary are used to measure the masses of the stars.

The range of stellar masses is not nearly as great as the range of stellar luminosities. The least massive stars have masses of about $0.08\, M_\odot$; the most massive stars probably have masses greater than $150\, M_\odot$. Why is there a limit to the mass of a star? These lower and upper limits are determined solely by the physical processes that go on deep in the interior of the star. A minimum stellar mass is necessary to ignite the nuclear furnace that keeps a star shining, but the furnace can run out of control if the stellar mass is too great.

Although the most luminous stars are 10^{10}, or 10 billion, times more luminous than the least luminous ones, the most massive stars are only about 10^3, or a 1,000, times more massive than the least massive stars.

13.4 The H-R Diagram Is the Key to Understanding Stars

The next step in understanding stars is to look for patterns in their properties. The first astronomers to take this step were Ejnar Hertzsprung (1873–1967) and Henry Norris Russell (1877–1957). In the early part of the 20th century (from 1906 to 1913), Hertzsprung and Russell were independently studying the properties of stars. Each scientist plotted the luminosities of stars versus a particular measure of their surface temperatures. The resulting plot is referred to as the *Hertzsprung-Russell diagram* or simply **H-R diagram**. ▶▶ NEBRASKA SIMULATION: HERTZSPRUNG-RUSSELL DIAGRAM EXPLORER

The H-R Diagram

The H-R diagram is one of the most used and useful diagrams in stellar astronomy. Everything about stars, from their formation to their old age and eventual death, is discussed using the H-R diagram (**Figure 13.16**). The surface temperature of stars is plotted on the horizontal axis (the *x*-axis), but it is plotted backward: temperature starts out hot on the left side of the diagram and *decreases* going to the right. Hot blue stars are on the left side of the H-R diagram; cool red stars are on the right.

The H-R diagram is extremely useful in astronomy.

Temperature is plotted *logarithmically*. A step along the axis from a point representing a star with a surface temperature of 30,000 K to one with a surface temperature of 10,000 K (that is, a temperature change by a factor of 3) is the same as a step between points representing a star with a temperature of 10,000 K and a star with a temperature of 3333 K (also a factor-of-3 change). Astronomers also often label the temperature axis with the spectral types or color indices of stars (**Process of Science Figure**).

The luminosity of stars is plotted along the vertical axis (the *y*-axis). More luminous stars are toward the top of the diagram, and less luminous stars are toward the bottom. As with the temperature axis, luminosities are plotted logarithmically. In this case each step along the axis corresponds to a multiplicative factor of 10 in the luminosity. To understand why the plotting is done this way, recall that the most luminous stars are 10 billion times more luminous

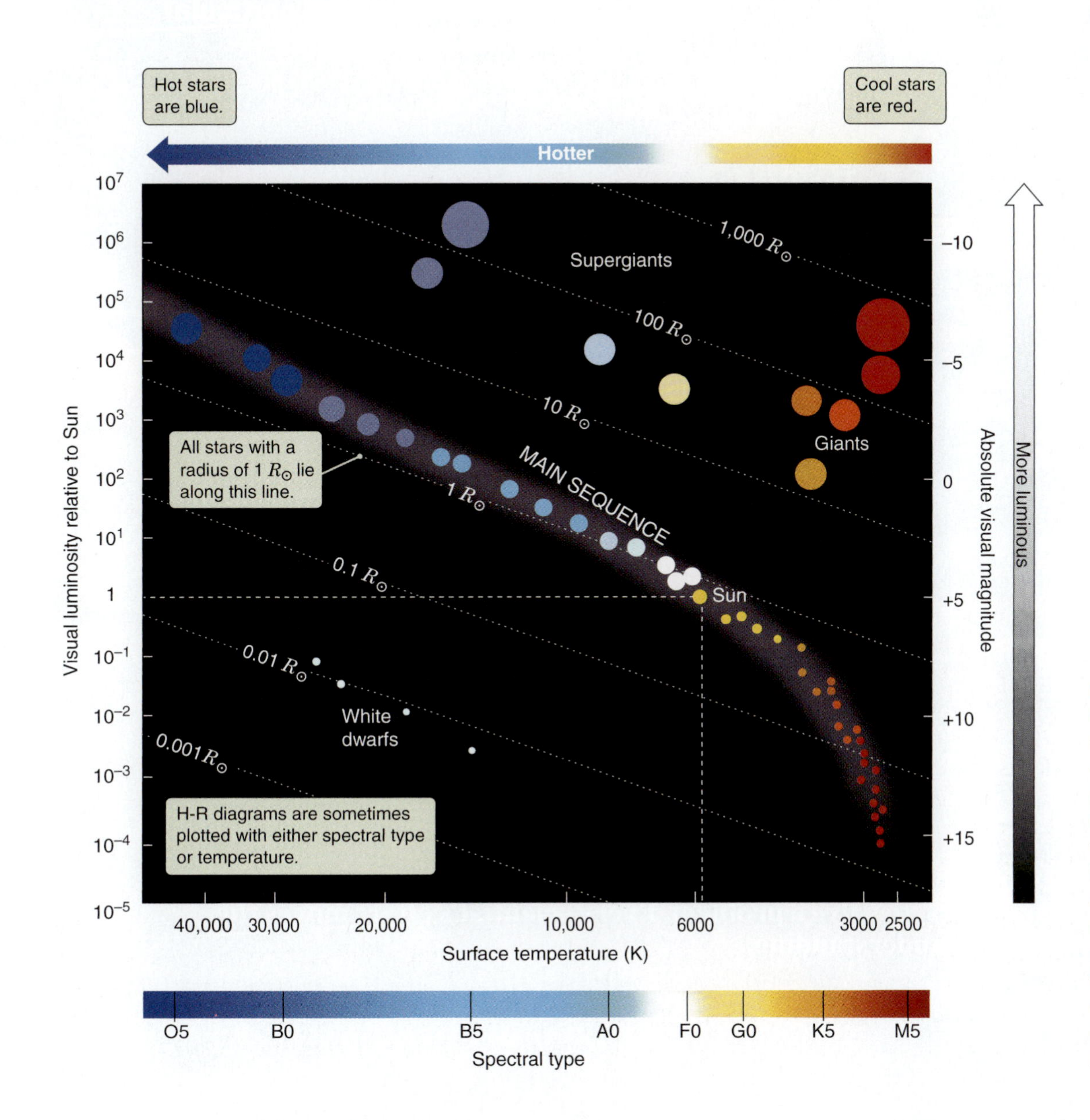

FIGURE 13.16 The layout of the H-R (Hertzsprung-Russell) diagram used to plot the properties of stars. More luminous stars are at the top of the diagram. Hotter stars are on the left. Stars of the same radius (R) lie along the dotted lines moving from upper left to lower right. Absolute magnitudes are discussed in Connections 13.1 and Appendix 6.

An understanding of the meaning behind stellar data took decades, and the contributions of dozens of people, all working toward a common goal.

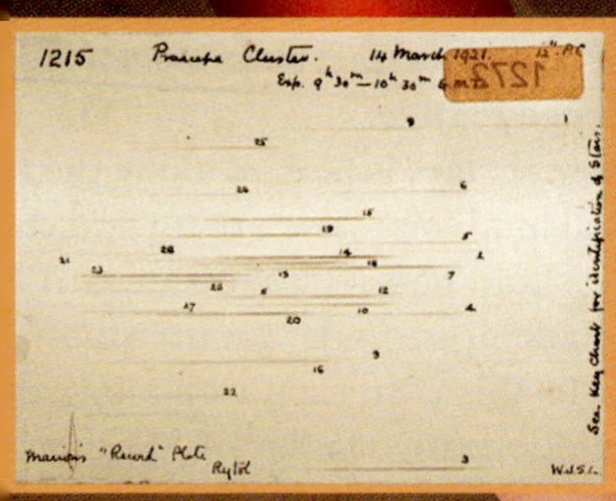

The Observations:
About 500,000 photographs of stellar spectra are obtained, by many astronomers at many telescopes.

The Classification:
Annie Jump Cannon leads a team that classifies all the spectra according to the strengths of particular absorption lines at particular wavelengths.

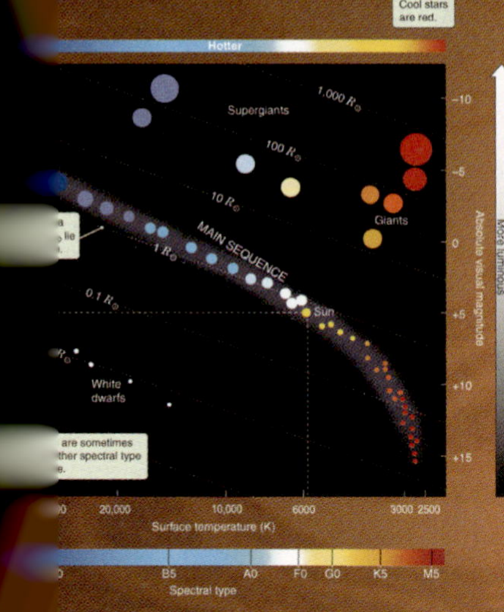

The Graph:
Hertzsprung and Russell independently develop what will later be called the H-R diagram. They do not understand why the x-axis, when ordered O-B-A-F-G-K-M, gives such a nice band across the middle. Russell hypothesizes there must be a single stellar attribute at work.

The Understanding:
Meghnad Saha shows that the stellar attribute in question is temperature. Cecilia Payne-Gaposchkin shows that stars are mostly composed of hydrogen and helium. Modern astrophysics is born; others go on to develop the understanding of stellar atmospheres.

c discoveries, while they sometime appear to happen all at once, take many vorking for many years, to solve a problem. Every scientist's effort moves the tle closer to a better understanding.

than the least luminous stars, yet all of these stars must fit on the same plot. Sometimes the luminosity axis is labeled with the absolute visual magnitude instead of luminosity.

The H-R diagram tells astronomers more than just the surface temperatures and luminosities of stars. Recall that if the temperature and the luminosity of a star are known, its radius can be calculated. Because each point in the H-R diagram is specified by a surface temperature and a luminosity, the luminosity-temperature-radius relationship will give the radius of a star at that point as well. For example, a star in the upper right corner of the H-R diagram is very cool, which, according to the Stefan-Boltzmann law, means that each square meter of its surface is radiating only a small amount of energy. But at the same time, this star is extremely luminous. Such a star must be huge to account for its high overall luminosity, despite the feeble radiation coming from each square meter of its surface. Conversely, a star in the lower left corner of the H-R diagram is very hot, which means that a large amount of energy is coming from each square meter of its surface. However, this star has a very low overall luminosity. Therefore, its surface area cannot be very large. Stars in the lower left corner of the H-R diagram are small. **ASTROTOUR: H-R DIAGRAM**

In the H-R diagram, moving up and to the right takes you to larger and larger stars. Moving down and to the left takes you to smaller and smaller stars. All stars of the same radius lie along lines running across the H-R diagram from the upper left to the lower right, as shown in Figure 13.16.

Like an experienced surveyor who "sees" the lay of the land when looking at a topographic map, or a classical musician who "hears" the rhythms and harmonies when looking at a piece of sheet music, you should get to the point where you automatically "see" the properties of a star—its temperature, color, size, and luminosity—from a glance at its position on the H-R diagram.

The Main Sequence Is a Grand Pattern in Stellar Properties

Figure 13.17 shows 16,600 nearby stars plotted on an H-R diagram. The data are based on observations obtained by the Hipparcos satellite. A quick look at this diagram immediately reveals a remarkable fact. Instead of being strewn all about the diagram, about 90 percent of the stars in the sky lie along a well-defined sequence running across the H-R diagram from lower right to upper left. This sequence of stars is called the **main sequence**. On the left end of the main sequence are the O stars: hotter, larger, and more luminous than the Sun. On the right end are the M stars: cooler, smaller, and fainter than the Sun. The location of a star on the main sequence gives its approximate luminosity, surface temperature, and size.

Most stars lie along the main sequence of the H-R diagram.

Astronomers determine whether a star is a main-sequence star by looking at the absorption lines in its spectrum and identifying its spectral type. The spectral type or color indicates the star's temperature. The location of such a main-sequence star on the H-R diagram is uniquely determined by its temperature. From that location, the luminosity can be read from the y-axis of the diagram. From its luminosity and its apparent brightness, the inverse square law of radiation gives the star's distance. (How far away must a star of a particular luminosity be to have the measured brightness?) This method of determining distances to stars is called **spectroscopic parallax**; it is discussed further in Appendix 6. (Despite the similarity

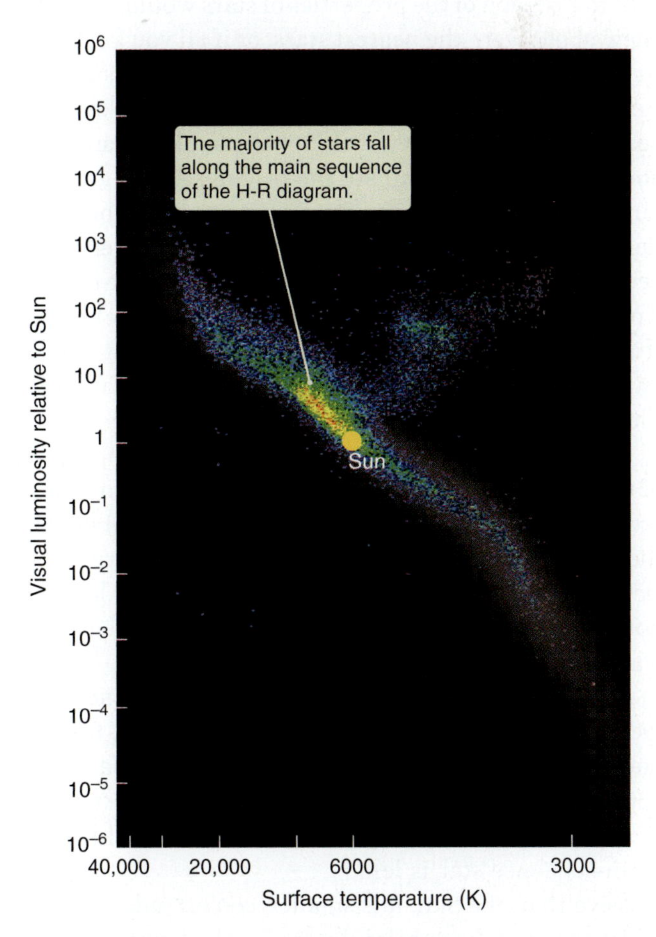

FIGURE 13.17 An H-R diagram for 16,600 stars obtained by the Hipparcos satellite. Most of the stars lie in a band running from the upper left of the diagram toward the lower right, called the *main sequence*.

between the names, this method is very different from the *parallax* method using geometry, discussed earlier in the chapter.) Spectroscopic parallax is used to measure the distances to stars that are farther away than a few hundred light-years. Parallax and spectroscopic parallax are the first two steps along a chain of reasoning that will build a knowledge of distances all the way to the edge of the observable universe. ▶▶ **NEBRASKA SIMULATION: SPECTROSCOPIC PARALLAX SIMULATOR**

Figure 13.18 makes another important point for astronomy, as well as for all other sciences—namely, that the sample studied can affect the conclusions. The red dots represent 46 of the stars nearest to the Sun; the blue symbols represent the 97 brightest stars as seen in Earth's sky. There are many more cool, low-luminosity stars in the sky than there are hot, high-luminosity stars. Only cool, low-luminosity stars are visible near the Sun, yet the sample of the brightest stars in the sky give a very different picture. Even though these stars are farther away, they are so luminous that they dominate the night sky. Think about how different your impression of the properties of stars would be if all you knew about were the nearest stars, or if all you knew about were the brightest stars. Neither of these groups alone accurately represents what stars as a whole are like, but the nearest stars provide a much fairer sample because luminous stars, which dominate the brightest-stars group, are rare. Whether you are an astronomer studying the properties of stars or a political pollster measuring the sense of the people, the validity of your results depends on the care with which you choose the sample (of stars or people) that you study. If astronomers were conducting a poll, they would be much better off interviewing the nearest stars rather than the brightest.

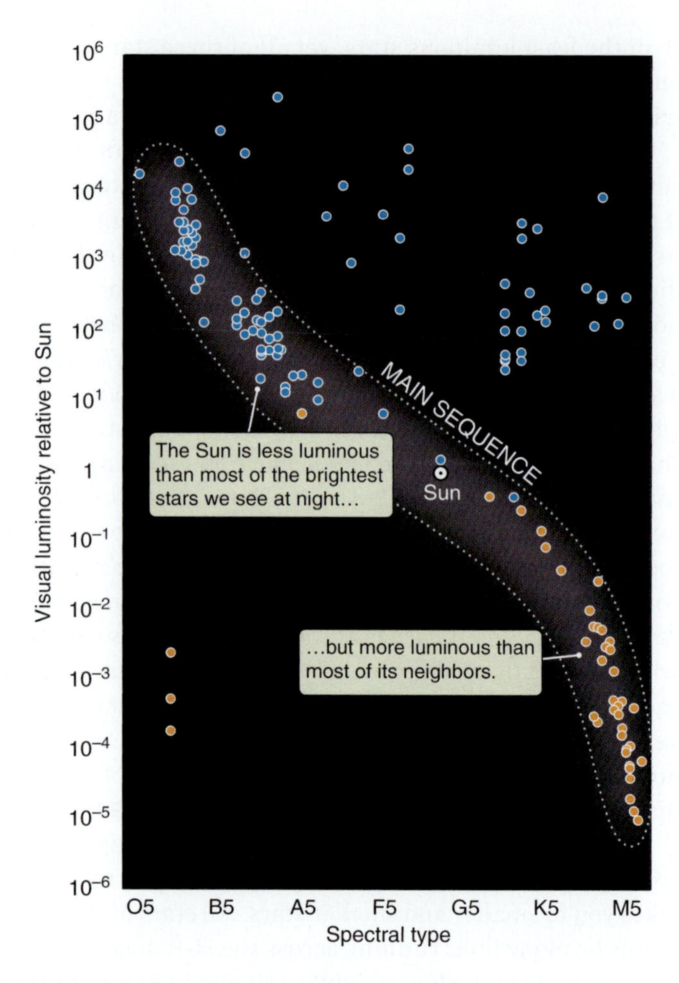

FIGURE 13.18 An H-R diagram for two different samples of stars. The red symbols show the H-R diagram for 46 stars that are especially close to the Sun. The blue symbols show the 97 brightest stars in Earth's sky. Note that because these are observational H-R diagrams, they are plotted against an observed quantity: the stars' spectral type.

Adding mass to the picture makes the main sequence of the H-R diagram even more interesting. Stellar mass increases smoothly from the lower right to the upper left along the main sequence. The faint, cool stars on the right-hand side of the main sequence have low masses; the luminous, hot stars on the left side are high-mass stars. Stated another way, the mass of a main-sequence star indicates its location on the main sequence, and thus the star's approximate temperature, size, and luminosity. If a main-sequence star is less massive than the Sun, it is smaller, cooler, redder, and less luminous than the Sun; it is located to the lower right of the Sun on the main sequence. If a main-sequence star is more massive than the Sun, it is larger, hotter, bluer, and more luminous than the Sun; it is located to the upper left of the Sun on the main sequence. *The mass of a star determines where on the main sequence the star will lie.*

Low-mass main-sequence stars are faint and cool. High-mass main-sequence stars are hot and luminous.

To see the relationship between a star's mass and its position on the main sequence directly, look at **Table 13.2** and **Figure 13.19**, which show the properties of main-sequence stars. If, for example, a main-sequence star has a mass of 17.5 $M_{\odot}$, it is a B0 star. It has a surface temperature of about 30,000 K, a radius of about 6.7 $R_{\odot}$, and a luminosity about 32,500 times that of the Sun. If a main-sequence star instead has a mass of 0.21 $M_{\odot}$, it is an M5 star. It has a surface temperature of 3,170 K, a radius of about 0.29 $R_{\odot}$, and a luminosity of about 0.008 $L_{\odot}$. All main-sequence stars with a mass of 1 $M_{\odot}$ are G2 stars like the Sun and have the same surface temperature, size, and luminosity as the Sun.

The relationship between the mass and the luminosity of stars is very sensitive. Relatively small differences

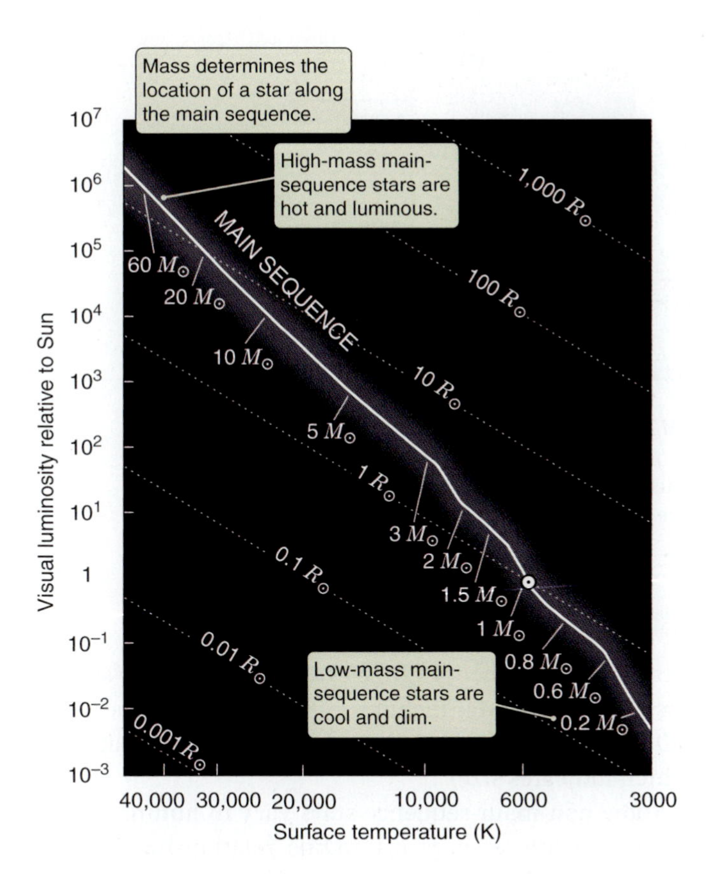

FIGURE 13.19 The main sequence of the H-R diagram is a sequence of masses.

TABLE 13.2 | The Main Sequence of Stars*

Spectral Type	Temperature (K)	Mass ($M_\odot$)	Radius ($R_\odot$)	Luminosity ($L_\odot$)
O5	42,000	60	13	500,000
B0	30,000	17.5	6.7	32,500
B5	15,200	5.9	3.2	480
A0	9800	2.9	2.0	39
A5	8200	2.0	1.8	12.3
F0	7300	1.6	1.4	5.2
F5	6650	1.4	1.2	2.6
G0	5940	1.05	1.06	1.25
G2 (Sun)	5780	1.00	1.00	1.0
G5	5560	0.92	0.93	0.8
K0	5150	0.79	0.93	0.55
K5	4410	0.67	0.80	0.32
M0	3840	0.51	0.63	0.08
M5	3170	0.21	0.29	0.008

*Approximately 90 percent of the stars in the sky fall on the main sequence of the H-R diagram. This table gives the properties of stars along the main sequence.

in the masses of stars result in large differences in their main-sequence luminosities. One method for estimating the masses of non-binary main-sequence stars is to use the **mass-luminosity relationship**, $L \propto M^{3.5}$, which is based on observed luminosities of stars of known mass (**Figure 13.20a**). The exponent varies for different ranges of stellar masses, but this method is useful for estimating masses of single stars.

It is difficult to overemphasize the importance of this result, so we state it again for clarity: for stars of similar chemical composition, *the mass of a main-sequence star alone determines all of its other characteristics*. The mass (and chemical composition) of a main-sequence star indicates how large it is, what its surface temperature is, and how luminous it is, as **Figure 13.20** illustrates—as well as what its internal structure is, how long it will live, how it will evolve, and what its final fate will be.

The mass of a main-sequence star determines its fate.

This relationship is possibly the most important and fundamental relationship in all of **astrophysics**. If there is a certain amount and type of material to make a star, there is only one kind of star that can be made. As we go on to discuss the physical processes that give a star its structure, this statement will make even more sense. A star is a "battle" between gravity (which is trying to pull the star together) and the energy released by nuclear reactions in the interior of the star (which is trying to blow it apart). The mass of the star determines the strength of its gravity, which in turn determines how much energy must be generated in its interior to prevent it from collapsing under its own weight. The mass of a star determines where the balance is struck.

Not All Stars Are Main-Sequence Stars

Before concluding our discussion of the observed properties of stars, we need to point out that although 90 percent of stars are main-sequence stars, some are not. Some stars are found in the upper right portion of the H-R diagram, well above the main sequence. This position indicates that

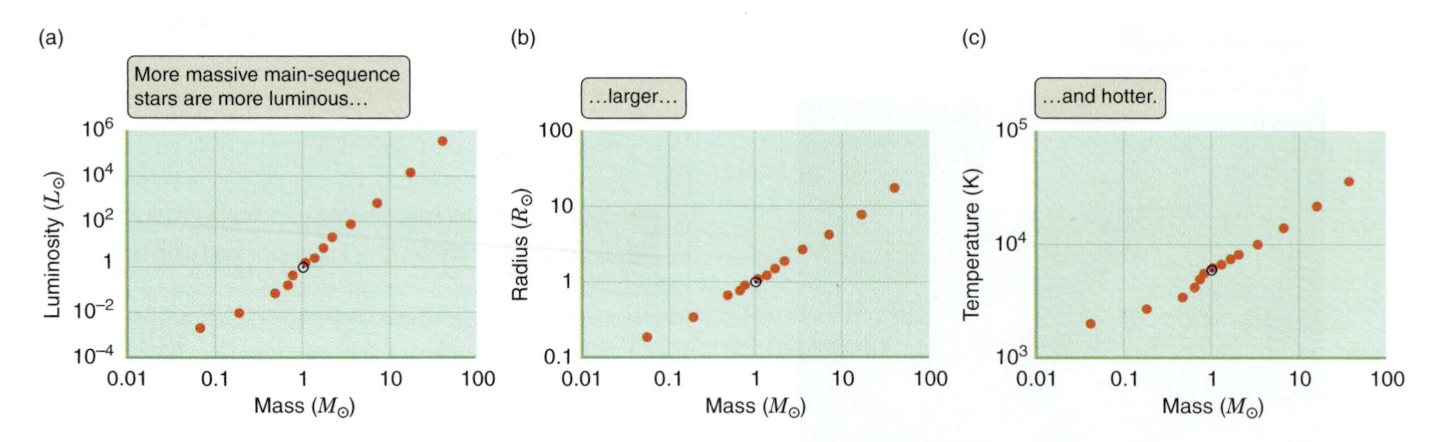

FIGURE 13.20 Plots of luminosity (a), radius (b), and temperature (c) versus mass for stars along the main sequence. The mass (and chemical composition) of a main-sequence star determines all of its other properties.

they must be large, luminous, cool giants, with radii hundreds or thousands of times the radius of the Sun. If the Sun were such a star, its atmosphere would swallow the orbits of the inner planets, including Earth. At the other extreme are stars found in the far lower left corner of the H-R diagram. These stars are tiny, comparable to the size of Earth. Their small surface areas explain why they have such low luminosities, despite having temperatures that can rival or even exceed the surface temperature of the hottest main-sequence O stars.

How can astronomers tell which stars are not members of the main sequence? Stars that lie off the main sequence on the H-R diagram can be identified by their luminosities (determined by their distance) or by slight differences within their spectral type. The width of a star's spectral lines is an indicator of the density and surface pressure of gas in the star's atmosphere. Those puffed-up stars above the main sequence have atmospheres with lower density and lower surface pressure and narrower absorption lines compared to main-sequence stars.

It is important for astronomers to know where stars fall on the H-R diagram. For example, when using the H-R diagram to estimate the distance to a star by the spectroscopic parallax method, they must know whether the star is on, above, or below the main sequence in order to get the star's luminosity. The spectral line widths from stars both on and off the main sequence indicate their **luminosity class**, which gives the *size* of the star. Luminosity classes are defined as follows: supergiant stars are luminosity class I, bright giants are class II, giants are class III, subgiants are class IV, main-sequence stars are class V, and white dwarfs are class WD. Luminosity classes I–IV lie above the main sequence, while class WD falls below and to the left of the main sequence (**Figure 13.21**). The complete spectral classification of a star includes both its spectral type (which indicates temperature and color) and its luminosity class (which indicates size).

Some non-main-sequence stars vary in luminosity, and the luminosity-temperature-radius relationship indicates that their temperature and size (radius) must be changing

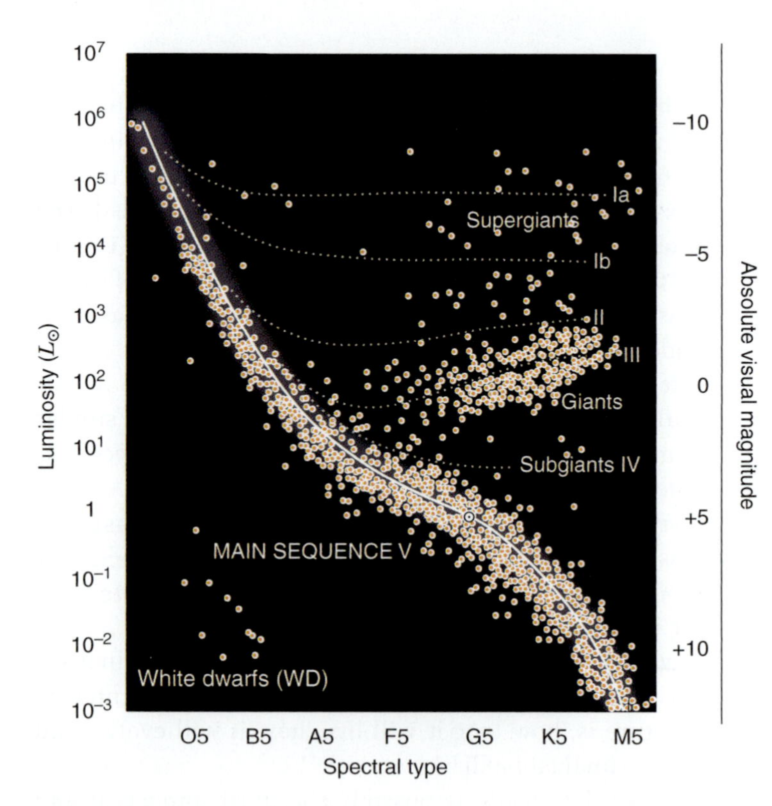

FIGURE 13.21 Stellar luminosity classes.

TABLE 13.3 | Taking the Measure of Stars*

Property	Methods
Luminosity	• For a star with a known distance, measure the brightness; then apply the inverse square law of radiation: $$\text{Luminosity} = 4\pi \times \text{Distance}^2 \times \text{Brightness}$$ • For a star without a known distance, take a spectrum of the star to determine its spectral and luminosity classes, plot them on an H-R diagram, and read the luminosity from the diagram.
Temperature	• Measure the color index of the star using blue and visual filters. Use Wien's law to relate the color to a temperature. • Take a spectrum of the star and estimate the temperature from its spectral class by noting which spectral lines are present.
Distance	• For a relatively nearby star (within a few hundred parsecs), measure the parallax shift of the star over the course of the year. • For a more distant star, find the luminosity using the H-R diagram as noted above, and then use the spectroscopic parallax method to relate luminosity, distance, and brightness.
Size	• For a few of the largest stars, measure the size directly or by the length of eclipse in eclipsing binary stars. • From the width of the star's spectral lines, estimate the luminosity class (supergiant, giant, or main-sequence). • For a star with known luminosity and temperature, use the Stefan-Boltzmann law to calculate the star's radius (the luminosity-temperature-radius relationship).
Mass	• Measure the motions of the stars in a binary system, and use these to determine the orbits of the stars; then apply Newton's form of Kepler's laws. • For a non-binary star, use the mass-luminosity relationship to estimate the mass from the luminosity.
Composition	• Analyze the lines in the star's spectrum to measure chemical composition.

*A brief summary of the methods used to measure basic properties of stars. Of the properties listed here, only temperature, distance, and composition can be *measured*. Luminosity must be *inferred* from the H-R diagram or calculated from distance and brightness, and size and mass must be *calculated*. Other properties that can be measured include brightness, color index, spectral type, and parallax shift.

as well. We will return to these stars in later chapters as we look at stellar evolution.

The existence of the main sequence, together with the fact that the mass of a main-sequence star determines where on the sequence it will lie, is a grand pattern that points to the possibility of a deep understanding of what stars are and how they "live." By the same token, the existence of stars that do *not* follow this grand pattern raises yet more questions. What is it about a star that determines whether it is part of the main sequence? In the decades that followed the discovery of the main sequence, few problems in astronomy attracted more attention than these questions. Their answers turned out to be as fundamental as stellar pioneers Annie Jump Cannon, Ejnar Hertzsprung, and Henry Norris Russell could ever have hoped. In the coming chapters you will learn that the main sequence holds the essential clue to what stars are and how they work, and that stars off the main sequence reveal how stars form, how they evolve, and how they die.

Table 13.3 summarizes the techniques that astronomers use to learn about certain types of stars, such as nearby stars and binary stars.

13.5 Origins: Habitable Zones

How might a basic property of a star, such as its luminosity, color, or surface temperature, affect the chance of there being a planet with life in orbit around that star? The only known life is that on planet Earth, where liquid water was essential for its formation and evolution. Of course, whether liquid water is an absolute requirement for life elsewhere is not known, but the presence of water is a good starting point for determining where to look. So astronomers look for planets that are at the right distance from their star to have a planetary temperature that permits water to exist in a liquid state. This range of distances

FIGURE 13.22 The distance and extent of a habitable zone (green) surrounding a star depends on the star's temperature. Regions too close to the star are too hot (red), and those too far are too cold (blue) to permit the existence of liquid water. The orbits of Mercury, Venus, Earth, and Mars have been drawn around these stars for scale.

is called the **habitable zone**. On planets that lie inside the habitable zone, water would exist only as a vapor—if at all. On planets that lie outside the habitable zone, water would be permanently frozen as ice.

Recall from Math Tools 5.4 that an important factor for estimating the temperature of a planet is the brightness of the sunlight that falls on that planet. This factor depends on the luminosity of the star and the planet's distance from the star. In the Solar System, the habitable zone ranges from about 0.9 to 1.4 AU, which includes Earth but just misses Venus and Mars. Main-sequence stars that are less luminous than the Sun are cooler and have narrower habitable zones, minimizing the chance that a habitable planet will form within that slender zone. Main-sequence stars that are more massive than the Sun are hotter and have larger habitable zones. **Figure 13.22** illustrates these zones around Sun-like, hotter, and cooler stars.

As of this writing, only a few planets have been found in the habitable zones of their respective stars. Methods of planet detection, as discussed in Chapter 7, work best when the planet is close to its star. In late 2011, scientists using the Kepler Mission telescope announced the discovery of a planet in a habitable zone. The planet Kepler-22b orbits a G5 star, which is slightly cooler and smaller than the Sun, at a distance of 0.85 AU. There are many other candidate planets in habitable zones, and it is likely that Kepler will confirm some of these in the near future. No one knows whether the habitable planets will all be found orbiting G stars like the Sun.

With an understanding of the basic physical properties of stars in place, we are now ready to ask much more fundamental questions about stars. How do stars work? How do they form? How do they evolve? How do they die? We will begin to address these questions by investigating the star that serves as the standard by which we measure other stars: the star we know best, the Sun. ■

Summary

13.1 The distance to a nearby star is measured by finding its parallax—that is, by measuring how the star's apparent position changes in the sky over the course of a year. The nearest star (other than the Sun) is about 4 light-years away. The brightness of a star can be measured directly. Combining brightness with distance gives the luminosity—how much light the star emits.

13.2 Astronomers determine the temperature, size, and composition of stars from stellar radiation. Emission and absorption lines in stellar spectra indicate the presence of specific elements and compounds in stars. Small, cool stars greatly outnumber large, hot stars. Blue stars are hotter and red stars are cooler.

13.3 Masses of stars are measured in binary star systems by observing the effects of the gravitational pull between the stars. Newton's universal law of gravitation and Kepler's laws connect the motion of the star to the forces they experience, and thus to their masses.

13.4 The H-R diagram is one of the most useful diagrams in stellar astronomy, showing the relationship among the various physical properties of stars. When stars are too remote to measure parallax using geometry, astronomers use the stars' spectra and the H-R diagram to estimate their distances. A star's luminosity class and temperature indicate its size. The mass and composition of a main-sequence star determine its luminosity, temperature, and size.

13.5 Stars of different luminosities and temperatures have habitable zones of different widths in different locations. The habitable zone is the distance from the star in which a planet could have the right temperature for liquid water.

Unanswered Questions

- What is the upper limit for stellar mass? Both theory and observation have shown that there is a lower limit for stellar mass, approximately 0.08 $M_{\odot}$. However, neither theory nor observation have provided a definitive value for the upper limit. Many astronomers believe that the upper limit lies somewhere in the range between 150 and 200 $M_{\odot}$, but the question remains unanswered.
- Are planets—and habitable planets in particular—likely to be found in binary star systems? Until recently, astronomers thought that planets would probably not be found in binary star systems, because they would not be gravitationally stable. But in 2011 the first "circumbinary" planet was discovered with the Kepler orbiting telescope. The planet Kepler-16b orbits the center of mass of the two stars in the binary system, and it may be at a great enough distance from the two stars that its orbit can remain stable. Sometimes there are two stars in its sky, sometimes none. The temperatures on Kepler-16b may vary greatly, and it may not be possible for the planet to stay in the temperature range for liquid water. Some astronomers think planets can be discovered in binary systems by observing variations in the timing of the eclipses. This method could be used to indicate the presence of a planet in binary systems.

Questions and Problems

Summary Self-Test

1. Star A and star B are two stars at nearly the same distance from Earth. Star A is half as bright as star B. Which of the following statements is true?
   a. Star B is farther away than star A.
   b. Star B is twice as luminous as star A.
   c. Star B is hotter than star A.
   d. Star B is larger than star A.
2. Star A and star B are two nearby stars. Star A is blue, and star B is red. Which of the following statements is true?
   a. Star A is hotter than star B.
   b. Star A is cooler than star B.
   c. Star A is farther away than star B.
   d. Star A is more luminous than star B.
3. Star A and star B are two stars nearly the same distance from Earth. Star A is blue, and star B is red, but they have equal brightness. Which of the following statements is true?
   a. Star A is more luminous than star B.
   b. Star A is larger than star B.
   c. Star A is smaller than star B.
   d. Star A is less luminous than star B.

4. Star A and star B are two stars exactly the same color. Star A's calcium absorption lines are deeper than star B's. Which of the following statements is true?
   a. Star A has more calcium than star B.
   b. Star A has less calcium than star B.
   c. There is a cloud of calcium atoms between Earth and star B.
   d. Star A is colder than star B.
5. Star A and star B are a binary system. The Doppler shift of star A's absorption lines is 3 times the Doppler shift of star B's absorption lines. Which of the following statements is true?
   a. Star A is 3 times as massive as star B.
   b. Star A is ⅓ as massive as star B.
   c. Star A is closer than star B.
   d. The binary pair is moving toward Earth, but star A is farther away.
6. Star A and star B are two red stars at nearly the same distance from Earth. Star A is many times brighter than star B. Which of the following statements is true?
   a. Star A is a main-sequence star, and star B is a red giant.
   b. Star A is a red giant, and star B is a main-sequence star.
   c. Star A is hotter than star B.
   d. Star A is a white dwarf, and star B is a red giant.
7. Star A and star B are two blue stars at nearly the same distance from Earth. Star A is many times brighter than star B. Which of the following statements is true?
   a. Star A is a main-sequence star, and star B is a red giant.
   b. Star A is a main-sequence star, and star B is a blue giant.
   c. Star A is a white dwarf, and star B is a blue giant.
   d. Star A is a blue giant, and star B is a white dwarf.
8. Which quantities are plotted on an H-R diagram?
   a. brightness on the horizontal axis; luminosity on the vertical axis
   b. brightness on the horizontal axis; temperature on the vertical axis
   c. luminosity on the vertical axis; temperature on the horizontal axis
   d. main sequence on the vertical axis; spectral type on the horizontal axis
9. Choose the two qualities that describe a star located in the lower right of the H-R diagram.
   a. hot c. bright
   b. cold d. faint
10. Star A is more massive than star B. Both are main-sequence stars. Therefore, star A is ________ than star B. (Choose all that apply.)
    a. brighter d. colder
    b. fainter e. larger
    c. hotter f. smaller

T/F and Multiple Choice

11. **T/F:** Brighter stars are always more luminous than fainter stars.
12. **T/F:** Hotter stars are always redder than cooler stars.
13. **T/F:** A star with no carbon absorption lines in its spectrum contains no carbon.
14. **T/F:** In a binary system, the more massive star moves more slowly than the less massive star.
15. **T/F:** The mass of a star is the primary indicator of its observable properties.
16. A telescope on Mars would be able to measure the distances to more stars than can be measured from Earth because
    a. the resolution of the telescope would be better.
    b. of Mars's thin atmosphere.
    c. it would be closer to the stars.
    d. the baseline would be longer.
17. Star A and star B are two nearby stars. Star A has a parallactic angle 4 times as large as star B's. Which of the following statements is true?
    a. Star A is ¼ as far away as star B.
    b. Star A is 4 times as far away as star B.
    c. Star A has moved through space ¼ as far as star B.
    d. Star A has moved through space 4 times as far as star B.
18. Which of these is the largest measure of an angle?
    a. arcsecond c. degree
    b. arcminute d. radian
19. Star A is twice as bright as star B, but also twice as far away. Star A is ________ as luminous as star B.
    a. 8 times c. twice
    b. 4 times d. half
20. O stars have relatively featureless visible spectra because O stars are
    a. too hot to have atoms.
    b. so hot that most atoms are ionized.
    c. not hot enough for atoms to be ionized.
    d. so cool that most atoms are in the ground state.
21. Table 13.1 shows two ways of reporting the amount of an element in the Sun. The percentage of hydrogen drops when changing from percentage by number to percentage by mass. But the percentage of helium grows. Why?
    a. Hydrogen is more massive than helium.
    b. Helium is more massive than hydrogen.
    c. Hydrogen is located in a different part of the Sun.
    d. It is difficult to measure the mass of hydrogen.
22. Star A and star B have the same luminosity, but star A is hotter than star B. Which of the following statements is true?
    a. Star A is smaller than star B.
    b. Star A is farther away than star B.
    c. Star A is larger than star B.
    d. Star A is closer than star B.
23. Capella (in the constellation Auriga) is the sixth-brightest star in the sky. When viewed with a high-power telescope, it is clear that Capella is actually two pairs of binary stars: the first pair are G-type giants; the second pair are M-type main-sequence stars. What color does Capella appear to be?
    a. red
    b. yellow
    c. blue
    d. Color cannot be determined from this information.

24. The H-R diagram is uniquely important because it shows the
    a. color of stars.
    b. temperature of stars.
    c. luminosity of stars.
    d. evolution of stars.
25. The reason most stars are main-sequence stars is that the main-sequence phase is the
    a. longest part of a star's lifetime.
    b. brightest part of a star's lifetime.
    c. hottest part of a star's lifetime.
    d. time when stars die.

Thinking about the Concepts

26. Distances to stars can be measured in inches, miles, kilometers (km), astronomical units (AU), light-years, and parsecs (pc). Why do many stellar astronomers prefer to use parsecs?
27. Even the best measurements always have experimental uncertainties. Given the measurement uncertainties described in this chapter, how accurately can the distance to a star be determined using trigonometric parallax?
28. The distances of nearby stars are determined by their parallaxes. Why is the uncertainty in astronomers' knowledge of a star's distance greater for stars that are farther from Earth?
29. To know certain properties of a star, you must first determine the star's distance. For other properties, knowledge of distance is not necessary. Into which of these two categories would you place each of the following properties: size, mass, temperature, color, spectral type, and chemical composition? In each case, state your reason(s).
30. The light from stars passes through dust in the Milky Way Galaxy before it reaches Earth, making stars appear dimmer than they actually are. How does this phenomenon affect stellar trigonometric parallax? How does it affect spectroscopic parallax?
31. In the constellation Cygnus, Albireo is a visual binary system whose two components can be seen easily with even a small, amateur telescope. Viewers describe the brighter star as "golden" and the fainter one as "sapphire blue."
    a. What does this description tell you about the relative temperatures of the two stars?
    b. What does it tell you about their respective sizes?
32. Very cool stars have temperatures around 2500 K and emit Planck spectra with peak wavelengths in the red part of the spectrum. Do these stars emit any blue light? Explain your answer.
33. The stars Betelgeuse and Rigel are both in the constellation Orion (pictured in the chapter-opening photo). Betelgeuse appears red, and Rigel is bluish white. To the eye, the two stars seem equally bright. If you can compare the temperature, luminosity, or size from just this information, do so. If not, explain why.
34. Explain why the stellar spectral types (O, B, A, F, G, K, M) are not in alphabetical order. What sequence of temperatures is defined by these spectral types?
35. Other than the Sun, the only stars whose mass astronomers can measure *directly* are those in eclipsing or visual binary systems. Why?
36. Once the mass of a certain spectral type of star located in a binary system has been determined, it can be assumed that all other stars of the same spectral type and luminosity class have the same mass. Why is this a safe assumption?
37. How do astronomers estimate the mass of stars that are not in eclipsing or visual binary systems?
38. Very old stars often have very few heavy elements, while very young stars have much more. What does this difference imply about the chemical evolution of the universe?
39. Could the spectral types of stars still be identified if there were no elements other than hydrogen and helium in their atmospheres? Explain your answer.
40. Explain why the Kepler Mission is finding eclipsing binary stars while it is searching for extrasolar planets using the transit method.

Applying the Concepts

41. Logarithmic (log) plots show major steps along an axis scaled to represent equal factors, most often factors of 10. Why do astronomers sometimes use a log plot instead of the more conventional linear plot? In the H-R diagram in Figure 13.16, how many times more luminous is the most luminous star than the least luminous?
42. Is the horizontal axis of the H-R diagram in Figure 13.16 logarithmic or linear?
43. Study Figure 13.20. Compared to the Sun, how luminous, large, and hot is a star that has 10 times the mass of the Sun?
44. Human eyes are typically 6 cm apart. Suppose your eye separation is average and you see an object jump from side to side by half a degree as you blink back and forth between your eyes. How far away is that object?
45. Sirius, the brightest star in the sky, has a parallax of 0.379 arcsec. What is its distance in parsecs? In light-years?
46. Sirius is actually a binary pair of two A-type stars. The brighter of the two stars is called the "Dog Star" and the fainter is called the "Pup Star" because Sirius is in the constellation Canis Major (meaning "Big Dog"). The Pup Star appears about 6,800 times fainter than the Dog Star. Compare the temperatures, luminosities, and sizes of the two stars.
47. Sirius and its companion orbit around a common center of mass with a period of 50 years. The mass of Sirius is 2.35 times the mass of the Sun.
    a. If the orbital velocity of the companion is 2.35 times greater than that of Sirius, what is the mass of the companion?
    b. What is the semimajor axis of the orbit?
48. Sirius is 22 times more luminous than the Sun, and Polaris (the "North Pole Star") is 2,350 times more luminous than the Sun. Sirius appears 23 times brighter than Polaris. How much farther away is Polaris than Sirius? What is the distance of Polaris in light-years?

49. Proxima Centauri, the star nearest to Earth other than the Sun, has a parallax of 0.772 arcsec. How long does it take light to reach Earth from Proxima Centauri?

50. Betelgeuse (in Orion) has a parallax of 0.00763 ± 0.00164 arcsec, as measured by the Hipparcos satellite. What is the distance to Betelgeuse, and the uncertainty in that measurement?

51. Rigel (also in Orion) has a Hipparcos parallax of 0.00412 arcsec. Given that Betelgeuse and Rigel appear equally bright in the sky, which star is actually more luminous? Knowing that Betelgeuse appears reddish while Rigel appears bluish white, which star would you say is larger and why?

52. The Sun is about 16 trillion (1.6×10^{13}) times brighter than the faintest stars visible to the naked eye.
    a. How far away (in astronomical units) would an identical solar-type star be if it were just barely visible to the naked eye?
    b. What would be its distance in light-years?

53. Our galaxy (the Milky Way) contains over 100 billion stars. If you assume that the average density of stars is the same as in the solar neighborhood, how much volume does the Milky Way take up? If the galaxy were a sphere, what would its radius be?

54. Find the peak wavelength of blackbody emission for a star with a temperature of about 10,000 K. In what region of the spectrum does this wavelength fall? What color is this star?

55. About 1,470 watts (W) of solar energy hits each square meter of Earth's surface. Use this value and the distance to the Sun to calculate the Sun's luminosity.

Using the Web

56. Go to the European Space Agency's Gaia (Global Astrometric Interferometer for Astrophysics) mission website (http://esa.int/science/gaia). Has Gaia been launched? How will it help astronomers determine the distances to more stars? Why will it map the stars from an L_2 orbit (see Connections 4.2)?

57. Go to the Eclipsing Binary Stars Lab website at http://astro.unl.edu/naap/ebs/ebs.html. Click on "Eclipsing Binary Simulator." Select preset Example 1, in which the two stars are identical. The animation will run with inclination 90° and show a 50 percent eclipse. What happens when you slowly change your viewing angle to the system—the inclination; how does this change the eclipse? At what value of inclination do you no longer see eclipses? What does the system look like at 0°? Reset the inclination to 90° and adjust the separation of the two stars. How does the light curve change when the separation is larger or smaller? Now make the two stars different. Change star 2 so that its radius is 3.0 $R_\odot$ and its temperature is 4000 K. At what value of inclination do you no longer see eclipses? What types of eclipsing binary systems do you think are the easiest to detect?

58. Go to the Kepler home page (http://kepler.nasa.gov) and mouse over "Confirmed Planets" on the upper right. How many eclipsing binary stars has Kepler found? Go to the Kepler Eclipsing Binary Catalog (http://keplerebs.villanova.edu) to see what new observations look like. Pick a few stars to study. What is the inclination ("sin i")? Look at the last column ("Figures"). The "raw" and "dtr" figures are pretty rough, but the "fit" figure shows a familiar light curve. How deep is the eclipse; that is, how much lower is the "normalized flux" during maximum eclipse?

59. Do a search for a photo of your favorite constellation (or go outside and take a picture yourself). Can you see different colors in the stars? What do the colors tell you about the surface temperatures of the stars? From your picture, can you tell which are the three brightest stars in the constellation? These stars will be named "alpha" (α), "beta" (β), and "gamma" (γ) for that constellation. Look up the constellation online and see if you chose the right stars. What are their temperatures and luminosities? What are their distances?

60. Go to the Spectral Types of Stars website (http://www.jb.man.ac.uk/distance/life/sample/java/spectype/specplot.htm) and run the applet provided. Note that the wavelengths are given in angstroms (Å), where 1 Å = 0.1 nm. On the right-hand side, use the menu to pull up (blue) spectra of stars from type O5 to type M5. "Fiducial wavelength" tells you where the applet will try to get the spectra to match with the (red) blackbody curve; it is set initially at 5,000 Å. Start with the O5 star and adjust the slider. What temperature blackbody curve gives you the best fit? Do the same for the other stars. Are the temperatures about what you expect? Explain why the blackbody curves seem to fit the longer wavelengths better than the shorter ones. If the observed (blue) spectra had come from space observations instead of from Earth observations, how might you expect the fit to look different?

Exploration | The H-R Diagram

Open the "HR Explorer" interactive simulation on the StudySpace website for this chapter. This simulation enables you to compare stars on the H-R diagram in two ways. You can compare an individual star (marked by a red *X*) to the Sun by varying its properties in the box in the left half of the window. Or you can compare groups of the nearest and brightest stars. Play around with the controls for a few minutes to familiarize yourself with the simulation.

Begin by exploring how changes to the properties of the individual star change its location on the H-R diagram. First, press the "Reset" button at the top right of the window.

Decrease the temperature of the star by dragging the temperature slider to the left. Notice that the luminosity remains the same. Since the temperature has decreased, each square meter of star surface must be emitting less light. What other property of the star changes in order to keep the total luminosity of the star constant?

Predict what will happen when you slide the temperature slider all the way to the right. Now do it. Did the star behave as you expected?

1 **As you move to the left across the H-R diagram, what happens to the radius?**

....

....

2 **What happens as you move to the right?**

....

....

Press "Reset" and experiment with the luminosity slider.

3 **As you move up on the H-R diagram, what happens to the radius?**

....

....

4 **What happens as you move down?**

....

....

Press "Reset" again and predict how you would have to move the slider bars to move your star into the red giant portion of the H-R diagram (upper right). Adjust the slider bars until the star is in that area. Were you correct?

5 **How would you adjust the slider bars to move the star into the white dwarf area of the H-R diagram?**

....

....

Press the "Reset" button and explore the right-hand side of the window. Add the nearest stars to the graph by clicking their radio button under "Plotted Stars." Using what you learned above, compare the temperatures and luminosities of these stars to the Sun (marked by the *X*).

6 **Are the nearest stars generally hotter or cooler than the Sun?**

....

7 **Are the nearest stars generally more or less luminous than the Sun?**

....

Press the radio button for the brightest stars. This action will remove the nearest stars and add the brightest stars in the sky to the plot. Compare these stars to the Sun.

8 **Are the brightest stars generally hotter or cooler than the Sun?**

....

9 **Are the brightest stars generally more or less luminous than the Sun?**

....

10 **How do the temperatures and luminosities of the brightest stars in the sky compare to the temperatures and luminosities of the nearest stars? Does this information support the claim in the chapter that there are more low-luminosity stars than high-luminosity stars? Explain.**

....

....

....

....

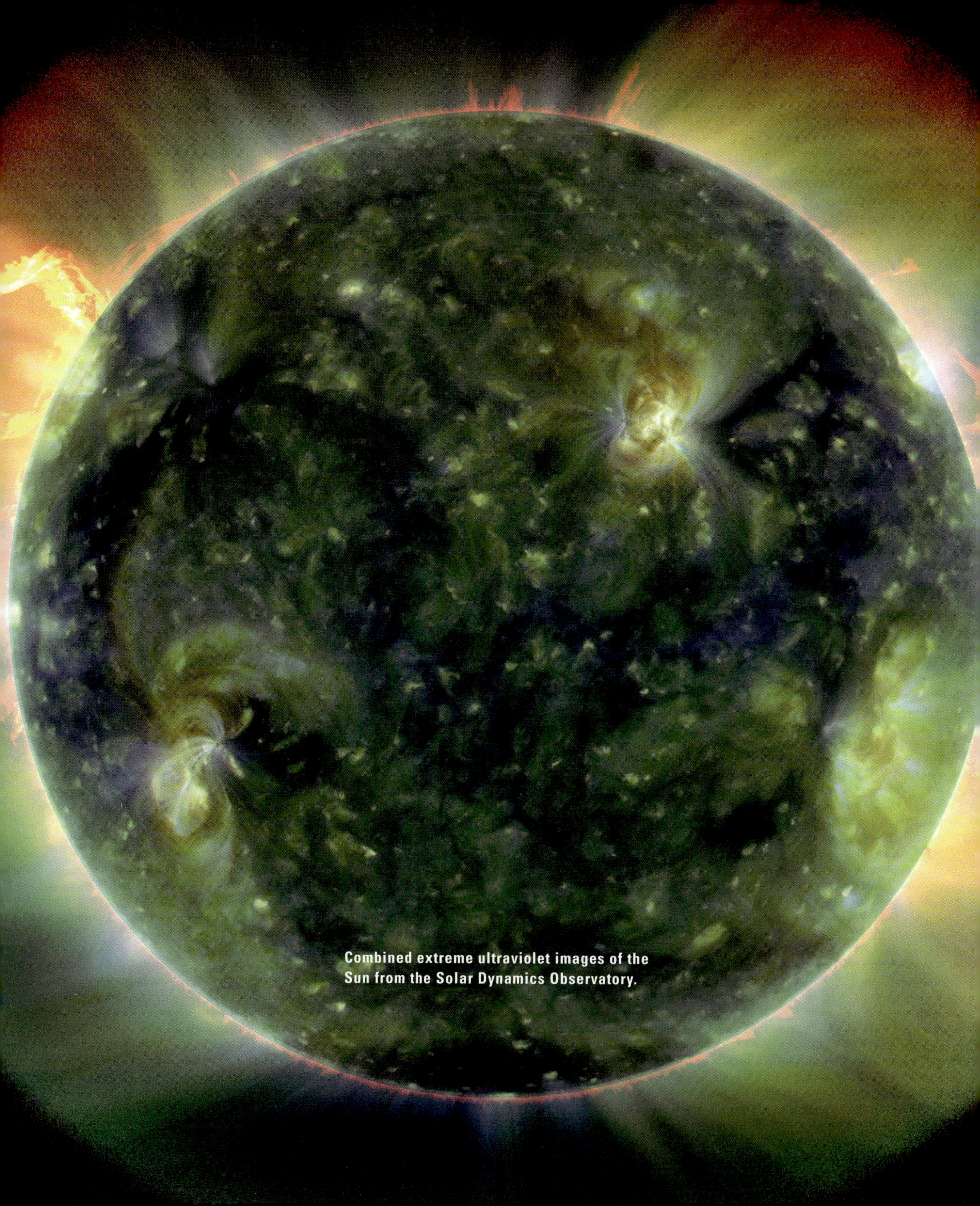

Combined extreme ultraviolet images of the Sun from the Solar Dynamics Observatory.

14 Our Star—The Sun

It is stern work, it is perilous work to thrust your hand in the sun
And pull out a spark of immortal flame to warm the hearts of men.

Joyce Kilmer (1886–1918)

LEARNING GOALS

To life on Earth the Sun is the most important object in the sky. It lights the days, warms the planet, and provides the energy for life. But to astronomers the Sun is simply a typical main-sequence star, located conveniently nearby for detailed study. In this chapter we take a closer look at Earth's star. By the conclusion of this chapter, you should be able to:

- Describe the balance between the forces that determine the structure of the Sun.
- Explain how mass is efficiently converted into energy in the Sun's core.
- Sketch a physical model of the Sun's interior, and describe how observations of solar neutrinos and seismic vibrations on the surface of the Sun test astronomers' models of the Sun.
- Describe how energy is transported through the Sun.
- Describe the solar activity cycles of 11 and 22 years, and explain how these cycles are related to the Sun's changing magnetic field.
- Explain how solar activity affects Earth.

14.1 The Structure of the Sun

Energy from the Sun is responsible for daylight, for Earth's weather and seasons, and for terrestrial life itself. No object in nature has been more revered or more worshipped than the Sun. With a spectral type of G2, the Sun is all but indistinguishable from billions of other stars in the Milky Way Galaxy. But it is the star against which all other stars are measured. The mass of the Sun, the size of the Sun, the luminosity of the Sun—these basic properties provide the yardsticks of modern astronomy.

How does the Sun produce the energy that is essential to life on Earth?

The Sun may be average as far as stars go, but that makes it no less impressive an object on a human scale. The mass of the Sun, 1.99×10^{30} kilograms (kg), is over 300,000 times that of Earth. The Sun's radius, 696,000 kilometers (km), is over 100 times that of Earth. At a luminosity of 3.85×10^{26} watts (W), the Sun produces more energy in a second than all of the electric power plants on Earth could generate in a half-million years. Because the Sun is also the only star at close range, much of the detailed knowledge about stars has come from studying the Sun.

In Chapter 13 we looked at the gross physical properties of distant stars, including their mass, luminosity, size, temperature, and chemical composition. In this chapter we ask fundamental questions about Earth's local star. How does the Sun work? Where does it get its energy? Why does it have the size, temperature, and luminosity that it has? How has it been able to remain so constant over the billions of years since the Solar System formed?

Geologists learn about the interior of Earth by using a combination of physical understanding, detailed computer models, and clever experiments that test the predictions of those models. The task of exploring the interior of the Sun is much the same. The structure of the Sun is governed by a number of physical processes and relationships just as Earth's structure is. With their understanding of physics, chemistry, and the properties of matter and radiation, astronomers can express these processes and relationships as mathematical equations. High-speed computers can be used to solve these equations and create a model of the Sun. One of the great successes of 20th century astronomy was the construction of a physical model of the Sun that agrees with observations of the mass, composition, size, temperature, and luminosity of the real thing.

The current model of the Sun's interior is the culmination of decades of work by thousands of physicists and astronomers. The essential ideas underlying what is known about the structure of the Sun emerge from a few key insights, summed up in a single statement: *The structure of the Sun is a matter of balance.*

The first key balance within the Sun is the balance between pressure and gravity (**Figure 14.1**). The Sun is a huge ball of hot gas. If gravity were stronger than pressure within the Sun, the Sun would collapse. Conversely, if pressure were stronger than gravity, the Sun would blow itself apart. The balance between the two is called **hydrostatic equilibrium**. Hydrostatic equilibrium establishes the pressure at each point within a planet and determines the atmospheric pressure at Earth's surface. According to hydrostatic equilibrium, the pressure at any point within the Sun's interior must be just enough to hold up the weight of all the layers above that point. If the Sun were not in hydrostatic equilibrium, forces within it would not be in balance, so the surface of the Sun would *move*. However, the size of the Sun does not change from one day to the next, so its interior must be in hydrostatic equilibrium.

Hydrostatic equilibrium becomes an even more powerful concept when combined with the way gases behave. Deeper in the interior of the Sun, the weight of the material above becomes greater, and hence the pressure must increase. In a gas, higher pressure means higher density and/or higher temperature. **Figure 14.2a** shows how conditions vary as distance from the center of the Sun changes. Deeper into the Sun, the pressure climbs; and as it does, the density and temperature of the gas climb as well.

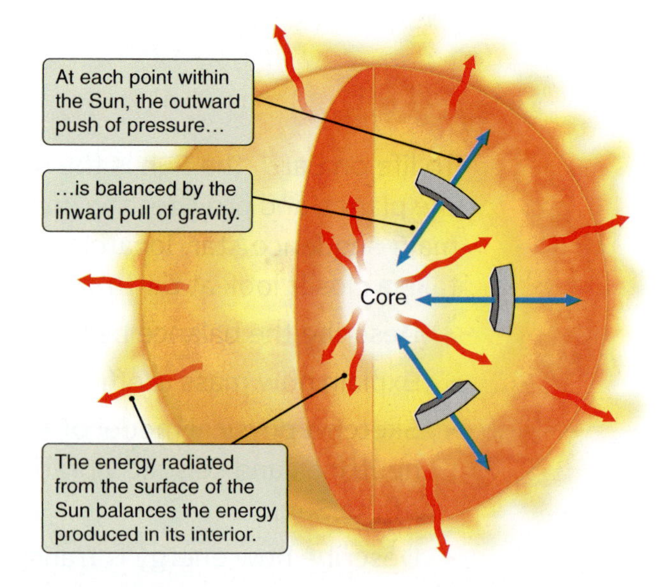

FIGURE 14.1 The structure of the Sun is determined by the balance between the forces of pressure and gravity, and the balance between the energy generated in its core and energy radiated from its surface.

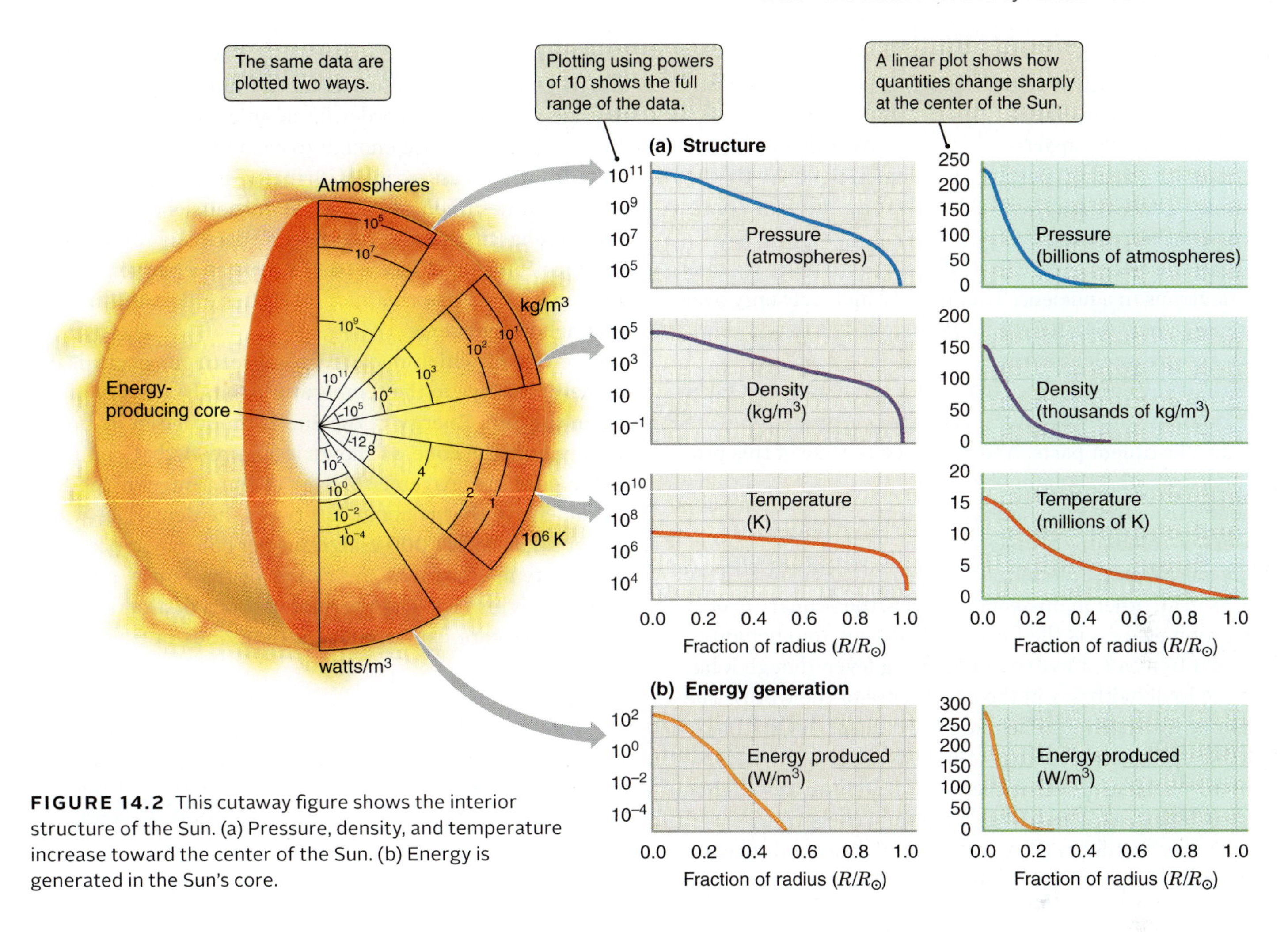

FIGURE 14.2 This cutaway figure shows the interior structure of the Sun. (a) Pressure, density, and temperature increase toward the center of the Sun. (b) Energy is generated in the Sun's core.

A second fundamental balance within the Sun is the balance of energy (see Figure 14.1). Stars such as the Sun are remarkably stable objects. Geological records show that the luminosity of the Sun has remained nearly constant for billions of years. Models of stellar evolution indicate that the luminosity of the Sun is increasing with time, but very, very slowly. The Sun's luminosity 4.5 billion years ago was about 70 percent of its current luminosity. The very existence of the main sequence means that stars do not change much over the main part of their lives. To remain in balance, the Sun must produce just enough energy in its interior each second to replace the energy that is radiated away by its surface each second. This energy balance depends on how energy is generated in the interior of the Sun (**Figure 14.2b**), and how that energy finds its way from the interior to the Sun's surface, where it is radiated away. We turn to this idea in Section 14.2.

Solar energy production must balance what is radiated away.

14.2 The Sun Is Powered by Nuclear Fusion

The amount of energy produced by the Sun each second is truly astronomical: 3.85×10^{26} W. One of the most basic questions facing the pioneers of stellar astrophysics was how the Sun and other stars get their energy. Theoretical studies and work in the laboratories of nuclear physicists revealed that the Sun's energy comes from nuclear reactions at its core. At the heart of the Sun lies a nuclear furnace capable of powering the star for billions of years.

Recall from Chapter 5 that the nucleus of most hydrogen atoms consists of a single proton. Nuclei of all other atoms are built from a mixture of protons and neutrons. Most helium nuclei, for example, consist of two protons and two neutrons. Most carbon nuclei consist of six protons and six neutrons. Protons have a positive electric charge, and neutrons have no net electric charge. Like charges repel, so all of the protons in an atomic nucleus are pushing away from each other with a tremendous force because they are so close to each other. If electric forces were all there was to it, the nuclei of atoms would rapidly fly apart—yet atomic nuclei hold together. There is another force in nature, even stronger than the electric force that "glues" together the protons and neutrons in a nucleus. That force, which acts only over extremely short distances, approximately 10^{-15} m, or meters, is the **strong nuclear force**.

Atomic nuclei are held together by the strong nuclear force.

The strong nuclear force is a very powerful force. It takes energy to pull apart the nucleus of an atom such as helium into its constituent parts. And when the reverse of this process occurs—when an atomic nucleus assembles from component parts—this same amount of energy is released. The process of combining two less massive atomic nuclei into a single more massive atomic nucleus is **nuclear fusion**. In the Sun, as in all other main-sequence stars, the primary energy generation process is the fusion of hydrogen into helium—a process often called **hydrogen burning** (even though it has nothing to do with fire in the usual sense of the word). The fusion of hydrogen into helium always takes several steps, but the net reaction is that four hydrogen atoms become the four particles in the nucleus of the helium atom. Thus nuclear fusion releases a large amount of energy.

The special theory of relativity (which we will discuss further in Chapter 18) explains that mass and energy are equivalent. This is the basis for the energy produced in nuclear fusion reactions in stars. Mass can be converted to energy, and energy can be converted to mass. Einstein's famous equation $E = mc^2$ provides the exchange rate between the two. By comparing the mass of the *products* of a reaction with the mass of the *reactants*, scientists can determine the fraction of the original mass that was turned into energy during the reaction. The mass of four separate hydrogen atoms is 1.007 times greater than the mass of a single helium atom; so when hydrogen fuses to make helium, 0.7 percent of the mass of the hydrogen is converted to energy (see **Math Tools 14.1**).

Nuclear fusion is a very efficient source of energy.

Whether they consist of a ball rolling downhill, a battery discharging itself through a lightbulb, or an atom falling to a lower state by emitting a photon, systems in nature tend to seek the lowest energy state available to them. The transition from hydrogen to helium is a big ride downhill in terms of energy, so it makes sense that hydrogen nuclei would naturally tend to fuse together to make helium. A major roadblock, however, stands in the way of nuclear fusion. As noted already, the strong nuclear force responsible for binding atomic nuclei together can act over only very short distances: 10^{-15} meter or so, or about a hundred-thousandth the size of an atom. In order for atomic nuclei to fuse, they must be brought close enough to each other for the strong nuclear force to assert itself. Bringing nuclei this close together is very difficult to do. All atomic nuclei have positive electric charges, so any two nuclei repel each other. This electric repulsion (**Figure 14.3**) serves as a barrier against nuclear fusion. Fusion cannot take place unless this barrier is somehow overcome.

The rate at which nuclear fusion reactions occur is extremely sensitive to the temperature and the density of the gas in the Sun. Energy in the Sun is produced in its innermost region, the **core**, as shown in Figure 14.2b. Conditions in the core are extreme. Matter at the center of the Sun has a density that is about 150 times the density of water (water's density is 1,000 kilograms per cubic meter, kg/m^3), and the temperature at the center of the Sun is about 15 million kelvins (K). The thermal motions of atomic nuclei in the Sun's core contain tens of thousands of times more

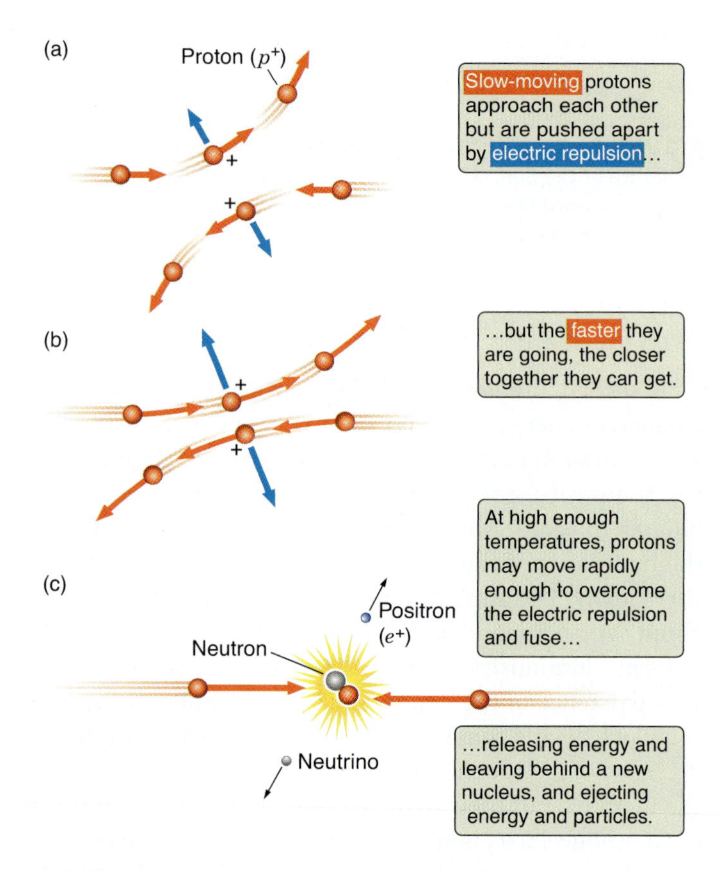

FIGURE 14.3 Atomic nuclei are positively charged and so repel each other. (a, b) If two nuclei are moving toward each other, the faster they are going, the closer they will get before veering away. (c) At the temperatures and densities found in the centers of stars, thermal motions of nuclei are energetic enough to overcome this electric repulsion, so fusion takes place.

Math Tools 14.1

The Source of the Sun's Energy

The conversion of four hydrogen nuclei (protons) into a single helium nucleus results in a loss of mass. The mass of a single proton is 1.6726×10^{-27} kg. So, four protons add up to a mass of 6.6904×10^{-27} kg. But the mass of a helium nucleus is 6.6447×10^{-27} kg. When we subtract the mass of the helium nucleus from the mass of the four protons:

$$\begin{aligned} m &= 6.6904 \times 10^{-27}\ \text{kg} - 6.6447 \times 10^{-27}\ \text{kg} \\ &= 0.0457 \times 10^{-27}\ \text{kg} = 4.57 \times 10^{-29}\ \text{kg} \end{aligned}$$

we see that this is 4.57×10^{-29} kg—a loss of about 0.7 percent. Conversion of 0.7 percent of the mass of the hydrogen into energy might not seem very efficient—until we compare it with other sources of energy and discover that it is millions of times more efficient than even the most efficient chemical reactions.

Using Einstein's equation $E = mc^2$, along with the definition of a joule, J ($1\ \text{J} = 1\ \text{kg m}^2/\text{s}^2$), we can calculate the energy released by this mass-to-energy conversion:

$$E = mc^2 = (4.57 \times 10^{-29}\ \text{kg})(3.00 \times 10^8\ \text{m/s})^2 = 4.11 \times 10^{-12}\ \text{J}$$

Each reaction that takes four hydrogen nuclei and turns them into a helium nucleus releases 4.11×10^{-12} J of energy. But there are many atoms in a gram, so fusing a single gram of hydrogen into helium releases about 6×10^{11} J of energy—about the equivalent of the chemical energy released in burning 100 barrels of oil. For the Sun to produce as much energy as it does, it must convert roughly 600 billion kg of hydrogen into helium every second (and about 4 billion kg of matter is converted to energy in the process). The sunlight falling on Earth may be responsible for powering almost everything that happens on the planet, but it amounts to only about a hundred-billionth of the energy that the Sun radiates.

The Sun has been burning hydrogen at this rate for at least the age of Earth and the Solar System, or 4.6 billion years. How much of its available fuel has already been spent, and how much longer will the Sun last? Astronomers estimate that only 10 percent of the Sun's total mass will ever be involved in fusion, because the other 90 percent will never get hot enough or dense enough for the strong nuclear force to make fusion happen.

Ten percent of the mass of the Sun is $(0.1) \times (2 \times 10^{30})$ kg, or 2×10^{29} kg. That is the amount of fuel the Sun has available. The Sun consumes hydrogen at a rate of 600 billion kilograms per second (kg/s), so each year the Sun consumes:

$$\begin{aligned} M_{\text{year}} &= (600 \times 10^9\ \text{kg/s}) \times (3.16 \times 10^7\ \text{s/yr}) \\ &= 1.90 \times 10^{19}\ \text{kg/yr} \end{aligned}$$

If we assume here that the rate has been constant throughout the Sun's age of 4.6 billion years, it has consumed:

$$M_{4.6\,BY} = (4.6 \times 10^9\ \text{yr}) \times (1.90 \times 10^{19}\ \text{kg/yr}) = 8.7 \times 10^{28}\ \text{kg}$$

This is about 44 percent of the Sun's original supply of available hydrogen fuel, confirming that the Sun is now in middle age. A favorite theme of science fiction is the fate that awaits Earth when the Sun uses up its hydrogen fuel, but you need not worry about that happening anytime soon. The Sun is only about halfway through its estimated 10-billion-year lifetime as a main-sequence star.

kinetic energy than the thermal motions of atoms at room temperature. As illustrated in **Figure 14.3c**, under these conditions atomic nuclei slam into each other hard enough to overcome the electric repulsion between them and allow short-range nuclear forces to act. The hotter and denser a gas is, the more of these energetic collisions will take place each second. For this reason, the rate at which nuclear fusion reactions occur is extremely sensitive to the temperature and the density of the gas. Half of the energy produced by the Sun is generated within the inner 9 percent of the Sun's radius, or less than 0.1 percent of the volume of the Sun (see Figure 14.2b).

Hydrogen fuses to helium in the core of the Sun.

There are several reasons why hydrogen burning is the most important source of energy in main-sequence stars. Hydrogen is the most abundant element in the universe, so it is the most abundant source of nuclear fuel at the beginning of a star's lifetime. Hydrogen burning is also the most efficient form of nuclear fusion, converting a larger fraction of mass into energy than does any other type of nuclear reaction. But the primary reason why hydrogen burning is the dominant process in main-sequence stars is that hydrogen atoms are also the easiest type of atoms to fuse. Hydrogen nuclei—protons—have an electric charge of +1. The electric barrier that must be overcome to fuse protons is the repulsion of a single proton against another. Compare this to the force required, for example, to get two carbon nuclei close enough that they can fuse. To fuse carbon, the repulsion of the six protons in one carbon nucleus pushing against the six protons in another carbon nucleus must be overcome. The resulting force is proportional to the product of the charges of the two atomic nuclei, making the repulsion between two carbon nuclei 36 times stronger than that between the two protons involved when hydrogen nuclei bond. For this reason, hydrogen fusion occurs at a much lower temperature than any other type of nuclear fusion. In the core of a low-mass star such as the Sun, hydrogen burns primarily through a process called the *proton-proton chain*. The dominant branch of the proton-proton chain is discussed in **Connections 14.1**.

Hydrogen burns mostly via the proton-proton chain.

ASTROTOUR: THE SOLAR CORE

Energy Produced in the Sun's Core Must Find Its Way to the Surface

Some of the energy released by hydrogen burning in the core of the Sun escapes directly into space in the form of neutrinos (extremely low-mass particles that interact very weakly with other forms of matter—see Connections 14.1), but most of the energy goes instead into heating the solar interior. The structure of the Sun is determined by the way thermal energy moves outward through the star. The na-

Connections 14.1

The Proton-Proton Chain

In the Sun and in other low-mass stars, hydrogen burning takes place in a series of nuclear reactions called the **proton-proton chain**, which has three different branches. The most important branch, responsible for about 85 percent of the energy generated in the Sun, consists of three steps (**Figure 14.4**). In the first step, two hydrogen nuclei (the nucleus of hydrogen consists of one proton) fuse. In that process, one of the protons is transformed into a neutron by emitting a positively charged particle called a **positron** and another type of elementary particle called a **neutrino**. This conversion is one variety of a process called **beta decay**.

The positron is expelled at a great velocity, carrying away some of the energy released in the reaction. Electrons and positrons have opposite electric charges, so they attract each other. As a result, the expelled positron soon collides with one of the many electrons moving freely about in the center of the Sun. But the positron is the **antiparticle** of the electron and, as we'll discuss in Chapter 22, when a particle and its antiparticle collide they annihilate each other, with their total mass being converted into energy. In this way, the annihilation of electrons and positrons in the Sun's core produces energy in the form of gamma-ray photons. These photons carry part of the energy released when the two protons fuse, thereby heating the surrounding gas. The neutrino, on the other hand, is a very elusive particle. Its interactions with matter are so feeble that its most likely fate is to escape from the Sun without further interactions with any other particles.

Follow along in Figure 14.4 as we step through the proton-proton chain. The new atomic nucleus formed by the first step in the chain consists of a proton and a neutron. This is the nucleus of a heavy isotope of hydrogen called *deuterium*, or ^{2}H. (To read this chemical shorthand, remember that the element symbol, H, indicates the number of protons; hydrogen always has one proton. The superscript indicates the sum of the number of neutrons and the number of protons. Deuterium has one proton because it is an isotope of hydrogen, $2 - 1 = 1$ neutron.) In the second step of the proton-proton chain, another proton slams into the deuterium nucleus, fusing with it to form the nucleus of a light isotope of helium, ^{3}He, consisting of two protons and a neutron. The energy released in this step is carried away as a gamma-ray photon. In the third and final step of the chain, two ^{3}He nuclei collide and fuse, producing an ordinary ^{4}He nucleus and ejecting two protons in the process. The energy released in this step is the kinetic energy of the helium nucleus and two ejected protons.

This dominant branch of the proton-proton chain can be written symbolically as:

$$^{1}\mathrm{H} + {}^{1}\mathrm{H} \rightarrow {}^{2}\mathrm{H} + e^{+} + \nu$$

followed by:

$$e^{+} + e^{-} \rightarrow \gamma + \gamma$$

$$^{2}\mathrm{H} + {}^{1}\mathrm{H} \rightarrow {}^{3}\mathrm{He} + \gamma$$

$$^{3}\mathrm{He} + {}^{3}\mathrm{He} \rightarrow {}^{4}\mathrm{He} + {}^{1}\mathrm{H} + {}^{1}\mathrm{H}$$

ture of **energy transport** within a star is one of the key factors determining the star's structure.

Thermal energy can be transported by a number of methods. Move a pot of boiling water from one side of the room to the other, and you've transported thermal energy. A common way in which energy is transported in everyday life is thermal **conduction**. For example, if you hold one end of a metal rod while putting the other end into a fire, soon the end of the rod that is in your hand becomes too hot to hold. Thermal conduction occurs as the energetic thermal vibrations of atoms and molecules in the hot end of the rod cause their cooler neighbors to vibrate more rapidly as well. Although thermal conduction is the most important way energy is transported in solid matter, it is typically ineffective in a gas because the atoms and molecules are too far apart to transmit vibrations to one another efficiently. Thermal conduction is unimportant in the transport of energy from the core of the Sun to its surface. Thermal energy is instead carried outward from the center of the Sun by two other mechanisms: *radiative transfer*, in which light carries the energy, and *convection*, in which energy is carried by rising and falling pockets of gas.

In **radiative transfer**, energy is transported from hotter to cooler regions by photons, which carry along the energy. Imagine a hotter region of the Sun adjacent to a cooler region (**Figure 14.5**). Recall from your study of radiation in Chapter 5 that the hotter region contains more (and more energetic) photons than the cooler region. More photons

Here the symbols are e^+ for a positron, e^- for an electron, ν (the Greek letter nu) for a neutrino, and γ (gamma) for a gamma-ray photon.

The rate of the proton-proton chain reaction depends on both temperature and pressure. At the temperature and pressure that exist within the Sun's core, the reaction rate is relatively slow—in fact, extremely slow compared to a nuclear bomb explosion. The Sun's slow nuclear fusion is fortunate for life on Earth; if its hydrogen burned quickly, the Sun would have exhausted its supply long ago, and life might not have had time to evolve. ▶▶ **NEBRASKA SIMULATION: PROTON-PROTON ANIMATION**

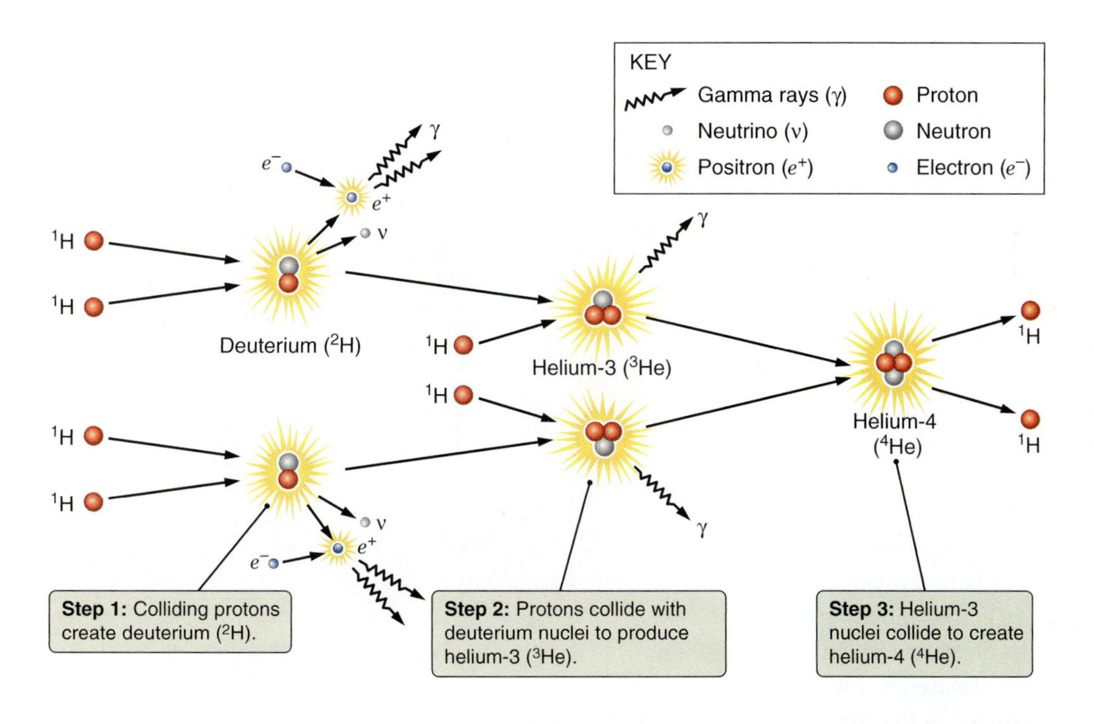

FIGURE 14.4 The Sun and all other main-sequence stars get their energy by fusing the nuclei of four hydrogen atoms together to make a single helium atom. In the Sun, about 85 percent of the energy produced comes from the branch of the proton-proton chain shown here.

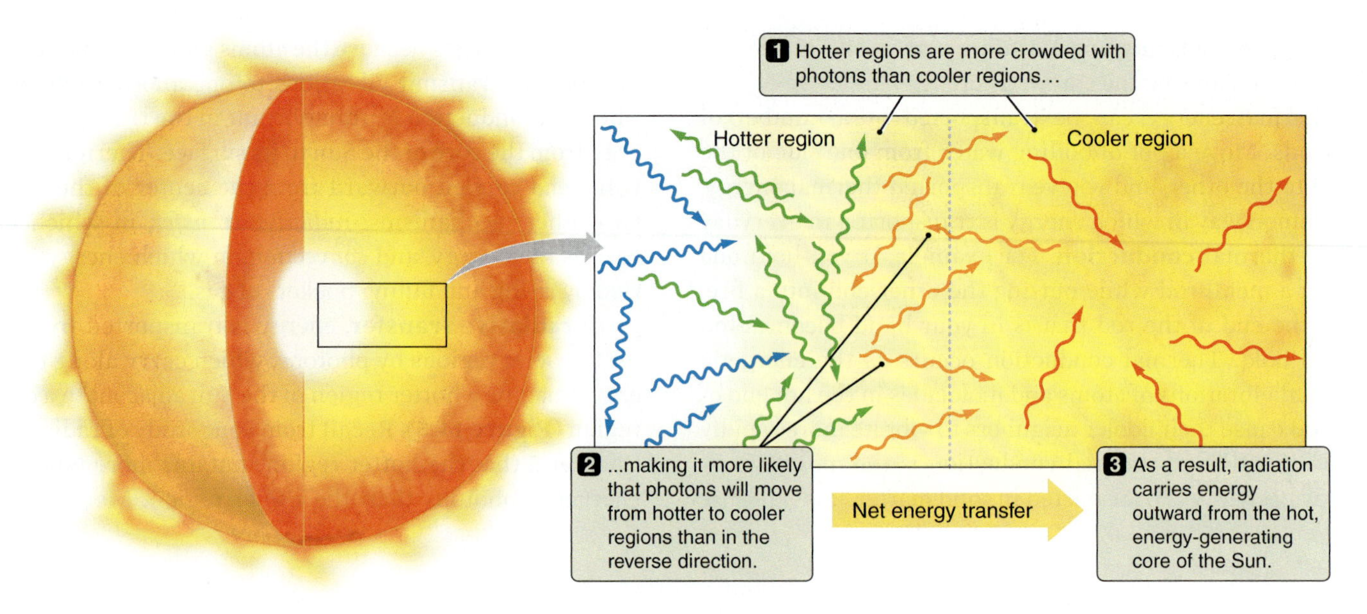

FIGURE 14.5 Higher-temperature regions deep within the Sun produce more radiation than do lower-temperature regions farther out. Although radiation flows in both directions, more radiation flows from the hotter regions to the cooler regions than from the cool regions to the hot regions. Therefore, radiation carries energy outward from the inner parts of the Sun.

move by chance from the hotter (more crowded) region to the cooler (less crowded) region than in the reverse direction. There is a net transfer of photons and photon energy from the hotter region to the cooler region, and in this way radiative transfer carries energy outward from the Sun's core.

Radiation carries energy from hotter regions to cooler regions.

If temperature varies by a large amount over a short distance, then the concentration of photons varies sharply as well. This difference favors rapid radiative energy transfer. The transfer of energy from one point to another by radiation also depends on how freely radiation can move from one point to another within a star. The degree to which matter impedes the flow of photons is called **opacity**. The opacity of a material depends on many things, including the density of the material, its composition, its temperature, and the wavelength of the photons moving through it.

Opacity impedes the outward flow of radiation.

Radiative transfer is most efficient in regions where opacity is low. In the inner part of the Sun, where temperatures are high and atoms are ionized, opacity comes mostly from the interaction between photons and free electrons (electrons not attached to any atom). Here opacity is relatively low, and radiation readily carries the energy produced in the core outward through the star. The region in which radiative transfer is responsible for energy transport extends 71 percent of the radius toward the surface of the Sun from the core. This region is the Sun's **radiative zone** (**Figure 14.6**). Even though the opacity of the radiative zone is relatively low, photons travel only a short distance before being absorbed, emitted, or deflected by matter, much like a beach ball being batted

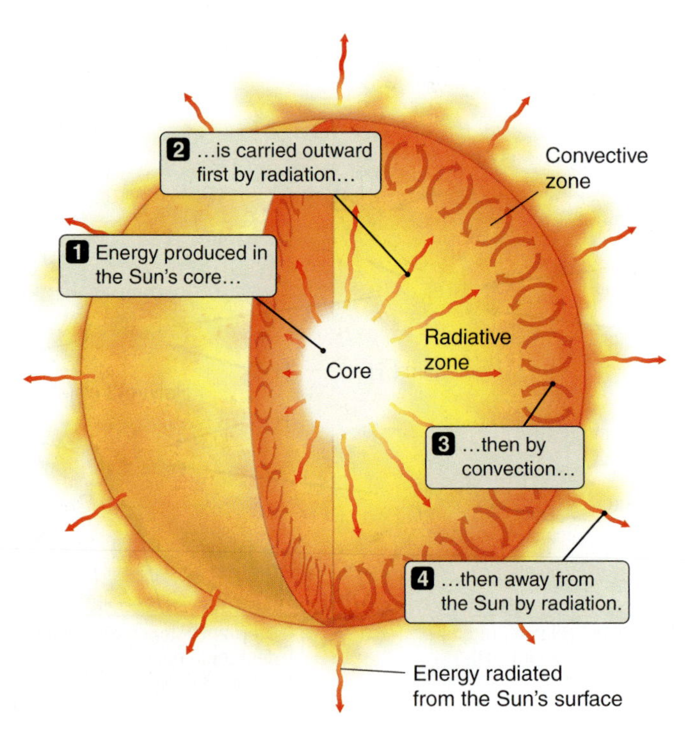

FIGURE 14.6 The interior structure of the Sun is divided into zones based on where energy is produced and how it is transported outward.

about by a crowd of people (**Figure 14.7**). Each interaction sends the photon in an unpredictable direction—not necessarily toward the surface of the star. The path that a photon follows is so convoluted that, on average, it takes the energy of a gamma-ray photon produced in the interior of the Sun about 100,000 years to find its way to the outer layers of the Sun. Opacity serves as a blanket, holding energy in the interior of the Sun and letting it seep away only slowly.

From a peak of 15 million K at the center of the Sun, the temperature falls to about 100,000 K at the outer margin of the radiative zone. At this temperature, atoms are no longer completely ionized, so there are fewer free electrons and the opacity is therefore higher. As the opacity increases, radiation becomes less efficient at carrying energy from one place to another. The energy that is flowing outward through the Sun "piles up." The physical sign that energy is accumulating is that the *temperature gradient* —how rapidly the temperature drops with increasing distance from the center of the Sun—becomes very steep. Radiative transfer carries energy from hotter regions to cooler regions, smoothing out temperature differences between them. As the opacity increases, radiation becomes less effective in smoothing out temperature differences, so temperature differences between one region and another become greater.

Nearer the surface of the Sun, radiative transfer becomes so inefficient (and the temperature gradient so steep) that a different way of transporting energy takes over: **convection**. As in a hot-air balloon, cells (or packets) of hot gas become buoyant and rise up through the lower-temperature gas above them, carrying energy with them. Just as convection carries energy from the interior of a planet to its surface, or from the Sun-heated surface of Earth upward through Earth's atmosphere, it also plays an important role in the transport of energy outward from the interiors of many stars, including the Sun. The solar **convective zone** (see Figure 14.6) extends from the outer boundary of the radiative zone outward to just below the visible surface of the Sun.

In the outer part of the Sun, energy is carried by convection.

In the outermost layers of stars, radiation again takes over as the primary mode of energy transport. Energy from the outermost layers of a star is transported into space through radiation. Even so, the effects of convection can be seen as a perpetual roiling of the visible surface of the Sun.

What If the Sun Were Different?

As noted earlier, a key point to take into account when calculating a model of the interior of the Sun is balance. The temperature and density at each point within the model Sun must be just right so that transport of energy away from the core by radiation and convection just balances the amount of energy produced by fusion in the core. The density, temperature, and pressure of the model Sun must vary from point to point in such a way that the outward push of pressure is everywhere balanced by the inward pull of gravity. Finally, the whole model must depend on only two things: the total mass of gas from which the star is made, and the chemical composition of that gas.

What if a hypothetical star had the same mass, surface temperature, and composition as the Sun, but was some-

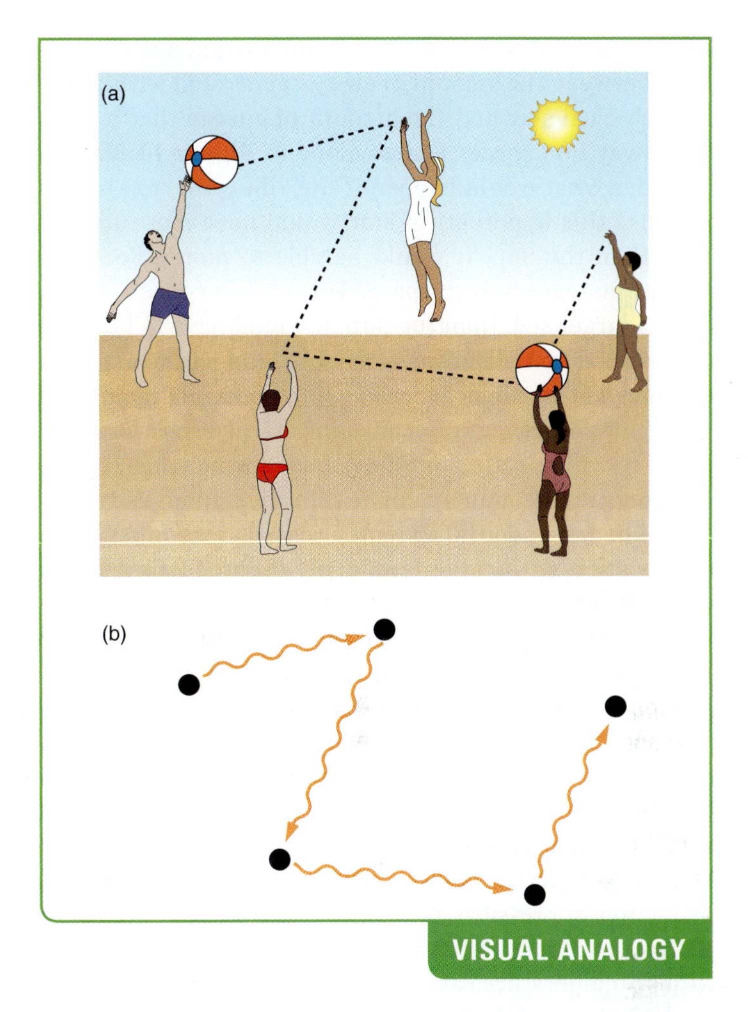

FIGURE 14.7 (a) When a crowd of people plays with a beach ball, the ball never travels very far before someone hits it, turning it in another direction. The ball moves randomly, sometimes toward the front of the crowd, sometimes toward the back. It often takes a ball a long time to make its way from one edge of the crowd to the other. (b) Similarly, when a photon travels through the Sun, it never travels very far before it interacts with an atom. The photon moves randomly, sometimes toward the center of the Sun, sometimes toward the outer edge. It takes a long time for a photon to make its way out of the Sun.

how larger than the Sun? What would happen to the balance between the amount of energy generated within this hypothetical star and the amount of energy that it radiates away into space? Follow along in **Figure 14.8** as we consider what would happen if the Sun were "too large." Because this hypothetical star would have more surface area than the Sun, it would be able to more effectively radiate its energy into space. For a 1-solar-mass star to have a larger size than the Sun, it would have to be more luminous than the Sun.

Now let's consider what is going on in the interior of this hypothetical star. Because the star is larger than the Sun but contains the same amount of mass as the Sun, the force of gravity at any point within the hypothetical star would be less than the force of gravity at the corresponding location within the Sun. (This difference is a result of the inverse square law of gravitation: if the radius R is larger in the hypothetical star, then $1/R^2$ must be smaller.) With weaker gravity, the weight of matter pushing down on the interior of the hypothetical star would be less than in the Sun. Because hydrostatic equilibrium means that the pressure at any point within a star is equal to the weight of overlying matter, the pressure at any point in the interior of this hypothetical star would be less than the pressure at the corresponding point in the Sun. This reduction in pressure would affect the amount of energy the star produced. The proton-proton chain runs faster at higher temperatures and densities, so the lower pressure in the interior of the hypothetical star means that less energy would be generated there than in the core of the Sun.

This hypothetical star would have to be more luminous than the Sun, but at the same time it would be producing less energy in its interior than the Sun does. This discrepancy violates the balance that must exist in any stable star between the amount of energy generated within the star and the amount of energy radiated into space. The hypothetical star cannot exist! Stated another way, even if the Sun were pumped up to a size larger than it actually is, it would not remain that way. Less energy would be generated in its core, while more energy would be radiated away at its surface. The Sun would be out of balance. As a result, the Sun would lose energy, the pressure in the interior of the Sun would decline, and the Sun would shrink back toward its original (true) size.

The same thought experiment could be done the other way around, asking what would happen if the Sun were smaller than it actually is. With less surface area, it would

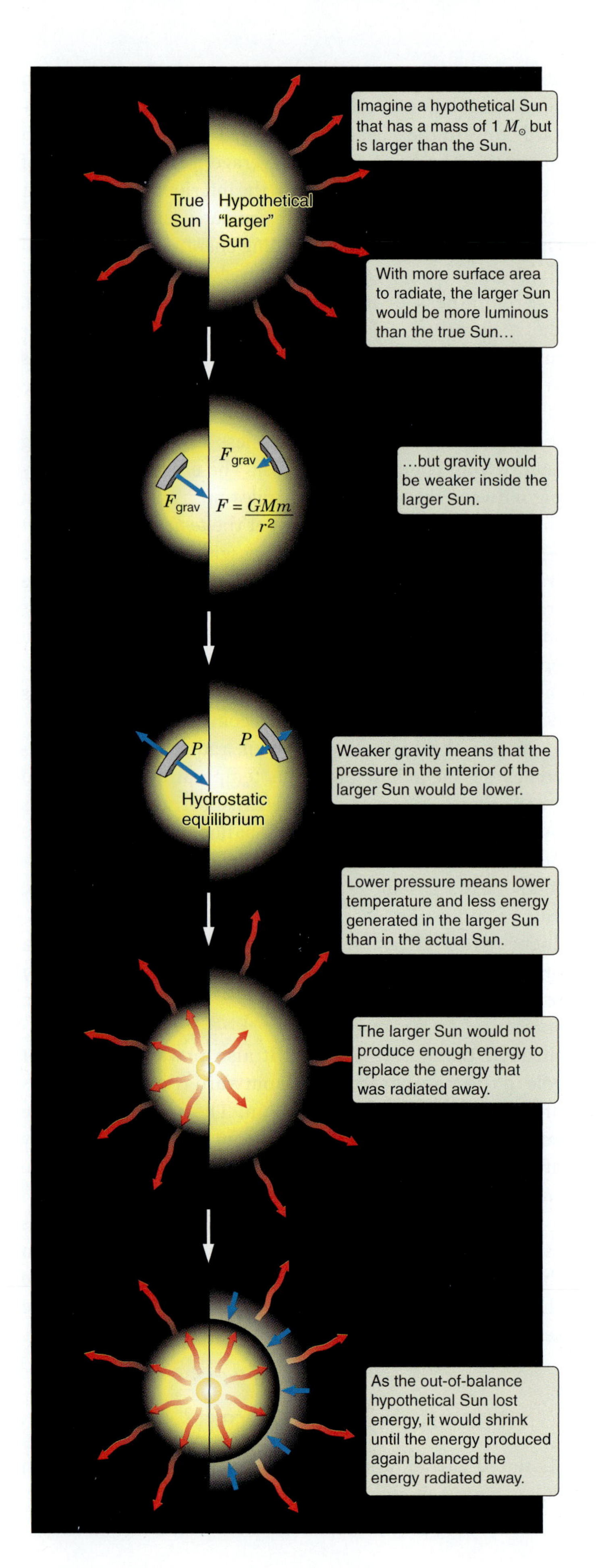

FIGURE 14.8 A star like the Sun can have only the structure that the Sun has. Here we imagine the fate of a Sun with too large a radius.

radiate less energy. At the same time the Sun's mass would be compacted into a smaller volume, driving up the strength of gravity and therefore the pressure in its interior. Higher pressure implies higher density and temperature, which in turn would cause the proton-proton chain to run faster, increasing the rate of energy generation. Again there would be a contradiction—an imbalance. This time, with more energy being generated in the interior than was being radiated away from the surface, pressure in the Sun would build up, causing it to expand toward its original (true) size.

14.3 The Interior of the Sun

The standard model of the Sun correctly matches observed global properties of the Sun such as its size, temperature, and luminosity. This is a remarkable feat, but the model predicts much more than these properties. In particular, the standard model of the Sun predicts exactly which nuclear reactions should be occurring in the core of the Sun, and at what rate. The nuclear reactions that make up the proton-proton chain produce a vast number of neutrinos. Since neutrinos barely interact with other ordinary matter, almost all of the neutrinos produced in the heart of the Sun travel freely through the outer parts of the Sun and on into space as if the outer layers of the Sun were not there. The core of the Sun lies buried beneath 700,000 km of dense, hot matter, yet the Sun is *transparent* to neutrinos.

Neutrinos escape freely from the core of the Sun.

It takes thermal energy produced in the heart of the Sun 100,000 years to find its way to the Sun's surface, but the solar neutrinos streaming through you as you read these words were produced by nuclear reactions in the heart of the Sun only 8⅓ minutes ago. (This is how far away the Sun is in light-minutes. Neutrinos travel very nearly at the speed of light.) In principle, neutrinos offer a direct window into the very heart of the Sun's nuclear furnace.

Astronomers Use Neutrinos to Observe the Heart of the Sun

Transforming the promise of neutrino astronomy into reality is a formidable technical challenge. The same property of neutrinos that makes them so exciting to astronomers—the fact that their interaction with matter is so feeble that they can escape unscathed from the interior of the Sun—also makes them notoriously difficult to observe. Suppose astronomers wanted to build a neutrino detector capable of stopping half of the neutrinos falling on it. This hypothetical detector would need the stopping power of a piece of lead a light-year thick. Yet despite the difficulties, neutrinos offer a unique window into the Sun so powerful that they are worth going to great lengths to try to detect.

Fortunately, the Sun produces an enormous number of neutrinos. As you lie in bed at night, about 400 trillion solar neutrinos pass through your body each second, having already passed through Earth. Because there are so many neutrinos about, a neutrino detector does not have to be very efficient to be useful. Several methods have been devised to measure neutrinos from the Sun and from other astronomical sources, and a number of such experiments are under way. These experiments have successfully detected neutrinos from the Sun, and in so doing they have provided crucial confirmation that nuclear fusion reactions are responsible for powering the Sun.

As with many good experiments, however, measurements of solar neutrinos raised new questions while answering others. After their initial joy at confirming that the Sun really is a nuclear furnace, astronomers became troubled that there seemed to be only about a third to a half as many solar neutrinos as predicted by solar models. The difference between the predicted and measured flux of solar neutrinos was referred to as the **solar neutrino problem** (**Process of Science Figure**).

One possible explanation of the solar neutrino problem was that the working model of the structure of the Sun was somehow wrong. This possibility seemed unlikely, however, because of the many other successes of the solar model. A second possibility was that an understanding of the neutrino itself was incomplete. The neutrino was long thought to have zero mass (like photons) and to travel at the speed of light. But if neutrinos actually do have a tiny amount of mass, then theories from particle physics predict that solar neutrinos should *oscillate* (alternate back and forth) among three different kinds, or "flavors"—the *electron*, *muon*, and *tau* neutrinos (**Figure 14.9**). Only one of these types, the electron neutrino, could interact with the atoms in the earlier neutrino detectors (described in **Connections 14.2** on page 440), so neutrino oscillations provided a convenient explanation for why only about a third of the expected number of neutrinos were detected. And, as seen in **Figure 14.9b**, electron neutrinos should also change flavor as they interact with solar material during their escape from the Sun.

After several decades of work on the solar neutrino problem, this last idea won out. Work currently under way at high-energy physics labs, nuclear reactors, and neutrino telescopes around the world is showing that neutrinos have a nonzero mass. This work has also uncovered evidence of neutrino oscillations.

Solving the solar neutrino problem is a good example of how science works—how a better model of the neutrino

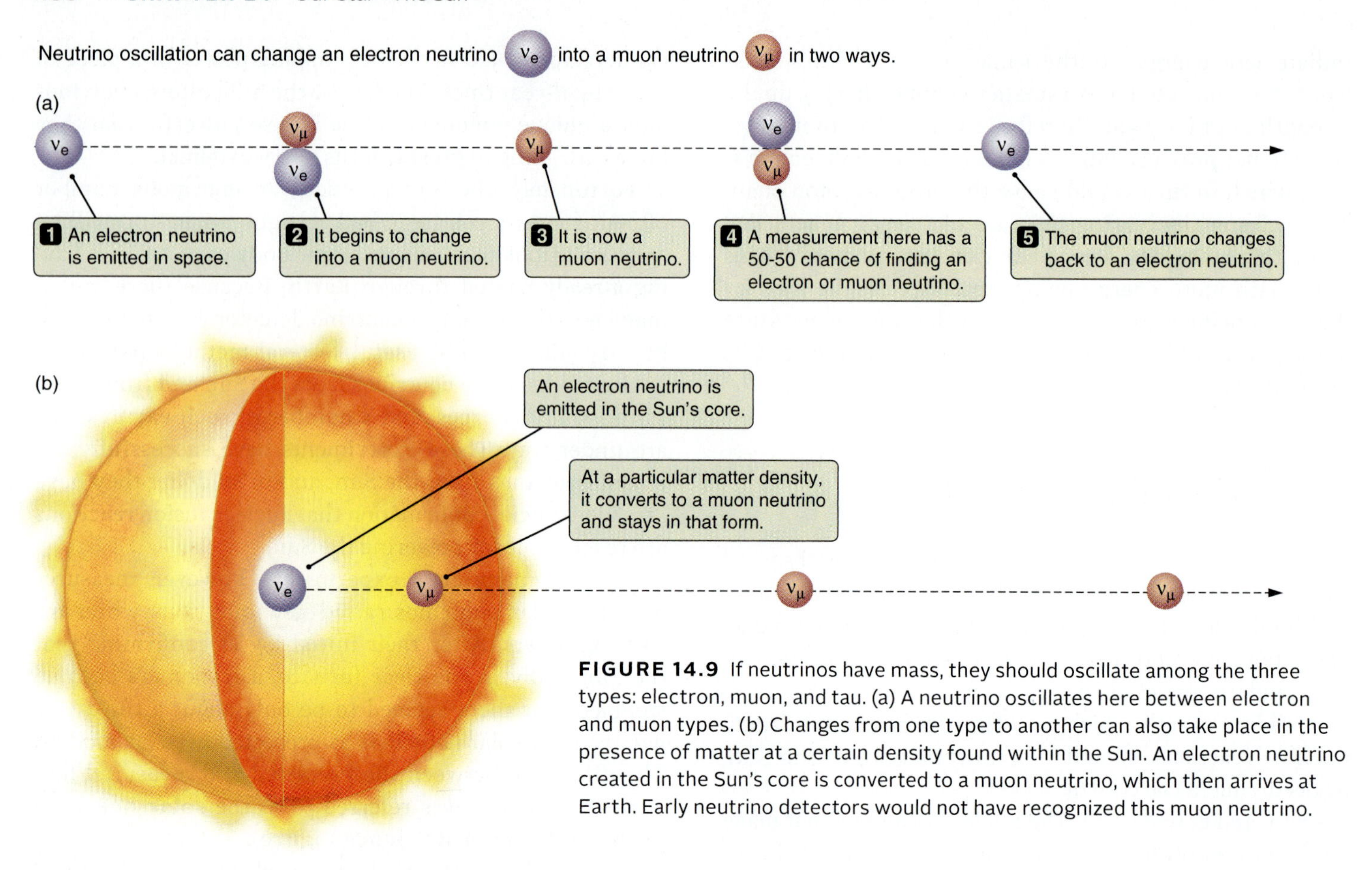

FIGURE 14.9 If neutrinos have mass, they should oscillate among the three types: electron, muon, and tau. (a) A neutrino oscillates here between electron and muon types. (b) Changes from one type to another can also take place in the presence of matter at a certain density found within the Sun. An electron neutrino created in the Sun's core is converted to a muon neutrino, which then arrives at Earth. Early neutrino detectors would not have recognized this muon neutrino.

showed that the solar neutrino problem was real and not merely an experimental mistake, and how a single set of anomalous observations was later confirmed by other, more sophisticated experiments. All of this effort has led to a better understanding of basic physics.

Helioseismology Can Be Used to Probe the Sun's Interior

Models of Earth's interior predict how density and temperature change from place to place within the planet. These density and temperature differences affect the way pressure waves travel through Earth, bending the paths of these waves. Geologists test models of Earth's interior by comparing measurements of seismic waves from earthquakes with model predictions of how seismic waves should travel through the planet.

The same basic idea has now been applied to the Sun. Detailed observations of motions of material from place to place across the surface of the Sun show that the Sun vibrates or "rings," something like a struck bell. Compared to a well-tuned bell, however, the vibrations of the Sun are very complex, with many different frequencies of vibrations occurring simultaneously. These motions are echoes of what lies below. Just as geologists use seismic waves from earthquakes to probe the interior of Earth, solar physicists use the surface oscillations of the Sun to test their understanding of the solar interior. This science is called **helioseismology** (**Figure 14.11**).

To detect the disturbances of helioseismic waves on the surface of the Sun, astronomers must measure Doppler shifts of less than 0.1 meter per second (m/s) while detecting changes in brightness of only a few parts per million at any given location on the Sun. Tens of millions of different wave motions are possible within the Sun. Some waves travel around the circumference of the Sun, providing information about the density of the upper convection zone. Other waves travel through the interior of the Sun, revealing the density structure of the Sun close to its core. Still others travel inward toward the center of the Sun, until they are bent by the changing solar density and return to the surface. All of these wave motions are going on at the same time. The Global Oscillation Network Group (GONG), a network of six solar observation stations spread around the world, enables astronomers to observe the surface of the Sun approximately 90 percent of the time.

To interpret helioseismology data, scientists compare the strength, frequency, and wavelengths of the data against predicted vibrations calculated from models of the

Science | The first detections of neutrinos raised more questions than they answered.

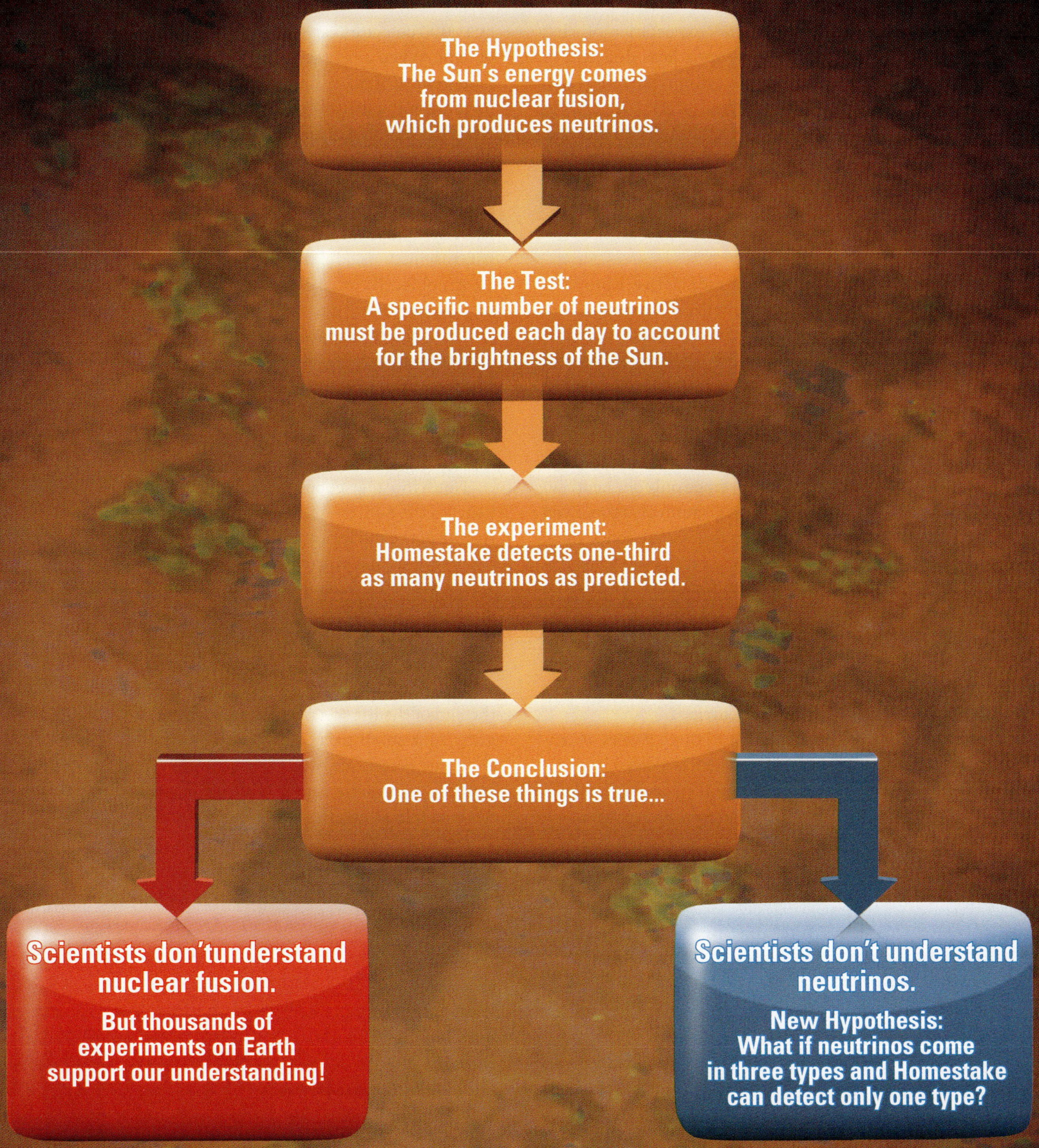

The Hypothesis:
The Sun's energy comes from nuclear fusion, which produces neutrinos.

The Test:
A specific number of neutrinos must be produced each day to account for the brightness of the Sun.

The experiment:
Homestake detects one-third as many neutrinos as predicted.

The Conclusion:
One of these things is true...

Scientists don'tunderstand nuclear fusion.
But thousands of experiments on Earth support our understanding!

Scientists don't understand neutrinos.
New Hypothesis:
What if neutrinos come in three types and Homestake can detect only one type?

Part of the "scientific attitude" is to find failure exciting. When experiments do not turn out as expected, good scientists get excited– there is something new to understand!

Connections 14.2

Neutrino Astronomy

A neutrino telescope hardly fits anyone's expectation of what a telescope should look like. The first apparatus designed to detect solar neutrinos consisted of a cylindrical tank filled with 100,000 gallons of dry-cleaning fluid—C_2Cl_4, or perchloroethylene—buried 1,500 meters deep within the Homestake Gold Mine in Lead, South Dakota. A tiny fraction of neutrinos passing through this fluid interact with chlorine atoms, causing this reaction:

$$^{37}\text{Cl} + \nu \rightarrow {}^{37}\text{Ar} + e^-$$

The ^{37}Ar formed in the reaction is a radioactive isotope of argon. The tank must be buried deep within Earth to shield the detector from the many other types of radiation capable of producing argon atoms. The argon is flushed out of the tank every few weeks and measured.

The Homestake detector (**Figure 14.10a**) operated from the late 1960s to the early 1990s. Over the course of 2 days, roughly 10^{22} (10 billion trillion) solar neutrinos passed through the Homestake detector. Of these, on average only *one* neutrino interacted with a chlorine atom to form an atom of argon. Even so, this interaction produced a measurable signal. Since then, many other neutrino detectors have been built, each using different reactions to detect neutrinos of different energies. In the 1990s, the Soviet-American Gallium Experiment (SAGE) and the European Gallium Experiment (GALLEX) and its successor the Gallium Neutrino Observatory used reactions involving the conversion of gallium atoms into germanium atoms ($^{71}\text{Ga} + \nu \rightarrow {}^{71}\text{Ge} + e^-$) to detect solar neutrinos.

Other detectors capable of detecting all three flavors of neutrinos include the Super-Kamiokande, which is located in an active zinc mine 2,700 meters under Mount Ikena, near Kamioka, Japan. It has a 50,000-ton tank of ultrapure water, surrounded by 13,000 detectors capable of registering extremely faint flashes of light. When a neutrino interacts with an atom in the tank, a faint conical flash of blue light is produced. This flash is seen by some of the detectors. The SNO+ experiment at the Sudbury Neutrino Observatory utilizes a chemical used in biodegradable detergents contained in a 12-meter sphere surrounded by light detectors (**Figure 14.10b**) and buried deep in a nickel mine near Sudbury, Ontario. The Double Chooz experiment uses neutrinos from the Chooz nuclear power plant in France to measure neutrino oscillations.

An objective of still-newer neutrino telescopes is to collect higher-energy neutrinos that originate from the most distant objects in space. **Figure 14.10c** shows the ANTARES experiment, which detects neutrinos passing through the Mediterranean Sea. The IceCube neutrino detector at the South Pole has optical sensors buried far beneath the surface, at depths of up to 2.5 km within the Antarctic ice (see Figure 6.33).

Neutrino telescopes observe neutrinos produced in the heart of the Sun, enabling astronomers to directly observe the results of the nuclear reactions going on there. Although these observations have provided crucial confirmation that stars are powered by nuclear reactions, they have also challenged models of the solar interior and led to changes in ideas about the nature of the neutrino itself. In addition to solar neutrinos, a number of experiments detected neutrinos from Supernova 1987A. As we will discuss in Chapter 17, this explosion marked the end of the life of a massive star located 160,000 light-years away in a small galaxy called the Large Magellanic Cloud. Neutrino astronomy was one of the great innovations of 20th century astronomy, and it will significantly benefit from the new neutrino detectors under construction.

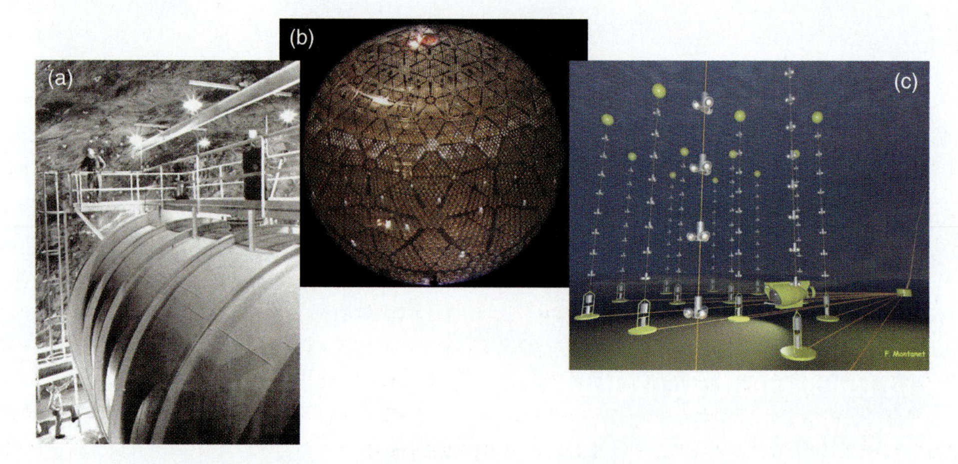

FIGURE 14.10 Neutrino "telescopes" do not look much like visible-light telescopes. (a) The first neutrino telescope, the Homestake neutrino detector, was a 100,000-gallon tank of dry-cleaning fluid located deep in a mine in South Dakota. (b) The SNO+ Sudbury Neutrino Observatory, buried 2 km deep in a Canadian nickel mine. (c) Artist's conception of the ANTARES neutrino observatory in the Mediterranean Sea.

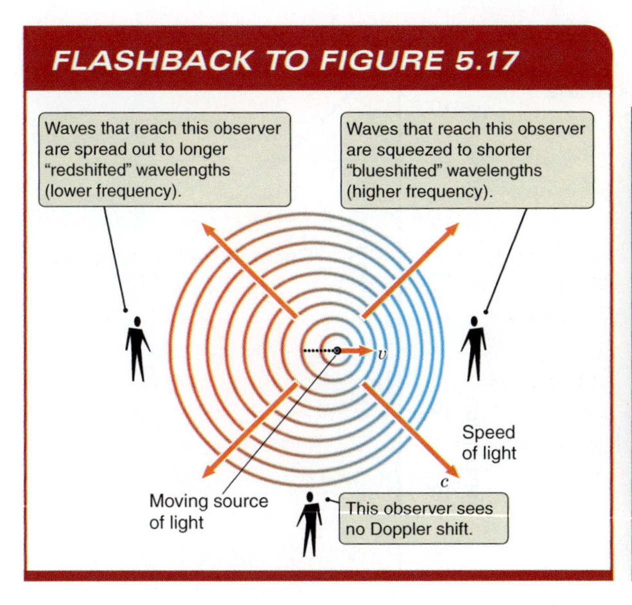

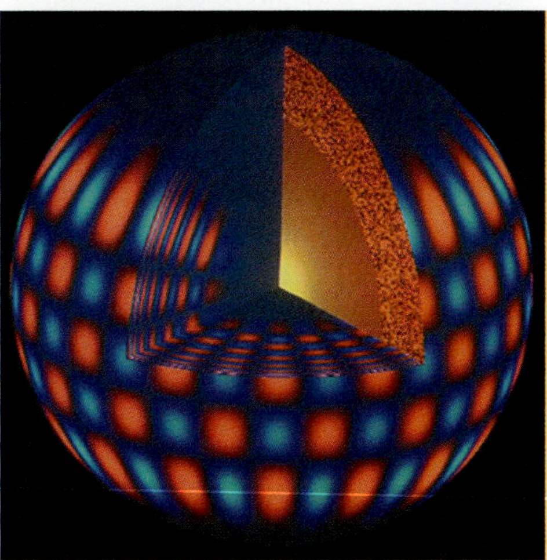

FIGURE 14.11 The interior of the Sun rings like a bell as helioseismic waves move through it. Waves with the right wavelength amplify and sustain the vibrations, while those with the wrong wavelength are damped out and disappear. In the particular "mode" of the Sun's vibration shown here, red indicates regions where gas is traveling inward; blue, where gas is traveling outward. Astronomers observe these motions by using Doppler shifts.

solar interior. This technique provides a powerful test of an understanding of the solar interior, and it has led both to some surprises and to improvements in the models. For example, some scientists had proposed that the solar neutrino problem might be solved if the models were found to have too much helium in the Sun—an explanation that was ruled out by analysis of the waves that penetrate to the core of the Sun. Helioseismology showed that the value for opacity used in early solar models was too low. This realization led astronomers to recalculate the location of the bottom of the convective zone. Both theory and observation now put the base of the convective zone at 71.3 percent of the way out from the center of the Sun, with an uncertainty in this number of less than half a percent.

Helioseismology confirms the predictions of solar models.

14.4 The Atmosphere of the Sun

The Sun is a large ball of gas, and so, unlike Earth, it has no solid surface. Instead, it has the kind of surface that a fog bank on Earth does; it is a gradual thing—an illusion really. Imagine watching some people walking into a fog bank. When they disappeared from view, you would say they were definitely inside the fog bank, even though they never passed through a definite boundary. The apparent surface of the Sun is defined by the same effect. Light from the Sun's surface can escape into space, so you can see it. Light from below the Sun's surface cannot escape directly into space, so you cannot see it.

An overview of the Sun's atmosphere shows that it is made of several layers that lie above the top of the convective zone (**Figure 14.12**). The Sun's atmosphere is where all visible solar phenomena take place. At the base of the atmosphere is the **photosphere**: the Sun's apparent surface. This is where features such as sunspots can be seen. Above this photosphere is the **chromosphere**, a region of strong emission lines. The top layer is the **corona**, which can be viewed during a solar eclipse as a halo around the Sun. Solar prominences, caused by the Sun's magnetic field, poke out into the corona. In the Sun's atmosphere the density of the gas drops very rapidly with increasing altitude. Figure 14.12 shows how pressure and temperature change across the atmosphere of the Sun. In this section we will explore each of these layers, beginning at the bottom, with the photosphere.

The Sun's apparent surface—the photosphere—has an **effective temperature** (the temperature of a blackbody that would emit at the same peak wavelength as the object) of 5780 K, ranging from 6600 K at the bottom to 4400 K at the top. It is a zone about 500 km thick, across which the density and opacity of the Sun increase sharply. The reason the Sun appears to have a well-defined surface and a sharp outline (but note that you should *never* look at the Sun directly) is that this zone is relatively shallow; 500 km does not look very thick when viewed from a distance of 150 million km.

The apparent surface of the Sun is called the photosphere.

Look at the photograph of the Sun in **Figure 14.13a** and notice that the Sun appears to be fainter near its edges than near its center. This effect, called **limb darkening**, is

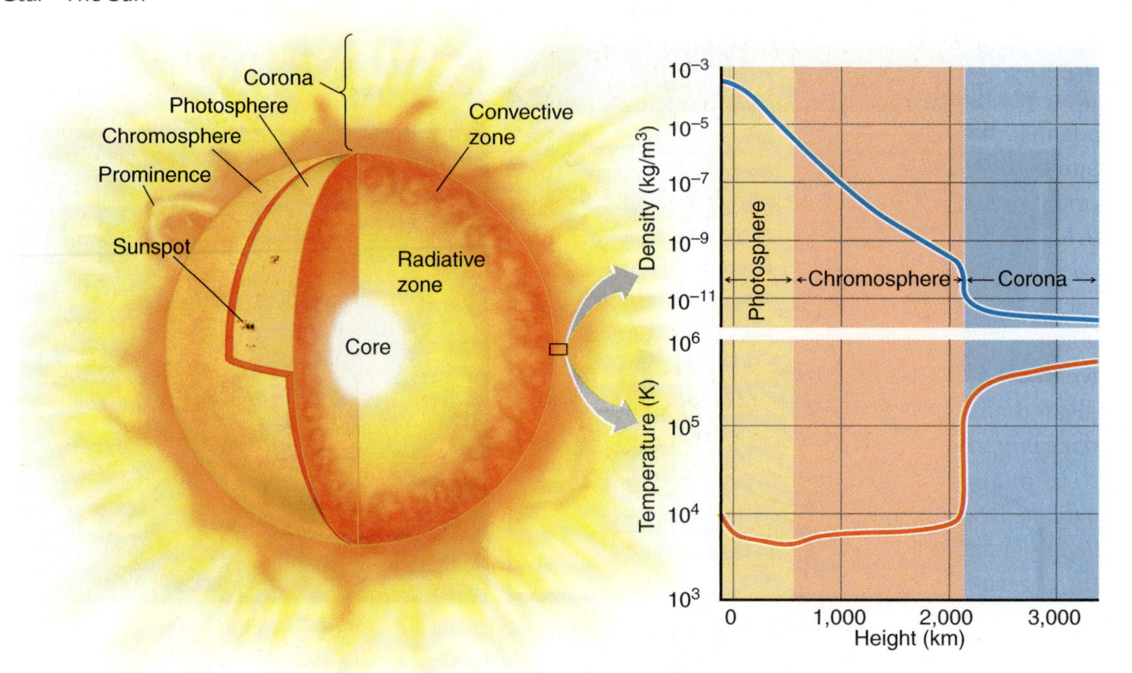

FIGURE 14.12 The components of the Sun's atmosphere, along with plots of how the temperature and density change with height at the base of the atmosphere. Note that the *y*-axes are logarithmic.

due to the structure of the Sun's photosphere. (The *limb* of a celestial body is the outer border of its visible disk.) **Figure 14.13b** illustrates the cause of limb darkening. Near the edge of the Sun you are looking through the photosphere at a steep angle. As a result, you do not see as deeply into the interior of the Sun as when you are looking directly down through the photosphere near the center of the Sun's disk. The light you see coming from near the limb of the Sun comes from a layer in the Sun that is shallower and hence cooler and dimmer.

Detailed observations from the ground and from space help astronomers understand the complex nature of the solar atmosphere. The Solar and Heliospheric Observatory (*SOHO*) spacecraft is a joint mission between NASA and the European Space Agency (ESA). By orbiting at the L_1 Lagrangian point of the Sun-Earth system (see Chapter 4), *SOHO* moves in lockstep with Earth at a location approximately 1,500,000 km from Earth that is almost directly in line between Earth and the Sun. *SOHO* carries a complement of 12 scientific instruments that monitor the Sun. In early 2010, NASA launched the Solar Dynamics Observatory (SDO) with the goal of improving scientists' understanding of the Sun to the point where they can *predict* when events will occur, rather than simply responding after they happen.

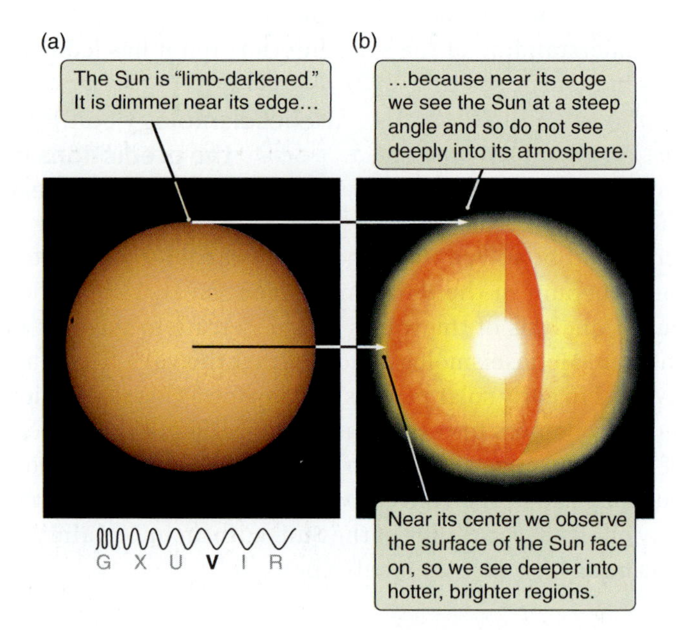

FIGURE 14.13 (a) When viewed in visible light, the Sun appears to have a sharp outline, even though it has no true surface. The center of the Sun appears brighter, while the limb of the Sun is darker—an effect known as limb darkening. (b) Looking at the middle of the Sun makes it possible to see deeper into the Sun's interior, where it is hotter, than looking at the edge of the Sun does. Because higher temperature means more luminous radiation, the middle of the Sun appears brighter than its limb.

The Solar Spectrum Is Complex

In one sense the surface of the Sun may be an illusion, but in another sense it is not. The transition between "inside" the Sun and "outside" the Sun is quite abrupt. Very nearly all the radiation from below the Sun's photosphere is absorbed by matter and cannot escape; it is trapped. This is exactly the definition from Chapter 5 for the conditions under which blackbody radiation forms. The radiation able to leak out of the Sun's interior has a spectrum very close to being a Planck (blackbody) spectrum. This is why, as you learned in Chapter 13, astronomers are able to understand much about the physical properties of stars by applying what they know of blackbody radiation.

This simple description of the spectra of stars is incomplete. Light from the solar photosphere must escape through the upper layers of the Sun's atmosphere, which affects the observed spectrum. In Chapter 13 we discussed the presence of absorption lines in the spectra of stars. Now we can take a closer look at how these absorption lines form. As photospheric light travels upward through the solar atmosphere, atoms in the solar atmosphere absorb the light at discrete wavelengths. Because the Sun appears so much brighter than any other star, its spectrum can be studied in far more detail, so specially designed telescopes and high-resolution spectrometers have been built specifically to study the Sun's light. The solar spectrum is shown in **Figure 14.14**. Absorption lines from over 70 elements have been identified. Analysis of these lines forms the basis for much of astronomers' knowledge of the solar atmosphere, including the composition of the Sun, and it is the starting point for an understanding of the atmospheres and spectra of other stars.

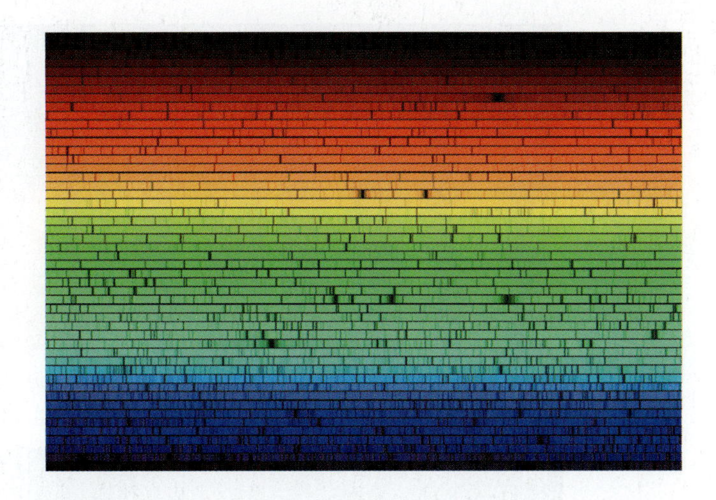

FIGURE 14.14 A high-resolution spectrum of the Sun, stretching from 400 nm (lower left corner) to 700 nm (upper right corner), showing a wealth of absorption lines. (The spectrum has been cut into strips and layered, like lines of text on a page, so that the entire visible part of the spectrum can be shown.)

The Sun's Outer Atmosphere: Chromosphere and Corona

Moving upward through the Sun's photosphere, the temperature continues to fall, reaching a minimum of about 4400 K at the top of the photosphere. At this point the trend reverses and the temperature slowly begins to climb, rising to about 6000 K at a height of 1500 km above the top of the photosphere. The region above the photosphere—the chromosphere (**Figure 14.15a**)—was discovered in the 19th century during observations of total solar eclipses (**Figure 14.15b**). The chromosphere is seen most strongly as a source of emission lines, especially the Hα line from hydrogen (the *hydrogen alpha line*). In fact, the deep red color of the Hα line is what gives the *chromosphere* ("the place where color comes from") its name. A spectrum of the Sun's chromosphere is also what led in 1868 to the discovery of the element helium. Helium is named after *helios*, the Greek word for "Sun." The reason for the chromosphere's temperature reversal with increasing height is not well understood, but it may be caused by magnetic waves propagating through the region.

The Sun's chromosphere lies above the photosphere.

At the top of the chromosphere, across a transition region that is only about 100 km thick, the temperature suddenly soars (see Figure 14.13). Above this transition lies the outermost region of the Sun's atmosphere, the corona, in which temperatures reach 1–2 million K. The corona is probably heated by magnetic waves and magnetic fields in much the same way the chromosphere is, but why the temperature changes so abruptly at the transition between the chromosphere and the corona is not at all clear. Since ancient times the Sun's corona has been known; it is visible during total solar eclipses as an eerie outer glow stretching a distance of several solar radii beyond the Sun's surface (**Figure 14.15c**). Because it is so hot, the solar corona is a strong source of X-rays. Atoms in the corona are also highly ionized.

The corona has a temperature of millions of kelvins.

Solar Activity Is Caused by Magnetic Effects

The Sun's magnetic field causes virtually all of the structure seen in the Sun's atmosphere. High-resolution images of the Sun show "coronal loops" that make the Sun look as though it were covered with matted, tangled hair (**Figure 14.16**). This fibrous or ropelike texture in the chromosphere is the result of magnetic structures called flux tubes. Magnetic fields are responsible for much of the structure of the corona as well. The corona is far too hot to be held in by the Sun's gravity. Instead, over most of the surface of the Sun coronal gas is confined by magnetic loops with both ends firmly anchored deep within the Sun. The magnetic field in the corona acts almost like a network of rubber bands

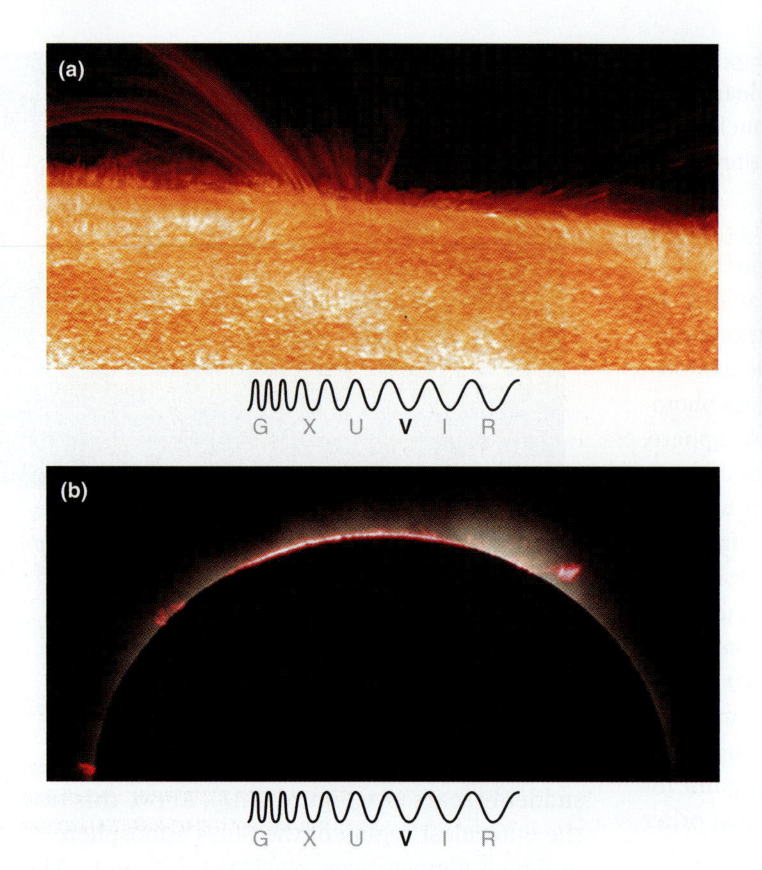

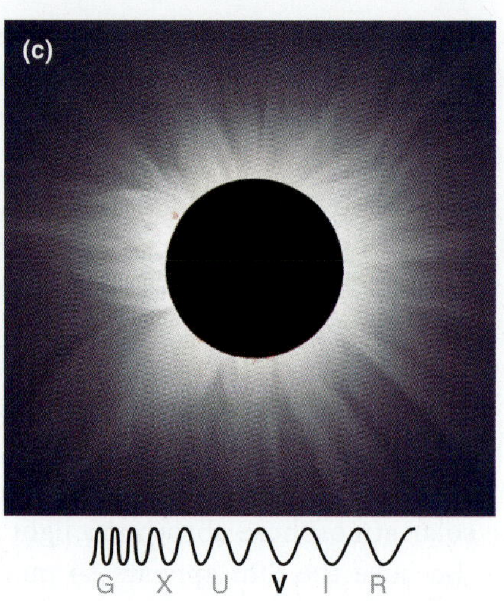

FIGURE 14.15 (a) Spacecraft image of the Sun showing fine structure in the chromosphere extending outward from the photosphere. (b) The chromosphere seen during a total eclipse. (c) This eclipse image shows the Sun's corona, consisting of million-kelvin gas that extends for millions of kilometers beyond the surface of the Sun.

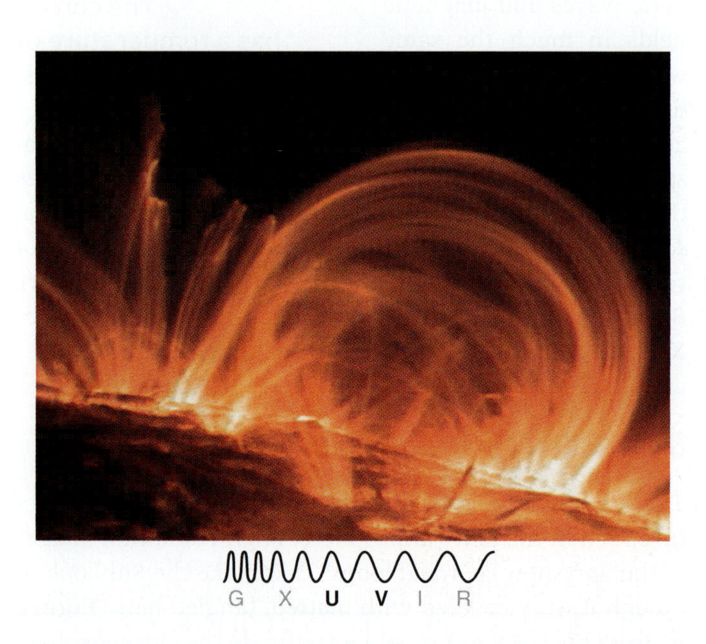

FIGURE 14.16 A close-up image of the Sun, showing the tangled structure of coronal loops.

that coronal gas is free to slide along but cannot cross. In contrast, about 20 percent of the surface of the Sun is covered by an ever-shifting pattern of **coronal holes**. These are visible in extreme ultraviolet images of the Sun as dark regions (as seen in the chapter-opener photo), indicating that they are cooler and lower in density than their surroundings. Coronal holes are large regions where the magnetic field points outward, away from the Sun, and where coronal material is free to stream away into interplanetary space as the **solar wind**.

A solar "wind" blows away from the Sun.

The relatively steady part of the solar wind consists of lower-speed flows, with velocities of about 350 km/s, and higher-speed flows, with velocities up to about 700 km/s. The higher-speed flows originate in coronal holes. Depending on their speed, particles in the solar wind take about 2–5 days to reach Earth. Frequently, 2–5 days after a coronal hole passes across the center of the face of the Sun, the speed and density of the solar wind reaching Earth increase. The solar wind drags the Sun's magnetic field along with it. The magnetic field in the solar wind gets "wound up" by the Sun's rotation (**Figure 14.17**). Consequently, the solar wind has a spiral structure resembling the stream of water from a rotating lawn sprinkler.

The effects of the solar wind are felt throughout the Solar System. The solar wind causes the tails of comets, shapes the magnetospheres of the planets, and provides the energetic particles that power Earth's spectacular auroral displays. Using space probes, astronomers have been able to observe the solar wind extending out to 100 AU from the Sun. But the solar wind does not go on forever. The farther it gets from the Sun, the more it has to spread out. Just like radiation, the density of the solar wind follows an inverse square law. At a distance of about 100 AU from the Sun, the solar wind is assumed no longer to be powerful enough to push the **interstellar medium** (the gas and dust that lie between stars in a galaxy and that surround the Sun) out of the way. There the solar wind stops abruptly, piling up against the pressure of the interstellar medium. **Figure 14.18** shows the region of space over which the wind from the Sun holds sway. The *Voyager 1* spacecraft is near the very outer edge of this boundary and beginning to send back the first direct measurements of true interstellar space. The *IBEX* (Interstellar Boundary Explorer) spacecraft, launched in 2008, is also exploring this region.

The best-known features on the surface of the Sun are relatively dark blemishes in the solar photosphere, called

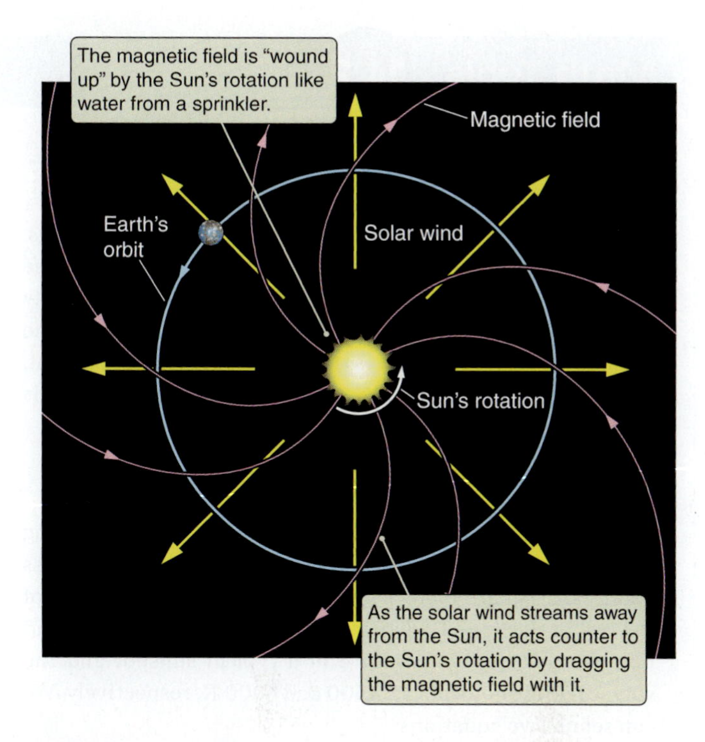

FIGURE 14.17 The solar wind streams away from active areas and coronal holes on the Sun. As the Sun rotates, the solar wind takes on a spiral structure, much like the spiral of water that streams away from a rotating lawn sprinkler.

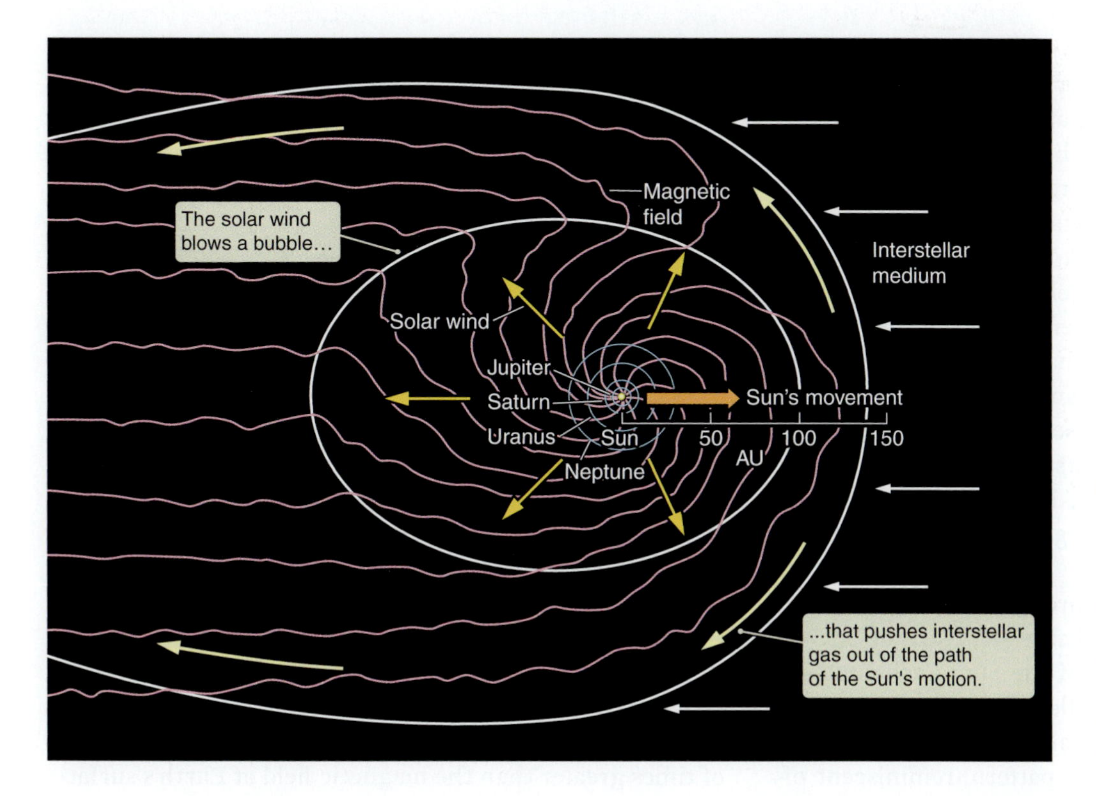

FIGURE 14.18 The solar wind streams away from the Sun for about 100 AU, until it finally piles up against the pressure of the interstellar medium through which the Sun is traveling. The *Voyager 1* spacecraft is now passing through this boundary and is expected to cross over into the true interstellar medium within the coming decade.

Math Tools 14.2

Sunspots and Temperature

Sunspots are about 1500 K cooler than their surroundings. What does this lower temperature indicate? Think back to the Stefan-Boltzmann law in Chapter 5. The flux from a blackbody is proportional to the fourth power of the temperature, *T*. The constant of proportionality is the Stefan-Boltzmann constant, σ, which has a value of 5.67×10^{-8} W/(m²K⁴). We write this relationship as:

$$\mathcal{F} = \sigma T^4$$

Remember that the flux is the amount of energy coming from a square meter of surface every second. How much less energy per square meter comes out of a sunspot than out of the rest of the Sun? To determine the answer, let's take round numbers for the temperature of a typical sunspot and the surrounding photosphere: 4500 and 6000 K, respectively. We can set up two equations:

$$\mathcal{F}_{\text{spot}} = \sigma T^4_{\text{spot}} \quad \text{and} \quad \mathcal{F}_{\text{surface}} = \sigma T^4_{\text{surface}}$$

We could solve each of these separately, and then divide the value of $\mathcal{F}_{\text{spot}}$ by $\mathcal{F}_{\text{surface}}$, to find out how much fainter it is, but it's much easier to solve for the *ratio* of the fluxes (recall the ratio example in Math Tools 1.1):

$$\frac{\mathcal{F}_{\text{spot}}}{\mathcal{F}_{\text{surface}}} = \frac{\sigma T^4_{\text{spot}}}{\sigma T^4_{\text{surface}}} = \frac{T^4_{\text{spot}}}{T^4_{\text{surface}}} = \left(\frac{T_{\text{spot}}}{T_{\text{surface}}}\right)^4$$

Plugging in our values for T_{spot} and T_{surface} gives:

$$\frac{\mathcal{F}_{\text{spot}}}{\mathcal{F}_{\text{surface}}} = \left(\frac{4500\ \text{K}}{6000\ \text{K}}\right)^4 = 0.32$$

and multiplying both sides by $\mathcal{F}_{\text{surface}}$ gives:

$$\mathcal{F}_{\text{spot}} = 0.32\mathcal{F}_{\text{surface}}$$

So the amount of energy coming from a square meter of sunspot every second is about one-third as much as the amount of energy coming from a square meter of surrounding surface every second. In other words, the sunspot is about one-third as bright as the surrounding photosphere. This is still extremely bright; if you could cut out the sunspot and place it elsewhere in the sky, it would be brighter than the full Moon.

sunspots, that come and go over time. These spots are places where material is trapped at the surface of the Sun by magnetic-field lines. When this material cools, convection cannot carry it downward, so it makes a cooler (therefore darker) spot on the surface of the Sun. Sunspots appear dark, but only in contrast to the brighter surface of the Sun (see **Math Tools 14.2**). Telescopic observations of sunspots made during the 17th century led to the discovery of the Sun's rotation, which has an average period of about 27 days as seen from Earth and 25 days relative to the stars. Observations of sunspots also show that the Sun (like Saturn), rotates more rapidly at its equator than it does at higher latitudes. This effect, called **differential rotation**, is possible because the Sun is a large ball of gas rather than a solid object.

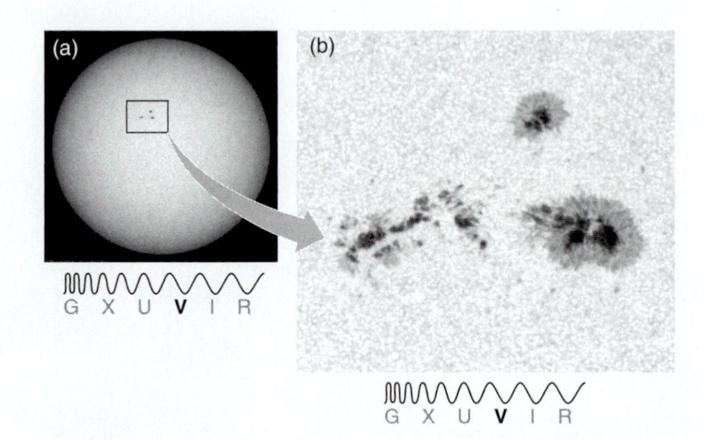

FIGURE 14.19 (a) This image from the Solar Dynamics Observatory (SDO), taken on October 26, 2010, shows a large sunspot group. Sunspots are magnetically active regions that are cooler than the surrounding surface of the Sun. (b) A high-resolution view of the sunspots in this group.

A large sunspot group is pictured in **Figure 14.19**, and **Figure 14.20** shows the remarkable structure of one of these blemishes on the surface of the Sun. Each sunspot consists of an inner dark core called the **umbra**, which is surrounded by a less dark region called the **penumbra**. The penumbra has an intricate radial pattern, reminiscent of the petals of a flower. Observations of sunspots show that they are magnetic in origin, with magnetic fields thousands of times greater than the magnetic field at Earth's surface. Sunspots occur in pairs that are connected by loops in the

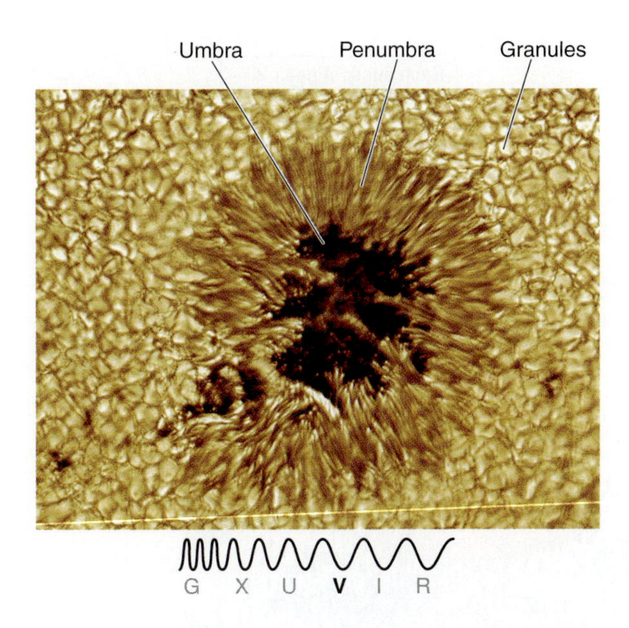

FIGURE 14.20 This very high-resolution view of a sunspot shows the dark umbra surrounded by the lighter penumbra. The solar surface around the sunspot bubbles with separate cells of hot gas called *granules*. The smallest features are about 100 km across.

magnetic field that resemble the shape of a horseshoe. Sunspots range in size from about 1,500 km across up to complex groups that may contain several dozen individual spots and span 50,000 km or more. The largest sunspot groups are so large that they can be seen with the naked eye and have been noted since antiquity. *But do not look directly at the Sun!* More than one casual observer of the Sun has paid the price of blindness for a glimpse of a naked-eye sunspot. Direct viewing through a commercial solar filter is safe, as is projecting the image through a telescope or binoculars onto a surface such as paper and looking only at the projection. Many websites have live images of the Sun viewed through ground and space telescopes; see the Web problems at chapter's end.

Magnetic fields cause sunspots.

Individual sunspots are ephemeral. Although sunspots occasionally last 100 days or longer, half of all sunspots come and go in about 2 days, and 90 percent are gone within 11 days. The number and distribution of sunspots change over time in a pronounced 11-year pattern called the **sunspot cycle** (**Figure 14.21a**). At the beginning of a cycle, sunspots begin to appear at solar latitudes of about 30° to the north and south of the solar equator. During the following years, the regions where most new sunspots are seen move toward the equator as the number of sunspots increases and then declines. As the last few sunspots near the equator are seen, new sunspots again begin appearing at middle latitudes, and the next cycle begins. A diagram showing the number of sunspots at a latitude plotted against time has the appearance of a series of opposing diagonal bands and is often referred to as the sunspot "butterfly diagram" (**Figure 14.21b**).

Sunspots come and go in an 11-year cycle.

In the early 20th century, solar astronomer George Ellery Hale (1868–1938) was the first to show that the 11-year sunspot cycle is actually half of a 22-year magnetic cycle during which the direction of the Sun's magnetic field reverses. In one 11-year sunspot cycle the leading sunspot in each pair tends to be a north magnetic pole, whereas the trailing sunspot tends to be a south magnetic pole. In the next 11-year sunspot cycle this polarity is reversed: the leading spot in each pair is a south magnetic pole. The transition between these two magnetic polarities occurs near the peak of each sunspot cycle (**Figure 14.21c**). The predominant theory of what causes this magnetic cycle involves a *dynamo* in the interior of the Sun, much like the dynamos that generate the magnetic fields of the planets.

The effects of magnetic activity on the Sun are felt throughout the Sun's photosphere, chromosphere, and corona. Sunspots are only one of a host of phenomena that follow the Sun's 22-year cycle of magnetic activity. The peaks of the cycle, called **solar maxima**, are times of intense activity.

Telescopic observations of sunspots date back almost 400 years, and there were naked-eye reports of sunspots even before that. **Figure 14.22** shows the historical record of sunspot activity. Although astronomers often speak of the 11-year sunspot cycle, the cycle is neither perfectly periodic nor especially reliable. The time between peaks in the number of sunspots actually varies between about 9.7 and 11.8 years. The number of spots seen during a given cycle fluctuates as well. Some cycles are real monsters. There have also been times when sunspot activity has disappeared almost entirely for extended periods. The most recent extended lull in solar activity, called the **Maunder Minimum**, lasted from 1645 to 1715. Normally there are about six peaks of solar activity in 70 years, virtually no sunspots were seen during the Maunder Minimum.

Although sunspots are darker-than-average features, they are often accompanied by a brightening of the solar chromosphere that is seen most clearly in the light of emission lines such as Hα. The magnificent loops arching through the solar corona are also anchored in solar active

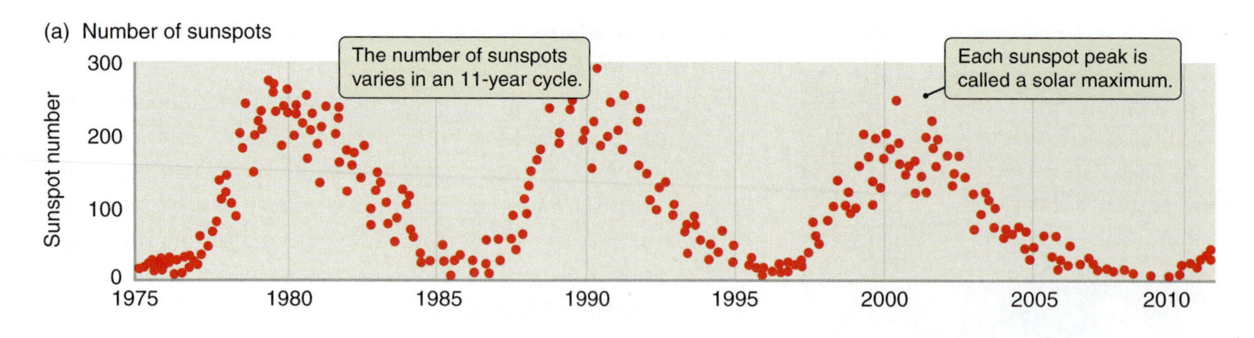

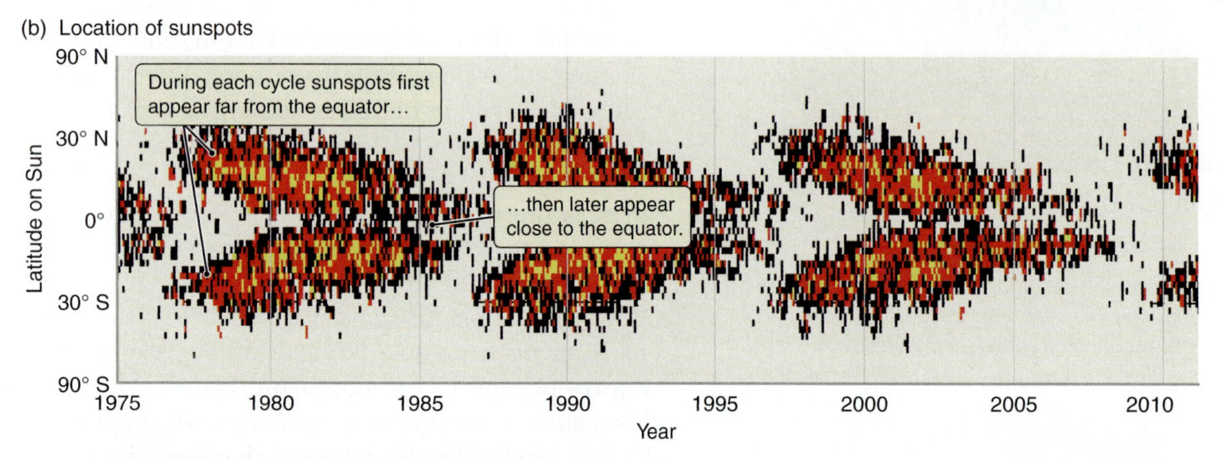

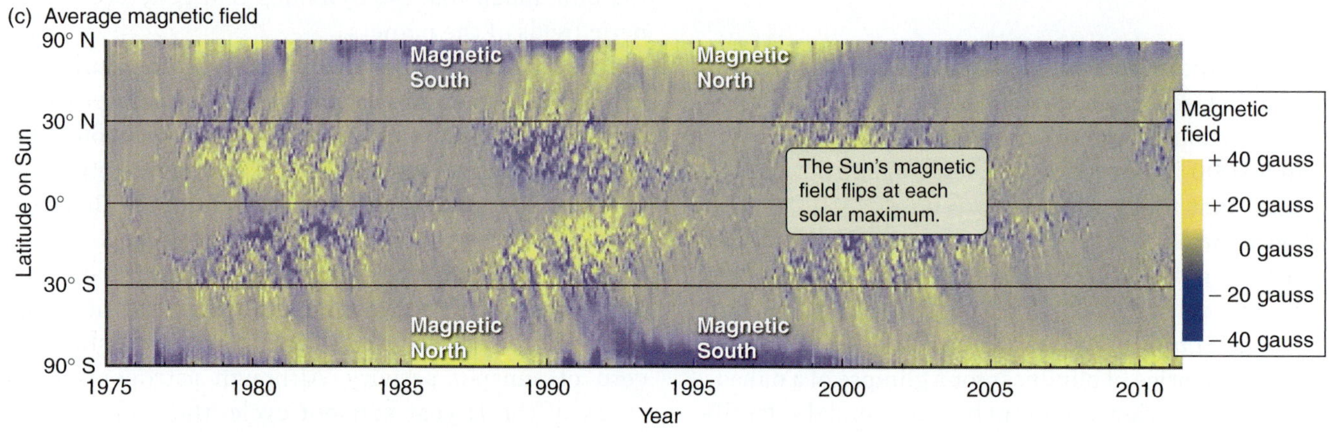

FIGURE 14.21 (a) The number of sunspots versus time for the last few solar cycles. (b) The solar butterfly diagram, showing the fraction of the Sun covered by sunspots at each latitude. (c) The Sun's magnetic poles flip every 11 years. Yellow indicates magnetic south polarity; blue indicates magnetic north.

regions. Solar **prominences** such as shown in **Figure 14.23** are magnetic flux tubes of relatively cool (5000–10,000 K) but dense gas extending through the million-kelvin gas of the corona. Although most prominences are relatively quiet, others can erupt out through the corona, towering a million kilometers or more over the surface of the Sun and ejecting material into the corona at velocities of 1,000 km/s.

The images in **Figure 14.24** show a **solar flare** erupting from a sunspot group. The two left-hand images (**Figures 14.24a and b**), taken in ultraviolet light, show material at very high temperatures. The spots in the visible-light image (**Figure 14.24c**) are at the base of the activity seen in Figures 14.24a and b. Solar flares are the most energetic form of solar activity. Flares are violent eruptions in which enormous amounts of energy are released over the course of a few minutes to a few hours. Solar flares can heat gas to temperatures of 20 million K, and they are the source of intense X-rays and gamma rays. Hot **plasma** (consisting of

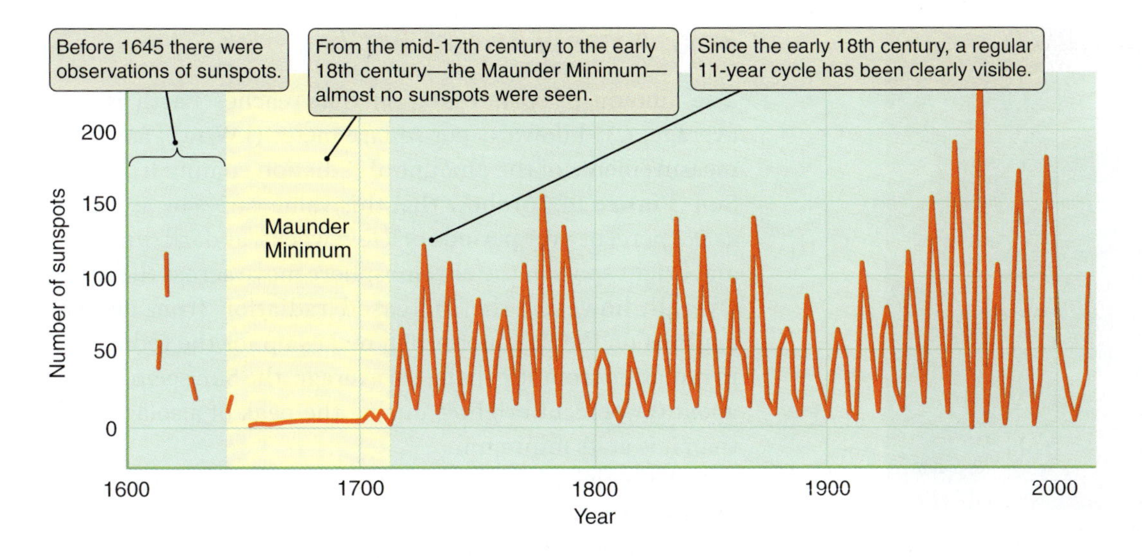

FIGURE 14.22 Sunspots have been observed for hundreds of years. In this plot the 11-year cycle in the number of sunspots (half of the 22-year solar magnetic cycle) is clearly visible. Sunspot activity varies greatly over time. The period from the middle of the 17th century to the early 18th century, when almost no sunspots were seen, is called the *Maunder Minimum*.

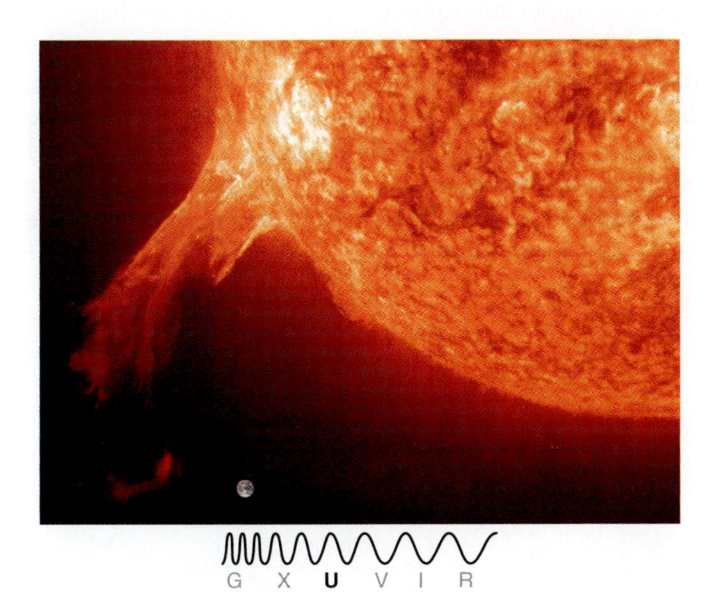

FIGURE 14.23 Solar prominences are magnetically supported arches of hot gas that rise high above active regions on the Sun. Here you can see a close-up view at the base of a large prominence. An image of Earth is included for scale (it is not actually that close to the Sun).

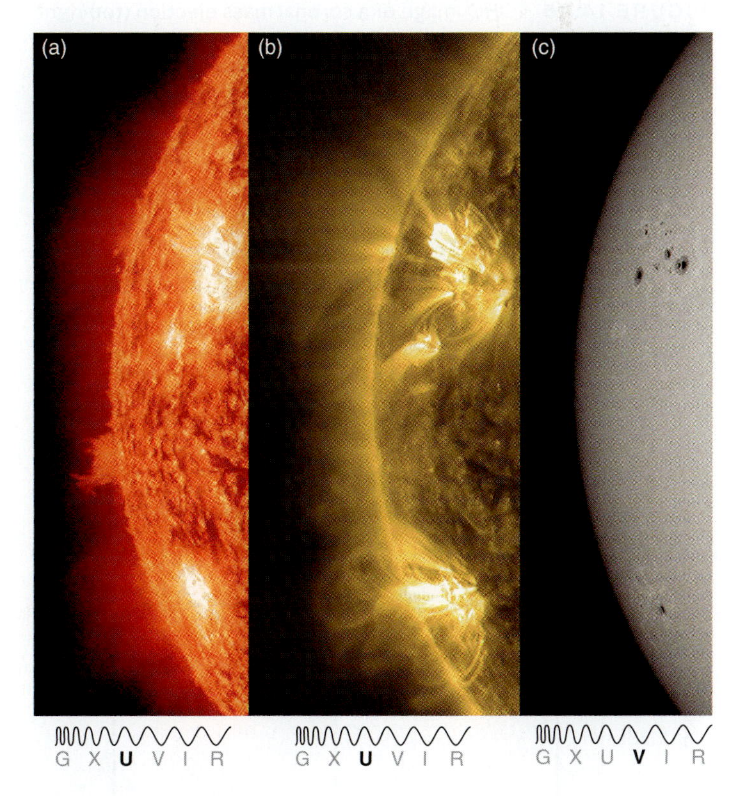

FIGURE 14.24 SDO observed these active regions of the Sun in August 2011. (a) Activity near the surface at 60,000 K in extreme ultraviolet light (along with a prominence rising up from the Sun's edge). (b) Many looping arcs, also in extreme UV light, and plasma heated to about 1 million K. (c) The magnetically intense sunspots that are the sources of all the activity.

atoms stripped of some of their electrons) moves outward from flares at speeds that can reach 1,500 km/s. Associated with the solar flares are **coronal mass ejections** (**Figure 14.25**), which are bursts of solar wind rising above the corona. Magnetic effects can then accelerate subatomic particles to high speeds and send powerful bursts of energetic particles outward through the Solar System. Coronal mass ejections can release up to 100 billion kg of material. They occur about once per week during the minimum of the solar activity cycle, but they can reach bursts of two or more per day near the maximum of the cycle.

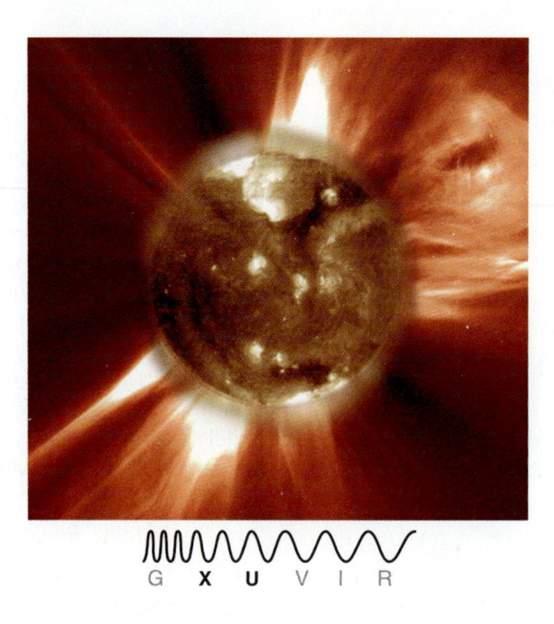

FIGURE 14.25 *SOHO* image of a coronal mass ejection (top right quadrant), with a simultaneously recorded ultraviolet image of the solar disk superimposed.

Solar Activity Affects Earth

The amount of solar radiation that reaches Earth is, on average, 1.35 kilowatts per square meter (kW/m²). Satellite measurements of the amount of radiation coming from the Sun (**Figure 14.26**) show that this value varies by as much as 0.2 percent over periods of a few weeks, as dark sunspots and bright spots in the chromosphere move across the disk. Overall, however, the increased radiation from **active regions** on the Sun more than makes up for the reduction in radiation from sunspots. On average, the Sun seems to be about 0.1 percent brighter during the peak of a solar cycle than it is at its minimum.

Solar activity has many effects on Earth. Solar active regions are the source of most of the Sun's extreme ultraviolet and X-ray emission. This energetic radiation heats Earth's upper atmosphere and, during periods of solar activity, causes the upper atmosphere to expand. When this happens, the swollen upper atmosphere can significantly increase the atmospheric drag on spacecraft orbiting near Earth, causing their orbits to decay. (Prior to the launch of the Hubble Space Telescope [HST], many people working on the project were concerned that, because it was to be launched right at a solar maximum, the telescope might not survive very long.) One reason for repeated shuttle visits to the HST was to boost it

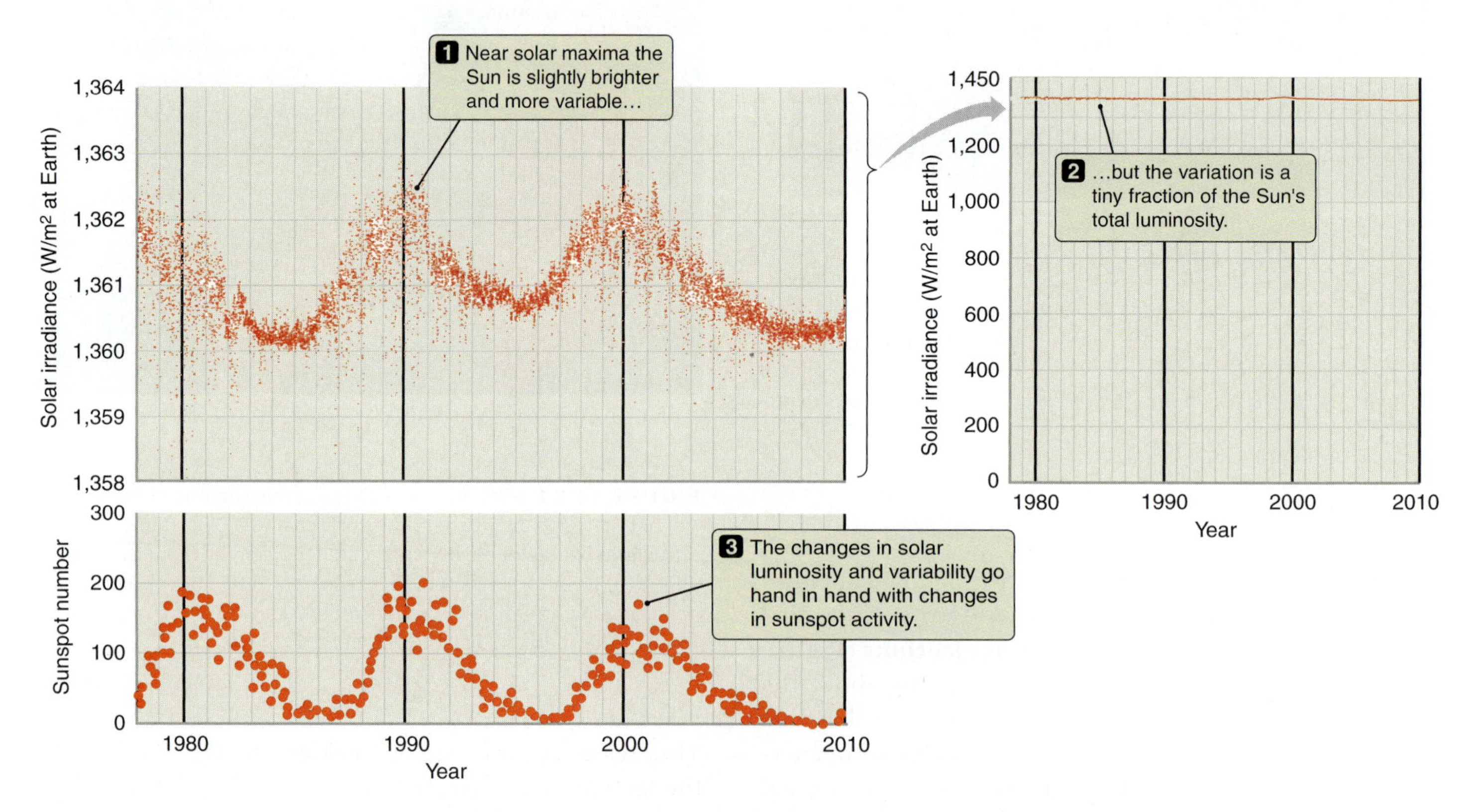

FIGURE 14.26 Measurements taken by satellites above Earth's atmosphere show that the amount of light from the Sun changes slightly over time.

back to its original orbit to make up for the slow orbital decay caused by drag in the rarefied outer parts of Earth's atmosphere. NASA gave a final shuttle boost to the HST in 2009, keeping the space telescope in operation until at least 2014. (The HST is scheduled to be replaced by the larger James Webb Space Telescope [JWST] in 2018.) Earth's atmospheric drag will eventually bring the HST back to Earth—although not in one piece.

Earth's magnetosphere is the result of the interaction between Earth's magnetic field and the solar wind. Increases in the solar wind accompanying solar activity, especially coronal mass ejections directed at Earth, can disrupt Earth's magnetosphere in ways that are obvious even to nonscientists. Spectacular auroras can accompany such events, as can magnetic storms that have been known to disrupt electric power grids, causing blackouts across large regions. Coronal mass ejections also have adverse effects on radio communication and navigation, and they can damage sensitive satellite electronics, including communication satellites.

Solar storms cause auroras and disrupt electric power grids on Earth.

14.5 Origins: The Solar Wind and Life

Solar flares and coronal mass ejections can affect the space around Earth. In fact, energetic particles accelerated in solar flares pose one of the greatest dangers to human exploration of space, and they need to be considered when astronauts are orbiting Earth in a space station or, someday, traveling to the Moon or farther. Earth's magnetic field protects life on the surface from these energetic particles. Thanks to this magnetic field, the particles travel along the magnetic field lines to the poles, creating the auroras, without bombarding the surface and causing danger to life on Earth. But the Moon does not have this protection, because its magnetic fields are very weak. Astronauts on its surfaces would be exposed just as astronauts traveling in space would be. The strength of the solar wind varies with the solar cycle, as noted in Section 14.4, so exposure danger varies as well.

The Solar System is surrounded by the **heliosphere** (**Figure 14.27**), in which the solar wind blows against the interstellar medium and clears out an area like the inside of a bubble. As the Sun and Solar System move through the Milky Way Galaxy, passing in and out of interstellar clouds, this heliosphere protects the entire Solar System from galactic high-energy particles known as cosmic rays that originate primarily in high-energy explosions of massive dying stars. When the Sun is in its lower-activity state, the heliosphere is weaker, so more galactic cosmic rays enter the Solar System. In addition, the intensity of these cosmic rays depends on where the Sun and Solar System are located in their orbit about the center of the Milky Way Galaxy.

Some scientists have theorized that at times when the Sun was quiet and the heliosphere was weaker than average, and the Solar System was passing through a particular part of the galaxy, the cosmic-ray flux in the

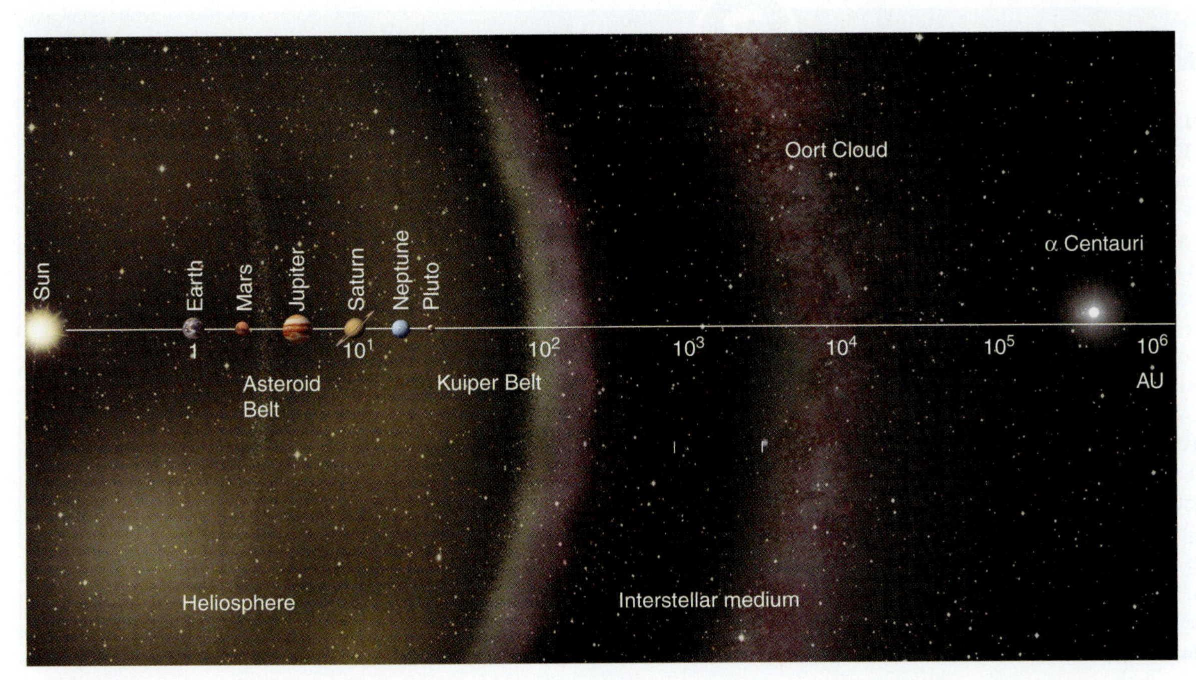

FIGURE 14.27 The heliosphere of the Sun, a bubble of charged particles surrounding the Solar System that is formed by the solar wind blowing against the interstellar medium.

Solar System—and on Earth—increased, possibly leading to a disruption in Earth's ozone layer, and possibly contributing to a mass extinction (in which many species died out) on Earth. The new satellite *IBEX* is studying the boundary between the Solar System and the interstellar medium.

Thus, in addition to the obvious contribution of the Sun to the heat and light on Earth, the extension of the Sun through the solar wind may have affected the evolution of life on Earth—and it may also affect humans' ability to live and work in space.

The physical properties of the Sun can be applied to all other stars. Starting with a successful model of the Sun, astronomers can now ask a host of other questions: What happens if the mass of a star is higher or lower than that of the Sun? What happens if its chemical abundances are different? What happens when all of the hydrogen at a star's center is gone? The study of the Sun has set the stage as we turn our attention to how stars form, how they differ from one another, how they change with time, and how they die. ■

Summary

14.1 The forces due to pressure and gravity balance each other in hydrostatic equilibrium, maintaining the Sun's structure.

14.2 Nuclear reactions converting hydrogen to helium are the source of the Sun's energy. In these reactions, some mass is converted directly to energy through the proton-proton chain. Energy created in the Sun's core moves outward to the surface, first by radiation and then by convection.

14.3 As hydrogen fuses to helium in the core of the Sun, neutrinos are emitted. Neutrinos are elusive, almost massless particles that interact only very weakly with other matter. Observations of neutrinos confirm that nuclear fusion is the Sun's primary energy source. Helioseismology probes the Sun's interior in much the same way that seismology probes Earth's interior.

14.4 The Sun's atmosphere has multiple layers, each with a characteristic density, temperature, and pressure. The apparent surface of the Sun is called the photosphere. The temperature of the Sun's atmosphere ranges from about 6000 K near the bottom to about 1 million K at the top. Material streaming away from the Sun's corona creates the solar wind. Sunspots are photospheric regions that are cooler than their surroundings, and they reveal the 11- and 22-year cycles in solar activity. These cycles are tied to the cycle of the Sun's changing magnetic field. Solar storms can produce terrestrial auroras and disrupt power grids.

14.5 The solar wind moves outward through the Solar System until it meets the interstellar medium. The wind may negatively affect astronauts in space or on planets that lack a protective magnetic field, but it also has protected the Solar System from galactic high-energy cosmic-ray particles.

Unanswered Questions

- Will nuclear fusion became a major source of energy production on Earth? Scientists have been working on controlled nuclear fusion for over 60 years, since the first hydrogen bombs were developed. But so far there are too many difficulties in replicating the conditions inside the Sun. It is not yet possible to confine the hydrogen isotopes with sufficient temperature, pressure, and density to start the reactions in a more controlled way than takes place during the explosion of a hydrogen bomb, and get more energy out than is put in. Several major experiments have attempted to fuse isotopes of hydrogen. An alternative approach is to fuse an isotope of helium, ^{3}He, which has only three particles in the nucleus (two protons and one neutron). On Earth, ^{3}He is found in very limited supply. But ^{3}He is in much greater abundance on the Moon, so some people propose setting up mining colonies on the Moon to extract ^{3}He for use in fusion reactions on Earth, or possibly even on the Moon (see question 62 at the end of the chapter).
- Are variations in Earth's climate related to solar activity? Solar activity affects Earth's upper atmosphere, and it may affect weather patterns as well. It has also been suggested that variations in the amount of radiation from the Sun might be responsible for past variations in Earth's climate. Current models indicate that observed variations in the Sun's luminosity could account for only about 0.1 K differences in Earth's average temperature—much less than the effects due to the ongoing buildup of carbon dioxide in Earth's atmosphere. Triggering the onset of an ice age may require a sustained drop in global temperatures of only about 0.2–0.5 K, so astronomers are continuing to investigate a possible link between solar variability and changes in Earth's climate.

Questions and Problems

Summary Self-Test

1. If the pressure inside a star increases slightly, the star will
   a. implode.
   b. explode.
   c. swell up.
   d. contract.

2. When hydrogen is fused into helium, energy is released from
   a. gravitational collapse.
   b. conversion of matter to light energy.
   c. the increase in pressure.
   d. the decrease in the gravitational field.

3. Which of the following energy transport mechanisms is active in the layer directly outside the core of the Sun?
   a. convection
   b. conduction
   c. radiation

4. Neutrinos are important to an understanding of the Sun because
   a. neutrinos from distant objects pass through the Sun, probing the interior.
   b. neutrinos from the Sun pass easily through Earth.
   c. neutrinos from the interior of the Sun easily escape.
   d. neutrinos change form on their way to Earth.

5. Place the following regions of the Sun in order of increasing radius.
   a. corona
   b. core
   c. radiative zone
   d. convective zone
   e. chromosphere
   f. photosphere
   g. a sunspot

6. Earth-bound observers can observe the solar wind by (select all that apply)
   a. observing auroras.
   b. mapping Earth's magnetic field.
   c. observing the stock market.
   d. tracking the behavior of power grids.

7. Coronal mass ejections
   a. carry away 1 percent of the mass of the Sun each year.
   b. are caused by breaking magnetic fields.
   c. are always emitted in the direction of Earth.
   d. are unimportant to life on Earth.

8. Which of the following describe sunspots? (Select all that apply.)
   a. brighter than the surrounding photosphere
   b. fainter than the surrounding photosphere
   c. hotter than the surrounding photosphere
   d. cooler than the surrounding photosphere
   e. round
   f. darker in the middle than at the edges
   g. caused by magnetic fields
   h. caused by convection

9. True or false: The Maunder Minimum shows a correlation between low solar activity and a cooler climate on Earth. This correlation means that the low solar activity definitely caused the cooler climate.

10. During periods of solar maximum, the Sun is ________ at solar minimum.
    a. much brighter than
    b. much fainter than
    c. nearly as bright as
    d. much hotter than
    e. much cooler than

True/False and Multiple Choice

11. **T/F:** In the Sun, a balance is maintained between pressure and energy.

12. **T/F:** The energy from the Sun comes from the fusion of helium into hydrogen.

13. **T/F:** Photons from the center of the Sun take a very, very long time to reach the surface.

14. **T/F:** Neutrinos from the center of the Sun take a very, very long time to reach the surface.

15. **T/F:** Solar activity is responsible for recent climate change on Earth.

16. Energy balance in the Sun means that
    a. the Sun does not change over time.
    b. the Sun absorbs and emits equal amounts of energy.
    c. radiation pressure balances the weight of overlying layers.
    d. energy produced in the core equals energy emitted at the surface.

17. The strong nuclear force
    a. holds atoms together.
    b. pulls protons together in nuclear fusion.
    c. is unimportant at scales bigger than the size of an atom.
    d. all of the above

18. The Sun's energy is supplied by
    a. nuclear fusion.
    b. gravitational collapse.
    c. energy balance.
    d. neutrinos.

19. Energy is produced primarily in the center of the Sun because
    a. the strong nuclear force is too weak elsewhere.
    b. that's where neutrinos are created.
    c. that's where most of the helium is.
    d. the outer parts have lower temperatures and pressures.
20. In the proton-proton chain, the mass of four protons is slightly greater than the mass of a helium nucleus. The "lost" mass
    a. binds the particles of the nucleus together.
    b. cancels the positive charge of the two protons that turned into neutrons.
    c. escapes as light, positrons, or neutrinos.
    d. is reabsorbed instantly, turning helium back into hydrogen.
21. Gas is not an efficient conductor of thermal energy because the particles
    a. are too slow.
    b. are too far apart.
    c. are too heavy.
    d. don't interact with light.
22. A high-energy photon and a neutrino are created in the Sun's core at the same instant. Which one will reach Earth first, and why?
    a. the neutrino, because neutrinos are much faster than light
    b. the photon, because nothing travels faster than light
    c. the neutrino, because neutrinos barely interact with matter
    d. the photon, because photons barely interact with matter
23. A solar flare releases ________ energy from the Sun.
    a. electric
    b. nuclear
    c. gravitational
    d. magnetic
24. The solar wind pushes on the magnetosphere of Earth, changing its shape, because
    a. the solar wind is so dense.
    b. the magnetosphere is so weak.
    c. the solar wind contains charged particles.
    d. the solar wind is so fast.
25. Sunspots peak every 11 years because the magnetic field of the Sun "resets" every ________ years.
    a. 5.5
    b. 11
    c. 16.5
    d. 22

Thinking about the Concepts

26. Explain how hydrostatic equilibrium acts as a safety valve to keep the Sun at its constant size, temperature, and luminosity.
27. Two of the three atoms in a molecule of water (H_2O) are hydrogen. Why are Earth's oceans not fusing hydrogen into helium and setting Earth ablaze?
28. Engineers and physicists dream of solving the world's energy supply problem by constructing power plants that would convert globally plentiful hydrogen into helium. The Sun seems to have solved this problem. On Earth, what is a major obstacle to this potentially environmentally clean solution?
29. Explain the proton-proton chain through which the Sun generates energy by converting hydrogen to helium.
30. On Earth, nuclear power plants use *fission* to generate electricity. In fission, a heavy element like uranium is broken into many atoms, where the total mass of the fragments is less than the original atom. Explain why fission could not be powering the Sun today.
31. Suppose an abnormally large amount of hydrogen suddenly burned in the core of the Sun. What would happen to the rest of the Sun? Would there be a change in the Sun as seen from Earth?
32. Why are neutrinos so difficult to detect?
33. Discuss the solar neutrino problem, and describe how it was solved.
34. The Sun's visible "surface" is not a true surface, but a feature called the photosphere. Explain why the photosphere is not a true surface.
35. The second most abundant element in the universe was not discovered here on Earth. What is it, and how was it discovered?
36. Describe the solar corona. Under what circumstances can it be seen without special instruments?
37. The solar corona has a temperature of millions of degrees; the photosphere has a temperature of only about 6000 K. Why isn't the corona much, much brighter than the photosphere?
38. What have sunspots revealed about the Sun's rotation?
39. How is the fate of the Hubble Space Telescope tied to solar activity?
40. Why are different parts of the Sun best studied at different wavelengths? Which parts are best studied from space?

Applying the Concepts

41. If a quantity decreases exponentially, it has a characteristic curve when plotted linearly and a different shape when plotted logarithmically. Which of the four quantities (pressure, density, temperature, and energy) decrease approximately exponentially with radius from the center of the Sun?

42. In Figure 14.12, density and temperature are both graphed versus height.
    a. Is the height axis linear or logarithmic? How do you know?
    b. Is the density axis linear or logarithmic? How do you know?
    c. Is the temperature axis linear or logarithmic? How do you know?

43. Study Figure 14.21, frames (b) and (c). Argue from these data that sunspots occur in regions of strong magnetic field.

44. Study the image in Figure 14.19 and the graph in Figure 14.22.
    a. Estimate the fraction of the Sun's surface that is covered by the large sunspot group in the image. (Remember to take into account that you are seeing only one hemisphere of the Sun.)
    b. From the graph, estimate the average number of sunspots that occurs at solar maximum.
    c. On average, what fraction of the Sun could be covered by sunspots at solar maximum? Is this a large fraction?
    d. Compare your conclusion to the graph of irradiance in Figure 14.26. Does this graph make sense to you?

45. The Sun has a radius equal to about 2.3 light-seconds. Explain why a gamma ray produced in the Sun's core does not emerge from the Sun's surface 2.3 seconds later.

46. Assume that the Sun's mass is about 300,000 Earth masses and that its radius is about 100 times that of Earth. The density of Earth is about 5,500 kg/m^3.
    a. What is the average density of the Sun?
    b. How does this compare with the density of water?

47. The Sun shines by converting mass into energy according to Einstein's well-known relationship $E = mc^2$. Show that if the Sun produces 3.85×10^{26} J of energy per second, it must convert 4.3 million metric tons (4.3×10^9 kg) of mass per second into energy.

48. Assume that the Sun has been producing energy at a constant rate over its lifetime of 4.6 billion years (1.4×10^{17} seconds).
    a. How much mass has it lost creating energy over its lifetime?
    b. The present mass of the Sun is 2×10^{30} kg. What fraction of its present mass has been converted into energy over the lifetime of the Sun?

49. Imagine that the source of energy in the interior of the Sun changed abruptly.
    a. How long would it take before a neutrino telescope detected the event?
    b. When would a visible-light telescope see evidence of the change?

50. On average, how long does it take particles in the solar wind to reach Earth from the Sun if they are traveling at an average speed of 400 km/s?

51. A sunspot appears only 70 percent as bright as the surrounding photosphere. The photosphere has a temperature of approximately 5780 K. What is the temperature of the sunspot?

52. The hydrogen bomb represents an effort to duplicate processes going on in the core of the Sun. The energy released by a 5-megaton hydrogen bomb is 2×10^{16} J.
    a. This textbook, *21st Century Astronomy*, has a mass of about 1.6 kg. If all of its mass were converted into energy, how many 5-megaton bombs would it take to equal that energy?
    b. How much mass did Earth lose each time a 5-megaton hydrogen bomb was exploded?

53. Verify the claim made at the start of this chapter that the Sun produces more energy per second than all the electric power plants on Earth could generate in a half-million years. Estimate or look up how many power plants there are on the planet, and how much energy an average power plant produces. Be sure to account for different kinds of power—for example, coal, nuclear, wind.

54. Let's examine the reason that the Sun cannot power itself by chemical reactions. Using Math Tools 14.1 and the fact that an average chemical reaction between two atoms releases 1.6×10^{-19} J, estimate how long the Sun could emit energy at its current luminosity. Compare that estimate to the known age of Earth.

55. How much distance must a photon actually cover in its convoluted path from the center of the Sun to the outer layers?

Using the Web

56. (a) Go to *QUEST*'s "Journey into the Sun" Web page (http://science.kqed.org/quest/video/journey-into-the-sun) to watch a short video on the Solar Dynamics Observatory (SDO), launched in 2010. Why is studying the magnetic field of the Sun so important? What is new and different in this observatory?

    (b) Go to the SDO website (http://sdo.gsfc.nasa.gov). Under "Data," select "The Sun Now" and view the Sun at many wavelengths. What activity do you observe in the images at the location of any sunspots seen in the "HMI Intensitygram" images? (You can download a free SDO app by Astra to get real-time images on your mobile device.) Look at a recent news story from SDO. What was observed, and why is it newsworthy?

57. (a) Go to the mission page for an older telescope, *SOHO* (Solar and Heliospheric Observatory—http://sohowww.nascom.nasa.gov) which was launched in 1995 by NASA and ESA. Click on "The Sun Now" to see today's images. The EIT images are in the far ultraviolet and show violent activity. How do these images differ from the ones of SDO in question 56?

(b) Go to the mission page for the Hinode telescope (http://hinode.nao.ac.jp/index_e.shtml), launched in 2005 by the Japanese Space Agency and NASA. Its primary instrument is an X-ray telescope. Why is it useful to observe the Sun in X-rays?

58. Go to the *STEREO* mission's website (http://stereo.gsfc.nasa.gov). What is *STEREO*? Where are the spacecraft located? How does this configuration enable observations of the entire Sun at once? (You can download the app "3-D Sun" to get the latest images on your mobile device.)

59. Go to the Space Weather website (http://spaceweather.com). Are there any solar flares today? What is the sunspot number? Is it about what you would expect for this year? (Click on "What is the sunspot number?" to see a current graph.) Are there any coronal holes today?

60. Go to the Solar Stormwatch website (http://solarstormwatch.com), a Zooniverse project from the Royal Observatory in Greenwich, England. Zooniverse projects offer an opportunity for people to contribute to science by analyzing pieces of data. Create an account for Zooniverse if you don't already have one (you will use it again in this course). Login and click on "Spot and Track Storms" and go through the Spot and Track training exercises. You are now ready to look at some real data. Click on an image to do the classification. Save a screen shot for your homework.

61. Go to the Advanced Technology Solar Telescope (ATST) website (http://atst.nso.edu). This adaptive-optics telescope under construction on Haleakala, Maui, will be the largest solar telescope. Click on "A Microscope for the Sun: the ATST Movie," and watch the movie. Why is it important to study the magnetic field of the Sun? What are some of the advantages of studying the Sun from a ground-based telescope instead of a space-based telescope? What wavelengths does the ATST observe? Why is Maui a good location? When is the telescope scheduled to be completed?

62. (a) Go to the National Ignition Facility (NIF) website (https://lasers.llnl.gov/about/nif). Under "Programs," click on "Internal Fusion Energy" and then "How to Make a Star." How are lasers used in experiments to develop controlled nuclear fusion on Earth? How does the fusion reaction here differ from that in the Sun?

(b) An alternative approach is to fuse ^{3}He + ^{3}He instead of the hydrogen isotopes. But on Earth, ^{3}He is found in limited supply. ^{3}He is in much greater abundance on the Moon, so some people propose setting up mining colonies on the Moon to extract ^{3}He for fusion reactions on Earth. Do a search on "helium 3 moon." Which countries are talking about going to the Moon for this purpose? What is the timeline for when this might happen? What are the difficulties?

STUDYSPACE is a free and open website that provides a Study Plan for each chapter of *21st Century Astronomy*. Study Plans include animations, reading outlines, vocabulary flashcards, and multiple-choice quizzes, plus links to premium content in SmartWork and the ebook. Visit **wwnorton.com/studyspace**.

SMARTWORK Norton's online homework system, includes algorithmically generated versions of these questions, plus additional conceptual exercises. If your instructor assigns questions in SmartWork, log in at **smartwork.wwnorton.com.**

Exploration | The Proton-Proton Chain

The proton-proton chain powers the Sun by fusing hydrogen into helium. This fusion process produces several different particles as by-products, as well as energy. In this Exploration we will explore the steps of the proton-proton chain in detail, with the intent of helping you keep them straight.

Visit StudySpace, and open the "Proton-Proton Animation" for this chapter.

Watch the animation all the way through once.

Play the animation again, pausing after the first collision. Two hydrogen nuclei (both positively charged) have collided to produce a new nucleus with only one positive charge.

1 **Which particle carried away the other positive charge?**

....

....

2 **What is a neutrino? Did the neutrino enter the reaction, or was the neutrino produced in the reaction?**

....

....

Compare the interaction on the top with the interaction on the bottom.

3 **Did the same reaction occur in each instance?**

....

....

Resume playing the animation, pausing it after the second collision.

4 **What two types of nuclei entered the collision? What type of nucleus resulted?**

....

....

5 **Was charge conserved in this reaction, or was it necessary for a particle to carry charge away?**

....

....

6 **What is a gamma ray? Did the gamma ray enter the reaction, or was it produced by the reaction?**

....

....

Resume the animation again, and allow it to run to the end.

7 **What nuclei enter the final collision? What nuclei are produced?**

....

....

8 **In chemistry, a catalyst facilitates the reaction but is not used up in the process. Do any nuclei act like catalysts in the proton-proton chain?**

....

....

Make a table of inputs and outputs. Which of the particles in the final frame of the animation were inputs to the reaction? Which were outputs? Fill in your table with these inputs and outputs.

9 **Which outputs are converted into energy that leaves the Sun as light?**

....

....

10 **Which outputs could become involved in another reaction immediately?**

....

....

11 **Which output is likely to stay in that form for a very long time?**

....

....

Herschel Space Observatory false-color far infrared image of the Rosette molecular cloud.

15 Star Formation and the Interstellar Medium

You must have chaos within you to give birth to a dancing star.

Friedrich Nietzsche (1844–1900)

LEARNING GOALS

The space between the stars is not empty; there are giant clouds of cool gas and dust. Hot gas fills the space between these clouds, pressing on them and helping to keep them together. Each of the atoms and molecules in a cloud is gravitationally attracted to every other particle. Some of these clouds collapse under this gravity, fragmenting to form multiple stars, and sometimes further fragmenting to form planets. In this chapter you will learn about how stars form from this interstellar material. By the conclusion of the chapter, you should be able to:

- Distinguish the types of material that exist in the space between the stars.
- Explain the conditions under which a cloud of gas can contract into a star, and the role that gravity and angular momentum play in the formation of stars and planets.
- Recognize how the evolution of a protostar is determined largely by its mass.
- Describe how astronomers study star formation using theoretical models and observations at many wavelengths.

15.1 The Interstellar Medium

Earth's local star, the Sun, has remained fairly constant long enough for life on Earth to arise, evolve, and begin to contemplate its own origin and fate. But the human perspective is one of a relatively brief existence; all of recorded human history amounts to less than a millionth of the age of the Sun and Earth. In Chapter 7 you learned how the Solar System formed some 4.6 billion years ago out of the swirling disk of gas and dust that surrounded the forming Sun. In this chapter we will return to that story, but with a different focus. First we will step back and look at the interstellar environment that gave birth to the Sun and the Solar System, and all the other stars, both large and small. Then we will focus our attention on the forming star—the **protostar**—and discuss how it becomes a star.

Does a star's birth environment affect its likelihood of having planets and possibly life?

The birth of a star—from a cloud of gas and dust to nuclear burning—is a process that can happen within tens of thousands of years (for the most massive stars) or can require hundreds of millions of years (for the least massive). Such evolutionary times are far too long to witness changes in any individual protostar; astronomers have pulled the larger picture together by observing many different stars at various stages of stellar development.

The volume of the Sun is about 1.3 million times the volume of Earth. Though this might seem enormous, it is almost incomprehensibly tiny compared with the volume of space. The Sun is the only star occupying a region of space that has a volume of over 300 cubic light-years. If the Sun were the size of a golf ball in Santa Fe, New Mexico, its nearest stellar neighbor would be a similar golf ball near Salt Lake City, Utah, 1,000 kilometers (km) away. The volume of the Milky Way Galaxy between stars is filled with the **interstellar medium**. It is here that the story of stars begins and ends. Stars are formed from the gas in the interstellar medium, they live their lives in the interstellar medium, and when they die they give some of their chemical and energetic legacy back to the interstellar medium.

The Sun formed from the interstellar medium, so it is not too surprising that the chemical composition of the interstellar medium in the Sun's region of the Milky Way is similar to the chemical composition of the Sun (see Table 13.1). In the interstellar medium, about 90 percent of the atomic nuclei are hydrogen, and the remaining 10 percent are almost all helium. The more massive elements account for only 0.1 percent of the atomic nuclei, or about 2 percent of the mass in the interstellar medium. Roughly 99 percent of that interstellar matter is gaseous, consisting of individual atoms or molecules moving about freely, as the molecules in the air do.

Stars and planets form from material in the space between stars.

However, interstellar gas is far less dense than the air that you breathe. Each cubic centimeter (cm^3) of the air around you contains about 2.7×10^{19} molecules. A good vacuum pump on Earth can reduce this density down to about 10^{10} molecules/cm^3—approximately a billionth as dense. By comparison, the interstellar medium has an average density of about 0.1 atom/cm^3—one 100-billionth as dense. To understand this concept more clearly, imagine a 1-meter-long hose, with an opening of 1 square centimeter (cm^2), filled with air and sealed at both ends. Imagine that you can stretch the hose without changing the diameter of the opening. As you stretch the hose, the air inside has to spread out to fill the larger volume and so is less dense. When the hose is stretched to a length of 10 meters, the air inside will be a tenth as dense as the air around you. When the hose is stretched to 100 meters, the air inside will be a hundredth as dense.

For the density of air in the hose to match the average density of interstellar gas—that is, for it to have the same number of gas particles per unit volume as the interstellar medium has—you would have to stretch it into a hose 26,000 light-years long! Stated another way, there is about as much material in a column of air between your eye and the floor beside you as there is interstellar gas in a column 1 cm in diameter and stretching from the Solar System to the center of the galaxy.

Interstellar gas is extremely tenuous.

The Interstellar Medium Is Dusty

About 1 percent of the material in the interstellar medium is in the form of solid grains, called **interstellar dust**. Ranging in size from little more than large molecules up to particles about 300 nanometers (nm) across, these solid grains more closely resemble the particles of soot from a candle flame than they do the dust that collects on a windowsill. (It would take several hundred "large" interstellar grains to span the thickness of a single human hair.) Interstellar dust begins to form when materials such as iron, silicon, and carbon stick together to form grains in dense, relatively cool environments like the outer atmospheres and "stellar

About 1 percent of the mass of the interstellar medium is in grains.

winds" of cool, red giant stars, or in dense material thrown into space by stellar explosions. Once these grains are in the interstellar medium, other atoms and molecules stick to them. This process is remarkably efficient: about half of all interstellar matter more massive than helium (1 percent of the total mass of the interstellar medium) is found in interstellar grains.

If there is only as much material between Earth and the center of the galaxy as there is between your eye and the floor, then you might expect the view of distant objects to be as clear as your view of your own big toe. This is not the case, however, because interstellar dust is extremely effective at blocking and diverting light. If the air around you contained as much dust as a comparable mass of interstellar material, it would be so dirty that you would be hard-pressed to see your hand 10 cm in front of your face. Go out on a dark summer night in the Northern Hemisphere (or a dark winter night in the Southern Hemisphere) and look closely at the Milky Way, visible as a faint band of diffuse light running through the constellation Sagittarius. You will see a dark "lane" running roughly down the middle of this bright band, splitting it in two. This dark band is a vast expanse of interstellar dust blocking astronomers' view of distant stars (see, for example, Figure 21.1).

Interstellar dust obscures the view of the Galaxy.

When interstellar dust gets in the way of radiation from distant objects, the effect is called **interstellar extinction**. Not all electromagnetic radiation suffers equally from interstellar extinction, however. **Figure 15.1** shows two images of the Milky Way Galaxy: one taken in visible light, the other taken in the infrared (IR). The dark clouds that block the shorter-wavelength visible light (**Figure 15.1a**) seem to have vanished in the longer-wavelength IR image (**Figure 15.1b**), enabling observations through the clouds to the center of the galaxy and beyond. These two images are *all-sky images*; they portray the entire sky surrounding Earth. They have been oriented so that the disk of the Milky Way, in which the Solar System is embedded, runs horizontally across the center of the image.

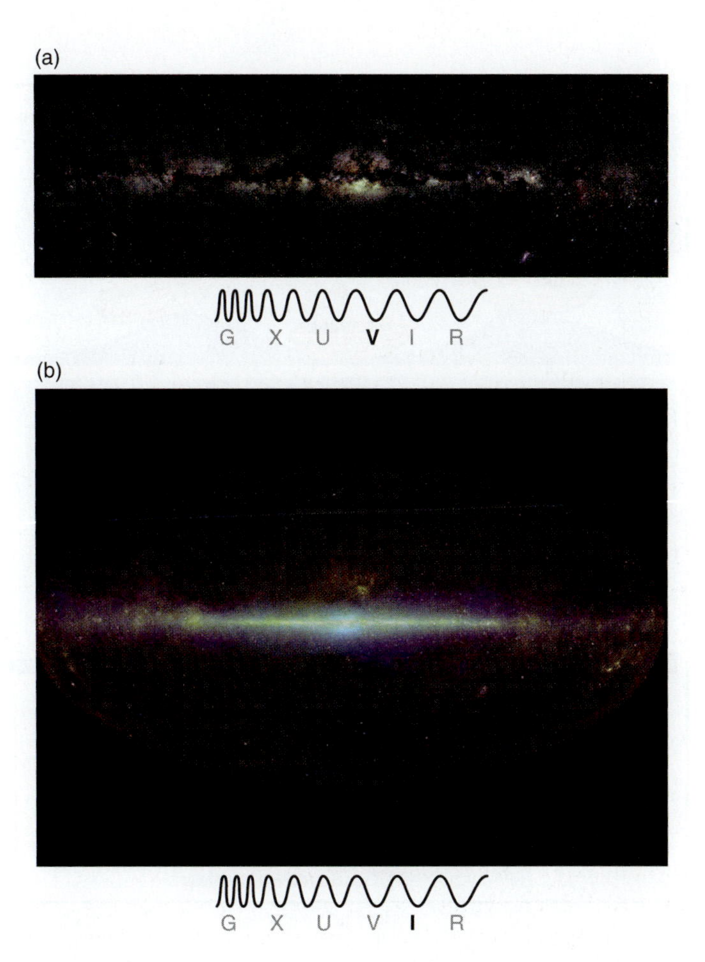

FIGURE 15.1 (a) An all-sky picture of the Milky Way taken in visible light. The dark splotches blocking the view are dusty interstellar clouds. The center of this (and all the other) all-sky images is the center of the Milky Way. (b) An all-sky picture as seen in the near infrared. Infrared radiation penetrates the interstellar dust, providing a clearer view of the stars in the Milky Way.

To understand why short-wavelength radiation is obscured by dust while long-wavelength radiation is not, think about a different kind of wave—a wave on the surface of the ocean (**Figure 15.2a**). Imagine you are on the ocean in a boat in a strong swell. If the waves are much bigger than your small boat, the swell causes you to bob gently up and down. But that is about all; there is no other interaction between the waves and the small boat. The story is quite different if you are on a larger boat that is about half as long as the wavelength of the ocean waves. Now the front of the boat may be on a wave crest while the back of the boat is in a trough, or vice versa. The boat tips wildly back and forth as the waves go by. If the size of the boat and the wavelength of the waves are the right match, even fairly modest waves will rock the boat. (You might have noticed this if you were ever in a canoe or rowboat when the wake from a speedboat came by.) Now imagine viewing these two situations from the perspective of the wave. The wave is hardly affected by the small boat, but it is strongly affected by the large boat. The energy to drive the wild motions of the boat comes from the wave, so the motion of the wave is affected by the interaction.

The interaction of electromagnetic waves with matter is more involved than that of a boat rocking on the ocean, but the same basic idea often applies (**Figure 15.2b**). Tiny interstellar dust grains interact most strongly with ultraviolet light and blue light, which have wavelengths

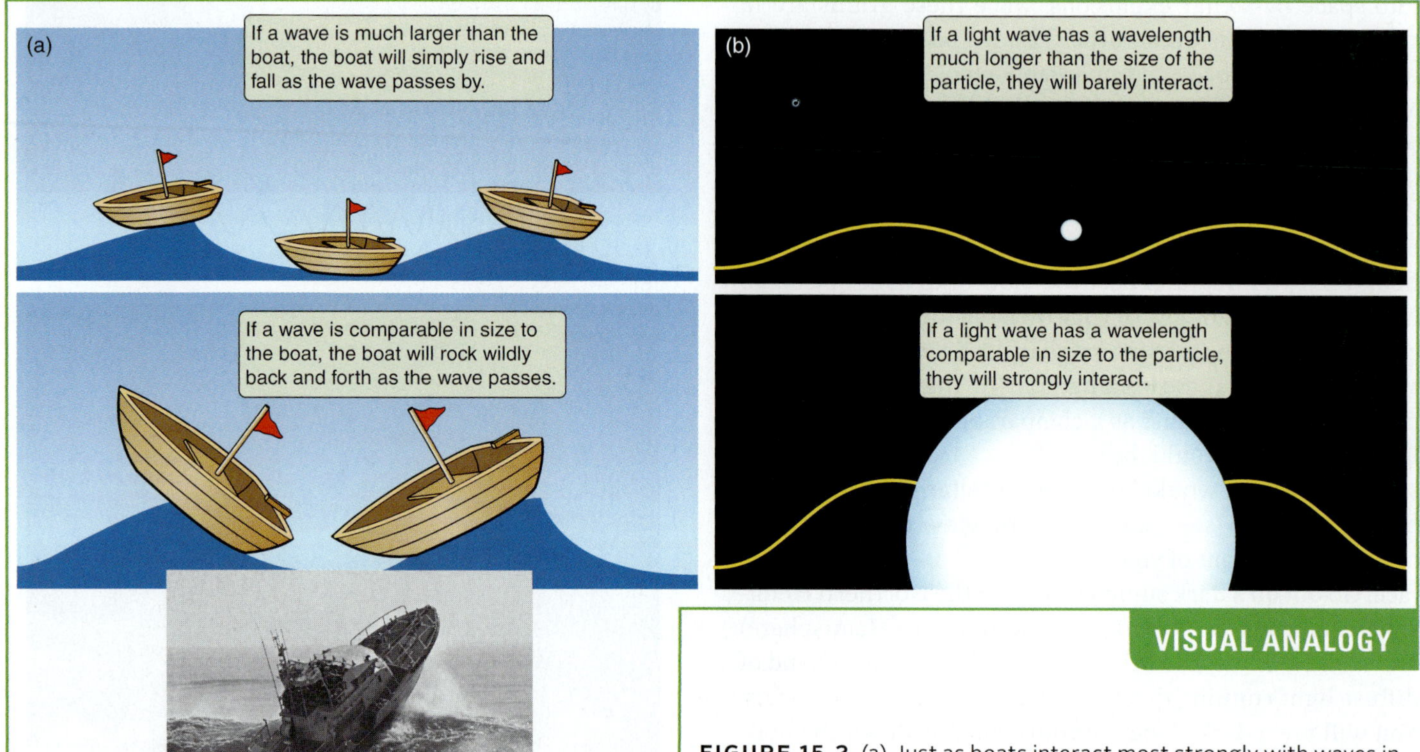

FIGURE 15.2 (a) Just as boats interact most strongly with waves in the ocean that are similar in size, (b) particles interact most strongly with wavelengths of light that are similar in size.

comparable to the typical size of dust grains. For this reason, ultraviolet light and blue light are effectively blocked by interstellar dust. Short wavelengths suffer heavily from interstellar extinction. Infrared and radio radiation, by contrast, have wavelengths that are too long to interact strongly with the tiny interstellar dust grains, so they travel largely unimpeded across great interstellar distances. To sum up, at visible and ultraviolet wavelengths, most of the galaxy is hidden from view by dust. In the infrared and radio portions of the spectrum, however, a far more complete view is possible.

Long-wavelength radiation penetrates interstellar dust.

In the visible part of the spectrum, short-wavelength blue light suffers more from extinction than does longer-wavelength red light. As a result, an object viewed through dust looks redder (actually less blue) than it really is (**Figure 15.3**). This effect is called **reddening**. Correcting for the fact that stars and other objects appear both fainter and redder than they would in the absence of dust can be one of the most difficult parts of interpreting astronomical observations, often adding to uncertainty in the measurement of an object's properties (**Process of Science Figure**).

Interstellar extinction is less of a concern at infrared wavelengths, but dust still plays an important role in what astronomers see when they look at the sky in the infrared. Like any other solid object, grains of dust glow at wavelengths determined by their temperature. In Chapter 5 you learned that the equilibrium between absorbed sunlight and emitted thermal radiation determines the temperatures of the terrestrial planets (see Figure 5.25). A similar equilibrium is at work in interstellar space, where dust is heated by starlight and by the gas in which it is immersed to temperatures of tens to hundreds of kelvins (**Figure 15.4**). At a temperature of 100 kelvins (K), Wien's law says that dust will glow most strongly at a wavelength of 29 microns (μm). Cooler dust—say, at a temperature of 10 K—glows most strongly at a longer wavelength (**Math Tools 15.1**).

Much of the light in infrared observations is thermal radiation from this dust. The first good look at the infrared sky at wavelengths out to 100 μm came in 1983 from images taken with the Infrared Astronomical Satellite (IRAS). IRAS showed the Milky Way's dark clouds glowing brilliantly in infrared radiation from dust. A more sensitive map from a 2010 survey by NASA's Wide-field Infrared Survey Explorer mission (WISE) shows the sky at 3.4, 12, and 22 μm (**Figure 15.5**).

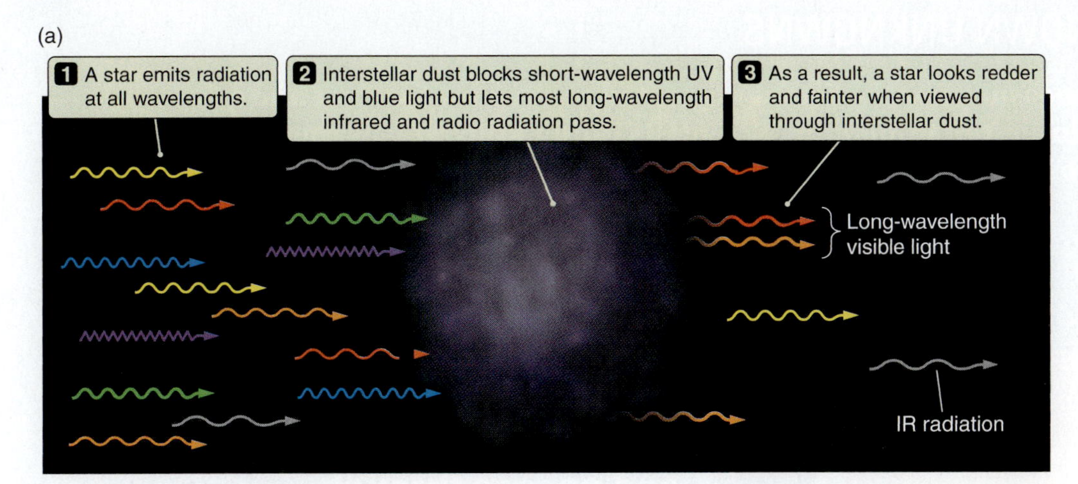

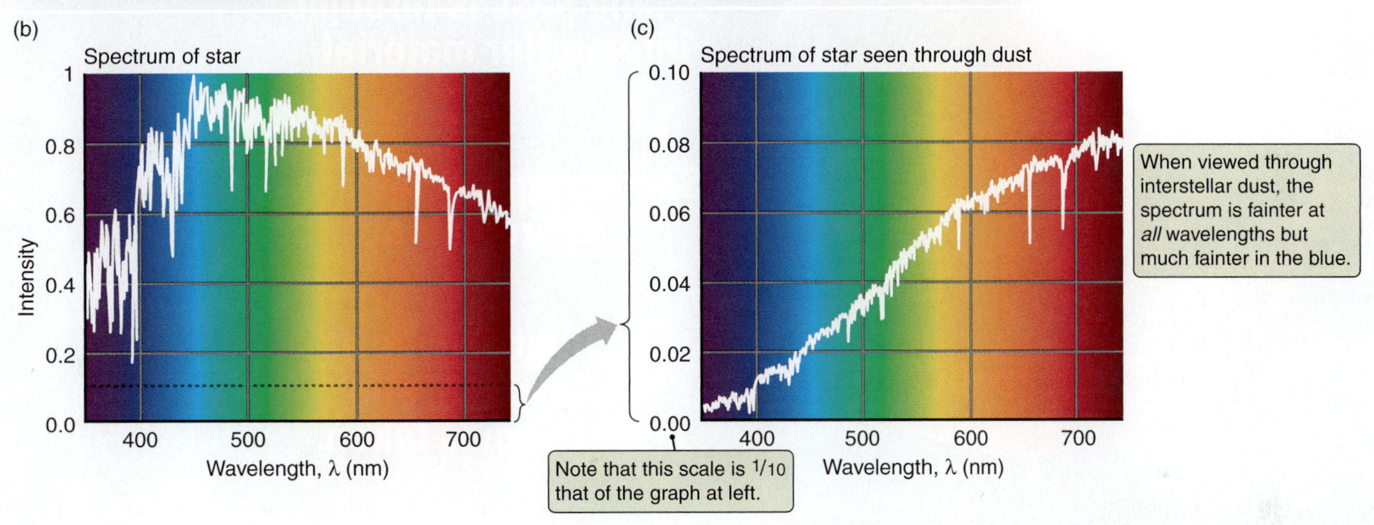

FIGURE 15.3 (a) The wavelengths of ultraviolet and blue light are close to the size of interstellar grains, so the grains effectively block this light. Grains are less effective at blocking longer-wavelength light. As a result, the spectrum of a star (b) when seen through an interstellar cloud (c) appears fainter and redder.

Interstellar Gas Has Different Temperatures and Densities

The gas and dust that fill interstellar space within the Milky Way Galaxy are not spread out evenly. About half of all interstellar gas is concentrated into only about 2 percent of the volume of interstellar space. These relatively dense regions are called **interstellar clouds**. Interstellar clouds are *not* like terrestrial clouds, which are locations where water vapor has condensed to form liquid droplets. In contrast, interstellar clouds are places where the interstellar gas is more concentrated than in surrounding regions. The other half of the interstellar gas is spread out through the remaining 98 percent of the volume of interstellar space. This gas that is found between the clouds is called **intercloud gas**.

> Most interstellar material is concentrated in relatively dense clouds.

The properties of intercloud gas vary from place to place (**Table 15.1**). Some intercloud gas is extremely hot, with temperatures in the millions of kelvins, close to those found in the centers of stars. Even so, were you to find yourself

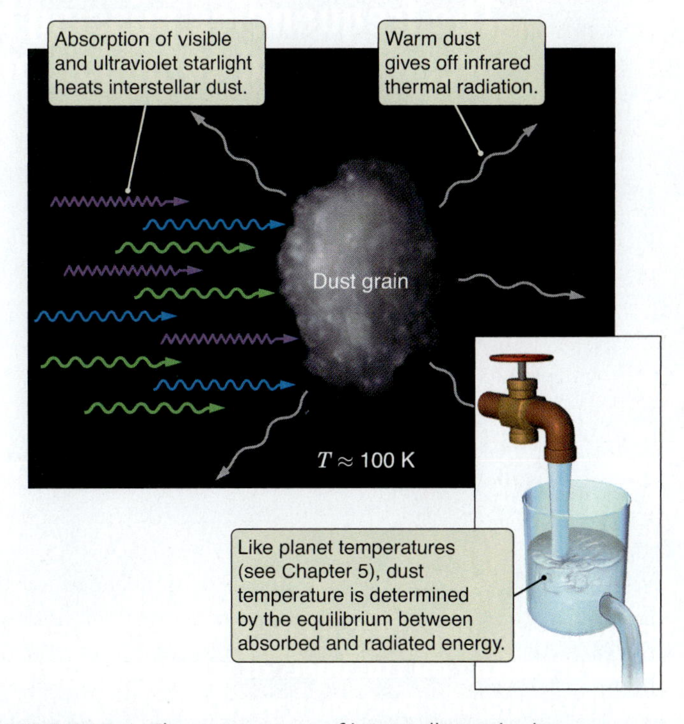

FIGURE 15.4 The temperature of interstellar grains is determined by the same type of equilibrium between absorbed and emitted radiation that determines the temperature of planets.

Process of Science

UNKNOWN UNKNOWNS

Early measurements of stars neglected to account for dust and gas in the interstellar medium. Scientists didn't know it was there until spectroscopy was developed.

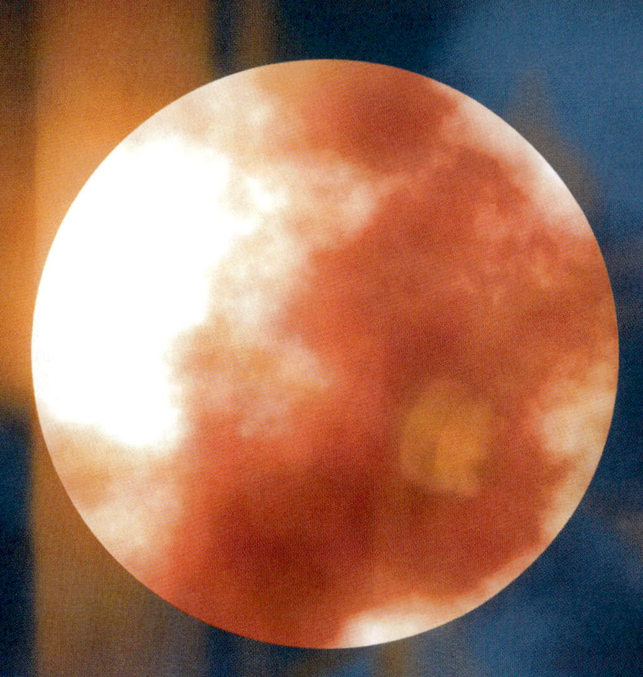

Light leaves a star. . .

. . .and passes through interstellar material.

Some frequencies are scattered away.

The light arrives at a telescope, and a spectrum is taken.

The frequencies missing from the spectrum reveal the presence and composition of the interstellar material.

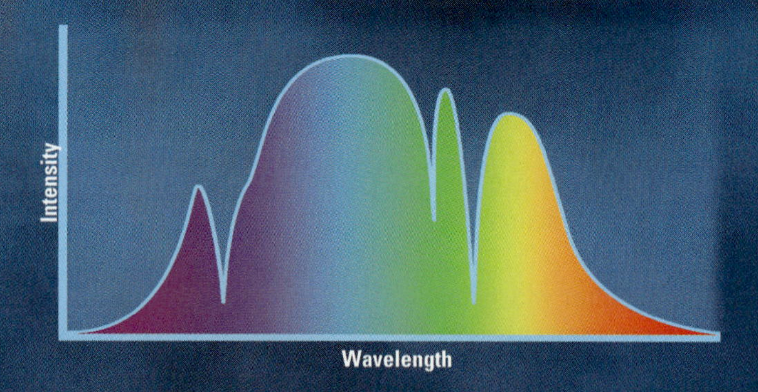

Scientists do not always know where the gaps in their knowledge lie. They must build their understanding on the basis of what is known but be prepared to change their minds when new information becomes available.

Math Tools 15.1

Dust Glows in the Infrared

Wien's law, discussed in Chapter 5, relates the temperature of an object to the peak wavelength (λ_{peak}) of its emitted radiation. For warm dust at a temperature of 100 K (recall that $1\ \mu m = 10^{-6}$ meter $= 1{,}000$ nm):

$$\lambda_{peak} = \frac{2{,}900\ \mu m\ K}{T} = \frac{2{,}900\ \mu m\ K}{100\ K} = 29\ \mu m$$

For cooler dust, at a temperature of 10 K:

$$\lambda_{peak} = \frac{2{,}900\ \mu m\ K}{T} = \frac{2{,}900\ \mu m\ K}{10\ K} = 290\ \mu m$$

The temperature and the peak wavelength are inversely proportional, so if the temperature drops, the peak wavelength gets longer. For the temperatures that are common for dust in the interstellar medium, the peak wavelength is in the far-infrared part of the electromagnetic spectrum.

adrift in an expanse of hot intercloud gas, your first concern would be freezing to death. The gas is hot, which means that the atoms that make it up are moving about very rapidly; so when one of these atoms runs into you, it hits you very hard. But the gas also has an extremely low density. Typically, you would have to search a liter (1,000 cm^3) or more of hot intercloud gas to find a single atom. To gather up 1 gram of hot intercloud gas—about the mass found in a liter of air—you would have to collect all of the gas in a cube that measured more than 8,000 km on an edge. There are so few atoms in a given volume of hot intercloud gas that atoms would very rarely run into you, so this million-kelvin gas would do little to keep you warm. You would radiate energy away and cool off much faster than the gas around you could replace the lost energy.

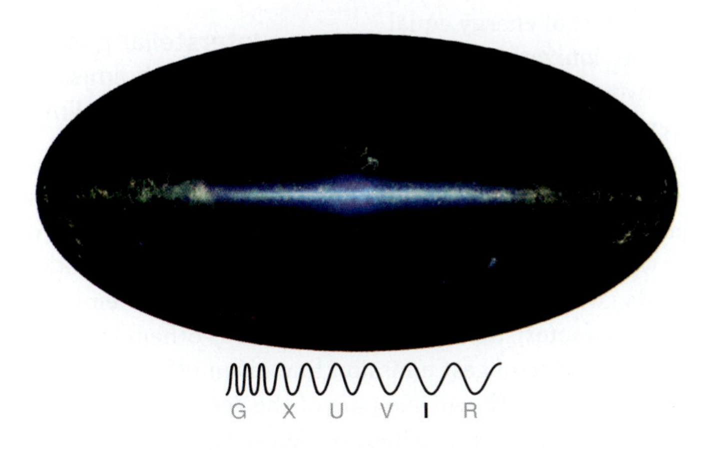

FIGURE 15.5 WISE galactic-plane image at 3.4, 12, and 22 μm is many times more sensitive than previous infrared images.

Extremely hot intercloud gas, heated primarily by the energy of tremendous stellar explosions called *supernovae*, occupies about half the volume of interstellar space. The Solar System is located inside a bubble of hot intercloud gas that has a density of about 0.005 hydrogen atom/cm^3 and is at least 650 light-years across. This may be the remnant of the hot bubble produced by a supernova explosion 300,000 years ago. Like the million-kelvin gas in the corona of the Sun, hot intercloud gas glows faintly in the energetic X-ray

TABLE 15.1 | Typical Properties of Components of the Interstellar Medium

Component	Temperature (K)	Number Density (atoms/cm^3)	State of Hydrogen
Hot intercloud gas	~1,000,000	~0.005	Ionized
Warm intercloud gas	~8000	0.01–1	Ionized or neutral
Cold intercloud gas	~100	1–100	Neutral
Interstellar clouds	~10	100–1,000	Molecular or neutral

portion of the electromagnetic spectrum. X-rays cannot penetrate Earth's atmosphere, so it is necessary to get above the atmosphere to observe the radiation coming from hot intercloud gas. Orbiting X-ray telescopes observe the entire sky aglow with faint X-rays coming from the bubble of million-kelvin hot gas in which the Solar System is immersed (**Figure 15.6**).

The Solar System is enveloped by a bubble of million-kelvin intercloud gas.

Not all intercloud gas is as hot as the local bubble. Most other intercloud gas has a temperature of about 8000 K and a density ranging from about 0.01 to 1 atom/cm^3. To gather up a gram of this more dense, warm intercloud gas would require all of the gas in a cube of interstellar space that was "only" 800 km on an edge.

Interstellar space is awash with the light from stars. Some of this is ultraviolet light with wavelengths shorter than 91.2 nm. Each photon of this ultraviolet light has enough energy that, if absorbed by a hydrogen atom, it will **ionize** the atom, kicking its electron free of the nucleus. If enough of these energetic photons are present, then atoms in the interstellar gas cannot remain neutral (because any neutral hydrogen atom would soon be ionized by starlight). In this way, about half of the volume of warm intercloud gas is kept ionized. But there is a large enough expanse of warm intercloud gas that the ionizing photons in the starlight get "used up." As a result, about half of the volume of warm intercloud gas is mostly neutral. This gas is usually surrounded by regions of ionized intercloud gas that shield the neutral gas from starlight, much as Earth's ozone layer shields the surface of Earth from harmful ultraviolet radiation from the Sun.

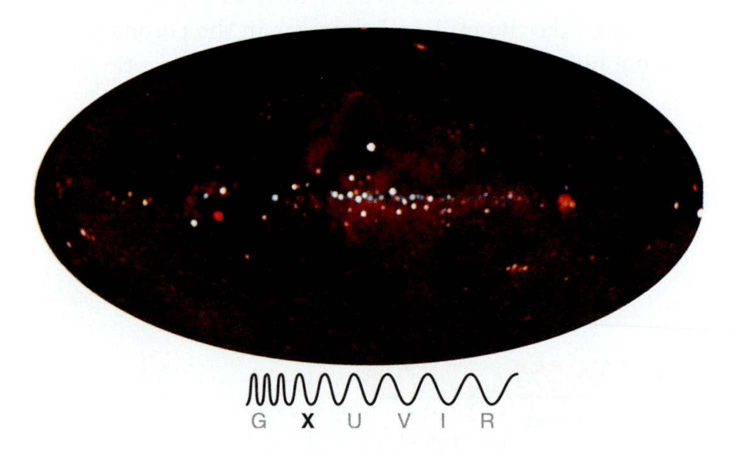

FIGURE 15.6 The faint X-ray glow that fills the sky is due largely to emission from the bubble of million-kelvin gas that surrounds the Sun. Bright spots are more distant sources, including objects such as bubbles of very hot, high-pressure gas surrounding the sites of recent supernova explosions.

One way to look for interstellar gas, including both warm and hot interstellar gas, is to study the spectra of distant stars. As starlight passes through interstellar gas, absorption lines form. Just as absorption lines that formed in the atmosphere of a star reveal much about the star's temperature, density, and chemical composition, interstellar absorption lines provide similar information about the gas the light has passed through. Astronomers can also study interstellar gas by looking for the radiation it emits. In regions of warm *ionized* gas, protons and electrons are constantly "finding each other" and recombining to make hydrogen atoms. It takes energy to tear a hydrogen atom apart, so according to the law of conservation of energy, when a proton and an electron combine to form a neutral hydrogen atom, this same amount of energy must be given up. This freed-up energy becomes electromagnetic radiation.

Typically, when a proton and an electron combine, the resulting hydrogen atom is left in an excited state (see Chapter 5). The atom then drops down to lower and lower energy states, emitting a photon at each step. Together, these photons carry away an amount of energy equal to the ionization energy of hydrogen plus whatever energy of motion the electron and proton had before recombining. The photons emitted in this process have wavelengths corresponding to the energy differences between the allowed energy states of hydrogen. In other words, warm ionized interstellar gas glows in emission lines characteristic of hydrogen (as well as other elements). Usually, the strongest emission line given off by warm interstellar gas in the visible part of the spectrum is the Hα (hydrogen alpha) line, which is seen in the red part of the spectrum at a wavelength of 656.3 nm.

Interstellar gas can produce both emission and absorption lines.

The image in **Figure 15.7** shows a portion of the sky taken in the light of Hα emission. The faint diffuse emission in the image comes mostly from warm, ionized intercloud gas. The especially bright spots, however, are quite different. These are regions where intense ultraviolet radiation from massive, hot, luminous O and B stars is able to ionize even relatively dense interstellar clouds. They are called **H II** ("H two") **regions**, because they contain large amounts of hydrogen. O stars live only a few million years, so they usually do not move very far from where they were born. The glowing clouds seen as H II regions are the very clouds from which these stars were born. Thus, H II regions are signposts to regions where *active star formation* is taking place.

H II regions are ionized by hot, luminous O and B stars.

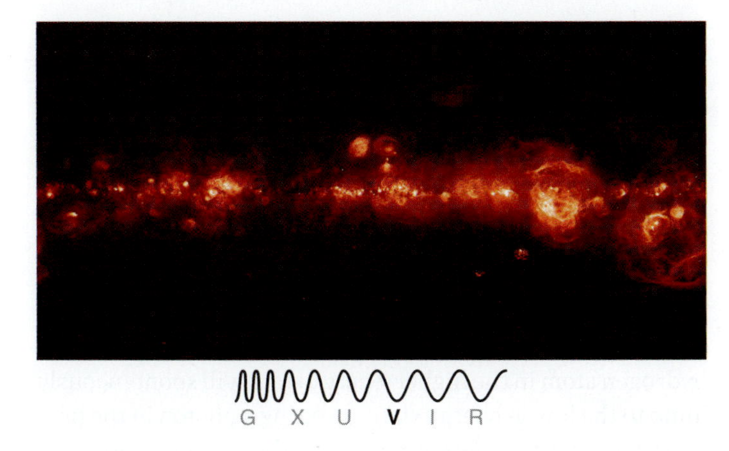

FIGURE 15.7 Warm (about 8000 K) interstellar gas glows in the Hα line of hydrogen. This image, showing the Hα emission from much of the northern sky, reveals how complex the structure of the interstellar medium is.

One of the H II regions that is closest to the Sun is the Orion Nebula, located 1,340 light-years from the Sun in the constellation Orion (**Figure 15.8**). Almost all of the ultraviolet light that powers this nebula comes from a single hot star, and only a few hundred stars are forming in its immediate vicinity. In contrast, **Figure 15.9** shows a giant H II region called 30 Doradus (also known as the Tarantula Nebula). The 30 Doradus H II region is located in the Large Magellanic Cloud, a small companion galaxy to the Milky Way, 160,000 light-years distant. This enormous cloud of ionized gas is powered by a dense star cluster containing thousands of hot, luminous stars. If 30 Doradus were as close as the Orion Nebula, it would be bright enough in the nighttime sky to cast shadows.

Warm, *neutral* hydrogen gas gives off radiation in a different way than does warm, ionized hydrogen. Many subatomic particles, including protons and electrons, have a property called *spin* that causes them to behave like tiny magnets—as if each particle had a bar magnet, with a north and a south pole, built into it (**Figure 15.10**). According to the rules of quantum mechanics, a hydrogen atom can exist in only two different configurations. Either the magnetic "poles" of the proton and electron point in the same

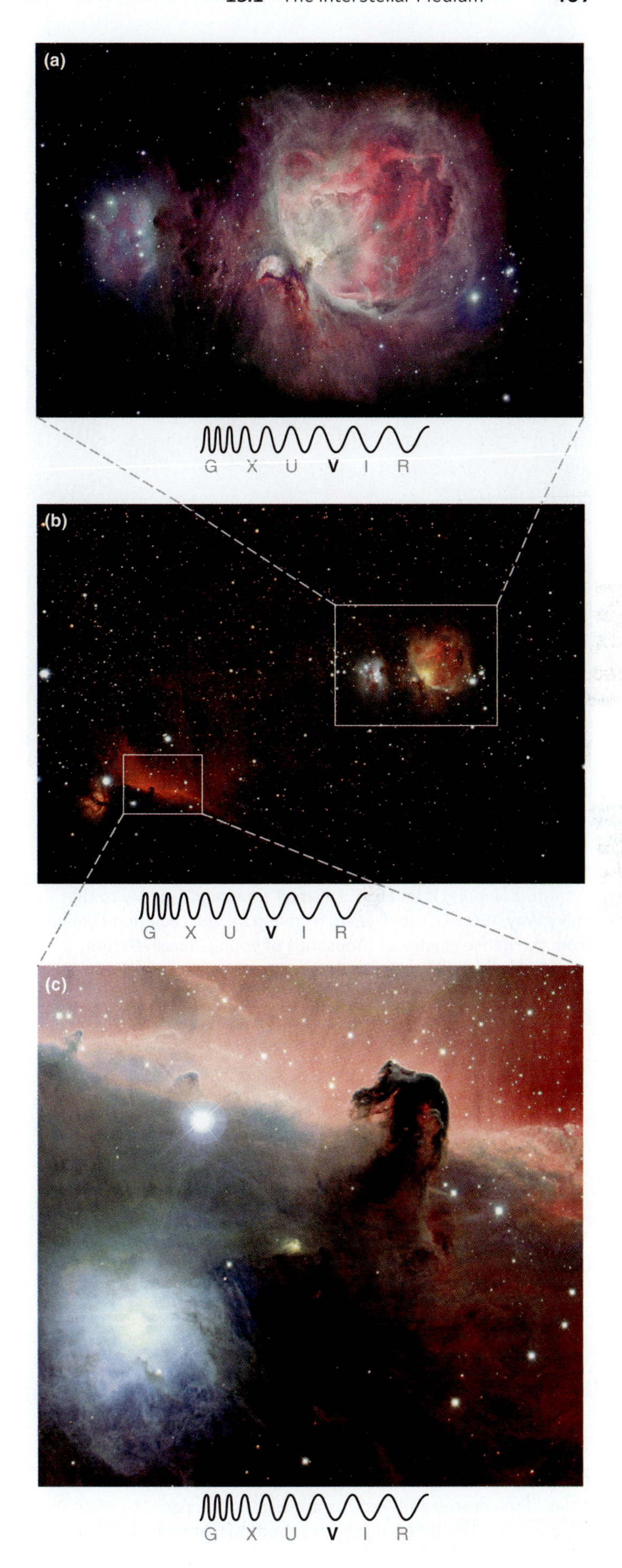

FIGURE 15.8 (a) The Orion Nebula is seen here as a glowing region of interstellar gas surrounding a cluster of young, hot stars. New stars are still forming in the dense clouds surrounding the nebula. (b) The Orion Nebula is only a small part of the larger Orion star-forming region. The dark Horsehead Nebula is seen at the lower left of image (b) and in (c). The circular halos around the bright stars in (c) are a photographic artifact.

direction (aligned), or they point in opposite directions (unaligned). The electron's "magnet" and the proton's "magnet" push on each other, so the way they line up affects the atom's energy. When the two "magnets" point in opposite directions, the atom has slightly less energy than when they point in the same direction.

FIGURE 15.9 The 30 Doradus H II region (also known as the Tarantula Nebula) is located in a small companion galaxy to the Milky Way, 160,000 light-years from Earth. The radiation coming from the dense cluster of thousands of young, massive stars causes the interstellar gas in this region of intense star formation to glow. Hot gas (blue) was detected by the Chandra X-ray Observatory; the surrounding cooler gas and dust were observed by the Spitzer Space Telescope in the infrared.

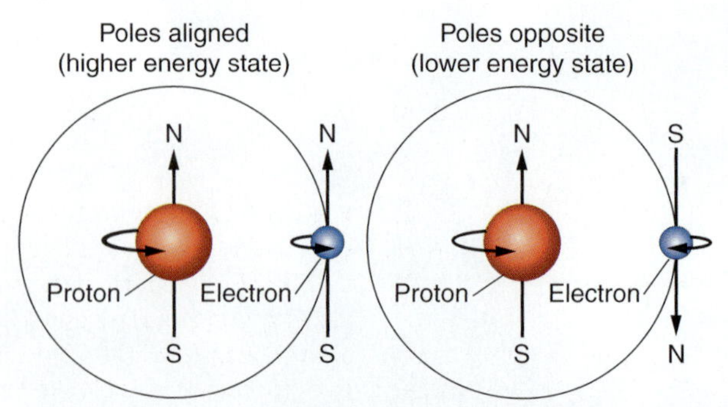

FIGURE 15.10 There is a slight difference in energy when the poles of the proton and electron are aligned compared to when they are opposite. This difference corresponds to a photon of 21 cm.

It takes energy to change a hydrogen atom from the lower-energy, "magnetically" unaligned spin state to the higher-energy, "magnetically" aligned spin state. In warm, neutral interstellar gas, interactions between atoms easily supply this energy. But once a hydrogen atom is in its higher-energy (magnetically aligned) spin state, getting back into its lower-energy (magnetically unaligned) spin state requires only time. If left undisturbed long enough, a hydrogen atom in the higher-energy state will spontaneously jump to the lower-energy state, emitting a photon in the process. The energy difference between the two magnetic spin states of a hydrogen atom is extremely small: less than half a millionth of the energy needed to tear the electron away from the proton completely. The radiation given off by these low-energy transitions has a wavelength of 21 cm, which lies in the radio portion of the spectrum.

Neutral hydrogen emits 21-cm radiation.

The tendency for hydrogen atoms to emit 21-cm radiation is extremely weak. On average, you would have to wait about 11 million years for a hydrogen atom in the higher-energy state to spontaneously jump to the lower-energy state and give off a photon. But there is a *lot* of hydrogen in the universe. In **Figure 15.11** the sky is aglow with 21-cm radiation from neutral hydrogen. Because of its long wavelength, 21-cm radiation freely penetrates dust in the interstellar medium, enabling astronomers to see neutral hydrogen throughout the galaxy, while measurements of the Doppler shift of the line indicate how fast the source of the radiation is moving. These two attributes make the 21-cm line of neutral hydro-

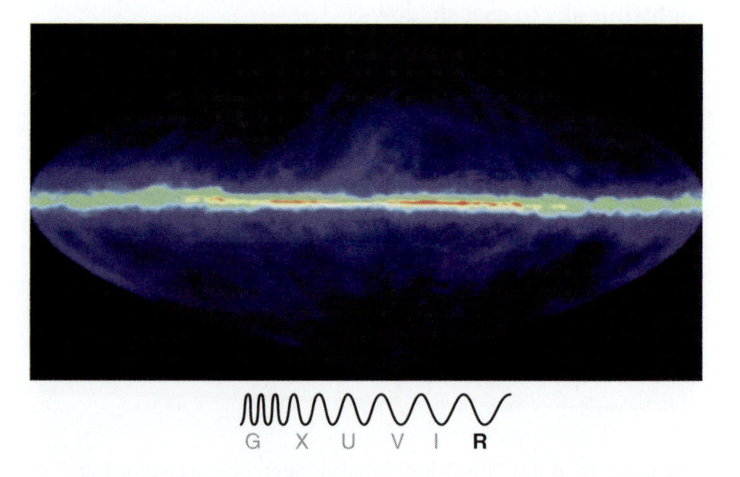

FIGURE 15.11 This radio image of the sky, a composite from many 21-cm surveys, shows the distribution of neutral hydrogen gas throughout the galaxy. Red indicates directions of high hydrogen density; blue and black show areas with little hydrogen. Because radio waves penetrate interstellar dust, 21-cm observations are a crucial probe of the structure of the galaxy.

gen perhaps the most important kind of radiation there is for understanding the structure of the Milky Way.

Clouds Are Regions of Cool, Dense Gas

As mentioned earlier, intercloud gas fills 98 percent of the volume of interstellar space but accounts for only half of the mass of interstellar gas. The remaining 50 percent of all interstellar gas is concentrated into much denser interstellar clouds that occupy only 2 percent of the volume of the galaxy. Most interstellar clouds are composed primarily of neutral atomic hydrogen (that is, isolated neutral hydrogen atoms). These clouds are much cooler and denser than the warm intercloud gas. They have temperatures of about 100 K and densities in the range of about 1–100 atoms/cm^3. But even at the high end of that range—100 atoms/cm^3—you would still have to gather up all the gas in a box measuring 180 km on an edge to collect a single gram of this material.

On Earth it is uncommon to find atoms in isolation; most atoms are tied up in molecules. (For example, in Earth's atmosphere, only nonreactive gases such as argon are typically found in their atomic form.) In most of interstellar space, however, including most interstellar clouds, molecules do not survive long. When interstellar gas is too hot, any molecules that do exist soon collide with other molecules or atoms that have enough energy to break the molecules back down into their constituent atoms. The temperature in a neutral hydrogen cloud may be low enough for some molecules to survive, but even in a neutral hydrogen cloud a different process will soon destroy any molecules that might form. Photons of starlight with enough energy to break molecules apart can penetrate neutral hydrogen clouds. Only in the hearts of the densest of interstellar clouds, where dust effectively blocks even the relatively low-energy photons required to shatter molecules, does interstellar chemistry get its chance. For this reason, these dark clouds are referred to as **molecular clouds**.

Molecular clouds are dense and cold.

In images such as those in **Figure 15.12**, molecular clouds are evident from their silhouettes against a background of stars. Inside such clouds it is dark and usually very cold, with a typical temperature of only about 10 K—that is, 10 kelvins above absolute zero. Most of these clouds have densities of about 100–1,000 molecules/cm^3, but densities as high as 10^{10} molecules/cm^3 have been observed. Even at 10^{10} molecules/cm^3 this gas is still less than a billionth as dense as the air around you, making it an extremely good vacuum by terrestrial standards. In this cold, relatively dense environment, atoms combine to form a wide variety of molecules.

By far the most common component of molecular clouds is molecular hydrogen. Molecular hydrogen (H_2) consists of two hydrogen atoms and is the smallest possible molecule. Molecules radiate mostly in the radio and infrared portions of the electromagnetic spectrum, as they make transitions between energy states determined by the way they rotate or vibrate, for example. Molecular emission lines are useful to astronomers in much the same way that the spectral lines from atoms are useful. Each type of molecule is unique in its properties, and the energies of its states are unique as well. The wavelengths of emission lines from molecules

(a)

(b)

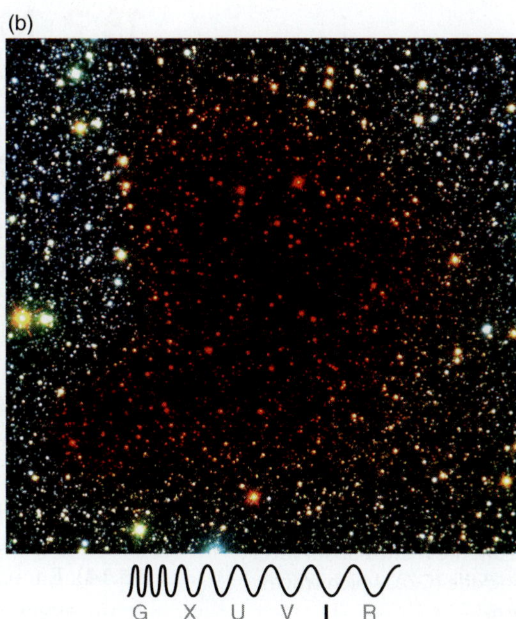

FIGURE 15.12 In visible light, interstellar molecular clouds are seen in silhouette against a background of stars and glowing gas. (a) Light from background stars is blocked by dust and gas in nearby Barnard 68, a dense, dark molecular cloud. (b) Infrared wavelengths can penetrate much of this gas and dust, as seen in this false-color image of Barnard 68.

are an unmistakable fingerprint of the kind of molecule responsible for them.

In addition to molecular hydrogen, approximately 150 other molecules have been observed in interstellar space. These molecules range from very simple structures such as carbon monoxide (CO), one of the more common molecules, to complex organic compounds such as methanol (CH_3OH) and acetone $(CH_3)_2CO$, to some with large carbon chains. Very large carbon molecules, made of hundreds of individual atoms, bridge the gap between large interstellar molecules and small interstellar grains. As mentioned earlier, visible light cannot escape from the molecular clouds where interstellar molecules are concentrated. The radio waves emitted by molecules, however, are unaffected by interstellar dust, so they escape easily from dark molecular clouds. Observations of radio waves from molecules give astronomers a remarkable look at the innermost workings of the densest and most opaque interstellar clouds. Among the more important molecules is CO (**Figure 15.13**). Many astronomers think that the ratio of CO to H_2 is relatively constant, and because molecular hydrogen is so difficult to detect even by infrared and radio observations, carbon monoxide is often used to reveal the amounts and distribution of molecular hydrogen.

Molecular clouds have masses ranging from a few times the mass of the Sun to 10 million solar masses. The smallest molecular clouds may be less than half a light-year across; the largest may be over a thousand light-years in size. The largest molecular clouds qualify for the title **giant molecular clouds**. These behemoths typically have masses a few hundred thousand times that of the Sun, and on average they are about 120 light-years across. The Milky Way contains about 4,000 giant molecular clouds and a much larger number of smaller ones. Molecular clouds fill only a tiny fraction of the volume of the galaxy—probably only about 0.1 percent of interstellar space. These clouds may be rare, but they are extremely important to our story because they are the cradles of star formation.

After looking at the many different phases of the interstellar medium, ranging from vast expanses of tenuous million-kelvin intercloud gas to the cold, dense interiors of molecular clouds, an obvious question to ask is: how does it all fit together? Discussing the structure of the interstellar medium is a lot like discussing the nature of weather on Earth: it is extremely complex, often difficult to understand, and all but impossible to predict precisely.

In the interstellar medium, the energy to power the "weather" comes from stars. Warm gas heated by ultraviolet radiation from massive, hot stars pushes outward into its surroundings. Blast waves from supernova explosions sweep out vast, hot "cavities," piling up everything in their path like snow in front of a snowplow. Interstellar clouds are destroyed under the onslaught of these violent events, but they are also formed by these same forces. Swept-up intercloud gas becomes the next generation of clouds. Hot bubbles of high-pressure interstellar gas crush molecular clouds, driving up their densities and perhaps triggering the formation of new generations of stars. The interstellar medium is so well stirred that, *on average*, interstellar clouds are moving around with random velocities of approximately 20 km per second (km/s). The fastest motions of interstellar material are those of very hot gas and are measured in thousands of kilometers per second.

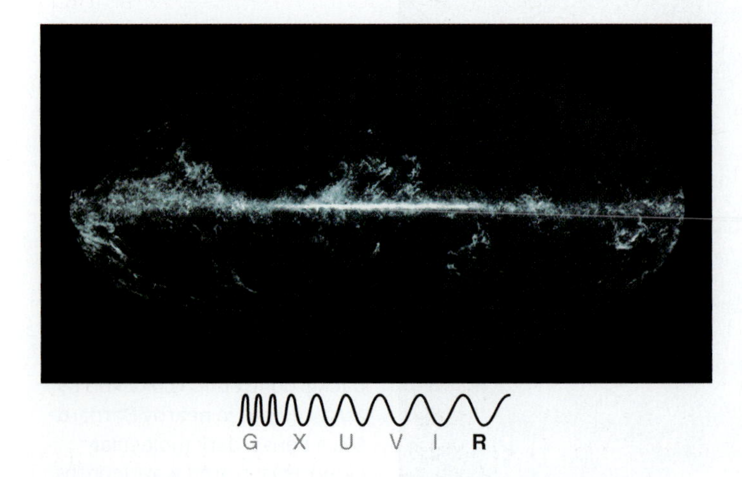

FIGURE 15.13 This all-sky image from the Planck observatory shows the distribution of carbon monoxide (CO), which traces molecular clouds where stars are born.

15.2 Molecular Clouds Are the Cradles of Star Formation

Objects such as planets and stars are held together by gravity. The mutual gravitational attraction between each part of a planet or star and every other part of the object results in a net inward force that pulls all parts of the object toward its center. If an object is stable, then this inward force of the object's self-gravity must be balanced by something—often the outward push of pressure within the object. Recall that in our discussion of the Sun (Chapter 14) we referred to the balance between gravity and pressure in a stable object as *hydrostatic equilibrium*.

These concepts also apply to clouds in the interstellar medium. Interstellar clouds also have self-gravity (**Figure 15.14**). Each part of the cloud feels a gravitational attraction from every other part of the cloud. However, interstellar

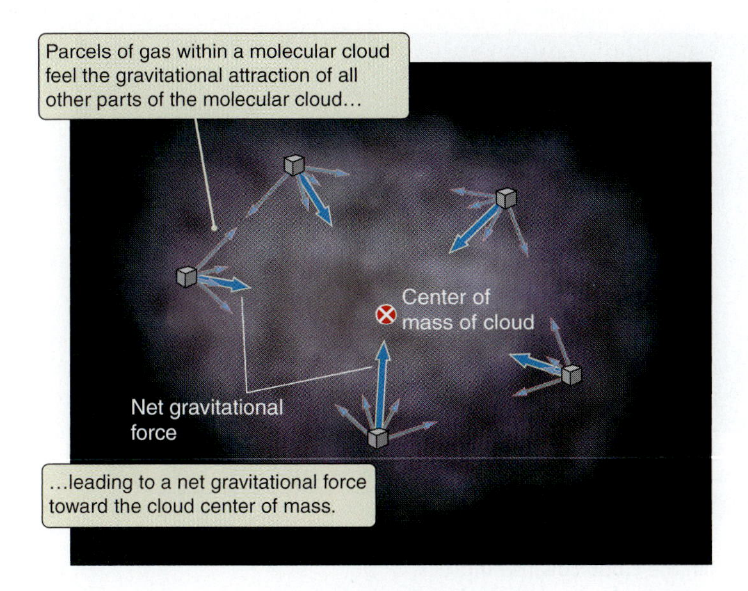

FIGURE 15.14 Self-gravity causes a molecular cloud to collapse.

clouds are not always in hydrostatic equilibrium. In most interstellar clouds, internal pressure is much stronger than self-gravity. (Because gravity follows an inverse square law, the more spread out an object's mass is, the weaker its self-gravity is.) The outward push of pressure in these clouds would cause them to expand if not for the opposing pressure of the surrounding intercloud medium. The intercloud medium is much less dense than interstellar clouds, but it is also much hotter, so its pressure is high enough to confine the clouds. (Pressure is proportional to both density and temperature.)

By contrast, some molecular clouds violate hydrostatic equilibrium in the other direction. These clouds are massive enough and dense enough—that is, there is enough mass packed together in a small enough volume of space—that their self-gravity becomes important. (More mass, along with smaller distances between different parts of the cloud, results in stronger gravity.) Furthermore, these clouds are cool enough that their internal pressure is relatively low despite their high density. In such clouds, self-gravity is much greater than pressure, so the clouds collapse under their own weight, beginning a chain of events that will culminate with the ignition of nuclear fires within a new generation of stars.

Some molecular clouds collapse under their own weight.

If self-gravity in a molecular cloud is much more significant than pressure, gravity should win outright, and the cloud should rapidly collapse toward its center. In practice, the process goes very slowly because several other effects stand in the way of the collapse. One such effect that slows the collapse of a cloud is conservation of angular momentum. (Recall the discussion of the effects of angular momentum and the flattening of a collapsing cloud in Chapter 7.) Other properties of molecular clouds that prevent them from collapsing rapidly are *turbulence* and the effects of magnetic fields. Even though these effects may slow the collapse of a molecular cloud, in the end gravity will win. Magnetic fields in the cloud can slowly die away. One part of the cloud can lose angular momentum to another part of the cloud, allowing the part of the cloud with less angular momentum to collapse further. Turbulence ultimately fades away. The details of these processes are complex and are the subject of much current research, but the crucial point is this: the effects preventing the collapse of a molecular cloud are temporary, and gravity is persistent. As the forces that oppose the cloud's self-gravity gradually fade away, the cloud slowly becomes smaller. ▶II **AstroTour: Star Formation**

Angular momentum, magnetic fields, and turbulence slow a collapsing molecular cloud, but ultimately gravity wins.

Molecular Clouds Fragment as They Collapse

Crucial to the process of star formation is the fact that as a cloud becomes smaller, the gravitational forces trying to pull it together grow stronger. This effect is a result of the inverse square law of gravitation discussed in Chapter 4. Suppose a cloud starts out being 4 light-years across. By the time the cloud has collapsed to 2 light-years across, the different parts of the cloud are, on average, only half as far away from each other as they were when they started. As a result, the gravitational attraction they feel toward each other will be 4 times stronger than it was when the cloud was 4 light-years across. When the cloud is ¼ as large as when it started out (or 1 light-year across), the force of gravity will be 16 times stronger. As a collapsing cloud becomes smaller, gravity becomes stronger. As gravity becomes stronger, the collapse picks up speed. And as the collapse picks up speed, gravity increases even more rapidly. In other words, the collapse "snowballs."

Molecular clouds are never uniform. Some regions within the cloud are invariably denser than others, so they collapse within themselves more rapidly than do surrounding regions. As these regions collapse, their self-gravity becomes stronger because they are more

Stars form in molecular-cloud cores.

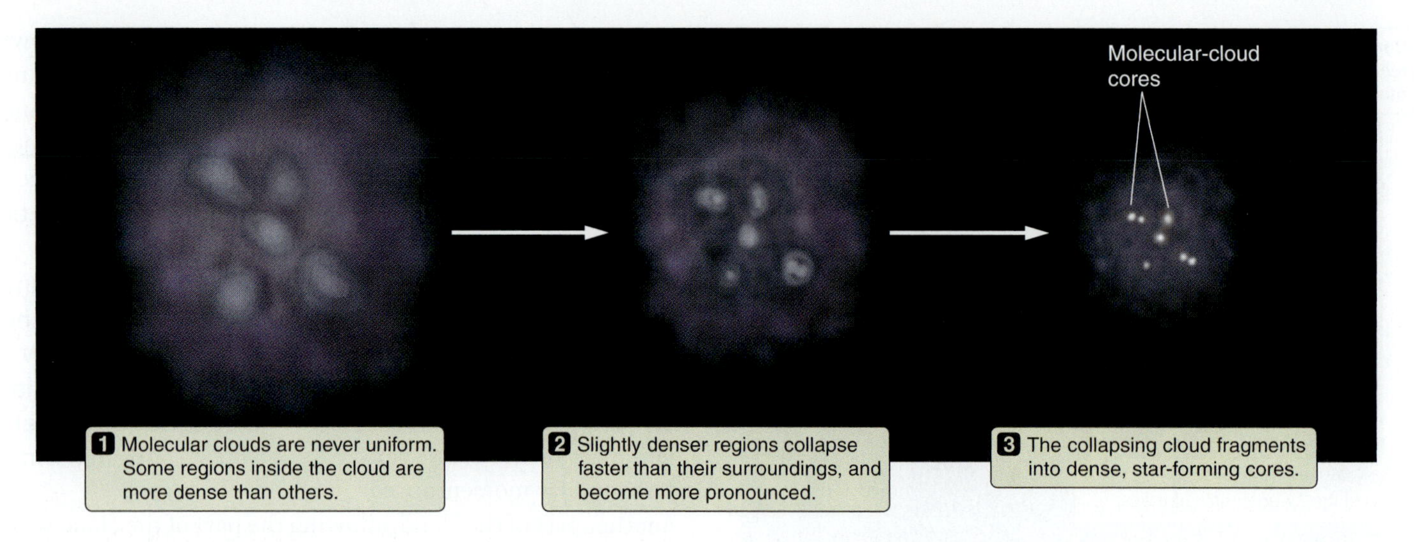

FIGURE 15.15 When a molecular cloud collapses, denser regions within the cloud collapse more rapidly than less dense regions. As this process continues, the cloud fragments into a number of very dense molecular-cloud cores that are embedded within the large cloud. These cloud cores may go on to form stars.

compact, so they collapse even faster. **Figure 15.15** shows the result of this process. What start out as slight variations in the density of the cloud grow to become very dense concentrations of gas. Instead of collapsing into a single object, the molecular cloud fragments into a number of very dense **molecular-cloud cores**. A single molecular cloud may form hundreds or thousands of molecular-cloud cores. It is from these dense cloud cores, typically a few light-months in size, that stars form.

As a molecular-cloud core becomes smaller, the gravitational forces that are trying to crush the cloud grow stronger. Eventually, gravity is able to overwhelm the forces due to pressure, magnetic fields, and turbulence that have been resisting it. This capitulation to gravity happens first near the center of the cloud because that's where the cloud material is most strongly concentrated. The pressure from the central part of the cloud core supports the weight of the layers above it. When the center

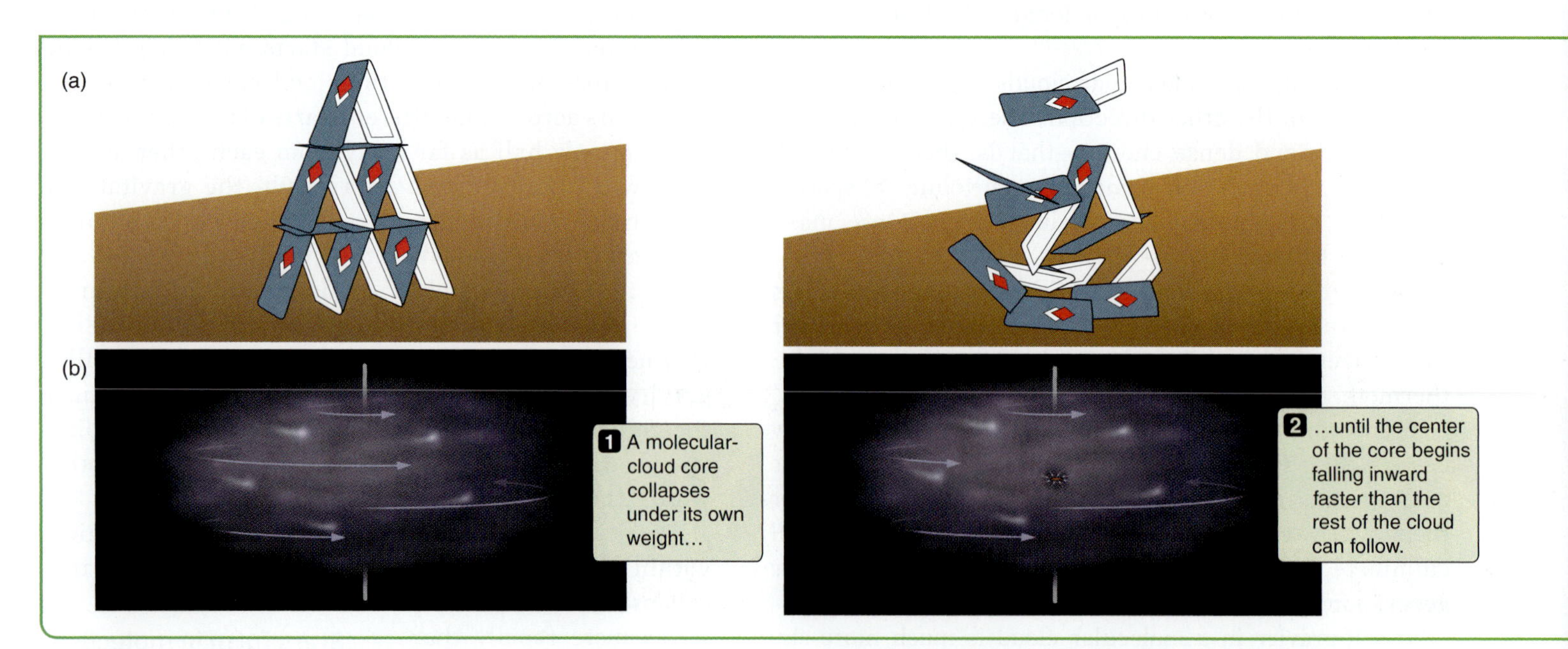

of the cloud collapses, this support is suddenly removed. Without this "pressure support," the next-outer layer begins to fall freely inward toward the center as well. But what about the layers farther out? Now there is nothing to hold them up either, so the process continues: each layer of the cloud core falls inward in turn, thereby removing support from the layers still farther out. As shown in **Figure 15.16**, the cloud core collapses like a house of cards when the bottom layer is knocked out. The whole structure comes crashing down.

It was at this point in the story of star formation that our discussion of the formation of the Solar System in Chapter 7 began. **Figure 15.17** shows an overview of the process. Material from the collapsing molecular-cloud core falls inward. Because of its angular momentum, this material lands on a flat, rotating accretion disk. (Figure 7.2 shows several such disks.) The dark bands in these images are the disks seen edge on; the bright regions are starlight reflected from the surfaces of the disks. Most of this material eventually finds its way inward onto the growing protostar at the center of the disk, but a small fraction of it remains to become the stuff planets are made of. In Chapter 7 we followed the evolution of the gas and dust left behind in the disk and saw how it leads naturally to a planetary system with the properties of the Solar System. This time through, we instead follow the story of the protostar as it becomes a star.

The Solar System began in a molecular cloud.

15.3 The Protostar Becomes a Star

We pick up the description of the protostar at a time when the cloud is still collapsing, and more and more material is falling onto the disk. At this point the surface of the protostar has been heated to a temperature of thousands of kelvins as the gravitational energy of the original molecular cloud has been converted into thermal energy. The protostar is also huge—hundreds of times larger than the Sun—which means that the surface of the protostar is tens of thousands of times larger than the surface of the Sun. Each square meter of its enormous surface is radiating energy away in accordance with the Stefan-Boltzmann law (see Chapter 5). As a result, the protostar is thousands of times more luminous than the Sun, even though the nuclear reactions that will power it through its life have yet to begin.

Protostars are large and luminous.

Although the protostar is extremely luminous at this stage of its life, it probably cannot yet be seen by astronomers in visible light, for two reasons: First, the protostar is relatively cool as stars go, so most of its radiation is in the infrared part of the spectrum. Second, and even more important, the protostar begins its life buried deep in the heart of a dense and dusty molecular cloud. The dust in the cloud blocks any visible light, so the view of the protostar is obscured. However, much of the longer-wavelength infrared light from the protostar *is* able to escape from the

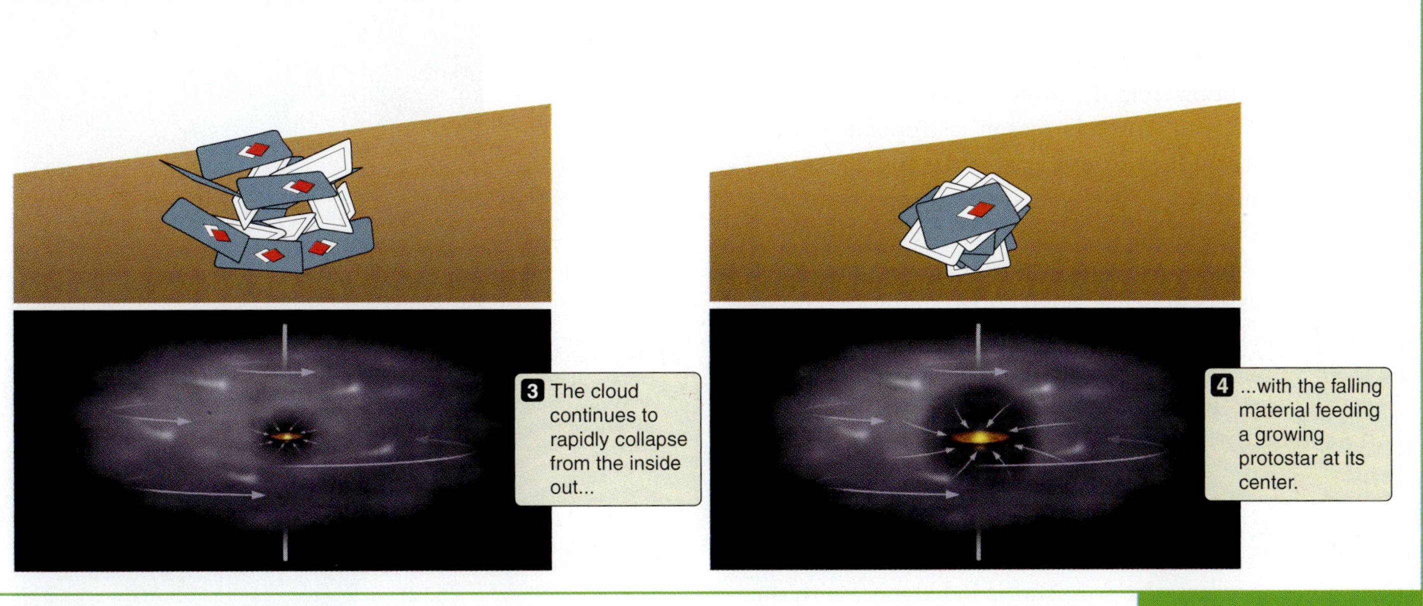

FIGURE 15.16 (a) When the bottom layer is knocked out from under a house of cards, the entire structure collapses from the ground up. (b) Similarly, when a molecular-cloud core gets very dense, it collapses from the inside out. Conservation of angular momentum causes the infalling material to form an accretion disk that feeds the growing protostar.

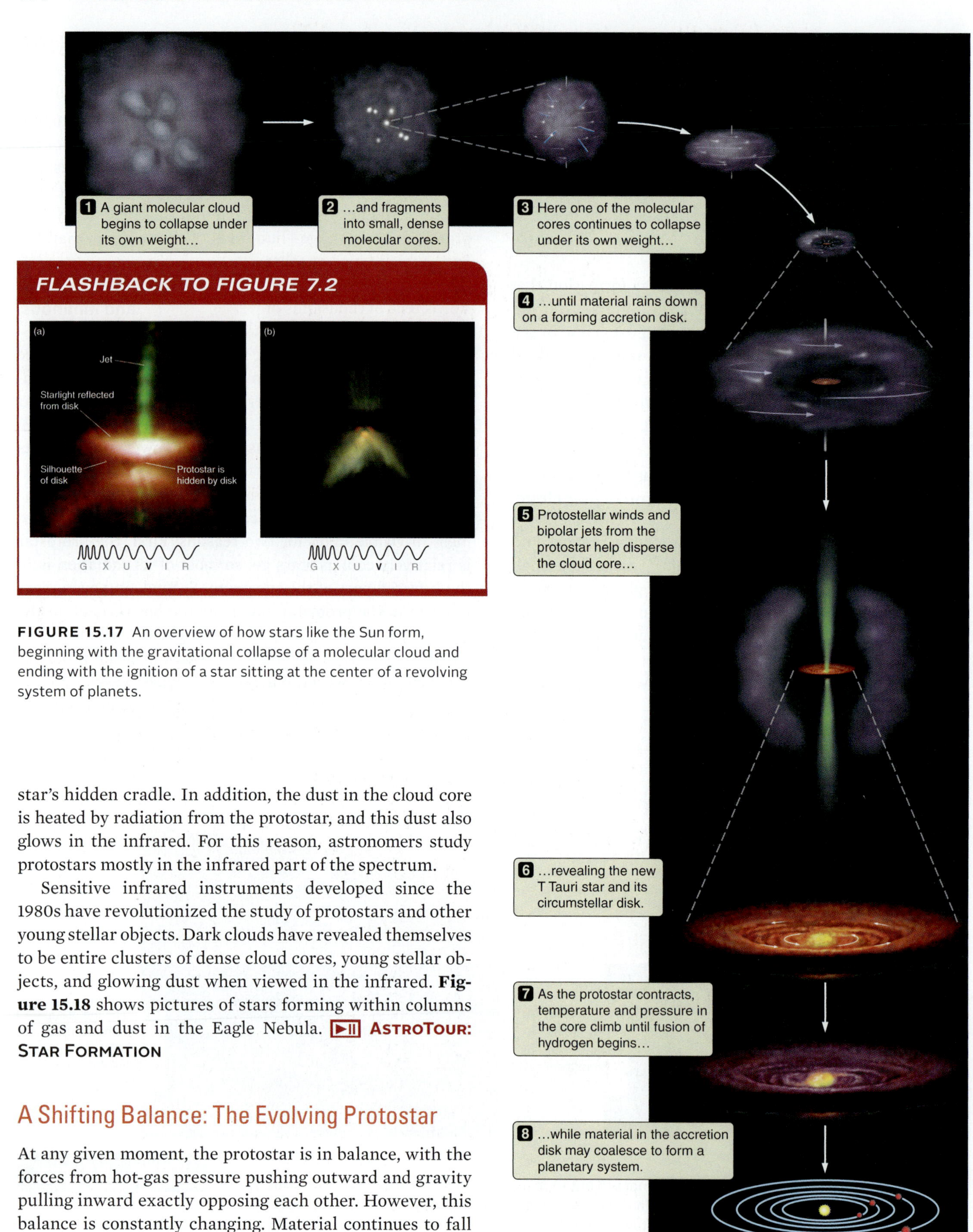

FIGURE 15.17 An overview of how stars like the Sun form, beginning with the gravitational collapse of a molecular cloud and ending with the ignition of a star sitting at the center of a revolving system of planets.

star's hidden cradle. In addition, the dust in the cloud core is heated by radiation from the protostar, and this dust also glows in the infrared. For this reason, astronomers study protostars mostly in the infrared part of the spectrum.

Sensitive infrared instruments developed since the 1980s have revolutionized the study of protostars and other young stellar objects. Dark clouds have revealed themselves to be entire clusters of dense cloud cores, young stellar objects, and glowing dust when viewed in the infrared. **Figure 15.18** shows pictures of stars forming within columns of gas and dust in the Eagle Nebula. **ASTROTOUR: STAR FORMATION**

A Shifting Balance: The Evolving Protostar

At any given moment, the protostar is in balance, with the forces from hot-gas pressure pushing outward and gravity pulling inward exactly opposing each other. However, this balance is constantly changing. Material continues to fall onto the protostar, adding to its mass and gravitational pull inward and therefore increasing the weight that underlying layers of the protostar must support. The protostar also slowly loses its internal thermal energy by radiating it away.

How can an object be in perfect balance and yet be changing at the same time? Consider a more everyday example. **Figure 15.19a** shows a simple spring balance, which works on the principle that the more a spring is compressed, the harder it pushes back. When sand is poured slowly onto the spring balance, at any point the downward weight of the sand is balanced by the upward force of the spring. As the weight of the sand increases, the spring is slowly compressed. The spring and the weight of the sand are always in balance, but this balance is *changing with time* as more and more sand is added. In just the same way, the protostar is always in balance, with the inward force of gravity matched by the outward force of pressure (**Figure 15.19b**). The protostar's balance, too, is changing with time. Just like sand that has fallen onto the spring balance, material that has fallen onto the protostar compresses the protostar more and more, and the protostar evolves.

As material continues to fall onto the protostar, it adds weight to the protostar's outer layers. This growing weight compresses the material in the interior of the protostar, and as it is compressed, the protostar's interior grows hotter. (Compressing a gas always causes it to heat up, and letting a gas expand always causes it to cool down. The cooling systems in your air conditioner and refrigerator work by compressing gas to make it hot, and then letting this hot gas cool so that when it expands again it gets really cold.) If the interior of the protostar is now denser and hotter, the pressure is also higher—just enough higher to balance the

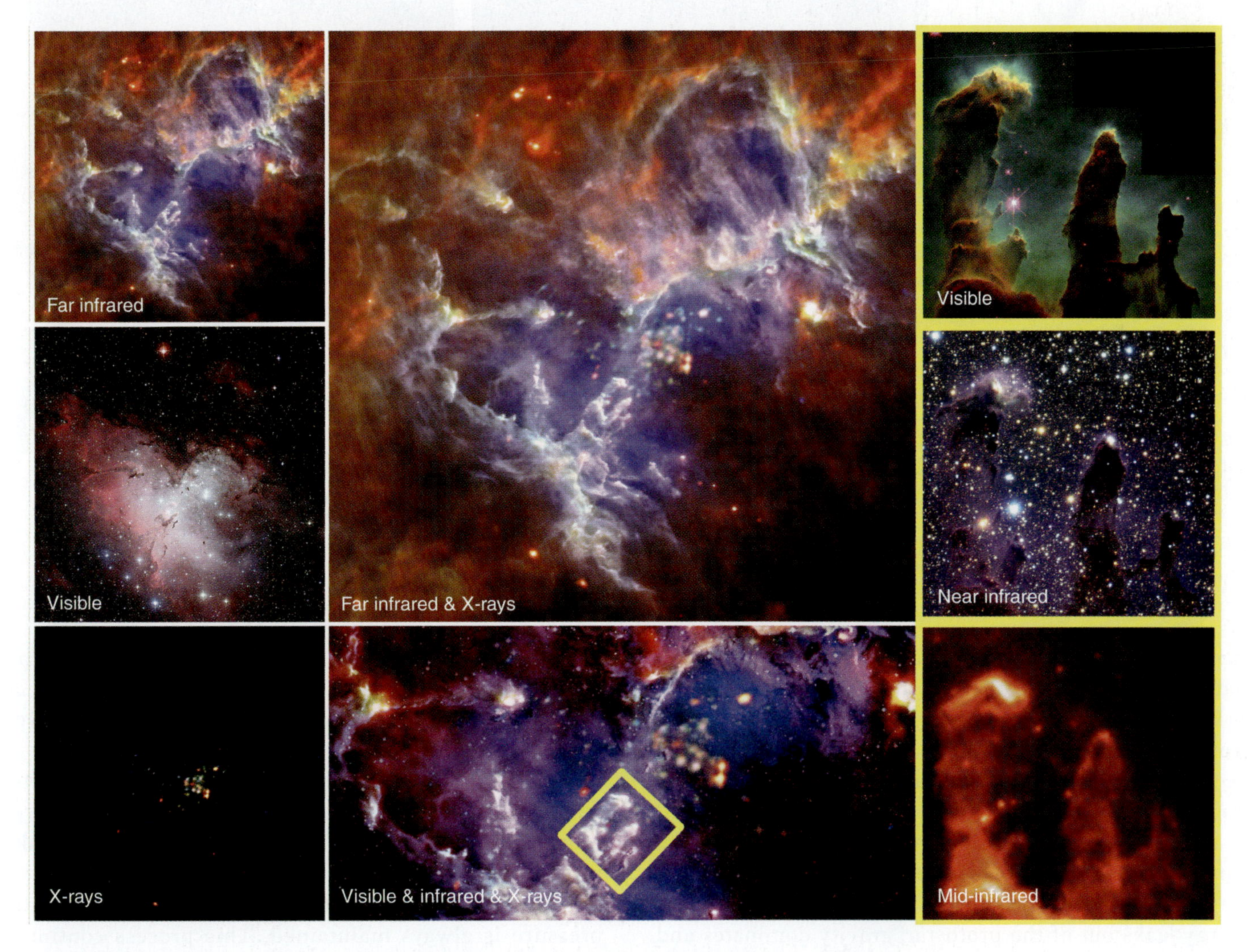

FIGURE 15.18 These multiwavelength images of the Eagle Nebula show dense columns of molecular gas and dust at the edge of an H II region. Infrared and X-ray images of the same field, taken with several telescopes, show young stars forming within these columns. The rightmost column of images zooms in to the area shown in the yellow box.

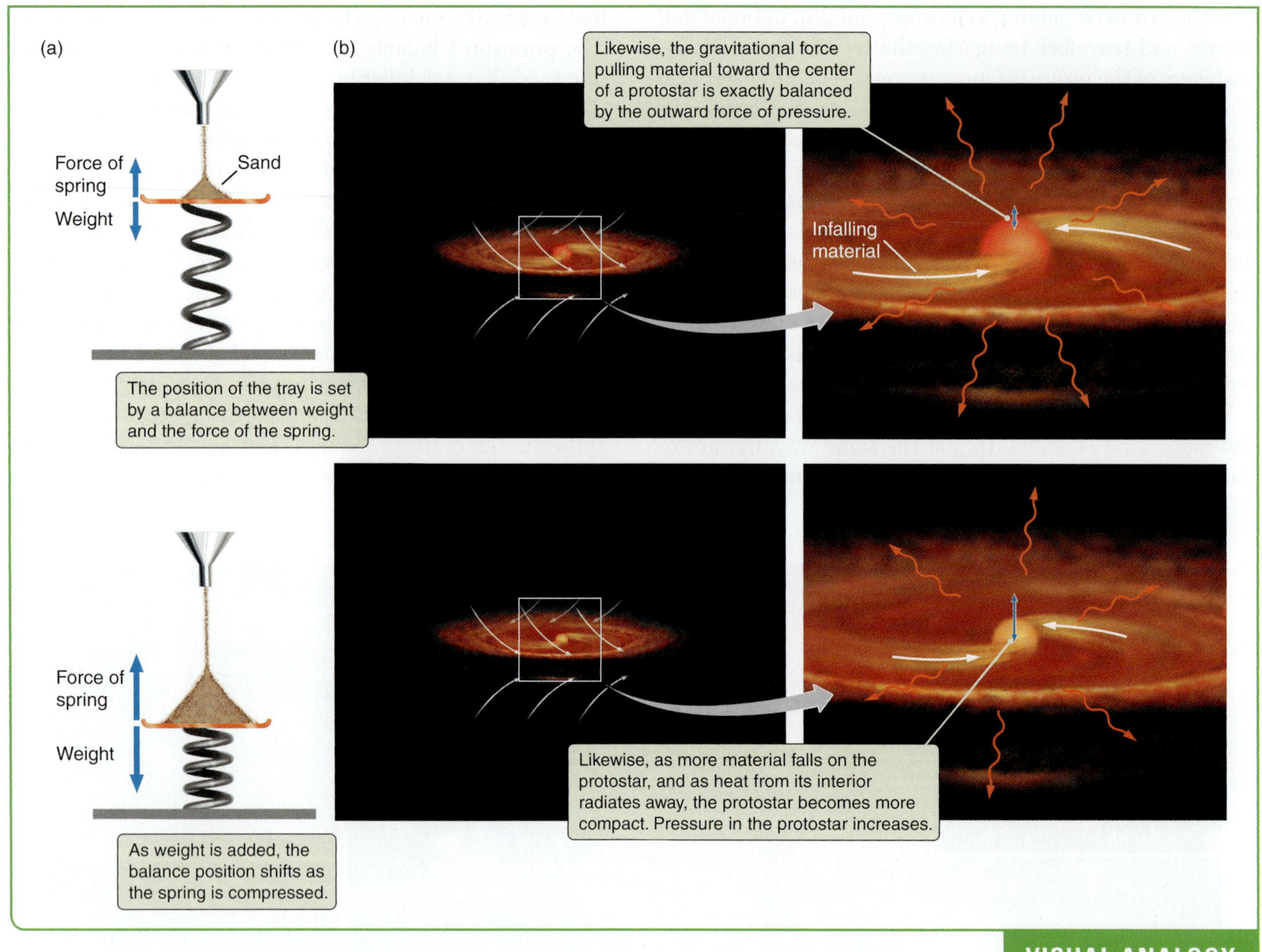

FIGURE 15.19 (a) A spring balance comes to rest at the point where the downward force of gravity is matched by the upward force of the compressed spring. As sand is added, the location of this balance point shifts. (b) Similarly, the structure of a protostar is determined by a balance between pressure and gravity. Like the spring balance, the structure of the protostar constantly shifts as the protostar radiates energy away and additional material falls onto its surface.

increased weight of the material above it. Balance is always maintained.

Even after material stops falling on the protostar, the protostar keeps evolving in much the same way. Earlier we stated that the protostar is radiating away many times more energy than the Sun does. Thermal energy trapped in the interior of the protostar is radiated away—the same thermal energy that is responsible for supporting the protostar against the forces of gravity. So, as thermal energy leaks out of the protostar, gravity dominates and the protostar slowly contracts. As the protostar shrinks, the forces of gravity become greater (because the parts of the protostar are closer together—remember the inverse square law of gravity). As the protostar shrinks, its interior becomes compressed, so it grows hotter. As the gas within the protostar radiates energy, the protostar actually heats up because the heating due to shrinking more than overcomes the cooling due to radiation. As density and temperature climb, so does the pressure—just enough to continue to counteract the growing force of gravity. Balance between gravity and pressure is always maintained. This process continues, with the protostar becoming smaller and smaller and its interior growing hotter and hotter, until the center of the protostar is finally hot enough for hydrogen to begin fusing into helium.

The protostar radiates away thermal energy and shrinks.

When an object radiates energy, it normally gets cooler. When you turn off the electric coil on your stove, for example, the coil cools as it loses the thermal energy within it. Yet we just concluded that as a protostar radiates thermal energy away, it actually grows hotter. How can that be? Once again the answer lies with the concept of conservation of energy. As the protostar contracts, every part of the protostar is slowly falling toward its center, which means that the protostar is *losing gravitational energy*. This gravitational energy has to show up in another form, which in this case is thermal energy. Some of this thermal energy is radiated away, but not all of it—so the interior of the protostar grows hotter.

Figure 15.20 illustrates this chain of events. The protostar radiates away thermal energy and, in the process, loses pressure support. Deprived of pressure support, the protostar contracts. As the protostar contracts, gravitational energy is converted to thermal energy, driving up the temperature in the protostar's interior. If the protostar is massive enough, its interior will eventually become so hot that nuclear fusion can begin. This is the point at which the transition from protostar to star takes place. The distinction between the two is that a *protostar* draws its energy from gravitational collapse, whereas a *star* draws its energy from thermonuclear reactions in its interior.

Fusion begins, and the protostar becomes a star.

The protostar's mass determines whether it will ever become a star. As the protostar slowly collapses, the temperature at its center gets higher and higher. If the

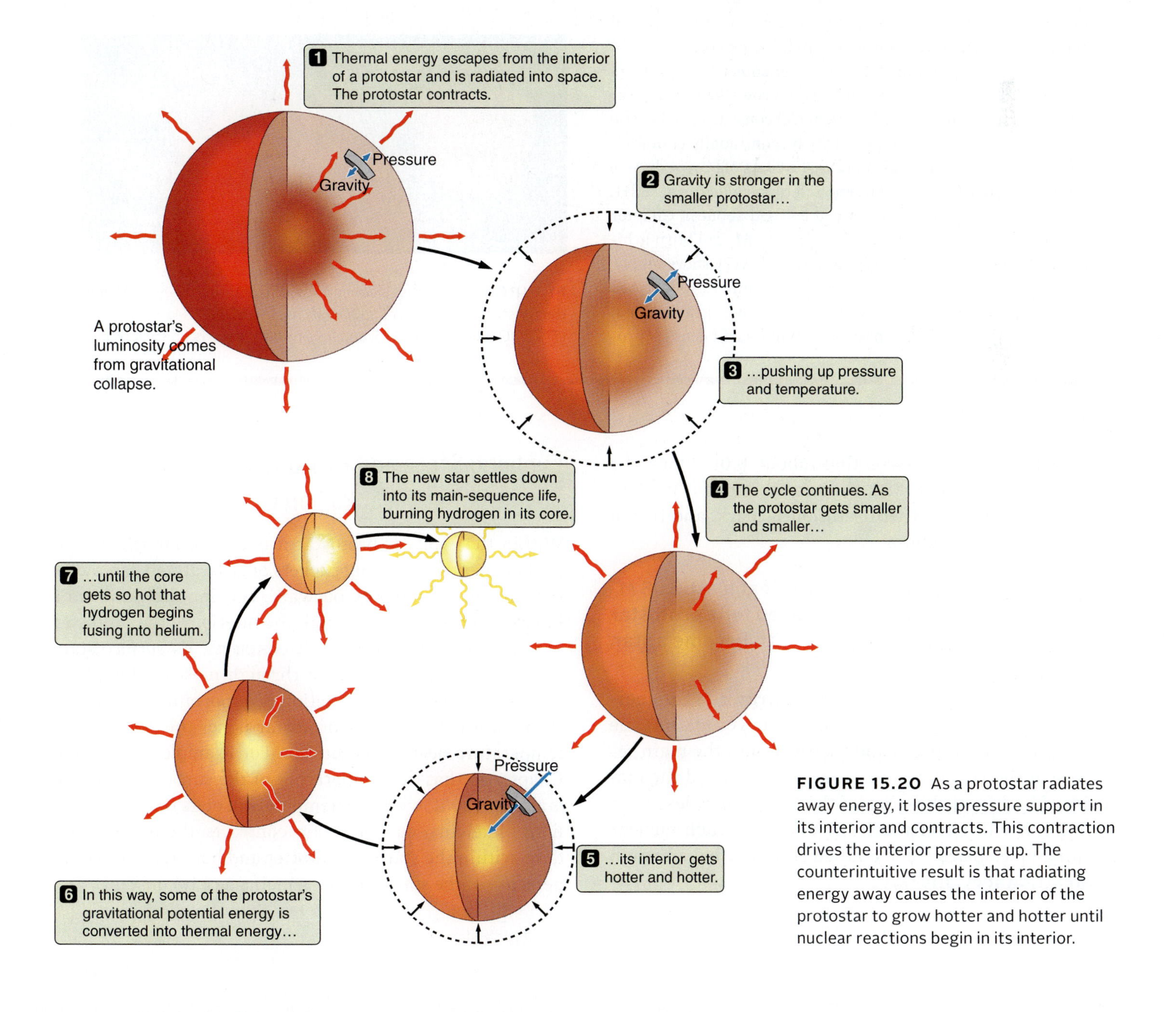

FIGURE 15.20 As a protostar radiates away energy, it loses pressure support in its interior and contracts. This contraction drives the interior pressure up. The counterintuitive result is that radiating energy away causes the interior of the protostar to grow hotter and hotter until nuclear reactions begin in its interior.

Connections 15.1

Brown Dwarfs

A brown dwarf is neither star nor planet, but something in between—sometimes called a *substellar object*. The International Astronomical Union (IAU) has somewhat arbitrarily set the boundary between a brown dwarf and a supermassive giant planet at 13 Jupiter masses (M_J), although there is still some debate as to whether this is really the appropriate criterion. Some astronomers think that 10 M_J would be more appropriate. That is, if the mass of the object is greater than 10 M_J, it must be a brown dwarf, not a supermassive planet. (The upper limit is about 75–80 M_J.)

A brown dwarf forms in the same way a star forms, yet in many respects it is like a giant Jupiter. Brown dwarf spectral types L, T, and Y have been added to the sequence of spectral classes to represent stars that are even cooler than M stars (**Figure 15.21**). A brown dwarf never grows hot enough to burn the most common hydrogen nuclei consisting of a single proton, but instead glows primarily by continually cannibalizing its own gravitational energy. The cores of brown dwarfs larger than 13 M_J can get hot enough to burn deuterium (^2H, the heavy isotope of hydrogen you learned about in Chapter 14); and those with a mass greater than 65 M_J can burn lithium. But both of these energy sources are in very short supply, and after a brief period of deuterium or lithium fusion, brown dwarfs shine only by the energy of their own gravitational contraction. As the years pass, a brown dwarf becomes progressively smaller and fainter. The coldest Y dwarfs observed with the WISE infrared satellite are colder than the human body, which radiates at 310 K.

Since the first brown dwarfs were identified in the mid-1990s, over a thousand have been found. The cooler among them have methane and ammonia in their atmospheres, similar to what is found in the atmospheres of the giant planets of the Solar System. Winds on brown dwarfs can be very high, producing weather far more violent than storms observed in the atmospheres of the giant planets.

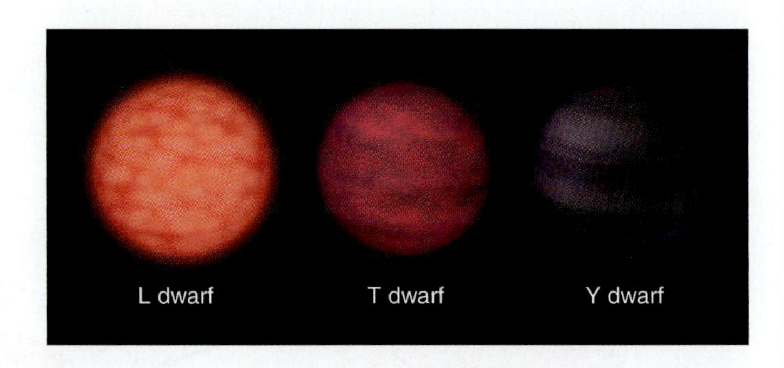

FIGURE 15.21 Artist's conception of the three types of brown dwarf stars: L dwarfs (T ~1700 K, mass ~65 M_J); T dwarfs (T ~1200 K, mass ~30 M_J); and low-mass Y dwarfs (T ~500 K).

protostar is more massive than about 0.08 times the mass of the Sun (0.08 $M_\odot$), the temperature in its core will eventually reach the 10-million-K mark, and fusion of hydrogen into helium will begin. At this onset of fusion, the newly born star will once again adjust its structure until it is radiating energy away from its surface at just the rate that energy is being liberated in its interior. As it does so, it achieves hydrostatic and thermal equilibrium and "settles" onto the main sequence of the H-R diagram, where it will spend the majority of its life. If the mass of the protostar is less than 0.08 $M_\odot$, it will never reach the point at which nuclear burning takes place. Such a failed star is called a **brown dwarf** (**Connections 15.1**).

A mass of at least 0.08 $M_\odot$ is needed for a protostar to become a star.

Evolving Stars and Protostars Follow "Evolutionary Tracks" on the H-R Diagram

Within the evolving protostar it is convection (the transport of energy by moving packets of gas) rather than radiation that carries energy outward, keeping the protostar's interior well stirred. Although the interior of the protostar grows hotter and hotter as it contracts, its *surface* stays about the same temperature through most of this phase of its evolution. This distinction is important. The surface temperature of a star or protostar is *not* the same as the temperature deep in its interior. For example, recall from Chapter 14 that the temperature of the surface of the Sun is about 5780 K, while the temperature of its interior is millions of kelvins. As a protostar contracts, the temperature deep within the star grows hotter and hotter, but the temperature of the star's *surface* remains nearly constant.

In the 1960s the theoretical physicist Chushiro Hayashi (1920–2010) explained why this is so. Hayashi pointed out that the atmospheres of stars and protostars contain a natural thermostat: the H^- ion. (An H^-, or "H minus," ion is a hydrogen atom that has acquired an extra electron and therefore has a negative charge.) The amount of H^- in the atmosphere of a protostar is highly sensitive to the temperature at the protostar's surface. The cooler the atmosphere of a star, the more slowly atoms and electrons are moving, and the easier it is for a hydrogen atom to hold on to an extra electron. As a result, the cooler the atmosphere of the star, the more H^- there is.

The H^- ion, in turn, helps control how much energy a star or protostar radiates away. The more H^- there is in the atmosphere of the star or protostar, the more opaque the atmosphere is, and the more effectively the thermal energy of the protostar is trapped in its interior. Imagine that the surface of the protostar is "too cool," meaning that extra H^- forms in the atmosphere and makes the atmosphere of the protostar more opaque. The atmosphere thus traps more of the radiation that is trying to escape, and the trapped energy heats up the star. As the temperature climbs, H^- ions are destroyed (that is, changed to neutral H atoms). Now imagine the other possibility—that the protostar is too hot. In this case, H^- in the protostar's atmosphere is destroyed, so the atmosphere becomes more transparent, allowing radiation to escape more freely from the interior. Because the protostar cannot hold on to enough of its energy to stay warm, the surface cools. In either case—too cold or too hot—H^- is formed or destroyed until the star's atmosphere once again traps just the right amount of escaping radiation. The H^- ion is basically doing the same thing that you do with your bedcovers at night. If you get too cold, you pile on extra covers to trap your body's thermal energy and keep warm (more H^- ions). If you get too hot, you kick off some covers to cool off (fewer H^- ions).

H^- is a stellar thermostat.

The amount of H^- in the atmosphere keeps the surface temperature of the protostar somewhere between about 3000 and 5000 K, depending on the protostar's mass and age. Because the surface temperature of the protostar is not changing much, the amount of energy per unit time (power) radiated away by each square meter of the surface of the protostar does not change much either. Recall the Stefan-Boltzmann law from Chapter 5, which says that the amount radiated by each square meter of the protostar's surface is determined by its temperature. But as the protostar shrinks, the area of its surface shrinks as well. There are fewer square meters of surface to radiate, so the luminosity of the protostar drops (**Math Tools 15.2**). As viewed from the outside, the protostar stays at nearly the

Math Tools 15.2

Luminosity, Surface Temperature, and Radius of Protostars

In Chapter 13 you learned how the luminosity, surface temperature, and radius of a star are related:

$$L = 4\pi R^2 \sigma T^4$$

What can this equation reveal about the changing properties of the protostar as it shrinks its radius?

Suppose that when the Sun was a protostar, it had a radius 100 times what it is now, and a surface temperature of 3300 K. What would its luminosity have been? The equations for each are:

$$L_{\text{protostar}} = 4\pi R^2_{\text{protostar}} \sigma T^4_{\text{protostar}}$$

and

$$L_\odot = 4\pi R^2_\odot \sigma T^4_\odot$$

We can set this up as a ratio, comparing the luminosity of the protostar Sun with its luminosity now, $L_\odot$:

$$\frac{L_{\text{protostar}}}{L_\odot} = \frac{4\pi R^2_{\text{protostar}} \sigma T^4_{\text{protostar}}}{4\pi R^2_\odot \sigma T^4_\odot}$$

We rewrite this as follows, grouping like terms together:

$$\frac{L_{\text{protostar}}}{L_\odot} = \frac{4\pi\sigma}{4\pi\sigma} \times \left(\frac{R_{\text{protostar}}}{R_\odot}\right)^2 \times \left(\frac{T_{\text{protostar}}}{T_\odot}\right)^4$$

Then we cancel out the constants, $4\pi\sigma$, and use the value for $T_\odot$ from Chapter 14: 5780 K. We know that the protostar's radius is 100 times that of the Sun, so $R_{\text{protostar}}/R_\odot = 100$. Then the equation becomes:

$$\frac{L_{\text{protostar}}}{L_\odot} = \left(\frac{100}{1}\right)^2 \times \left(\frac{3300}{5780}\right)^4 = 100^2 \times (0.57)^4 = 1{,}060$$

So the Sun was about 1,000 times more luminous as a protostar than it is now. We see this on the H-R diagram of protostars (see Figure 15.23). As the star approaches the main sequence on the diagram, it moves down (toward lower luminosity) and to the left (toward higher surface temperature).

same temperature and color but gradually gets fainter as it evolves toward its eventual life as a main-sequence star.

In Chapter 13 we introduced the H-R diagram and used it to help explain how the properties of stars differ. For the next several chapters we will use the H-R diagram to keep track of how stars change as they evolve through their lifetimes. The path across the H-R diagram that a star follows as it goes through the different stages of its life is called the star's **evolutionary track** (**Figure 15.22a**). The particular path that an evolving protostar follows as it approaches the main sequence is called its **Hayashi track** (**Figure 15.22b**). The protostar is brighter than it will be as a true star on the main sequence, so a protostar's Hayashi track is located above the main sequence on the H-R diagram. Because the surface temperature of the protostar stays nearly constant as the protostar contracts, the protostar's Hayashi track prior to the start of hydrogen burning is an almost vertical line on the H-R diagram. Figure 15.22 shows the pre-main-sequence evolutionary tracks of stars of several different masses. Astronomers say that an evolving protostar "descends the Hayashi track."

15.4 Not All Stars Are Created Equal

In Chapter 13 we found that stars can have a wide range of masses, varying from less than 1/10 the mass of the Sun up to about 100 times the mass of the Sun. The mass of a star determines its path through its life, so astronomers want to know what determines how massive a star will be. One obvious possibility is that a forming star grows until it uses up all of the gas around it; it becomes no larger, simply because it has run out of material. This scenario is not consistent with observations of how stars actually form. Only a small

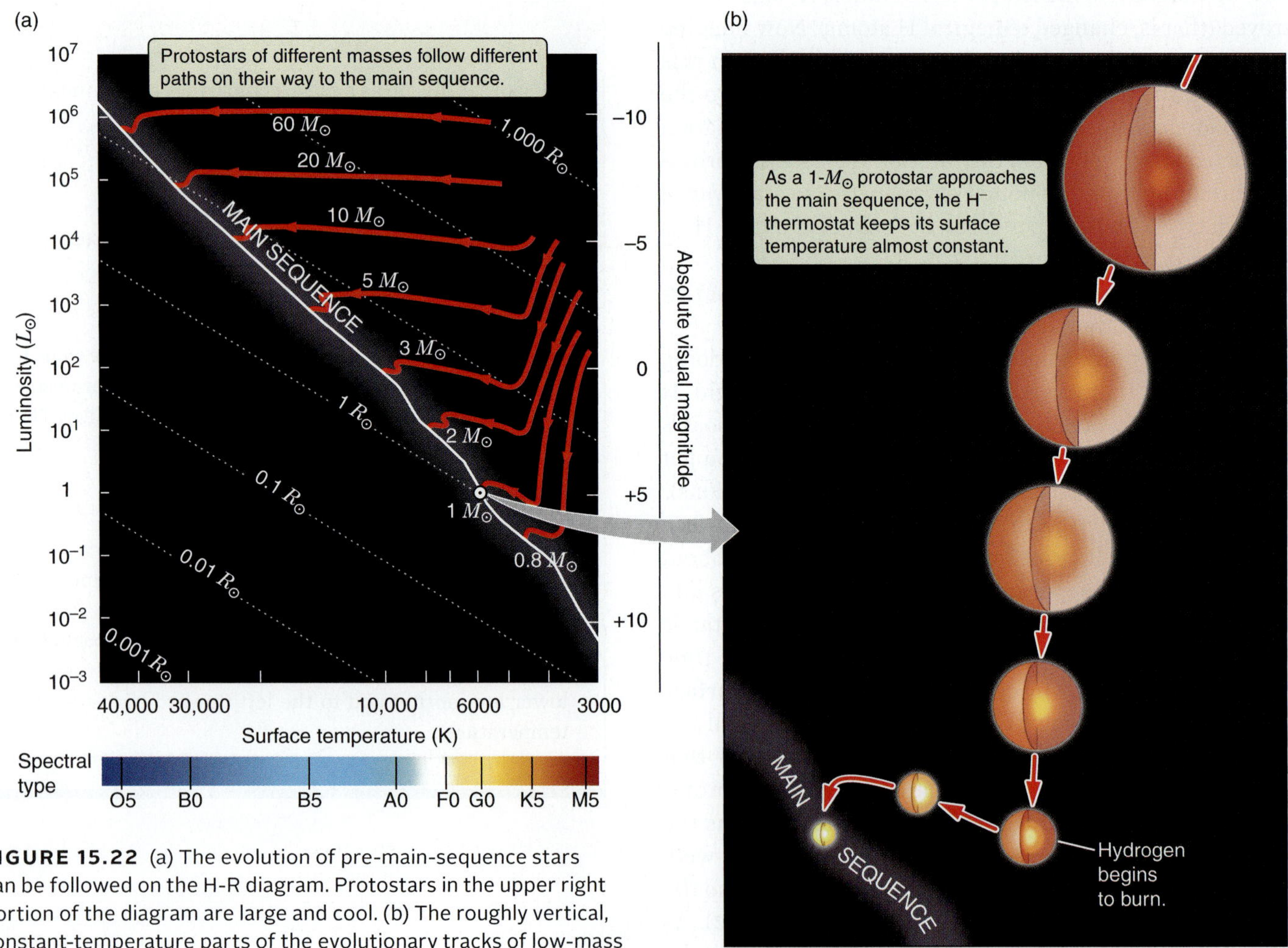

FIGURE 15.22 (a) The evolution of pre-main-sequence stars can be followed on the H-R diagram. Protostars in the upper right portion of the diagram are large and cool. (b) The roughly vertical, constant-temperature parts of the evolutionary tracks of low-mass protostars are called Hayashi tracks.

fraction of the material in a molecular cloud—perhaps a few percent—ends up in the stars forming within it. Something must prevent most of the material in a molecular cloud from ever actually falling onto protostars.

Material falls onto the accretion disk around a young stellar object and moves inward toward the equator of the star, while at the same time other material is blown away from the protostar and disk in two opposite directions away from the plane of the disk (**Figure 15.23a**). The resulting stream of material away from the protostar is called a **bipolar outflow**. Powerful outflows can disrupt the cloud core and accretion disk from which the protostar formed, shutting down the flow of material onto the protostar.

Some bipolar outflows from young stellar objects are slow and fairly disordered, but others produce remarkable **jets** of material that move away from the central protostar and disk at velocities of hundreds of kilometers per second (**Figure 15.23b**). The material in these jets flows out into the interstellar medium, where it heats, compresses, and pushes away surrounding interstellar gas. Knots of glowing gas accelerated by jets are referred to as **Herbig-Haro objects** (or **HH objects** for short), named after the two astronomers who first identified them and associated them with star formation. **Figure 15.24** shows Hubble Space Telescope images of jets in HH objects.

Protostars drive powerful bipolar outflows.

The origin of outflows from protostars is not well understood, but current models suggest that they are the result of magnetic interactions between the protostar and the disk. The interior of a protostar on its Hayashi track is convective. Great cells of hot gas are rising from this interior, while other cells of cooler gas are falling toward the center. This convection, coupled with the protostar's rapid rotation, can lead to the formation of a dynamo, similar to the dynamo that drives the Sun's magnetic field. The

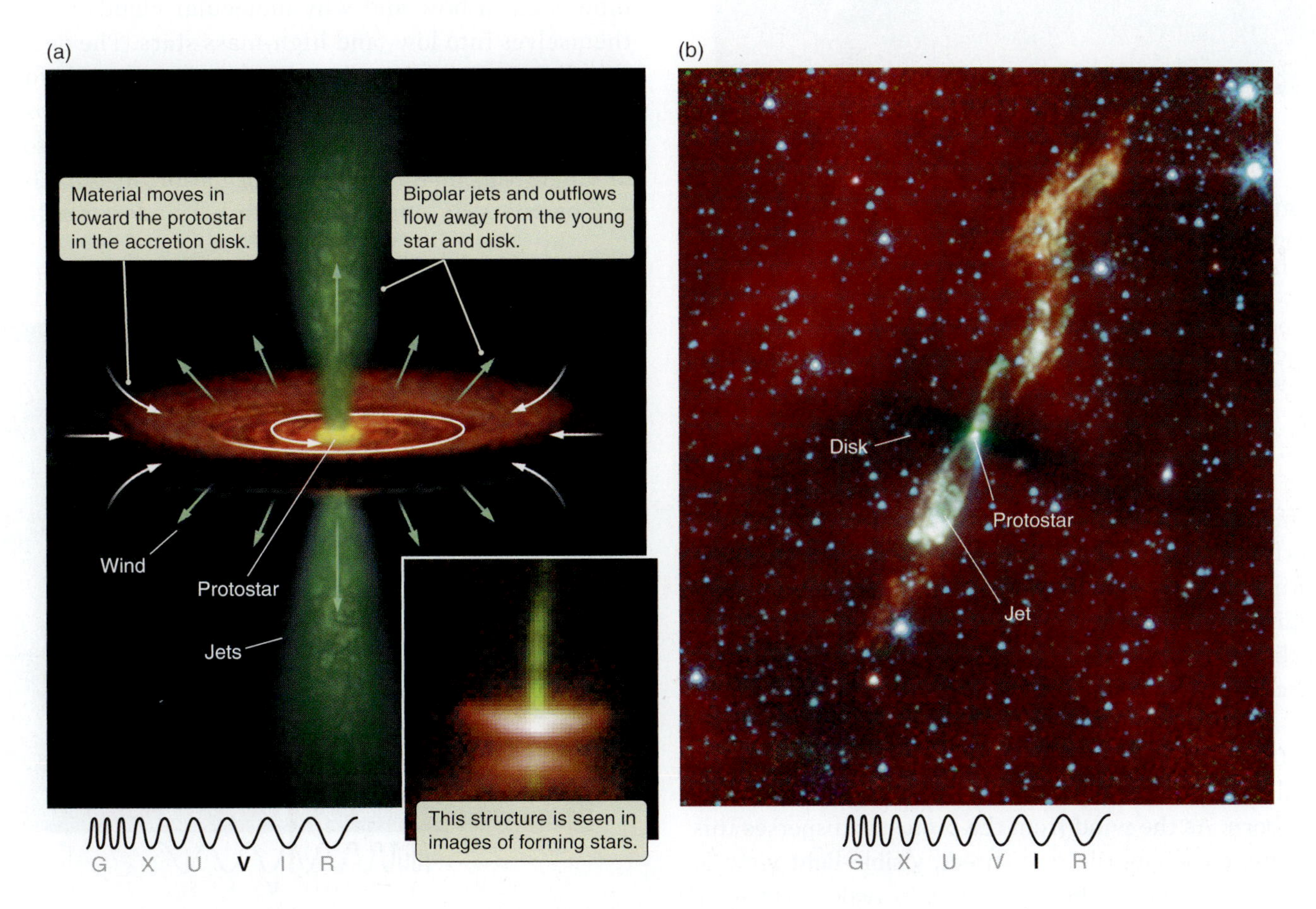

FIGURE 15.23 (a) Material falls onto an accretion disk around a protostar and then moves inward, eventually falling onto the protostar. In the process, some of this material is driven away in powerful jets that stream perpendicular to the disk. (b) This infrared Spitzer Space Telescope image shows jets streaming outward from a young, developing star. Note the nearly edge-on, dark accretion disk surrounding the young star.

FIGURE 15.24 Bipolar jets of material stream outward perpendicular to the disk around a forming star. When the objects hit interstellar material, they form glowing bow shocks called HH objects. (a) HST observations of HH 47 show a jet coming from a dark cloud hiding the new star, and the white bow shock at the end of the jet on the right. (b) HST image of HH 110; several bow shocks can be seen in this geyser of hot gas flowing from a new star.

dynamo in the center of a protostar would be much more powerful than the Sun's dynamo, however. The protostar's resulting strong magnetic field might cause the protostar to begin blowing a powerful wind. It might also act something like the blade in a blender, tearing at the inner edge of the accretion disk and flinging material off into interstellar space.

Until the protostellar wind begins, the protostar is enshrouded in the dusty molecular-cloud core from which it was born. As the wind from the protostar disperses this obscuring envelope, the first direct, visible-light view of the protostar emerges: the protostar is "revealed." Once the contracting protostar makes its appearance, it is referred to as a **T Tauri star**. This name comes from the first recognized member of this class of objects, the star labeled T in the constellation Taurus.

Astronomers have long known that stars are often found together in close collections called **star clusters**. **Figure 15.25** shows one such star cluster: a group called the Pleiades or "Seven Sisters." Star clusters gave astronomers their first evidence that many stars of all different masses can form together at the same place and at about the same time. A look around the Milky Way Galaxy at large reveals a hodgepodge of stars—some very old and others very young. If these were the only stars available to study, it would be extremely difficult to learn much about how stars evolve. Star clusters, by contrast, are large collections of stars that all formed *at the same time, in the same place, and from the same material*. They provide extremely useful samples for studying star formation.

Even though the few brightest and most massive stars in a cluster dominate what is visible, the vast majority of stars in a cluster are much less massive, like the Sun. In fact, some star-forming regions do not seem to form any especially massive stars at all. Astronomers are very interested in how and why molecular clouds subdivide themselves into low- and high-mass stars. The details of this division—specifically, what fraction of newly formed stars will have which masses—are crucial if observations of the stars in the Sun's vicinity today are to help untangle the history of star formation in our Galaxy. Astronomers do not have a detailed understanding yet

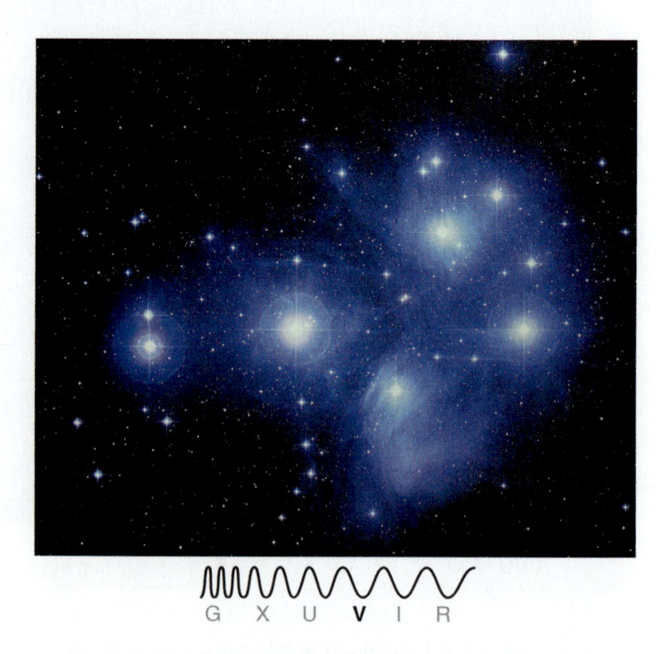

FIGURE 15.25 On a clear winter night you can see the brightest of the stars in this cluster—a tight bunch in the constellation Taurus called the Pleiades or "Seven Sisters." The diffuse blue light around the stars is starlight scattered by interstellar dust.

of why some cloud cores become 1-$M_{\odot}$ stars while others become 10-$M_{\odot}$ stars.

After a cloud core collapse, the evolution of a protostar is determined largely by its mass. Calculations suggest that a star with the mass of the Sun probably takes about 10 million years or so to descend its Hayashi track and become a star on the main sequence. The entire history, including the collapse and fragmentation of the molecular cloud itself, suggests that the total time is more like 30 million years. Because the self-gravity of a more massive core is stronger, more massive cores collapse to form stars more quickly. A 10-$M_{\odot}$ star might go from the stage of being a molecular-cloud core to burning hydrogen in its interior in only 100,000 years. A 100-$M_{\odot}$ star might take less than 10,000 years. By comparison, a 0.1-$M_{\odot}$ star might take 100 million years to finally reach the main sequence.

Star formation may take millions of years.

The 30 million years or so that it took for the Sun to form is a long time, but it is a tiny fraction of the 10 billion years during which the Sun will steadily fuse hydrogen into helium as a main-sequence star. It is no wonder that so few among the many stars visible in the sky are young. But every star was young at one time, including the Sun.

Not surprisingly, many questions about star formation remain. For example, how must astronomers modify their theories to account for the formation of binary stars or other multiple-star systems? When they observe the sky, they find that about half of the stars they see are part of multiple-star systems. At what point during star formation is it determined that a collapsing cloud core will form several stars instead of just one? Some models suggest that this split may happen early in the process, during the fragmentation and collapse of the molecular cloud. The advantage of these ideas is that they provide a natural way of dealing with much of the angular momentum of the cloud core: it goes into the orbital angular momentum of the stars about each other. Other models suggest that additional stars may form from the accretion disk around an initially single protostar.

Collapsing cloud cores often form multiple stars.

The picture of star formation presented here is remarkably complete, considering that astronomers have never visited a protostar or watched a star form. Instead, they have observed many different stars at different stages in their formation and evolution at different wavelengths, and they have used their knowledge of physical laws to tie these observations together into a coherent, consistent description of how, why, and where stars form. Astronomers use observations to see what things exist in the universe, physics to understand how they work and the relationships between them, and computer models to simulate the physical changes over time. Then they compare their results with observations. This multipronged attack is how all astronomy (and in a certain sense all physical science) works.

15.5 Origins: Star Formation, Planets, and Life

When astronomers consider the possibility of other life in the universe, one of the first things they think about is the rate of formation of stars and planets. Life probably needs planets, and planets form along with stars. The conditions under which a star is born, and the mass and chemical composition that it has when it begins its nuclear fusion, set the stage for the rest of its life. In Chapter 7 you saw that planet formation can be a common consequence of star formation; thus, if a star is going to have planets, they form at about the same time as the star. If the star is going to have rocky planets with hard surfaces (such as the planets of the inner Solar System), or gaseous planets with rocky cores and rocky moons (such as the planets of the outer Solar System), then the material from which the star and planets form must be "enriched" with the heavy elements that make up these rocky surfaces.

These enriched clouds would also provide elements that are essential to life on Earth. In addition to the presence of organic molecules mentioned earlier in the chapter, astronomers have detected water in star-forming regions such as W3 IRS5 (**Figure 15.26**). Water exists as ice mixed with dust grains in the cool molecular clouds, or as vapor when it's closer to a protostar and the dust grains and ice evaporate. In 2011 the Herschel Space Observatory detected oxygen molecules (O_2, the type we breathe) in a star-forming complex in Orion (**Figure 15.27**). Oxygen is the third-most-common element in the universe, yet it had not been decisively observed before in molecular form. This oxygen also may have come from the melting and evaporation of water ice on the tiny dust grains.

As noted in Chapter 13, astronomers had doubted that planets could exist in stable orbits in binary star systems, but now a few such circumbinary systems have been found. Planets that form within associations of O and B stars may be too unstable to last very long. Isolated planets that appear to be unattached to any star move through the Milky Way. Perhaps the isolated planets were gravitationally ejected soon after their formation in a multiple system. But these planets do not have a

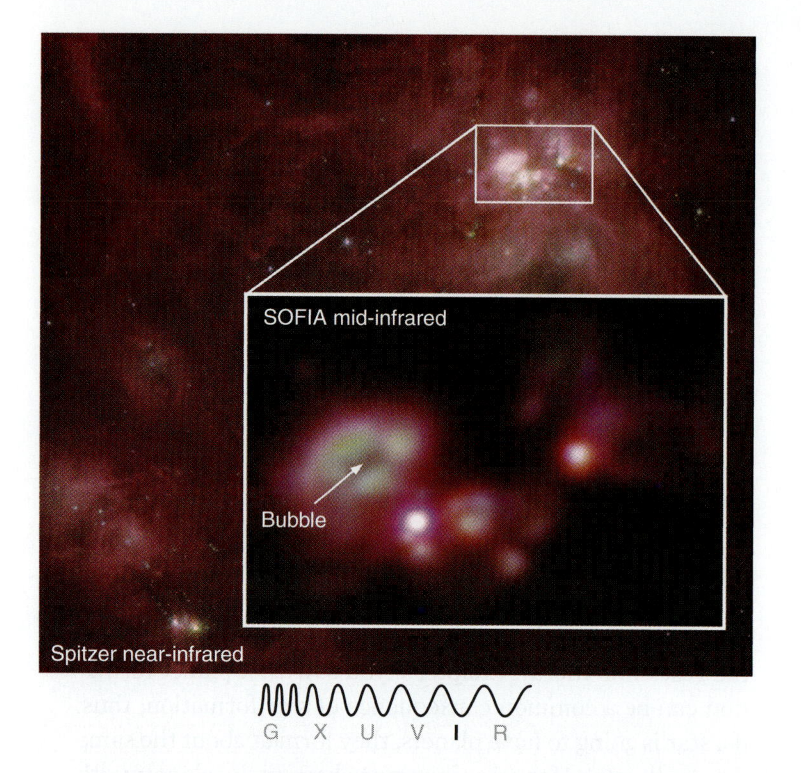

FIGURE 15.26 Water was detected in the W3 IRS5 star-forming complex. W3 IRS5 observed with the Spitzer Space Telescope in near infrared and with the Stratospheric Observatory for Infrared Astronomy (SOFIA) telescope in mid-infrared. A massive star has cleared the dust and gas from a small bubble, sweeping it into a dense shell (green).

source of energy like Earth's Sun. Astronomers theorize that only planets that orbit stars will be able to support life. So when they try to estimate the possibility of life in the galaxy, astronomers include estimations of the rate of formation of stars in the galaxy, along with the fraction of stars that have planets. Advances in the study of star formation and planet detection help astronomers to better understand the conditions under which life might develop elsewhere.

Stars are temporal objects. They are born from interstellar gas, they shine by the light of nuclear reactions deep within their cores, and when they exhaust their fuel they die. The changing balance of the protostar is only the first chapter in a process of evolution that continues throughout the star's life. It is convenient to follow the changes taking place within an evolving protostar by tracking its progress across the face of the H-R diagram. We will continue to use the H-R diagram as we explore the other stages of a star's life. ■

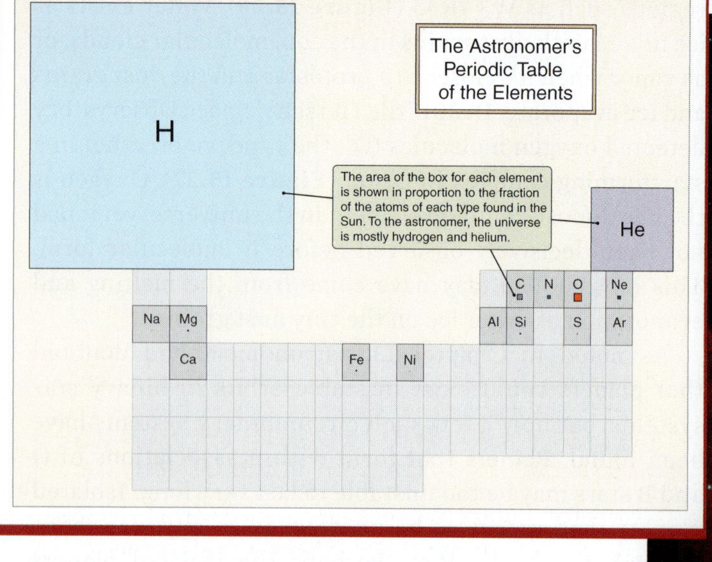

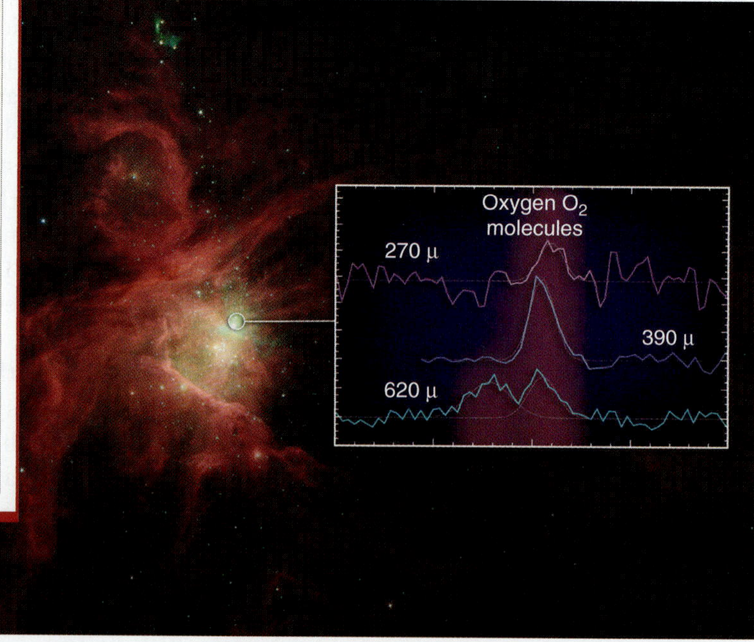

FIGURE 15.27 The Herschel Space Observatory discovered molecular oxygen (inset) in this star-forming region in Orion.

Summary

15.1 The interstellar medium is complex, ranging from cold, relatively dense molecular clouds to hot, tenuous intercloud gas heated and ionized by energy from stars and stellar explosions. Dust and gas in the interstellar medium blocks visible light but becomes more transparent at longer, infrared wavelengths. Different phases of the interstellar medium emit various types of radiation and can be observed at different wavelengths, ranging from radio waves to X-rays. Neutral hydrogen cannot be detected at visible and infrared wavelengths, but it is revealed by its 21-cm emission.

15.2 Star formation begins when the self-gravity of dense clouds exceeds outward pressure. The clouds collapse, heat up, and fragment to form stars. The conservation of angular momentum is important to the formation of disks during the collapse. Forming stars are detected from their infrared emission and from the effects they have on their surroundings.

15.3 Protostars collapse, radiating away their gravitational energy until fusion starts in their cores. When they reach hydrostatic and thermal equilibrium, they settle onto the main sequence. Stars form in clusters from dense cores buried within giant molecular clouds. A protostar must have a mass of at least 0.08 $M_{\odot}$ to become a true star. Brown dwarfs are neither stars nor planets, but something in between.

15.4 Because star formation takes tens of thousands to millions of years, what astronomers know about the evolution of the birth of stars comes from observations of many protostars at various stages of their development. Protostars drive powerful winds and jets that impact the space around them. Multiple stars may form from a single cloud core.

15.5 Observations and theory help astronomers learn about star formation and planetary formation. The conditions under which a star is born determine whether it will have planets and the chemical elements required by life as it exists on Earth.

Unanswered Questions

- Do high-mass and low-mass stars form in very different ways? The smallest stars, spectral type M, are most likely to form as single stars, but a high fraction of medium-mass stars are formed in binary pairs. One theory is that these binaries actually start out as triple systems, from which the smallest star is gravitationally ejected, leading to a remaining pair and a single star. The highest-mass stars are less likely to form alone; many form in OB associations–larger groups of massive stars in which the formation of one large star may stimulate the formation of another nearby in the molecular cloud.
- How common are brown dwarfs? As with extrasolar planets, they have been observed only recently, so their space density is not well known. The *Kepler* mission, which has been finding extrasolar planets (see Chapter 7) and eclipsing binaries (see Chapter 13), also detects brown dwarfs, so there may be an answer to this question in a few years.

Questions and Problems

Summary Self-Test

1. Phases of the interstellar medium include (choose all that apply)
   a. hot, low-density gas.
   b. cold, high-density gas.
   c. hot, high-density gas.
   d. cold, low-density gas.

2. Dust in the interstellar medium can be observed in
   a. visible light.
   b. infrared radiation.
   c. radio waves.
   d. X-rays.

3. The interstellar medium in the Sun's region of the galaxy is closest in composition to
   a. the Sun.
   b. Jupiter.
   c. Earth.
   d. comets in the Oort Cloud.

4. Interstellar dust is effective at blocking visible light because
   a. the dust is so dense.
   b. dust grains are so few.
   c. dust grains are so small.
   d. dust grains are so large.

5. Hot intercloud gas is heated primarily by
   a. starlight.
   b. the cosmic microwave background radiation.
   c. supernova explosions.
   d. neutrinos.

6. Astronomers determined the composition of the interstellar medium from
   a. observing its emission and absorption lines.
   b. measuring the composition of the Sun and other stars that form from it.
   c. taking samples with spacecraft.
   d. all of the above

7. Molecular clouds fragment as they collapse because
   a. the rotation of the cloud throws some mass to the outer regions.
   b. the density increases fastest in the center of the cloud.
   c. density variations from place to place grow larger as the cloud collapses.
   d. the interstellar wind is stronger in some places than others.

8. The energy required to begin nuclear fusion originally came from
   a. the gravitational potential energy of the protostar.
   b. the kinetic energy of the protostar.
   c. the wind from nearby stars.
   d. the pressure from the interstellar medium.

9. In astronomy, the term *bipolar* refers to outflows that
   a. point in opposite directions.
   b. alternate between expanding and collapsing.
   c. rotate about a polar axis.
   d. show spiral structure.

10. Astronomers understand the process of star formation because
    a. they have observed star formation as it happens for a small number of stars.
    b. they have observed star formation as it happens for a large number of stars.
    c. they have observed many different stars at each step of the process.
    d. theoretical models predict that this must be the way stars form.

True/False and Multiple Choice

11. **T/F:** The interstellar medium can be observed only in the infrared.

12. **T/F:** A molecular cloud is held together by gravity.

13. **T/F:** Only very simple molecules exist in the interstellar medium.

14. **T/F:** Most of the material in the interstellar medium is in the form of dust grains.

15. **T/F:** Star formation takes longer for more massive stars.

16. Cold neutral hydrogen can be detected because
    a. it emits light when electrons drop through energy levels.
    b. it blocks the light from more distant stars.
    c. it is always hot enough to glow in the radio and infrared wavelengths.
    d. the atoms change spin states.

17. The Hayashi track is a nearly vertical evolutionary track on the H-R diagram. What does the vertical nature of this track tell you about a protostar as it moves along it?
    a. The star remains the same brightness.
    b. The star remains the same luminosity.
    c. The star remains the same color.
    d. The star remains the same size.

18. What distinguishes a T Tauri star from a protostar?
    a. It can be seen in the visible wavelengths.
    b. It is more massive than a protostar.
    c. It emits in the visible wavelengths.
    d. It comes from more massive protostars.

19. What two forces establish hydrostatic equilibrium in an evolving protostar?
    a. the force from radiation pressure and gravity
    b. the force from radiation pressure and the strong nuclear force
    c. gravity and the strong nuclear force
    d. energy emitted and energy produced

20. Why are so few of the many stars that astronomers see in the sky protostars?
    a. Protostars are hidden in giant molecular clouds.
    b. Protostars are small.
    c. Protostars are dim.
    d. Protostars are short-lived.

21. Suppose you are studying a visible-light image of a distant galaxy, and you see a dark line cutting across the bright disk. This dark line is most likely caused by
    a. gravitational instabilities that clear the area of stars.
    b. dust in the Milky Way blocking the view of the distant galaxy.
    c. dust in the distant galaxy blocking the view of stars in the disk.
    d. a flaw in the instrumentation.

22. What causes a hydrogen atom to radiate a photon of 21-cm radio emission?
    a. The electron drops down one energy level.
    b. The (formerly free) electron is captured by the proton.
    c. The electron flips to an aligned spin state.
    d. The electron flips to an unaligned spin state.

23. What determines whether a protostar will become a true star or a brown dwarf?
    a. the protostar's composition
    b. the protostar's temperature
    c. the protostar's companion
    d. the protostar's mass

24. Astronomers know that there are dusty accretion disks around protostars because
    a. there is often a dark band across the protostar.
    b. there is often a bright band across the protostar.
    c. theory says accretion disks should be there.
    d. there are planets in the Solar System.

25. What is the single most important property of a star that will determine its evolution?
    a. temperature
    b. composition
    c. mass
    d. radius

Thinking about the Concepts

26. The interstellar medium is approximately 99 percent gas and 1 percent dust. Why is it the dust and not the gas that blocks a visible-light view of the galactic center?

27. Explain why observations in the infrared are necessary for astronomers to study the detailed processes of star formation.

28. How does the material in interstellar clouds and intercloud gas differ in density and distribution?

29. When a star forms inside a molecular cloud, what happens to the cloud? Is it possible for a molecular cloud to remain cold and dark with one or more stars inside it? Explain your answer.

30. If you placed your hand in boiling water (100°C) for even one second, you would get a very serious burn. If you placed your hand in a hot oven (200°C) for a second or two, you would hardly feel the heat. Explain this difference and how it relates to million-kelvin regions of the interstellar medium.

31. How do astronomers know that the Sun is located in a "local bubble" formed by a supernova?

32. Interstellar gas atoms typically cool by colliding with other gas atoms or grains of dust; during the collision, the gas atom loses energy and hence its temperature is lowered. How does this explain why very low-density gases are generally so hot, while dense gases tend to be so cold?

33. Explain how 21-cm radio emission has enabled astronomers to detect interstellar clouds of neutral hydrogen (H I), even when large amounts of interstellar dust are in the way.

34. Molecular hydrogen is very difficult to detect from the ground, but astronomers can easily detect carbon monoxide (CO) by observing its 2.6-cm microwave emission. Describe how observations of CO might help astronomers infer the amounts and distribution of molecular hydrogen within giant molecular clouds.

35. The Milky Way contains several thousand giant molecular clouds. Describe a giant molecular cloud and the role it plays in star formation.

36. As a cloud collapses to form a protostar, the forces of gravity felt by all parts of the cloud (which follow an inverse square law) become stronger and stronger. One might argue that under these conditions, the cloud should keep collapsing until it becomes a single massive object. Why doesn't this happen?

37. The internal structure of a protostar maintains hydrostatic equilibrium even as more material is falling onto it. Explain how this can be.

38. You can think of a brown dwarf as a failed star—that is, one lacking sufficient mass for nuclear reactions to begin. What are the similarities and differences between a brown dwarf and a giant planet such as Jupiter? Would you classify a brown dwarf as a supergiant planet? Explain your answer.

39. The H^- ion acts as a thermostat in controlling the surface temperature of a protostar. Explain the process.

40. Describe a Herbig-Haro object.

Applying the Concepts

41. In Chapter 13 you learned that astronomers can measure the temperature of a star by comparing its brightness in blue and yellow light. Does reddening by interstellar dust affect a star's temperature measurement? If so, how?

42. When a hydrogen atom is ionized, it splits into two components.
    a. Identify the two components.
    b. If both components have the same kinetic energy, which moves faster?

43. Estimate the typical density of dust grains (grains per cubic centimeter) in the interstellar medium. A typical grain has a mass of about 10^{-17} kilogram (kg). (Hint: You know the typical density of gas, and the fraction of the interstellar medium, by mass, that is made of dust.)

44. Referring to Figure 15.3, estimate the blackbody temperature of the star as shown in part (b) (without dust) and part (c) (with dust). How significant are the effects of interstellar dust when observed data are used to determine the properties of a star?

45. A typical temperature of intercloud gas is 8000 K. Using Wien's law (see Math Tools 15.1 and Chapter 5), calculate the wavelength at which this gas would radiate.

46. Some parts of the Orion Nebula have a blackbody peak wavelength of 0.29 μm. What is the temperature of these parts of the nebula?

47. Stellar radiation can convert atomic hydrogen (H I) to ionized hydrogen (H II).
    a. Why does a B8 main-sequence star ionize far more interstellar hydrogen in its vicinity than does a K0 giant of the same luminosity?
    b. What properties of a star are important in determining whether it can ionize large amounts of nearby interstellar hydrogen?

48. The mass of a proton is 1,850 times the mass of an electron. If a proton and an electron have the same kinetic energy ($E_K = \frac{1}{2}mv^2$), how many times greater is the velocity of the electron than that of the proton?

49. If a typical hydrogen atom in a collapsing molecular-cloud core starts at a distance of 1.5×10^{12} km (10,000 AU) from the core's center and falls inward at an average velocity of 1.5 km/s, how many years does it take to reach the newly forming protostar? Assume that a year is 3×10^7 seconds.

50. Table 13.1 indicates that the ratio of hydrogen atoms (H) to carbon atoms (C) in the Sun's atmosphere is approximately 2,400:1. It would be reasonable to assume that this ratio also applies to molecular clouds. If 2.6-cm radio observations indicate 100 $M_\odot$ of carbon monoxide (CO) in a giant molecular cloud, what is the implied mass of molecular hydrogen (H_2) in the cloud? (Carbon represents $^3/_7$ of the mass of a CO molecule.)

51. Neutral hydrogen emits radiation at a radio wavelength of 21 cm when an atom drops from a higher-energy spin state to a

lower-energy spin state. On average, each atom remains in the higher-energy state for 11 million years (3.5×10^{14} seconds).
a. What is the probability that any given atom will make the transition in 1 second?
b. If there are 6×10^{59} atoms of neutral hydrogen in a 500-$M_\odot$ cloud, how many photons of 21-cm radiation will the cloud emit each second?
c. How does this number compare with the 1.8×10^{45} photons emitted each second by a solar-type star?

52. The Sun took 30 million years to evolve from a collapsing cloud core to a star, with 10 million of those years spent on its Hayashi track. It will spend a total of 10 billion years on the main sequence. Suppose the Sun's main-sequence lifetime were compressed into a single day.
a. How long would the total collapse phase last?
b. How long would the Sun spend on its Hayashi track?

53. A protostar with the mass of the Sun starts out with a temperature of about 3500 K and a luminosity about 200 times larger than the Sun's current value. Estimate this protostar's size and compare it to the size of the Sun today.

54. The star-forming region 30 Doradus is 160,000 light-years away in the nearby galaxy called the Large Magellanic Cloud and appears about ⅙ as bright as the faintest stars visible to the naked eye. If it were located at the distance of the Orion Nebula (1,300 light-years away) how much brighter than the faintest visible stars would it appear?

55. Assume a brown dwarf has a surface temperature of 1000 K and approximately the same radius as Jupiter. What is its luminosity compared to that of the Sun? How many brown dwarfs like this one would be needed to produce the luminosity of a star like the Sun?

Using the Web

56. Go to the Astronomy Picture of the Day (APOD) website (http://apod.nasa.gov/apod), do a search on "molecular clouds," and pick out a few images. Were these pictures obtained from space or on the ground, and at what wavelengths? With which telescopes? What wavelengths do the colors in the images represent? Are they "real" or "false-color" images? Explain your answers.

57. Space infrared telescopes:
a. Go to NASA's Spitzer Space Telescope website (http://spitzer.caltech.edu). Click on "News" and find a recent story about star formation. What did Spitzer observe? What wavelengths do the colors in the picture represent? How does this "false color" help astronomers to analyze these images? Why is it better to study star formation in the infrared than in the visual part of the spectrum?
b. Go to the website for ESA's Herschel Space Observatory (http://sci.esa.int/science-e/www/area/index.cfm?fareaid=16), which has a 3.5-meter primary mirror and is located at Lagrangian point L_2. Compare this telescope with Spitzer, which has a 0.85-meter primary mirror. How much more light-gathering power does Herschel have than Spitzer? Why do astronomers put infrared telescopes in space? What is new from Herschel?

58. Infrared astronomy:
a. The Wide-field Infrared Survey Explorer (WISE) surveyed the entire sky in four infrared wavelength bands between January 2010 and February 2011. Why do astronomers want to see the whole sky in the infrared? Go to the WISE website (http://wise.ssl.berkeley.edu/news.html). An all-sky map and catalog was released in March 2012 (see Figure 15.5). What types of objects were detected with this mission? Have many new brown dwarfs been detected?
b. The Stratospheric Observatory for Infrared Astronomy (SOFIA) is a 2.5-meter telescope on a modified Boeing 747 aircraft. Go to the SOFIA website (http://sofia.usra.edu). Why would astronomers put an infrared telescope on an airplane? What has been detected with this telescope?

59. Go to the website for Stardust (http://stardustathome.ssl.berkeley.edu), a Citizen Science project that asks Internet users to use a virtual microscope to analyze digital scans of particles collected by the *Stardust* mission in 2006. The goal is to identify tiny interstellar dust grains. Follow the four steps under "Get Started" (you need to create a log-in account) and help search for stardust. Click on "News." What has been learned from this project? Remember to save the images for your homework, if required.

60. Do a Google (or other) news search for a story about brown dwarfs. Is this story from an observatory? A NASA mission? A press release? What is new, and why is it interesting?

Exploration | The Stellar Thermostat

In this Exploration you will see how the H^- thermostat works in forming stars. You will need about 20 coins (they do not have to be all the same type).

Place your coins on a sheet of paper and draw a circle around them—the smallest possible circle that will fit all the coins. Then divide the circle into three parts as shown in **Figure 15.28**. This circle represents a star with a changing temperature. The coins represent H^- ions. Removing a coin from the circle means that the H^- ion has turned into a neutral hydrogen atom. Placing a coin in the circle means that the neutral hydrogen atom has become an H^- ion.

Place all the coins back on the circle.

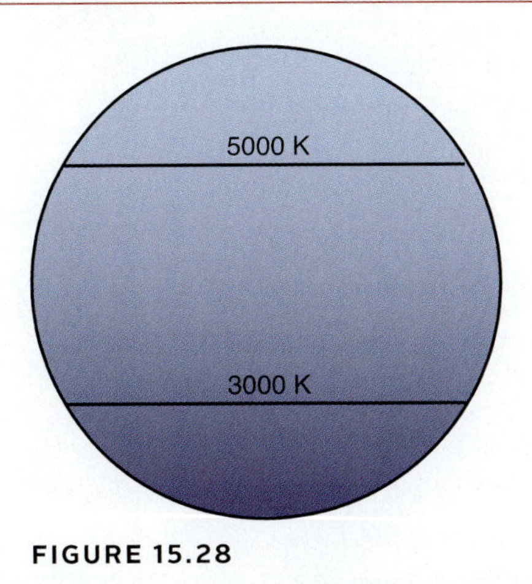

FIGURE 15.28

1 **How many "H^- ions" are now in the star?**

....

....

The "blanket" of H^- ions holds heat in the star, so the star begins to heat up until it reaches about 5000 K. At that surface temperature, the H^- ions begin to be destroyed. Now that the star is hot, begin removing coins one at a time, starting from the top of the circle and working downward. When you see the line marking 3000 K, stop removing coins.

2 **How many "H^- ions" are now in the star?**

....

....

3 **What will happen to the surface temperature of the star, now that there are fewer ions?**

....

....

When the star cools off to about 3000 K, H^- ions begin to form. Place the coins back on the circle, starting from the bottom and working your way up to the 5000-K line.

4 **How many "H^- ions" are now in the star?**

....

....

5 **What will happen to the surface temperature of the star, now that there are more ions?**

....

....

Now that the star is hot, begin removing coins one at a time, starting from the top of the circle and working downward. When you see the line marking 3000 K, stop removing coins.

6 **What should happen next?**

....

....

7 **Make a circular flowchart that includes the following steps, in the proper order: the star heats up; the star cools down; H^- is formed; H^- is destroyed.**

....

....

Optical and X-ray image of SNR 0509-67.5, a remnant of a Type Ia supernova located in the Large Magellenic Cloud.

16 Evolution of Low-Mass Stars

It is said an Eastern monarch once charged his wise men to invent him a sentence to be ever in view, and which should be true and appropriate in all times and situations. They presented him the words "And this, too, shall pass away."

Abraham Lincoln (1809–1865), September 30, 1859

LEARNING GOALS

Within its core, the Sun fuses more than 4 billion kilograms of hydrogen to helium each second; and although the Sun may seem immortal by human standards, eventually it will run out of fuel. When it does, some 5 billion years from now, the Sun's time on the main sequence will come to an end. In this chapter we examine what happens when a low-mass star like the Sun nears the end of its life. By the conclusion of this chapter, you should be able to:

- Estimate the lifetime of a star from its mass.
- Explain why low-mass stars grow larger and more luminous as they run out of fuel.
- Sketch post-main-sequence evolutionary tracks on an H-R diagram.
- Make a flowchart of the stages of evolution for low-mass stars.
- Describe how planetary nebulae and white dwarfs form.

16.1 The Life of a Main-Sequence Star

Suppose you had 1 minute to observe all of the people in a crowded stadium, and from that minute of observation, draw a conclusion about the life cycle of humans. It is possible, but highly unlikely, that you might observe a significant life change such as a birth or a death. More probably, you would observe people of different ages and note some properties of people of various ages that would give you clues to the fact that some are young, some are old, and most are in between. But you wouldn't see individual people change over the course of the minute, since a minute is a very small fraction of a typical human lifetime.

Similarly, astronomers are able to observe only a very brief fraction of each star's lifetime. They would have to observe a star like the Sun for several hundred years to be watching it for the equivalent of 1 minute in a human life span. The vast majority of the stars that are observed in any particular year will not change noticeably in that amount of time. Astronomers do not see individual stars age. But because there are so many stars—billions more than the number of people in a stadium—they can observe enough stars at different stages to piece together an evolutionary picture. And sometimes when they are fortunate, astronomers have a chance to observe a star undergoing a dramatic change.

To understand what is happening at the core of the star, theorists use the most powerful computers available to model the nuclear reactions that take place there. These models make predictions about how a star of a given mass and chemical composition will change over its lifetime. These predictions then must be connected to what astronomers observe. The study of stellar evolution involves back-and-forth between observation and theory, which has led to a general understanding of how and when stars die.

A star cannot remain on the main sequence forever. It will eventually exhaust the hydrogen fuel in its core, and when it does, its structure will begin to change dramatically. Just as the balance between pressure and gravity within a protostar constantly changes as the star descends the Hayashi track toward the main sequence, new balances are constantly found as a star evolves beyond the main sequence. The mass and the composition of a star determine the star's life on the main sequence, and these two qualities remain at center stage as the star begins to die.

How long do stars and their planets "live"?

Each star is unique. Relatively minor differences in the masses and chemical compositions of two stars can sometimes result in significant, and possibly even dramatic, differences in their fates. The course followed by a star with a mass that is 1.1 times the mass of the Sun (1.1 $M_{\odot}$) is not identical to the fate of a star with a mass of 0.9 $M_{\odot}$. Nevertheless, stars can be divided roughly into two broad categories whose members evolve in qualitatively different ways. Massive, luminous O and B stars follow a course fundamentally different from that of the cooler, fainter, less massive stars found toward the lower right end of the main sequence (see Figure 13.16). These stars, which have masses less than about 3 $M_{\odot}$, are considered **low-mass stars** and are typified by the Sun. Stars with masses between 3 and 8 $M_{\odot}$ are **intermediate-mass** stars, and stars with masses above 8 $M_{\odot}$ are **high-mass stars**. In this chapter we begin our discussion of stellar evolution by examining the stages in the evolution of low-mass stars.

Low-mass and high-mass stars evolve differently.

Recall from Chapter 14 that the structure of the Sun is determined by a balance between the inward force of gravity and the outward force of pressure. The pressure within the Sun is, in turn, maintained by energy released by nuclear fusion in the heart of the star. If the mass of the Sun were to increase, the weight of material pushing down on the inner regions of the star would increase. Gravity would gain an advantage. The inner parts of the Sun would be compressed by the added weight, driving up the temperature and density there. This increase in temperature and density would in turn accelerate the pace of the nuclear reactions occurring in the core.

Mass determines a star's fate.

When the temperature and pressure at the center of a star increase, more atomic nuclei are packed together into a smaller volume and it is more likely that they will run into each other and fuse. Higher density means more frequent collisions between atomic nuclei, and therefore faster burning. Higher temperature also drives up the rate of nuclear reactions because the atomic nuclei are moving faster, so they are more likely to collide with one another more violently, increasing the chances that they will overcome the electric repulsion that pushes the positively charged nuclei apart. As a result of the combined effects of temperature and density, modest increases in pressure within a star can sometimes lead to dramatic increases in the amount of energy released by nuclear burning.

Increases in temperature and density speed up nuclear burning.

This is the key to understanding why the main sequence is primarily a sequence of masses, with low-mass stars on the faint end and high-mass stars on the luminous end. More mass means stronger gravity, stronger gravity means higher temperature and pressure in the star's interior, higher temperature and pressure mean faster nuclear reactions, and faster nuclear reactions mean a more luminous star. If the Sun were more massive, it would necessarily have a different balance between gravity and pressure—a balance in which the Sun would burn its nuclear fuel more rapidly and thus be more luminous. In other words, if the Sun were more massive, it would be located at a different position on the H-R diagram: farther up and to the left on the main sequence. Mass determines the structure of a star and its place on the main sequence.

Higher Mass Means a Shorter Lifetime

A main-sequence star will eventually run out of fuel. Think about how long you can drive your car before it runs out of gas; it depends on how much gas your tank holds and on the size and efficiency of your engine. A gas-guzzling eight-cylinder SUV drinks fuel a lot faster than a subcompact. The amount of time your car will run is determined by a competition between these two effects. The larger car might run out of gas first, even if it is attached to a much larger tank.

The competition between these two effects—tank size and engine size—is most readily expressed as a ratio. How long your car runs is given by the amount of gas in the tank, divided by how quickly the car uses it:

$$\text{Lifetime of tank of gas (hours)} = \frac{\text{Amount of fuel (gallons)}}{\text{Rate at which fuel is used (gallons/hour)}}$$

For example, if you have a 15-gallon tank and your engine is burning fuel at a rate of 3 gallons each hour, then your car will use up all of the gas in just 5 hours.

The same principle works for main-sequence stars. The amount of fuel is determined by the mass of the star. The more massive the star is, the more hydrogen is available to power nuclear burning. The rate at which fuel is used is measured by the luminosity of the star. Main-sequence stars are "in balance," so energy is radiated into space from the surface of the star at the same rate at which energy is being generated in its core. This balance between energy generation and luminosity remains true at almost every stage of a star's evolution. If one main-sequence star has twice the luminosity of another, then it must be burning hydrogen at twice the rate of the other star.

How quickly a star runs out of fuel depends on its mass and luminosity.

An expression for the **main-sequence lifetime** of the star looks very similar to the expression for the time it takes your car to run out of fuel:

$$\text{Lifetime of star} = \frac{\text{Amount of fuel } (\propto \text{ mass of star})}{\text{Rate fuel is used } (\propto \text{ luminosity of star})}$$

At first glance, you might expect that since a more massive star has more mass, it will live longer. A more massive star, however, not only has more fuel, but also burns that fuel faster, and this higher rate of fuel use is the key factor. Stars with higher masses live shorter lives, not longer ones, because they burn their fuel faster. Lower-mass stars have much longer lives. This concept is developed further in **Math Tools 16.1**.

The Structure of a Star Changes as It Uses Its Fuel

Although stable, a main-sequence star slowly changes. When the Sun formed, about 90 percent of its atoms were hydrogen atoms. Since then, the Sun has produced its energy by converting hydrogen into helium via the proton-proton chain. As the composition of a star changes, so must its structure. When we discussed the collapse of a protostar toward the main sequence in Chapter 15, we considered the idea of a changing balance between gravity and pressure. The protostar is always in balance as it contracts; but as it radiates away thermal energy, this balance constantly shifts toward that of a smaller and denser object.

The same concept applies here. As a main-sequence star uses the fuel in its core, its structure must continually shift in response to the changing core composition. At any given point in its lifetime, a main-sequence star like the Sun is in balance, but the balance in the Sun today is slightly different from the balance the Sun had billions of years ago, and slightly different from the balance it will have billions of years from now. Between the time the Sun was born and the time it will leave the main sequence, its luminosity will roughly double, with most of this change occurring during the last billion years of its life on the main sequence. Stars evolve even as they "sit" on the main sequence, although this evolution is slow and modest in comparison with the events that follow.

Helium Ash Builds Up in the Center of the Star

As you saw in Chapter 14, at the end of the proton-proton chain in main-sequence stars, two ^{3}He nuclei fuse together to form ^{4}He (and 2 ^{1}H). But at the temperatures found at the centers of main-sequence stars, collisions are not energetic or frequent enough for ^{4}He nuclei to fuse into more massive

Math Tools 16.1

Estimating Main-Sequence Lifetimes

Astronomers can estimate the lifetime of main-sequence stars either observationally or by modeling the evolution of stars of a given composition. Using what is known about how much hydrogen must be converted into helium each second to produce a given amount of energy, as well as the fraction of its hydrogen that a star burns, we can state that the main-sequence lifetime of a star, $\text{Lifetime}_{\text{MS}}$, can be expressed as:

$$\text{Lifetime}_{\text{MS}} \propto \frac{M_{\text{MS}}}{L_{\text{MS}}}$$

where M is mass (the amount of fuel) and L is luminosity (the rate the fuel is used). The same equation would apply for the Sun:

$$\text{Lifetime}_{\odot} \propto \frac{M_{\odot}}{L_{\odot}}$$

We can express the lifetime as a ratio, adding in that the computed lifetime of a 1-$M_{\odot}$ star like the Sun is 10 billion (1.0×10^{10}) years:

$$\frac{\text{Lifetime}_{\text{MS}}}{\text{Lifetime}_{\odot}} = \frac{\text{Lifetime}_{\text{MS}}}{10^{10}\text{ years}} = \frac{M_{\text{MS}}/L_{\text{MS}}}{M_{\odot}/L_{\odot}}$$

Multiplying through by 10^{10} years and rearranging the fractions yields:

$$\text{Lifetime}_{\text{MS}} = 10^{10}\text{ yr} \times \frac{M_{\text{MS}}/L_{\text{MS}}}{M_{\odot}/L_{\odot}} = 10^{10} \times \frac{M_{\text{MS}}/M_{\odot}}{L_{\text{MS}}/L_{\odot}}\text{yr}$$

Now let's compare the lifetime of a star with that of the Sun. The relationship between the mass and the luminosity of stars is

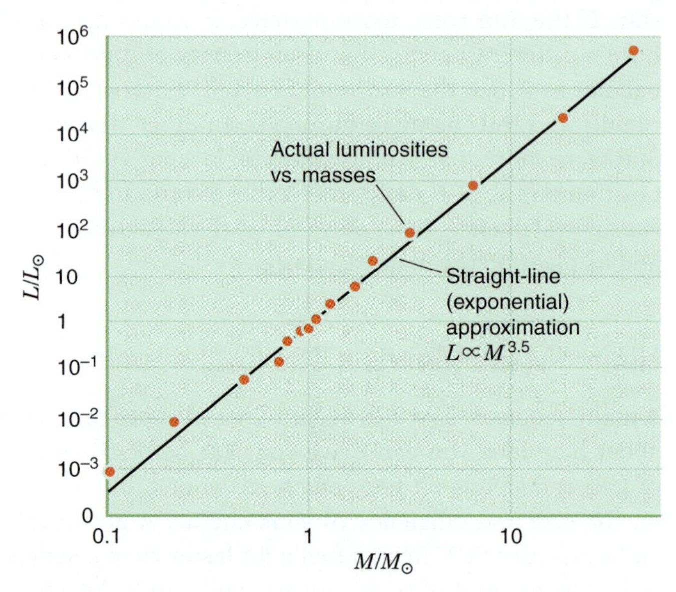

FIGURE 16.1 The mass-luminosity relationship for main-sequence stars: $L \propto M^{3.5}$. The exponent (3.5) is an average value over the wide range of main-sequence star masses. Observational data show that the deviation of stars from the average relationship depends on their composition.

elements. As hydrogen burns in the core of a low-mass star, the resulting helium collects there, building up like the nonburning ash in the bottom of a fireplace.

The temperature and pressure are highest at the center of a main-sequence star, and hydrogen burns most rapidly there as well. As a result, nonburning helium "ash" accumulates most rapidly at the center of the star. As a star evolves, its chemical composition changes most rapidly at its center and less rapidly farther out in the star. **Figure 16.2** illustrates how the chemical composition inside a star like the Sun changes throughout its main-sequence lifetime. When the Sun formed, it had a uniform composition of about 70 percent hydrogen and 30 percent helium by mass. As hydrogen fused into helium, the helium fraction in the core of the Sun climbed. Today, roughly 5 billion years later, only about 35 percent of the mass *in the core* of the Sun is hydrogen.

Helium ash builds up in the core of a main-sequence star.

16.2 A Star Runs Out of Hydrogen and Leaves the Main Sequence

Hydrogen burning in the core of a star cannot continue forever. Eventually—about 5 billion years from now in the case of the Sun—a star exhausts all of the hydrogen fuel available at its center. At this point, the innermost core of the star is composed entirely of helium ash. As thermal energy leaks out of the helium core into the surrounding layers of the star, no more energy is generated within the core to replace it. The balance that has maintained the structure of the star throughout its life is now broken. The star's life on the main sequence has come to an end.

The Helium Core Is Degenerate

All of the matter you directly experience is mostly *empty space*. An atom is mostly empty, except for the tiny bit of very sensitive. Relatively small differences in the masses of stars result in large differences in their main-sequence luminosities. Recall from Chapter 13 that one method for estimating luminosities of main-sequence stars is known as the **mass-luminosity relationship**, $L \propto M^{3.5}$, which is based on observed luminosities of stars of known mass such as binary stars. The exponent can vary from 2.5 to 5.0, depending on the mass of the star, as illustrated in **Figure 16.1**. As we did in the preceding example, we can express this relationship relative to the Sun's mass and luminosity:

$$\frac{L_{MS}}{L_\odot} = \left(\frac{M_{MS}}{L_\odot}\right)^{3.5}$$

Substituting the mass-luminosity relationship into the lifetime equation gives us:

$$\text{Lifetime}_{MS} = 10^{10} \times \frac{M_{MS}/M_\odot}{(M_{MS}/M_\odot)^{3.5}}\ \text{yr} = 10^{10} \times \left(\frac{M_{MS}}{M_\odot}\right)^{-2.5} \text{yr}$$

Note that (something)$^{-2.5}$ = (1/something)$^{2.5}$.

For example, let's look at a K5 main-sequence star. According to **Table 16.1**, a K5 star has a mass that is equal to about 0.67 $M_\odot$:

$$\text{Lifetime}_{K5} = 10^{10} \times (0.67)^{-2.5}\ \text{yr} = 10^{10} \times \frac{1}{(0.67)^{2.5\ \text{yr}}}$$

$$= 2.7 \times 10^{10}\ \text{years}$$

Instead of the 10-billion-year life span of the Sun, a K5 star has a main-sequence lifetime 2.7 times as long as the Sun's. Even though the K5 star starts out with *less* fuel than the Sun, it burns that fuel more slowly, so it lives longer.

TABLE 16.1 | Main-Sequence Lifetimes

Spectral Type	Mass ($M_\odot$)	Luminosity ($L_\odot$)	Main-Sequence Lifetime (years)
O5	60	500,000	3.6×10^5
B0	17.5	32,500	7.8×10^6
B5	5.9	480	1.2×10^8
A0	2.9	39	7×10^8
A5	2.0	12.3	1.8×10^9
F0	1.6	5.2	3.1×10^9
F5	1.4	2.6	4.3×10^9
G0	1.05	1.25	8.9×10^9
G2 (the Sun)	1.0	1.0	1.0×10^{10}
G5	0.92	0.8	1.2×10^{10}
K0	0.79	0.55	1.8×10^{10}
K5	0.67	0.32	2.7×10^{10}
M0	0.51	0.08	5.4×10^{10}
M5	0.21	0.008	4.9×10^{11}
M8	0.06	0.0012	1.1×10^{12}

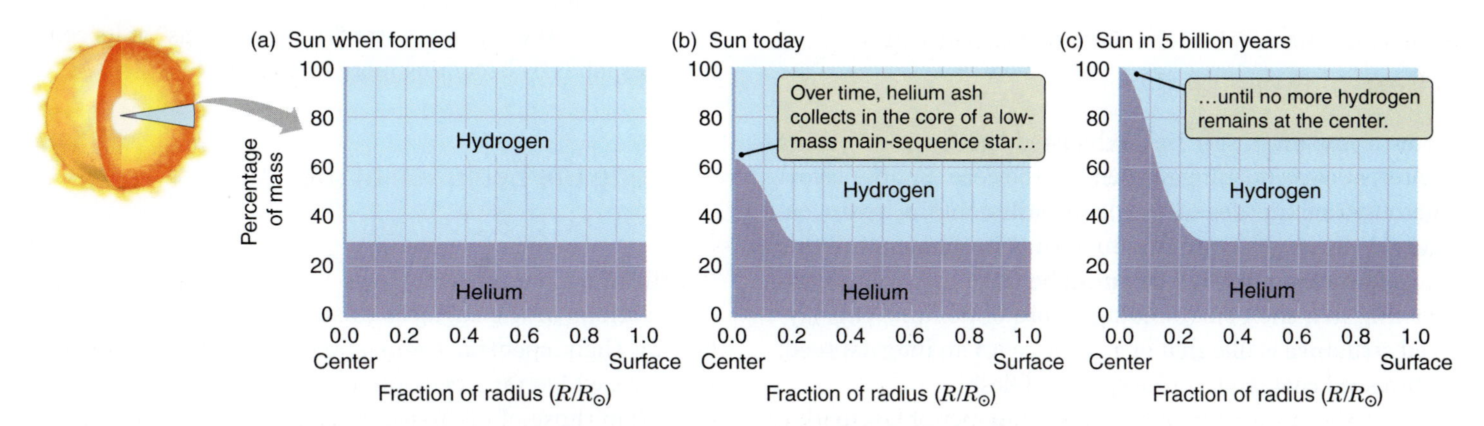

FIGURE 16.2 Chemical composition of the Sun is plotted here as a percentage of mass against distance from the center of the Sun. (a) When the Sun formed some 5 billion years ago, about 30 percent of its mass was helium and 70 percent was hydrogen throughout. (b) Today the material at the center of the Sun is about 65 percent helium and 35 percent hydrogen. (c) The Sun's main-sequence life will end in about 5 billion years, when all of the hydrogen at the center of the Sun will be gone.

space occupied by the nucleus and the electrons. The same is true for the matter within the Sun. At the enormous internal temperatures of the Sun, all of the electrons have been stripped away from their nuclei by energetic collisions. (In other words, the gas is completely ionized.) So the gas inside the Sun is a mixture of electrons and atomic nuclei all flying about freely. Because the size of atomic nuclei is very much less than the distance between nuclei in this gas, most of the space inside the Sun is empty, with electrons and atomic nuclei filling only a tiny fraction of the Sun's volume.

When a low-mass star like the Sun exhausts the hydrogen at its center, the situation changes. As gravity begins to win its shoving match against pressure, the helium core is crushed to an ever-smaller size and an ever-greater density. But there is a limit to how dense the core can get. The rules of quantum mechanics (the same rules that say atoms can have only certain discrete amounts of energy and that light comes in packets called photons) limit the number of electrons that can be packed into a given volume of space at a given pressure. As the matter in the core of the star is compressed further and further, it finally bumps up against this limit. The space occupied by the core of the star is no longer mostly empty, but is now effectively "filled" with electrons that are smashed tightly together. Matter at the center of the star is now so dense that a single cubic centimeter of this material can have a mass of 1,000 kilograms (kg) or more. Matter that has been compressed to this point is said to be **electron-degenerate** matter.

The crushed helium core is electron-degenerate.

Hydrogen Burns in a Shell Surrounding a Core of Helium Ash

After a low-mass star has exhausted the hydrogen at its center, nuclear burning may end in the core. But the layers surrounding the degenerate core still contain hydrogen, and this hydrogen continues to burn. Astronomers call this process **hydrogen shell burning** because the hydrogen is burning in a shell surrounding a core of helium. This layered structure is like that of a peach, with an internal seed, a thin seed coat, and a large sphere of flesh.

Electron-degenerate matter has a number of fascinating properties. For example, as more and more helium ash piles up on the electron degenerate core, the core *shrinks* in size. (This is one of the rules that degenerate matter breaks: the more massive it is, the smaller it is.) The reason the core shrinks is that the added mass increases the strength of gravity, and therefore the weight bearing down on the core, so the electrons are smashed together into a smaller volume. The presence of the electron degenerate core triggers the following chain of events that dominates the evolution of a 1-$M_\odot$ star for the next 50 million years:

1. Just outside the star's degenerate helium core, the gravitational acceleration g_{core} is proportional to the core's mass (M_{core}) divided by the square of its radius (R_{core}). As the helium core grows, both its larger mass (bigger M_{core}) and its shrinking size (smaller R_{core}) cause the gravitational acceleration to increase. As more helium is added to the core, the gravitational acceleration at the core increases dramatically.
2. Increasing the gravitational acceleration around the core increases the weight of the overlying material pushing down on the hydrogen-burning shell surrounding the core.
3. This increase in weight must be balanced by an increase in pressure in the inner parts of the star. In particular, the pressure in the hydrogen-burning shell must increase.
4. Increasing the pressure and temperature in the hydrogen-burning shell drives up the rate of the nuclear reactions occurring in the shell.
5. Faster nuclear reactions release more energy, so the luminosity of the star *increases*.

This is a very counterintuitive result. When a star like the Sun uses up the nuclear fuel at its center, it becomes more, not less, luminous. A degenerate core means stronger gravity, stronger gravity means higher pressure, and higher pressure means faster nuclear burning, producing greater and greater amounts of energy. This faster nuclear burning means a more luminous star. So, when the low-mass star "runs out of gas" at its center, it responds by becoming more luminous.

Running out of fuel makes the star grow more luminous.

Tracking the Evolution of the Star on the H-R Diagram

Recall from Chapter 13 that the H-R diagram is based on observations of the surfaces of stars. Stars can be identified by their spectra. Giant stars have a lower surface pressure, which results in narrower widths in the spectral lines than those of main-sequence stars of the same spectral class.

The changes that occur in the heart of a star with a degenerate helium core are reflected in changes in the overall structure of the star. With time, the mass of the degenerate helium core grows as more and more hydrogen is converted into helium ash in the surrounding shell. And as the mass of the degenerate helium core grows, so does the rate of energy generation in the surrounding hydrogen-burning shell. This increase in energy generation heats the overlying layers of the star, causing them to expand. The star becomes a bloated, luminous giant. As illustrated in **Figure 16.3**, the internal structure of the star is now fundamentally different from the structure when the star was on the main sequence. The giant can grow to have a luminosity hundreds of times the luminosity of the Sun and a radius of over 50 solar radii (50 $R_\odot$). Yet at the same time, the core of the giant star is compact; much of the star's mass becomes concentrated into a volume that is only a few times the size of Earth.

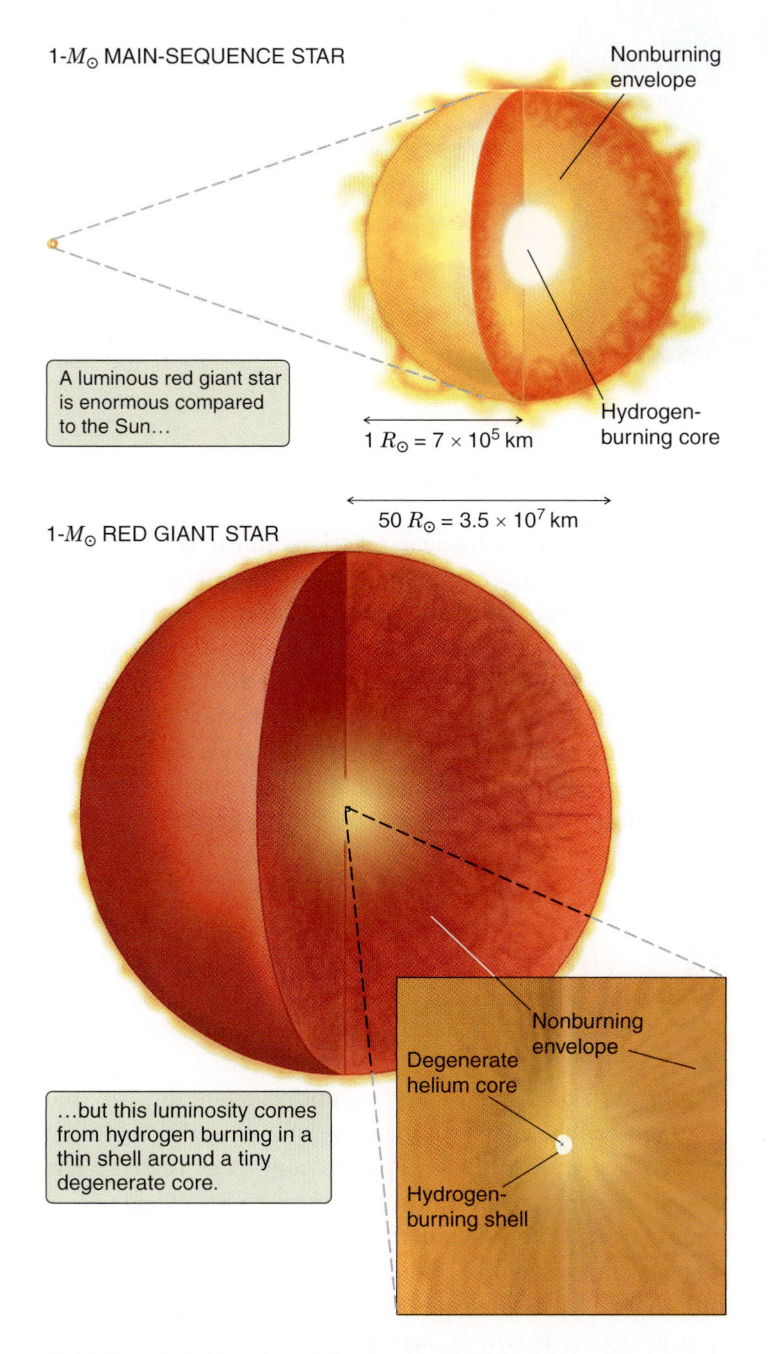

FIGURE 16.3 The size of the Sun (top left) is compared with the size of a star near the top of the red giant branch (bottom). The structure of the Sun is compared with the core of the red giant in the 50-times larger views identified by the dashed lines.

Changes taking place deep within a star's interior are hidden; the star becomes larger and more luminous, and *cooler and redder* as well. The enormous expanse of the star's surface allows it to cool very efficiently. Even though its *interior* grows hotter and its luminosity increases, the *surface* temperature of the star actually begins to drop. The relation among radius, temperature, and luminosity that we discussed in Math Tools 13.2 and Chapter 15 still applies ($L = 4\pi\sigma R^2T^4$), so a change in one of these variables will lead to a change in the other two.

A bloated luminous giant surrounds a tiny degenerate core.

Just as the H-R diagram shows the changes in a protostar on its way to the main sequence, it is a handy device for keeping track of the changing luminosity and surface temperature of the star as it evolves away from the main sequence (**Figure 16.4**). As soon as the star exhausts the hydrogen in its core, it leaves the main sequence and begins to move upward and to the right on the H-R diagram, growing more luminous but cooler. Such a star, which is somewhat brighter and larger than it was on the main sequence, is called a **subgiant** star. As the subgiant continues to evolve, it grows larger and cooler, but after a time its progress to the right on the H-R diagram hits a roadblock: the H^- thermostat. When the surface temperature of the subgiant star has dropped by about 1000 kelvins (K) relative to its temperature on the main sequence, H^- ions start to form in great abundance in its atmosphere. Recall from Chapter 15 that it was the H^- ion in the protostar's atmosphere that acted as a thermostat, regulating the star's temperature. The H^- ion serves exactly the same role here, regulating how much radiation can escape from the star and preventing it from becoming any cooler.

Because the star can cool no further, it begins to move almost vertically upward on the H-R diagram, growing larger and more luminous but remaining about the same temperature. The star has become a **red giant**—an obvious name for a star that is now both redder and larger than it was on the main sequence. You can think of the path that a star follows on the H-R diagram as it leaves the main sequence as being a tree "branch" growing out of the "trunk" of the main sequence. Astronomers refer to these tracks as the **subgiant branch** and the **red giant branch** of the H-R diagram. The path that a red giant follows on the H-R diagram closely parallels the path that it followed earlier as a collapsing protostar on its way toward the main sequence—except in reverse: this time the star is moving up that path rather than coming down it. This similarity is not a coincidence. The same physical processes (such as the H^- thermostat) that give rise to the Hayashi track followed by a collapsing protostar also control the relationship of luminosity, size, and surface temperature in an expanding red giant.

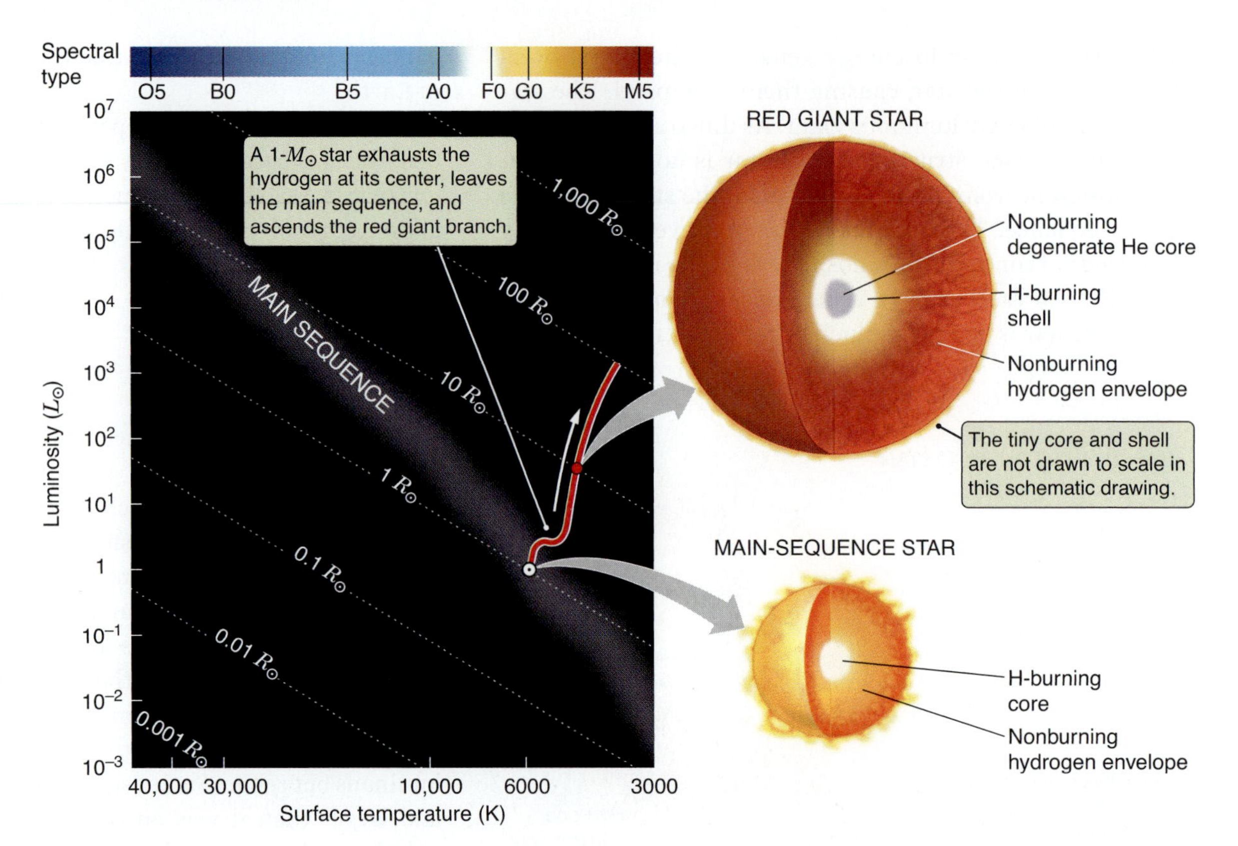

FIGURE 16.4 The evolution of a red giant star on the H-R diagram. A red giant star consists of a degenerate core of helium ash surrounded by a hydrogen-burning shell. As the star moves up the red giant branch, it comes close to retracing the Hayashi track that it followed when it was a protostar collapsing toward the main sequence.

As the star leaves the main sequence, the changes in its structure occur slowly at first, but then the star moves up the red giant branch faster and faster. It takes 200 million years or so for a star like the Sun to go from the main sequence to the top of the red giant branch. Roughly the first half of this period is spent on the subgiant branch as the star's luminosity increases to about 10 times the luminosity of the Sun ($L_\odot$). During the second half of this period, the star's luminosity skyrockets from 10 $L_\odot$ to almost 1,000 $L_\odot$. The evolution of the star, illustrated in **Figure 16.5**, is reminiscent of the growth of a snowball rolling downhill. The larger the snowball becomes, the faster it grows; and the faster it grows, the larger it becomes. Growth and size feed off each other, and what began as a bit of snow at the top of the mountain soon becomes a huge ball.

An evolving low-mass star moves up and to the right on the H-R diagram.

The analogy between the evolution of a red giant star and the growth of a snowball is not perfect. The helium core of the star grows in mass—but not in radius as the rest of the star does—as hydrogen is converted to helium in the hydrogen-burning shell. The increasing mass of the ever-more-compact helium core increases the force of gravity in the heart of the star. Stronger gravity means higher pressure, and higher pressure accelerates nuclear burning in the shell. But faster nuclear reactions in the shell convert hydrogen into helium more quickly, so the part of the star's mass that is in its core grows more rapidly. The cycle feeds on itself: increasing core mass leads to ever-faster burning in the shell; and the faster hydrogen burns in the shell, the faster the core mass grows. As a result, the star's luminosity climbs at an ever-higher rate. Note that since the luminosity has increased but the mass has not, the main-sequence mass-luminosity relation discussed at the beginning of this chapter no longer applies.

16.3 Helium Begins to Burn in the Degenerate Core

The growth of the red giant cannot continue forever, and the next crucial question for understanding the evolution of stars becomes, What will be the next thing to give? The answer lies in another unusual property of the degenerate helium core. The core of the red giant star is electron-degenerate, which means that as many *electrons* are packed into that space as the rules of quantum mechanics allow at that pressure. However, *atomic nuclei* in the core are still able to move freely about as shown in **Figure 16.6**, just as they are throughout the rest of the star.

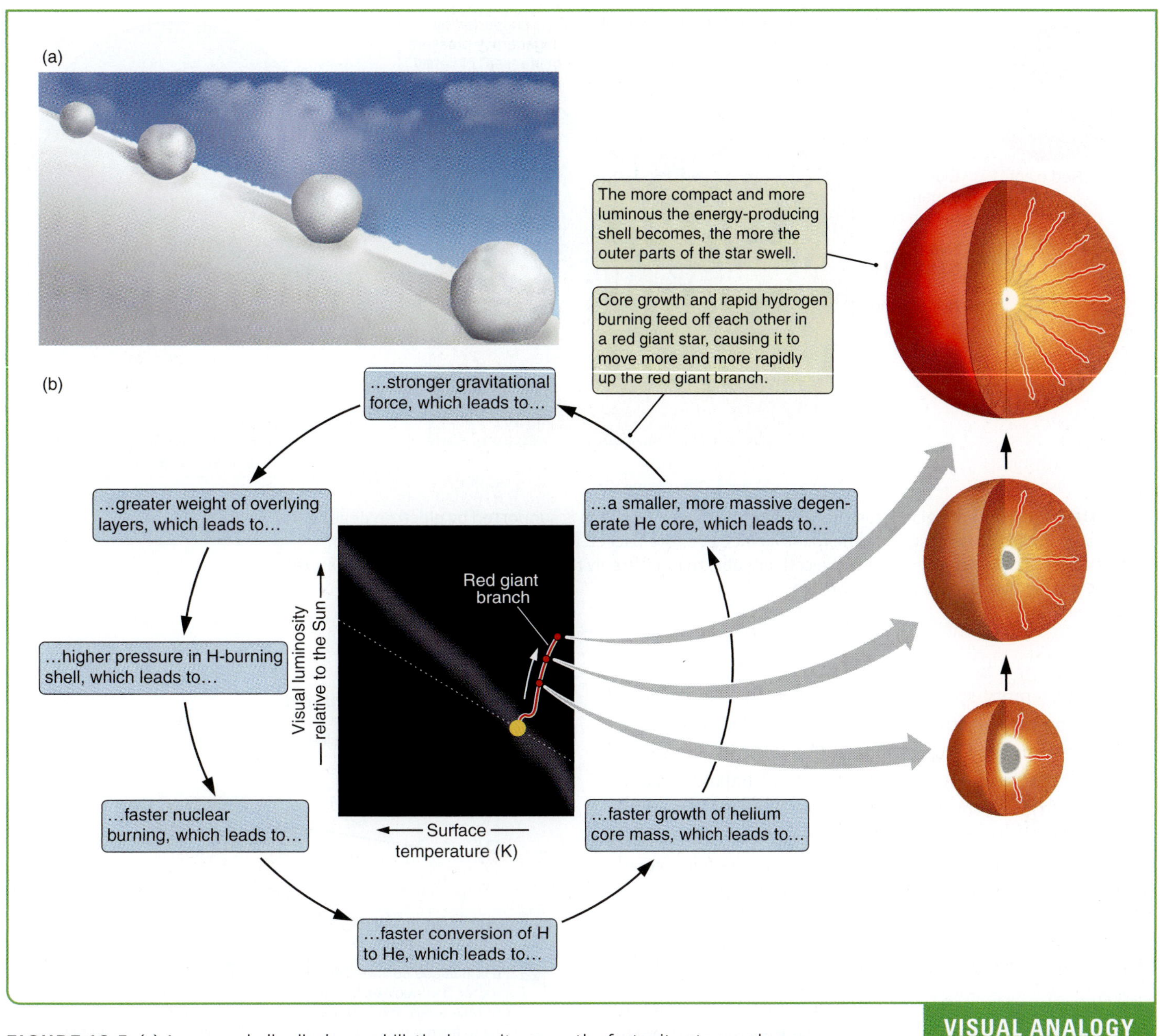

FIGURE 16.5 (a) As a snowball rolls down a hill, the larger it grows, the faster it gets even larger. (b) Similarly, as a star moves up the red giant branch, burning hydrogen to helium in a shell surrounding a degenerate helium core, its evolution feeds on itself. The luminosity of the star grows faster and faster.

You are used to thinking about all material as being equal: if a room were packed as tightly as possible with cats, it would be surprising if people could move freely through the room. Similarly, you normally think of atomic nuclei as being larger than electrons, so if the electrons are packed as tightly as possible into the core of the star, then you might expect atomic nuclei to be packed as tightly as possible into the core as well. But the rules of quantum mechanics say that particles are also waves, and electron waves are larger than nuclear waves. When packed together tightly, electrons effectively take up more space than do more massive particles like atomic nuclei. As the core of the star is compressed, electrons become degenerate much sooner than the atomic nuclei do.

The laws of quantum mechanics place few restrictions on electrons and atomic nuclei occupying the same physical space (unlike people and cats). As far as the atomic nuclei are concerned, the electron-degenerate core of the star is

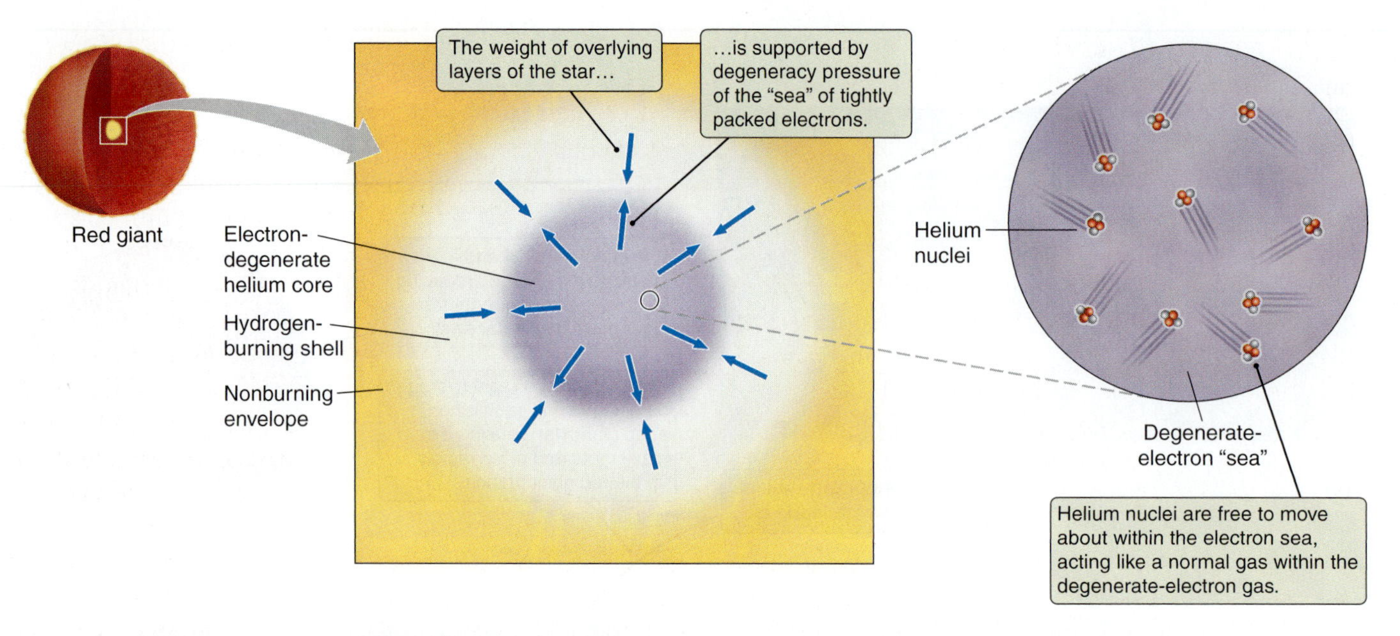

FIGURE 16.6 In a red giant star, the weight of the overlying layers is supported by electron degeneracy pressure in the core arising from the fact that electrons are packed together as tightly as quantum mechanics allows. Atomic nuclei in the core are able to move freely about within the sea of degenerate electrons, so they behave as a normal gas.

still mostly empty space. The nuclei behave like a normal gas, moving through the sea of degenerate electrons almost as if the electrons were not there. The negative charges of the electrons are important because they balance out the positive charges of the nuclei; but apart from that, the atomic nuclei in the electron-degenerate core of a star are a perfectly normal "gas within a gas," behaving just as the (electrically neutral) atoms and molecules in the air on Earth do.

Helium Burning and the Triple-Alpha Process

As the star evolves up the red giant branch, its helium core grows not only smaller and more massive, but hotter as well. This increase in temperature is due partly to the gravitational energy released as the core shrinks (just as the protostar's core grew hotter when it was collapsing) and to the energy released by the ever-faster pace of hydrogen burning in the surrounding shell. The climbing temperature of the core means that the thermal motions of the atomic nuclei in the core become more and more energetic. Eventually, at a temperature of about 10^8 K (a hundred million kelvins), helium burning begins.

Helium burns in a two-stage process referred to as the **triple-alpha process** (**Figure 16.7**). First, two helium-4 nuclei (^{4}He) fuse to form a beryllium-8 nucleus (^{8}Be) consisting of four protons and four neutrons. The ^{8}Be nucleus is extremely unstable. Left on its own, it would break apart again after only about a trillionth (10^{-12}) of a second. But if, in that short time, it collides with a ^{4}He nucleus, the two nuclei will fuse into a stable nucleus of carbon-12 (^{12}C) consisting of six protons and six neutrons. (The reaction rate is very temperature-dependent, with higher temperatures enabling more reactions and increasing the number of ^{8}Be

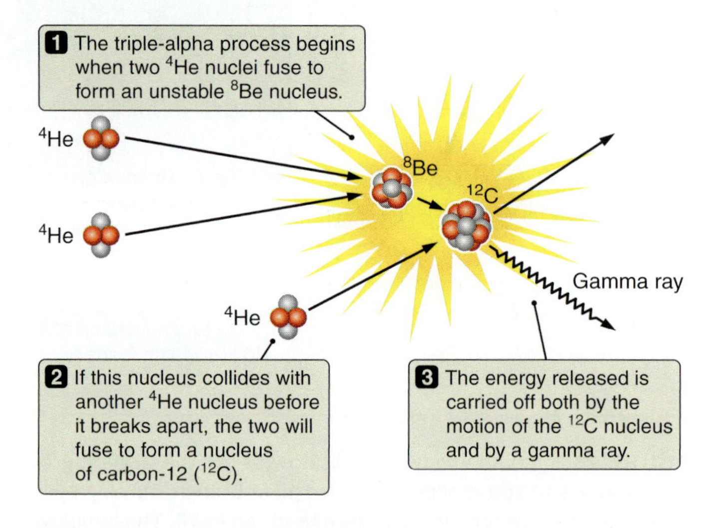

FIGURE 16.7 The triple-alpha process. Two helium-4 (^{4}He) nuclei fuse to form an unstable beryllium-8 (^{8}Be) nucleus. If this nucleus collides with another ^{4}He nucleus before it breaks apart, the two will fuse to form a stable nucleus of carbon-12 (^{12}C). The energy produced is carried off both by the motion of the ^{12}C nucleus and by a high-energy gamma ray emitted in the second step of the process.

nuclei that collide with a ^{4}He nucleus). The triple-alpha process takes its name from the fact that it involves the fusion of three ^{4}He nuclei, which are traditionally referred to as **alpha particles**.

The Helium Core Ignites in a Helium Flash

In the next phase of the star's evolution, the helium in the core begins burning. Degenerate material is a very good conductor of thermal energy, so any differences in temperature within the core rapidly even out. As a result, when helium burning begins at the center of the core, the energy released quickly heats the entire core. Within a few minutes the entire core is burning helium into carbon by the triple-alpha process.

In a normal gas like the air around you, the pressure of the gas comes from the random thermal motions of the atoms. Increasing the temperature of such a gas means that the motions of the atoms become more energetic, so the pressure of the gas increases. If the helium core of a red giant star were a normal gas, the increase in temperature that accompanies the onset of helium burning would lead to an increase in pressure. The core of the star would expand; the temperature, density, and pressure would decrease; nuclear reactions would slow; and the star would settle into a new balance between gravity and pressure. These are exactly the sorts of changes that are steadily occurring within the core of a main-sequence star like the Sun, as the structure of the star steadily and smoothly shifts in response to the changing composition in the star's core.

However, the degenerate core of a red giant is not a normal gas. The pressure in a red giant's degenerate core comes from how tightly the electrons in the core are packed together. Heating the core does not change the number of electrons that can be packed into its volume, so the core's pressure does not respond to changes in temperature. And if the pressure does not increase, the core does not expand.

When helium begins to burn in the degenerate core, the temperature of the core goes up but the pressure does not. So the onset of helium burning in the degenerate core of a red giant does not cause the core to expand. Yet even though the higher temperature does not change the pressure, it does cause the helium nuclei to be slammed together with more frequency and greater force, so the nuclear reactions become more vigorous. The process begins to snowball again. More vigorous reactions mean higher temperature, and higher temperature means even more vigorous reactions. Thermonuclear burning in the degenerate core runs away with itself, wildly out of control as increasing temperature and increasing reaction rates feed each other. This is the **helium flash**.

The helium flash is a thermonuclear runaway—an explosion within the star.

Helium burning begins at a temperature of about 100 million K. By the time the temperature has climbed by just 10 percent, to 110 million K, the rate of helium burning has increased to 40 times what it was at 100 million K. By the time the core's temperature reaches 200 million K, the core is burning helium 460 million times faster than it was at 100 million K. As the temperature in the core grows higher and higher, the thermal motions of the electrons and nuclei become more energetic, and the pressure due to these thermal motions becomes greater and greater. Within seconds of ignition, the thermal pressure increases until it is no longer smaller than the degeneracy pressure, at which point the core literally explodes. Because the explosion is contained within the star, however, it cannot be seen. The energy released in this runaway thermonuclear explosion lifts the overlying layers of the star, and as the core expands, the electrons are able to spread out. The drama is over within a few hours. The expanded helium-burning core is no longer degenerate, and the star is on its way toward a new equilibrium.

Following the helium flash, the star once again does something counterintuitive. Helium burning in the core does not cause the star to grow more luminous. The tremendous energy released during the helium flash goes into fighting gravity and puffing up the core. After the helium flash, the core (which is no longer degenerate) is much larger, so the forces of gravity within it and the surrounding shell are much smaller. (Again, a larger core radius means smaller values of gravitational acceleration, *g*.) Weaker gravity means less weight pushing down on the core and the shell, which means lower pressure. Lower pressure, in turn, slows the nuclear reactions. The net result is that after the helium flash, core helium burning keeps the core of the star puffed up, and the star becomes less luminous than it was as a red giant.

The star spends the next 100,000 years or so settling into a stable structure in which helium burns to carbon in a normal, nondegenerate core while hydrogen burns to helium in a surrounding shell. The star is now about one-hundredth as luminous as it was at the time of the helium flash. The lower luminosity means that the outer layers of the star are not as puffed up as they were as a red giant. The star shrinks, and as it does so its surface temperature climbs. (This is just the reverse of the sequence of events that caused the red giant to become larger and redder as it grew more luminous.)

At this point in their evolution, low-mass stars with chemical compositions similar to that of the Sun will bunch up on the H-R diagram just to the left of the red giant branch. Stars that contain much less iron than the Sun tend to distribute themselves away from the red giant branch

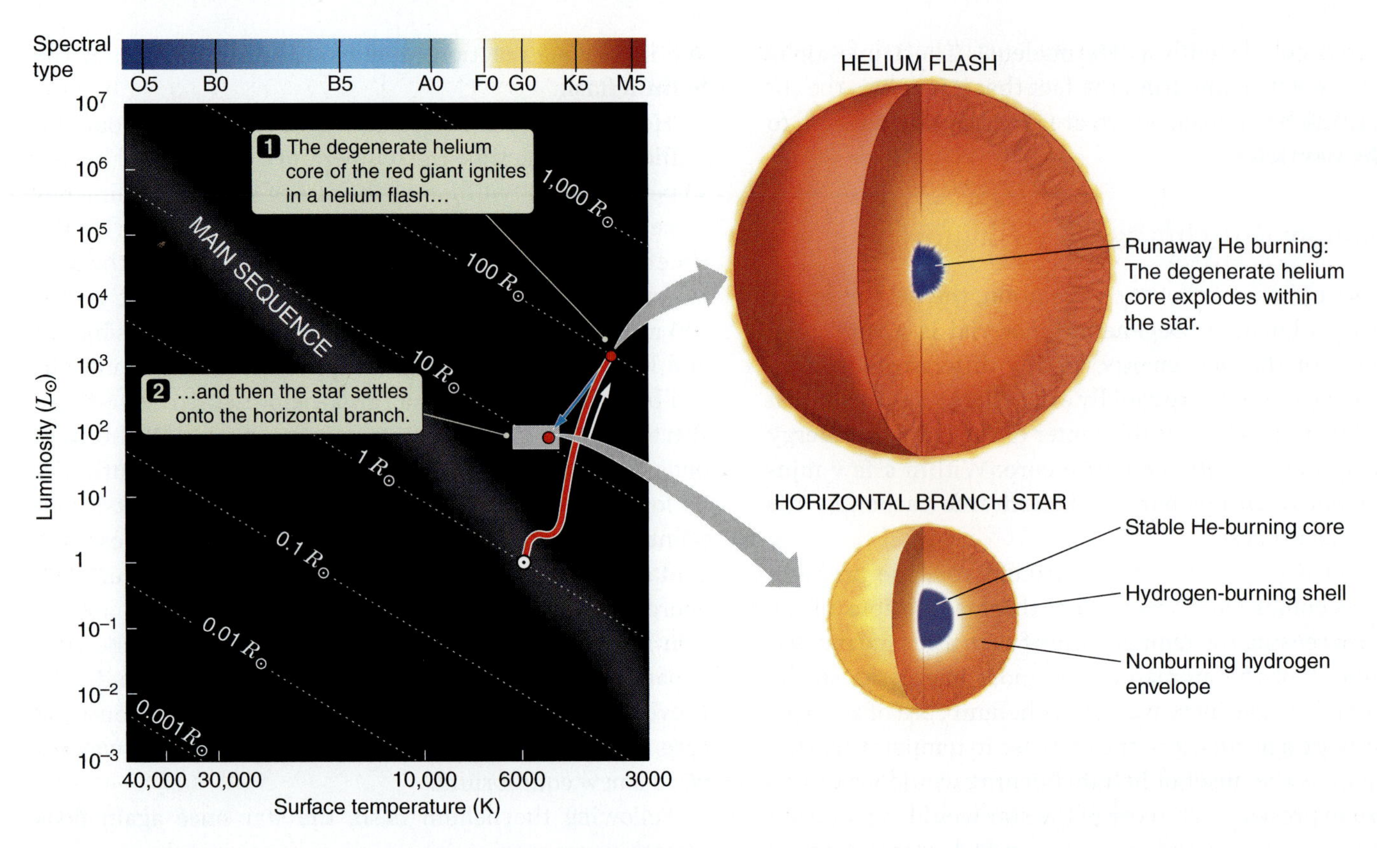

FIGURE 16.8 When the core temperature of a red giant reaches about 10^8 K, helium begins to burn explosively in the degenerate core, leading to a helium flash. After a few hours the core of the star begins to inflate, ending the helium flash. Over about the next 100,000 years (a relatively short time), the star settles onto the *horizontal branch*, where it burns helium in its core and hydrogen in a surrounding shell.

along a nearly horizontal line on the H-R diagram. This stage of stellar evolution takes its name from this horizontal band. The star is now called a **horizontal branch** star (**Figure 16.8**).

16.4 The Low-Mass Star Enters the Last Stages of Its Evolution

The evolution of a star like the Sun from the main sequence through its helium flash and on to the horizontal branch is fairly well understood. Just as an understanding of the interior of the Sun comes from computer models of the physical conditions within it, an understanding of the evolution of a red giant comes from computer models that look at the changes in structure as the star's degenerate helium core grows. These models show that any star with a mass of about 1 $M_\odot$ will follow the march from main sequence to helium flash, and then drop down onto the horizontal branch. But computer models are less clear about what happens next. We already noted that differences in chemical composition between stars significantly affect where they fall on the horizontal branch. From this point on, small changes in the properties of a star—mass, chemical composition, strength of the star's magnetic field, or even the rate at which the star is rotating—can lead to noticeable differences in how the star evolves. The more time that passes beyond the helium flash, the more possible it is that divergent evolutionary paths are open to a low-mass star.

With this caveat in mind, we continue our story of the evolution of a 1-$M_\odot$ star with solar composition, presenting the most likely sequence of events.

The Star Moves Up the Asymptotic Giant Branch

The structure and behavior of a star on the horizontal branch are remarkably similar to those of a star on the main sequence in many respects. The biggest difference is that now, instead of burning hydrogen into helium in a stable, nondegenerate core, the horizontal branch star is burning helium into carbon in a stable, nondegenerate core. (The other difference is that hydrogen is continuing to burn in a shell surrounding the core.)

The star's time on the horizontal branch, however, is much shorter than its span on the main sequence. There is now less fuel to burn in its core. In addition, the star is more luminous, so it is consuming fuel more rapidly. Finally, helium is a much less efficient nuclear fuel than hydrogen. Even so, for 100 million years the horizontal branch star remains stable, burning helium to carbon in its core, and hydrogen to helium in a shell.

The temperature at the center of a horizontal branch star is not high enough for carbon to burn, so carbon ash builds up in the heart of the star. When the horizontal branch star has burned all of the helium at its core, gravity once again begins to win. The nonburning carbon ash core is crushed by the weight of the layers of the star above it until once again the electrons in the core are packed together as tightly as the laws of quantum mechanics allow at its pressure. The carbon core is now electron-degenerate, with physical properties much like those of the degenerate helium core at the center of a red giant.

The star forms a degenerate carbon core as it leaves the horizontal branch.

The small, dense electron-degenerate carbon core drives up the strength of gravity in the inner parts of the star, which in turn drives up the pressure, which speeds up the nuclear reactions, which causes the degenerate core to grow more rapidly. The internal changes occurring within the star are similar to the changes that took place at the end of the star's main-sequence lifetime, and the path the star follows as it leaves the horizontal branch echoes that earlier phase of evolution as well. Just as the star accelerated up the red giant branch as its degenerate helium core grew, the star now leaves the horizontal branch and once again begins to grow larger, redder, and more luminous as its degenerate carbon core grows. The path that the star follows on the H-R diagram (**Figure 16.9**) closely parallels the path it followed as a red giant, getting closer to the red giant branch as the star grows more luminous. That is why this phase of evolution is called the **asymptotic giant branch** (**AGB**) of the H-R diagram. An AGB star is burning helium and hydrogen in nested concentric shells surrounding a degenerate carbon core.

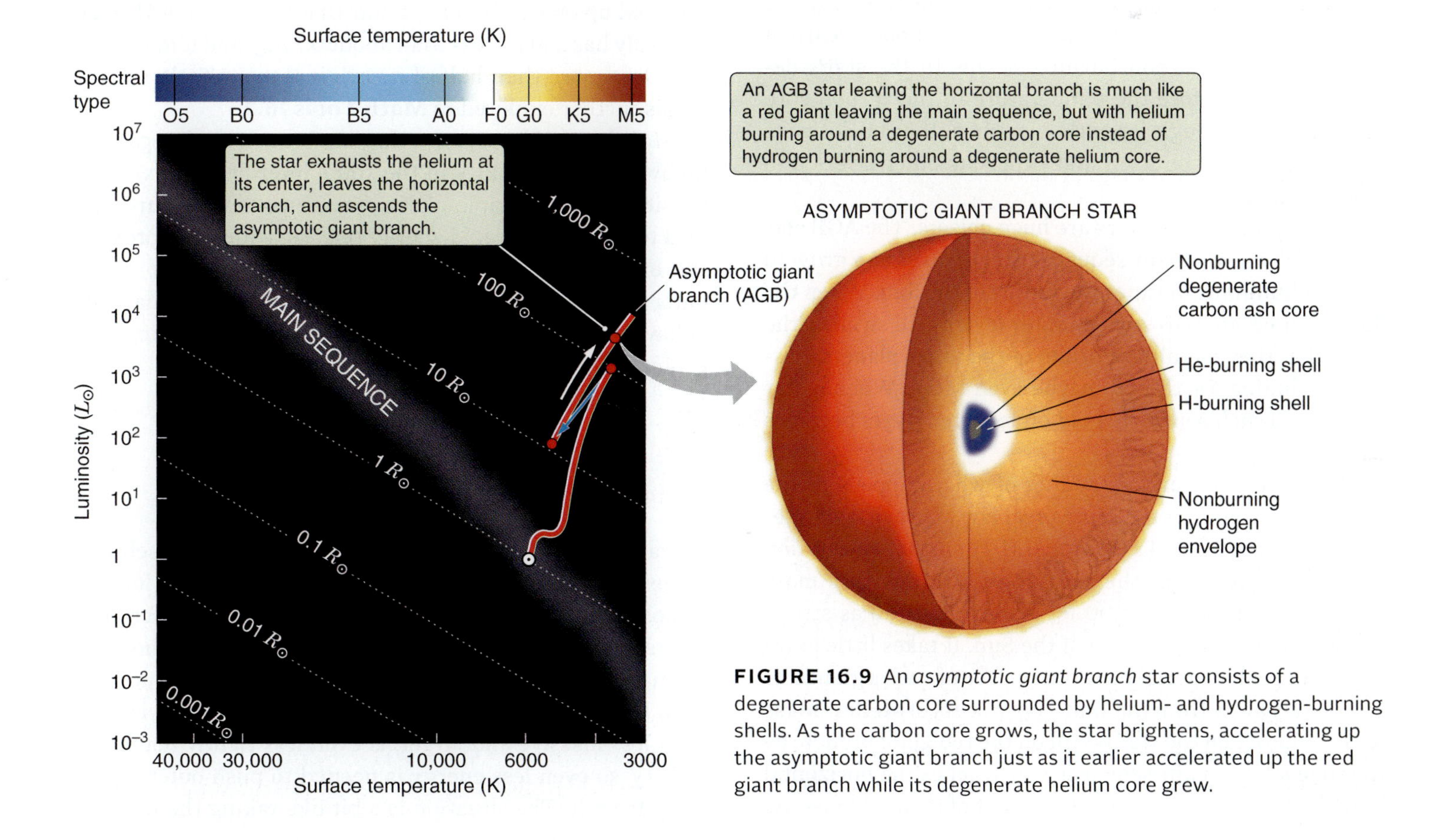

FIGURE 16.9 An *asymptotic giant branch* star consists of a degenerate carbon core surrounded by helium- and hydrogen-burning shells. As the carbon core grows, the star brightens, accelerating up the asymptotic giant branch just as it earlier accelerated up the red giant branch while its degenerate helium core grew.

Math Tools 16.2

Escaping the Surface of an Evolved Star

Why are giant stars likely to lose mass? The escape velocity from the surface of a star is given by:

$$v_{esc} = \sqrt{2GM/R}$$

How does v_{esc} change when a star becomes a red giant? Let's look at the Sun as an example. When the Sun is on the main sequence, its escape velocity (recall Chapter 4) can be calculated (using $M_\odot = 1.99 \times 10^{30}$ kg and $R_\odot = 6.96 \times 10^8$ m) as follows:

$$v_{esc} = \sqrt{2 \times (6.67 \times 10^{-11}\ \text{m}^3\text{kg}^{-1}\text{s}^{-2}) \times (1.99 \times 10^{30}\ \text{kg}) \Big/ 6.96 \times 10^8\ \text{m}}$$

$$= \sqrt{2.65 \times 10^{20}\ \text{m}^3\text{s}^{-2} \Big/ 6.96 \times 10^8} = \sqrt{3.81 \times 10^{11}\ \text{m}^2\text{s}^{-2}}$$

$$= 617{,}000\ \text{m/s} = 617\ \text{km/s}$$

What will the escape velocity be when the Sun becomes a red giant, with a radius 50 times greater than the radius it has today and a mass 0.9 times its current mass:

$$v_{esc} = \sqrt{2 \times (6.67 \times 10^{-11}\ \text{m}^3\text{kg}^{-1}\text{s}^{-2}) \times 0.9 \times (1.99 \times 10^{30}\ \text{kg}) \Big/ 50 \times (6.96 \times 10^8\ \text{m})}$$

$$= \sqrt{2.4 \times 10^{20}\ \text{m}^3\text{s}^{-2} \Big/ 3.48 \times 10^{10}\ \text{m}} = \sqrt{6.9 \times 10^9\ \text{m}^2\text{s}^{-2}}$$

$$= 83{,}000\ \text{m/s} = 83\ \text{km/s}$$

The escape velocity from the surface of a red giant star is only 13 percent that of a main-sequence star. This is part of the reason that red giant and AGB stars lose mass. The magnetic fields and stellar pulsations in these stars also contribute to their loss of mass. The Sun may eventually lose as much as 30 percent of its mass.

Giant Stars Lose Mass

AGB stars are analogous to red giants in some ways. But the next step in the evolution of an AGB star is not a "carbon flash," in which carbon burning begins in the star's degenerate core. Before the temperature in the carbon core becomes high enough for carbon to burn, the star loses its gravitational grip on itself and expels its outer layers into interstellar space.

Red giant and AGB stars are huge objects. The AGB star into which a 1-$M_\odot$ main-sequence star evolves can grow to a radius hundreds of times the radius of the Sun. When the Sun becomes an AGB star, its outer layers will swell to the point that they engulf the orbits of the inner planets, possibly including Earth and maybe even Mars. When a star expands to such a size, its hold on its outer layers becomes tenuous.

Once again, an understanding of what happens to the star comes from Newton's universal law of gravitation. Recall from Math Tools 4.1 that $g = GM/r^2$; so, the acceleration due to gravity (g) at the surface of a star with a mass of 1 $M_\odot$ and a radius of 100 $R_\odot$ is only 1/10,000 as strong as the gravity at the surface of the Sun. It takes little extra energy to push material near the surface of such a giant star completely away from the star. The process of **stellar-mass loss** begins when the star is still on the red giant branch; by the time a 1-$M_\odot$ main-sequence star reaches the horizontal branch, it may have lost 10–20 percent of its total mass. As the star ascends the asymptotic giant branch, it loses another 20 percent or even more of its total mass. By the time it is well up on this branch, a star that began as a 1-$M_\odot$ star probably has a mass less than about 0.7 $M_\odot$, and it may even have lost more than half of its original mass. Stellar-mass loss is further explored in **Math Tools 16.2**.

Mass loss on the asymptotic giant branch can be spurred on by a lack of stability in the star's interior. The extreme sensitivity of the triple-alpha process to temperature can lead to episodes of rapid energy release, which can provide the extra kick needed to expel material from the star's outer layers. Even stars that are initially quite similar can behave very differently when they reach this stage in their evolution.

The Post-AGB Star May Cause a Planetary Nebula to Glow

Toward the end of an AGB star's life, mass loss itself becomes a runaway process. When a star loses a bit of its outermost layers, the weight pushing down on the underlying layers of the star is reduced. Without this weight holding them down, the remaining outer layers of the star puff up even larger than they were before. The star, which is now both less massive and larger, is even less tightly bound by gravity, so even less energy is needed to push outer layers away from it. The situation is a bit like taking the lid off a pressure cooker. Mass loss leads to weaker gravity, which leads to faster mass loss, which leads to weaker gravity, and so on. When the end comes, much of the remaining mass of the star is ejected into space, typically at speeds of 20–30 kilometers per second (km/s).

After ejection of its outer layers, all that is left of the low-mass star itself is a tiny, very hot electron-degenerate carbon core, surrounded by a thin envelope in which hydrogen and helium are still burning. This star is now somewhat less luminous than when it was at the top of the asymptotic giant branch, but it is still much more luminous than a horizontal branch star. The remaining hydrogen and helium in the star rapidly burn to carbon, and as more and more of the mass of the star ends up in the carbon core, the star itself shrinks and becomes hotter and hotter. Over the course of only 30,000 years or so following the beginning of runaway mass loss, the star moves very rapidly from right to left across the top of the H-R diagram (**Figure 16.10**).

The surface temperature of the star may eventually reach 100,000 K or hotter. Wien's law says that at such temperatures, most of the light from the star is in the high-energy ultraviolet part of the spectrum. The intense UV light from what remains of the star heats and ionizes the expanding shell of gas that was recently ejected by the star. The ultraviolet light causes the shell of gas to glow in the same way that UV light from an O star causes an H II region to glow.

If conditions are right, the mass ejected by the AGB star will pile up in a dense, expanding shell. If you were to look at such a shell through a telescope, you would see it as a round or oblong patch of light, perhaps with a hole and a dot in the middle. When these glowing shells were first observed in small telescopes, they looked like disks of planets, which is why such objects were named **planetary nebulae**. But there is nothing

Planetary nebulae may form around dying low-mass stars.

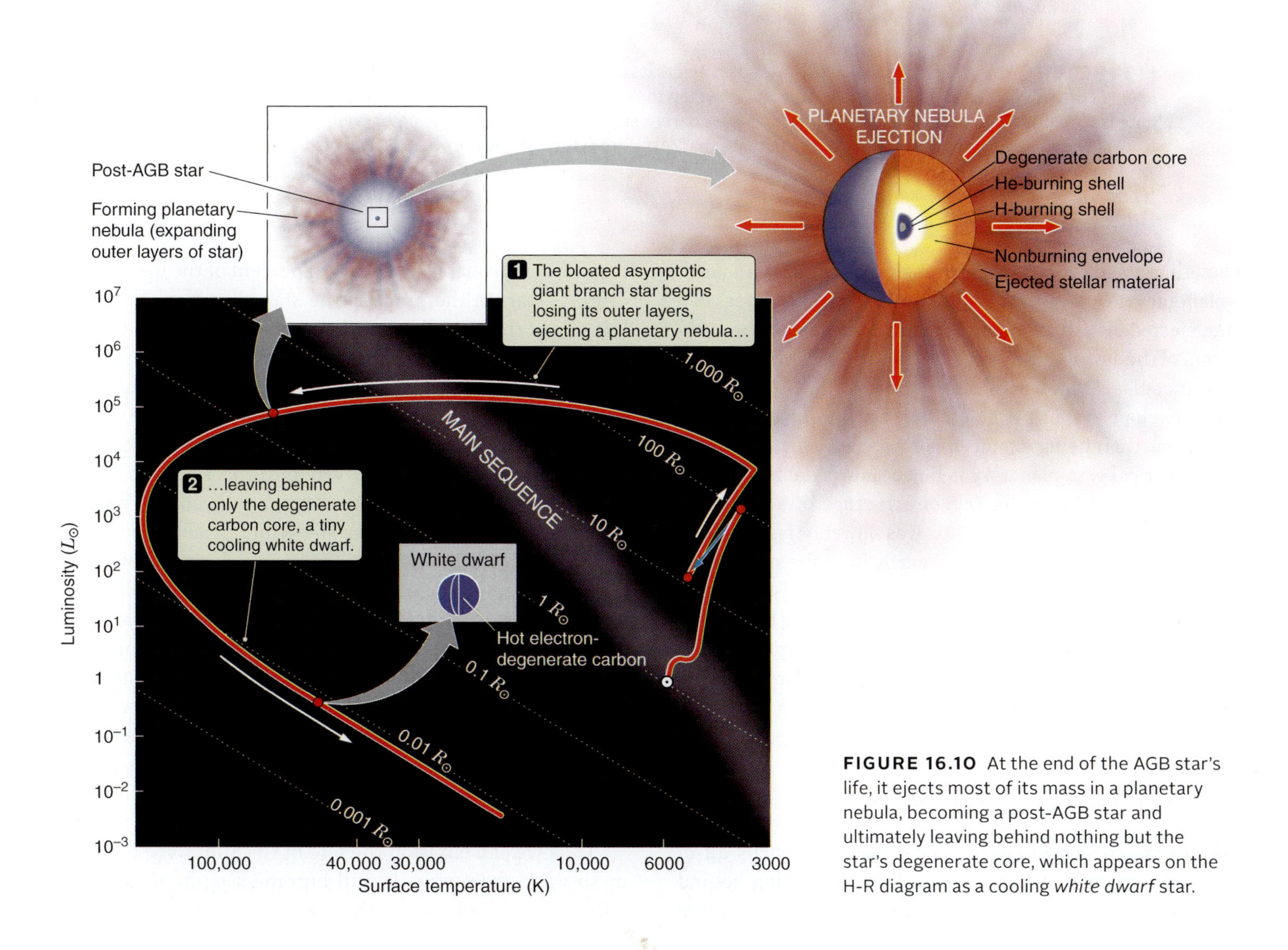

FIGURE 16.10 At the end of the AGB star's life, it ejects most of its mass in a planetary nebula, becoming a post-AGB star and ultimately leaving behind nothing but the star's degenerate core, which appears on the H-R diagram as a cooling *white dwarf* star.

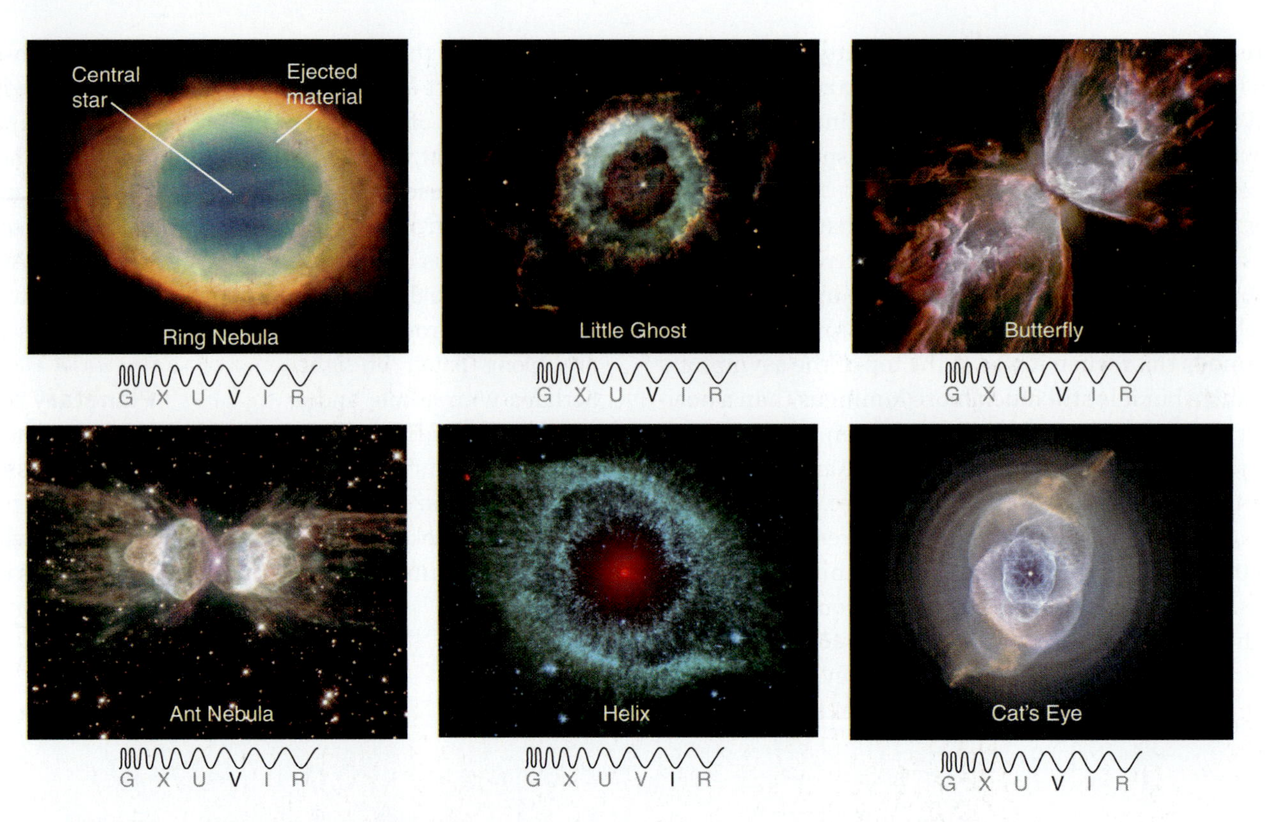

FIGURE 16.11 At the end of its life, a low-mass star ejects its outer layers and may form a *planetary nebula* consisting of an expanding shell of gas surrounding the white-hot remnant of the star. Planetary nebulae are not all simple spherical shells around their parent stars. These images of planetary nebulae from HST and *Spitzer* show the wealth of structures that result from the complex processes by which low-mass stars eject their outer layers.

planetlike about them, as is apparent in **Figure 16.11**. Instead, a planetary nebula consists of the remaining outer layers of a star, ejected into space as a dying gasp at the end of the star's ascent of the asymptotic giant branch. Not all stars form planetary nebulae. Stars with insufficient mass take too long, so their envelope evaporates before they can illuminate it.

Rather than being simple spherical shells surrounding a nice spherical star, planetary nebulae show a dazzling range in appearance (Figure 16.11). The structure of a planetary nebula tells of eras when mass loss was slower or faster, and of times when mass was ejected primarily from the star's equator or its poles. The gas in the planetary nebula also contains chemical elements that were produced by nuclear burning in the star, offering the first look at the processes responsible for the chemical evolution of the universe. A planetary nebula is visible for 50,000 years or so before the gas ejected by the star disperses so far that the nebula is too faint to be seen.

The Star Becomes a White Dwarf

Within about 50,000 years, a post-AGB star burns all of the fuel remaining on its surface, leaving nothing behind but a cinder—a nonburning ball of carbon with a remaining mass that is probably less than 70 percent of the mass of the original star. In the process, the star plummets down the left side of the H-R diagram, becoming smaller and fainter. Within a few thousand years the burned-out core shrinks to about the size of Earth, at which point it has become fully electron-degenerate and can shrink no further. The degenerate stellar cinder is now a **white dwarf**. The white dwarf continues to radiate energy away into space, and as it does so it cools, just as the filament of a lightbulb cools when the switch is turned off. Because the white dwarf is electron-degenerate, its size does not change much as it cools, so it moves down and to the right on the H-R diagram, following a line of constant radius. Even though the white dwarf may remain very hot for 10 million years or so, its luminosity may now be only 1/1,000 that of a main-sequence star like the Sun. Many white dwarfs are known, but all are much too faint to be seen without the aid of a telescope. Yet when you look at Sirius, you are also looking at a white dwarf. Sirius, the brightest star in Earth's sky has a faint white dwarf as a binary companion.

Thus we can envision the fate of the Sun, some 6 billion or so years from now. It will become a white dwarf that will fade as it radiates its thermal energy away into space.

Recall that this superdense ball—with a density of a ton per teaspoonful—actually began its life billions of years earlier as a cloud of interstellar gas billions of times more tenuous than the vacuum in the best vacuum chamber on Earth.

A newly formed white dwarf is hot but tiny.

Figure 16.12 illustrates the evolution of a solar-type 1-$M_\odot$ main-sequence star through to its final existence as a 0.6-$M_\odot$ white dwarf. This process is representative of the fate of low-mass stars. Although every low-mass star forms a white dwarf at the end point of its evolution, the exact path a low-mass star follows from core hydrogen burning on the main sequence to white dwarf depends on many details particular to the star. Some stars less massive than the Sun may become white dwarfs composed largely of helium rather than carbon. Temperatures in the cores of evolved 2- to 3-$M_\odot$ stars are high enough to allow additional nuclear reactions to occur, leading to the formation of somewhat more massive white dwarfs composed of materials such as oxygen, neon, and magnesium. Differences in chemical composition of a star can also lead to dramatic differences in its post-main-sequence evolution. Finally, it is important to note that stars whose main-sequence masses are less than about 0.8 $M_\odot$ are able to burn hydrogen in their cores for longer than the known age of the universe, so these have not been observed in a post-main-sequence phase.

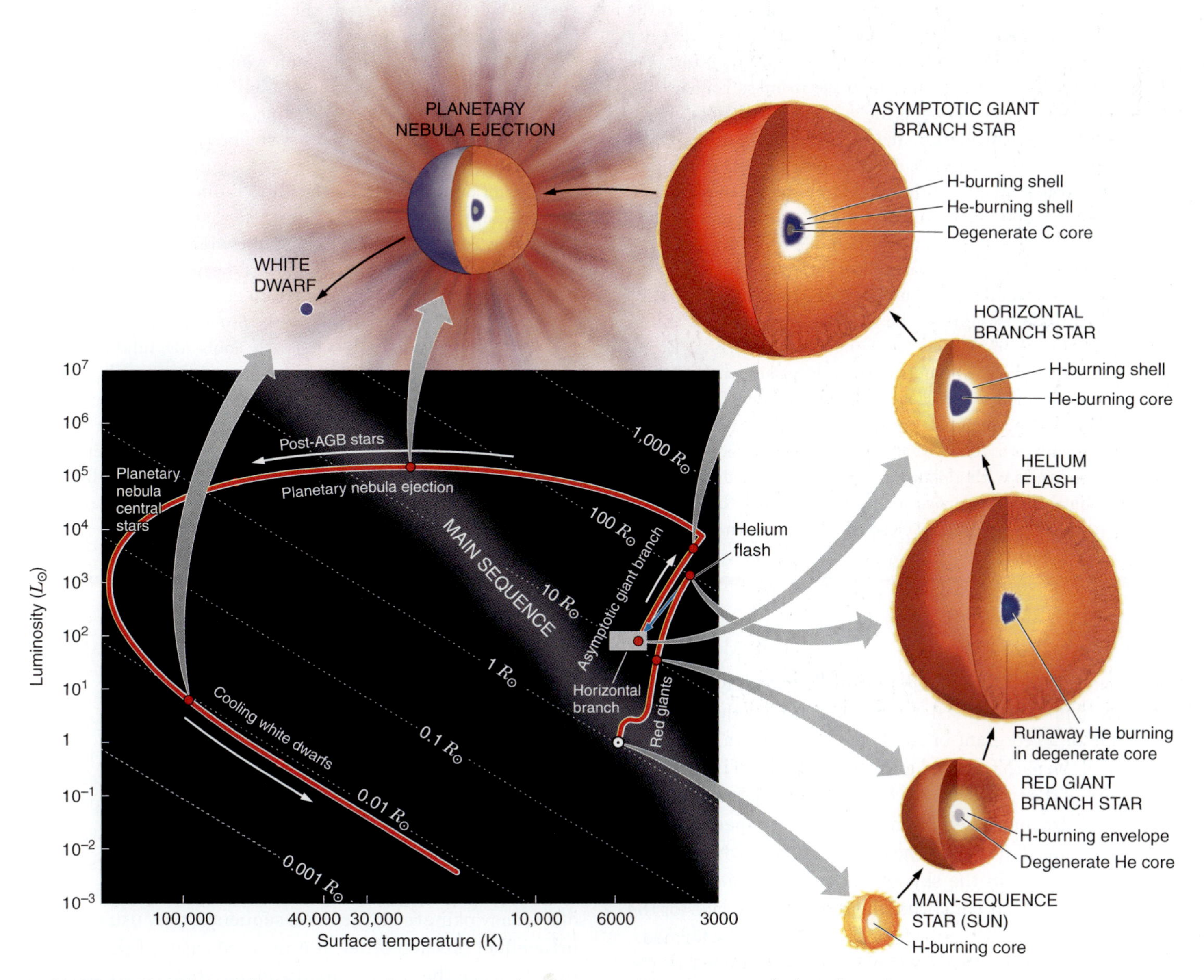

FIGURE 16.12 This H-R diagram summarizes the stages in the post-main-sequence evolution of a 1-$M_\odot$ star.

Connections 16.1

What Happens to the Planets?

The first decade of extrasolar planet discoveries yielded primarily planets with orbits that are very close to their star, because of the observational selection effects discussed in Chapter 7. Astronomers assumed that any planets closer than 1–2 astronomical units (AU) would not survive the post-main-sequence expansion of the star, and therefore did not expect to find planets in orbit around evolved stars.

But extrasolar planets are full of surprises. Planets have been observed in orbit around red giants, AGB stars, and horizontal branch stars. Astronomers cannot be sure whether the planets are in the same orbital location they were in when the star was on the main sequence, or if the **planets migrated** to different orbits. The surface gravity of a red giant is low, and some of the mass in its outermost layers will blow away. This decrease in mass reduces the gravitational force between the star and its planets, which by conservation of energy can lead to planetary orbits evolving *outward* away from the star. In some models the migration of large planets and the tidal forces between planets or between the star and a planet are significant factors too, and as a result, some planetary orbits could evolve *inward*.

In addition, there is some observational evidence of the remains of planetary systems around white dwarf stars. As a star loses mass during the evolutionary process, stellar or planetary companions may change their orbits. Rocky asteroids or smaller planets could migrate inward past the Roche limit (see Chapter 4) of the white dwarf and break up. Some of the material could remain in orbit around the white dwarf as a debris disk, similar to those seen in main-sequence stars with a planetary system. These dusty debris disks have been observed around many white dwarf stars with the Spitzer Space Telescope. Some of this dusty or rocky material may also fall onto the white dwarf, "polluting" its spectrum with heavy elements that were not produced in the stellar core.

And what about Earth? The age of the Sun from radioactive dating of meteorites indicates that the Solar System formed 4.6 billion years ago. In Chapter 14 we estimated how long the Sun might last by calculating its rate of hydrogen burning. The Sun's luminosity is about 30 percent higher now than it was early in the history of the Solar System, and it will continue to increase at a steady rate over the rest of the Sun's main-sequence lifetime of another 5 billion years. Models estimate that the Sun's luminosity may increase enough, even while the Sun is a main-sequence star, that Earth will heat up to the point where the oceans would evaporate, perhaps as soon as 1–2 billion years from now. By the time the Sun leaves the main sequence, the habitable zone may have moved out to Mars and no longer encompass the Earth.

It is not certain whether the Sun's radius, when the Sun is a red giant, will extend past 1 AU, so that Earth becomes completely engulfed. The red giant Sun will have low surface gravity and will lose mass, which could cause Earth's orbit to expand, thereby enabling Earth to escape the encroaching solar surface. Alternatively, as the Sun expands in radius and its rotation rate slows (see Math Tools 7.1), tidal forces might pull Earth inward. The habitable zone of a red giant Sun might be close to Jupiter or Saturn. Eventually, depending on how much mass the Sun loses in the red giant and AGB stages, the outer layers may or may not form a planetary nebula. The solar core will become a white dwarf, perhaps with a dusty disk and a "polluted" atmosphere as the only remaining evidence that a rocky planet once was here.

In our discussion of stellar evolution we have focused on what happens after a star leaves the main sequence. The spectacle of a red giant or AGB star is ephemeral. The Sun will travel the path from red giant to white dwarf in less than one-tenth of the time it spends on the main sequence steadily burning hydrogen to helium in its core. Stars spend most of their luminous lifetimes on the main sequence, which is why most of the stars in the sky are main-sequence stars. The fainter white dwarfs constitute the final resting place for the vast majority of stars that have been or ever will be formed.

In **Connections 16.1** we discuss what happens to the planets of an evolving star.

16.5 Binary Star Evolution

Possibly the most significant complication in this picture of the evolution of low-mass stars arises from the fact that many stars are members of binary systems. Recall from Chapter 13 that binary stars are very common and are quite helpful when Kepler's laws are applied to the orbits of the two stars to determine their masses. While both members of a binary pair are on the main sequence, they usually have little effect on each other. But in some cases, if the separation between the stars is small and one star is more massive than the other, their evolution may become linked.

Mass Flows from an Evolving Star onto Its Companion

Think for a moment about what would happen if you were to travel in a spacecraft from Earth toward the Moon. As you move away from Earth and closer to the Moon, the gravitational attraction of Earth weakens, and the gravitational attraction of the Moon becomes stronger. You eventually reach an intermediate zone where neither body has the stronger pull. If you continue beyond this point, the lunar gravity begins to successfully assert itself until you find yourself firmly in the gravitational grip of the Moon.

Exactly the same situation exists between two stars. Gas near each star clearly belongs to that star. When one star leaves the main sequence and swells up, its outer layers may cross that gravitational dividing line separating the star from its companion. Any material that crosses this line no longer belongs to the first star, but instead can be pulled toward the companion. A star reaches this point when it fills up its portion of an imaginary figure eight–shaped volume of space (**Figure 16.13**). These regions surrounding the two stars—their gravitational domains—are called the **Roche lobes** of the system. Once the first star has expanded to fill its Roche lobe, material begins to pour through the "neck" of the figure eight and fall toward the other star. This exchange of material between the two stars is called **mass transfer**.

Recall from Chapter 4 the Roche limit, which describes how close a small object might come to a planet before being torn apart by the planet's tides. Similarly, mass transfer between stars in a binary can be thought of as the "tidal stripping" of one star by the other. The physical principles at work in the two situations are much the same, and Édouard A. Roche is responsible for early calculations of both.

Evolution of a Binary System

The best way to understand how mass transfer affects the evolution of stars in a binary system is to apply what is known from studying the evolution of single low-mass stars. **Figure 16.13a** shows a binary system consisting of two low-mass stars. In the figure, the more massive of the two stars is "star 1," and the less massive of the two is "star 2." This is an ordinary binary system, and each of these stars is an ordinary main-sequence star for most of the system's lifetime. When one of the stars begins to evolve, however, things start to get interesting.

More massive main-sequence stars evolve more rapidly than less massive main-sequence stars. Therefore, star 1 will be the first to use up the hydrogen at its center and begin to evolve off the main sequence (**Figure 16.13b**). If the two stars are close enough together, star 1 will eventually grow to fill its Roche lobe, and material will transfer

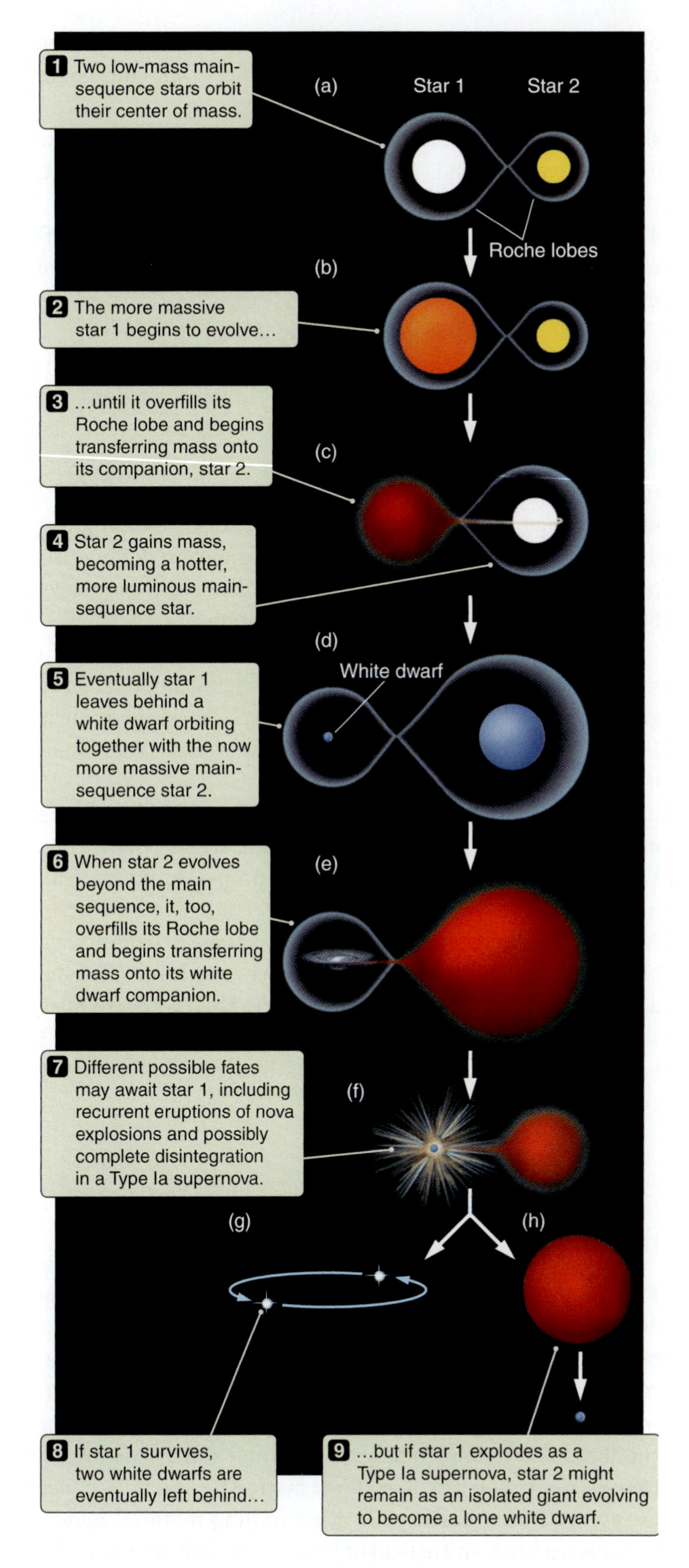

FIGURE 16.13 A compact binary system consisting of two low-mass stars passes through a sequence of stages as the stars evolve and mass is transferred back and forth. The evolution of the binary system can lead to novae or even to a Type Ia supernova that destroys the initially more massive star 1.

onto star 2 (**Figure 16.13c**). The transfer of mass between the two stars can result in a sort of "drag" that causes the orbits of the two stars to shrink, bringing the stars closer together and further enhancing mass loss. The two stars can even reach the point where they are effectively two cores sharing the same extended envelope of material.

The more massive star evolves first.

Despite these complexities, star 2 probably remains a basically normal main-sequence star throughout this process, burning hydrogen in its core. However, it does not remain the same main-sequence star that it started out as. The mass of star 2 increases because of the accumulation from its companion. As it does so, the structure of star 2 must change to accommodate its new status as a higher-mass star. Star 2's position on the H-R diagram during this period moves up and to the left along the main sequence, becoming larger, hotter, and more luminous.

While star 2 gains from the interaction, star 1 suffers. Star 1 never gets to experience the full glory of being an isolated red giant or AGB star, because star 1 can never grow larger than its Roche lobe. The presence of star 2 prevents star 1 from swelling to the size of the behemoths that populate the top of the H-R diagram. Yet star 1 continues to evolve, burning helium in its core on the horizontal branch, proceeding through a stage of helium shell burning, and finally losing its outer layers and leaving behind a white dwarf. **Figure 16.13d** shows the binary system after star 1 has completed its evolution. All that remains of star 1 is a white dwarf, orbiting about its bloated main-sequence companion.

The first star becomes a white dwarf orbiting a main-sequence star.

Fireworks Occur When the Second Star Evolves

Figure 16.13e picks up the evolution of the binary system as star 2 begins to evolve off the main sequence. Like star 1 before it, star 2 grows to fill its Roche lobe; as it does, material from star 2 begins to pour through the "neck" connecting the Roche lobes of the two stars. However, this time the mass is not being added to a normal star but is drawn toward the tiny white dwarf left behind by star 1. Because the white dwarf is so small, the infalling material generally misses the star. Instead of landing directly on the white dwarf, the infalling mass forms an **accretion disk** around the white dwarf, similar in some ways to the accretion disk that forms around a protostar. The accretion disk serves as a way station for material that is destined to find its way onto the white dwarf but that starts out with too much angular momentum to hit the white dwarf directly.

You have already seen that a white dwarf has a mass comparable to that of the Sun, but a size comparable to that of Earth. A large mass and a small radius mean strong gravity. In Chapter 7 you saw how the gravitational energy of material falling toward a forming protostar is converted into thermal energy when the material hits the accretion disk around the protostar. The same principle applies here. A kilogram of material falling from space onto the surface of a white dwarf releases 100 times more energy than a kilogram of material falling from the outer Solar System onto the surface of the Sun. The material streaming toward the white dwarf in the binary system falls into an incredibly deep gravitational "well." All of this energy has to go somewhere, and that somewhere is thermal energy. The spot where the stream of material from star 2 hits the accretion disk can be heated to millions of kelvins, where it glows in the far ultraviolet and X-ray parts of the electromagnetic spectrum.

Material from the second star gets hot as it falls onto the white dwarf.

Eventually, the infalling material accumulates on the surface of the white dwarf (**Figure 16.14a**), where it is compressed by the enormous gravitational pull of the white dwarf to a density close to that of the white dwarf itself. As more and more material builds up on the surface of the white dwarf, the white dwarf shrinks (just as the core of a red giant shrinks as it grows more massive). The density increases more and more, and at the same time the release of gravitational energy drives the temperature of the white dwarf higher and higher. Recall that the infalling material comes from the outer, unburned layers of star 2, so it is composed mostly of hydrogen. Hydrogen, the best nuclear fuel around, is being compressed to higher and higher densities and heated to higher and higher temperatures on the surface of the white dwarf.

The mental picture you should have at this point is of gasoline pooling on the floor of a match factory. Once the temperature at the base of the layer of hydrogen reaches about 10 million K, hydrogen begins to burn. But this is not the contained hydrogen burning that takes place in the center of the Sun. Rather, this is explosive hydrogen burning in a degenerate gas. Energy released by hydrogen burning drives up the temperature, and the higher temperature drives up

Runaway burning of hydrogen on a white dwarf causes a nova.

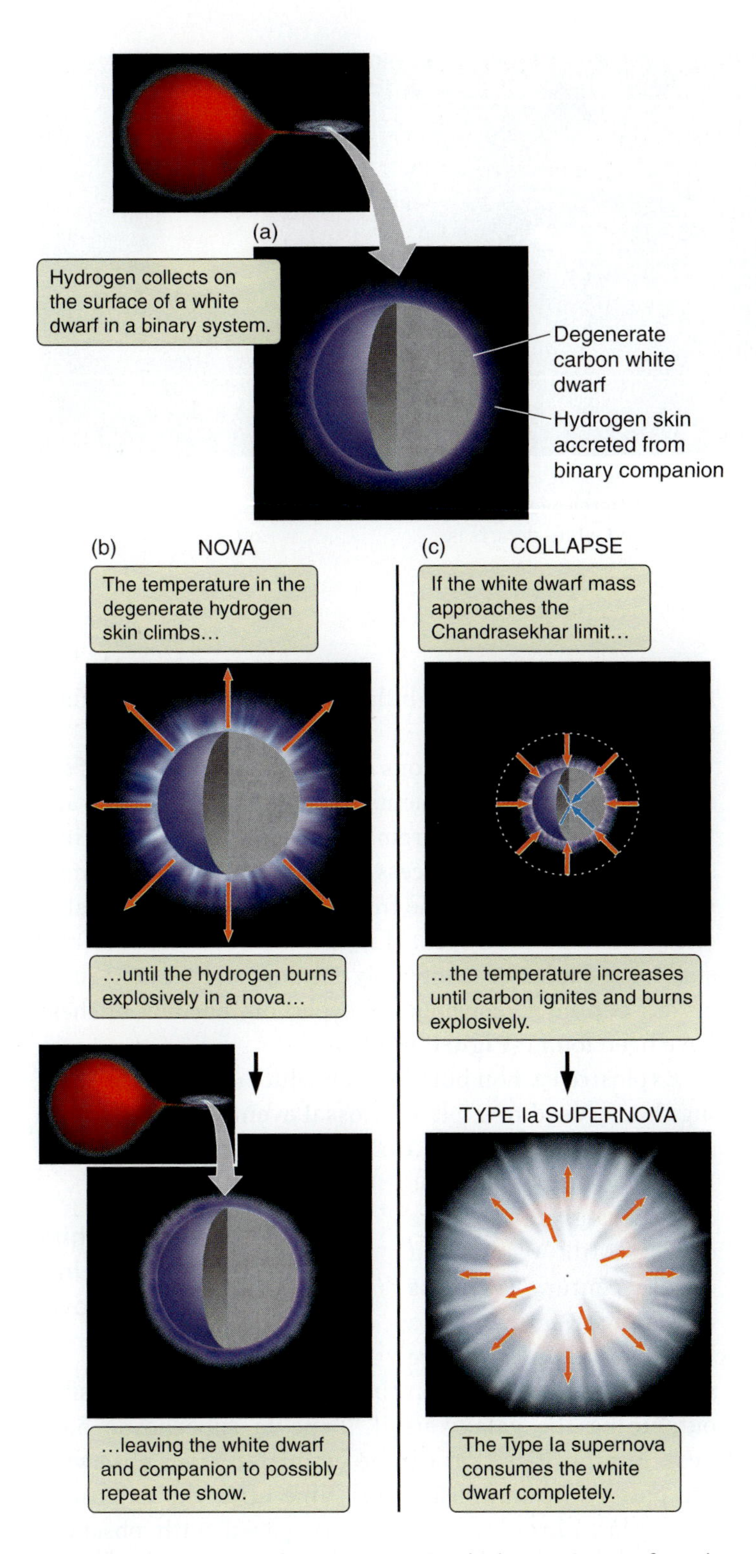

FIGURE 16.14 (a) In a binary system in which mass is transferred onto a white dwarf, a skin of hydrogen builds up on the surface of the degenerate white dwarf. (b) If hydrogen burning ignites on the surface of the white dwarf, the result is a nova. (c) If enough hydrogen accumulates to raise the core temperature high enough, carbon ignites and the result is a Type Ia supernova.

the rate of hydrogen burning. This runaway thermonuclear reaction is much like the runaway helium burning that takes place during the helium flash, except now there are no overlying layers of a star to keep things contained. The result is a tremendous explosion—a **nova** (plural: *novae*)—that blows part of the layer covering the white dwarf out into space at speeds of thousands of kilometers per second (**Figures 16.13f** and **16.14b**).

About 50 novae are thought to occur in the Milky Way Galaxy each year, but because of the dust and gas that obscure the view from Earth, only a few of these are observed, typically two or three each year. Novae get bright in a hurry—typically reaching their peak brightness in only a few hours—and for a brief time they can be almost half a million times more luminous than the Sun. Although the brightness of a nova sharply declines in the weeks following the outburst, it can sometimes still be seen for years. During this time the glow from the expanding cloud of material ejected by the explosion is powered by the decay of radioactive isotopes created in the explosion.

The explosion of a nova does not destroy the underlying white dwarf star. In fact, much of the material that had built up on the white dwarf may remain behind after the explosion. The nova leaves the binary system in much the same configuration it was in previously—the configuration shown in Figure 16.15a—with material from star 2 still pouring onto the white dwarf. This cycle can repeat many times, with material building up and igniting over and over again on the surface of the white dwarf. In most cases, outbursts are separated by thousands of years, so most novae have been seen only once in historical times. Some novae, however, are known to be recurrent, erupting anew every decade or so.

A Stellar Cataclysm May Await the White Dwarf in a Binary System

Novae may be spectacular events, but they pale in comparison with an alternate fate that may await the white dwarf in a binary system. One possibility is that eventually star 2 will simply go on to form a white dwarf too, leaving behind a binary system consisting of two degenerate white dwarfs (**Figure 16.13g**). The two white dwarfs might orbit around each other quickly, getting closer and closer. Eventually, tidal forces disrupt the smaller one and material falls onto the larger one until they merge, creating one object with a larger mass (**Figure 16.15**). Another possibility is that over millions of years, mass is transferred from star 2 onto the white dwarf, and through

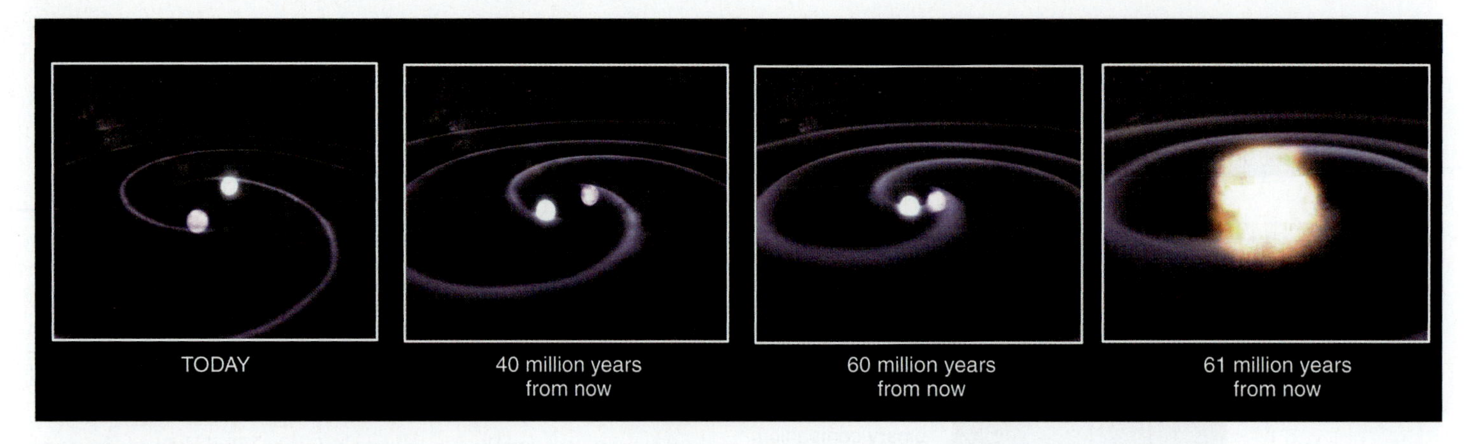

FIGURE 16.15 Two white dwarfs in close orbit about each other get closer and faster over time, eventually merging into one more massive white dwarf. If the mass of the combined white dwarfs is above the Chandrasekhar limit, it will explode as a Type Ia supernova.

countless nova outbursts, the mass of the white dwarf slowly increases. As it does, the white dwarf shrinks, and the pull of self-gravity on this already extremely dense object gets stronger and stronger.

In either of these two possibilities, the mass of a white dwarf can be pushed only so far. There comes a point when even the pressure supplied by degenerate electrons is no longer enough to balance gravity. This precipice occurs at a mass of about 1.4 $M_{\odot}$, which is called the **Chandrasekhar limit**, named for Subrahmanyan Chandrasekhar (1910–1995), the astrophysicist who derived it. It is impossible for a star to increase in mass above 1.4 $M_{\odot}$ and remain a white dwarf. But astronomers think that a white dwarf that is accreting mass like this does not reach the Chandrasekhar limit (**Figure 16.14c**).

The mass of a white dwarf cannot exceed 1.4 $M_{\odot}$.

As a white dwarf in a binary system approaches this limit, the increase in pressure and density lead to an increase in the temperature of the core. When the core temperature reaches about 6×10^8 K, carbon nuclei can slam together with enough force to overcome their electric repulsion and begin to fuse. The star might have a "simmering" phase with a growing central convective region that prevents thermonuclear runaway for awhile. Whereas the thermonuclear runaway hydrogen burning in a nova involves only a thin shell of material on the surface of the white dwarf, runaway carbon burning involves the entire white dwarf. After about 1,000 years, when the temperature reaches about 8×10^8 K, the star explodes and the white dwarf is consumed within about a second. About as much energy is liberated as will be given off by the Sun over its entire 10-billion-year lifetime on the main sequence.

Runaway fusion reactions convert a large fraction of the mass of the star into elements such as iron and nickel, and the explosion blasts the remains of the white dwarf into space at top speeds in excess of 20,000 km/s. If this more massive white dwarf came from the merger of two smaller ones, it will be destroyed. If it grew out of accretion from star 2, the explosion completely destroys star 1. Star 2 could be left behind to continue its evolution, but few of these have been found (**Figure 16.13h**).

Explosive carbon burning in a white dwarf is the leading theory used to explain colossal events called **Type Ia supernovae** (**Figure 16.16**; see also Figures 16.14c and 16.15). Type Ia supernovae occur in a galaxy the size of the MilkyWay about once a century (**Process of Science Figure**). For a brief time they can shine with luminosity 10 billion times that of the Sun, possibly outshining the galaxy itself. Type Ia supernovae have peak luminosities that can be estimated from observing the pattern of luminosity over time after the explosion. This peak luminosity can be compared with observed brightness to estimate a distance to the supernova. Type Ia supernovae are so luminous that they can be seen at great distances, so they are useful for estimating distances to faraway galaxies.

The carbon in a white dwarf can ignite in a Type Ia supernova.

Other types of **supernovae** can be distinguished by their spectra and by the pattern of their change in brightness. We will discuss these in more detail in Chapter 17.

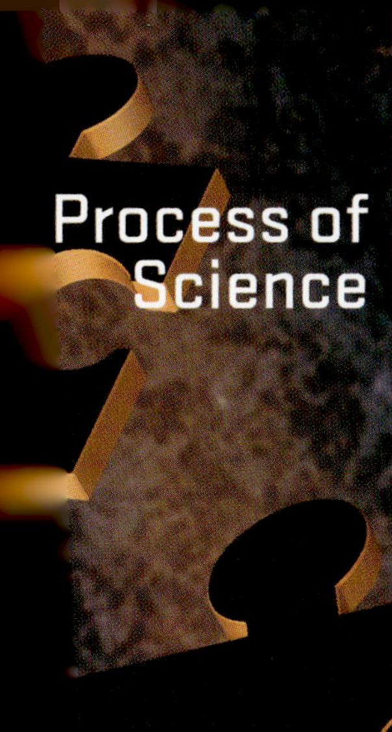

Process of Science

SCIENCE IS NOT FINISHED

Type Ia supernovae are fundamental to an understanding of the universe as a whole. Yet the observational evidence remains incomplete.

observational evidence is inconclusive, each scientist adds a piece to the puzzle. will never see a completed picture. Exploring nature is a project of humanity, individual humans.

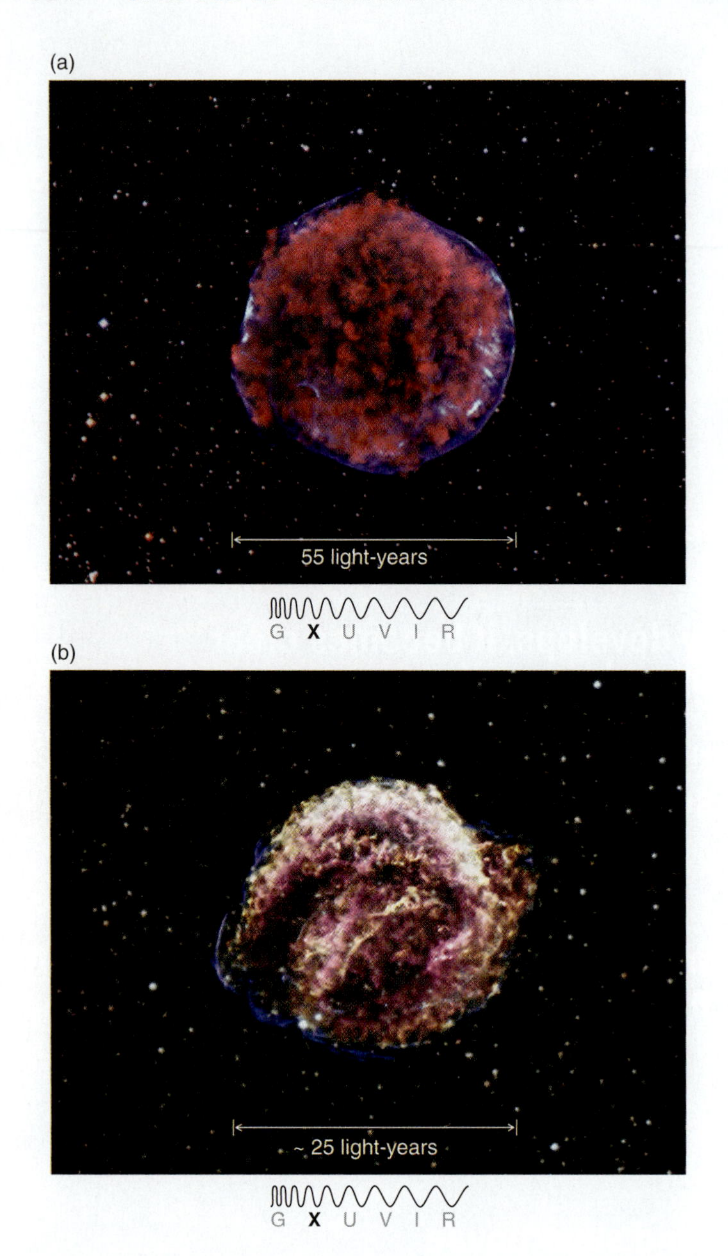

FIGURE 16.16 Type Ia supernovae. The material heated by the expanding blast wave from a supernova glows in X-rays. (a) Tycho's supernova was observed in 1572. Low-energy X-rays are red; high-energy X-rays are blue. (b) Kepler's supernova, observed in 1604, is shown in five different X-ray wavelengths.

16.6 Origins: Stellar Lifetimes and Biological Evolution

From fossil records and DNA analysis, scientists estimate that life appears to have taken hold on Earth within 400 million–1 billion years after the formation of the Solar System, 4.6 billion years ago. It took another 1.5 billion years for more complex cells to develop, and another billion years to develop multicellular life. The first animals didn't appear on Earth until 600 million years ago, 4 billion years after the formation of the Sun and the Solar System.

The only known example of biology is the life on Earth. It is always risky to extrapolate from one data point, and it is not known whether biology is widespread in the universe, or whether Earth's biological timeline is "typical." Still, reasoning from this one example is the only way to begin thinking about life in the universe. How does the preceding timeline of the evolution of life on Earth compare with the lifetimes of main-sequence stars? Table 16.1 indicates that the lifetime of an O5 star is less than a million years; of a B5 star, about 80 million years; and of an A0 star, about 400 million years. These stars would have run out of hydrogen in the core and started post-main-sequence evolution in less time than it took for Earth to settle down after its periods of heavy bombardment by debris early in the Solar System.

The 1-billion-year main-sequence lifetime of an A5 star corresponds to the amount of time it took for the simplest life-forms to have developed on Earth. The 2-billion-year lifetime for F0 stars corresponds to the amount of time it took for photosynthetic bacteria to have developed on Earth. Only F5 and cooler, less massive stars have main-sequence lifetimes longer than the 4 billion years it took for life to evolve into animals on Earth. Thus, projects searching for extrasolar planets that are inhabited by intelligent life aim their telescopes at stars that are F5 or cooler—on the theory that hotter and more massive stars don't live long enough on the main sequence for advanced life to develop.

After a star leaves the main sequence, the helium-burning red giant stage is estimated to last for about 1/10 of the main-sequence lifetime, so this doesn't help much with the biological timescale for stars with short lifetimes. Could life survive the transition of its star to a red giant? As noted in Connections 16.1, even if a planet is not destroyed, its orbit, temperature, and atmospheric conditions drastically change, and life might have to relocate if it is to survive.

What we have discussed in this chapter is representative of the fate awaiting the vast majority of stars in the universe. In the next chapter we turn our attention to the evolution of the small fraction of stars that fall into the category of high-mass stars. ■

Summary

16.1 Low-mass and high-mass stars evolve differently. The more massive a star is, the shorter will be its lifetime.

16.2 All stars eventually exhaust their nuclear fuel. Helium accumulates like ash in the cores of main-sequence stars. After exhausting its hydrogen, a low-mass star leaves the main sequence and swells to become a red giant.

16.3 A red giant burns helium via the triple-alpha process. In their dying stages, some stars form planetary nebulae.

16.4 All low-mass stars eventually become white dwarfs. A white dwarf is very hot but very small, and cannot exceed 1.4 $M_{\odot}$. White dwarfs cool to become cold, dark cinders—the stellar graveyard.

16.5 Transfer of mass within a binary system can lead to novae or supernovae. Type Ia supernovae are used as standard candles to measure the distances to the most distant objects in the universe.

16.6 Low-mass stars have main-sequence lifetimes comparable to the evolutionary timescales of life on Earth. If life on Earth is typical (a big if), then more massive stars with shorter lifetimes won't be stable long enough for advanced life to evolve.

Unanswered Questions

- Could Earth be moved farther from the Sun to accommodate the Sun's inevitable changes in luminosity, temperature, and radius? One proposal suggests that Earth could capture energy from a passing asteroid and migrate outward, thus staying in the habitable zone while moving farther from the Sun as the Sun ages. Or a huge, thin "solar sail" could be constructed so that radiation pressure from the Sun would slowly push Earth into a larger orbit. These feats of "astronomical engineering" are not feasible anytime in the near future, but perhaps in millions of years (when they will be needed) this could be accomplished.
- Why do planetary nebula have different shapes? Some are not simply chaotic but are well organized, with varying types of symmetry. Some are spherically symmetric (like a ball), some have bipolar symmetry (like a long hollow tube, pinched in the middle), and some are even point-symmetric (like the letter *S*). How can an essentially spherically symmetric object like a star produce such beautifully organized outflows? Because these are three-dimensional objects and astronomers can view each object from only one direction, it is difficult to determine how much of this variation is due to orientation, and how much is due to actual differences in the shape of the object. For example, a bipolar nebula, viewed from one end, would appear spherically symmetric. This orientation effect, among other problems, complicates efforts to understand how these shapes are formed. No single explanation has yet satisfactorily covered all the object types.

Questions and Problems

Summary Self-Test

1. Place the main-sequence lifetimes of the following stars in order.
   a. the Sun: mass 1 $M_{\odot}$, luminosity 1 $L_{\odot}$
   b. Capella Aa: mass 3 $M_{\odot}$, luminosity 76 $L_{\odot}$
   c. Rigel: mass 24 $M_{\odot}$, luminosity 85,000 $L_{\odot}$
   d. Sirius A: mass 2 $M_{\odot}$, luminosity 25 $L_{\odot}$
   e. Canopus: mass 8.5 $M_{\odot}$, luminosity 13,600 $L_{\odot}$
   f. Achernar: mass 7 $M_{\odot}$, luminosity 3,150 $L_{\odot}$
2. When the Sun runs out of fuel in its core, the core will be
   a. empty.
   b. filled with hydrogen.
   c. filled with helium.
   d. filled with carbon.
3. When a star runs out of fuel in its center, it becomes a
   a. smaller, cooler star.
   b. larger, hotter star.
   c. smaller, less luminous star.
   d. larger, more luminous star.
4. Place the following steps in the evolution of a low-mass star in order.
   a. main-sequence star
   b. planetary nebula ejection
   c. horizontal branch
   d. helium flash
   e. red giant branch
   f. asymptotic giant branch
   g. white dwarf

5. If a star follows a horizontal path across the H-R diagram, the star
   a. maintains the same temperature.
   b. stays the same color.
   c. maintains the same luminosity.
   d. keeps the same spectral type.

6. Degenerate matter is different from normal matter because as the mass goes up,
   a. the size goes down.
   b. the temperature goes down.
   c. the density goes down.
   d. the luminosity goes down.

7. Stars begin burning helium to carbon when the temperature rises in the core. This temperature increase is caused by
   a. gravitational collapse.
   b. fusing hydrogen into helium in the core.
   c. fusing hydrogen into helium in a shell around the core.
   d. degenerate-electron pressure.

8. A planetary nebula comes from
   a. the ejection of mass from a low-mass star.
   b. the collision of planets around a dying star.
   c. the collapse of the magnetosphere of a high-mass star.
   d. the remainders of the original star-forming nebula.

9. A solitary white dwarf will eventually
   a. brighten and explode.
   b. accrete mass from other objects.
   c. implode to become a black hole.
   d. become cold and dim.

10. A white dwarf will become a nova if
    a. the original star was more than 1.4 $M_{\odot}$.
    b. it accretes an additional 1.4 $M_{\odot}$ from a companion.
    c. even a small amount of mass falls on it from a companion.
    d. enough mass accretes from a companion to give the white dwarf a total mass of 1.4 $M_{\odot}$.

T/F and Multiple Choice

11. **T/F:** All stars eventually die.

12. **T/F:** All stars on the main sequence are fusing hydrogen into helium in their cores.

13. **T/F:** A 5-$M_{\odot}$ star has a longer lifetime than a 1-$M_{\odot}$ star.

14. **T/F:** When a star first runs out of fuel in the core, it becomes brighter.

15. **T/F:** A 1.2-$M_{\odot}$ white dwarf has a smaller radius than a 1.0-$M_{\odot}$ white dwarf.

16. Why are most nearby stars low-mass, low-luminosity stars?
    a. Those stars are easier to see.
    b. The more distant low-mass, low-luminosity stars are not visible from Earth.
    c. Most stars are low-mass, low-luminosity stars.
    d. Most nearby stars formed in the same region as the Sun, and thus are similar.

17. The most massive stars have the shortest lifetimes because
    a. the temperature is higher in the core, so they burn their fuel faster.
    b. they have less fuel in the core.
    c. their fuel is located farther from the core.
    d. the temperatures are lower in the core, so they burn their fuel slower.

18. If a main-sequence star suddenly started burning hydrogen at a faster rate in its core, it would become
    a. larger, hotter, and more luminous.
    b. larger, cooler, and more luminous.
    c. smaller, hotter, and more luminous.
    d. smaller, cooler, and more luminous.

19. It is rare to see a helium flash, because
    a. few stars go through this stage.
    b. stars that go through this stage are all far away.
    c. the flash is not very bright.
    d. the flash does not take very long.

20. Giant stars lose mass because
    a. the radiation pressure from fusion becomes so high.
    b. the mass of the star drops because of mass loss from fusion.
    c. the magnetic field causes increasing numbers of coronal mass ejections.
    d. the star swells until the surface gravity is too low to hold material.

21. A planetary nebula glows because
    a. it is hot.
    b. fusion is happening in the nebula.
    c. it is heating up the interstellar medium around it.
    d. light from the central star causes emission lines.

22. As an AGB star evolves into a white dwarf, it runs out of nuclear fuel, and one might guess that the star should cool off and move to the right on the H-R diagram. Why does the star move instead to the left?
    a. It becomes larger.
    b. More of the star is involved in fusion.
    c. As outer layers are lost, deeper layers are exposed.
    d. The star gets hotter.

23. When compressed, ordinary gas heats up but degenerate gas does not. Why, then, does a degenerate core heat up as the star continues shell burning around it?
    a. It is heated by the radiation from fusion.
    b. It is heated by the gravitational collapse of the shell.
    c. It is heated by the weight of helium ash falling on it.
    d. It is insulated by the shell.

24. All Type Ia supernovae
    a. are at the same distance from Earth.
    b. always involve a white dwarf of the same mass.
    c. involve degenerate matter.
    d. always release the same amount of energy in fusion.

25. In Latin, *nova* means "new." This word is used for novae and supernovae because they are
    a. newly formed stars.
    b. newly dead stars.
    c. newly visible stars.
    d. newly expanded stars.

Thinking about the Concepts

26. Is it possible for a star to skip the main sequence and immediately begin burning helium in its core? Explain your answer.

27. Consider a hypothetical star with the same mass as the Sun being one of the first stars that formed when the universe was still very young. Would you expect it to be surrounded by planets? Explain your answer.

28. Describe some possible ways in which a star might increase the temperature within its core while at the same time lowering its density.

29. Astronomers typically say that the mass of a newly formed star determines its destiny from birth to death. However, there is a frequent environmental circumstance for which this statement is not true. Identify this circumstance and explain why the birth mass of a star might not fully account for the star's destiny.

30. What physical effect or explanation do you think could be responsible for the "break" in slope in the mass-luminosity relationship between low- and high-mass stars, as shown in Figure 16.1? Justify your answer.

31. Do stars change their structure while on the main sequence? Why or why not?

32. Suppose Jupiter were not a planet, but instead were a G5 main-sequence star with a mass of 0.8 $M_\odot$.
    a. How do you think life on Earth would be affected, if at all?
    b. How would the Sun be affected as it came to the end of its life?

33. Why are the paths along the H-R diagram that a star follows as it forms (from a protostar) and as it leaves the main sequence (climbing the red giant branch) so similar?

34. Why is a horizontal branch star (which burns helium at a high temperature) less luminous than a red giant branch star (which burns hydrogen at a lower temperature)?

35. Suppose a star is able to heat its core temperature high enough to begin fusing oxygen. Predict how the star will continue to evolve, including how you think the star will evolve on the H-R diagram.

36. Why does a white dwarf move down and to the right along the H-R diagram?

37. Why does fusion in degenerate material always lead to a runaway reaction?

38. The intersection of the Roche lobes in a binary system is the equilibrium point between the two stars where the gravitational attraction from both stars is equally strong and opposite in direction. Is this an example of stable or unstable equilibrium? Explain.

39. Suppose the more massive red giant star in a binary system engulfs its less massive main-sequence companion, and their nuclear cores combine. What structure do you think the new star will have? Where will the star lie on the H-R diagram?

40. T Coronae Borealis is a well-known recurrent nova.
    a. Is it a single star or a binary system? Explain.
    b. What mechanism causes a nova to flare up?
    c. How can a nova flare-up happen more than once?

Applying the Concepts

41. Figure 16.1 contains the text "Straight-line (exponential) approximation..." What does this text tell you about the axes on the graph; are they linear or logarithmic? Explain.

42. Use Figure 16.2 to estimate the percentage of the Sun's mass that is turned from hydrogen into helium over its lifetime.

43. Figure 16.11 shows a gallery of planetary nebulae. What fraction of these show spherical symmetry (being symmetric in every direction, like a circle)? What fraction show bipolar symmetry (having an axis about which they are symmetric, like a person's face)? What fraction have point symmetry (being symmetric about a point, like the letter *S*)? The cause of this wild variation in the shapes of planetary nebulae is an active area of study for astronomers.

44. Study Figure 16.12. How many times brighter is a star at the top of the giant branch than the same star (a) when it was on the main sequence? and (b) when it was on the horizontal branch?

45. Study Figure 16.12. Make a graph of surface temperature versus time for the evolutionary track shown—from the time the star leaves the main sequence until it arrives at the dot showing that it is a white dwarf. Your time axis may be approximate, but it should show that the star spends different amounts of time in the different phases.

46. For most stars on the main sequence, luminosity scales with mass as $M^{3.5}$ (see Math Tools 16.1). What luminosity does this relationship predict for (a) 0.5-$M_\odot$ stars, (b) 6-$M_\odot$ stars, and (c) 60-$M_\odot$ stars? Compare these numbers to values given in Table 16.1.

47. Calculate the main-sequence lifetimes for (a) 0.5-$M_\odot$ stars, (b) 6-$M_\odot$ stars, and (c) 60-$M_\odot$ stars. Compare them to the values given in Table 16.1.

48. What will the escape velocity be when the Sun becomes an AGB star with a radius 200 times greater and a mass only 0.7 times that of today? How will these changes in escape velocity affect mass loss from the surface of the Sun as an AGB star?

49. Each form of energy generation in stars depends on temperature.
    a. The rate of hydrogen fusion (proton-proton chain) near 10^7 K increases with temperature as T^4. If the temperature of the hydrogen-burning core is raised by 10 percent, how much does the hydrogen fusion energy increase?
    b. Helium fusion (the triple-alpha process) at 10^8 K increases with an increase in temperature at a rate of T^{40}. If the temperature of the helium-burning core is raised by 10 percent, how much does the helium fusion energy increase?

50. Roughly how large does a planetary nebula grow before it disperses? Use an expansion rate of 20 km/s and a lifetime of 50,000 years.

51. In typical binary systems, a red giant can transfer mass onto a white dwarf at rates of about 10^{-9} $M_{\odot}$ per year. Roughly how long after mass transfer begins will the white dwarf explode as a Type Ia supernova? How does this length of time compare to the typical lifetime of a low-mass star? Assume that a typical white dwarf starts with a mass of 0.6 $M_{\odot}$.

52. How fast does material in an accretion disk orbit around a white dwarf? Use Kepler's third law.

53. A white dwarf has a density of approximately 10^9 kilograms per cubic meter (kg/m^3). Earth has an average density of 5,500 kg/m^3 and a diameter of 12,700 km. If Earth were compressed to the same density as a white dwarf, how large would it be?

54. What is the density of degenerate material? Calculate how large the Sun would be if all of its mass were degenerate.

55. Recall from Chapter 4 that the luminosity of a spherical object at temperature T is given by $L = 4\pi R^2 \sigma T^4$, where R is the object's radius. If the Sun became a white dwarf with a radius of 10^7 meters, what would its luminosity be at the following temperatures: (a) 10^8 K; (b) 10^6 K; (c) 10^4 K; (d) 10^2 K?

Using the Web

56. Go to the website for the Katzman Automatic Imaging Telescope at Lick Observatory (http://astro.berkeley.edu/bait/public_html/kait.html). What is this project? Why can a search for supernovae be automated? Pick a recent year. How many supernovae were discovered? Look at some of the images. How bright do the supernovae look compared to their galaxies?

57. Go to the American Association of Variable Star Observers (AAVSO) website (www.aavso.org). What does this 100-year-old organization do? Read about the types of intrinsic variable stars. Click on "Getting Started." If you have access to dark skies, you can contribute to the study of variable stars. Go to the page for observers (www.aavso.org/observers) and click on each item in the "For New Observers" list, including the list of stars that are easy to observe. Assemble a group and observe a variable star from this list. (Another option is to join AAVSO's project searching for novae, at www.aavso.org/nova-search-section.)

58. Go to the University of Washington's "Properties of Planetary Nebulae" Web page (www.astro.washington.edu/courses/labs/clearinghouse/labs/ProppnShort/proppn.html) and do the lab exercise.

59. Go to the Hubble Space Telescope's planetary nebula gallery (http://hubblesite.org/gallery/album/nebula/planetary). Find an example of a nebula that shows clearly each type of symmetry: spherical, bipolar, point-symmetric (see question 43 for an explanation of these types of symmetry). Print each of the three images you choose, and label the type of symmetry each one represents. For all three nebulae, identify the location of the central star. For bipolar symmetry, draw a line that shows the axis about which the nebula is symmetric. For point symmetry, identify several features that are symmetric across the location of the central star.

60. In the Hubble telescope news archive, look up press releases on planetary nebulae (http://hubblesite.org/newscenter/archive/releases/nebula/planetary) and white dwarf stars (http://hubblesite.org/newscenter/archive/releases/star/white-dwarf). Pick a story for each. What observations were reported, and why were they important?

StudySpace is a free and open website that provides a Study Plan for each chapter of *21st Century Astronomy*. Study Plans include animations, reading outlines, vocabulary flashcards, and multiple-choice quizzes, plus links to premium content in SmartWork and the ebook. Visit **wwnorton.com/studyspace**.

SmartWork Norton's online homework system, includes algorithmically generated versions of these questions, plus additional conceptual exercises. If your instructor assigns questions in SmartWork, log in at **smartwork.wwnorton.com.**

Exploration | Low-Mass Stellar Evolution

The evolution of a low-mass star, as discussed in this chapter, corresponds to many twists and turns on the H-R diagram. In this Exploration we return to the "HR Explorer" interactive simulation to investigate how these twists and turns affect the appearance of the star.

Open the "H-R Explorer" simulation, linked from the Chapter 16 area on StudySpace. The box labeled "Size Comparison" shows an image of both the Sun and the test star. Initially, these two stars have identical properties: the same temperature, the same luminosity, and the same size.

Examine the box labeled "Cursor Properties." This box shows the temperature, luminosity, and radius of a test star located at the "X" in the H-R diagram. Before you change anything, answer these questions:

1 **What is the temperature of the test star?**

....

2 **What is the luminosity of the test star?**

....

3 **What is the radius of the test star?**

....

As a star leaves the main sequence, it moves up and to the right on the H-R diagram. Move the cursor (the X on the diagram) up and to the right.

4 **What changes about the image of the test star next to the Sun?**

....

....

5 **What is the test star's temperature? What property of the image of the test star indicates that its temperature has changed?**

....

....

6 **What is the test star's luminosity?**

....

7 **What is the test star's radius?**

....

8 **Ordinarily, the hotter an object is, the more luminous it is. In this case, the temperature has gone down, but the luminosity has gone up. How can this be?**

....

....

The star then moves around quite a lot in that part of the H-R diagram. Look at Figure 16.12, and then use the cursor to approximate the motion of the star as it moves up the red giant branch, back down and onto the horizontal branch, and then back to the right and up the asymptotic giant branch.

9 **Are the changes you observe in the image of the star as dramatic as the ones you observed for question 4?**

....

....

10 **What is the most noticeable change in the star as it moves through this portion of its evolution?**

....

....

Next, the star begins moving across the H-R diagram to the left, maintaining almost the same luminosity. Drag the cursor across the top of the H-R diagram to the left, and study what happens to the image of the star in the "Size Comparison" box.

11 **What changed about the star as you dragged it across the H-R diagram?**

....

....

12 **How does the star's size now compare to that of the Sun?**

....

....

Finally, the star drops to the bottom of the H-R diagram and then begins moving to the right. Move the cursor toward the bottom of the H-R diagram, where the star becomes a white dwarf.

13 **What changed about the star as you dragged it down the H-R diagram?**

....

....

14 **How does its size now compare to that of the Sun?**

....

....

To thoroughly cement your understanding of stellar evolution, press the "Reset" button and then move the star from main sequence to white dwarf several times. This exercise will help you remember how this part of a star's life appears on the H-R diagram.

Shells of gas blown into space by the oldest documented supernova, RCW 86, witnessed by Chinese astronomers in 186 CE.

17 Evolution of High-Mass Stars

It does not do to leave a live dragon out of your calculations, if you live near him.

J. R. R. Tolkien (1892–1973), The Hobbit

LEARNING GOALS

The vast majority of stars are smaller and less massive than the Sun, and live a relatively long time. The most massive stars, the O and B stars, are rarer and live a much shorter time. When these massive stars die, the result is far more spectacular than when lower-mass stars die. By the conclusion of this chapter, you should be able to:

- Make a flowchart of the sequence of the stages of evolution for high-mass stars.
- Describe how the death of high-mass stars differs from that of low-mass stars.
- Explain the origin of chemical elements up to and heavier than iron.
- Identify how H-R diagrams enable astronomers to measure the ages of stars and test theories of stellar evolution.

17.1 High-Mass Stars Follow Their Own Path

So far in our discussion of the lives of stars, we have concentrated on what happens to low-mass stars like the Sun. Low-mass stars steadily burn hydrogen to helium in their cores and produce a fairly constant output of energy for billions of years. High-mass stars—stars with masses greater than about 8 solar masses ($M_{\odot}$)—are very different. These objects have luminosities thousands or even millions of times as great as the Sun, but they use up their higher allotments of nuclear fuel much faster. High-mass stars live only hundreds of thousands to millions of years—less time than humans have walked on Earth.

High-mass stars and low-mass stars evolve differently because of the relationships among gravity, pressure, and the rate of nuclear burning. More mass means stronger gravity and therefore more force bearing down on the inner parts of the star. The increase in force means higher pressure, higher pressure means faster reaction rates, and faster reaction rates mean greater luminosity. There are many differences in the evolution of low- and high-mass stars, but in the end they all trace back to the greater gravitational force bearing down on the interior of a high-mass star.

How do stars create the elements that constitute life?

Recall from Chapter 14 that the first step in the hydrogen-burning proton-proton chain is the collision and fusion of two protons. The lone proton in a hydrogen nucleus has only a single positive charge, so hydrogen requires less energy to fuse than does any other atomic nucleus. This is a great advantage for low-mass stars, in which hydrogen burns at temperatures as low as a few million kelvins. However, the proton-proton chain has a large disadvantage as well. Even when two protons are slammed together hard enough for the strong nuclear force to act on them, the probability that they will fuse is low. The lack of vigor of this first step in the proton-proton chain limits how rapidly the entire process can move forward.

At the much higher temperatures at the center of a high-mass star, additional nuclear reactions become possible. In particular, hydrogen nuclei are able to interact with the nuclei of more massive elements such as carbon. It takes a lot of energy to get a hydrogen nucleus past the electric barrier set up by a carbon nucleus, with its six protons. But if this barrier can be overcome, fusion is much more probable than in the interaction between two hydrogen nuclei. This reaction, in which a carbon-12 (^{12}C) nucleus and a proton combine to form a nitrogen-13 (^{13}N) nucleus consisting of seven protons and six neutrons, is the first reaction in the **carbon-nitrogen-oxygen (CNO) cycle (Figure 17.1)**. Note that carbon is not consumed by the CNO cycle but instead is a **catalyst**. (A catalyst is a reaction helper. It facilitates the reaction but is not used up in the process.) The ^{12}C nucleus that started the cycle is present again at the end of the cycle, but now instead of the four hydrogen

The CNO cycle burns hydrogen in massive stars.

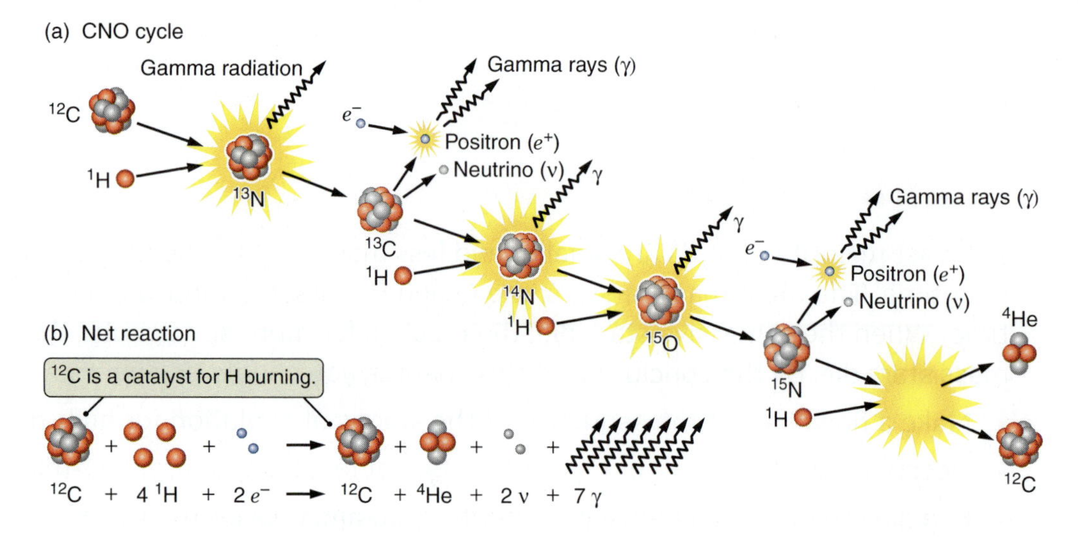

FIGURE 17.1 In high-mass stars, carbon serves as a catalyst for the fusion of hydrogen to helium. This process is the carbon-nitrogen-oxygen (CNO) cycle.

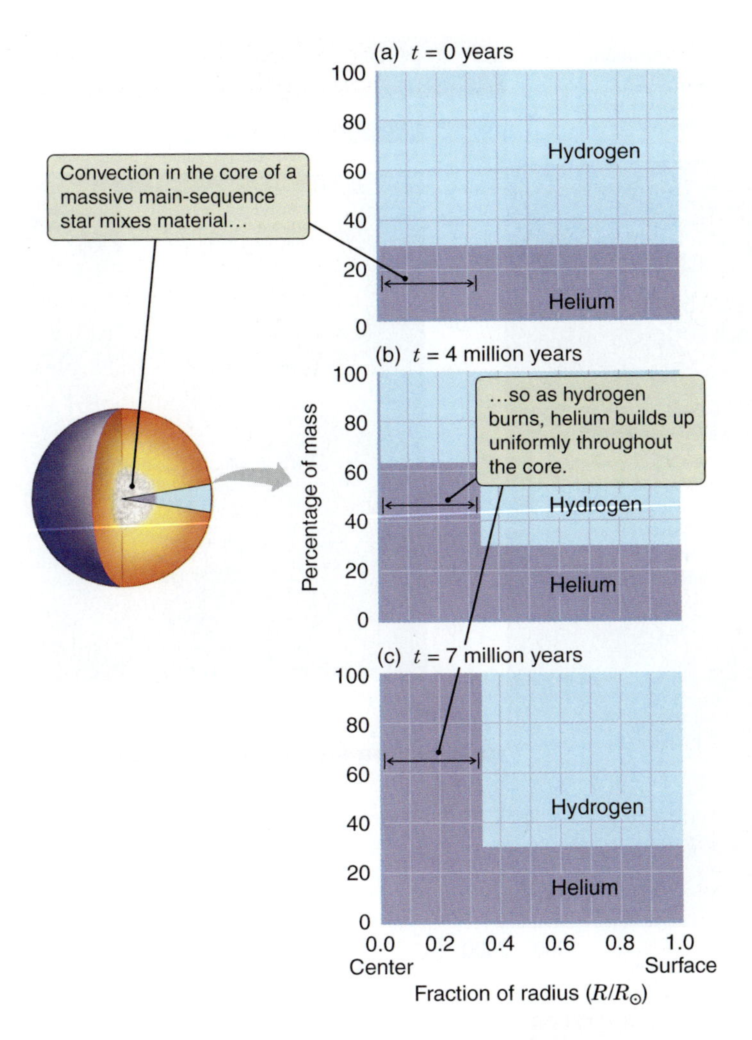

FIGURE 17.2 Convection keeps the core of a high-mass main-sequence star well mixed, so the composition remains uniform throughout the evolving core. (Evolution times are for a 25-$M_{\odot}$ star.)

nuclei that took part in the chain of reactions, there is a helium nucleus. The CNO cycle is far more efficient than the proton-proton chain in stars more massive than about 1.3–1.5 $M_{\odot}$. ▶▶ **NEBRASKA SIMULATION:** CNO CYCLE ANIMATION

The different ways that hydrogen burning takes place in high- and low-mass stars are reflected in the different core structures of the two types of stars. The temperature gradient in the core of a high-mass star is so steep that convection sets in within the core itself, "stirring" the core like the water in a boiling pot. Compare **Figure 17.2** with Figure 16.2. Rather than building up from the center outward as it does in low-mass stars, helium ash spreads uniformly throughout the core of a high-mass star as the star consumes its hydrogen.

Convection stirs the core of a high-mass star.

The High-Mass Star Leaves the Main Sequence

As the high-mass star's life on the main sequence comes to an end, the visible differences in its structure and evolution become far more pronounced. When the high-mass star runs out of hydrogen in its core, the weight of the overlying star compresses the core, just as it does in a low-mass star. Yet long before the core of the high-mass star becomes electron-degenerate, the pressure and temperature in the core reach the 10^8-kelvin (K) point needed for helium burning to begin. This rapid increase in pressure and temperature prevents the growth of a degenerate core in the high-mass star. The star makes a fairly smooth transition from hydrogen burning to helium burning. The star's overall structure responds to the changes taking place in its interior, but its luminosity changes relatively little.

High-mass stars live fast and die young.

Recall that when a low-mass star leaves the main sequence, the path it follows on the H-R diagram is largely vertical (see Figure 16.4), going to higher and higher luminosities. But as a high-mass star leaves the main sequence, it grows in size while its surface temperature falls, so it moves mostly off to the right on the H-R diagram (**Figure 17.3**). When this happens, the massive star has the same structure as a low-mass horizontal branch star, burning helium in its core and hydrogen in a surrounding shell. Stars more massive than 10 $M_{\odot}$ become red supergiants during their helium-burning phase. They have very cool surface temperatures (about 4000 K) and radii as big as 1,500 times that of the Sun.

The next stage in the evolution of a high-mass star has no analog in low-mass stars. When the high-mass star exhausts the helium in its core, the core again begins to collapse. However, this time as the core collapses it reaches temperatures of 8×10^8 K or higher, and carbon begins to burn (see Table 17.1). Carbon burning produces a number of more massive elements, including, neon, sodium, magnesium, and some oxygen. The star at this time consists of a carbon-burning core surrounded by a helium-burning shell, which in turn is surrounded by the outward-moving hydrogen-burning shell. The sequence does not end here. When carbon is exhausted as a nuclear fuel at the center of the star, neon breaks down to oxygen and helium or burns to magnesium; and when neon is exhausted, oxygen begins to burn. The structure of the

Progressively more massive elements burn as massive stars evolve.

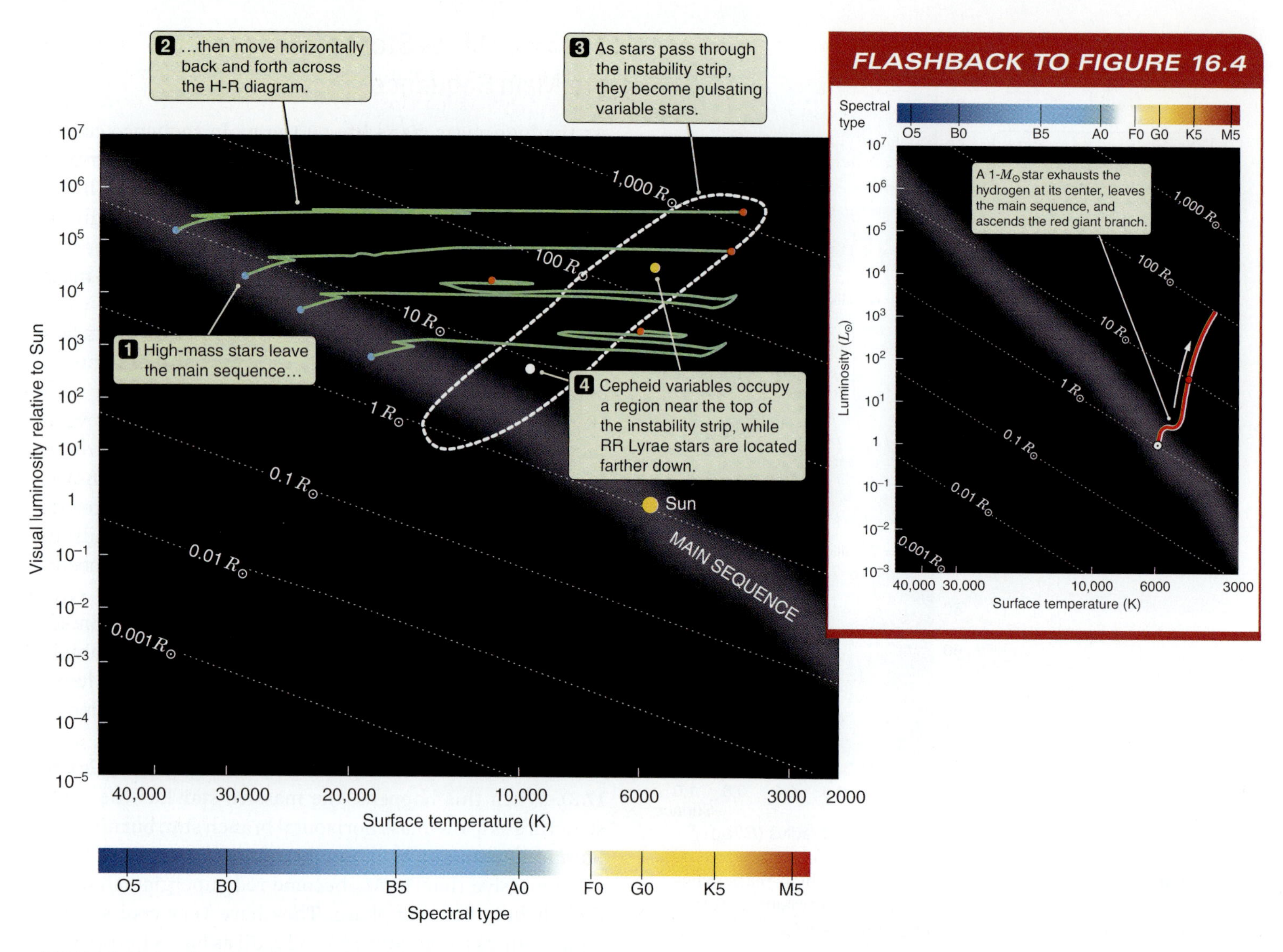

FIGURE 17.3 When massive stars leave the main sequence, they move horizontally across the H-R diagram.

evolving high-mass star is reminiscent of an onion, with many concentric layers (**Figure 17.4**). Inside the layer of hydrogen burning is a layer of helium burning, then a layer of carbon burning, and so on.

Not All Stars Are Stable

As a star undergoes post-main-sequence evolution, it may make one or more passes through a region of the H-R diagram known as the **instability strip** (see Figure 17.3). Rather than achieving a steady balance between pressure and gravity, stars in the instability strip pulsate, alternately growing larger and smaller. Such stars are called **pulsating variable stars**.

> Evolved stars pulsate while passing through the instability strip.

Thermal energy powers the pulsations of stars lying within the instability strip of the H-R diagram. They do not settle at a constant radius like the Sun; instead the stars alternately expand and shrink. At each change the star overshoots the equilibrium point where forces due to pressure and gravity balance each other, shrinking too far or expanding too far. The pulsations in the outer parts of the star have very little effect on the nuclear burning in the star's interior. However, the pulsations do affect the light escaping from the star. From Chapter 13, recall the **luminosity-temperature-radius relationship** for stars: both the luminosity and the color of the star change as the star expands and shrinks. The star is at its brightest and bluest while it expands through its

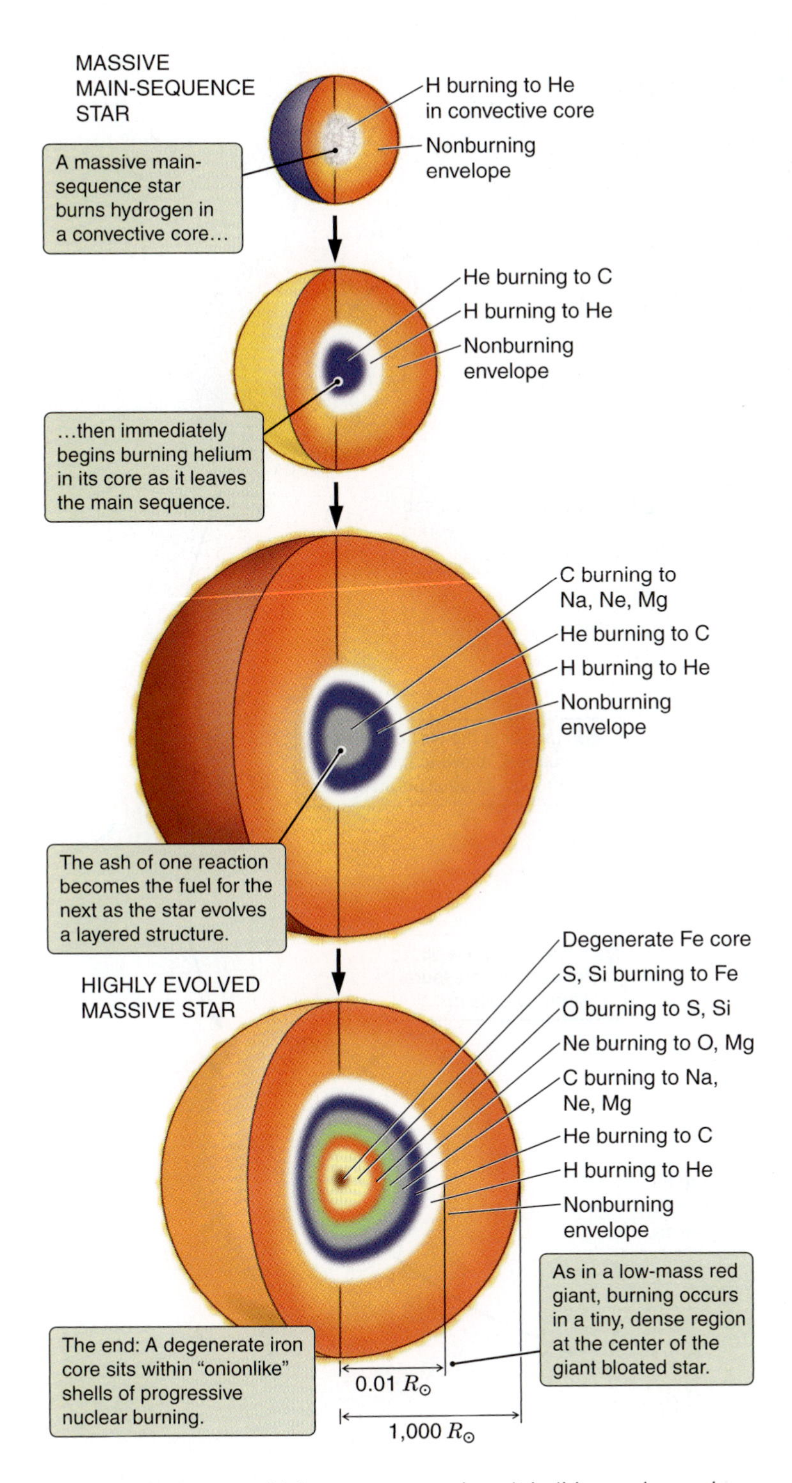

FIGURE 17.4 As a high-mass star evolves, it builds up a layered structure like that of an onion, with progressively more advanced stages of nuclear burning found deeper and deeper within the star. Note the change in scale for the bottom image.

equilibrium size, and at its faintest and reddest while it falls back inward.

One type of these pulsating variable stars are **Cepheid variables**, named after Delta Cephei, the first recognized member of this class of stars (**Figure 17.5a**). Classical, or Type I, Cepheids are massive and luminous yellow supergiants. A Cepheid variable completes one cycle of its pulsation in anywhere from about 1 to 100 days, depending on its luminosity. The more luminous the star, the longer it takes it to complete its cycle. This **period-luminosity relationship** for Cepheid variables, first discovered experimentally by Henrietta Leavitt (1868–1921) in 1912, is the basis for the use of Cepheid variables as indicators of the distances to galaxies beyond the Milky Way. Cepheid variables pulsate because of the changing ionization state of helium atoms inside the star as the gas heats, expands, and cools (**Figure 17.6**).

Classical (Type I) Cepheid variables are not the only type of **variable star**. The horizontal branch of the evolutionary tracks of low-mass stars (see Figure 16.8) may pass through the instability strip as well. Old low-mass (0.8-$M_\odot$) stars can be Type II Cepheid or **RR Lyrae variables** (with periods of less than a day). These are further subdivided by their luminosity and the length of their brightness cycle. These unstable horizontal branch stars pulsate by the same mechanism as Cepheid variables but are typically hundreds of times less luminous. They, too, follow a period-luminosity relationship (**Figure 17.5b**). **NEBRASKA SIMULATION: H-R EXPLORER**

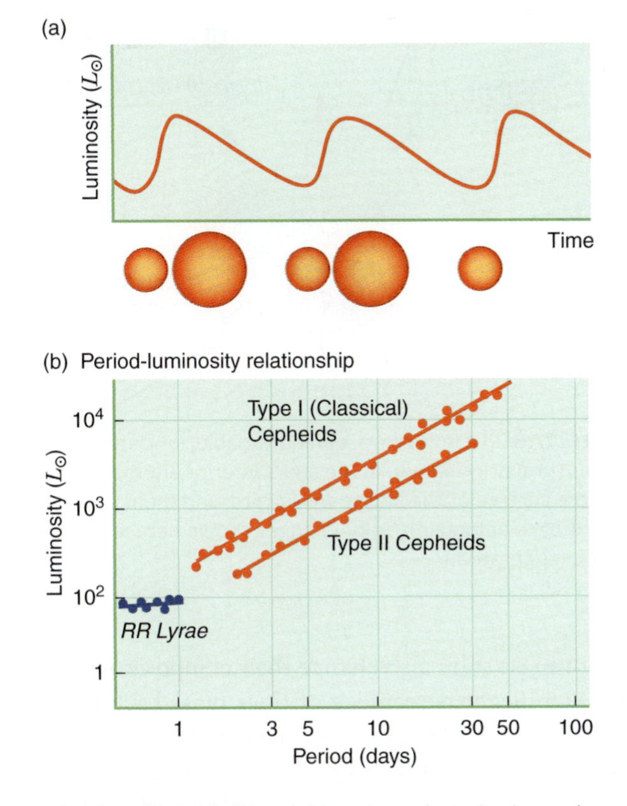

FIGURE 17.5 (a) Cepheid variable stars pulsate in size and therefore luminosity. (b) The length of the period of pulsation is related to the star's luminosity.

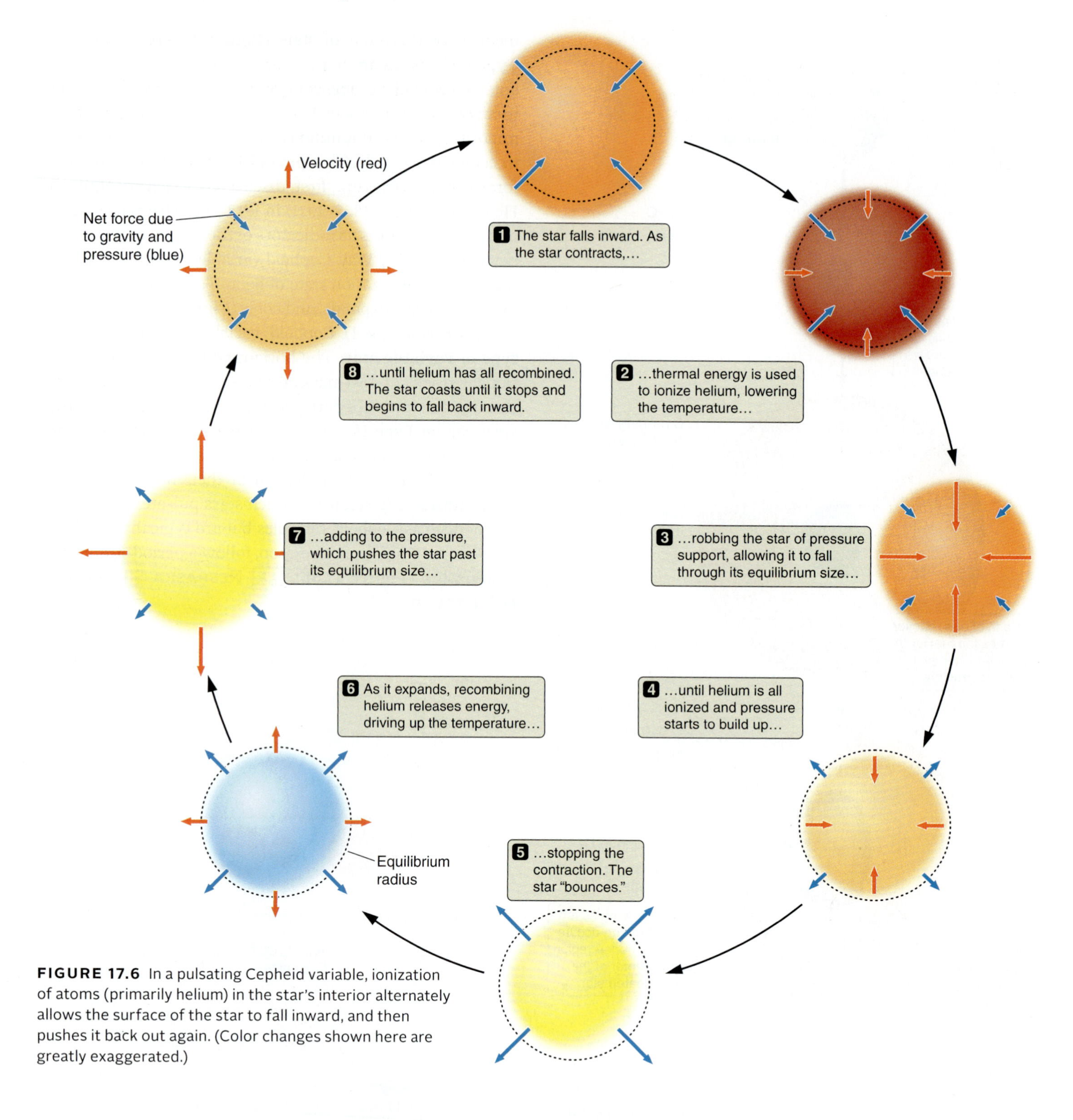

FIGURE 17.6 In a pulsating Cepheid variable, ionization of atoms (primarily helium) in the star's interior alternately allows the surface of the star to fall inward, and then pushes it back out again. (Color changes shown here are greatly exaggerated.)

High-mass stars also change their composition by expelling a significant percentage of their mass back into space throughout their lifetimes. Even while on the main sequence, massive O and B stars have low-density winds with velocities as high as 3,000 kilometers per second (km/s). These winds are pushed outward by the pressure of the radiation from the star. You may not normally think of light as "pushing" on something, but the pressure of the intense radiation at the surface of a massive star overcomes the star's gravity and drives away material in its outermost layers. Main-sequence O and B stars lose mass at rates ranging from about 10^{-7} to 10^{-5} $M_{\odot}$ of

Massive stars lose mass as they generate high-velocity winds.

material per year. The greatest mass loss occurs in the most massive stars.

These numbers may sound tiny, but over millions of years mass loss plays a prominent role in the evolution of high-mass stars. O stars with masses of 20 $M_{\odot}$ or more may lose about 20 percent of their mass while on the main sequence, and possibly more than 50 percent of their mass over their entire lifetimes. Even an 8-$M_{\odot}$ star may lose 5–10 percent of its mass.

17.2 High-Mass Stars Go Out with a Bang

You have already seen how a low-mass star approaches the end of its life—relatively slowly and gently, ejecting its outer parts into nearby space (sometimes forming a planetary nebula) and leaving behind a degenerate core. In stark contrast, for a high-mass star the end comes suddenly and amid considerable fury. An evolving high-mass star builds up its onionlike structure (see Figure 17.4) as nuclear burning in its interior proceeds to more and more advanced stages. Hydrogen burns to helium, helium burns to carbon and oxygen, carbon burns to magnesium, oxygen burns to sulfur and silicon, and then silicon and sulfur burn to iron. Many different types of nuclear reactions occur up to this point, forming almost all of the different stable isotopes of elements less massive than iron. But the essential point is this: the chain of nuclear fusion stops with iron.

Iron is the most massive element formed by fusion.

Gasoline burns because energy is released when the fuel chemically combines with oxygen. This released energy increases the temperature, which speeds up the chemical reaction. The reaction is *self-sustaining*, meaning that the reaction itself is the source of thermal energy needed to cause the reaction to go. The same is true of nuclear fusion reactions in the interiors of stars. Recall from Chapter 14 that when four hydrogen atoms combine to form a helium atom, the resulting helium atom has less mass than the sum of the four individual hydrogen atoms. This difference in mass has been converted to energy, which maintains the temperature of the gas at the high levels needed to sustain the reaction.

In the triple-alpha process (see Figure 16.7), three helium nuclei combine to form a ^{12}C nucleus. The energy available from this nuclear reaction is the energy it takes to break down each of the three helium nuclei into its constituent six neutrons and six protons, and how much energy is released if these six protons and neutrons combine to form a ^{12}C nucleus. The net energy produced by the reaction is the difference between these two amounts.

The **binding energy** of an atomic nucleus is the energy required to break the nucleus into its constituent parts. A nuclear reaction that increases a nucleus's binding energy releases energy. Conversely, decreasing the binding energy absorbs energy. **Figure 17.7** shows the binding energy per nucleon (that is, per each proton or neutron in the nucleus) for different atomic nuclei. Moving up on the plot from helium to carbon increases the binding energy, so fusing helium to carbon releases energy. Iron is at the peak of the binding-energy curve, so moving up from lighter elements to iron (Fe) also releases energy; conversely, moving from iron down to heavier elements *absorbs* energy. Iron fusion is not self-sustaining, because it absorbs energy in the reaction (**Math Tools 17.1**).

Different nuclei have different binding energies.

The Final Days in the Life of a Massive Star

The nuclear reactions following hydrogen burning are energetically much less favorable than conversion of hydrogen to helium. For example, conversion of helium to carbon produces less than 1/10 as much energy as conversion of hydrogen to helium. In order to support the star against gravity, this less efficient nuclear fuel must be

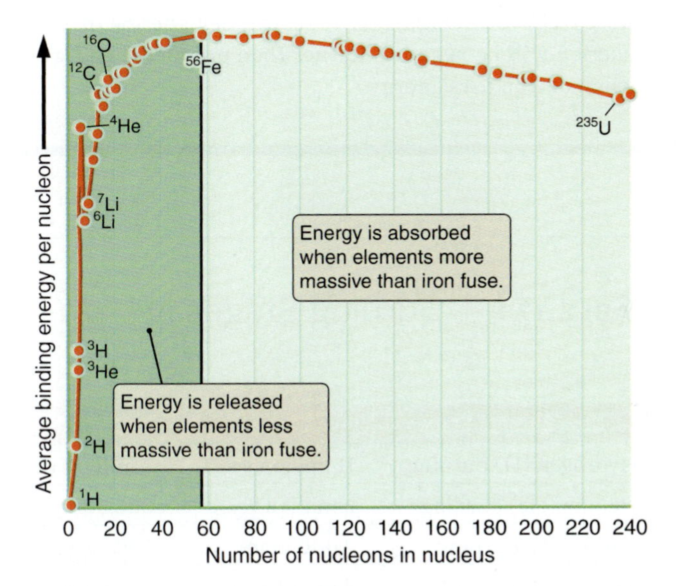

FIGURE 17.7 The binding energy per nucleon is plotted against the number of nucleons for each element. This is the energy it would take to break the atomic nucleus apart into protons and neutrons. Energy is released by nuclear fusion only if the resulting element is higher on the curve.

Math Tools 17.1

Binding Energy of Atomic Nuclei

The net energy released by a nuclear reaction is the difference between the binding energy of the products and the binding energy of the reactants:

$$\text{Net energy} = \begin{pmatrix}\text{Binding energy}\\ \text{of products}\end{pmatrix} - \begin{pmatrix}\text{Binding energy}\\ \text{of reactants}\end{pmatrix}$$

In the example of the triple-alpha process, the binding energy of a helium nucleus is 6.824×10^{14} joules (J) per kilogram (kg) of helium, and the binding energy of the produced ^{12}C nucleus is 7.402×10^{14}. The amount of energy available from fusing 1 kg of helium nuclei into carbon is given by:

$$\begin{pmatrix}\text{Net energy from}\\ \text{burning 1 kg He}\end{pmatrix} = \begin{pmatrix}\text{Binding energy}\\ \text{of C formed}\end{pmatrix} - \begin{pmatrix}\text{Binding energy}\\ \text{of He burned}\end{pmatrix}$$
$$= (7.402 \times 10^{14}\ \text{J}) - (6.824 \times 10^{14}\ \text{J})$$
$$= 5.780 \times 10^{13}\ \text{J}$$

This release of net energy indicates that helium is a good nuclear fuel, as Figure 17.7 shows.

What about fusing iron into more massive elements? Because iron is at the peak of the binding-energy curve, the products of iron burning will have *less* binding energy than the initial reactants. Going from iron to more massive elements means moving down on the binding-energy curve in Figure 17.7, so the net energy in the reaction will be *negative*. Rather than producing energy, fusion of iron *uses* energy.

consumed more rapidly. Although conversion of hydrogen into helium can provide the energy needed to support the high-mass star against the force of gravity for millions of years, helium burning can support the star for only a few hundred thousand years.

Following helium burning, the nature of the balance within the star becomes qualitatively different. There is almost as much energy available from burning a kilogram of carbon, neon, oxygen, or silicon as there is from burning a kilogram of helium. **Table 17.1** shows that the star proceeds from helium burning to the end of its life at a much faster pace.

The balance in a star is analogous to trying to keep a leaky balloon inflated. The larger the leak, the more rapidly air must be pumped into the balloon. A star that is burning hydrogen or helium is like a balloon with a slow leak (**Figure 17.8**). At the temperatures generated by hydrogen or helium burning, energy leaks out of the interior of a star primarily by radiation and convection. Neither of these processes is very efficient, because the outer layers of the star act like a thick, warm blanket. Much of the energy is kept in the star, so nuclear fuels need to burn at a relatively modest rate to support the weight of the outer layers of the star while keeping up with the energy escaping outward.

Beginning with carbon burning, this balance shifts in a dramatic and fundamental way. Rather than being carried away entirely by radiation and convection, energy begins to escape from the core primarily in the form of neutrinos produced by the many nuclear reactions occurring there. Like air pouring out through a huge hole in the side of a balloon, neutrinos produced in the interior of the star stream through the overlying layers of the star as if they were not even there, carrying the energy from the stellar interior out into space. As thermal energy pours out of the interior of

TABLE 17.1 | Burning Stages in High-Mass Stars

Core Burning Stage	15-$M_\odot$ Star	25-$M_\odot$ Star	Typical Core Temperatures
Hydrogen (H) burning	11 million years	7 million years	$(3–10) \times 10^7$ K
Helium (He) burning	2 million years	800,000 years	$(1–7.5) \times 10^8$ K
Carbon (C) burning	2,000 years	500 years	$(0.8–1.4) \times 10^9$ K
Neon (Ne) burning	8 months	11 months	$(1.4–1.7) \times 10^9$ K
Oxygen (O) burning	2.6 years	5 months	$(1.8–2.8) \times 10^9$ K
Silicon (Si) burning	18 days	0.7 day	$(2.8–4) \times 10^9$ K

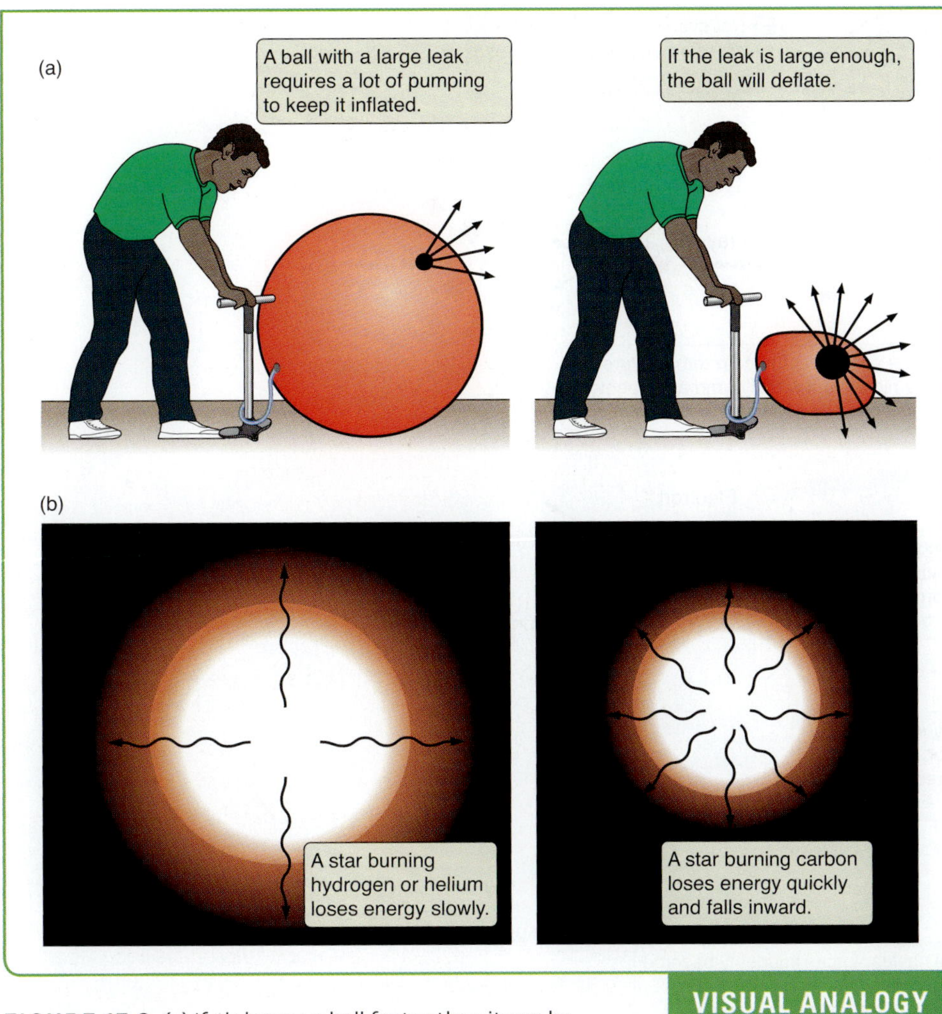

FIGURE 17.8 (a) If air leaves a ball faster than it can be replaced, the ball will deflate. (b) Similarly, if energy leaves a star faster than it can be replaced, the energy balance is disrupted, and the star begins to contract.

the star, the outer layers of the star push inward, driving up the density and temperature, and forcing nuclear reactions to run at a furious rate to replace the energy escaping in the form of neutrinos.

Once this process of **neutrino cooling** becomes significant, the star begins evolving much more rapidly. Carbon burning is capable of supporting the star for less than about a thousand years. Oxygen burning holds the star up for only about a year. Silicon burning lasts only a few days. A silicon-burning star is not much more luminous than it was while burning helium. But because of neutrino cooling, the silicon-burning star actually releases about 200 million times more energy per second than did its former self.

Stages of burning after helium burning are progressively shorter-lived.

The Core Collapses and the Star Explodes

After silicon burns to form an iron core in the star, no source of nuclear energy remains to replenish the energy that is being taken away by escaping neutrinos. The high-mass star's life balancing gravity and controlled nuclear energy production is over. No longer supported by thermonuclear fusion, the iron core of the massive star begins to collapse (**Figure 17.9**).

The early stages of collapse of the iron ash core of an evolved massive star are much the same as in the collapse of a nonburning core in a low-mass star. As the core collapses, its density and temperature skyrocket, and the force of gravity grows even stronger. The gas in the core becomes electron-degenerate when it reaches the approximate size of Earth. Unlike the electron-degenerate core of a low-mass red giant, however, the weight bearing down on the interior of the iron ash core is too great to be held up by electron degeneracy. As the collapse continues, the core reaches temperatures of 10 billion K (10^{10} K) and higher, while the density exceeds 10^{10} kilograms per cubic meter (kg/m^3)—10 times the density of an electron-degenerate white dwarf.

These phenomenal temperatures and pressures trigger fundamental changes in the core. The laws describing thermal radiation say that at these temperatures the nucleus of the star is awash in extremely energetic thermal radiation. This radiation is so energetic that thermal gamma-ray photons are produced with enough energy to break iron nuclei apart into helium nuclei. This process, called photodisintegration, absorbs thermal energy and begins reversing the results of nuclear fusion. At the same time, the density of the core is so great that electrons are squeezed into atomic nuclei, where they combine with protons to produce neutrons (and neutron-rich isotopes in the core of the star). This process consumes thermal energy as well, robbing the core of more of its pressure support. Neutrinos continue to pour out of the core of the dying star, taking more energy with them. The collapse of the core accelerates, reaching

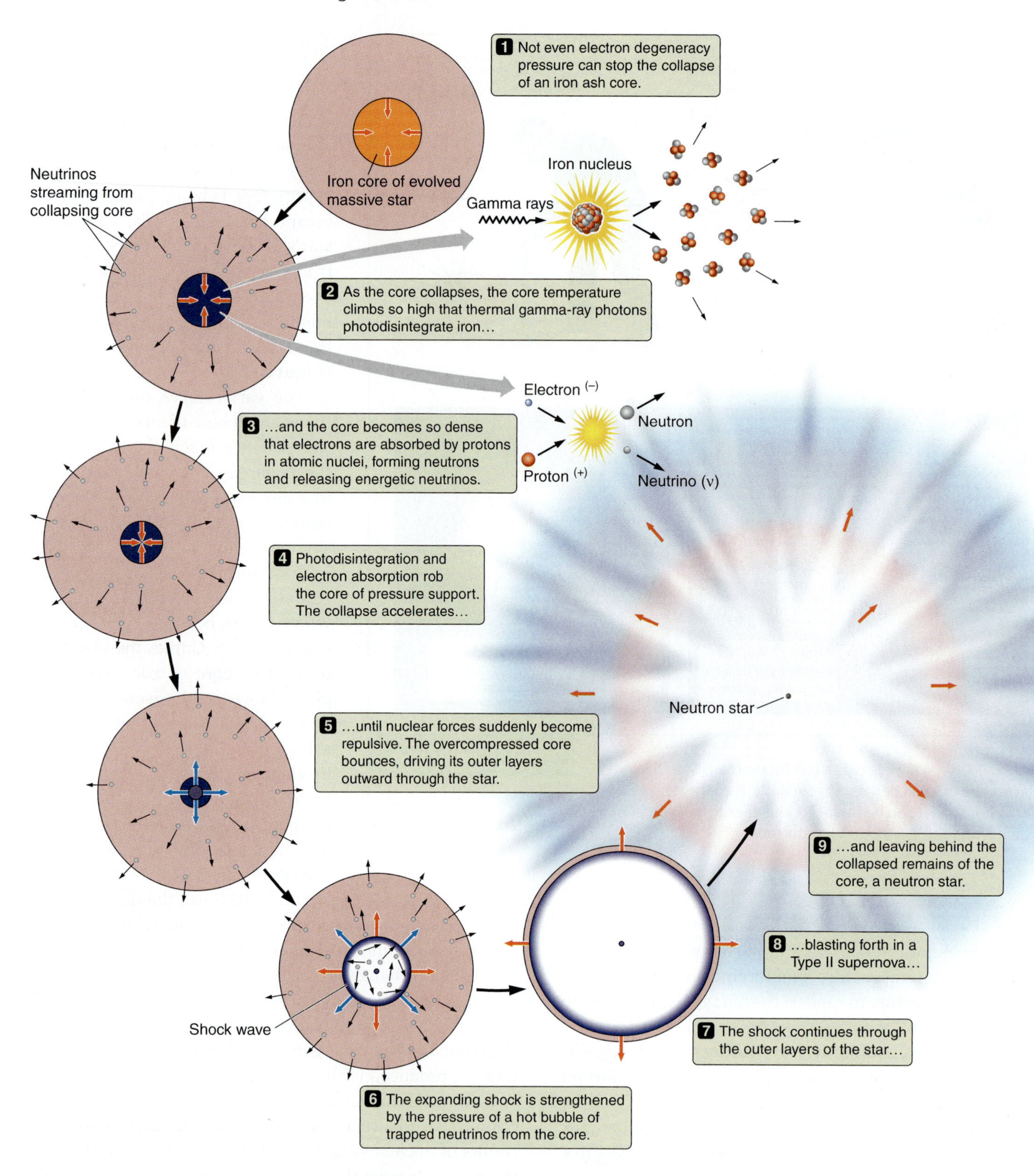

FIGURE 17.9 The stages that a high-mass star goes through at the end of its life as its core collapses and the star explodes as a Type II supernova.

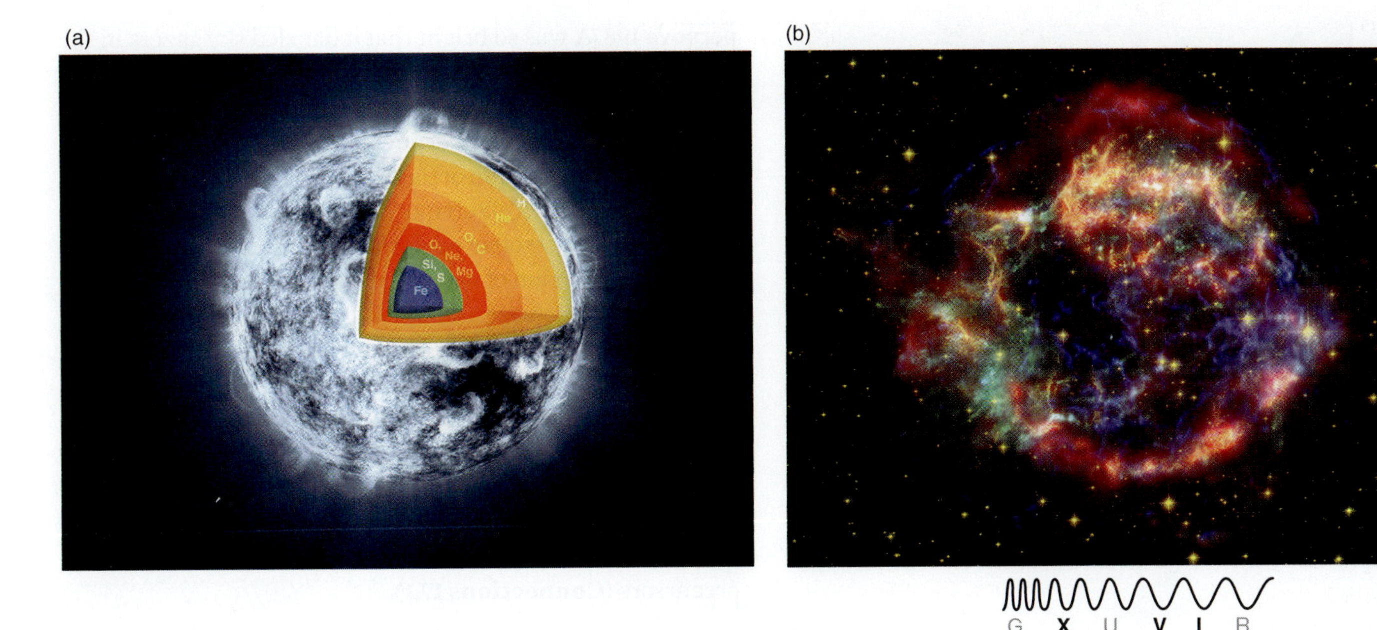

FIGURE 17.10 (a) An evolved massive star has a layered structure, shown here (not to scale). When the evolved star explodes, it forms a supernova remnant like Cassiopeia A. (b) X-ray observations such as these suggest that the star may have turned itself inside out as it spit out elements from its core.

velocities of 70,000 km/s (almost one-fourth the speed of light) on its inward fall. These events take place within a second.

The next hurdle for the collapsing core is the force that holds atomic nuclei together. As material in the collapsing core reaches and exceeds the density of an atomic nucleus, the strong nuclear force actually becomes repulsive. The collapsing core suddenly slows its inward fall. The remaining half slams into the innermost part of the star at a fraction of the speed of light and "bounces," sending a tremendous shock wave back out through the star.

Under the extreme conditions in the center of the star, neutrinos are produced at an enormous rate. Over the next second or so, almost a fifth of the core mass is converted into neutrinos. Most of these neutrinos pour outward through the star; but at the extreme densities found in the collapsing core of the massive star, not even neutrinos pass with complete freedom. A few tenths of a percent of the energy of the neutrinos streaming out of the core of the dying star is trapped by the dense material behind the expanding shock wave. The energy of these trapped neutrinos drives the pressure and temperature in this region ever higher, inflating a bubble of extremely hot gas and intense radiation around the core of the star. The pressure of this bubble adds to the strength of the shock wave moving outward through the star. Within about a minute the shock wave has pushed its way out through the helium shell within the star. Within a few hours it reaches the surface of the star itself, heating the stellar surface to 500,000 K and blasting material outward at velocities of up to about 30,000 km/s. The evolved massive star (**Figure 17.10a**) has exploded in a **Type II supernova**, leaving behind a cloud of dust and gas (**Figure 17.10b**).

The high-mass star explodes as a tremendous Type II supernova.

17.3 The Spectacle and Legacy of Supernovae

For a brief time a Type II supernova shines with the light of a billion Suns. Like Type Ia supernovae, Type II supernovae can be so bright for a few weeks that they can be seen easily in distant galaxies. Astronomers determine the difference between the two types of supernovae by studying both spectra and light curves from these suddenly bright objects (**Figure 17.11**).

The energy carried away by light from a supernova represents only about 1 percent of the kinetic energy being carried away by the outer parts of the star. This kinetic

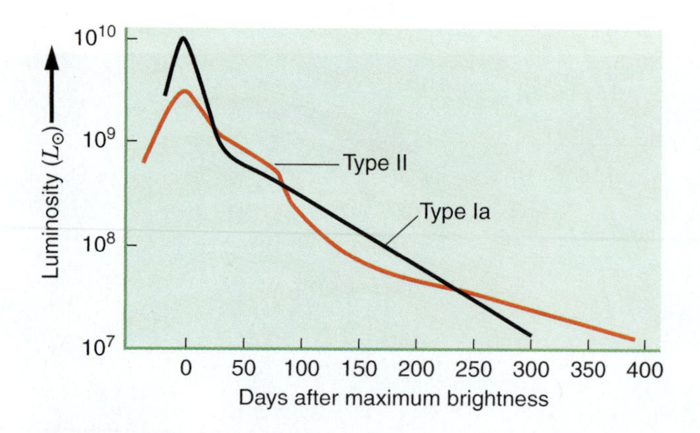

FIGURE 17.11 Light curves showing the changes in brightness of average Type Ia and Type II supernovae. The shape of the curve indicates the type.

energy, in turn, is only about 1 percent of the energy carried away by neutrinos. This ejected material contains approximately 10^{47} J of kinetic energy—enough energy to accelerate the entire Sun to a speed of 10,000 km/s. The kinetic energy of the material ejected from both Type Ia and Type II supernovae heats the hottest phases of the interstellar medium and pushes around the clouds in the interstellar medium. Yet even this amount of energy is small by comparison with the energy carried away from the supernova explosion by neutrinos—an amount of energy at least 100 times larger.

In 1987, astronomers detected a massive star exploding in the Large Magellanic Cloud (LMC, a companion galaxy to the Milky Way). Although 160,000 light-years away, Supernova 1987A was so bright that it dazzled sky gazers in the Southern Hemisphere (**Figure 17.12**). Astronomers working in all parts of the electromagnetic spectrum pointed their telescopes at the new supernova—the first naked-eye supernova since the invention of the telescope. Neutrino telescopes recorded a burst of neutrinos passing through Earth from this tremendous stellar explosion that had occurred in the LMC. The detection of neutrinos from SN 1987A provided astronomers with a rare and crucial glimpse of the very heart of a massive star at the moment of its death, confirming a fundamental prediction of theories about the collapse of the core and its effects. Astronomers were ultimately surprised to discover from looking at old photographs that the star that blew up was not a red supergiant, but a B3 I blue supergiant. The 20-$M_\odot$ star Sanduleak −69° 202 is now classified as a luminous blue variable star, one of several types of supernova precursors (**Connections 17.1**).

The Energetic and Chemical Legacy of Supernovae

Supernova explosions leave a rich and varied legacy in the universe. Huge expanding bubbles of million-kelvin gas glow in X-rays (**Figure 17.14a**) and ultraviolet radiation (**Figure 17.14b**) and drive visible shock waves (**Figure 17.14c**) into the surrounding interstellar medium (the dust and gas between the stars). These bubbles are the still-powerful blast waves of supernova explosions that took place thousands of years

Supernovae eject newly formed massive elements into interstellar space.

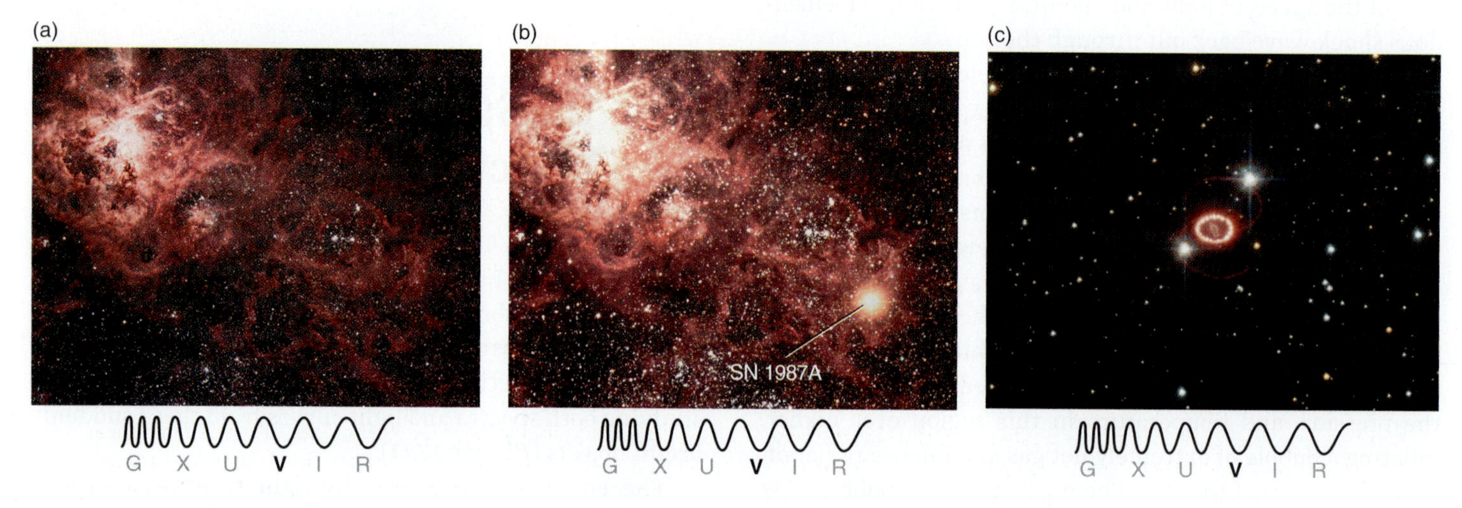

FIGURE 17.12 SN 1987A was a supernova that exploded in a small companion galaxy of the Milky Way called the Large Magellanic Cloud (LMC). These images show the LMC before the explosion (a) and while the supernova was near its peak (b). Notice the "new" bright star at lower right. Image (c) from 2011 zooms in to show rings that have developed around the exploded star.

Connections 17.1

Variations by Stellar Mass

Stars with masses between 3 and 8 $M_{\odot}$ exist in something of a gray area between the low-mass stars discussed in Chapter 16 and the high-mass stars discussed in this chapter. Like massive stars, these **intermediate-mass stars** burn hydrogen via the CNO cycle. They also leave the main sequence as massive stars do, burning helium in their cores immediately after their hydrogen is exhausted and skipping the helium flash phase of low-mass star evolution. When helium burning in the core is complete, however, the temperature at the center of an intermediate-mass star is too low for carbon to burn. From this point on, the star evolves more like a low-mass star, ascending the asymptotic giant branch, burning helium and hydrogen in shells around a degenerate core, and then ejecting its outer layers and leaving behind a white dwarf. However, the chemical compositions of planetary nebulae and white dwarfs left behind by intermediate-mass stars can be quite distinct from those of truly low-mass stars.

Stars that are massive enough to burn heavier elements display much variation before they explode. What happens to the star when it leaves the main sequence is determined primarily by its mass, although chemical composition and rotation also play a role. Chemical composition varies because new stars form from material that has been enriched by the deaths of previous stars; so new stars are born with different amounts of heavy elements. But in general, stars with masses below 15 $M_{\odot}$ go through a red supergiant stage, perhaps with a Cepheid variable phase. Stars under 30 $M_{\odot}$ move back and forth on the H-R diagram—becoming red, then blue, then red supergiants again (sometimes with a brief period as a yellow supergiant), depending on what is burning in their core (see Figure 17.4).

Higher-mass stars go through Wolf-Rayet (WR) and or luminous blue variable (LBV) phases. WR stars are identified from strong emission lines of helium (He), carbon (C), nitrogen (N), and oxygen (O) in their spectra. These stars continually eject gas into space (**Figure 17.13a**), and most of them eventually become supernovae. LBV stars are hot, luminous, extremely rare stars that may be as massive as 150 $M_{\odot}$. An example is Eta Carinae (**Figure 17.13b**), a binary system with a 120-$M_{\odot}$ star and a luminosity (summed over all wavelengths) of 5 million Suns (5 million $L_{\odot}$). Currently, Eta Carinae is losing mass at a rate of about 10^{-3} $M_{\odot}$ per year (or 1 $M_{\odot}$ every 1,000 years). However, during a 19th century eruption, when Eta Carinae became the second-brightest star in the sky, its mass loss must have reached the rate of 0.1 $M_{\odot}$ per year, shedding 2 $M_{\odot}$ of material over a mere 20 years. Eta Carinae is expected to explode in the astronomically near future.

(a)

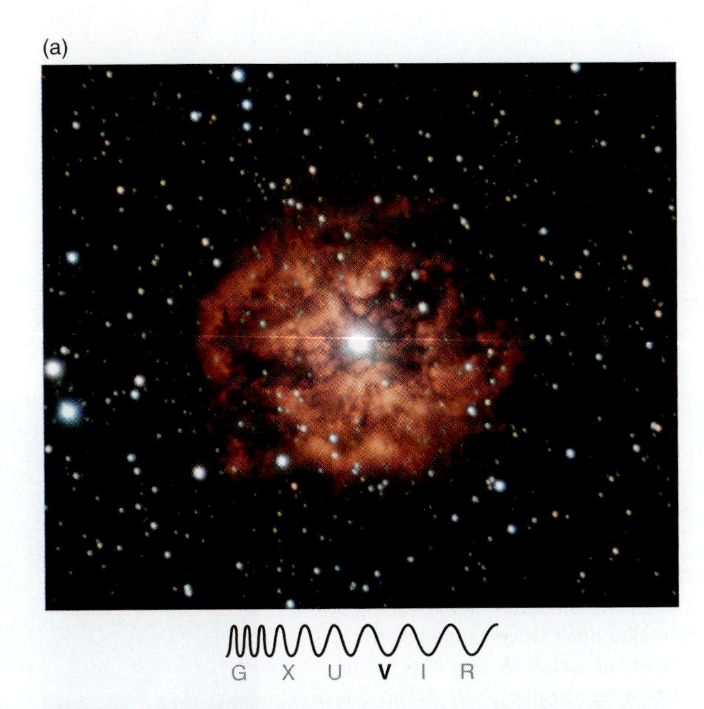

(b)

FIGURE 17.13 (a) A Wolf-Rayet (WR) star (center) is shedding outer layers, forming a nebula of dust and gas. (b) In this image of the luminous blue variable (LBV) star Eta Carinae, an expanding cloud of ejected dusty material is seen in optical (blue) and X-ray (yellow) light. The star itself, which is largely hidden by the surrounding dust, has a luminosity of 5 million $L_{\odot}$ and a mass probably in excess of 120 $M_{\odot}$. Dust is created when volatile material ejected from the star condenses.

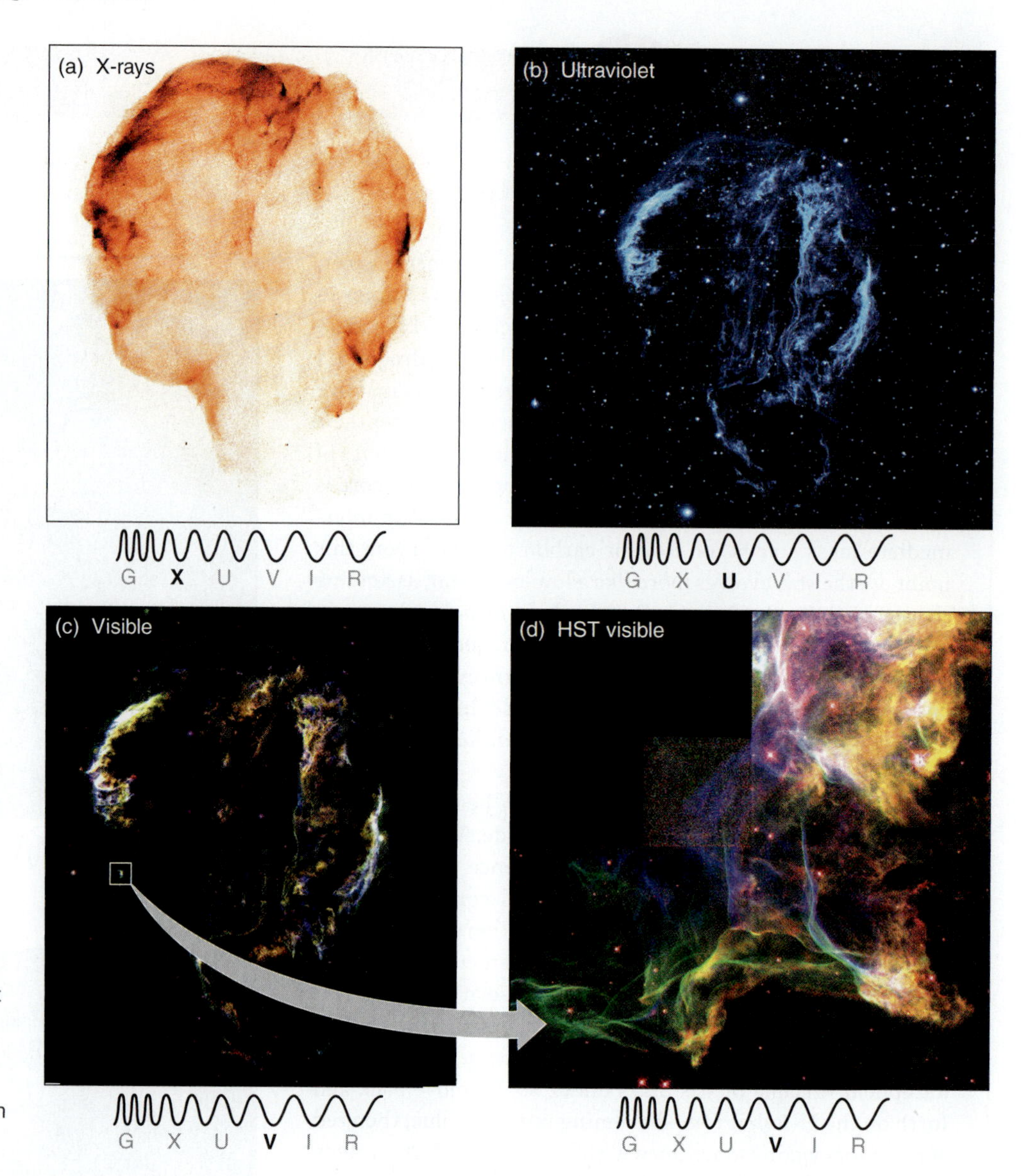

FIGURE 17.14 The Cygnus Loop is a supernova remnant—an expanding interstellar blast wave caused by the explosion of a massive star. Gas in the interior with a temperature of millions of kelvins glows in X-rays (a) and ultraviolet radiation (b), while visible light (c) comes from locations where the expanding blast wave pushes through denser gas in the interstellar medium. (d) A Hubble Space Telescope image zooms in to show a location where the blast wave is hitting an interstellar cloud.

ago. Supernova explosions compress nearby clouds (**Figure 17.14d**), triggering the initial collapse that begins star formation.

Perhaps even more important is the chemical legacy left behind by supernova explosions. Only the least massive chemical elements formed at the beginning of the universe: hydrogen, helium, and trace amounts of lithium, beryllium, and boron. All the rest of the chemical elements formed in the stars through nuclear reactions and then returned to the interstellar medium. This process, which is responsible for the progressive chemical enrichment of the universe, is **nucleosynthesis**. Types I and II supernovae are the true champions of nucleosynthesis, because the highest-mass elements are fused in their cores.

Elements up to carbon and oxygen form from nuclear fusion in the cores of low-mass stars and travel to the interstellar medium in asymptotic giant branch (AGB) winds and planetary nebulae. High-mass stars produce elements as massive as iron before their supernova explosions (see Figure 17.4), and the explosions enrich the interstellar medium. A look at a table of elements that occur in nature (see Appendix 3) shows that many elements, up through uranium (the most massive among the naturally occurring elements), are far more massive than iron. Recall that fusion up to iron creates energy, but fusion beyond iron absorbs energy. So elements heavier than iron fuse only under conditions in which energy can be absorbed—such as the enormous energies of supernova explosions.

Normally, electric repulsion keeps positively charged atomic nuclei far apart. Extreme temperatures are needed to slam nuclei together hard enough to overcome this electric repulsion. Free neutrons, however, are not subject to these rules: they have no net electric charge, so there is no electric repulsion to prevent them from simply colliding with an atomic nucleus, regardless of how many protons that nucleus contains. Under normal circumstances in nature, free neutrons are very rare. In the interiors of evolved stars, however, a number of nuclear reactions produce free neutrons, and under some circumstances—including those shortly before and during a supernova—free neutrons are produced in very large numbers. Free neutrons are easily captured by atomic nuclei and later decay to become protons. (Recall that the number of protons—that is, the atomic number—is what distinguishes one element from another.) In this way, elements with atomic numbers and masses higher than those of iron are formed.

The abundances of the elements measured on Earth, in meteorite material, and in the atmospheres of stars and their remains have been predicted by nuclear physicists (**Figure 17.15**). Less massive elements are far more abundant than more massive elements, because more massive elements are progressively built up from less massive elements. An exception to this pattern is the dip in the abundances of the light elements lithium (Li), beryllium (Be), and boron (B). These elements are destroyed in nuclear burning. Conversely, carbon (C), nitrogen (N), and oxygen (O) are produced in quantity in the triple-alpha process, so they are very abundant. Even the sawtooth pattern in the abundances of even- and odd-numbered elements is a consequence of the formation of atomic nuclei in stars. By comparing the predictions of nuclear physics with observations of elemental abundances, astronomers repeatedly test the theory of stellar evolution.

Neutron Stars and Pulsars

In the explosion of a Type II supernova, the outer parts of the star are blasted back into interstellar space—but what remains of the core that was left behind? The matter at the center of the massive star has collapsed to the point where it has about the same density as the nucleus of an atom. As long as the mass of the core left behind by the explosion is no more than about 3 $M_{\odot}$, this collapse is halted by quantum mechanical effects similar to those responsible for supporting a white dwarf. But in this case, instead of electrons, it is neutrons that are forced together as tightly as the rules of quantum mechanics allow. The neutron-degenerate

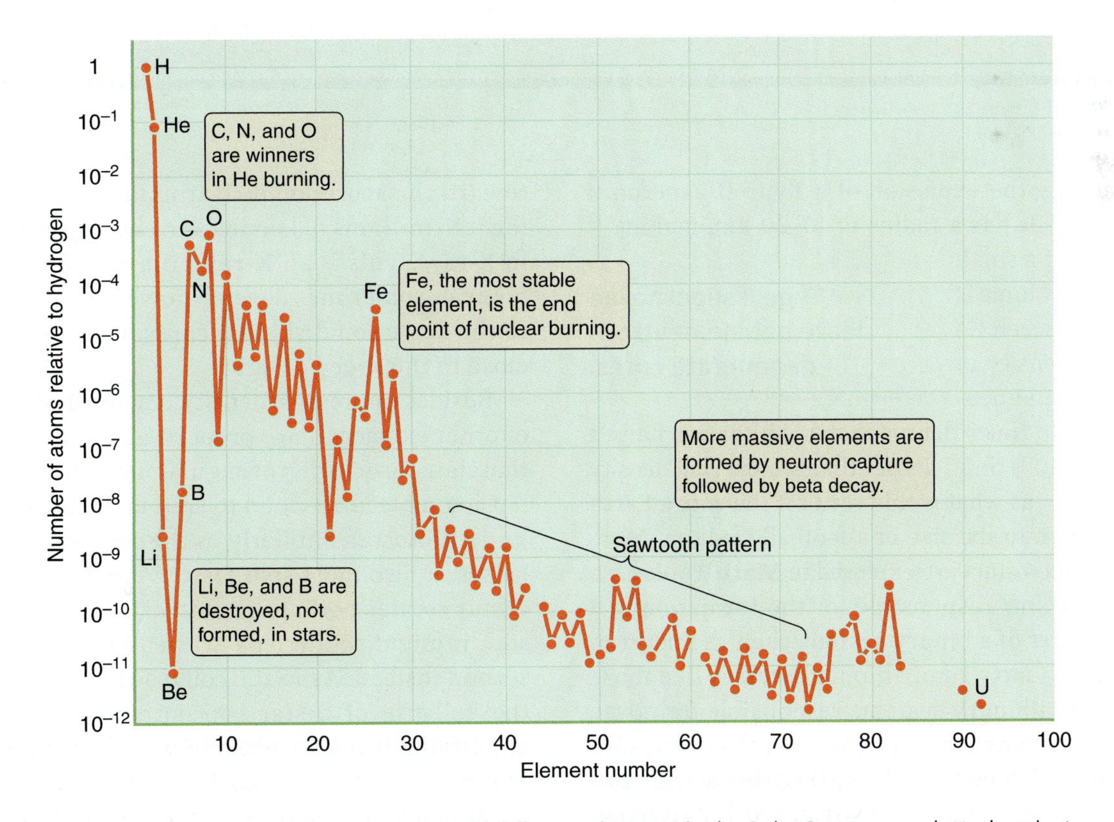

FIGURE 17.15 The relative abundances of different elements in the Solar System are plotted against the element number of the nucleus. This pattern can be understood as a result of nucleosynthesis in stars. The periodic table of elements in Appendix 3 identifies individual elements by their element number (the number of protons).

Math Tools 17.2

Gravity on a Neutron Star

Neutron stars are incredibly dense objects. As a result, the surface gravity and escape velocity of neutron stars are very high. For example, let's look at a typical neutron star with a radius of 15 km and a mass of 2 $M_\odot$.

Recall from Math Tools 4.1 that the acceleration due to gravity on the surface—in this case the surface of a neutron star (NS)—is given by:

$$g = \frac{GM_{NS}}{R_{NS}^2}$$

$$= 6.67 \times 10^{-11} \frac{m^3}{kg\ s^2} \times \frac{2.0 \times (1.99 \times 10^{30}\ kg)}{(15 \times 10^3\ m)^2}$$

$$= 1.2 \times 10^{12} \frac{m}{s^2}$$

Dividing this number by the gravitational acceleration on Earth, 9.8 m/s^2, shows that the gravitational acceleration on a neutron star is over 100 billion times as large as that on Earth.

What about the escape velocity from a neutron star? From Math Tools 16.2, we know that the escape velocity is given by:

$$v_{esc} = \sqrt{2GM/R}$$

Putting in the above numbers for a typical neutron star yields:

$$v_{esc} = \sqrt{2 \times (6.67 \times 10^{-11}\ m^3 kg^{-1} s^{-2}) \times 2.0 \times (1.99 \times 10^{30}\ kg) / (15 \times 10^3\ m)}$$

$$= 1.9 \times 10^8\ m/s = 190{,}000\ km/s$$

Dividing this result by the speed of light gives:

$$\frac{v_{esc}}{c} = \frac{190{,}000\ km/s}{300{,}000\ km/s} = 0.63$$

The escape velocity from a neutron star is over 60 percent of the speed of light, and almost 17,000 times greater than the escape velocity from Earth (which is 11.2 km/s). The physicist Albert Einstein showed that strange things happen at velocities near the speed of light (including modifications to Newton's equations), which we will discuss in detail in Chapter 18.

core left behind by the explosion of a Type II supernova is a **neutron star**. It has a radius of 10–20 km, making it roughly the size of a small city; but into that volume is packed a mass between 1.4 and 3 $M_\odot$. At a density of about 10^{18} kg/m^3, the neutron star is a billion times denser than a white dwarf and a thousand trillion (10^{15}) times denser than water. That density is roughly the same as what would result if the entire Earth were crushed down to the size of a football stadium. More properties of neutron stars are explored in **Math Tools 17.2**.

Type II supernovae leave behind neutron-degenerate cores.

If the massive star responsible for the formation of a neutron star is part of a binary system, then the neutron star is left with a binary companion. Processes like those in the white dwarf binary systems responsible for novae and Type Ia supernovae (discussed in Chapter 16) are possible. As the lower-mass star in such a binary system evolves and overfills its Roche lobe (Figure 16.13), matter falls toward the accretion disk around the neutron star, heating it to millions of kelvins and causing it to glow brightly in X-rays. This is an **X-ray binary** (**Figure 17.16**). X-ray binaries sometimes develop powerful jets, perpendicular to the accretion disk, that carry material away at speeds close to the speed of light.

X-ray binaries arise from the accretion of mass onto neutron stars.

Besides impressive density, a neutron star has a number of other extraordinary properties. Recall from Chapter 15 that the conservation of angular momentum requires a collapsing molecular cloud to spin faster as it grows smaller into a protostar. Similarly, as the core of a massive star collapses, it also must spin faster. A massive main-sequence O star rotates perhaps once every few days. As a neutron star, it might rotate tens or even hundreds of times each second instead. As in the collapse of an interstellar cloud, the collapse of a star concentrates the magnetic field, amplifying it to strengths trillions of times greater than the magnetic field at Earth's surface. A neutron star has a magnetosphere, so it is surrounded by intense magnetic fields and plasmas, (just as on Earth and several other planets). But the neutron star's magnetosphere is much stronger and is whipped around many times a second by

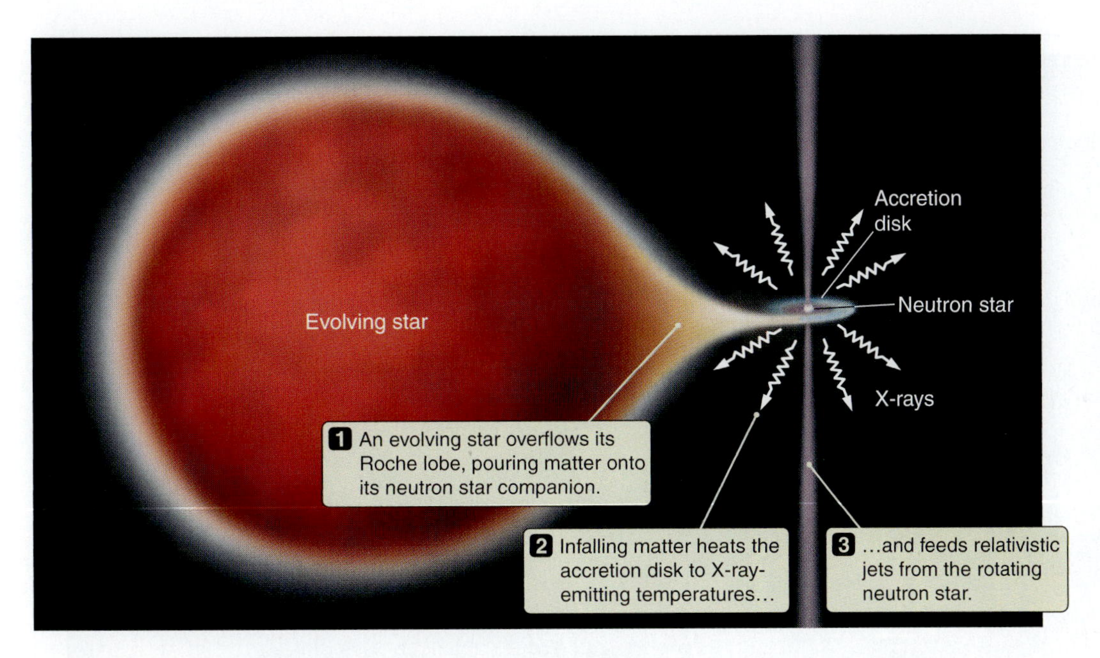

FIGURE 17.16 X-ray binaries are systems consisting of a normal evolving star and a white dwarf, a neutron star, or a black hole. As the evolving star overflows its Roche lobe, mass falls toward the collapsed object. The gravitational well of the collapsed object is so deep that when the material hits the accretion disk, it is heated to such high temperatures that it radiates away most of its energy as X-rays.

the spinning star. As in planets, in stars the magnetic axis is often not aligned with the rotation axis.

Energetic subatomic particles such as electrons and positrons move along the magnetic-field lines of the neutron star and are funneled by the field toward the magnetic poles of the system. Conditions there produce intense electromagnetic radiation, which is beamed away from the magnetic poles of the neutron star (**Figure 17.17**). As the neutron star rotates, these beams of radiation sweep through space like the rotating beams of a lighthouse. When Earth is located in the paths of these beams, telescopes detect a pattern resembling what the beams from a lighthouse look like to sailors entering a harbor at night. The neutron star appears to flash on and off with a regular period equal to the period of rotation of the star (or half the rotation period, if both beams are seen).

Many of the unusual objects discussed in this and the previous chapter—such as pulsating stars, supernovae, and planetary nebulae—puzzled astronomers when they were first observed, but later they were understood to be associated with the end points of stellar evolution. In contrast, neutron stars were *predicted* in 1934, not long after neutrons themselves were discovered. Astronomers Walter Baade and Fritz Zwicky proposed that supernova explosions could lead to the formation of a neutron star. But neutron stars were not actually observed for another 30 years. In 1967, rapidly pulsing objects were first discovered by observers working with radio wavelengths (**Process of Science Figure**). These objects, which blinked like very fast, regularly ticking clocks, puzzled astronomers. Today these rotating neutron stars are called **pulsars**. Over 2,000 pulsars are known, and more are being discovered all the time (**Figure 17.18**).

Pulsars are rapidly spinning, magnetized neutron stars.

The Crab Nebula—Remains of a Stellar Cataclysm

In 1054 CE, Chinese astronomers recorded the presence of a "guest star" newly observed in the direction of the constellation Taurus. The new star was so bright that it could be seen during the daytime for 3 weeks, and it did not fade away altogether for many months. From the Chinese description of the changing brightness and color of the object, modern-day astronomers concluded that the guest star of 1054 was a fairly typical Type II supernova. Today the expanding cloud of debris from this explosion is called the Crab Nebula (**Figure 17.19**).

The Crab Nebula has several components. Images of the Crab Nebula taken in the light of specific emission lines show filaments of glowing gas. Doppler shift measurements of these filaments reveal an expanding shell much like that seen in planetary nebulae. But whereas planetary nebulae are expanding at 20–30 km/s, the shell of the Crab is expanding at closer to 1,500 km/s. Studies of the spectra of these filaments show that they contain anomalously high abundances of helium and other more massive chemical elements—the products of the nucleosynthesis that took place in the supernova and its progenitor star.

The pulsar at the center of the Crab Nebula was the first to be seen at visible wavelengths. The Crab pulsar flashes 60 times a second—first with a main pulse associated with

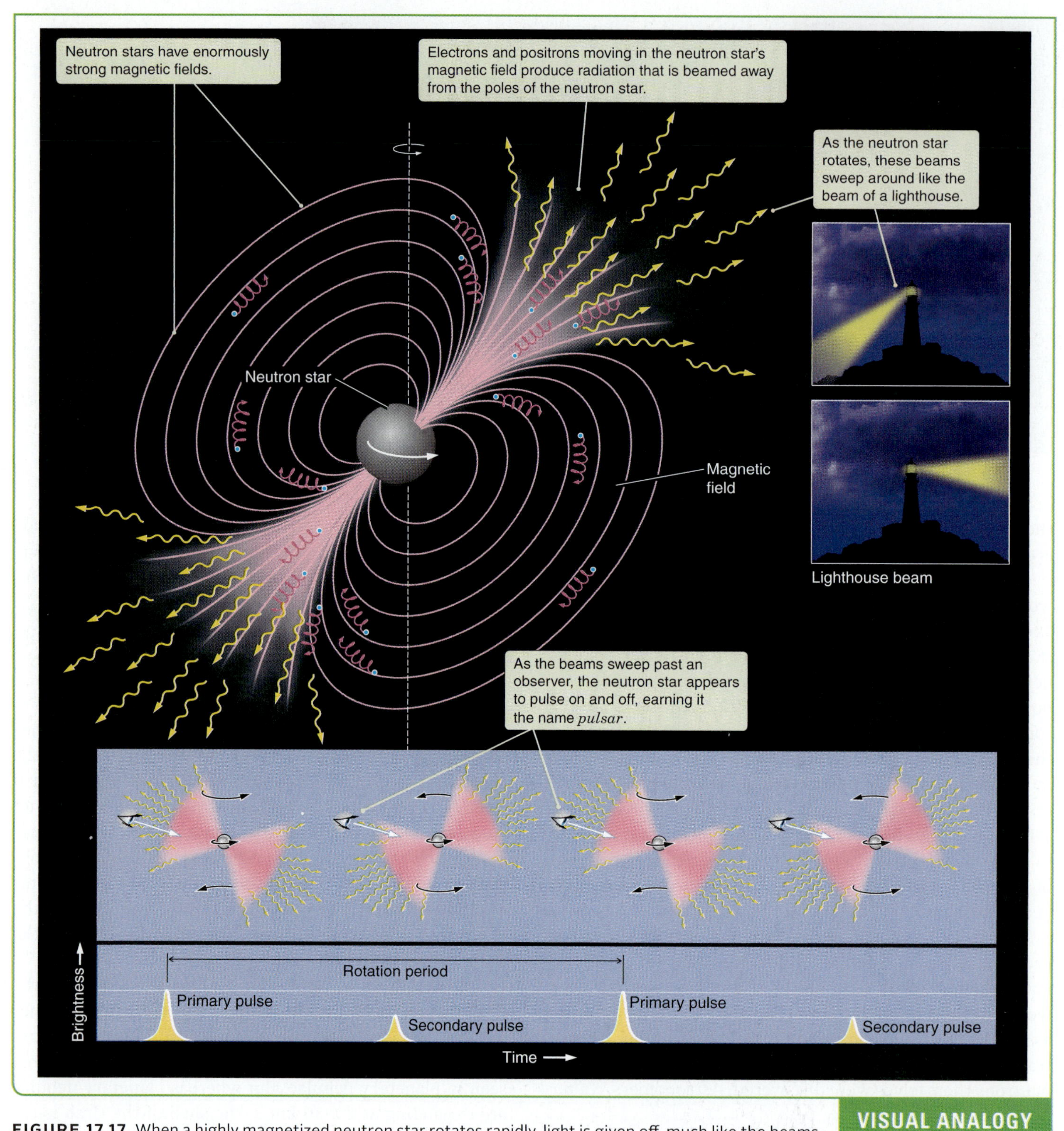

FIGURE 17.17 When a highly magnetized neutron star rotates rapidly, light is given off, much like the beams from a rotating lighthouse lamp. As these beams sweep past Earth, the star will appear to pulse on and off, earning it the name *pulsar*.

one of the "lighthouse" beams, then with a fainter secondary pulse associated with the other beam. The Crab pulsar spins 30 times a second, whipping its powerful magnetosphere around with it. A few thousand kilometers from the pulsar, material in its magnetosphere must move at almost the speed of light to keep up with this rotation. Like a tremendous slingshot, the rotating pulsar magnetosphere flings particles away from the neutron star in a powerful wind moving at

Occam's razor is a guiding principle in science: when scientists consider two hypotheses that explain a phenomenon equally well, they should adopt the *simpler* of the two. Simpler does not mean that the math is easier, or even that the concept is easy to understand. It means that the fewest number of other, new assumptions need to be made.

Jocelyn Bell, a student at Cambridge, has a "mystery signal" in her data.

Her adviser, Anthony Hewish, half-jokingly suggests "little green men" as the cause of the signal.

Bell and Hewish find three more such signals. It is unlikely that the same "little green men" would be sending the same signal from three separate locations in the sky.

They suggest the signals are from pulsating white dwarfs or neutron stars.

Franco Pacini and Thomas Gold each develop a detailed explanation involving rotating neutron stars.

This explanation relies entirely on previously understood physical phenomena: rotation, magnetic fields, and neutron stars.

It does not require assumptions about the existence of extra-terrestrials.

The neutron star explanation is "simpler."

If more than one explanation exists, scientists choose the simplest one that explains all the data. This is always the best place to start.

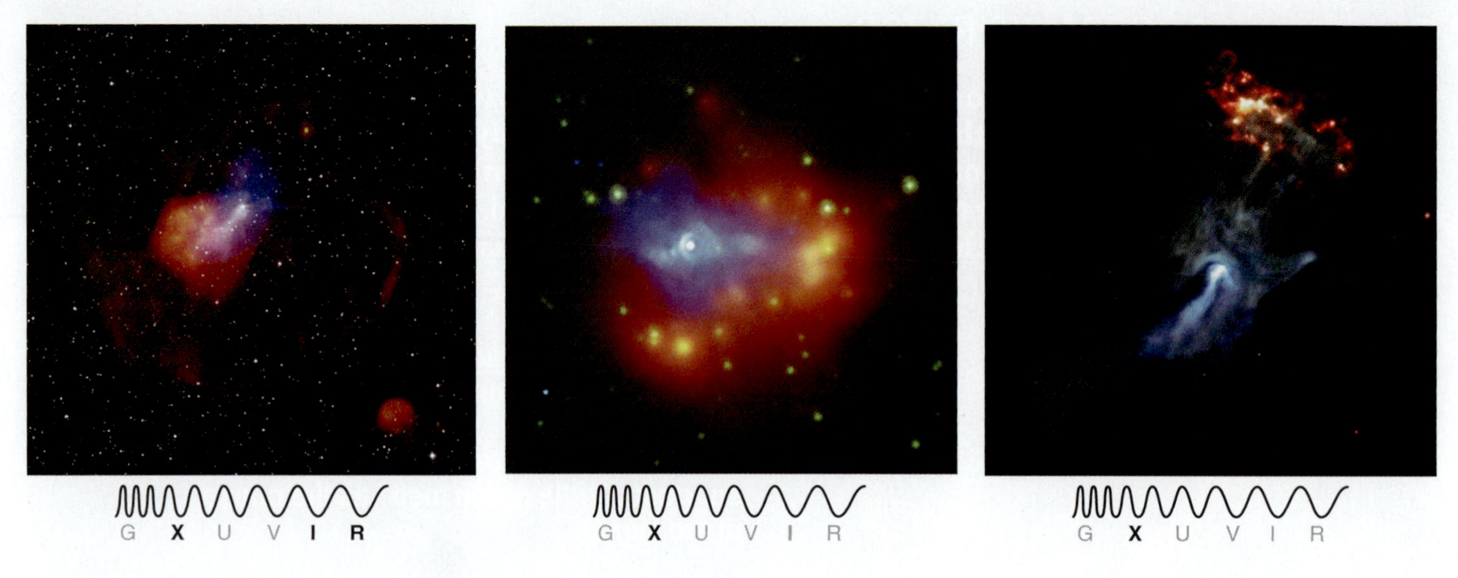

FIGURE 17.18 Three pulsars observed with the Chandra X-ray telescope.

FIGURE 17.19 The Crab Nebula is the remnant of a supernova explosion witnessed by Chinese astronomers in 1054 CE. The X-ray image of the pulsar is in the center; the outer part of the nebula was observed with the Hubble and Spitzer space telescopes.

nearly the speed of light. Material from this wind fills the space between the pulsar and the expanding shell. The Crab Nebula is almost like a big balloon; but instead of being filled with hot air, it is filled with a mix of very fast particles and strong magnetic fields. Images of the Crab Nebula show this bubble as a glow from *synchrotron radiation*—a type of beamed radiation that is emitted as very fast moving particles spiral around the magnetic field.

A pulsar powers the Crab Nebula's glow.

17.4 Star Clusters Are Snapshots of Stellar Evolution

Observations of star clusters provide evidence for the evolution of stars of different masses. Recall from Chapter 15 that when an interstellar cloud collapses, it breaks into pieces, forming not one star but many stars of different masses. Two types of star clusters that have been observed since the earliest telescopes are globular clusters and open clusters. **Globular clusters** are densely packed collections of hundreds of thousands to millions of stars (**Figure 17.20a**). **Open clusters** are much less tightly bound collections of a few dozen to a few thousand stars (**Figure 17.20b**).

Cluster Ages

In the 1920s, astronomers started plotting the observed brightness versus the spectral type for as many stars as possible in each cluster. The resulting cluster H-R diagrams showed stars in all the categories in the "textbook" H-R diagram (see Figure 13.16). Since all of the stars in a

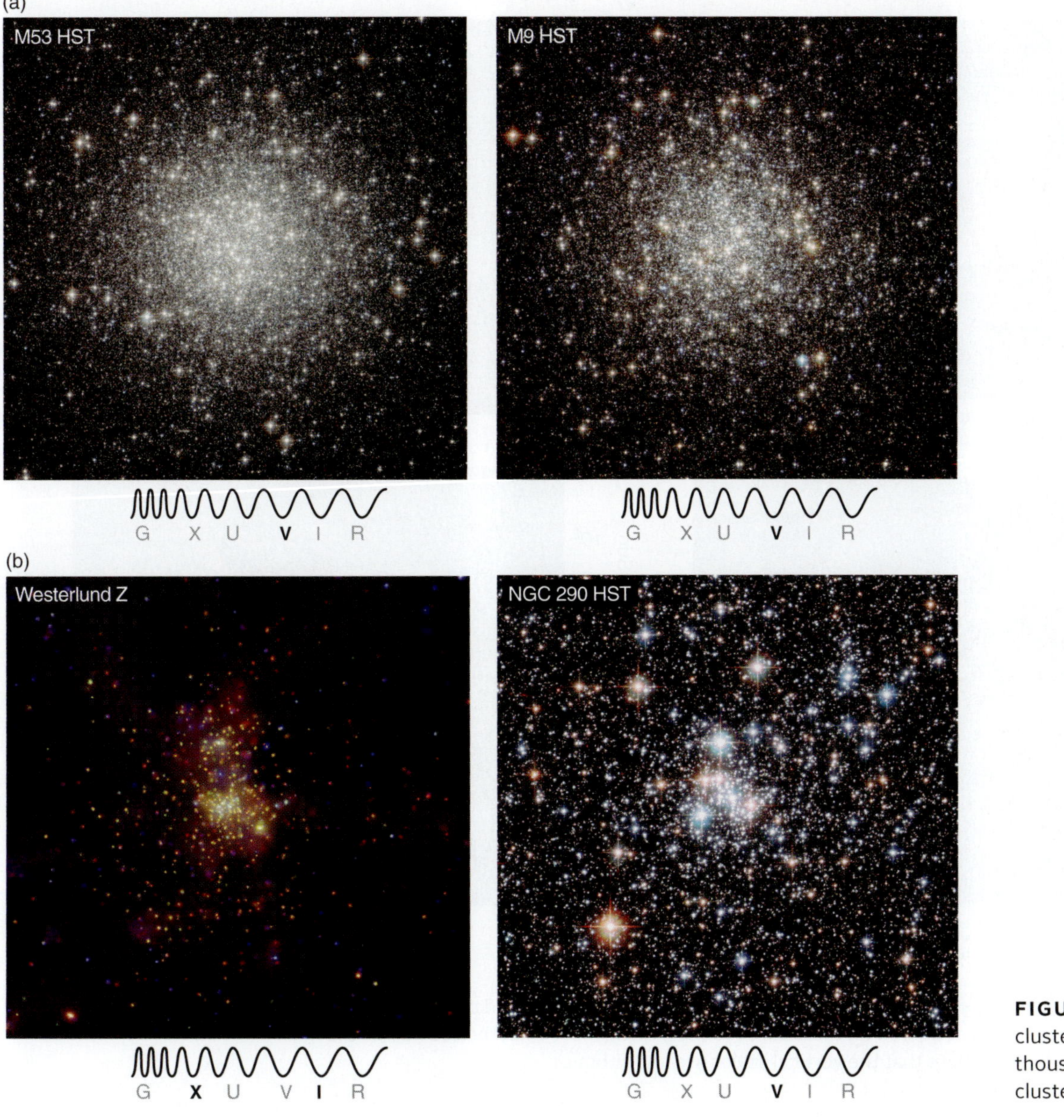

FIGURE 17.20 (a) Globular clusters can have hundreds of thousands of stars. (b) Open clusters have hundreds.

cluster are at approximately the same distance from Earth, the difference between the observed brightness and the actual luminosity is the same for all of them. By matching the main sequence on the observed H-R diagram to the main sequence on the standard H-R diagram, astronomers could estimate the distance to a cluster.

Astronomers also realized that the cluster H-R diagrams offered clues to the newly developing theories of stellar evolution. All of the stars in a cluster formed together at nearly the same time, so a look at a cluster that is 10 million years old shows what the stars of all different masses evolve into during the first 10 million years after they form. In other words, a look at a cluster 10 billion years after it formed shows what becomes of stars of different masses after 10 billion years. **Figure 17.21a** shows an H-R diagram of a very young cluster, NGC 6530. O, B, and A stars are on the main sequence; F through M stars with lower masses are still evolving *to* the main sequence. There are no red giants or white dwarfs. In contrast, **Figure 17.21b** shows the H-R diagram of a very old cluster, M55. There are no high-mass stars on the main sequence, because they have evolved off of it, but there are stars on the horizontal, red giant, and asymptotic giant branches, and in the lower part of the main sequence. This globular cluster is about 12 billion years old; globular clusters are among the oldest astronomical objects known.

Of course, astronomers cannot watch an individual cluster age over millions of years, but they can observe different clusters of different ages. They explore cluster

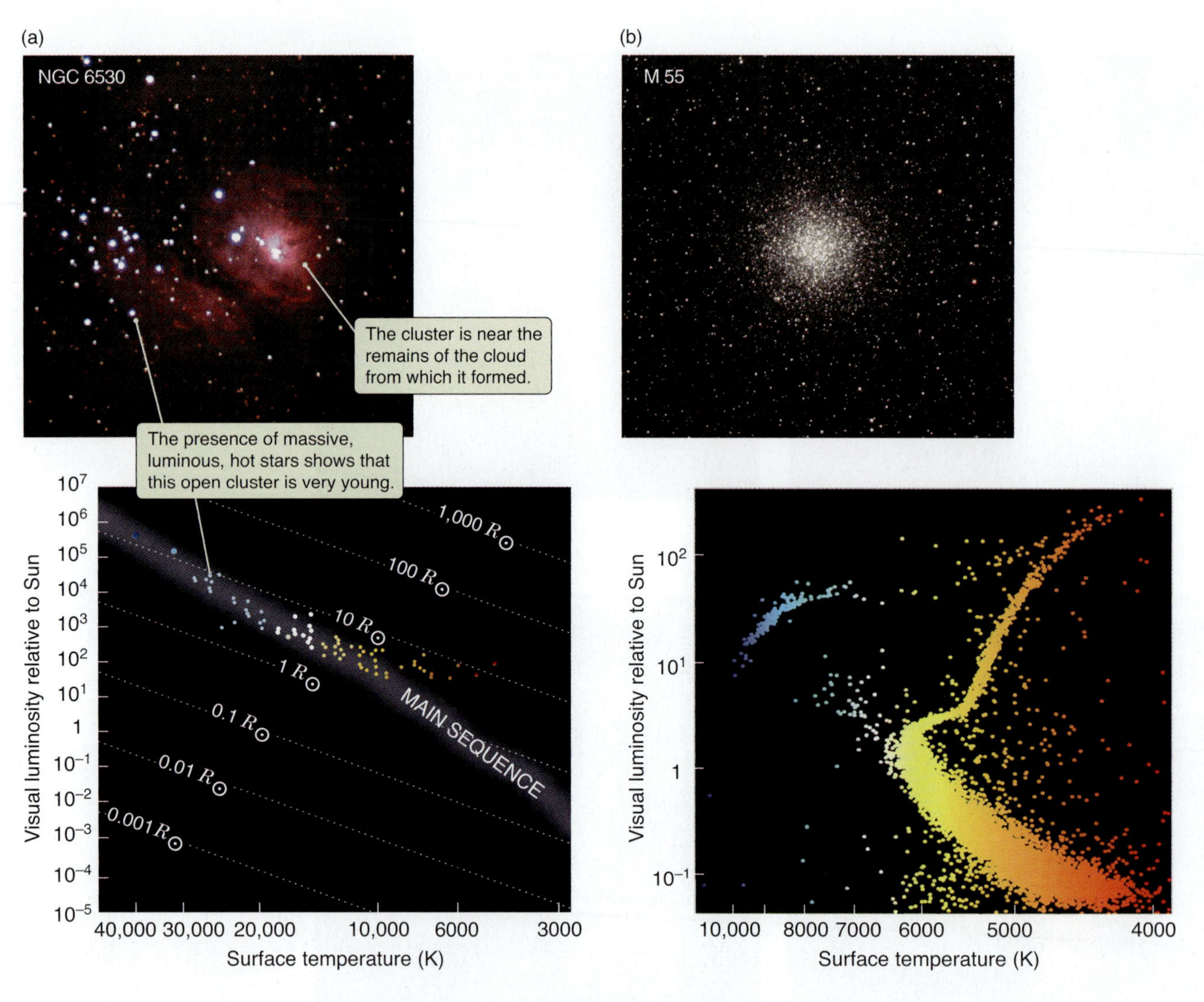

FIGURE 17.21 H-R diagrams of a very young cluster, NGC 6530 (a) and a very old cluster, M55 (b). In (a), some of the stars aren't yet arrived on the main sequence. In (b), more than the top half of the main sequence has already evolved. Note that the vertical scales are logarithmic.

evolution more systematically by examining an H-R diagram of a *simulated* cluster (**Figure 17.22**) of 40,000 stars as it would appear at several different ages, and then compare it to observed H-R diagrams of actual clusters. In **Figure 17.22a**, stars of all masses are located on the zero-age main sequence, showing where they begin their lives as main-sequence stars. The increasing masses of stars along the main sequence are indicated. An actual cluster H-R diagram will never look like the one in Figure 17.22a, however, since the stars in a cluster do not all reach the main sequence at exactly the same time. Star formation in a molecular cloud is spread out over several million years, and it takes considerable time for lower-mass stars to contract to reach the main sequence. The H-R diagram of a very young cluster typically shows many lower-mass stars located well above the main sequence.

Stars in clusters formed together at about the same time.

The more massive a star is, the shorter its life on the main sequence will be. After only 4 million years (**Figure 17.22b**), all stars with masses greater than about 20 $M_{\odot}$ have evolved off the main sequence and are now spread out across the top of the H-R diagram. The most massive stars have already disappeared from the H-R diagram entirely, having vanished in supernovae. As time goes on, stars of lower and lower mass evolve off the main sequence, and the turnoff point moves toward the bottom right in the H-R diagram. By the time the cluster is 10 million years old (**Figure 17.22c**), only stars with masses less than about 15 $M_{\odot}$ remain on the main sequence. The location of the most massive star that remains on the main sequence is called

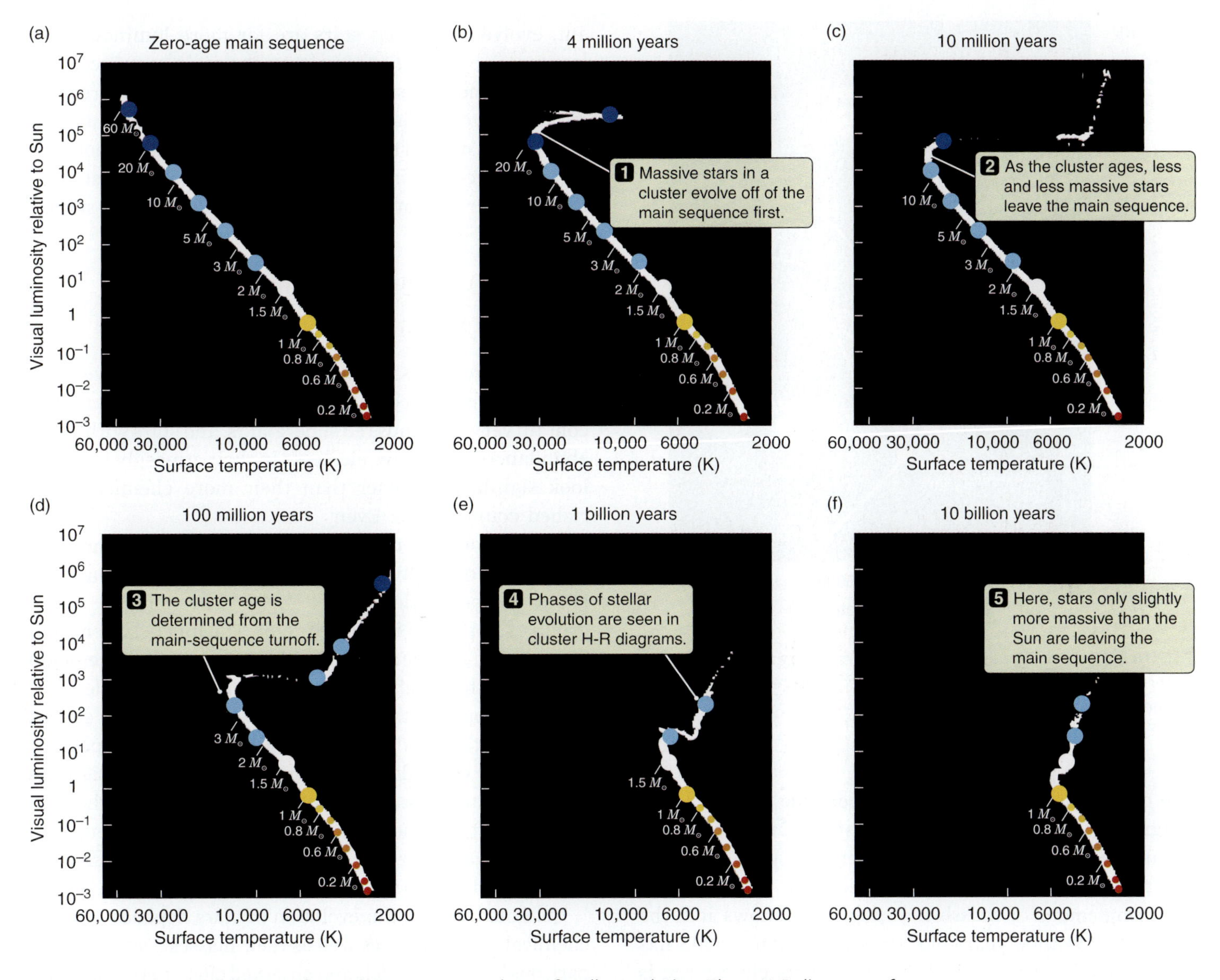

FIGURE 17.22 H-R diagrams of star clusters are snapshots of stellar evolution. These H-R diagrams of a *simulated* cluster of 40,000 stars of solar composition are shown at different times following the birth of the cluster. Note the progression of the *main-sequence turnoff* to lower and lower masses.

the **main-sequence turnoff**. As the cluster ages, the main-sequence turnoff moves farther and farther down the main sequence to stars of lower and lower mass.

As a cluster ages (**Figures 17.22d and e**), the details of all stages of stellar evolution appear. By the time the star cluster is 10 *billion* years old (**Figure 17.22f**), stars with masses of only 1 $M_\odot$ are beginning to die. Stars slightly more massive than this are seen as giant stars of various types. Note how few supergiant and giant stars are present in any of the cluster H-R diagrams. The supergiant, giant, horizontal, and asymptotic giant branch phases in the evolution of stars pass so quickly in comparison with a star's main-sequence lifetime that even though this simulated cluster started with 40,000 stars, only a handful of stars are seen in these phases of evolution at any given time. Similarly, it takes a newly formed white dwarf only a few tens of millions of years to cool to the point that it disappears off the bottom of these figures. Even though the majority of evolved stars in an old cluster are white dwarfs, all but a few of these stars will have cooled and faded into obscurity at any given time. More of these evolved stars are seen in the larger globular cluster M55 in Figure 17.21b.

The main-sequence turnoff shifts to lower-mass stars as a cluster ages.

Cluster evolution models show the sequence of ages of a cluster, and thus are a powerful tool for studying the history of a cluster. To astronomers observing a star cluster, the location of the main-sequence turnoff immediately indicates the

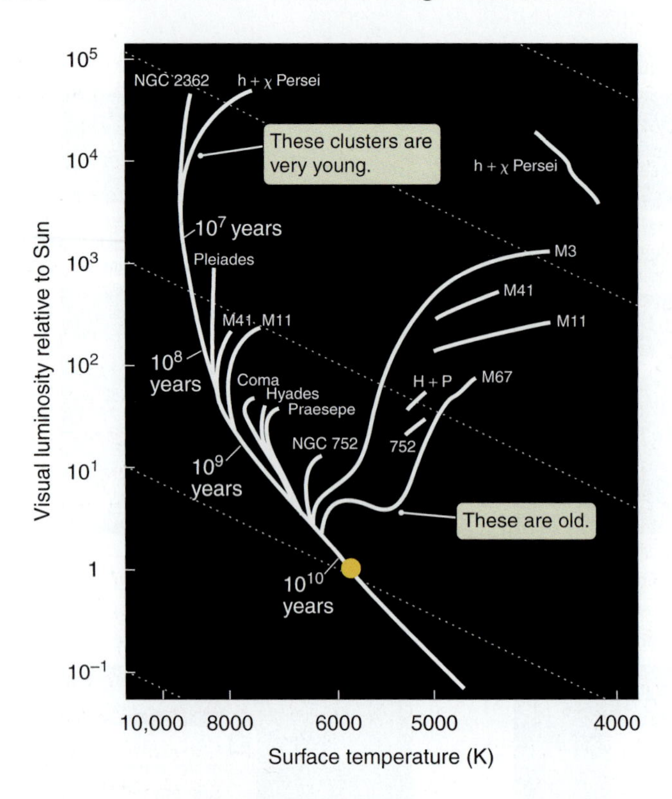

FIGURE 17.23 H-R diagrams for clusters having a range of different ages. The ages associated with the different main-sequence turnoffs are indicated.

age of the cluster. **Figure 17.23** traces the observed H-R diagrams for several real star clusters. Once you know what to look for, the difference between young and old clusters is obvious. NGC 2362 is clearly a young cluster. Its complement of massive, young stars shows it to be only a few million years old. In contrast, cluster M3 has a main-sequence turnoff that indicates its cluster age is about 11 billion years. When the H-R diagrams of open clusters are studied, a wide range of ages is observed. Some open clusters contain the short-lived O and B stars and are therefore very young. Other open clusters contain stars that are somewhat older than the Sun. But even the youngest globular clusters are several billion years older than the oldest open clusters. ▶▶ **NEBRASKA SIMULATION: H-R DIAGRAM STAR CLUSTER FITTING EXPLORER**

This understanding of stellar evolution applies even when groups of stars are so far away that individual stars cannot be seen. Although many fewer high-mass stars form in a cluster than do low-mass stars, higher-mass stars are *far* more luminous than lower-mass stars. Likewise, giant, evolved low-mass stars are far more luminous than less massive stars that remain on the main sequence. As a result, the most massive, most luminous stars that are present dominate the light from a star cluster. If the cluster is young, then most of the light comes from luminous hot, blue stars. If the cluster is old, then the light from the cluster has the color of red giants and relatively cool, low-mass stars.

Stars of Different Age and Different Chemical Composition

As always, general statements of the sort that we have made in this discussion come with caveats. Young clusters may contain very luminous red supergiants, and stars with lower abundances of massive elements in their atmospheres often look significantly bluer than their more chemically enriched counterparts. Even so, astronomers usually can figure out something about the properties of a group of stars from its overall color. The link between color and specific characteristics will be useful when we begin to discuss the much larger collections of stars called galaxies. A group of stars with similar ages and other shared characteristics is called a **stellar population**. An especially bluish color to a galaxy or a part of a galaxy often signifies that the galaxy contains a young stellar population that still includes hot, luminous, blue stars that formed recently. In contrast, a galaxy or part of a galaxy that has a reddish color is composed primarily of an old stellar population.

Colors reveal the ages of stellar populations.

As noted earlier, the evolution of stars depends on their chemical composition as well as their masses. Detailed comparisons between models and observed cluster H-R diagrams must account for the abundances of massive elements in the atmospheres of cluster stars, as well as the cluster age. In addition to grouping stars according to their ages, why focus on the chemical composition of stars? Recall that all elements more massive than boron are formed by nucleosynthesis in stars. For this reason, the abundance of massive elements in the interstellar medium provides a record of the cumulative amount of star formation that has taken place up to the present time. Gas that shows large abundances of massive elements has gone through a great deal of stellar

Abundances of massive elements record the cumulative history of star formation.

processing—so it contains more "recycled" material. Gas with low abundances of massive elements is more pristine.

Thus, the abundances of massive elements in the atmosphere of a star provide a snapshot of the chemical composition of the interstellar medium *at the time the star formed.* (In main-sequence stars, material from the core does not mix with material in the atmosphere, so the abundances of chemical elements inferred from the spectra of a star are the same as the abundances in the interstellar gas from which the star formed.) The chemical composition of a star's atmosphere reflects the cumulative amount of star formation that has occurred up to that moment. Stars in globular clusters contain only very small amounts of massive elements; some globular-cluster stars contain only 0.5 percent as much of these massive elements as the Sun does, indicating that they were among the earliest stars to form. Open clusters are younger and contain stars that formed from a more enriched interstellar medium, and therefore they have higher amounts of the more massive elements.

Young stars typically have higher massive-element abundances than do older stars.

Even the very oldest globular-cluster stars contain *some* amount of massive chemical elements. There must have been at least one generation of massive stars that lived and died, ejecting newly synthesized massive elements into space, before even the oldest globular clusters formed. Further, every star less massive than about 0.8 $M_{\odot}$ that ever formed is still around today. Note, though, that even a chemically rich star like the Sun, which is made of gas processed through approximately 9 billion years of previous generations of stars, is still composed of less than 2 percent massive elements. Luminous matter in the universe is still dominated by hydrogen and helium formed before the first stars. In upcoming chapters you will learn that these variations in the chemical content of stars indicate a lot about the chemical evolution of galaxies.

17.5 Origins: Seeding the Universe with New Chemical Elements

On Earth, massive elements are everywhere. The surfaces of rocky planets contain silicon, oxygen, magnesium, and sodium. The iron-and-nickel solid inner and liquid outer core of Earth are responsible for Earth's magnetic field. The most common chemical elements in biological molecules are carbon (C), hydrogen (H), nitrogen (N), oxygen (O), phosphorus (P), and sulfur (S)—all but hydrogen created in a dying star. The fact that these elements are here means that the Sun is not a first-generation star; it formed from material in the interstellar medium enriched by material from dying massive stars.

In supergiant, giant, and AGB stars, more massive chemical elements that formed from nuclear fusion deep within their interiors are carried upward and mixed with material in the outer parts of the star. As a star ages, its core grows hotter and hotter, and the temperature gradient within the star grows steeper. Under certain circumstances, convection can spread so deep into a star that chemical elements formed by nuclear burning within the star are dredged up and carried to the star's surface. For example, carbon stars (a type of AGB star) show an overabundance of carbon and other by-products of nuclear burning in their spectra. This extra carbon originated in each star's helium-burning shell and was carried to the surface by convection. For stars with lower masses, stellar winds and planetary nebulae carry the enriched outer layers off into interstellar space. The nuclear burning that occurs in supergiant stars goes well beyond the formation of elements such as carbon. Supernova explosions seed the universe with much more massive atoms, from iron and nickel up to uranium.

The oxygen atoms in the air you breathe and the water you drink were created by nucleosynthesis in dying stars. The iron atoms that are a key element of hemoglobin, which makes up the red blood cells that carry oxygen from your lungs to the rest of your body, formed in the cores of massive stars just before they exploded as supernovae. The nickel, copper, and zinc atoms in the coins in your pocket, and the rare earth atoms in your electronics, were created in massive stars. The Sun, the planets (including Earth), and all life on Earth are made of recycled stars. Supernovae are in you.

Our discussion of the death of massive stars in this chapter ended with neutron stars and pulsars. But the most massive stars can continue to collapse to the most extreme objects of all: black holes. To study these objects in the next chapter, we will go beyond Newtonian physics into the theories of Albert Einstein. ■

Summary

17.1 The CNO cycle burns hydrogen in massive stars. As these stars evolve, they burn and exhaust heavier elements in their core, creating concentric shells of progressive nuclear burning. If a high-mass star passes through the instability strip on the H-R diagram, it will become a pulsating variable star.

17.2 The chain of nuclear fusion consists of increasingly shorter stages of burning resulting in more massive elements, up to iron. High-mass stars eventually explode as Type II supernovae. Elements more massive than iron are created during the explosion.

17.3 Type II supernovae eject newly formed massive elements into interstellar space. Some leave behind neutron stars, which contain between 1.4 and 3 $M_{\odot}$ of neutron-degenerate matter packed into a sphere 10–20 km in diameter. Accretion of mass onto neutron stars produces X-rays in some binary systems. Pulsars are rapidly spinning, magnetized neutron stars.

17.4 Clusters are groups of stars that were born together and are all at about the same distance from Earth. H-R diagrams of clusters show stars leaving the main sequence in a progression from the highest-mass stars to the lowest-mass stars, confirming theories of stellar evolution.

17.5 The Sun, the Solar System, Earth, and all life on Earth contain heavy elements created inside of earlier generations of short-lived massive stars.

Unanswered Questions

- Blue straggler stars, found in clusters, are bluer and brighter than the stars at the main-sequence turnoff point. How do they fit into the picture of stellar evolution? They may have resulted from mass transfer in a binary pair or from the merger of two single or two binary stars, either of which could have resulted in a more massive star than what might be expected from the age of the cluster. Astronomers study the environments of these stars by estimating the likelihood of collisions and the number of binary systems, which may be different in clusters where the density of stars is high.
- What creates magnetars? There is a class of pulsars called magnetars, which are characterized by extremely large magnetic fields. These objects are observed to produce bursts of soft gamma rays. The origin of their huge magnetic fields is not well understood. These fields may originate from a dynamo in the interior of a superconducting region of the neutron star, but we do not know whether ordinary pulsars go through a magnetar phase.

Questions and Problems

Summary Self-Test

1. Why does the interior of an evolved high-mass star have layers like an onion?
   a. Heavier atoms sink to the bottom because stars are not solid.
   b. Before the star formed, heavier atoms accumulated in the centers of clouds, because of gravity.
   c. Heavier atoms fuse closer to the center, because the temperature and pressure are higher there.
   d. Different energy transport mechanisms occur at different densities.
2. True or false: When iron fuses into heavier elements, it produces energy.
3. Arrange the following elements in the order they burn inside the nucleus of a high-mass star during the star's evolution.
   a. helium c. oxygen e. hydrogen
   b. neon d. silicon f. carbon
4. Elements heavier than iron originated
   a. in the Big Bang.
   b. in the cores of low-mass stars.
   c. in the cores of high-mass stars.
   d. in the explosions of high-mass stars.
5. A pulsar pulses because
   a. its spin axis crosses Earth's line of sight.
   b. it spins.
   c. it has a strong magnetic field.
   d. its magnetic axis crosses Earth's line of sight.
6. True or false: A pulsar changes in brightness because its size pulsates.
7. Study Figure 17.3. If it were possible to watch a high-mass star move to the right, along one of these post-main-sequence lines, what would you observe happening to the star in terms of its color?
   a. It would become redder. c. It would become more yellow.
   b. It would become bluer. d. It would remain the same.
8. Study Figure 17.3. If it were possible to watch a high-mass star move to the right, along the topmost of these post-main-sequence lines, what would you observe happening to the star in terms of its size?
   a. It would become much larger. c. It would remain the same.
   b. It would become much smaller.
9. Study Figure 17.3. If it were possible to watch a high-mass star move along the topmost of these post-main-sequence lines,

what would you observe happening to the star in terms of its luminosity?
   a. It would become much more luminous.
   b. It would become much less luminous.
   c. It would remain about the same.

10. Study Figure 17.22. If the Sun were a member of a globular cluster, that cluster's H-R diagram would fall between
   a. (a) and (b)
   b. (b) and (c)
   c. (c) and (d)
   d. (d) and (e)
   e. (e) and (f)

True/False and Multiple Choice

11. **T/F:** The end result of the CNO cycle is that four hydrogen nuclei become one helium nucleus.
12. **T/F:** Stars in the instability strip are pulsating variable stars.
13. **T/F:** Electrons and protons can combine to become neutrons.
14. **T/F:** Uranium forms in the core of a star.
15. **T/F:** A supernova can be as bright as its entire host galaxy.
16. In a high-mass star, hydrogen fusion occurs via
   a. the proton-proton chain.
   b. the CNO cycle.
   c. gravitational collapse.
   d. spin-spin interaction.
17. The layers in a high-mass star occur roughly in order of
   a. atomic number.
   b. decay rate.
   c. atomic abundance.
   d. spin state.
18. Pulsations in a Cepheid variable star are controlled by
   a. the spin.
   b. the magnetic field.
   c. the ionization state of helium.
   d. the gravitational field.
19. Eta Carinae is an extreme example of
   a. a massive star.
   b. a rotating star.
   c. a magnetized star.
   d. a high-temperature star.
20. Iron fusion cannot support a star, because
   a. iron oxidizes too quickly.
   b. iron absorbs energy when it fuses.
   c. iron emits energy when it fuses.
   d. iron is not dense enough to hold up the layers.
21. The start of photo disintegration of iron in a star sets off a process that *always* results in a
   a. supernova.
   b. neutron star.
   c. black hole.
   d. pulsar.
22. Supernova remnants
   a. can be viewed at all wavelengths.
   b. can be viewed at only a few emission lines.
   c. are never seen in radio waves.
   d. have colors because the moving gas emits Doppler-shifted emission lines.
23. X-ray binaries are similar to another type of system you have studied:
   a. the Solar System.
   b. a system consisting of progenitors of Type Ia supernovae.
   c. a system consisting of progenitors of Type II supernovae.
   d. a system consisting of progenitors of planetary nebulae.
24. The Type II supernova that created the Crab Nebula was seen by Chinese and Arab astronomers in the year 1054 CE. The radius of the nebula is now 5.5 light-years, so the average expansion has been about:
   a. 15 km/s.
   b. 150 km/s.
   c. 1,500 km/s.
   d. 15,000 km/s.
25. Very young star clusters have main-sequence turnoffs
   a. nowhere. All the stars have already turned off in a young cluster.
   b. at the top left of the sequence.
   c. at the bottom right of the sequence.
   d. in the middle of the sequence.

Thinking about the Concepts

26. Explain the differences between the ways that hydrogen is converted to helium in a low-mass star (proton-proton chain) and in a high-mass star (CNO cycle). What is the catalyst in the CNO cycle, and how does it take part in the reaction?
27. How does a low-mass star begin burning helium in its core? What about a high-mass star? How are these processes different or similar?
28. Why does the core of a high-mass star not become degenerate, as the core of a low-mass star does?
29. List the two reasons why each post-helium-burning cycle for high-mass stars (carbon, neon, oxygen, silicon, and sulfur) becomes shorter than the preceding cycle.
30. Cepheids are highly luminous, variable stars in which the period of variability is directly related to luminosity. Why are Cepheids good indicators for determining stellar distances that lie beyond the limits of accurate parallax measurements?
31. Identify and explain two important ways in which supernovae influence the formation and evolution of new stars.
32. Is a Type II supernova an explosion or an implosion? Explain your answer.
33. Describe what an observer on Earth will witness when Eta Carinae explodes.
34. Recordings show that SN 1987A was detected by neutrinos on February 23, 1987. About 3 hours later it was detected in optical light. What was the reason for the time delay?
35. Why can the accretion disk around a neutron star release so much more energy than the accretion disk around a white dwarf, even though the two stars have approximately the same mass?
36. In Section 17.2, you learned that Type II supernovae blast material outward at 30,000 km/s. The material in the Crab Nebula (see Section 17.3) is expanding at only 1,500 km/s. What explains the difference?
37. An experienced astronomer can take one look at the H-R diagram of a star cluster and immediately estimate its age. How is this possible?
38. Explain how astronomers know that there was an even earlier generation of stars before the oldest observed stars.
39. Does the main-sequence turnoff indicate anything besides the age of a cluster? If so, what?

40. What is the *binding energy* of an atomic nucleus? How does this quantity help astronomers calculate the energy given off in nuclear fusion reactions?

Applying the Concepts

41. Study Figure 17.2. What fraction of the star is helium at time $t = 0$ and at time $t = 7$ million years?
42. Study Figure 17.4. Are the radius of the core and the radius of the star represented to scale in this figure? What fraction of the star's radius is the core's radius?
43. Study Figure 17.3. How much hotter, larger, and more luminous than the Sun is the uppermost main-sequence star on this H-R diagram?
44. Study Figure 17.6. When the velocity (red) and the force (blue) arrows point in the same direction, is the star's expansion accelerating or slowing down? When the velocity and force arrows point in opposite directions, is the star's expansion accelerating or slowing down?
45. Suppose you observe a classical Cepheid variable with a period of 10 days. What is the luminosity of this star? What other piece of information would you need to find out how far away this star is?
46. Suppose you had a random sample of universe in which you found one beryllium (Be) atom. Using the graph of chemical abundances in Figure 17.15, estimate roughly how many hydrogen atoms you would expect to find in your sample.
47. Follow Math Tools 17.2 to find the surface gravity on a neutron star with radius 10 km and mass 2.8 $M_\odot$.
48. Follow Math Tools 17.2 to find the escape velocity from a neutron star with radius 10 km and mass 2.8 $M_\odot$.
49. The Milky Way has about 50,000 stars of average mass (0.5 $M_\odot$) for every main-sequence star of 20 $M_\odot$. But 20-$M_\odot$ stars are about 10,000 times as luminous as the Sun, and 0.5-$M_\odot$ stars are only 0.08 times as luminous as the Sun.
    a. How much more luminous is a single massive star than the total luminosity of the 50,000 less massive stars?
    b. How much mass is in the lower-mass stars compared to the single high-mass star?
    c. Which stars—lower-mass or higher-mass stars—contain more mass in the galaxy, and which produce more light?
50. In a large outburst in 1841, the 120-$M_\odot$ star Eta Carinae was losing mass at the rate of 0.1 $M_\odot$ per year. Let's put that into perspective.
    a. The mass of the Sun is 2×10^{30} kg. How much mass (in kilograms) was Eta Carinae losing each minute?
    b. The mass of the Moon is 7.35×10^{22} kg. How does Eta Carinae's mass loss per minute compare with the mass of the Moon?
51. Using values given in Section 17.1, show that an O star can lose 20 percent of its mass during its main-sequence lifetime.
52. The approximate relationship between the luminosity and the period of Cepheid variables is L_{star} (in $L_\odot$) = 335 P (in days). Delta Cephei has a cycle period of 5.4 days and a parallax of 0.0033 arcsecond (arcsec). A more distant Cepheid variable appears 1/1,000 as bright as Delta Cephei and has a period of 54 days.
    a. How far away (in parsecs) is the more distant Cepheid variable?
    b. Could the distance of the more distant Cepheid variable be measured by parallax? Explain.
53. If the Crab Nebula has been expanding at an average velocity of 3,000 km/s since 1054 CE, what was its average radius in the year 2013? (Note: There are approximately 3×10^7 seconds in a year.)
54. Pulsars are rotating neutron stars. For a pulsar that rotates 30 times per second, at what radius in the pulsar's equatorial plane would a co-rotating satellite (rotating about the pulsar 30 times per second) have to be positioned to be moving at the speed of light? Compare this to the pulsar radius of 1 km.
55. Verify the claim in Section 17.3 that Earth would be roughly the size of a football stadium if it were as dense as a neutron star.

Using the Web

56. Go to the Chandra X-ray Observatory's "Variable Stars" Web page (http://chandra.harvard.edu/edu/formal/variable_stars/index.html#). Do the two exercises on Cepheid variable stars, which ask you to estimate their changes in brightness. You might want to look at Appendix 6 to review apparent magnitudes before you do the projects.
57. The International Astronomical Union's "List of Recent Supernovae" (http://cbat.eps.harvard.edu/lists/RecentSupernovae.html) includes all recently discovered supernovae. Pick a few of the most recent ones. What type of supernova is each one? How bright is it? Why are these so much fainter than the novae you looked at in Chapter 16? Are Type Ia or Type II supernovae more common?
58. The Palomar Transient Factory (PTF) survey (www.astro.caltech.edu/ptf), which looks at the same patch of sky every 5 days, is another automated experiment. What kinds of supernovae has this study found? Anything new?
59. Go to the website for the *Gaia* mission (http://esa.int/science/gaia). How will this mission contribute to the study of variable stars? How will it contribute to the study of novae and supernovae?
60. Go to the Einstein@Home website (http://einsteinathome.org). In this distributed computing project, volunteers use their spare computer processing power to help search for new pulsars. Look over the "News" section on the right. Have any pulsars been found lately? Join the project, create an account, download BOINC, and follow directions to look for pulsars.

STUDYSPACE is a free and open website that provides a Study Plan for each chapter of *21st Century Astronomy*. Study Plans include animations, reading outlines, vocabulary flashcards, and multiple-choice quizzes, plus links to premium content in SmartWork and the ebook. Visit **wwnorton.com/studyspace**.

SMARTWORK Norton's online homework system, includes algorithmically generated versions of these questions, plus additional conceptual exercises. If your instructor assigns questions in SmartWork, log in at **smartwork.wwnorton.com.**

Exploration | The CNO Cycle

Nuclear reactions usually involve many steps. In the Exploration for Chapter 14, you investigated the proton-proton chain. In this Exploration you will study the CNO cycle, which is even more complex. Visit StudySpace, and open the "CNO Cycle" interactive simulation in Chapter 17.

First, press "Play" and watch the animation all the way through. Press "Stop" to clear the screen, and then press "Play" again, allowing the animation to proceed past the first collision before pressing "Pause."

1 **Which atomic nuclei are involved in this first collision?**

....

2 **What color is used to represent the proton (hydrogen nucleus)?**

....

3 **What does the blue squiggle represent?**

....

4 **What atomic nucleus is created in the collision?**

....

5 **The resulting nucleus is not the same type of element as either of the two that entered the collision. Why not?**

....

Press "Play" again, and then press "Pause" as soon as the yellow ball and the dashed line appear.

6 **Is this a collision or a spontaneous decay?**

....

7 **What does the yellow ball represent?**

....

8 **What does the dashed line represent?**

....

9 **The resulting nucleus has the same number of nucleons (13), but it is a different element. What happened to the proton that was in the nitrogen nucleus but is not in the carbon nucleus?**

....

Proceed past the next two collisions, to "^{15}O."

10 **Study the pattern that's forming. When a blue ball comes in, what happens to the number of nucleons and the type of the nucleus (that is, what happens to the "12" and the "C," or the "14" and the "N")?**

....

11 **What is emitted in these collisions?**

....

Proceed until "^{15}N" appears.

12 **Is this a collision or a spontaneous decay?**

....

13 **Which previous reaction is this most like?**

....

Now proceed to the end of the animation.

14 **After the final collision, a line is drawn back to the beginning, telling you what type of nucleus the upper red ball represents. What is this nucleus?**

....

15 **How many nucleons are not accounted for by the upper red ball? (Hint: Don't forget the ^{1}H that came into the collision.) These nucleons must be in the nucleus represented by the bottom red ball.**

....

16 **Carbon has six protons. Nitrogen has seven. How many protons are in the nucleus represented by the bottom red ball?**

....

17 **How many neutrons are in the nucleus represented by the bottom red ball?**

....

18 **What element does the bottom red ball represent?**

....

19 **What is the net reaction of the CNO cycle? That is, which nuclei are combined and turned into the resulting nucleus?**

....

20 **Why is ^{12}C not considered to be part of the net reaction?**

....

An Einstein ring created by the gravity of a luminous red galaxy gravitationally lensing the light from a much more distant blue galaxy.

18 Relativity and Black Holes

Everything should be made as simple as possible, but not simpler.

Albert Einstein (1879–1955)

LEARNING GOALS

So far, our discussions of stars and planets have been based on Galilean and Newtonian views of space and gravity. The beginning of the 20th century brought about a revolution in understanding these fundamental concepts, which established the idea that space and time are inexorably linked. In this chapter we will move beyond Newtonian ideas of space and time. By the conclusion of this chapter, you should be able to:

- Describe why the speed of light is a universal constant.
- List the key implications and predictions of special and general relativity.
- Explain how space and time are related.
- Recognize that gravity is a consequence of the way mass distorts the very shape of spacetime.
- Describe the key features of stellar black holes.

18.1 Beyond Newtonian Physics

Let's return to the stellar remnants left behind by the evolution of the most massive stars. The neutron star does not represent the final extreme of stellar evolution; there is another fate that awaits some massive stars. The physics of a neutron star is much like the physics of a white dwarf, except that neutrons rather than electrons are what cause a neutron star to be degenerate. A white dwarf can have a mass of no more than about 1.4 Solar masses ($M_{\odot}$). This is the *Chandrasekhar limit*, discussed in Chapter 16. If the mass of the object exceeds this limit, then gravity is able to overcome electron degeneracy pressure, and the white dwarf will begin collapsing again.

Just as there is a Chandrasekhar limit for white dwarfs, there is an upper limit for the mass of neutron stars. If the mass of a neutron star exceeds about 3 $M_{\odot}$, gravity begins to win out over pressure once again. The neutron star grows smaller, and gravity at the star's surface becomes stronger and stronger at an ever-accelerating pace. In this case, however, there is no force in nature powerful enough to prevent gravity's final victory. The collapsing object quickly crosses a threshold where the escape velocity from its surface exceeds the speed of light, and not even light can escape its gravity. From this point on,

What are the origins of black holes?

Connections 18.1

Aberration of Starlight

It is extremely difficult to sense the motion of Earth around the Sun. The first direct measurement of the effect of Earth's motion was made in the 18th century. Imagine that you are sitting in a car in a windless rainstorm, as shown in **Figure 18.1**. If the car is sitting still and the rain is falling vertically, when you look out your side window you see raindrops falling straight down. If you were to hold a vertical tube out the window, raindrops would fall straight through the tube. When the car is moving forward, however, the situation is different. Between the time a raindrop appears at the top of your window and the time it disappears beneath the bottom of your window, the car has moved forward. The raindrop disappears beneath the window *behind* the point at which it appeared, which means the raindrop *looks as if* it falls at an angle, even though in reality it is falling straight down. For raindrops to fall directly through a tube held out the window now, you would have to tilt the top of the tube forward. As you go faster, the apparent front-to-back motion of the raindrops increases, and their apparent paths become more slanted. An observer by the side of the road would say the raindrops are coming from directly overhead, but to you in the moving car they are coming from a direction in front of the car (see Figure 18.1). You are observing this apparent motion of the raindrops from within your own unique frame of reference.

The same phenomenon occurs with starlight, as shown in **Figure 18.2**. The light from a distant star arrives at Earth from the direction of this star. Because Earth moves, however, to an observer on Earth the starlight seems to be coming from a slightly different direction, just as the raindrops appeared to be coming from in front of the moving car in Figure 18.1b. Because the direction of Earth's motion around the Sun continually changes during the year, the apparent position of a star in the sky moves in a small loop. Called the **aberration of starlight**, this shift in apparent position was

(a)

From the vantage point of a person outside, the rain falls vertically, even when the car is moving.

(b)

From the frame of reference of a person inside the car…

…the rain falls vertically if the car is stationary…

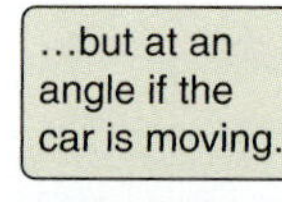

FIGURE 18.1 On a windless day the direction in which rain falls depends on the frame of reference in which it is viewed. (a) From outside the car, the rain is seen to fall vertically downward whether the car is stationary or moving. (b) From inside the car, the rain is seen to fall vertically downward if the car is stationary; but if the car is moving, the rain is seen to fall at an angle determined by the speed and direction of the car's motion.

nothing can escape from the collapsing object and find its way back to where it originated. The object is now a **black hole**.

A black hole will form if the stellar core left behind by a massive-star supernova exceeds about 3 $M_\odot$. Alternatively, a neutron star in a binary system will collapse to become a black hole if it accretes enough matter from its companion to push it over the 3-$M_\odot$ limit. Regardless of how it forms, any collapsed object with a mass greater than 3 $M_\odot$ must be a black hole. Black holes are so strange that the laws of Newtonian physics (explored in Chapter 4) are inadequate to describe them. To discuss them we must first look at how Einstein questioned intuitive assumptions about the very nature of space and time.

If a neutron star's mass exceeds about 3 $M_\odot$, it will collapse to form a black hole.

18.2 Special Relativity

The Galilean view of frames of reference led to many successes in explaining relative motion (see Chapter 2). One such example of relative motion has the effect of changing the apparent position of stars as Earth orbits the Sun—a phenomenon called the *aberration of starlight,* discussed in **Connections 18.1**. As relative velocities approach the speed of light, however, the story becomes quite different.

first detected in the 1720s by two astronomers—Samuel Molyneux and James Bradley. Measurement of the aberration of starlight shows that Earth moves on a roughly (but not exactly) circular path about the Sun with an average speed of just under 30 kilometers per second (km/s). Because distance equals speed multiplied by time, the distance around this near-circle—its circumference—is the speed of Earth (29.8 km/s) multiplied by the length of one year (3.16×10^7 seconds), which equals 9.42×10^8 km.

From the circumference of Earth's nearly circular orbit we can find the radius, which is 1.50×10^8 km (150 million km). Recall that this is the value of the astronomical unit (AU). Modern measurements of the size of the astronomical unit are made in very different ways, such as by bouncing radar signals off Venus and measuring the time it takes for the photons to return to Earth. However, the aberration of starlight provided a simple and compelling demonstration that Earth orbits about the Sun—and a pretty good value for the size of Earth's orbit as well.

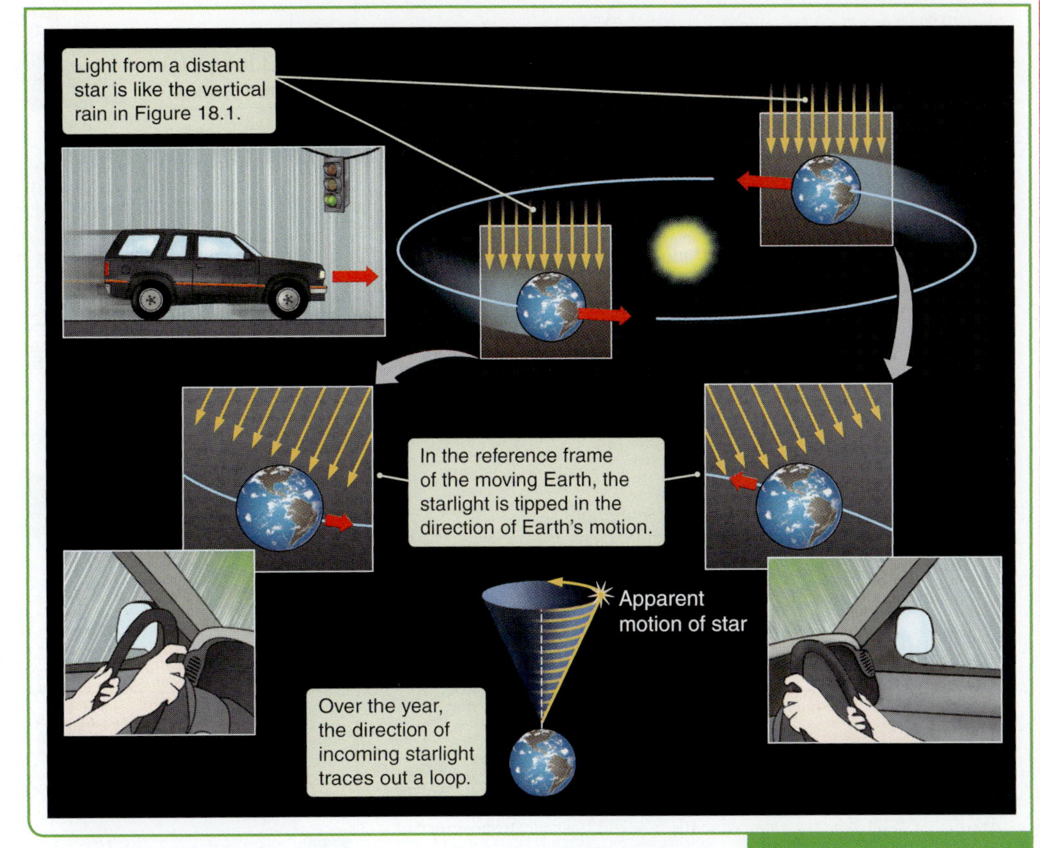

FIGURE 18.2 The apparent positions of stars are deflected slightly toward the direction in which Earth is moving. As Earth orbits the Sun, stars appear to trace out small loops in the sky. This effect is called *aberration of starlight*.

A moving car is a useful example of a moving frame of reference (as in Figure 2.3). Imagine that you are sitting in a moving car and there is a ball on the seat beside you. In your frame of reference the ball is at rest. But if the car is moving at 50 kilometers per hour (km/h) down the highway, someone standing by the road will say that the ball is moving at 50 km/h. To someone in oncoming traffic moving at 50 km/h, the relative speed of both your car and the ball is 100 km/h. There really is no difference among these three perspectives. The laws of physics are the same in *any* **inertial frame of reference** (a frame of reference in which observers feel no net force).

As a variant of the ball-in-the-car scenario, imagine two cars approaching one another, each traveling at 50 km/h, as in **Figure 18.3a**. As your car (the red car) moves down the highway at 50 km/h, you pitch a ball forward at 100 km/h. In *your* frame of reference (top panel of Figure 18.3a), you are stationary, the oncoming (green) car is approaching you at 100 km/h, an observer standing still by the side of

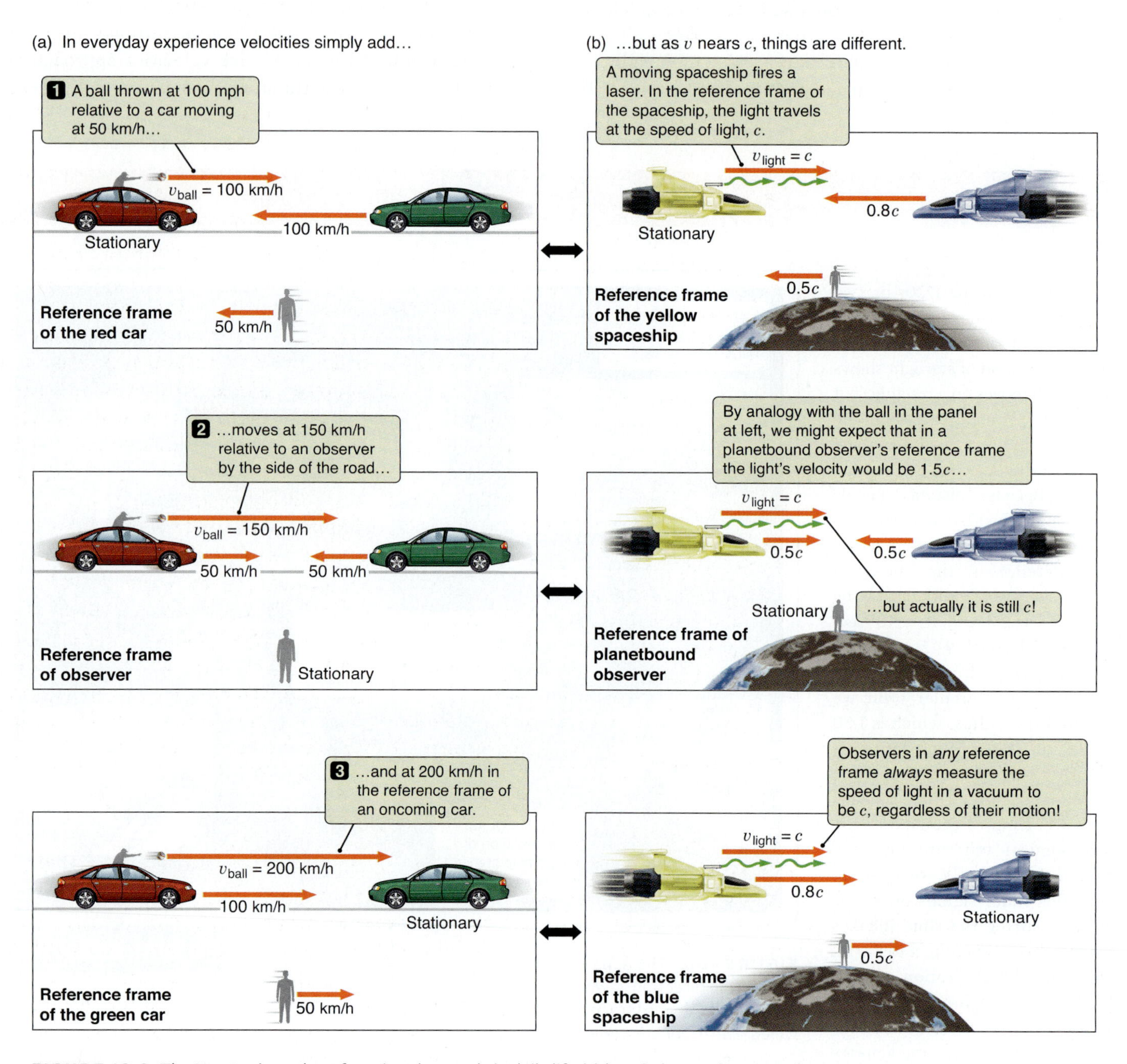

FIGURE 18.3 The Newtonian rules of motion that apply in daily life (a) break down when speeds approach the speed of light. (b) The fact that light itself always travels at the same speed for any observer is the basis of special relativity. (Note that relativity also affects the relative speeds of the two spacecraft.)

the road is moving toward you at 50 km/h, and the ball is moving at 100 km/h. Now consider the frame of reference of the observer (middle panel of Figure 18.3a). Both cars are approaching at 50 km/h, and the ball is moving at 150 km/h. (The ball has the original 50-km/h speed of your car plus the additional 100 km/h that you gave it with your throw.) In the frame of reference of the oncoming green car (bottom panel of Figure 18.3a) traveling at 50 km/h, the ball is moving at 200 km/h. (This is the 150 km/h that the ball is moving relative to the ground plus the 50-km/h motion of the oncoming car.) In everyday experience, velocities simply add. This is also how Newton's laws say the universe should behave.

The Speed of Light Is a Special Value

During the closing years of the 19th century and the early years of the 20th century, physicists conducted laboratory experiments in an attempt to better understand light. What they found puzzled them greatly. Rather than the speed of a beam of light differing from one observer to the next, as expected on the basis of Newton's physics and "common sense," they found instead that *all observers measure exactly the same value for the speed of a beam of light, regardless of the motion of their frame of reference.*

Instead of cars traveling at 50 km/h, now imagine two spacecraft approaching each other, both traveling at half the speed of light (0.5*c*) as shown in **Figure 18.3b**. You are in the yellow spaceship traveling at 0.5*c* and, instead of throwing a baseball, you shine a beam of light forward. If you replace the 100-km/h speed of the ball in the previous example with *c*, in your frame of reference (top panel of Figure 18.3b) you might expect that you would be stationary, the oncoming (blue) spacecraft would be approaching at 1.0*c*, an observer on a nearby planet would be moving toward you at 0.5*c*, and the beam of light would be moving away from you at the speed of light, *c*. In the frame of reference of an observer on a nearby planet (middle panel of Figure 18.3b), the light should travel by at a speed equal to the speed of your spacecraft plus the speed of light, or 1.5*c*. Similarly, in the reference frame of the oncoming blue spacecraft (bottom panel of Figure 18.3b) traveling at 0.5*c*, the light should appear to travel at a speed of 2.0*c*. (The 1.5*c* that the light is moving relative to the nearby planet plus the 0.5*c* motion of the oncoming spacecraft.) But something is wrong here. The beam of light appears to be traveling at 1.5*c* and 2.0*c* relative to, respectively, observers on the nearby planet and the approaching blue spacecraft, and the two spacecraft appear to be approaching one another at the speed of light. You already learned in Chapter 5 that the speed of light has a constant value, *c*; and, as you will see later in this chapter, objects cannot move at a speed equal to or greater than the speed of light. In the relativistic world, it turns out, speeds do *not* simply add. The relative speed between the two spacecraft in the top and bottom panels of Figure 18.3b ($0.5c + 0.5c$) adds to 0.8*c*, not 1.0*c*! At relativistic speeds, everyday experience no longer holds true.

As you ride in your spaceship, you measure the speed of the beam of light to be *c*, or 3×10^8 meters per second (m/s). That is as expected because you are holding the source of the light. But the observer on the planet *also* measures the speed of the passing beam of light to be 3×10^8 m/s. Even the passenger in the oncoming spacecraft finds that the beam from your light is traveling at exactly *c* in her own frame of reference. *Every observer always finds that light in a vacuum travels at exactly the same speed,* c, *regardless of his or her own motion or the motion of the light source.*

Experiments show that the speed of light is the same for all observers.

Albert Einstein's first scientific paper, written when he was a 16-year-old student, was about traveling along with a light wave, and he continued to think about this topic for the next decade as he earned a doctorate in physics. Light travels in a straight line at a constant speed. Einstein reasoned that according to Newton's laws of motion, there should be an inertial frame of reference that moves along with the light and in which the light is stationary. That is, it should be possible to "keep up" with light so that you are moving right along with it. But if that were possible, then the light would be an oscillating electric and magnetic wave that does not move. This was impossible according to James Clerk Maxwell's equations for electromagnetic waves (discussed in Chapter 5). There was a contradiction here. Either Maxwell was wrong in his understanding of electricity and magnetism, or Newtonian physics did not apply at the very large velocities traveled by light. As the experimental results came in on measurements of the speed of light, it became clear that it was Newtonian physics, and not Maxwell's equations, that needed revision.

Time Is Relative

Einstein resolved the contradiction between Maxwell and Newton and ushered in a scientific revolution with his **special theory of relativity**, published in 1905. Special relativity was Einstein's answer to the question, What must the universe be like if every observer always measures the same value for the speed of light in a vacuum? Einstein focused his thinking on pairs of *events*. In relativity, an **event** is something that happens at a particular location in space at a particular time. Snapping your fingers is an event, because this action has both a time and a place. Everyday experience indicates that the distance between any two events depends on the frame of reference of the person observing them. Suppose you are sitting in a car that is traveling down the highway in a straight line at a constant 60 km/h. You snap your fingers (event 1), and a minute later you snap your fingers again (event 2). In your frame of reference *you* are stationary and the two events happened at exactly the same place. They are separated by a minute in *time*, but there is no separation between the two events in *space*. This is very different from what happens in the frame of reference of an observer sitting by the road. This observer agrees that the second snap of your fingers (event 2) occurred a minute after the first snap of your fingers (event 1), but to this observer the two events were separated from each other in space by a kilometer, the distance your car traveled in the minute. In this Newtonian view, the *distance* between two events depends on the motion of the observer, but the *time* between the two events does not.

Special relativity describes the relationship between events in space and time.

Einstein questioned the distinction between the way Newton treated space and the way Newton treated time. Einstein realized that the *only* way the speed of light can be the same for all observers is if *the passage of time is different from one observer to the next.* To Newton, and in everyday life, the march of time seems immutable and constant. But in reality, the only thing that is truly constant is the speed of light, and even time itself flows differently for different observers.

This is the heart of Einstein's special theory of relativity. In the everyday Newtonian view of the world, you live in a three-dimensional space through which time marches steadily onward. Events occur in space at a certain time. By the time Einstein finished working out the implications of his insight, he had reshaped this three-dimensional universe into a four-dimensional **spacetime**. Events occur at specific locations within this four-dimensional spacetime, but how this spacetime translates into what you perceive as "space" and what you perceive as "time" depends on your frame of reference.

Einstein did not "disprove" Newtonian physics; we were not wasting our time in Chapter 4 when we discussed Newton's laws of motion. Instead, Einstein found that Newtonian physics is contained within special relativity. In everyday experience you could never encounter speeds that approach the speed of light. Even the fastest human-made object (the *Helios II* spacecraft) traveled at only about 0.0002*c*. At speeds much less than the speed of light, Einstein's equations become the same equations that describe Newtonian physics. In your everyday life you experience a Newtonian world. Only when relative velocities approach that of light do things begin to depart measurably from the predictions of Newtonian physics. When great velocities cause an effect different from what Newtonian physics predicts, this is called a **relativistic** effect (discussed further in **Math Tools 18.1**).

Newtonian physics is contained within special relativity.

The Implications of Relativity Are Far-Ranging

Special relativity is a useful case study of how science works. Newton's laws had proven for a long time to be an extraordinarily powerful way of viewing the world; however, as scientists studied a different phenomenon—light—difficulties arose. Newton's theory of motion, Maxwell's theory of electromagnetic radiation, and empirical measurements of the speed of light conflicted in some aspects. Einstein was able to step in and reconcile this conflict, and in the process he changed scientific ideas about the universe. Einstein's ideas remained controversial well into the 20th century, and his 1921 Nobel Prize in Physics was awarded for his work on the photoelectric effect (see Chapter 5) and not for his work on relativity. But as one experiment after another confirmed the strange and counterintuitive predictions of relativity, scientists came to accept its validity.

Today, special relativity is an integral and indispensable part of all physics, shaping ideas about the motions of both the tiniest subatomic particles and the most distant galaxies. Here we discuss only a few of the essential insights that come from Einstein's work.

1. **Mass and energy are two manifestations of the same thing.** The energy of an object depends on its speed; the faster it moves, the more energy it has. But Einstein's famous equation $E = mc^2$ says that even a *stationary* object has an intrinsic "rest" energy that equals the mass (m) of the object multiplied by the speed of light (c) squared. The speed of light is a very large number. This relationship between mass and energy says that a single tablespoon of water has a rest energy equal to the energy released in the explosion of over 300,000 tons of TNT. All nuclear reactions that produce energy do so by converting some of the mass of the reactants into other forms of energy. But even the most efficient chemical or nuclear reactions release only tiny fractions of the total energy available. Exploding TNT, for example, converts less than a trillionth of its mass into energy. Even the explosion of a hydrogen bomb releases far less than 1 percent of the energy contained in the mass of the bomb. And as we discussed in the earlier chapters on stars, the

Math Tools 18.1

The Boxcar Experiment

The special theory of relativity is a *very* counterintuitive idea, but it is central to a modern scientific understanding of the universe. To measure time, the first thing we need is a clock. Consider the thought experiment known as the boxcar experiment. In this experiment, observer 1 is in a boxcar of a train moving to the right. Observer 1 has a lamp, a mirror, and a clock. Observer 2 is sitting on the ground outside. The clock is based on a value that everyone can agree on—such as the speed of light.

Figure 18.4a shows the experimental setup as seen by observer 1, who is stationary with respect to the clock. At time t_1 event 1 happens: the lamp gives off a pulse of light. The light bounces off a mirror at a distance l meters away, and then heads back toward its source. At time t_2 event 2 happens: the light arrives at the clock and is recorded by a photon detector. The time between events 1 and 2 is just the distance the light travels ($2l$ meters), divided by the speed of light: $t_2 - t_1 = 2l/c$.

Now let's look from the perspective of observer 2 sitting on the ground outside the train, in a frame of reference in which the clock is moving (**Figure 18.4b**). In *this* observer's reference frame, he is stationary and the boxcar is moving at speed v. (Recall that because any inertial frame of reference is as good as any other, this observer's perspective is as valid as the first observer's perspective.) In observer 2's reference frame, the clock *moves* to the right between the two events, so the light has *farther to go* because of the horizontal distance. (If you do not see this right away, use a ruler to measure the total length of the light path in Figure 18.4b and compare it with the total length of the light path in Figure 18.4a.) The time between the two events is still the distance traveled divided by the speed of light, but now that distance is *longer* than $2l$ meters. Because the speed of light is the same for all observers, the time between the two events must be longer as well.

The two events are the *same two events*, regardless of the frame of reference from which they are observed. The question is, how much time passed between the two events? Because the speed of light is the same for all observers, there must be more time between the two events when viewed by observer 2 from a frame of reference in which the clock is moving. It takes a moving clock more time than a stationary clock to complete one "tick." The seconds of a moving clock are stretched. Moving clocks must run slowly, and the passage of time must depend on an observer's frame of reference. Because both frames of reference are equally good places to do physics, both time measurements are equally valid, even though they differ from one another.

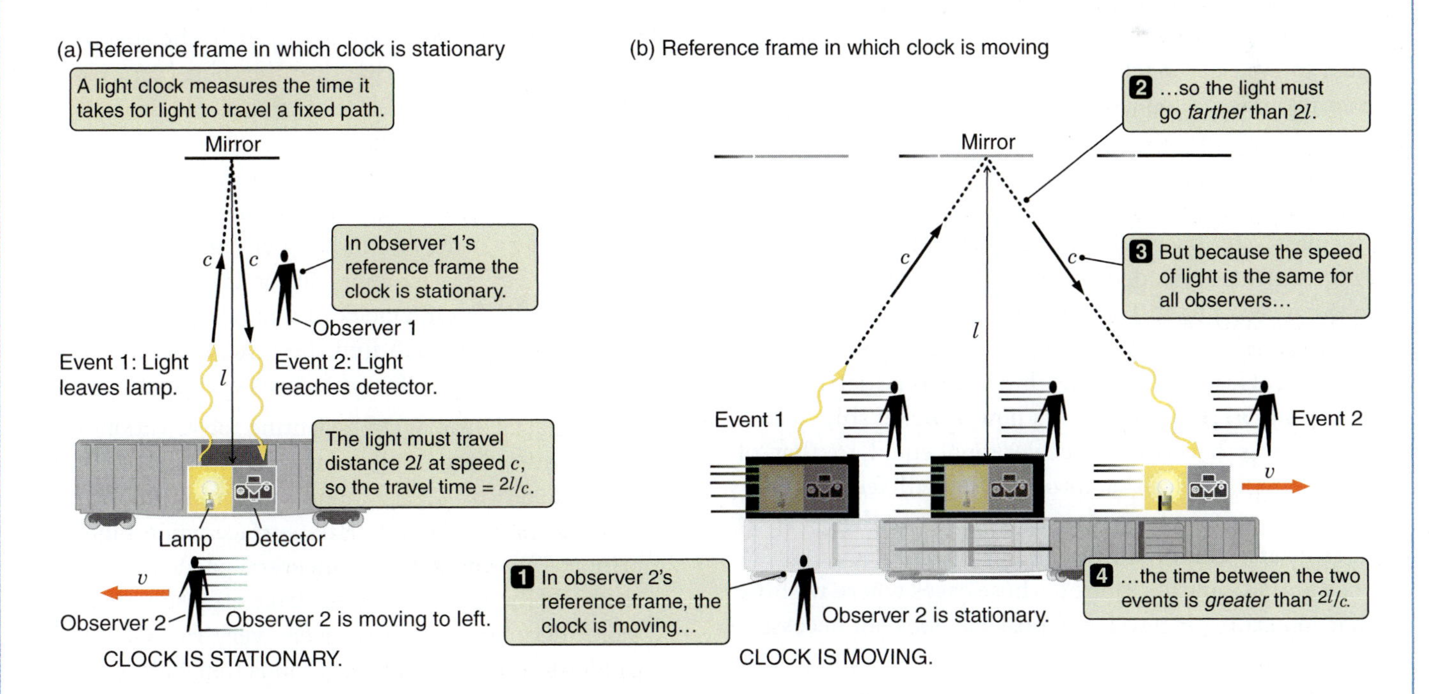

FIGURE 18.4 The "tick" of a light clock as seen in two different reference frames: stationary (in the boxcar) (a) and moving (observer 2 on the tracks) (b). As Einstein's thought experiment demonstrates, if the speed of light is the same for every observer, then moving clocks *must* run slowly compared to stationary clocks.

thermonuclear energy released in the cores of stars originates in the conversion of some of the mass of atomic nuclei into energy.

The equivalence between mass and energy points both ways. In Chapter 3 we defined *mass* as the property of matter that resists changes in motion. The energy of an object increases this resistance. Even adding to the energy of motion of an object increases its inertia. For example, a proton in a high-energy particle accelerator may approach the speed of light so closely that its total energy is 1,000 times greater than its rest energy. Such an energetic proton is, indeed, harder to "push around" (in other words, it has more inertia) than a proton at rest. It also more strongly attracts other masses through gravity.

2. **The speed of light is the ultimate speed limit.** We already discussed the insight that led Einstein to relativity in the first place. If it were possible to travel at the speed of light, then in that frame of reference light would cease to be a traveling wave, and all of the laws of physics would fall apart. You can also think about this limit in terms of the equivalence of mass and energy just discussed. As the speed of an object gets closer and closer to the speed of light, the energy of that object, and therefore its mass, become greater and greater, so it becomes increasingly resistant to further changes in its motion.

The situation is like trying to get from 0 to 1 by halving the remaining distance again and again. The resulting sequence—0, $\frac{1}{2}$, $\frac{3}{4}$, $\frac{7}{8}$, $\frac{15}{16}$, $\frac{31}{32}$, $\frac{63}{64}$, . . .—gets arbitrarily close to 1 but never actually reaches it. In the same way, a continuous force applied to an object will cause its velocity to get closer and closer to the speed of light, but it will never actually reach the speed of light. It would take an *infinite* amount of energy to accelerate an object with a nonzero rest mass to the speed of light. In short, all the energy in the entire universe is inadequate to accelerate a single electron to the speed of light. The electron can get arbitrarily close to that number—$0.99999999999999999999999\ldots \times c$ is possible, at least in principle—but there is no getting over the hump. **Figure 18.5** shows how a rocket ship, which experiences a constant acceleration equal to that of gravity on Earth (so that its occupants will feel at home), moves faster and faster but never reaches the speed of light. Faster-than-light travel may be a mainstay of science fiction, but this is one of those cases where wishing that something is physically possible does not necessarily make it so.

3. **"At the same time" is a relative concept.** Two events that occur at the same time for one observer may occur at different times for another observer. Hold out your arms and snap the fingers on both hands at the same time. For you, the two snaps were simultaneous. But to an observer moving by you from your left to your right at nearly the speed of light, you snapped the fingers of your left hand first and the fingers of your right hand later.

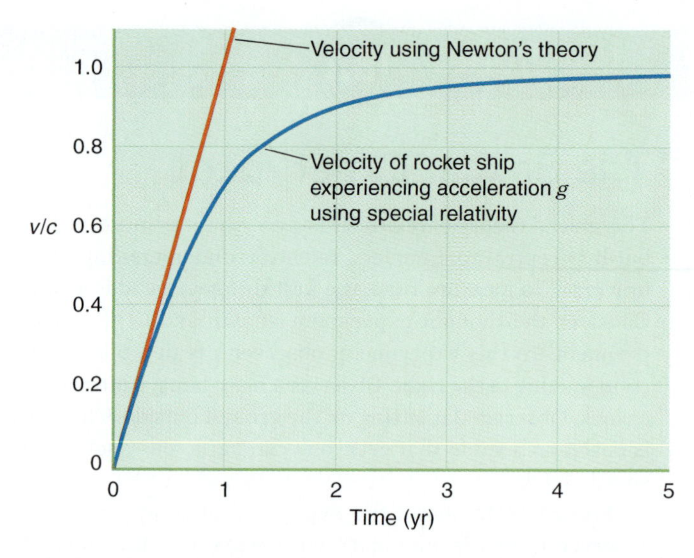

FIGURE 18.5 The speed of a rocket ship experiencing an acceleration equal to Earth's acceleration of gravity (g). The rocket ship approaches the speed of light but never gets there.

4. **Time passes more slowly in a moving reference frame.** This phenomenon is referred to as **time dilation** because time is "stretched out" in the moving reference frame. Were you to compare your clock with that of another observer moving at $\frac{9}{10}$ the speed of light ($0.9c$), you would find that the other observer's clock was running 0.44 times as fast as your clock. From this, you might guess that to the observer your clock would be fast, but actually the other observer would find instead that it is *your* clock that was running slowly. To you, the other observer may be moving at $0.9c$, but to the other observer, *you* are moving. Either frame of reference is equally valid, so if a clock in a moving reference frame runs slowly, you each find the other's clock to be slow. This time dilation effect increases with speed.

A straightforward scientific observation demonstrates this effect. As illustrated in **Figure 18.6**, fast particles called *cosmic-ray muons* are produced at a height of about 15 km above Earth's surface when high-energy primary **cosmic rays**—elementary particles accelerated to nearly the speed of light—strike atmospheric atoms or molecules. Muons at rest decay very rapidly into other particles. Within 2.2 microseconds (μs), half of all muons will have decayed into other particles. This decay happens so quickly that, even if they *could* move at the speed of light, virtually all muons would have decayed long before traveling the 15 km to reach Earth's surface. However, the time dilation effect causes the muons' clocks to run slower,

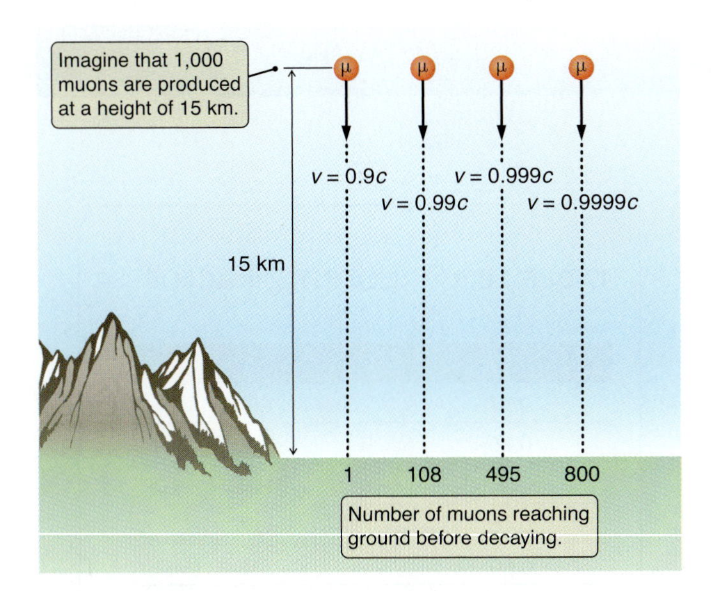

FIGURE 18.6 Muons created by cosmic rays high in Earth's atmosphere would decay long before reaching the ground if they were not traveling at nearly the speed of light. An observer on the ground sees a longer muon lifetime because of time dilation. Here we show what happens to four different sets of 1,000 muons produced at an altitude of 15 km for a variety of speeds.

so the particles live longer and can travel farther. That is why cosmic-ray muons can be detected on the ground. In fact, the faster muons move, the slower their clocks run and the more of them are able to reach the ground.

The same general principle is observed in particle accelerators, where particles that are traveling at speeds near the speed of light live longer before decaying.

5. **An object is shorter in motion than it is at rest.** More specifically, moving objects are compressed in the direction of their motion. A meter stick moving at 0.9*c* is only 43.6 centimeters (cm)—less than half a meter—long. This also explains the muon experiment we just discussed from the perspective of the muons themselves. In the frame of reference of the fast-moving muon produced at a height of 15 km, Earth's atmosphere appears to be much shorter than 15 km; indeed, it is so compressed, from the muon's perspective, that the muon may be able to reach the ground before decaying. This effect, called length contraction, also increases with speed.

These consequences of relativity can be combined in what is often called the *twin paradox*. Suppose you take a trip into space. Your spectacularly powerful star drive accelerates your ship to nearly the speed of light. To you, the distance you travel will seem shorter than the distance as measured on Earth. After you arrive at your destination, you take a picture and then turn around to head home to Earth, again traveling at a speed close to *c*. To your twin on Earth, who is in the reference frame of Earth, you were in a moving reference frame, so clocks (including biological clocks) ran much slower, and you should return younger than your twin. However, from your perspective your twin is the one who traveled at just under the speed of light, and your twin's biological clock (and all clocks on Earth), ran slowly compared to yours, so your twin has aged much less. Both twins cannot be correct, and that is the nature of the paradox, which at first glance appears quite puzzling.

The twin paradox illustrates many aspects of relativity.

Both on the way out into space and on the way back, in your reference frame it is the clocks on Earth that are running slowly, so you are aging *faster* than your twin. Yet when you return, more time has passed for your twin on Earth than for you, and your twin has aged more than you have. How can this be? The answer is that, unlike your twin, you *changed reference frames* during your trip. Event 1 is when you left Earth, and event 2 is when you returned to Earth. Your twin went from one event to the other while riding along in Earth's frame of reference. You, however, changed reference frames when you left Earth, changed again when you stopped at your destination, changed a third time when you left your destination to return home, and changed reference frames one final time when you arrived back at Earth.

Another way to view this is that the key difference between you and your twin is that you experienced acceleration during your trip, while your twin did not. When two observers are in *uniform motion* relative to one another, *neither* of them can lay claim to being in a unique frame of reference. However, *acceleration is a real phenomenon*. You feel acceleration when you are riding in a car, and you would surely feel the acceleration of the spaceship in this example. It is the fact that you experienced an acceleration that enabled you to be younger than your Earth-bound twin. The twin paradox is explored further in **Math Tools 18.2**.

Space Travel

Imaginary travel in spaceships is one thing, but what is the current reality of human travel in space? Some American astronauts have been to the Moon and back, and a number of robotic spacecraft have been sent to explore objects throughout the Solar System. But how realistic would it be for humans to visit and explore other planetary systems in the Milky Way Galaxy? Traveling to other planetary systems turns out to be very difficult because of the constraints of energy and the ultimate speed limit of light.

Math Tools 18.2

The Twin Paradox

The physicist Hendrik Lorentz (1853–1928) derived the equation for how much time is dilated and how much space is contracted when something is traveling at velocities near the speed of light. This *Lorentz factor* (abbreviated γ) is given by:

$$\gamma = \frac{1}{\sqrt{1 - \frac{v^2}{c^2}}}$$

Figure 18.7 shows a plot of this Lorentz factor versus velocity, and **Table 18.1** shows the calculated value of γ for different values of velocities v/c. You can see that for something moving at half the speed of light the Lorentz factor is 1.15. But for something moving at 90 percent of c the factor is 2.3, and it goes up quickly from there, becoming arbitrarily large as the velocity approaches but never quite reaches the speed of light.

Let's conduct a thought experiment of how this would work for the twin paradox. Suppose you could take a trip to the star 82 G. Eridani at a distance of 20 light-years in order to study the "super-Earth" planets in its orbit. Your spaceship accelerates to a velocity of 0.995c. To you, moving through space at this speed, the distance of 20 light-years is compressed by the Lorentz factor, which Table 18.1 indicates is equal to 10:

$$d_{\text{moving}} = \frac{d_{\text{rest}}}{\gamma} = \frac{20 \text{ ly}}{10} = 2 \text{ ly}$$

To you on the spaceship, it would seem as if the distance were 2 light-years. At your speed, you would cross this distance in slightly more than 2 years. You take a picture of the planets orbiting the star 82 G. Eridani and then head home. Again, the return trip would take 2 years. So you would have traveled to this star and back again in slightly more than 4 years. When you return, how much time would have passed on Earth? The equation for time dilation also uses the Lorentz factor, but here we will solve it for the time passing on Earth:

$$t_{\text{Earth}} = \gamma \times t_{\text{moving}} = 10 \times 4 \text{ yr} = 40 \text{ yr}$$

Four years would have passed for you in the spaceship, but *40* years would have passed on Earth. Your twin on Earth would be 36 years older than you are!

TABLE 18.1 | Lorentz Factor

v/c	γ
0.10	1.005
0.20	1.02
0.30	1.05
0.40	1.09
0.50	1.15
0.60	1.25
0.70	1.40
0.80	1.67
0.90	2.29
0.95	3.20
0.99	7.09
0.995	10.01
0.999	22.37
0.9999	70.71

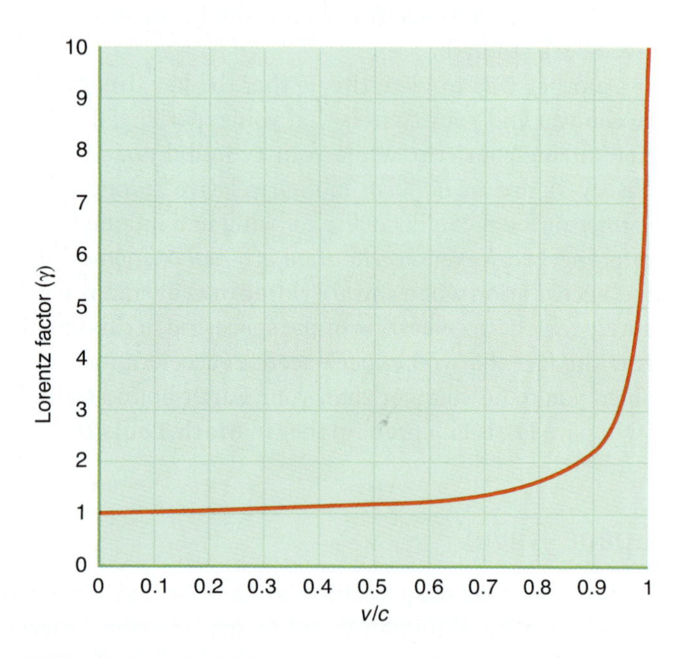

FIGURE 18.7 The Lorentz factor γ plotted against v/c. This factor doesn't become significant until velocities are about 50 percent the speed of light.

With current technology, engineers can construct rockets able to travel at speeds of up to 20,000 m/s. At such a speed, a one-way trip to Earth's nearest neighbor star, Proxima Centauri, at a distance of 4.2 light-years, would take well over 50,000 years. Travel to more distant stars would take even longer. It would take thousands of human lifetimes to make a round-trip to even the closest neighboring star, Proxima Centauri, using current technology.

In principle, travel just under the cosmic speed limit c is possible. The ability to move at this speed would provide two advantages: one could travel nearly as fast as is theoretically possible, and one could take advantage of the relativistic time dilation to make such adventures well within the lifetime of a space traveler. In the example of the twin paradox from Math Tools 18.2, an astronaut could travel to the planets orbiting 82 G. Eridani and back in just 4 years (in the astronaut's frame of reference) by traveling at $0.995c$. Or an astronaut could travel to the center of the Milky Way Galaxy and back in just 2 years, by traveling at a speed of $0.9999999992c$. Although this is theoretically possible, the energy needs are enormous. If M is the mass of the astronauts, rocket ship, and fuel, it would take γMc^2 of energy just to accelerate the rocket up to such a high speed, or $10Mc^2$ in the first example to 82 G. Eridani, and $25{,}000Mc^2$ in the second example to the center of the Milky Way. Such enormous amounts of energy may never be attainable in a spaceship that can sustain life. So, while not theoretically impossible, visits to other stars in our galaxy will not take place anytime soon.

18.3 Gravity Is a Distortion of Spacetime

Our exploration of special relativity began with the observation that the speed of light is always the same (regardless of the motion of an observer or the source), and ended by shattering everyday notions of space and time. As we discuss the properties of black holes—indeed, of all massive objects in the universe—the concepts of space and time will be pulled even further from the absolutes of Newtonian physics. First you learned that what are traditionally called (three-dimensional) space and time are actually just a result of a particular, limited perspective on a four-dimensional spacetime that is different for each observer. Next you will learn that this four-dimensional spacetime is warped and distorted by the masses it contains. One of the consequences of this deformation is the gravity that holds you to Earth. The **general theory of relativity** describes how mass distorts spacetime. This theory is another of Einstein's great contributions to science.

Mass warps the fabric of spacetime.

A crucial clue to the fundamental connection between gravity and spacetime was noted in Chapter 4, where we showed that the *inertial mass* of an object—the mass appearing in Newton's $F = ma$—is *exactly* the same as the object's gravitational mass. Another clue is that, left on their own, any two objects at the same location and moving with the same velocity will follow the same path through spacetime, regardless of their masses. The space shuttle astronaut falls around Earth, moving in lockstep with the space shuttle itself. A feather dropped by an *Apollo* astronaut standing on the Moon falls toward the surface of the Moon at exactly the same rate as a dropped hammer does. Rather than thinking of gravity as a "force" that "acts on" objects, it is more accurate to think of it as a consequence of the warping of spacetime in the presence of a mass. *Gravitation is the result of the shape of spacetime that objects move through.* This is one of the key insights of general relativity.

Inertial mass and gravitational mass are the same.

Free Fall Is the Same as Free Float

The essence of special relativity is that any inertial reference frame is as good as any other. No experiment can distinguish between sitting in an enclosed spaceship floating stationary in deep space (**Figure 18.8a**) and sitting in an enclosed spaceship traveling at 0.9999 times the speed of light (**Figure 18.8b**). You cannot tell any difference between these two cases—they do not feel any different—because there *is* no difference between them. Each of these reference frames is an equally valid inertial reference frame. As long as nothing acts to *change* the motion—that is, nothing accelerates either spaceship—the laws of physics are exactly the same inside both spacecraft.

General relativity begins by the application of this same idea to an astronaut inside a space shuttle orbiting Earth (**Figure 18.8c**). As long as we restrict our discussion to a small enough volume of space and a short enough period of time that we can ignore changes in the strength and direction of gravity from place to place, the astronaut again has no way to tell the difference between being inside the space shuttle as it falls around Earth and being inside a spaceship coasting through interstellar space. If you close your eyes and jump off a diving board, for the brief time that you are falling

A freely falling object defines an inertial reference frame.

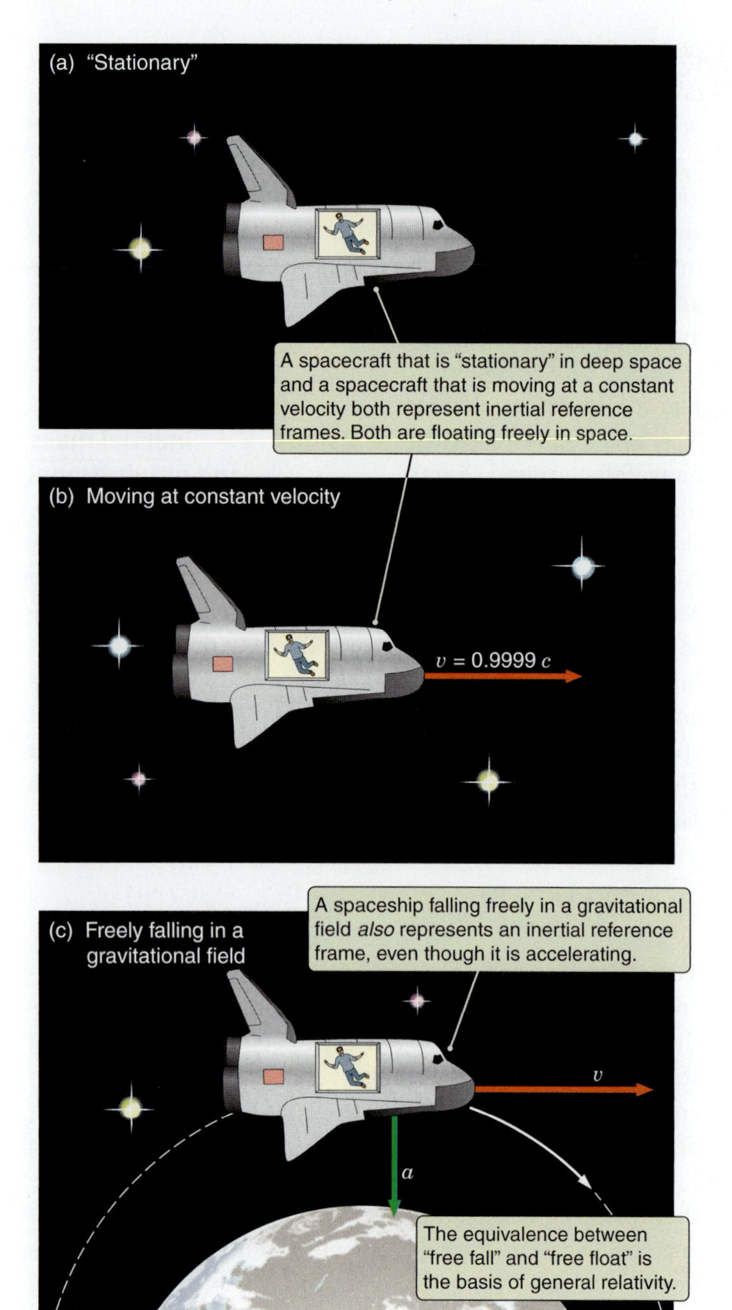

FIGURE 18.8 Special relativity says that there is no difference between (a) a reference frame that is floating "stationary" in space and (b) one that is moving through the galaxy at constant velocity. General relativity adds that there is no difference between these inertial reference frames and (c) an inertial reference frame that is falling freely in a gravitational field. Free fall is the same as free float, as far as the laws of physics are concerned.

freely through Earth's gravitational field, the sensation you feel is exactly the same as the sensation you would feel floating in interstellar space. Even though its velocity is constantly changing as it falls, the inside of a space station orbiting Earth is as good an inertial frame of reference as that of an object drifting along a straight line through interstellar space. This principle—which can be simply stated as "free fall is the same as free float"—is the **equivalence principle**.

The equivalence principle says that a falling object is simply following its "natural" path through spacetime—it is going where its inertia carries it—just like an object that drifts along a straight line at a constant speed through deep space. The natural path that an object will follow through spacetime in the absence of other forces is the object's **geodesic**. In the absence of a gravitational field, the geodesic of an object is a straight line—hence Newton's statement of inertia: an object, unless acted on by an unbalanced external force, will move at a constant speed in a constant direction. In the presence of mass, however, the shape of spacetime becomes distorted, so an object's geodesic becomes curved.

The equivalence principle explains why gravitational mass and inertial mass are one and the same thing. When we discussed Newton's law, $F = ma$, we found it useful to state it as $a = F/m$. When a force F is applied to an object, it moves away from its natural path, and its inertial mass m indicates how strongly it resists the change. General relativity says that when you are standing still on the surface of Earth, your natural path through spacetime—your geodesic—is actually a path falling inward toward the center of Earth. In the absence of any external forces, this is what you would do. Of course, the surface of Earth gets in your way. Put another way, the surface of Earth exerts an external force on your feet, and that force causes you to accelerate continuously away from your natural path through spacetime.

Falling objects follow curved paths through curved spacetime.

This idea leads to a different, equally valid, way of stating the equivalence principle. Another thought experiment (**Figure 18.9a**) demonstrates the point. Imagine you are in a box inside a spaceship that is accelerating through deep space at a rate of 9.8 meters per second per second (m/s^2) in the direction of the arrow shown in **Figure 18.9b**. The floor of the box exerts enough of a force on you to overcome your inertia and cause you to accelerate at 9.8 m/s^2, so you feel as though you are being pushed into the floor of the box. Now imagine instead that you are sitting in a closed box on the surface of Earth. Again the floor of the box exerts enough upward force on you to overcome your inertia, causing you to accelerate at 9.8 m/s^2. You feel as though you are being pushed into the floor of the box.

According to the equivalence principle, *the two cases are identical.* There is no difference between sitting in an armchair in a spaceship traveling with an acceleration of 9.8 m/s^2 and sitting in an armchair on the surface of Earth reading this book. In the first case, the force of the spaceship is

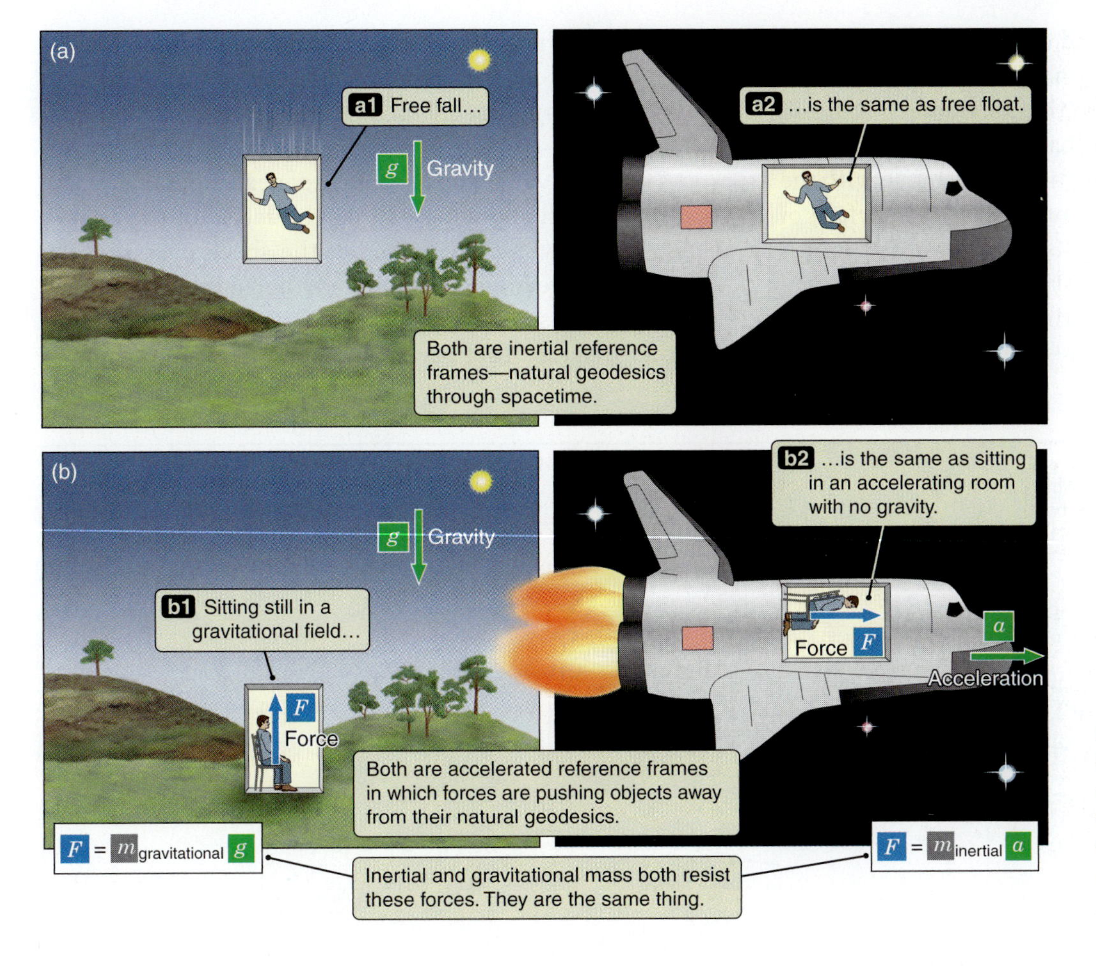

FIGURE 18.9 According to the equivalence principle, an object falling freely in a gravitational field is in an inertial reference frame (a), and an object at rest in a gravitational field is in an accelerated frame of reference (b).

pushing you away from your "floating" straight-line geodesic through spacetime. In the second case, Earth's surface is pushing you away from your curved "falling" geodesic through a spacetime that has been distorted by the mass of Earth. Whether you are being accelerated off a straight-line geodesic through deep space or being accelerated off a "falling" geodesic in the gravitational field of Earth, they are both accelerations. And in all cases, it is the same mass—the mass that gives an object inertia—that resists the change. Gravitational mass and inertial mass are the same thing.

Being stationary in a gravitational field is the same as being in an accelerated reference frame.

There is an important caveat to the equivalence principle. In an accelerated reference frame such as an accelerating spaceship, the *same* acceleration is experienced *everywhere*. By contrast, the curvature of space by a massive object changes from place to place. Tides are one result of changes in the curvature of space from one place to another. Another way to say it is that on Earth the force of gravity always points to the center of Earth, so in a huge enough box, the force is not exactly parallel throughout the box. This contrasts with the accelerating spaceship, where the acceleration is always parallel throughout the ship. A more careful statement of the equivalence principle is that the effects of gravity and acceleration are indistinguishable *locally*—that is, as long as you restrict your attention to small enough regions of space and time, changes in gravity can be ignored.

Spacetime as a Rubber Sheet

The general theory of relativity describes how mass distorts the *geometry* of spacetime. Imagine the surface of a tightly stretched flat rubber sheet. A marble will roll in a straight line across the sheet. Euclidean geometry, the geometry of everyday life, applies on the surface of the sheet: if you draw a circle, its circumference is equal to 2π multiplied by its radius, r; if you draw a triangle, the angles add up to 180°; lines that are parallel anywhere are parallel everywhere.

Now place a bowling ball in the middle of the rubber sheet, creating a deep depression, or "well." The surface of the sheet is stretched and distorted. If you roll a marble across the sheet, its path dips and curves (**Figure 18.10a**). You can roll the marble so that it moves around and around the bowling ball, like a planet orbiting about the Sun. If you draw a circle around the bowling ball, measure the distance around that circle, and then compare that distance with the distance from the circle to its center along the surface of the sheet (**Figure 18.10b**), you will find that the circumference of the circle is less than $2\pi r$. If you draw a triangle on the surface of the sheet, connecting three points around the bowling ball with the straightest and shortest lines you can draw on the surface of the sheet (**Figure 18.10c**), the angles add up to more than 180°. The surface of the sheet is no longer flat, and Euclid's geometry ("plane geometry") no longer applies.

Mass has an effect on the fabric of spacetime that is like the effect of the bowling ball on the fabric of the rubber sheet. The bowling ball stretches the sheet, changing the distances between any two points on the surface of the sheet. Similarly, mass distorts the shape of spacetime, changing the "distance" between any two locations or events in

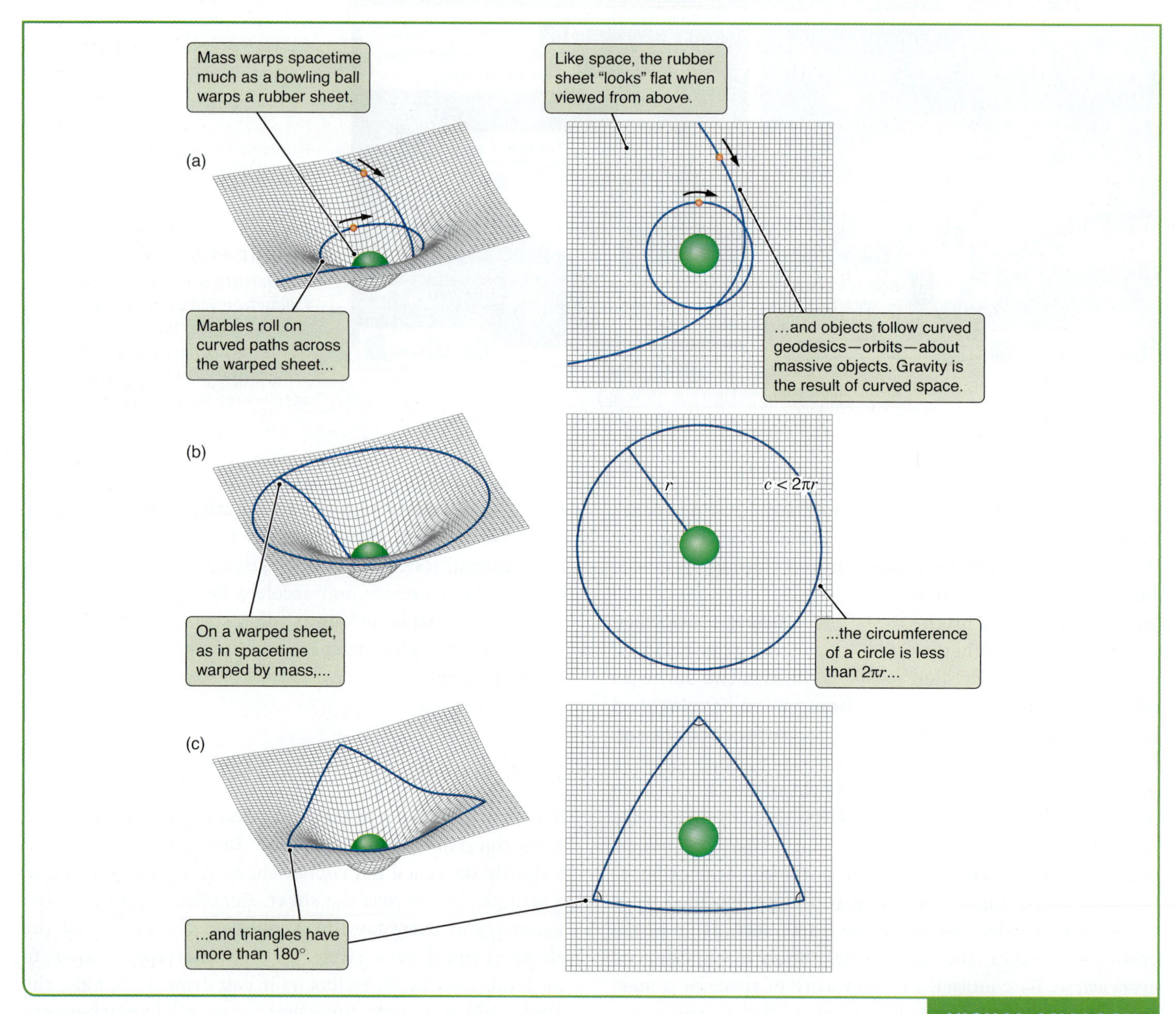

FIGURE 18.10 Mass warps the geometry of spacetime in much the same way that a bowling ball warps the surface of a stretched rubber sheet. This distortion of spacetime has many consequences; for example, objects follow curved paths or geodesics through curved spacetime (a), the circumference of a circle around a massive object is less than 2π times the radius of the circle (b), and angles in triangles need not add to exactly 180° (c).

Connections 18.2

When One Physical Law Supplants Another

So far in this book, we have described gravity as a force that obeys Newton's universal law of gravitation: $F = Gm_1m_2/r^2$. In this chapter we have introduced the ideas of general relativity and asked you to totally change the way you think about gravity. If general relativity is right, does that not imply that Newton's formulation of gravity is wrong? And if it does, then why does Newton's law continue to be used?

The answers to these questions go to the heart of how science progresses and how scientists' conceptions of the universe evolve. Under most circumstances there is virtually no difference between the predictions made using general relativity and the predictions made using Newton's law. As long as a gravitational field is not too strong, Newton's law of gravitation is a very close *approximation* to the results of a calculation using general relativity. The meaning of *too strong* in this context is relative. For example, in most ways the enormous gravitational field near the core of a massive main-sequence star would be considered "weak." Using a general relativistic formulation of gravity rather than Newton's laws to calculate the structure of a main-sequence star would have made virtually no difference in the results of the calculation. Similarly, even though spacetime is curved by the presence of mass, this curvature near Earth is very slight, so over small regions it can be ignored entirely. The flat Euclidean geometry is a good "local" approximation even to curved spacetime—and it is a lot easier to use. This is exactly the kind of approximation people use when they navigate with a flat road map, despite the curvature of Earth.

Similarly, Newton's laws of motion are actually *approximations* of the more generally applicable rules of special relativity and quantum mechanics. In fact, Newton's laws can be derived from special relativity and quantum mechanics using the "everyday" assumptions that speeds of objects are much less than the speed of light and that objects are much larger than the subatomic particles from which atoms are made. Newton's laws of motion and gravitation are used most of the time because they are far easier to apply than the relativistically and quantum mechanically "correct" laws, and because any inaccuracies introduced by using Newtonian approximations are usually far too tiny to matter. Only when conditions are very different from those of everyday life (such as the behavior of an electron in an atom or the gravitational field of a black hole), or in special cases when very high accuracy is needed (such as the precise timing used by the global positioning system [GPS] satellite network), must the relativistic and quantum mechanical laws be used.

This is a general feature of new scientific theories. If a new theory is to replace an earlier, highly successful scientific theory, the new theory must normally hold the old theory within it—it must be able to reproduce the successes of the earlier theory—just as special relativity contains Newton's laws of motion, and general relativity holds within it the successful Newtonian description of gravity that we have relied on throughout the book.

that spacetime. More mass produces even more spacetime distortion. It is easy to visualize how a rubber sheet with a bowling ball on it is stretched through a third spatial dimension, but it is impossible for most people to visualize what a curved four-dimensional spacetime would "look like." Yet experiments verify that the geometry of four-dimensional spacetime is distorted much like the rubber sheet, whether or not it can be easily pictured.

Mass distorts the geometry of spacetime.

As with special relativity and Newton's laws of motion, it is important to point out that general relativity does not mean that Newton's law of gravitation is "wrong" (**Process of Science Figure**). See **Connections 18.2** for a discussion of what happens when one physical law supplants another.

The Observable Consequences of General Relativity

Curved spacetime does have observable consequences. You can imagine stretching a rope all the way around the circumference of Earth's orbit about the Sun, and then comparing the length of that rope with the length of a rope taken from the orbit of Earth to the center of the Sun. You might expect to find that the circumference of Earth's orbit is equal to 2π times the radius of Earth's orbit, just like a circle drawn on a flat piece of paper. If you could carry out this experiment, however, you would find instead that *the rope around the circumference of Earth's orbit is shorter by 10 km than 2π times the length of the rope stretched from Earth to the center of the Sun*—just as the circumference of the circle on the stretched rubber sheet is less than 2π times the radius of the circle.

Process of Science

NEW SCIENCE INCLUDES THE OLD

General relativity is more accurate than Newton's laws: it also explains why gravity acts as it does. Still, for objects like Earth or the Solar System, the two calculations agree.

General relativity is needed when masses are large and distances are small, so the pull of gravity is large.

But far from a mass, where gravity is weak, general relativity gives the same result that Newton found.

One way that scientists check new theories is by considering the limits. What happens at infinity? What happens if the mass is very small? In these limits, new, more complete theories must be compatible with old ones.

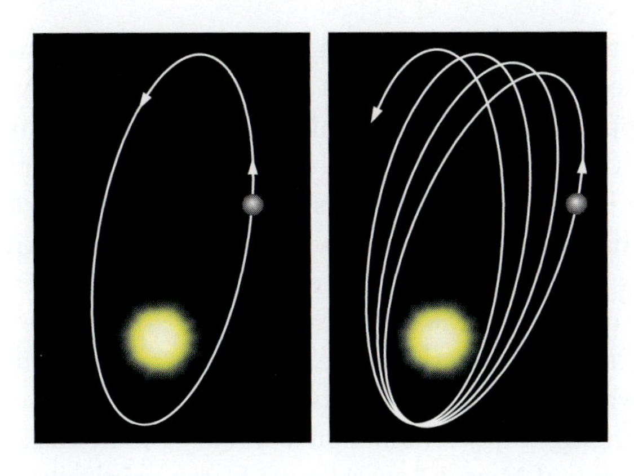

FIGURE 18.11 The precession of Mercury's orbit.

It is not practical to stretch a rope from Earth to the Sun, but there are observations that are conceptually very similar. The long axis of Mercury's elliptical orbit about the Sun is slowly *precessing*—that is, slowly changing its direction (**Figure 18.11**). Newton's law of gravity predicts the major component of Mercury's precession; but even after taking this into account, there remains a very small component, equal to 43 arcseconds (arcsec) per century, that cannot be explained by Newton's laws alone. This observed precession puzzled astronomers as far back as the mid-19th century. Einstein showed that this observed component of Mercury's precession is predicted by general relativity for the path of a planet that is moving alternately deeper and outward within the non-Euclidean spacetime warped by the mass of the Sun.

The consequences of curved spacetime include precession of Mercury's orbit.

The real-life equivalent of the triangle with more than 180° is easier to visualize. A straight line in space is defined by the path followed by a beam of light. This is the shortest distance between any two points. A beam of light moving through the distorted spacetime around a massive object is bent by gravity, just as the lines in Figure 18.10c are bent by the curvature of the sheet. This phenomenon is called **gravitational lensing** because the way the curvature of spacetime bends the path of light resembles the way the lenses bend light in a pair of eyeglasses.

The first measurement of gravitational lensing came during the solar eclipse of 1919. Prior to the eclipse, astrophysicist Sir Arthur Stanley Eddington (1882–1944) measured the positions of a number of stars in the direction of the sky where the eclipse would occur. Eddington then repeated his measurement during the eclipse and found that the apparent positions of the stars had been deflected outward by the presence of the Sun. The light from the stars followed a bent path through the curved spacetime around the Sun, causing the stars to appear farther apart in Eddington's measurement (**Figure 18.12**). During the eclipse, the triangle formed by Earth and the two stars contained more

Gravitational lensing can displace and distort an object's image.

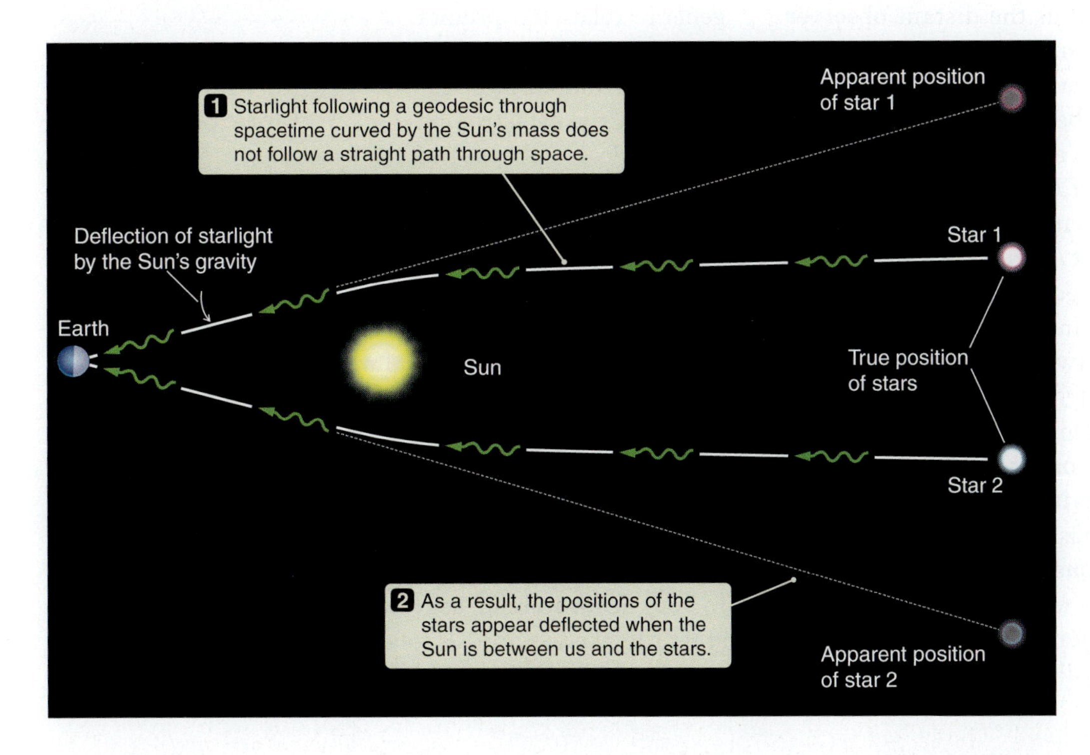

FIGURE 18.12 Measurements obtained by Sir Arthur Eddington during the total solar eclipse of 1919 found that the gravity of the Sun bends the light from distant stars by the amount predicted by Einstein's general theory of relativity. This is an example of gravitational lensing. Note that the "triangle" formed by Earth and the two stars contains more than 180°, just like the triangle in Figure 18.10c.

than 180°—just like the triangle on the surface of our rubber sheet. The results of Eddington's measurements were just as predicted by Einstein's theory. Eddington's result was the first experimental test of a prediction of general relativity and is considered to be one of the landmark experiments of 20th century physics.

More recently, gravitational lensing has been used to search for unseen massive objects in space. These objects do not give off much light, but their gravity can distort the light from background stars that they happen to pass in front of. Similarly, images of galaxies can be distorted by gravitational lensing by other galaxies or clusters of galaxies. In an extreme case, the lensed image can be distorted into a ring called an **Einstein ring**, as illustrated in the chapter-opening photo.

Mass distorts not only the geometry of space, but also the geometry of time. The deeper one descends into the gravitational field of a massive object, the more slowly clocks appear to run from the perspective of a distant observer. This effect is called **general relativistic time dilation**. Suppose a light is attached to a clock sitting on the surface of a neutron star. The light is timed so that it flashes once a second. Because time near the surface of the star is dilated, an observer far from the neutron star perceives the light to be pulsing with a lower frequency—less than once a second. Now suppose there is an emission line source on the surface of the neutron star. Because time is running slowly on the surface of the neutron star, the light that reaches the distant observer will have a lower frequency than when it was emitted. Recall that a lower frequency means a longer wavelength, so the light from the source will be seen at a longer, redder wavelength than the wavelength at which it was emitted. This shift in the wavelengths of light from objects deep within a gravitational well is what gives the phenomenon its name: **gravitational redshift** (**Figure 18.13**).

Time runs more slowly near massive objects.

The effect of gravitational redshift is similar to the Doppler redshift discussed in Chapter 5. In fact, there is no way to tell the difference between light that has been redshifted by gravity and light from an object moving away from you that has been Doppler-shifted. Astronomers often describe the gravitational redshift of an object as an *equivalent velocity*. The gravitational redshift of lines formed on the surface of the Sun is equivalent to a Doppler shift of 0.6 km/s. The gravitational redshift of light from the surface of a white dwarf is equivalent to a Doppler shift of about 50 km/s. The gravitational redshift from the surface of a neutron star is equivalent to a Doppler shift of about a tenth the speed of light. Sometimes astronomers talk about the gravitational redshift as if it truly were a Doppler shift. They might say, for example, that the "gravitational redshift of the surface of a particular white dwarf is 57.1 km/s." This does not mean that the surface of the white dwarf is moving away from Earth at 57.1 km/s. It means that time is running so slowly on the surface of the white dwarf that the light reaching Earth from the white dwarf *looks as though* it were coming from an object moving away from Earth at 57.1 km/s.

Bringing this phenomenon a bit closer to home, a clock on the top of Mount Everest runs faster, gaining about 80 nanoseconds (ns—billionths of a second) a day compared with a clock at sea level. The difference between an object on the surface of Earth and an object in orbit is even greater. A GPS receiver uses the results of sophisticated calculations of the effects of general relativistic gravitational redshift to help you accurately find your position on the surface of Earth. Even after allowing for slowing due to special relativity, the clocks on the satellites that make up the GPS run faster than clocks on the surface of Earth. If the satellite clocks *and* your GPS receiver did not correct for this disparity and other effects of general relativity, then the position your GPS receiver reported would be in error by up to half a kilometer. The fact that the GPS works is actually strong experimental confirmation of several predictions of general relativity.

If you thump the surface of the rubber sheet, waves move away from where you thump it, like ripples spreading out over the surface of a pond. Similarly, the equations of general relativity predict that if the fabric of spacetime is "thumped" (for example, by the catastrophic assymetrical collapse of a high-mass star), then ripples in spacetime, or **gravitational waves**, move outward at the speed of light. These gravitational waves are like electromagnetic waves in some respects. Accelerating an electrically charged particle gives rise to an electromagnetic wave. Accelerating a massive object gives rise to gravitational waves.

Gravitational waves travel through the fabric of spacetime.

Gravitational waves have never been observed in a laboratory or anywhere else, but there is strong circumstantial evidence for their existence. In 1974, astronomers discovered a binary system consisting of two neutron stars, one of which is an observable pulsar. By using the pulsar as a precise clock, astronomers have been able to very accurately measure the orbits of both stars. The stars themselves are 2.8 solar radii apart, but their orbits are gradually decaying, which means that they are losing

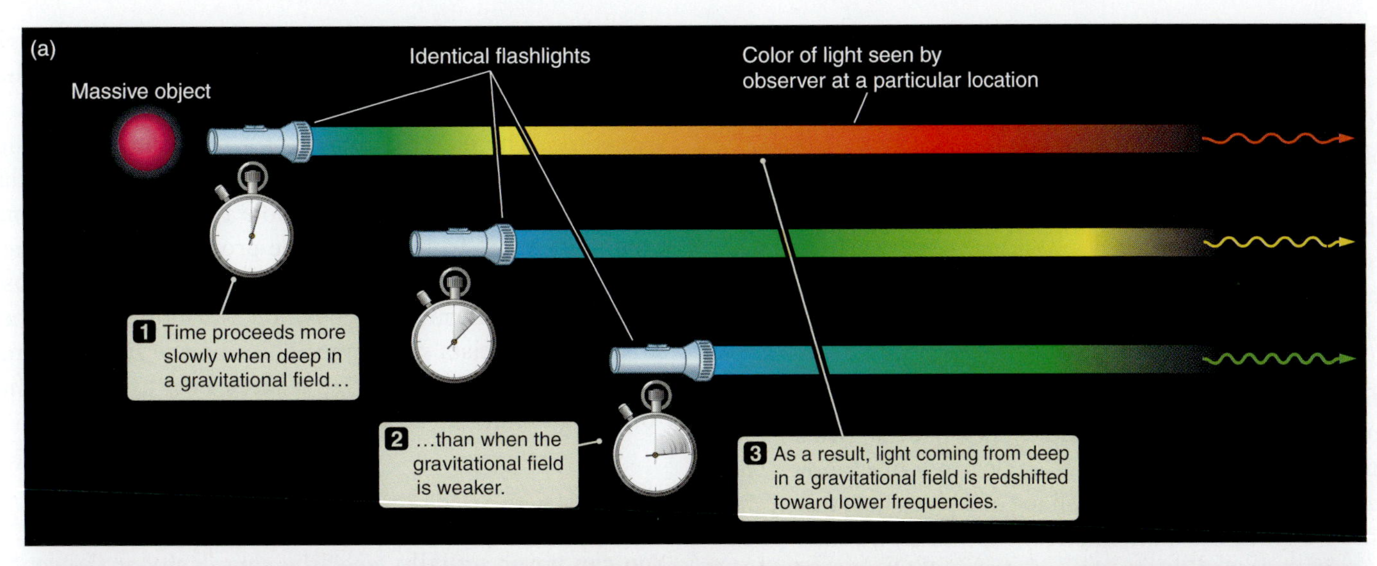

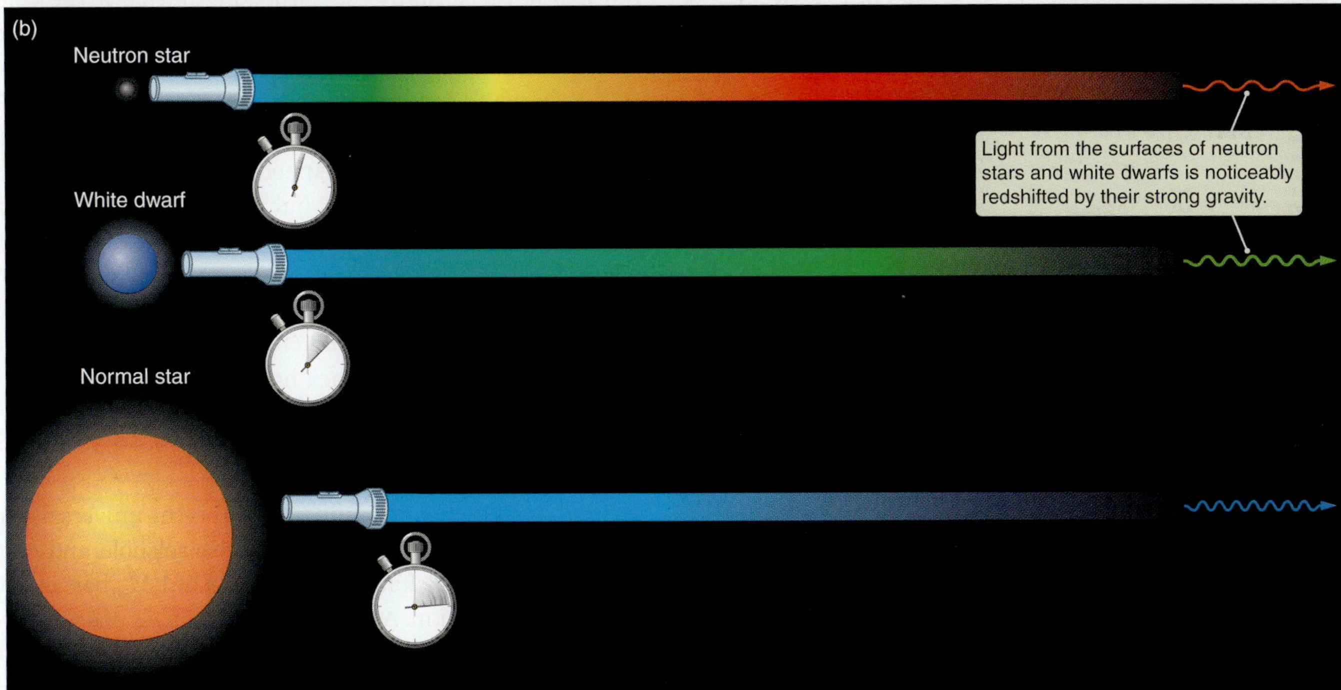

FIGURE 18.13 Time passes more slowly near massive objects because of the curvature of spacetime. As a result, to a distant observer light from near a massive object will have a lower frequency and longer wavelength. The closer the source of radiation is to the object (a) or the more massive and compact the object is (b), the greater the gravitational redshift will be.

energy *somewhere*. Calculations show that the energy being lost by the system is just what general relativity predicts the system should be losing in the form of gravitational waves. Other similar binary pairs have been found with an orbital energy loss consistent with the radiation of gravitational waves. These systems provide indirect evidence for the existence of gravitational waves, but they do not prove it. As we discussed in Chapter 6, astrophysicists have a new kind of "telescope," called the Laser Interferometer Gravitational-Wave Observatory (LIGO), that might be able to detect the predicted gravitational waves emanating from such events. Either they will be seen or they will not. The theory that predicts gravitational waves can be tested because it is falsifiable.

18.4 Black Holes

We began our digression into relativity because of black holes, and we return to the nature of black holes now. When an object is placed on the surface of the rubber sheet, it causes a funnel-shaped distortion that is analogous to the distortion of spacetime by a mass. Now imagine such a funnel-shaped distortion in the rubber sheet that is *infinitely* deep—a funnel that keeps getting narrower and narrower as it goes deeper and deeper, but that has no bottom. This is the rubber-sheet analog to a black hole. The mathematics describing the shape of a black hole fail in the same way that the mathematical expression $1/x$ fails when $x = 0$. Such a mathematical anomaly is called a **singularity**. Black holes are singularities in spacetime (**Figure 18.14**).

Black holes are bottomless pits in spacetime.

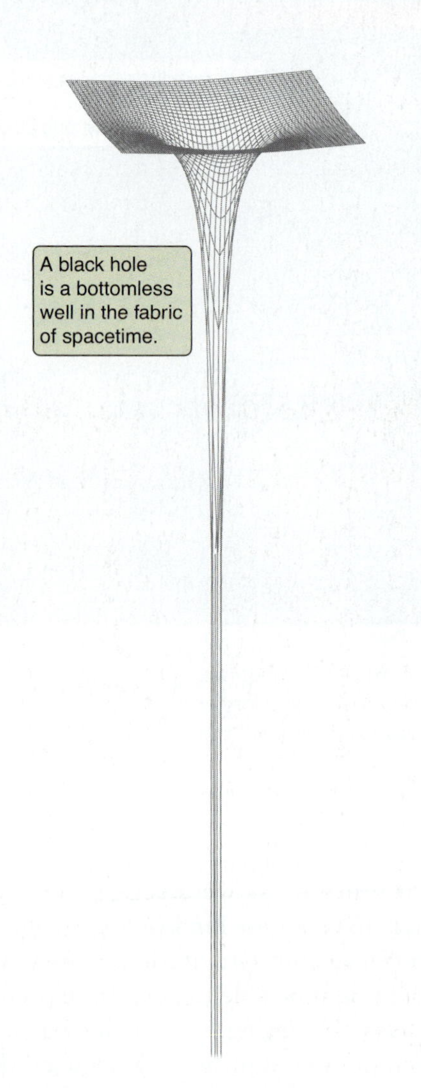

FIGURE 18.14 A black hole is a singularity in the curvature of spacetime. It is a gravitational well with no bottom.

A black hole has only three properties: mass, electric charge, and angular momentum. The amount of mass that falls into a black hole determines how much it distorts spacetime. The electric charge of a black hole is the net electric charge of the matter that fell into it. The angular momentum of a rotating black hole twists the spacetime around it. Apart from these three properties, all information about the material that fell into the black hole is lost. Nothing of its former composition, structure, or history survives. Physicists sometimes describe the fact that only three properties characterize a black hole by saying that "black holes have no hair."

Event Horizons

You can never actually "see" the singularity at the center of a black hole. The closer an object is to a black hole, the greater is its escape velocity (the speed at which it would have to move to escape from the gravity of the black hole). There is a radial distance from the black hole at which the escape velocity reaches the speed of light. This point of no return, beyond which even light is trapped by the black hole, is its **event horizon**. The radius of the event horizon of a black hole that has neither angular momentum nor charge is called the **Schwarzschild radius**, named for the physicist Karl Schwarzschild (1873–1916), and it is proportional to the mass of the black hole:

$$R_S = \frac{2GM_{BH}}{c^2} = 3\ \text{km} \times \frac{M_{BH}}{M_\odot}$$

where R_S is the Schwarzschild radius, G is the universal gravitational constant, M_{BH} is the mass of the black hole, and c is the speed of light. A black hole with a mass of 1 $M_\odot$ has an event horizon of about 3 km. A black hole with a mass of 5 $M_\odot$ has an event horizon 5 times that, or about 15 km. A black hole with a mass equivalent to that of Earth would have an event horizon of only about a centimeter—the mass of Earth compressed into a volume equal to that of a marble. All the mass of a black hole is concentrated at its very center, but this fact is unobservable from outside the black hole.

The event horizon is the boundary of no return.

Let's consider what would happen if an adventurer were willing to journey into a black hole (**Figure 18.15**). From the perspective outside the black hole, the adventurer would appear to fall toward the event horizon. As she fell, her watch would run more and more slowly and her progress toward the event horizon would become slower and slower as well. Even though she would get closer and closer to the event horizon, she would never quite make it, from the external observer's perspective. The event horizon

FIGURE 18.15 Hypothetical journey into a black hole.

is where the gravitational redshift becomes infinite and where clocks stop altogether. Yet the adventurer's experience would be very different. From her perspective, there would be nothing special about the event horizon at all. She would fall past the event horizon and continue deeper into the black hole's gravitational well. She would now have entered a region of spacetime cut off from the rest of the universe. The event horizon is like a one-way door: after the adventurer has passed through, she can never again pass back into the larger universe she once belonged to.

Actually, we have overlooked a rather important detail—namely, that the adventurer would have been torn to shreds long before she reached the black hole. Near the event horizon of a 3-$M_{\odot}$ black hole, the difference in gravitational force between the explorer's feet and her head—the tidal "force" pulling her apart—would be about a billion times her weight on the surface of Earth. This is not an experiment that anyone would ever want to perform, of course. Although scientific theories must produce testable predictions, not all individual predictions have to be tested directly.

"Seeing" Black Holes

In 1974 the physicist Stephen Hawking (1942–) realized that black holes should actually be *sources* of radiation. In the ordinary vacuum of empty space, quantum theory says that particles and their antiparticles spontaneously spring into existence and then, within about 10^{-21} second, annihilate each other and disappear. If such a pair of **virtual particles** comes into existence near the event horizon of a very small black hole (**Figure 18.16**), one of the particles might wind up falling into the black hole while the other particle is able to escape. Some of the gravitational energy of the black hole will have been used up in making one of the pair of virtual particles real. Hawking was able to show that a black hole should actually emit a Planck spectrum, and that the effective temperature of this spectrum would increase as the black hole became smaller. Although this phenomenon, called **Hawking radiation**, is of considerable interest to physicists, in a practical sense it is usually negligible and not a likely way to see a black hole.

The strongest evidence for black holes comes from X-ray binary stars in the Milky Way. In 1972, the brightest X-ray source in the constellation Cygnus, called Cygnus X-1, was found to be rapidly flickering—changing in as little as 0.01 second. This means that the source of the X-rays must be smaller than the distance that light travels in 0.01 second, or 3,000 km. Thus, the source of X-rays in Cygnus X-1 must be smaller than Earth.

When astronomers began to study this object in other parts of the electromagnetic spectrum, Cygnus X-1 was identified both with a star that had radio emission and with an already cataloged optical star called HD 226868. The spectrum of HD 226868 shows that it is a normal O9.7 I supergiant star with a mass of about 20 $M_{\odot}$, far too cool to produce X-ray emission. The wavelengths of absorption lines in the spectrum of HD 226868 are Doppler-shifted back and forth with a period of 5.6 days, indicating that HD 226868 is part of a binary system. Using the same techniques we discussed to measure the masses of stars in Chapter 13 (namely, analyzing the orbits of spectroscopic binaries), astronomers found that the mass of the unseen compact companion of HD 226868 must be about 13–17 $M_{\odot}$ (see **Math Tools 18.3**). The companion to HD 226868 is too compact to be a normal star, yet it is much more massive than the Chandrasekhar limit for a white dwarf or the upper mass limit of a neutron star. The laws of physics suggest that such an object can only be a black hole. Astronomers believe that the X-ray emission from Cygnus X-1 arises when material from the O9.7 I supergiant falls onto an accretion disk surrounding the black hole (**Figure 18.17**).

In some similar systems, winds have been observed blowing off the disk around the black hole. These winds are likely caused by magnetic fields in the disk. The fastest

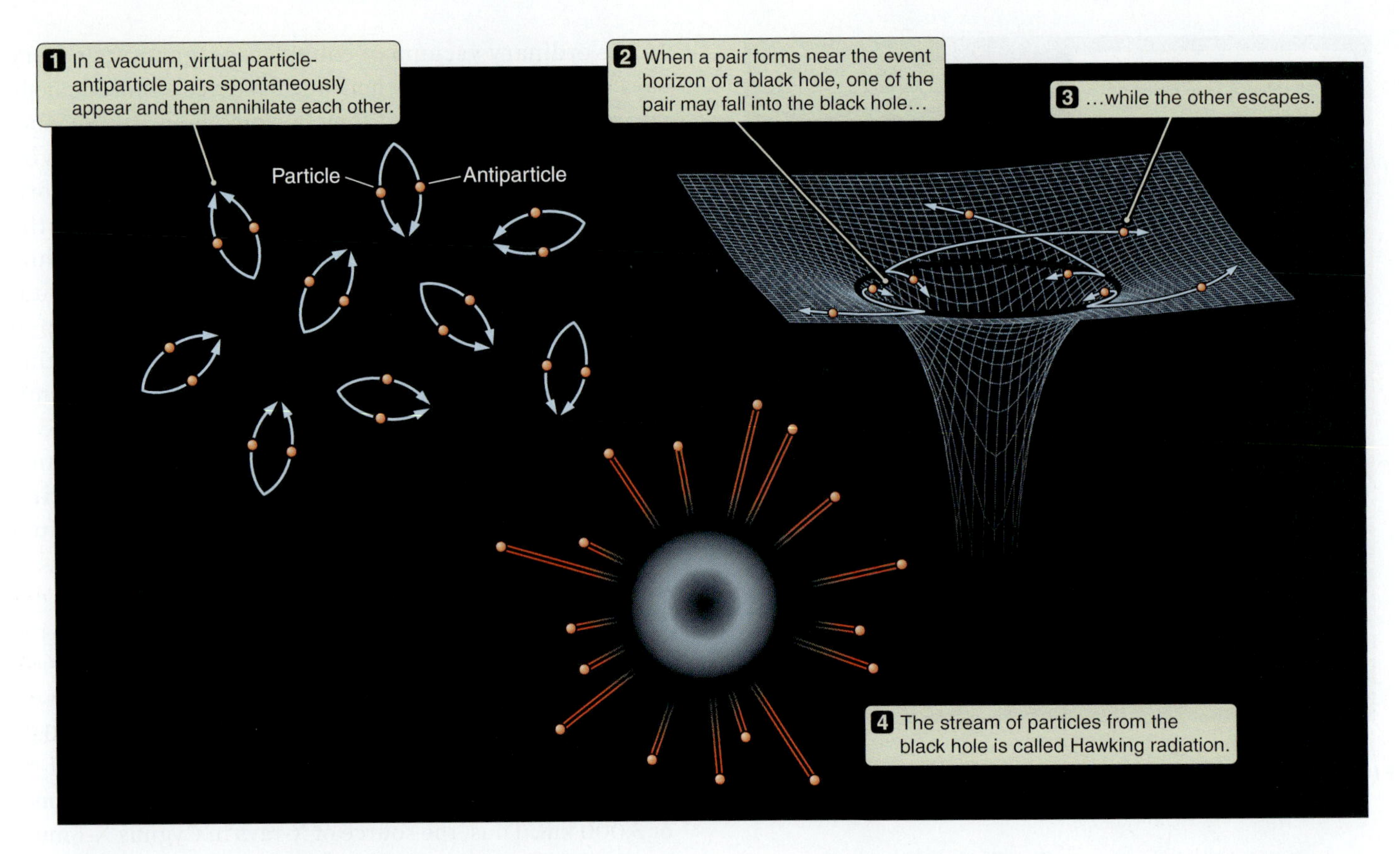

FIGURE 18.16 In the vacuum of empty space, particles and antiparticles are constantly being created and then annihilating each other. Near the event horizon of a black hole, however, one particle may cross the horizon before it recombines with its partner. The remaining particle exits the black hole as *Hawking radiation*.

Math Tools 18.3

Masses in X-Ray Binaries

Cygnus X-1 is part of a binary system with a blue supergiant star O9.7 I (very close to B0 I), and an unseen compact object located about 0.2 AU away from it. The blue supergiant and the compact object orbit a common center of mass every 5.6 days. We can use the simple formula from Math Tools 13.3 to calculate the sum of the masses:

$$\frac{M_{\text{blue}}}{M_{\odot}} + \frac{M_{\text{compact}}}{M_{\odot}} = \frac{A^3}{P^2}$$

with A in AU and P in years. In this case, $A = 0.2$ AU and $P =$ 5.6 days = 5.6 days/365.25 days per year = 0.015 year, so:

$$\frac{M_{\text{blue}}}{M_{\odot}} + \frac{M_{\text{compact}}}{M_{\odot}} = \frac{0.2^3}{0.015^2} = 36$$

Thus, the sum of the masses of the two stars is 36 $M_{\odot}$.

To find the values of the two individual masses from their orbits, we need to know the velocities of the two stars, or the distances of each star to the center of mass and the orbital inclination of the system. Obtaining such information is difficult when one star is compact and not observed separately. However, the mass of the blue supergiant star can be estimated from spectroscopic and photometric data at many wavelengths, assuming the distance to the system is known. When this is done, the mass of the supergiant is estimated at 19–23 $M_{\odot}$. Subtracting from 36, the mass of the compact object is 13–17 $M_{\odot}$—well over the mass limit for a neutron star. Therefore, Cygnus X-1 is assumed to be a black hole. A recent study with data from several X-ray telescopes concluded that the black-hole mass of Cygnus X-1 is 14.8 $M_{\odot}$.

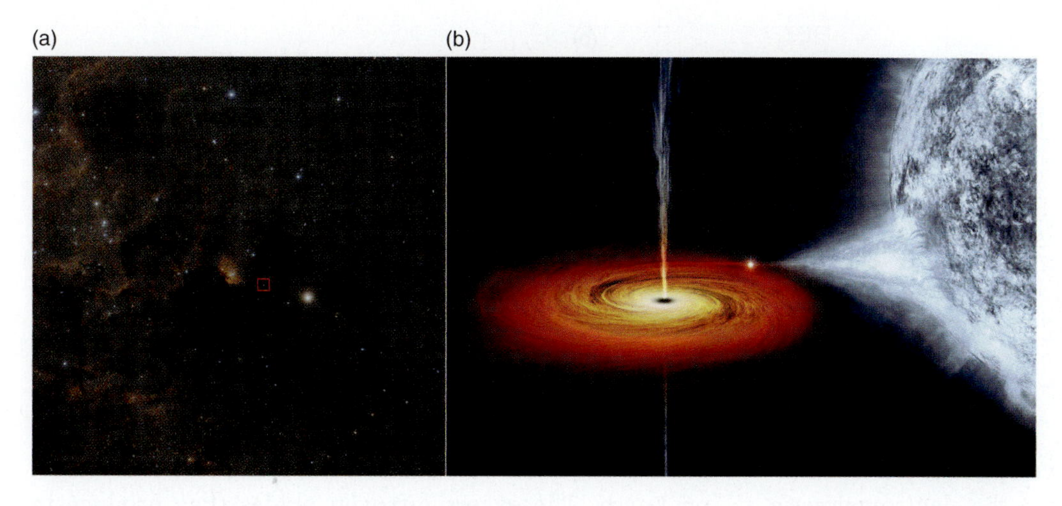

FIGURE 18.17 Black hole in the Cygnus X-1 binary system. (a) Optical image. (b) Artist's model illustrating how material from the O9.7 supergiant is pulled off and falls onto an accretion disk surrounding the black hole, thereby producing X-ray emission.

winds observed are in the binary system IGR J17091, where the winds are as high as 32 million km/h (about 3 percent of the speed of light). This wind is blowing in many directions, and it may be carrying away more mass than is being captured by the black hole (**Figure 18.18**).

Black holes are found through the effects of their gravity.

Astronomers have modeled the observational data of two dozen good candidates for stellar-mass black holes in X-ray binary systems in the Milky Way. They have found that the black-hole masses are greater than 4.5–5 $M_{\odot}$—that is, not very close to the limit of 2.5–3 $M_{\odot}$ for a neutron star. This gap in mass between the most massive neutron stars and the least massive stellar black holes is not yet understood, and it is assumed to be a result of the mass transfer processes between the stars.

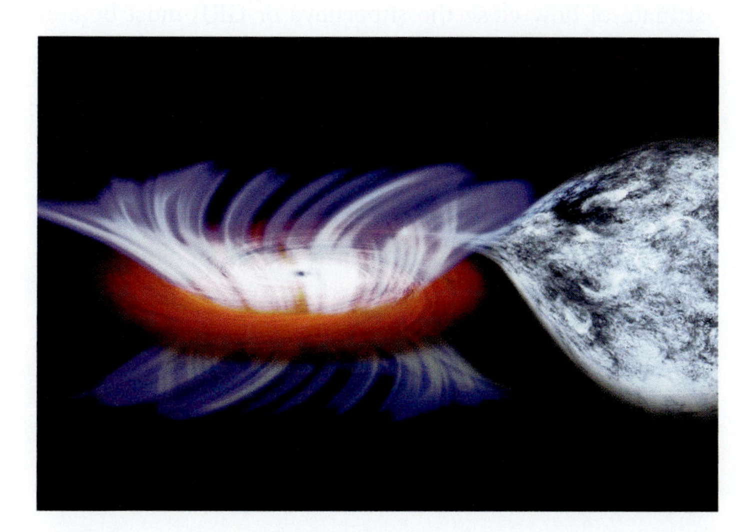

FIGURE 18.18 Artist's model of strong winds being emitted from the disk around a stellar black hole. These winds can remove more material than the amount that actually falls into the hole.

The black holes we discussed in this chapter came from collapsing massive stars, but this is not the only type of black hole. In Chapters 20 and 21 you will learn that supermassive black holes can be found at the centers of galaxies, including the Milky Way.

18.5 Origins: Gamma-Ray Bursts

The most energetic explosions in the universe are likely related to stellar black holes. **Gamma-ray bursts**, or **GRBs**, are intense bursts of gamma rays. The bursts are followed by a weaker "afterglow" that is observed at many wavelengths. GRBs were first observed in the 1960s by satellites designed to look for radiation from nuclear weapons being tested in space after such tests were banned on Earth. In the 1990s, gamma-ray astronomy satellites discovered that these bursts were coming from all directions in the sky, and that they might be associated with supernova explosions in distant galaxies. Short-duration GRBs that last less than 2 seconds probably originate from the merging of two neutron stars or a neutron star and a black hole in a close binary system that collapses into a single black hole. The more common long-duration GRBs are easier to study because they have a longer afterglow. Astronomers think that they originate in the collapse of a very high-mass rapidly spinning star to a neutron star or, more likely, to a black hole, following a supernova explosion. (Supernovae from very high-mass stars are sometimes called **hypernovae**, or Type Ib or Ic supernovae).

Unlike regular supernovae that radiate equally in all directions, GRBs are beamed events, so their enormous

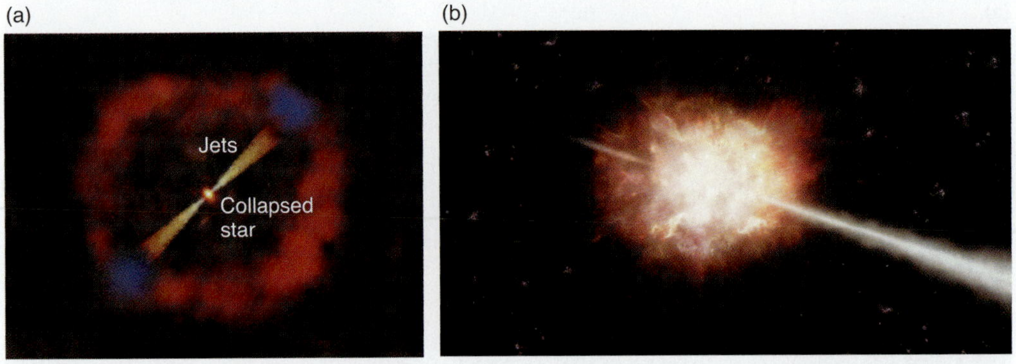

FIGURE 18.19 Artist's model of a gamma-ray burst (GRB). (a) Narrow beams of intense energy are sent in two opposite directions. (b) If one beam is pointed toward the observer, the GRB will appear bright.

energies are concentrated into two opposite jets of emission (**Figure 18.19a**). In addition to the electromagnetic radiation, there are relativistic jets of cosmic rays. Astronomers have not observed any GRBs in the Milky Way; there has not been a massive supernova here for at least a century. But the energy of GRBs is so intense that people have wondered what might happen to Earth if one went off nearby and was beamed in Earth's direction (**Figure 18.19b**). A leading candidate for a future GRB in our galaxy is the massive star Eta Carinae, discussed in Chapter 17. This star is "only" 7,500 light-years away—a neighbor, astronomically speaking. However, its rotation axis is such that is unlikely to form a GRB that beams toward Earth.

Some scientists wonder whether past supernova and GRB events could have affected the history of life on Earth. Supernova "archaeologists" may have found evidence on Earth of past supernovae. In one study, rocks deep in the Pacific Ocean were found to have amounts of a radioactive isotope of iron that is too short-lived to be left over from the formation of Earth. This iron-60 could have been deposited 2.8 million years ago on Earth after a supernova explosion. In another study, high concentrations of nitrates were found in some layers in Antarctic ice cores. Gamma radiation from supernovae can produce excess nitrogen oxides in the atmosphere, which then become converted to nitrates that are trapped in snowfall. Nitrate spikes correlate to 1006 and 1054 CE—two years when there were known to be bright supernovae in the Milky Way.

What about more drastic effects on Earth from a nearby supernova or a more distant but beamed GRB? Normally, Earth is protected from cosmic radiation and cosmic-ray particles by its ozone and its magnetic field layer. Cosmic-ray particles might not be a major problem unless they were really close, but the high-energy gamma-ray radiation could have a more serious effect on Earth. The excess nitrogen oxides they produce in the atmosphere can absorb sunlight, thus cooling Earth. The gamma radiation could ionize Earth's atmosphere, reducing or destroying the ozone layer that protects life from ultraviolet radiation. Even a short burst of a few seconds could lead to ozone damage lasting for decades. The gamma rays could trigger a burst of solar UV radiation at Earth's surface, which could damage the DNA of phytoplankton a few hundred meters deep in the ocean, affecting their ability to photosynthesize. Phytoplankton are the base of Earth's food chain, so a drastic reduction in phytoplankton could upset the entire biosphere. It has been hypothesized that a GRB may have been responsible for the Ordovician mass extinction event 450 million years ago.

Statistically speaking, GRBs that beam to Earth may be quite rare, and some astronomers have argued that they are less likely to be produced in a Milky Way–type galaxy than in other types of galaxies. There is a lot of uncertainty in any estimate of how close the supernova or GRB must be and how often these explosive events must occur to have a serious effect on Earth. One estimate is that there is a supernova or GRB explosion close enough to alter Earth's biosphere a few times every billion years, possibly leading to "mass extinction" events. In Chapter 17 we noted that the chemical elements that make up life were created in supernova explosions. Here you see that supernovae may have had some effect on the *evolution* of life on Earth as well.

In the next chapter we will describe how Einstein's theory of relativity led to theoretical predictions about the evolution of the universe, at about the same time that astronomers were first gathering the data that showed how vast the universe is. ■

Summary

18.1 There is an upper limit to the mass for both white dwarfs and neutron stars. Dense stellar remnants more massive than about 3 $M_{\odot}$ collapse to form black holes.

18.2 Special relativity concerns the relationship between events in space and time. The speed of light, *c*, is the ultimate speed limit. As observers approach that speed, they observe moving objects to be contracted and moving clocks appearing to run more slowly than their own clocks.

18.3 Inertial mass and gravitational mass are the same, leading to the principle of equivalence, in which acceleration cannot be distinguished from gravity if the region in question is small enough. In general relativity, mass warps the fabric of spacetime so that objects move on the shortest path in this warped geometry. Gravity is a consequence of the spacetime warping. Time runs more slowly near massive objects, and sources of radiation in a gravitational well appear to be redshifted.

18.4 Black holes are bottomless pits in spacetime that are surrounded by a one-way surface called the event horizon. Nothing that falls into a black hole can emerge.

18.5 Gamma-ray bursts are beamed high-energy explosions that result from the merger of two compact objects or the rapid collapse of a high-mass star to a black hole. The radiation from these bursts could affect life on Earth.

Unanswered Questions

- What happens to the information that falls into a black hole? We said earlier that a black hole is characterized by only three properties: mass, angular momentum, and electric charge. Where did all the other information go? To a distant observer, it takes an infinitely long time for material to fall into a black hole, so the properties of the material seem to be observable for all time. But from the perspective of the infalling material, it takes a finite time to cross the event horizon, so information other than mass, angular momentum, and charge can no longer be shared with the outside world.
- Do wormholes exist in spacetime, connecting one region with another, perhaps through black holes? Wormholes are a mathematical solution to the equations of general relativity. The idea is that when something goes into a black hole, it travels through a wormhole and emerges in a different part of the universe. In this way a wormhole acts as a shortcut through spacetime. In science fiction, wormholes are a popular means of traveling large distances by exploiting the strange geometry of spacetime. But many scientists doubt that wormholes can exist in nature.

Questions and Problems

Summary Self-Test

1. Rank the following in terms of the mass of the star that produces each.
   a. neutron star
   b. black hole
   c. white dwarf

2. The maximum mass of a neutron star is determined by
   a. electron degeneracy pressure.
   b. neutron degeneracy pressure.
   c. radiation pressure.
   d. none of the above

3. A car approaches you at 50 km/h. A fly inside the car is flying toward the back of the car at 7 km/h. From your point of view, the fly is moving at ________ km/h.
   a. 7
   b. 28.5
   c. 43
   d. 57

4. A car approaches you at 50 km/h. The driver turns on the headlights. From your point of view, the light from the headlights is moving at
   a. $c + 50$ km/h.
   b. $c - 50$ km/h.
   c. $(c + 50 \text{ km/h})/2$.
   d. c.

5. Imagine that you are on a spaceship. A second spaceship rockets past yours at $0.5c$. You start a stopwatch and stop it 10 seconds later. For an astronaut in the other spaceship, the number of seconds that have ticked by during the 10 seconds on your stopwatch is
   a. more than 10 seconds.
   b. equal to 10 seconds.
   c. less than 10 seconds.
   d. there is not enough information

6. The International Space Station flies overhead. Using a telescope, you take a picture and measure its length to be ________ than its length as it would be measured if it were sitting on the ground.
   a. much greater
   b. slightly greater
   c. slightly less
   d. much less

7. Astronauts in the International Space Station
   a. have no mass.
   b. have no weight.
   c. are outside of Earth's gravitational field.
   d. are in free fall.

8. Einstein's formulation of gravity
   a. is approximately equal to Newton's universal law of gravitation for small gravitation fields.
   b. is always used to calculate gravitational effects in modern times.
   c. explained why Newton's universal law of gravitation describes the motions of masses.
   d. both a and c

9. List the three properties of a black hole.

10. As the mass of a black hole increases, its Schwarzschild radius
    a. increases as the square of the mass.
    b. increases proportionally.
    c. stays the same.
    d. decreases proportionally.
    e. decreases as the square of the mass.

True/False and Multiple Choice

11. **T/F:** Any star with an initial mass greater than 3 $M_\odot$ will collapse to form a black hole.

12. **T/F:** A clock ticks more quickly on a fast-moving airplane.

13. **T/F:** Fast-moving objects are shorter (in the direction of travel) than when they are at rest.

14. **T/F:** Near a massive body, the angles of a triangle always add up to more than 180°.

15. **T/F:** Gravitational redshift is the same as the Doppler shift.

16. If a neutron star is more than 3 times as massive as the Sun, it collapses because
    a. the force of electron degeneracy is stronger than gravity.
    b. gravity overpowers the force of electron degeneracy.
    c. gravity overpowers the force of neutron degeneracy.
    d. the force of neutron degeneracy is stronger than gravity.

17. Relative motion between two objects is apparent
    a. even at everyday speeds, such as 10 km/h.
    b. only at very large speeds, such as $0.8c$.
    c. only near very large masses.
    d. only when both objects are in the same reference frame.

18. If a spaceship approaches you at $0.5c$, and a light on the spaceship is turned on pointing in your direction, how fast will the light be traveling when it reaches you?
    a. $1.5c$
    b. between $1.0c$ and $1.5c$
    c. exactly c
    d. between $0.5c$ and $1.0c$

19. Imagine two protoms traveling past each other at a distance d, with relative speed $0.9c$. Compared with two *stationary* protons a distance d apart, the gravitational force between these two protons will be
    a. smaller, because they interact for less time.
    b. smaller, because the moving proton acts as if it has less mass.
    c. the same, because in both cases, the mass of the two particles are identical.
    d. larger, because the moving proton acts as if it has more mass.

20. Suppose Earth had a perfectly circular orbit around the Sun, with radius r. The circumference of this orbit would be
    a. less than $2\pi r$.
    b. equal to $2\pi r$.
    c. equal to πr^2.
    d. greater than $2\pi r$.

21. If two spaceships approach each other, each traveling at $0.5c$, spaceship 1 will measure spaceship 2 to be traveling
    a. much faster than c.
    b. slightly faster than c.
    c. at c.
    d. more slowly than c.

22. If two events at the same point are separated by an elapsed time, t, an observer in another reference frame will measure the elapsed time to be
    a. shorter than t.
    b. t.
    c. longer than t.
    d. You can't tell from the information given.

23. Consider a lemon located on the surface of a white dwarf. The lemon is observed both by an observer on the surface of the white dwarf and by a person in orbit around the white dwarf. In this unusual situation,
    a. the lemon will look redder to the observer on the surface than to the observer in orbit.
    b. the lemon will look the same color to both observers.
    c. the lemon will look redder to the observer in orbit than to the observer on the surface.
    d. neither observer will be able to see the lemon, because light will not escape from the white dwarf.

24. The strongest evidence for black holes comes from rapidly flickering X-ray sources. This observation indicates that
    a. X-rays are not light, because even light cannot escape from a black hole.
    b. the X-rays are coming from a very small source—so small that it is a black hole.
    c. black holes are (at least sometimes) surrounded by hot gas.
    d. black holes have a very high temperature.

25. Short gamma-ray bursts (GRBs) and their afterglows are difficult to observe because
    a. it is difficult to build cameras that can take pictures in less than 2 seconds.
    b. they are faint, so exposure times must be longer than 2 seconds to detect them.
    c. by the time they are detected, the light has traveled so far that astronomers do not know where they are.
    d. it is difficult to move other telescopes fast enough to observe them once they've been detected.

Thinking about the Concepts

26. An astronomer sees a redshift in the spectrum of an object. Without any other information, can she determine whether this is an extremely dense object (exhibiting gravitational redshift) or one that is receding from her (exhibiting Doppler redshift)? Explain your answer.
27. Imagine you are traveling in a spacecraft at 0.9999999*c*. You point your laser pointer out the back window of the spacecraft. At what speed does the light from the laser pointer travel away from the spacecraft? What speed would be observed by someone on a planet traveling at 0.000001*c*?
28. Einstein's special theory of relativity says that no object can travel faster than, or even at, the speed of light. Recall that light is both an electromagnetic wave and a particle called a photon. If it acts as a particle, how can a photon travel at the speed of light?
29. Twin A takes a long trip in a spacecraft and returns younger than twin B, who stayed behind. Could twin A ever return before twin B was born? Explain.
30. In one frame of reference, event A occurs before event B. Is it possible, in another frame of reference, for the two events to be reversed, so that B occurs before A? Explain.
31. Imagine you are watching someone whizzing by at very high speed in a spacecraft. Will that person's pulse rate appear to be extremely fast or extremely slow?
32. Suppose you had a density meter that could instantly measure the density of an object. You point the meter at a person in a spacecraft zipping by at very high speed. Is that person's mass density larger than an average person's, or smaller? Explain.
33. Imagine a future astronaut traveling in a spaceship at 0.866 times the speed of light. Special relativity says that the length of the spaceship along the direction of flight is only half of what it was when it was at rest on Earth. The astronaut checks this prediction with a meter stick that he brought with him. Will his measurement confirm the contracted length of his spaceship? Explain your answer.
34. Looking into a speeding spaceship, you observe that the travelers are playing soccer with a perfectly round soccer ball. What is the shape of the ball according to observers on the spacecraft?
35. You observe a meter stick traveling past you at 0.9999*c*. You measure the meter stick to be 1 meter long. How is the meter stick oriented relative to you?
36. Of the four forces in nature (strong nuclear, electromagnetic, weak nuclear, and gravity), gravity is by far the weakest. Why, then, is gravity such a dominant force in stellar evolution? (Note: The weak nuclear force, which has not been discussed in this text yet, is involved in certain decay processes within the nucleus.)
37. Suppose astronomers discover a 3-$M_\odot$ black hole located a few light-years from Earth. Should they be concerned that its tremendous gravitational pull will lead to Earth's untimely demise?
38. If you could watch a star falling into a black hole, how would the color of the star change as it approached the event horizon?
39. Why don't people detect the effects of special and general relativity in their everyday lives here on Earth?
40. Many movies and TV programs (like *Star Wars*, *Star Trek*, and *Battlestar Galactica*) are premised on faster-than-light travel. How likely is it that such technology will be developed in the near future?

Applying the Concepts

41. Study Figure 18.1b. Does the angle of the rain falling outside the car depend on the speed of the car? Knowing only this angle and the information on your speedometer, how could you determine the speed of the falling rain?
42. Compare Figure 18.1 with Figure 18.2. If you knew the speed of Earth in its orbit from a prior experiment, how could you determine the speed of light from the angle of the aberration of starlight?
43. According to Einstein, mass and energy are equivalent. So, which weighs more on Earth: a cup of hot coffee or a cup of iced coffee? Why? Do you think the difference is measurable?
44. Study Figure 18.5, which compares two imaginary rocket ships experiencing the same acceleration, *g*. The blue line correctly accounts for relativistic effects. The red line does not.
    a. Approximately how fast are the two spaceships going when the effects of relativity begin to be significant?
    b. Approximately how long did it take for the spaceships to reach the speed in (a)?
45. Figure 18.7 shows how the Lorentz factor depends on speed. At about what speed (in terms of *c*) does the Lorentz factor begin to differ noticeably from 1? What happens to the Lorentz factor as the speed of an object approaches the speed of light?
46. The perihelion of Mercury advances 2° per century. How many arcseconds does the perihelion advance in a year? (Recall that there are 60 arcseconds in an arcminute and 60 arcminutes in a degree.) Is it possible to measure Mercury's position well enough to measure the advance of perihelion in 1 year?
47. Study Figure 18.12. If the Sun were twice as massive, would the distance between the apparent positions of stars 1 and 2 increase or decrease during an eclipse?
48. Follow Math Tools 18.2 to find out how much younger you would be than your twin if you made the journey described there at 0.5*c*.

49. Use Math Tools 18.2 to predict how much younger you would be than your twin if you traveled at 0.999*c* instead of 0.5*c* as in question 48. Then calculate the difference in ages and compare the calculated result to your prediction.

50. What is the Schwarzschild radius of a black hole that has a mass equal to the average mass of a person (~70 kilograms)?

51. What is the mass of a black hole with a Schwarzschild radius of 1.5 km?

52. The Moon has a mass equal to $3.7 \times 10^{-8}\ M_{\odot}$. Suppose the Moon suddenly collapsed into a black hole.
    a. What would be the radius of the event horizon (the "point of no return") around the black-hole Moon?
    b. What effect would this collapse have on tides raised by the Moon on Earth? Explain.
    c. Do you think this event would generate gravitational waves? Explain.

53. If a spaceship approaching Earth at 0.9 times the speed of light shines a laser beam at Earth, how fast will the photons in the beam be moving when they arrive at Earth?

54. Suppose you discover signals from an alien civilization coming from a star that is 25 light-years away, and you go to visit it using the spaceship described in the discussion of the twin paradox in Math Tools 18.2.
    a. How long will it take you to reach that planet, according to your clock? According to a clock on Earth? According to the aliens on the other planet?
    b. How likely is it that someone you know will be here to greet you when you return to Earth?

55. Math Tools 18.3 relates the mass of a binary pair to the period and the size of the orbit. Suppose that a spaceship orbited a black hole at a distance of 1 AU, with a period of 0.5 year. What assumptions could you make that would allow you to calculate the mass of the black hole from this information? Make those assumptions, and calculate the mass.

Using the Web

56. Go to the "Through Einstein's Eyes" website (http://anu.edu.au/Physics/Savage/TEE/site/tee/home.html), click on "Start Here," and "Take the tour" and "Movie explained." Take the ride on the "relativistic rollercoaster." Why do colors look different on the relativistic rollercoaster? Click on "Continue tour" and view the cube, tram, and desert road. How do things look different? Why do you get rainbow when driving down the desert at high gamma? Continue for the Solar System tours (or return to the home page). What do you see when you approach the Sun or a planet at a relativistic speed?

57. NASA missions:
    a. Go to NASA's Swift Gamma-Ray Burst Mission website (http://nasa.gov/mission_pages/swift/bursts/index.html). Locate a recent result related to supernovae, gamma-ray bursts, or stellar black holes. Why would two merging neutron stars likely form a black hole?
    b. NASA's Fermi Gamma-ray Space Telescope (http://nasa.gov/mission_pages/GLAST/main/index.html and http://fermi.gsfc.nasa.gov) is exploring the gamma-ray universe. What objectives of this mission relate to the study of black holes? What is a recent news story related to black holes?

58. Go to the LIGO website (http://ligo.org/science.php) and read about gravitational waves. Click on "Sources of Gravitational Waves" and listen to the example. What are the differences among the four listed sources of gravitational waves? Click on "Advanced LIGO." What's new with the project?

59. The newest NASA mission to study black holes, gamma-ray bursts, and neutron stars is named NuSTAR (Nuclear Spectroscopic Telescopic Array). Go to the NuSTAR website (www.nustar.caltech.edu/news-updates). What wavelengths and energies will this telescope observe? What has been observed? What new science has been learned?

60. Go to the "Inside Black Holes" website (http://jila.colorado.edu/~ajsh/insidebh/index.html), enter, and click on "Schwarzschild." Work your way down the page, watching the videos. What does it look like when you go into a black hole? Why is there gravitational lensing when Earth is in orbit? What happens when you fall through the horizon; is everything black? Click on "Reissner-Nordström" to see an electrically charged black hole. What is a wormhole? Why is there a warning at the top of the page? Click on "4D perspective." What does it look like if you move toward the Sun at the speed of light?

STUDYSPACE is a free and open website that provides a Study Plan for each chapter of *21st Century Astronomy*. Study Plans include animations, reading outlines, vocabulary flashcards, and multiple-choice quizzes, plus links to premium content in SmartWork and the ebook. Visit **wwnorton.com/studyspace**.

SMARTWORK Norton's online homework system, includes algorithmically generated versions of these questions, plus additional conceptual exercises. If your instructor assigns questions in SmartWork, log in at **smartwork.wwnorton.com.**

Exploration | Black Holes

Because it's not possible to go grab a black hole and bring it into the lab, and because Earth has never actually been close to one, astronomers can only conduct "thought experiments" to explore the properties of black holes. The following are a few thought experiments to help you think about what's happening near and around a black hole:

Imagine a big rubber sheet. It is very stiff and not easily stretched, but it does have some "give" to it. At the moment, it is perfectly flat. Imagine rolling some golf balls across it.

1 **Describe the path of the golf balls across the sheet.**

....

....

Now imagine putting a bowling ball (very much heavier than a golf ball) in the middle of the sheet, so that it makes a big, slope-sided pit. Roll some more golf balls.

2 **What happens to the path of the golf balls when they are very far from the bowling ball?**

....

....

3 **What happens to the path of the golf balls when they come just inside the edge of the dip?**

....

....

4 **What happens to the path of the golf balls when they go directly toward the bowling ball?**

....

....

5 **How do each of the three cases in questions 2–4 change if the golf balls are moving very, very fast? What if they are moving very slowly?**

....

....

6 **What happens to the depth and width of the pit as the golf balls fall into the center near the bowling ball? (Imagine putting *lots* of golf balls in.)**

....

....

All of the preceding relates to ordinary stuff. Stars, people, planets—everything interacts in this way because of gravity. With black holes, things are a bit different. In this case it is more accurate to think of the bowling balls as holes in the sheet that pull it down, rather than as objects that sit on it. But they still affect the sheet in the same way. The hole is a good analogy for the event horizon of a black hole. Objects outside the event horizon will know that the black hole is there, because the sheet is sloping, but they won't be captured unless they come within the event horizon. Think about light for a moment, as though it were, say, grains of sand rolling across the sheet.

7 **What happens to the light as it passes far from the pit? What happens if it reaches the hole?**

....

....

Now suppose you roll another bowling ball across the sheet.

8 **What happens to the sheet when the second bowling ball falls in after the first? Would this change affect your golf balls and grains of sand? How? What happens to the hole? What happens to the size of the pit?**

....

....

None of these thought experiments take into account relativistic effects (length contraction and time dilation). Imagine for a moment that you are traveling close to the black hole.

9 **Look out into the galaxy and describe what you see. Consider the lifetimes of stars, the distances between them, their motions in your sky, and how they die. Add anything else that occurs to you.**

....

....

A deep image of a section of dark sky made with HST in many exposures over the past 10 years. Nearly every object in this image is a galaxy; the faintest smudges are galaxies forming more than 13 billion years ago.

19 The Expanding Universe

A man said to the universe: "Sir, I exist!"
"However," replied the universe,
"The fact has not created in me a sense of obligation."

Stephen Crane (1871–1900)

LEARNING GOALS

In this final part of the book, we move to a discussion of the largest groups of stars—the galaxies. As we did in the other parts of this book, we begin this chapter with the big picture: the discovery of galaxies and their distribution in space. In later chapters we will explore the types of galaxies, including the Milky Way, and see what these observations indicate about the ultimate fate of the universe. By the conclusion of this chapter, you should be able to:

- Describe the distance ladder and how distances to galaxies are estimated.
- Sketch a graph of Hubble's law, and explain why this relation implies that the universe is expanding.
- Describe how Hubble's law is used to map the universe and to look back in time
- Recount the major predictions made by the Big Bang theory.
- Link observations of the cosmic microwave background radiation to the properties of the young universe.

19.1 Twentieth Century Astronomers Discovered the Universe of Galaxies

Look at a deep-space image of a piece of sky, such as the Hubble Space Telescope (HST) image shown in the chapter-opening photo. It reveals a myriad of faint smudges of light filling the gaps among a sparse smattering of nearby stars. Observers have long known that the sky contains faint, misty patches of light. These objects were originally called **nebulae** (singular: *nebula*) because of their nebulous (fuzzy) appearance. Prior to the 1780s, only about 100 of these smudges of light had been found with and without telescopes. In 1784, astronomer Charles Messier (1730–1817) published a catalog of 103 nebulous objects. Twenty years later, because of the remarkable observations by the astronomers William Herschel and his sister Caroline Herschel, that number jumped to 2,500. From this time on, astronomers were aware of systematic differences in the appearance of nebulae. Although some of the Herschels' nebulae looked diffuse and amorphous, most were round or elliptical or resembled spiraling whirlpools. These distinctions were the basis for the original three categories—diffuse, elliptical, and spiral—used to classify nebulae.

Speculations about the nature of these objects abounded for the next 140 years. It was suggested that spiral nebulae might be relatively nearby planetary systems in various stages of formation. The influential 18th century philosopher Immanuel Kant (1724–1804) had a very different idea. He speculated that spiral nebulae were instead "island universes"—separate from the Milky Way Galaxy. Herschel himself shared this belief but realized that telescopes at that time would never be able to resolve the issue. Not until the first third of the 20th century did the technological tools became available to turn Kant's philosophical musing into scientific knowledge.

Today astronomers know that Kant was correct. The Milky Way is only one of many island universes, which were renamed *galaxies* (from the Greek *gala*, meaning "milky"). Most diffuse nebulae are clouds of gas and dust near Earth in the Milky Way Galaxy, but what Messier and others thought of as elliptical and spiral nebulae are instead galaxies located far beyond the bounds of the Milky Way. A **galaxy** is a gravitationally bound collection of dust, gas, and millions to hundreds of billions of stars. Each tiny smudge in the chapter-opening photo is such a galaxy. The universe contains billions upon billions of galaxies—more galaxies than there are stars in the Milky Way Galaxy. Most of these galaxies are located at such astonishing distances that they appear too small and faint to see with any but the most powerful telescopes. (We will discuss distant galaxies and the Milky Way in Chapters 20 and 21.)

What chemical elements formed in the early universe?

The Milky Way is only one of billions and billions of galaxies.

These questions about the nature of galaxies were, in essence, questions about size and distance. Early attempts to understand the size of the Milky Way were confounded by interstellar dust, which blocks the passage of visible light and thus limits the view. Early astronomers, not knowing of the existence or consequences of this dust, assumed that what they could see in visible light was all there is. Unable to see past this obscuring shroud, they concluded that the Milky Way is a system of stars some 1,800 parsecs (pc) across (recall from Chapter 13 that 1 parsec = 3.26 light-years). Not until the beginning of the 20th century did astronomer Harlow Shapley (1885–1972), of the Harvard College Observatory, estimate that instead, the Milky Way

FIGURE 19.1 Edwin Hubble, shown on the cover of the February 9, 1948, issue of *Time* magazine.

is over 50 times larger—92,000 pc (300,000 light-years) in size. Shapley based his estimate on observations of globular clusters, and these observations greatly expanded the previously held views about the extent of space.

Shapley thought his far larger estimate for the size of the Milky Way meant that it was big enough to encompass everything in the universe, and therefore he thought that the spiral and elliptical nebulae were inside of the Milky Way. Astronomer Heber D. Curtis (1872–1942) from California's Lick Observatory preferred the earlier, smaller model of the Milky Way. He also favored the idea that the spiral nebulae were in fact galaxies separate from the Milky Way and that therefore the whole universe was larger than the Milky Way. In 1920, Shapley and Curtis met in Washington DC to publicly debate these two issues. Historians call this meeting astronomy's *Great Debate*.

The Great Debate focused attention on the size and distance of nebulae.

Unlike questions of politics and law, scientific questions are not resolved by the rhetorical skills of partisans. Instead, they are settled by the results of well-crafted and carefully conducted experiments and observations. However, scientific debates do help bring issues into sharper focus, leading scientists to concentrate their attention and efforts on key questions. While this "Great Debate" did not resolve the issue at the time, it set the stage and pointed the direction for the subsequent work of Edwin P. Hubble (1889–1953), whose discoveries fundamentally changed the modern understanding of the universe (**Figure 19.1**).

Using the newly finished 100-inch telescope on Mount Wilson, high above the then small city of Los Angeles, Hubble was able to find some variable stars in the spiral nebula Andromeda (**Figure 19.2**). He recognized that these stars

FIGURE 19.2 The Andromeda Galaxy, the nearest large galactic neighbor to the Milky Way, is about 2.5 million light-years (780,000 pc) away. The arrow points to the Cepheid variable star V1, a standard candle that Hubble used to estimate the distance to Andromeda. The insets show V1's variability. His measurement provided the first observational evidence of the vastness of the universe.

were very similar to but fainter than the Cepheid variable stars studied by Henrietta Leavitt in the Milky Way Galaxy and the nearby Magellanic Cloud nebula. Using the period-luminosity relation for Cepheid variable stars discussed in Chapter 17, Hubble turned his observations of these stars into measurements of the distances to these objects. The results showed that the distances to these nebulae are far greater than Shapley's size of the Milky Way (100,000 pc). Spiral and elliptical nebulae are really separate galaxies, similar in size to the Milky Way but located at truly immense distances.

Hubble settled the Great Debate by measuring the distances to other galaxies.

19.2 The Cosmological Principle

In the 1920s, around the same time that astronomers were first measuring distances to galaxies, theoretical physicists were applying Einstein's general theory of relativity to **cosmology**—the study of space and time and the dynamics of the universe as a whole. The cosmologist Alexander Friedmann (1888–1925) produced mathematical models of the universe, which assumed that Earth is not in a special place in the universe, and that on large scales the physical properties of the universe were the same everywhere, and in every direction.

Recall from Chapter 1 that the **cosmological principle** states that *the physical laws that apply to one part of the universe apply everywhere.* We have used this principle throughout this book when applying laws of physics and chemistry to objects near and far in the universe. The cosmological principle is a testable scientific theory. An important prediction of the principle is that the conclusions that scientists reach about the universe should be more or less the same, regardless of whether they are in the Milky Way or in a galaxy billions of parsecs away. In other words, if the cosmological principle is correct, then the universe is **homogeneous**, having the same composition and properties at all places. Clearly the universe is not homogeneous in an absolute sense, since the conditions on Earth are very different from those in deep space or in the center of the Sun. In cosmology, homogeneity of the universe means that stars and galaxies in Earth's part of the universe are much the same, and behave in the same manner, as stars and galaxies in remote corners of the universe. It also means that stars and galaxies everywhere are distributed in space in much the same way that they are distributed in Earth's cosmic neighborhood, and that observers in those galaxies see the same properties for the universe that astronomers see from here.

It is not easy to verify the prediction of homogeneity directly. Scientists cannot travel from the Milky Way to a galaxy in the remote universe to see whether conditions are the same. However, they can compare light arriving from closer and farther locations in the distant universe and see the ways in which features look the same or different. For example, they can look at the way galaxies are distributed in distant space and see whether that distribution is similar (that is, homogeneous) to the distribution nearby.

Observers everywhere should see the same universe.

In addition to predicting that the universe is homogeneous, the cosmological principle requires that all observers (including those on Earth) have the same impression of the universe, regardless of the *direction* in which they are looking. If something is the same in all directions, then it is **isotropic**. This prediction of the cosmological principle is much easier to test directly than is homogeneity. For example, if galaxies were lined up in bands—a violation of the cosmological principle—astronomers would get very different impressions, depending on the direction they looked. In most instances, isotropy goes together with homogeneity, and the cosmological principle requires both. **Figure 19.3** shows examples of how the universe could have violated the cosmological principle by being inhomogeneous or anisotropic, as well as examples of how the universe might satisfy the cosmological principle.

The distribution of galaxies is isotropic and homogeneous.

The isotropy and homogeneity of the distribution of galaxies in the universe are predictions of the cosmological principle that astronomers can test directly. All observations show that the properties of the universe are basically the same, regardless of the direction observers are looking. On very large scales, thousands of millions of parsecs, the universe appears homogeneous as well. The cosmological principle is now a fundamental tenet of modern cosmology.

19.3 The Universe Is Expanding

In the 1920s, Hubble and his coworkers were studying the properties of a large collection of galaxies. Vesto Slipher (1875–1969), one of Hubble's colleagues, was obtaining spectra of these galaxies at Lowell Observatory in Flagstaff, Arizona. Slipher's galaxy spectra looked like the spectra of ensembles of stars with a bit of glowing interstellar gas mixed in. But he was surprised to find that the emission and absorption lines in the spectra of

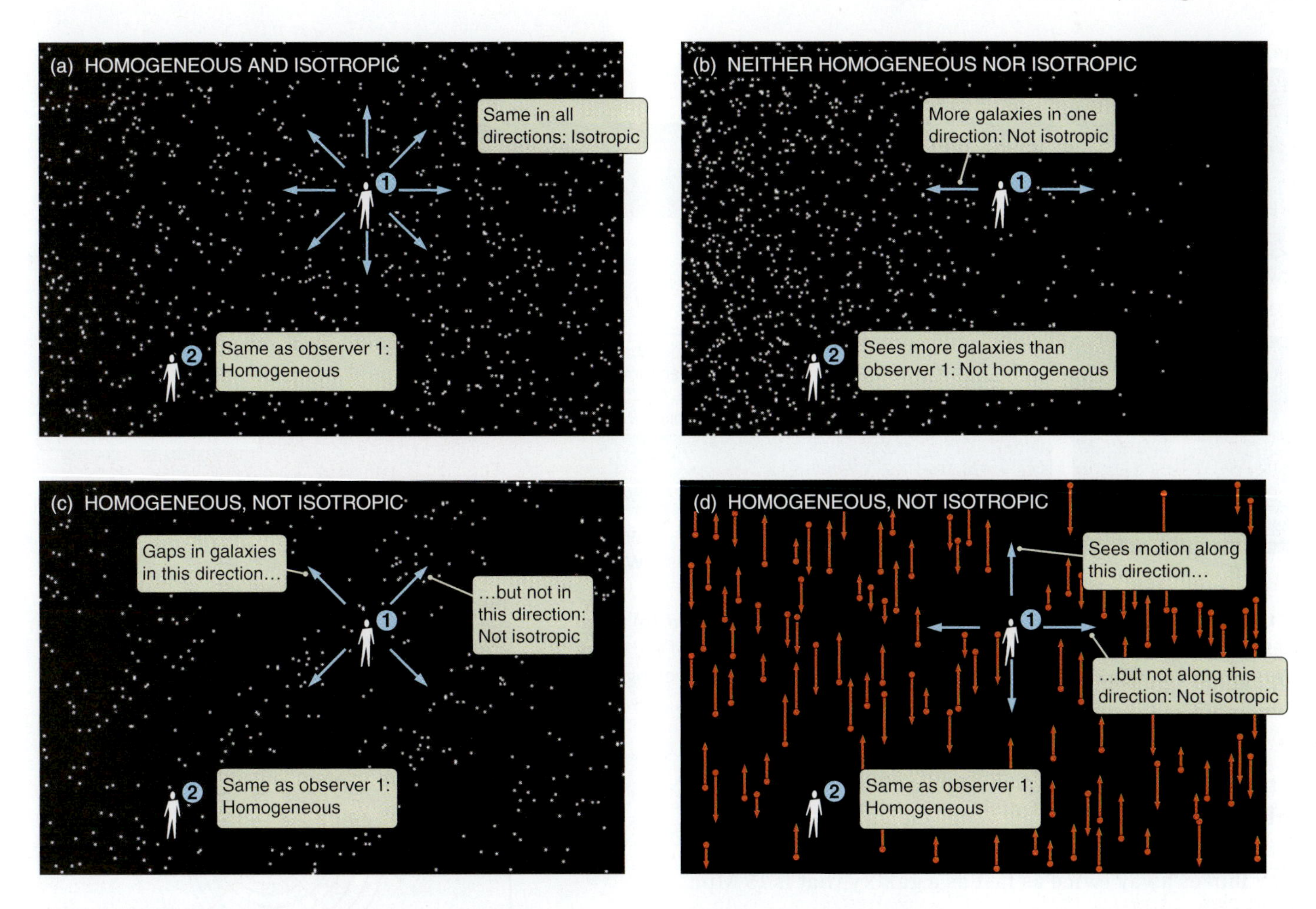

FIGURE 19.3 Homogeneity and isotropy in four different theoretical models of a universe. Blue arrows indicate the direction of view. (a) The distribution of galaxies is uniform, so this universe is both homogeneous and isotropic. (b) The density of galaxies is decreasing in one direction, so this universe is neither homogeneous nor isotropic. (c) The bands of galaxies lie along a unique axis, making this universe anisotropic. (d) The distribution of galaxies is uniform, but galaxies move along only one direction, so this universe also is not isotropic.

these galaxies were seldom seen at the same wavelengths as in the spectra of stars observed in the Milky Way Galaxy. The lines were almost always shifted to longer wavelengths (**Figure 19.4**).

Slipher characterized most of the observed shifts in galaxy spectra as **redshifts** because the light from these galaxies is shifted to longer, or redder, wavelengths. Recall from Chapter 5 that the Doppler shift causes the observed wavelengths of objects moving away to shift toward the red end of the spectrum. The wavelength at which a line is observed in an object that is stationary relative to the observer is called the **rest wavelength** of the line, written λ_{rest}. (If you took a spectrum of a gas in the lab, the spectral lines would be at the rest wavelength.) The redshift of a galaxy, called z, is the difference between the observed wavelength (λ_{obs}) and the rest wavelength, divided by the rest wavelength:

Light from distant galaxies is redshifted.

$$z = \frac{\lambda_{obs} - \lambda_{rest}}{\lambda_{rest}}$$

▶▶ NEBRASKA SIMULATION: GALACTIC REDSHIFT SIMULATOR

Hubble interpreted Slipher's redshifts as Doppler shifts, and he concluded that almost all of the galaxies in the universe are moving away from the Milky Way. When he combined these measurements of galaxy recession velocities with his own estimates of the distances to these galaxies, Hubble made one of the greatest discoveries in the history of astronomy. He found that distant galaxies are moving away from Earth more rapidly than are nearby galaxies. Specifically, *the velocity at which a galaxy is moving away from an observer is proportional to the distance of that galaxy*. This relationship between

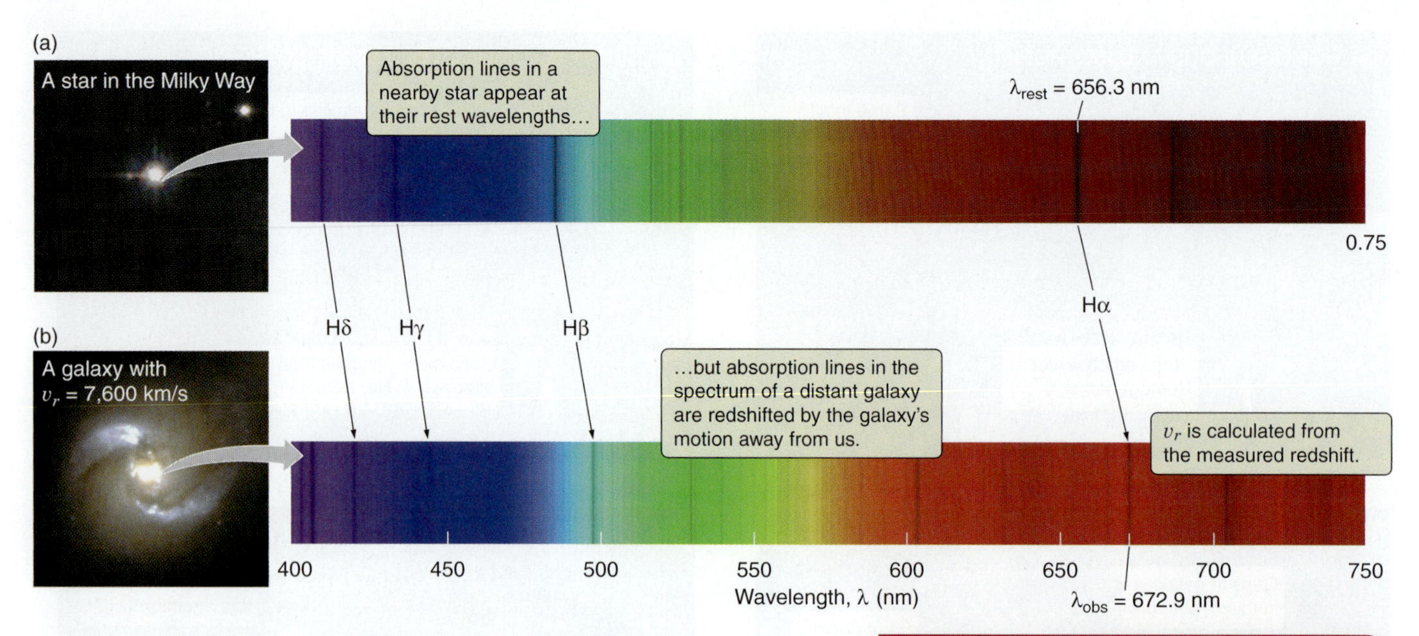

FIGURE 19.4 (a) A star in the Milky Way, shown with its spectrum. (b) A distant galaxy, shown with its spectrum at the same scale as that of the star. Note that lines in the galaxy spectrum are redshifted to longer wavelengths.

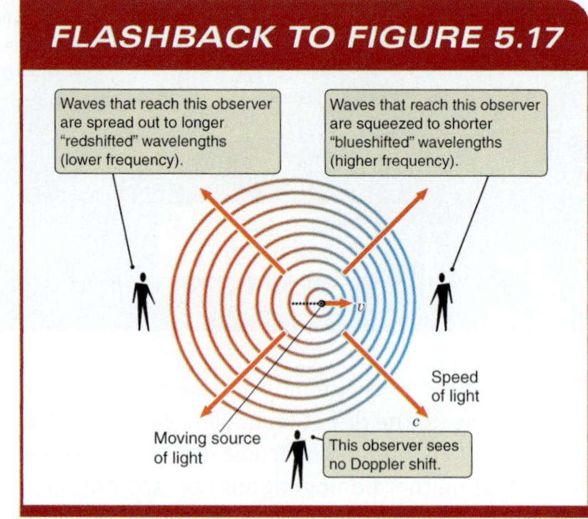

distance and recessional velocity has become known as **Hubble's law**.

Hubble's law says that a galaxy 30 **megaparsecs (Mpc)** away moves away twice as fast as a galaxy that is 15 Mpc distant. (A megaparsec is 3.26 million light-years.) H_0 is the constant of proportionality between the distance and the speed, and it is called the **Hubble constant**. This constant is one of the most important numbers in cosmology, and many astronomical careers have been dedicated to precisely and accurately determining its value. Using this constant, Hubble's law is written as $v_r = H_0 \times d_G$, where d_G is the distance to the galaxy and v_r is the galaxy's recessional velocity. (See example in **Math Tools 19.1**). ▶II **ASTROTOUR: HUBBLE'S LAW**

Hubble's law says that a galaxy's recessional velocity is proportional to its distance.

Hubble's constant relates a galaxy's recessional velocity to its distance.

All Observers See the Same Hubble Expansion

Hubble's law is a remarkable observation about the universe that has far-reaching implications. For one thing, Hubble's law helps astronomers investigate whether the universe is homogeneous and isotropic. When they look at galaxies in one direction in the sky, they find that these galaxies obey the same Hubble law that galaxies observed in other directions in the sky obey. Although Hubble's law corroborates the prediction that Earth's view of the universe is isotropic, the law appears at first glance to contradict the prediction of the cosmological principle that the universe is homogeneous. Hubble's law might seem to imply that Earth is in a very special place—at the *center* of a tremendous expansion of space, with everything else in the universe streaming away. This initial impression, however, is incorrect. *Hubble's law actually says that Earth is located in a uniformly expanding universe and that the expansion looks the same, regardless of the location of the observer* (**Process of Science Figure**). To help you visualize this, we now turn to a useful model

Hubble's law says that the universe is expanding uniformly.

Math Tools 19.1

Redshift—Calculating the Recessional Velocity and Distance of Galaxies

Recall from Math Tools 5.2 that the Doppler equation for spectral lines showed that:

$$v_r = \frac{\lambda_{obs} - \lambda_{rest}}{\lambda_{rest}} \times c$$

The fraction in front of the c is equal to z, the redshift. Substituting for the fraction, we get:

$$v_r = z \times c$$

(Note: This correspondence works only for velocities much slower than the speed of light; see Connections 19.1.)

Because lines from distant galaxies have wavelengths shifted to the red, the galaxies must be moving away from Earth. Suppose astronomers observe a spectral line with a rest wavelength of 373 nanometers (nm) in the spectrum of a distant galaxy. If the observed wavelength of the spectral line is 379 nm, then its redshift (z) is:

$$z = \frac{\lambda_{obs} - \lambda_{rest}}{\lambda_{rest}}$$

$$= \frac{379 \text{ nm} - 373 \text{ nm}}{373 \text{ nm}}$$

$$= 0.0161$$

Note that the value of the redshift of a galaxy is independent of the wavelength of the line used to measure it; the same result would have been calculated if a different line had been observed.

We can now calculate the recessional velocity from this redshift as follows:

$$v_r = z \times c = 0.0161 \times 300{,}000 \text{ km/s} = 4{,}830 \text{ km/s}$$

How far away is the distant galaxy? This is where *Hubble's law* and the *Hubble constant* apply. (In this book we use $H_0 = 70$ kilometers per second per megaparsec, or km/s/Mpc.) Hubble's law relates a galaxy's recessional velocity to its distance as:

$$v_r = H_0 \times d_G$$

where d_G is the distance to a galaxy measured in megaparsecs. Dividing through by H_0 yields:

$$d_G = \frac{v_r}{H_0} = \frac{4{,}830 \text{ km/s}}{70 \text{ km/s/Mpc}} = 69 \text{ Mpc}$$

From a simple measurement of the wavelength of a spectral line, we have learned that the distant galaxy is approximately 69 million pc (Mpc) away.

that you can build for yourself with materials you probably have in your desk.

Figure 19.5 shows a long rubber band with paper clips attached along its length. If you stretch the rubber band, the paper clips, which represent galaxies in an expanding universe, get farther and farther apart. To get a sense of what this expansion would look like up close, imagine yourself an ant riding on paper clip A. As the rubber band is stretched, you notice that all of the paper clips are moving away from you. Clip B, the closest one, is moving away slowly. Clip C, located twice as far from you as B, is moving away twice as fast as B. Clip E, located 4 times as far away as B, is moving away 4 times as fast as B. From your perspective as an ant riding on clip A, all of the other paper clips on the rubber band are moving away with a velocity proportional to their distance. The paper clips located along the rubber band obey a Hubble-like law.

The key insight to the analogy comes from realizing that there is nothing special about the perspective of paper clip A. If you were riding on clip E, clip D would be the one moving away slowly and clip A would be moving away 4 times as fast. For an ant on any paper clip along the rubber band, the speed at which other clips are moving away from the ant is proportional to their distance. The stretching rubber band, like the universe, is "homogeneous." The same Hubble-like law applies, regardless of where the ant is located.

The observation that nearby paper clips move away slowly and distant paper clips move away more rapidly does not say that the paper clip selected as a vantage point is at the center of anything. Instead, it says that the rubber band is being stretched uniformly along its length. In like fashion, Hubble's law for galaxies means that the Milky Way is not at the center of an expanding universe. *Hubble's law means that the universe is expanding uniformly.* Any observer in any galaxy sees nearby galaxies moving away slowly and more distant galaxies moving away more rapidly. The expansion of the universe is homogeneous.

Process of Science

AUTHORITY IS IRRELEVANT

Einstein is one of the most famous scientists of all time. His genius is recognized by everyone. But even Einstein had to change his mind in the face of new data.

Einstein develops general relativity, which predicts the universe is expanding or contracting.

He "fudges" his equations to make the universe stationary because he believes it should be.

Hubble takes data that prove Einstein wrong. The universe is expanding!

Einstein changes his mind.

Data always trump preconceived notions. No matter who thinks otherwise.

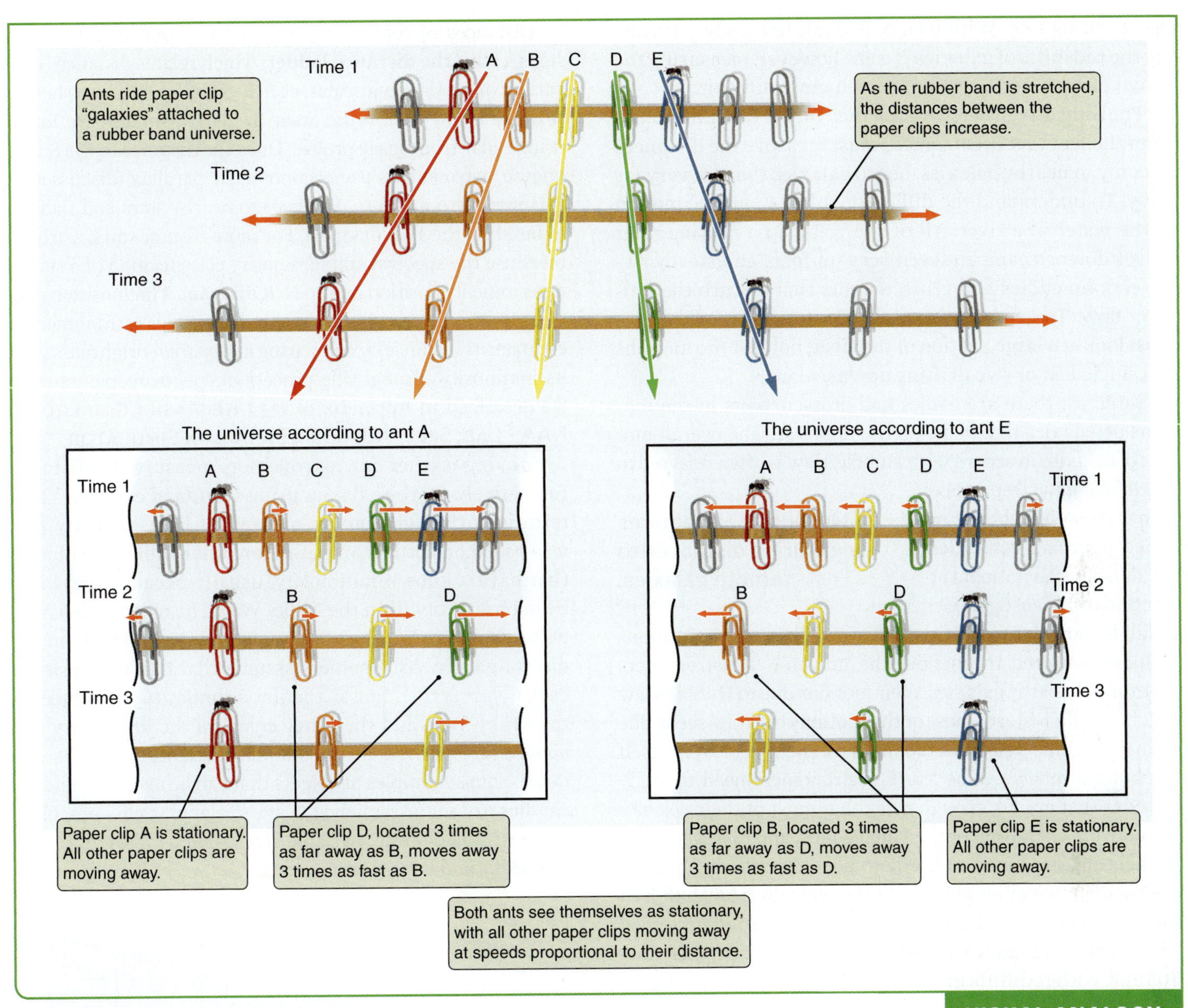

FIGURE 19.5 In this one dimensional analogy of Hubble's law, a rubber band with paper clips evenly spaced along its length is stretched. As the rubber band stretches, an ant riding on clip A observes clip C moving away twice as fast as clip B. Similarly, an ant riding on clip E sees clip C moving away twice as fast as clip D. Any ant sees itself as stationary, regardless of which paper clip it is riding, and it sees the other clips moving away with speed proportional to distance.

The only exception to this rule is the case of galaxies that are close together, in which case gravitational attraction dominates over the expansion of space. For example, the Andromeda Galaxy and the Milky Way are being pulled together by gravity. The Andromeda Galaxy is approaching the Milky Way at about 110 kilometers per second (km/s), so the light from the Andromeda Galaxy is blueshifted, not redshifted. The fact that gravitational or electromagnetic forces can overwhelm the expansion of space also explains why the Solar System is not expanding, and neither are you.

Astronomers Build a Distance Ladder to Measure the Hubble Constant

Hubble's law indicates that the universe is expanding. But to know the present *rate* of the expansion, astronomers need a good value for the Hubble constant, H_0. To obtain this value requires knowing both the recessional velocity and the distance of a large number of galaxies, including galaxies that are very far away. When these velocities and distances are plotted against each other, H_0 is the slope of the resulting line. With a large enough telescope, measuring the redshifts of galaxies is easy; however, measuring the actual distances to galaxies is much more difficult.

The difficulty in determining the Hubble constant comes from the fact that astronomers must measure the distances not only to nearby galaxies, but to galaxies that are very far away. To understand the difficulty, think about the motion of the water in a river. All of the water in a flowing river moves downstream, but even very uniform and steady rivers contain eddies and crosscurrents that disturb the uniform flow. To get a good overall picture of river flow, you must look at a large portion of the river, not just the motions of a single leaf or two drifting downstream.

Similarly, there are eddies and crosscurrents in the motions of galaxies that make up the universe. The overall motion of galaxies in accord with Hubble's law is often referred to as *Hubble flow*. Departures from a smooth Hubble flow, called **peculiar velocities**, result from gravitational attractions that cause galaxies to fall toward their neighbors or toward large concentrations of mass scattered throughout the universe. If astronomers look only at nearby galaxies, their motions due to Hubble's law will be small; instead, most of the motion that they see is due to their peculiar velocities. To measure the Hubble flow itself to obtain a reliable value for H_0, astronomers need to study galaxies that are far enough away that most of their velocity comes from expansion or the Hubble flow, and relatively little of their observed motion is due to peculiar velocity or gravity. Peculiar velocities of galaxies are typically a few hundred kilometers per second, so to determine H_0, astronomers need to measure accurately the distances to galaxies farther than 50 Mpc, with redshifts greater than 0.01.

Measuring H_0 requires measuring distances to remote galaxies.

Distances of remote objects are measured in a series of steps, called the **distance ladder**, which relates distances on a variety of overlapping scales, each method building on the last (**Figure 19.6**). Within the Solar System, distances are found using radar from space probes. Once the distance to the Sun is known, astronomers use trigonometric parallax (discussed in Chapter 13) to measure distances to nearby stars and thereby to build up the H-R diagram. For more distant stars, astronomers use the spectral and luminosity classification of a star to determine its position on the H-R diagram. That position provides a star's luminosity, which in turn enables astronomers to estimate its distance by comparing its *apparent* brightness with its luminosity through the process of spectroscopic parallax (as described in Appendix 6). **NEBRASKA SIMULATION: NAAP LAB: SPECTROSCOPIC PARALLAX SIMULATOR**

Moving farther out, astronomers measure the distance to relatively nearby galaxies using **standard candles**. (This term is borrowed from an old unit of light intensity that was based on actual candles.) Standard candles are objects that have a known luminosity, usually because they have been observed within the Milky Way. The objects must also be bright enough to be observable and recognizable in the distant galaxy. Astronomers assume that the luminosity of each object is the same as that for a similar type of object in the Milky Way, and then they compare the luminosity and apparent brightness of the standard candle to find its distance. Some examples of objects that can be used as standard candles are main-sequence O stars, globular clusters, planetary nebulae, novae, variable stars such as RR Lyraes and Cepheids, and supernovae.

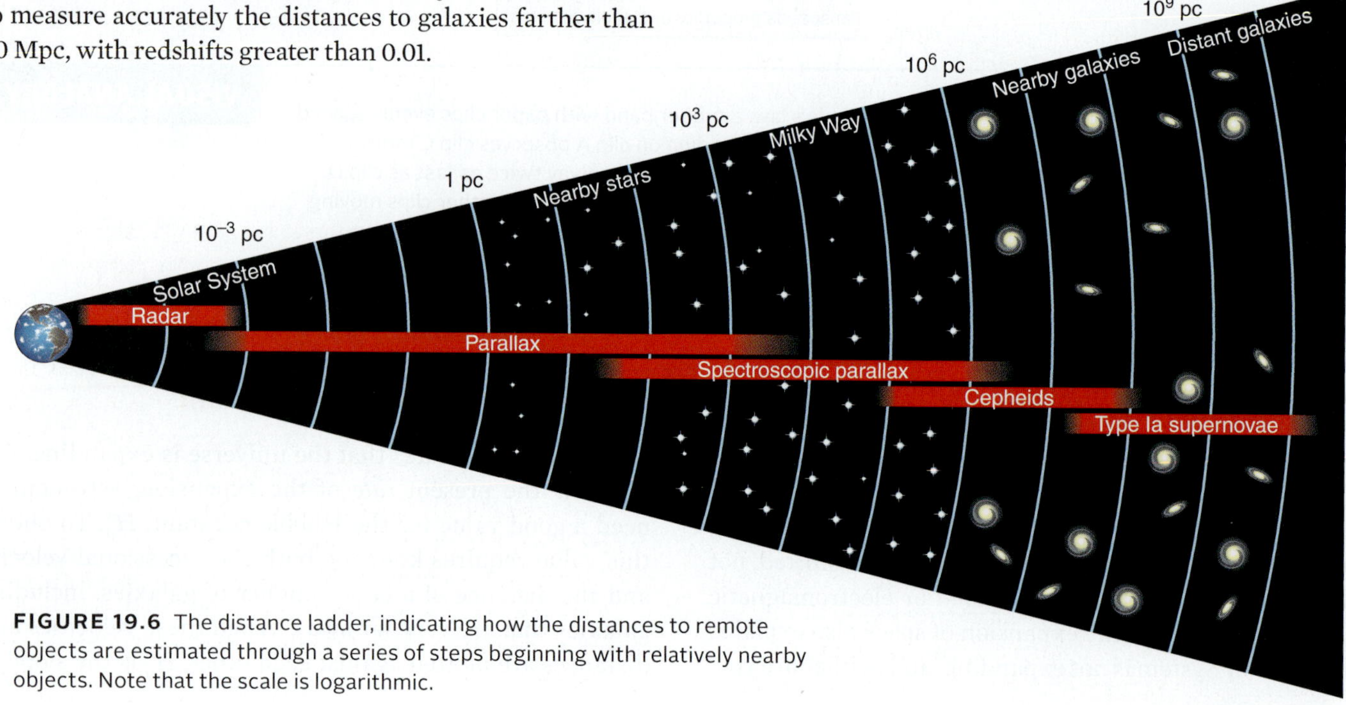

FIGURE 19.6 The distance ladder, indicating how the distances to remote objects are estimated through a series of steps beginning with relatively nearby objects. Note that the scale is logarithmic.

(a)

(b)

FIGURE 19.7 Type Ia supernovae are extremely luminous standard candles. (a) Supernova 2011fe in the Pinwheel Galaxy, 6.4 Mpc away. (b) The galaxy in ultraviolet light before and after the explosion of the supernova.

For example, HST observations of Cepheid variables enable astronomers to accurately measure distances to galaxies as far away as 30 Mpc. Even this distance is not far enough to determine a final value for the Hubble constant, but within that volume of space are many galaxies that can be studied for yet more powerful distance indicators. Among the best of these are Type Ia supernovae.

Recall from Chapter 16 that Type Ia supernovae can occur when gas flows from an evolved star onto its white dwarf companion, so that the white dwarf approaches the Chandrasekhar limit for the mass of an electron-degenerate object. When this happens, the white dwarf burns carbon and then explodes. These Type Ia supernovae occur in white dwarfs of the same mass, so astronomers expect all such explosions to have similar luminosity, with some calibration adjustment for the rate at which the brightness declines after it peaks. An alternate possibility is that both stars are white dwarfs that merge and then explode. In this case there might be a mass difference for the double white dwarf supernova, which could mean a difference in luminosity (because different Type Ia supernovae may have different origins, they are sometimes called "standardizable candles"). To test the prediction that they all have about the same luminosity, astronomers observe Type Ia supernovae in galaxies with distances determined from other methods, such as measuring light curves of Cepheid variables. With a peak luminosity that outshines a billion Suns (**Figure 19.7**),

Math Tools 19.2

Finding the Distance from a Type Ia Supernova

Let's see how astronomers use a standard candle to estimate distance. Figure 19.7 shows M101 (the Pinwheel Galaxy), with a supernova that was observed in 2011. Astronomers can compare the maximum observed brightness of this supernova with the maximum luminosity for this type of supernova to compute the distance.

In Section 13.1 we gave the equation relating brightness, luminosity and distance:

$$\text{Brightness} = \frac{\text{Luminosity}}{4\pi d^2}$$

Rearranging to solve for distance:

$$d = \sqrt{\frac{\text{Luminosity}}{4\pi \times \text{Brightness}}}$$

The maximum observed brightness of this supernova is 7.5×10^{-12} watts per square meter (W/m^2). The graph in Figure 17.11 shows that a typical maximum luminosity of a Type Ia supernova is 9.5×10^9 times the luminosity of the Sun:

$$\text{Luminosity L} = 9.5 \times 10^9 \times L_\odot$$

$$= 9.5 \times 10^9 \times (3.9 \times 10^{26}\ \text{W}) = 3.7 \times 10^{36}\ \text{W}$$

Thus, we can solve the equation:

$$d = \sqrt{\frac{3.7 \times 10^{36}\ \text{W}}{4\pi \times 7.5 \times 10^{-12}\ \text{W/m}^2}} = 2.0 \times 10^{23}\ \text{m}$$

To put this into megaparsecs:

$$d = \frac{2.0 \times 10^{23}\ \text{m}}{3.1 \times 10^{22}\ \text{m/Mpc}} = 6.4\ \text{Mpc}$$

The distance is 6.4 Mpc. This is the closest Type Ia supernova observed since 1986, and it was caught early, so it is one of the youngest supernovae ever observed. Because the Pinwheel Galaxy is relatively close, other standard candles can be observed in this galaxy to help calibrate the distance. Also note that 6.4 Mpc = 21 million light-years, so this supernova explosion took place 21 million years ago.

Type Ia supernovae can be seen and measured with modern telescopes at very high redshifts (**Math Tools 19.2**). ▶▶ NEBRASKA SIMULATION: NAAP LAB: SUPERNOVA LIGHT CURVE FITTING EXPLORER

Figure 19.8 plots the measured recession velocities of galaxies against their measured distances. Because the velocity and distance are proportional to each other, the points lie along a line on the graph with a slope equal to the proportionality constant H_0. Notice how well the data line up along the line. This strong correlation indicates that the universe follows Hubble's law. Hubble's original value was 8 times too large, which led to inconsistencies, but the problem was resolved when astronomers realized there that there are two types of Cepheids with slightly different period-luminosity relationships. Today, astronomers have measured the Hubble constant by several different methods, using observations from the *WMAP*, Hubble, and Spitzer observatories. These yield a value of 70–74 km/s/Mpc. The value is likely to be further refined in the years to come. In this text we use 70 km/s/Mpc for simplicity.

Current measurements put H_0 at 70 km/s/Mpc.

Hubble's Law Maps the Universe in Space and Time

Hubble's law gives astronomers a practical tool for measuring distances to remote objects. Once they know the value of H_0, they can use a straightforward measurement of the redshift of a galaxy to find its distance. In other words, once H_0 is known, Hubble's law makes the once-difficult task of measuring distances in the universe relatively easy, providing astronomers with a tool to map the structure of the observable universe. This may seem like a logical impossibility, because they are using redshifts and distances to find H_0 and then using H_0 to find the distances. But astronomers find H_0 from one set of galaxies and then use that value to find the distance to a different, more distant set of galaxies (see Math Tools 19.1).

Redshift indicates a galaxy's distance.

Since the distances to the galaxies are so large, galaxies are placed in time as well as in space. Light travels at a huge but finite speed. Recall from Chapter 1 that when you look at the Sun, you see it as it existed 8⅓ minutes ago. When you look at Alpha Centauri, the nearest stellar system beyond the Sun, you see it as it existed 4.3 years ago. If you look at a picture

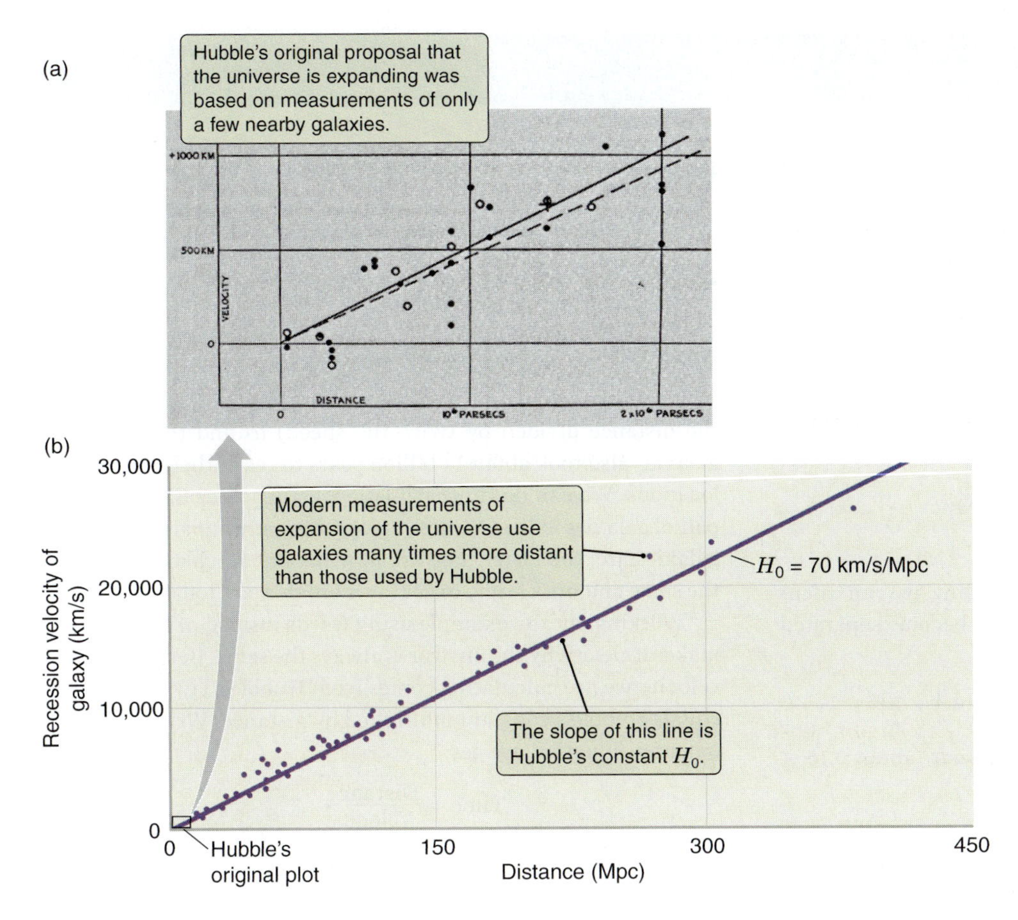

FIGURE 19.8 (a) Hubble's original figure illustrating that more distant galaxies are receding faster than less distant galaxies. (b) Modern data on galaxies many times farther away than those studied by Hubble show that recessional velocity is proportional to distance.

of the center of the Milky Way, the light you see is 27,000 years old. The **look-back time** of a distant object is the time it has taken for the light from that object to reach a telescope on Earth. As astronomers look into the distant universe, look-back times become very great indeed. The distance to a galaxy whose redshift $z = 0.1$ is 1.4 billion light-years (assuming $H_0 = 70$ km/s/Mpc), so the look-back time to that galaxy is 1.4 billion years. The look-back time to a galaxy where $z = 0.2$ is 2.7 billion years. The look-back time of the most distant galaxies observed, at about $z = 10$, is 13.2 billion years. As astronomers observe objects with greater and greater redshifts, they are seeing increasingly younger stages of the universe.

When observing the universe, elsewhere is elsewhen.

19.4 The Universe Began in the Big Bang

Hubble's law provides a very powerful and practical tool for mapping the distribution of galaxies throughout the universe. Yet the most significant aspect of Hubble's law is what it indicates about the structure and origin of the universe itself. All galaxies in the universe are moving away from each other, so if you could imagine running the "movie" of this expansion backward in time, you would find the galaxies getting closer and closer together. According to Hubble's law, about 6.8 billion years ago, when the universe was half its present age, all of the galaxies in the universe were separated from each other by half of their present distances. Twelve billion years ago, all of the galaxies in the universe must have been separated from each other by about a fifth of their present distances. Assuming that galaxies have been moving apart all that time at the same speed as they do today, then 13.7 billion years ago (a time equal to $1/H_0$, where $1/H_0$ is known as the **Hubble time**), *all the stars and galaxies that make up today's universe must have been concentrated together at the same location* (see **Math Tools 19.3**). Today's universe is hurtling outward from a tremendous expansion of space that started approximately 13–14 billion years ago. This colossal event that marked the

Expansion started with the Big Bang.

Math Tools 19.3

Expansion and the Age of the Universe

We can use Hubble's law to estimate the age of the universe. Consider two galaxies located 30 Mpc ($d_G = 9.3 \times 10^{20}$ km) away from each other (**Figure 19.9**). If these two galaxies are moving apart, then at some time in the past they must have been together in the same place at the same time. According to Hubble's law, and assuming that $H_0 = 70$ km/s/Mpc, the distance between these two galaxies is increasing at the following rate:

$$\begin{aligned} v_r &= H_0 \times d_G \\ &= 70 \text{ km/s/Mpc} \times 30 \text{ Mpc} \\ &= 2{,}100 \text{ km/s} \end{aligned}$$

Knowing the speed at which they are traveling, we can calculate the time it took for the two galaxies to become separated by 30 Mpc:

$$\text{Time} = \frac{\text{Distance}}{\text{Speed}} = \frac{9.3 \times 10^{20} \text{ km}}{2{,}100 \text{ km/s}} = 4.4 \times 10^{17} \text{ s}$$

Dividing by the number of seconds in a year (about 3.16×10^7 s/yr) gives:

$$\text{Time} = 1.4 \times 10^{10} \text{ yr}$$

In other words, *if* expansion of the universe has been constant, two galaxies that today are 30 Mpc apart started out at the same place 14 billion years ago.

Now let's do the same calculation with two galaxies that are 60 Mpc, or 18.6×10^{20} km, apart. These two galaxies are twice as far apart, but the distance between them is increasing twice as rapidly:

$$v_r = H_0 \times d_G = 70 \text{ km/s/Mpc} \times 60 \text{ Mpc} = 4{,}200 \text{ km/s}$$

Therefore:

$$\text{Time} = \frac{18.6 \times 10^{20} \text{ km}}{4{,}200 \text{ km/s}} = 4.4 \times 10^{17} \text{ s} = 1.4 \times 10^{10} \text{ yr}$$

Again, we calculate time as distance divided by speed (twice the distance divided by twice the speed) to find that these galaxies also took about 14 billion years to reach their current locations. We can do this calculation again and again for any pair of galaxies in the universe today. The farther apart the two galaxies are, the faster they are moving. But all galaxies took the same amount of time to get to where they are today.

Working out the example using words instead of numbers makes it clear why the answer is always the same. Because the velocity we are calculating comes from Hubble's law, velocity equals Hubble's constant multiplied by distance. Writing this out as an equation, we get:

$$\begin{aligned} \text{Time} &= \frac{\text{Distance}}{\text{Velocity}} \\ &= \frac{\text{Distance}}{H_0 \times \text{Distance}} \end{aligned}$$

Distance divides out to give:

$$\text{Time} = \frac{1}{H_0}$$

where $1/H_0$ is the Hubble time. This is one way of estimating the age of the universe.

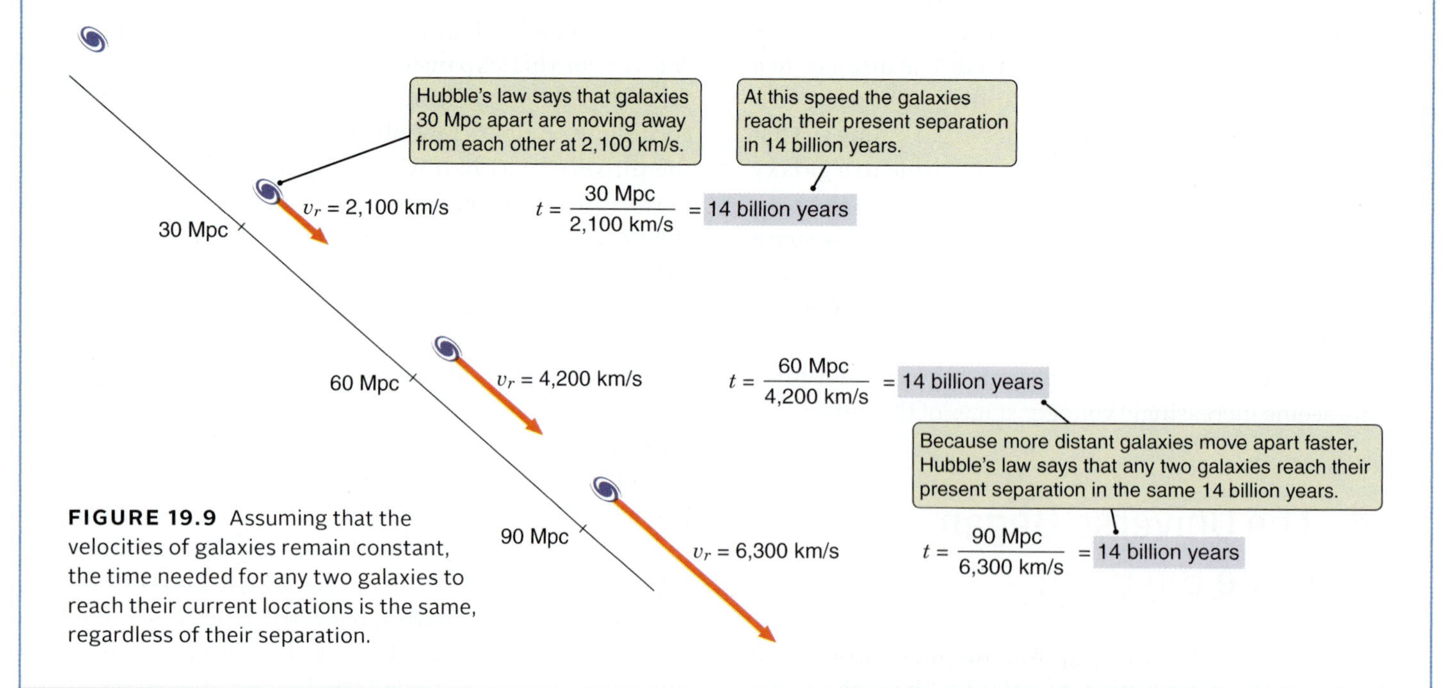

FIGURE 19.9 Assuming that the velocities of galaxies remain constant, the time needed for any two galaxies to reach their current locations is the same, regardless of their separation.

beginning of the universe is called the **Big Bang** (**Figure 19.10**). ▶II **AstroTour: Hubble's Law**

Georges Lemaître (1894–1966), a professor of physics, cosmologist, and priest at the Catholic University of Leuven in Belgium, was the first to propose the theory of the Big Bang. This idea greatly troubled many astronomers in the early and middle years of the 20th century. Several different suggestions were put forward to explain the observed fact of Hubble expansion without resorting to the idea that the universe came into existence in an extraordinarily dense "fireball" billions of years ago. However, as more and more distant galaxies have been observed, and as more discoveries about the structure of the universe have been made, the Big Bang theory has grown stronger. Virtually all the major predictions of the Big Bang theory have proven to be correct. The Big Bang theory for the origin of the universe is now such a well-corroborated theory that most astronomers would say it has crossed into the realm of scientific fact.

The implications of Hubble's law are striking. This single discovery forever changed the scientific concepts of the origin, history, and possible future of the universe. At the same time, Hubble's law has pointed to many new questions about the universe. To address them, we next need to consider exactly what is meant by the term *expanding universe*.

Galaxies Are *Not* Flying Apart through Space

At this point in our discussion, you may be picturing the expanding universe as a cloud of debris from an explosion flying outward through surrounding space. However, this is not an explosion in the usual sense of the word, and in fact there is no surrounding space.

A common question about the Big Bang is, Where did it take place? The answer is that the Big Bang took place *everywhere*. Wherever anything is in the universe today, it is at the site of the Big Bang. The reason is that galaxies are not flying apart through space at all. Rather, *space itself* is expanding, carrying the stars and galaxies that populate the universe along with it.

The Big Bang happened everywhere.

We have already dealt with the basic ideas that explain the expansion of space. In our discussion of black holes in Chapter 18, you encountered Einstein's general theory of relativity. General relativity says that space is distorted by the presence

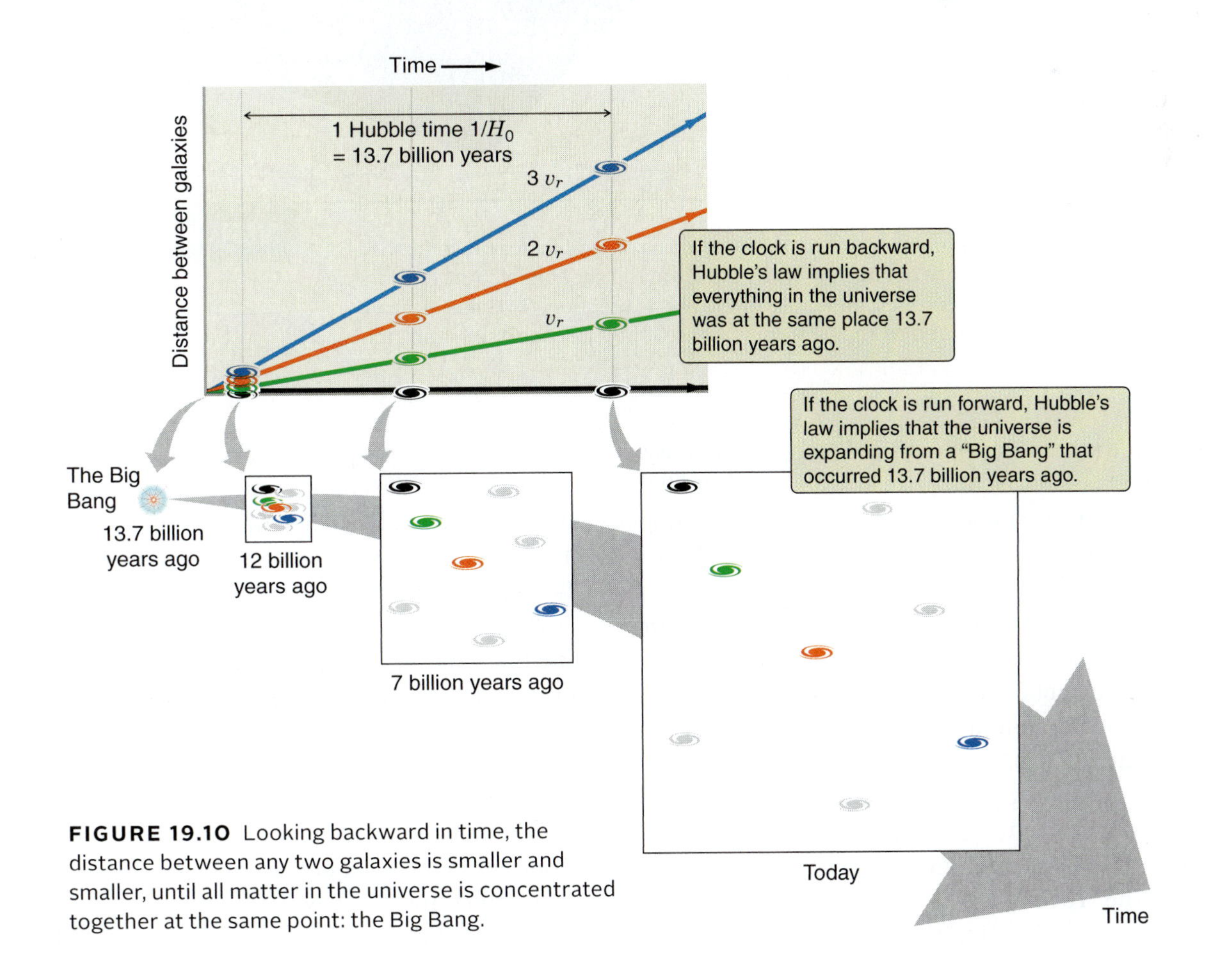

FIGURE 19.10 Looking backward in time, the distance between any two galaxies is smaller and smaller, until all matter in the universe is concentrated together at the same point: the Big Bang.

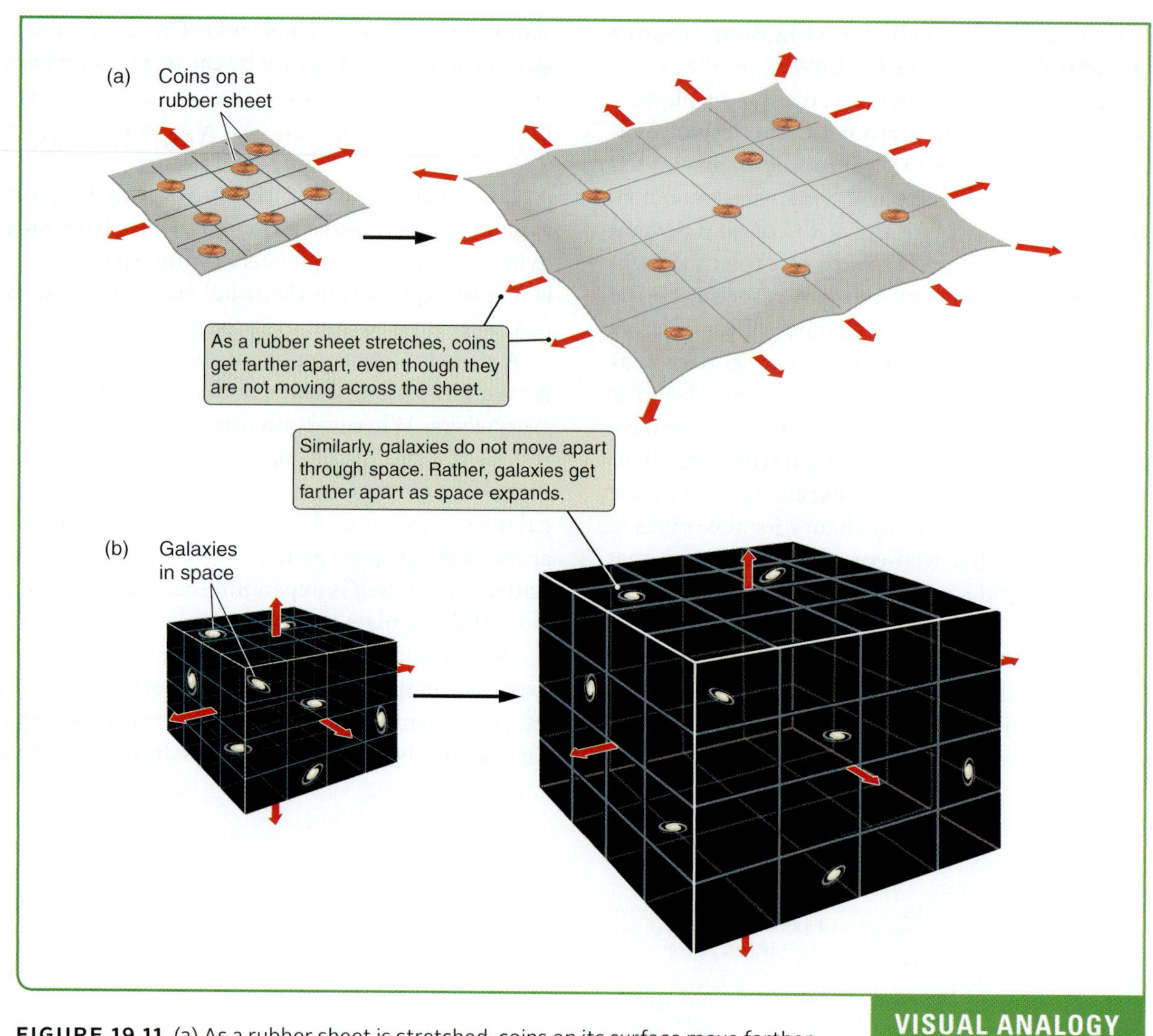

FIGURE 19.11 (a) As a rubber sheet is stretched, coins on its surface move farther apart, even though they are not moving with respect to the sheet itself. Any coin on the surface of the sheet observes a Hubble-like law in every direction. (b) Galaxies in an expanding universe are not flying apart through space. Rather, space itself is stretching.

of mass, and that the consequence of this distortion is gravity. For example, the mass of the Sun, like any other object, distorts the geometry of spacetime around it; so Earth, coasting along in its inertial frame of reference, follows a curved path around the Sun. We illustrated this phenomenon in Figure 18.10 with the analogy of a ball placed on a stretched rubber sheet, showing how the ball distorted the surface of the sheet.

The surface of a rubber sheet can be distorted in other ways as well. Imagine a number of coins placed on a rubber sheet (**Figure 19.11**). Then imagine grabbing the edges of the sheet and beginning to pull them outward. As the rubber sheet stretches, each coin remains at the same location on the surface of the sheet, but the distances between the coins increase. Two coins sitting close to each other move apart only slowly, while coins farther apart move away from each other more rapidly. The distances and relative motions of the coins on the surface of a rubber sheet obey a Hubble-like relationship as the sheet is stretched.

The expansion of space is similar to stretching a two-dimensional rubber sheet.

This movement is analogous to what is happening in the universe, with galaxies taking the place of the coins and space itself taking the place of the rubber sheet. In the case of coins on a rubber sheet, there is a limit to how far the sheet can be stretched before it breaks. With space and the real universe, there is no such limit. The fabric of space can, in principle, go on expanding forever. Hubble's law is the observational consequence of the fact that the space making up the universe is expanding.

How will this expansion of the universe behave in the future? In the next few chapters you will learn that most of the mass in the universe consists of invisible *dark matter*, and the gravity of the dark matter has the effect of slowing down the expansion of the universe. But you will also learn that there is another unseen constituent of the universe, called *dark energy*, and this constituent causes the expansion of the universe to *accelerate*. At the current stage in the expansion of the universe, the accelerating effect of dark energy dominates over the slowing effect of dark matter, and therefore the universe will continue to expand at an ever-faster rate.

Expansion Is Described with a Scale Factor

The expansion of the universe is sometimes discussed in terms of the **scale factor** of the universe. To understand this concept, let's return to the analogy of the rubber sheet. Suppose you place a ruler on the surface of the rubber sheet and draw a tick mark every centimeter (**Figure 19.12a**). To measure the distance between two points on the sheet, you can count the marks between the two points and multiply by 1 centimeter (cm) per tick mark.

As the sheet is stretched, however, the distance between the tick marks does not remain 1 cm. When the sheet is stretched to 150 percent of the size it had when the ruler was drawn, each tick mark is separated from its neighbors by 1½ times the original distance, or 1.5 cm. The distance between two points can still come from counting the marks, but you need to scale up the distance between tick marks by 1.5 to find the distance in centimeters. The scale factor of the sheet is now 1.5. If the sheet were twice the size it was when the ruler was drawn (**Figure 19.12b**), each mark would correspond to 2 cm of actual distance; the scale factor of the sheet would now be 2. The scale factor indicates the size of the sheet relative to its size at the time when the ruler was drawn. The scale factor also indicates how much the distance between points on the sheet has changed.

Suppose astronomers choose today to lay out a "cosmic ruler" on the fabric of space, placing an imaginary tick mark every 10 Mpc. The scale factor of the universe at this time is defined to be 1. In the past, when the universe was smaller, distances between the points in space marked by this cosmic ruler would have been less than 10 Mpc. The scale factor of that younger, smaller universe would have been less than 1 compared to the scale factor today. In the future, as the universe continues to expand, the distances between the tick marks on this cosmic ruler will grow to more than 10 Mpc, and the scale factor of the universe will be greater than 1. Astronomers use the scale factor, usually written as R_U, to keep track of the changing scale of the universe.

The scale factor, R_U, increases as the universe expands.

It is important to remember that the laws of physics are themselves unchanged by the expansion of the universe, just as stretching a rubber sheet does not change the properties of the coins on its surface. At noncosmological scales, the nuclear and electromagnetic forces within and between atoms, as well as the gravitational forces between relatively nearby objects, dominate over the expansion. As the universe expands, the sizes and other physical properties of atoms, stars, and galaxies also remain unchanged.

Expansion does not affect local physics—stars, atoms, or any material object.

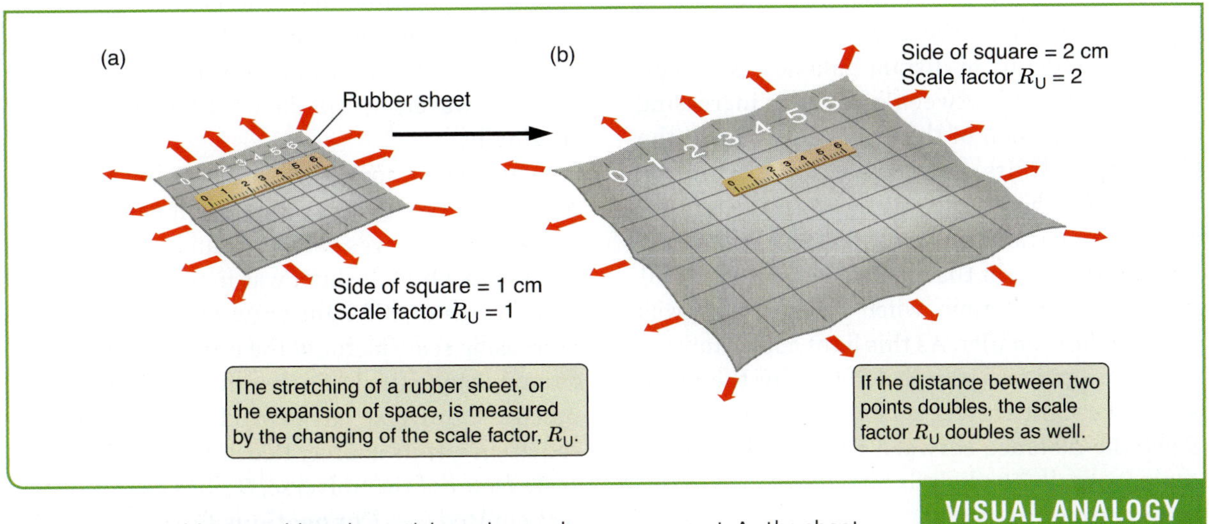

FIGURE 19.12 (a) On a rubber sheet, tick marks are drawn 1 cm apart. As the sheet is stretched, the tick marks move farther apart. (b) When the spacing between the tick marks is 2 cm, or twice the original value, the scale factor of the sheet, R_U, is said to have doubled. A similar scale factor, R_U, is used to describe the expansion of the universe.

Let's return now to the question of locating the center of expansion. Looking back in time, the scale factor of the universe gets smaller and smaller, approaching zero as it comes closer and closer to the Big Bang. The fabric of space that today spans billions of parsecs spanned much smaller distances when the universe was young. When the universe was only a day old, all of the space visible today amounted to a region only a few times the size of the Solar System. When the universe was 1/50 of a second old, the vast expanse of space that makes up today's observable universe (and all the matter in it) occupied a volume only the size of today's Earth. Going backward in time approaching the Big Bang itself, the space that makes up today's observable universe becomes smaller and smaller—the size of a grapefruit, a marble, an atom, a proton. Every point in the fabric of space that makes up today's universe was right there at the beginning, a part of that unimaginably tiny, dense universe that emerged from the Big Bang.

These points bear repetition: *Where is the center of the Big Bang?* There is no center. The Big Bang did not occur at a specific point in space, because space itself came into existence with the Big Bang. *Where did the Big Bang happen?* It happened everywhere, including right where you are sitting. This is an important concept. If a particular point in today's universe marked the site of the Big Bang, that would be a very special point indeed. But there is no such point. The Big Bang happened everywhere. A Big Bang universe is homogeneous and isotropic, consistent with the cosmological principle.

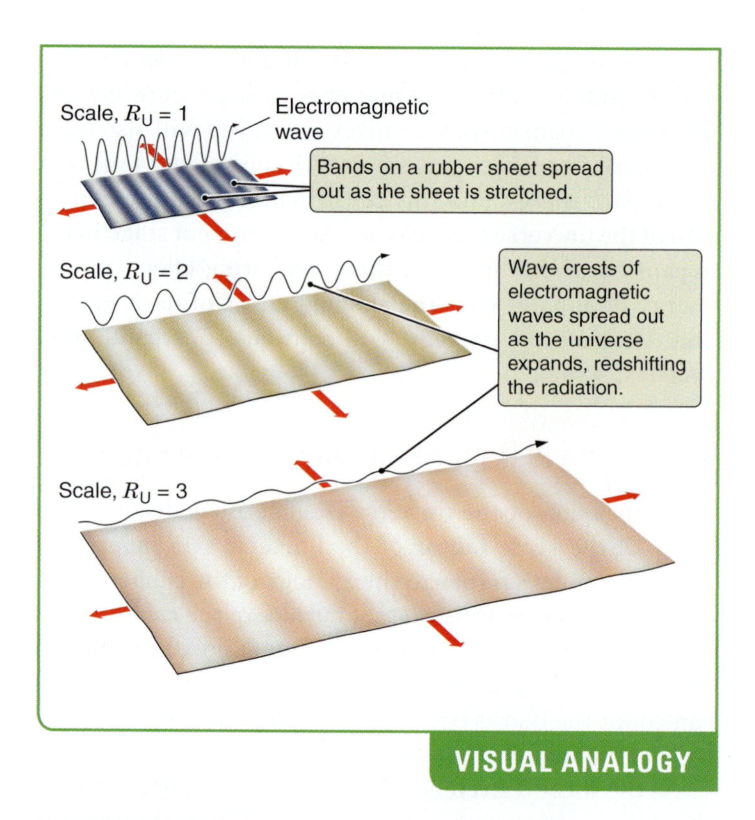

FIGURE 19.13 Bands drawn on a rubber sheet represent the positions of the crests of an electromagnetic wave in space. As the rubber sheet is stretched—that is, as the universe expands—the wave crests get farther apart. The light is redshifted.

Redshift Is Due to the Changing Scale Factor of the Universe

General relativity is a powerful tool for interpreting Hubble's great discovery of the expanding universe. It also relates to the redshift of distant galaxies. Although it is true that the distance between galaxies is increasing as a result of the expansion of the universe, and that the *equation* for Doppler shifts can be used to measure the redshifts of galaxies, these redshifts are not due to Doppler shifts in the same way that we described for a moving star in the Milky Way. Light that comes from very distant objects was emitted at a time when the universe was younger and therefore smaller. As this light comes toward Earth from distant galaxies, the scale factor of the space through which the light travels is constantly increasing; and as it does, the distance between adjacent light wave crests increases as well. The light is "stretched out" as the space it travels through expands.

Let's return to the rubber sheet analogy. If you draw a series of bands on the rubber sheet to represent the crests of an electromagnetic wave, as in **Figure 19.13**, you can watch what happens to the wave as the sheet is stretched out. By the time the sheet is stretched to twice its original size—that is, by the time the scale factor of the sheet is 2—the distance between wave crests has doubled. When the sheet has been stretched to 3 times its original size (a scale factor of 3), the wavelength of the wave will be 3 times what it was originally.

This idea is applied to light coming from a distant galaxy. When the light left the distant galaxy, the scale factor of the universe was smaller than it is today. The universe expanded while the light was in transit, and as it did so, the wavelength of the light grew longer in proportion to the increasing scale factor of the universe. The redshift of light from distant galaxies is therefore a direct measure of how much the universe has expanded since the time when the radiation left its source. Redshift measures how much the scale factor of the universe, R_U, has changed since the light was emitted (see **Connections 19.1**).

Connections 19.1

When Redshift Exceeds 1

In our discussion of the Doppler shift in Chapter 5, we noted that $(\lambda_{obs} - \lambda_{rest})/\lambda_{rest}$ is equal to the velocity of an object moving away, divided by the speed of light. In this chapter you have seen how Edwin Hubble used this result to interpret the observed redshifts of galaxies as evidence that galaxies throughout the universe are moving away from the Milky Way. Einstein's special theory of relativity says that nothing can move faster than the speed of light. Hubble initially assumed that redshifts are due to the Doppler effect. The resulting relation, $z = v_r/c$, would then seem to imply that no object can have a redshift (z) greater than 1. Yet that is not the case. Astronomers routinely observe redshifts significantly in excess of 1. As of this writing, the most distant objects known have redshifts as large as 9 or 10. How can redshifts exceed 1?

To arrive at the expression for the Doppler effect, $v_r/c = (\lambda_{obs} - \lambda_{rest})/\lambda_{rest}$, we have to *assume* that v_r is much less than c. If v_r was close to c, we would have to consider more than just the fact that the waves from an object are stretched out by the object's motion away. We would also have to consider relativistic effects, including the fact that moving clocks run slowly (see Math Tools 18.1). When combining these effects, we would find that as the speed of an object approaches the speed of light, its redshift approaches infinity (**Figure 19.14**).

A second source of redshift is the gravitational redshift discussed in Chapter 18. As light escapes from deep within a gravitational well, it loses energy, so photons are shifted to longer and longer wavelengths. If the gravitational well is deep enough, then the observed redshift of this radiation can be boundlessly large. In fact, the event horizon of a black hole—that is, the surface around the black hole from which not even light can escape—is where the gravitational redshift becomes infinite.

Cosmological redshift, which is most relevant to this chapter, results from the amount of "stretching" that space has undergone during the time the light from its original source has been en route to Earth. The amount of stretching is given by the factor $1 + z$. When astronomers observe light from a distant galaxy whose redshift $z = 1$, then the wavelength of this light is twice as long as when it left the galaxy. When the light left its source, the universe was half the size that it is today. When they see light from a galaxy with $z = 2$, the wavelength of the radiation is 3 times its original wavelength, and they are seeing the universe when it was one-third its current size. This direct relationship enables astronomers to use the observed redshift of the galaxy to calculate the size of the universe at the look-back time to that galaxy. Nearby, this means that distance and look-back time are proportional to z. As they look back closer and closer to the Big Bang, however, redshift climbs more and more rapidly.

Written as an equation, the scale factor of the universe that astronomers see when looking at a distant galaxy is equal to 1, divided by 1 plus the redshift of the galaxy:

$$R_U = \frac{1}{1+z}$$

Doppler's original formula is essentially correct—for relatively nearby objects whose measured velocities are far less than the speed of light. In this case, $v_r/c = (\lambda_{obs} - \lambda_{rest})/\lambda_{rest}$. When astronomers look at the motions of orbiting binary stars or the peculiar velocities of galaxies relative to the Hubble flow, this equation works just fine. But anytime there is a redshift of 0.4 or greater, relativity must be taken into account.

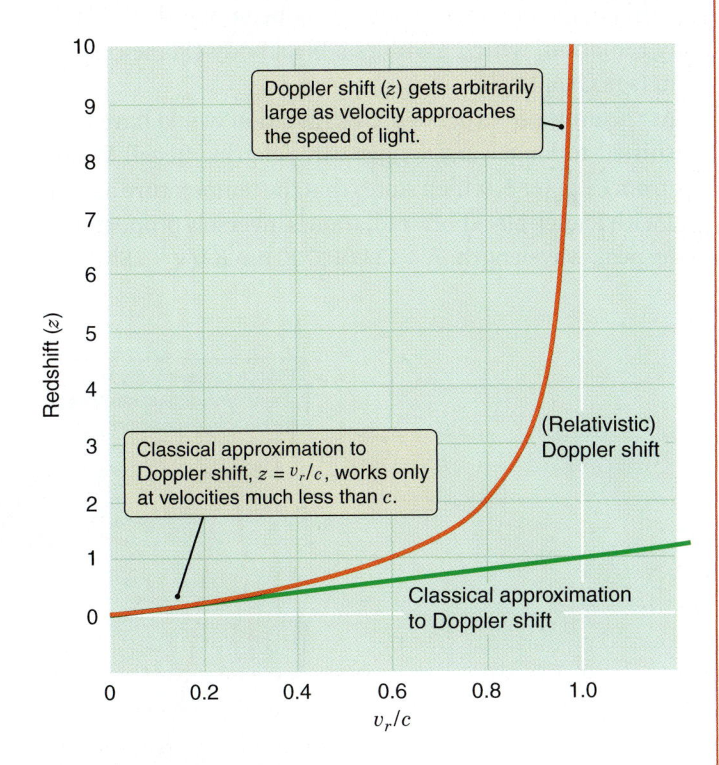

FIGURE 19.14 Plot of the redshift (z) of an object versus its recessional velocity (v_r) as a fraction of the speed of light. According to special relativity, as v_r approaches c, the redshift becomes large without limit.

19.5 Astronomers Observe Radiation Left Over from the Big Bang

Cosmological questions about the origin of the universe are some of the most fundamental questions humankind can ask. Throughout human history, answers to these questions have been the highly prized goals of philosophers and theologians. It is quite remarkable to live in a time when astronomers are finding real, testable answers to these questions by using the empirical methods of science. It is essential, then, that the evidence supporting the theory of the Big Bang be of extraordinary quality. What is the evidence that the Big Bang actually took place?

The expansion of the universe can be related to changes in temperature: when a gas is compressed it grows hotter, but as it expands it becomes cooler. Since the universe is expanding, it must also be cooling. Consequently, when the universe was very young and small, it must have consisted of an extraordinarily hot, dense gas. As with any hot, dense gas, this early universe would have been awash in blackbody radiation, which exhibits a blackbody (Planck) spectrum (see Chapter 5).

As the universe expanded, this radiation would have been redshifted to longer and longer wavelengths. Recall Wien's law from Chapter 5, which states that the temperature associated with Planck blackbody radiation is inversely proportional to the peak wavelength: $T = (2{,}900{,}000 \text{ nm K})/\lambda_{peak}$. Shifting the Planck radiation to longer and longer wavelengths is therefore the equivalent of shifting the temperature of the radiation to lower and lower values. As illustrated in **Figure 19.15**, doubling the wavelength of the photons in a Planck spectrum by doubling the scale factor of the universe is equivalent to cutting the temperature of the Planck spectrum in half. In the late 1940s, cosmologists Ralph Alpher (1921–2007), Robert Herman (1914–1997), and George Gamow (1904–1968) concluded that this radiation should still be visible today and should have a Planck blackbody spectrum with a temperature of about 5–50 kelvins (K).

The existence of the glow from a young hot universe was predicted in 1948.

Measuring the Temperature of the CMB

This prediction remained untested until the early 1960s, when, as we noted in Chapter 6, two physicists at Bell Laboratories—Arno Penzias and Robert Wilson (**Figure 19.16**)—detected a faint microwave signal in all parts of the sky. Physicist Robert Dicke (1916–1997) and his colleagues at Princeton University also predicted a hot early universe, arriving independently at the same basic conclusions that Alpher and Gamow had reached two decades earlier. When Dicke and colleagues heard of the signal that Penzias and Wilson had found, they interpreted

Penzias and Wilson discovered the cosmic background radiation.

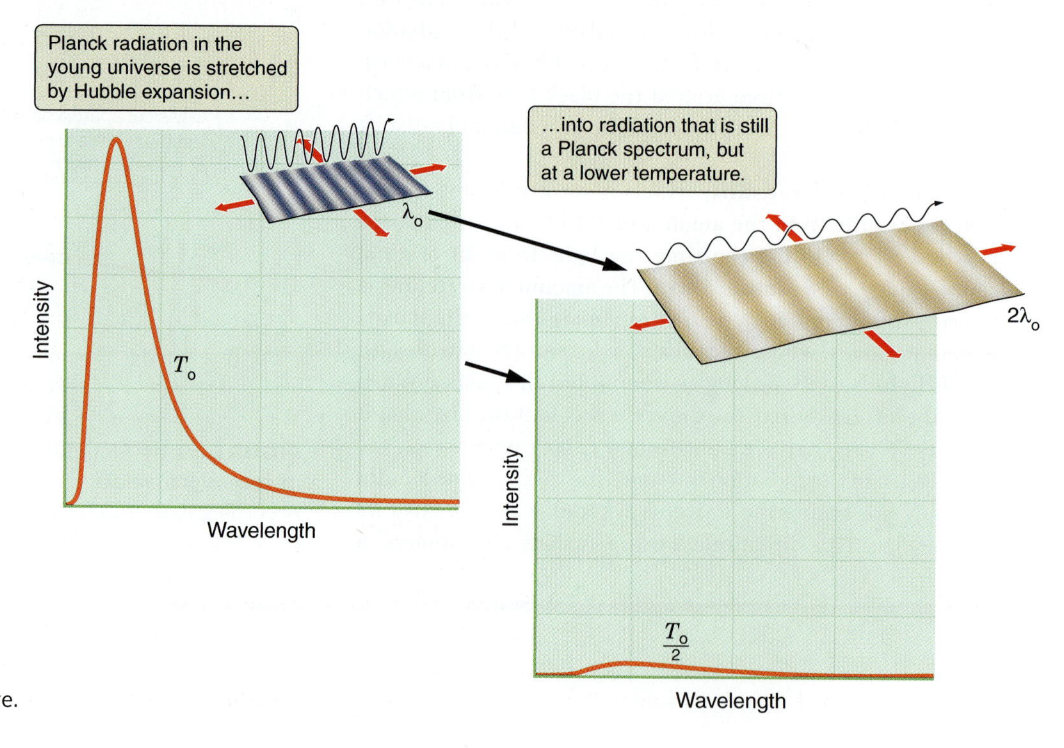

FIGURE 19.15 As the universe expanded, Planck radiation left over from the hot young universe was redshifted to longer wavelengths. Redshifting a Planck spectrum is equivalent to lowering its temperature.

FIGURE 19.16 Robert Wilson (left) and Arno Penzias next to the Bell Labs radio telescope antenna with which they discovered the cosmic background radiation. This antenna is now a U.S. National Historic Landmark.

it as the radiation left behind by the hot early universe. The strength of the detected signal was consistent with the glow from a blackbody with a temperature of about 3 K, very close to the predicted value. Their results, published in 1965, reported the discovery of the "glow" left behind by the Big Bang.

It is worth noting that timing can be everything in science. Alpher, who first predicted the existence of a faint glow from the Big Bang, had searched unsuccessfully for the signal 10 years before Penzias and Wilson made their discovery. Unfortunately, however, the technology of the late 1940s and early 1950s was simply not up to the task.

This radiation left over from the early universe is called the **cosmic microwave background radiation (CMB)**. Today, the CMB and the conditions in the early universe are much better understood than they were in the early 1960s. When the universe was young, it was hot enough that all atoms were ionized, so that the electrons were separate from the atomic nuclei. Recall from our discussion of the structure of the Sun and stars in Chapter 14 that radiation does not travel well through ionized plasma. The free electrons in the plasma interact strongly with the radiation, blocking its progress.

The CMB is thermal radiation that arose when the universe was hot and ionized.

The conditions within the early universe were much like the conditions within a star: the universe was an opaque blackbody (**Figure 19.17a**). As the universe expanded, the gas filling it cooled. By the time the universe was about a thousandth of its current size, the temperature had dropped to a few thousand kelvins, so protons and electrons were able to combine to form hydrogen atoms. This event, called the **recombination** of the universe, occurred when the universe was several hundred thousand years old (**Figure 19.17b**).

Hydrogen atoms are much less effective at blocking radiation than free electrons are, so when recombination occurred, the universe suddenly became transparent to radiation. Since that time, the radiation left behind from the Big Bang has been able to travel largely unimpeded throughout the universe. At the time of recombination, when the temperature of the universe was a few thousand kelvins, the wavelength of this radiation peaked at about 1 micron (μm), according to Wien's law (see Chapter 5). As the universe expanded, this radiation was redshifted to longer and longer wavelengths—and therefore cooler temperatures (see Figure 19.15). Today, the scale of the universe has increased a thousandfold since recombination, and the peak wavelength of the cosmic background radiation has increased by a thousandfold as well, to a value close to 1 millimeter (mm). The spectrum of the CMB still has the shape of a Planck blackbody spectrum, but with a characteristic temperature of 2.73 K—only a thousandth what it was at the time of recombination.

Since recombination, the CMB has traveled freely and cooled by a factor of 1,000.

The presence of cosmic background radiation with a Planck blackbody spectrum is a very strong prediction of the Big Bang theory. Penzias and Wilson had confirmed that a signal with the correct strength was there, but they could not say for certain whether the signal they saw had the spectral shape of a Planck spectrum. From the late 1960s to the 1980s, most experiments at different wavelengths supported these same conclusions. Instruments carried on the COBE satellite launched in 1989 made extremely precise measurements of the CMB at many wavelengths, from a few microns out to 1 cm. Recall from Figure 6.36 that the COBE spectrum of the CMB is a Planck blackbody spectrum with a temperature of 2.73 K. The agreement between theoretical prediction and observation is truly remarkable. The observed spectrum so perfectly matches the one predicted by Big Bang cosmology that there can be no real doubt that this is the residual radiation left behind from the primordial fireball of the early universe.

COBE unambiguously showed the CMB to have a Planck spectrum at 2.73 K.

The CMB Measures Earth's Motion Relative to the Universe Itself

COBE data included much more than a measurement of the spectrum of the cosmic background radiation. **Figure 19.18a** shows a map obtained by COBE of the CMB from

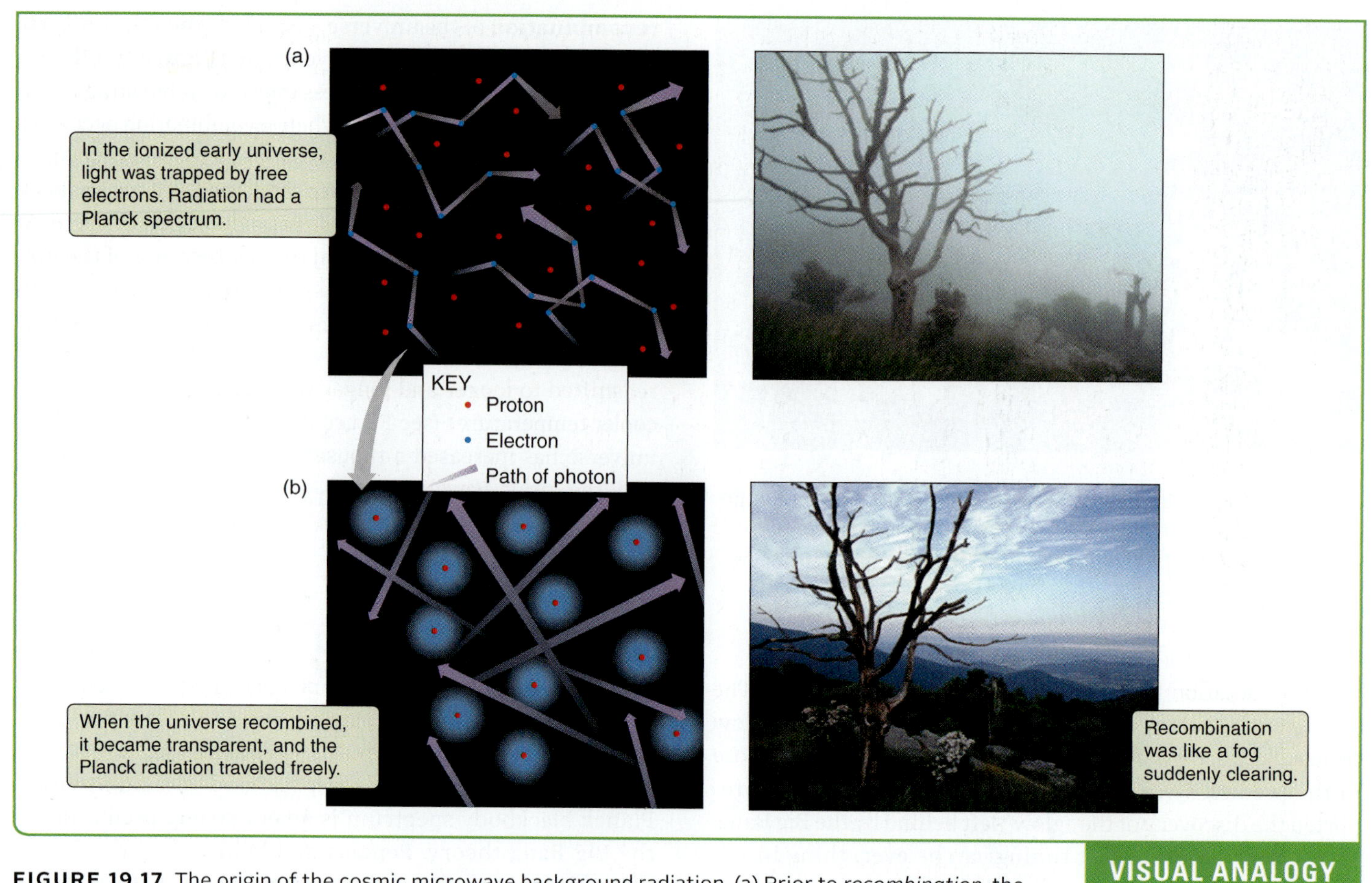

FIGURE 19.17 The origin of the cosmic microwave background radiation. (a) Prior to *recombination*, the universe was like a foggy day, except that the "fog" was a sea of electrons and protons. Radiation interacted strongly with free electrons and so could not travel far. The trapped radiation had a Planck spectrum. (b) When the universe recombined to form neutral hydrogen atoms, the fog cleared and this radiation was free to travel unimpeded.

the entire sky. The different colors in the map correspond to variations of about 0.1 percent in the temperature of the CMB. Most of this range of temperature is present because one side of the sky looks slightly warmer than the opposite side of the sky. This difference has nothing to do with the large-scale structure of the universe itself, but rather is the result of the motion of Earth with respect to the CMB.

Some CMB asymmetry is caused by the motion of Earth.

We have emphasized that there is no preferred frame of reference. The laws of physics are the same in *any* inertial reference frame, so none is better than any other. Yet in a certain sense there is a preferred frame of reference at every point in the universe. This is the frame of reference that is at rest with respect to the expansion of the universe and in which the CMB is isotropic, or the same in all directions. The COBE map shows that one side of the sky is slightly hotter than the other because Earth and the Sun are moving at a velocity of 368 km/s in the direction of the constellation of Crater, relative to this cosmic reference frame. Radiation coming from the direction in which Earth is moving is slightly blueshifted (shifted to a higher characteristic temperature) by this motion, whereas radiation coming from the opposite direction is Doppler-shifted toward the red (or cooler temperatures). Earth's motion is due to a combination of factors, including the motion of the Sun around the center of the Milky Way Galaxy and the motion of the Milky Way relative to the CMB.

When this asymmetry in the CMB caused by the motion of Earth is subtracted from the COBE map, only slight variations in the CMB remain (**Figure 19.18b**), with amplitudes only about 1/100,000 the brightness of the CMB. This means that the brighter parts of this image are only about 1.00001 times brighter than the fainter parts. These slight variations might not seem like much, but they are actually of

Observations show tiny CMB variations caused by the early formation of structure.

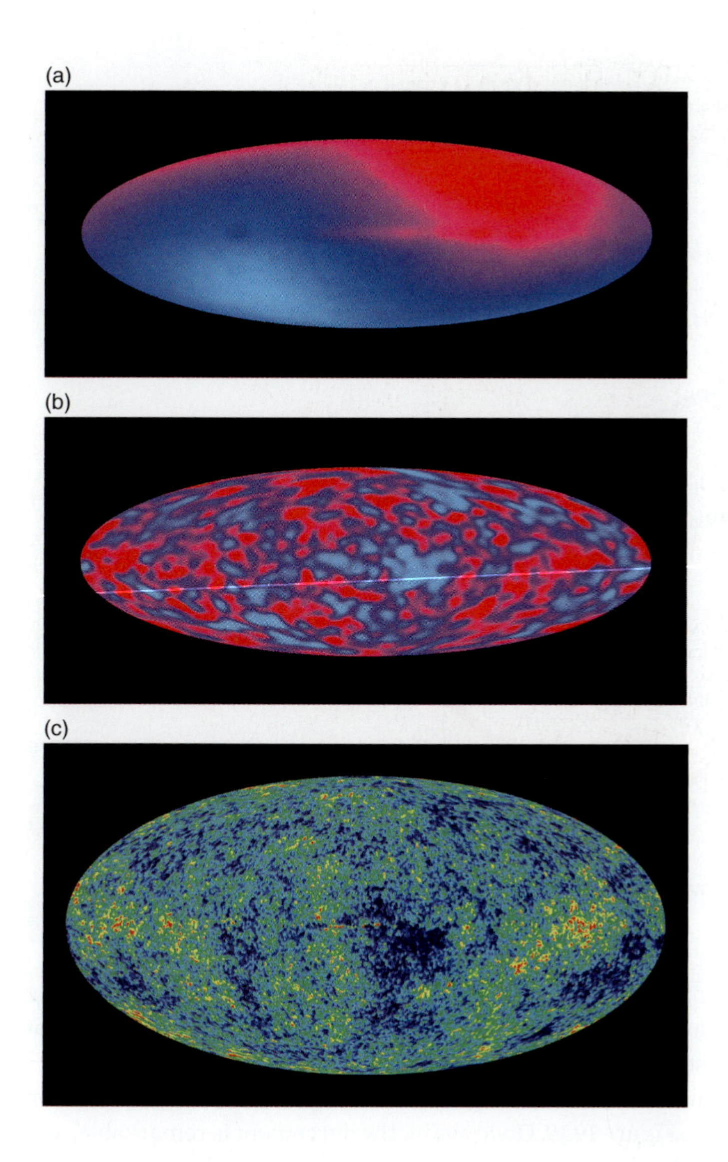

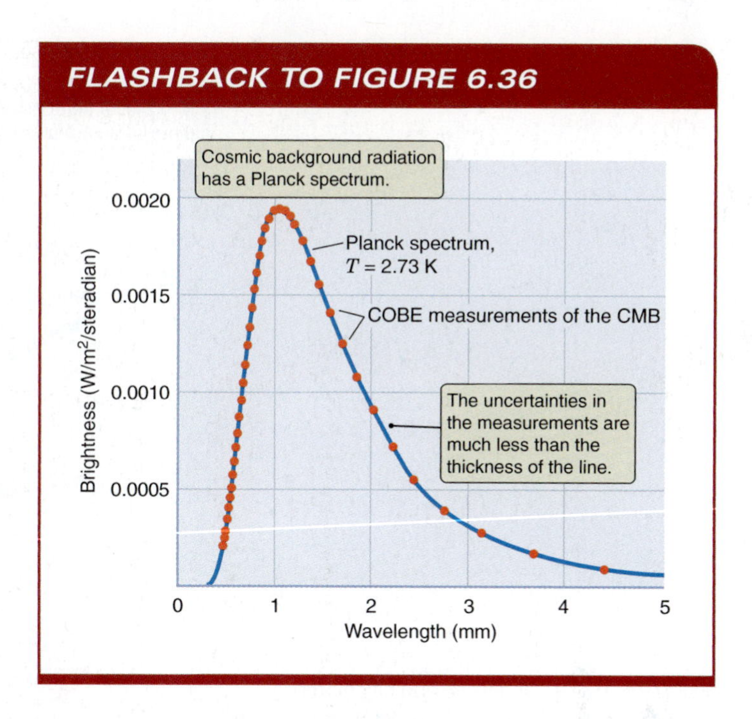

FIGURE 19.18 (a) The COBE map of the cosmic background radiation. The CMB is slightly hotter (by about 0.003 K) in one direction in the sky than in the other direction. This difference is due to Earth's motion relative to the CMB. (b) The COBE map with Earth's motion removed, showing tiny ripples remaining in the CMB. (c) A *WMAP* image of the CMB. The radiation seen here was emitted less than 400,000 years after the Big Bang.

crucial importance in the history of the universe. Recall from Chapter 18 that gravity itself can create a redshift. These tiny fluctuations in the cosmic background radiation are the result of gravitational redshifts caused by concentrations of mass that existed in the early universe. These concentrations later gave rise to galaxies and the rest of the structure that is evident in the universe today.

Subsequent observations from instruments carried aloft by balloons in Antarctica supported the COBE findings. Since 2001, more precise measurements of the CMB variations have been carried out by a NASA satellite named the Wilkinson Microwave Anisotropy Probe, or *WMAP*. **Figure 19.18c** shows the ripples measured by *WMAP* with much higher resolution than could be detected by COBE. The European Space Agency's Planck satellite collected data of even higher resolution from 2009 to 2012. The much higher-resolution maps obtained by *WMAP* and Planck enable astronomers to refine their ideas about the development of structure in the early universe, which we will discuss in Chapter 23.

19.6 Origins: Big Bang Nucleosynthesis

Another key piece of evidence supporting the Big Bang theory came from observations of the number and types of chemical elements in the universe. When the universe was only a few minutes old, its temperature and density were high enough for nuclear reactions to take place. Collisions between protons in the early universe built up low-mass nuclei, including deuterium (heavy hydrogen) and isotopes of helium, lithium, beryllium, and boron. This process, called **Big Bang nucleosynthesis**, determined the final chemical composition of the matter that emerged from the hot phase of the Big Bang.

ASTROTOUR: BIG BANG NUCLEOSYNTHESIS

No elements more massive than boron formed in the Big Bang, because at that time the density in the universe was too low for reactions such as the triple-alpha process, which forms carbon in the interiors of stars (see Chapter 16). Therefore, all

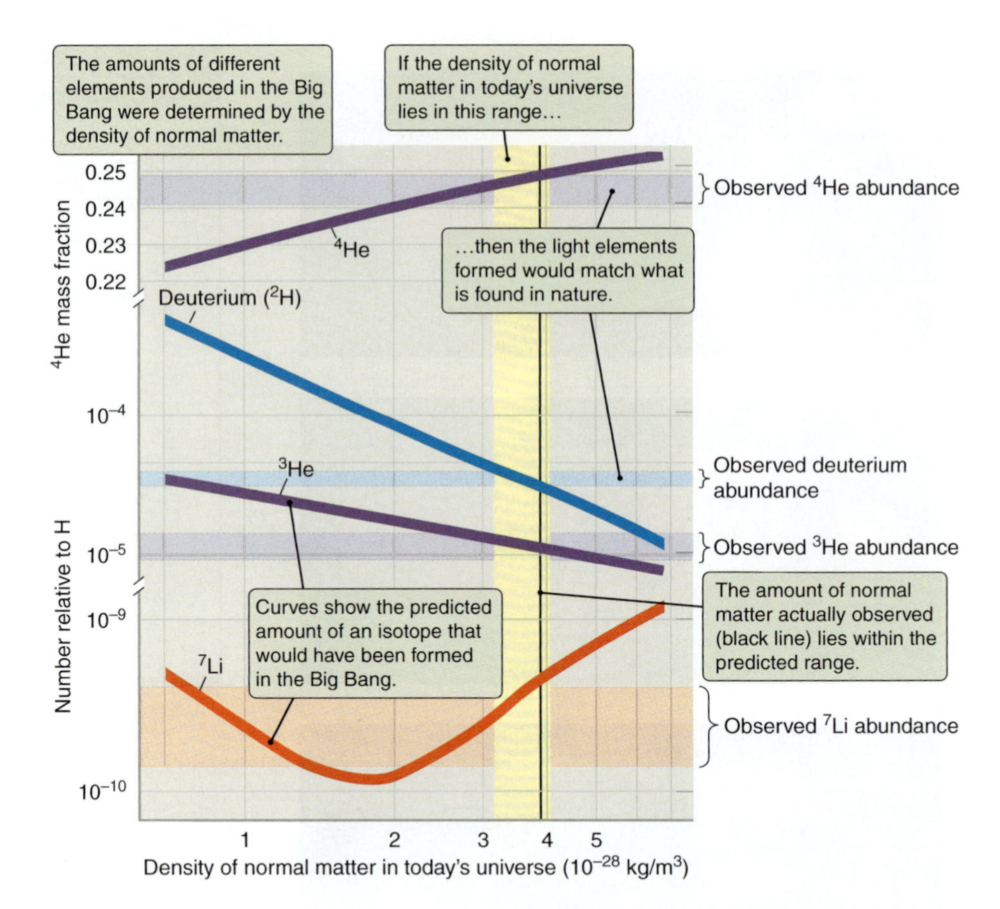

FIGURE 19.19 Observed and calculated abundances of the products of Big Bang nucleosynthesis, plotted against the density of normal matter in today's universe. Big Bang nucleosynthesis correctly predicts the amounts of these isotopes found in the universe today.

of the more massive elements in the universe, including the atoms making up the bulk of Earth and its life, must have formed in subsequent generations of stars.

The calculated predictions of abundances of deuterium, helium, and lithium from Big Bang nucleosynthesis can be plotted as a function of the present-day density of normal (luminous) matter in the universe, as in **Figure 19.19**. The first thing to note in Figure 19.19 is that about 24 percent of the mass of the normal matter formed in the early universe should have ended up in the form of the very stable isotope ^{4}He, regardless of the total density of matter in the universe. Indeed, about 24 percent of the mass of normal matter in the universe today is in the form of ^{4}He, in complete agreement with the prediction of Big Bang nucleosynthesis.

The other isotope abundances formed in the Big Bang depended more sensitively on the density of normal matter in the universe. Comparing current abundances with models of isotope formation in the Big Bang helps pin down the density of the early universe. Beginning with the abundances of isotopes such as ^{2}H (deuterium) and ^{3}He found in the universe today (shown as the roughly horizontal pastel bands in Figure 19.19), and comparing them to predictions in different models of how abundant isotopes *should* be when formed at different densities (saturated colors in Figure 19.19), cosmologists can find the density of normal matter (the vertical yellow band in Figure 19.19). The best current measurements give a value of about 3.9×10^{-28} kilogram per cubic meter (kg/m^3) for the average density of normal matter in the universe today (at sea level, the density of air is about 1.2 kg/m^3). This value lies well within the range predicted by the observations shown in Figure 19.19. Once again, the agreement is remarkable, and it holds for many different isotopes. Turning this around, cosmologists can begin with an observation of the amount of normal matter in and around galaxies and then compare with calculations of what the chemical composition emerging from the Big Bang should have been. The observations agree remarkably well with the amounts of these elements actually found in nature. The major predictions of the Big Bang theory are resoundingly confirmed.

This agreement also provides a powerful constraint on the nature of dark matter, which dominates the mass in the universe. Dark matter cannot consist of normal matter made up of neutrons and protons; if it did, the density of neutrons and protons in the early universe would have been much higher, and the resulting abundances of light elements in the universe would have been much different from what is actually observed.

In the next chapter we will examine galaxies—the immense collections of stars, gas, and dust that were once called "island universes"—and the evidence that they are composed mostly of dark matter. ■

Summary

19.1 Fuzzy objects observed in earlier (and smaller) telescopes were eventually shown to be distant galaxies, collections of stars similar to the Milky Way. The universe contains hundreds of billions of galaxies, each with millions to hundreds of billions of stars.

19.2 Observations suggest that the universe is homogeneous and isotropic, in agreement with the cosmological principle.

19.3 Astronomers build a distance ladder of overlapping techniques for measuring distances to faraway stars and galaxies. Hubble found that all galaxies are moving away, with a recessional velocity proportional to distance, indicating that the universe is expanding. The local physics and structure of objects are not affected by the expansion.

19.4 The universe is expanding uniformly from a Big Bang, which occurred nearly 14 billion years ago. The Big Bang happened everywhere; it is not an explosion spreading out from a single point. Redshifts are observed from distant galaxies because space itself is expanding, increasing the scale factor.

19.5 Astronomers observe thermal radiation at 2.73 K, which is the cooled remnant of the radiation present in the universe when it was much smaller and hotter. The cosmic microwave background radiation enables a measurement of Earth's velocity with respect to it, and it also shows evidence of the ripples that grew to become large-scale structure in the universe.

19.6 The amounts of helium and trace amounts of other light elements measured today agree with what would be expected from nuclear reactions of normal matter in the hot early universe.

Unanswered Questions

- How standard are "standard candles"? For example, Cepheid variable light curves are slightly different, depending on the amount of heavy elements in the stars, so this variation must be calibrated. Some percentage of Type Ia supernovae may originate from the merging of two compact objects rather than one white dwarf exploding at nearly 1.4 $M_{\odot}$. Astronomers try to address these difficulties by using multiple methods to find distances to galaxies—for example, observing numerous Cepheids, bright O stars, and Type Ia supernovae in the same galaxy, to check that their calculated distances agree. In one recent study, astronomers used HST to observe more than 600 Cepheid variable stars in eight galaxies that had Type Ia supernova detections, and they were able to reduce the uncertainty in their distances (and thus their calculated value of the Hubble constant). The Type Ia supernovae calibrated in this way are consistent with earlier-calibrated maximum luminosities. But it is possible that those supernovae in galaxies far enough away that there are no other standard candles may have slightly different maximum luminosities, leading to an incorrect distance estimate (and value of H_0).
- What existed before the Big Bang? The usual (and somewhat unsatisfactory) answer is that the Big Bang was the beginning of space and *time*, so there could be no time before it happened. A more recent answer is that this universe may be just one of many universes, and the Big Bang was the beginning of this universe only. We will return to this topic in Chapter 22.

Questions and Problems

Summary Self-Test

1. In astronomy, *isotropy* means that the universe is the same
   a. in all locations.
   b. in all directions.
   c. at all times.
   d. at all size scales.
2. In astronomy, *homogeneity* means that the universe is the same
   a. in all locations.
   b. in all directions.
   c. at all times.
   d. at all size scales.
3. Cosmological redshifts are calculated from observations of spectral lines from
   a. individual stars in distant galaxies.
   b. clouds of dust and gas in distant galaxies.
   c. spectra of entire galaxies.
   d. rotations of the disks of distant galaxies.

4. When they look into the universe, astronomers observe that nearly all galaxies are moving away from the Milky Way. This observation suggests that
   a. the Milky Way is at the center of the universe.
   b. the Milky Way must be at the center of the expansion.
   c. the Big Bang occurred at the location of the Milky Way.
   d. an observer in a distant galaxy would make the same observation.
5. Hubble's law requires the measurement of two properties of a galaxy: ________ and ________.
   a. size; mass
   b. distance; rotation speed
   c. distance; recessional velocity
   d. size; recessional velocity
6. Some galaxies have redshifts z that if equated to v/c correspond to velocities greater than the speed of light. Special relativity is not violated in this case
   a. because of relativistic beaming.
   b. because of superluminal motion.
   c. because redshifts carry no information.
   d. because these velocities do not measure motion *through* space.
7. The Big Bang theory predicted (select all that apply)
   a. the Hubble law.
   b. the cosmic microwave background radiation.
   c. the cosmological principle.
   d. the abundance of helium.
   e. that the sky should be dark at night.
   f. the period-luminosity relation of Cepheid variables.
8. The simplest way to estimate the age of the universe is from
   a. using the slope of the Hubble law.
   b. the age of Moon rocks.
   c. models of stellar evolution.
   d. measurements of the abundances of elements.
9. The CMB includes information about (select all that apply)
   a. the age of the universe.
   b. the temperature of the early universe.
   c. the density of the early universe.
   d. density fluctuations in the early universe.
   e. the motion of Earth around the center of the Milky Way.
10. Repeated measurements showing that the current helium abundance is much less than the value predicted by the Big Bang would imply that
    a. some part of the Big Bang theory is incorrect or incomplete.
    b. the current helium abundance is wrong.
    c. scientists don't know how to measure helium abundances.

True/False and Multiple Choice

11. **T/F:** Observers in a very distant galaxy would observe that galaxies far from them are more redshifted than are galaxies nearby.
12. **T/F:** The distribution of galaxies is both homogeneous and isotropic.
13. **T/F:** Peculiar velocities of distant galaxies are often larger than the velocities from the Hubble flow.
14. **T/F:** The expansion of space means that galaxies are becoming larger over time.
15. **T/F:** Cosmological redshift is caused by the movement of galaxies through space.
16. If you found a galaxy with an Hα emission line that had a wavelength of 756.3 nm, what would be the galaxy's redshift? Note that the rest wavelength of the Hα emission line is 656.3 nm.
    a. 0.01
    b. 0.05
    c. 0.10
    d. 0.15
17. The cosmological principle says that
    a. the universe is expanding.
    b. the universe began in the Big Bang.
    c. the rules that govern the universe are the same everywhere.
    d. all of the above
18. Peculiar velocities are
    a. the velocities of galaxies that are moving toward the Milky Way Galaxy.
    b. the velocities of galaxies that result from something other than the expansion of space.
    c. the velocities of galaxies toward their neighbors.
    d. the random velocities of galaxies from when they formed.
19. Why is the Milky Way Galaxy not expanding together with the rest of the universe?
    a. It is not expanding because it is at the center of the expansion.
    b. It is expanding, but the expansion is too small to measure.
    c. The Milky Way is a special location in the universe.
    d. Local gravity dominates over the expansion of the universe.
20. As astronomers extend their distance ladder beyond 30 Mpc, they change their measuring standard from Cepheid variable stars to Type Ia supernovae. Why is this change necessary?
    a. Type Ia supernovae are more luminous than Cepheid variables.
    b. Type Ia supernovae are less luminous than Cepheid variables.
    c. Type Ia supernovae vary more slowly than do Cepheid variables.
    d. Type Ia supernovae vary more quickly than do Cepheid variables.
21. The scale factor keeps track of
    a. the movement of galaxies through space.
    b. the current distances between many galaxies.
    c. the changing distance between any two galaxies.
    d. the location of the center of the universe.
22. The Big Bang is
    a. the giant supernova explosion that triggered the formation of the Solar System.
    b. the explosion of a supermassive black hole.
    c. the eventual demise of the Sun.
    d. the beginning of space and time.

23. The CMB is essentially uniform in all directions in the sky. This is an example of
   a. anisotropy.
   b. isotropy.
   c. thermal fluctuations.
   d. none of the above

24. The CMB comes from
   a. the moment when the universe became transparent.
   b. the outer Solar System.
   c. the edge of the universe.
   d. the instant of the Big Bang.

25. According to the definitions in Chapter 1, the Big Bang is
   a. an idea.
   b. a hypothesis.
   c. a law.
   d. a theory.

Thinking about the Concepts

26. What was the subject of the Great Debate, and why was it important to astronomers' understanding of the scale of the universe?

27. How did observations of Cepheid variable stars finally settle the Great Debate?

28. Why is it better to observe more than one type of standard candle in a distant galaxy?

29. Imagine that you are standing in the middle of a dense fog.
   a. Would you describe your environment as isotropic? Why or why not?
   b. Would you describe it as homogeneous? Why or why not?

30. Early in the 20th century, astronomers discovered that most galaxies are moving away from the Milky Way (that is, they are redshifted.)
   a. What was the significance of this discovery?
   b. Edwin Hubble later made an even more important discovery: that the speed at which galaxies are receding is proportional to their distance. Why was this among the more important scientific discoveries of the 20th century?

31. Why is it not possible to use the measured radial velocities of nearby galaxies, such as the Andromeda Galaxy, to evaluate the Hubble constant (H_0)?

32. Explain what astronomers mean by *distance ladder.*

33. Why is it important to know the type of progenitor of a Type Ia supernova in a distant galaxy?

34. As the universe expands from the Big Bang, galaxies are not actually flying apart from one another. What is really happening?

35. Knowing that you are studying astronomy, a curious friend asks where the center of the universe is located. You answer, "Right here and everywhere." Explain in detail why you give this answer.

36. The general relationship between recessional velocity (v_r) and redshift (z) is $v_r = cz$. This simple relationship fails, however, for very distant galaxies with large redshifts. Explain why.

37. Why is it significant that the CMB displays a Planck spectrum?

38. What is the significance of the tiny brightness variations that are observed in the CMB?

39. What important characteristics of the early universe are revealed by today's observed abundances of various isotopes, such as ^{2}H and ^{3}He?

40. Study Figure 19.6.
   a. Why is it important that the different "rungs" of the distance ladder overlap in the distances that they measure?
   b. Why does the figure end at the right edge—because there are no more ways to measure distance or because there is no more universe to measure? How do you know?

Applying the Concepts

41. Sketch what the graph in Figure 19.8b would look like for each of the following cases:
   a. a universe in which galaxies moved randomly through space
   b. a contracting universe
   c. a universe that started out expanding but is now contracting

42. Study Figure 19.19. Error bars have not been plotted in this figure. Why not? Was this a very precise measurement or a very imprecise measurement? How does the precision of the measurement affect your confidence in the conclusions drawn from it?

43. Study Figure 19.19. This figure includes both predictions and observations.
   a. What do the vertical yellow bar and the slanted lines and curves represent: theory or observation?
   b. What do the pastel horizontal lines and the vertical black line represent: predictions or observations?
   c. Do the predictions and observations match? Choose one example, and explain how you know.

44. The Hubble time ($1/H_0$) represents the age of a universe that has been expanding at a constant rate since the Big Bang. Assuming an H_0 value of 70 km/s/Mpc and a constant rate of expansion, calculate the age of the universe in years. How is the age different if $H_0 = 75$ km/s/Mpc? (Note: 1 year $= 3.16 \times 10^7$ seconds, and 1 Mpc $= 3.09 \times 10^{19}$ km.)

45. Throughout the latter half of the 20th century, estimates of H_0 ranged from 55 to 110 km/s/Mpc. Calculate the age of the universe in years for each of these estimated values of H_0.

46. A distant galaxy has a redshift $z = 5.82$ and a recessional velocity $v_r = 287{,}000$ km/s (about 96 percent of the speed of light).
    a. If $H_0 = 70$ km/s/Mpc and if Hubble's law remains valid out to such a large distance, then how far away is this galaxy?
    b. Assuming a Hubble time of 13.7 billion years, how old was the universe at the look-back time of this galaxy?
    c. What was the scale factor of the universe at that time?

47. The spectrum of a distant galaxy shows the Hα line of hydrogen ($\lambda_{rest} = 656.28$ nm) at a wavelength of 750 nm. Assume that $H_0 = 70$ km/s/Mpc.
    a. What is the redshift (z) of this galaxy?
    b. What is its recessional velocity (v_r) in kilometers per second?
    c. What is the distance of the galaxy in megaparsecs?

48. A distant galaxy has a redshift $z = 7.6$.
    a. What would be the observed wavelength of the Hα line ($\lambda_{rest} = 656.28$ nm)?
    b. In what region of the spectrum would this line be located?

49. Suppose you observe two galaxies: one at a distance of 10.7 Mpc with a recessional velocity of 580 km/s, and another at a distance of 337 Mpc with a radial velocity of 25,400 km/s.
    a. Calculate the Hubble constant (H_0) for each of these two observations.
    b. Which of the two calculations would you consider to be more trustworthy? Why?
    c. Estimate the peculiar velocity of the closer galaxy.
    d. If the more distant galaxy had this same peculiar velocity, how would your calculated value of the Hubble constant change?

50. The temperature of the CMB is 2.73 K. What is the peak wavelength of its Planck blackbody spectrum expressed both in microns and in millimeters?

51. COBE observations show that the Solar System is moving in the direction of the constellation Crater at a speed of 368 km/s relative to the cosmic reference frame. What is the blueshift (negative value of z) associated with this motion?

52. The average density of normal matter in the universe is 4×10^{-28} kg/m^3. The mass of a hydrogen atom is 1.66×10^{-27} kg. On average, how many hydrogen atoms are there in each cubic meter in the universe?

53. To get a feeling for the emptiness of the universe, compare its density (4×10^{-28} kg/m^3) with that of Earth's atmosphere at sea level (1.2 kg/m^3). How much denser is Earth's atmosphere? Write this ratio using standard notation.

54. Assume that the most distant galaxies have a redshift $z = 10$. The average density of normal matter in the universe today is 4×10^{-28} kg/m^3. What was its density when light was leaving those distant galaxies? (Hint: Keep in mind that volume is proportional to the cube of the scale factor.)

55. Suppose a Type Ia supernova is found in a distant galaxy. The measured supernova brightness is 10^{-17} W/m^2. What is the distance of the galaxy?

Using the Web

56. Go to the Goddard Multimedia website and view the animation of a Cepheid variable star in a spiral galaxy (http://svs.gsfc.nasa.gov/goto?10145). Explain how astronomers use data like these to estimate the distance to the galaxy. What is actually observed, what is assumed, and what is calculated? (Review the discussion in Chapter 16 as needed.)

57. For more details on the history of the discovery of the expanding universe, go to the American Institute of Physics' "Cosmic Journey: A History of Scientific Cosmology" website (http://aip.org/history/cosmology). Read through the sections titled "Island Universes," "The Expanding Universe," and "Big Bang or Steady State?" Why was Albert Einstein "irritated" by the idea of an expanding universe? What was the contribution of Belgian astrophysicist (and Catholic priest) Georges Lemaître? What is the steady-state theory, and what was the main piece of evidence against it?

58. Go to the University of Washington Astronomy Department's "Hubble's Law: An Introductory Astronomy Lab" Web page (www.astro.washington.edu/courses/labs/clearinghouse/labs/HubbleLaw/hubbletitle), and do the lab exercise, which uses real data from galaxies to calculate Hubble's law. Your instructor will indicate whether you should use the regular or the shorter version.

59. Go to the "Astronomy" page of the Phys.Org website (http://phys.org/space-news/astronomy), click on "Search," and enter "Type Ia supernova" in the Search box. Find a recent story about one of these supernovae. What is its distance and brightness? What type of star produced the explosion?

60. Observations of the CMB from the Planck space telescope are expected to be released in 2013. Go to ESA's Planck website (http://esa.int/SPECIALS/Planck) and view the all-sky CMB map. Compare this with the *WMAP* data in Figure 19.18c; what is different? What has been learned from this mission?

STUDYSPACE is a free and open website that provides a Study Plan for each chapter of *21st Century Astronomy.* Study Plans include animations, reading outlines, vocabulary flashcards, and multiple-choice quizzes, plus links to premium content in SmartWork and the ebook. Visit **wwnorton.com/studyspace**.

SMARTWORK Norton's online homework system, includes algorithmically generated versions of these questions, plus additional conceptual exercises. If your instructor assigns questions in SmartWork, log in at **smartwork.wwnorton.com.**

Exploration | Hubble's Law for Balloons

The expansion of the universe is extremely difficult to visualize, even for professional astronomers. In this Exploration you will use the surface of a balloon to get a feel for how an "expansion" changes distances between objects. Throughout this exploration, remember to think of the surface of the balloon as a two-dimensional object, much as the surface of Earth is a two-dimensional object for most people. The average person can move east or west, or north or south, but into Earth and out to space are not options. For this Exploration you will need a balloon, 11 small stickers, a piece of string, and a ruler. A partner is helpful as well. **Figure 19.20** shows some of the steps involved.

Blow up the balloon partially and hold it closed, but *do not tie it shut*. Stick the 11 stickers on the balloon (these represent galaxies) and number them. Galaxy 1 is the reference galaxy.

Measure the distance between the reference galaxy and each of the galaxies numbered 2–10. The easiest way to do this is to use your piece of string. Lay it along the balloon between the two galaxies and then measure the string. Record these data in the "Distance 1" column of a table like the one shown below.

Simulate the expansion of your balloon universe by *slowly* blowing up the balloon the rest of the way. Have your partner count the number of seconds it takes you to do this, and record this number in the "Time Elapsed" column of the table (each row has the same time elapsed, because the expansion occurred for the same amount of time for each galaxy). Tie the balloon shut. Measure the distance between the reference galaxy and each numbered galaxy again. Record these data under "Distance 2."

Subtract the first measurement from the second. Record the difference in the table.

Divide this difference, which represents the distance traveled by the galaxy, by the time it took to blow up the balloon. Distance divided by time gives an average speed.

Make a graph with velocity on the *y*-axis and distance 2 on the *x*-axis to get "Hubble's law for balloons." You may wish to roughly fit a line to these data to clarify the trend.

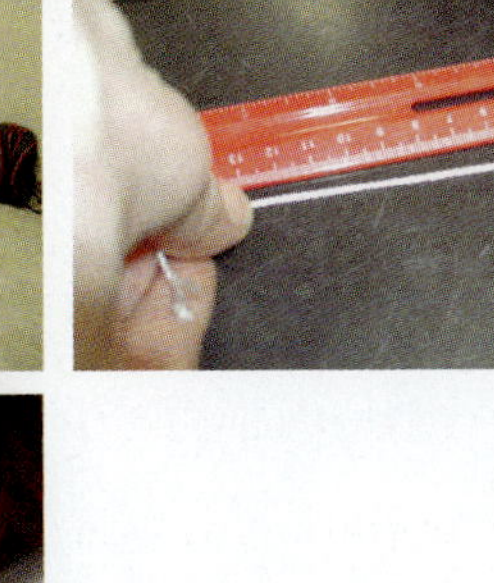

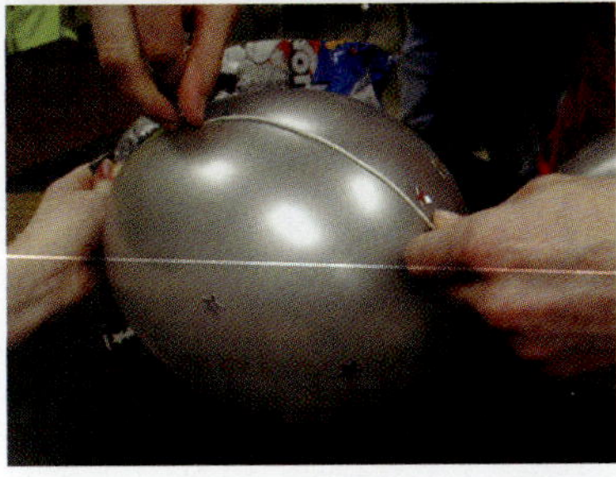

FIGURE 19.20 These students are using balloons to develop an understanding of the expansion of space.

Galaxy Number	Distance 1	Distance 2	Difference	Time Elapsed	Velocity
1 (reference)	0	0	0		0
2					
3					
4					
5					
6					
7					
8					
9					
10					
11					

1 **Describe your data. If you fit a line to them, is it horizontal or does it trend upward or downward?**

2 **Is there anything special about your reference galaxy? Is it different in any way from the others?**

3 **If you had picked a different reference galaxy, would the trend of your line be different? If you are not sure of the answer, get another balloon and try it.**

4 **The expansion of the universe behaves similarly to the movement of the galaxies on the balloon. We don't want to carry the analogy too far, but there is one more thing to think about. In your balloon, some areas probably expanded less than others because the material was thicker; there was more "balloon stuff" holding it together. How is this similar to some places in the actual universe?**

The large, barred spiral galaxy NGC 1300 is about 20 million parsecs away, in the constellation Eridanus.

20 Galaxies

It may not be amiss to point out some other very remarkable Nebulae which cannot well be less, but are probably much larger than our own system; and being also extended, the inhabitants of the planets that attend the stars which compose them must likewise perceive the same phenomena. For which reason they may also be called milky ways.

Sir William Herschel (1738–1822)

LEARNING GOALS

As you learned in Chapter 19, it has been less than a century since astronomers realized that the universe is filled with huge collections of stars and dust called galaxies. Just as stars vary in their mass or their stage of evolution, galaxies come in many forms. As we did for stars, we begin our discussion of galaxies with a survey of the types of galaxies and their basic properties in order to better understand the differences among them. In the next chapter we will focus on the Milky Way, and in the following chapters we'll look at the evolution of galaxies and the universe itself. By the conclusion of this chapter, you should be able to:

- Determine a galaxy's type from its appearance, and describe the motions of its stars.
- Explain how the arms of spiral galaxies form and why they are sites of star formation.
- Describe the evidence suggesting that galaxies are composed mostly of dark matter.
- Discuss the evidence indicating that most—perhaps all—large galaxies have supermassive black holes at their centers.

20.1 Galaxies Come in Many Types

Imagine taking a handful of coins and throwing them in the air. You know that all of these objects are very much the same: dimes, pennies, nickels, and quarters—all flat and circular. When you look at the objects falling through the air, however, they do not appear all the same. Some coins appear face on and look circular. Some coins are seen edge on and look like thin lines. Most coins are seen from an angle between these two extremes and appear with various degrees of "ellipticity," or flattening. In principle, you can learn a lot about the properties of coins by looking at a picture like the one in **Figure 20.1a**. If you began by assuming that coins have a particular three-dimensional shape, you could predict what you should see. You could then compare your prediction with what you actually observed. Even if this one image was the only information you had, you could use it to figure out the three-dimensional shape of a coin—flat and circular.

Do some types of galaxies have conditions more suitable for life to develop?

Astronomers play exactly this game in their efforts to discover the true three-dimensional shapes of galaxies. **Figure 20.1b** shows a set of galaxies seen from various viewing angles, from face on to edge on. You can infer from images of the sky that, just like the coins in Figure 20.1a, galaxies often have disk-like shapes and are randomly oriented on the sky.

A quick look at a group of galaxies imaged by the Gemini telescope (**Figure 20.2**) shows that galaxies come in a wide range of apparent sizes and shapes. The classifications for galaxies used today date back to the 1930s, when Edwin Hubble sorted the different shapes into categories like those shown in **Figure 20.3**. Hubble grouped all galaxies according to appearance and positioned them on a diagram that resembles the tuning fork used in the tuning of a musical instrument.

At the bottom (or "handle") of this **tuning fork diagram** are objects that are generally elliptical in three dimensions.

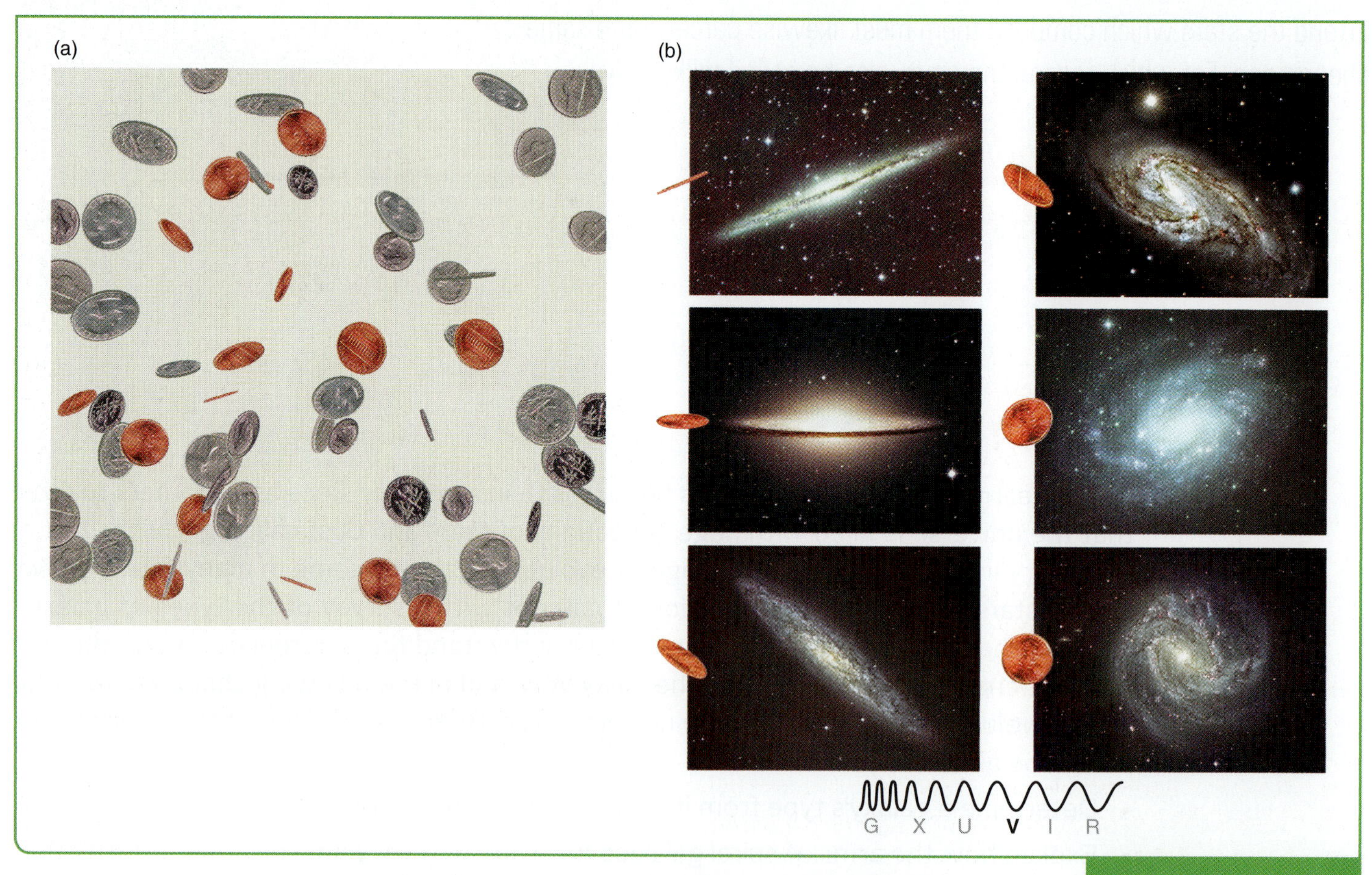

FIGURE 20.1 (a) A handful of coins thrown in the air provides an analogy for the difficulties in identifying the shapes of certain types of galaxies, which, like the coins in this picture, are seen in various orientations—some face on, some edge on, and most somewhere in between. (b) Disk-shaped galaxies seen from various perspectives or angles. The variety of angles for galaxies corresponds to the range of perspectives for coins.

FIGURE 20.2 This image of a small group of galaxies (at approximately the same distance) called Hickson Compact Group 87 illustrates something of the range of shapes and sizes found among galaxies.

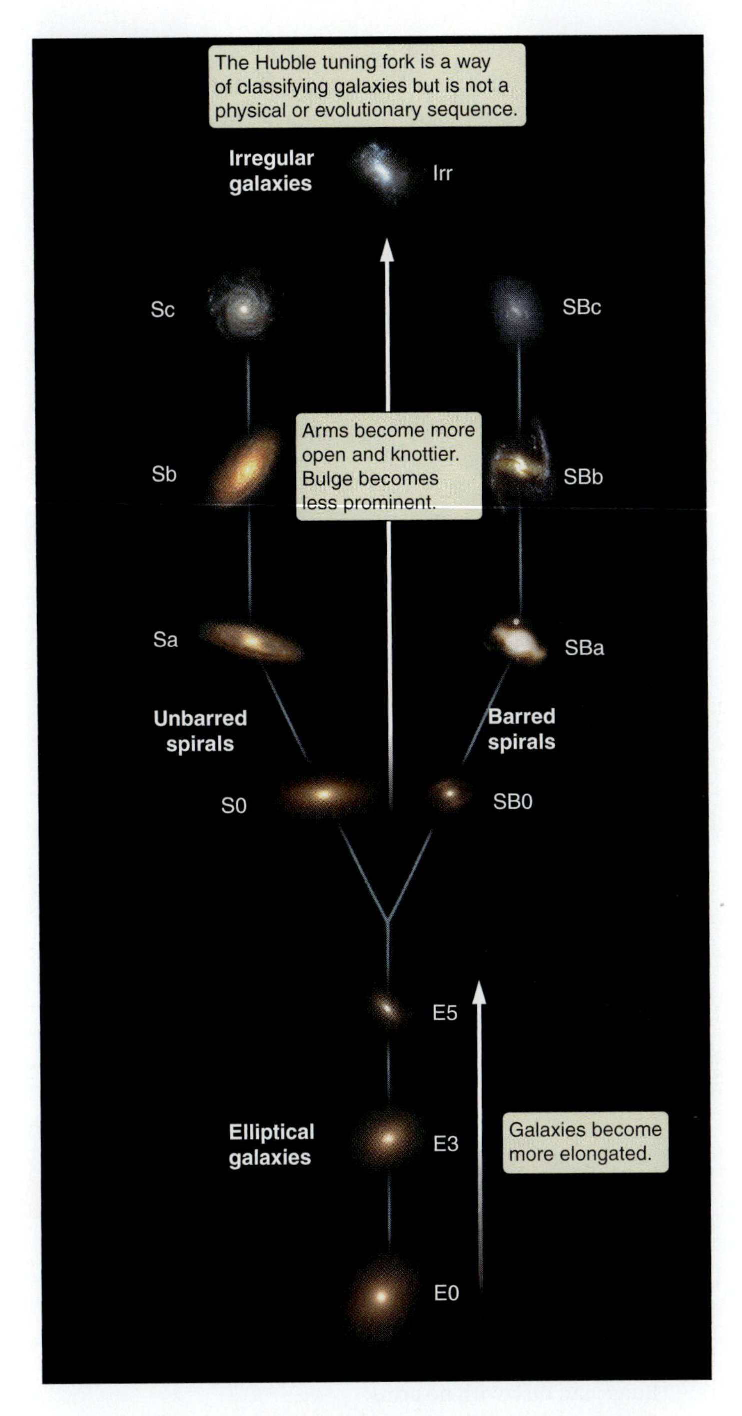

FIGURE 20.3 Tuning fork diagram showing Edwin Hubble's scheme for classifying elliptical (E), spiral (S, SB), and SO galaxies according to their appearance only. Irregular (Irr) galaxies were not placed on the tuning fork. Hubble's scheme does not show an evolutionary sequence of galaxies.

These **elliptical galaxies** (labeled "E" on the diagram) can have either spherical or ellipsoidal shapes, and they show little evidence of the flat, round disk seen in other types of galaxies. On the two "tines" of the fork are the **spiral galaxies**, designated with an initial "S." The defining feature of a spiral galaxy is a flattened, rotating disk. Spiral galaxies also contain spiral arms that give these galaxies their name. In addition to disks and arms, spiral galaxies have central **bulges**, which look like elliptical galaxies.

Hubble noticed that the bulges of about half of the spiral galaxies are bar-shaped (see the chapter-opening photo). He called these galaxies **barred spirals** ("SB") and placed them along the right-hand tine of the tuning fork (see Figure 20.3). Spirals that lack a barlike bulge ("S") lie along the left-hand tine. Notice that Hubble positioned the spiral galaxies vertically along the tines of the fork according to the prominence of the central bulge and how tightly the spiral arms are wound. For example, Sa and SBa galaxies have the largest bulges and display tightly wound and smooth spiral arms. Sc and SBc galaxies have small central bulges and more loosely woven spiral arms that are often very knotty in appearance.

Hubble recognized that the distinction between spiral and elliptical galaxies is not always clear-cut. Some galaxies seem to be a cross between the two types, having stellar disks but no spiral arms. Hubble called these intermediate types of galaxies **S0 galaxies**—either barred (SB0) or unbarred (S0)—and placed them near the junction of his tuning fork. Today, the distinction between elliptical and

S0 galaxies is even more blurred. Modern telescopic observations have revealed that many, if not most, elliptical galaxies contain small rotating disks at their centers. Elliptical and S0 galaxies share another similarity: at the present time, neither produces many new stars.

Galaxies that fall into none of these classes are **irregular galaxies** ("Irr" in Figure 20.3). As their name implies, irregular galaxies often lack symmetry in shape or structure, and they do not fit neatly on Hubble's tuning fork. About 25 percent of galaxies are irregular, and now astronomers think that most of them once were spirals or ellipticals that became distorted by the gravity of another galaxy.

Originally, Hubble thought that his tuning fork diagram might do for galaxies what the H-R diagram had done for stars. It did not, but his classification scheme has held up for classifying galaxies by their appearance in visible light (**Process of Science Figure**). **Table 20.1** summarizes the criteria that Hubble used to classify galaxies.

Stellar Motions Give Galaxies Their Shapes

A galaxy is not a solid object like a coin, but a collection of stars, gas, and dust orbiting under the influence of the galaxy's overall gravitational field. The shapes of elliptical galaxies are determined from the almost random orbits of their stars. For example, in any region in an elliptical galaxy, some stars are falling in while others are climbing out. In fact, stars are moving in all possible directions. Unlike planets in the Solar System, which move on simple elliptical orbits about the Sun, stars in an elliptical galaxy follow orbits with a wide range of different shapes (**Figure 20.4**). These orbits are more complex than the orbits of planets because the gravitational field within an elliptical galaxy does not come from a single central object. Taken together, all of these stellar orbits are what give an elliptical galaxy its shape.

The collective orbits of all its stars give an elliptical galaxy its shape.

Orbital speeds are also a factor. The faster the stars are moving, the more spread out the galaxy is. (If the stars were not moving at all, they would all clump together at the center of the galaxy.) If the stars in an elliptical galaxy are moving in truly random directions, the galaxy has a spherical shape. However, if stars tend to move faster in one direction than in others, the galaxy is more spread out in that direction, giving it an elongated shape. Differences in stellar orbits are responsible for the variety of shapes that Hubble noted. Some elliptical galaxies (those at the bottom of the tuning fork handle—see Figure 20.3) are spherical, while others (those at the top of the handle) are elongated.

TABLE 20.1 | The Hubble Sequence of Galaxies

A Morphological Classification Scheme Based on the Properties of Galaxies

Category/Criteria	Abbreviation	Range of Features		
Ellipticals Mostly bulge Old, red stellar population Smooth-appearing	E0 ↕ E7	More spherical ↕ More elongated		
S0 (unbarred/barred) Bulge and disk with no arms Bulge and disk with mostly old, red stars	S0/SB0	Smooth disk and bulge		
Spirals (unbarred/barred) Bulge and disk with arms Bulge has old, red stars Disk has both old, red stars and young, blue stars Spirals (S) have roundish bulges Barred spirals (SB) have elongated or barred bulges	Sa/Sba Sb/SBb Sc/SBc	More bulge ↕ Little bulge	Tightly wound arms ↕ Open arms	Smooth arms ↕ Knotty arms
Irregulars No arms, no bulge Some old stars, but mostly young stars, gas and dust, giving a knotty appearance	Irr			

Process of Science

WRONG IDEAS ARE SOMETIMES USEFUL

Hubble created the "tuning fork" diagram because he was looking for an evolutionary sequence. The idea has since been falsified, but the figure remains useful.

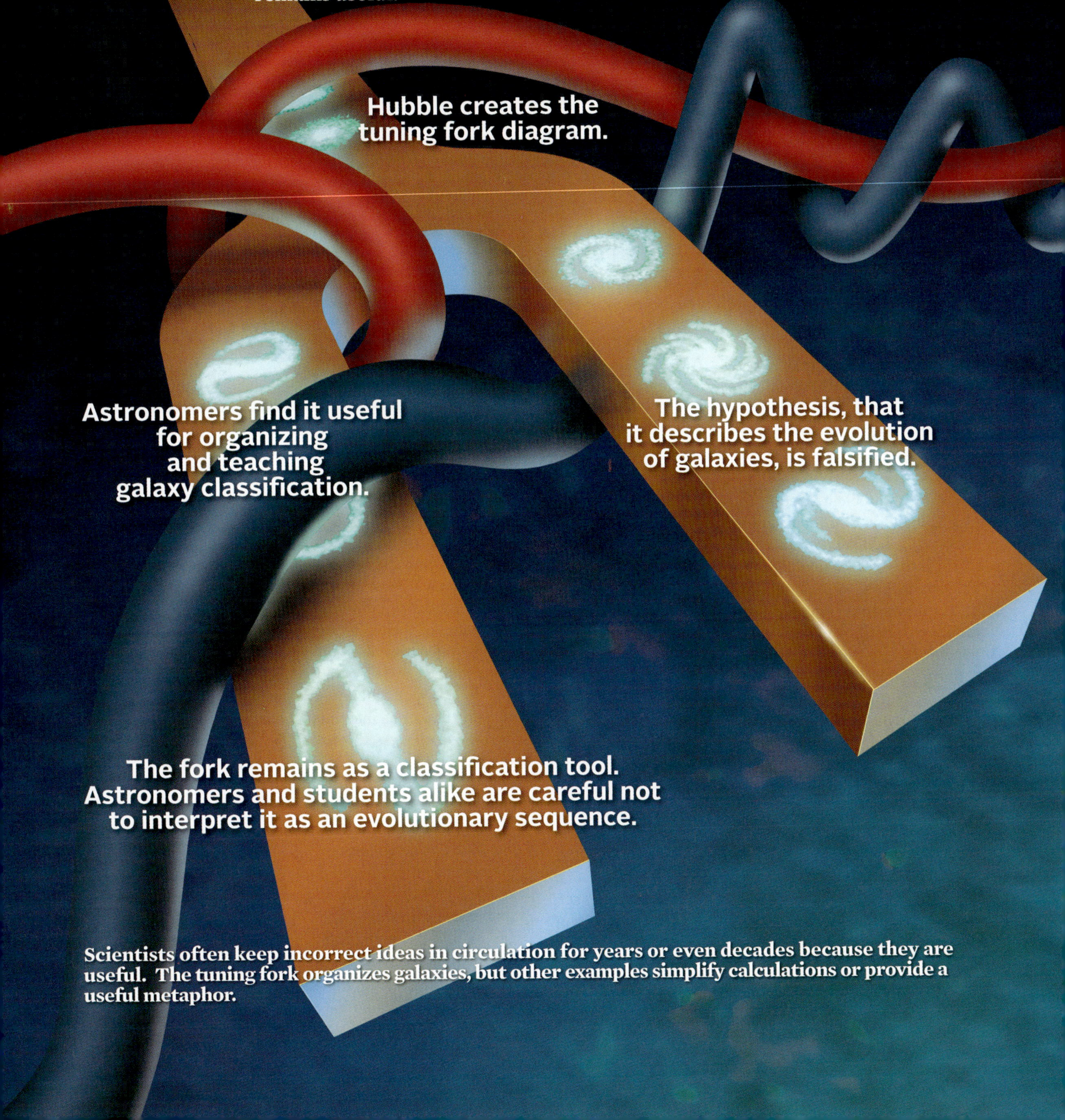

Hubble creates the tuning fork diagram.

Astronomers find it useful for organizing and teaching galaxy classification.

The hypothesis, that it describes the evolution of galaxies, is falsified.

The fork remains as a classification tool. Astronomers and students alike are careful not to interpret it as an evolutionary sequence.

Scientists often keep incorrect ideas in circulation for years or even decades because they are useful. The tuning fork organizes galaxies, but other examples simplify calculations or provide a useful metaphor.

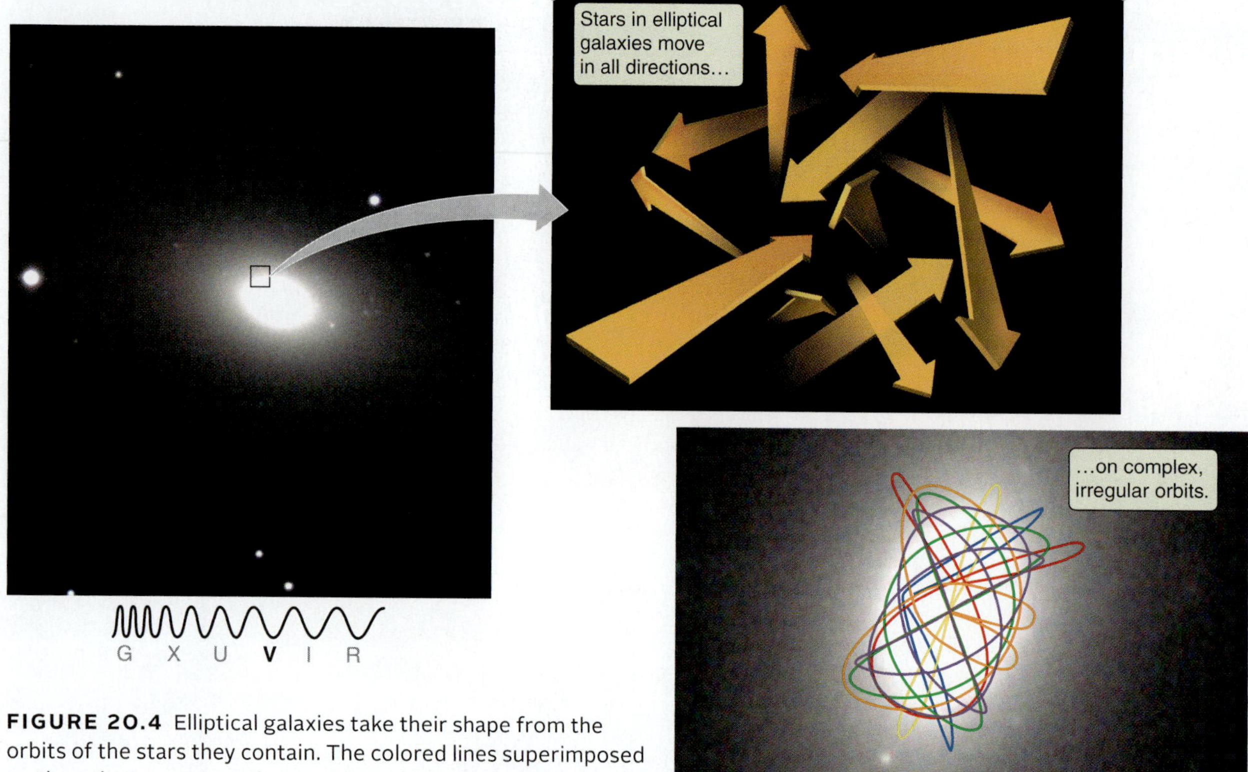

FIGURE 20.4 Elliptical galaxies take their shape from the orbits of the stars they contain. The colored lines superimposed on the galaxy represent the complex orbits of its stars.

One difficult problem faced by astronomers studying elliptical galaxies is that their appearance in the sky does not necessarily indicate their true shape. For example, a galaxy might actually be shaped like a football,[1] but if viewed end on, it looks round like a soccer ball instead.

The orbits of stars in the disks of spiral galaxies are quite different from those of stars in elliptical galaxies. The components of a spiral galaxy are shown in **Figure 20.5**. The defining feature of a spiral galaxy is that it has a flattened, rotating disk. Like the planets of the Solar System, most of the stars in the disk of a spiral galaxy follow nearly circular orbits and travel in the same direction about a concentration of mass at the center of the galaxy. But the stellar orbits in a spiral galaxy's central bulge are quite different from those in the galaxy's disk. As with elliptical galaxies, the gravitational field within the bulge does not come from a single object, and the stars therefore follow random orbits. The bulges of spiral galaxies are thus roughly spherical in shape.

Spiral galaxies have a rotating disk and a central bulge.

1 More like a rugby ball or an Australian football, which have rounded ends, than like an American football, which has more pointed ends.

Other Differences among Galaxies

In addition to the differences in their stellar orbits, there is another important distinction between spiral and elliptical galaxies. Most spiral galaxies contain large amounts of dust and cold, dense molecular gas concentrated in the midplanes of their disks. Just as the dust in the disk of the Milky Way can be seen on a clear summer night as a dark band slicing the galaxy in two (see Figure 15.1), the dust in an edge-on spiral galaxy appears as a dark, obscuring band running down the midplane of the disk (**Figure 20.6**). The cold molecular gas that accompanies the dust can also be seen in radio observations of spiral galaxies. By contrast, elliptical galaxies are extremely gas-poor.

Some of the giant ellipticals contain very hot gas that astronomers see primarily by observing the X-rays that the gas emits. The difference in shape between elliptical and spiral galaxies offers some insight into why the gas in giant ellipticals is hot, while spirals contain large amounts of cold, dense gas. Just as gas settles into a disk around a forming star, cold gas settles into the disk of a spiral galaxy as a result of conser-

Ellipticals contain mostly hot gas; spirals contain mostly cold gas.

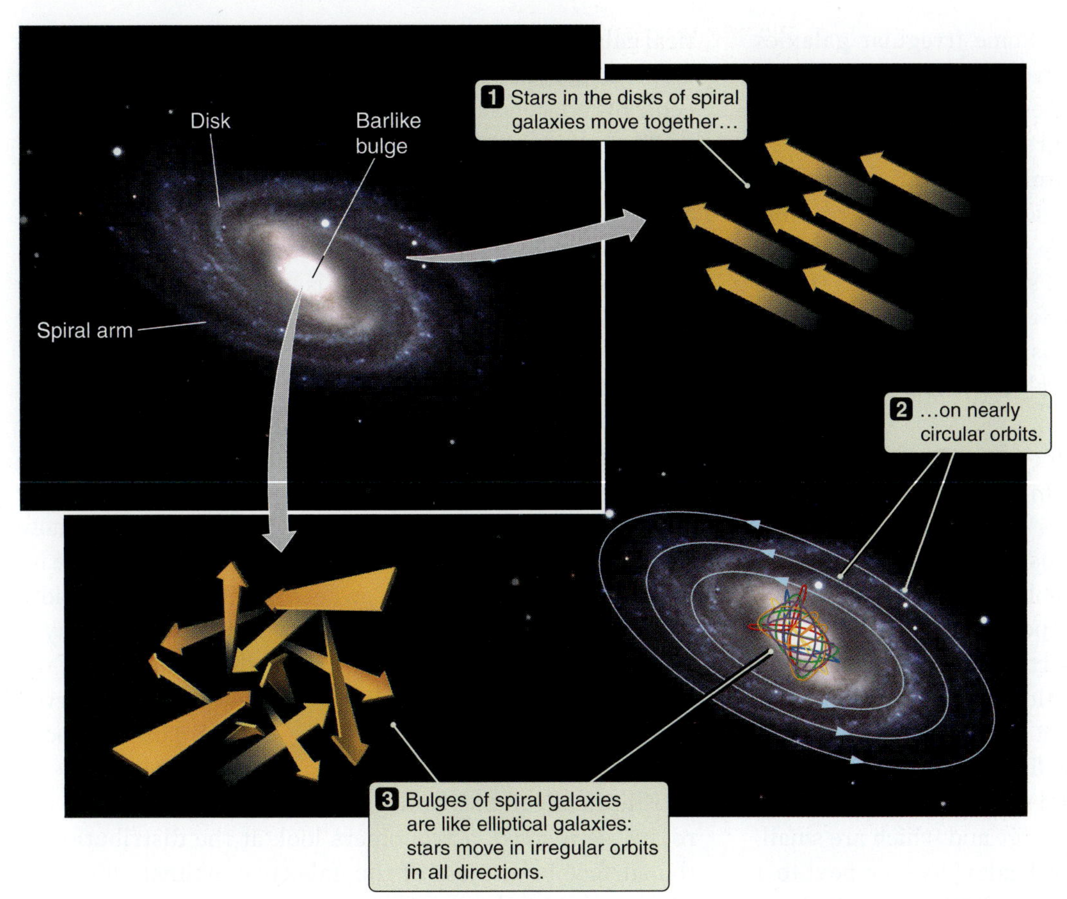

FIGURE 20.5 The components of a barred spiral galaxy. The orbits of stars in the rotating disk and the ellipse-like bulge are indicated.

vation of angular momentum. The only place in an elliptical galaxy where cold gas could collect is at the center. However, the density of stars in elliptical galaxies is so high that Type Ia supernovae continually reheat this gas, preventing most of it from cooling off and forming cold clouds.

The colors of spiral and elliptical galaxies reveal a great deal about their star formation histories. Recall from Chapter 15 that stars form from dense clouds of cold molecular gas. Because the gas observable in elliptical galaxies is very hot, astronomers know that active star formation is not taking place in those galaxies today. The reddish colors of elliptical and S0 galaxies confirm that little or no star formation has occurred there for quite some time. The stars in these galaxies are an older population of lower-mass stars. The bluish colors of the disks of spiral galaxies, by contrast, indicate that massive, young, hot stars are forming in the cold molecular clouds contained within the disk. Even though most of the stars in a spiral disk are old, the massive, young stars are so luminous that their blue light dominates. When it comes to star formation, most irregular

Stars are still forming in spiral galaxies but not in elliptical galaxies.

FIGURE 20.6 HST image of the nearly edge-on spiral galaxy M104 (the Sombrero Galaxy, type Sa). The dust in the plane is seen as a dark, obscuring band in the midplane of the galaxy. Note the bright halo made up of stars and globular clusters. Compare this image with Figure 15.1, which shows the dust in the plane of the Milky Way.

galaxies are like spiral galaxies. Some irregular galaxies form stars at prodigious rates, given their relatively small sizes. Irregular and disk galaxies that undergo intense bursts of star formation are called starburst galaxies.

The relationship between luminosity and size among the different types of galaxies is not straightforward. Normal galaxies range in luminosity from tens of thousands up to a trillion (a million million) solar luminosities (10^4 to 10^{12} $L_{\odot}$) and in size from about a few hundred to hundreds of thousands of parsecs. There is no distinct size difference between elliptical and spiral galaxies; about half of both types of galaxies fall within a similar range of sizes. Although it is true that the most luminous elliptical galaxies are more luminous than the most luminous spiral galaxies, there is considerable overlap in the range of luminosities among all Hubble types.

All types of galaxies come in a wide range of sizes.

Mass is the single most important parameter in determining the properties and evolution of a star. In contrast, differences in mass and size do not lead to such obvious differences in galaxies. Only subtle differences in color and concentration exist between large and small galaxies, making it difficult to distinguish which are large and which are small. Even when a smaller, nearby spiral galaxy is seen next to a larger, more distant spiral, as in **Figure 20.7**, it can be hard to tell which is which. Still, galaxies that have relatively low luminosity (less than 1 billion $L_{\odot}$) are called **dwarf galaxies**, and those that are more luminous than this are called giant galaxies. Only elliptical and irregular galaxies come in both types; among spiral and S0 galaxies, there are only giants. It is relatively easy to tell the difference between a dwarf elliptical galaxy and a giant elliptical galaxy (**Figure 20.8**). Giant elliptical galaxies have a much higher density of stars, more centrally concentrated than in dwarf ellipticals.

The motions of the stars in galaxies give rise to two "secondary" methods for estimating the distance to a galaxy (in addition to the methods discussed in Chapter 19). In rotating spiral galaxies, some of the light is approaching Earth and thus blueshifted; some light is moving away from Earth and thus redshifted. These redshifts and blueshifts together broaden a spectral line in the galaxy, and the amount of broadening indicates how fast the galaxy is rotating. Astronomers can measure the broadening from the Doppler shifts of the 21-centimeter (21-cm) radio emission line of hydrogen, which tells them the speed of rotation, which then relates to the galaxy's mass by Newton's version of Kepler's law. The more massive galaxies have more stars and are therefore more luminous. This empirical relation between the measured width of the 21-cm line and the luminosity of the spiral galaxy is called the Tully-Fisher relation. Once the luminosity of the galaxy is known, it can be compared to the galaxy's observed apparent brightness to estimate its distance. This method is thought to work out to about 100 megaparsecs (Mpc).

Elliptical galaxies and the bulges of S0 galaxies do not rotate, so instead, astronomers look at the distribution of the surface brightness of the galaxy to estimate distance. Closer galaxies show more variations in the surface brightness because the distribution of stars throughout the galaxy isn't perfectly uniform. For more distant galaxies, these differences are less noticeable, and the surface brightness appears more uniform across the galaxy. This method is less precise than the Tully-Fisher method for spirals, but generally it is thought also to work to about 100 Mpc.

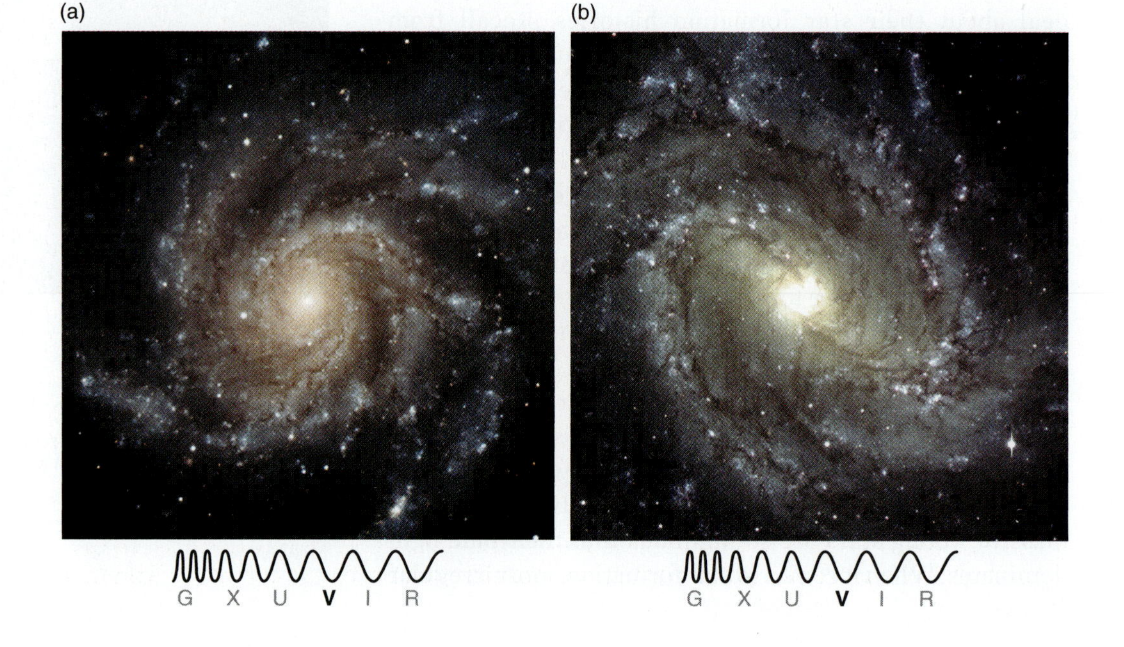

FIGURE 20.7 The mass or size of a spiral galaxy does not determine its appearance. Even though these galaxies appear to be similar in size and luminosity, the larger galaxy (a) is 4 times more distant and 10 times more luminous than the smaller galaxy (b).

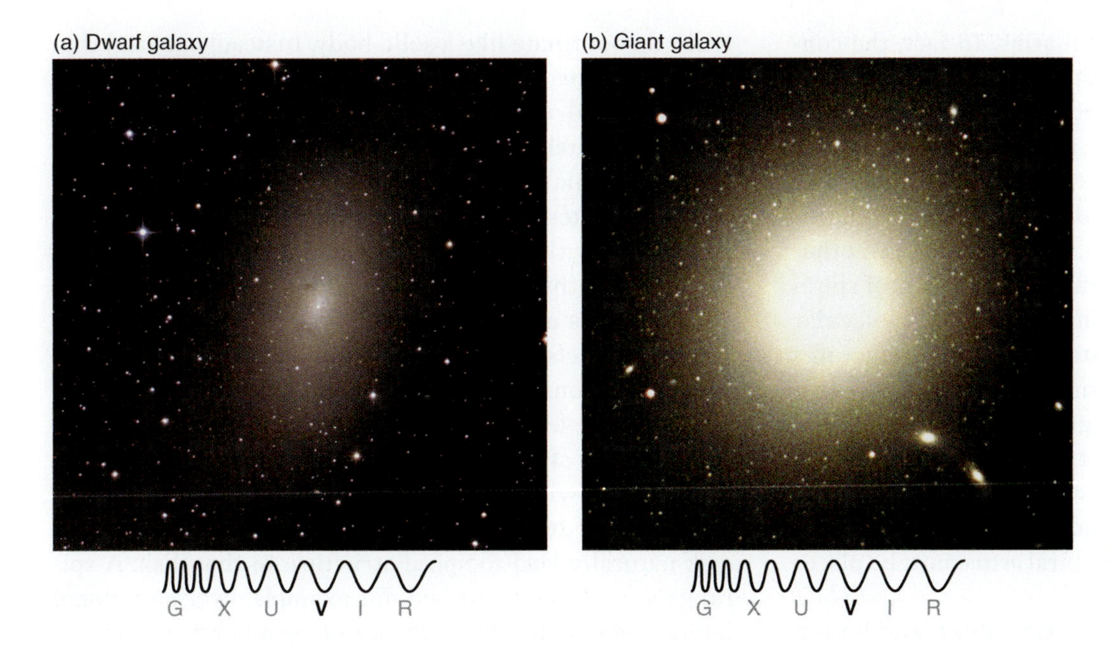

FIGURE 20.8 Dwarf elliptical galaxies (a) differ in appearance from giant elliptical galaxies (b).

20.2 In Spiral Galaxies, Stars Form in the Spiral Arms

Spiral galaxies take their name from the spiral arms they contain, but what is a spiral arm? From pictures of spiral galaxies it might appear that stars in the disk of a spiral galaxy are concentrated in the spiral arms. This turns out not to be the case. **Figure 20.9** shows images of the Andromeda Galaxy taken in ultraviolet and visible light. Notice that the spiral arms are relatively prominent in the UV image (**Figure 20.9a**) and less prominent when viewed in visible light (**Figure 20.9b**). If astronomers counted the actual numbers of stars rather than just their brightness, they would find that although stars are slightly concentrated in spiral arms, this concentration is not strong enough to account for the

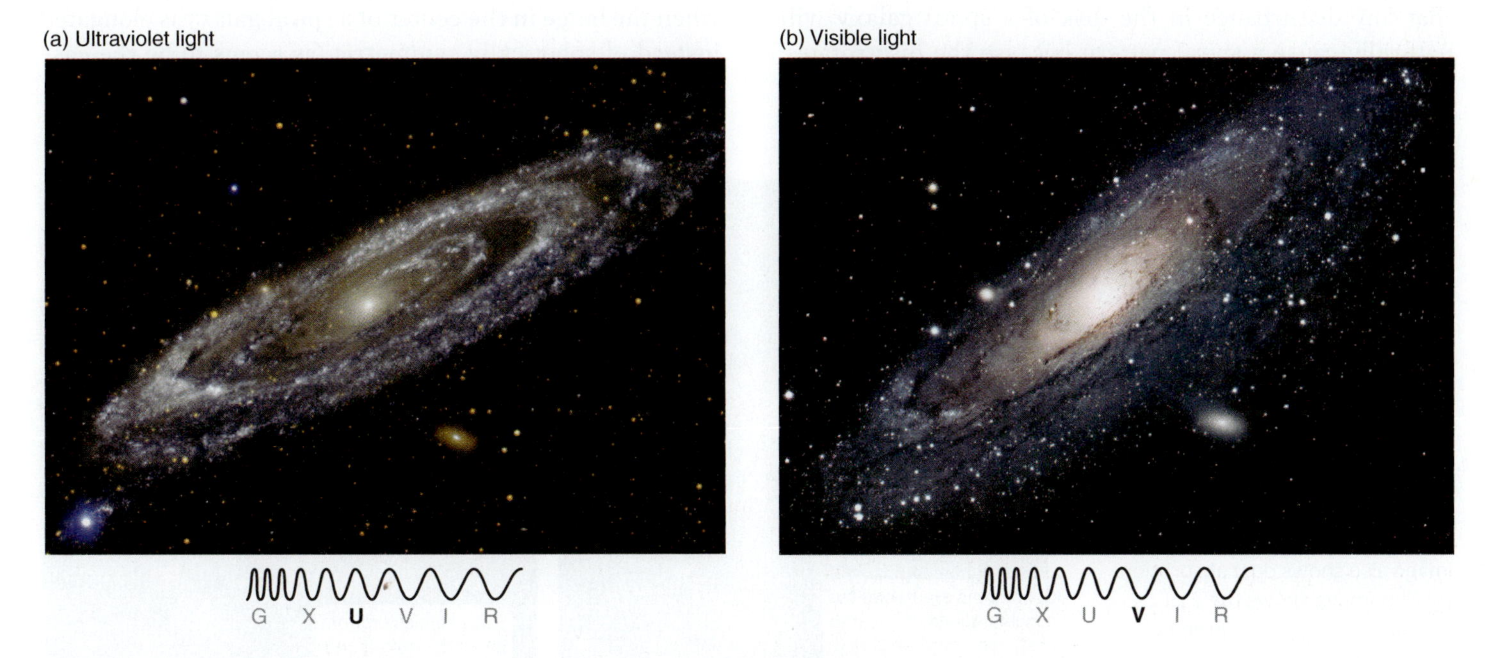

FIGURE 20.9 The Andromeda Galaxy in ultraviolet light (a) and visible light (b). Note that the spiral arms, which are dominated by hot young stars, are most prominent in ultraviolet light. The spiral arms are less prominent in visible light.

prominence of the observed spiral arms. In fact, the concentration of stars in the disks of spiral galaxies varies quite smoothly as it decreases outward from the center of the disk to the visible edge of the galaxy.

Spiral arms look so prominent when viewed in blue or UV light because the arms are co-located with significant concentrations of young, massive, luminous stars. In other words, what is strongly concentrated in the arms of spiral galaxies is ongoing star formation. H II regions, molecular clouds, associations of O and B stars, and other structures that are associated with star formation are all found predominantly in the spiral arms of galaxies.

Recall from Chapter 15 that stars form when dense interstellar clouds become so massive and concentrated that they begin to collapse under the force of their own gravity. If stars form in spiral arms, then spiral arms must be places where clouds of interstellar gas pile up and are compressed. Such is indeed the case. There are many ways to trace the presence of gas in the spiral arms of galaxies. Pictures of face-on spiral galaxies, such as the one featured in **Figure 20.10a**, show dark lanes where clouds of dust block starlight. These lanes provide one of the best tracers of spiral arms. Spiral arms also show up in other tracers of concentrations of gas, such as radio emission from neutral hydrogen or from carbon monoxide (**Figure 20.10b**).

Gas, dust, and young stars are concentrated in spiral arms.

Spiral arms are concentrations of gas where stars form, but why do spiral arms exist at all? Part of the answer is that any disturbance in the disk of a spiral galaxy will naturally cause a spiral pattern because the disk rotates. Disks do not rotate like a solid body. Instead, material that is closer to the center takes less time to complete a revolution around the galaxy than does material farther out in the galaxy. **Figure 20.11** illustrates the point. Beginning with a single linear "arm" through the center of a model galaxy, watch what happens as the model galaxy rotates. In the time it takes for objects in the inner part of the galaxy to complete several rotations, objects in the outer part of the galaxy may not have completed even a single revolution. In the process, the originally straight arms are slowly made into the spiral structure shown.

Rotation in a disk galaxy naturally produces spiral structure.

As Figure 20.11 shows, any disturbances to a galaxy's disk naturally lead to spiral structure in that disk. A spiral galaxy can be disturbed, for example, by gravitational interactions with other galaxies or by a burst of star formation. However, a single disturbance will not produce a *stable* spiral-arm pattern. Spiral arms produced from one disturbance will wind themselves up completely in two or three rotations of the disk and then disappear. Thus, this mechanism can create only short-term spiral patterns.

Gravitational interactions and star formation are processes that disturb disks.

Other types of disturbances are repetitive, so they are capable of sustaining spiral structure indefinitely. Some disturbances come from within a galaxy itself. For example, when the bulge in the center of a spiral galaxy is elongated instead of spherically symmetric (as seems to be the case

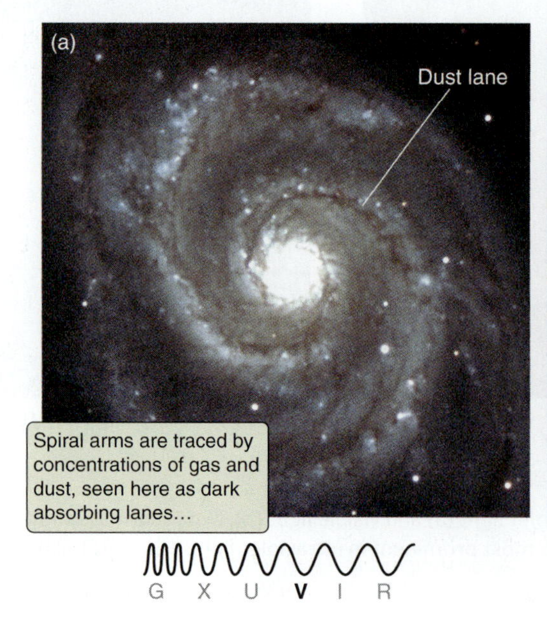

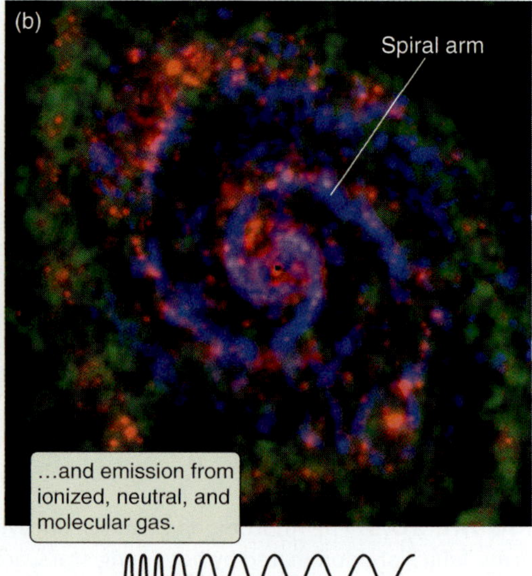

FIGURE 20.10 Two images of a face-on spiral galaxy showing the spiral arms. (a) This visible-light image also shows dust absorption. (b) This image shows the distribution of neutral interstellar hydrogen (green), carbon monoxide (CO) emission from cold molecular clouds (blue), and Hα emission from ionized gas (red).

for most spiral galaxies), then the bulge produces a gravitational disturbance in the disk. As the disk rotates through this disturbance, repeated episodes of star formation occur and more stable spiral arms form.

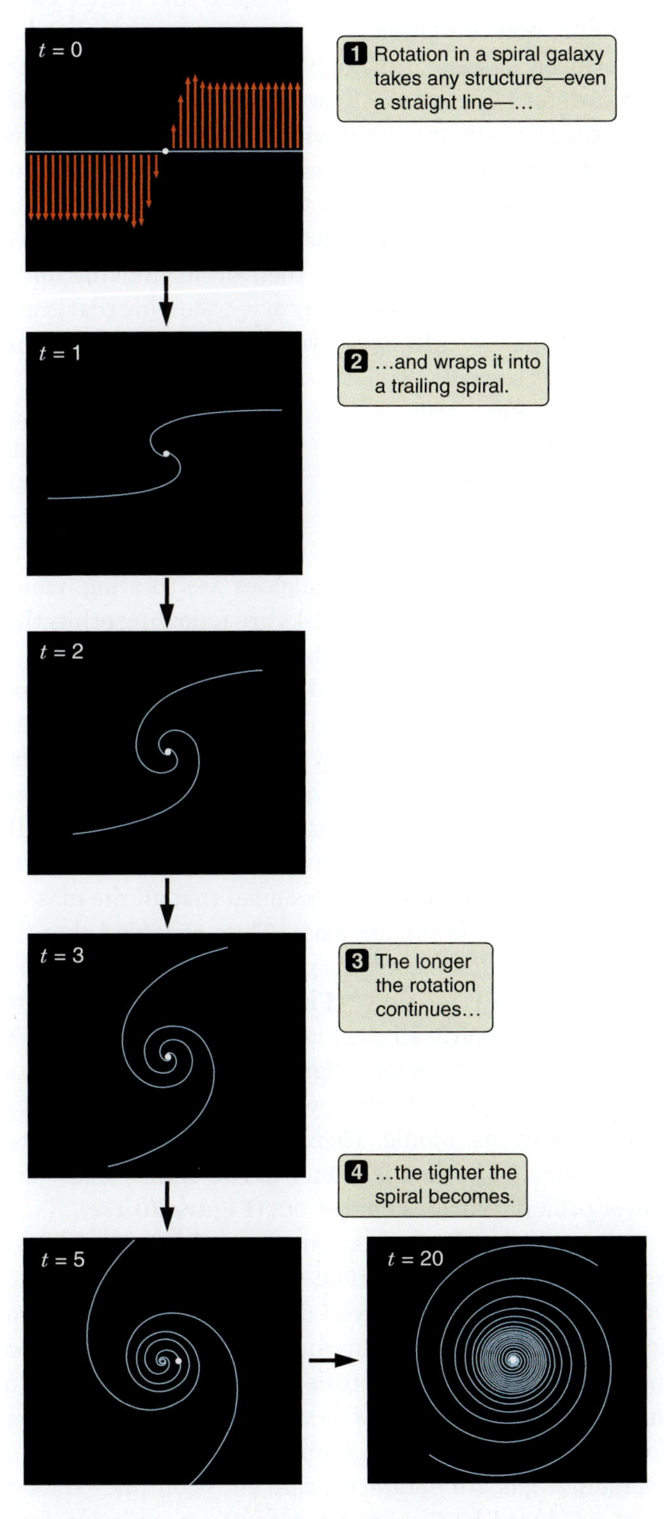

FIGURE 20.11 The differential rotation of a spiral galaxy will naturally take even an originally linear structure ($t = 0$) and wrap it into a progressively tighter spiral as time (t) goes by.

The process of star formation itself can also create spiral structure. Regions of star formation release considerable energy into their surroundings in the form of UV radiation, stellar winds, and supernova explosions. This energy drives up the pressure in the region, compressing clouds of gas and triggering more star formation. Typically, many massive stars form in the same region at about the same time; their combined mass outflows and supernova explosions occur one after another in the same region of space over the course of only a few million years. The result can be large, expanding bubbles of hot gas that sweep out cavities in the interstellar medium and concentrate the swept-up gas into dense, star-forming clouds, much like the snow that piles up in front of a snowplow. In this way, star formation can actually propagate through the disk of a galaxy. Rotation bends the resulting strings of star-forming regions into spiral structures.

Many galaxies show clear evidence of a relationship between the shapes of their bulges and the structure of their spiral arms. Barred spirals, for example, have a characteristic two-armed spiral pattern that is connected to the elongated bulge, as seen in the chapter-opening photograph. Even the bulges of galaxies that are not obviously barred may be nonspherical enough to contribute to the formation of two-armed spiral structure. Smaller galaxies in orbit about larger galaxies can also give rise to a periodic gravitational disturbance, triggering the same sort of two-armed structure.

Regular disturbances in the disks of spiral galaxies are called **spiral density waves** because they are regions of greater mass density and increased pressure in the galaxy's interstellar medium. These waves move around a disk in the pattern of a two-armed spiral. Spiral density waves act like the spiral-shaped blade in a blender slicing through the liquid. Models of how spiral density waves form indicate that this spiral-shaped, two-armed wave pattern does not necessarily rotate at the same rate as the rest of the galaxy. Consequently, as material in the disk orbits about, it passes through the spiral density waves. Because they are waves, it is the disturbance (the wave) that moves. The stars in an arm today are not the same stars that were in the arm 20 million years ago.

Regular disturbances lead to two-armed spirals.

A spiral density wave has very little effect on the motions of stars as they pass through it, but it does compress the gas that flows through it. As an analogy for this process, consider what happens when you turn on the tap in your kitchen sink (**Figure 20.12**). The water hits the bottom of the sink and spreads out in a thin, rapidly moving layer. A few centimeters out, depending on the rate at which water is flowing, there is a sudden increase in the

FIGURE 20.12 Water from a tap flows in a thin layer along the bottom of a kitchen sink. The sudden increase in the depth of the water is called a hydrostatic jump.

depth of the water, called a hydrostatic jump. Spiral arms in galaxies work in much the same way. Gas flows into the spiral density wave and piles up like the water in the sink. Stars form in the resulting compressed gas. Massive stars have such short lives (typically 10 million years or so) that they never have the chance to drift far from the spiral arms where they were born, so that is where they are seen. Less massive stars, however, have plenty of time to move away from their places of birth, so they form and fill in a smooth underlying disk.

Spiral density waves compress gas, triggering star formation.

NEBRASKA SIMULATION: NEBRASKA SIMULATION: TRAFFIC DENSITY ANALOGY

20.3 Galaxies Are Mostly Dark Matter

We now turn to another important property of a galaxy: its mass. Recall from Part III (Chapters 13–18) that mass plays the dominant role in determining the properties of stars. Mass does not play such a clear-cut role for galaxies. Even so, efforts to measure the masses of galaxies during the last decades of the 20th century led to some of the most remarkable and surprising findings in the history of astronomy. To measure the mass of a galaxy, astronomers add up the mass of the stars, dust, and gas that they observe. Because a galaxy's spectrum is composed primarily of starlight, once they know what types of stars are in the galaxy they can use stellar evolution to estimate the total stellar mass from the galaxy's luminosity. They estimate the mass of the dust and gas by using the physics of radiation from interstellar gas at X-ray, infrared, and radio wavelengths. Together, the stars, gas, and dust in a galaxy are called **luminous matter** (or simply **normal matter**) because this matter emits electromagnetic radiation.

However, looking at the light from a galaxy is not sufficient to determine a galaxy's total mass. Imagine, for example, if the Sun were replaced with a black hole of the same mass. No light would be coming from this black hole, and thus its mass would not be included in an estimate based on the luminosity of starlight from the Milky Way Galaxy. Yet the planets would continue on their orbits, moving under the influence of the black hole's gravity, showing that gravitational attraction still exists, even if no light is coming from the center of the Solar System. Fortunately, there is a method for determining mass that does not involve luminosity: the effect of gravity on an object's motion can be used to determine its mass. The disks of spiral galaxies are rotating, which means that the stars in those disks are following orbits that are much like the Keplerian orbits of planets around a parent star and binary stars around each other (see Math Tools 13.3). To measure the mass of a spiral galaxy, astronomers apply Kepler's laws, just as they do for those other systems.

Astronomers use Kepler's laws to measure galaxy mass.

Applying Kepler's laws requires having some idea of how the mass is distributed in a galaxy. Astronomers originally hypothesized that the mass and the light are distributed in the same way; that is, they assumed that all the mass in these galaxies is luminous mass. They observed that the light of all galaxies, including spiral galaxies, is highly concentrated toward the center (**Figure 20.13a**) and therefore predicted that nearly all the mass of a spiral galaxy would be contained in its center (**Figure 20.13b**). This situation is much like the Solar System, where nearly all the mass is in the Sun, in the middle. Therefore, they predicted faster orbital velocities near the center of the spiral galaxy and slower orbital velocities farther out (**Figure 20.13c**).

To test this prediction, astronomers used the Doppler effect to measure orbital motions of stars, gas, and dust. Not until the mid-1970s, however, did telescope instrumentation provide reliable measurements of orbital velocities of stars and interstellar gas outside the inner, bright regions of galaxies. The velocities of stars are obtained from observations of absorption lines in their spectra. The velocities of interstellar gas are obtained using emission lines such as those produced by Hα emission or 21-cm emission from neutral hydrogen (see Chapter 15). Once the velocities have been found, astronomers can create a graph that shows

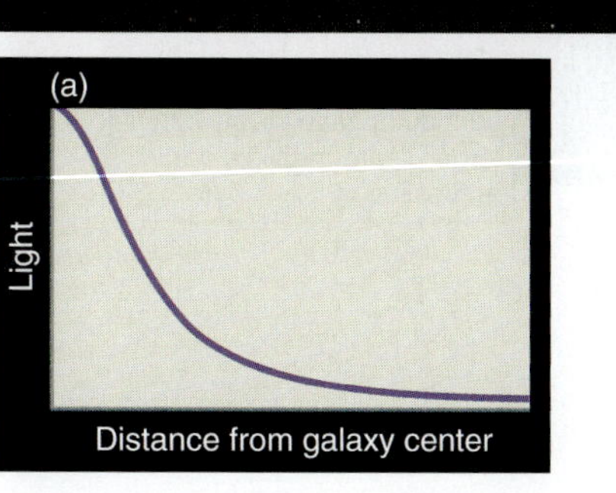

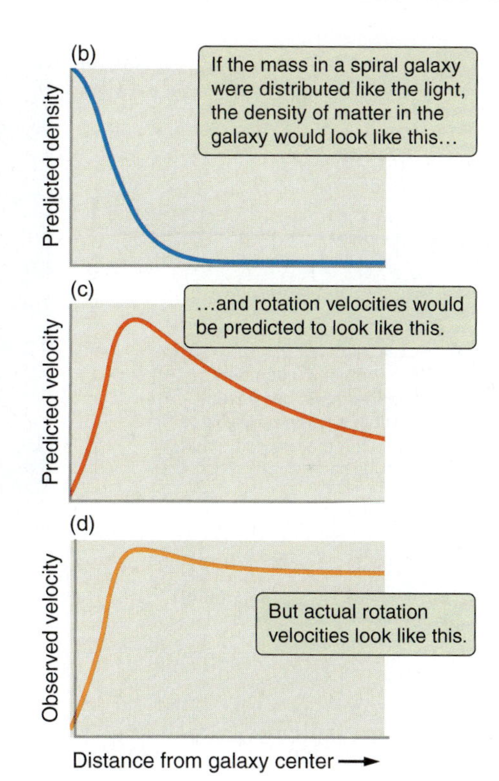

FIGURE 20.13 (a) The profile of visible light in a typical spiral galaxy. (b) The predicted mass density of stars and gas located at a given distance from the galaxy's center. If stars and gas accounted for all of the mass of the galaxy, then the galaxy's rotation curve would be as shown in (c). However, the rotation curves actually observed look more like the one shown in (d).

how orbital velocity in a galaxy varies with distance from the galaxy's center. This kind of graph is called a **rotation curve** (see Figures 20.13c and d).

Vera Rubin (1928–) pioneered work on galaxy rotation rates. She discovered that, contrary to the earlier prediction, the rotation velocities of spiral galaxies remain about the same out to the most distant measured parts of the galaxies (**Figure 20.13d**). Observations of 21-cm radiation from neutral hydrogen show that the rotation curves appear level, or "flat," in their outer parts, rather than sloping downward as predicted, even well outside the extent of the visible disks. These observations indicated that the hypothesis that mass and light are distributed in the same way is wrong.

Rotation curves of spiral galaxies are remarkably flat.

The rotation curve of a spiral galaxy enables astronomers to directly determine how the mass in that galaxy is distributed, by applying Kepler's laws to the rotation curves and asking, What mass distribution would cause this unexpected rotation curve? Recall from Chapter 4 that only the mass inside a given radius contributes to the net gravitational force felt by an object. (Strictly speaking, this is true only for spherically symmetric objects, but a spiral galaxy is symmetric enough for this to be a good approximation.) In addition to the centrally concentrated luminous matter, these galaxies must have a second component consisting of matter that does not show up in the census of stars, gas, and dust. This material, which reveals itself only by the influence of its gravity, is called **dark matter**. In the graph in **Figure 20.14a**, the red line shows how much luminous mass is inside a particular radius; the blue line shows how much dark matter is inside a particular radius; the black line shows the speed of rotation *at* a particular radius.
AstroTour: Dark Matter

The rotation curves of the inner parts of spiral galaxies match predictions based on their luminous matter, indicating that the inner parts of spiral galaxies are mostly luminous matter. Within the entire *visual* image of a galaxy, the mix of dark and luminous matter is about half and half. However, rotation curves measured with 21-cm radiation from neutral hydrogen indicate that the outer parts of spiral galaxies are mostly dark matter. Astronomers currently estimate that as much as 95 percent of the total mass in some spiral galaxies consists of a **dark matter halo** (**Figure 20.14b**), which can extend up to 10 times farther than the visible spiral portion of the galaxy located at the galaxy's center. This is a startling statement. The luminous part of a spiral galaxy is only a part of a much larger distribution of mass that is dominated by some type of invisible dark matter.

Most of the mass in spiral galaxies is dark matter.

What about elliptical galaxies? Again, astronomers need to compare the luminous mass, measured from the light they can see, with the gravitational mass, measured

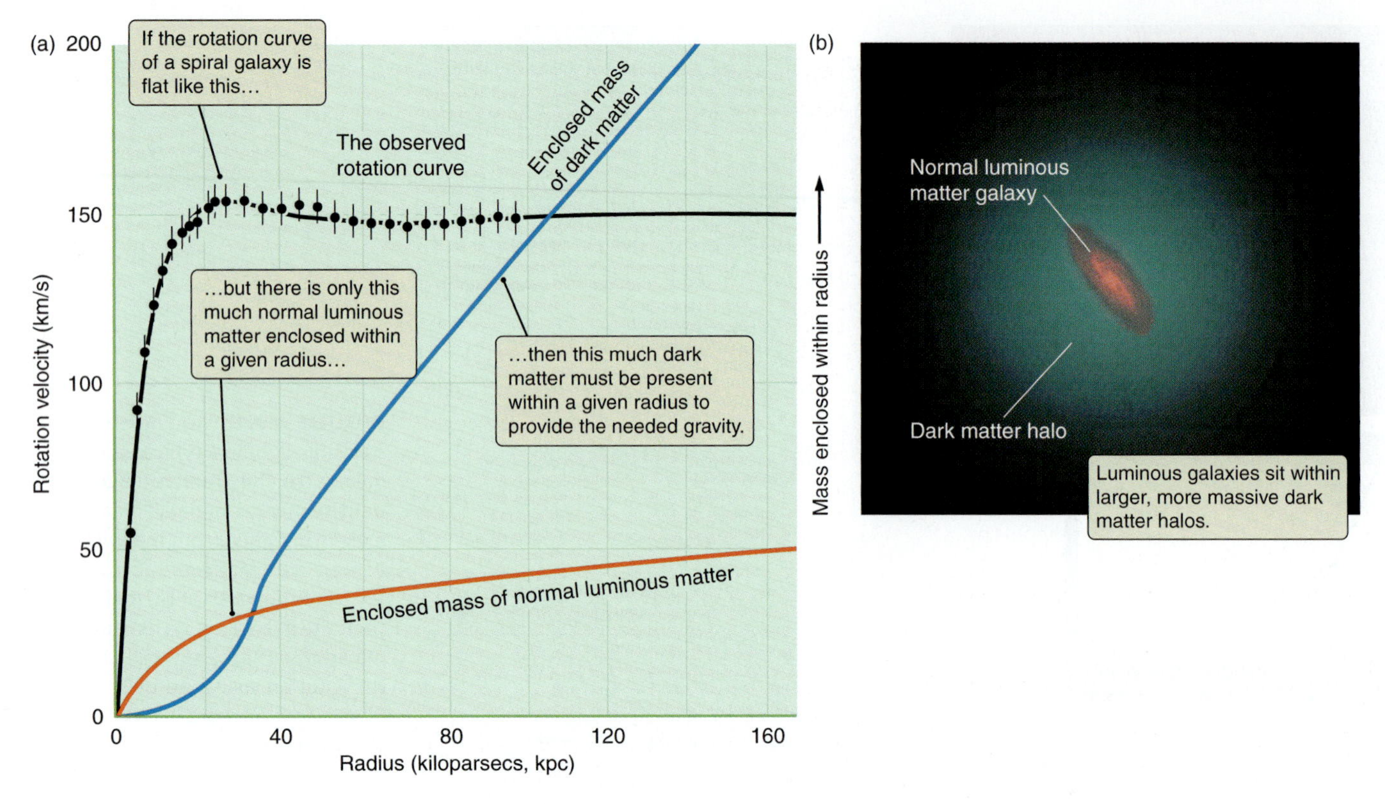

FIGURE 20.14 (a) The flat rotation curve of the spiral galaxy NGC 3198 can be used to determine the total mass within a given radius. Notice that the normal mass that can be accounted for by stars and gas accounts for only part of the needed gravity. Extra dark matter is needed to explain the rotation curve. (b) In addition to the matter that is visible, galaxies must be surrounded by halos containing a large amount of dark matter.

from the effects of gravity. Since elliptical galaxies do not rotate, astronomers cannot use Kepler's laws to measure the gravitational mass. Instead, they noticed that an elliptical galaxy's ability to hold on to its hot, X-ray-emitting gas depends on its mass (just as a planet's ability to hold on to its atmosphere depends on its mass). If the galaxy is not massive enough, the hot atoms and molecules will escape into intergalactic space. To find the mass of an elliptical galaxy, astronomers first infer the total amount of gas from X-ray images, such as the blue and purple halo seen in **Figure 20.15**. Then they calculate the mass that is needed to hold on to the gas, and compare that gravitational mass with the luminous mass. The amount of dark matter is the difference between what is needed to hold on to the inferred amount of gas and the observed amount of luminous matter.

The dark matter of elliptical galaxies enables them to hold on to their hot gases.

Some elliptical galaxies contain up to 20 times as much mass as can be accounted for by their stars and gas alone, so they must be dominated by dark matter, just as spiral galaxies are. As with spirals, the luminous matter in ellipticals is more centrally concentrated than is the dark matter. The

FIGURE 20.15 Combined visible-light and X-ray images of elliptical galaxy NGC 1132. The false-color blue/purple halo is X-ray emission from hot gas surrounding the galaxy. The hot gas extends well beyond the visible light from stars.

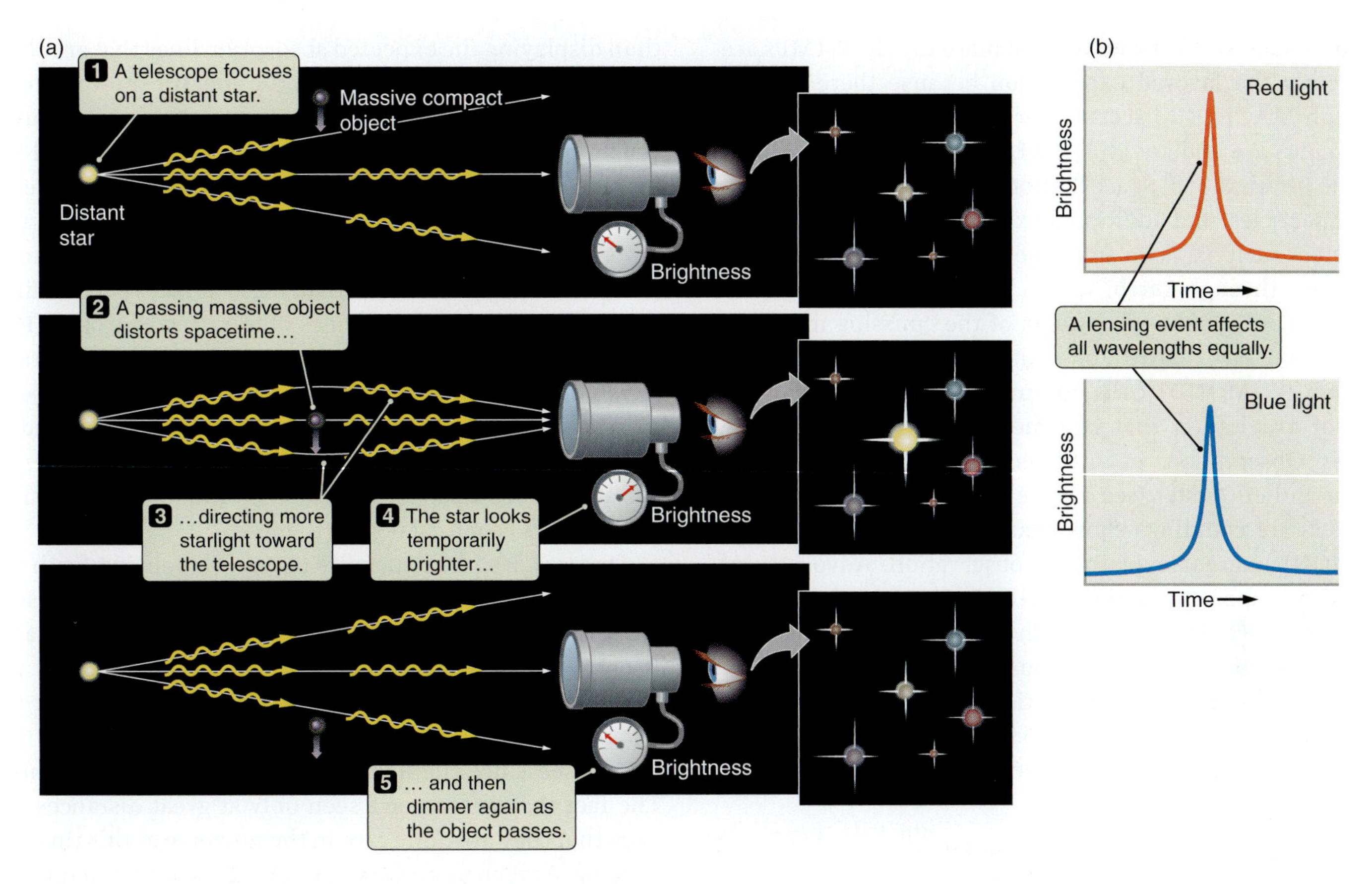

FIGURE 20.16 (a) The light from a distant star is affected by a compact object crossing the observer's line of sight. (b) The observed light curves of a stellar lensing event.

transition from the inner parts of galaxies (where luminous matter dominates) to the outer parts (which are dominated by dark matter) is remarkably smooth. Some galaxies may contain less dark matter than others, but about 90–95 percent of the total mass in a typical galaxy is in the form of dark matter. The same is true for the smaller dwarf galaxies, and this high percentage of dark matter distinguishes smaller dwarf galaxies from globular clusters, which do not have dark matter. This is an important observation that will need to be explained in the context of the evolution of galaxies.

What is the dark matter that makes up most of a galaxy? A number of suggestions are under investigation. Some candidates are astronomical objects such as large planets, compact stars, and exotic unknown elementary particles. These candidates can be lumped into two groups: MACHOs and WIMPs.

Dark matter candidates such as small main-sequence M stars, Jupiter-sized planets, white dwarfs, neutron stars, or black holes are collectively referred to as **MACHOs**, which stands for *massive compact halo objects*. If the dark matter in a galaxy's halo consists of MACHOs, there must be a lot of these objects, and they must each exert gravitational force but not emit much light. Because they have mass, MACHOs gravitationally deflect light according to Einstein's general theory of relativity—a phenomenon called gravitational lensing (see Chapter 18). If astronomers were observing a distant star and a MACHO passed between Earth and the star, the star's light would be deflected and, if the geometry were just right, focused by the intervening MACHO as it passed across their line of sight (**Figure 20.16a**). Because gravity affects all wavelengths equally, such lensing events should look the same in all colors, ruling out other causes of variability.

Astronomers would be remarkably lucky if such an event occurred just as they were observing a single distant star. When they monitored the stars in the Large and Small Magellanic Clouds (two of the small companion galaxies of the Milky Way), observing tens of millions of stars for several years, they saw numerous examples of events like those represented in the graphs of **Figure 20.16b**. However, these were not nearly enough to account for the amount of dark matter in the halo of the Milky Way. Thus, it was concluded that the dark matter in this galaxy is not composed primarily of MACHOs.

The other dark matter candidates are the exotic unknown elementary particles commonly known as **WIMPs**, which stands for *weakly interacting massive particles*. WIMPs are predicted to be similar to neutrinos (see Chapter 14) in that they would barely interact with ordinary matter, yet would be more massive and would move more slowly. WIMPs are currently the favored explanation because there are not enough MACHOs to account for the observed effects. Experiments are under way at the Large Hadron Collider and on the International Space Station to detect the existence of such particles, and additional experiments are being done to detect such particles from the halo of the Milky Way as they pass through Earth.

Some proposed explanations of the "missing mass" do not rely on dark matter. For example, Modified Newtonian Dynamics (MOND) calls for modifications to Newton's law of gravitation that become apparent only on large scales. Observations at intermediate scales—for example, in the Bullet Cluster (see Figure 23.23), where two galaxy clusters are colliding—show that the original formulation of MOND cannot explain the observation. Advocates of the idea propose to further modify MOND. As the science proceeds, MOND, as well as other proposed explanations, will continue to be tested as possible alternatives to dark matter. Eventually, all explanations but one will be ruled out, and the surviving explanation will become the theory that explains the problem of galaxy rotation curves.

20.4 Most Galaxies Have a Supermassive Black Hole at the Center

Galaxies are remarkable objects, each shining with the light of hundreds of billions of stars. But galaxies are not as luminous as the most brilliant beacons of all: **quasars**. *Quasar* is short for *quasi-stellar radio source*, so named because astronomers first observed these mysterious objects as unresolved points at radio wavelengths. The story of the discovery of quasars provides an interesting insight into the discovery of new phenomena in astronomy and how conventional thinking can sometimes hinder progress.

Quasars are phenomenally luminous.

In the late 1950s, radio surveys had detected a number of bright, compact objects that at first seemed to have no optical counterparts. Eventually, astronomers found the optical counterparts when more accurate radio positions revealed that the radio sources coincided with faint, very blue, stellar-like objects. Unaware of the true nature of these objects, astronomers called them "radio stars." Obtaining spectra of the first two radio stars was a laborious task, requiring 10-hour exposures with the only recording technique available at the time: slow photographic plates. Astronomers were greatly puzzled by the results. Rather than displaying the expected absorption lines that are characteristic of blue stellar objects, the spectra showed only a single pair of emission lines that were broad—indicating very rapid motions within these objects—and that did not seem to correspond to the lines of any known substances.

For several years astronomers believed they had discovered a new type of star, until one astronomer, Maarten Schmidt (1929–), realized that these broad spectral lines were the highly redshifted lines of ordinary hydrogen (**Figure 20.17**). The implications were surprising: these "stars" were not stars. They were extraordinarily luminous objects at enormous distances. Other "quasars," were soon found by the same techniques. Many were relatively easy to identify because of their unusual blue color. As still more were found, astronomers began cataloging them.

Quasars are phenomenally powerful, pouring forth the luminosity of a trillion to a thousand trillion (10^{12} to 10^{15}) Suns. They are also very distant; the quasar nearest to Earth is approximately 300 Mpc away. Billions of galaxies are closer to Earth than is the nearest quasar. Recall that the distance to an object also indicates the amount of time that has passed since the light from that object left its source. The fact that quasars are seen only at great distances implies that they are quite rare in the universe at this time but were once much more common. The discovery that quasars existed in the distant and therefore earlier universe provided one of the first pieces of evidence demonstrating that the universe has evolved over time.

Quasars are not the isolated beacons they were once thought to be. Instead, they are centers of violent activ-

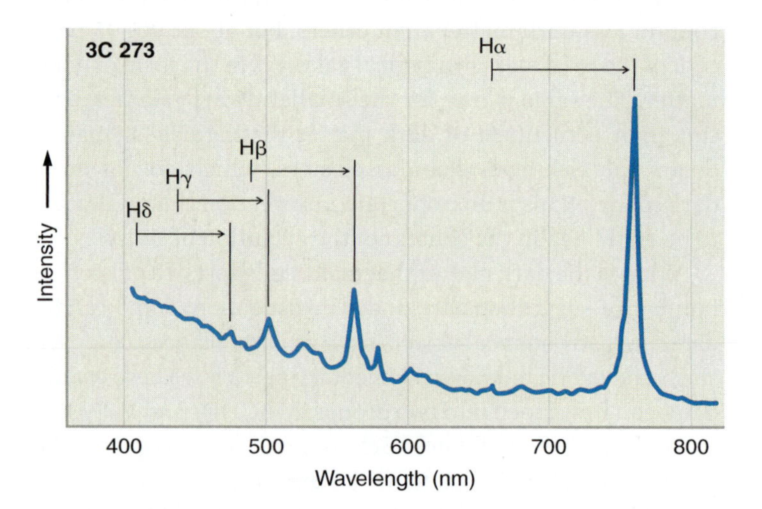

FIGURE 20.17 Spectrum of quasar 3C 273, one of the closest and most luminous known quasars. Active galactic nuclei are identified by the emission lines in their spectra, which distinguishes them from normal galaxies, which show mostly only absorption lines. The emission lines are redshifted by $z = 0.16$, indicating that the quasar is at a distance of about 750 Mpc.

ity in the hearts of large galaxies (**Figure 20.18**). Astronomers now recognize that quasars result from the most extreme form of activity that can occur in the nuclei of galaxies. Together, quasars and their less luminous cousins are called **active galactic nuclei**, or simply **AGNs**. Several distinct types of active nuclei can exist within galaxies. These are identified from the spectrum of the galaxy. A "normal" galaxy has an absorption spectrum that is a composite of the light from its billions of stars. A galaxy with an AGN exhibits emission lines in addition to the stellar absorption spectrum.

There are several types of active galactic nuclei.

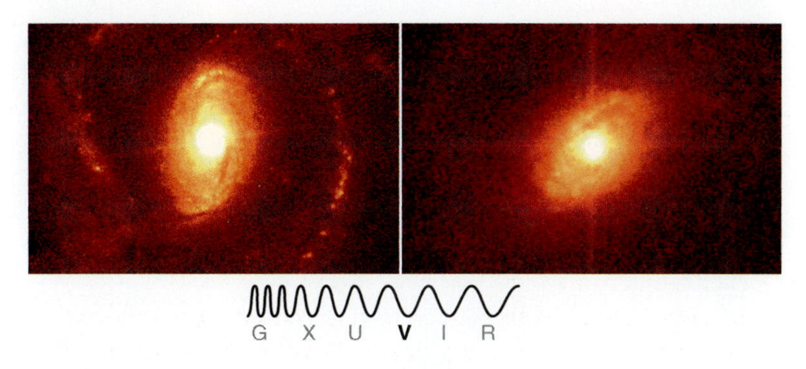

FIGURE 20.18 HST images showing quasars embedded in the centers of galaxies.

Seyfert galaxies, named after Carl Seyfert (1911–1960), who discovered them in 1943, are spiral galaxies whose centers contain AGNs. The luminosity of a typical Seyfert nucleus can be 10–100 billion $L_{\odot}$, comparable to the luminosity of the rest of the galaxy as a whole. The luminosities of AGNs found in elliptical galaxies are similar to those of Seyfert nuclei. Unlike Seyfert nuclei, however, AGNs in elliptical galaxies are usually most prominent in the radio portion of the electromagnetic spectrum, earning those galaxies the name **radio galaxies**. Radio galaxies, and their distant, extremely luminous cousins the quasars, are often the sources of slender jets that extend outward millions of light-years from the galaxy, powering twin lobes of radio emission (**Figure 20.19**).

Much of the light from AGNs is synchrotron radiation. This is the same type of radiation that comes from extreme environments such as the Crab Nebula supernova remnant (see Chapter 17). Synchrotron radiation comes from relativistic charged particles spiraling around in the direction of a magnetic field. The fact that AGNs accelerate large amounts of material to nearly the speed of light indicates that they are very violent objects. In addition to the continuous spectrum of synchrotron emission, the spectra of many quasars and Seyfert nuclei also show emission lines that are smeared out by the Doppler effect across a wide range of wavelengths. This observation implies that gas in AGNs is swirling around the centers of these galaxies at speeds of thousands or even tens of thousands of kilometers per second.

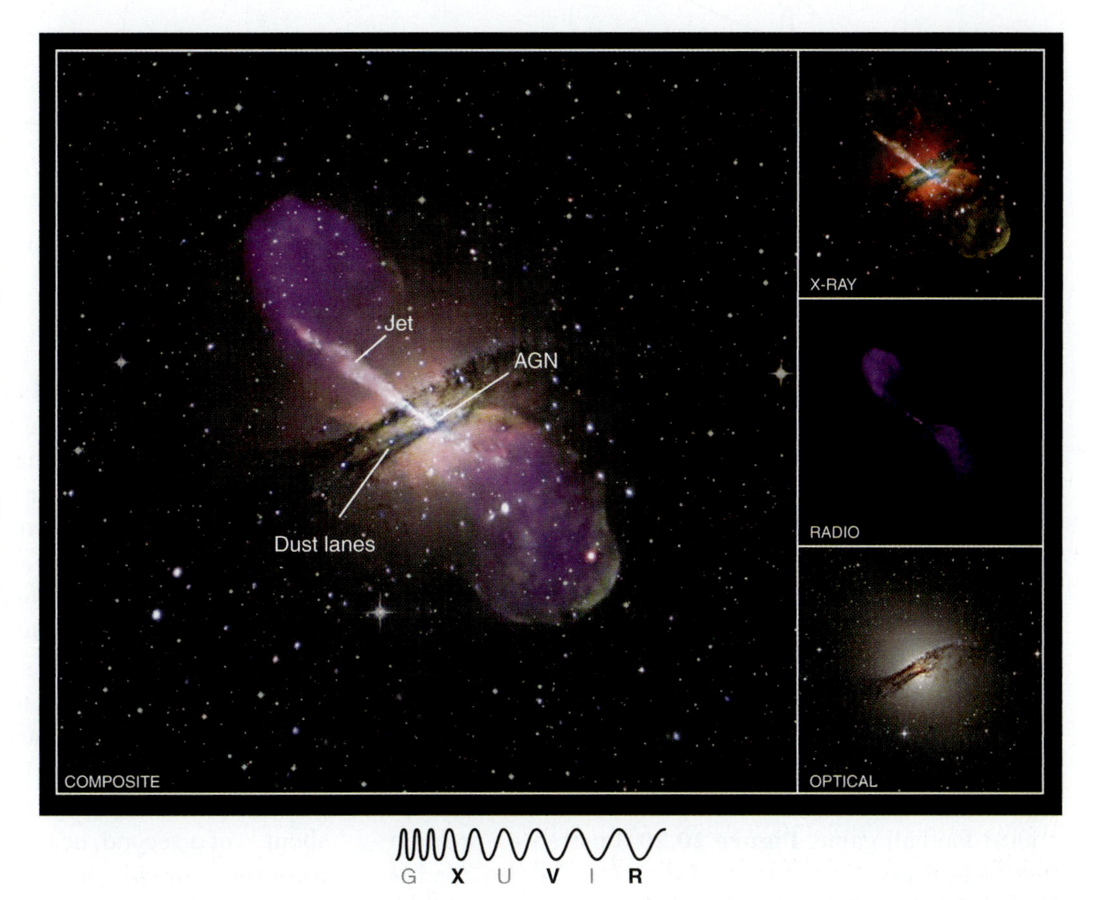

FIGURE 20.19 Radio galaxy Centaurus A. The visible-light (optical) image shows the galaxy, the X-ray image shows the hot gas, and the radio image shows the jets and lobes.

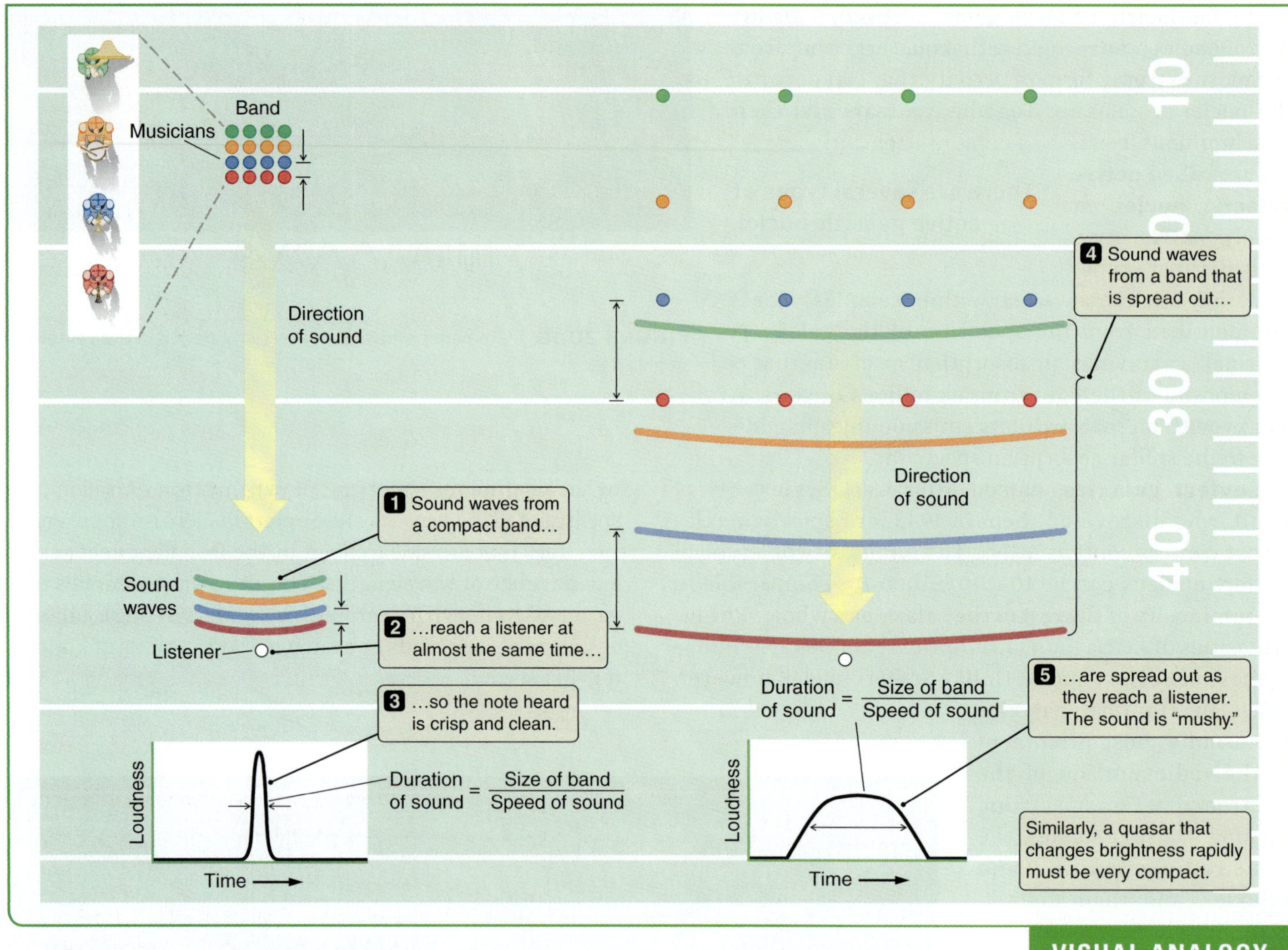

FIGURE 20.20 A marching band spread out across a field cannot play a clean note. Similarly, AGNs must be very compact to explain their rapid variability (see the text).

AGNs Are the Size of the Solar System

The enormous radiated power and mechanical energy of active galactic nuclei are made even more spectacular by the fact that all of this power emerges from a region that can be no larger than a light-day or so across—a region comparable in size to the Solar System. Although the Hubble Space Telescope (HST) and large ground-based telescopes show faint fuzz—light from the surrounding galaxy—around the images of some quasars and AGNs, the quasars and AGNs themselves remain as unresolved points of light.

Why do astronomers think that AGNs are compact objects? As an analogy, think about the halftime show at a local football game. **Figure 20.20** illustrates a problem that faces every marching-band director. When a band is all together in a tight formation at the center of the field, the notes you hear in the stand are clear and crisp; the band plays together beautifully. But as the band spreads out across the field, its sound begins to get mushy. This is not because the marchers are poor musicians. Instead, it is a consequence of the fact that sound travels at a finite speed. On a cold, dry, December day sound travels at a speed of about 330 meters per second (m/s). At this speed, it takes sound approximately ⅓ of a second to travel from one end of the football field to the other. Even if every musician on the field plays a note at exactly the same instant in response to the director's cue, in the stands you hear the instruments close to you first but have to wait longer for the sound from the far end of the field to arrive.

If the band is spread from one end of the field to the other, then the beginning of a note will be smeared out over about ⅓ of a second, or the difference in sound travel time from the near side of the field to the far side. If the band were spread out over two football fields, it would take about ⅔ of a second for the sound from the most distant musicians to arrive at your ear. If our marching band were spread out over a kilometer, then it would take roughly 3 seconds—the time it takes sound to travel a kilometer—for you to hear a crisply played note start and stop. Even with your eyes closed, it would be easy to tell whether the band was in a tight group or spread out across the field.

Exactly the same principle applies to the light observed from active galactic nuclei. Astronomers observe that quasars and other AGNs can change their brightness dramatically over the course of only a day or two—and in some cases as briefly as in a few hours. This rapid variability sets an upper limit on the size of the AGN, just as hearing clear music from the band indicates that the band musicians are close together. The AGN powerhouse must therefore be no more than a light-day or so across because if it were larger, what astronomers see could not possibly change in a day or two. An AGN has the light of 10,000 galaxies pouring out of a region of space that would come close to fitting within the orbit of Neptune. ▶II **ASTROTOUR: ACTIVE GALACTIC NUCLEI**

AGNs vary rapidly and so must be relatively small.

Supermassive Black Holes and Accretion Disks

When astronomers first discovered AGNs, they put forward a variety of ideas to explain them. But as observations revealed the tiny sizes and incredible energy output of AGNs, only one answer seemed to make sense. Violent accretion disks surrounding **supermassive black holes**—black holes with masses from thousands to tens of billions of solar masses—power AGNs. Recall that you have already encountered accretion disks several times in this book. Accretion disks surround young stars, providing the raw material for planetary systems. Accretion disks around white dwarfs, fueled by material torn from their bloated evolving companions, lead to novae and Type Ia supernovae. Accretion disks around neutron stars and stellar-mass black holes a few kilometers across are seen as X-ray binary stars. If you scale these examples up to a black hole with a mass of a billion solar masses and a radius comparable in size to the orbit of Neptune (**Math Tools 20.1**), and then imagine an accretion disk fed by substantial amounts of mass rather than the small amounts of material being siphoned off a star, *that* is an active galactic nucleus.

AGNs are powered by accretion onto supermassive black holes.

Astronomers have developed this basic picture of a supermassive black hole surrounded by an accretion disk into a more complete AGN model. This model attempts to explain all types of AGNs—quasars, Seyfert galaxies, and radio galaxies. **Figure 20.21** shows the various components of this AGN model, in which an accretion disk surrounds a supermassive black hole. Much farther out lies a large torus (doughnut) of gas and dust consisting of material that is feeding the central engine. This model is discussed further in **Connections 20.1** (on page 632).

Math Tools 20.1

Supermassive Black Holes

Recall from Chapter 18 our discussion of the Schwarzschild radius, where you saw that stellar-mass black holes are kilometers in size. What are the sizes of supermassive black holes? The formula for the Schwarzschild radius is given by:

$$R_S = \frac{2GM_{BH}}{c^2}$$

where G is the gravitational constant and c is the speed of light.

The largest supermassive black holes observed have about 10 billion solar masses ($M_\odot$). For example, the black hole at the center of the galaxy M87 is 6.6 billion $M_\odot$. To compute its size, recall that $M_\odot = 1.99 \times 10^{30}$ kilograms (kg), $c = 3 \times 10^8$ m/s, and $G = 6.67 \times 10^{-11}$ cubic meters per kilogram per second squared (m³/kg s²). Then a 6.6-billion-$M_\odot$ black hole has a Schwarzschild radius of:

$$R_S = \frac{2 \times (6.67 \times 10^{-11} \text{m}^3/\text{kg s}^2) \times (6.6 \times 10^9 \times 1.99 \times 10^{30} \text{ kg})}{(3 \times 10^8 \text{ m/s})^2}$$
$$= 2.0 \times 10^{13} \text{ m}$$

We can convert this value into astronomical units. Recall that 1 astronomical unit (AU) = 1.5×10^{11} meters. Therefore, this supermassive black hole has a radius of 130 AU—somewhat larger than the Solar System. We know that light takes 8⅓ minutes to reach Earth from the Sun at a distance of 1 AU, so this 130 AU corresponds to a distance of 1,083 light-minutes, or 18 light-hours.

What is the density of this object? The mass of the black hole divided by the volume within the Schwarzschild radius is:

$$\text{Density} = \frac{\text{Mass}}{\text{Volume}} = \frac{(6.6 \times 10^9) \times (1.99 \times 10^{30} \text{ kg})}{\frac{4}{3} \times \pi \times (2.0 \times 10^{13} \text{ m})^3}$$
$$= 0.4 \text{ kg/m}^3$$

This is much less than the density of water. Supermassive black holes do not have the extremely high mean densities of stellar-mass black holes.

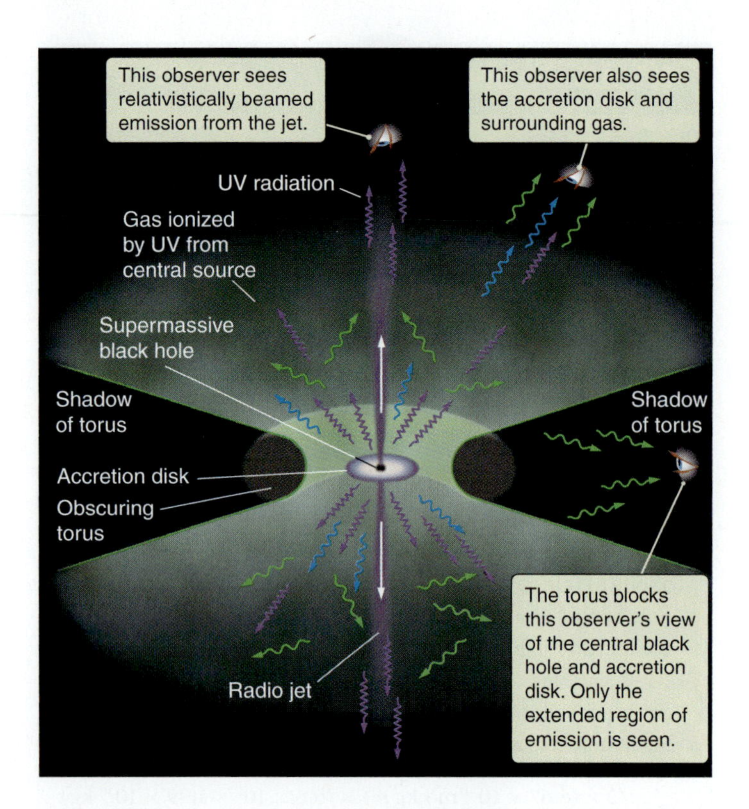

FIGURE 20.21 The basic model of active galactic nuclei, with a supermassive black hole surrounded by an accretion disk at the center. A larger, dusty torus sometimes blocks the view of the black hole. The mass of the central black hole, the rate at which it is being fed, and the viewing angle determine the observational properties of an AGN.

In the previous discussion of star formation (see Chapter 15), you learned that gravitational energy is converted to thermal energy as material moves inward toward the growing protostar. Here, as material moves inward toward a supermassive black hole, conversion of gravitational energy heats the accretion disk to hundreds of thousands of kelvins, causing it to glow brightly in visible, ultraviolet, and X-ray light. Conversion of gravitational energy to thermal energy as material falls onto the accretion disk is also a source of energetic emission. When we discussed the Sun, we noted the efficiency of fusion, which converts 0.7 percent of the mass of hydrogen into energy. In contrast, as much as 20 percent of the mass of infalling material around a supermassive black hole is converted to luminous energy (**Math Tools 20.2**). The rest of that mass is pulled into the black hole itself, causing it to grow even more massive.

AGN accretion disks are extremely efficient at converting mass to energy.

The interaction of the accretion disk with the black hole gives rise to powerful radio jets—superpowerful analogs to the jets in Herbig-Haro objects formed by accretion disks around young stellar objects (see Chapter 15). Throughout, twisted magnetic fields accelerate charged particles such as electrons and protons to relativistic speeds, accounting for the observed synchrotron emission. Gas in the accretion disk or in nearby clouds orbiting the central black hole at high speeds gives off emission lines that are smeared out by the Doppler effect into the broad lines seen in AGN spectra. This accretion disk surrounding a supermassive black hole is the "central engine" that leads to AGNs.

Normal Galaxies and AGNs

The essential elements of an AGN are a central engine (an accretion disk surrounding a supermassive black hole) and a source of fuel (gas and stars flowing onto the accretion disk). Without a source of matter falling onto the black hole, an AGN would no longer be an active nucleus. Astronomers looking at such an object would see a normal galaxy with a supermassive black hole sitting in its center.

Only about 3 percent of present-day galaxies contain AGNs as luminous as the host galaxy. But when astronomers look at more distant galaxies (and therefore look back in time), the percentage of galaxies with AGNs is much larger. These observations show that when the universe was younger, there were many more AGNs than there are today. If astronomers' understanding of AGNs is correct, then all the supermassive black holes that powered those dead AGNs should still be around. If they combine what they know of the number of AGNs in the past with ideas about how long a given galaxy remains in an active AGN phase, they are led to predict that many—perhaps even *most*—normal galaxies today contain supermassive black holes.

If supermassive black holes are present in the centers of normal galaxies, these black holes should reveal themselves in a number of ways. For one thing, such a concentration of mass at the center of a galaxy should draw surrounding stars close to it. The central region of such a galaxy should be much brighter than could be explained if stars alone were responsible for the gravitational field in the inner part of the galaxy. Stars feeling the gravitational pull of a supermassive black hole in the center of a galaxy should also orbit at very high velocities. Astronomers should therefore see large Doppler shifts in the light from stars near the centers of normal gal-

All large galaxies probably contain supermassive black holes.

Math Tools 20.2

Feeding an AGN

Power for an AGN is created when some infalling matter around the central supermassive black hole is converted to energy, which is quantified by Einstein's mass-energy equation: $E = mc^2$. About how much material has to be converted to energy to produce the observed luminosities? Astronomers estimate the efficiency of the accretion to be about 10–20 percent. Here, we'll assume that 15 percent of the infalling matter is converted to energy, or:

$$E = 0.15mc^2$$

Astronomers can measure how much energy is produced by the infalling material and radiated to space. For a relatively weak AGN like the earlier example of the galaxy M87, $L = 5 \times 10^{35}$ joules per second (J/s), or 5×10^{35} kg $\times$ m^2/s^2 each second. Dividing both sides of Einstein's equation by $0.15c^2$ gives us the mass consumed each second:

$$m = \frac{E}{0.15c^2}$$

$$= \frac{5 \times 10^{35}\ \text{kg m}^2/\text{s}^2}{0.15 \times (3 \times 10^8\ \text{m/s})^2}$$

$$= 3.7 \times 10^{19}\ \text{kg}$$

Multiplying this result (the mass consumed each second) by 3.2×10^7 seconds per year yields an accreted mass of 10^{27} kg, or about half the mass of Jupiter.

Quasars, the most powerful AGNs, have a luminosity $L = 10^{39}$ J/s $= 10^{39}$ kg $\times$ m^2/s^2 each second. The mass consumed each second is then:

$$m = \frac{10^{39}\ \text{kg m}^2/\text{s}^2}{0.15 \times (3 \times 10^8\ \text{m/s})^2}$$

$$= 7.4 \times 10^{22}\ \text{kg}$$

Multiplying by 3.2×10^7 seconds per year yields a mass of 2.4×10^{30} kg per year. Recall that the mass of the Sun is 2×10^{30} kg. Therefore, this quasar supermassive black hole is accreting about 1.2 $M_\odot$ each year.

axies. They have, in fact, found evidence of this sort in every normal galaxy with a substantial bulge in which a careful search has been conducted. The masses inferred for these black holes range from 10,000 $M_\odot$ (for a "small" black hole) to 20 billion $M_\odot$ (for a "gargantuan" black hole). The mass of the supermassive black hole seems to be related to the mass of the elliptical-galaxy or spiral-galaxy bulge in which it is found. All large galaxies probably contain supermassive black holes. These observations reveal something remarkable about the structure and history of normal galaxies.

Apparently the only difference between a normal galaxy and an active galaxy is whether the supermassive black hole at its center is currently being fed. The low percentage of present-day galaxies with very luminous AGNs does not indicate which galaxies have the potential for AGN activity. Rather, it indicates which galaxy centers are being lit up at the moment. If a large amount of gas and dust were dropped directly into the center of any large galaxy, this material would fall inward toward the central black hole, forming an accretion disk and a surrounding torus. The predicted result would be that the nucleus of this galaxy would change into an AGN.

In Chapter 23 we will discuss galaxy evolution and note that many of the observed properties of galaxies discussed in this chapter, including the formation of galaxy type, spiral structure, star formation, and AGN, depend on the interactions and mergers between galaxies. To account for the many large galaxies visible today, interactions and mergers must have been much more prevalent in the past when the universe was younger; this is one explanation for the larger number of AGNs that existed in the past. Computer models show that galaxy-galaxy interactions can cause gas located thousands of parsecs from the center of a galaxy to fall inward toward the galaxy's center, where it can provide fuel for an AGN. During mergers, a significant fraction of a galaxy might wind up being cannibalized. HST images of quasars often show that quasar host galaxies are tidally distorted or are surrounded by other visible matter that is probably still falling into the galaxies. Galaxies that show evidence of recent interactions with other galaxies are more likely to house AGNs

Galaxy-galaxy interactions fuel AGN activity.

Connections 20.1

Unified Model of AGN

In the **unified model of AGN**, the different types of AGNs observed from Earth are partly explained by astronomers' view of the central engine (see Figure 20.21). The outer torus obscures this view in different ways, depending on the viewing angle. Variation in this angle, in the mass of the black hole, and in the rate at which it is being fed accounts for a wide range of AGN properties. When the AGN is viewed edge on, astronomers see emission lines from the surrounding torus and other surrounding gas. They can also sometimes see the torus in absorption against the background of the galaxy. From this nearly edge-on orientation, they cannot see the accretion disk itself, so they do not expect to see the Doppler-smeared lines that originate closer to the supermassive black hole. If jets are present in the AGN, however, these should be visible emerging from the center of the galaxy.

If astronomers look at the inner accretion disk somewhat more face on, they can see over the edge of the torus and thus get a more direct look at the accretion disk and the location of the black hole. In this case they see more of the synchrotron emission from the region around the black hole and the Doppler-broadened lines produced in and around the accretion disk. **Figure 20.22** shows an HST image of one such object, called M87, at an intermediate inclination. M87 is a source of powerful jets that continue outward for 100,000 light-years but originate in the tiny engine at the heart of the galaxy. Spectra of the disk at the center of this galaxy show the rapid rotation of material around a central black hole with a mass of 3 billion (3×10^9) $M_\odot$. (See **Figure 20.23** and **Figure 20.24**.)

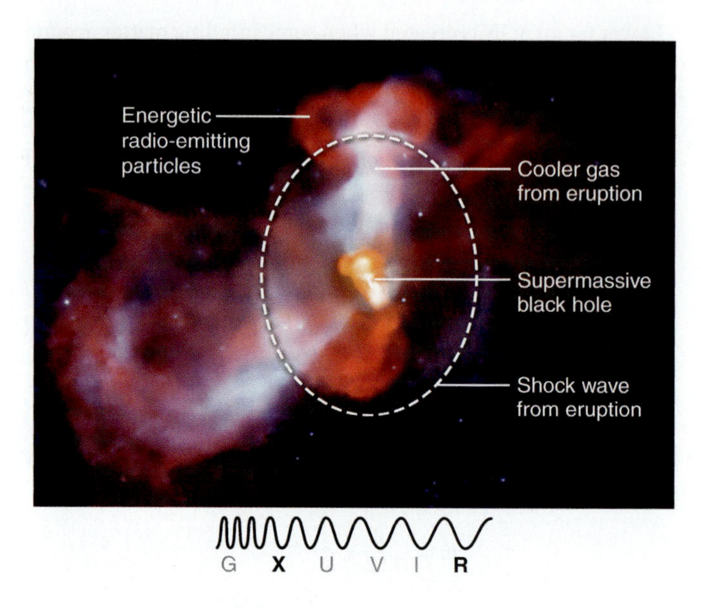

FIGURE 20.22 M87 in radio and X-rays, showing the location of the supermassive black hole.

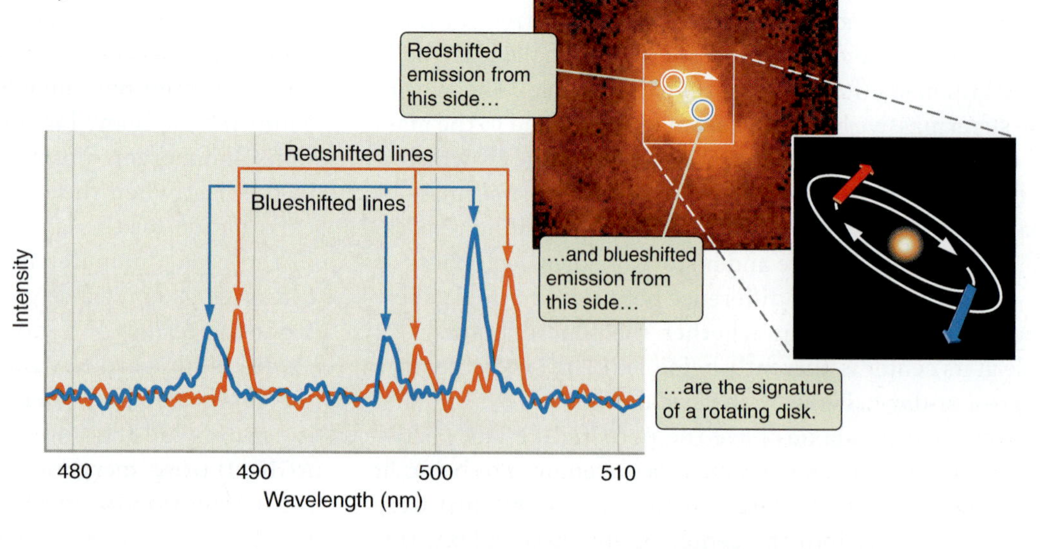

FIGURE 20.23 An HST image of the nearby radio galaxy M87. The high velocities observed provide direct evidence of rotation about a supermassive black hole at this galaxy's center.

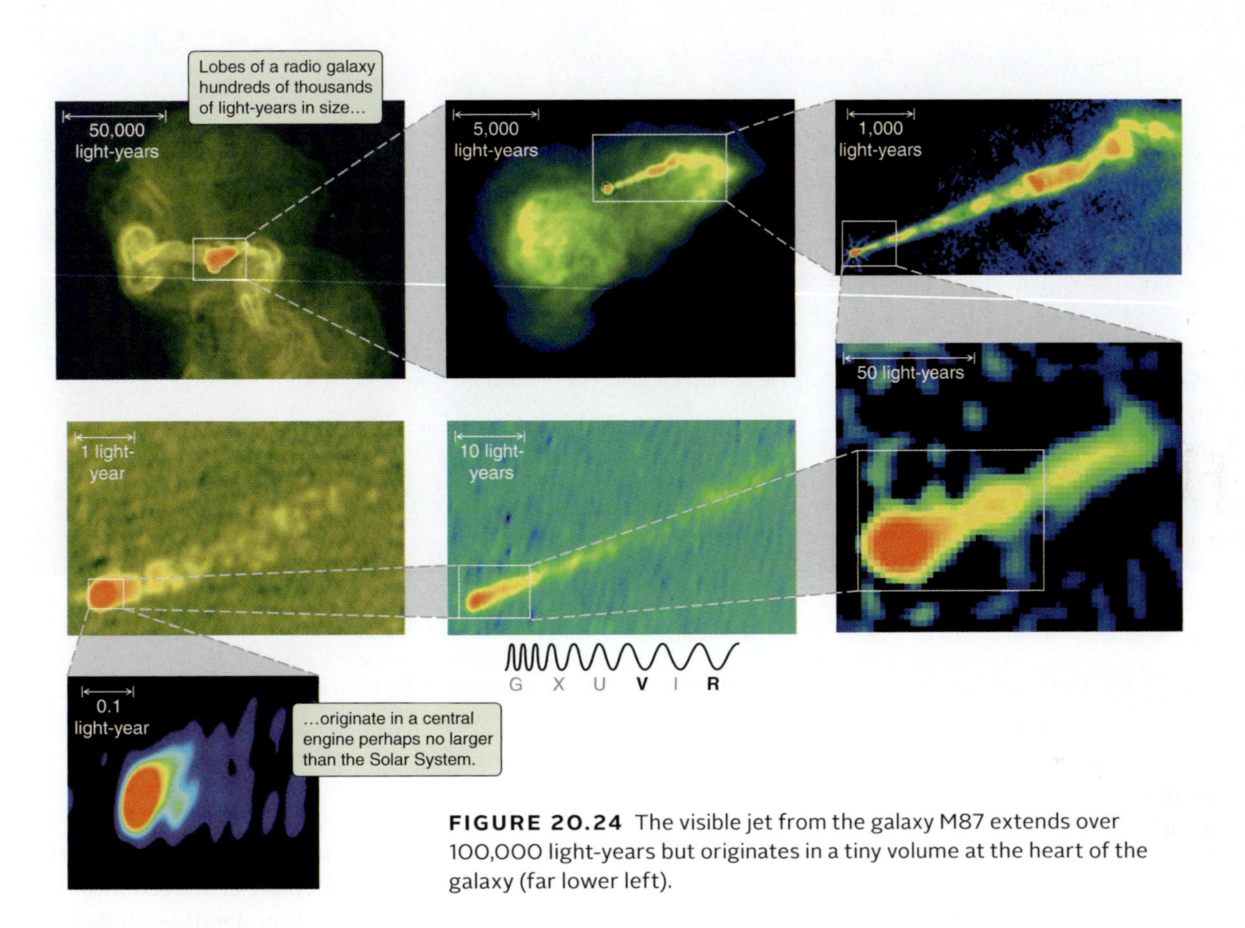

FIGURE 20.24 The visible jet from the galaxy M87 extends over 100,000 light-years but originates in a tiny volume at the heart of the galaxy (far lower left).

The material in an AGN jet travels very close to the speed of light. As a result, what astronomers see is strongly influenced by relativistic effects. One of these is an extreme form of the Doppler effect called **relativistic beaming**: matter traveling at close to the speed of light concentrates any radiation it emits into a tight beam pointed in the direction in which it is moving. So astronomers often observe only one side of the jets from AGNs, even though the radio lobes of radio galaxies are always two-sided. The jet moving away is just too faint to observe.

In rare instances when the accretion disk in a quasar or radio galaxy is viewed almost directly face on, relativistic beaming dominates the observations (see Figure 20.21). In these *blazars*, emission lines and other light coming from hot gas in the accretion disk are overwhelmed by the bright glare of jet emission beamed directly at Earth.

Relativistic beaming is not the only thing that complicates what astronomers see when they look down the barrel of an AGN jet. The material in an AGN jet is moving so close to the speed of light that the radiation it emits is barely able to outrun its source. Astronomers see all of the light that was emitted by the jet over thousands of years arrive at their telescopes over the course of only a few years. Time appears to be compressed. From their perspective, the jet seems to travel great distances in brief periods of time. In extreme cases, such as the jet in M87, features in the jet *appear* to be moving across the sky faster than the speed of light. We stress the word *appear* because this phenomenon, referred to as **superluminal motion**, is an optical illusion. Despite the name, nothing in these jets is actually traveling through space faster than the speed of light. Einstein's special theory of relativity remains safe.

FIGURE 20.25 The Swift Gamma-Ray observatory has detected active black holes (circles) in these merging galaxies.

in their centers (**Figure 20.25**). Any large galaxy might be only an encounter away from becoming an AGN.

20.5 Origins: Habitability in Galaxies

In this chapter we discussed the different types of galaxies that have been observed. Can we say anything about their potential for life? The short answer is that there is no solid information. These galaxies are too far away for astronomers to have detected any planets around their stars. So all we can do is speculate about the habitability of other galaxies. Two key requirements are the presence of heavy elements to form planets (and life), and an environment without too much radiation that might be damaging to life.

A study of the host stars of Kepler exoplanet candidates suggests that stars with a higher percentage of heavier elements may be likely to have planets. (This finding fits with the core accretion models of planet formation discussed in Chapter 7.) The first generation of stars made from the hydrogen and helium of Big Bang nucleosynthesis do not have heavy elements. Recall from the chapters on stellar evolution that elements heavier than helium are created in the cores of dying stars, and then are scattered into the galactic environment through planetary nebulae, stellar winds, and supernova explosions. So the amount of heavier elements in a star depends on the cosmic history of the material from which the star formed. Therefore, astronomers must consider the galactic environment of the star, which varies among different types of galaxies and different locations within the galaxies.

Spiral galaxies have had more continuous star formation in their disks throughout their history. They contain more stars born from recycled material, and therefore more stars with a higher fraction of heavy elements. Elliptical (and S0) galaxies have older, redder populations of stars and very little star formation. Old, massive ellipticals have a larger percentage of lower-mass stars than smaller ellipticals or spirals have. Astronomers had previously thought that this difference meant that large elliptical galaxies would not be good environments for planet formation. But the Kepler telescope has found many planets around small, red, main-sequence stars like the ones that populate elliptical galaxies. One study of two elliptical galaxies showed that both had some fraction of stars with a heavy-element fraction similar to that of the stars hosting Kepler exoplanets in the Milky Way.

Another issue considered by astronomers is the presence of radiation that might be hazardous to life. This radiation is most likely to come from the center of the galaxy, especially if AGN jets are present. Galaxies that are in an active AGN state might have too much radiation in regions close to their centers to be conducive to life. Stars whose orbits cross spiral arms many times might also be exposed to higher-than-average levels of radiation, but this is not the case for the majority of stars in a galaxy.

The conditions in these galaxies may also change as the galaxies evolve. Galaxy mergers can shake up stellar orbits and relocate stars and their planets to different locations. Mergers may also affect the growth and activity level of supermassive black holes and thus the presence of radiation. Some galactic environments just may not remain habitable for the length of time—billions of years—that it took life to evolve from bacteria to intelligence on Earth.

It is unlikely that there will ever be a simple scheme that does for galaxies what the H-R diagram did for stars. However, astronomers are beginning to better understand what the "ecology" within galaxies is like and to piece together something of how galaxies form and evolve. In the next chapter we will look in more detail at Earth's own Milky Way, and see what it reveals about galaxy evolution. ■

Summary

20.1 The shapes of galaxies and the types of orbits of their stars determine their Hubble classification as elliptical (E), spiral (S), barred spiral (SB), or irregular (Irr). Stars are currently forming in the disks of spiral galaxies but not in ellipticals or S0 galaxies.

20.2 Spiral arms of galaxies are regions of intense star formation. The arms are visible because of the concentration of bright young stars. Any significant disturbance in the rotating disk of a galaxy will lead to spiral structure. Regular disturbances called spiral density waves trigger star formation.

20.3 Most of the mass in galaxies does not reside in gas, dust, or stars; rather, about 90 percent of a galaxy's mass is in the form of dark matter, which does not emit or absorb light to any significant degree. The two main groups of candidates for the composition of dark matter are MACHOs (astronomical objects such as planets, stars, and black holes) and WIMPs (massive elementary particles).

20.4 Most—perhaps all—large galaxies have supermassive black holes at the center. When gas accretes onto one of these supermassive black holes, the center of the galaxy becomes an active galactic nucleus (AGN). AGNs can emit as much as 1,000 times the light of the whole galaxy, all coming from a region the size of the Solar System.

20.5 In thinking about the potential habitability of galaxies, astronomers consider the activity state of the galaxy, including AGNs and mergers, and the amount of heavy elements in the stars in the galaxy, which is related to the star formation rate and galaxy type.

Unanswered Questions

- How long do spiral arms and bars in spiral galaxies last? In contrast to the spiral arms, the bars in spiral galaxies are not density waves. The stars remain in the bar, and star formation can sometimes be seen at the bar ends. The bars may drive the spiral density waves, but the extent to which this happens is not yet known. This question is studied theoretically by running computer models that simulate the contents of a disk galaxy and gravitational interaction over many years, to see how long the patterns will last. Observationally large samples of spiral galaxies are studied to look for the connection between the spiral arms and the bar.
- Where do supermassive black holes come from? To explore this question, astronomers are studying computer models in which the supermassive black hole grew from a large number of stellar-mass black holes, or grew along with the galaxy by swallowing large amounts of central gas, or increased after the merger of two or more galaxies. We will return to this question when we discuss galaxy evolution in Chapter 23.

Questions and Problems

Summary Self-Test

1. Galaxies are classified according to
   a. mass.
   b. color.
   c. density.
   d. shape.

2. Spiral arms have ________ stars than the rest of the disk has.
   a. more
   b. younger
   c. older
   d. fewer

3. Which of the following contributes the largest percentage of total mass of a spiral galaxy?
   a. dark matter
   b. central black hole
   c. stars
   d. dust and gas

4. Currently, stars form in
   a. ellipticals.
   b. S0 galaxies.
   c. spiral galaxies.
   d. all of the above

5. A rotation curve plots ________ versus distance from the galaxy center.
   a. orbital speed
   b. radial speed
   c. luminosity
   d. luminous mass

6. Spiral structure in a disk can be created by
   a. star formation.
   b. a bar-shaped bulge.
   c. gravitational interaction with other galaxies.
   d. all of the above

7. In the context of spiral galaxies, Kepler's laws could be used to estimate
   a. P, the period.
   b. A, the semimajor axis.
   c. M, the mass of the galaxy.
   d. v, the rotation speed of the galaxy.

8. The existence of supermassive black holes at the centers of massive galaxies is
   a. an idea.
   b. a fact.
   c. a law.
   d. a theory.

9. The reason all quasars are very distant is that
   a. they are so luminous.
   b. they are so faint.
   c. they are so large.
   d. the universe has evolved.

10. Astronomers determine the radius of an AGN by measuring
    a. how much light comes from it.
    b. how hard it pulls on stars nearby.
    c. how quickly its light varies.
    d. how quickly it rotates.

True/False and Multiple Choice

11. **T/F:** An E7 galaxy is more spherical than an E0 galaxy.

12. **T/F:** A spiral galaxy's dark matter is distributed spherically.

13. **T/F:** Dust does not shine in visible light, so it is considered dark matter.

14. **T/F:** Most of the mass of a galaxy is in stars.

15. **T/F:** AGNs are small but extremely luminous.

16. Which of the following galaxy types has a spherical bulge and a well-defined disk?
    a. spiral
    b. barred spiral
    c. elliptical
    d. irregular

17. Which of the following galaxy types describes a galaxy shaped like a rugby ball?
    a. Sb
    b. SBb
    c. E0
    d. E5

18. For a galaxy, the term *morphology* refers to
    a. its shape.
    b. its evolution over time.
    c. the motion of its stars.
    d. its overall density.

19. If all the stars in an elliptical galaxy traveled in random directions in their orbits, the elliptical galaxy would be type
    a. E0.
    b. E2.
    c. E5.
    d. E7.

20. In spiral galaxies, stars form primarily in the spiral arms
    a. because that's where most of the mass of the galaxy is.
    b. because of the sudden increase in density.
    c. because that's where rotation is the fastest.
    d. because of conservation of angular momentum.

21. Dark matter is different from normal matter because
    a. it doesn't emit light.
    b. it doesn't absorb light.
    c. it doesn't scatter light.
    d. all of the above

22. Astronomers know that dark matter is present in galactic halos because the speeds of orbiting stars ________ far from the center of the galaxy.
    a. decrease
    b. increase
    c. remain about constant
    d. fluctuate dramatically

23. Luminous matter includes
    a. stars.
    b. dust.
    c. gas.
    d. all of the above

24. The differences among various types of AGNs are caused by
    a. the type of the host galaxy.
    b. the size of the central black hole.
    c. the amount of dark matter in the galaxy's halo.
    d. the viewing angle.
25. If a Seyfert galaxy's nucleus varies in brightness on the timescale of 10 hours, then approximately what is the size of the emitting region?
    a. 20 AU
    b. 70 AU
    c. 90 AU
    d. 140 AU

Thinking about the Concepts

26. Name and describe a common, everyday object that appears so dissimilar when viewed from different angles that from these different views you might think it was actually a different object.
27. How does molecular-gas temperature differ between elliptical and spiral galaxies?
28. Some galaxies have regions that are relatively blue; other regions appear redder. What does this variation indicate about the differences between these regions?
29. Describe the characteristics of irregular galaxies.
30. Describe how elliptical galaxies and spiral bulges are similar.
31. Describe the spiral arms in a galaxy and explain at least one of the mechanisms that create them.
32. Name some of the candidates for the composition of dark matter.
33. How would you explain a quasar to a relative or friend?
34. Which is more luminous: a quasar or a galaxy with 100 billion solar-type stars? Explain your answer.
35. The nearest quasar is about a 300 Mpc away. Why don't astronomers observe any that are closer?
36. What distinguishes a normal galaxy from one that contains an AGN?
37. Contrast the size of a typical AGN with the size of the Solar System. How do astronomers know how big an AGN is?
38. Describe what astronomers think is happening at the centers of galaxies that contain AGNs.
39. The material in some AGN jets appears to be moving faster than the speed of light (superluminal motion). Why is it impossible to exceed, and what contributes to this illusion?
40. It is likely that most galaxies contain supermassive black holes, yet in many galaxies there is no obvious evidence for their existence. Why do some black holes reveal their presence while others do not?

Applying the Concepts

41. Study Figure 20.14. The small vertical bars (known as error bars) on the data points indicate the size of the measurement error.
    a. At a radius of 25,000 parsecs (pc), what is the approximate measurement error in the rotation velocity?
    b. What is this value as a percentage of the measured velocity?
    c. Error bars are important because they show how wrong the measurement could possibly be. One way to think about this is that the black line could be as high as the top of the error bars, or as low as the bottom of the error bars. In either case, would shifting the black line change the overall conclusion? Why or why not?
42. Assume that there are 1 trillion (10^{12}) galaxies in the universe, that the average galaxy has a mass equivalent to 100 billion (10^{11}) average stars, and that an average star has a mass of 2×10^{30} kg.
    a. Ignoring dark matter, how much mass (in kilograms) does the universe contain?
    b. If the mass of an average particle of normal matter is 10^{-27} kg, how many particles are there in the entire universe?
43. Suppose the number density of galaxies in the universe is, on average, 3×10^{-68} galaxy/m^3. If astronomers could observe all galaxies out to a distance of 10^{10} parsecs, how many galaxies would they find?
44. If astronomers observe a quasar with a redshift of 0.5, what can they say about the size of the universe when the light from that quasar was emitted?
45. The nearest known quasar is 3C 273. It is located in the constellation of Virgo and is bright enough to be seen in a medium-sized amateur telescope. With a redshift of 0.158, what is the distance to 3C 273 in parsecs?
46. The quasar 3C 273 has a luminosity of $10^{12}\ L_{\odot}$. Assuming that the total luminosity of a large galaxy, such as the Andromeda Galaxy, is 10 billion times that of the Sun, compare the luminosity of 3C 273 with that of the entire Andromeda Galaxy.
47. A quasar has the same brightness as a galaxy that is seen in the foreground 2 Mpc distant. If the quasar is 1 million times more luminous than the galaxy, what is the distance of the quasar?
48. Estimate the Schwarzschild radius for a supermassive black hole with a mass of 26 billion $M_{\odot}$.
49. You read in the newspaper that astronomers have discovered a "new" cosmological object that appears to be flickering with a period of 83 minutes. Because you have read *21st Century Astronomy*, you are able to quickly estimate the maximum size of this object. How large can it be?
50. A quasar has a luminosity of 10^{41} watts (W), or J/s, and $10^{8}\ M_{\odot}$ to feed it. Assuming constant luminosity and 20 percent conversion efficiency, what is your estimate of the quasar's lifetime?

51. A solar-type star ($M_{\odot} = 2 \times 10^{30}$ kg), approaches a supermassive black hole. As it crosses the event horizon, half of its mass falls into the black hole while the other half is completely converted to energy in the form of light. How much energy does this dying star send out to the rest of the universe?

52. Suppose that an object with the mass of Earth ($M_{\oplus} = 5.97 \times 10^{24}$ kg) fell into a supermassive black hole with a 10 percent energy conversion.
    a. How much energy (in joules) would be radiated by the black hole?
    b. Compare your answer with the energy radiated by the Sun each second: 3.85×10^{26} J.

53. If a luminous quasar has a luminosity of 2×10^{41} W, or J/s, how many solar masses ($M_{\odot} = 2 \times 10^{30}$ kg) per year does this quasar consume to maintain its average energy output?

54. Material ejected from the supermassive black hole at the center of galaxy M87 extends outward from the galaxy to a distance of approximately 30,000 pc. M87 is approximately 17 Mpc away.
    a. If this material were visible to the naked eye, how large would it appear in the nighttime sky? Give your answer in degrees, recalling from Chapter 13 that 1 radian = 57.3°.
    b. Compare this size with the angular size of the Moon.

55. A lobe in a visible jet from galaxy M87 is observed at a distance of 1,530 pc (5,000 light-years) from the galaxy's center moving outward at a speed of 0.99 times the speed of light (0.99*c*). Assuming constant speed, how long ago was the lobe expelled from the supermassive black hole at the galaxy's center?

Using the Web

56. Go to Astronomy Picture of the Day app or website (http://apod.nasa.gov/apod) and look at some recent pictures of galaxies. In each case, consider the following questions: Was the picture taken from a large or small telescope, and from the ground or from space? Are galaxies in the image face on, edge on, or at an angle? What wavelengths were used for making the image? Are any of the colors "false colors"? If the picture is a combination of images from several telescopes, what do the different colors indicate?

57. Go to the website for Galaxy Zoo (http://galaxyzoo.org), the original Zooniverse citizen science project. (Log in with your Zooniverse password.) The specific project in action at any given time depends on the real data that need to be examined. One of the projects is likely a classification project. Click on and read, "Story," "Science," and "Classify," and then classify some galaxies. Save a copy of your classifications for your homework if necessary.

58. a. Go to the Hubble Space Telescope website's "News Release Archive: Galaxy" page (http://hubblesite.org/newscenter/archive/releases/galaxy) and look for a news release on the subject of galaxies. Describe a recent story. What has been observed, and what is its importance? Do the observations support or contradict anything you read in this chapter?
    b. Go to the website for NuSTAR (Nuclear Spectroscopic Telescope Array—www.nustar.caltech.edu), a space telescope launched by NASA in 2012. This mission is studying active galaxies hosting supermassive black holes. What type of telescope is this (wavelengths observed, general design)? What has been discovered?

59. a. Go to the website for the Fermi Gamma-ray Space Telescope (http://fermi.gsfc.nasa.gov). Scroll down to click on "Full News Archive" and look for a story about dark matter. What has this telescope discovered about dark matter?
    b. Go to the website for the NASA Swift Gamma-Ray observatory (http://heasarc.nasa.gov/docs/swift/swiftsc.html), which studies gamma-ray bursts. Click on "Latest Swift News" and look for a story about supermassive black holes.

60. Go to the website for the Alpha Magnetic Spectrometer (http://ams02.org), a particle physics detector located on the International Space Station to search for dark matter, including WIMPs. Has it found anything?

StudySpace is a free and open website that provides a Study Plan for each chapter of *21st Century Astronomy*. Study Plans include animations, reading outlines, vocabulary flashcards, and multiple-choice quizzes, plus links to premium content in SmartWork and the ebook. Visit **wwnorton.com/studyspace**.

SmartWork Norton's online homework system, includes algorithmically generated versions of these questions, plus additional conceptual exercises. If your instructor assigns questions in SmartWork, log in at **smartwork.wwnorton.com.**

Exploration | Galaxy Classification

Galaxy classification sounds simple, but it can become complicated when you actually attempt it. The image in **Figure 20.26**, taken by the Hubble Space Telescope, shows a small portion of the Coma Cluster of galaxies. The Coma Cluster is made up of thousands of galaxies, each containing billions of stars. Some of the objects in this image (the ones with a bright cross) are foreground stars, in the Milky Way. Some of the galaxies in this image are far behind the Coma Cluster. Working with a partner, in this Exploration you will classify the 20 or so brightest galaxies in this cluster.

First, make a map by laying a piece of paper over the image and numbering the 20 or so brightest (or largest) galaxies in the image (label them "galaxy 1," "galaxy 2," and so on). Copy this map so that you and your partner each have a list of the same galaxies.

Separately, classify each galaxy by type. If it is a spiral galaxy, what is its subtype: a, b, or c? If it is an elliptical, how elliptical is it? Make a table that contains the galaxy number, the type you have assigned it, and any comments that will help you remember why you made that choice. When you are done classifying, compare your list with your partner's. Now comes the fun part! Argue about the classifications until you agree—or until you agree to disagree.

G X U **V** I R

FIGURE 20.26 An HST image of the Coma Cluster.

1 **Which galaxy type was easiest to classify?**

....

....

2 **Which galaxy type was hardest to classify?**

....

....

3 **What makes it hard to classify some of the galaxies?**

....

....

4 **Which galaxy type did you and your partner agree about most often?**

....

....

5 **Which galaxy type did you and your partner disagree about most often?**

....

....

6 **How might you improve your classification technique?**

....

....

If you found this activity interesting and rewarding, astronomers can use your help: visit http://galaxyzoo.org to get involved in a citizen science project to classify galaxies, some of which have never been viewed before by human eyes.

The star fields and dark dust clouds of the Milky Way Galaxy, with a Lyrid meteor in the foreground.

21 The Milky Way—A Normal Spiral Galaxy

Milky Way, sister in whiteness
To Canaan's rivers and the bright
Bodies of lovers drowned,
Can we follow toilsomely
Your path to other nebulae?

Guillaume Apollinaire (1880–1918)

LEARNING GOALS

Of the more than hundreds of billions of galaxies in the universe, the Milky Way is the only one that astronomers can study at close range. In this chapter we focus our attention on the Milky Way, and how it offers clues to understanding all galaxies. By the conclusion of this chapter, you should be able to:

- Explain how astronomers measured the size and structure of the Milky Way.
- Describe the environment within the disk of the Milky Way, and the halo of stars, globular clusters, and dark matter that surrounds and permeates the galaxy.
- Chart how the chemical composition of the Milky Way has evolved with time.
- Explain the evidence for a supermassive black hole at the center of the Milky Way.
- Describe the Local Group of galaxies.

21.1 Measuring the Shape and Size of the Milky Way

As you learned in Chapters 19 and 20, the universe is full of galaxies of many sizes and types. These galaxies are visible in astronomers' most powerful telescopes all the way to the edge of the observable universe. Yet when you go outside at night away from city lights and look up, it is not this universe of galaxies that you see. Rather, the night sky is filled with a single galaxy—the galaxy called the Milky Way. With what you know of galaxies from the previous chapter, you can learn a great deal about the local galaxy by looking at the night sky. **Figure 21.1a** shows what the Milky Way Galaxy looks like as seen in Earth's skies. For comparison, **Figure 21.1b** is an image of an edge-on spiral galaxy, suggesting that the Milky Way is flat like an edge-on galaxy. From a dark location at night you can even see dark bands where clouds of interstellar gas and dust obscure much of the central plane of the Milky Way. This view of the Milky Way from inside offers a different and much closer perspective of a galaxy than can be obtained by viewing external galaxies. Astronomers can merge this perspective from inside our galaxy with what is known about other galaxies to better understand the Milky Way as a spiral galaxy.

How might a planet's location in the Milky Way affect its chances for having life?

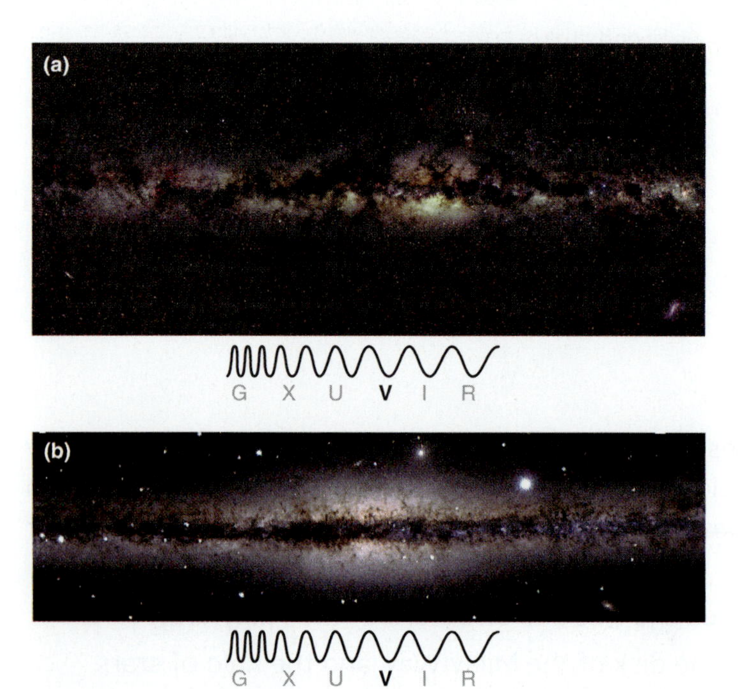

FIGURE 21.1 (a) The Milky Way is seen as a luminous band stretching across the night sky. Note the prominent dark lanes caused by interstellar dust that obscures the light from more distant stars. (b) The edge-on spiral galaxy NGC 891, whose disk greatly resembles the Milky Way.

Spiral Structure in the Milky Way

Because the Sun and its planets are buried within the disk of the Milky Way, however, astronomers lack a bird's-eye view of Earth's galactic home. Dust in the surrounding interstellar medium further complicates the situation by limiting the view. **Figure 21.2a** shows a model of the structure of the Milky Way proposed in the late 1700s by William Herschel. Bearing little resemblance to modern ideas about the Milky Way, Herschel's model was based simply on counting the number and brightness of stars seen in different directions. Herschel did not know about the interstellar extinction of starlight, the properties of stars, or the relationship between the Milky Way and the "spiral nebulae" that he studied throughout his life. Before multiwavelength astronomy, it was easier for observers to see the prominence of the spiral arms, the presence of a bar, and the size of the bulge in distant galaxies than to see these features in the Milky Way.

Dust obscures the view of the Milky Way's center.

It took modern infrared and radio observations to form the current model of the Milky Way Galaxy, and this picture is still being revised. Recall from Chapter 15 that neutral hydrogen emits radiation at wavelength 21 centimeters (Figures 15.10 and 15.11). This radiation was predicted in the 1940s and then detected in the early 1950s. By 1952, astronomers had the first maps of the neutral hydrogen in the Milky Way and other galaxies. The maps showed spiral structure in the other galaxies and suggested spiral structure in the Milky Way. At about the same time, observations of ionized hydrogen gas in visible light showed two spiral arms with concentrations of young, hot O and B stars. The Milky Way was confirmed to be a spiral galaxy.

However, new discoveries are still being made about the shape of our galaxy. In the 1990s, there were some hints that the Milky Way has a bar, and this was confirmed in 2005 with Spitzer infrared observations of the distribution and motions of stars toward the center of the galaxy. **Figure 21.2b** shows an artist's rendering of the major features of the Milky Way. The Milky Way has a substantial bar with a modest bulge at its center. Two major spiral arms—Scutum-Centaurus and Perseus—connect to the ends of the central bar and sweep through the galaxy's disk, just like the arms observed in external spiral galaxies. There are several smaller arm segments, including the Orion

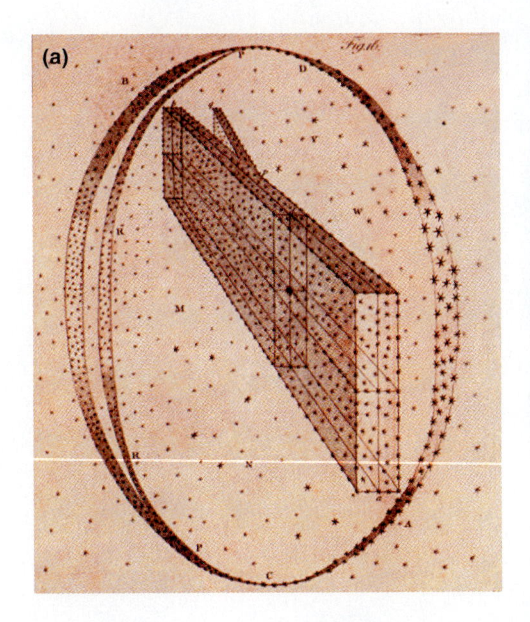

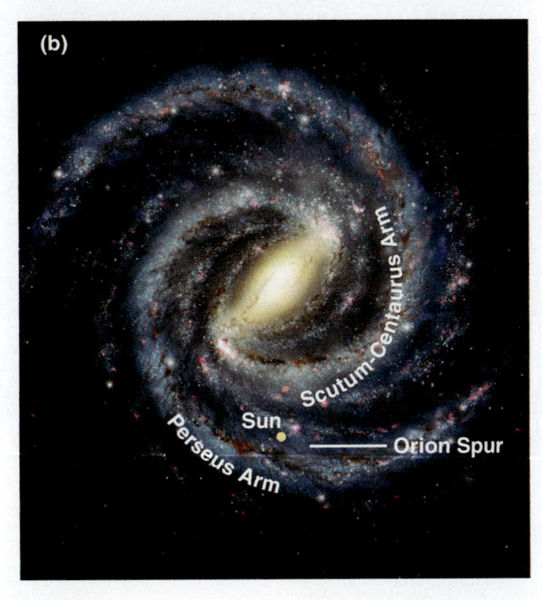

FIGURE 21.2 (a) An early model proposed by William Herschel in the late 1700s depicting the Milky Way as a dense slab of stars. (b) Infrared and radio observations contribute to an artist's model of the Milky Way Galaxy. The galaxy's two major arms (Scutum-Centaurus and Perseus) are seen attached to the ends of a thick central bar.

Spur that contains the Sun and Solar System. Astronomers conclude that the Milky Way is a giant barred spiral and is more luminous than an average spiral. From outside, it probably looks much like the galaxy shown in **Figure 21.3**, with a Hubble classification of SBbc.

Globular Clusters and the Size of the Milky Way Galaxy

In Chapter 19 we mentioned that Harlow Shapley made a first determination of the size of the Milky Way and the Sun's offset from the center (**Proces of Science Figure**). We also discussed Hubble's discovery of Cepheid variables in other galaxies, and how distances to galaxies are estimated through the use of standard candles. Similarly, the key to determining the size of the Milky Way Galaxy was finding standard candles whose distances could be measured throughout the galaxy. Recall from Chapter 17 that globular clusters (**Figure 21.4**) are large, spheroidal groups of stars held together by gravity. Many clusters can be seen through small telescopes. At first glance they look something like dwarf elliptical galaxies, and the motions of stars within a globular cluster are much like the motions of

FIGURE 21.3 From the outside, the Milky Way would look much like this barred spiral galaxy, M109.

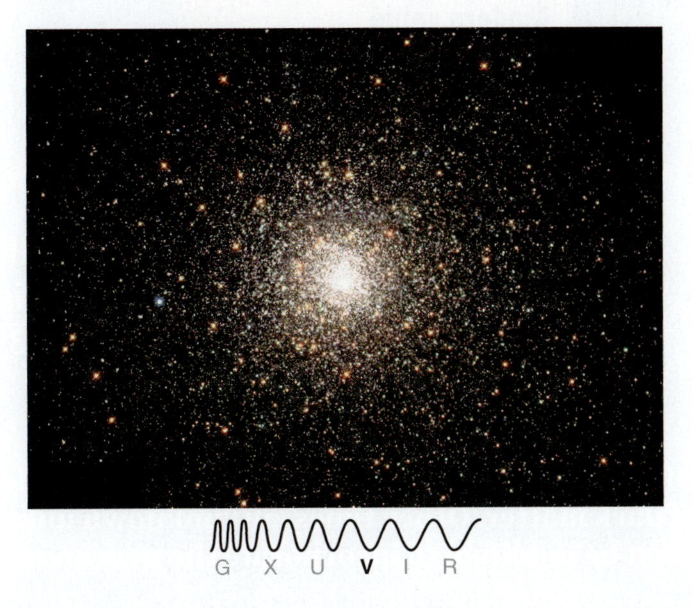

FIGURE 21.4 A Hubble Space Telescope image of the globular cluster M80.

Process of Science

UNKNOWN UNKNOWNS

Shapley's initial efforts to measure the Milky Way did not include dust and gas, because he did not know about them.

Shapley measures the disk of the Milky Way to be roughly 90 kiloparsecs (kpc) across, with the Sun 50 kpc (50,000 light-years) from the center.

Dust and gas are discovered in the disk. Reddening caused Shapley to overestimate the distance.

Today, astronomers routinely account for reddening. The Milky Way's disk is about 30 kpc in diameter and the Sun is 8.3 kpc from the center.

Often, scientists are very aware of what they don't know—for example, the composition of dark matter. Other times, an "unknown unknown" is later discovered, and prior results must be modified to incorporate the new knowledge.

Connections 21.1

Nightfall

Isaac Asimov (1920–1992), a writer of science and science fiction, imagined what might happen to a civilization on a planet orbiting within a system of six stars located in the heart of a giant globular cluster. His short story "Nightfall" has become one of the more famous works of science fiction. On Asimov's fictional planet Lagash, at least one of its six stars is almost always above the horizon. "Nightfall" occurs on Lagash only once every 2,049 years. The story tells of the great madness that afflicts the planet's inhabitants on this one night as they recoil in fear from a sky filled with hundreds of thousands of bright stars.

Asimov's story compels readers to consider how human perspectives are formed by the circumstances in which they live. Yet science fiction stories are not the only way to ask these kinds of questions. Human perspective on the universe can change with time. At the moment, the Solar System is traversing an open, rather dust-free part of the Milky Way. On a moonless night you can see a dark sky and gaze with telescopes into a universe of galaxies, but that will not always be the case. In the future, the Solar System may encounter dark interstellar clouds and star-forming regions filled with glowing gas and obscuring dust through which the Sun and its entourage of planets will occasionally pass. How different would your view of the universe be if, instead of a dark sky, you looked up each night and saw a sky filled with a soft green glow, punctuated by a few points of intense light? How much different would history be if, at some point during the rise of human civilization, the Solar System had suddenly emerged over the course of just a few years from within a molecular cloud, giving people their first look at the larger universe?

---

stars within an elliptical galaxy. However, globular clusters differ from dwarf elliptical galaxies in their concentration of stars and in their lack of a supermassive black hole or dark matter.

The Milky Way contains over 150 cataloged globular clusters (and very likely more that are hidden by dust in the disk). The known globular clusters have luminosities ranging from a low of 400 solar luminosities ($L_\odot$) to a high of about 1 million $L_\odot$. A typical globular cluster consists of 500,000 stars packed into a volume of space with a radius of only 5 parsecs (pc). To put this density into perspective, consider that there are only about 50 stars within that same distance out from the Sun. Globular clusters are much denser concentrations of stars than occur on average throughout the Milky Way, but they contain only about 0.1 percent of the Milky Way's stars. (It is interesting to imagine how Earth's sky would appear if the Solar System were located at the center of a globular cluster, as portrayed in **Connections 21.1**.)

Globular clusters are very luminous and easy to identify at great distances.

About one-fourth of the globular clusters in the Milky Way reside in or near the disk of the galaxy. The rest of them lie in a halo surrounding the disk and bulge. Globular clusters are very luminous, so the ones that lie outside of the dusty disk can be easily seen at great distances. To find the clusters' actual distance, astronomers look at the properties of the stars within each globular cluster that can be used as standard candles. The best candidates are primarily RR Lyrae variables, and some Cepheid variables. These stars have a period-luminosity relationship (see Figure 17.5b), so by measuring their period astronomers can estimate their luminosity, compare that with the observed brightness, and calculate their distance.

Also recall from Chapter 17 that the H-R diagrams of globular clusters show that they are old, with main-sequence turnoff points less than 1 solar mass ($M_\odot$), so astronomers would not expect to find bright blue or red supergiant stars (see Figure 17.21b). The main-sequence turnoff in the H-R diagrams of Milky Way globular clusters occurs for stars with masses of about 0.8 $M_\odot$, which corresponds to a main-sequence lifetime of close to 13 billion years. Globular clusters are the *oldest* objects known in the Milky Way or in any nearby galaxy. Globular clusters must have formed when the universe and the galaxy were very young. Compared to globular-cluster stars, the Sun, at about 5 billion years old, is a relatively young member of the Milky Way.

In an H-R diagram of an old cluster, the horizontal branch crosses the instability strip. These RR Lyrae stars, which you learned about in Chapter 17, are easy to spot in globular clusters because they are relatively luminous and have a distinctive light curve. Astronomer Henrietta Leavitt (1868–1921) determined that there is a relationship between the period and the luminosity for RR Lyrae stars. As with Cepheid variables, the time it takes for an RR Lyrae star to undergo one

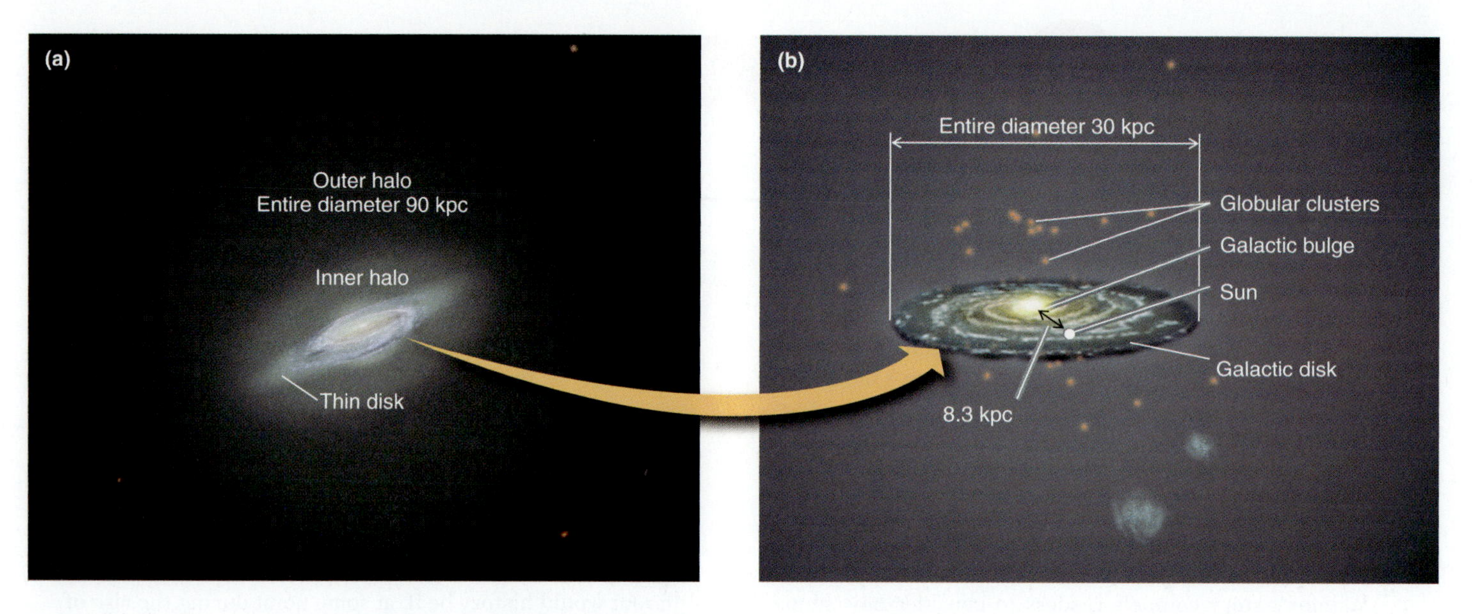

FIGURE 21.5 Parts of the Milky Way Galaxy: (a) the disk and the inner and outer halos; (b) the galactic bulge and disk, and the location of the Sun within the disk (1 kpc = 1,000 pc).

pulsation is related to the star's luminosity. Harlow Shapley used this period-luminosity relationship to find the luminosities of RR Lyrae stars in globular clusters. He then used the inverse square law of radiation to combine these luminosities with measured brightnesses to determine the distances to globular clusters. Finally, Shapley cross-checked his results by noting that more distant clusters (as measured with his standard candle) also tended to appear smaller in the sky, as expected.

RR Lyrae variable stars in globular clusters are used to measure the clusters' distances.

Shapley made a three-dimensional map of globular clusters from the distances he had determined and the locations of the clusters in the sky. In this map, globular clusters occupy a roughly spherical region of space with a diameter of about 90 kiloparsecs. These globular clusters trace out the luminous part of the halo of the Milky Way Galaxy, which reflects the modern view of the globular-cluster distribution.

The globular clusters around the Milky Way are moving about under the gravitational influence of the galaxy just as stars in an elliptical galaxy move about under the galaxy's gravity. Therefore, the center of the distribution of globular clusters coincides with the gravitational center of the galaxy. Shapley realized that, because he could determine the distance to the center of this distribution, he had actually determined the Sun's distance from the center of the Milky Way, as well as the size of the galaxy itself. A modern determination indicates that the Sun is located about 8,300 pc (27,000 light-years) from the center of the galaxy, or roughly halfway out toward the edge of the disk. **Figure 21.5** identifies the disk, bulge, and inner and outer halos of the Milky Way.

Shapley used globular clusters to determine the size of the Milky Way.

21.2 Dark Matter in the Milky Way

Because the Solar System is inside the dusty disk of the galaxy itself, the visible-light view of the Milky Way is badly obscured. If you go out on a dark night, away from any streetlights, and look in the direction of the center of the Milky Way (located in the constellation Sagittarius), instead of a bright spot you will see the dark lane of dusty clouds shown in Figure 21.1a. To probe the structure of the Milky Way, astronomers must use long-wavelength infrared and radio radiation that can penetrate the disk without being affected much by dust. The most powerful tool for this work is the same 21-cm line from neutral interstellar hydrogen, which (as we described in Chapter 20) is used to measure the rotation of other galaxies.

The velocities of interstellar hydrogen measured from 21-cm radiation are plotted in **Figure 21.6** as a function of the direction in which astronomers pointed their telescopes. Looking in the region around the center of the galaxy, on one side hydrogen clouds are moving toward Earth while on the other side clouds are moving away from Earth. This is a pattern that you have seen before; it is the pattern

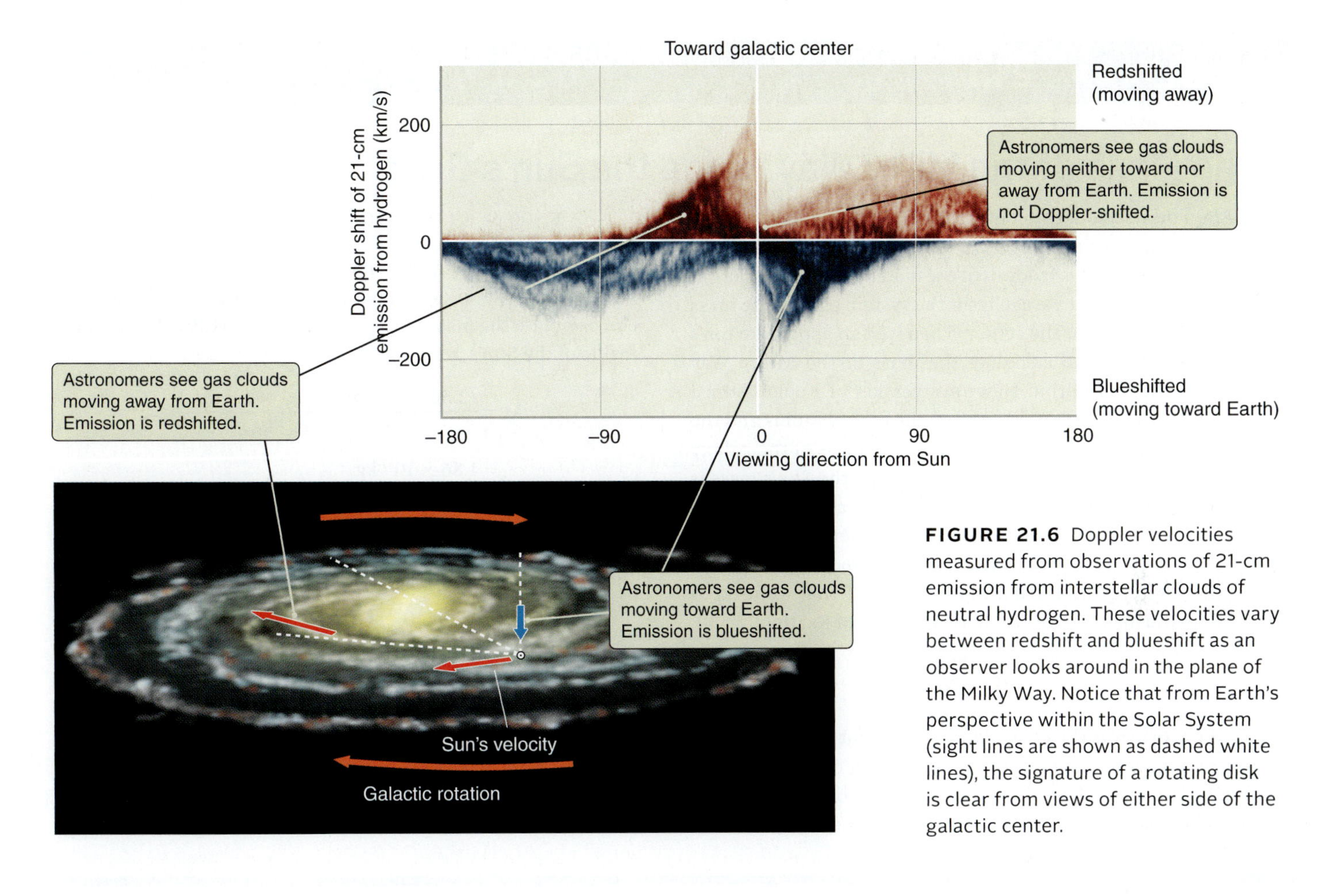

FIGURE 21.6 Doppler velocities measured from observations of 21-cm emission from interstellar clouds of neutral hydrogen. These velocities vary between redshift and blueshift as an observer looks around in the plane of the Milky Way. Notice that from Earth's perspective within the Solar System (sight lines are shown as dashed white lines), the signature of a rotating disk is clear from views of either side of the galactic center.

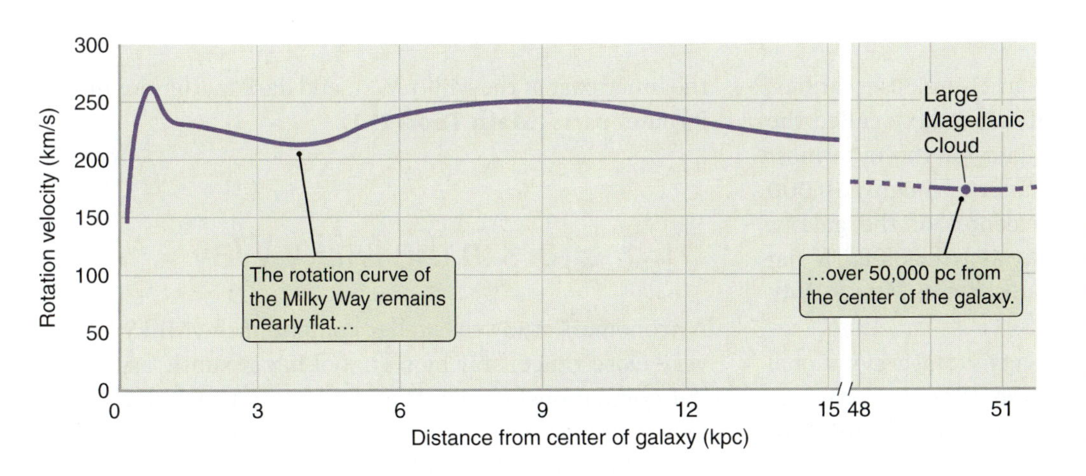

FIGURE 21.7 Rotation velocity is plotted against distance from the center of the Milky Way. The most distant point comes from measurements of the orbit of the Large Magellanic Cloud. The nearly flat rotation curve indicates that dark matter dominates the outer parts of the Milky Way.

of the rotation velocity of gas in a disk (see Figures 20.13 and 20.14). The only difference is that instead of looking at it from outside, you see the Milky Way Galaxy's rotation curve from a vantage point located within—and rotating with—the galaxy. In other directions, the measured velocities are complicated by Earth's moving vantage point within the disk and so are more difficult to interpret at a glance. Even so, 21-cm radio data enable astronomers to measure the Milky Way Galaxy's rotation curve and even determine the structure present throughout its disk.

Recall from the previous chapter that observations of rotation curves led astronomers to conclude that the masses of spiral galaxies consist mostly of dark matter. **Figure 21.7** shows the rotation curve of the

The Milky Way Galaxy has a flat rotation curve.

Math Tools 21.1

The Mass of the Milky Way inside the Sun's Orbit

The Sun orbits about the center of the Milky Way Galaxy. Recall from Connections 4.1 that even though the gravitational pull on the Sun comes from all of the material inside this orbit, Newton showed that we can treat the system as if all the mass were concentrated at the center. Then we can apply Newton's and Kepler's laws to calculate the mass of the Milky Way inside of the Sun's orbit. Newton's version of Kepler's third law relates the period of the orbit to the orbital radius and the masses of the objects. But in this case the mass of the galaxy is much larger than the mass of the Sun, so the Sun's mass is negligible by comparison. In addition, what astronomers can *measure* is the orbital speed of the Sun or other stars about the galactic center, rather than the orbital period. Thus, we can use the same equation from Math Tools 4.2 that we used to estimate the mass of the Sun from the orbit of Earth:

$$M = \frac{rv_{\text{circ}}^2}{G}$$

The Sun orbits the center of the galaxy at 220 kilometers per second, which equals 220,000 meters per second (m/s), or 2.2×10^5 m/s. The distance of the Sun from the center of the galaxy is 8,300 pc, which converts to meters as:

$$8{,}300 \text{ pc} \times (3.09 \times 10^{16} \text{ m/pc}) = 2.56 \times 10^{20} \text{ m}$$

Since we know that the gravitational constant $G = 6.67 \times 10^{-11}$ cubic meters per kilogram per second squared ($\text{m}^3/\text{kg s}^2$), we can calculate the mass of the portion of the Milky Way that is inside of the Sun's orbit:

$$M = \frac{(2.56 \times 10^{20} \text{ m}) \times (2.2 \times 10^5 \text{ m/s})^2}{6.67 \times 10^{-11} \text{ m}^3/\text{kg s}^2}$$

$$= 1.86 \times 10^{41} \text{ kg}$$

To put this in units of the Sun's mass, we divide this answer by $M_\odot = 1.99 \times 10^{30}$ kg, yielding:

$$M = \frac{1.86 \times 10^{41} \text{ kg}}{1.99 \times 10^{30} \text{ kg}/M_\odot} = 9.35 \times 10^{10} M_\odot$$

The mass of the Milky Way inside of the Sun's orbit is about 94 billion times the mass of the Sun. The mass *outside* of the Sun's orbit does not greatly affect the Sun's orbit. The total mass of the Milky Way Galaxy is currently estimated to be about 1.0–1.5 trillion (10^{12}) times the mass of the Sun, and it is mostly dark matter.

Milky Way as inferred primarily from 21-cm observations. The orbital motion of the nearby dwarf galaxy called the Large Magellanic Cloud provides data for the outermost point in the rotation curve, at a distance of roughly 50,000 pc (160,000 light-years) from the center of the galaxy. Like other spiral galaxies, the Milky Way has a fairly flat rotation curve. ▶▶ **NEBRASKA SIMULATION: MILKY WAY ROTATIONAL VELOCITY**

Astronomers estimate that the galaxy's total gravitational mass must be about 1.0–1.5 $\times$ 10^{12} $M_\odot$. But the luminous mass, estimated by measuring the light from the stars, dust, and gas and knowing how much stellar mass is needed to produce that much light (observing in the infrared to see through the dust), is only about one-tenth as much. Astronomers can therefore infer that, like other spiral galaxies, the Milky Way's mass consists mainly of dark matter. The spatial distribution of dark and normal matter within the Milky Way is also much like what astronomers have observed in other galaxies. Visible matter dominates the inner part of the Milky Way, and dark matter dominates its outer parts (**Math Tools 21.1**).

The Milky Way is mostly dark matter.

21.3 Stars in the Milky Way

Astronomers study the stellar content of the Milky Way at very close range, star by star, looking at subtle aspects of the populations of stars that give them direct clues about how spiral galaxies form. For example, stellar orbits determine the shapes of the different parts of the galaxy, and it is much easier to measure stellar orbits in the Milky Way than in other galaxies. The stars in the Milky Way's disk rotate about the center of the galaxy, as do the gas and dust in the disk. The stars in the halo move in orbits similar to those of stars in elliptical galaxies, sometimes at high velocities. The bar in the bulge of the Milky Way is shaped primarily by stars and gas moving both in highly elongated orbits up and down the long axis of the bar and in short orbits aligned perpendicular to the bar.

The Sun is a middle-aged disk star located among other middle-aged stars that orbit around the galaxy within the disk. Yet near the Sun are other stars, usually much older, that are a part of the galactic halo and whose orbits are carrying them *through* the disk. Using the ages, chemical abundances, and motions of nearby stars, astronomers can differentiate between disk and halo stars to learn more about the galaxy's structure.

Stars Have Different Ages and Chemical Compositions

Stellar ages and chemical abundances provide the most fundamental categories into which populations of stars can be grouped. Conveniently, some stars fit into distinct groups that split up along just these two lines. In addition to the globular clusters in the halo, younger open clusters orbit in the disk of the Milky Way. These open clusters are less tightly bound collections of a few dozen to several thousand stars (see Figure 17.20b). As with globular clusters, the stars in an open cluster all formed in the same region at about the same time. Open clusters have a wider range of ages. Some open clusters contain the very youngest stars known; others contain stars that are somewhat older than the Sun. Because open clusters are loosely bound together, they are easily disrupted by the gravitational tug from nearby objects, so they do not survive long in the disk of the galaxy. The oldest open clusters in the disk are several billion years younger than the youngest globular clusters in the halo. The differences in ages between globular and open clusters indicate that stars in the halo formed first, but this epoch of star formation did not last long. No young globular clusters are seen. Star formation in the disk started later but has been continuing ever since.

Recall from Chapter 19 that when the universe was very young, only the least massive of elements existed. All elements more massive than boron were formed by nucleosynthesis in stars. For this reason, the abundance of more massive elements in the interstellar medium provides a record of the cumulative amount of star formation that has taken place up to the present time. Gas that shows higher fractions of massive (heavy) elements must have gone through a great deal of stellar processing, whereas gas with low abundances of massive elements must not.

In turn, the relative abundance of massive elements in the atmosphere of a star provides a snapshot of the chemical composition of the interstellar medium *at the time the star formed.* (In main-sequence stars, material from the core does not mix with material in the atmosphere, so the relative abundances of chemical elements inferred from the spectra of a star are the same as the relative abundances in the interstellar gas from which the star formed.) The chemical composition of a star's atmosphere reflects the cumulative amount of star formation that has occurred up to that moment (**Figure 21.8**).

Abundances of heavier elements record the cumulative history of star formation.

If these ideas about the chemical evolution of the universe are correct, then astronomers would expect to see large differences in massive-element abundances between globular and open clusters. Stars in globular clusters, being among the earliest stars to form, should contain only very small amounts of massive elements. And that is exactly what is observed. Some globular-cluster stars contain only 0.5 percent as much of these massive elements as the Sun has. This relationship between age and abundances of massive elements is evident throughout much of the galaxy. The chemical evolution of the Milky Way has continued within the disk as generation after generation of disk stars have further enriched the interstellar medium with the products of their nucleosynthesis. Within the disk, younger stars typically have higher abundances of massive elements than do older stars. Similarly, older stars in the outer parts of the galaxy's bulge have lower massive-element abundances than do young stars in the disk. Such lower abundances of heavy elements characterize not only globular-cluster stars, but also all of the stars in the galaxy's halo, where globular-cluster stars constitute only a minority among the total number of stars in the galactic halo.

Within the galaxy's disk, astronomers can even see differences in abundances of massive elements from place to place that are related to the rate of star formation in different regions. Star formation is generally more active in the inner part of the Milky Way than in the outer parts. This higher level of activity is a result of the denser concentrations of gas that are found in the inner galaxy. If such activity has continued throughout the history of the galaxy, then massive elements should be more abundant in the inner part of the galaxy than in the outer parts. Observations of chemical abundances in the interstellar medium, based both on interstellar absorption lines in the spectra of stars and on emission lines in glowing H II regions, confirm this prediction. There is a smooth decline in abundances of massive elements from the inner to the outer parts of the disk. Astronomers have documented similar trends in other galaxies. These trends can be seen with stars as well. Within a galactic disk,

Younger stars typically have higher fractions of heavier elements than older stars have.

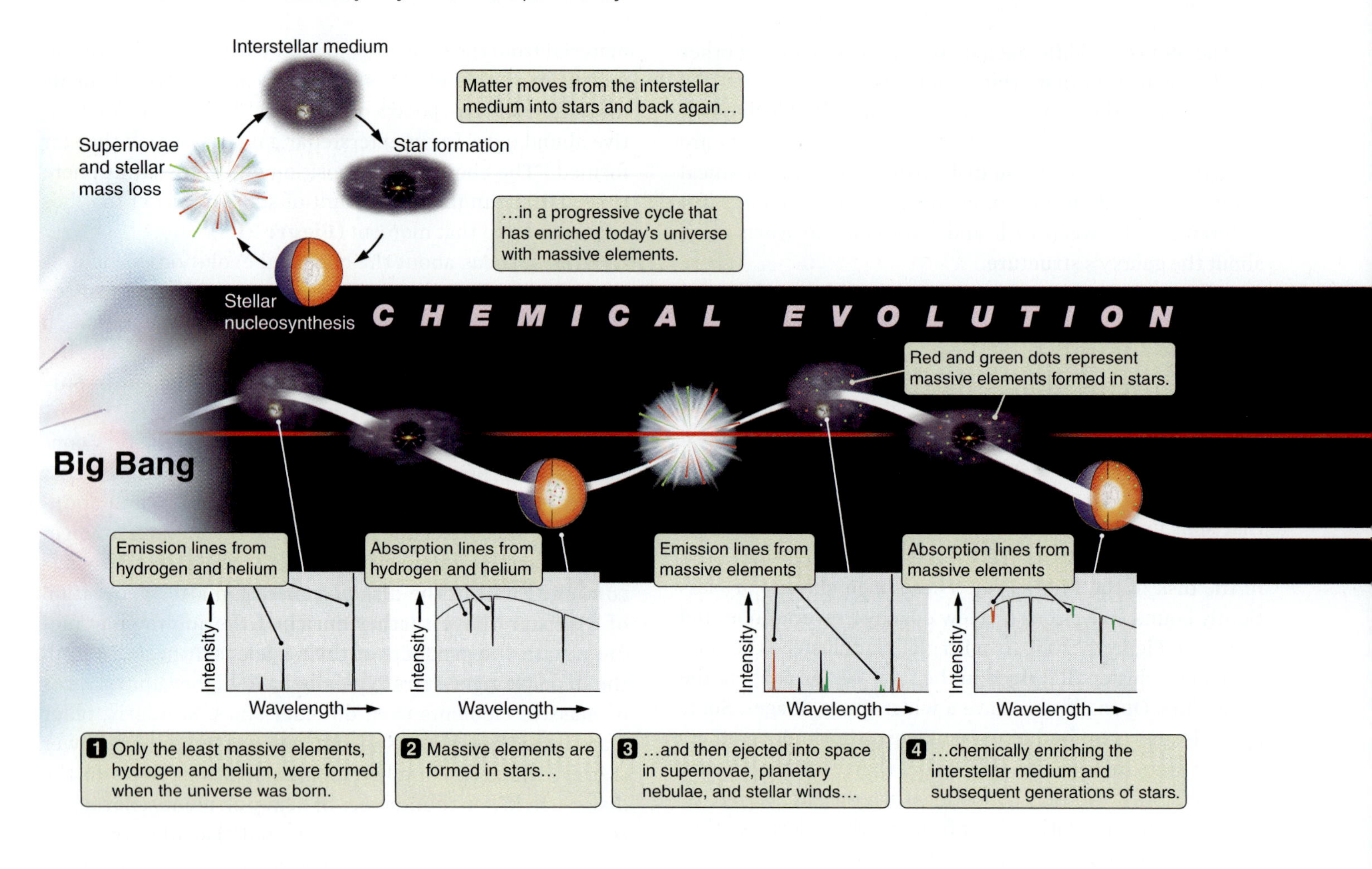

relatively old stars near the center of a galaxy often have greater massive-element abundances than do young stars in the outer parts of the disk.

The basic idea of higher massive-element abundances following the more prodigious star formation in the inner galaxy seems correct, but as always, the full picture is not this simple. The chemical composition of the interstellar medium at any location depends on a wealth of factors. New material falling into the galaxy might affect interstellar chemical abundances. Chemical elements produced in the inner disk might be blasted into the halo in great "fountains" powered by the energy of massive stars (**Figure 21.9**) and ultimately fall back onto the disk elsewhere. Past interactions with other galaxies might have stirred the Milky Way's interstellar medium, mixing gas from those other galaxies with gas already there. Surveys of hundreds of thousands of stars are under way to better map stellar abundances throughout the Milky Way. These surveys will help astronomers obtain a deeper understanding of what the variation of chemical abundances within the Milky Way and other galaxies indicates about the history of star formation and nucleosynthesis.

Although the details are complex, several clear and important lessons can be learned from the observed patterns in the abundances of more massive elements in the galaxy. The first is that even the very oldest globular-cluster stars contain some chemical elements fused in previous generations of more massive stars. This observation implies that globular-cluster stars and other halo stars were not the first stars to form. At least one generation of massive stars lived and died, ejecting newly synthesized massive elements into space, before even the oldest globular clusters formed. (We will return to these first stars in Chapter 23.) Every star less massive than about 0.8 $M_{\odot}$ that ever formed is still around as a mian-sequence star today. Even so, astronomers find *no* disk stars with exceptionally low massive-element abundances. Therefore, the gas that wound up in the plane of the Milky Way must have seen a significant amount of star formation *before* it settled into the disk of the galaxy and made stars.

In this discussion we have focused mainly on the variations in chemical abundances from place to place. These variations indicate a lot about the history of the galaxy and the origin of the material that makes up the Sun, So-

FIGURE 21.8 As subsequent generations of stars form, live, and die, they enrich the interstellar medium with massive elements—the products of stellar nucleosynthesis. The chemical evolution of the Milky Way and other galaxies can be traced in many ways, including by the strength of interstellar emission lines and stellar absorption lines.

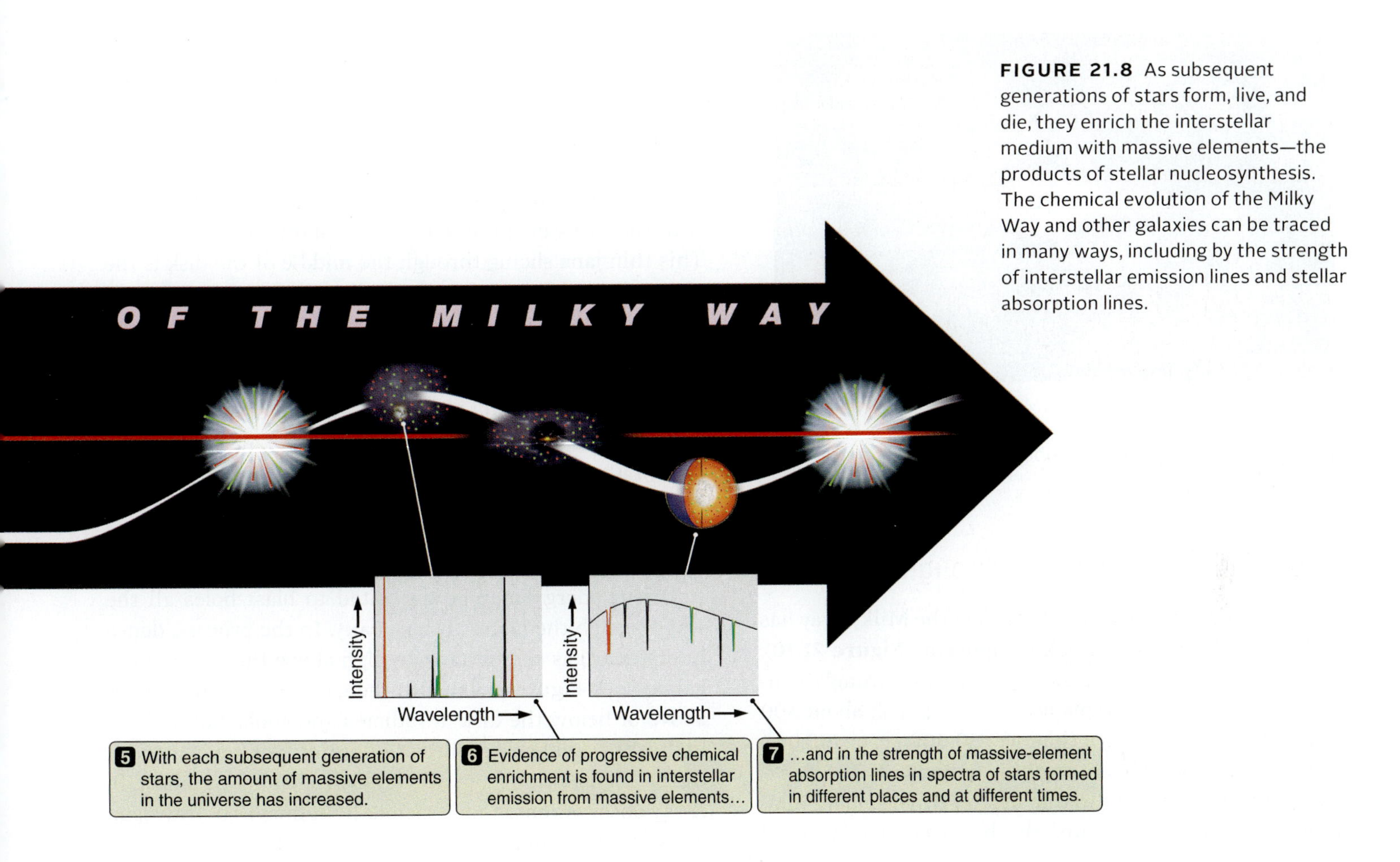

lar System, planets, and us. It is important to remember, however, that even a chemically "rich" star like the Sun, which is made of gas processed through approximately 9 billion years of previous generations of stars, is composed of less than 2 percent massive elements. Luminous matter in the universe is still dominated by hydrogen and helium formed just after the Big Bang, long before the first stars.

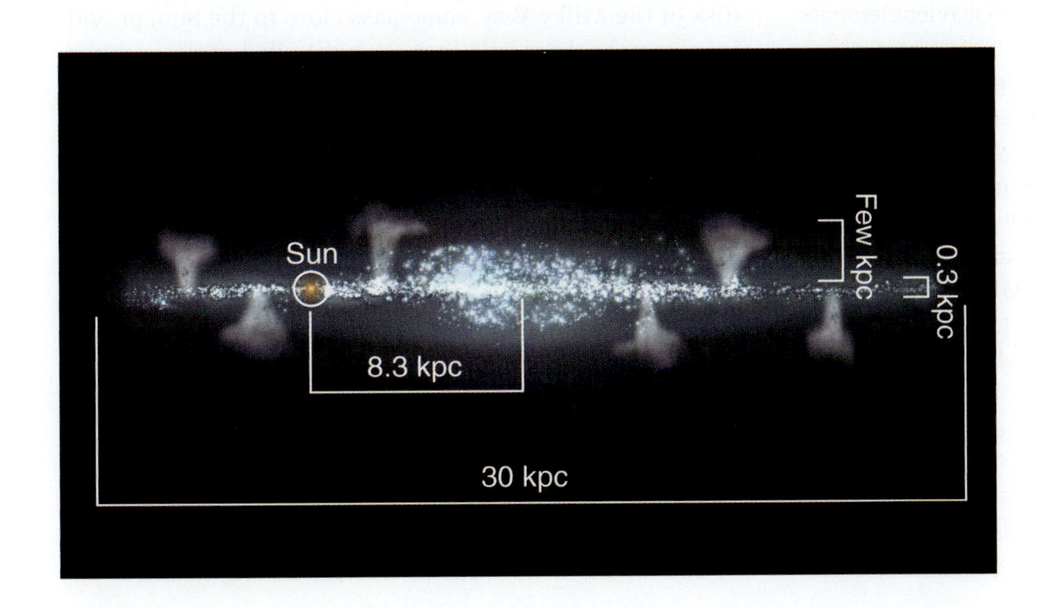

FIGURE 21.9 In the "galactic fountain" model of the disk of a spiral galaxy, gas is pushed away from the plane of the galaxy by energy released by young stars and supernovae and then falls back onto the disk.

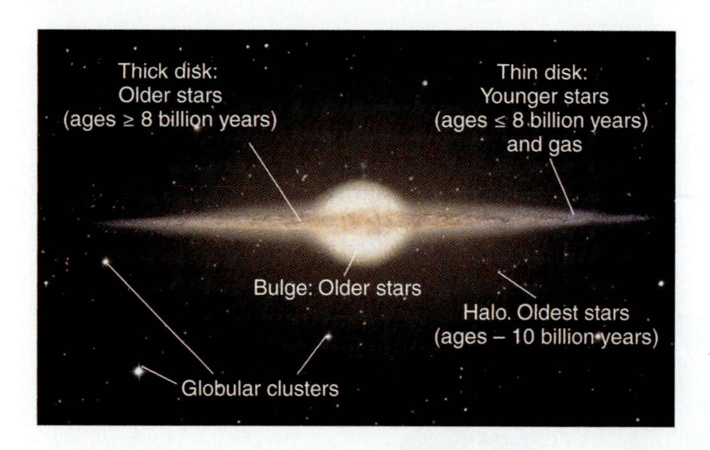

FIGURE 21.10 The disks, bulge, and halo of the Milky Way Galaxy, showing the location of globular clusters.

Looking at a Cross Section through the Disk

Most astronomers think that the disk of the Milky Way has a thin component and a thick component (**Figure 21.10**). The youngest stars in the galaxy are most strongly concentrated in the galactic plane, defining a disk about 300 pc (1,000 light-years) thick (but over 30,000 pc, or 100,000 light-years, across). This ratio of the diameter to the thickness of the disk is similar to that of a DVD. The older population of disk stars, distinguishable by lower abundances of massive elements, has a much thicker distribution, about 3,700 pc (12,000 light-years) across. The population of stars changes with distance perpendicular to the galactic plane. The youngest stars are concentrated closest to the plane of the galaxy because this is where the molecular clouds are. These stars show a decrease in heavy-element content as the distance from the galactic center increases. Older stars make up the thicker parts of the disk. The stars farther from the galactic plane have similar amounts of heavier elements at all distances from the galactic center.

There are two hypotheses for the origin of this thicker disk. One suggests that these stars formed in the midplane of the disk long ago but were affected by gravitational interactions with massive molecular clouds in the spiral arms and then kicked up out of the plane of the galaxy. The other hypothesis suggests that these stars were acquired from the merging processes that formed the Milky Way Galaxy.

As we discussed in Chapters 7 and 15, accretion disks form from a rotating cloud of gas that collapses into a thin disk as a consequence of gas falling from one direction (above the disk) running into gas falling from the other direction (below the disk.) Clouds of gas cannot pass through each other, so the colliding gas clouds instead settle into a disk. The same process applies to clouds of gas that are pulled by gravity toward the midplane of the disk of a spiral galaxy. Although stars are free to move back and forth from one side of the disk to the other, cold, dense clouds of interstellar gas settle down into the central plane of the disk. These clouds appear as the concentrated dust lanes that slice the disks of spiral galaxies (as shown in Figure 21.1). This thin lane slicing through the middle of the disk is the place where new stars form and where new stars are found.

Gas tends to concentrate into thin disks.

The interstellar medium is a dynamic place: energy from star-forming regions can shape it into impressively large structures. As we mentioned earlier, energy from regions of star formation could impose interesting structure on the interstellar medium, clearing out large regions of gas in the disk of a galaxy. Many massive stars forming in the same region can blow "fountains" of hot gas out through the disk of the galaxy via a combination of supernova explosions and strong stellar winds. If enough massive stars form together, sufficient energy may be deposited to blast holes all the way through the plane of the galaxy. In the process, dense interstellar gas can be thrown high above the plane of the galaxy (see Figure 21.9). Once the gas is a few kiloparsecs above or below the disk, it radiates and cools, falling back to the disk. Evidence of these "fountains" is seen in maps of the 21-cm emission from neutral hydrogen in the galaxy, in X-ray observations, and in visible-light images of hydrogen emission from some edge-on external galaxies.

The Halo Is More Than Globular Clusters

The globular clusters in the galactic halo tell astronomers a great deal about the history of star formation in the halo. Yet globular clusters account for only about 1 percent of the total mass of stars in the halo. As halo stars fall through the disk of the Milky Way, some pass close to the Sun, providing a sample of the halo that can be studied at closer range.

Most of the stars near the Sun are disk stars like the Sun, but astronomers can distinguish nearby halo stars in two ways. First, most halo stars have much lower abundances of massive elements than disk stars have. Second, halo stars appear to be whizzing by at high velocities. Actually, it is often not the halo stars that are moving fast, but the Sun that is moving fast relative to them. Halo stars do not rotate about the center of the galaxy in the same way that disk stars do. Instead they are in the same kind of odd-shaped orbits as stars in elliptical galaxies. In contrast, the disk stars near the Sun are moving, mostly together, in a 220-km/s rotation about the center of the galaxy. Just as it is easy to tell the difference between a person sitting beside you on a bus and the people who (in your frame of reference) are whizzing by outside the window, astronomers can tell the difference between disk stars that share Earth's motion and halo stars that are just passing through.

Astronomers call the halo stars high-velocity stars (even though it is usually the disk stars that are moving faster relative to the galaxy as a whole). Enough halo stars are near the Sun that astronomers can measure their distances and map out their orbits, presenting a very detailed look at the kinds of orbits that stars in a halo have. The picture is complex. The halo has two separate components: an inner halo that includes stars up to about 15 kiloparsecs (kpc) from the center, and an outer halo beyond about 15 kpc (see Figure 21.5a). The stars in the outer halo have lower fractions of heavier elements, and many of them are moving in the *opposite* direction to the rotation of the galaxy, suggesting that the outer halo may have originated years ago in a merger with a small dwarf galaxy. The complete orbits of the halo stars near the Sun can be deduced by measuring their motions. Such orbits suggest that halo stars fill a volume of space similar to that occupied by the globular clusters in the halo.

Recent observations with several X-ray telescopes suggest that there is a halo of hot gas surrounding the Milky Way. The gas halo may extend for 100 kpc from the galactic center and contain as much mass as that of all the stars in the galaxy (**Figure 21.11**). Its temperature is estimated to be 1.0–2.5 million kelvins (K), meaning that the gas particles are moving very quickly. But the gas is extremely diffuse, so the particles are not colliding with each other and transferring energy. The gas wouldn't feel hot, much like the solar corona that we discussed in Chapter 14.

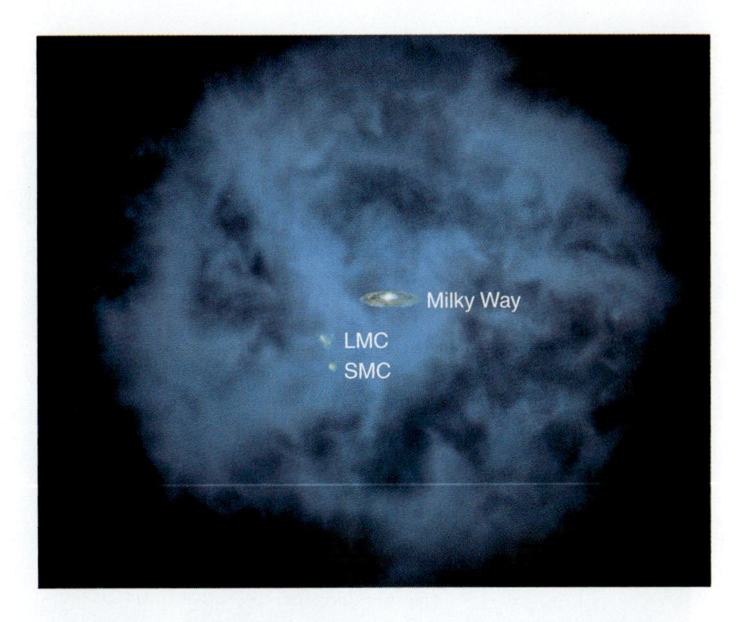

FIGURE 21.11 Artist's model of a hot gas halo surrounding the Milky Way, which may contain as much mass as that of all of the stars in the galaxy combined. The Large Magellanic Cloud (LMC) and Small Magellanic Cloud (SMC) are nearby dwarf galaxies.

Magnetic Fields and Cosmic Rays Fill the Galaxy

The interstellar medium of the Milky Way is laced with magnetic fields that are wound up and strengthened by the rotation of the galaxy's disk. The total interstellar magnetic field, however, is a million times weaker than Earth's magnetic field. In general, charged particles spiral around magnetic fields, causing a net motion along the field rather than across it. In addition, magnetic fields cannot freely escape from a cloud of gas containing charged particles. The dense clouds of interstellar gas in the midplane of the Milky Way anchor the galaxy's magnetic field to the disk (**Figure 21.12**). These magnetic fields trap **cosmic rays** within the galaxy. Despite their name, cosmic rays are not a form of electromagnetic radiation; they are charged particles moving at nearly the speed of light. (They were named before their true nature was known.)

Cosmic rays are continually hitting Earth. Most cosmic-ray particles are protons, but some are nuclei of helium, carbon, and other elements produced by nucleosynthesis. A few are high-energy electrons and other subatomic particles. Cosmic rays span an enormous range in particle energy. Astronomers can observe the lowest-energy cosmic rays by using interplanetary spacecraft. These cosmic rays have energies as low as about 10^{-11} joule (J), which corresponds to the energy of a proton moving at a velocity of a few tenths the speed of light.

In contrast, the most energetic cosmic rays are 10 trillion (10^{13}) times as energetic as the lowest-energy cosmic rays, and they move very close to the speed of light, 0.99999*c*. These high-energy cosmic rays are detected from the showers of elementary particles that they cause when crashing through Earth's atmosphere. Astronomers hypothesize that cosmic rays are accelerated to these incredible energies by the shock waves produced in supernova explosions. The very highest-energy cosmic rays must have a different origin and are much more difficult to explain, but they could be from active galactic nuclei (AGNs), gamma-ray bursts, or even dark matter particles.

Some cosmic rays have enormously high energies.

The disk of the galaxy glows from synchrotron radiation (see Chapter 17) produced by cosmic rays (mostly electrons) spiraling around the galaxy's magnetic field. Such synchrotron emission is seen in the disks of other spiral galaxies as well, indicating that they, too, have magnetic fields and populations

Cosmic rays are trapped by the Milky Way's magnetic field.

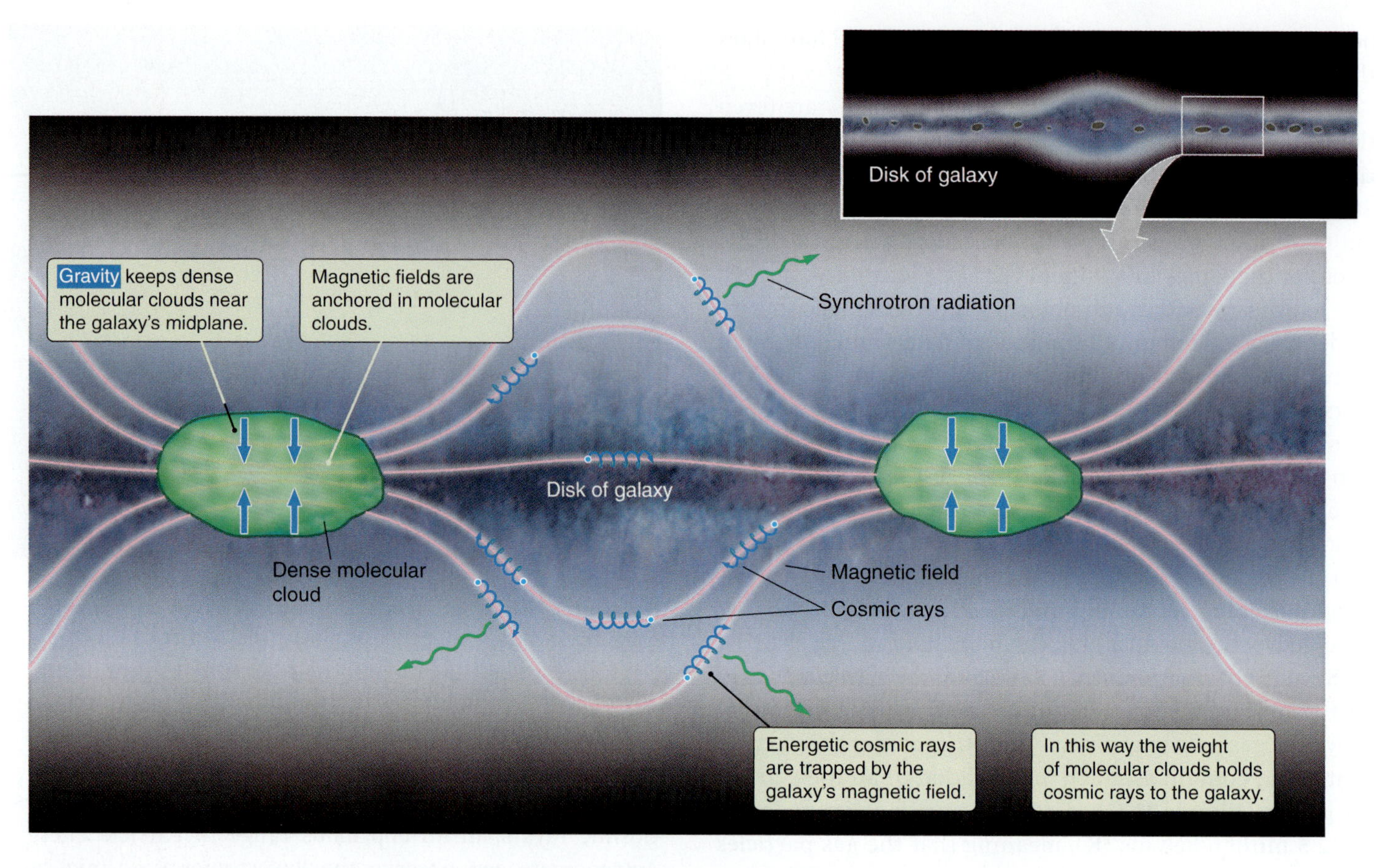

FIGURE 21.12 The weight of interstellar clouds anchors the magnetic field of the Milky Way to the disk of the galaxy. The magnetic field, in turn, traps the galaxy's cosmic rays.

of energetic cosmic rays. Even so, the very highest-energy cosmic rays are moving much too fast to be confined by the gravitational force of their originating galaxy. Any such cosmic rays that formed in the Milky Way would soon stream away from the galaxy into intergalactic space. Thus, it is likely that some of the energetic cosmic rays reaching Earth originated in energetic events outside of the Milky Way Galaxy.

The total energy of all the cosmic rays in the galactic disk can be estimated from the energy of the cosmic rays reaching Earth. The strength of the interstellar magnetic field can be measured in a variety of ways, including by observing the effect that it has on the properties of radio waves passing through the interstellar medium. These measurements indicate that in the Milky Way Galaxy, the magnetic-field energy and the cosmic-ray energy are about equal to each other. Both are comparable to the energy present in other energetic components of the galaxy, including the motions of interstellar gas and the total energy of electromagnetic radiation within the galaxy. This similarity in energy level suggests that they are all connected. One possible scenario is that supernova explosions accelerate the cosmic rays, stirring the interstellar medium and generating turbulent motions and the magnetic field. The pressure from all of these components supports the disk and keeps it in vertical equilibrium (see Figure 21.12).

21.4 The Milky Way Hosts a Supermassive Black Hole

Dense clouds of dust and gas hide astronomers' visible-light view of the Milky Way's center. However, infrared, radio, and some X-ray radiation passes through dust. **Figure 21.13** shows images of the Milky Way's center taken with the Chandra X-ray Observatory and the Spitzer Space Telescope. The X-ray view (**Figure 21.13a**) shows the location of a strong radio source called Sagittarius A* (abbreviated Sgr A*), which astronomers believe lies at the exact center of the Milky Way. The infrared image (**Figure 21.13b**) cuts through the dust to reveal the crowded, dense core of the galaxy containing hundreds of thousands of stars.

Studies of the motions of stars closest to the Sgr A* source suggest a central mass very much greater than that of the few hundred stars orbiting there. Furthermore,

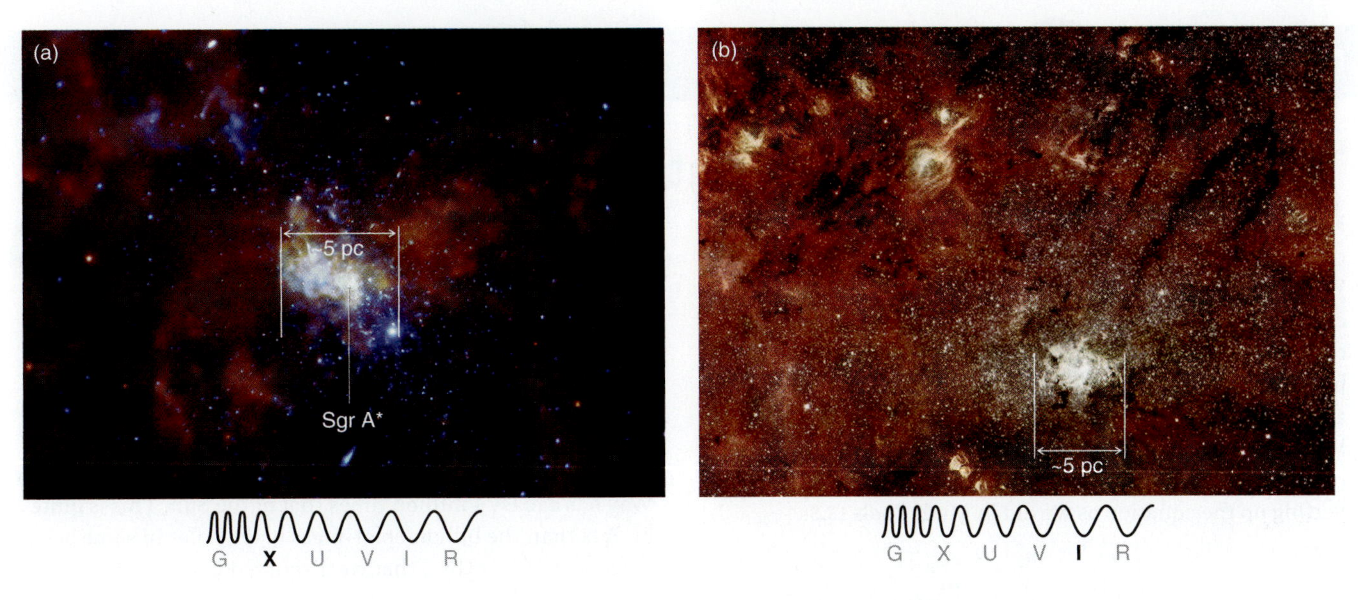

FIGURE 21.13 (a) An X-ray view of the Milky Way's central region showing the active source, Sgr A*, as the brightest spot at the middle of the image. Lobes of superheated gas (shown in red) are evidence of recent, violent explosions happening near Sgr A*. (b) This infrared view of the central core of the Milky Way shows hundreds of thousands of stars. The bright white spot at the lower right marks the galaxy's center, home of a supermassive black hole.

observations of the galaxy's rotation curve show rapid rotation velocities very close to the galactic center. Stars closer than 0.1 light-year from the galactic center follow Kepler's laws, indicating that their motion is dominated by mass within their orbit. The closest stars studied are only about 0.01 light-year from the center of the galaxy—so close that their orbital periods are only about a dozen years. The positions of these stars change noticeably over time, and astronomers can see them speed up as they whip around what can only be a supermassive black hole at the focus of their elliptical orbits (**Figure 21.14**). Using Newton's version of Kepler's third law, we can then estimate that the black hole at the center of the Milky Way Galaxy is a relative lightweight, having a mass of "only" 4 million times the mass of the Sun (**Math Tools 21.2**).

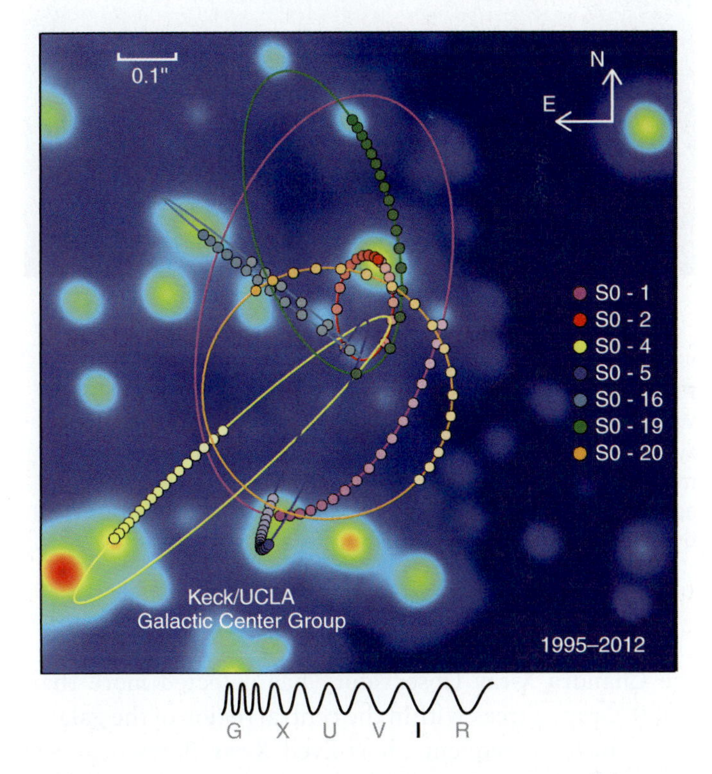

FIGURE 21.14 Orbits of seven stars within 0.03 pc (0.1 light-year, or about 6,000 astronomical units [AU]) of the Milky Way's center. The Keplerian motions of these stars reveal the presence of a 4-million-$M_{\odot}$ supermassive black hole at the galaxy's center. Colored dots show the measured positions of each of the stars over many years; the dots go from lighter in 1995 to darker in 2012.

The central region of the Milky Way Galaxy is awash in radio, infrared, X-ray, and gamma radiation. Radio observations (**Figure 21.15**) reveal synchrotron emission from wisps and loops of material distributed throughout the region. This is similar to the synchrotron emission seen from AGNs, except that this emission is at far lower levels.

The galactic center shows rapid rotation and AGN-like emission.

Clouds of interstellar gas at the galaxy's center are heated to millions of degrees by shock waves from supernova explosions and colliding stellar winds blown outward by young massive stars. Superheated gas produces X-rays, and

Math Tools 21.2

The Mass of the Milky Way's Central Black Hole

Figure 21.14 illustrates data points for the stars in the central region orbiting closely to the central black hole of the Milky Way Galaxy. These stars have highly elliptical orbits with changing speeds, but the orbital periods are short enough that they can be observed and measured. Star S0-2 in the figure has a measured orbital period of 15.8 years. The semimajor axis of its orbit is estimated to be 1.5×10^{11} km = 1,000 AU. With this information we can use Newton's version of Kepler's third law to estimate the mass inside of S0-2's orbit. Setting up the equation as we did in Math Tools 13.3:

$$\frac{m_{\rm BH}}{M_\odot} + \frac{m_{\rm S0\text{-}2}}{M_\odot} = \frac{A^3_{\rm AU}}{P^2_{\rm years}}$$

The mass of star S0-2 is much less than the mass of the black hole, so the sum of the two is very close to the mass of the black hole. Therefore, we can write:

$$\frac{m_{\rm BH}}{M_\odot} = \frac{A^3_{\rm AU}}{P^2_{\rm years}} = \frac{1{,}000^3}{15.8^2} = 4.0 \times 10^6$$

$$m_{\rm BH} = 4.0 \times 10^6\, M_\odot$$

The supermassive black hole at the center of the Milky Way has a mass 4 million times that of the Sun. This is quite a bit less than the billion-solar-mass black holes in some active galactic nuclei (AGNs) that we discussed in Chapter 20.

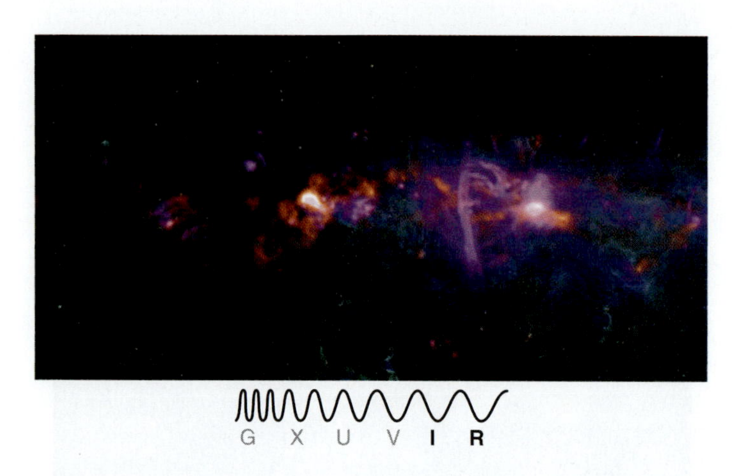

FIGURE 21.15 Radio observations of the center of the Milky Way reveal wispy molecular clouds (purple) glowing from strong synchrotron emission. Cold dust (20–30 K) associated with molecular clouds is shown in orange. Diffuse infrared emission appears in blue-green. The galactic center (Sgr A*) lies within the bright area to the right of center.

the Chandra X-ray Observatory has detected more than 9,000 X-ray sources within the central region of the galaxy. These include frequent, short-lived X-ray flares near Sgr A* (see Figure 21.13a), which provide direct evidence that matter falling toward the supermassive black hole fuels the energetic activity at the galaxy's center.

The Fermi Gamma-ray Space Telescope has observed gamma-ray-emitting bubbles that extend 8 kpc above and below the galactic plane. The bubbles may have formed after a burst of star formation a few million years ago produced massive star clusters near the center of the galaxy. If some of the gas formed stars and about 2,000 $M_\odot$ of material fell into the supermassive black hole, enough energy could have been released to power the bubbles. More recently, faint gamma-ray signals were observed that look like jets coming from the center, within the bubbles (**Figure 21.16**). If these jets are originating from material falling into the supermassive black hole, activity might be

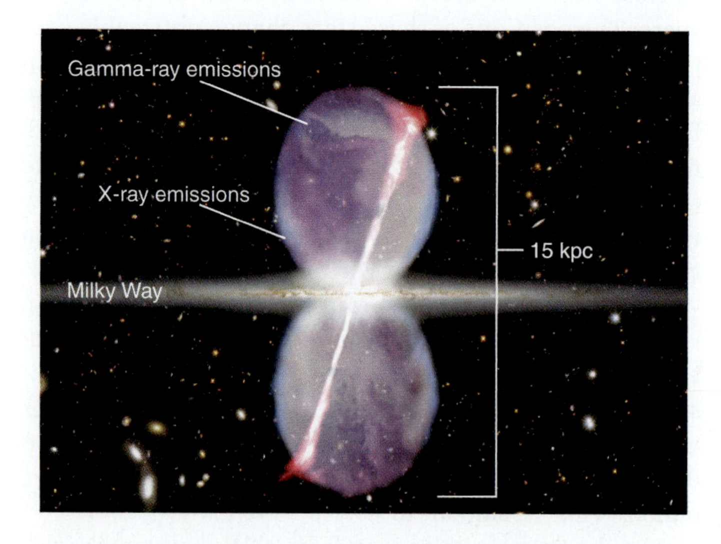

FIGURE 21.16 The Fermi Gamma-ray Space Telescope observed gamma-ray bubbles (purple) extending 8 kpc above and below the galactic plane. Hints of the edges of the bubbles were first observed in X-rays (blue) in the 1990s. In this artist's conceptual view from outside of the galaxy, the gamma-ray jets (pink) are tilted with respect to the bubbles, which might suggest that the accretion disk around the black hole is tilted as well.

even more recent, maybe 20,000 years ago. Some astronomers predict that gas clouds are heading toward the center and will soon be accreted by the black hole. At this time the observed activity is not as intense as that seen in active galaxies with central supermassive black holes. The inner Milky Way is a reminder that it was almost certainly "active" in the past and could become active once again.

21.5 The Milky Way Offers Clues about How Galaxies Form

One of the fundamental goals of stellar astronomy is to understand the life cycle of stars, including how stars form from clouds of interstellar gas. In Chapter 15 we told a fairly complete story of this process, at least as it occurs today, and tied this story strongly to observations of Earth's galactic neighborhood. Galactic astronomy has a similar basic goal. Astronomers would like very much to have a complete and well-tested theory of how the Milky Way Galaxy formed. The distribution of stars of different ages with different amounts of heavy elements is one clue. Additional clues come from studying other galaxies at different distances (and therefore of different ages), their supermassive black holes, and their merger history.

Important among these clues are the properties of globular clusters and high-velocity stars in the halo of the galaxy. For reasons we discussed earlier in this chapter, these objects must have been among the first stars formed that still exist today. The fact that they are not concentrated in the disk or bulge of the galaxy says that they formed from clouds of gas well before those clouds settled into the galaxy's disk. The observations that globular clusters are very old and that the youngest globular cluster is older than the oldest disk stars agree with this hypothesis. The presence of extremely small amounts of massive elements in the atmospheres of halo stars also indicate that at least one generation of stars must have lived and died *before* the formation of the halo stars visible today.

At least one generation of stars had to exist before today's halo stars formed.

Galaxies do not exist in utter isolation. The vast majority of galaxies are parts of gravitationally bound collections of galaxies. The smallest and most common of these are called **galaxy groups**. A galaxy group contains as many as several dozen galaxies, most of them dwarf galaxies. The Milky Way is a member of the **Local Group**, first identified by Edwin Hubble in 1936. Hubble labeled 12 galaxies as part of the Local Group, but now astronomers count at least 50. The group includes the two giant barred spirals (the Milky Way and the Andromeda Galaxy), along with a few ellipticals and irregulars, and at least 30 smaller dwarf galaxies in a volume of space about 3 million parsecs (10 million light-years) in diameter (**Figure 21.17**). Almost 98 percent of all the galaxy mass in the Local Group resides in just its two giant galaxies. The third largest galaxy, Triangulum, is an unbarred spiral with about one-tenth the mass of the Milky Way or

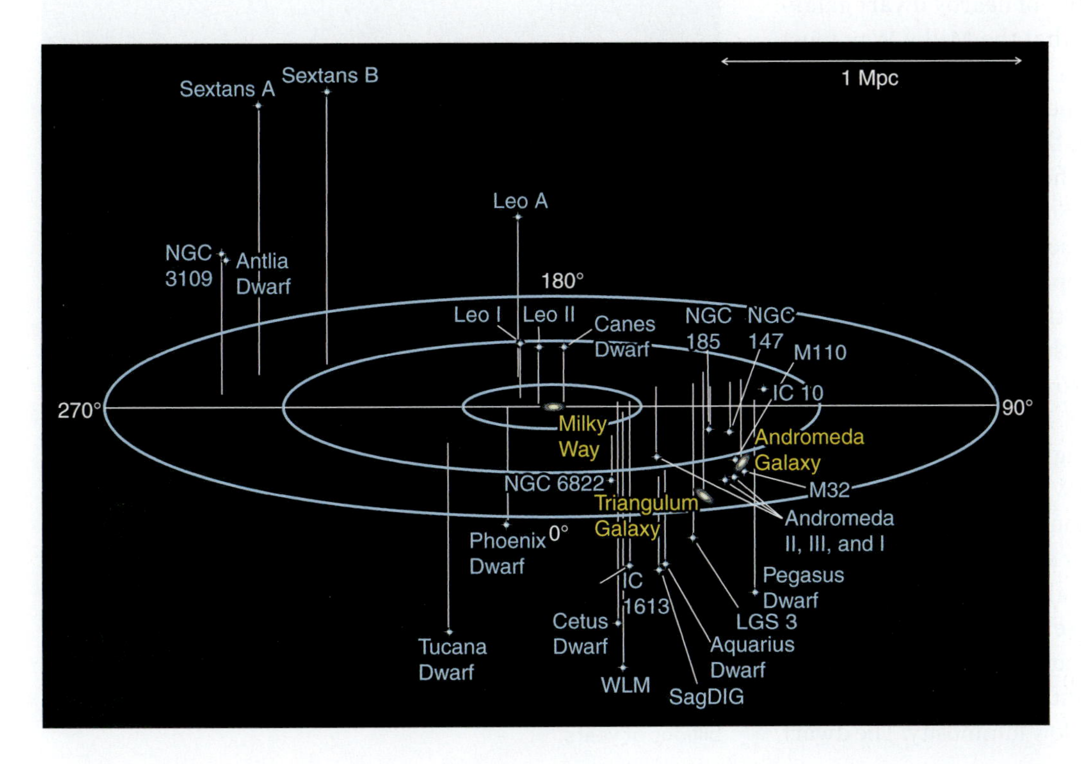

FIGURE 21.17 A graphical map showing the members of the Local Group of galaxies. Most are dwarf galaxies. Spiral galaxies are shown in yellow. The closest galaxies to the Milky Way are not seen on this scale; 1 Mpc = 1 megaparsec = 1 million parsecs.

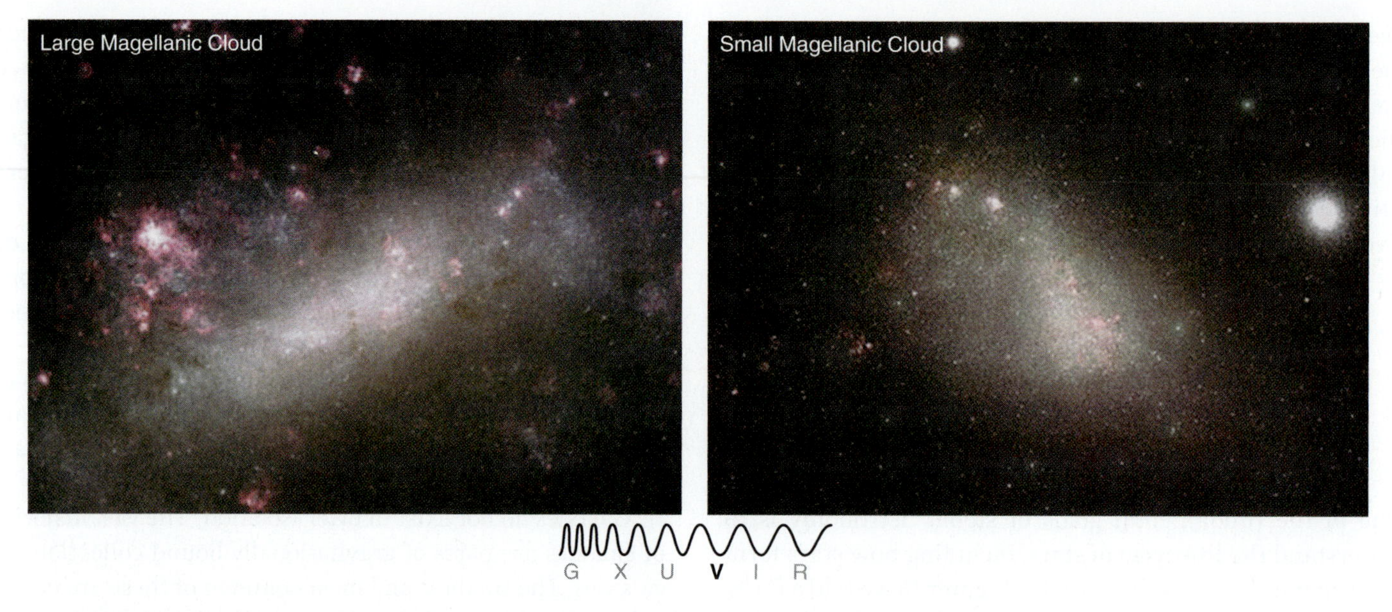

FIGURE 21.18 The Milky Way is surrounded by more than 20 dwarf companion galaxies; the largest among them are the Large (LMC) and Small (SMC) Magellanic Clouds (see Figure 21.11). The Magellanic Clouds were named for Ferdinand Magellan (c. 1480–1521), who headed the first European expedition that ventured far enough into the Southern Hemisphere to see them.

Andromeda. Most but not all of the dwarf elliptical and dwarf spheroidal galaxies in the group are satellites of the Milky Way or Andromeda. The Local Group interacts with a few nearby groups, and in Chapter 23 you will learn that the Local Group is part of the local supercluster called the Virgo Supercluster, a collection of many groups of galaxies.

From the properties of globular-cluster and high-velocity stars in the halo of the galaxy, the presence of the central massive black hole, and the number of nearby dwarf galaxies, astronomers have concluded that the Milky Way must have formed when the gas within a huge "clump" of dark matter collapsed into a large number of small *protogalaxies*. Some of these smaller clumps are the small, satellite dwarf galaxies in the Local Group near the Milky Way. The larger among them are the Large and Small Magellanic Clouds (**Figure 21.18**), which are easily seen by the naked eye in the Southern Hemisphere and appear much like detached pieces of the Milky Way. Among the closest is Sagittarius Dwarf, which is plowing through the disk of the Milky Way on the other side of the bulge. Astronomers have observed streams of stars from Sagittarius Dwarf and some of the other dwarf galaxies that are being tidally disrupted by the Milky Way. These dwarf galaxies will become incorporated into the Milky Way—an indication that the galaxy is still growing (**Figure 21.19**). Computer simulations suggest that such mergers could cause the spiral-arm structure.

There are more than 20 of these satellite dwarf galaxies, although it's not certain that all are gravitationally bound to the Milky Way. Some of the fainter ones were discovered only very recently, because of their low luminosity. The dwarf galaxies are the lowest-mass galaxies observed, and they are dominated by an even greater fraction of invisible dark matter than are other known galaxies. They also contain stars very low in elements more massive than helium. These ultra-faint dwarf galaxies offer clues to the formation of the Local

FIGURE 21.19 Artist's impression of tidal tails of stars from the Sagittarius Dwarf Elliptical Galaxy (reddish-orange). These stars have been stripped from the dwarf galaxy by the much more massive Milky Way, and the two galaxies will eventually merge, in billions of years.

Connections 21.2

Will Andromeda and the Milky Way Collide?

Recall from Chapter 19 that the Andromeda Galaxy appears to violate Hubble's law because its spectrum shows blueshifts, indicating that the galaxy is moving toward, not away from, the Milky Way at 110 km/sec (400,00 km/hr). Andromeda and the Milky Way are the two largest galaxies in the Local Group, about 770 Mpc apart. (Think of the two galaxies as quarters an arm's length apart.) Since Hubble noticed over 100 years ago that Andromeda's spectrum was blueshifted, indicating that it was approaching the Milky Way, astronomers have wondered if and when Andromeda and the Milky Way would collide and form a merged galaxy.

Recently, astronomers used the Hubble Space Telescope to measure the perpendicular motion of Andromeda and determine whether a collision will be head-on, partial, or a total miss. They concluded that the two galaxies will "collide" head-on in about 4 billion years, although because most of a galaxy is space between the stars, actual collisions between the stars themselves are quite unlikely; it is more that the material of one galaxy will pass through the other. It will take another 2 billion years for the two galaxies to merge completely and form one giant elliptical galaxy. The third spiral in the Local Group, the Triangulum Galaxy, may merge first with the Milky Way, or with Andromeda, or be ejected from the Local Group temporarily.

Note the timing here. This first close encounter with Andromeda in 4 billion years will occur *before* the Sun runs out of hydrogen in its core, although it is likely that the Sun will have increased its luminosity enough by then that Earth's habitability will have been affected. In 6 billion years, the Sun and its planets could end up near the center of the merged galaxy, or more likely they will have a new location farther from the center with a different orbit and in a different stellar neighborhood (**Figure 21.20**). If by chance the encounter leads to a star passing close to the Sun, the orbits of the Solar System planets could be disrupted, causing them to be at different distances from the Sun. **Figure 21.21** shows what this merged galaxy might look like in Earth's sky—if anyone is still around to see it! (For links to the merger simulation, see question 60 at the end of the chapter.)

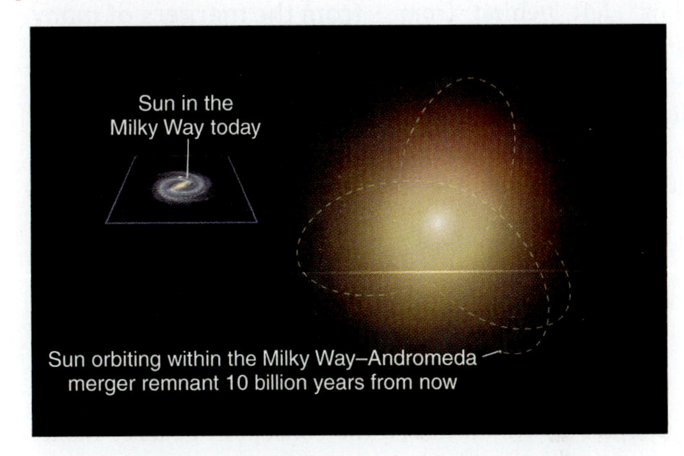

FIGURE 21.20 Collision scenario for a Milky Way and Andromeda Galaxy encounter. Triangulum, the third spiral galaxy, might also join in.

FIGURE 21.21 Computer simulation of the view in the sky on Earth when the Andromeda and Milky Way galaxies collide. The spiral Andromeda appears larger in the sky as it gets closer, and then the sky becomes increasingly bright as the two collide.

Group. In addition, observations of the motions and speeds of the dwarf galaxies about the Milky Way may lead to new estimates of the dark matter mass within the Milky Way itself.

Protogalaxies merged to form the large barred spiral galaxies in the Local Group. Mergers and collisions of Local Group galaxies continue today (**Connections 21.2**).

Astronomers seeking to understand how the Milky Way formed must consider the distribution of stars of different ages with different amounts of heavy elements, the presence of dark matter and the supermassive black hole, and evidence that indicates when and where mergers took place. The Milky Way and the Local Group offer many clues about the way galaxies form, but much of what astronomers know of the process comes from looking beyond the local system. Images of distant galaxies (seen as they existed *billions* of years ago), as well as observations of the "glow" left behind from the earliest hot phases of the universe itself, provide equally important pieces of the puzzle. In the coming chapters, we will return to the story of galaxy formation along with the development of structure in the universe.

The Milky Way formed from the mergers of many smaller protogalaxies.

21.6 Origins: The Galactic Habitable Zone

In Chapter 20 we discussed the concept of galactic habitable zones in general; in this section we will focus more specifically on ideas about the habitable zone of the Milky Way. Stars that are situated too far from the galactic center may have protoplanetary disks with insufficient quantities of heavy elements—such as oxygen, silicon, iron, and nickel (to make up rocky planets like Earth) or carbon, nitrogen, and oxygen (to make up the molecules of life). Stars that are too close to the galactic center may have planets that are strongly affected by its high-energy radiation environment (X-rays and gamma rays from the supermassive black hole) and by supernova explosions and gamma-ray bursts (GRBs). The bulge has a higher density of stars creating a strong radiation field, and the halo and thick disk have stars with lower amounts of heavy elements, so perhaps only stars in the thin disk of the galaxy are candidates for residence in a galactic habitable zone.

Astronomers must also consider stellar lifetimes versus the 4 billion years after the formation of Earth that it took for life to evolve into land animals (see Chapter 16)—so only stars with masses low enough that they will live at least 4 billion years on the main sequence are considered potential hosts of more complex life. In this simple model based on the evolution times for life on Earth, the galactic habitable zone could be a doughnut-shaped region around the galactic center. In one version of the model, this zone is estimated to contain stars born 4–8 billion years ago located between 7 and 9 kpc from the galactic center, with the Sun exactly in the middle of the doughnut. The doughnut would grow larger over time as heavier-element formation spread outward from the galactic center.

Some have conducted additional studies and proposed more complex models. For example, one group initiated a search for the molecule formaldehyde (H_2CO)—a key "prebiotic" molecule—by observing molecular clouds in the outer parts of the Milky Way, 12–23.5 kpc from the galactic center. Formaldehyde was detected in two-thirds of their sample of 69 molecular clouds, suggesting that at least one important prebiotic molecule is available far from the galactic center.

Another computer model took into account details of the evolution of individual stars within the Milky Way, including birth rates, locations, distribution within the galaxy, abundances of heavy elements, stellar masses, main-sequence lifetimes, and the likelihood that stars became or will become supernovae. The model also assumed 4 billion years to the development of complex life. One result of this model is that stars in the inner part of the Milky Way are more likely to be affected by supernova explosions, but these stars are even more likely to have the heavier elements for the formation of planets. So in this model the *inner* part of the galaxy, about 2.5–4 kpc from the center, in and near the midplane of the thin disk, is the most likely place for habitable planets. In this case the Sun is *not* in the middle of the most probable zone for habitable planets. As more extrasolar planets are discovered, astronomers will have a better idea of their distribution throughout the Milky Way. ▶▶ **NEBRASKA SIMULATION: MILKY WAY HABITABILITY EXPLORER**

As noted in Connections 21.2, mergers with other galaxies could cause stars to migrate into or out of the galactic habitable zone. The uncertainties increase as the assumptions in these models move from the *astronomical* to the *biological*. For example, maybe life evolved faster on other planets, so stars of higher mass and shorter lifetimes should be included. Or maybe life evolved slower elsewhere, in which case the older stars would be the best candidates. It is unknown whether intense radiation from a supernova or a gamma-ray burst would permanently sterilize a planet or only affect evolution for a while. For example, if Earth's ozone layer was temporarily destroyed, life on land might die out but life in the oceans would continue. There are many, many uncertainties involved in thinking about habitability in the Milky Way Galaxy.

The Milky Way offers many clues about the way galaxies form, but much of what astronomers know of the process comes from looking beyond the Local Group. Images of distant galaxies (seen as they existed billions of years ago), as well as observations of the glow left behind by the formation of the universe itself, provide equally important pieces of the puzzle. We will return to the story of galaxy formation in Chapter 23, but first we will turn our attention to the structure of the early universe as a whole. ■

Summary

21.1 The Sun is located in the disk of a barred spiral SBbc galaxy called the Milky Way, which is 30,000 pc (100,000 light-years) across, and about 8,300 pc (27,000 light-years) from the Milky Way's center. Astronomers use variable stars of known luminosity to find the distances to globular clusters, which enable them to measure the size of the luminous part of the Milky Way's extended halo.

21.2 The rotation curve of the Milky Way is flat, like those of other galaxies, indicating that its mass is mostly in the form of dark matter.

21.3 The abundance of heavy elements in the Milky Way has increased with time as each generation of stars has produced more of these elements during the final phases of the stars' life. There must have been a generation of stars before the oldest halo and globular-cluster stars visible today formed, because there are heavier elements than those produced in the Big Bang. Star formation is actively occurring in the disk of the galaxy, leading to complex structures within the disk.

21.4 At the center of the galaxy is a supermassive black hole. Evidence for the black hole includes rapid orbital velocities of nearby stars, and symmetric X-ray and gamma-ray outflows of material.

21.5 The Milky Way is part of the Local Group of galaxies, which consists of two large barred spirals and several dozen smaller galaxies. Collisions and mergers between these galaxies likely happened in the past, and a merger with the Andromeda Galaxy may be part of the Milky Way's future.

21.6 The idea of a galactic habitable zone is that certain parts of the Milky Way may be more suitable for the existence of habitable planets. This zone would have enough heavy elements for the formation of rocky planets and organic molecules, but not too much radiation that it would damage any life.

Unanswered Questions

- Are there many more ultrafaint dwarf galaxies than have been detected so far? These types of dwarf galaxies are so faint that they are hard to detect even when close to the Milky Way. As we will discuss in Chapter 23, the large giant spirals such as the Milky Way were built up by mergers of these small faint galaxies, so models predict that there should be hundreds or even thousands of them. They could be so dominated by dark matter that they are not at all visible, or too small ever to have formed stars, or they might have merged with the Milky Way or other Local Group members long ago.
- Will the merger of the Milky Way and Andromeda form a quasar at the center of the new elliptical galaxy? If the dust and gas at the center of the galaxy did not completely block the view from Earth, such a quasar could be brighter than the full Moon. The supermassive black hole at the center of Andromeda is thought to be 25–50 times larger than the one in the Milky Way, but both black holes combined still would be on the lower end of the masses of the black holes in AGNs discussed in Chapter 20.

Questions and Problems

Summary Self-Test

1. The size of the Milky Way is determined from studying ________ stars in globular clusters.
   a. Cepheid variable
   b. blue supergiant
   c. RR Lyrae
   d. Sun-like
2. Rank, in increasing order of age, globular clusters with main-sequence turnoffs at the following temperatures.
   a. 10,000 K
   b. 3000 K
   c. 6000 K
3. Detailed observations of the structure of the Milky Way are difficult because
   a. the Solar System is embedded in the dust and gas of the disk.
   b. the Milky Way is mostly dark matter.
   c. there are too many stars in the way.
   d. the galaxy is rotating too fast (about 200 km/s).

4. In general, what does the abundance of a star's heavy elements indicate about the age of the star?
   a. Stars with higher abundance are older.
   b. Stars with higher abundance are younger.

5. Older stars are found farther from the midplane of a galactic disk because
   a. the disk used to be thicker.
   b. the stars have lived long enough to move there.
   c. the younger stars in the thick disk were more massive and have already died.
   d. none of the above

6. The magnetic field of the Milky Way has been detected by
   a. synchrotron radiation from cosmic rays.
   b. direct observation of the field.
   c. its interaction with Earth's magnetic field.
   d. studying molecular clouds.

7. Evidence of a supermassive black hole at the center of the Milky Way comes from
   a. direct observations of stars that orbit it.
   b. X-rays from material that is falling in.
   c. strong radio emission from the region of the accretion disk.
   d. all of the above

8. A most likely fate of the Large and Small Magellanic Clouds is that they will
   a. become part of the Milky Way.
   b. remain orbiting forever.
   c. become attached to another passing galaxy.
   d. escape from the gravity of the Milky Way.

9. Rank, in increasing order, the following in terms of the fraction of the Milky Way's mass that they represent.
   a. stars in the bulge
   b. stars in the halo
   c. stars in the disk
   d. dark matter

10. Globular clusters are important because
    a. they provide information about how the Milky Way formed and evolved.
    b. they provide information about how stars evolve.
    c. they reveal the size of the Milky Way, and Earth's location in it.
    d. all of the above

True/False and Multiple Choice

11. **T/F:** Most of the Milky Way is visible because it is so close.

12. **T/F:** A globular cluster spends most of its time in the galaxy's halo.

13. **T/F:** The disk of the Milky Way contains most of its mass.

14. **T/F:** The Milky Way rotates like a solid disk.

15. **T/F:** Most of the stars in the Milky Way formed at the same time.

16. Why are globular clusters so useful in determining the size of the galaxy?
    a. They are large.
    b. They are bright.
    c. They are evenly distributed.
    d. all of the above

17. In order for a variable star to be useful as a standard candle, its luminosity must be related to its
    a. period of variation.
    b. mass.
    c. temperature.
    d. radius.

18. The best evidence for dark matter in the Milky Way comes from the observation that the rotation curve
    a. is quite flat at great distances from the center.
    b. rises swiftly in the interior.
    c. falls off and then rises again.
    d. has a peak at about 2,000 light-years from the center.

19. Cosmic rays are
    a. a form of electromagnetic radiation.
    b. high-energy particles.
    c. high-energy dark matter.
    d. high-energy photons.

20. In the Hubble scheme for classifying galaxies, what kind of galaxy is the Milky Way?
    a. elliptical
    b. spiral
    c. barred spiral
    d. irregular

21. Where are the youngest stars in the Milky Way Galaxy?
    a. in the core
    b. in the bulge
    c. in the disk
    d. in the halo

22. Halo stars are found in the vicinity of the Sun. What observational evidence distinguishes them from disk stars?
    a. the direction of their motion
    b. their speed
    c. their composition
    d. their temperature

23. How does the mass of the supermassive black hole at the center of the Milky Way compare with that found in most other spiral galaxies?
    a. It is comparable in mass.
    b. It is much less massive.
    c. It is much more massive.
    d. Astronomers don't yet know the masses well enough to tell.

24. Why are most of the Milky Way's satellite galaxies so difficult to detect?
    a. They are very small.
    b. They are very far away.
    c. The halo of the Milky Way is in the way.
    d. They are very faint.

25. The concept of a galactic habitable zone does not consider
    a. the radiation field.
    b. the ages of stars.
    c. the amount of heavy elements.
    d. the distance of a planet from its central star.

Thinking about the Concepts

26. Why is it so difficult for astronomers to get an overall picture of the structure of the Milky Way Galaxy?
27. What do astronomers mean by *standard candle*?
28. How do astronomers know that the stars in globular clusters are among the oldest stars in the Milky Way Galaxy?
29. Compare and contrast globular and open clusters.
    a. What are the main differences in the gas out of which globular and open clusters formed?
    b. Why do globular clusters have such high masses while open clusters have low masses?
    c. Are the characteristics of interest in parts (a) and (b) related? Explain.
30. Old stars in the inner disk of the Milky Way have higher abundances of massive elements than do young stars in the outer disk. Explain how this difference might have developed.
31. How do 21-cm radio observations reveal the rotation of the Milky Way Galaxy?
32. What does the rotation curve of the Milky Way indicate about the presence of dark matter in the galaxy?
33. Explain the observational evidence showing that Earth is located in a spiral galaxy, not an elliptical galaxy.
34. What is one source of synchrotron radiation in the Milky Way, and where is it found?
35. Why must astronomers use X-ray, infrared, and 21-cm radio observations to probe the center of the galaxy?
36. What is Sgr A* and how was it detected?
37. Explain the evidence for a supermassive black hole at the center of the Milky Way.
38. To observers in Earth's Southern Hemisphere, the Large and Small Magellanic Clouds look like detached pieces of the Milky Way. What are these "clouds," and why is it not surprising that they look so much like pieces of the Milky Way?
39. What has been the fate of most of the Milky Way's original satellite galaxies?
40. Use your imagination to describe how Earth's skies might appear if the Sun and Solar System were located
    a. near the center of the galaxy.
    b. near the center of a large globular cluster.
    c. near the center of a large, dense molecular cloud.

Applying the Concepts

41. From Figure 21.5 estimate the ratio between the radius of the Milky Way's halo and the radius of the disk.
42. Study Figure 21.7.
    a. What is the rotation velocity of a disk star located 6,000 pc from the center of the Milky Way?
    b. Assuming a circular orbit, how long does it take that star to orbit once?
43. From the data in Figure 21.7, estimate the time it would take the Large Magellanic Cloud to orbit the Milky Way, if it were on a circular orbit.
44. The Sun completes one trip around the center of the galaxy in approximately 230 million years. How many times has the Solar System made the circuit since its formation 4.6 billion years ago?
45. The Sun is located about 8,300 pc from the center of the galaxy, and the galaxy's disk probably extends another 9,000 pc farther out from the center. Assume that the Sun's orbit takes 230 million years to complete.
    a. With a truly flat rotation curve, how long would it take a globular cluster located near the edge of the disk to complete one trip around the center of the galaxy?
    b. How many times has that globular cluster made the circuit since its formation about 13 billion years ago?
46. Parallax measurements of the variable star RR Lyrae indicate that it is located 230 pc from the Sun. A similar star observed in a globular cluster located far above the galactic plane appears 160,000 times fainter than RR Lyrae.
    a. How far from the Sun is this globular cluster?
    b. What does your answer to part (a) tell you about the size of the galaxy's halo compared to the size of its disk?
47. Although the flat rotation curve indicates that the total mass of the Milky Way is approximately $1 \times 10^{12}\ M_{\odot}$, electromagnetic radiation associated with normal matter suggests a total mass of only $9 \times 10^{10}\ M_{\odot}$. Given this information, calculate the fraction of the Milky Way mass that is made up of dark matter.
48. Given what you have learned about the distribution of massive elements in the Milky Way, and what you know about the terrestrial planets, where do you think such planets are most likely and least likely to form?
49. A cosmic-ray proton is traveling at nearly the speed of light (3×10^8 m/s).
    a. Using Einstein's familiar relationship between mass and energy ($E = mc^2$), show how much energy (in joules) the cosmic-ray proton would have if m were based only on the proton's rest mass (1.7×10^{-27} kg).
    b. The actual measured energy of the cosmic-ray proton is 100 J. What, then, is the relativistic mass of the cosmic-ray proton?
    c. How much greater is the relativistic mass of this cosmic-ray proton than the mass of a proton at rest?

50. One of the fastest cosmic rays ever observed had a speed of (1.0 minus 10^{-24}) $\times$ *c*. Assume that the cosmic ray and a photon left a source at the same instant. To a stationary observer, how far behind the photon would the cosmic ray be after traveling for 100 million years?

51. Consider a black hole with a mass of 5 million $M_{\odot}$ ($M_{\odot} = 2 \times 10^{30}$ kg). A star's orbit about the black hole has a semi-major axis of 0.02 light-year (1.9×10^{14} meters). Calculate the star's orbital period. (Hint: For this and the questions that follow, you may want to refer back to Chapter 4.)

52. A star in a circular orbit about the black hole at the center of the Milky Way (whose mass $M_{BH} = 8 \times 10^{36}$ kg) has an orbital radius of 0.0131 light-year (1.24×10^{14} meters). What is the average speed of this star in its orbit?

53. What is the Schwarzschild radius of the black hole at the center of the Milky Way? What is its density? How does this compare with the density of a stellar black hole?

54. A star is observed in a circular orbit about a black hole with an orbital radius of 1.5×10^{11} kilometers and an average speed of 2,000 km/s. What is the mass of this black hole in solar masses?

55. One model of the galactic habitable zone contains stars in a doughnut between 7 and 9 kpc from the center of the galaxy, in the disk. Assuming that this doughnut is as thick as the disk itself, what fraction of the disk of the Milky Way lies in this habitable zone?

Using the Web

56. Go to the "Night Sky" Web page of the National Park Service (NPS—http://nature.nps.gov/night). What is a "natural lightscape"? Click on and read "Light Pollution" and "Measuring Lightscapes." Why is it becoming more rare for people to see the Milky Way? Why does the NPS consider viewing the Milky Way an important part of the parks experience?

57. a. Go to the Astronomy Picture of the Day (APOD) app or website for July 2, 2012 and watch the video clip "Zooming into the Center of the Milky Way" (http://apod.nasa.gov/apod/ap120702.html). Why is there a shift in wavelengths of the selected pictures? On APOD, run a search for "Milky Way" to look at some of the best photographs of it. Where were the pictures taken from? Can you see the Milky Way from *your* location on a clear night?

    b. Why does the galactic center have to be observed in infrared and X-ray wavelengths? Go to the "Milky Way Galaxy" page of the Chandra X-ray telescope website (http://chandra.harvard.edu/photo/category/milkyway.html) and to the Spitzer infrared telescope website (http://spitzer.caltech.edu). Are there any new images of the galactic center? What has been learned?

58. Go to the websites of the two main groups studying the supermassive black hole at the galactic center: the UCLA Galactic Center Group (http://astro.ucla.edu/~ghezgroup/gc) and the Galactic Center Research group at the Max Planck Institute for Extraterrestrial Physics (http://mpe.mpg.de/ir/GC). Watch some of the time-lapse animations of the stars orbiting something unseen. Why is it assumed that the unseen object is a black hole? What new results are these groups reporting?

59. Go to the Milky Way Project website (http://milkywayproject.org). This citizen science project aims to sort Spitzer telescope observations of the dusty material in the galaxy. Log in with your Zooniverse account name, and read the information under "About" and "Tutorial." Participants in this project have already discovered some of these bubbles (http://spitzer.caltech.edu/images/4938-sig12-002-Finding-Bubbles-in-the-Milky-Way). Start looking.

60. Go to the HST website and watch the videos about the possible collision of the Milky Way and Andromeda (http://hubblesite.org/newscenter/archive/releases/galaxy/2012/20/video). Read the report under "The Full Story." Now do a Web search on this story, which received a lot of press in 2012. Did other astronomers dispute this study later? Is it still receiving a lot of press attention?

STUDYSPACE is a free and open website that provides a Study Plan for each chapter of *21st Century Astronomy*. Study Plans include animations, reading outlines, vocabulary flashcards, and multiple-choice quizzes, plus links to premium content in SmartWork and the ebook. Visit **wwnorton.com/studyspace**.

SMARTWORK Norton's online homework system, includes algorithmically generated versions of these questions, plus additional conceptual exercises. If your instructor assigns questions in SmartWork, log in at **smartwork.wwnorton.com.**

Exploration | The Center of the Milky Way

Adapted from *Learning Astronomy by Doing Astronomy*, by Ana Larson

Astronomers once thought that the Sun was at the center of the Milky Way. In this Exploration you will repeat Harlow Shapley's globular cluster experiment that led to a more accurate picture of the size and shape of the Milky Way.

Imagine that the disk of the Milky Way is a flat, round plane, like a pizza. Globular clusters are arranged in a rough sphere around this plane. To map globular clusters on **Figure 21.22**, imagine that a line is drawn straight "down" from a globular cluster to the plane of the Milky Way. The "projected distance" in kiloparsecs is the distance from the Sun to the place where the line hits the plane. The galactic longitude indicates the direction toward that point; it is marked around the outside of the graph, along with the several constellations.

Make a dot at the location of each globular cluster by finding the galactic longitude indicated outside the circle and then coming in toward the center to the projected distance. The two globular clusters in bold in **Table 21.1** have been plotted for you as examples. After plotting all of the globular clusters, estimate the center of their distribution and mark it with an X. This is the center of the Milky Way.

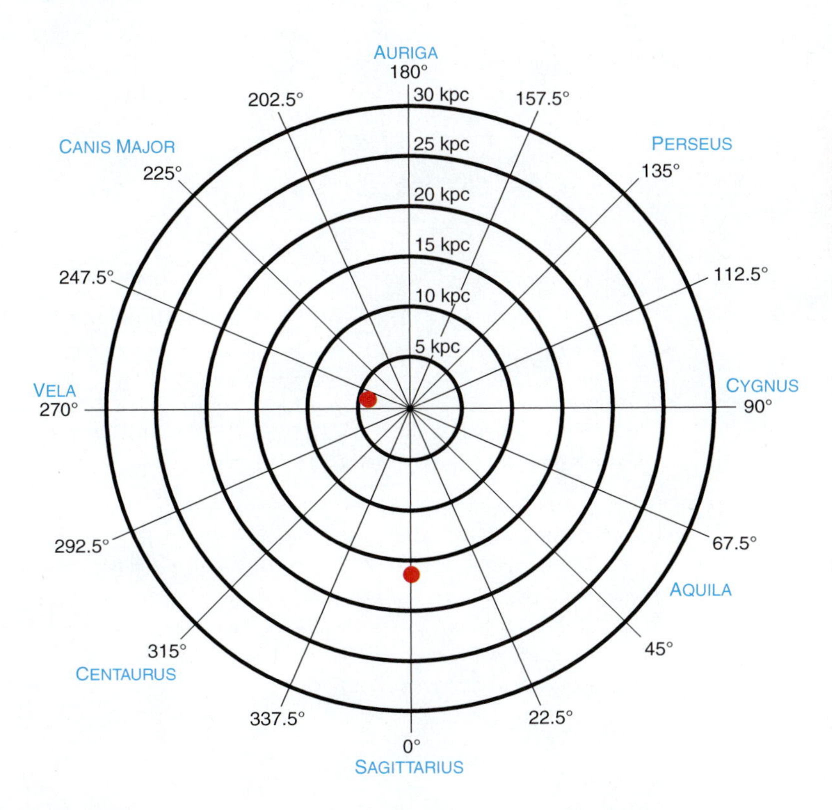

FIGURE 21.22 A polar graph to plot distance and direction.

TABLE 21.1 | Globular Cluster Data

Cluster	Galactic Longitude	Projected Distance (kpc)	Cluster	Galactic Longitude	Projected Distance (kpc)
104	306	3.5	6273	357	7
362	302	6.6	**6287**	**0**	**16.6**
2808	283	8.9	6333	5	12.6
4147	**251**	**4.2**	6356	7	18.8
5024	333	3.4	6397	339	2.8
5139	309	5	6535	27	15.3
5634	342	17.6	6712	27	5.7
Pal 5	1	24.8	6723	0	7
5904	4	5.5	6760	36	8.4
6121	351	4.1	Pal 10	53	8.3
O 1276	22	25	Pal 11	32	27.2
6638	8	15.1	6864	20	31.5
6171	3	15.7	6981	35	17.7
6218	15	6.7	7089	54	9.9
6235	359	18.9	Pal 12	31	25.4
6266	353	11.6	288	147	0.3
6284	358	16.1	1904	228	14.4
6293	357	9.7	Pal 4	202	30.9
6341	68	6.5	4590	299	11.2
6366	18	16.7	5053	335	3.1
6402	21	14.1	5272	42	2.2
6656	9	3	5694	331	27.4
6717	13	14.4	5897	343	12.6
6752	337	4.8	6093	353	11.9
6779	62	10.4	6541	349	3.9
6809	9	5.5	6626	7	4.8
6838	56	2.6	6144	352	16.3
6934	52	17.3	6205	59	4.8
7078	65	9.4	6229	73	18.9
7099	27	9.1	6254	15	5.7

1 **What is the approximate distance from the Sun to the center of the Milky Way?**

....

2 **What is the galactic longitude of the center of the Milky Way?**

....

3 **How do astronomers know that the Sun is not at the center of the Milky Way?**

....

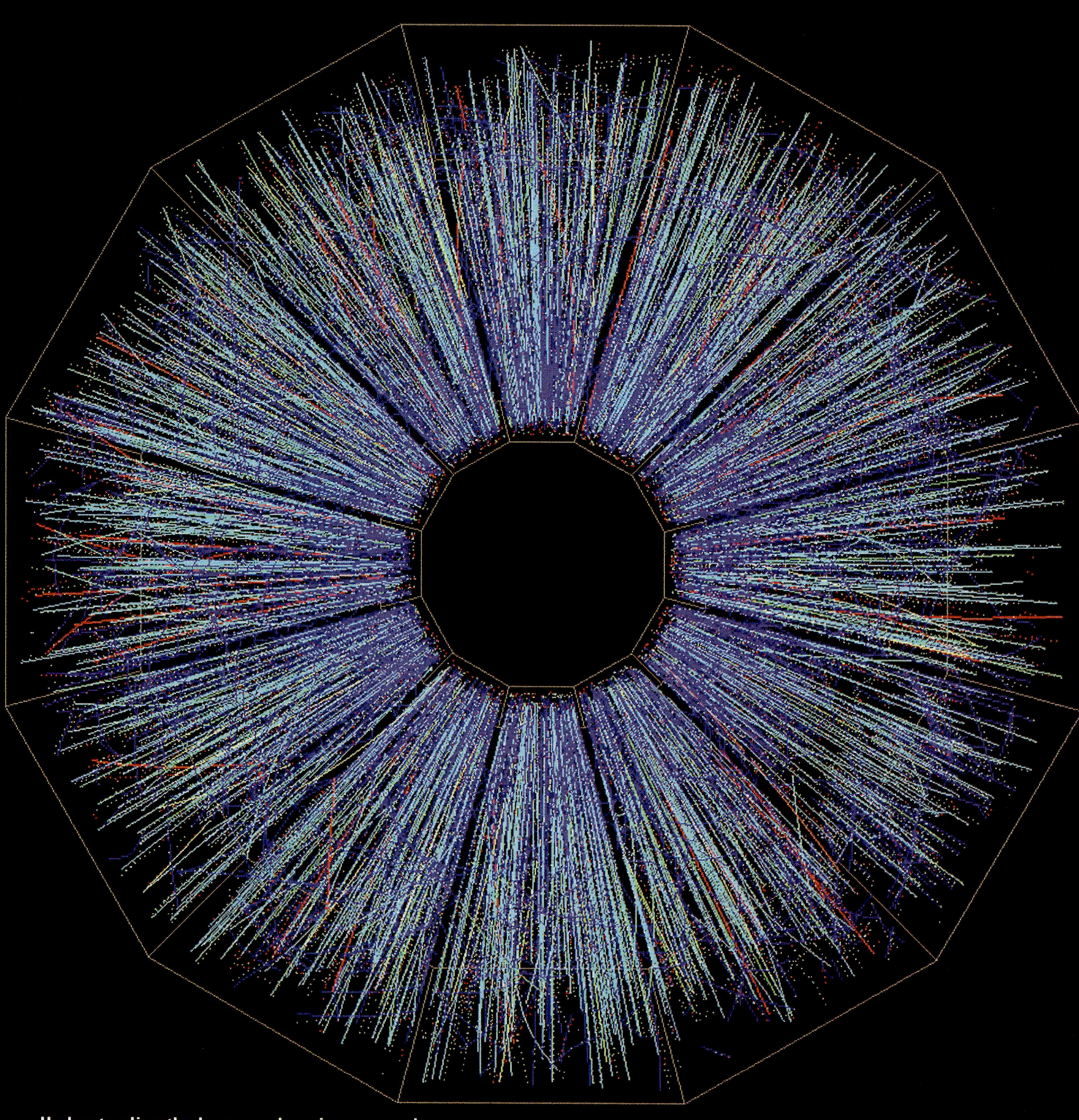

Understanding the large-scale universe requires knowledge of everything from the smallest subatomic particles to the largest structures. The STAR particle detector at Brookhaven National Lab has studied the form of matter that existed moments after the Big Bang.

22 Modern Cosmology

There is a theory which states that if ever anyone discovers exactly what the universe is for and why it is here, it will instantly disappear and be replaced by something even more bizarre and inexplicable. There is another which states that this has already happened.

Douglas Adams (1952–2001)

LEARNING GOALS

Cosmology is the study of the large-scale universe, including its nature, origin, evolution, and ultimate destiny. In this chapter we will discuss the early stages of the universe. By the conclusion of this chapter, you should be able to:

- Explain how the history, shape, and fate of the universe are affected by gravity from its mass.
- Connect the accelerating expansion of the universe to dark energy.
- Describe the reasoning that leads astronomers to hypothesize an early period of rapid expansion known as inflation.
- List the fundamental forces of nature and discuss when they became distinct.
- Explain the ideas about multiverses.

22.1 The Universe Has a Destiny and a Shape

Cosmology is the study of the universe on the very grandest of scales. In Chapter 19 you learned that the universe is expanding, after originating in a Big Bang 13.7 billion years ago; that it was once filled with very hot Planck blackbody radiation that has now cooled to a temperature of 2.7 kelvins (K); and that light elements in the universe were produced within the first few minutes after the Big Bang. In this chapter we take a closer look at the nature of the universe, how it has evolved over time, and its ultimate fate. We also discuss the physics of the smallest particles, which is necessary for describing the very smallest pieces of the universe.

How did the universe evolve from its smooth beginning to its "lumpy" state today?

At first glance, particle physics and cosmology might seem to have almost nothing in common. Whereas particle physics is the study of the quantum mechanical world that exists on the tiniest scales imaginable, cosmology is the study of the changing structure of a universe that extends for billions of parsecs and probably much farther. Yet the last quarter of the 20th century saw the boundary between these two fields fade as cosmologists and particle physicists came to realize that the structure of the universe and the fundamental nature of matter are related.

What is the fate of the universe? This is clearly one of the key questions of modern cosmology. The simplest answer depends in part on the amount of mass distributed across the universe on very large scales. The gravitational effect of this distributed matter is only one of the factors—the first one we will discuss—that determines how the universe evolves.

To see how gravity affects the expansion of the universe, recall gravity's effects on the motion of projectiles and the discussion of escape velocity from Chapter 4. For example, the fate of a projectile fired straight up from the surface of Earth depends on its speed. As long as the speed is less than Earth's escape velocity (11.2 kilometers per second, km/s), gravity will eventually stop the rise of the projectile and pull it back to Earth's surface. But if the speed of the projectile is greater than Earth's escape velocity, gravity will lose. In this case, although the projectile will slow down, it will never stop, and it will escape from Earth entirely.

Just as Earth's mass gravitationally pulls on a projectile to slow its climb, the mass distributed across the universe gravitationally slows the universe's expansion. If there is enough mass in the universe, then gravity will be strong enough to stop the expansion. And in that case, the universe will slow its expansion, stop, and eventually collapse in on itself. If there is not enough mass, the expansion of the universe may slow, but it will never stop. The universe will expand forever.

Gravity slows the expansion of the universe.

A planet's mass and radius determine the escape velocity from its surface. The "escape velocity" of the universe is also determined by its mass and size—specifically, its average density. If the universe is denser on average than a particular value, called the **critical density**, then gravity will be strong enough to eventually stop and reverse the expansion. If the universe is less dense than the critical density, gravity will be too weak and the universe will expand forever.

The faster the universe is expanding, the more mass is needed to turn that expansion around. For this reason, the critical density depends on the value of several constants, including the gravitational constant G and the Hubble constant, H_0. Assuming that gravity is the only factor affecting the expansion, the critical density of the universe is fewer than six hydrogen atoms in every cubic meter (**Math Tools 22.1**). Astronomers use the ratio of the actual density of the universe to its critical density, called Ω_{mass} (which is pronounced "omega sub mass"). Because it is a ratio of two densities, Ω_{mass} has no units.

The expansion of the universe changes over time. **Figure 22.1** shows the scale factor R_U (see Chapter 19) versus time for a universe dominated by mass and controlled by gravity alone. The colored lines show different possible values of Ω_{mass}. If Ω_{mass} is greater than 1, then gravity is strong enough to turn the expansion around. The expansion will slow and eventually stop, and the universe will then fall back in on itself. Conversely, if Ω_{mass} is less than 1, that universe will be slowed by gravity, but it will still expand forever. The dividing line, where Ω_{mass} equals 1, corresponds to a universe that expands more and more slowly—continuing forever, but never quite stopping. Such a universe is expanding at exactly the "escape velocity."

Ω_{mass} determines the fate of a universe that is governed exclusively by gravity.

Look again at the three plots for different values of Ω_{mass} in Figure 22.1 and you will see that they are not straight lines. The curvature in these plotted lines shows that gravity is slowing the expansion. As the value of Ω_{mass} increases (corresponding to greater density in the universe), the amount of curvature also increases. When Ω_{mass} is greater than 1, the scale factor begins to decrease and gravity wins.

Until the closing years of the 20th century, most astronomers thought that this straightforward application of

Math Tools 22.1

Critical Density

Using the assumption of the cosmological principle—that the universe is homogeneous and isotropic—and using a set of equations from Einstein on gravitation, Alexander Friedmann (1888–1925) derived equations for cosmology in an expanding universe. The equations are complicated, but he discussed this idea of critical density. When he assumed that the energy of empty space was zero, making gravity from the total mass of the universe the only factor, he derived the following equation for the critical density, ρ_c:

$$\rho_c = \frac{3H_0^2}{8\pi G}$$

Using the value 70 kilometers per second per megaparsec (km/s/Mpc) for H_0, we will put everything in terms of meters to match the units in G. Using 1 km = 10^3 meters, and 1 Mpc = 3.1×10^{22} meters, we can rewrite H_0 as:

$$H_0 = \frac{70 \times 10^3 \text{ m/s}}{3.1 \times 10^{22} \text{ m}} = 2.3 \times 10^{-18}/\text{s}$$

Using $G = 6.67 \times 10^{-11}$ m^3/kg s^2, the current value of the critical density is given by:

$$\rho_c = \frac{3 \times (2.3 \times 10^{-18}/\text{s})^2}{8 \times \pi \times (6.67 \times 10^{-11} \text{ m}^3/\text{kg s}^2)}$$

$$= 9.5 \times 10^{-27} \text{ kg/m}^3$$

Dividing by the mass of the proton, 1.67×10^{-27} kg, we find that this equals about 5.7 hydrogen atoms per cubic meter, which is equivalent to about 140 billion (1.4×10^{11}) solar masses per cubic megaparsec ($M_\odot$/Mpc^3). The observed mass density of the universe is much less than this value of the critical density.

The Hubble constant measures the rate of expansion that has changed with time, so the critical density changes with time as well (which makes sense in an expanding universe). Thus, 5 billion years ago the universe was smaller, and the critical density would have been larger.

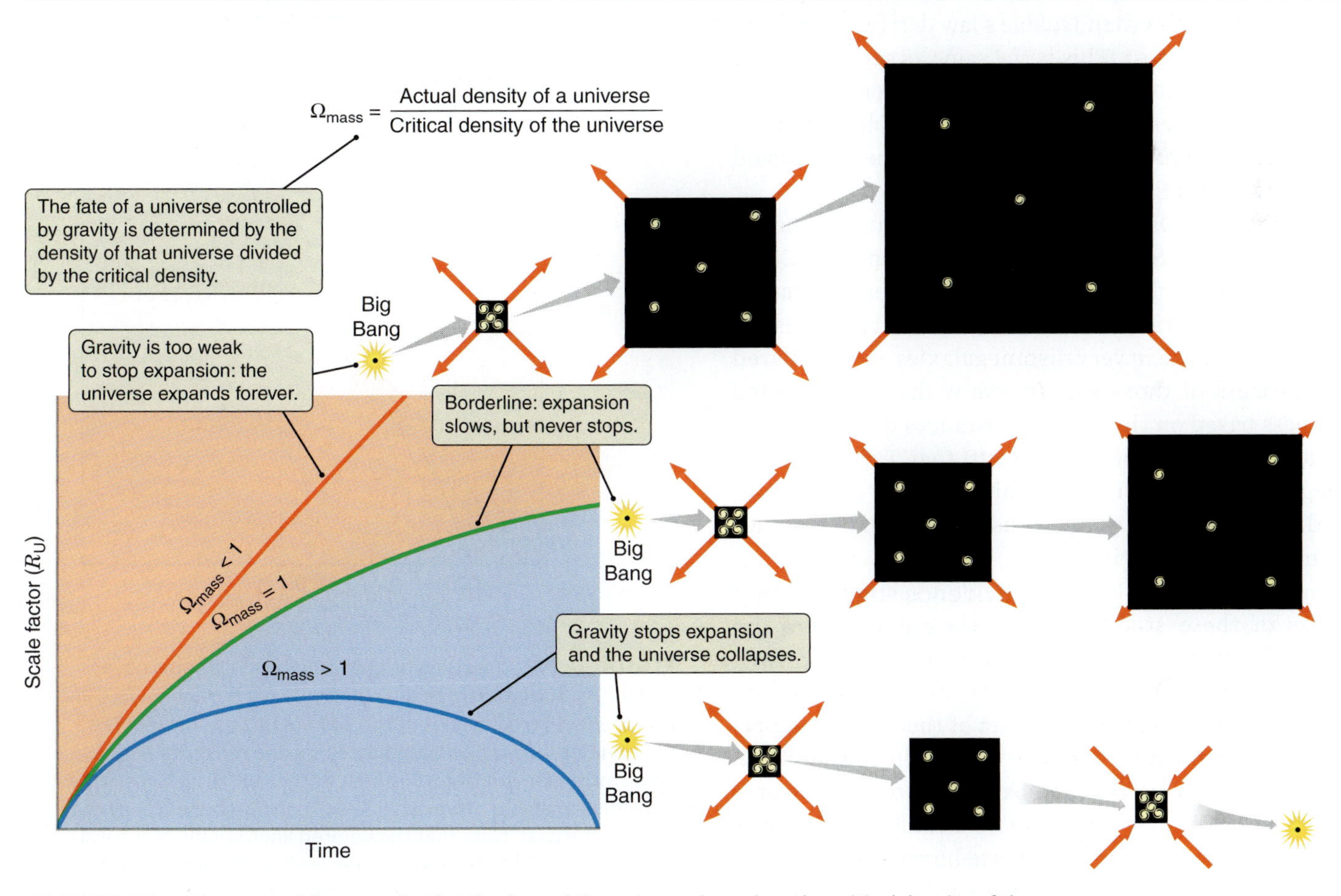

FIGURE 22.1 Three possible scenarios for the fate of the universe based on the critical density of the universe, but ignoring any cosmological constant.

gravity to the universe was all there was to the question of expansion and collapse. Researchers carefully measured the masses of galaxies and assemblages of galaxies in the expectation that these data would reveal the density and therefore the fate of the universe. Measuring masses of objects in the universe is tricky. The luminous matter that is seen in galaxies and groups of galaxies gives an Ω_{mass} value of about 0.02. But galaxies contain about 10 times as much dark matter as normal luminous matter, so adding in the dark matter in galaxies pushes the value of Ω_{mass} up to about 0.2. When the mass of dark matter *between* galaxies is included (a subject we will return to in the next chapter), Ω_{mass} could increase to 0.3 or higher. By this accounting there is only at most a third as much mass in the universe as is needed to stop the universe's expansion.

22.2 The Accelerating Universe

If the expansion of the universe is in fact slowing with time, as these simple models using gravity predict, then when the universe was young it must have been expanding more rapidly than it is today. Objects that are very far away should have larger velocities than Hubble's law derived from local galaxies would suggest. This is the same as saying that the plots in Figure 22.1 are all curving away from a straight line. (If gravity were not present, the Hubble expansion [see Chapter 19] would not be changing, and the plots would instead be straight lines.)

During the 1990s, some groups of astronomers began using the Hubble Space Telescope, telescopes in Chile, and the giant Keck Observatory telescopes in Hawaii to test this prediction. They measured the brightness of Type Ia supernovae in very distant galaxies and compared the brightness of those supernovae with their expected brightness based on the redshift distances of those galaxies. (Recall from Chapters 19 and 20 that Type Ia supernovae have a very high peak luminosity that can be calibrated, and therefore they are used as "standardizable" candles for measuring distances to galaxies.) The findings of these studies sent a wave of excitement through the astronomical community. The observational data of the Type Ia supernovae at different distances are plotted in **Figure 22.2**. Rather than showing that the expansion of the universe has slowed down over time, the data indicated that it is *speeding up*—accelerating. For this to be true, a force must be pushing the *entire universe* outward in opposition to gravity. In 2011, the Nobel Prize in Physics was awarded to Saul Perlmutter, Brian P. Schmidt, and Adam G. Riess for their observations of Type Ia supernovae and the discovery of the accelerating universe. Results obtained early in the 21st century by the *WMAP* experiment (see Chapter 19) provided independent confirmation of this increasing rate of expansion of the universe.

Evidence suggests that the expansion of the universe is accelerating.

The idea of a repulsive force opposing the attractive force of gravity is not new. In the early 20th century, Einstein used his newly formulated equations of general relativity to calculate the structure of spacetime in the universe. The equations clearly indicated that any universe containing mass could not be static, any more than a ball can hang motionless in the air. He found the same result that Figure 22.1 illustrates—namely, that gravity always makes the universe move toward slower expansion or even collapse. However, Einstein's formulation of spacetime came more than a decade before Hubble discovered the expansion of the universe (see Chapter 19), and the conventional wisdom

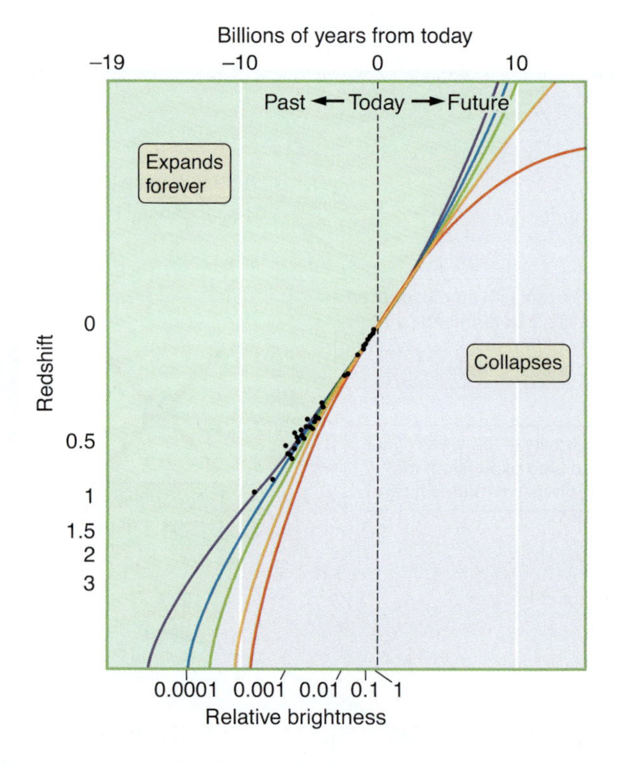

FIGURE 22.2 Observed brightness of Type Ia supernovae, plotted as a function of their redshift. The different colored lines represent different values of the cosmological constant. The observations (data points) indicate that the redshifts are too small for their distances; they best fit the line for an accelerating universe—that is, a universe that is expanding faster today than it did in the past. Note that the colored lines are more separate at higher redshifts, so astronomers look for higher-redshift supernovae to better differentiate the data.

at the time was that the universe was indeed static—that it neither expands nor collapses.

To force his new general theory of relativity to allow for a static universe, Einstein inserted a "fudge factor" called the **cosmological constant** into his equations. Einstein's cosmological constant acts as a repulsive force in the equations, opposing gravity and allowing galaxies to remain stationary despite their mutual gravitational attraction.

Einstein invented a cosmological constant to oppose gravity in a static universe.

When Hubble announced his discovery that the universe is expanding, Einstein realized his mistake. The unmodified equations of general relativity demand that the structure of the universe be dynamic. Instead of inventing the cosmological constant, Einstein could have *predicted* that the universe must either be expanding or contracting with time. What a coup it would have been for his new theory to successfully predict such an amazing and previously unsuspected result. He called the introduction of his fudge factor, the cosmological constant, the "biggest blunder" of his career as a scientist. With the results on the brightness of Type Ia supernovae, Einstein's cosmological constant turned out to be critical to understanding the expansion of the universe. The repulsive force represented by the cosmological constant in Einstein's equations is just what is needed to describe a universe that is expanding at an ever-accelerating rate. Today, this constant is written as Ω_Λ, pronounced "omega sub lambda" (**Process of Science Figure**).

How does the possibility of a nonzero value for Ω_Λ affect the possible fate of the universe? If Ω_Λ is not zero, the fate of the universe is no longer controlled exclusively by Ω_{mass}, but rather is affected by both Ω_Λ and Ω_{mass}. If something is effectively pushing outward from within the universe, adding to its expansion, then gravity will have a harder time turning the expansion around. In that case, the mass needed to halt the expansion of the universe will be greater than the critical mass we already discussed. **Figure 22.3** shows plots of the scale factor R_U versus time that are similar to those in Figure 22.1, but now the effects of a nonzero cosmological constant are included. If Ω_{mass} is very large, gravity will stop the expansion and the universe will collapse back on itself, regardless of whether Ω_Λ is zero. In contrast, if Ω_{mass} is less than one, and therefore not large enough to stop the expansion and thus the universe expands forever, whether the expansion of the universe will accelerate or decelerate will depend on whether Ω_Λ is zero.

When the universe was young and compact, gravity was strong enough to dominate the effect of the cosmological constant. As the universe expanded, gravity grew weaker and weaker because the mass was more and more spread out. Unlike gravity, however, the effect of the cosmological constant has become increasingly *greater*. Unless gravity is able to turn the expansion around, the cosmological constant will win in the end, causing the expansion to continue accelerating forever. Indeed, even for a universe with Ω_{mass} greater than one, so that otherwise it would have collapsed back on itself because of gravity, a large enough cosmological constant could make that same universe expand forever (dashed blue line in Figure 22.3). One of the first scientists to work with the cosmological constant was Willem de Sitter (1872–1934). He provided an early cartoon (**Figure 22.4**) showing that it is the cosmological constant that drives the expansion of the universe.

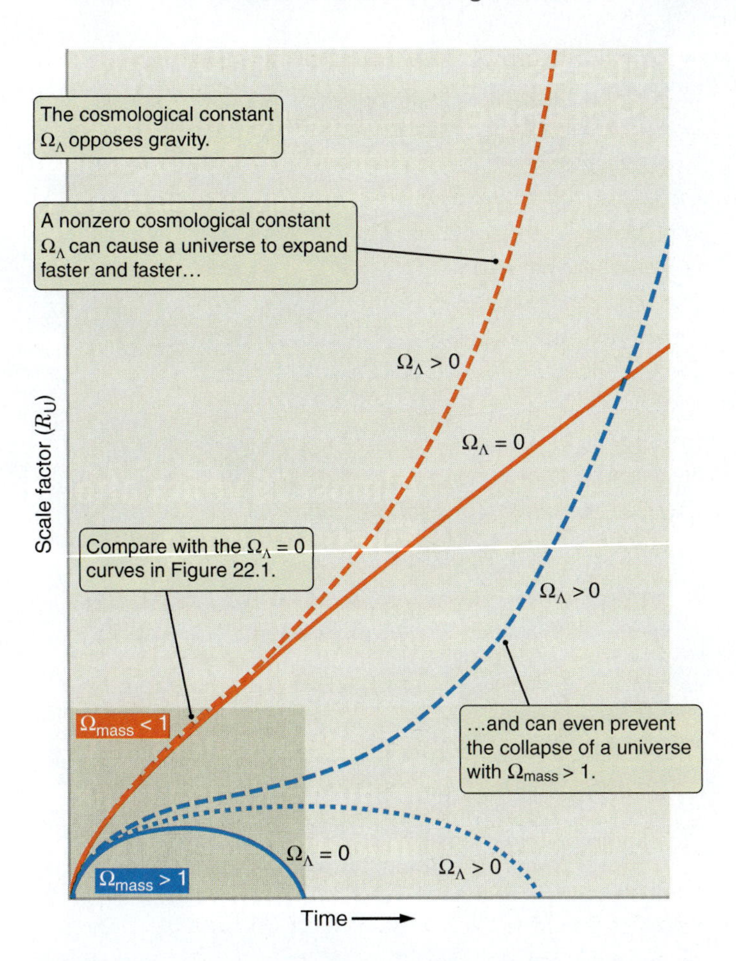

FIGURE 22.3 Plots of scale factor R_U versus time for a possible universe with or without a cosmological constant, Ω_Λ. If there is enough mass in a universe, gravity could still overcome the cosmological constant and cause that universe to collapse. In the presence of a cosmological constant, any universe without enough mass to eventually collapse will instead end up expanding at an ever-increasing rate.

Process of Science

NEVER THROW ANYTHING AWAY

The accelerating universe is a good example of a scientist being right for the wrong reason. Einstein's famous mathematical device has become useful in a way he never would have imagined.

Almost 75 years after Hubble proves Einstein wrong . . .

. . . astronomers have been working hard to measure the expansion rate of the universe. They discover the rate is accelerating!

Einstein's "fudge factor" is revived as the cosmological constant, which accounts for the acceleration.

Science has a memory that revisits old ideas in the light of new data, often saving a great deal of effort, even when the interpretation has changed.

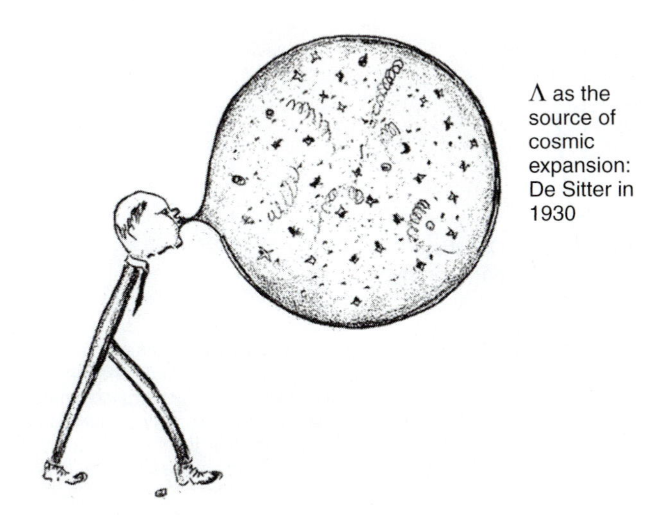

FIGURE 22.4 Cartoon from physicist Willem de Sitter showing that the cosmological constant is the force blowing up the balloon that is the universe. Note that the cartoon figure looks like lambda.

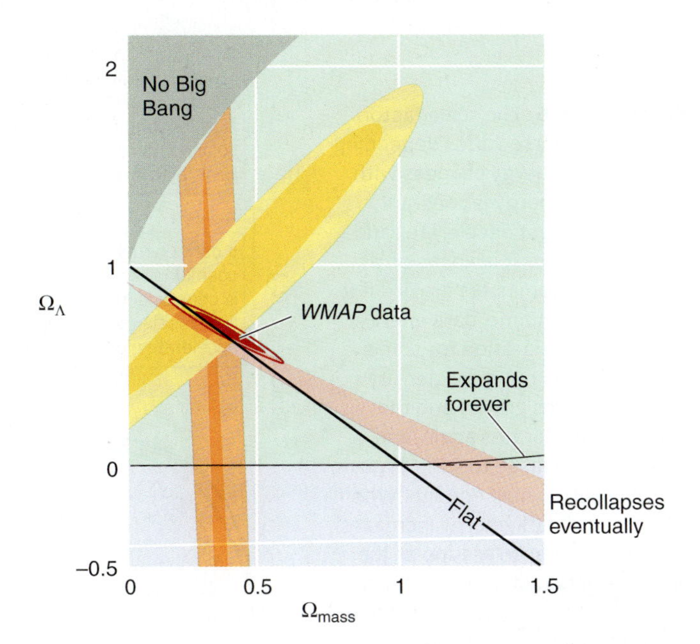

FIGURE 22.5 Current observations from different sources—Type Ia supernovae (yellow), measurements of mass in galaxies and clusters of galaxies (orange), and detailed observations of the structure of the cosmic background radiation (pink and bright red)—suggest that the best current estimates are about 0.7 for Ω_Λ and about 0.3 for Ω_{mass}, which means that the expansion of the universe is accelerating

When Einstein added the cosmological constant to his equations of general relativity, he introduced what he considered a new fundamental constant, in many ways similar to Newton's universal gravitational constant *G*. Physicists call "empty space" the **vacuum**, and it has some distinct physical properties. For example, the vacuum can have energy even in the total absence of matter. Called **dark energy**, this energy produces exactly the same kind of repulsive force that Einstein's cosmological constant does. To astronomers, therefore, the terms *dark energy* and *cosmological constant* mean the same thing.

What fate, then, actually awaits the universe? **Figure 22.5** shows the range of possible values for Ω_{mass} and Ω_Λ that are allowed by current observations. Each colored region represents the data from a different experiment. Because the values of Ω_{mass} and Ω_Λ outside of these regions are ruled out by these experiments, the allowed values must lie within the area on the graph where all of these regions overlap. The data from Type Ia supernovae, from *WMAP*, and from clusters of galaxies (which will be discussed in Chapter 23) are all consistent with values for Ω_{mass} and Ω_Λ of about 0.3 and 0.7, respectively. Thus, it appears that the expansion of the universe is accelerating under the dominant effect of dark energy, and has been doing so for 5–6 billion years.

The universe appears to be accelerating and will expand forever.

Before we conclude this discussion, there is one more speculative idea to consider: What if the cosmological constant is not constant with time? Einstein introduced the cosmological constant as a true constant. But since scientists do not yet understand the origin of dark energy, it is possible that it is not really a constant of nature and instead could be either increasing or decreasing with time.

A changing cosmological constant would significantly change the future of the universe, as illustrated in **Figure 22.6**. For example, if dark energy were to decrease rapidly enough with time, the accelerating expansion of the universe that is observed now would change to a *deceleration* as the mass once again dominates over dark energy. In fact, if the universe were much denser than measured values indicate, the expansion could reverse and the universe could collapse to what astronomers call the **Big Crunch**. By contrast, if the effect of dark energy were to increase with time, the universe would accelerate its expansion at an ever-increasing rate. Ultimately, expansion could be so rapid that the scale factor would become infinite within a finite period of time—a phenomenon called the **Big Rip**. In the Big Rip, the repulsive force of dark energy would become so dominant that the entire universe would come apart. First, gravity would no longer keep groups of galaxies together; then gravity would no longer be able to hold individual galaxies together; and so on. Just before the end, the Solar System would come apart, and even atoms would be ripped into their constituent components. But don't worry too much about the Big Crunch or the Big Rip; the best observational data seem consistent with constant dark energy.

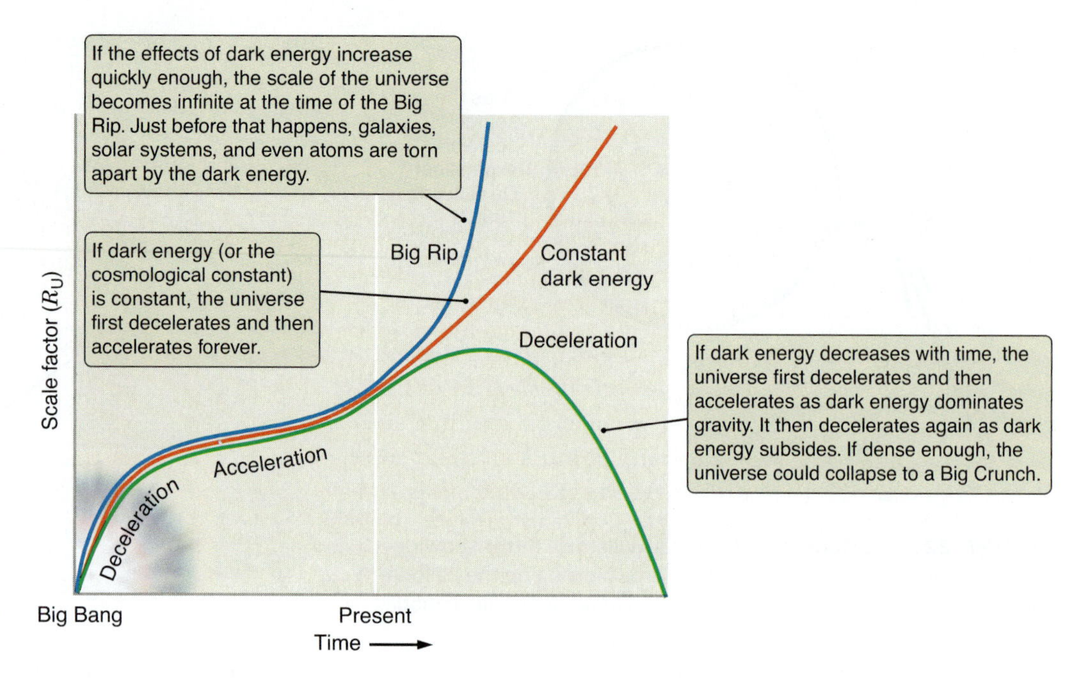

FIGURE 22.6 The scale factor R_U of the universe varies depending on how dark energy changes with time. If dark energy is constant, the universe first decelerates and then accelerates its expansion as dark energy dominates gravity. If it decreases with time, dark energy can cause acceleration for some period of time, but mass density later dominates gravity and the universe decelerates again, and could even collapse to a Big Crunch if the density is high enough. When the effect of dark energy increases with time, the acceleration of the universe gets faster and faster, and the scale factor can become infinite at a time called the Big Rip.

The Age of the Universe

The values for Ω_{mass} and Ω_Λ not only affect predictions for the future of the universe; they also influence how astronomers interpret the past. **Figure 22.7** shows plots of the scale factor of the universe versus time. Measurement of the Hubble constant (H_0) indicates how fast the universe is expanding *today*. That is, it indicates the slope of the curves in Figure 22.7 at the current time. If the expansion of the universe has not changed with time, then the plot of R_U versus time is the straight red line in Figure 22.7. The age of the universe in this case is equal to the Hubble time: $1/H_0$. If the expansion of the universe has been slowing down with time (green line in Figure 22.7), then the universe is actually younger than the Hubble time. (The curve crosses $R_U = 0$ at a point more recent than $1/H_0$.) If the expansion of the universe has been speeding up with time (blue line), then the true age of the universe is greater than the Hubble time.

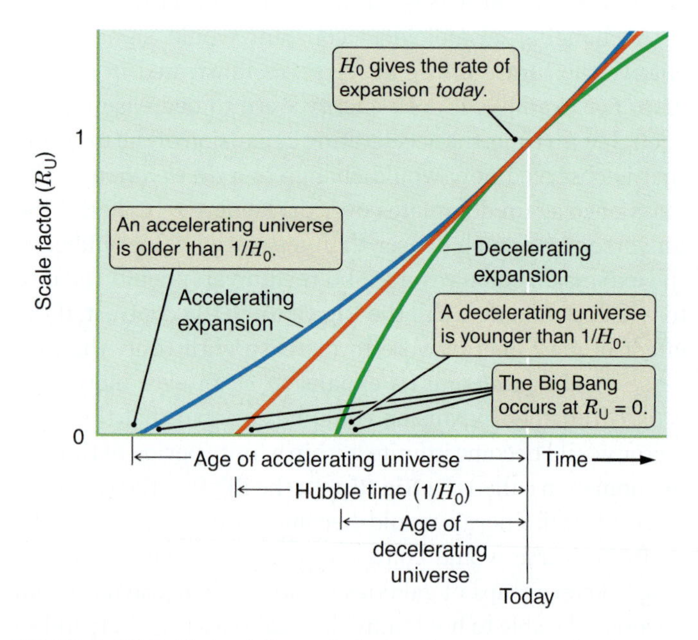

FIGURE 22.7 Plots of the scale factor R_U versus time for three possible universes. If the universe has expanded at a constant rate, then its age is equal to the Hubble time, $1/H_0$. If the expansion of the universe has slowed with time, then the universe is younger than the Hubble time. If the expansion has sped up with time, the universe is older than $1/H_0$.

A Hubble constant of $H_0 = 70$ km/s/Mpc corresponds to a Hubble time ($1/H_0$) of about 13.7 billion years (see Math Tools 19.3). If expansion of the universe has slowed over time, the universe is actually younger than 13.7 billion years. Having a younger universe is a problem if the measured ages of globular clusters—13 billion years—is correct. (Globular clusters clearly cannot be older than the universe that contains them.) But if the expansion of the universe has sped up with time, as suggested by the observations of Type Ia supernovae and by *WMAP*, then the universe is about 13.7 billion years old—comfortably older than globular clusters. The effects from gravity and dark energy have nearly canceled out at the present time.

The Universe Has a Shape

We have already discussed such properties of the universe as density, dark energy, and age. The universe also has another key property: its shape in spacetime. Recall the

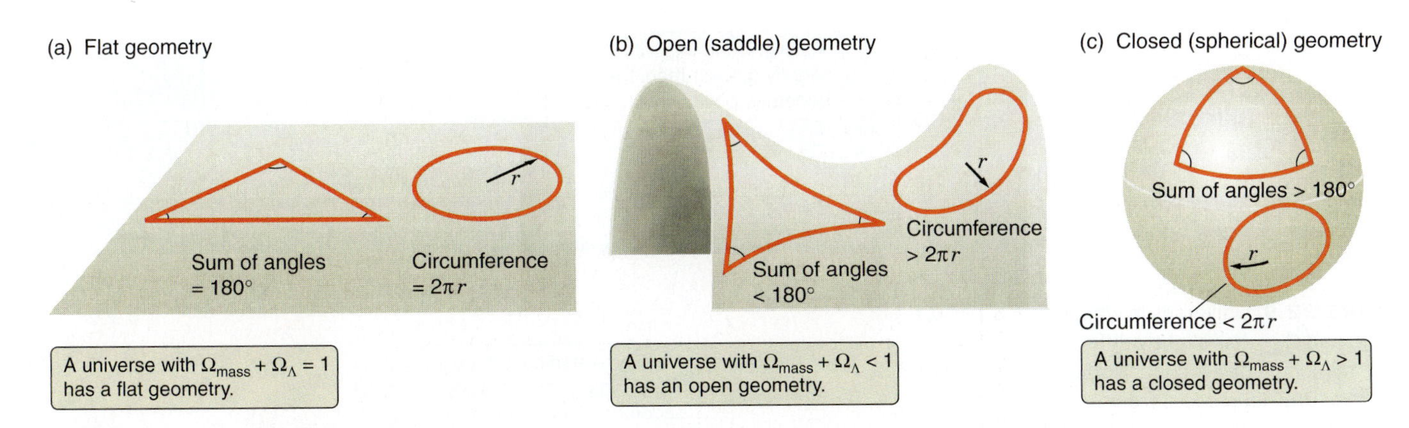

FIGURE 22.8 Two-dimensional representations of the possible geometries that space can have in a universe. In a flat universe (a), Euclidean geometry holds: triangles have angles that sum to 180°, and the circumference of a circle equals 2π times the radius. In an open universe (b) or a closed universe (c), these relationships are no longer correct over very large distances.

concept of spacetime as described by general relativity (see Chapter 18). Space is a "rubber sheet" that has stretched outward from the Big Bang. In Chapter 18 you saw that the rubber sheet of space is also curved by the presence of mass. You saw how the shape of space around a massive object is detected through changes in geometric relationships, such as the ratio of the circumference of a circle to its radius, or the sum of the angles in a triangle. Recall that the mass of a star, planet, or black hole causes a distortion in the shape of space; similarly, the mass of everything in the universe—including galaxies, dark matter, and dark energy—distorts the shape of the universe *as a whole*.

Three basic shapes are possible for the universe. Which shape actually describes the universe is determined by the total amount of mass and energy—in other words, the sum of Ω_{mass} and Ω_Λ. Continuing with the rubber-sheet analogy, the first possibility, corresponding to $\Omega_{mass} + \Omega_\Lambda = 1$, is that the universe is a **flat universe**. A flat universe is described overall by the rules of the basic Euclidean geometry that you learned in high school. As shown in **Figure 22.8a**, circles in a flat universe have a circumference of 2π times their radius ($2\pi r$), and triangles contain angles whose sum is 180°. A flat universe stretches on forever.

The second possibility is that the universe is shaped something like the surface of a saddle (**Figure 22.8b**). This type of universe, in which $\Omega_{mass} + \Omega_\Lambda < 1$, is also infinite and is called an **open universe**. In an open universe the circumference of a circle is greater than $2\pi r$, and triangles contain less than 180°.

The third possibility, in which $\Omega_{mass} + \Omega_\Lambda > 1$, is a universe shaped like the surface of a sphere (**Figure 22.8c**). The geometric relationships on a sphere are similar to those in the vicinity of a massive object, as discussed in Chapter 18. The circumference of a circle on a sphere is less than ($2\pi r$), and triangles contain more than 180°. This possibility is called a **closed universe** because space is finite and closes back on itself.

The measurements to directly estimate which of these shapes describes the universe are difficult. Even so, as seen in Figure 22.5, $\Omega_{mass} + \Omega_\Lambda$ is close to 1 (0.3 + 0.7), meaning that the universe is very nearly flat.

Evidence suggests that the universe is remarkably flat.

22.3 Inflation

A century ago, astronomers were struggling just to understand the size of the universe. Today scientists have a comprehensive theory that ties together many diverse facts about nature: the constancy of the speed of light, the properties of gravity, the motions of galaxies, and even the origins of the atoms that make up planets and life. The case for the Big Bang is compelling. Even so, improved observations of the cosmic background radiation and measurements of the expansion of the universe have revealed a number of puzzles. To address these observations, cosmologists have had to consider some remarkable ideas about how the universe expanded when it was very young.

The Universe Is Much Too Flat

The first puzzle that astronomers had to deal with is the observation that the universe is so close to being exactly

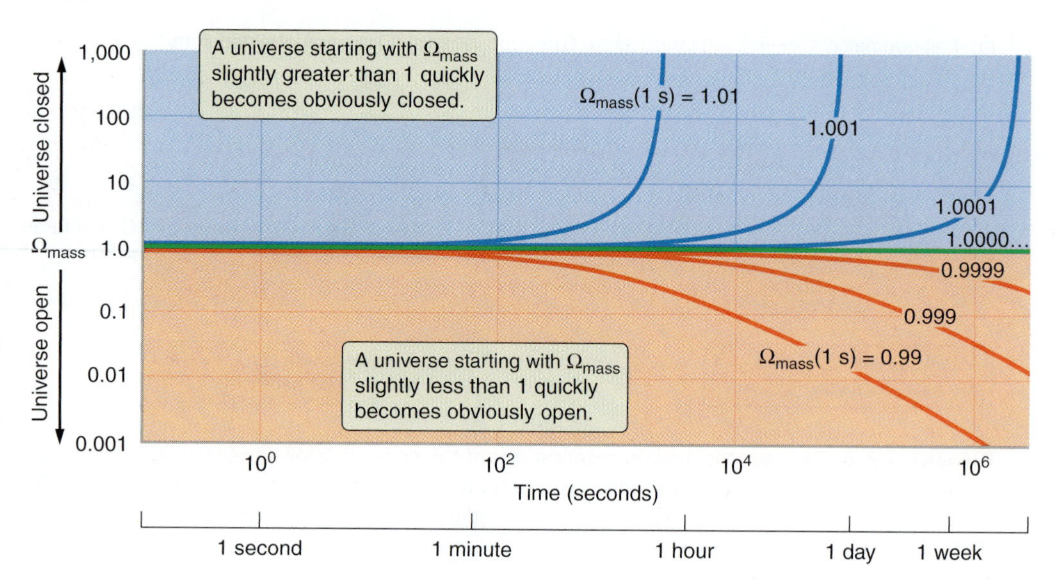

FIGURE 22.9 Universes in which $\Omega_\Lambda = 0$ and Ω_{mass} has slightly different values at an age of 1 second. Notice that even tiny differences from $\Omega_{mass} = 1$ are rapidly amplified as the model universe expands. To have a value of Ω_{mass} so close to 1 today, the universe must have started out with a value of Ω_{mass} exquisitely close to 1.

flat that it is too improbable to have happened by chance. To see why this is a problem, imagine a universe in which $\Omega_\Lambda = 0$. (This is a reasonable approximation for the very early days of *any* possible universe. When a universe is very young, it is also very dense, and Ω_{mass} is all that matters.) As this universe expands out of its own version of the Big Bang, its density falls. At the same time, because this universe is expanding more slowly, the critical density needed to eventually stop the expansion falls as well. If Ω_{mass} is *exactly* 1, the decline in the actual density and the decline in the critical density go together: the ratio between the two, Ω_{mass}, remains 1 for all time, as shown by the green middle curve in **Figure 22.9**. A universe that starts out perfectly flat *stays* perfectly flat.

A universe that does *not* start out perfectly flat has a very different fate. If a universe started out with Ω_{mass} even slightly greater than 1, its expansion would slow more rapidly than that of the flat universe, meaning that less and less density would be required to stop the expansion. At the same time, the actual density would be falling less rapidly than in the flat universe. This disparity between the actual density of the universe and the critical density would increase, causing the ratio between the two, Ω_{mass}, to skyrocket. This condition is illustrated by the blue curves that climb toward the top of Figure 22.9. A universe that starts out even slightly closed rapidly becomes obviously closed, and would collapse long before stars could form.

Conversely, if a universe started with Ω_{mass} even a tiny bit less than 1, the expansion would slow less rapidly than in a flat universe. As time passed, more and more mass would be required for gravity to stop the too-rapidly-expanding universe. At the same time, the actual density of the universe would be dropping faster than in a flat universe. In this case, Ω_{mass} (the ratio between the actual density and the critical density) would plummet, leading to the red curves that dive toward the bottom of Figure 22.9.

Adding Ω_Λ to the picture makes the math a bit more complex, but it does not change the basic results. Try balancing a razor blade on its edge. If the blade is tipped just a tiny bit in one direction, it quickly falls that way. If the blade is tipped just a tiny bit in the other direction, it quickly falls in the other direction instead. It would seem that the actual universe should be either obviously open or obviously closed—analogous to the tipped razor blade. Instead the universe has $\Omega_{mass} + \Omega_\Lambda$ so close to 1 that it is difficult to tell which way the razor blade is tipped at all. Discovering that $\Omega_{mass} + \Omega_\Lambda$ is extremely close to 1 after more than 13 billion years is like balancing a razor blade on its edge and coming back 10 years later to find that it still has not tipped over.

For the present-day value of $\Omega_{mass} + \Omega_\Lambda$ to be as close to 1 as it is, $\Omega_{mass} + \Omega_\Lambda$ could not have differed from 1 by more than one part in 100,000 when the universe was 2,000 years old. When the universe was 1 second old, it had to be flat by at least one part in 10 billion. At even earlier times, it had to be much flatter still. In cosmology this is called the **flatness problem**, because this is too special a situation to be the result of chance. Something about the early universe must have *forced* $\Omega_{mass} + \Omega_\Lambda$ to have a value incredibly close to 1.

The Cosmic Background Radiation Is Much Too Smooth

The second problem faced by cosmological models is that the cosmic microwave background radiation (CMB—see Chapter 19) is surprisingly smooth. Following the discovery of the CMB in the 1960s, many observational cosmologists turned their attention to mapping this background glow.

At first, result after result showed that the temperature of the CMB is remarkably constant, regardless of where one looks in the sky. Yet over time this strong confirmation of Big Bang cosmology turned instead into a puzzle that challenged cosmologists' view of the early universe. Once Earth's motion relative to the CMB is removed from the picture, the CMB is not just smooth—it is *too* smooth.

The early universe was subject to the quantum mechanical **uncertainty principle**. In Chapter 5 we discussed the bizarre world of quantum mechanics that shapes the world of atoms, light, and elementary particles. When the universe was extremely young, it was so small that quantum mechanical effects played a role in shaping the structure of the universe as a whole. The uncertainty principle says that as a system is studied at extremely small scales, the properties of that system become less and less well determined. This principle applies to the properties of an electron in an atomic orbital, or to the entire universe at the time when it would have fit within the size of an atom.

The early universe was nonuniform.

Consider a simple analogy of how the uncertainty principle applies to the universe. Imagine sitting on the beach looking out across the ocean. Off in the distance, you see more total ocean and average out over larger scales; therefore, the surface of the ocean appears smooth and flat. The horizon looks almost like a geometric straight line. Yet the apparent smoothness of the ocean as a whole hides the tumultuous structure present at smaller scales, where waves and ripples fluctuate dramatically from place to place. Similarly, quantum mechanics says that at smaller and smaller scales in the universe, conditions *must* fluctuate in unpredictable ways. In particular, quantum mechanics says that the smaller the universe is—that is, the earlier in the history of the universe—the more dramatic those fluctuations are. When the universe was young, it could *not* have been smooth. There must have been dramatic variations ("ripples") in the density and temperature of the universe from place to place.

If the universe had expanded slowly, those ripples would have smoothed themselves out. But the universe expanded much too rapidly for such smoothing to be possible. Different parts of the universe could not have "communicated" with each other rapidly enough. There just wasn't enough time after the Big Bang for a smoothing signal to travel from one region to the other. So when cosmologists look at the universe today, they should see the fingerprint of those early ripples imprinted on the cosmic background radiation—but they do not. The fact that the CMB is so smooth is called the **horizon problem** in cosmology. The horizon problem states that different parts of the universe are too much like other parts of the universe that should have been "over their horizon" and beyond the reach of any signals that might have smoothed out the early quantum fluctuations. Basically, the horizon problem is this: How can different parts of the universe that underwent different fluctuations and were never able to communicate with one another still show the same temperature in the cosmic background radiation to an accuracy of better than one part in 100,000?

The CMB is smoother than the early universe should have been.

Inflation Solves the Problems

In the early 1980s, physicist Alan Guth (1947–) offered a solution to the flatness and horizon problems of cosmology. Guth suggested that for a brief time the young universe expanded at a rate *far* in excess of the speed of light in a very short period of time. This rapid expansion is called **inflation**. In the first 10^{-33} second of the universe, the scale factor R_U increased by a factor of at least 10^{30} and perhaps very much more. In that incomprehensibly brief instant, the size of the observable universe grew from ten-trillionths the size of the nucleus of an atom to a region about 3 meters across. That is like a grain of very fine sand growing to the size of today's *entire universe*—all in a billionth the time it takes light to cross the nucleus of an atom. During inflation, space itself expanded so rapidly that the distances between points in space increased faster than the speed of light. Inflation does not violate the rule that no signal can travel through space at greater than the speed of light because the space itself was expanding.

Inflation was a period of intense expansion of the universe.

To understand how inflation solves the flatness and horizon problems of cosmology, imagine that you are an ant living in the two-dimensional universe defined by the surface of a golf ball (**Figure 22.10**). Your universe would have two very apparent characteristics: First, it would be obviously positively curved. If you were to walk around the circumference of a circle in your two-dimensional universe and then measure the radius of the circle, you would find the circumference to be less than $2\pi r$. If you were to draw a triangle in your universe, the sum of its angles would be greater than 180°. The second obvious characteristic would be the dimples, approximately a half millimeter deep, on the surface of the golf ball.

Now imagine that the golf-ball universe suddenly grew to the size of Earth. First, the curvature of the universe would no longer be apparent. An ant walking along the surface of Earth would be hard-pressed to tell that Earth

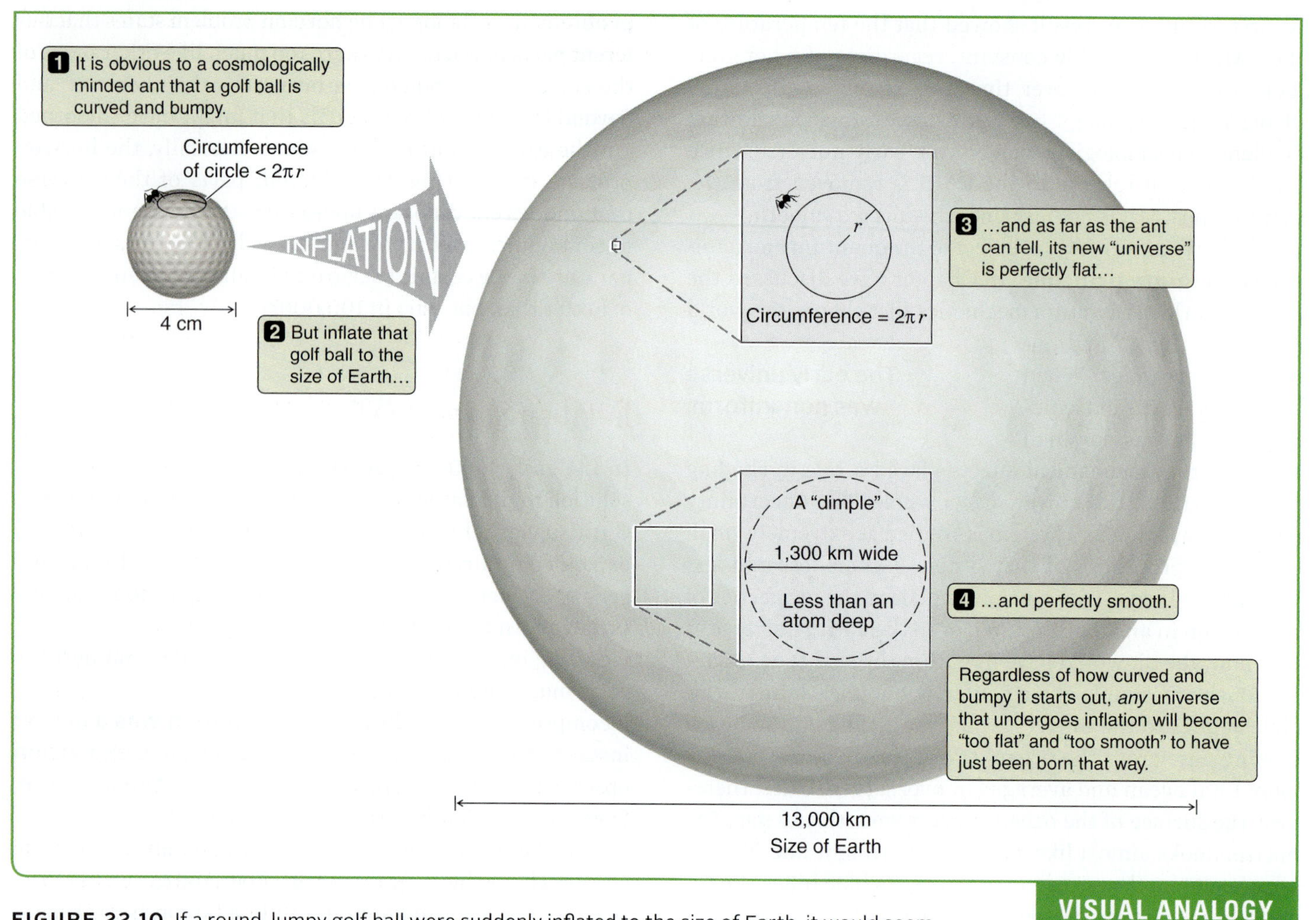

FIGURE 22.10 If a round, lumpy golf ball were suddenly inflated to the size of Earth, it would seem extraordinarily flat and smooth to an ant on its surface. Similarly, following inflation, any universe would seem both extremely flat and extremely smooth, regardless of the exact geometry and irregularities it started with.

is not flat. The circumference of a circle would be $2\pi r$, and there would be 180° in a triangle. In fact, it took most of human history to realize that Earth is a sphere. In the case of inflationary cosmology, the universe after inflation would be extraordinarily flat (that is, having $\Omega_{mass} + \Omega_{\Lambda}$ extraordinarily close to 1) *regardless* of what the geometry of the universe was before inflation. Because the universe was inflated by a factor of at least 10^{30}, $\Omega_{mass} + \Omega_{\Lambda}$ immediately after inflation must have been equal to 1 within one part in 10^{60}, which is flat enough for $\Omega_{mass} + \Omega_{\Lambda}$ to remain close to 1 today. Today's universe is not flat by chance. It is flat because *any* universe that underwent inflation would become flat.

What about the horizon problem? When the golf-ball universe inflates to the size of Earth, the dimples that covered the surface of the golf ball stretch out as well. Instead of being a half millimeter or so deep and a few millimeters across, these dimples now are only an atom deep but are hundreds of kilometers across. The ant would be hard-pressed to detect any dimples at all. In the case of the real universe, inflation took the large fluctuations in conditions caused by quantum uncertainty in the preinflationary universe and stretched them out so much that they are not measurable in today's postinflationary local universe. The slight irregularities observed in the CMB (see Figure 19.18c) are the faint ghosts of quantum fluctuations that occurred as the universe inflated.

Huge expansion smooths out inhomogeneities.

An early era of inflation in the history of the universe offers a way of solving the horizon and flatness problems, but it seems quite remarkable that the universe should undergo a period during which it expanded at such a high rate. The cause of inflation lies in the fundamental physics that governed the behavior of matter and energy at the earliest moments of the universe.

22.4 The Earliest Moments

To understand the universe requires an understanding of the forces that govern the behavior of all matter and energy in the universe. There are four fundamental forces in nature, and everything in the universe is a result of their action (**Table 22.1**). Chemistry and light are products of the **electromagnetic force** acting between protons and electrons in atoms and molecules. The energy produced in fusion reactions in the heart of the Sun comes from the **strong nuclear force** that binds together the protons and neutrons in the nuclei of atoms. Beta decay of nuclei, in which a neutron decays into a proton, an electron, and an antineutrino, is governed by the **weak nuclear force**. Finally, there is **gravity**, which has played such a major role throughout astronomy. How these forces—these physical laws—originated and evolved is part of the history of the universe.

There are four fundamental forces in nature.

The Forces of Nature

Recall from Chapter 5 the description of light as an electromagnetic wave resulting from electric and magnetic fields, and the quantum mechanical description of light as a stream of particles called photons. These descriptions of electromagnetism have to coexist. The branch of physics that deals with this reconciliation is called **quantum electrodynamics**, or **QED**.

QED treats charged particles almost as if they were baseball players engaged in an endless game of catch. As baseball players throw and catch baseballs, they experience forces. Similarly, in QED, charged particles "throw" and "catch" an endless stream of "virtual photons" (**Figure 22.11**). Earlier we discussed the idea that quantum mechanics is a science of probabilities rather than certainties. The QED description of the electromagnetic interaction between two charged particles is an average of all the possible ways that the particles could throw photons back and forth. The resulting force acts, over large scales, like the classical electric and magnetic fields described by Maxwell's equations. Physicists describe this as the electromagnetic force being mediated by the exchange of photons. As with the quantum mechanics we discussed in Chapter 5, the world described by QED is hard to picture; but QED is one of the most accurate, well-tested, and precise branches of physics. Not even the tiniest measurable difference between the predictions of the theory and the outcome of an actual experiment has been found.

In QED, photons carry the electromagnetic force.

The central idea of QED—forces mediated by the exchange of carrier particles—provides a template for understanding two of the other three fundamental forces in nature. The electromagnetic and weak nuclear forces have been combined into a single theory called **electroweak theory**. This theory predicts the existence of three particles—labeled W^+, W^-, and Z^0—that mediate the weak nuclear force. Sheldon Glashow, Abdus Salam, and Steven Weinberg received the 1979 Nobel Prize in Physics for their work on the theory of the unified weak and electromagnetic forces. In the 1980s, physicists identified these particles in laboratory experiments and confirmed the essential predictions of electroweak theory.

The weak nuclear and electromagnetic forces combine in electroweak theory.

The strong nuclear force is described by a third theory, called **quantum chromodynamics**, or **QCD**. This theory

TABLE 22.1 | The Four Fundamental Forces of Nature

Force	Relative Strength	Range of Force	Particles That Can Carry the Force	Example of What the Force Does
Strong nuclear	1	10^{-15} m	Gluons	Holds protons and neutrons together in atomic nuclei.
Electromagnetic	10^{-2}	Infinite	Photons	Binds the electrons in an atom to the nucleus.
Weak nuclear	10^{-4}	10^{-16} m	W^+, W^-, and Z^0	Responsible for beta decay.
Gravitational	10^{-38}	Infinite	Gravitons	Holds you to Earth; binds together planetary systems, stars, galaxies, clusters of galaxies, etc.

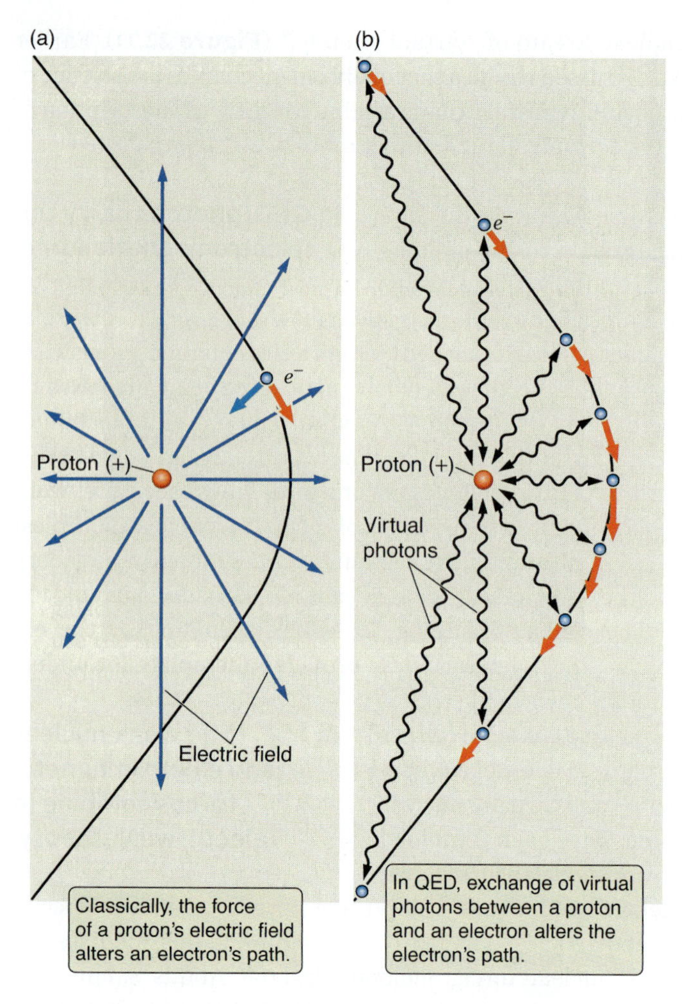

FIGURE 22.11 (a) The classical view of an electron being deflected from its course (black line) by the electric field from a proton. (b) According to quantum electrodynamics, the interaction is properly viewed as an ongoing exchange of virtual photons between the two particles.

states that particles such as protons and neutrons are composed of more fundamental building blocks called **quarks**, which are bound together by the exchange of another type of carrier particle, dubbed **gluons**. Together, electroweak theory and QCD comprise the **standard model** of particle physics. A deeper investigation of the standard model is beyond the scope of this textbook. Here we conclude the discussion by pointing out that, excluding gravity, the standard model is able to explain all the currently observed interactions of matter and has made many predictions that were subsequently confirmed by laboratory experiments. However, the standard model leaves many questions unanswered, such as whether neutrinos have mass, or why strong interactions are so much stronger than weak interactions.

Electroweak theory + QCD = the standard model of particle physics.

A Universe of Particles and Antiparticles

Every type of particle in nature has an **antiparticle** that is its opposite. A positron is identical to an electron except that it has a positive charge instead of a negative charge. (You saw the electron's antiparticle, the positron, in nuclear reactions within the Sun in Chapter 14.) For the proton there is the antiproton; for the neutron, the antineutron; and so on down the list. Collectively these antiparticles are called **antimatter**.

One property of these particle-antiparticle pairs is that if you bring such a pair together, the two particles annihilate each other. When a particle-antiparticle pair annihilates, the mass of the two particles is converted into energy in accord with Einstein's special theory of relativity ($E = mc^2$). For example, in **Figure 22.12a** an electron and a positron annihilate each other, and the energy is carried away by a pair of gamma-ray photons. (This is the idea behind *Star Trek*'s "antimatter" engines.) Particle-antiparticle pairs were produced when two high-energy photons collided with each other, as in **Figure 22.12b**, creating in their place an electron-positron pair. This example of how an energetic event can create a particle and its corresponding antiparticle—a process called **pair production**—has been observed in particle accelerators.

In principle, *any* type of particle and its antiparticle can be created in this way. The only limitation comes when there is not enough energy available to supply the mass of the particles being created (**Math Tools 22.2**). If two gamma-ray photons with a combined energy greater than the rest mass energy of an electron-positron pair collide, then the two photons may disappear and leave an electron-positron pair behind in their place. If the photons have more than

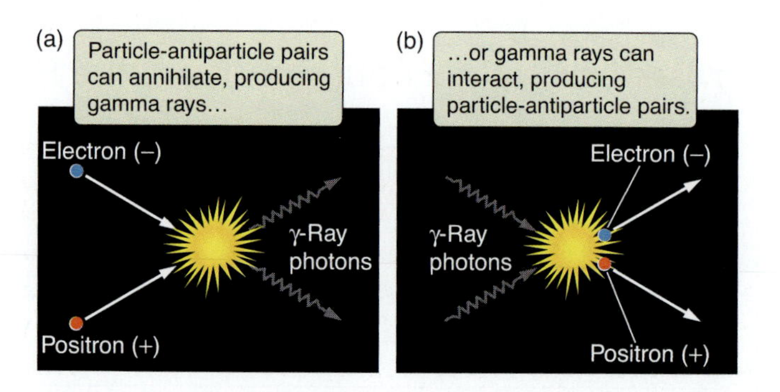

FIGURE 22.12 (a) An electron and a positron annihilate, creating two gamma-ray photons that carry away the energy of the particles. (b) In the reverse process, pair production, two gamma-ray photons collide to create an electron-positron pair.

Math Tools 

Pair Production in the Early Universe

The early universe was hot and awash in a bath of Planck blackbody radiation. How did this radiation become particles? Recall Einstein's equation for the conversion of mass to energy and energy to mass:

$$E = mc^2$$

To produce a pair consisting of a particle plus an antiparticle of a certain rest mass, a minimum amount of energy is required. The formula shows that producing a higher-mass particle plus antiparticle requires a greater amount of energy than does producing a lower-mass pair. For example, a proton has 1,836 times the mass of an electron, so it will require 1,836 times as much energy to produce a proton-antiproton pair of particles than to produce an electron-antielectron pair.

We can relate this amount of energy to the average energy of particles at a given temperature through the following equation:

$$E = 3/2\ kT$$

where k is Boltzmann's constant: 1.38×10^{-23} joule per kelvin (J/K) or, substituting 1 kg m^2/s^2 for each joule, 1.38×10^{-23} kg m^2/s^2/K. Equating the two energies yields:

$$mc^2 = 3/2\ kT$$

or

$$T = \frac{2mc^2}{3k}$$

Let's look at some numbers. The proton and antiproton each have a mass of 1.67×10^{-27} kg. What temperature would the radiation have to be to produce this proton-antiproton pair?

$$T = \frac{2 \times (2 \times 1.67 \times 10^{-27}\ \text{kg}) \times (3.0 \times 10^8\ \text{m/s})^2}{3 \times (1.38 \times 10^{-23}\ \text{kg m}^2/\text{s}^2/\text{K})}$$

$$= 1.45 \times 10^{13}\ \text{K}$$

Such high temperatures are thought to have existed only during the first few seconds after the Big Bang.

Similarly, to produce an electron-positron pair, each of which has a mass of 9.1×10^{-31} kg, the temperature of the CMB would have to be:

$$T = \frac{2 \times (2 \times 9.11 \times 10^{-31}\ \text{kg}) \times (3.0 \times 10^8\ \text{m/s})^2}{3 \times (1.38 \times 10^{-23}\ \text{kg m}^2/\text{s}^2/\text{K})}$$

$$= 7.92 \times 10^9\ \text{K}$$

This is still very hot, but less than the temperature required to form a proton-antiproton pair, so electron-positron production lasted longer after the Big Bang than did proton-antiproton production.

We can also think of this in terms of the energy of the photons involved in creating these particles. From Chapter 5, recall that the energy of a photon is related to its wavelength by:

$$E = \frac{hc}{\lambda} \quad \text{or} \quad \lambda = \frac{hc}{E}$$

where h = Planck's constant = 6.63×10^{-34} kg m^2/s. We then use $E = 3/2\ kT$ with our value of T for the electron-positron production above, yielding:

$$\lambda = \frac{2hc}{3kT} = \frac{2 \times (6.63 \times 10^{-34}\ \text{kg m}^2/\text{s}) \times (3.0 \times 10^8\ \text{m/s})}{3 \times (1.38 \times 10^{-23}\ \text{kg m}^2/\text{s}^2/\text{K}) \times (7.92 \times 10^9\ \text{K})}$$

$$= 1.21 \times 10^{-12}\ \text{m}$$

The electromagnetic spectrum pictured in Figure 5.7 shows that the photons involved in the pair production of electrons and positrons are high-energy gamma-ray photons.

the necessary energy, the extra energy goes into the kinetic energy of the two newly formed particles.

Now we apply this idea to a hot universe awash in a bath of blackbody radiation. When the universe was less than about 100 seconds old and had a temperature greater than a billion kelvins, it was filled with energetic photons that were constantly colliding, creating electron-positron pairs; and these electron-positron pairs were constantly annihilating each other, creating pairs of gamma-ray photons. The whole process reached an equilibrium, determined strictly by temperature, in which pair production and pair annihilation exactly balanced each other. Rather than being filled only with a swarm of photons, at this time the universe was filled with a swarm of photons, electrons, and positrons (**Figure 22.13a**). Earlier, when the universe was even hotter, photons would have produced a swarm of protons and antiprotons. Still earlier, there was a swarm of quarks/antiquarks and gluons called a "quark-gluon plasma," as has been observed in some heavy-nucleus accelerators on Earth.

The early universe was filled with photons, particles, and antiparticles.

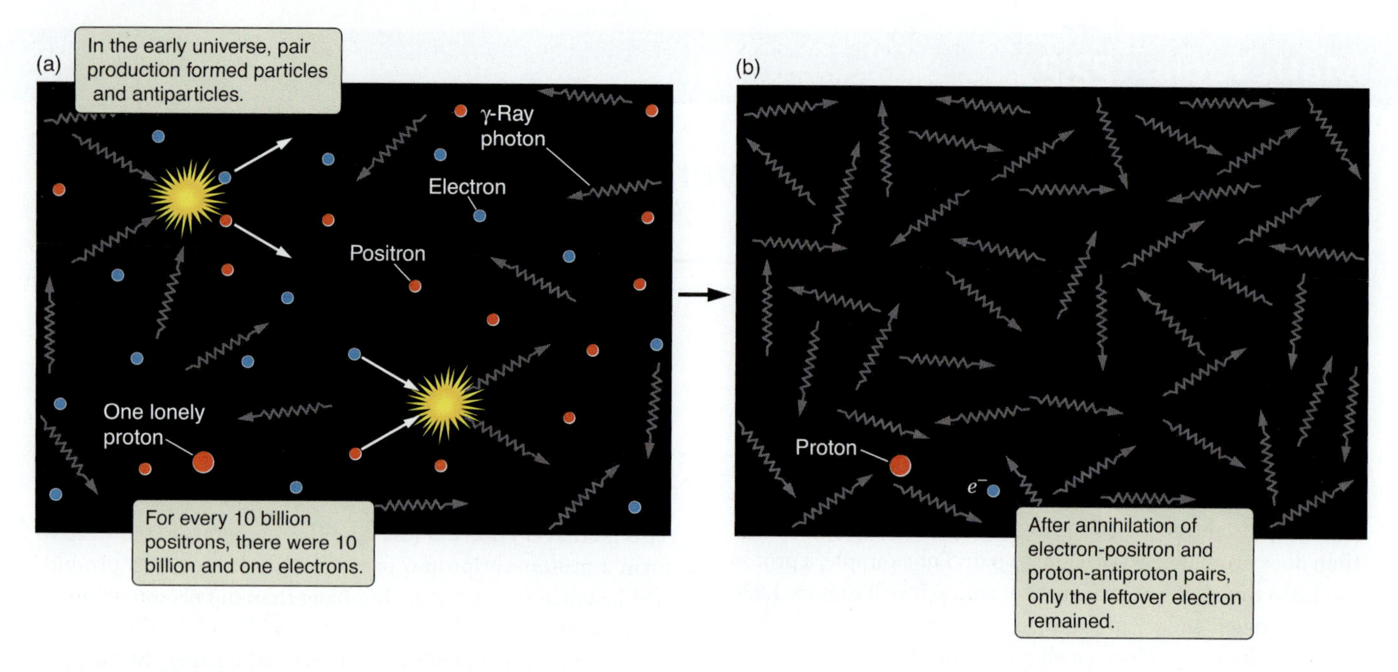

FIGURE 22.13 (a) A swarm of electrons, positrons, and photons in the very early universe. For every 10 billion positrons, there were 10 billion and one electrons. (b) After these particles annihilated, only the one electron was left.

The Frontiers of Physics

In the process of pair production there is a symmetry between matter and antimatter: for every particle created, its antiparticle is created as well. As the universe cooled, there was no longer enough energy to support the production of particle pairs, so the particles and antiparticles forming the swarm that filled the early universe annihilated each other and were not replaced. When this cooling happened, first every proton should have been annihilated by an antiproton. Then at still cooler temperatures, every electron should have been annihilated by a positron. This was almost the case, but not quite. For every electron in the universe today, there were 10 billion and one electrons in the early universe, but only 10 billion positrons. This one-part-in-10-billion excess of electrons over positrons meant that when electron-positron pairs finished annihilating each other, some electrons were left over—enough to account for all the electrons in all the atoms in the universe today (**Figure 22.13b**). Similarly, there was an excess of protons over antiprotons in the early universe, and the protons observed today are all that is left from the annihilation of proton-antiproton pairs.

For every 10 billion positrons there were 10 billion and one electrons in the early universe.

If the standard model of particle physics were a complete description of nature, then the imbalance of one part in 10 billion between matter and antimatter would not have been present in the early universe. The symmetry between matter and antimatter would have been complete. No matter at all would have survived into today's universe, and galaxies, stars, and planets would not exist. The fact that you are reading this page demonstrates that something more needs to be added.

The symmetry between matter and antimatter may be broken in a theory that joins the electroweak and strong nuclear forces in much the same way that electroweak theory unified the electromagnetic and weak nuclear forces. There are theories that combine three of the four fundamental forces into a single grand, unified force are called **grand unified theories**, or **GUTs**. Grand unified theories break the particle/antiparticle symmetry and explain why the universe is composed of matter rather than antimatter. There are several competing GUTs, and only the very simplest of GUTs have been ruled out.

Many possible GUTs exist, and they make many predictions about the universe. Unfortunately, most of those predictions are impossible to test with even the largest of today's particle colliders. The problem is that the particles carrying the forces are so massive that it takes enormous amounts of energy to bring them into existence—roughly a trillion times as much energy as can be achieved in today's particle accelerators. Even so, some predictions of GUTs can be tested with current technology. For example, GUTs predict that protons should be unstable particles that, given enough time, will decay into other types of elementary particles. This is a *very* slow process. Over the course of 100 years, GUTs predict that there may be as much as a 1 percent chance that *one* of the 10^{28} or so protons in your body will decay. Seawater experiments have given a lower limit on the lifetime of a proton of 10^{34} years, but as of this writing, proton decay has yet to be observed.

One of the unresolved questions, the existence of the Higgs boson, may have been answered in 2012 when scientists at the European Organization for Nuclear Research (CERN) announced the discovery of a new particle that seems to be consistent with the Higgs boson. The Higgs boson is the particle that all other particles must interact with to gain their masses. In the standard model, all particles are created without mass. When the electroweak symmetry breaks (as expansion and cooling continue) this "special" particle is created throughout the universe. All existing particles interact with the Higgs boson and gain their mass in this process. The more they interact, the more massive (heavier) they become.

The particles that mediate GUTs may be beyond the reach of today's high-energy physics labs. But when the universe was *very* young (younger than about 10^{-35} second) and *very* hot (hotter than about 10^{27} K), enough energy was available for these particles to be freely created. During this time, the electromagnetic, weak nuclear, and strong nuclear forces had not yet established their unique identities; there was only the one grand, unified force. During this era of GUTs, the apparent size of the entire observable universe was less than a trillionth the size of a single proton. How does gravity fit into this scheme?

General relativity provides a beautifully successful description of gravity that correctly predicts the orbits of planets, describes the collapse of stars down to the radius of the singularity, and enables astronomers to calculate the structure of the universe. Yet general relativity's description of gravity "looks" very different from the theories of the other three forces. Rather than talking about the exchange of photons or gluons or other mediating particles, general relativity talks about the large, smooth, continuous canvas of spacetime that events are painted on. The era of GUTs is described perfectly if gravity is treated as a separate force. Even closer to the moment of the Big Bang, the universe was extremely small, and then both the quantum mechanical and the general relativistic descriptions had to apply.

GUTs unify three forces but not the fourth, gravity.

Toward a Theory of Everything

When the universe was younger than about 10^{-43} second old, its density was incomprehensibly high. The observable universe was so small that 10^{60} universes would have fit into the volume of a single proton. Under these extreme conditions, quantum physics is needed to describe not just particles, but spacetime itself. Rather than a smooth sheet, spacetime was a quantum mechanical "foam." The failure of general relativity to describe this early universe is much like the failure of Newtonian mechanics to describe the structure of atoms. An electron in an atom must be thought of in terms of probabilities rather than certainties. Similarly, there is no certainty for the earliest moments after the Big Bang. This era in the history of the universe is called the **Planck era**, signifying that physicists can understand the structure of the universe during this period only by using the ideas of quantum mechanics.

In the Planck era, the whole universe was a quantum mechanical "foam."

The conflict between general relativity and quantum mechanics is at the current limits of human knowledge. Physics can explain things back to a time when the universe was a millionth of a trillionth of a trillionth of a trillionth of a second old, but to push back any further, something new is needed. To understand the earliest moments of the universe, physicists need a theory that combines general relativity and quantum mechanics into a single theoretical framework unifying all *four* of the fundamental forces. Such a theory is called a **theory of everything** (**TOE**).

A successful theory of everything would do more than unify general relativity with quantum mechanics. It would suggest which of the possible GUTs is correct and would provide an answer for the nature of dark matter. A successful theory of everything would also necessarily answer several outstanding issues in cosmology, including the how, when, and why of inflation, and the underlying physics of the dark energy that is accelerating the expansion of the universe. Physicists are currently grappling with what a TOE might look like. One leading contender is superstring theory, discussed in **Connections 22.1**.

The Forces Separated in the Cooling Universe

To understand the very earliest moments in the history of the universe, physicists look backward to higher and higher energies and correspondingly to earlier and earlier times. The universe started out with one TOE with all four forces united, and as the universe expanded and cooled the

Connections 22.1

Superstring Theory

One leading idea for a theory of everything is **superstring theory**, in which elementary particles are viewed not as points but as tiny loops called "strings." A guitar string vibrates in one way to play an F, another way to play a G, and yet another way to play an A. According to superstring theory, different types of elementary particles are like different "notes" played by vibrating loops of string.

In principle, superstring theory provides a way to reconcile general relativity and quantum mechanics. To make superstring theory work, physicists imagine that these tiny loops of string are vibrating in a universe with 10 spatial dimensions instead of 3. (Adding time to the list would make the universe 11-dimensional.) Whereas the usual 3 spatial dimensions spread out across the vastness of the universe, the other 7 dimensions predicted by superstring theory wrap tightly around themselves (**Figure 22.14**), extending no further today than they did a brief instant after the Big Bang.

To better appreciate this bizarre notion, imagine what it would be like to live in a three-dimensional universe (like the one you experience) in which one of those dimensions extended for only a tiny distance. Living in such a universe would be like living within a thin sheet of paper that extended billions of parsecs in two directions but was far smaller than an atom in the third. In such a universe you would easily be aware of length and width; you could move in those directions at will. In contrast, you would have no freedom to move in the third dimension at all, and you might not even recognize that the third dimension existed. Perhaps your only inkling of the true nature of space would come from the fact that in order to explain the results of particle physics experiments, you would have to assume that particles extended into a third, unseen dimension. If superstring theory is correct, everyone now sees three spatial dimensions extending on possibly forever, but is unaware of the fact that each point in this three-dimensional space actually has a tiny but finite extent in seven other dimensions at the same time.

Superstring theory is only a pale shadow of the sort of well-tested theories that have been discussed throughout this book. In some respects, superstring theory is no more than a promising idea providing direction to theorists searching for a TOE. Physicists will probably never be able to build particle accelerators that enable them to directly search for the most fundamental particles predicted by a TOE; the energies required are simply too high. Fortunately, however, nature has provided the ultimate particle accelerator: the Big Bang itself. The structure of the universe that you see today is the observable result of that grand experiment.

FIGURE 22.14 It is virtually impossible to visualize what six spatial dimensions wrapped up into structures far smaller than the nucleus of an atom would be like. Here such geometries are projected onto the two-dimensional plane of the paper.

various forces emerged separately. **Figure 22.15** illustrates how the four fundamental forces emerged in the evolving universe. In the first 10^{-43} second after the Big Bang, as described by the TOE, the physics of elementary particles and the physics of spacetime were one and the same. As the universe expanded and cooled, gravity broke away and separated from the forces described by the GUT. Spacetime took on the properties described by general relativity. Inflation may also have been taking place at this time.

As the universe continued to expand and its temperature dropped further, less and less energy was available for the creation of particle-antiparticle pairs. When the particles responsible for mediating GUT interactions could no longer form, the strong force split off from the electroweak force. During this time, as the unity of the original TOE was lost, the symmetry between matter and antimatter was broken. As a result, the universe ended up with more matter than antimatter.

The next big change took place when the electromagnetic and weak nuclear forces separated, leaving these two forces independent of each other. All four fundamental forces of nature that govern today's universe were then separate. At one 10-trillionth of a second, the temperature of the universe had fallen to 10^{16} K. It was a full minute or two

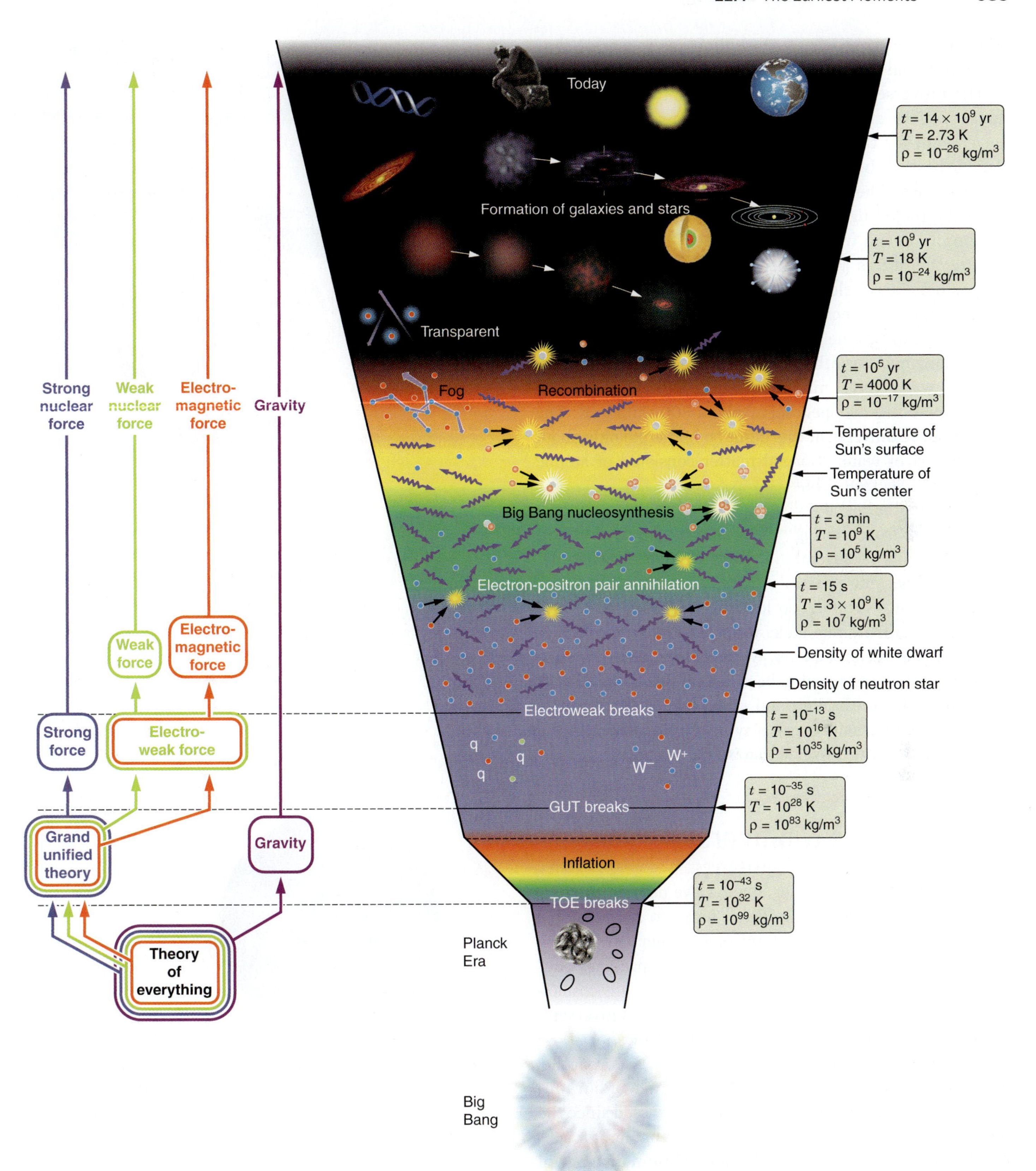

FIGURE 22.15 Eras in the evolution of the universe. The left side shows the four forces separating in stages after the Big Bang. The right side shows the temperature and density of the universe at different key times. As the universe expanded and cooled following the Big Bang, it went through a series of phases determined by what types of particles could be created freely at that temperature. Later, the structure of the universe was set by the gravitational collapse of material to form galaxies and stars, and by the chemistry made possible by elements formed in stars.

before the universe cooled to the billion-kelvin mark, below which not even pairs of electrons and positrons could form.

The universe at that point was too cool to form additional particles and their antiparticles. However, it was still hot enough for the fast-moving protons to overcome the electric barriers between them, allowing nuclear reactions to take place. These reactions formed the least massive elements, including helium and some lithium, beryllium, and boron. Recall from stellar evolution that increasingly high temperatures are needed for the nucleosynthesis of increasingly heavy elements. So as the universe continued to expand, it soon became too cool for the nucleosynthesis of more massive elements.

Big Bang nucleosynthesis came to an end by the time the universe was about 5 minutes old and the temperature of the universe had dropped below about 800 million K. The density of the universe at this point had fallen to only about a tenth that of water. Normal matter consisted of atomic nuclei and electrons, but it was still dominated by the radiation from the Big Bang. After another several hundred thousand years, the temperature dropped so low that electrons were able to combine with atomic nuclei to form neutral atoms. This was the era of recombination, which is seen directly when astronomers observe cosmic background radiation. At this stage the radiation background could no longer dominate over matter, and gravity began playing its role in forming the vast structure of the universe that is now observed.

As the universe cooled, atomic nuclei formed first, followed by atoms.

22.5 Multiple Multiverses

Is our universe the only one? Since we've defined the universe as "everything," what does it mean to say "multiple universes"? Are there parallel universes either separated in space or even occupying exactly the same space as "our" universe? These ideas are quite speculative, but many cosmologists think seriously about the idea of multiple universes—or **multiverses**—collections of parallel universes.

Let's begin with the simplest example of such parallel universes, illustrated in **Figure 22.16**. The age of our universe—that is, the amount of time that has passed since the Big Bang—is 13.7 billion years. That means that light reaching Earth today can have traveled a maximum distance of about 13.7 billion light-years (4,200 Mpc). Therefore, the observable universe—everything astronomers can possibly observe today—must be within a sphere of radius 13.7 billion light-years. Anything farther than that is outside of the observable universe and cannot be seen.

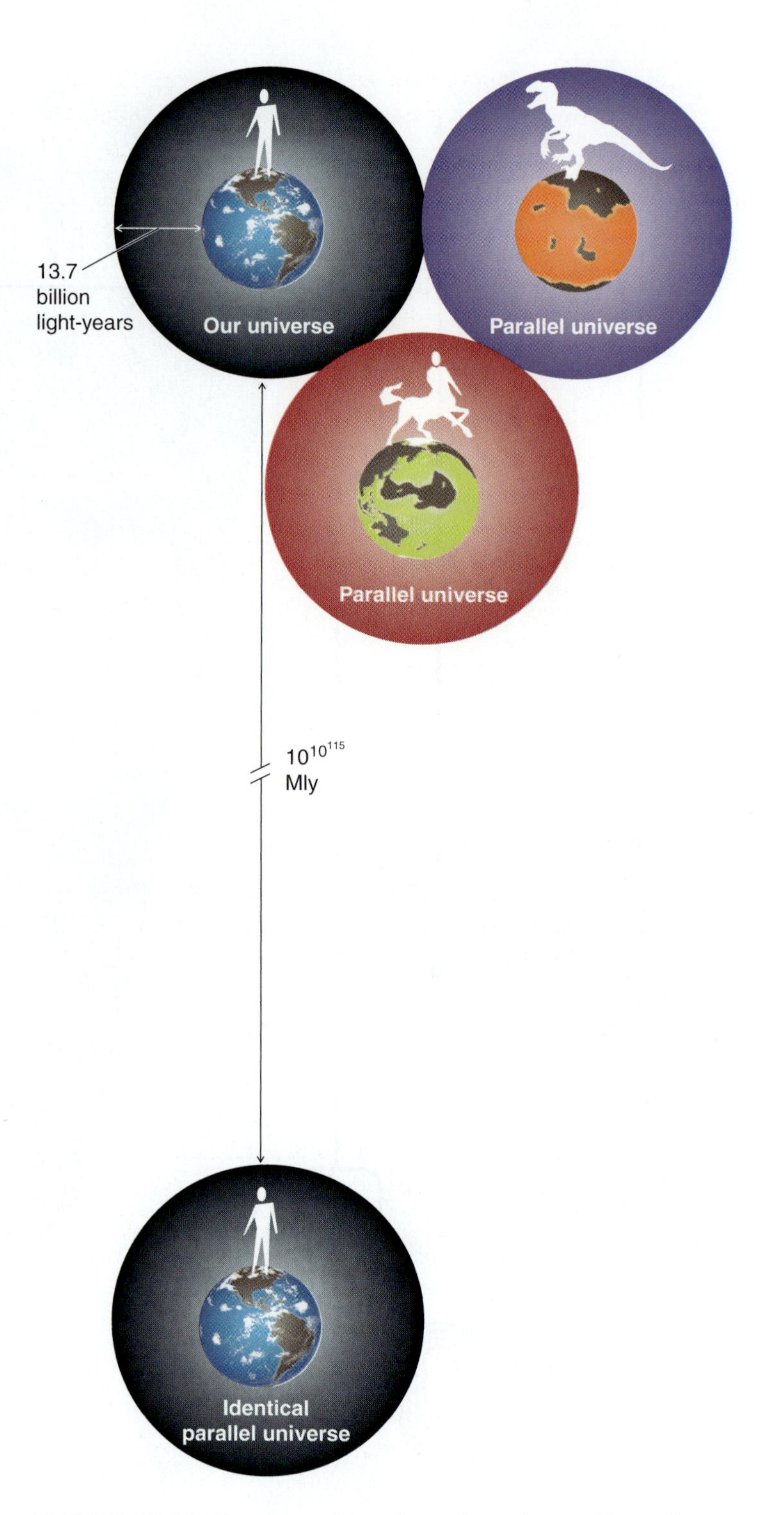

FIGURE 22.16 The observable universe is a sphere with a radius equal to the distance light has traveled since the Big Bang (13.7 billion light-years). Since the universe is infinite, there must be an infinite number of similar spheres. The rules of probability dictate that some of these are *exactly* like our own.

As we discussed earlier in the chapter, the observational evidence suggests that the geometry of space is flat. A flat universe is infinite in size and must therefore contain an infinite number of similar spheres. As dark energy causes the universe to expand faster and faster, the separate observable universes move farther apart and will never overlap. These parallel universes are simply too far away to ever be observed from Earth.

What are these other parallel universes like? Physicists suggest several things based on what has been learned about the observable universe. First, if the cosmological principle holds, then on large scales each of these observable universes should look pretty much like our own, although the details could be very different. In a truly infinite universe there must be an infinite number of observable universes exactly like this one, with an exact copy of you reading an identical version of *21st Century Astronomy*. The argument is that if our own observable universe is cooler than about 10^8 K everywhere, there can be no more than 10^{115} particles in the observable universe, and only so many ways that those particles can be distributed. If you then ask how far you must go before you are sure to find an observable universe *identical to* our own, the answer is about $10^{10^{115}}$ Mpc. Yes, that's 10 raised to the power 10^{115}. So, in an infinite universe—as enormous as it might be—the identical copy of you must be at about that distance.

The observable part of our universe will become isolated from all others.

The collection of parallel universes we just described represents the first of four types of multiverses theorized by cosmologists (**Figure 22.17a**). The inflationary universe model forms the basis of the second type of multiverse. Imagine a universe that undergoes **eternal inflation**, with no beginning or end to the inflation. This idea was hypothesized by physicist Andrei Linde (1948–), who realized that if such a universe exists, then quantum fluctuations may cause some regions to expand more slowly than the rest of the universe. As a result, such a region may form a bubble whose inflating phase will soon end (**Figure 22.17b**). In this

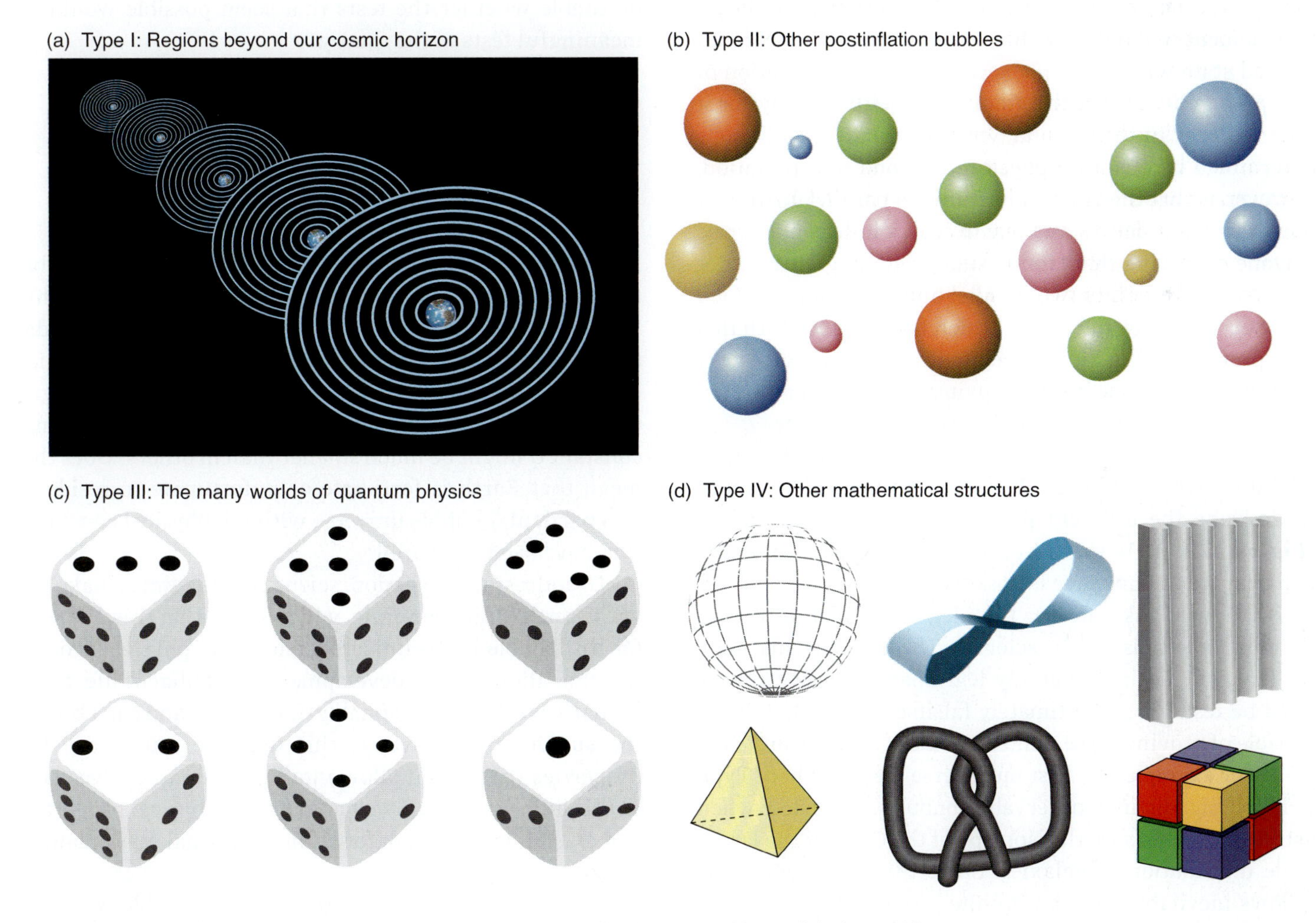

FIGURE 22.17 Cosmologists have proposed four types of multiverses: (a) parallel observable universes; (b) the inflationary model; (c) the quantum mechanical model; (d) parallel universes with different physics.

scenario, Earth is inside such a region, and "our" Big Bang would just be the condensation of our bubble within the eternally inflating universe.

One nice feature of eternal inflation is that it neatly answers the question of what there was before the Big Bang. Since the universe has been inflating and will continue to inflate forever, there is no beginning or end. "Our" own bubble or parallel universe separated from the rest of the universe at a time called the Big Bang, but other bubbles are constantly separating and becoming their own parallel-universe big bangs. Given that these bubble universes are finite, could Earth be near the edge of such a bubble? This scenario is possible, but the multiverses may be so huge that it is very unlikely Earth is anywhere near the edge.

If eternal inflation is happening, new big bangs are forever forming.

The third type of multiverse has its origins purely in quantum mechanics (**Figure 22.17c**). Recall from Chapter 6 that the quantum description of the world is a probabilistic description. There is a certain probability that a radioactive element will decay within a day or that an atom in an excited state will decay in a second. One interpretation of this probability is that the atomic state either decays or it doesn't decay in that second, with the relative likelihoods determined by quantum physics. A second interpretation, however, is that the atom splits into two parallel universes: one in which it decays and one in which it does not decay. Anyone observing that atomic state, watching for it to decay, will also split into two parallel universes: one in which they see the decay and one in which they do not. With this interpretation of quantum mechanics, new universes are forming all the time and occupying the same space, even though they cannot communicate with one another.

The fourth type of multiverse is characterized by parallel universes that have different mathematical structure to describe the different physics within these universes (**Figure 22.17d**). In this all-encompassing case, almost any behavior for the universe is possible.

It is fair to ask whether the idea of parallel universes, or multiverses, is really science. Throughout this book we have emphasized that any legitimate scientific theory must be testable and ultimately falsifiable. Are there tests capable of proving these multiverse ideas to be wrong? Possibly. For example, the first multiverse we described, Type I, involving distinct observable bubbles, is tested when astronomers measure the isotropy of the CMB or the large-scale distribution of galaxies. Since this multiverse type follows inevitably from an infinite, expanding universe, it makes predictions that can be tested. The eternal inflation model, Type II, is harder to test, because it is impossible to directly observe parallel universes. But if physicists obtain a theory of everything that predicts eternal inflation, and if that theory of everything is itself falsifiable, then there is a connection between eternal inflation and observation. In addition, if the bubbles in eternal inflation have very different properties from one another, astronomers can also test whether it is theoretically possible for such bubbles to expand into parallel universes that do not evolve too fast or do not produce stars and planets, in which case our very existence would be unlikely and such models would be highly suspect. Types III and IV might also be modeled with future computers and with an ultimate TOE.

There is considerable debate within the scientific community as to whether the multiverse hypotheses, except for Type I, can be truly falsified. Even if there are some tests, the concern is that the tests are not truly meaningful. For example, you might want to test Newton's theory of gravity by releasing an apple and watching whether it falls upward or downward. But this test is not very discerning; it might not always distinguish between two or more sensible theories. Similarly, for the multiverse hypothesis it is still debatable whether the tests that seem possible would be meaningful tests of the theory.

22.6 Origins: Our Own Universe Must Support Life

If eternal inflation describes our universe, then it is possible (and indeed modern particle theories even suggest) that each bubble could contain different values of the fundamental constants of physics. In some bubble universes, for example, the strength of the nuclear force might be larger, the electric charge might be smaller, and the gravitational constant G might be much smaller than in others. Does this mean that Earth is fortunate or unfortunate to reside in this particular bubble universe with the physical constants that have been measured?

To address this question, scientists sometimes make use of the **anthropic principle**, which states that this universe (or this bubble in the universe) must have physical properties that allow for the development of intelligent life. Since humans exist, are (presumably) intelligent, and can observe the surrounding universe, this universe must have the properties that would allow intelligent life to evolve. That is, this universe must have had the right physical properties and existed long enough for atoms, stars, galaxies, planets, and life to have formed.

In the case of a multiverse that contains bubbles with different physical constants in each of them, the anthropic principle provides information about the values of those physical constants. Consider a few examples. In a bubble universe where the gravitational constant G was much bigger than G as measured in our universe, stellar evolution would occur much faster, and there would not be time for intelligent life to evolve on a planet before its star burned itself out. Similarly, the anthropic principle provides a relatively narrow range for the strength of the strong nuclear force that holds nuclei together. If that force were much weaker than what physicists now measure, nuclei (which are all positively charged) would not be able to overcome their electric repulsion in order to fuse. Without nuclear fusion, stars could not shine, heavy elements would not form, and planets and life could not evolve. Alternatively, if the nuclear force were stronger, it would be easier for two protons to fuse together in the early universe, thus depleting the universe entirely of hydrogen, and therefore also of the water and organic molecules that are necessary for life as we know it.

As another example of the anthropic principle applied to multiverses, consider the cosmological constant Ω_Λ. If Ω_Λ were 20 times larger than what is currently observed, the universe would have begun accelerating much earlier, and galaxies would not have had time to form. But without galaxies, there would be little star formation. It follows that stellar nucleosynthesis would not have taken place, and rocky planets and life could not have evolved. Alternatively, if the cosmological constant were negative, then this entire bubble universe would have reached a maximum size and begun to collapse even before galaxies and stars could have evolved. A universe with an intermediate value might last only a few billion years—enough time for stars and galaxies to be established, but perhaps not enough time for sufficient amounts of heavy elements to form or for life to evolve to intelligence. As these examples illustrate, the cosmological constant must be within a particular range of values to allow for the intelligent life that exists on Earth to evolve.

The physics in the very early universe sets the stage for what comes next: the formation of structure. In the next chapter we'll look at how the universe evolved from its smooth beginning to one that is filled with galaxies, stars, and planets. ■

Summary

22.1 The shape and the age of the universe depend on the amount of mass and therefore gravity in the universe.

22.2 Observations suggest that the expansion of the universe is accelerating. Both gravity and the cosmological constant (or dark energy) determine the fate of the universe.

22.3 The very early universe may have gone through a brief but dramatic period of exceptionally rapid expansion, called inflation. If true, inflation would explain both the flatness and the homogeneity of the universe observed today.

22.4 The earliest moments in the universe were determined by the four fundamental forces of nature, which were all unified into one basic phenomenon at the moment the universe began.

22.5 The observable universe may be only one of an infinite number of simultaneously existing universes. Scientists disagree about whether the theories of multiverses are testable and falsifiable.

22.6 Our observable universe must be one in which physics can support the formation of life.

Unanswered Questions

- What mechanism leads to an extra electron for every 10 billion electron-positron pairs in the early universe? In this chapter we mentioned that grand unified theories predict an asymmetry between particles and antiparticles, as well as proton decay. But physicists do not currently know which GUT is the correct one, and so far they have not actually observed proton decay. Measuring the actual lifetime of a proton would enable physicists to home in on the correct GUT and therefore the mechanism leading to particle-antiparticle asymmetry. In addition, if the correct TOE was really understood, that theory could predict which GUT describes the universe, and therefore the mechanism and amount of asymmetry.
- What is the origin of dark energy, and why is Ω_Λ so close to Ω_{mass} at the present time? Is that a coincidence? We have already said that dark energy is a form of vacuum energy, and it is one of the grand challenges of any successful theory of everything to explain how big Ω_Λ really is. The simplest estimates for the size of Ω_Λ yield results that are much larger than the observed value—by a factor of about 10^{120}. So there must be an as yet undetermined mechanism that affects the size and evolution of Ω_Λ, and that is one of the bigger questions in modern cosmology.

Questions and Problems

Summary Self-Test

1. The force that acts to slow the expansion of the universe is the ________ force.
   a. gravitational
   b. electromagnetic
   c. strong nuclear
   d. weak nuclear
2. In an open universe, the angles of a triangle add to ________ 180° and the circumference of a circle is ________ $2\pi r$.
3. If astronomers ignored any cosmological constant, the future of the universe could be determined solely from
   a. the mass of the universe.
   b. the volume of the universe.
   c. the amount of light in the universe.
   d. the density of the universe.
4. The cosmological constant accounts for the effects of
   a. dark matter.
   b. the Big Bang.
   c. dark energy.
   d. gravity.
5. The universe is accelerating. Therefore, the age of the universe is probably ________ than previously thought.
6. The cosmic microwave background radiation indicates that the early universe
   a. was quite uniform.
   b. varied greatly in density from one place to another.
   c. varied greatly in temperature from one place to another.
   d. was shaped differently from the modern universe.
7. Identify the two problems of cosmology that are solved by inflation.
8. According to the definitions in Chapter 1, superstring theory is
   a. a hypothesis.
   b. an idea.
   c. a theory.
   d. a law.
   e. a principle.
9. Place in order the following events in the history of the universe.
   a. Planck era
   b. grand unified theory breaks
   c. today
   d. Big Bang nucleosynthesis
   e. electroweak breaks
   f. theory of everything breaks
   g. electron-positron pair annihilation
   h. formation of galaxies and stars
   i. recombination
   j. inflation
10. Place the following forces in order of their separation in the first moments after the Big Bang.
    a. gravity
    b. strong nuclear force
    c. weak nuclear force
    d. electromagnetic force

True/False and Multiple Choice

11. **T/F:** The sum of the angles of a triangle is always 180°.
12. **T/F:** Gravity acts to accelerate the universe.
13. **T/F:** The future of the universe is determined only by the mass it contains.
14. **T/F:** The light in the cosmic microwave background radiation is the oldest light in the universe.
15. **T/F:** Grand unified theories combine all four fundamental forces in the universe.
16. As applied to the universe, what is the meaning of *critical density*?
    a. Above this density, nebulae collapse to form stars.
    b. Above this density, dark matter becomes important.
    c. Above this density, the universe will eventually collapse.
    d. Above this density, matter becomes degenerate.
17. Of the four fundamental forces in nature, which one depends on electric charge?
    a. gravitational force
    b. electromagnetic force
    c. strong nuclear force
    d. weak nuclear force
18. The principal difference between normal matter and dark matter is that
    a. normal matter interacts with light, while dark matter does not.
    b. normal matter has gravity, while dark matter does not.
    c. things made of normal matter are larger when they are more massive; things made of dark matter are smaller.
    d. there is no difference; dark matter was just discovered later.
19. Suppose you measure the angles of a triangle and find that they add to 185°. From this you can determine that the space the triangle occupies is
    a. flat.
    b. positively curved.
    c. negatively curved.
    d. filled with dark matter.
20. Quarks are
    a. virtual particles.
    b. massless particles.
    c. candidates for dark matter.
    d. building blocks of larger particles.

21. Current understanding indicates that the universe is
    a. closed.
    b. flat.
    c. open.
    d. inflating.

22. When a particle and an antiparticle come together, they
    a. annihilate each other, releasing photons.
    b. create a black hole.
    c. release astronomical amounts of energy.
    d. create new particles.

23. The vast majority of antimatter in the early universe
    a. is still around today, filling the space between galaxies.
    b. became dark matter.
    c. formed antimatter galaxies and stars.
    d. annihilated with matter.

24. Astronomers will never directly observe the first few minutes of the universe, because
    a. the universe was opaque at that time.
    b. the universe is too large now.
    c. there were no particles or other matter to see.
    d. there were no photons.

25. The anthropic principle states that
    a. the universe was created so that life exists.
    b. life exists, so the universe must be such that life can exist.
    c. if the universe were otherwise, life would not exist.
    d. life has made the universe the way it is.

Thinking about the Concepts

26. What set of circumstances would cause an expanding universe to reverse its expansion and end up in a "Big Crunch"?

27. Describe the observational evidence suggesting that Einstein's cosmological constant (a repulsive force) may be needed to explain the historical expansion of the universe.

28. What do astronomers mean by *dark energy*?

29. If the universe is being forced apart by dark energy, why isn't the Milky Way Galaxy, the Solar System, or the planet Earth being torn apart?

30. Describe the cause and consequences of a Big Rip.

31. In Chapter 19 we said we could estimate the age of the universe with Hubble time ($1/H_0$). Why does that method not give the best answer?

32. What is the flatness problem, and why has it created difficulties for cosmologists?

33. During the period of inflation, the universe may have briefly expanded at 10^{30} (a million trillion trillion) or more times the speed of light. Why did this ultra-rapid expansion not violate Einstein's special theory of relativity, which says that neither matter nor communication can travel faster than the speed of light?

34. Why is particle physics important for understanding the early universe?

35. The fundamental forces of the universe are generally assumed not to change.
    a. How would the fate of the universe be affected if Newton's gravitational constant changed with time?
    b. What if, instead, the electric force between charged particles changed with time?

36. The standard model cannot explain why neutrinos have mass, or why electron-positron asymmetry existed in the early universe. Do these failings make it an incomplete theory? Should all of its predictions be ignored until the theory can resolve these remaining issues?

37. Explain the process of pair production.

38. Describe the Planck era.

39. What are the basic differences between a grand unified theory (GUT) and a theory of everything (TOE)?

40. Consider the term *superstring theory* in light of the discussion in Chapter 1. Many scientists object to using the word *theory* to describe superstring theory. Why?

Applying the Concepts

41. Study Figure 22.2.
    a. Is the vertical axis linear or logarithmic?
    b. There are two labels for the horizontal axis. The top label is measured in billions of years. Is this axis linear or logarithmic?
    c. The bottom label for the horizontal axis is measured in relative brightness. Is this axis linear or logarithmic?
    d. What is the relationship between billions of years and relative brightness?

42. Figure 22.2 shows a blue curve that indicates a model in which the universe first decelerated and then accelerated, as well as a red curve indicating continuous deceleration.
    a. How are the two curves different?
    b. What is it about one of these curves that indicates deceleration? What indicates acceleration?
    c. If a straight line were plotted on this graph, what would the model that the new line represents indicate about the expansion of the universe?

43. Study Figure 22.2. On this graph, the colored lines represent various models, and the black dots represent data taken in the actual universe.
    a. Why are there no data points on the right-hand side of the graph?
    b. Which models are excluded by the data?
    c. Roughly how far back in time do the data go?
    d. What fraction of the age of the universe is the answer to part (c) (assuming an age of 13.7 billion years)?

44. Compare Figure 22.3, which shows several predictions for the future of possible universes and Figure 22.5, which displays data taken about our own universe. Given the data in Figure 22.5, which lines in Figure 22.3 must be rejected as possibilities for describing the future of the universe?

45. Study Figure 22.5. From the plotted data, determine the following:
    a. Is the universe flat, open, or closed?
    b. Will the universe expand forever, coast to a stop, or recollapse eventually?
    c. Was there a Big Bang?
    d. What is the most probable value of omega sub lambda (Ω_Λ)? Of omega sub mass (Ω_{mass})?

46. Suppose that new data coming in from a new instrument give the value 0.1 for both Ω_Λ and Ω_{mass}. How would astronomers probably respond to these new data?

47. Study Figure 22.15.
    a. Is the time axis (the vertical dimension of the figure) approximately linear or approximately logarithmic?
    b. By how many orders of magnitude (factors of 10) has the density ρ of the universe dropped since earliest time?
    c. By how many orders of magnitude has the temperature dropped since earliest time?

48. Currently, the Hubble constant has an uncertainty of about 4 percent. What are the corresponding maximum and minimum ages allowed for the universe?

49. How many hydrogen atoms need to be in 1 cubic meter (m^3) of space to equal the critical density of the universe?

50. The universe today has an average density $\rho_0 = 3 \times 10^{-28}$ kg/m^3. Assuming that the average density depends on the scale factor, as $\rho = \rho_0/R_U^3$, what was the scale factor of the universe when its average density was about the same as Earth's atmosphere at sea level ($\rho = 1.23$ kg/m^3)?

51. The proton and antiproton each have the same mass, $m_p = 1.67 \times 10^{-27}$ kg. What is the energy (in joules) of each of the two gamma rays created in a proton-antiproton annihilation?

52. There are about 500 million CMB photons in the universe for every hydrogen atom. What is the equivalent mass of these photons? Is it large enough to factor into the overall density of the universe?

53. Suppose you brought together a gram of ordinary-matter hydrogen atoms (each composed of a proton and an electron) and a gram of antimatter hydrogen atoms (each composed of an antiproton and a positron). Keeping in mind that 2 grams is less than the mass of a dime,
    a. Calculate how much energy (in joules) would be released as the ordinary-matter and antimatter hydrogen atoms annihilated one another.
    b. Compare this amount of energy with the energy released by a 1-megaton hydrogen bomb (1.6×10^{14} J).

54. One GUT theory predicts that a proton will decay in about 10^{31} years, which means if you have 10^{31} protons, you should see one decay per year. The Super-Kamiokande observatory in Japan holds about 20 million kg of water in its main detector, and it did not see any decays in 5 years of continuous operation. What limit does this observation place on proton decay and on the GUT theory described here?

55. Assume a planet's orbit is perfectly circular as it travels in the gravitational well of its star. If this were true, would the orbit's circumference be greater than, less than, or equal to 2π times the radius of the orbit?

Using the Web

56. Animations of the stages after the Big Bang shown in Figure 22.15 can be found at these websites: http://expositions.bnf.fr/ciel/elf/1big/big.htm (ignore the text in French) and http://superstringtheory.com/cosmo/cosmo3.html. What are the major stages? What is the evidence for each of these stages?

57. Go to the website for the Dark Energy Survey, an international project beginning in 2012 (https://www.darkenergysurvey.org/index.shtml). What observations will be made for this project? What will it tell scientists about dark energy? Click on "News." What is the status of this project? Are there any results yet?

58. Go to the website of the European Organization for Nuclear Research (CERN—http://public.web.cern.ch/public/en/Science/Recipe-en.html) and read through the pages indexed on the left. What was the role of the Higgs boson after the Big Bang? Go to CERN's press release page (http://press.web.cern.ch/press) to see what's new in the search for the Higgs boson. (*Note*: The World Wide Web was invented at CERN.)

59. Scientists debate whether there ever can be scientific evidence for a multiverse. A good example is a discussion in the journal *Scientific American*. Read the article "Does the Multiverse Really Exist?" in the August 2011 issue (it is probably accessible online through your school library) and the response at http://scientificamerican.com/article.cfm?id=multiverse-the-case-for-parallel-universe. What are some of the arguments for and against multiverses?

60. Clips from a four-part episode of the public television series *NOVA* called "The Fabric of the Cosmos" can be accessed on PBS's website (http://pbs.org/wgbh/nova/physics/fabric-of-cosmos.html), and complete episodes can be viewed on the Top Documentary Films website (http://topdocumentary-films.com/the-fabric-of-the-cosmos). Watch the clips or at least one of the episodes. Are the arguments made in these programs compelling? Is the science explained in a way that makes sense to a general audience?

STUDYSPACE is a free and open website that provides a Study Plan for each chapter of *21st Century Astronomy*. Study Plans include animations, reading outlines, vocabulary flashcards, and multiple-choice quizzes, plus links to premium content in SmartWork and the ebook. Visit **wwnorton.com/studyspace**.

SMARTWORK Norton's online homework system, includes algorithmically generated versions of these questions, plus additional conceptual exercises. If your instructor assigns questions in SmartWork, log in at **smartwork.wwnorton.com.**

Exploration | Understanding Orders of Infinity

Astronomers often grapple with the concept of infinity. Not just "really, really big," but truly infinite. For example, the universe is likely infinite in size. Yet it is also expanding. This can be a very difficult concept to wrap your head around. In this Exploration you will develop your understanding of infinity.

First, picture a number line extending to your right and left, with 0 in front of you.

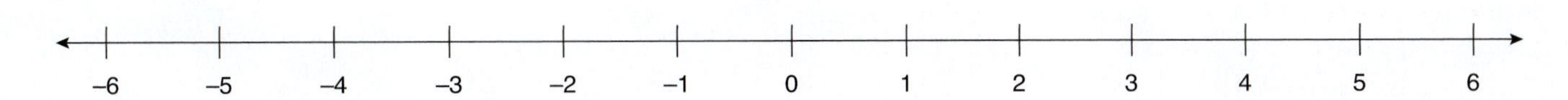

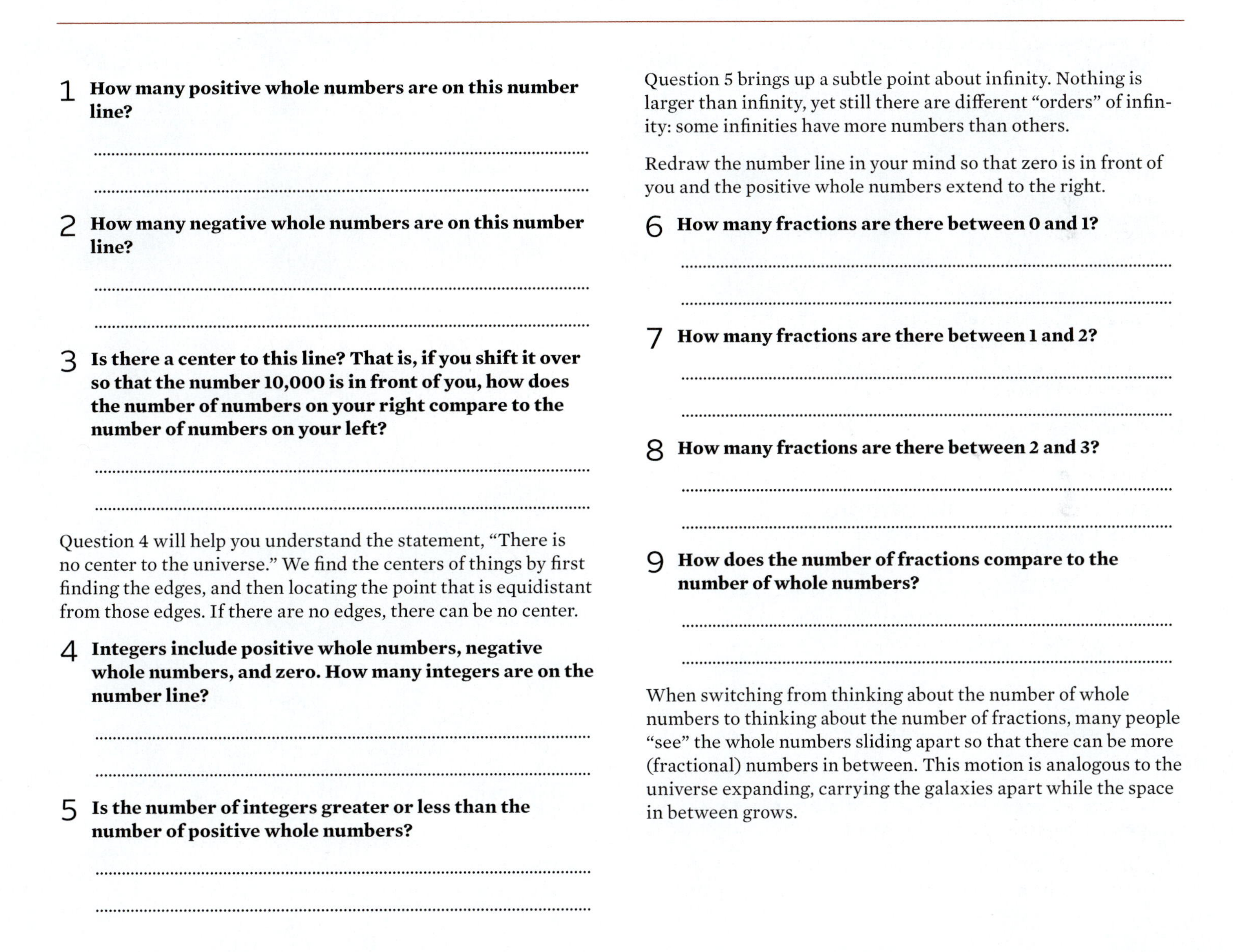

1 **How many positive whole numbers are on this number line?**

....

....

2 **How many negative whole numbers are on this number line?**

....

....

3 **Is there a center to this line? That is, if you shift it over so that the number 10,000 is in front of you, how does the number of numbers on your right compare to the number of numbers on your left?**

....

....

Question 4 will help you understand the statement, "There is no center to the universe." We find the centers of things by first finding the edges, and then locating the point that is equidistant from those edges. If there are no edges, there can be no center.

4 **Integers include positive whole numbers, negative whole numbers, and zero. How many integers are on the number line?**

....

....

5 **Is the number of integers greater or less than the number of positive whole numbers?**

....

....

Question 5 brings up a subtle point about infinity. Nothing is larger than infinity, yet still there are different "orders" of infinity: some infinities have more numbers than others.

Redraw the number line in your mind so that zero is in front of you and the positive whole numbers extend to the right.

6 **How many fractions are there between 0 and 1?**

....

....

7 **How many fractions are there between 1 and 2?**

....

....

8 **How many fractions are there between 2 and 3?**

....

....

9 **How does the number of fractions compare to the number of whole numbers?**

....

....

When switching from thinking about the number of whole numbers to thinking about the number of fractions, many people "see" the whole numbers sliding apart so that there can be more (fractional) numbers in between. This motion is analogous to the universe expanding, carrying the galaxies apart while the space in between grows.

In this example of galactic cannibalism, the NGC 1532/1531 galaxy pair shows how a large galaxy grows by accreting and subsuming a small galaxy that strays too close.

23 Large-Scale Structure in the Universe

Cosmologists are often wrong but never in doubt.

Lev Landau (1908–1968)

LEARNING GOALS

The universe that emerged from the Big Bang was incredibly uniform, wholly unlike today's universe of galaxies, stars, and planets. In this chapter we investigate the origin of the current structure of the universe and find that complex structure is a natural consequence of the action of physical law in an evolving universe. By the conclusion of this chapter, you should be able to:

- Identify the structures in the universe that have larger scales than individual galaxies have.
- Describe how astronomers use observations, theory, and simulations on supercomputers to connect the formation of structure following the Big Bang with the large-scale structure of our universe today.
- Discuss the formation of the first stars and the first galaxies.
- Explain how observations of galaxies at many different redshifts contribute to understanding the evolution of the large-scale structure of the universe.

23.1 Galaxies Form Groups, Clusters, and Larger Structures

The early universe was an extraordinary place—an expanding "fireball" that was far more uniform than the blue of the bluest sky on the clearest day. That early universe was very different from the universe that astronomers observe today. Today's universe is not smooth and uniform; it contains galaxies and stars. How did the universe progress from its relatively smooth beginning to the structure that is observed today?

What is the origin of structure?

Just as stars and clouds of glowing gas reveal the structure of the Milky Way, the distribution of galaxies indicates the structure of the universe. And just as it is gravity that holds galaxies together, giving them their shape, it is gravity that shapes larger structures. Most galaxies are gravitationally bound in groups (**Figure 23.1**). Larger gravitationally bound systems of galaxies, called **galaxy clusters**, can consist of thousands of galaxies, often with a more regular structure than is found in galaxy groups. Galaxy clusters are larger than groups, typically occupying a volume of space 2–10 megaparsecs (Mpc) across. The Local Group, which includes the Milky Way and Andromeda galaxies, is about 3 Mpc in diameter. It is located near two large clusters: the Virgo Cluster and the Coma Cluster.

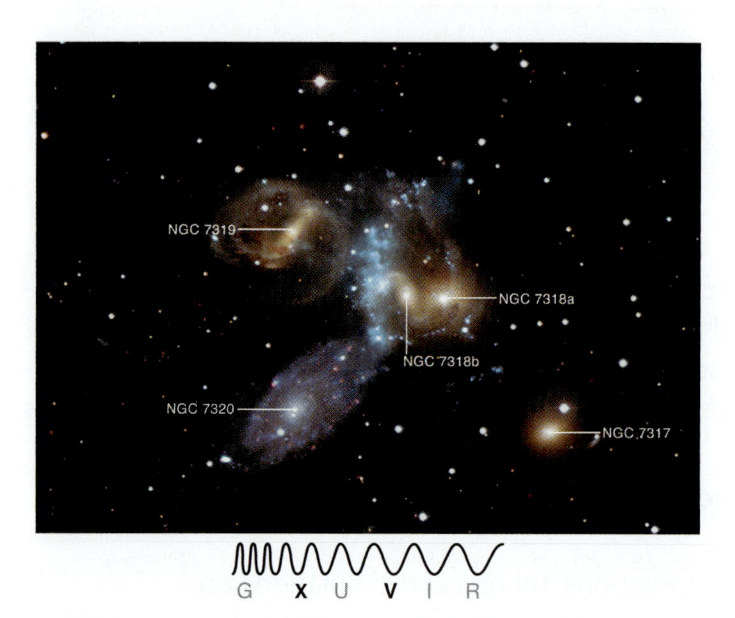

FIGURE 23.1 Stephen's Quintet, a compact group of galaxies 86 Mpc away. NGC 7320 is in the foreground, not a member of the group. NGC 7318b is passing though the core of the others, generating a shock wave that heats the gas, producing X-rays (light blue). Other colors are optical light from the Canada-France-Hawaii Telescope.

Like groups, galaxy clusters contain far more dwarf galaxies than giant galaxies. However, most of the stellar *mass* in galaxy clusters resides in the giant galaxies. In addition, although spiral galaxies are common in most clusters, elliptical galaxies are prevalent in about one-fourth of galaxy clusters. The Virgo Cluster (**Figure 23.2**), located 16.5 Mpc from the Local Group, is an example of a cluster that contains mostly spiral galaxies. The more distant Coma Cluster (see Figure 20.26) is dominated by giant elliptical and S0 galaxies.

Clusters and groups of galaxies bunch together to form enormous **superclusters**, which contain tens of thousands or even hundreds of thousands of galaxies and span regions of space typically more than 30 Mpc in size. The Local Group is part of the Virgo Supercluster, which also includes the Virgo Cluster.

Mapping the Universe

Hubble's law is a powerful tool for mapping the distribution of galaxies, groups, clusters, and superclusters in space. Using this law, astronomers estimate the distance to a galaxy by measuring the redshift in the galaxy's spectrum. The first redshifts were measured from spectra recorded on photographic plates, which required exposures of several hours to capture the faint signal. By 1975, astronomers had documented redshifts for only about a thousand of the several hundred billion observable galaxies. Now astronomers use larger telescopes with electronic detectors and spectrographs capable of observing many galaxies at once, along with more powerful computers for collecting and analyzing the data, to measure the redshifts of millions of galaxies.

Galaxy redshift surveys measure distances to large numbers of galaxies.

The first large redshift survey looked only at local space, but later surveys extended much farther. These surveys were used to illustrate a slice of the universe (**Figure 23.3**). Observations show that clusters and superclusters of galaxies, rather than being scattered randomly through space, are linked in an intricate web of filaments and "walls." The concentrations of galaxies, in turn, surround large **voids**, regions of space that are largely empty of galaxies. These voids represent some of the largest "structures" seen in the universe; they are mainly empty of observable galaxies, but they might contain dark matter, as well as normal matter that has never been seen. Clusters and superclusters

Large-scale structure fills the universe.

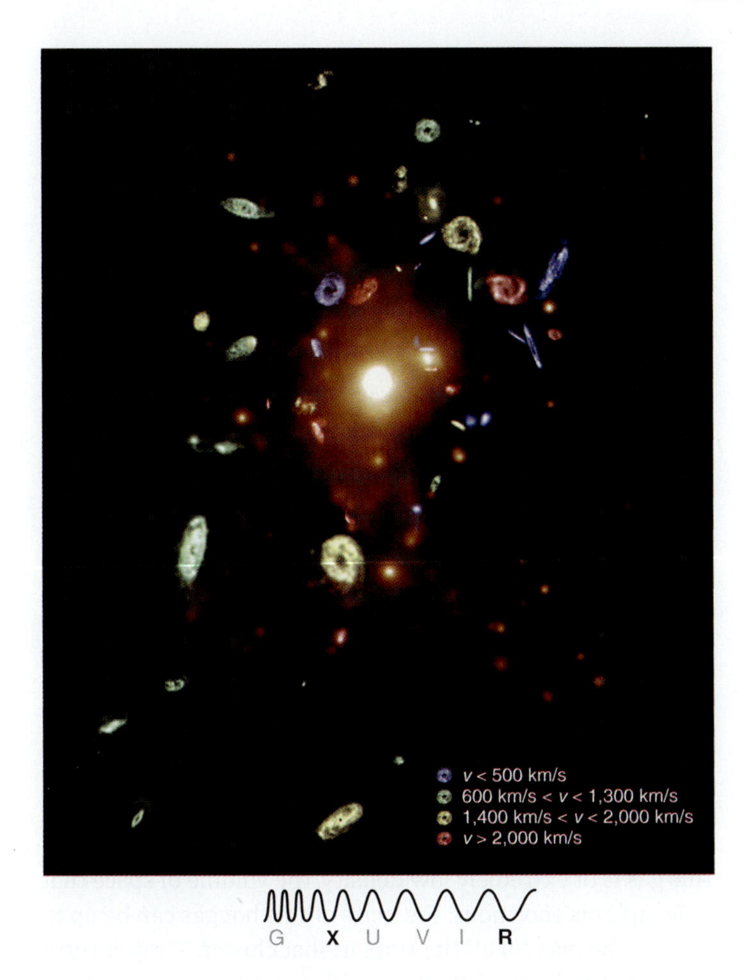

FIGURE 23.2 Composite X-ray and radio image of the Virgo Cluster of galaxies. The X-ray emission (orange) is centered on three giant elliptical galaxies. The radio images are magnified 10 times to show more detail in the spiral structures. The colors indicate distances, with the blue ones closest and the red ones farthest. The cluster environment affects the evolution of the galaxies, as they accrete or lose gas, or merge with each other.

are located within the walls and filaments. The Sloan Digital Sky Survey (SDSS) includes the Sloan Great Wall, a string of galaxies 400 million Mpc long (**Figure 23.3b**). For as far out as observations measure, the universe has a porous structure much like a sponge. Together, galaxies and the larger groupings in which they are found are called **large-scale structure**. **Figure 23.4** shows some of the nearest of these largest structures: superclusters and walls.

The **peculiar velocity** of a galaxy is its motion relative to the CMB (see Chapter 19). Observations of the peculiar velocity reveal the distribution of mass near that galaxy. For example, the peculiar velocity of the Local Group was originally attributed to the Great Attractor, which has a mass of several thousand times the mass of the Milky Way and is located about 75 Mpc away. More recent X-ray observations led to the discovery of the Shapley Supercluster,

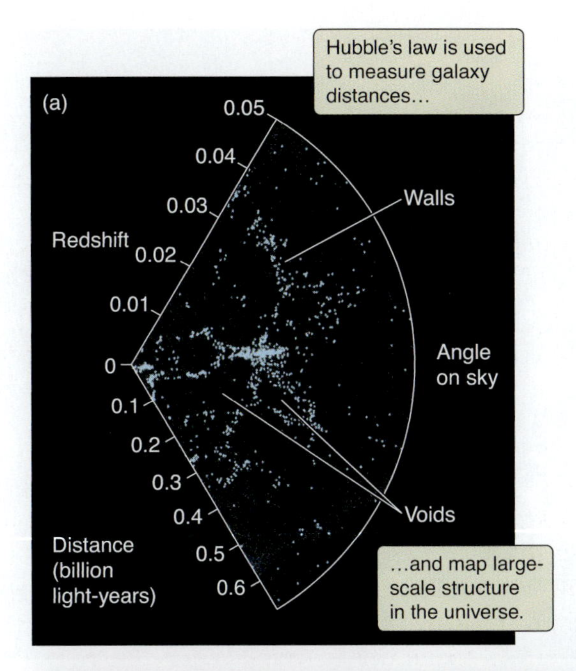

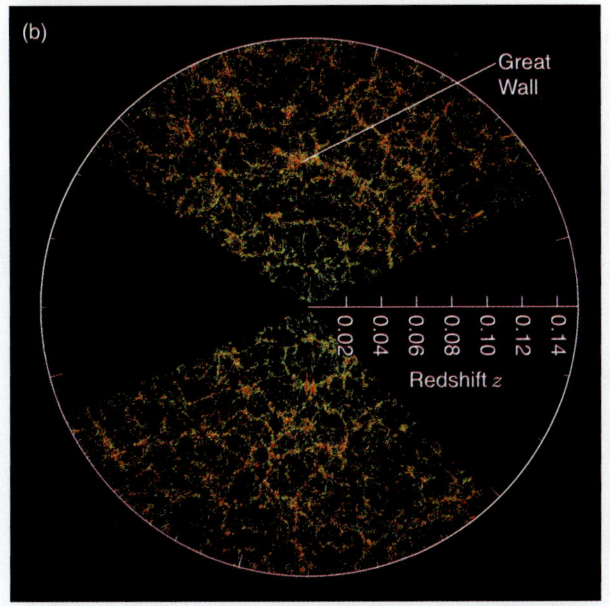

FIGURE 23.3 Redshift surveys use Hubble's law to map the universe. (a) In 1986, the Harvard-Smithsonian Center for Astrophysics redshift survey, called "A Slice of the Universe," was the first to show that clusters and superclusters of galaxies are part of even larger-scale structures. (b) The 2008 Sloan Digital Sky Survey map of the universe extends outward to a distance of about 2 billion light-years. Shown here is a sample of 67,000 galaxies colored according to the ages of their stars, with the redder, more strongly clustered points showing galaxies that are made of older stars.

a more massive supercluster located 125 Mpc beyond the Great Attractor. These two large structures gravitationally tug on the Local Group, accelerating the galaxies within it, including the Milky Way (**Figure 23.5**).

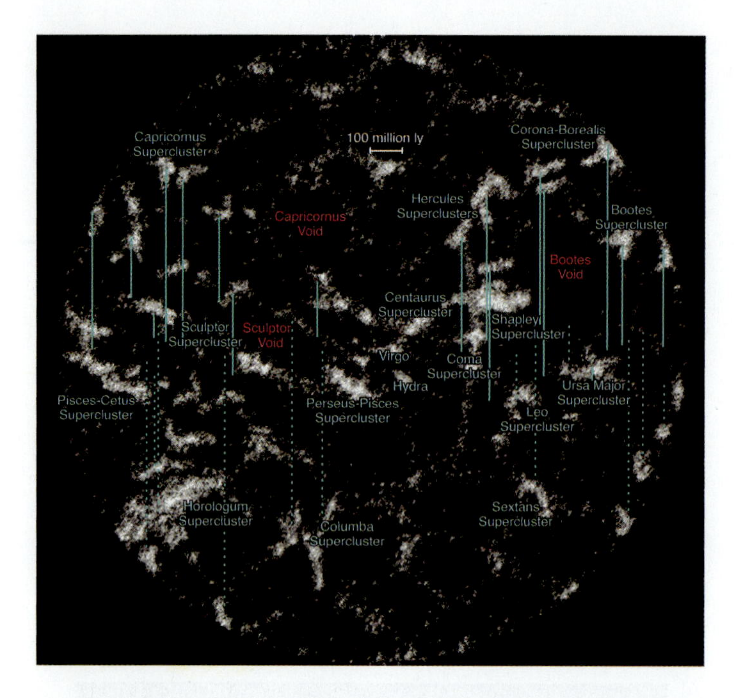

FIGURE 23.4 The superclusters and walls that are closest to the Milky Way.

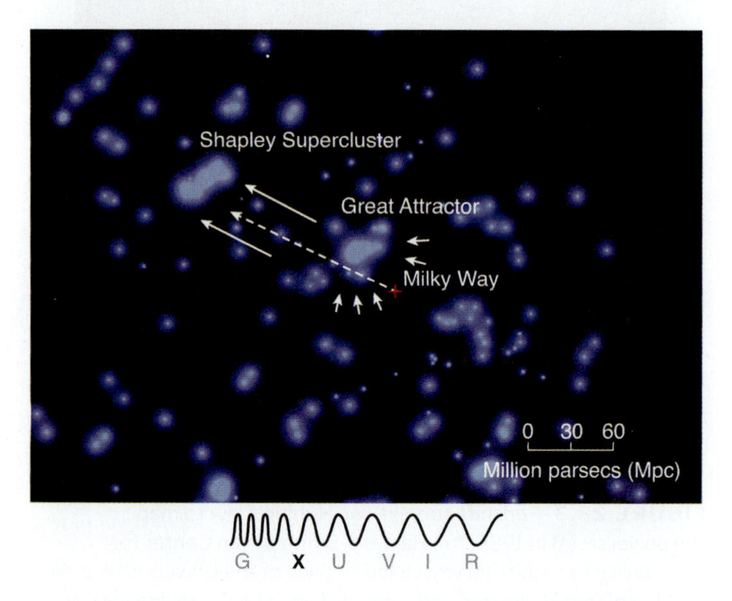

FIGURE 23.5 The blue halos represent clusters of galaxies within 250 Mpc of the Milky Way. Multiple halos grouped together are superclusters. The gravitational pull of both the Great Attractor (small arrows) and the Shapley Supercluster (large arrows) produces a large-scale flow affecting the motion through space of local clusters and the Milky Way.

Peculiar velocities in groups, clusters, and superclusters map out the local distribution of mass in those structures. To separate the peculiar velocity from the velocity caused by the expansion of space, an astronomer must obtain the distance to the galaxy by a method independent of the Hubble law. The Hubble law is then used to find the expansion velocity that should correspond to this distance. Then this expansion velocity is compared to the measured velocity of the galaxy. The difference between the two measurements is the peculiar velocity.

Dark Matter Dominates the Mass of Galaxy Groups and Clusters

Just as dark matter dominates galaxies, it also dominates galaxy groups and clusters. Astronomers can infer this in several ways. They can look at the motion of a small satellite galaxy orbiting the central dominant galaxy of the cluster and estimate the mass of the cluster inside of this orbit (**Math Tools 23.1**), similar to the way they measure dark matter in spiral galaxies. Or they can look at the motions of all of a cluster's galaxies and calculate how strong gravity must be to hold the cluster together. And again, the conclusion is that the total mass of the clusters, including dark matter, must be about 8–10 times greater than the normal matter they contain.

Another piece of evidence for dark matter is that the space between galaxies in a cluster is filled with extremely hot gas that is 10–100 million kelvins, K (**Figure 23.6**). Even though this gas is of extremely low density, the volume of space that it occupies is enormous: the mass of this hot gas can be up to 5 times the mass of all the stars in that cluster. X-ray spectra show that the gas contains significant amounts of massive elements that must have formed in stars. This chemically enriched gas has been either blown out of galaxies in winds driven by the energy of massive stars, or stripped from galaxies during encounters with neighboring galaxies. This hot gas would have dispersed long ago, were it not for the gravity of the dark matter filling the volume of the cluster.

An additional way to look for dark matter relies on the predictions of Einstein's general theory of relativity (Chapter 18), which states that mass distorts the geometry of spacetime, causing even light to bend near a massive object. In particular, light from a distant object is bent by the gravity of a galaxy or cluster of galaxies, so that images of the distant object can be seen magnified on either side of the intervening galaxy or cluster. The result is a gravitational lens (**Figure 23.7a**).

Recall that we mentioned gravitational lensing in the discussion of MACHOs in Chapter 20, where we noted that lenses can make background objects appear brighter. Lenses can also show multiple images of background objects, and that these magnified images are often drawn out into arcs. The greater the gravitational lensing, the greater the mass that must be in the cluster. **Figure 23.7b** shows an image of a galaxy cluster that is acting as a gravitational lens for

Gravitational lenses are one way to measure the masses of galaxy clusters.

Math Tools 23.1

Mass of a Cluster of Galaxies

Recall from Math Tools 21.1 that astronomers use the orbit of a star about the center of a galaxy to estimate the mass of the galaxy within the star's orbit. A similar estimation is made with clusters of galaxies. These clusters are often dominated by one giant galaxy at the center, with smaller ones orbiting around it. The orbital velocities of the smaller galaxies are measured from the Doppler shifts of the lines in their spectra. The distance between the central and orbiting galaxies (the radius of a circular orbit) is estimated. The equation then looks like the one from Math Tools 4.2 and Math Tools 21.1:

$$M = \frac{rv^2_{\text{circ}}}{G}$$

Consider a typical cluster. Suppose a smaller galaxy is orbiting the large galaxy at the center of the cluster at a speed of 1,000 km/s at a distance of about 3 Mpc. The gravitational constant is $G = 6.67 \times 10^{-11}$ m^3/kg s^2, so it's best to put everything else into those same units. Converting 1,000 km/s, we get 1,000,000 m/s, or 10^6 m/s. As in Chapter 21, we must convert the orbital radius (3 Mpc) to meters too:

$$(3 \times 10^6 \text{ pc}) \times (3.09 \times 10^{16} \text{ m/pc}) = 9.3 \times 10^{22} \text{ m}$$

The mass of the cluster is given by:

$$M = \frac{(9.3 \times 10^{22} \text{ m}) \times (10^6 \text{ m/s})^2}{6.67 \times 10^{-11} \text{ m}^3/\text{kg s}^2}$$

$$= 1.4 \times 10^{45} \text{ kg}$$

We can divide this by the mass of the Sun, 2.0×10^{30} kg, to get a cluster mass of $7.0 \times 10^{14}\ M_\odot$. If we divide this cluster mass by the mass of the Milky Way Galaxy, $10^{12}\ M_\odot$, we see that the cluster has the mass of about 700 Milky Way galaxies.

Astronomer Fritz Zwicky (1898–1974) made a calculation like this one in 1933, and measured more mass than was expected from the visible light. He concluded that there must be dark matter within clusters of galaxies. But his work was ignored for decades, until Vera Rubin (1928–) and colleagues made rotation curves of individual galaxies and discovered dark matter in these galaxies too.

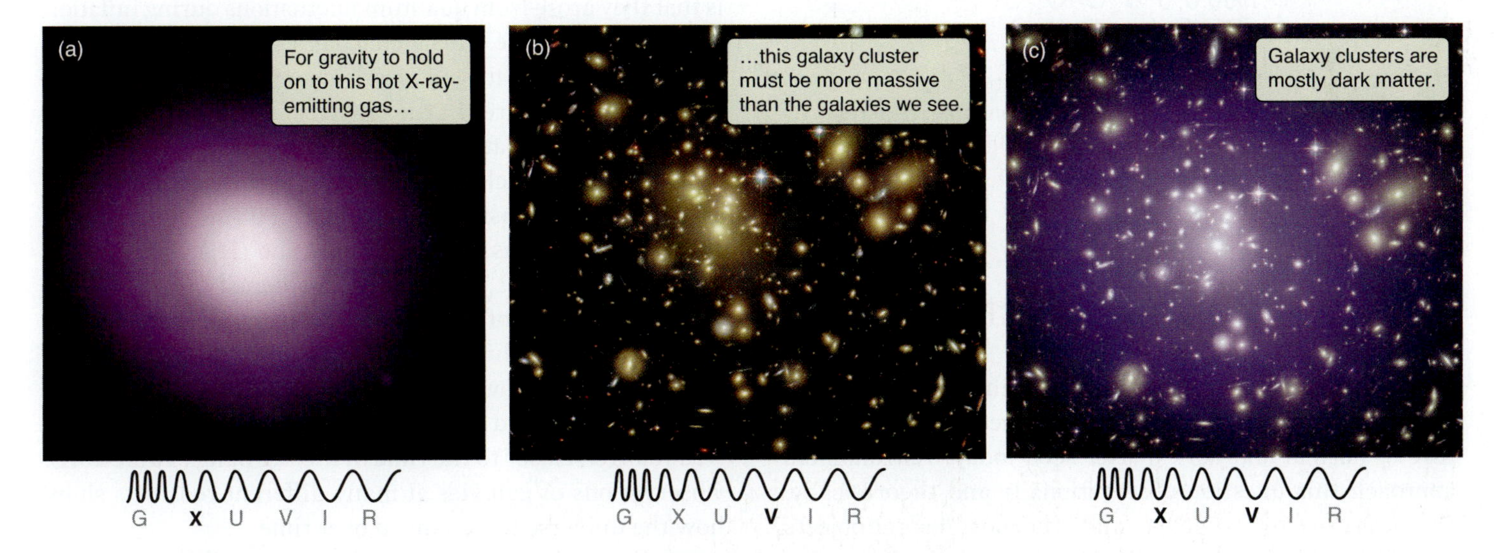

FIGURE 23.6 The massive galaxy cluster Abell 1689, located about 2.3 billion light-years away. (a) Chandra X-ray image of the cluster. (b) The individual galaxies are seen in a Hubble Space Telescope image in visible light. (c) The combined image shows that the 100-million-K gas fills the cluster.

a number of background galaxies. Analysis of such images reveals the mass of the lensing cluster.

Regardless of how astronomers measure the masses of galaxy clusters—by looking at the motions of their galaxies, by measuring their hot gas, or by using them as gravitational lenses—the results are the same. Dark matter dominates the mass of galaxy clusters and superclusters.

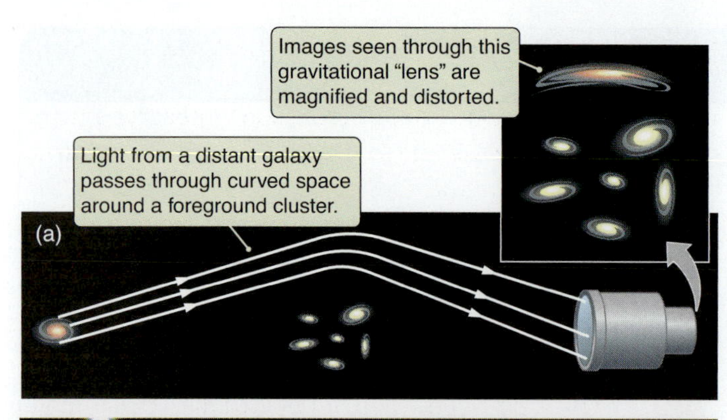

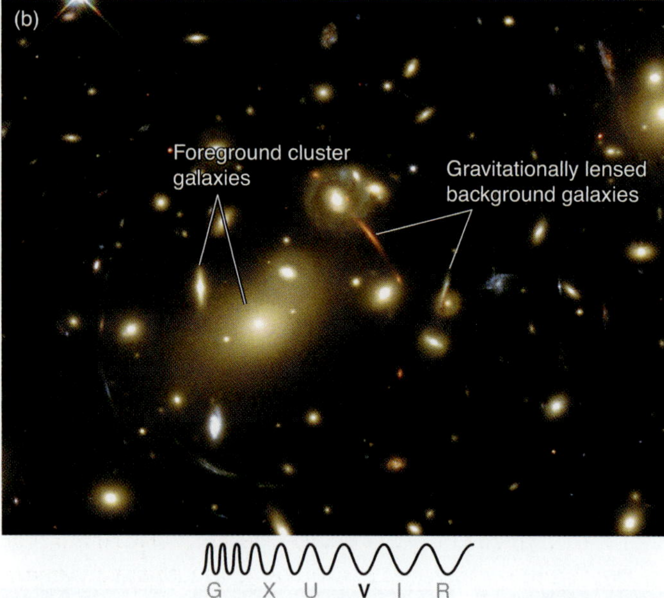

FIGURE 23.7 (a) The geometry of a gravitational lens. A mass can gravitationally focus the light from a distant object, thereby magnifying and distorting the image. (b) A Hubble Space Telescope image of the cluster Abell 2218, showing many gravitationally lensed galaxies, seen as arcs.

23.2 The Origin of Structure

How did the universe evolve from the initially smooth radiation field following the Big Bang to the large-scale structure of normal and dark matter seen today? Astronomers approach this question observationally and theoretically: Observers use telescopes to study the most distant objects and the CMB. Theorists utilize the largest supercomputers to simulate the growth of small- and large-scale structure.

Theorists think about gravity, inflation, and dark energy, and consider the balance among radiation, normal matter, and dark matter in the early universe. Models of inflation are especially important because they tie together the large-scale structure of today's universe and the structure of the universe immediately after the Big Bang. The early structure included clumpiness produced by quantum effects. Over time, gravity amplified these clumps (sometimes called "seeds," since structure grows from them). Smaller structures (such as subgalactic clumps and dwarf galaxies) formed first, whereas larger structures took more time to form. This concept is called **hierarchical clustering** because the structure forms in a "bottom-up" hierarchy. Hierarchical clustering is supported by observations and is fundamental to how structure formed in the universe.

Smaller structures form first. Larger structures form later.

It is worth taking a moment to reflect on the meaning of these seeds. When the universe is not 100 percent uniform but has small density variations from place to place, gravity will amplify those variations over time. For example, on the scale of a galaxy, the early universe may have regions that are more dense than average by perhaps 1 percent. Over time, gravity can amplify this difference in density so that it becomes greater than 100 percent different, and a galaxy can collapse and form. But the seeds are essential, since gravity cannot produce structure in a perfectly uniform universe where there is nothing to amplify. Currently, one of the few, if not the only, viable model for these seeds is that they arose from quantum fluctuations during inflation in the early universe. This model has a profound implication: the seeds leading to galaxies, clusters, and superclusters (the largest structures in the universe) arose from the same quantum physics that describes the smallest structures in the universe (atoms, nuclei, and elementary particles).

Ripples in the early universe were the seeds of galaxies and large-scale structure.

Observational astronomers measure the constants of cosmology, such as Hubble's constant (H_0), the cosmological constant (Ω_Λ), and the mass density (Ω_{mass}). They estimate the ratio of normal and dark matter; and look for galaxies with the highest redshifts. Remember that more distant galaxies have higher measured values of redshift (z), so the observed light was emitted when the universe was *younger*, closer to the time of the Big Bang (**Table 23.1**). Observations of galaxies at many different redshifts show how the universe has changed over time.

However, there is a gap of about 400 million years between the CMB maps of the universe at age 400,000 years and the highest-redshift galaxies observed. New telescopes such as the international Atacama Large Millimeter/submillimeter Array (ALMA) in Chile and the future James Webb Space Telescope (JWST) in space may be able to observe even younger galaxies closer to the time when the CMB became observable.

TABLE 23.1 | Redshift and Age

Observed z	Age of Universe (years)
1,100	380,000 (recombination)
30	100 million
20	200 million
15	270 million
10	480 million
9	560 million
8	650 million
7	750 million
6	900 million
5	1.2 billion
4	1.6 billion
3	2.2 billion
2	3.3 billion
1	5.9 billion
0.5	8.6 billion
0.25	10.5 billion
0	13.7 billion

The details of what happened in the very early universe affected the growth of the large-scale structure seen today. The values of Ω_Λ and Ω_{mass} are important in part because they determine how rapidly the universe expanded and therefore how difficult it is for gravity to overcome this expansion in a particular region. The more rapidly the universe is expanding, or the less mass it contains, the more difficult it will be for gravity to pull material together into galaxies and larger-scale structures.

The available observations and measured constants of cosmology become inputs to theoretical models and supercomputer simulations of how the large-scale structure formed. By the beginning of the 21st century, a "standard model" of Big Bang cosmology had broad support for explaining the large-scale structure and accelerated expansion of the universe, as well as the CMB and the amounts of the light elements from Big Bang nucleosynthesis (discussed in Chapters 19 and 22). This model is called **Lambda-CDM** (*lambda* for the cosmological constant Ω_Λ, and *CDM* for the "cold dark matter" that played a vital role in structure formation).

Galaxies Formed Because of Dark Matter

When the cosmic background radiation is observed using such satellites as COBE and *WMAP* (see Chapter 19), variations in the background radiation are found at a level of about one part in 100,000. The theoretical models clearly show that such tiny variations at the time of recombination (when the universe was about 400,000 years old) are far too small to explain the structure observed in today's universe. Gravity is not strong enough to grow galaxies and clusters of galaxies from such small clumps. These models indicate that for today's galaxies to have formed, the density of those clumps must have been a few tenths of a percent greater than the average density of the universe at the time of recombination. But if normal luminous matter in the early universe had clumps with this higher density, the variations in the CMB today would be at least 30 times larger than they are. How do astronomers reconcile this problem?

There is much more dark matter than normal luminous matter in the universe, and dark matter is an essential ingredient in forming the observed structure. The amount of normal matter seen in the universe predicts just the right abundances of light elements from Big Bang nucleosynthesis. Therefore, the much greater amount of dark matter cannot be made of normal matter consisting of neutrons, protons, and electrons. If it were, it would have affected the formation of chemical elements in the early universe, and the abundances of several isotopes of the least massive elements would be quite different from what is found in nature.

Dark matter must be something else—something that has no electric charge (so it does not interact with electromagnetic radiation) and that interacts only weakly with normal matter. Clumps of such dark matter in the early universe would not have interacted with radiation or normal matter, so astronomers would not see them directly when looking at the CMB. Dark matter clumpiness can be large enough to form galaxies without producing too much variation in the CMB, as long as the clumpiness in the ordinary matter is much smaller. This unseen dark matter solves the problems of modeling the formation of galaxies and clusters of galaxies.

Dark matter and normal matter behaved differently in the early universe. At that time, pressure waves and radiation smoothed out ripples in the distribution of normal matter. Feebly interacting dark matter is immune to these processes, so clumps of dark matter survived long after clumps in the normal matter had been smoothed out (**Figure 23.8**). In addition, the

Observed structure forms when ordinary matter falls onto dark matter clumps.

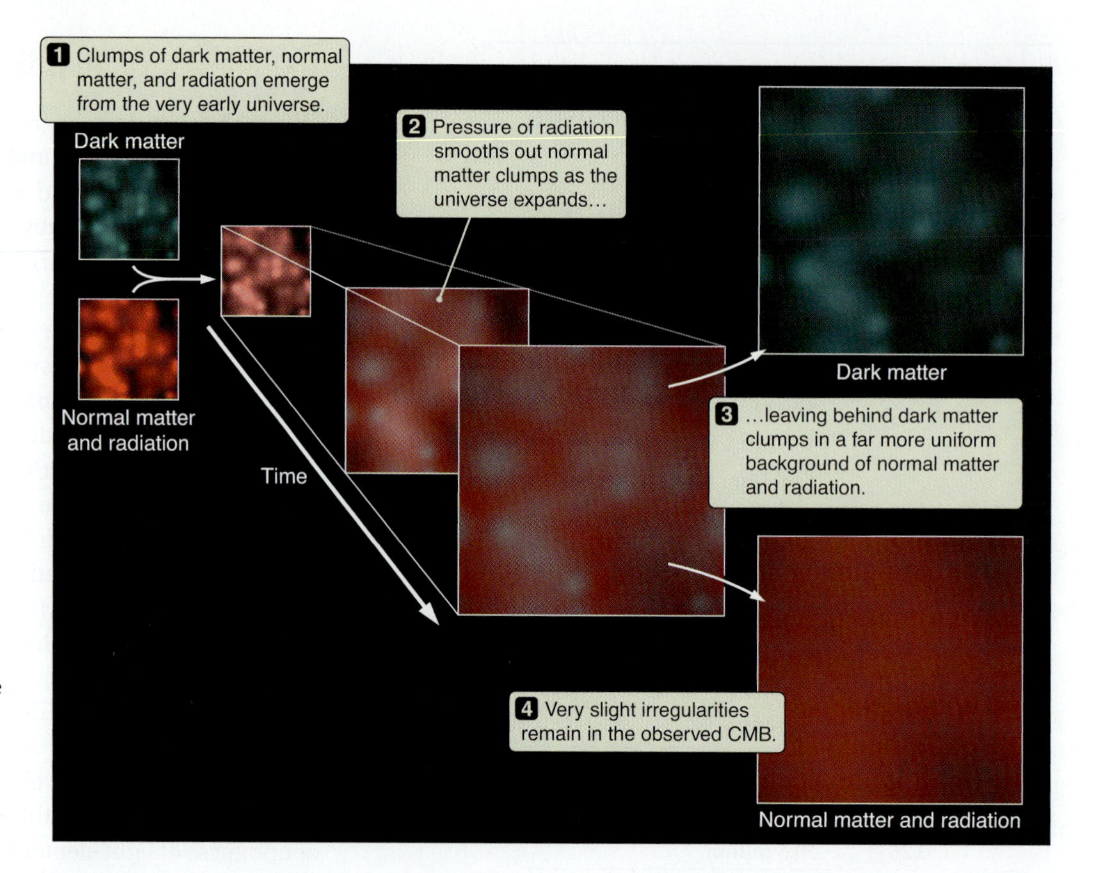

FIGURE 23.8 Radiation pressure and other processes in the early universe smoothed out variations in normal matter, but clumps in the dark matter survived and pulled in normal matter to form galaxies.

dark matter in these clumps does not glow, so astronomers do not see it in the cosmic microwave background radiation. Although these clumps of dark matter do cause slight gravitational redshifts in the light coming from normal matter, the resulting variations fit well with current observations of the CMB. Analogous to how dark matter dominates the gravitational fields of today's galaxies and clusters, the mass of dark matter controlled the growth of gravitational instabilities in the early universe.

Hot and Cold Dark Matter

Dark matter in the early universe was much more strongly clumped than normal matter. Within a few million years after recombination, these dark matter clumps pulled in the surrounding normal matter. Later, gravitational instabilities caused these clumps to collapse. The normal matter in the clumps went on to form visible galaxies. The details of how this happened depend greatly on the properties of the dark matter itself.

One possibility is that dark matter consists of weakly interacting particles that are moving about relatively slowly, like the slow-moving atoms and molecules in a cold gas. This type of dark matter is called **cold dark matter**. There are several candidates for cold dark matter. One possibility is that it consists of tiny black holes that might have been produced in the early universe, but few physicists and cosmologists favor this idea. Most think instead that cold dark matter consists of an unknown **elementary particle**. One candidate is the **axion**, a hypothetical particle first proposed to explain some observed properties of neutrons. Even though axions would have very low mass, they would have been produced in great abundance in the Big Bang. Another candidate is the **photino**, an elementary particle related to the photon. Some theories of particle physics predict that the photino exists and has a mass about 10,000 times that of the proton. Physicists are looking for these types of particles using existing particle accelerators such as the Large Hadron Collider, and several experiments are under way to search for axions and photinos that are trapped in the dark matter halo of the Milky Way.

> Cold dark matter consists of relatively massive, slowly moving particles.

The other class of dark matter is known as **hot dark matter**. Hot dark matter consists of particles that are moving very rapidly. One example is neutrinos, which interact with matter so feebly that they are able to flow freely outward from the center of the Sun, passing through the overlying layers of matter as if they were not there. The universe is filled with neutrinos; calculations indicate that about 300 million of these relics of the Big Bang fill each cubic meter of space. Physicists have conducted laboratory experiments to find the mass of the neutrino, but it has not yet been measured accurately. Astronomers have been using the universe itself as a particle detector to measure the neutrino mass, and early results indicate that it's a few million times less than the mass of the electron. Neutrinos account for only a few percent of the mass of the universe, but because they are light and fast, they might have affected the formation and distribution of galaxies in the universe.

Hot dark matter consists of less massive, rapidly moving particles.

Cold and hot dark matter have different effects on structure formation because of the way they respond to a gravitational field. Slow-moving particles are more easily corralled by gravity than are fast-moving particles, so particles of cold dark matter clump together more easily into galaxy-sized structures than do particles of hot dark matter. As a result, theoretical models show that on the largest scales of massive superclusters, both hot dark matter and cold dark matter can form the kinds of structures observed; but on much smaller scales, only cold dark matter can clump enough to produce structures like the galaxies filling the universe. If instead hot dark matter dominated, then structure formation would be top-down, galaxy clusters would be older than galaxies, and the CMB would look different. None of these are observed.

At the time of recombination, the distribution of matter was remarkably uniform, with dark matter slightly clumpier than normal matter. By a few million years after recombination, the universe had expanded drastically, but the clumps of dark matter did not expand as rapidly as their surroundings did, because their self-gravity slowed down its expansion. The clumps of dark matter stood out more with respect to their surroundings. The gravity of the dark matter clumps began to pull in normal matter. Eventually, the clumps of dark matter stopped expanding when their own self-gravity slowed and then stopped their initial expansion. Unlike dark matter (which cannot emit radiation), the normal matter in the clumps radiated away energy and cooled, collapsing toward the center of the dark matter clumps (**Figure 23.9**).

Normal matter cools and falls toward the center of the dark matter halo.

These clumps did not exist in isolation; they were tugged on by the gravity of neighboring clumps and were pushed

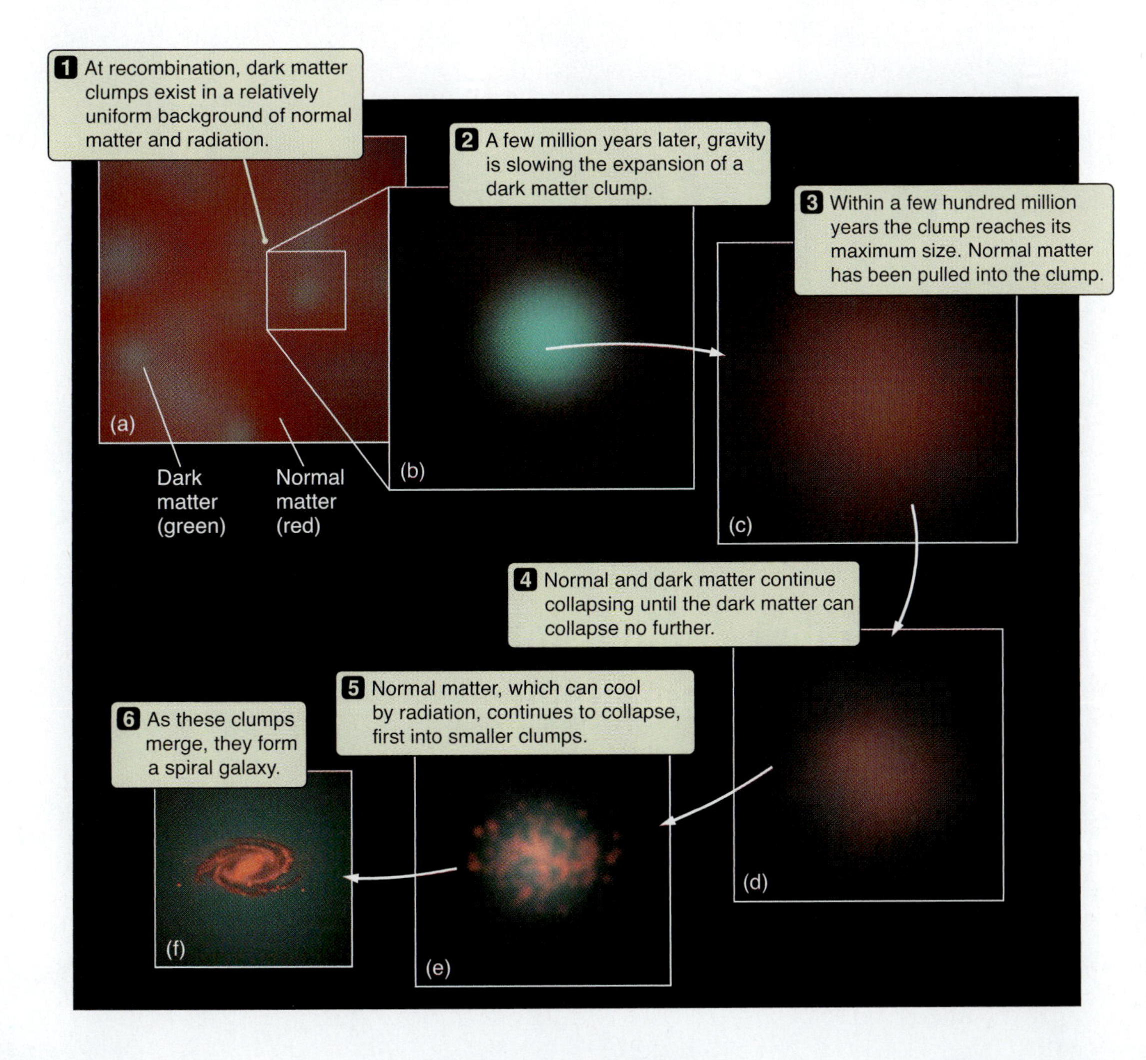

FIGURE 23.9 Stages in the formation of a galaxy from the collapse of a clump of cold dark matter.

Process of Science

NATURE DOES WHAT NATURE DOES

Roughly 90 percent of the mass in the universe is in an unknown form or composition, called dark matter. So many observations, models, and experiments provide supporting evidence for dark matter that scientists have concluded that it exists.

DARK MATTER

Galaxy rotation curves

Motions in galaxy clusters

Cosmic microwave background

Gravitational lensing

Confined hot gas in clusters

Models of cluster formation

Models of galaxy formation

Models of the Big Bang

Scientists do not have the luxury of ignoring evidence that does not fit current theories. They must see "being wrong" as an opportunity to learn and as a challenge to try harder.

around by the pressure waves that ran through the young universe, smoothing out its structure. As a result, each clump had a little bit of rotation when it began its collapse (Figure 23.9). As normal matter fell inward toward the center of the dark matter clump, this rotation forced much of the gas to settle into a rotating disk (just as the collapsing clouds of protostars settled first into an accretion disk), which later became the disk of a spiral galaxy (**Process of Science Figure**).

Cold dark matter forms galaxies.

23.3 First Light

Recall from Chapter 19 that recombination occurred approximately 400,000 years after the Big Bang, when the CMB cooled enough for electrons and protons to combine to form hydrogen and helium atoms. Before this time the universe was opaque; after recombination it became transparent, and the CMB could be observed. This epoch in the history of the universe is called the **Dark Ages** because there was no visible "light" from astronomical objects. The only light available at this time was from the cooling—and darkening—CMB, and perhaps some amount of 21-centimeter (21-cm) radio radiation from hydrogen. The Dark Ages lasted from about 200 to about 600 million years after the Big Bang. During this time objects such as the first stars, were forming in the early universe. As they formed, they heated up until they emitted UV photons with enough light to reionize neutral hydrogen. During this **reionization** stage, the hydrogen began to glow at visible wavelengths. The reionization started at about 270–480 million years ago, with photons from the first stars, and continued with star formation in the first low-luminosity galaxies and with radiation from the first supermassive black holes. Reionization was completed by about 750–900 million years ago.

Table 23.1 shows how these times compare with observed redshifts. Only within the last few years have astronomers been detecting objects at $z > 6$, representing light from the first billion years of the universe. Many astronomers were surprised by the identification of galaxies, quasars, and gamma-ray bursts at such high redshifts, as it had been thought that these objects did not form until after the universe was at least a billion years old. As we discussed in Chapter 18, gamma-ray bursts (GRBs) are extremely luminous and arise from the explosive deaths of massive stars. For GRBs to be detected at $z = 8$, there must have been massive stars that had already died by 650 million years after the Big Bang. Similarly, the detection of quasars at $z = 7$ indicates that supermassive black holes already existed 750 million years after the Big Bang. The study of the highest-redshift objects and what they tell astronomers about the early universe is one of the most dynamic topics in astronomy today. New telescopes and new instruments are regularly detecting objects with higher and higher redshifts. By the time you read this book, the highest known redshifts likely will be higher.

The First Stars

Astronomers utilize what has been learned in stellar evolution theory (Part III of this book) to study galaxy evolution and cosmology. The very first stars must have formed from the elements created in Big Bang nucleosynthesis: hydrogen, helium, and a very small amount of lithium. Observational astronomers look for stars with only these elements, but so far they have not been detected. Instead, computer simulations using the conditions in the early universe on one end, and data for the stars with the lowest amount of heavier elements that have been detected in the halo of the Milky Way on the other end, inform theoretical predictions for what happened in between (**Figure 23.10**).

The formation of the first stars would have been different from what we discussed in Chapter 15 for another reason besides their chemical content. That there were no heavy elements meant also that there was no dust, and there were no molecular clouds with cold, dense gas for the stars to form in. Instead, these stars are thought to have formed inside of dark matter minihalos, of about 0.5–1.0 million

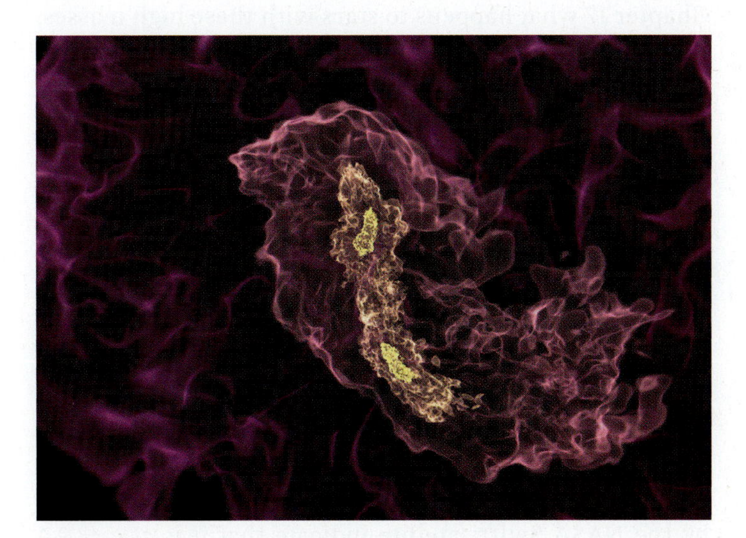

FIGURE 23.10 Supercomputer simulation of first star formation from primordial gas in a dark matter minihalo a few hundred million years after the Big Bang. In this image, two massive stars are forming a few hundred astronomical units apart. The higher-density regions are brighter.

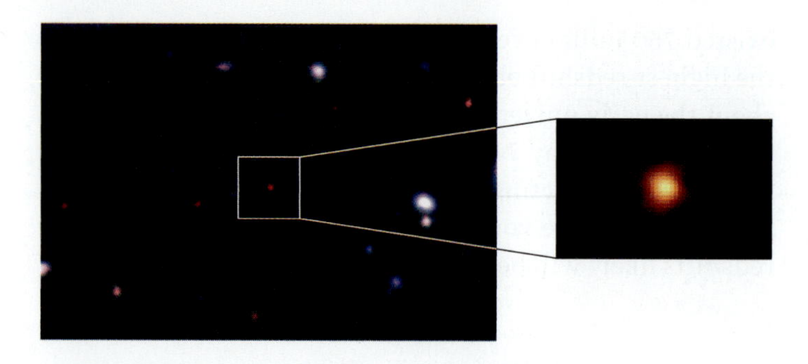

FIGURE 23.11 The NASA Swift satellite observed this very distant gamma-ray burst, which exploded 500 million years after the Big Bang ($z = 9.4$).

solar masses ($M_\odot$) and 100 pc across, that themselves had formed a few hundred million years after the Big Bang ($z \approx 20$–30). Primordial gas clouds within these minihalos contained neutral hydrogen, and over time, some small amounts of molecular hydrogen (H_2) formed. The molecular hydrogen cooled the gas; and as it cooled, it lost pressure and collapsed to the center of the minihalo. A tiny protostar grew in this gas cloud, accreting more gas to become a star. The models and computer simulations indicate that the stars that formed in this way were likely to be hot, massive, and single, double, or in small multiples (Figure 23.10). Estimates of the masses of the first stars range from 10 to over 100 $M_\odot$ for single stars and 10–40 $M_\odot$ for double stars. These stars had high luminosity, peaking in the ultraviolet, which ionized the gas near the star.

These first stars were much more massive than the average star observed today. Recall from Table 16.1 and Chapter 17 what happens to stars with these high masses: they have very short lifetimes, burning hydrogen in their cores for 10 million years or less. Today, massive stars utilize the CNO cycle for more efficient hydrogen burning, but carbon, nitrogen and oxygen were not available to these first stars. These first stars ended their brief lives in supernova explosions or as black holes. As supernovae, they scattered some heavy elements into nearby space. If the core of such a star had rapid rotation at the time of the supernova explosion, it might have emitted a gamma-ray burst of extremely high luminosity. Follow-up ground observations of a GRB detected by the NASA Swift satellite indicate that it is at $z = 9.4$ (**Figure 23.11**) and emitted the light we see 500 million years after the Big Bang.

The first stars may have been much more massive than a typical star formed today.

Some of these stars were massive enough to have become black holes after their short lifetimes. If they were in binary or multiple systems, they could have become energetic X-ray binary systems. If two or more of the resulting black holes that formed in this way were close enough that they eventually merged, they might have emitted gravitational waves in the process. These gravitational waves might be detectable in future experiments. Some theorists think that these merged black holes might have become the seeds for the supermassive black holes found in galaxies, but other models suggest it would take too long for these stellar black holes to have built up to a mass of 1 million–1 billion $M_\odot$.

The explosions of these massive first stars scattered heavy elements produced in stellar nucleosynthesis. Carbon, oxygen, and other elements might have mixed in and cooled nearby star-forming gas clouds. Some of these elements condensed into dust grains, which substantially added to the cooling of the clouds, so that the next generation of stars could have formed in a manner similar to the way stars form in today's cold molecular clouds form. These "second-generation stars" would still have had very low amounts of heavy elements, but measurably more than in the first stars. Since they formed in a cooler environment than the first stars, these stars could have had lower masses. Any stars less massive than 0.8–0.9 $M_\odot$ are still burning hydrogen on the main sequence today. These stars are not very luminous, but a few of them have been found in the halo of the Milky Way. Such stars have very low amounts of heavy elements, but their spectra show small amounts of many of the elements on the periodic table—including uranium. Astronomers are very interested in studying these second-generation stars, because they offer strong clues about the nature of the first stars and the conditions early in the history of the Milky Way.

The First Galaxies

What about the formation of the first galaxies? From a theoretical perspective, the minihalos just discussed are not considered galaxies. A galaxy is supposed to have stellar systems

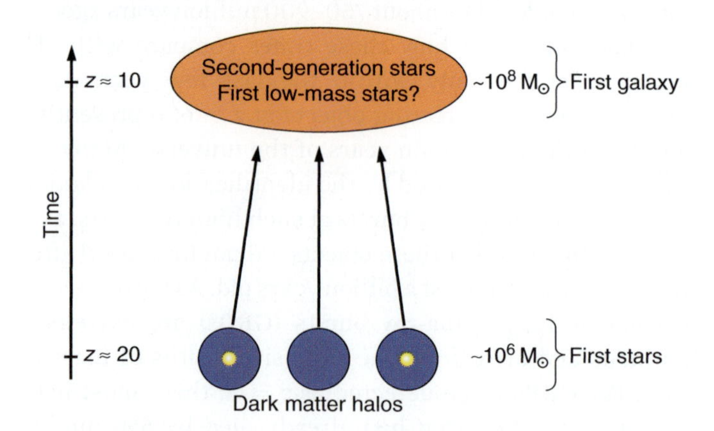

FIGURE 23.12 In "bottom-up" (hierarchical) growth, the first stars formed in minihalos and then came together to form larger halos, which became the first small dwarf galaxies.

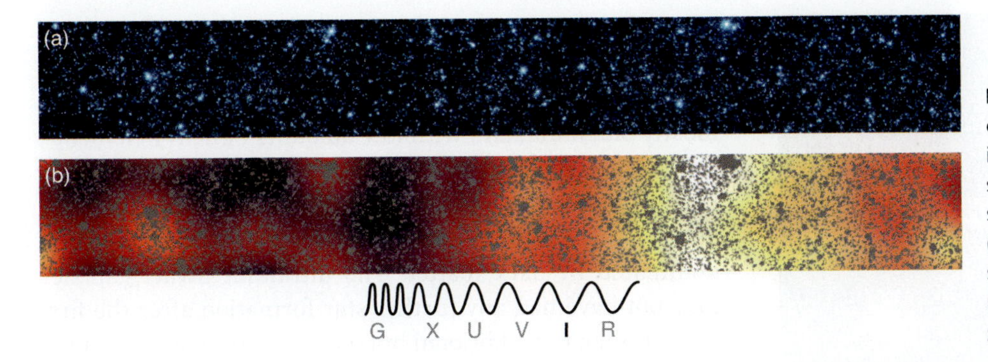

FIGURE 23.13 Spitzer telescope deep observations of a strip of sky. (a) A standard image of the nearby universe shows stars and some galaxies. (b) In this image, the nearby stars and galaxies have been subtracted out (gray) and the remaining glow enhanced, showing some structure from the early universe at a time when the earliest stars and galaxies were forming.

that live a long time and are able to retain gas heated by the ultraviolet light of stars or by supernova explosions. After the first stars died, the energy from supernovae, GRBs, or X-ray binaries may have heated any remaining gas in the minihalo too much for further star formation, or it may have escaped the minihalo because of its relatively low gravitational pull. Thus, the minihalos may have produced one generation of short-lived massive stars and may not have been able to hang on to heated or blasted gas with their small mass.

One theory is that the first galaxy was made up of the first system of stars that were gravitationally bound in a dark matter halo. These stars may have been first stars or chemically enriched second stars. The properties of the first galaxies were shaped by the radiation, the presence of some heavier elements from first-star nucleosynthesis, and the black holes produced by the first stars. The masses of these galaxies are thought to have been about $10^8\ M_\odot$, and they were built up hierarchically from the merging of minihalos (**Figure 23.12**).

From an observational perspective, one piece of evidence for this early structure is the cosmic infrared background (**Figure 23.13**). **Figure 23.13a** includes the usual nearby stars and galaxies, but when these are all subtracted, a glow remains. This remaining structure, seen in **Figure 23.13b**, likely arose from the first stars and galaxies, maybe 500 million years after the Big Bang.

Another piece of evidence comes from the discoveries of galaxies and quasars at higher and higher redshifts (**Math Tools 23.2**). These observations constrain the timeline; they indicate how soon the first galaxies—and first supermassive black holes—formed after the Big Bang. The peak light of these early galaxies has been redshifted into the infrared (Math Tools 23.2), so the 2009 near-infrared instruments

Math Tools 23.2

Observing High-Redshift Objects

New instruments and new telescopes constructed in the past few years enable astronomers to detect galaxies at higher and higher redshifts. Recall from Connections 19.1 that at high redshifts, the observed wavelengths of radiation are very different from the wavelengths emitted. The equation is as follows:

$$1 + z = \frac{\lambda_{\text{observed}}}{\lambda_{\text{emitted}}}$$

where z is the redshift. For $z = 1$, the observed wavelength is twice that emitted; for $z = 2$, three times that emitted; and so on. For the highest-redshift galaxies discovered in the last few years, with $z = 8$–10, the observed wavelengths are 9–11 times that emitted.

Neutral hydrogen in gas clouds absorbs light at wavelengths shorter (bluer) than 121.6 nanometers (nm). As a result of this absorption, a galaxy spectrum is brighter at emitted wavelengths longer than 121.6 nm, and the spectrum is fainter, or "drops out," at emitted wavelengths shorter than 121.6 nm. This dropout is noticeable even in a weak spectrum from a very faint galaxy, and even if other spectral details cannot be discerned. For a nearby galaxy at $z = 0$, this dropout occurs in the far ultraviolet, but for distant galaxies the dropout is red-shifted. For example, at $z = 9$:

$$1 + 9 = \frac{\lambda_{\text{observed}}}{121.6\ \text{nm}}$$

Solving for the observed wavelength of the dropout:

$$\lambda_{\text{observed}} = 10 \times 121.6\ \text{nm} = 1{,}216\ \text{nm} = 1.2\ \mu\text{m}$$

In this way, galaxies at the highest redshifts are detected in observations at wavelengths longer than 1.2 μm (microns) in the near-infrared, but they do not show up in optical images.

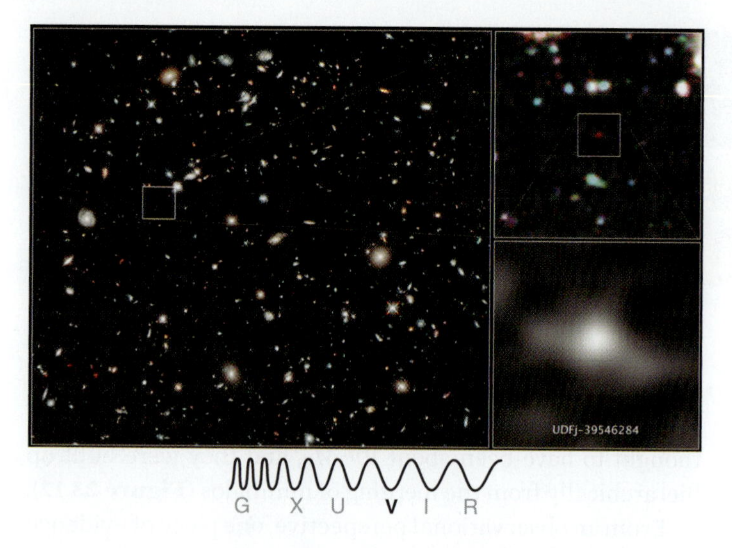

FIGURE 23.14 The youngest, highest-redshift galaxies are identified in infrared light. This HST image shows a small, faint, compact galaxy at $z \approx 10$, or 480 million years after the Big Bang. At least 100 of these small galaxies would be needed to build up a giant spiral galaxy like the Milky Way.

installed on the Hubble Space Telescope (HST) and instruments on the Spitzer Space Telescope and the Herschel Space Observatory have provided infrared images of these very young, highly redshifted objects (**Figure 23.14**). The images of the highest-redshift objects look like small, faint dots, with none of the detail seen in the pictures of closer galaxies in this text. Astronomers are excited by these images anyway, because just the detection of these objects contributes to an understanding of when and how galaxies formed.

The first galaxies are thought to have formed by about $z \approx 10$–15, or by about 400–500 million years after the Big Bang. For the second generation of stars, the physics of formation was more complicated than for the first: heavy elements and dust from the first generation mixed into the halo, along with magnetic fields and turbulence. The heavier elements carbon, oxygen, and iron cooled the gas, which then collapsed to the center of this larger dark matter halo, probably to a disk, and stars formed in the dark matter halo. These highest-redshift—that is, youngest—galaxies appear to be small, 20 times smaller than the Milky Way, which adds support for the bottom-up models of galaxy formation. Only in the past few years have astronomers been identifying galaxies at $z = 7$–10, so there are not so many yet to compare with somewhat older galaxies, at $z = 2$–6.

Galaxy formation began about 500 million years after the Big Bang.

Recall from Chapter 21 that the Local Group has small, faint dwarf galaxies orbiting the Milky Way, and streams from the dwarfs indicate that the Milky Way is still accreting matter. Recently, the number of these galaxies identified in the Local Group has doubled. About a dozen of them are called **ultrafaint dwarf galaxies** because they are dim, only 1,000–100,000 times the Sun's luminosity. They contain mostly old, faint stars with low amounts of heavy elements; about one-third of the stars with the lowest abundances of heavy elements are found in these galaxies. The mass of these galaxies compared to their luminosity suggests that they are dominated by dark matter. The ultrafaint dwarf galaxies may not have had any further star formation after the first stars, unlike the additional bursts of star formation that occurred in regular dwarf galaxies. The ultrafaint dwarfs may have contributed to building the Milky Way's halo, and they may be the oldest galaxies around that have been involved in galaxy mergers. They may even be the fossil remains of the first galaxies or of the first minihalos.

23.4 Galaxy Evolution

Galaxies continued to evolve hierarchically, with smaller protogalactic fragments merging to form larger ones. The earlier universe was smaller; recall from Chapter 19 that the light from a galaxy currently at redshift z was emitted when the universe was $z + 1$ times smaller than it is now, with a volume $(1 + z)^3$ times smaller than its volume today. The early fragments and galaxies were closer together because the universe was smaller, and therefore mergers were more

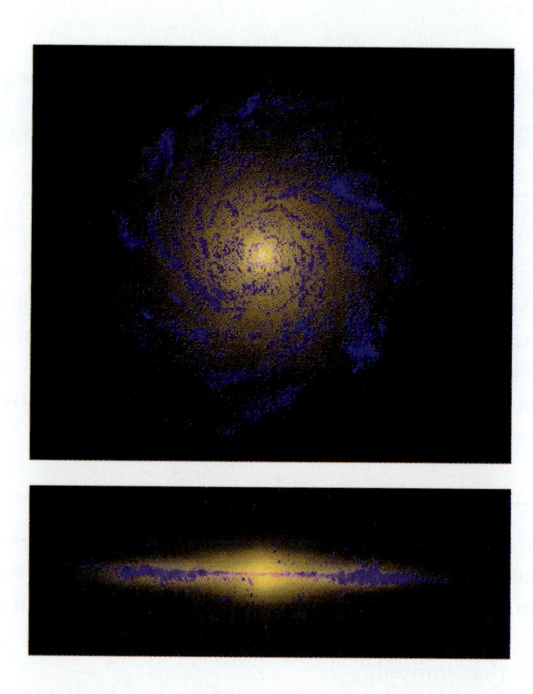

FIGURE 23.15 Supercomputer simulation of the formation of a Milky Way–sized spiral galaxy with a small bulge and a big disk, using the Lambda-CDM model. At the bottom is an edge-on view of the galaxy. Blue colors indicate recent star formation, while older stars are redder.

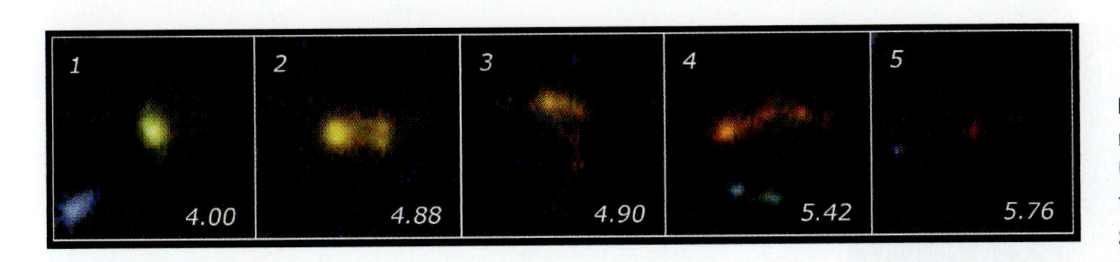

FIGURE 23.16 Young galaxies at redshifts 4–6 from the 2004 Hubble Ultra Deep Field. They are "messier" than nearby evolved spiral and elliptical galaxies.

likely. Simulations indicate that small concentrations of normal matter within the dark matter could have clumped and collapsed under their own gravity as they radiated and cooled, forming clumps of normal matter that ranged from the size of globular clusters to the size of dwarf galaxies.

In a large spiral galaxy such as the Milky Way, faint dwarf spheroidal galaxies (with dark matter) and the oldest globular clusters (without much dark matter) may be leftover protogalactic fragments. The gas collapsed to form a disk as it cooled. A recent 9-month-long supercomputer simulation that included dark matter, gravity, star formation, and supernova explosions was able to reproduce a Milky Way–like galaxy with a large disk and a small bulge (**Figure 23.15**). Observationally, astronomers conduct "stellar archaeology" on the oldest parts of the Milky Way to understand better how they are all assembled into a galaxy. For example, the oldest globular clusters are about 13.5 billion years old, whereas the halo may be as much as 2 billion years younger. The thick and thin components of the disk have taken even longer to form.

The Most Distant Galaxies

The galaxies observed at $z = 7$ are so faint that astronomers cannot see any structure. By $z = 4$–6, some structure becomes more evident (**Figure 23.16**), and young galaxies were much messier than the ones seen today. Even at $z = 0.4$–0.8, about 6 billion years ago, galaxies were much messier. **Figure 23.17** shows two numerically representative samples

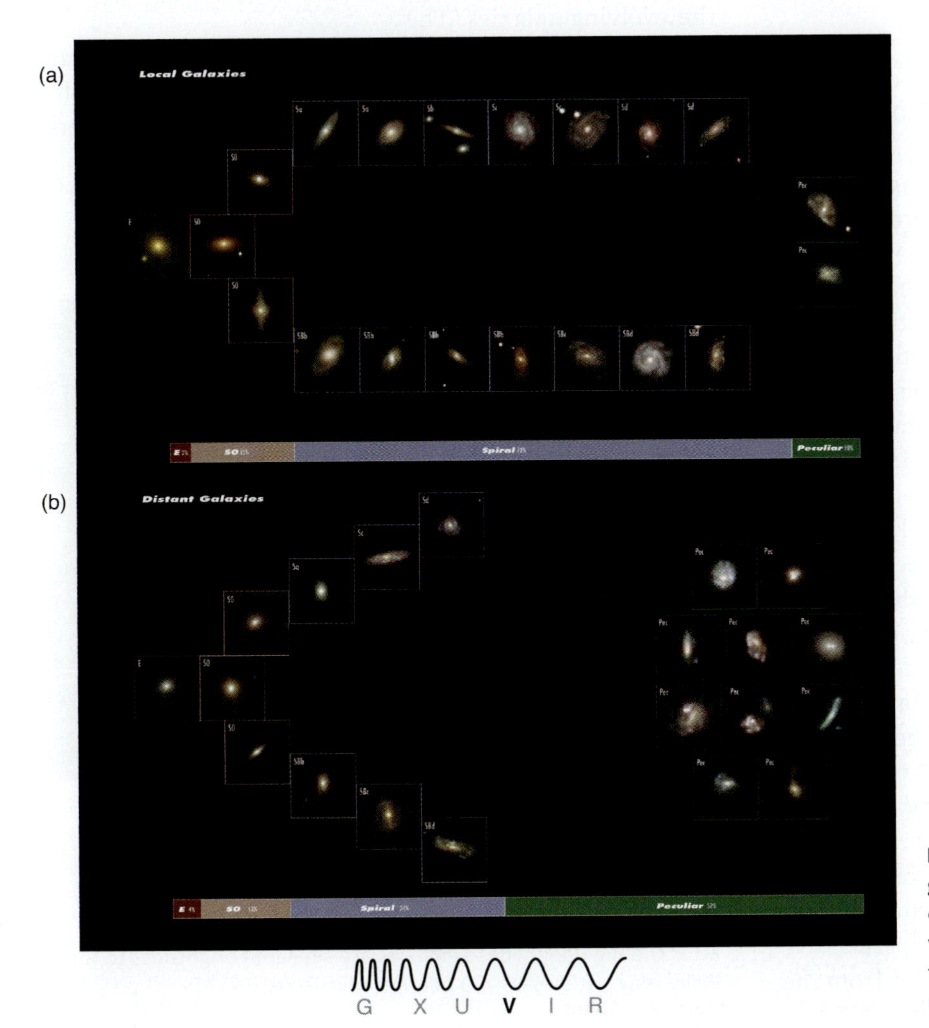

FIGURE 23.17 Hubble classification of nearby galaxies today (a), compared with the Hubble classification of galaxies 6 billion years ago (b). There were more "peculiar" galaxies and fewer spirals in the past; these peculiar galaxies must have eventually merged to become spirals.

FIGURE 23.18 The galaxy NGC 6240, as observed by Chandra in X-rays (red, orange, and yellow) and the HST (blue and white), shows that it has two black holes only 3,000 light-years apart. The black holes will likely merge in 100 million years or so.

of galaxies. The nearby galaxies ($z = 0.02–0.03$) conform to the Hubble tuning fork discussed in Chapter 20, and only about 10 percent of the galaxies are labeled "peculiar" (**Figure 23.17a**). But 6 billion years ago, over half of the galaxies were peculiar, and there were many fewer spirals (**Figure 23.17b**). (The percentages of elliptical and S0 galaxies are about the same at the two different times.) This difference suggests that it took time for many spiral galaxies to build up to their current form. The messy state of early galaxies shows that there were many mergers in the past, when galaxies were closer together.

Young galaxies are messier.

The bottom-up hierarchical merging also likely triggered the growth of the supermassive black holes at the centers of galaxies. The first supermassive black holes, which power the quasars seen at $z = 6–7$ with masses of $10^9\ M_\odot$, could have grown from the merging of minihalos with stellar black holes left after the first stars. Or they could have formed through the accretion of gas from the material between the galaxies during mergers of the first galaxies, or through rapid collapse from a hot, dense gas at the center of the first galaxies. Nearby galaxies show a correlation between the mass of the supermassive black hole and the bulge properties of the galaxies, suggesting that their growth might have been linked when they were younger. Supermassive black holes could have grown even more massive from the mergers of large galaxies too. **Figure 23.18** shows a nearby galaxy with two supermassive black holes 3,000 light-years apart and in the process of merging.

The hierarchical merging and growth of the supermassive black hole also affected the rates of star formation in the evolving galaxies. The tidal interactions between the galaxies and the collisions between gas clouds in the galaxies probably triggered many regions of star formation throughout the

Galaxy bulges and supermassive black holes grow in mass together.

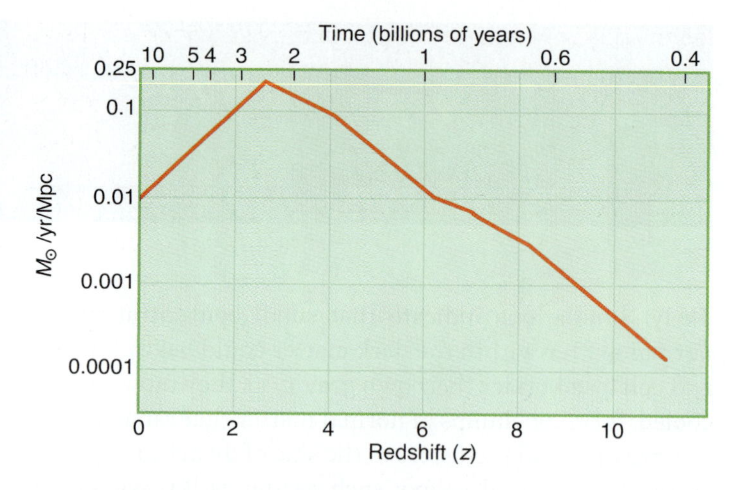

FIGURE 23.19 The rate of star formation has changed over time in the universe, peaking at about 2–3 billion years ago.

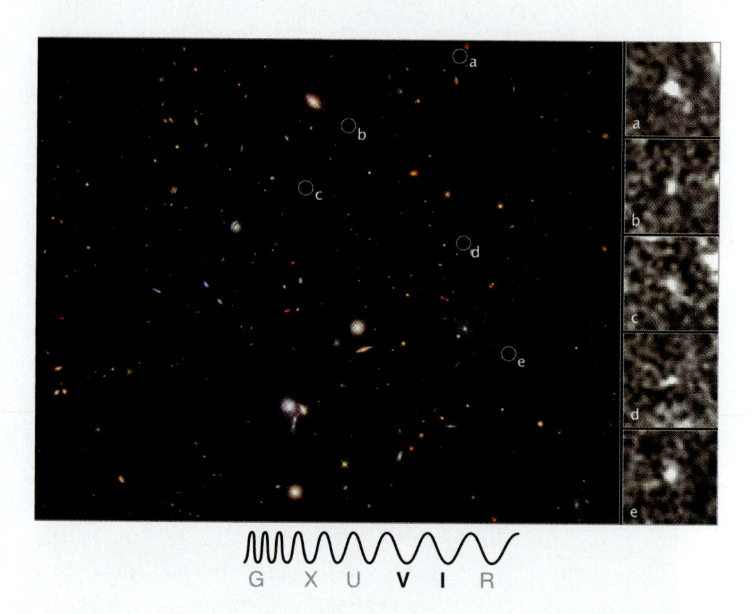

FIGURE 23.20 The circled galaxies in this HST image are tiny young galaxies forming a protocluster, at only 650 million years after the Big Bang. These galaxies are similar in brightness to the Milky Way but only 10–50 percent of its size.

combined system. Star formation generally increased sporadically over time, including a rapid increase in the 200 million years between $z = 10$ and $z = 8$. The star formation rate seems to have peaked around $z = 3$ (2.5–3 billion years after the Big Bang), before decreasing again to the current star formation rate (**Figure 23.19**). Some parallels between galaxy formation and star formation are discussed in **Connections 23.1**.

Clusters of galaxies also evolve hierarchically. **Figure 23.20** shows a young protocluster at $z = 8$,650 million years after the Big Bang. As in the images of single galaxies, these early clusters are messier than the older, closer ones. These observations offer additional evidence that bottom-up processes led to the formation of structures larger than galaxies.

Young galaxies in the process of merging are seen in HST images (**Figure 23.21**). By observing galaxy mergers at different distances, astronomers can see how they differ at various times in the history of the universe. Ellipticals are now thought to result from the merger of two or more

Connections 23.1

Parallels between Galaxy and Star Formation

As you read about galaxy formation, think back to the discussion of star formation in Chapter 15. Both processes involve the gravitational collapse of vast clouds to form denser, more concentrated structures. To help you with the comparison, we list here a few of the similarities and differences between the two.

Gravitational instability. In both star and galaxy formation, the collapse begins with a gravitational instability. Regions only slightly denser than their surroundings are pulled together by their own self-gravity. As the matter in these regions becomes more compact, gravity becomes stronger, and the collapse process snowballs. One key difference between galaxy and star formation is that for a galaxy to form, the dark matter clump must collapse rapidly enough to counteract the overall expansion of the universe itself.

Fragmentation. In both star and galaxy formation, the original cloud separates into smaller pieces as a result of the gravitational instability. However, the order of fragmentation differs between star and galaxy formation. In molecular clouds, first large regions begin to collapse, and then they fragment further to form individual stars. In contrast to this "top-down" star formation process, galaxy formation is "bottom-up": smaller structures collapse first and then merge to form galaxies and, eventually, assemblages of galaxies.

Compression, heating, and thermal support. As an interstellar molecular cloud collapses, its temperature climbs and the pressure in the cloud increases. The higher pressure would eventually be enough to prevent further collapse, except that the cloud core is able to radiate away thermal energy. That energy is the bright infrared radiation that enables astronomers to see star-forming cores. Compare this process with galaxy formation: As a dark matter clump collapses, its temperature also climbs, and it, too, reaches a point at which there is a balance between gravity and the thermal motions of the dark matter. However, dark matter is *not* able to radiate away energy, so once this balance is reached, the collapse of the dark matter is over. Only the normal matter within the cloud of dark matter is able to radiate away thermal energy and continue collapsing. That's why normal matter collapses to form galaxies, while dark matter remains in much larger dark matter halos.

As galaxies form, dark matter remains in extended halos. Dark matter is moving too fast to settle into galaxy disks, or to become concentrated into even smaller structures such as molecular clouds, or to take part in the formation of stars. Dark matter may be the dominant form of matter in the universe, and it may determine the structure of galaxies; but dark matter can never collapse enough to play a role in the processes that shape stars, planets, or the interstellar medium.

Angular momentum and the formation of disks. Conservation of angular momentum is responsible for the formation of disk galaxies, just as it is responsible for the formation of the accretion disks around young stars and for the flatness of both the Milky Way and the Solar System. The origin of the angular momentum is different, though. Whereas turbulent motions within star-forming molecular clouds produce the net angular momentum for stellar disks, gravitational interactions with nearby clumps are responsible for the angular momentum of the Milky Way.

The end product. Once a stellar accretion disk forms, most of the matter moves inward and is collected into a star. In contrast, much of the matter in a spiral galaxy remains in the disk, as discussed in Chapter 21.

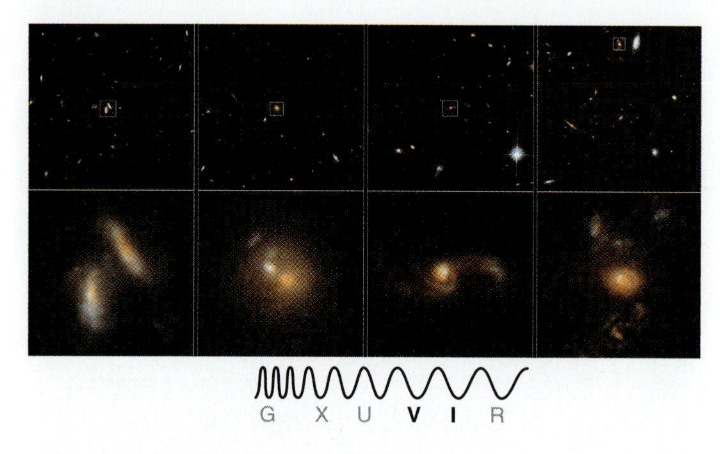

FIGURE 23.21 HST images of young, merging galaxies. From left to right, galaxies at 2.4–6.2 billion light-years from Earth merging at 11–7.5 billion years after the Big Bang.

spiral galaxies. Compare these young mergers with those of closer, older galaxies (**Figure 23.22**). (A computer simulation of such a merger is shown in Figure 6.35.) The dark matter halos of the galaxies merge, and the stars eventually settle down into the blob-like shape of an elliptical galaxy. Elliptical galaxies are known to be more common in dense clusters, where mergers are likely to have been more frequent.

Ellipticals form from the mergers of spirals.

Observations reveal *clusters* of galaxies merging as well. **Figure 23.23** shows the high-speed collision and merging of two galaxy clusters. Optical images show the individual galaxies. Ordinary matter, mostly hot gas, is seen in X-rays, and the distribution of the total mass was deduced from the gravitational lensing produced by the clusters. The ordinary matter slowed down when it collided, but the dark matter did not. This separation provides evidence for dark matter in galaxy clusters. **Figure 23.24** shows the collision of four galaxy clusters.

Simulating Structure

Astronomers use the most powerful supercomputers available to simulate the universe. These simulations start with billions of particles of dark matter and use the most recent *WMAP* data from observing the CMB. The simulations model the formation and evolution of dark matter clumps and halos, filaments and voids, small and large galaxies, and galaxy groups and clusters. The simulations seek to approximate what the universe should look like at different times (different redshifts) and can then be compared to the observations.

The Bolshoi simulation discussed here was run on NASA supercomputers in 2009. **Figure 23.25** shows how the slight variations in the CMB after inflation led to higher-density regions that became the seeds for the growth of structure. During the first few billion years, dark matter fell together into structures comparable in size to today's clusters of galaxies. The spongelike filaments, walls, and voids became well defined later (**Figure 23.26**). Zooming

FIGURE 23.22 These tidally interacting galaxies in the more nearby universe show severe distortions, including stars and gas drawn into long tidal tails.

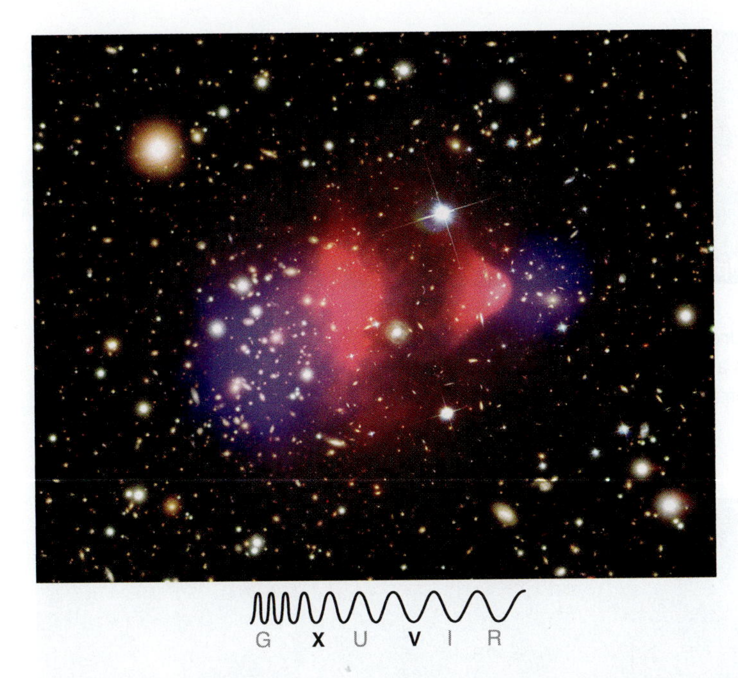

FIGURE 23.23 The Bullet Cluster of galaxies, at a redshift of $z = 0.3$, represents a later stage in the collision of two giant clusters of galaxies in the process of merging. The smaller cluster on the right seems to have moved through the larger cluster like a bullet.

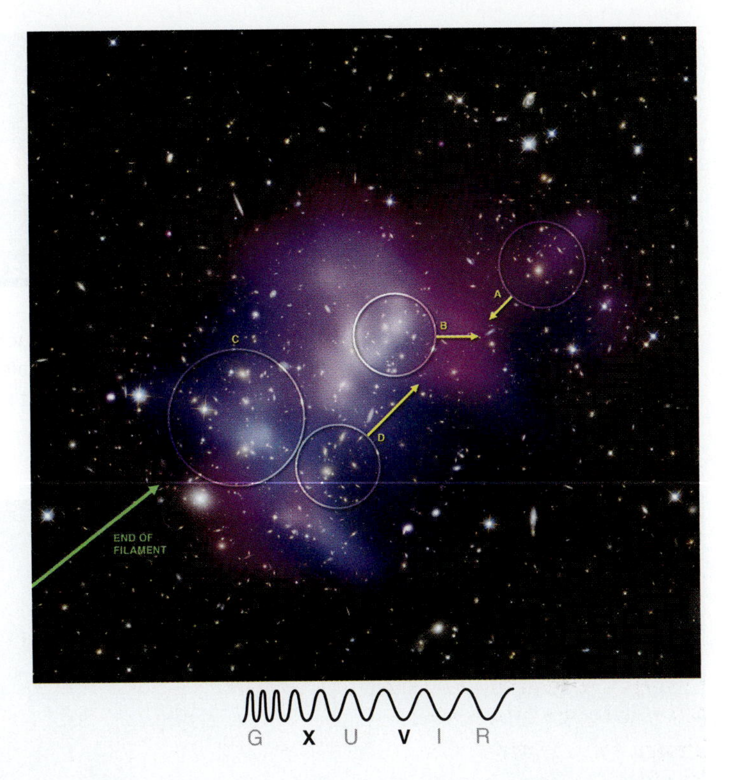

FIGURE 23.24 Four galaxy (A, B, C, D) clusters in the process of merging. In the Chandra X-ray image, the cooler gas is magenta and the hotter gas is blue. This is one of the most complex galaxy clusters observed.

in on some simulated filaments and voids shows a cluster of galaxies. The similarities between the results of the models and observations of large-scale structure are quite remarkable. **Figure 23.27** compares the simulated view with the observed slice of the universe from the Sloan Digital Survey that was pictured in Figure 23.3b. The comparison is striking. Only simulations with certain combinations of mass, CMB variations, types of dark matter, dark matter halos, and values for the cosmological constant will produce structure similar to what is actually observed. This is a very important result. Models contain assumptions consistent with observational and theoretical knowledge of the early universe, and they predict the formation of large-scale structure that is similar to what is actually seen in today's universe.

The precision cosmology developed over the past two decades has given astronomers a detailed model of the universe in space and time, so that galaxy evolution can now be reliably sequenced. **Figure 23.28** neatly summarizes the galaxy formation process, in which smaller objects form first and merge into ever-larger structure,

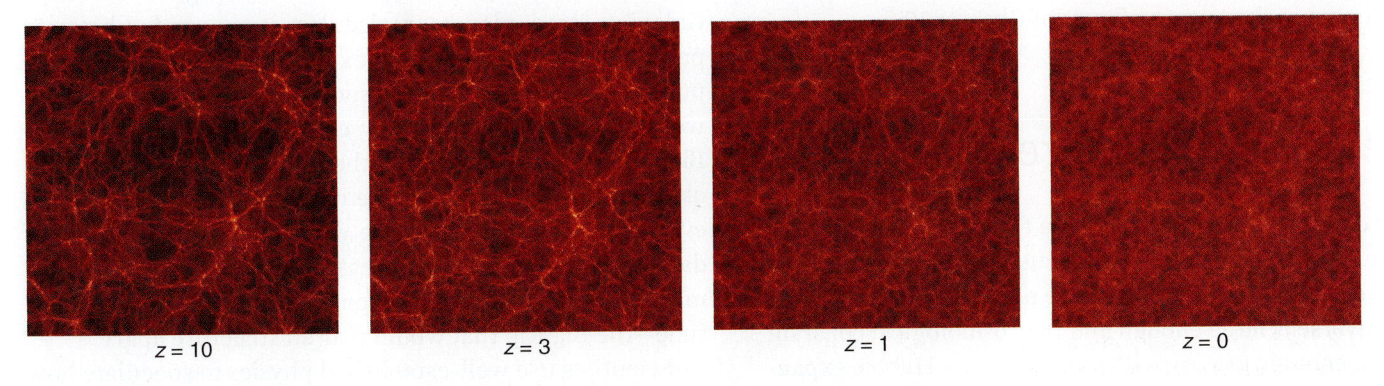

FIGURE 23.25 The Bolshoi supercomputer simulation of the formation of very large-scale structure in a universe filled with cold dark matter. The growth of filaments and voids is seen in these images.

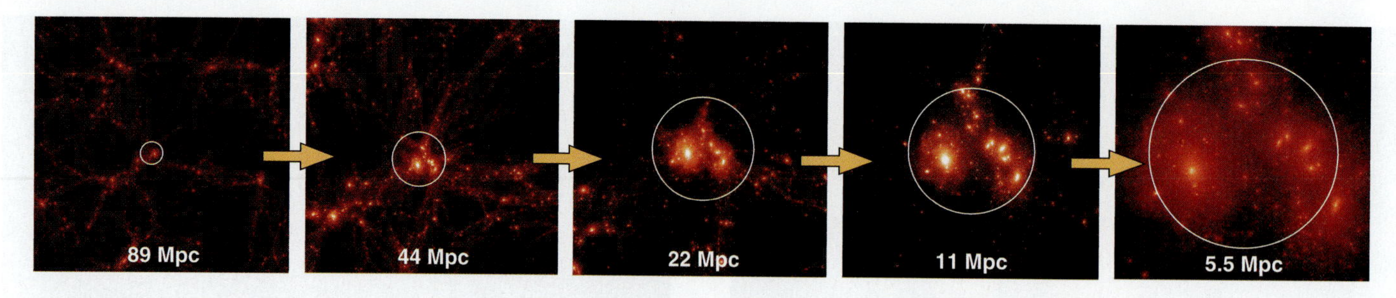

FIGURE 23.26 Simulations enable astronomers to model structure at different size scales. Each image zooms in more to show smaller structures. Dark matter halos are seen as the dense blobs in the images. The smallest blob in the last image could become a giant spiral galaxy like the Milky Way.

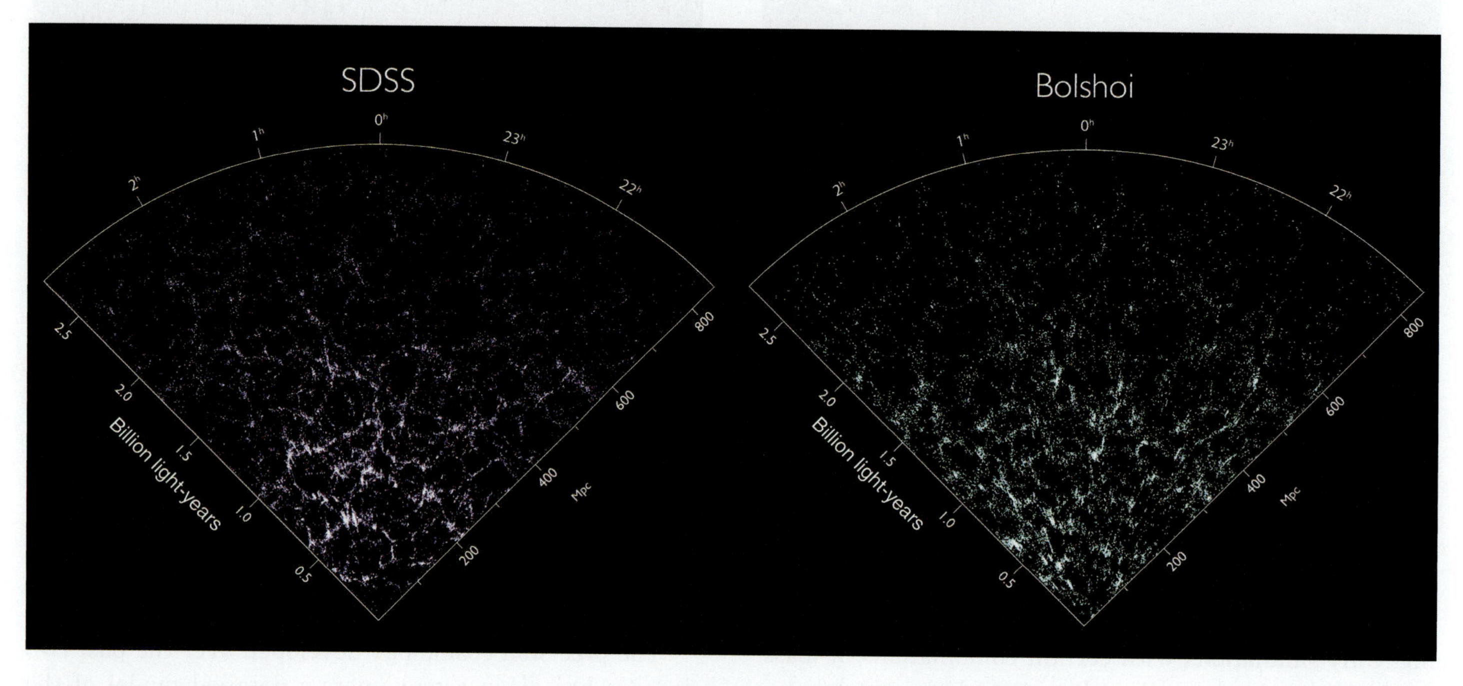

FIGURE 23.27 The large-scale structure of dark matter halos produced by the Bolshoi simulation (right) is remarkably similar to the distribution of galaxies observed in the Sloan Digital Sky Survey (SDSS, left).

leading ultimately to the Hubble Extreme Deep Field picture shown in the opening figure of Chapter 19.

23.5 The Deep Future

What does the future hold in store for the universe and its structure? In Chapter 22 we discussed how the universe evolved to its current state, and we noted that if the mass of the universe is large enough and the cosmological constant is small enough, gravity will win in the end. Hubble expansion will eventually reverse, and the universe will collapse back into a state resembling that of the young, hot universe. Galaxies, stars, planets, molecules, atoms, and subatomic particles—all might cease to exist as matter is replaced by pure energy. Perhaps from such a state, a new universe would emerge. However, few cosmologists predict such a Big Crunch for the future of the universe. Instead, current observations indicate that the universe is expanding at an ever-accelerating rate. As we discussed in Chapter 22, if dark energy's effects are increasing with time, the universe may expand so fast that it becomes infinite in size in a finite time—the Big Rip that would tear all structure apart.

Scientists use well-established physics to speculate how the existing structures in the universe will evolve over a very

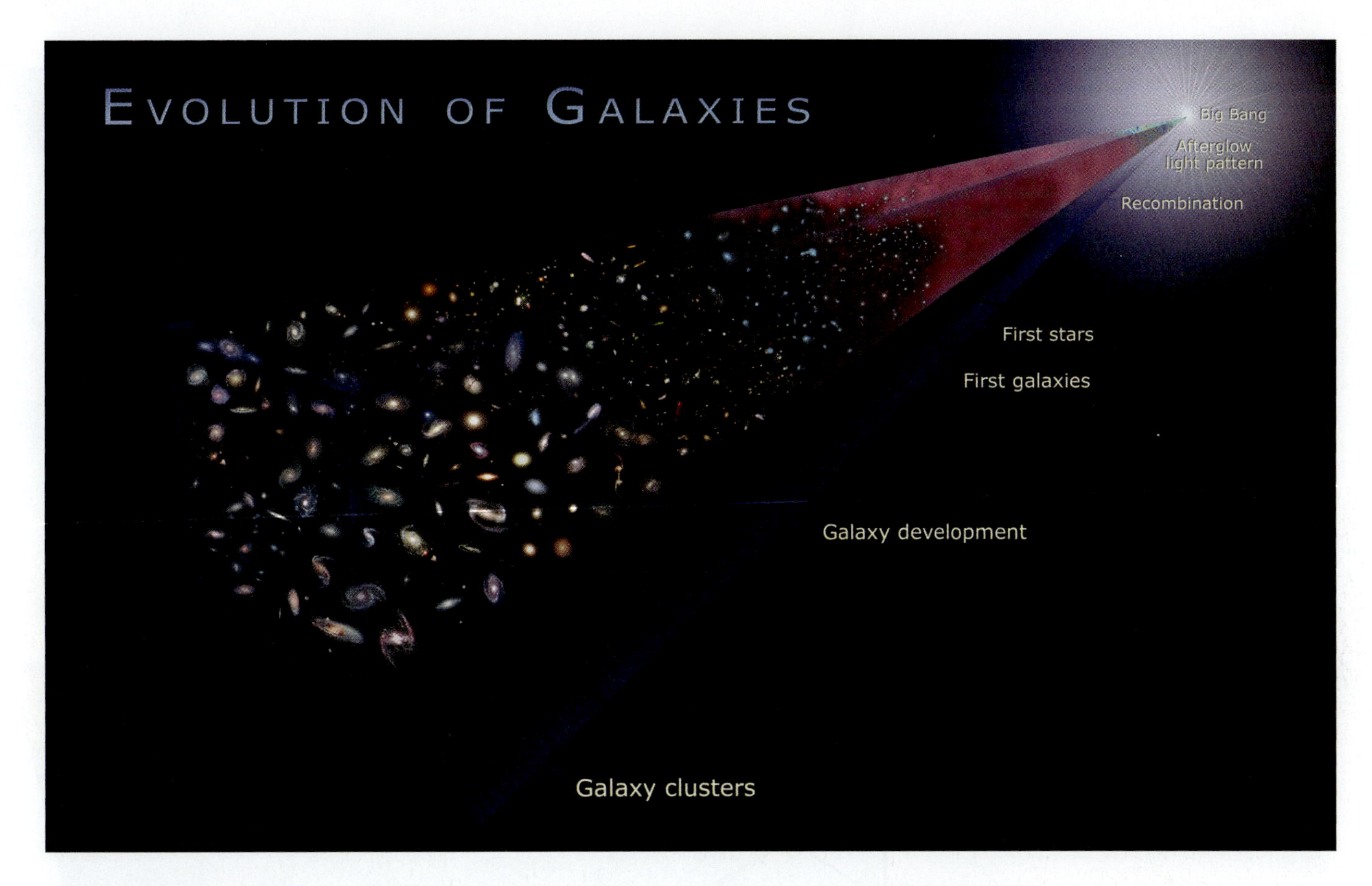

FIGURE 23.28 A schematic view of how structure formed in the universe, from smaller systems to larger ones, leading ultimately to the kind of Hubble Deep Field picture shown in the opening figure of Chapter 19.

long time. Astrophysicist Avi Loeb and others have argued that the study of galaxies and cosmology in the distant future will be much more difficult, if not impossible. In 90 billion years or so, the only visible galaxy will be the one that resulted from the merger of the Milky Way and Andromeda (see Chapter 21). Because of the acceleration of the universe, the other galaxies will be too far away to be detectable from here. In a trillion years, even the wavelength of the CMB will be greater than the size of the observable universe, and the CMB won't be visible either.

In another example, astrophysicists Fred Adams and Gregory Laughlin published their calculations of the great eras, past and future, in the history of the universe (**Figure 23.29**). During the first era, the Primordial Era—the first 400,000 years after the Big Bang and before recombination—the universe was a swarm of radiation and elementary particles. The current time is the second era, the Stelliferous Era ("era of stars"); this era, too, will end. Some 100 trillion (10^{14}) years from now the last molecular cloud will collapse to form stars, and a mere 10 trillion years later the least massive of these stars will evolve to form white dwarfs.

Following the Stelliferous Era, most of the normal matter in the universe will be locked up in degenerate stellar objects: brown dwarfs, white dwarfs, and neutron stars. During this Degenerate Era, the occasional star will still flare up as ancient substellar brown dwarfs collide and merge to form low-mass stars that burn out in a trillion years or so. However, the main source of energy during this era will come from the decay of protons and neutrons and the annihilation of particles of dark matter. Even these processes will eventually run out of fuel. In 10^{39} years, white dwarfs will have been destroyed by proton decay, and neutron stars will have been destroyed by the decay of neutrons.

As the Degenerate Era comes to an end, the only significant remaining concentrations of mass will be black holes. These will range from black holes with the masses of single stars to supermassive black holes that grew during the Degenerate Era to the size of galaxy clusters. During the period that follows—the Black Hole Era—these

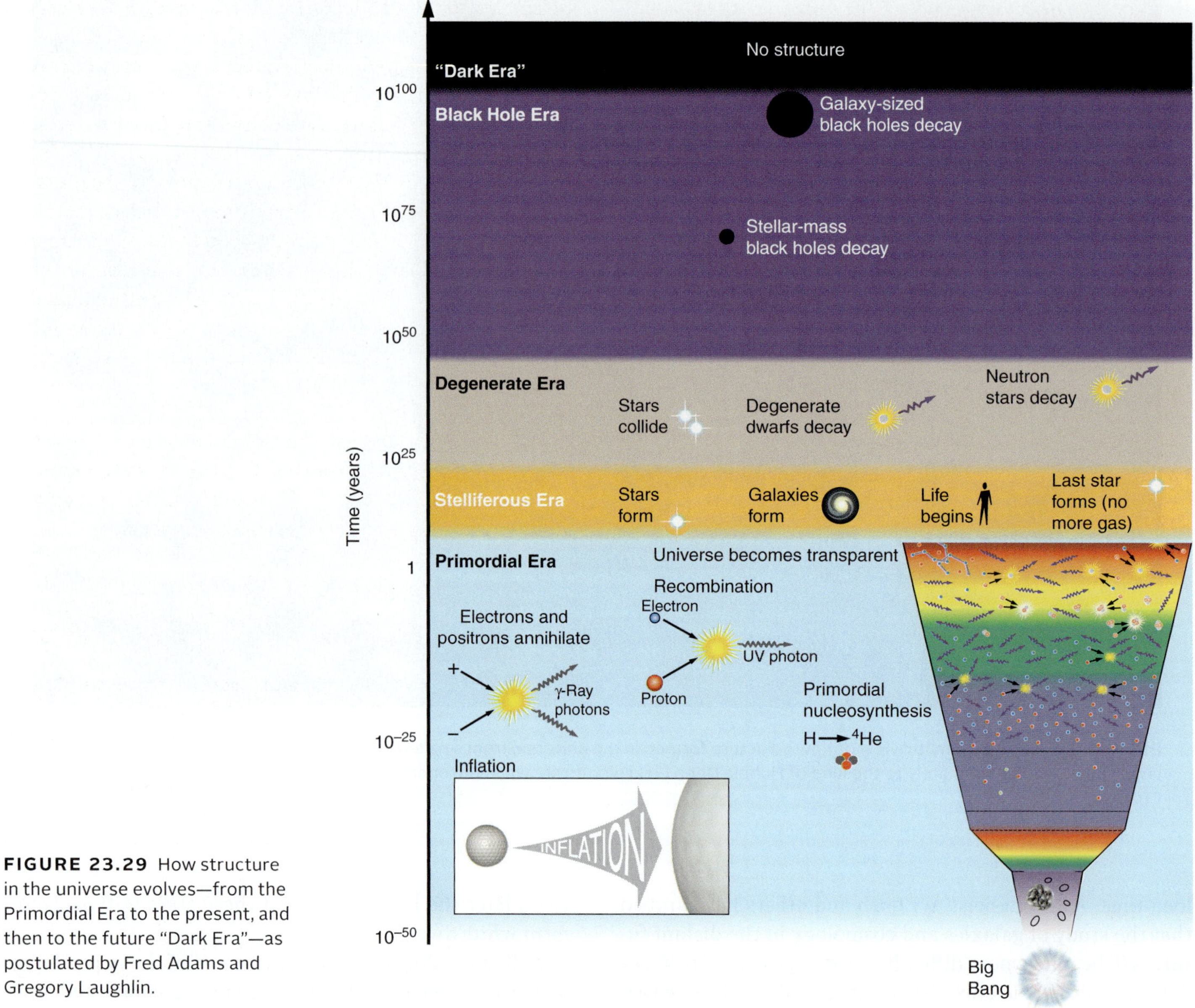

FIGURE 23.29 How structure in the universe evolves—from the Primordial Era to the present, and then to the future "Dark Era"—as postulated by Fred Adams and Gregory Laughlin.

black holes will slowly evaporate into elementary particles through the emission of Hawking radiation (see Chapter 18). A black hole with a mass of a few solar masses will evaporate into elementary particles in 10^{65} years, and galaxy-sized black holes will evaporate in about 10^{98} years. By the time the universe reaches an age of 10^{100} years, even the largest of the black holes will be gone. A universe vastly larger than our current one will contain little but photons with colossal wavelengths, neutrinos, electrons, positrons, and other waste products of black hole evaporation. The "Dark Era" will have arrived as the universe continues to expand forever—into the long, cold, dark night of eternity. This structureless state may be the final victory of **entropy**—the **heat death** of the universe.

Ultimately, all structure in the universe will decay.

23.6 Origins: We Are the 4 or 5 Percent

The very first stars initiated post–Big Bang nucleosynthesis, which chemically enriched the universe with elements heavier than lithium or boron, including the elements necessary to form planets and life on Earth. The first stars and galaxies "lit" up the universe and brought it out of the Dark Ages, thanks to the glow of their ordinary matter. But this ordinary matter is not the primary constituent of the universe. In their study of the largest structures in the universe—the galaxies, and groups and clusters of galaxies—astronomers have realized that these objects are not composed of the same stuff as stars and planets; instead, they are dominated by dark matter.

As you have seen in these last few chapters, several types of independent observations suggest that dark matter accounts for about 83 percent of all the matter in the universe. But matter is only about 26 percent of the universe, so dark matter is about 22 percent of today's universe. Dark matter dominated over normal matter at the time the first stars and galaxies formed. Dark matter dominated the evolution of the galaxies as they went through mergers to become the larger systems seen today. But despite decades of study, astronomers don't know exactly what dark matter is.

As the universe evolved, dark energy became more important. When the universe was younger and galaxies were forming, it was dominated by matter. The expansion of the universe was slowing because of the pull of gravity from all of this matter. But about 5–7 billion years ago, dark energy began to dominate over matter in the universe, and the expansion of the universe accelerated. Dark energy currently makes up about 74 percent of the mass and energy of the universe—nearly 3 times as much as the dark matter—and astronomers don't know exactly what it is either.

Ordinary matter is only about 4.6 percent of the mass energy of the present-day universe. Most of what astronomers have studied since people first looked at the sky is this 4.6 percent. The parts of the local universe that are important to life on Earth—the Sun, its planets and their moons, and the local environment in the Milky Way—are composed of this 4.6 percent. The matter that constitutes us is a surprisingly small part of the universe.

One type of structure that does not form through the action of gravity is the structure called life. In the next chapter we will discuss life on Earth and see that humans have existed for only a small part of the time of the universe. ■

Summary

23.1 Galaxies reside in groups, clusters, and larger structures, which generally formed after galaxies did. On very large scales, the structures in the small percentage of the matter content of the universe that astronomers can currently study reside mainly on filaments and walls surrounding large regions devoid of galaxies. Dark matter dominates the mass of galaxy groups and clusters.

23.2 Structure formed in the universe as slight variations in the density of the dark matter emerging from the Big Bang grew. These "seeds" then collapsed under the force of gravity, pulling in normal matter as well. Structure formed "bottom-up."

23.3 Minihalos of dark matter collapsed to form the first stars. Radiation from the first stars and their supernovae affected the growth of the first galaxies. Radiation from the first stars, galaxies, and black holes ended the Dark Ages.

23.4 Larger structures form from smaller structures. Distant young galaxies look very different from the nearby galaxies in the present-day universe; they are smaller, fainter, and more likely to be merging.

23.5 Structure will continue to form in our universe for at least another 10^{14} years, but current knowledge of quantum effects suggests that ultimately all structure, from planets to black holes, will decay.

23.6 Dark matter and dark energy are responsible for the formation of structure and the future of the universe. Ordinary matter, which makes up stars, planets, and us, is only 4 or 5 percent of the mass energy of the universe. Dark energy is playing an increasingly dominant role in the universe and will shape its future.

Unanswered Questions

- Do astronomers correctly understand how galaxy colors change over time as the constituent stars evolve off the main sequence? Stellar evolution is reasonably well established and will contribute to an understanding of galaxy evolution. Much of the work on identifying the highest-redshift galaxies is based on an understanding of their colors, how they change over billions of years, and what they were like when their light was emitted.
- How can astrophysicists gain a better understanding of dark matter? Will there be a way to detect dark matter by means other than gravity? If it's a particle, is it stable against decay? Experiments in particle accelerators on the ground and observations of Local Group halos in space have put some limits on the type of dark matter particles that might be out there, but there is still not an answer.

Questions and Problems

Summary Self-Test

1. Place the following in order of size, from smallest to largest.
   a. a galaxy
   b. star clusters
   c. the Local Group
   d. a wall
   e. Virgo cluster
   f. Virgo supercluster
   g. a star

2. The dominant factor in the formation of galaxies is the distribution of ________ in the early universe.
   a. ordinary matter
   b. dark matter
   c. energy
   d. dark energy

3. The dominant force in the formation of galaxies is
   a. gravity.
   b. angular momentum.
   c. the electromagnetic force.
   d. the strong nuclear force.

4. Larger galaxies form from the merging of small protogalaxies. This process is similar to the formation of
   a. stars.
   b. planets.
   c. molecular clouds.
   d. asteroids.

5. Large groups of galaxies are formed and held together by
   a. gravity.
   b. dark matter.
   c. photons.
   d. structural strength.

6. The difference between cold and hot dark matter is
   a. temperature.
   b. composition.
   c. the way they clump under the influence of gravity.
   d. all of the above

7. Gravitational lenses can be used to find
   a. planets in the interstellar medium.
   b. the presence of dark matter around galaxies.
   c. the masses of galaxy clusters.
   d. all of the above

8. In the far distant future, the universe
   a. will be cold and dark.
   b. will be bright and hot.
   c. will collapse and re-form.
   d. will be the same as it is now, on large scales.

9. Place the following in order of occurrence.
   a. Degenerate Era
   b. Primordial Era
   c. Black Hole Era
   d. Stelliferous Era
   e. the Big Bang

10. Place the following in increasing order.
    a. the fraction of the universe that is stars, planets, dust and gas
    b. the fraction of the universe that is dark energy
    c. the fraction of the universe that is dark matter

True/False and Multiple Choice

11. **T/F:** Nearly all the galaxy mass in the Local Group is located in just two galaxies.

12. **T/F:** Galaxies are never found in voids.

13. **T/F:** Gravity is primarily responsible for large-scale structure.

14. **T/F:** Hot dark matter cannot account for the formation of today's galaxies.

15. **T/F:** Most of the mass in the Universe is just like the mass found on Earth.

16. What is the primary difference between galaxy groups and galaxy clusters?
    a. the volume they occupy
    b. the number of galaxies
    c. the total mass of the galaxies
    d. all of the above

17. Once the redshift of a galaxy has been found, its ________ also known.
    a. mass
    b. velocity
    c. distance
    d. both b and c

18. Galaxy formation is similar to star formation because both
    a. are the result of gravitational instabilities.
    b. are dominated by the influence of dark matter.
    c. end with the release of energy through fusion.
    d. result in the formation of a disk.

19. Dark matter cannot consist of neutrons, protons, and electrons, because if it did,
    a. the abundances of isotopes would not be the same as those observed.
    b. it would have interacted with light in the early universe.
    c. stars and galaxies would be much more massive.
    d. both a and b

20. Neutrinos are an example of
    a. hot dark matter.
    b. cold dark matter.
    c. charged particles.
    d. both a and c

21. Dark matter clumps stop collapsing because
    a. angular momentum must be conserved.
    b. they are not affected by normal gravity.
    c. fusion begins, and radiation pressure stops the collapse.
    d. the particles are moving too fast to collapse any further.

22. According to our definitions in Chapter 1, dark matter is classified as
    a. an idea.
    b. a law.
    c. a theory.
    d. a principle.

23. Elliptical galaxies come from
    a. the gravitational collapse of clouds of normal and dark matter.
    b. the collision of smaller elliptical galaxies.
    c. the fragmentation of large clouds of normal and dark matter.
    d. the merging of two or more spiral galaxies.

24. Astronomers have never observed a star that has no elements heavier than boron in its atmosphere. What does this fact imply about the history of star formation in the early universe?
    a. It occurred even before galaxies were fully formed.
    b. It did not begin until after galaxies were fully formed.
    c. The earliest stars must have had very low masses.
    d. The earliest stars must have been very enriched.

25. Which observation reveals the most about how the universe will end?
    a. The universe is expanding.
    b. The universe is accelerating.
    c. The universe contains large amounts of dark matter.
    d. The universe contains large amounts of normal matter.

Thinking about the Concepts

26. Suppose you were able to view the early universe at a time when galaxies were first forming. How would it be different from today's universe?

27. Is it likely that voids are filled with dark matter? Why or why not?

28. Imagine that there are galaxies in the universe composed mostly of dark matter, with relatively few stars or other luminous normal matter. If this were true, how might you learn of the existence of such galaxies?

29. How are the processes of star formation and galaxy formation similar? How do they differ?

30. What is the origin of large-scale structure?

31. Why is dark matter essential to the galaxy formation process?

32. Which of the following is the correct evolutionary sequence? (a) Small star clusters formed first, which were bound together into galaxies, which were later bound together in clusters and superclusters; or (b) supercluster-sized regions collapsed to form clusters, which then later collapsed to form galaxies, which formed small clusters of stars. Justify your answer.

33. Why does the current model of large-scale structure require dark matter?

34. Why do scientists think that some hot dark matter exists?

35. How does a roughly spherical cloud of gas collapse to form a disk-like, rotating spiral galaxy?

36. What are some of the observational signs that dark matter exists?

37. Using the current model of galaxy formation, describe how galaxies should appear as you look further back in time. Are the features you described observed?

38. Why do astronomers think that the dark matter in the universe must be mostly cold, rather than hot, dark matter?

39. Using broad strokes, describe the process of structure formation in the universe, starting at recombination (half a billion years after the Big Bang) and ending today.

40. Why do scientists think that gravity, and not the other fundamental forces, is responsible for large-scale structure?

Applying the Concepts

41. Figure 23.3a shows the redshifts and velocities for a large number of galaxies. Find the average recession velocity for the galaxies in the wall indicated by the line labeled "Walls."

42. The theory of cosmology assumes that on large scales, the structure in the universe is uniform no matter where you look. Maps of structure, like the ones shown in Figure 23.3, support this assumption. Does the presence of large masses like the Great Attractor violate this principle? Explain your answer.

43. As clumps containing cold dark matter and normal matter collapse, they heat up. When a clump collapses to about half its maximum size, the increased thermal motion of particles tends to inhibit further collapse. Whereas normal matter can overcome this effect and continue to collapse, dark matter cannot. Explain the reason for this difference.

44. In previous chapters we painted a fairly comprehensive picture of how and why stars form. Why, then, is it difficult to model the star formation history of a young galaxy? Is this difficulty a failure of scientific theories?

45. If 300 million neutrinos fill each cubic meter of space, and if neutrinos account for only 5 percent of the mass density (including dark energy) of the universe, estimate the mass of a neutrino.

46. What are the approximate masses of (a) an average group of galaxies, (b) an average cluster, and (c) an average supercluster?

47. The lifetime of a black hole varies in direct proportion to the cube of the black hole's mass. How much longer does it take a supermassive black hole of $10^{30}\ M_{\odot}$ to decay compared to a stellar black hole of $3\ M_{\odot}$?

48. Knowing what elliptical galaxies are made of, estimate how old they must be. Knowing that ellipticals form via mergers of spirals, and knowing when galaxies first formed, estimate how long it took to complete the merging events that formed the elliptical galaxies seen today.

49. Currently, the Milky Way and Andromeda galaxies are separated by about 2.3 million light-years and are moving toward each other at about 110 kilometers per second (km/s). Estimate how long it might take for the two to collide. Why do you think your answer may or may not be a good estimate of how long it will take these galaxies to fully merge?

50. Compare the timescales in questions 48 and 49. Is the timescale of the impending merger between the Milky Way and Andromeda galaxies typical of past galaxy mergers? Can this timescale place any constraints on the distances or ages of galaxies that merged to form the elliptical galaxies seen today?

51. It is likely that the initial fluctuations leading to large-scale structure arose from quantum fluctuations in the early universe. How would the universe look different today if those fluctuations were 10 times bigger? What about 10 times smaller?

52. From Figure 23.5, estimate the distance from the Milky Way to the Great Attractor (pointed to by the arows).

53. From Figure 23.5, estimate the distance from the Shapley Supercluster to the Great Attractor.

54. Examine Figure 23.19. Is the early universe on the left or the right in this graph? (Alternatively, you could wonder if "now" was on the left or right in the graph.) Compare the star formation rate today with the star formation rate at the peak. How much more star formation occurred during the peak than occurs now?

55. Figure 23.27 shows real data in the left panel and simulated data in the right panel. These two panels are not in exact agreement. Do these differences indicate a significant problem in the simulation's ability to represent reality? Why or why not?

Using the Web

56. a. Go to the website for the Sloan Digital Sky Survey III, SDSS-III (http://sdss3.org/index.php), which has made a three-dimensional map of the sky. The 2012 video fly-through can be accessed at http://sdss3.org/press/dr9.php, and new ones may be posted as the project acquires more data. Why are the SDSS-III scientists making this map? What are the goals of the SDSS-III project? What do astronomers learn from this fly-through?

    b. Go to the website for the Galaxy and Mass Assembly (GAMA) survey (http://gama-survey.org). What are the science goals of this project? What are the observations? Are any results posted on the site?

57. Go to the website for the Bolshoi simulation (http://hipacc.ucsc.edu/Bolshoi). What is this simulation? Click on "Videos" to watch the Bolshoi and some other videos that compare observed and simulated universes. What do astronomers learn from these simulations? Are there any results yet from "Big Bolshoi"?

58. Go to the website for Galaxy Crash (http://burro.cwru.edu/JavaLab/GalCrashWeb), a Java applet that lets you run simple models of galaxy collisions. Read the sections under "Background" and then click on "Lab." Pick an exercise from the list (or go to the one suggested by your instructor), and work through the questions.

59. Use your favorite search engine to find the highest-redshift galaxy, quasar, and GRB observed so far. Why are astronomers interested in finding objects at higher and higher redshifts? Why is it also important for astronomers to estimate the relative frequency of such objects, compared to the frequency at $z = 6$ or $z = 2$?

60. Go to the website for the new ALMA telescope (http://almaobservatory.org/en). What is unique about this telescope? How will it study the Dark Ages? Why is this telescope also going to do a "Deep Field" project? How will this project be different from the Deep Field observations with the Hubble Space Telescope? Look at the items under "ALMA Latest News." Are there any reports about galaxy formation?

Exploration | The Story of a Proton

Now that you have surveyed the current astronomical understanding of the universe, you are prepared to put the pieces together to make a story of how you came to be sitting in your chair and reading these pages. It is valuable to take a moment to work your way backward through the book, from the Big Bang through all the intervening steps that had to occur, to the beginning of the book, which started with looking at the sky.

1 **In the Big Bang, how did a proton form?**

....

....

2 **How might that proton have become part of one of the first stars?**

....

....

3 **Suppose that proton later became part of a carbon atom in a 4-$M_{\odot}$ star. Through what type of nebula would it have passed before returning to the interstellar medium?**

....

....

4 **Suppose that carbon atom then became part of the molecular-cloud core forming the Sun and the Solar System. What two physical processes dominated the core's collapse as the Solar System formed and that carbon atom became part of a planet?**

....

....

5 **Beginning with the Big Bang, create a time line that traces the full history of a proton that becomes a part of the nucleus of a carbon atom in you.**

....

....

The Allen Telescope Array, shown against the background of the Milky Way Galaxy.

24 Life

An extraordinary claim requires extraordinary proof.

Marcello Truzzi (1935–2003)

LEARNING GOALS

In the "Origins" sections of each chapter throughout this book, we have discussed how astronomers think about topics in astrobiology. In this chapter we expand on some of these topics and provide a more systematic overview. By the conclusion of this chapter, you should be able to:

- Explain how, life, like planets, stars, and galaxies, is a structure that has evolved through the action of the physical and chemical processes that shape the universe.
- Present the general time line of when scientists think life began on Earth.
- Describe the concept and attributes of a habitable zone.
- Describe some of the methods used to search for extraterrestrial life.
- Explain why all life on Earth must eventually come to an end.

24.1 Life's Beginnings on Earth

Like planets, stars, and galaxies, **life** is a type of structure that evolved in the universe. Before delving into the questions of how life first appeared on Earth and how it has evolved, we first ask, How do scientists define life? Many scientists suggest that there is no universal definition of life. With the one example we know of here on Earth, we can compose a definition of terrestrial (that is, Earth) life, but a true universal definition would have to encompass different forms of life that may exist elsewhere in the universe. A complete definition would have to take into account life-forms that scientists know nothing about.

A Definition of Life

Defined on the basis of terrestrial experience, life is a set of interconnected complex biochemical processes that draw energy from the environment in order to survive and reproduce. With the assistance of specific biological molecules such as ribonucleic acid (RNA) and deoxyribonucleic acid (DNA), organisms on Earth are able to reproduce and evolve. All terrestrial life involves carbon-based chemistry and employs liquid water as its biochemical solvent. (We will take a closer look at the chemistry of life in Section 24.2.)

Are we alone in the universe?

With this basic idea of what we mean by *life*, we turn to the greater question of how life got its start here on Earth. Recall from Chapter 7 that Earth's secondary atmosphere was formed in part by carbon dioxide and water vapor that poured out of volcanoes. Then, a heavy bombardment of comets and asteroids likely added large quantities of water, methane, and ammonia to the mix. These are all simple molecules, incapable of carrying out life's complex chemistry. However, early Earth had abundant sources of energy, such as lightning and ultraviolet solar radiation, that had the ability to tear these relatively simple molecules apart, creating fragments that could subsequently reassemble into molecules of greater mass and complexity. Rain carried the molecules out of the atmosphere, and these heavier **organic** molecules (compounds that contain carbon) ended up in Earth's oceans, forming a "primordial soup" (a liquid rich in organic compounds and the environmental conditions able to support the emergence and growth of life-forms.)

In 1952, chemists Harold Urey (1893–1981) and Stanley Miller (1930–2007) attempted to create conditions similar to what they thought had existed on early Earth. To a sterilized laboratory jar containing liquid water as an "ocean," they added methane, ammonia, and hydrogen as a primitive atmosphere; electric sparks simulated lightning as a source of energy (**Figure 24.1**). Within a week, the Urey-Miller experiment yielded 11 of the 20 basic amino acids that link together to form proteins, the structural molecules of life. Other organic molecules that are components of nucleic acids, the precursors of RNA and DNA, also appeared in the mix.

Fifty years later, scientists discovered additional sealed samples from this old experiment. The samples were similar to those of the original experiment but had been used in investigations conducted with methane, ammonia, carbon dioxide, and hydrogen sulfide as the primitive atmosphere. When the samples were analyzed with modern instrumentation, 23 amino acids were found, suggesting that hydrogen sulfide, which would have come from volcanic plumes in the early Earth, was an important factor. More recent experiments with carbon dioxide and nitrogen as the primitive atmosphere have produced results similar to those of Urey and Miller. The conclusion is that a feasible atmospheric composition with an energy source can produce significant quantities of amino acids and other substances important to life.

Life is a chemical process.

From laboratory experiments such as these, scientists have developed various models to explain how life was able to get its start in locations rich in prebiotic organic molecules. The details of where and how these prebiotic molecules evolved into the molecules of life are not so clear. Some biologists think life began in the ocean depths, where volcanic vents provided the hydrothermal energy needed to create the highly organized

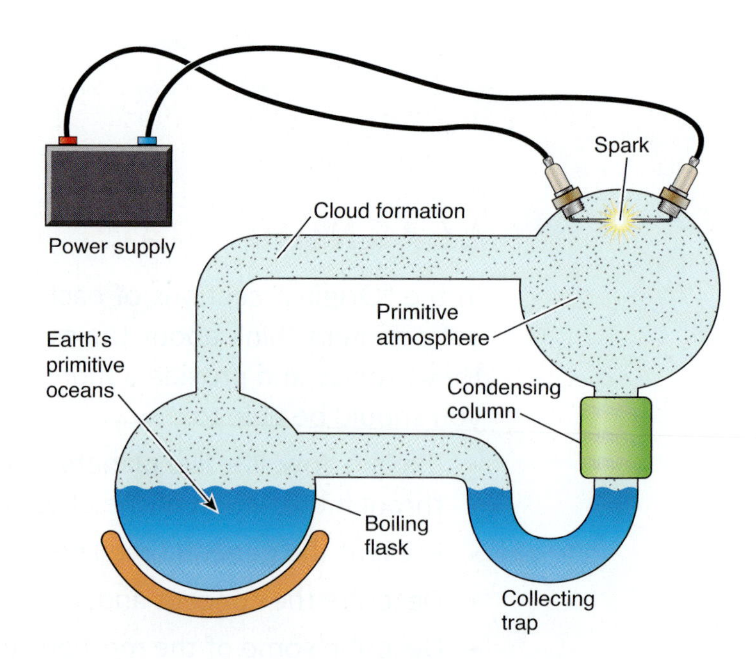

FIGURE 24.1 The Urey-Miller experiment was designed to simulate conditions in an early-Earth atmosphere.

(a)

(b)

FIGURE 24.2 (a) Life on Earth may have arisen near ocean hydrothermal vents like this one. Similar environments might exist elsewhere in the Solar System. (b) Living organisms around hydrothermal vents, such as the giant tube worms shown here, rely on hydrothermal rather than solar energy for their survival.

molecules responsible for biochemistry (**Figure 24.2**). Others think that life originated in tide pools, where lightning and ultraviolet radiation supplied the energy (**Figure 24.3**). In either case, short strands of self-replicating molecules may have formed first, later evolving into RNA, and finally into DNA, the huge molecule that serves as the biological "blueprint" for self-replicating organisms.

Terrestrial life probably began in Earth's oceans.

Finally, we should briefly mention the suggestion by a few scientists that life on Earth may have been "seeded" from space in the form of microbes (microorganisms) brought here by meteoroids or comets. Although this hypothesis, called *panspermia,* might explain how life began on Earth, it does not explain how life began elsewhere in the Solar System or beyond. However, there is no scientific evidence at this time to support the "seeding" hypothesis.

FIGURE 24.3 An alternative theory is that life began in tide pools.

The First Life

If life did indeed get its start in Earth's oceans, when did it happen? Recall from Chapter 7 that the young Earth suffered severe bombardment by Solar System debris for several hundred million years following its formation roughly 5 billion years ago. These conditions might have been too harsh for life to form and evolve on Earth. However, once the bombardment had abated and Earth's oceans appeared, the opportunities for living organisms to evolve greatly improved. Scientists are debating whether there is *indirect* evidence of terrestrial life in carbonized material found in Greenland rocks dating back 3.65–3.85 billion years. Stronger and more direct evidence for early life appears in the form of fossilized masses of simple microbes called **stromatolites**, which date back about 3.5 billion years. Fossilized stromatolites have been found in western Australia and southern Africa, and living examples still exist today (**Figure 24.4**). Scientists may never know precisely when life first appeared on Earth, but it seems that terrestrial life quickly took advantage of the favorable environment that followed the cessation of pummeling by planetesimal fragments. The current evidence suggests that the earliest life-forms appeared within a billion years after the Solar

Life on Earth may have begun more than 3.6 billion years ago.

FIGURE 24.4 Modern-day stromatolites growing in colonies along an Australian shore.

FIGURE 24.5 Thermophiles in the Grand Prismatic Spring in Yellowstone National Park. The different colors result from different amounts of chlorophyll.

System formed, and within 500 million years of the end of the young Earth's late heavy bombardment.

Analysis of DNA sequences suggests that the earliest organisms were extremophiles—life-forms that not only survive, but thrive, under extreme environmental conditions. Extremophiles include organisms such as thermophiles, which flourish in water temperatures as high as 120°C, occurring in the vicinity of deep-ocean hydrothermal vents (see Figure 24.2). Other extremophiles are found under the severe conditions of extraordinary cold, salinity, pressure, dryness, acidity, or alkalinity. Scientists today study extremophiles in boiling hot sulfur springs in Yellowstone National Park, in salt crystals beneath the Atacama Desert in Chile, at the bottoms of glaciers, in ice fields in the Arctic, and in other extreme environments (**Figure 24.5**).

Among the early life-forms was an ancestral form of **cyanobacteria**, single-celled organisms otherwise known as *blue-green algae* (**Figure 24.6**). Cyanobacteria were responsible for creating oxygen in Earth's atmosphere by photosynthesizing carbon dioxide and releasing oxygen as a waste product. (In photosynthesis, organisms use sunlight and carbon dioxide as food, and generate oxygen as a by-product.) Oxygen, however, is a highly reactive gas, and the newly released oxygen was quickly removed from Earth's atmosphere by the rusting (oxidation) of surface minerals. Only when the exposed minerals could no longer absorb more oxygen did atmospheric levels of the gas begin to rise. Oxygenation of Earth's atmosphere and oceans began about 2 billion years ago, and the current level was reached only about 250 million years ago. Without cyanobacteria and other photosynthesizing organisms, Earth's

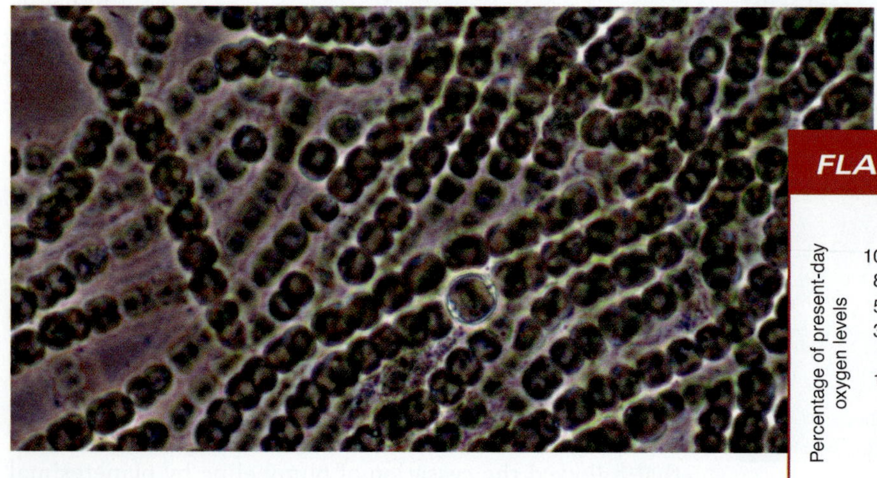

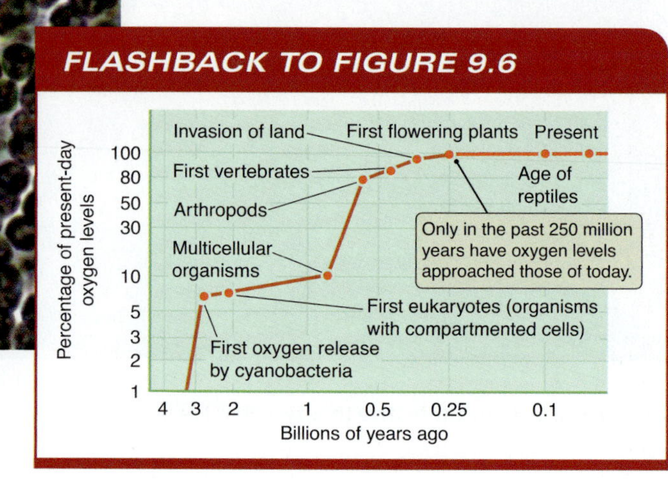

FIGURE 24.6 Present-day cyanobacteria, similar to the ones that put oxygen into Earth's atmosphere. The Flashback figure shows oxygen levels on Earth over time.

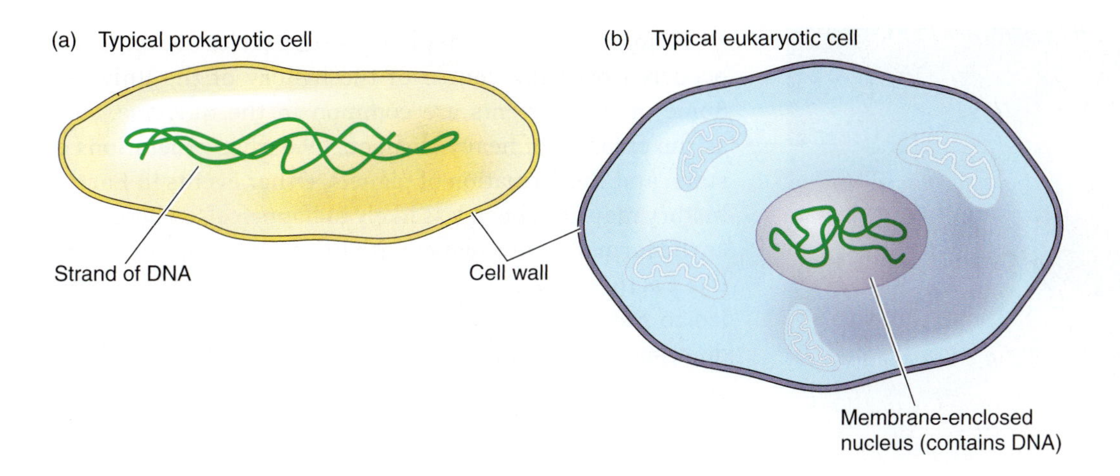

FIGURE 24.7 (a) A simple prokaryotic cell contains little more than the cell's genetic material. (b) A eukaryotic cell contains several membrane-enclosed structures, including a nucleus that houses the cell's genetic material.

atmosphere would be as oxygen-free as the atmospheres of Venus and Mars.

Biologists comparing genetic (DNA) sequences find that terrestrial life is divided into three major domains: Bacteria, Archaea, and Eukarya (consisting of bacteria, archaeans, and eukaryotes, respectively). Bacteria and archaeans are organisms, known as *prokaryotes*, that consist of free-floating DNA inside a cell wall. Prokaryotes are simple single-celled organisms, without a nucleus (**Figure 24.7a**). Even today, the largest component of Earth's total biomass is in the form of simple microbes. Eighty percent of the history of terrestrial life has been microbial. Eukaryotes have a more complex form of DNA contained within each eukaryotic cell's membrane-enclosed nucleus (**Figure 24.7b**). The first eukaryote fossils date back about 2 billion years, coincident with the rise of free oxygen in the oceans and atmosphere, although the first *multicellular* eukaryotes did not appear until a billion years later.

Life Becomes More Complex

All life on Earth, whether prokaryotic or eukaryotic, shares a similar genetic code that originated from a common ancestor. Close analysis of DNA enables biologists to trace backward to the time when different types of life first appeared on Earth, and to identify the species from which these life-forms evolved. Scientists have used DNA analysis to establish the evolutionary tree of life, which describes the interconnectivity of all species (**Figure 24.8**). The tree has revealed some interesting relationships; for example, it indicates that animals (including humans) are most similar to fungi. The evolutionary tree for primates suggests that the earliest primates branched off from other mammals about 70 million years ago, and the great apes (gorillas, chimpanzees, bonobos, and orangutans) split off from the lesser apes about 20 million years ago. DNA analysis shows that humans and chimpanzees share about 98 percent of their DNA; scientists think the two groups evolved from a common ancestor about 6 million years ago.

The pace of evolution proceeded very slowly over the eons that followed the first appearance of terrestrial life. Living creatures in Earth's oceans remained much the same—a mixture of single-celled and relatively primitive multicellular organisms—for more than 3 billion years. Then, between 540 and 500 million years ago, the number and diversity of biological species increased spectacularly.

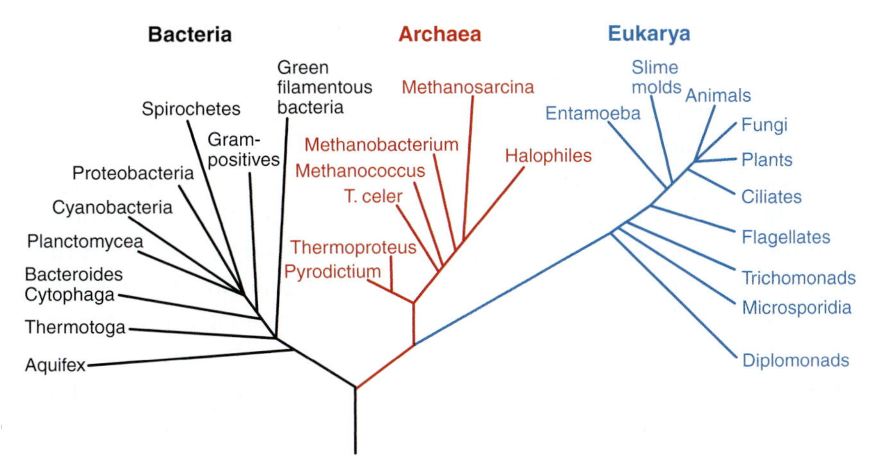

FIGURE 24.8 A basic tree of life, constructed from analysis of the DNA strands of different life-forms.

FIGURE 24.9 Artist's reconstruction from a fossil found in the Canadian Arctic of *Tiktaalik*, a fish with limblike fins and ribs. *Tiktaalik* is an animal in a midevolutionary step of leaving the water for dry land.

Biologists call this event the **Cambrian explosion**. The trigger of this sudden surge in biodiversity remains unknown, but possibilities include rising oxygen levels, an increase in genetic complexity, major climate change, a change in the Solar System's location in the Milky Way Galaxy, or a combination of these. The "Snowball Earth" hypothesis suggests that Earth was in a period of extreme cold between about 750 and 550 million years ago, before the Cambrian explosion, and was covered almost entirely by ice. During this period of extreme cold, predatory animals may have died out, setting the stage for new species to adapt and thrive. Another possibility is that a marked increase in atmospheric oxygen, O_2 (see Figure 24.6), would have been accompanied by a corresponding increase in stratospheric ozone (O_3), which shields Earth's surface from deadly solar ultraviolet radiation. With this protective ozone layer in place, life was free to leave the oceans and move to land (**Figure 24.9**).

The number of new species exploded about a half-billion years ago.

The first plants appeared on land about 475 million years ago, and large forests and insects go back 360 million years. The age of dinosaurs began 230 million years ago and ended abruptly 65 million years ago, when a small asteroid or comet collided with Earth and led to the extermination of more than half of all existing plant and animal species, according to leading theories. Mammals were the big winners in the aftermath; the subsequent evolutionary pathway led to the earliest human ancestors a few million years ago. The first civilizations occurred a mere 10,000 years ago. Present-day industrial society, barely more than two centuries old, is but a footnote in the history of life on Earth.

Humans are here today because of a series of events that occurred over the course of the history of the universe. Some of these events are common in the universe, such as the creation of heavy elements by earlier generations of stars, and the formation of planets. Other events in Earth's history may have been less likely to happen elsewhere, such as the formation of a planet with life-supporting conditions like Earth, or the creation of self-replicating molecules that led to Earth's earliest life. A few events stand out as random, such as the impact of a piece of space debris 65 million years ago that led to the extinction of many species. This event made possible the evolution of advanced forms of mammalian life and, ultimately, human beings.

Connections 24.1 shows the relative timing of the major events in the history of the universe scaled down to fit into a single 24-hour day that began with the Big Bang and ended today.

Evolution Is a Means of Change and Advancement

Imagine that just once during the first few hundred million years after the formation of Earth, a single molecule formed by chance somewhere in Earth's oceans. That molecule had a very special property: as a result of chemical reactions between that molecule and other molecules in the surrounding water, the molecule made a copy of itself. After that event there were two such molecules. Chemical reactions then produced copies of each of these molecules as well, making 4. Four became 8, 8 became 16, 16 became 32, and so on. By the time the original molecule had copied itself just 100 times, over a *million trillion trillion* (10^{30}) of these molecules existed. That is about 100 million times more of these molecules than there are stars in the observable universe. Such unconstrained replication would be highly unlikely for a number of reasons, such as the limited availability of raw materials needed for reproduction and survival.

To create life, a self-replicating molecule needed to form by chance.

Chemical reactions are never perfect. Sometimes when a molecule replicates, the new molecule is not an exact duplicate of the old one. The likelihood that a copying error will occur while a molecule is replicating increases significantly with the number of copies being made. An imperfection in an attempted copy is called a **mutation**. Most of the time such an error is devastating, leading to a molecule that can no longer reproduce at all. But occasionally a mutation is actually helpful, resulting in a molecule that is better at duplicating itself than the original was. Even if imperfections in the copying process are rare, and even if only a

Connections 24.1

Forever in a Day

The events that ultimately led to the existence of human inhabitants on planet Earth stretched out over many billions of years—intervals so large that even astronomers have difficulty visualizing them. Sometimes enormous spans of time like this are easier to grasp if they are compressed into much shorter intervals that are more familiar. For example, try to imagine the age of the universe and the important events associated with human origins as if they took place within a single day. This cosmic day begins at the stroke of midnight.

12:00:00 A.M. The embryonic universe is a mixed broth—minute specks of matter suspended in a vast soup of radiation. The entire universe exists only as an extraordinarily hot bath of photons and elementary particles.

12:00:02 A.M. At just 2 seconds after midnight, the early era in the history of the universe has passed. The fundamental forces have separated, Big Bang nucleosynthesis has formed the universe's original complement of hydrogen and helium and traces of a few other atomic nuclei, and things have cooled down enough for these atomic nuclei to combine with electrons to produce neutral atoms. Both normal and dark matter are now available to create galaxies and stars, but that process will take some time.

12:10 A.M. The first stars and then the first galaxies appear. At some point the Milky Way Galaxy becomes visible as star formation begins. Throughout the cosmic day, stars will continue to form. The more massive stars go through their brief life cycles in only 5–10 seconds of this imaginary 24-hour clock. Each massive star shines briefly, creates its heavy elements, and then disperses its material throughout interstellar space as it dies in a violent supernova explosion. Stars similar to the Sun go through less dramatic life cycles, each lasting about 16 hours in this imaginary day. Stars of less than 0.8 solar mass ($M_{\odot}$) last several dozen cosmic hours and will survive beyond the end of this cosmic day.

4:00 P.M. The Solar System forms out of a giant cloud of gas and dust. Collapse of the cloud's protostellar core and then the appearance of the Sun and the planets—including Earth—all take place within the span of a single cosmic minute.

4:05 P.M. A Mars-sized planetesimal crashes into Earth, forming the Moon.

5:20 P.M. The first primitive life appears on Earth. It evolves into the simplest life-forms: unicellular organisms such as bacteria and cyanobacteria.

8:40 P.M. More complex single-celled organisms appear, making it possible for multicellular life to develop.

9:30 P.M. The first multicellular organisms (fungi) appear on dry land.

11:00 P.M. Multicellular organisms become abundant. Their evolution paves the way for still larger and more complex life-forms.

11:20 P.M. The first animals (fish) make the transition from ocean to dry land—a major step toward human existence.

11:35 P.M. The first dinosaurs make their appearance. Various small mammals appear as well, but they remain dominated by larger and more powerful life-forms.

11:53:10 P.M. A large asteroid crashes into Mexico's Yucatán Peninsula. More than half of all species on Earth (including the dinosaurs) suddenly vanish. In the minutes that follow, the mammals, who are more adaptable to the changed environment, gain prominence.

11:59:20 P.M. The earliest human ancestors finally appear on the plains of Africa just 40 seconds before the end of this cosmic day.

11:59:58.5 P.M. *Homo sapiens* first appears 1.5 seconds before the day's end.

11:59:59.8 P.M. Modern humans arrive with only a fifth of a second to spare—a fraction of a heartbeat before the day's end.

small fraction of these errors turns out to be beneficial, after just 100 generations there are trillions of errors that, by luck, might improve on the original molecule (**Math Tools 24.1**). Copies of each of these improved molecules inherit the change. These molecules will have the property of **heredity**—the ability of one generation of structure to pass on its characteristics to future generations.

As molecules of the early Earth continued to interact with their surroundings and make copies of themselves, they split into many different varieties. Eventually, the descendants of the original molecule became so numerous that the building blocks they needed in order to reproduce became scarcer. In the face of this scarcity of resources, varieties of molecules that were more successful than others

Math Tools 24.1

Exponential Growth

Self-replication is an example of exponential growth. The doubling time, n, for exponential growth is given by the ratio of the original and final amounts:

$$\frac{P_F}{P_O} = 2^n$$

Assume a hypothetical self-replicating molecule that makes one copy of itself each minute. How many molecules will exist after an hour? Here the number of generations is given by $n = 60$, for 60 minutes in an hour:

$$\frac{P_F}{P_O} = 2^{60} = 1.2 \times 10^{18}$$

There will be a billion billion of these molecules after 1 hour.

Now suppose an imperfection in the copying process occurs only once every 50,000 times that a molecule reproduces itself, and one out of 200,000 of these errors turns out to be beneficial. After 100 generations, how many molecules with these beneficial errors might exist? This equation is similar to the previous equation, but in this case $n = 100$:

$$\frac{P_F}{P_O} = 2^{100} = 1.3 \times 10^{30}$$

The total number of molecules is 1.3×10^{30}. The number of imperfections is this number divided by 50,000, or 2.6×10^{25} imperfect molecules. The number of beneficially imperfect molecules is this number divided by 200,000, or 1.3×10^{20} molecules. So there will be 100 million trillion (10^{20}) errors that, by luck, might improve the survivability of the original molecule.

in reproducing themselves became more numerous. Varieties that could break down other varieties of self-replicating molecules and use them as raw material were especially successful in this world of limited resources. After a few generations, certain molecules dominated the mix, while less successful varieties became less and less common. This process, in which better-adapted molecules thrive and less well-adapted molecules die out, is called **natural selection**.

Mutations can breed success.

Four billion years is a long time—enough time for the combined effects of heredity and natural selection to shape the descendants of that early self-copying molecule into a huge variety of complex, competitive, successful structures. Geological processes on Earth have preserved a fossil record of the history of these structures (**Figure 24.10**). Among these descendants are human "structures" capable of thinking about their own existence and unraveling the mysteries of the stars.

The molecules of DNA (**Figure 24.11**) in the nuclei of the cells of all advanced life today are direct descendants of

FIGURE 24.10 A variety of fossils and human remains record the history of the evolution of life on Earth.

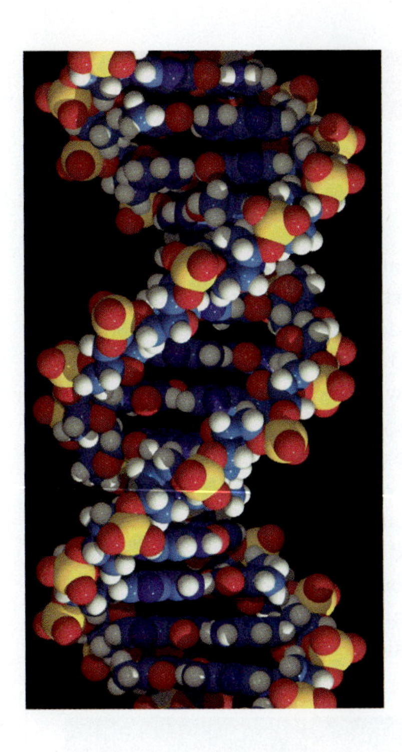

FIGURE 24.11 DNA molecules form the blueprint for life.

those early self-duplicating molecules that flourished in the oceans of a young Earth. Although the game played by the molecules of DNA in your body is far more elaborate than the game played by those early molecules in Earth's oceans, the fundamental rules remain the same. This process is inevitable: any system that combines the elements of heredity, mutation, and natural selection will evolve.

Evolution is inevitable.

24.2 The Chemistry of Life

The story of the formation and evolution of life cannot be separated from the narrative of astronomy. The universe is full of stars, and systems of planets orbit many—probably most—of those stars. The evolution of life on Earth is but one of many examples of the emergence of structure in an evolving universe (**Connections 24.2**). This point leads naturally to one of the more profound questions that scientists can ask about the universe: Has life arisen elsewhere? Unlike the study of planets, stars, and galaxies, there is only one known case for the study of life—Earth—and scientists do not know how much can be generalized to other places (**Process of Science Figure**).

Recall that the infant universe was composed basically of hydrogen and helium and very little else. But 9 billion years later, all the heavier chemical elements essential to life were present and available in the molecular cloud that gave birth to the Solar System. Those heavier elements were formed in nuclear fusion in the cores of earlier generations of stars and were then dispersed into space. When low-mass stars such as the Sun die, the carbon and oxygen created in their cores are blown off into space, eventually finding their way into molecular clouds. High-mass stars produce even heavier elements through nucleosynthesis in their cores—up to and including iron. But some of the trace elements essential to biology on Earth are even more massive than iron. They are produced within a matter of minutes during the violent supernova explosions that mark the death of high-mass stars, and then are thrown into the chemical mix found in molecular clouds.

The chemistry of life means life itself; all *known* organisms are composed of a more or less common suite of complex chemicals. Approximately two-thirds of the atoms in human bodies are hydrogen (H), about one-fourth are oxygen (O), a tenth are carbon (C), and a few hundredths are nitrogen (N); together these four are called CHON. Note that C, N, and O are the three most abundant products of stellar nucleosynthesis after helium. The several dozen remaining atomic elements in your body make up only 0.2 percent. All living creatures that we know of are assemblages of molecules composed almost entirely of the four CHON elements, along with small amounts of phosphorus and sulfur. Some of these molecules are enormous. Consider DNA, which is responsible for genetic codes. DNA is made up entirely of only five atomic elements: CHON and phosphorus. But the DNA in each cell of our bodies is composed of combinations of *tens of billions* of atoms of these same five elements. Consider also proteins, the huge molecules responsible for the structure and function of living organisms. Proteins are long chains of the smaller molecules we discussed in Section 24.1 called amino acids. Recall from the discussion of the Urey-Miller experiment that terrestrial life contains 20 specific amino acids, which also consist of no more than five atomic elements—in this case CHON plus sulfur instead of phosphorus.

All life on Earth is composed primarily of only six elements.

The chemistry of life is far too complex to have only a half dozen atomic elements. Many of the other elements, which are present in smaller amounts, are essential to the complicated chemical processes that make living organisms. These elements include sodium, chlorine, potassium, calcium, magnesium, iron, manganese, and iodine. Trace elements such as copper, zinc, selenium, and cobalt also play a crucial role in life chemistry.

Life on Earth is based on carbon, which has the important chemical property of allowing as many as four

Connections 24.2

Life, the Universe, and Everything

Popular discussions of the origin of life frequently grapple with the question of how complex, highly ordered structures such as living things could have emerged from simpler, less ordered antecedents. Physicists discuss the degree of order of a system using the concept of **entropy**, which is a measure of the number of different ways a system could be rearranged and still appear the same. For example, place a drop of ink in a glass of water and watch what happens. The ink spreads out, diffusing through the water until the only sign that the ink is there is the fact that the water is a different color. The order represented by the discrete drop of ink naturally fades away as the ink spreads out through the water. Yet no matter how long you watch, you will never see that drop of ink spontaneously reassemble itself.

Initially, the drop of ink and the glass of clear water form a neatly ordered system with low entropy. By contrast, the glass of inky water, which can be stirred or turned and still look the same, is a disordered system with higher entropy. The **second law of thermodynamics** says that, left on its own, an isolated system will always move toward higher entropy—that is, from order toward disorder.

In light of the inescapable march toward disorder that is dictated by the second law, how can ordered structure emerge spontaneously? Unscientific claims that the origin of life violates the second law of thermodynamics make a crucial mistake: they focus attention on one player while ignoring the rest of the game.

When a glass of ice water is set out on a hot, humid day, water vapor from the surrounding air condenses into drops of liquid water on the surface of the cold glass (**Figure 24.12**). Ordered structure (drops of water) spontaneously emerges from disorder (water vapor in the air), almost as if the drop of ink were to reassemble itself. This simple, everyday event appears to violate the second law of thermodynamics—but in reality it does not. When the drops condense, they release a small amount of thermal energy that slightly heats both the glass and the surrounding air. Heating something increases its entropy. The decrease in entropy due to the formation of the drops is more than made up for by the increase in the entropy of the surroundings. Ordered structure spontaneously emerges in the form of the drops, but overall, disorder increases.

A living thing, whether an amoeba or a human being, represents a huge local increase in order. However, no violation of the second law of thermodynamics is involved. The evolution of life on Earth was powered primarily by energy striking Earth in the form of sunlight, which was absorbed by photosynthetic organisms that produced oxygen and food. A local increase in order on Earth (such as you) is "paid for" in the end by the much greater decrease in order that accompanies thermonuclear fusion in the heart of the Sun. Order may emerge in localized regions within a system, but the second law of thermodynamics is obeyed overall.

The unifying theme of the evolution of the universe is the origin of structure. Whether we are discussing matter and the fundamental forces in the young universe; the gravitational collapse of stars, planets, and galaxies; the evolution of life; or water beading up on the outside of a cold glass—ordered structure does emerge spontaneously as an unavoidable consequence of the action of physical law.

FIGURE 24.12 On a humid day, water condenses into droplets on the surface of a cold glass. The second law of thermodynamics does not mean that ordered structure cannot spontaneously emerge.

Process of Science

ALL OF SCIENCE IS INTERCONNECTED

More than many other subjects, astrobiology draws on all of science and makes clear that all the fields are interconnected.

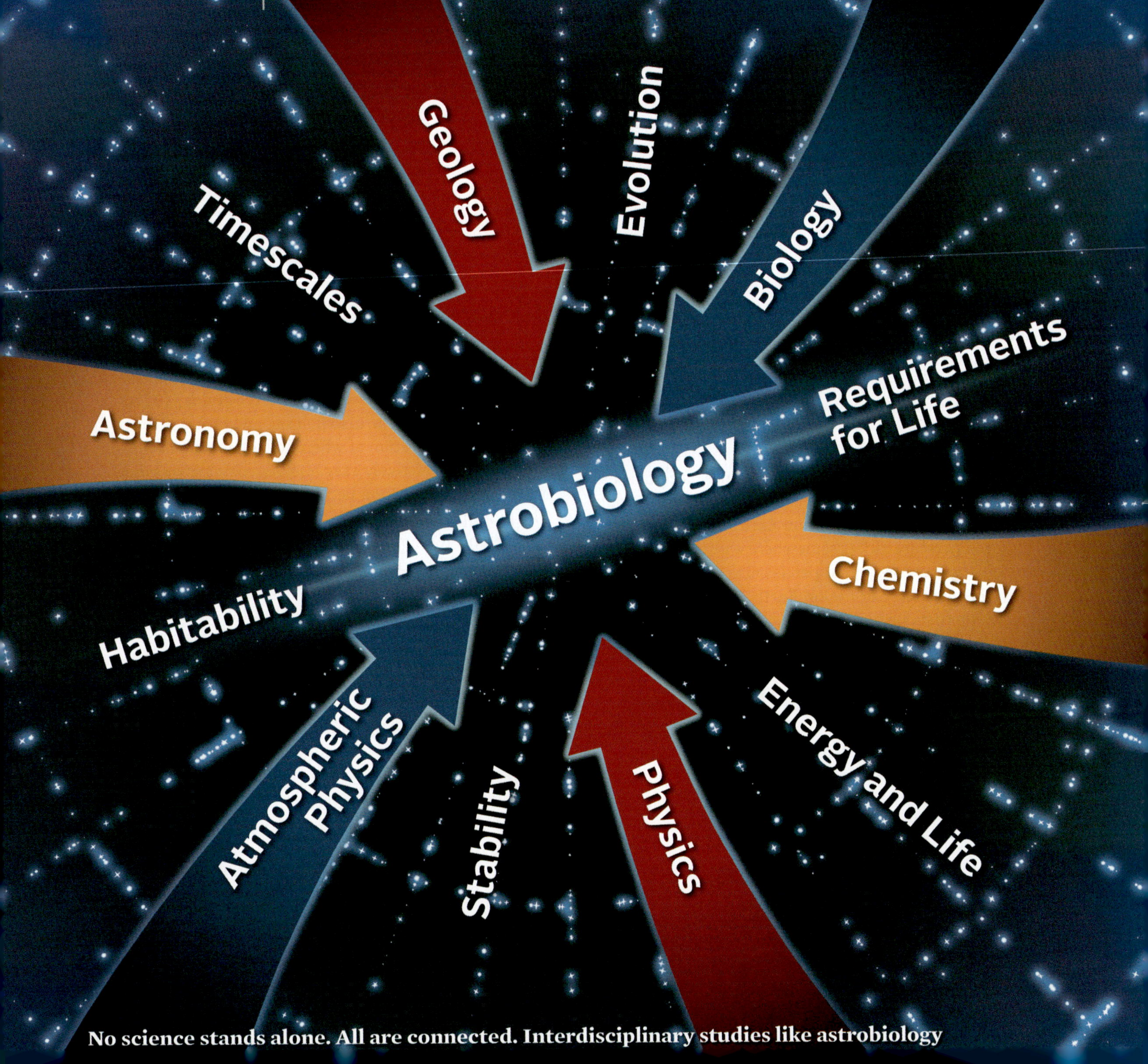

Geology

Evolution

Biology

Timescales

Requirements for Life

Astronomy

Astrobiology

Chemistry

Habitability

Atmospheric Physics

Stability

Physics

Energy and Life

No science stands alone. All are connected. Interdisciplinary studies like astrobiology

other atoms or molecules to bind themselves to each carbon atom. If the attached molecules also contain carbon, the result can be an enormous variety of long-chain molecules. This great versatility enables carbon to form the complex molecules that provide the basis for terrestrial life's chemistry.

Terrestrial life is based on carbon.

There could be forms of extraterrestrial life that are also carbon based but have chemistries quite different from that of life on Earth. For example, there are countless varieties of amino acids besides the 20 employed by terrestrial life. Furthermore, molecules other than RNA and DNA may be capable of self-replication.

Science fiction writers often speculate about silicon-based life-forms because silicon, like carbon, can bind to four other atoms. As a potential life-enabling atom, silicon has both advantages and disadvantages when compared to carbon. An advantage is that silicon-based molecules remain stable at much higher temperatures than do carbon-based molecules, possibly enabling silicon-based life to thrive in high-temperature environments, such as on planets that orbit close to their parent star. But silicon has a serious disadvantage: it is a larger and more massive atom than carbon, and it cannot form long chains of molecules as complex as those based on carbon. Any silicon-based life likely would be simpler than life-forms here on Earth, but it might exist in high-temperature niches somewhere within the universe. Although carbon's unique properties make it readily adaptable to the chemistry of life on Earth, scientists just don't know what other chemistries life might adopt elsewhere.

24.3 Life beyond Earth

The logical place to start searching for evidence of extraterrestrial life is the Solar System. People have long speculated about the possibility of life beyond Earth. Some early conjectures seem ridiculous, considering what is now known about the Solar System. Two centuries ago, the eminent astronomer Sir William Herschel, discoverer of Uranus, proclaimed, "We need not hesitate to admit that the Sun is richly stored with inhabitants." In 1877, astronomer Giovanni Schiaparelli (1835–1910) observed what appeared to be linear features on Mars and dubbed them *canali* ("channels" in Italian). In one of astronomy's great ironies, the famous observer of Mars, Percival Lowell (1855–1916), misinterpreted Schiaparelli's *canali* as "canals," suggesting that they were constructed by intelligent beings. Initial public fascination with Martians turned to hysteria in 1938 when Orson Welles aired H. G. Wells's fictitious *War of the Worlds* in the form of "live" radio news coverage of militant Martians invading Earth. Panic ensued when many listeners believed that the "invasion" was actually happening.

Exploration of the Solar System

During the mid-20th century, astronomers using ground-based telescopes discovered that Mars possesses an atmosphere, water ice, and carbon dioxide ice. Liquid water is considered essential for any terrestrial-type life to get its start and evolve. During the 1960s, the United States and the Soviet Union sent reconnaissance spacecraft to the Moon, Venus, and Mars, but the instrumentation carried aboard these spacecraft was more suited to learning about the physical and geological properties of these bodies than to searching for life. Serious efforts to look for signs of life—past or present—had to await advanced spacecraft with specialized bioinstrumentation.

In the meantime, astronomers and biologists alike were discussing where to look and what to look for. From this discussion was born the science of **astrobiology**—the study of the origin, evolution, distribution, and future of life in the universe. The question of where to look for life in the Solar System was a relatively easy one. Astrobiologists ruled out Mercury and the Moon because they lacked atmospheres. The giant planets and their moons were thought to be too remote and too cold to sustain life. The surface of Venus was far too hot, but Mars seemed more promising. In the mid-1970s, two American *Viking* spacecraft were sent to Mars with detachable landers containing a suite of instruments designed to find evidence of a terrestrial type of life. (Some scientists were critical of the specific sites chosen, claiming that higher-latitude locations, where water ice might exist, would have been preferable.) When the *Viking* landers failed to find any evidence of life on Mars, hopes faded for finding life on any other body orbiting the Sun.

Since that time, however, a better understanding of the Solar System has generated renewed optimism. Observations of Mars suggest that the planet was wetter and warmer at some time in the past—although it is not known for how long. One possibility is that there is water below the surface, which led some scientists to propose that fossil life or even living microbes might be buried under the planet's surface.

Mars missions beginning in the 1990s mapped the planet's surface from the ground and from space, but even those on the ground were not designed to explore far beneath the surface. In 2008, NASA's *Phoenix* spacecraft landed at a far-northern latitude, inside the planet's arctic circle. Specialized instruments dug into and analyzed the martian water-ice permafrost. *Phoenix* found that the martian arctic soil has a chemistry similar to the Antarctic dry valleys

FIGURE 24.13 The Mars *Curiosity* rover detected evidence that Mars had a watery past. The rounded gravel surrounding the bedrock suggests there was an ancient flowing stream.

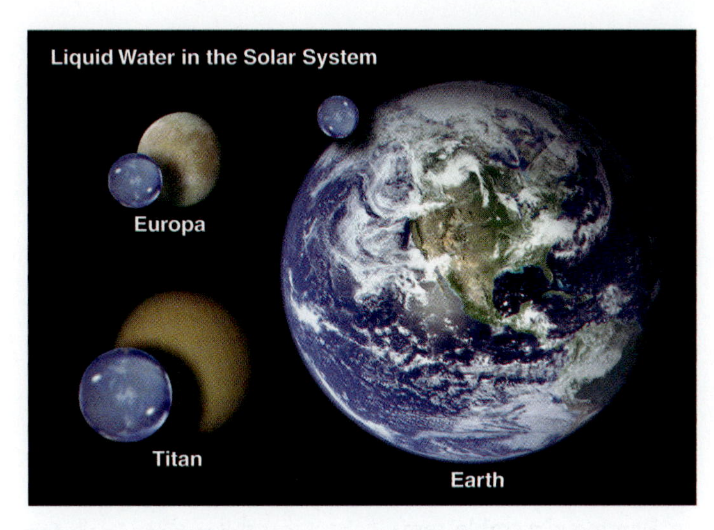

FIGURE 24.14 The total amount of liquid water (blue ball) on Jupiter's moon Europa and Saturn's moon Titan compared to the amount on Earth. All figures are drawn to scale and assume average ocean depths of 4 km (Earth), 100 km (Europa), and 200 km (Titan).

on Earth, where life exists deep below the surface at the ice-soil boundary. Aqueous minerals, such as calcium carbonate, suggested that oceans existed in the past on Mars. However, *Phoenix* did not find direct evidence of life.

The most recent Mars mission, the *Mars Science Laboratory Curiosity*, landed in Gale Crater on Mars in 2012. This large rover studies the rocks and soil of Mars in order to provide data for a better understanding of the history of the planet's climate and geology. A month after landing, *Curiosity* found evidence that a stream of liquid water had once flowed in the crater. The rover observed rounded, gravelly pebbles stuck together, which have been interpreted as coming from an ankle-to-hip-deep stream moving at 1 meter per second (**Figure 24.13**).

The next mission, the *Mars Atmosphere and Volatile Evolution Mission* (*MAVEN*), is scheduled for launch in 2013. It will study the upper atmosphere in order to learn more about the escape of carbon dioxide, hydrogen, and nitrogen, from the planet's atmosphere, and how the loss of those gases affected surface pressure and the existence of liquid water.

NASA's instrumented robots reached the outer Solar System starting in the 1980s, and many astrobiologists were surprised by the findings. Although the outer planets themselves did not appear to be habitats for life, some of their moons became objects of special interest. Jupiter's moon Europa is covered with a layer of water ice that appears to overlie a great ocean of briny liquid water (**Figure 24.14**). The water remains liquid because of high pressure and tidal heating by Jupiter. Impacts by comet nuclei may have added a mix of organic material, another essential ingredient for life. Once thought to be a frozen, inhospitable world, Europa is now a candidate for biological exploration.

Mars and the icy moons in the outer Solar System are the best candidates for life.

Saturn's moon Titan has an atmosphere that is rich in organic chemicals, many of which are thought to be precursor molecules of a type that existed on prebiotic Earth. The *Cassini* mission currently orbiting Saturn found additional evidence for a variety of prebiotic molecules in Titan's atmosphere, as well as a liquid lake of methane on the surface and probably a liquid-water ocean under the surface. In addition, the *Cassini* spacecraft detected water-ice crystals spouting from cryovolcanoes (which erupt ice crystals instead of rocks) near the south pole of Saturn's tiny moon Enceladus, suggesting that liquid water lies beneath its icy surface. Some outer-planet moons are possible habitats of extremophile life, perhaps similar to the environment found near hydrothermal vents deep within Earth's oceans.

The discovery of life on even one Solar System body beyond Earth would be exciting: if life arose independently *twice* in the same planetary system, astronomers think that would increase the probability that life exists throughout the universe.

Habitable Zones

Recall from Chapter 7 that there are hundreds of confirmed and thousands of candidate extrasolar planets within the Milky Way Galaxy. To decide which planets to focus on for further study, astronomers consider issues such as each planet's orbit, its inferred temperature, its distance from its star, and its location in the galaxy. As noted in Chapters 2, 3, and 4, astronomers think about the effects of a planet's rotation and orbit. Planets in nearly circular orbits have relatively uniform climatological environments, whereas planets in more elliptical orbits or with a large axial tilt experience more intense temperature swings that could be detrimental to the survival of life.

In Chapter 7 we discussed the idea of the **habitable zone**, the location of a planet relative to its parent star that provides a range of temperature that permits the existence of liquid water. Liquid water was essential for the formation and evolution of life on Earth, so a stable temperature that maintains the existence of water in a liquid state might be important to the development of life elsewhere. On planets that are too close to their parent star, water would exist only as a vapor—if at all. On planets that are too far from their star, water would be permanently frozen as ice. Yet another consideration is planet size: Large gas giants retain most of their light gases during formation and do not have a surface. Small planets may be rocky or a mix of water, rock, and ice; measurement of the mass and radius enables scientists to estimate the density. Most astrobiologists estimate the habitable zone of the Solar System as starting at about 0.7–0.9 astronomical units (AU) and ending at 1.2–1.4 AU (recall that 1 AU is Earth's average distance from the Sun), which obviously includes Earth and may have included Venus and Mars at the very edges of the zone at some time in the planet's history. However, this limit may be too narrow, because it does not account for the possibility of extremophile organisms that may be thriving in liquid water beneath the surfaces of some icy moons of Jupiter and Saturn.

Scientists look for planets that could have liquid water.

Astronomers must also think about the type of star they are observing when searching for planets that could have liquid water (see Figure 7.25). Stars that have less mass than the Sun ($M_\odot$), and thus are cooler, will have narrower habitable zones, and if the planet is close to its star, it might be tidally locked to the star so there is no day/night cycle. Stars that are more massive than the Sun are hotter and will have a larger habitable zone, but these stars have shorter lifetimes and, depending on their mass, might not last long enough for evolution on terrestrial timescales to take place (**Figure 24.15**). A 3-$M_\odot$ star has a lifetime of only a few hundred million years; a 1.5-$M_\odot$ star has a lifetime of a couple of billion years. On Earth, a billion years was long enough for bacterial life to form and cover the planet, but it took 3.5 billion years to reach Earth's Cambrian "explosion." Of course, evolution might happen at a faster pace elsewhere; the only known example is that of life on Earth. Still, stellar lifetime is a sufficiently strong consideration that astronomers focus their efforts on stars with longer lifetimes—specifically, stars with 0.6–1.4 $M_\odot$.

Another factor to consider is a planet's atmosphere. The ability of a planet to keep an atmosphere depends on its mass and radius (and therefore its escape velocity) and its temper-

FLASHBACK TO FIGURE 13.22

Hotter stars
Sun-like stars
Cooler stars

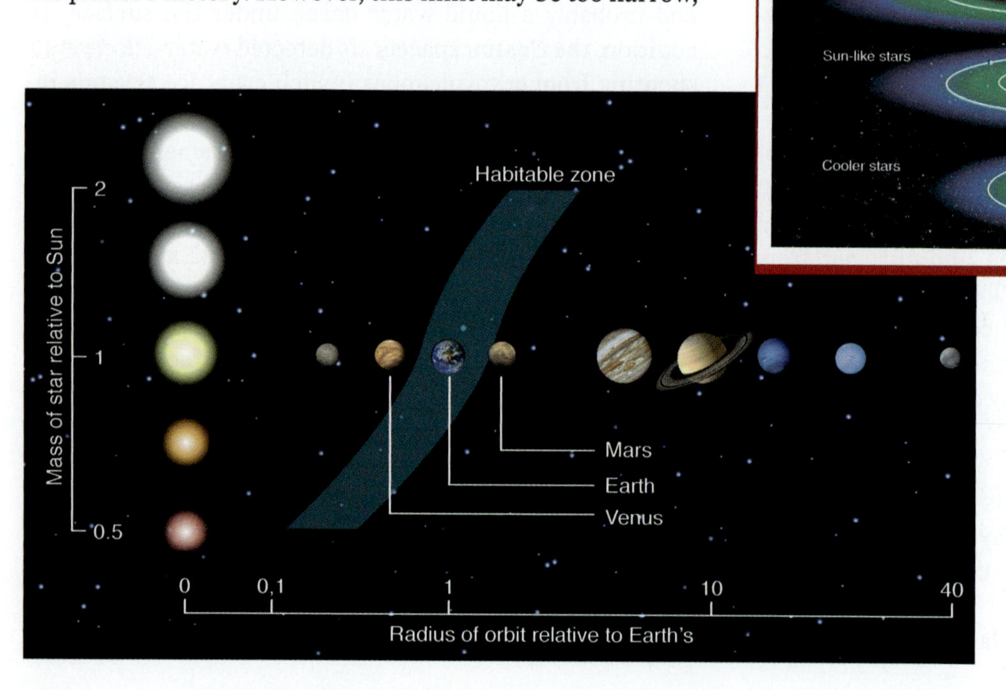

FIGURE 24.15 The habitable zone changes with the mass and temperature of a star. Habitable zones around hot, high-mass stars are more distant than the zones around cooler, lower-mass stars. The Sun and Solar system are shown for comparison.

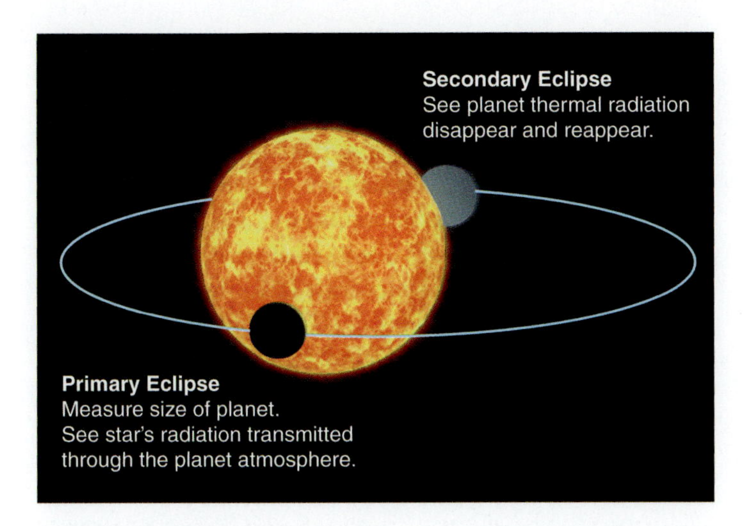

FIGURE 24.16 Artist's conception of a Jupiter-sized planet HD 189733 b observed with the Hubble and Spitzer space telescopes. To isolate the spectrum of the planet's atmosphere, astronomers observe the system when the planet and star are seen together just before an eclipse, and again when the star eclipses the planet as viewed from Earth. Subtraction of the stellar spectrum from the combined spectrum reveals water vapor, methane, carbon monoxide, and carbon dioxide on this planet.

ature. Planets that are very small may have insufficient surface gravity to retain their atmospheric gases. In the inner Solar System the Moon and Mercury were too small to keep any atmosphere. Mars lost its atmosphere over time, but the larger Earth and Venus were able to keep a thick atmosphere. Another important consideration is the greenhouse effect, which traps heat underneath an atmosphere and raises the temperature on a planetary surface. This has happened on Venus, Earth, and Mars, each of which has a higher surface temperature because of its atmosphere. The thickness and chemical content of the atmosphere affect the strength of the greenhouse effect, so that, for example, Venus is much hotter than its distance from the Sun would suggest, because of its thick atmosphere of CO_2 (see Figure 5.25 and Section 5.7). The total amount of atmosphere affects the atmospheric pressure at the surface, which, along with the temperature, determines whether water (or other molecules) can be in a liquid state on the surface. Mars's current thin atmosphere does not permit standing liquid water on its surface.

In the next few years there likely will be many observations from planet-finding projects that identify atmospheres on extrasolar planets. Water vapor, methane, and carbon dioxide may have been found on a few extrasolar planets already (**Figure 24.16**). In particular, the discovery of oxygen in the atmosphere of an extrasolar planetary atmosphere would be exciting. The oxygen in Earth's atmosphere makes it stand out from the rest of the planets and moons in the Solar System, and we know that terrestrial oxygen was created by photosynthetic life.

Recently, some astrobiologists began using two different indices to quantify the habitability of extrasolar planets. The first one, the **Earth Similarity Index (ESI)** uses the currently available data on an extrasolar planet to estimate how much it is like Earth. Factors include the radius, density, escape velocity, and surface temperature. The ESI ranges from 0 to 1. An ESI of 0.8–1 is used for rocky planets that can retain an atmosphere at temperatures suitable for liquid water—that is, that are Earth-like (**Figure 24.17**). This is the Earth-centric approach, based on the experience of life on Earth.

The **Planetary Habitability Index (PHI)** aims to be less Earth-centric and to broaden the options for habitability, but it depends on factors not yet measured or measurable for most extrasolar planets. The PHI depends on whether the planet has a surface on which organisms can grow, as well as the right kind of chemistry, a source of energy, and the ability to hold a liquid solvent. Saturn's moon Titan or Jupiter's moon Europa might satisfy these conditions, and Mars might have done so the past. Over the next decade, improvements in observations will likely lead to enough information that at least some of the extrasolar planets can be classified by their PHI.

Astronomers also consider the *galactic habitable zone*—the idea that there may be some locations within the Milky Way Galaxy where planets might have a higher probability for hosting life. Stars that are situated too far from the galactic center may have protoplanetary disks with insufficient quantities of heavy elements—such as oxygen, silicon (silicates), and iron—that make up small rocky planets and the molecules of life seen on Earth. Stars that are too close to the galactic

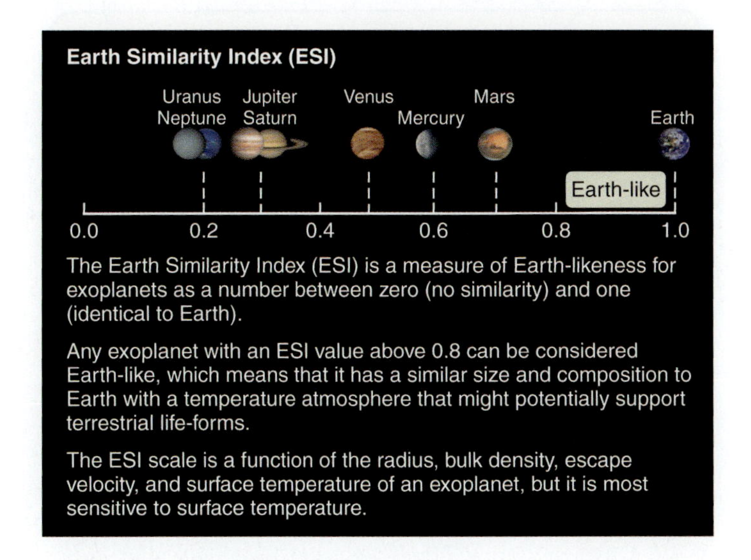

FIGURE 24.17 The Earth Similarity Index (ESI) estimates how much an extrasolar planet resembles Earth, using Solar-System planets as examples.

center may be affected by the high-energy radiation environment (X-rays and gamma rays from supermassive black holes or gamma-ray bursts), which can damage self-replicating molecules. Stars that migrate within the galaxy and change their distance from the galactic center over time may move in and out of any galactic habitable zone. ▶▶ **NEBRASKA SIMULATION: MILKY WAY HABITABILITY EXPLORER**

To summarize, astronomers try to narrow down the vast number of stars as they conduct their search. But they acknowledge that the criteria they use might not be completely applicable when looking for life in other planetary systems.

24.4 The Search for Signs of Intelligent Life

During the 1970s, scientists made preliminary efforts to send evidence of Earth civilization to space. The *Pioneer 11* spacecraft, which will probably spend eternity drifting through interstellar space, carries the plaque shown in **Figure 24.18**. It pictures humans and the location of Earth for any future interstellar traveler who might happen to find it and understand its content. Another message to the cosmos accompanied the two *Voyager* spacecraft on identical phonograph records that contained greetings from planet Earth in 60 languages, samples of music, animal sounds, and a message from then-President Jimmy Carter. These messages created concern among some politicians, who felt that scientists were dangerously advertising our location in the galaxy, even though radio signals had been broadcast into space for nearly 80 years at the time. The messages were also criticized by some philosophers, who claimed that they contained ridiculous anthropomorphic assumptions about the aliens being sufficiently like Earthlings to decode them.

Sending messages on spacecraft is an inefficient way to make contact with extraterrestrial life, but it was a significant gesture nonetheless. A somewhat more practical effort was made in 1974, when astronomers used the 300-meter-wide dish of the Arecibo radio telescope to beam a message in binary code (**Figure 24.19**) toward the star cluster M13, located 25,000 light-years away. (If someone from M13 answers, the message will arrive in 50,000 years.) In 2008, a radio telescope in Ukraine sent a message to the exoplanet Gliese 581 c. The message was composed of 501 digitized images and text messages selected by users on a social networking site and will arrive at Gliese 581 c in 2029.

The Drake Equation

The first serious effort to search for intelligent extraterrestrial life was made by astronomer Frank Drake (1930–) in 1960. He developed the **Drake equation**, which seeks to estimate

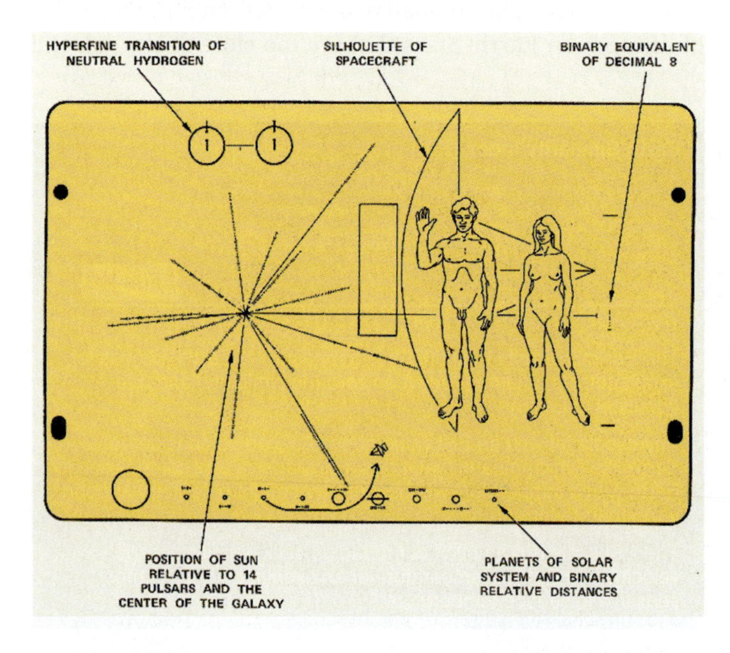

FIGURE 24.18 The plaque included with the *Pioneer 11* probe, which was launched in 1973 and will eventually leave the Solar System to travel in interstellar space.

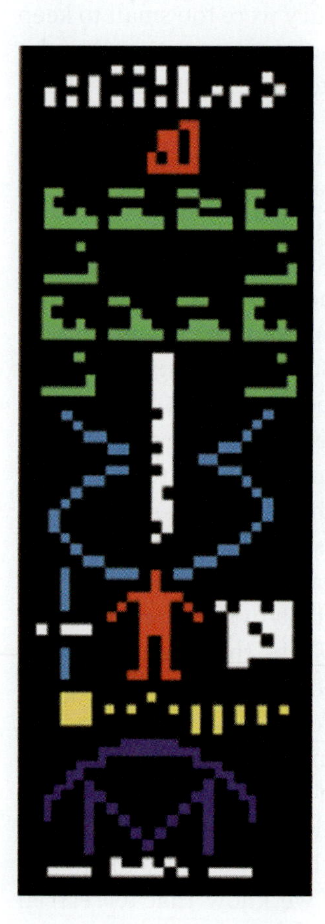

FIGURE 24.19 The message beamed toward the star cluster M13 in 1974. This binary-encoded message contains the numbers 1–10, hydrogen and carbon atoms, some interesting molecules, DNA, a human with description, basics of the Solar System, and a depiction of the Arecibo telescope. A reply may be forthcoming in 50,000 years.

the likely number (N) of intelligent civilizations currently existing in the Milky Way Galaxy. This is a way to sort out some of the factors that relate to the conditions that Drake thought must be met for a civilization to exist. The Drake equation is different from the other equations in this book because the values for each variable are uncertain. In addition, there is no way of knowing whether there are some effects that greatly change the values of these variables or whether there are some other factors that have been completely missed. The equation is discussed further in **Math Tools 24.2**.

As illustrated in **Figure 24.20**, any conclusions drawn using the Drake equation depend a great deal on the assumptions made and therefore the numbers entered into the equation. For the most pessimistic estimates, the Drake equation sets the likelihood of finding a technological civilization in our galaxy at about 1, in which case we are the *only* technological civilization in the Milky Way. Such a universe would still be full of intelligent life. Because the observable universe contains 100 billion galaxies, even these pessimistic assumptions mean that there could be 100 billion technological civilizations out there somewhere. However, if the nearest neighbors are in another galaxy, they are *very* far away—millions of light-years on average.

The Drake equation estimates the number of technologically advanced civilizations in the galaxy.

Math Tools 24.2

Putting Numbers into The Drake Equation

The Drake equation states that the number, N, of extraterrestrial civilizations in the Milky Way Galaxy that can communicate by electromagnetic radiation is given by:

$$N = R^* \times f_p \times n_e \times f_l \times f_i \times f_c \times L$$

where the factors on the right-hand side of the equation are explained as follows:

1. R^* is the number of stars that form in the Milky Way Galaxy each year—roughly 5–7.
2. f_p is the fraction of stars that form planetary systems. The discoveries of extrasolar planets over the past two decades have shown that planets form as a natural by-product of star formation and that many—perhaps most—stars have planets. For this calculation astronomers assume that f_p is between 0.5 and 1.
3. n_e is the number of planets and moons in each planetary system with an environment suitable for life. In the Solar System, this number is at least 1 (for Earth), but it could be more if Mars or an outer-planet moon has suitability for life.
4. f_l is the fraction of suitable planets and moons on which life actually arises. Remember that just a single self-replicating molecule may be enough to get the ball rolling. Some biochemists think that if the right chemical and environmental conditions are present, then life *will* develop, but others disagree. Values of f_l range from 100 percent (life always develops) to 1 percent (life is more rare). Astronomers use a range of 0.01 to 1.
5. f_i is the fraction of those planets harboring life that eventually develop intelligent life. Intelligence is certainly the kind of survival trait that might often be strongly favored by natural selection. Yet on Earth it took about 4 billion years—roughly half the expected lifetime of the Sun—to evolve tool-building intelligence. The correct value for f_i might be close to 0.01, or it might be closer to 1. The truth is, no one knows.
6. f_c is the fraction of intelligent life-forms that develop technologically advanced civilizations—that is, civilizations that send communications into space. With only one example of a technological civilization to work with, f_c is unknown. Astronomers estimate f_c to be between 0.1 and 1.
7. L is the number of years that technologically advanced civilizations exist. This factor is certainly the most difficult of all to estimate, because it depends on the long-term stability of these civilizations. There has been a technological civilization capable of sending communications into space on Earth for about 70 years, and during that time people have developed, deployed, and used weapons with the potential to eradicate this civilization and render Earth hostile to life for many years to come. People have also so degraded the planet's ecosystem that many respectable biologists and climatologists wonder whether Earth is in danger of reducing or losing its habitability. Do all technological civilizations destroy themselves within a thousand years? Or do most technological civilizations learn to use their technology for survival rather than self-destruction, and instead survive for a million years? Astronomers usually put a value between 1,000 years and 1 million years in their estimates.

At the other extreme—taking the most optimistic numbers, which assume that intelligent life arises and survives everywhere it gets the chance—the Drake equation says that there could be 40 million technological civilizations in the Milky Way alone! In this case the nearest neighbors may be "only" 40 or 50 light-years away. If scientists in that nearest alien civilization are listening to the universe with their own radio telescopes, hoping to answer the question of other life in the universe for themselves, then as you read this page those scientists may be puzzling over an episode of the original *Star Trek* television series, which began in 1966.

If humans did meet a technologically advanced civilization, what would it be like? The Drake equation suggests that it's highly unlikely there are neighbors nearby, unless civilizations typically live for many thousands or even millions of years. If that were the case, any civilization we encountered would almost certainly have been around for much longer than we have. Having survived that long, would its members have learned the value of peace, or would they have developed strategies for controlling their pesky neighbors? Fictional movies and books tell amusing and thoughtful stories that explore what life in the universe might be like (**Figure 24.21**). How do scientists search for intelligent life?

Technologically Advanced Civilizations

During lunch with colleagues, the physicist Enrico Fermi (1901–1954), a firm believer in extraterrestrial life, is reported to have asked, "If the universe is teeming with aliens ... where is everybody?" Fermi's question—first posed in 1950 and sometimes called the *Fermi paradox*—remains unanswered. If intelligent life-forms are common but interstellar travel is difficult or impossible, would the aliens send out messages—perhaps by electromagnetic waves instead? And if they did, why haven't astronomers detected these messages?

Drake used what was then astronomy's most powerful radio telescope to listen for signals from intelligent life around two nearby stars, but he found nothing unusual. His original project has grown over the years into a much

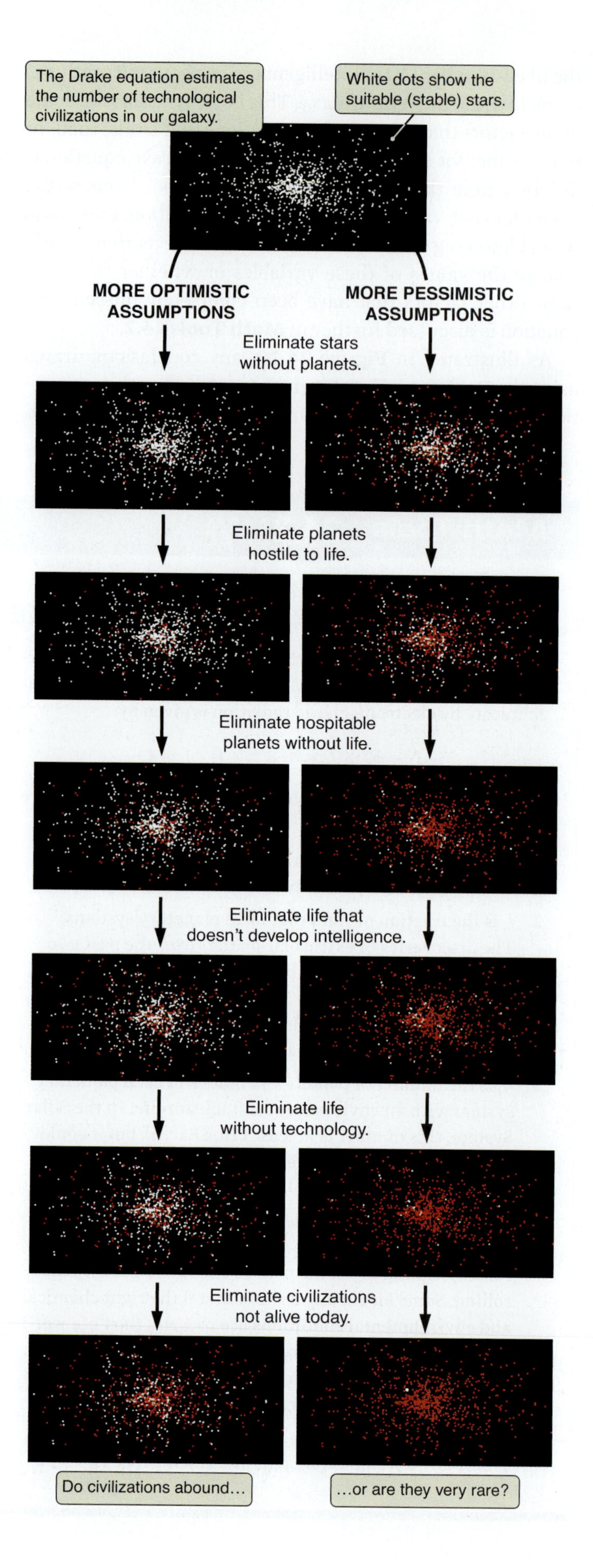

FIGURE 24.20 Estimates of the existence of intelligent, communicative civilizations in the Milky Way Galaxy using the Drake equation. White dots represent stars with possible civilizations. Notice how widely these estimates vary, given optimistic and pessimistic assumptions about the seven factors in the equation (see Math Tools 24.2).

FIGURE 24.21 The classic 1951 film *The Day the Earth Stood Still* portrayed intelligent extraterrestrials who threatened to destroy Earth if Earthlings carried their violent ways into space.

more elaborate program called the Search for Extraterrestrial Intelligence, or **SETI**. Scientists from around the world have thought carefully about what strategies might be useful for finding life in the universe. Most of these have focused on the idea of using radio telescopes to listen for signals from space that bear an unambiguous signature of an intelligent source. Some have listened intently at certain key frequencies, such as the frequency of the interstellar 21-centimeter (21-cm) line from hydrogen gas. The assumption behind this approach is that if a civilization wants to be heard, its members will tune their broadcasts to a channel that astronomers throughout the galaxy should be listening to. More recent searches have made use of advances in technology to record as broad a range of radio signals from space as possible. Analysts use computers to search these databases for types of regularity in the signals that might suggest they are intelligent in origin.

SETI listens for radio signals from other civilizations.

Unlike much astronomical research, SETI receives its funding from private rather than government sources, and SETI researchers have found ingenious ways to continue the search for extraterrestrial civilizations with limited resources. SETI@home was one of the first—and largest—distributed computing projects that uses hundreds of thousands of otherwise idle personal computers around the world from volunteers to analyze radio observations from the SETI Institute (see question 60 at the end of the chapter).

The SETI Institute's Allen Telescope Array (ATA) received much of its initial financing from Microsoft cofounder Paul Allen. The ATA consists of a "farm" of small, inexpensive radio dishes like those used to capture signals from orbiting TV broadcasting satellites (see the chapter-opener photo). One of the key projects of the ATA is to observe the planets discovered by the *Kepler* mission in search of signs of intelligent life. Each dish has a diameter of 6.1 meters, which collectively add to a large total signal-receiving area. Just as your brain can sort out sounds coming from different directions, this array of radio telescopes is able to determine the direction a signal is coming from, allowing it to listen to many stars at the same time. Over several years' time, astronomers using the ATA are expected to survey as many as a million stars, hoping to find a civilization that has sent a signal toward Earth.

As we stated earlier in this chapter, finding even one nearby civilization in the Milky Way Galaxy—that is, a *second* technological civilization in Earth's small corner of the universe—will make scientists optimistic that the universe as a whole is teeming with intelligent life. SETI may not be in the mainstream of astronomy, and the likelihood of its success is difficult to predict, but its potential payoff is enormous. Few discoveries would have a more profound impact than the certain knowledge that we on Earth are not alone.

Science fiction is filled with tales of humans who leave Earth to "seek out new life and new civilizations." Unfortunately, these scenarios are not scientifically realistic. The distances to the stars and their planets are enormous; to explore a significant sample of stars would require extending the physical search over tens or hundreds of light-years. Special relativity limits how fast one can travel. The speed of light is the limit, and even if we could travel that fast, it would take over 4 years to reach the *nearest* star. The relativistic effect of time dilation would favor astronauts traveling at very high speeds, and they would return to Earth younger than if they had stayed at home. For example, suppose astronauts visited a star 15 light-years distant. Even if they traveled at speeds close to the speed of light, by the time they returned to Earth, 30 years would have passed. Some science fiction writers get around this problem by invoking "warp speed" or "hyperdrive," which enables travel faster than the speed of light, or by using wormholes as shortcuts across the galaxy—but there is absolutely no evidence that any of these options are possible.

Some people claim that aliens have already visited Earth: tabloid newspapers, books, and websites are filled with tales of UFO sightings, government conspiracies and cover-ups, alleged alien abductions, and UFO religious cults. However, none of these reports meet the basic standards of science, and the only possible conclusion is that there is no scientific evidence for any alien visitations.

24.5 Origins: The Fate of Life on Earth

About 5 billion years from now, the Sun will end its long period of relative stability. It will expand to become a "red giant" star, swelling to hundreds of times bigger than it is at present, and thousands of times more luminous. The giant planets, orbiting outside the extended red giant atmosphere, will probably survive. But at least some of the planets of the inner Solar System will not. Just as an artificial satellite is slowed by drag in Earth's tenuous outer atmosphere and eventually falls to the ground, so, too, will a planet caught in the Sun's atmosphere be engulfed by the expanding Sun. If this is what happens to Earth, no trace of this planet will remain other than a slight increase in the amount of massive elements in the Sun's atmosphere.

Another possibility is that the red giant Sun will lose mass in a powerful wind, its gravitational grasp on the planets will weaken, and the orbits of both the inner and outer planets will spiral outward. If Earth moves out far enough, it may survive as a seared cinder, orbiting the small, hot "white dwarf" star that the Sun will become. Barely larger than Earth and with its nuclear fuel exhausted, the white dwarf Sun will slowly cool, eventually becoming a cold, inert sphere of densely packed carbon, orbited by what remains of its planets. The ultimate outcome for Earth—consumed in the heart of the Sun or left behind as a cold, burned rock orbiting a long-dead white dwarf—is not yet certain. In either case, however, Earth's status as a garden spot in the habitable zone will be at an end. If the Sun does not expand too far, or Mars also migrates outward, Mars could become the habitable planet in the Solar System, at least for a while. As the dying Sun loses more and more of its atmosphere in a stellar wind, Earth's atoms might be expelled back into the reaches of interstellar space from which they came, perhaps to be recycled into new generations of stars, planets, and even life itself.

Far-future Earth will be consumed by the Sun or left as an icy cinder.

But even before the Sun's change into a red giant star, the Sun's luminosity will begin to rise. As solar luminosity increases, so will temperatures on all the planets. Eventually, Earth's temperatures will climb so high that all animal and plant life will perish. Even the extremophiles that inhabit the oceanic depths will die, as the oceans themselves boil away. Models of the Sun's evolution are still not precise enough for astronomers to predict with certainty when that fatal event will occur, but the end of all terrestrial life may be only 1 or 2 billion years away.

It is far from certain, however, that the descendants of today's humanity will even be around a billion years from now. Some of the threats to life come from beyond Earth. For the remainder of the Sun's life, the terrestrial planets, including Earth, will continue to be bombarded by asteroids and comets. Perhaps a hundred or more of these impacts will involve kilometer-sized objects, capable of causing the kind of devastation that eradicated the dinosaurs and most other species 65 million years ago. Although these events may create new surface scars and may be harmful to human life as we know it, they will have little effect on the integrity of Earth itself. Earth's geological record is filled with such events, and each time they happen, life manages to recover and reorganize.

It seems likely, then, that some form of life will survive to see the Sun begin its march toward the red giant stage. However, individual species do not necessarily fare so well when faced with cosmic cataclysm. If humanity survives, it will be because it developed technology to detect the most threatening asteroids and to modify asteroid orbits well before they strike Earth. Comets are more difficult to guard against because long-period comets appear from the outer Solar System with little warning. Although impacts from kilometer-sized objects are infrequent, objects only a few dozen meters in size, carrying the punch of a several-megaton bomb, strike Earth about once every 100 years.

Humans might protect themselves from the fate of the dinosaurs, but in the long run humanity will either leave this world or die out. Planetary systems are common to other stars, and many other Earth-like planets may well exist throughout the Milky Way Galaxy. Colonizing other planets is currently the stuff of science fiction, but if the descendants of modern-day humans are ultimately to survive the death of the home planet, off-Earth colonization must become science fact at some point in the future.

Although humanity may soon develop ways of protecting Earth from life-threatening comet and asteroid impacts, in some ways it is its own worst enemy. Humans are poisoning the atmosphere, the water, and the land that form the habitat for all terrestrial life. As the human population grows, more and more of Earth's land is occupied, and more and more of its resources are consumed. Thousands of species of plants and animals become extinct each year. At the same time, human activities are dramatically affecting the balances of atmospheric gases. The climate and ecosystem of Earth constitute a finely balanced, complex system capable of exhibiting chaotic behavior. The fossil record shows that Earth has undergone sudden and dramatic climatic changes in response to even minor perturbations. Such drastic changes in the overall balance of nature will certainly have consequences for human survival. Humans possess the means to unleash nuclear, chemical, or biological disasters that could threaten the very survival of the species. In the end, the fate

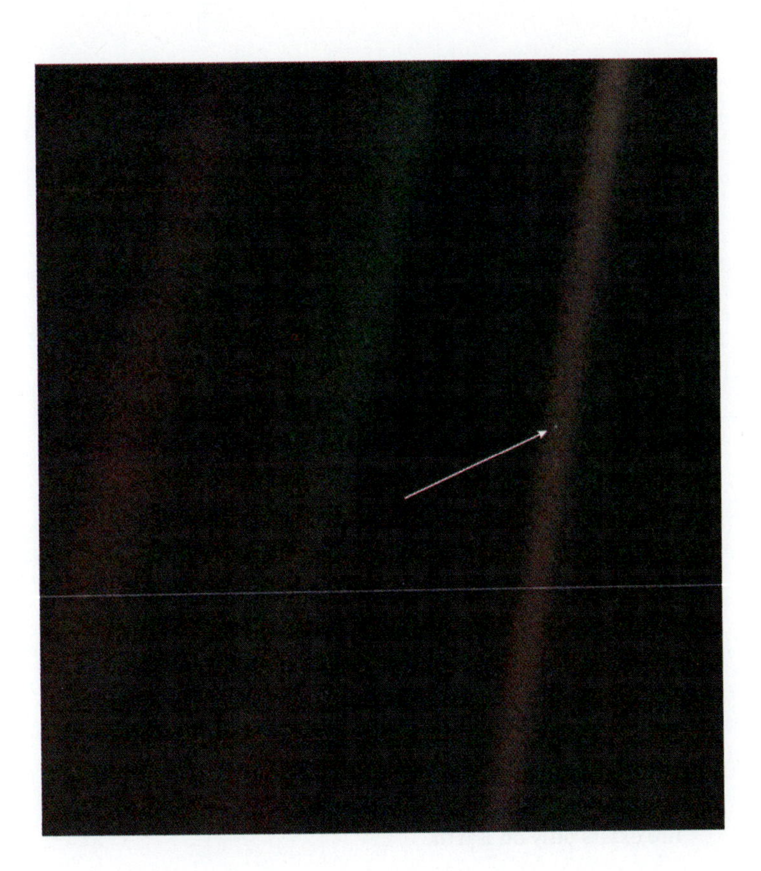

FIGURE 24.22 This image from the *Voyager 1* spacecraft shows Earth from a distance of 3.8 billion miles, well past the orbit of Neptune. The streaks are scattered sunlight. The "pale blue dot" in the rightmost streak is Earth.

of humanity will depend more than anything on whether we accept stewardship of planet Earth—and ourselves.

Figure 24.22 is the famous "pale blue dot" picture, taken by the *Voyager 1* spacecraft from beyond the orbit of Neptune at a distance of more than 40 AU. The beams in the picture are sunlight scattered off the spacecraft. The arrow points to a dot, which is Earth—the only place in the entire universe where life is confirmed to exist. Compare the size of that dot to the size of the universe. Compare the history of life on Earth to the history of the universe. Compare Earth's future with the fate of the universe. Astronomy is humbling. We occupy a tiny part of space and time. Yet we are unique, as far as we know. Think for a moment about what that means to you. This may be the most important lesson the universe has to offer. ■

Summary

24.1 Life likely formed in Earth's oceans, evolving chemically from an organic "soup" of prebiotic molecules into self-replicating organisms.

24.2 Life is a form of complex carbon-based chemistry, made possible by certain molecules capable of reproducing themselves. All terrestrial life is composed primarily of only six elements: carbon, hydrogen, oxygen, nitrogen, sulfur, and phosphorus. Life-forms that are very different from those on Earth, including those based on silicon chemistry, cannot be ruled out.

24.3 Within the Solar System, Mars and some moons of Jupiter and Saturn are the most promising candidates for life. Outside of the Solar System, astronomers study extrasolar planets, especially those that are the most similar to Earth and orbit in habitable zones surrounding solar-type stars.

24.4 The Drake equation includes different factors that astronomers consider when thinking about the possibility of life in the universe. Astronomers use radio telescopes to search for signals from extraterrestrial life. None have yet been detected.

24.5 Even before the Sun expands in size and evolves to a red giant star, its luminosity will increase enough to alter the location of its habitable zone. When that happens, Earth may no longer be the planet at the right location for keeping liquid water.

Unanswered Questions

- Will humans spread life into space? Some scientists have suggested that seeding from Earth may have already happened, as Earth microbes scattered into space following giant impacts. More intentional methods of seeding include sending microorganisms from Earth to other planets or moons to try to jump-start evolution, "terraforming" Mars or a moon to change conditions on it to make it more habitable for humanity, or sending humans in spaceships to colonize the galaxy.
- Does life exist elsewhere in the universe?

Questions and Problems

Summary Self-Test

1. The study of life and the study of astronomy are connected because (select all that apply)
   a. life may be quite commonplace in the universe.
   b. studying other planets may help explain why there is life on Earth.
   c. explorations of extreme environments on Earth suggest where to look for life elsewhere.
   d. life is a structure that evolved through physical processes, and life on Earth may not be unique.
   e. life elsewhere is most likely to be found by astronomers.

2. Scientists look for "life as they know it" because
   a. they might be able to recognize it when they see it.
   b. they think it's the only possibility.
   c. they believe life on Earth is special.
   d. no other chemistry could create life.

3. Extremophiles are organisms that
   a. are extremely reactive.
   b. are extremely rare.
   c. have an extreme quality, such as mass or size.
   d. live in extreme conditions.

4. After periods of mass extinction, there is usually a period of rapid evolution, possibly because
   a. creatures that used to fill ecological niches have vanished, leaving an opportunity for those that remain.
   b. the predatory pressure has been removed from the creatures that remain.
   c. changes in the environment make it easier for new creatures to thrive.
   d. all of the above

5. The replication of molecules results in a rapid rise in the number of copies. This is an example of
   a. linear growth.
   b. exponential growth.
   c. limitless growth.
   d. all of the above

6. The Drake equation enables astronomers to
   a. calculate precisely the number of alien civilizations.
   b. organize their thoughts about probabilities.
   c. locate the stars they should study to find life.
   d. find new kinds of life.

7. Scientists think that terrestrial life probably originated in Earth's oceans because
   a. all the chemical pieces were there.
   b. energy was available there.
   c. earliest evidence for life on Earth comes in ocean-dwelling forms.
   d. all of the above

8. Any system characterized by heredity, mutation, and natural selection will
   a. change over time.
   b. improve over time.
   c. decay over time.
   d. develop intelligence.

9. The fact that no alien civilizations have yet been detected indicates that
   a. they are not there.
   b. they are rare.
   c. Earth is in a "blackout," and they are not talking to us.
   d. not enough is known yet to draw any conclusions.

10. All life on Earth *must* eventually come to an end, because
    a. humans will make the planet uninhabitable.
    b. asteroids will make the planet uninhabitable.
    c. new species will arise that won't be recognized as life.
    d. the Sun will make the planet uninhabitable.

True/False and Multiple Choice

11. **T/F:** Evolution always leads to more complexity.

12. **T/F:** All life on Earth branches from the same evolutionary tree.

13. **T/F:** Scientists know with certainty that in the Solar System, life exists only on Earth.

14. **T/F:** It is impossible that humans will exist on Earth 10 billion years from now.

15. **T/F:** Water once flowed on Mars.

16. The Cambrian explosion
    a. killed the dinosaurs.
    b. produced the carbon that is now here on Earth.
    c. was a sudden increase in biodiversity.
    d. released a lot of CO_2 into the atmosphere.

17. The habitable zone is the place around a star where
    a. life has been found.
    b. atmospheres can contain oxygen.
    c. water exists.
    d. liquid water can exist on the surface of a planet.

18. Carbon is a favorable base for life because
    a. it can bond to many other atoms in long chains.
    b. it is nonreactive.
    c. it forms weak bonds that can be readily reorganized as needed.
    d. it is organic.

19. The difference between a prokaryote and a eukaryote is that prokaryotes
    a. have no DNA.
    b. have no cell wall.
    c. have no nucleus.
    d. do not exist today.

20. A thermophile is an organism that lives in extremely ________ environments.
    a. salty
    b. hot
    c. cold
    d. dry

21. The search for life elsewhere in the Solar System is carried out primarily by
    a. astronauts.
    b. robots.
    c. astronomers at Earth-bound telescopes.
    d. none of the above
22. Contact with an alien species (should one be found) might be problematic, because
    a. stars are so far apart.
    b. the aliens may be much more advanced.
    c. the aliens will be aggressive.
    d. humans will be aggressive.
23. The Drake equation depends on the number of stars in the Milky Way because
    a. that is a proxy for the size of the galaxy.
    b. life probably needs a star to evolve.
    c. if life is not around a star, it will never be found.
    d. stars live longer than civilizations, so the information about the civilization does not die with the civilization.
24. SETI listens for transmissions from other civilizations in the radio part of the spectrum because
    a. Earth's atmosphere is transparent in the radio.
    b. radio broadcasting is an essential step along the way to becoming a communicating civilization.
    c. no one is interested in radio astronomy, so it is easy to get time on telescopes.
    d. it can analyze the data for free using screen savers on other people's computers.
25. Life first appeared on Earth
    a. billions of years ago.
    b. millions of years ago.
    c. hundreds of thousands of years ago.
    d. thousands of years ago.

Thinking about the Concepts

26. Why do scientists generally talk about molecules such as DNA or RNA when discussing life on Earth?
27. How do scientists think that the building blocks of DNA first formed on Earth?
28. If processes on Earth were energetic enough to form DNA, shouldn't they have been able to destroy it as well? Why, then, does life exist today?
29. Today, most known life enjoys moderate climates and temperatures. Compare this environment to some of the conditions in which early life developed.
30. How was Earth's carbon dioxide atmosphere changed into today's oxygen-rich atmosphere? How long did that transformation take?
31. What was the Cambrian explosion, and what might have caused it?
32. Why do you suppose plants and forests appeared in high numbers before large animals did?
33. Why was evolution inevitable on Earth?
34. Which general conditions were needed on Earth for life to arise?
35. Is evolution underway on Earth today? If so, how might humans continue to evolve?
36. Where did all the atoms in your body come from?
37. The *Viking* spacecraft did not find evidence of life on Mars when it visited that planet in the late 1970s, nor did the *Phoenix* lander when it examined the martian soil in 2009. Do these results imply that life never existed on the planet? Why or why not?
38. Some scientists believe that humans may be the only advanced life in the galaxy today. If this is indeed the case, which factors in the Drake equation must be extremely small?
39. The second law of thermodynamics says that the entropy (a measure of disorder) of the universe is always increasing. Yet living organisms exist. Why does this not violate the second law of thermodynamics?
40. Why is it likely that life on Earth as we know it will end long before the Sun runs out of nuclear fuel?

Applying the Concepts

41. The *Kepler* mission is currently searching for planets in the habitable zones of stars. Explain which factors in the Drake equation are affected by this search, and how the final number N will be impacted if *Kepler* finds that most stars have planets in their habitable zones.
42. Study Figure 24.19. The four rows of white blocks at the top represent the numbers 1–10. The bottom row is a placeholder, and the top three rows are the actual counting. Explain the "rule" for the kind of counting shown here. (For example, how do three white blocks represent the number 7?)
43. Connections 24.1 ("Forever in a Day") takes events spread out over enormous intervals of time and compresses them into the more comprehensible interval of a single 24-hour day. Make your own "Life in a Day" by compressing all the important events of your lifetime into a single day, starting with your birth at the stroke of midnight and continuing to the present at the end of the day.
44. As noted in Section 24.1, some scientists suspect the early Earth was "seeded" with primitive life stored in comets and meteoroids. Knowing when and how our Solar System, galaxy, and universe formed, what time line is required for such seeding to be possible?
45. As discussed in Math Tools 24.1, the doubling time for exponential growth is given by $P_F/P_0 = 2^n$. Assume a self-replicating molecule that makes one copy of itself each second. Make a graph of the number of molecules versus time for the first 60 seconds after the molecule begins replicating.
46. If a self-replicating molecule has begun replicating and seven doubling times have passed, how many molecules are there?
47. The doubling time for *E. coli* is 20 minutes, and you start getting sick when just 10 bacteria enter your system. How many bacteria are in your body after 12 hours?

48. If the chance that a given molecule will make a copying error is one in 100,000, how many generations are needed before, on average, at least one mutation has occurred?

49. Consider an organism Beta that, because of a genetic mutation, has a 5 percent greater probability of survival than its non-mutated form, Alpha. Alpha has only a 95 percent probability (p_r) of reproducing itself compared to Beta. After *n* generations, Alpha's population within the species will be $S_p = (p_r)^n$ compared to Beta's. Calculate Alpha's relative population after 100 generations. (Note: You may need a scientific calculator or help from your instructor to evaluate the quantity 0.95^{100}.)

50. To fully appreciate the power of heredity, mutation, and natural selection, consider Alpha's relative population (from question 49) after 5,000 generations if Beta has a mere 0.1 percent survivability advantage over Alpha.

51. Study the Drake equation, in Math Tools 24.2. Make your own optimistic and pessimistic assumptions for each of the variables in the equation. What values do you find for *N*?

52. Look back at Figure 5.25 in Chapter 5. Trace (or photocopy) this graph, and then add horizontal lines for the temperatures at which water freezes and boils. Which planets in the Solar System have measured temperatures that fall within those lines? Which planets have predicted temperatures (based on the equilibrium model) that fall within those lines? What does this tell you about assumptions about the habitable zone?

53. Figure 5.25 shows that Mercury's measured range of temperatures overlaps the temperatures at which water is a liquid. Is Mercury in the habitable zone? Why or why not?

54. Suppose astronomers announce the discovery of a new planet around a star with a mass equal to the Sun's. This planet has an orbital period of 87 days. Is this planet in the habitable zone for a Sun-like star?

55. Why do you think astronomers sent a coded radio signal to the globular cluster M13 in 1974—rather than, say, to a nearby star?

Using the Web

56. a. Go to the online *Astrobiology Magazine* (http://astrobio.net), which covers many topics included in this chapter. Under "Hot Topics" and then "Origins," click on "Extreme Life." What are some new findings about extremophiles? Why is this a hot topic for astrobiologists?
    b. Go to the "Life, Unbounded" blog (http://blogs.scientificamerican.com/life-unbounded). What is a recent topic of interest? Is the discussion based on some new data? A new theory?

57. Solar System space missions:
    a. Go to the website for the *Cassini* mission to Saturn (http://saturn.jpl.nasa.gov). Is it still collecting data? What has been found recently on one of the moons that would be of interest to astrobiologists?
    b. Go to the website for the *Mars Science Laboratory* (http://mars.jpl.nasa.gov/msl/mission). The rover *Curiosity* landed on the surface in August 2012. What are the science goals of this mission, and how do they relate to astrobiology? What are some recent results?
    c. The next Mars mission, *MAVEN* (http://lasp.colorado.edu/home/maven), is scheduled for launch in late 2013, with arrival in 2014. What are the science goals of this mission? Why are astrobiologists interested in the history of Mars's climate and atmosphere? Are there any results?

58. Go to the Habitable Exoplanets Catalog website (http://phl.upr.edu/projects/habitable-exoplanets-catalog). Click on "Methods" to read about the different criteria for evaluating a planet's habitability. How many confirmed habitable exoplanets are there? How many candidates? How might the number of confirmed exoplanets affect the terms in the Drake equation?

59. a. Go to the website for the Kepler Space telescope (http://kepler.nasa.gov); click on "News" and then "Planet-finding News." What is a recent discovery of a planet in the habitable zone? What is a recent discovery of a planet with an interesting atmosphere?
    b. Go to the website for the European Space Agency (ESA) *Gaia* mission (http://sci.esa.int/science-e/www/area/index.cfm?fareaid=26), scheduled for launch in 2013. What are the science objectives of this mission? Click on "Extra-solar Planets" on the left. How will *Gaia* identify new planets? What properties of the planet will it be able to measure? Has this mission been launched? What's new?

60. a. Go to the website for the Zooniverse citizen science project SETILive (http://setilive.org), which uses Allen Telescope Array observations of Kepler exoplanet candidates to look for interesting signals. How were the targets selected for observation? Click on "Classify" and then "Signals." What types of signals are picked up that are not from extraterrestrial aliens? Watch the tutorial and video under "Classify." If the ATA is observing, look at some data. What happens if several citizen scientists report finding something interesting?
    b. Go to the website for SETI@home (http://setiathome.berkeley.edu). What is this project? What are the advantages of using many online computers instead of one supercomputer? Under "About," read "About SETI@home" and "Science status." Then click on "Participate" (under "...more"). To get started, you will need to download, install, and run the BOINC software. What do you think will happen if someone detects a positive signal?

Exploration | Fermi Problems and the Drake Equation

The Drake equation is a way of organizing ideas about other intelligent, communicating civilizations in the galaxy. This type of thinking is very useful for estimating a value, particularly when analyzing systems for which counting is not possible. The types of problems that can be solved in this way are often called Fermi problems after Enrico Fermi, who was mentioned in this chapter. For example, we might ask, What is the circumference of Earth?

You could Google this question, or you could already "know" the answer, or you might look it up in this textbook. Alternatively, you could very carefully measure the shadow of a stick in two locations at the same time on the same day. Or you could drive around the planet.

Or you could reason this way:

How many time zones are between New York and Los Angeles?
3 time zones. You know this from traveling, or from television.

How many miles is it from New York to Los Angeles?
3,000 miles. You know this from traveling, or from living in the world.

So, how many miles per time zone?
$3{,}000/3 = 1{,}000$

How many time zones in the world?
24, because there are 24 hours in a day, and each time zone marks an hour.

So, what is the circumference of Earth?
24,000 miles, because there are 24 time zones, each 1,000 miles wide.

The measured circumference is 24,900 miles, which agrees with our estimate to within 4 percent.

Here we list several Fermi problems. Time yourself for an hour, and work as many of them as possible. (You don't have to do them in order!)

1. **How much has the mass of the human population on Earth increased in the last year?**

   ....

2. **How much energy does a horse consume in its lifetime?**

   ....

3. **How many pounds of potatoes are consumed in the United States annually?**

   ....

4. **How many cells are there in the human body?**

   ....

5. **If your life earnings were given to you by the hour, how much is your time worth per hour?**

   ....

6. **What is the weight of solid garbage thrown out by American families each year?**

   ....

7. **How fast does human hair grow (in feet per hour)?**

   ....

8. **If all the people on Earth were crowded together, how much area would be covered?**

   ....

9. **How many people could fit on Earth if every person occupied 1 square meter of land?**

   ....

10. **How much carbon dioxide (CO_2) does an automobile emit each year?**

    ....

11. **What is the mass of Earth?**

    ....

12. **What is the average annual cost of an automobile, including overhead (maintenance, looking for parking, insurance, cleaning, and so on)?**

    ....

13. **How much ink was used printing all the newspapers in the United States today?**

    ....

APPENDIX 1

Mathematical Tools

Working with Proportionalities

Most of the mathematics in *21st Century Astronomy* involves proportionalities—statements about how one physical quantity changes when another quantity changes. We began a discussion of proportionality in Math Tools 1.1 and 3.3; here we offer a few more examples of working with proportionalities.

To use a statement of proportionality to compare two objects, begin by turning the proportionality into a ratio. For example, the price of a bag of apples is **proportional** to the weight of the bag:

$$\text{Price} \propto \text{Weight}$$

Here the symbol $\propto$ is pronounced "is proportional to." What this means is that the ratio of the prices of two bags of apples is equal to the ratio of the weights of the two bags:

$$\text{Price} \propto \text{Weight} \quad \text{means} \quad \frac{\text{Price of A}}{\text{Price of B}} = \frac{\text{Weight of A}}{\text{Weight of B}}$$

Let's work a specific example. Suppose bag A weighs 2 pounds, and bag B weighs 1 pound. That means bag A will cost twice as much as bag B. The price per pound is an example of a **constant of proportionality**. We can turn this proportionality into the following equation:

$$\frac{\text{Price of A}}{\text{Price of B}} = \frac{\text{Weight of A}}{\text{Weight of B}} = \frac{2\text{ lb}}{1\text{ lb}} = 2$$

In other words, the price of bag A is 2 times the price of bag B.

Now let's work another, more complicated example. In Chapter 13 we discuss how the luminosity, brightness, and distance of stars are related. The luminosity of a star—the total energy that the star radiates each second—is proportional to the star's brightness multiplied by the square of its distance:

$$\text{Luminosity} \propto \text{Brightness} \times \text{Distance}^2$$

What this proportionality means is that for any two stars—call them A and B:

$$\frac{\text{Luminosity of A}}{\text{Luminosity of B}} = \frac{\text{Brightness of A}}{\text{Brightness of B}} \times \left(\frac{\text{Distance of A}}{\text{Distance of B}}\right)^2$$

If we use the symbols L, b, and d to represent luminosity, brightness, and distance, respectively, this equation becomes

$$\frac{L_A}{L_B} = \frac{b_A}{b_B} \times \left(\frac{d_A}{b_B}\right)^2$$

As an example, suppose that star A appears twice as bright in the sky as star B, but star A is located 10 times as far away as star B. Compare the luminosities of the two stars. Because we know that:

$$\text{Luminosity} \propto \text{Brightness} \times \text{Distance}^2$$

we write:

$$\frac{\text{Luminosity of A}}{\text{Luminosity of B}} = \frac{\text{Brightness of A}}{\text{Brightness of B}} \times \left(\frac{\text{Distance of A}}{\text{Distance of B}}\right)^2$$

$$= \frac{2}{1} \times \left(\frac{10}{1}\right)^2 = 200$$

In other words, star A is 200 times as luminous as star B.

Proportionalities are used to compare one object to another. Constants of proportionality are used to calculate actual values. In *21st Century Astronomy*, it is usually the proportionality itself that is important.

Scientific Notation

Astronomy is a science of both the very large and the very small. The mass of an electron, for example, is:

0.0000000000000000000000000000009109 kilograms (kg)

whereas the distance to a galaxy far, far away might be about:

100,000,000,000,000,000,000,000,000 meters

All it takes is a quick glance at these two numbers to see why astronomers, like most other scientists, make heavy use of **scientific notation** to express numbers.

Powers of 10

Our number system is based on powers of 10. Going to the left of the decimal place,

$$10 = 10 \times 1$$

$$100 = 10 \times 10 \times 1$$

$$1{,}000 = 10 \times 10 \times 10 \times 1$$

and so on. Going to the right of the decimal place,

$$0.1 = \frac{1}{10} \times 1,$$

$$0.01 = \frac{1}{10} \times \frac{1}{10} \times 1,$$

$$0.001 = \frac{1}{10} \times \frac{1}{10} \times \frac{1}{10} \times 1$$

and so on. In other words, each place to the right or left of the decimal place in a number represents a power of 10. For example, 1 million can be written as:

$$1 \text{ million} = 1{,}000{,}000 = 1 \times 10 \times 10 \times 10 \times 10 \times 10 \times 10$$

That is, 1 million is "1 multiplied by six factors of 10." Scientific notation combines these factors of 10 in convenient shorthand. Rather than all being written out, the six factors of 10 are combined into shorthand using an exponent:

$$1 \text{ million} = 1 \times 10^6$$

which also means "1 multiplied by six factors of 10."

Moving to the right of the decimal place, each step *removes* a power of 10 from the number. One-millionth can be written:

$$1 \text{ millionth} = 1 \times \frac{1}{10} \times \frac{1}{10} \times \frac{1}{10} \times \frac{1}{10} \times \frac{1}{10} \times \frac{1}{10}$$

This removes powers of 10, so this number is written using a negative exponent, as:

$$\frac{1}{10} = 10^{-1}$$

and

$$1 \text{ millionth} = 1 \times 10^{-1} \times 10^{-1} \times 10^{-1} \times 10^{-1} \times 10^{-1} \times 10^{-1}$$

$$= 1 \times 10^{-6}$$

Returning to our earlier examples, the mass of an electron is 9.109×10^{-31} kg, and the distant galaxy is located 1×10^{26} meters away. These are much more convenient ways of writing these values. Notice that *the exponent in scientific notation gives you a feel for the size of a number at a glance.* The exponent of 10 in the electron mass is −31, which quickly indicates that it is a very small number. The exponent of 10 in the distance to a remote galaxy, +26, quickly indicates that it is a very large number. This exponent is often called the *order of magnitude* of a number. When you see a number written in scientific notation while reading *21st Century Astronomy* (or elsewhere), just remember to look at the exponent to better understand what the number is telling you.

Scientific notation is also convenient because it makes multiplying and dividing numbers easier. For example, two billion multiplied by eight-thousandths can be written:

$$2{,}000{,}000{,}000 \times 0.008$$

but it is more convenient to write these two numbers using scientific notation, as:

$$(2 \times 10^9) \times (8 \times 10^{-3})$$

We can regroup these expressions in the following form:

$$(2 \times 8) \times (10^9 \times 10^{-3})$$

The first part of the problem is just $2 \times 8 = 16$. The more interesting part of the problem is the multiplication in the right-hand parentheses. The first number, 10^9, is just shorthand for $10 \times 10 \times 10 \ldots$ nine times. That is, it represents nine factors of 10. The second number stands for three factors of 1/10—or removing three factors of 10 if you prefer to think of it that way. Altogether, that makes $9 - 3 = 6$ factors of 10. In other words,

$$10^9 \times 10^{-3} = 10^{9-3} = 10^6$$

Putting the problem together, we get:

$$(2 \times 10^9) \times (8 \times 10^{-3}) = (2 \times 8) \times (10^9 \times 10^{-3})$$

$$= 16 \times 10^6$$

By convention, when a number is written in scientific notation, only one digit is placed to the left of the decimal point. In this case, there are two. However, 16 is 1.6×10, so we can add this additional factor of 10 to the exponent at right, making the final answer:

$$1.6 \times 10^7$$

Dividing is just the inverse of multiplication. Dividing by 10^3 means removing three factors of 10 from a number. Using the previous number,

$$(1.6 \times 10^7) \div (2 \times 10^3) = (1.6 \div 2) \times (10^7 \div 10^3)$$
$$= 0.8 \times 10^{7-3}$$
$$= 0.8 \times 10^4$$

This time we have only a zero to the left of the decimal point. To get the number into proper form, we can substitute 8×10^{-1} for 0.8, giving:

$$0.8 \times 10^4 = (8 \times 10^{-1}) \times 10^4 = 8 \times 10^3$$

Calculator hint: Adding and subtracting numbers in scientific notation is somewhat more difficult, because all numbers must be written as values multiplied by the *same* power of 10 before they can be added or subtracted. Therefore, you will need to use a calculator that has scientific notation. Most scientific calculators have a button that says EXP or EE. These mean "times 10 to the." So for 4×10^{12}, you would type [4][EXP][1][2] or [4][EE][1][2] into your calculator. Usually this number shows up in the window on your calculator either just as you see it written in this book, or as a 4 with a smaller 12 all the way over in the right of the window.

Significant Figures

In the previous example, we actually broke some rules in the interest of explaining how powers of 10 are treated in scientific notation. The rules we broke involve the *precision* of the numbers. When expressing quantities in science, it is extremely important to know not only the value of a number, but also how precise that value is.

The most complete way to keep track of the precision of numbers is to actually write down the uncertainty in the number. For example, suppose you know that the distance to a store (call it d) is between 0.8 and 1.2 km, then you can write:

$$d = 1.0 \pm 0.2 \text{ km}$$

where the symbol $\pm$ is pronounced "plus or minus." In this example, d is between $1.0 - 0.2 = 0.8$ km and $1.0 + 0.2 = 1.2$ km. This is an unambiguous statement about the limitations on knowing the value of d, but carrying along the formal errors with every number written would be cumbersome at best. Instead, you keep track of the approximate precision of a number by using *significant figures*.

The convention for significant figures is this: Assume the written number has been rounded from a number that had one additional digit to the right of the decimal point. If a quantity d, which might represent the distance to the store, is "1.", then d is close to 1. It is likely not as small as "0.", and it is likely not as large as "2.". If instead it is written as:

$$d = 1.0$$

then d is likely not 0.9 and is likely not 1.1. It is roughly 1.0 to the nearest tenth. The greater the number of significant figures, the more precisely the number is being specified. For example, 1.00000 is not the same number as 1.00. The first number, 1.00000, represents a value that is probably not as small as 0.99999 and probably not as large as 1.00001. The second number, 1.00, represents a value that is probably not as small as 0.99 nor as large as 1.01. The number 1.00000 is much more precise than the number 1.00.

In mathematical operations, significant figures are important. For example, $2.0 \times 1.6 = 3.2$. It does *not* equal 3.20000000000. *The product of two numbers cannot be known to any greater accuracy than the numbers themselves.* As a general rule, when you multiply and divide, the answer should have the same number of significant figures as the less precise of the numbers being multiplied or divided. In other words, $2.0 \times 1.602583475 = 3.2$. Because all you know is that the first factor is probably closer to 2.0 than to 1.9 or 2.1, all you know about the product is that it is between about 3.0 and 3.4. It is 3.2. It is not 3.205166950 (*even if that is the answer on your calculator*). The rest of the digits to the right of 3.2 just do not mean anything.

When two numbers are added or subtracted, if one number has a significant figure with a particular place value but another number does not, their sum or difference cannot have a significant figure in that place value. For example,

$$\begin{array}{r} 1{,}045. \\ +1.34567 \\ \hline 1{,}046. \end{array}$$

The answer is "1,046.", *not* "1,046.34567". Again, the extra digits to the right of the decimal place have no meaning, because "1,045." is not known to that accuracy.

What is the precision of the number 1,000,000? As it is written, the answer is unclear. Are all those zeros really significant, or are they placeholders? If the number is written in scientific notation, however, there is never a question. Instead of 1,000,000, you write 1.0×10^6 for a number that is known to the nearest hundred thousand or so; or you write 1.00000×10^6 for a number that is known to the nearest 10.

So the earlier example would have been more correct if written as:

$$(2.0 \times 10^9) \times (8.0 \times 10^{-3}) = 1.6 \times 10^7$$

Algebra

There are many branches of mathematics. The branch that focuses on the relationships between quantities is called **algebra**. If you are reading this book, you have almost certainly taken an algebra class, but you are not alone if you feel a little review is in order. Basically, algebra begins by using symbols to represent quantities.

For example, you could write the distance you travel in a day as d. As it stands, d has no value. It might be 10,000 miles. It might be 30 feet. It does, however, have **units**—in this case, the units of distance. The average speed at which you travel is equal to the distance you travel divided by the time you take. By using the symbol v to represent your average speed and the symbol t to represent the time you take, instead of writing out "Your average speed is equal to the distance you travel divided by the time you take," you can write:

$$v = \frac{d}{t}$$

The meaning of this algebraic expression is exactly the same as the sentence quoted before it, but it is much more concise. As it stands, v, d, and t still have no specific values. There are no numbers assigned to them yet. However, this expression indicates what the relationship between those numbers will be when you look at a specific example. For example, if you go 500 km ($d = 500$ km) in 10 hours ($t = 10$ hours), this expression tells you that your average speed is:

$$v = \frac{d}{t} = \frac{500\ \text{km}}{10\ \text{h}} = 50\ \text{km/h}$$

Notice that the units in this expression are multiplicative factors that act exactly like the numerical values. Dividing the two shows that the units of v are kilometers divided by hours, or km/h (pronounced "kilometers per hour").

We introduced algebra as shorthand for expressing relations between quantities, but it is far more powerful than that. Algebra provides rules for manipulating the symbols used to represent quantities. We begin with a bit of notation for *powers* and *roots*. Raising a quantity to a power means multiplying the quantity by itself some number of times. For example, if S is a symbol for something (anything), then S^2 (pronounced "S squared" or "S to the second power") means $S \times S$, and S^3 (pronounced "S cubed" or "S to the third power") means $S \times S \times S$. Suppose S represents the length of the side of a square. The area of the square is given by:

$$\text{Area} = S \times S = S^2$$

If $S = 3$ meters (m), then the area of the square is:

$$S^2 = 3\ \text{m} \times 3\ \text{m} = 9\ \text{m}^2$$

(pronounced "9 square meters"). It should be obvious why raising a quantity to the second power is called *squaring* the quantity. We could have done the same thing for the sides of a cube and found that the volume of the cube is:

$$\text{Volume} = S \times S \times S = S^3$$

If $S = 3$ meters, then the volume of the cube is:

$$S^3 = 3\ \text{m} \times 3\ \text{m} \times 3\ \text{m} = 27\ \text{m}^3$$

(pronounced "27 cubic meters"). Again, it is clear why raising a quantity to the third power is called *cubing* the quantity.

Roots are the reverse of this process. The square root of a quantity is the value that, when squared, gives the original quantity. The square root of 4 is 2, which means that $2 \times 2 = 4$. The square root of 9 is 3, which means that $3 \times 3 = 9$. Similarly, the cube root of a quantity is the value that, when cubed, gives the original quantity. The cube root of 8 is 2, which means that $2 \times 2 \times 2 = 8$. Roots are written with the symbol $\sqrt{\ }$. For example, the square root of 9 is written as:

$$\sqrt{9} = 3$$

and the cube root of 8 as:

$$\sqrt[3]{8} = 2$$

If the volume of a cube is $V = S^3$, then:

$$S = \sqrt[3]{V} = \sqrt[3]{S^3}$$

Roots can also be written as powers. Powers and roots behave exactly like the exponents of 10 in our discussion of scientific notation. (The exponents used in scientific notation are just powers of 10.) For example, if a, n, and m are all algebraic quantities, then

$$a^n \times a^m = a^{n+m} \quad \text{and} \quad \frac{a^n}{a^m} = a^{n-m}$$

(To see whether you understand all this, explain why the square root of a can also be written $a^{1/2}$ and the cube root of a can be written $a^{1/3}$.)

Some of the rules of algebra are listed next. These are the rules of arithmetic applied to the symbolic quantities of algebra. The important thing is this: as long as the rules of algebra are applied properly, then the relationships among symbols arrived at through algebraic manipulation remain true for the physical quantities that those symbols represent.

Here we summarize a few algebraic rules and relationships. In this summary, a, b, c, m, n, r, x, and y are all algebraic quantities:

Associative rule:

$$a \times b \times c = (a \times b) \times c = a \times (b \times c)$$

Commutative rule:

$$a \times b = b \times a$$

Distributive rule:

$$a \times (b + c) = (a \times b) + (a \times c)$$

Cross-multiplication:

$$\text{If } \frac{a}{b} = \frac{c}{d}, \text{ then } ad = bc.$$

Working with exponents:

$$\frac{1}{a^n} = a^{-n} \quad a^n a^m = a^{n+m}$$

$$\frac{a^n}{a^m} = a^{n-m} \quad (a^n)^m = a^{n \times m} \quad \left(\frac{a}{b}\right)^n = \frac{a^n}{b^n}$$

Equation of a line with slope *m* and *y*-intercept *b*:

$$y = mx + b$$

Equation of a circle with radius *r* centered at *x* = 0, *y* = 0:

$$x^2 + y^2 = r^2$$

Angles and Distances

The farther away something is, the smaller it appears. This is common sense and everyday experience. Since astronomers cannot walk up to the object they are studying and measure it with a meterstick, their knowledge about the sizes of things usually depends on relating the size of an object, its distance, and the angle it covers in the sky.

The natural way to measure angles is to use a unit called the **radian**. As shown in **Figure A1.1a**, the size of an angle in radians is the length of the arc subtending the angle, divided by the radius of the circle. In the figure, the angle $x = S/r$ radians.

Because the circumference of a circle is 2π multiplied by the radius ($C = 2\pi r$), a complete circle has an angular measure of $(2\pi r)/r = 2\pi$ radians. In more conventional angular measure, a complete circle is 360°, so:

$$360° = 2\pi \text{ radians}$$

or:

$$1 \text{ radian} = \frac{360°}{2\pi} = 57.2958°$$

Often seconds of arc (**arcseconds**) are used to measure angles for stars and galaxies. A degree is divided into 60 minutes of arc (**arcminutes**), each of which is divided into 60 seconds of arc—so there are 3,600 seconds of arc in a degree. Therefore,

$$3{,}600 \frac{\text{arcseconds}}{\text{degree}} \times 57.2958 \frac{\text{degree}}{\text{radian}} = 206{,}265 \frac{\text{arcseconds}}{\text{radian}}.$$

If the angle is small enough (which it usually is in astronomy), there is very little difference between the pie slice just described and a long skinny triangle with a short side of length *S*, as **Figure A1.1b** illustrates. So, if you know the distance *d* to an object and you can measure the angular size *x* of the object, then the size of the object is given by:

$$S = x(\text{in radians}) \times d = \frac{x(\text{in degrees})}{57.2958 \text{ degrees/radian}} \times d$$

$$= \frac{x(\text{in arcseconds})}{206{,}265 \text{ arcseconds/radian}} \times d$$

which is how astronomers relate an object's angular size, distance, and physical size.

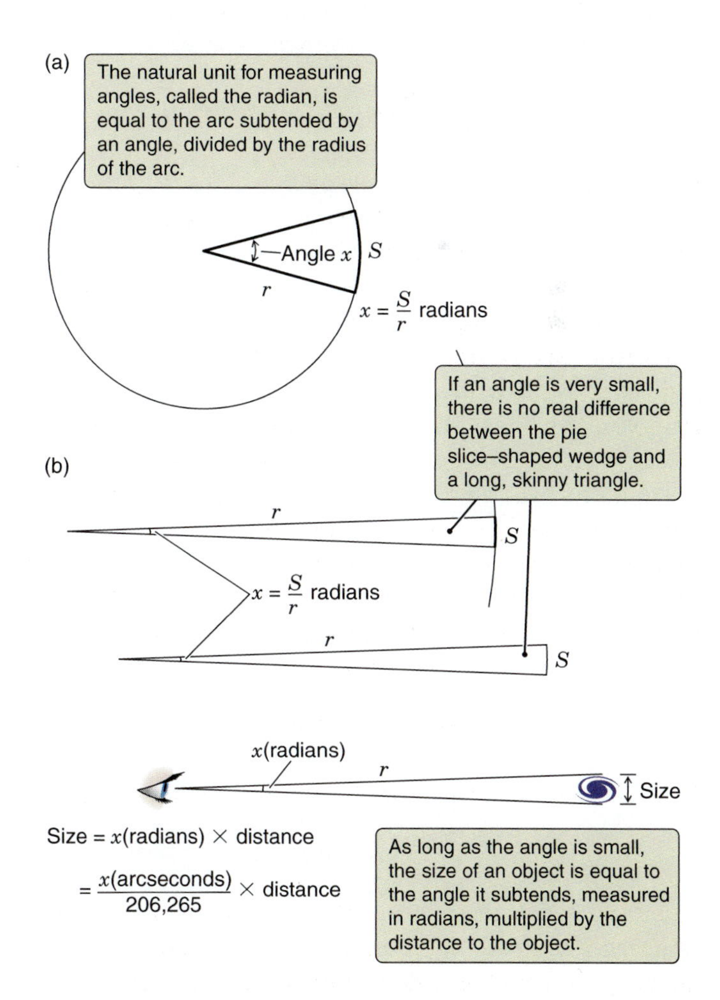

FIGURE A1.1 Measuring angles.

Circles and Spheres

To round out these mathematical tools, here are a few useful formulas for circles and spheres. The circle or sphere in each case has a radius r.

$$\text{Circumference}_{\text{circle}} = 2\pi r$$

$$\text{Area}_{\text{circle}} = \pi r^2$$

$$\text{Surface area}_{\text{sphere}} = 4\pi r^2$$

$$\text{Volume}_{\text{sphere}} = \frac{4}{3}\pi r^3$$

APPENDIX 2

Physical Constants and Units

Fundamental Physical Constants

Constant	Symbol	Value
Speed of light in a vacuum	c	2.99792×10^{8} m/s
Universal gravitational constant	G	6.673×10^{-11} $m^3/(kg\ s^2)$
Planck constant	h	6.62607×10^{-34} J-s
Boltzmann constant	k	1.38065×10^{-23} J/K
Stefan-Boltzmann constant	σ	5.67040×10^{-8} $W/(m^2\ K^4)$
Mass of electron	m_e	9.10938×10^{-31} kg
Mass of proton	m_p	1.67262×10^{-27} kg
Electric charge of electron or proton	e	1.60218×10^{-19} C

Source: Data from the Particle Data Group (http://pdg.lbl.gov).

Unit Prefixes

Prefix*	Name	Factor†
n	nano-	10^{-9}
μ	micro-	10^{-6}
m	milli-	10^{-3}
k	kilo-	10^{3}
M	mega-	10^{6}
G	giga-	10^{9}
T	tera-	10^{12}

These prefixes (*), when appended to a unit, change the size of the unit by the factor (†) given. For example, 1 km (kilometer) is 10^3 meters (m).

Units and Values

Quantity	Fundamental Unit	Values
Length	meters (m)	radius of Sun ($R_\odot$) = 6.9551×10^8 m astronomical unit (AU) = 1.49598×10^{11} m 1 AU = 149,598,000 km light-year (ly) = 9.4605×10^{15} m 1 ly = 6.324×10^4 AU 1 parsec (pc) = 3.262 ly = 3.0857×10^{16} m 1 m = 3.281 feet
Volume	meters^3 (m^3)	1 m^3 = 1,000 liters = 264.2 gallons
Mass	kilograms (kg)	1 kg = 1,000 grams mass of Earth ($M_\oplus$) = 5.9726×10^{24} kg mass of Sun ($M_\odot$) = 1.9885×10^{30} kg
Time	seconds (s)	1 hour (h) = 60 minutes (min) = 3,600 s solar day (noon to noon) = 86,400 s sidereal day (Earth rotation period) = 86,164.1 s tropical year (equinox to equinox) = 365.24219 days = 3.15569×10^7 s sidereal year (Earth orbital period) = 365.25636 days = 3.15581×10^7 s
Speed	meters/second (m/s)	1 m/s = 2.237 miles/h 1 km/s = 1,000 m/s = 3,600 km/h $c = 2.99792 \times 10^8$ m/s = 299,792 km/s
Acceleration	meters/second^2 (m/s^2)	g = gravitational acceleration on Earth = 9.81 m/s^2
Energy	joules (J)	1 J = 1 kg m^2/s^2 1 megaton = 4.18×10^{15} J
Power	watts (W)	1 W = 1 J/s solar luminosity ($L_\odot$) = 3.828×10^{26} W
Force	newtons (N)	1 N = 1 kg m/s^2 1 pound (lb) = 4.448 N 1 N = 0.22481 lb
Pressure	newtons/meter^2 (N/m^2)	atmospheric pressure at sea level = 1.013×10^5 N/m^2 = 1.013 bar
Temperature	kelvins (K)	absolute zero = 0 K = −273.15°C = −459.67°F

Source: Data from the Particle Data Group (http://pdg.lbl.gov).

APPENDIX 3

Periodic Table of the Elements

1 — Atomic number; H — Symbol; Hydrogen — Name; 1.00794 — Average atomic mass

Metals | Metalloids | Nonmetals

Period	1 1A	2 2A	3 3B	4 4B	5 5B	6 6B	7 7B	8 8B	9 8B	10 8B	11 1B	12 2B	13 3A	14 4A	15 5A	16 6A	17 7A	18 8A
1	1 **H** Hydrogen 1.00794																	2 **He** Helium 4.002602
2	3 **Li** Lithium 6.941	4 **Be** Beryllium 9.012182											5 **B** Boron 10.811	6 **C** Carbon 12.0107	7 **N** Nitrogen 14.0067	8 **O** Oxygen 15.9994	9 **F** Fluorine 18.9984032	10 **Ne** Neon 20.1797
3	11 **Na** Sodium 22.98976928	12 **Mg** Magnesium 24.3050											13 **Al** Aluminum 26.9815386	14 **Si** Silicon 28.0855	15 **P** Phosphorus 30.973762	16 **S** Sulfur 32.065	17 **Cl** Chlorine 35.453	18 **Ar** Argon 39.948
4	19 **K** Potassium 39.0983	20 **Ca** Calcium 40.078	21 **Sc** Scandium 44.955912	22 **Ti** Titanium 47.867	23 **V** Vanadium 50.9415	24 **Cr** Chromium 51.9961	25 **Mn** Manganese 54.938045	26 **Fe** Iron 55.845	27 **Co** Cobalt 58.933195	28 **Ni** Nickel 58.6934	29 **Cu** Copper 63.546	30 **Zn** Zinc 65.38	31 **Ga** Gallium 69.723	32 **Ge** Germanium 72.64	33 **As** Arsenic 74.92160	34 **Se** Selenium 78.96	35 **Br** Bromine 79.904	36 **Kr** Krypton 83.798
5	37 **Rb** Rubidium 85.4678	38 **Sr** Strontium 87.62	39 **Y** Yttrium 88.90585	40 **Zr** Zirconium 91.224	41 **Nb** Niobium 92.90638	42 **Mo** Molybdenum 95.96	43 **Tc** Technetium [98]	44 **Ru** Ruthenium 101.07	45 **Rh** Rhodium 102.90550	46 **Pd** Palladium 106.42	47 **Ag** Silver 107.8682	48 **Cd** Cadmium 112.411	49 **In** Indium 114.818	50 **Sn** Tin 118.710	51 **Sb** Antimony 121.760	52 **Te** Tellurium 127.60	53 **I** Iodine 126.90447	54 **Xe** Xenon 131.293
6	55 **Cs** Cesium 132.9054519	56 **Ba** Barium 137.327	57 **La** Lanthanum 138.90547	72 **Hf** Hafnium 178.49	73 **Ta** Tantalum 180.94788	74 **W** Tungsten 183.84	75 **Re** Rhenium 186.207	76 **Os** Osmium 190.23	77 **Ir** Iridium 192.217	78 **Pt** Platinum 195.084	79 **Au** Gold 196.966569	80 **Hg** Mercury 200.59	81 **Tl** Thallium 204.3833	82 **Pb** Lead 207.2	83 **Bi** Bismuth 208.98040	84 **Po** Polonium [209]	85 **At** Astatine [210]	86 **Rn** Radon [222]
7	87 **Fr** Francium [223]	88 **Ra** Radium [226]	89 **Ac** Actinium [227]	104 **Rf** Rutherfordium [261]	105 **Db** Dubnium [262]	106 **Sg** Seaborgium [266]	107 **Bh** Bohrium [264]	108 **Hs** Hassium [277]	109 **Mt** Meitnerium [268]	110 **Ds** Darmstadtium [271]	111 **Rg** Roentgenium [272]	112 **Cn** Copernicium [285]		114 **Uuq** Ununquadium [289]		116 **Uuh** Ununhexium [292]		

6 Lanthanides	58 **Ce** Cerium 140.116	59 **Pr** Praseodymium 140.90765	60 **Nd** Neodymium 144.242	61 **Pm** Promethium [145]	62 **Sm** Samarium 150.36	63 **Eu** Europium 151.964	64 **Gd** Gadolinium 157.25	65 **Tb** Terbium 158.92535	66 **Dy** Dysprosium 162.500	67 **Ho** Holmium 164.93032	68 **Er** Erbium 167.259	69 **Tm** Thulium 168.93421	70 **Yb** Ytterbium 173.05	71 **Lu** Lutetium 174.967
7 Actinides	90 **Th** Thorium 232.03806	91 **Pa** Protactinium 231.03588	92 **U** Uranium 238.02891	93 **Np** Neptunium [237]	94 **Pu** Plutonium [244]	95 **Am** Americium [243]	96 **Cm** Curium [247]	97 **Bk** Berkelium [247]	98 **Cf** Californium [251]	99 **Es** Einsteinium [252]	100 **Fm** Fermium [257]	101 **Md** Mendelevium [258]	102 **No** Nobelium [259]	103 **Lr** Lawrencium [262]

We have used the U.S. system as well as the system recommended by the International Union of Pure and Applied Chemistry (IUPAC) to label the groups in this periodic table. The system used in the United States includes a letter and a number (1A, 2A, 3B, 4B, etc.), which is close to the system developed by Mendeleev. The IUPAC system uses numbers 1–18 and has been recommended by the American Chemical Society (ACS). While we show both numbering systems here, we use the IUPAC system exclusively in the book. Elements with atomic numbers higher than 112 have been reported but not yet fully authenticated.

APPENDIX 4

Properties of Planets, Dwarf Planets, and Moons

Physical Data for Planets and Dwarf Planets

Planet	Equatorial Radius (km)	Equatorial Radius ($R/R_{\oplus}$)	Mass (kg)	Mass ($M/M_{\oplus}$)	Average Density (relative to water*)	Rotation Period (days)	Tilt of Rotation Axis (degrees, relative to orbit)	Equatorial Surface Gravity (relative to Earth†)	Escape Velocity (km/s)	Average Surface Temperature (K)
Mercury	2,440	0.383	3.30×10^{23}	0.055	5.427	58.65	0.01	0.378	4.3	340 (100, 700)§
Venus	6,052	0.949	4.87×10^{24}	0.815	5.243	243.02‡	177.3	0.907	10.36	735
Earth	6,378	1.000	5.97×10^{24}	1.000	5.513	1.000	23.44	1.000	11.19	288 (185, 331)§
Mars	3,396	0.532	6.42×10^{23}	0.107	3.934	1.026	25.19	0.377	5.03	227 (186, 268)§
Ceres	487.3	0.075	9.47×10^{20}	0.0002	2.09	0.378	10.59	0.28	1.86	170
Jupiter	71,492	11.209	1.90×10^{27}	317.8	1.326	0.4135	3.13	2.528	6.02	165
Saturn	60,268	9.449	5.68×10^{26}	95.16	0.687	0.4440	26.73	1.065	36.1	134
Uranus	25,559	4.007	8.68×10^{25}	14.54	1.270	0.7183‡	97.77	0.887	21.4	76
Neptune	24,764	3.883	1.02×10^{26}	17.148	1.638	0.6713	28.32	1.14	23.6	58
Pluto	1,195	0.187	1.25×10^{22}	0.0021	1.750	6.387‡	122.53	0.083	1.23	44
Haumea	~650	0.11	4.0×10^{21}	0.0007	~3	0.163	?	0.045	0.84	<50
Makemake	750	0.12	3×10^{21}	0.0005	~2	0.32	?	0.041	0.75	~30
Eris	1,163	0.182	1.67×10^{22}	0.0028	2.5	1.08	?	0.084	1.38	43

*The density of water is 1,000 kg/m^3.
†The surface gravity of Earth is 9.81 m/s^2.
‡Venus, Uranus, and Pluto rotate opposite to the directions of their orbits. Their north poles are south of their orbital planes.
§Where provided, values in parentheses give extremes of recorded temperatures.

Orbital Data for Planets and Dwarf Planets

Planet	Mean Distance from Sun (A*) (10^6 km)	(AU)	Orbital Period (P) (sidereal years)	Eccentricity	Inclination (degrees, relative to ecliptic)	Average Speed (km/s)
Mercury	57.9	0.387	0.241	0.2056	7.0	47.36
Venus	108.2	0.723	0.615	0.0068	3.39	35.02
Earth	149.6	1.000	1.000	0.0167	0.000	29.78
Mars	227.9	1.524	1.881	0.0934	1.85	24.08
Ceres	413.7	2.765	4.603	0.079	10.59	17.88
Jupiter	778.3	5.203	11.863	0.0484	1.30	13.06
Saturn	1,426.7	9.537	29.447	0.0539	2.49	9.64
Uranus	2,870.7	19.189	84.017	0.0473	0.77	6.80
Neptune	4,495.1	30.047	164.79	0.0113	1.769	5.43
Pluto	5,906.38	39.48	247.68	0.2488	17.16	4.72
Haumea	6,432.0	43.0	281.9	0.198	28.22	4.50
Makemake	6,783.3	45.5	306.8	0.161	29.01	4.42
Eris	10,123	68.0	561.4	0.436	43.84	3.44

*A is the semimajor axis of the planet's elliptical orbit.

Properties of Selected Moons*

		Orbital Properties		Physical Properties		
Planet	Moon	P (days)	A (10^3 km)	R (km)	M (10^{20} kg)	Relative Density† (g/cm²) (water = 1.00)
Earth (1 moon)	Moon	27.32	384.4	1,737.5	735	3.34
Mars (2 moons)	Phobos	0.32	9.38	13.4 × 11.2 × 9.2	0.0001	1.9
	Deimos	1.26	23.46	7.5 × 6.1 × 5.2	0.00002	1.5
Jupiter (66 known moons)	Metis	0.30	128	21.5	0.00012	3
	Amalthea	0.50	181.4	131 × 73 × 67	0.0207	0.8
	Io	1.77	421.8	1,822	893	3.53
	Europa	3.55	671.1	1,561	480	3.01
	Ganymede	7.15	1,070	2,631	1,482	1.94
	Callisto	16.69	1,883	2,410	1,076	1.83
	Himalia	250.56	11,461	85	0.067	2.6
	Pasiphae	744‡	23,624	30	0.0030	2.6
	Callirrhoe	759‡	24,102	4.3	0.00001	2.6

(continued)

Properties of Selected Moons*

(continued)

Planet	Moon	Orbital Properties: P (days)	Orbital Properties: A (10^3 km)	Physical Properties: R (km)	Physical Properties: M (10^{20} kg)	Physical Properties: Relative Density† (g/cm²) (water = 1.00)
Saturn (62 known moons)	Pan	0.58	133.58	14	0.00005	0.42
	Prometheus	0.61	139.38	74 × 50 × 34	0.0016	0.48
	Pandora	0.63	141.70	57 × 42 × 31	0.0014	0.49
	Mimas	0.94	185.54	198	0.38	1.15
	Enceladus	1.37	238.04	252	1.08	1.6
	Tethys	1.89	294.67	533	6.18	0.97
	Dione	2.74	377.42	562	11.0	1.48
	Rhea	4.52	527.07	764	23.1	1.23
	Titan	15.95	1,222	2,575	1,346	1.88
	Hyperion	21.28	1,501	205 × 130 × 110	0.0559	0.54
	Iapetus	79.33	3,561	736	18.1	1.08
	Phoebe	550.3†	12,948	107	0.08	1.6
	Paaliaq	687.5	15,024	11	0.0001	2.3
Uranus (27 known moons)	Cordelia	0.34	49.8	20	0.0004	1.3
	Miranda	1.41	129.9	236	0.66	1.21
	Ariel	2.52	190.9	579	12.9	1.59
	Umbriel	4.14	266.0	585	12.2	1.46
	Titania	8.71	436.3	789	34.2	1.66
	Oberon	13.46	583.5	761	28.8	1.56
	Setebos	2,225†	17,420	24	0.0009	1.5
Neptune (13 known moons)	Naiad	0.29	48.2	48 × 30 × 26	0.002	1.3
	Larissa	0.56	73.5	108 × 102 × 84	0.05	1.3
	Proteus	1.12	117.6	220 × 208 × 202	0.5	1.3
	Triton	5.88†	354.8	1,353	214	2.06
	Nereid	360.13	5,513.8	170	0.3	1.5
Pluto (5 moons)	Charon	6.39	17.54	604	15.5	1.65
Haumea (2 moons)	Namaka	18	25.66	85	0.018	~1
	Hi'iaka	49	49.88	170	0.179	~1
Eris	Dysnomia	15.8	37.4	50–125?	?	?

*Innermost, outermost, largest, and/or a few other moons for each planet.
†Irregular moon (has retrograde orbit).

APPENDIX 5

Nearest and Brightest Stars

Stars within 12 Light-Years of Earth

Name*	Distance (ly)	Spectral Type†	Relative Visual Luminosity‡ (Sun = 1.000)	Apparent Magnitude	Absolute Magnitude
Sun	1.55×10^{-5}	G2V	1.000	−26.74	4.83
Alpha Centauri C (Proxima Centauri)	4.24	M5.0V	0.000052	11.05	15.48
Alpha Centauri A	4.36	G2V	1.5	0.01	4.38
Alpha Centauri B	4.36	K0V	0.44	1.34	5.71
Barnard's star	5.96	M4Ve	0.00043	9.57	13.25
CN Leonis	7.78	M5.5	0.000019	13.53	16.64
BD +36-2147	8.29	M2.0V	0.0057	7.47	10.44
Sirius A	8.58	A1V	22.1	−1.43	1.47
Sirius B	8.58	DA2	0.0025	8.44	11.34
BL Ceti	8.73	M5.5V	0.000059	12.61	15.40
UV Ceti	8.73	M6.0	0.000039	13.06	15.85
V1216 Sagittarii	9.68	M3.5V	0.00050	10.44	13.08
HH Andromedae	10.32	M5.5V	0.00010	12.29	14.79
Epsilon Eridani	10.52	K2V	0.28	3.73	6.20
Lacaille 9352	10.74	M1.0V	0.011	7.34	9.76
FI Virginis	10.92	M4.0V	0.00033	11.16	13.53
EZ Aquarii A	11.26	M5.0V	0.000063	13.03	15.33
EZ Aquarii B	11.26	M5e	0.000050	13.27	15.58
EZ Aquarii C	11.26	—	0.000010	15.07	17.37
Procyon A	11.40	F5IV-V	7.38	0.38	2.66
Procyon B	11.40	DA	0.00055	10.70	12.98
61 Cygni A	11.40	K5.0V	0.087	5.20	7.48

(continued)

Stars within 12 Light-Years of Earth

(continued)

Name*	Distance (ly)	Spectral Type†	Relative Visual Luminosity‡ (Sun = 1.000)	Apparent Magnitude	Absolute Magnitude
61 Cygni B	11.40	K7.0V	0.041	6.03	8.31
Gliese 725 A	11.52	M3.0V	0.0029	8.90	11.17
Gliese 725 B	11.52	M3.5V	0.0014	9.69	11.96
Andromedae GX	11.62	M1.5V	0.0064	8.08	10.32
Andromedae GQ	11.62	M3.5V	0.00041	11.06	13.30
Epsilon Indi A	11.82	K5Ve	0.15	4.68	6.89
Epsilon Indi B (brown dwarf)	11.82	T1.0	–	–	–
Epsilon Indi C (brown dwarf)	11.82	T6.0	–	–	–
DX Cancri	11.82	M6.0V	0.000012	14.90	17.10
Tau Ceti	11.88	G8.5V	0.46	3.49	5.68
Gliese 1061	11.99	M5.5V	0.000067	13.09	15.26

Source: From the Research Consortium on Nearby Stars (http://chara.gsu.edu/RECONS).
*Stars may carry many names, including common names (such as Sirius), names based on their prominence within a constellation (such as Alpha Canis Majoris, another name for Sirius), or names based on their inclusion in a catalog (such as BD +36-2147). Addition of letters A, B, and so on, or superscripts indicates membership in a multiple-star system.
†Spectral types such as M3 are discussed in Chapter 13. Other letters or numbers provide additional information. For example, *V* after the spectral type indicates a main-sequence star, and III indicates a giant star. Stars of spectral type T are brown dwarfs.
‡Luminosity in this table refers only to radiation in "visual" light.

The 25 Brightest Stars in the Sky

Name	Common Name	Distance (ly)	Spectral Type	Relative Visual Luminosity* (Sun = 1.000)	Apparent Visual Magnitude	Absolute Visual Magnitude
Sun	Sun	1.58×10^{-5}	G2V	1.000	−26.74	4.83
Alpha Canis Majoris	Sirius	8.60	A1V	22.9	−1.46	1.43
Alpha Carinae	Canopus	313	F0II	14,900	−0.72	−5.60
Alpha[1] Centauri	Rigil Kentaurus A	4.36	G2V	1.51	−0.01	4.38
Alpha[2] Centauri	Rigil Kentaurus B	4.36	K1V	0.44	1.33	5.71
Alpha Bootis	Arcturus	36.7	K1.5III	113	−0.04	−0.30
Alpha Lyrae	Vega	25.3	A0Va	49.2	0.03	0.60
Alpha Aurigae	Capella	43	G5IIIe+G0III	137	0.08	−0.51
Beta Orionis	Rigel	860	B8Iab	54,000	0.12	−7.0
Alpha Canis Minoris	Procyon	11.5	F5IV-V	7.73	0.34	2.61
Alpha Eridani	Achernar	140	B3Vpe	1,030	0.46	−2.70
Beta Centauri	Hadar	392	B1III	7,180	0.61	−4.81
Alpha Orionis	Betelgeuse	570	M1-2Iab	13,600	0.7	−5.5
Alpha Aquilae	Altair	16.7	A7V	11.1	0.77	2.22
Alpha Crucis	Acrux	325	B0.5IV+B1V	3,100	1.3	−3.9
Alpha Tauri	Aldebaran	67	K5III	163	0.85	−0.70
Alpha Scorpii	Antares	550	M1.5Ib	16,300	0.96	−5.7
Alpha Virginis	Spica	250	B1IV+B4V	1,920	1.04	−3.38
Beta Geminorum	Pollux	34	K0III	32.2	1.14	1.06
Alpha Piscis	Fomalhaut	25	A3V	17.4	1.16	1.73
Beta Crucis	Mimosa	280	B0.5III	1,980	1.25	−3.41
Alpha Cygni	Deneb	1,425	A2Ia	58,600	1.25	−7.09
Alpha Leonis	Regulus	79	B7V	146	1.35	−0.58
Epsilon Canis Majoris	Adhara	405	B2II	3,400	1.50	−4.0
Alpha Gemini	Castor	51	A1V+A5Vm	49	1.58	0.61
Gamma Crucis	Gacrux	88	M3.5III	138	1.63	−0.52

Sources: Data from Jim Kaler's *STARS* page (http://stars.astro.illinois.edu/sow/bright.html); SIMBAD Astronomical Database (http://simbad.u-strasbg.fr/simbad).

*Luminosity in this table refers only to radiation in "visual" light.

APPENDIX 6

Observing the Sky

The purpose of this appendix is to provide enough information so that you can make sense of a star chart or list of astronomical objects, as well as find a few objects in the sky.

Celestial Coordinates

In Chapter 2 we discuss the **celestial sphere**—the imaginary sphere with Earth at its center upon which celestial objects appear to lie. A number of different coordinate systems are used to specify the positions of objects on the celestial sphere. The simplest of these is the *altitude-azimuth coordinate system*. The altitude-azimuth coordinate system is based on the "map" direction to an object (the object's azimuth, with north = 0°, east = 90°, south = 180°, and west = 270°) combined with how high the object is above the horizon (the object's altitude, with the horizon at 0° and the zenith at 90°). For example, an object that is 10° above the eastern horizon has an altitude of 10° and an azimuth of 90°. An object that is 45° above the horizon in the southwest is at altitude 45°, azimuth 225°.

The altitude-azimuth coordinate system is the simplest way to tell someone where in the sky to look at the moment, but it is not a good coordinate system for cataloging the positions of objects. The altitude and azimuth of an object are different for each observer, depending on the observer's position on Earth, and they are constantly changing as Earth rotates on its axis. To specify the direction to an object in a way that is the same for everyone requires a coordinate system that is fixed relative to the celestial sphere. The most common such coordinates are called *celestial coordinates*.

Celestial coordinates are illustrated in **Figure A6.1**. Celestial coordinates are much like the traditional system of latitude and longitude used on the surface of Earth. On Earth, latitude specifies how far you are from Earth's equator, as discussed in Chapter 2. If you are on Earth's equator, your latitude is 0°. If you are at Earth's North Pole, your latitude is 90° north. If you are at Earth's South Pole, your latitude is 90° south.

The latitude-like coordinate on the celestial sphere is called **declination**, often signified with the Greek letter δ (delta). The celestial equator has $\delta = 0°$. The north celestial pole has $\delta = +90°$. The south celestial pole has $\delta = -90°$. (See Chapter 2 if you need to refresh your memory about the celestial equator or celestial poles.) Declination is usually expressed in degrees, minutes of arc, and seconds of arc. For example, Sirius, the brightest star in the sky, has $\delta = -16°42'58''$, meaning that it is located not quite 17° south of the celestial equator.

On Earth, east–west position is specified by longitude. Lines of constant longitude run north–south from one pole to the other. Unlike latitude, for which the equator provides a natural place to call "zero," there is no natural starting point for longitude, so one was invented. By arbitrary convention, the Royal Observatory in Greenwich, England, is defined to lie at a longitude of 0°. On the celestial sphere the longitude-like coordinate is called **right ascension**, often signified with the Greek letter α (alpha). Unlike the case with longitude, there *is* a natural point on the celestial sphere to use as the starting point for right ascension: the vernal equinox, or the point at which the ecliptic crosses the celestial equator with the Sun moving from the southern sky into the northern sky. The (Northern Hemisphere) vernal equinox defines the line of right ascension at which $\alpha = 0°$. The (Northern Hemisphere) autumnal equinox, located on the opposite side of the sky, is at $\alpha = 180°$.

Normally, right ascension is measured in units of time rather than degrees. It takes Earth 24 hours (of sidereal time) to rotate on its axis, so the celestial sphere is divided into 24 hours of right ascension, with each hour of right ascension corresponding to 15°. Hours of right ascension are then subdivided into minutes and seconds of time. Right ascension increases going to the east. The right ascension of Sirius, for example, is $\alpha = 06^h45^m08.9^s$, meaning that Sirius is about 101° (that is, 06^h45^m) east of the vernal equinox. Time is a natural unit for measuring right ascension because time naturally tracks the motion of objects due to Earth's rotation on its axis. If stars on the meridian at a certain time have $\alpha = 06^h$, then an hour later the stars on the meridian will have 07^h,

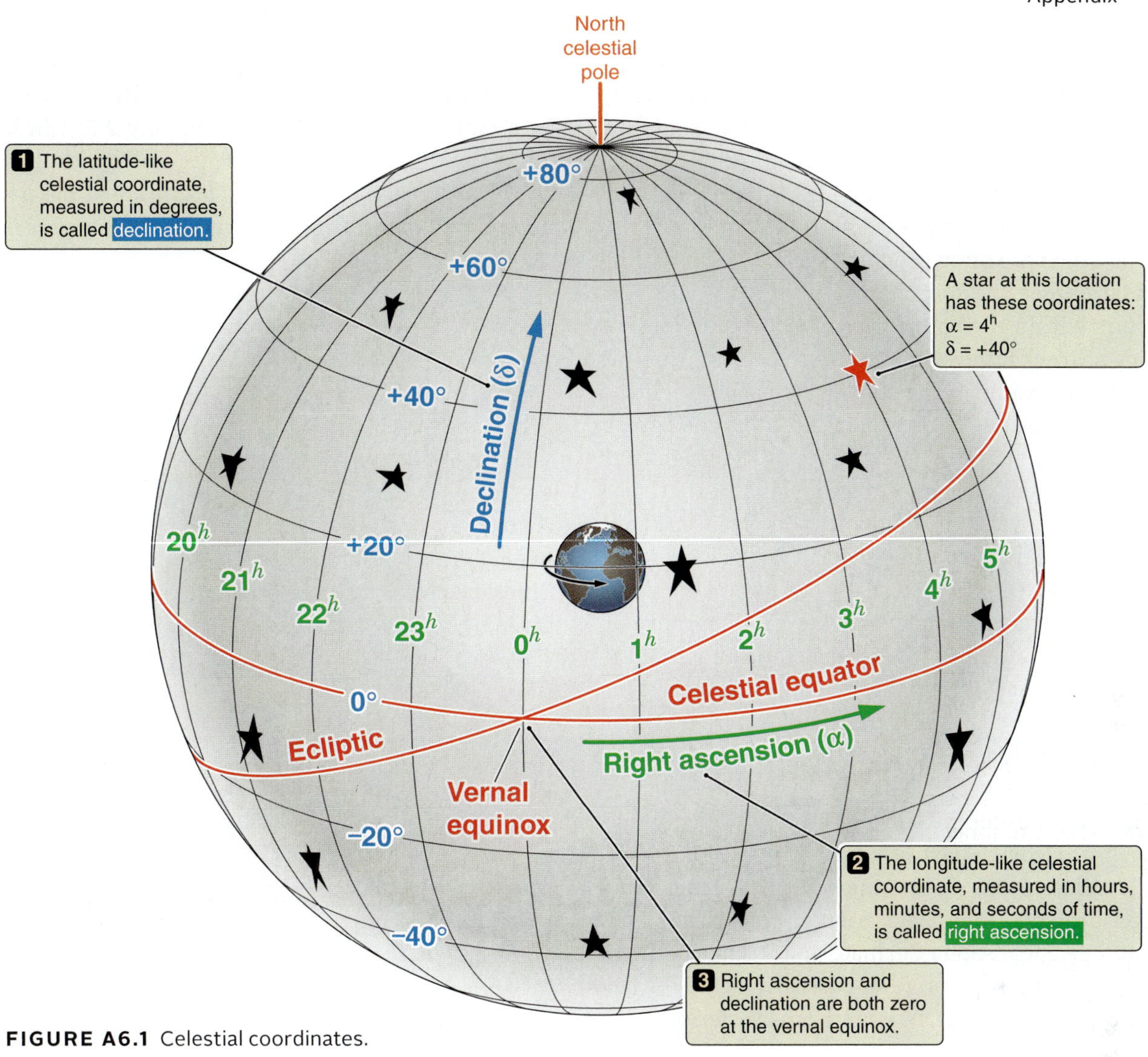

FIGURE A6.1 Celestial coordinates.

and an hour after that they will have 08^h. The *local sidereal time*, or *star time*, at your location right now is equal to the right ascension of the stars that are on your meridian at the moment. Because of Earth's motion around the Sun, a sidereal day is about 4 minutes shorter than a solar day, so local sidereal time constantly gains on solar time. At midnight on September 22, the local sidereal time is 0^h. By midnight on December 21, local sidereal time has advanced to 06^h. On March 20, local sidereal time at midnight is 12^h. And at midnight on June 20, local sidereal time is 18^h.

Putting this all together, right ascension and declination provide a convenient way to specify the location of any object on the celestial sphere. Sirius is located at $\alpha = 06^h45^m08.9^s$, or $-16°42'58''$, which means that at midnight on December 21 (local sidereal time $= 06^h$) you will find Sirius about 45^m east of the meridian, not quite 17° south of the celestial equator.

There is just one final caveat. As we discussed in Chapter 2, the directions of the celestial equator, celestial poles, and vernal equinox are constantly changing as Earth's axis wobbles like the axis of a spinning top. In Chapter 2 we called this 26,000-year wobble the **precession of the equinoxes**, meaning that the location of the equinoxes is slowly advancing along the ecliptic. So when we specify the celestial coordinates of an object, we need to specify the date at which the positions of the vernal equinox and celestial poles were measured. By convention, coordinates are usually referred to with the position of the vernal equinox on January 1, 2000. A complete, formal specification of the coordinates of Sirius would then be $\alpha(2000) = 06^h45^m08.9^s$, $\delta(2000) = -16°42'58''$, where the "2000" in parentheses refers to the equinox of the coordinates.

Constellations and Names

Although it is certainly possible to exactly specify any location on the surface of Earth by giving its latitude and longitude, it is usually convenient to use a more descriptive address. We might say, for example, that one of the coauthors of this book works near latitude 37° north, longitude 122° west; but it would probably mean a lot more to you if we said that George Blumenthal works in Santa Cruz, California.

Just as the surface of Earth is divided into nations and states, the celestial sphere is divided into 88 **constellations**, the names of which are often used to refer to objects within their boundaries (see the star charts in **Figure A6.2**). The brightest stars within the boundaries of a constellation are named using a Greek letter combined with the name of the constellation. For example, the star Sirius is the brightest star in the constellation Canis Major (literally, the "great dog"), so it is called "Alpha Canis Majoris." The bright red star in the northeastern corner of the constellation of Orion is called "Alpha Orionis," also known as Betelgeuse. Rigel, the bright blue star in the southwest corner of Orion, is also called "Beta Orionis."

Astronomical objects can take on a bewildering range of names. For example, the bright southern star Canopus, also known as "Alpha Carinae" (the brightest star in the constellation of Carina), has no fewer than 34 different names, most of which are about as memorable as "SAO 234480" (number 234,480 in the Smithsonian Astrophysical Observatory catalog of stars).

There is a slight difference in the way a constellation is spelled when it becomes part of a star's name. For example, Sirius is called "Alpha Canis Majoris," not "Alpha Canis Major"; Rigel is referred to as "Beta Orionis," not "Beta Orion"; and Canopus becomes "Alpha Carinae," not "Alpha Carina." This is because the Latin genitive, or possessive, case is used with star names; for example, *Orionis* means "of Orion."

Astronomical Magnitudes

Apparent Magnitudes

We first introduced magnitudes in Connections 13.1; here we provide some additional information. You are most likely to see this system if you take a lab course in astronomy or if you use a star catalog. Astronomers use the logarithmic system of **apparent magnitudes** to compare the apparent brightness of objects in the sky. Other common systems of logarithmic measurements that you may have encountered include decibels for measuring sound levels, and the Richter scale for measuring the strength of earthquakes. For example, an earthquake of magnitude 6 is not just a little stronger than an earthquake of magnitude 5; it is, in fact, 10 times stronger.

As discussed in Connections 13.1, a difference of five magnitudes between the apparent brightness of two stars (say, a star with $m = 6$ and a star with $m = 1$), corresponds to 100 times difference in brightness, and *the greater the magnitude, the fainter the object*. If five steps in magnitude correspond to a factor of 100 in brightness, then one step in magnitude must correspond to the fifth root of 100—that is, a factor of $100^{1/5}$ = approximately 2.512 in brightness ($100^{1/5} \times 100^{1/5} \times 100^{1/5} \times 100^{1/5} \times 100^{1/5} = 100$).

If star 1 has a brightness of b_1 and star 2 has a brightness of b_2, then the ratio of the brightness of the stars is given by:

$$\frac{b_1}{b_2} = (2.512)^{m_2 - m_1} = 100^{\frac{(m_2 - m_1)}{5}}$$

We can put this into the more common base 10 by noting that $100 = 10^2$, so this becomes:

$$\frac{b_1}{b_2} = 10^{2 \times \frac{(m_2 - m_1)}{5}} = 10^{0.4(m_2 - m_1)}$$

After taking the log of both sides and dividing by 0.4, the difference in magnitude $(m_2 - m_1)$ between the two stars is given by:

$$m_2 - m_1 = 2.5 \log_{10} \frac{b_1}{b_2}$$

The following table shows some examples using the preceding equations.

Apparent Magnitude Difference $(m_2 - m_1)$	Ratio of Apparent Brightness (b_1/b_2)
1	2.512
2	$2.512^2 = 6.3$
3	$2.512^3 = 15.8$
4	$2.512^4 = 39.8$
5	$2.512^5 = 100$
10	$2.512^{10} = 100^2 = 10{,}000$
15	$2.512^{15} = 100^3 = 1{,}000{,}000$
20	$2.512^{20} = 100^4 = 10^8$
25	$2.512^{25} = 100^5 = 10^{10}$

Absolute Magnitudes

Recall that stars differ in their brightness for two reasons: the amount of light they are actually emitting, and their distance from Earth. The magnitude system is also used for **luminosity**, with the same scale as for brightness: a difference of five magnitudes corresponds to 100 times difference in luminosity. Astronomers call these **absolute magnitudes** (M_{abs}), and the idea is to imagine how bright the star would be if it were at a distance of 10 parsecs (pc). Absolute magnitudes enable comparison of how luminous two stars really are, without the factor of distance. The Sun is very bright because it is so close (apparent visual magnitude = −27); but if the Sun were at a distance of 10 pc, its magnitude would be only about 5.[1] Thus the absolute magnitude of the Sun is $M_{abs} = 5$. Recall that the luminosity of a star is usually expressed by comparing it with the luminosity of the Sun. As with apparent magnitudes, higher magnitude numbers corresponding to lower luminosity. Thus, a star that is 100 times less luminous than the Sun will be 5 absolute magnitudes fainter, or $M_{abs} = 10$. A star that is 10,000 times more luminous than the Sun will be 10 absolute magnitudes brighter, or $M_{abs} = -5$.

Absolute magnitudes and luminosities follow the same equations as those we provided already, using L instead of b and M_{abs} instead of m:

$$\frac{L_1}{L_2} = 10^{2 \times \frac{(M_{abs(2)} - M_{abs(1)})}{5}} = 10^{0.4(M_{abs(2)} - M_{abs(1)})}.$$

and

$$M_{abs(2)} - M_{abs(1)} = 2.5 \log_{10} \frac{L_1}{L_2}$$

Most often, astronomers think about the luminosity of a star compared to the luminosity of the Sun. In this case $L_1 = L_{star}$ and $L_2 = L_\odot$. The following table compares luminosity (where $L_\odot = 1$) with absolute magnitude of a star.

$L_{star}/L_\odot$	M_{abs}
1,000,000	−10
10,000	−5
100	0
1	5
1/100	10
1/10,000	15

Distance Modulus

The difference between the apparent magnitude and the absolute magnitude depends on the star's distance. By definition, a star at a distance of exactly 10 pc will have an apparent magnitude equal to its absolute magnitude. Astronomers can always measure the brightness of a star and thus its apparent magnitude, and can estimate the luminosity of a star and thus its absolute magnitude using the H-R diagram. This is the way the distances to most stars are found.

Using the preceding equations and the definition of absolute magnitude, we can get to the following relatively simple expression:

$$m - M_{abs} = 5 \log_{10} d - 5$$

where distance d is in parsecs.

We can rewrite this equation to solve for distance as follows:

$$d = 10^{\left(\frac{m - M_{abs} + 5}{5}\right)}$$

The following table shows how the difference between an object's apparent and absolute magnitudes leads to its distance in parsecs.

$m - M_{abs}$	Distance (pc)
−3	2.5
−2	4.0
−1	6.3
0	10
1	16
2	25
3	40
4	63
5	100
10	1,000
15	10,000
20	100,000

Although the system of astronomical magnitudes is convenient in many ways—which is why astronomers continue to use it—it can also be confusing to new students. Just remember three things and you will probably get by:

1. The greater the magnitude, the fainter the object.
2. One magnitude *smaller* means about two and a half times *brighter.*
3. The brightest stars in the sky have magnitudes of less than 1, and the faintest stars that can be seen with the naked eye on a dark night have magnitudes of about 6.

A final note: Astronomers sometimes use "colors" based on the ratio of the brightness of a star as seen in two different parts of the spectrum. The "b_B/b_V color," for example, is the ratio of the brightness of a star seen through a blue filter, divided by the brightness of a star seen through a yellow-green (visual) filter. Normally, astronomers instead discuss the "$B - V$ color" of a star, which is equal to the difference between a star's blue magnitude and its visual magnitude. We can use the previous expression for a magnitude difference to write:

$$B - V \text{ color} = m_B - m_V = -2.5 \log_{10} (b_B/b_V)$$

Thus, a star with a b_B/b_V color of 1.0 has a $B - V$ color of 0.0, and a star with a b_B/b_V color of 1.4 has a $B - V$ color of −0.37. Notice that, as with magnitudes, $B - V$ colors are "backward": the bluer a star, the greater its b_B/b_V color but the less its $B - V$ color.

1 The apparent and absolute magnitudes of the Sun are −26.74 and +4.83, respectively. We use +5 for the Sun's absolute magnitude as an approximation.

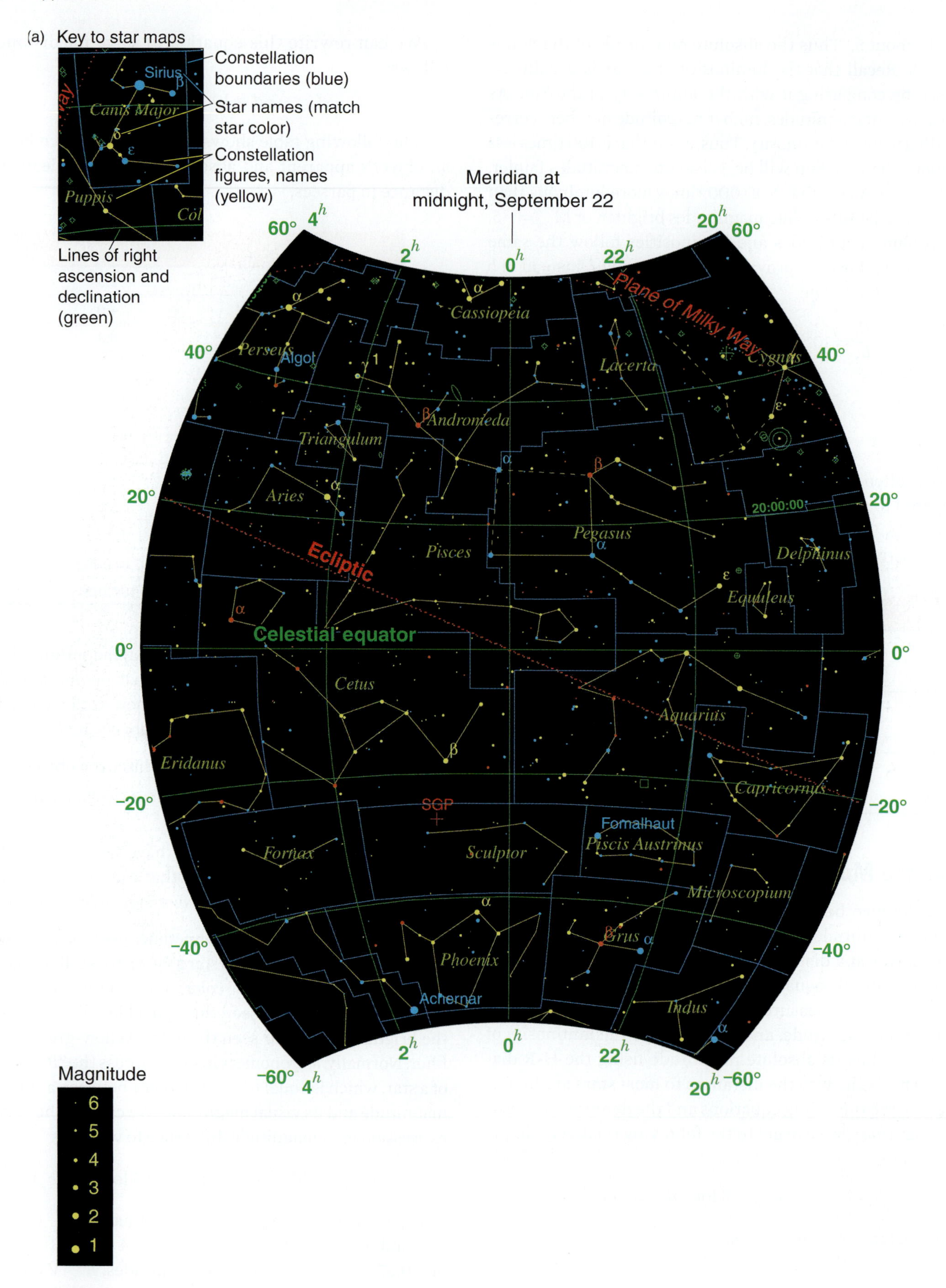

FIGURE A6.2A The sky from right ascension 20ʰ to 04ʰ and declination −60° to +60°.

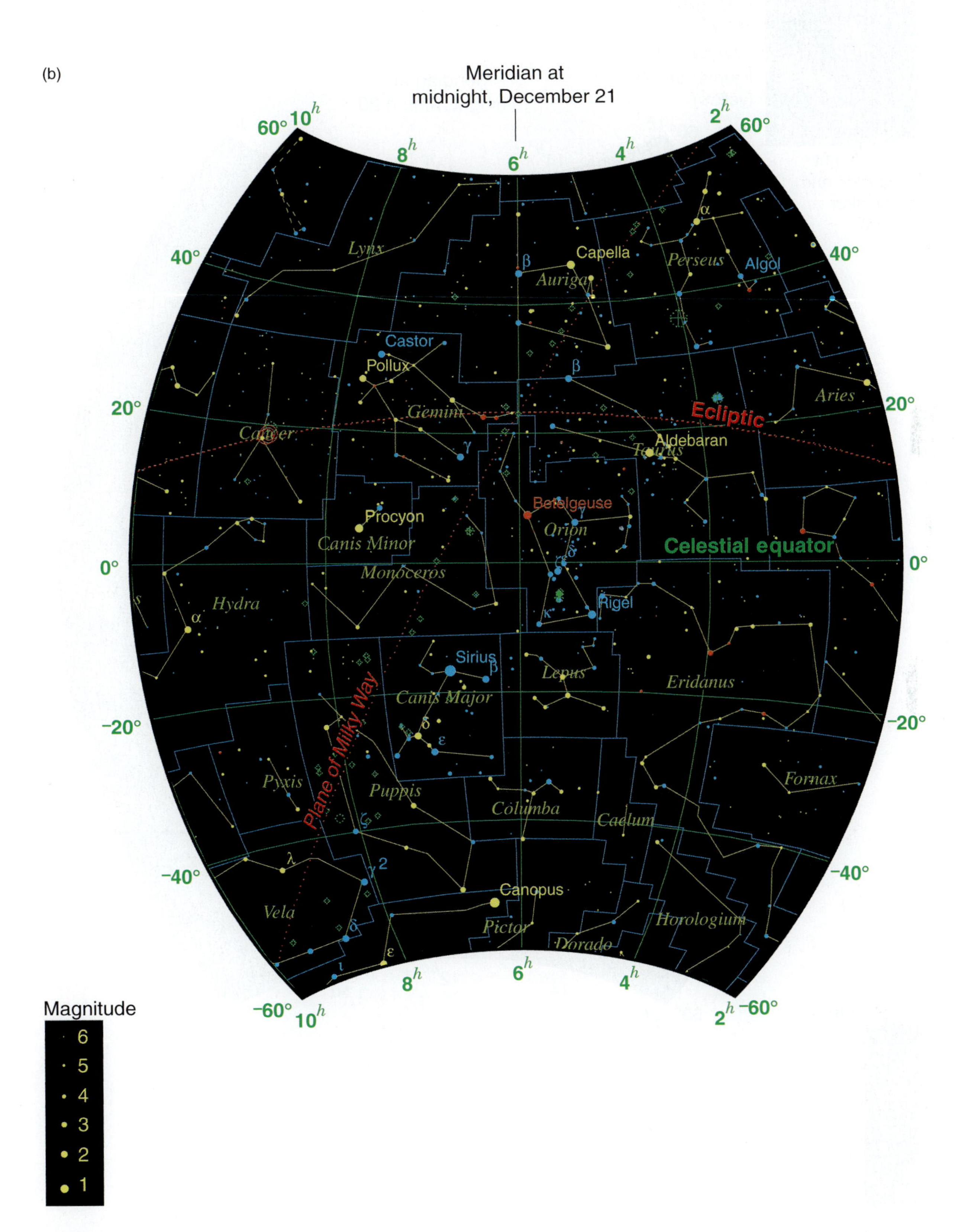

FIGURE A6.2B The sky from right ascension 02ʰ to 10ʰ and declination −60° to +60°.

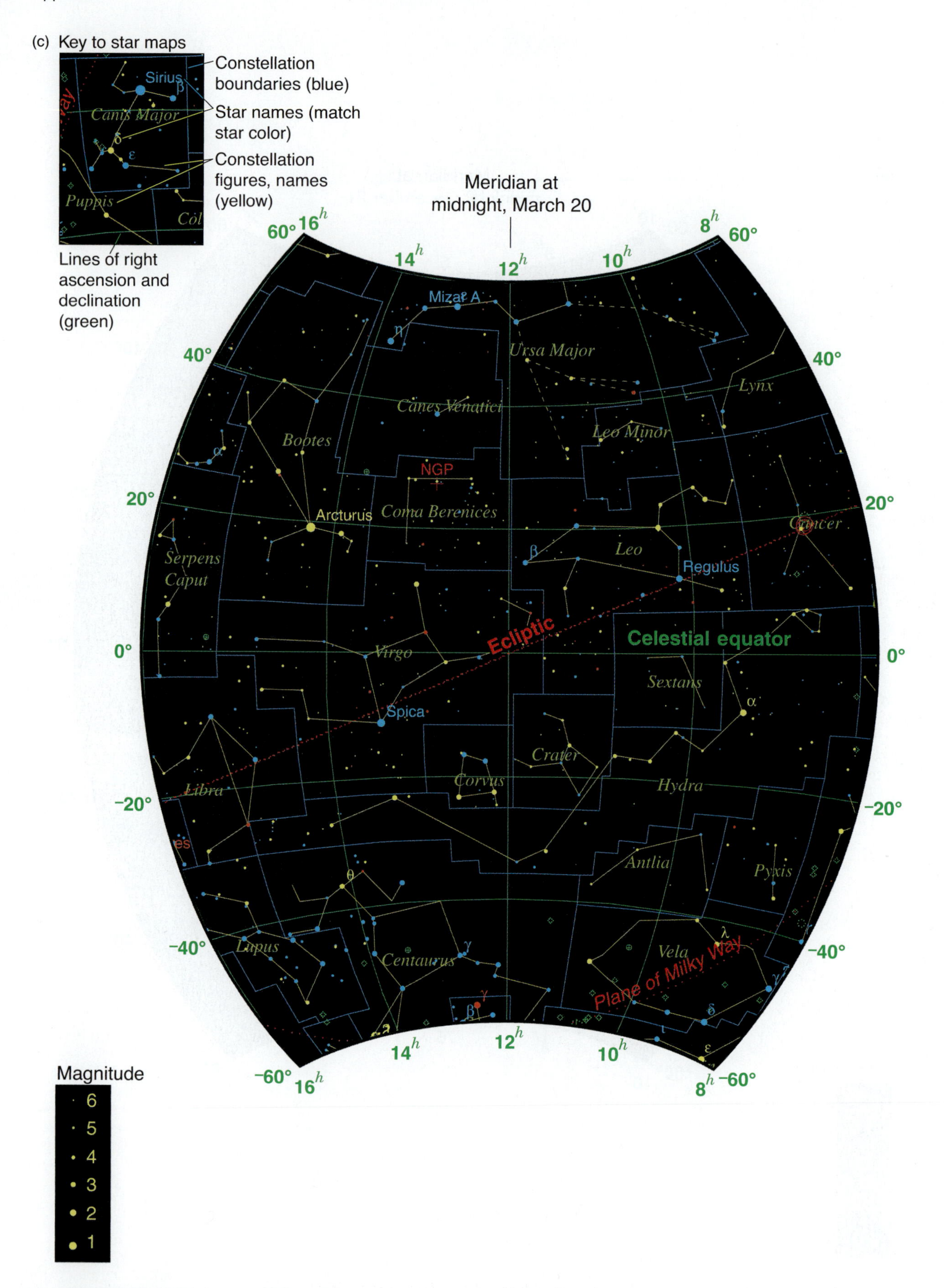

FIGURE A6.2C The sky from right ascension 08^h to 16^h and declination −60° to +60°.

(d)

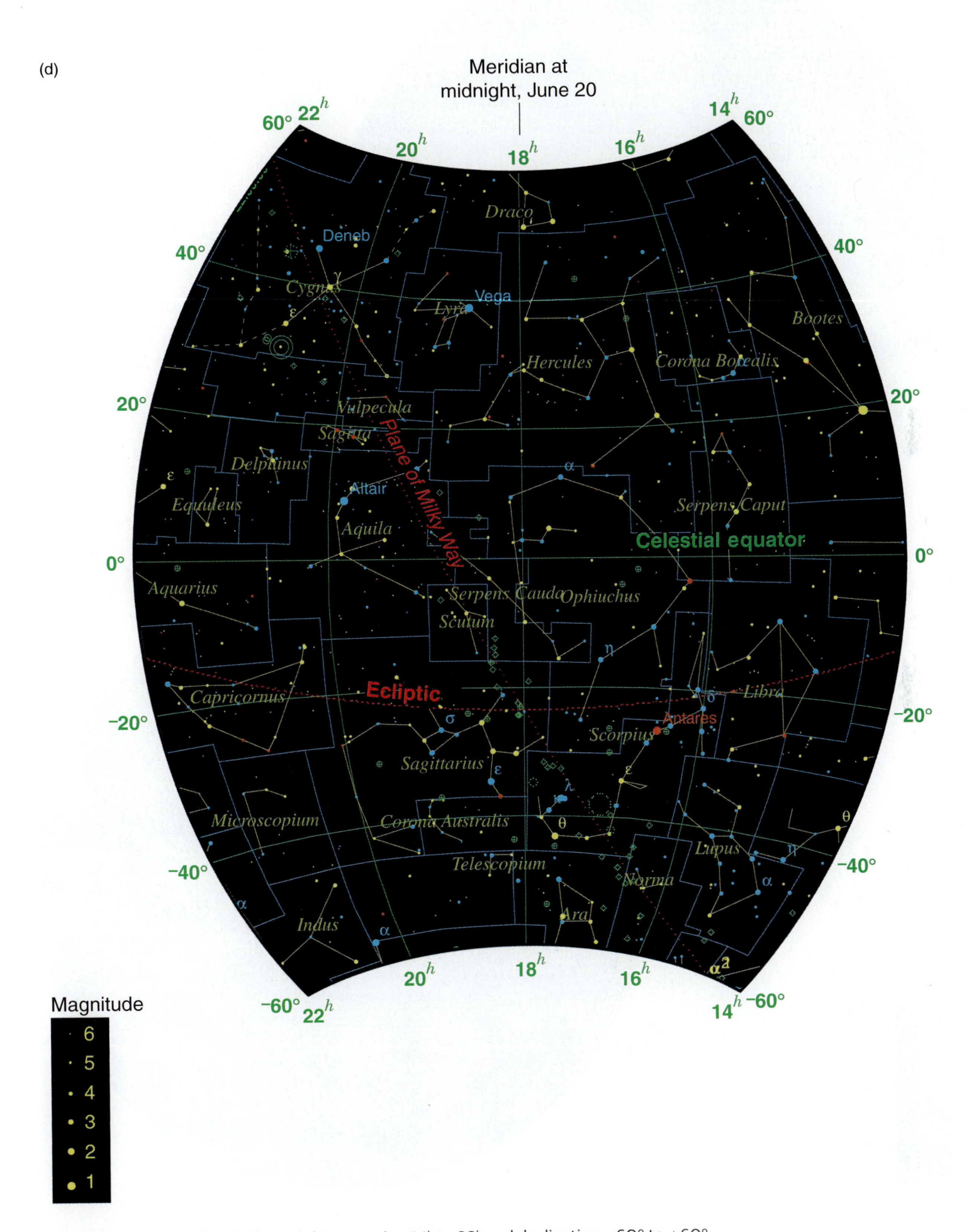

FIGURE A6.2D The sky from right ascension 14^h to 22^h and declination −60° to +60°.

(e)

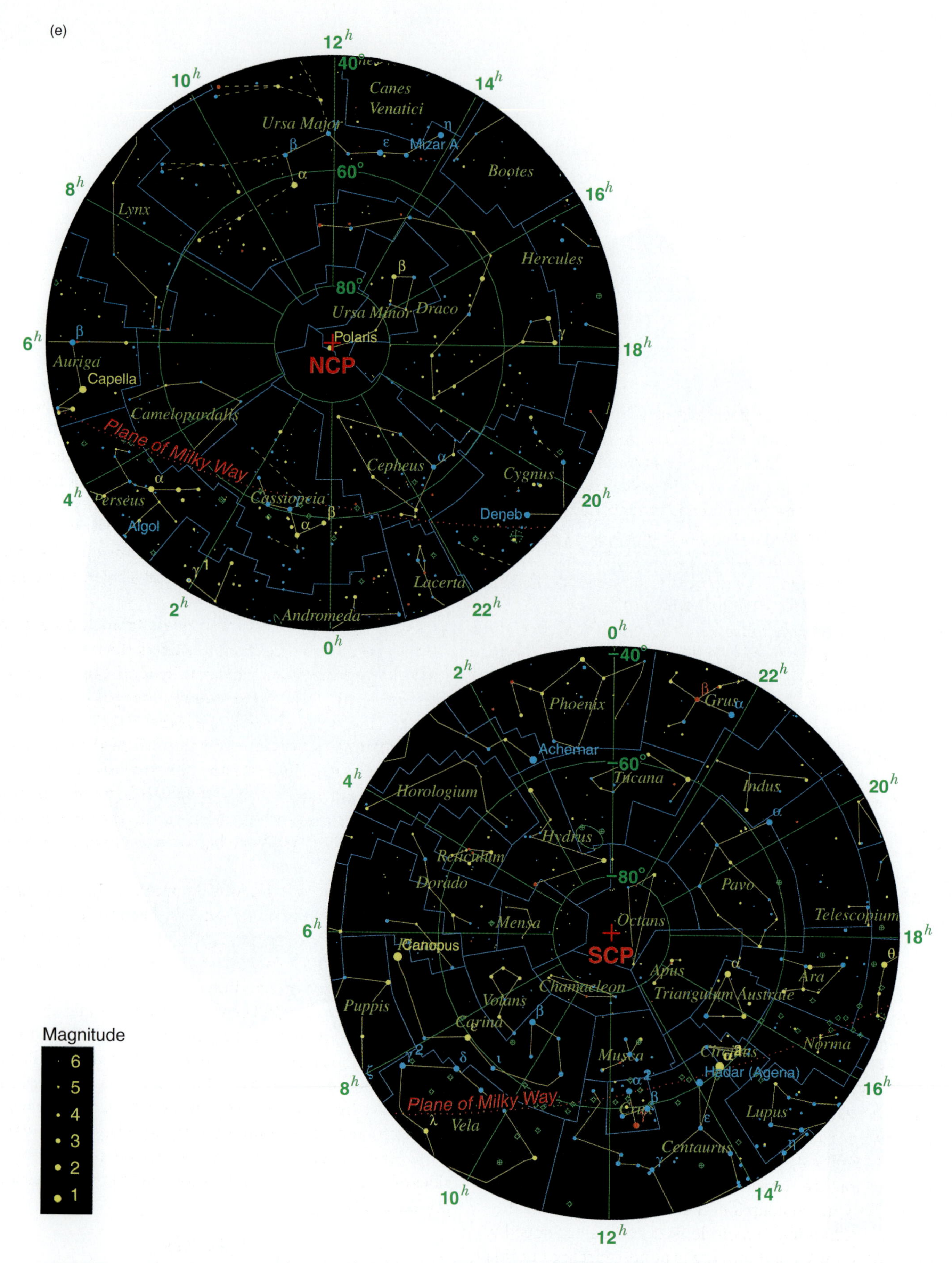

FIGURE A6.2E The regions of the sky north of declination +40° and south of declination −40°. NCP = north celestial pole; SCP = south celestial pole.

APPENDIX 7

Uniform Circular Motion and Circular Orbits

Uniform Circular Motion

In Chapter 4 (see Section 4.2 and Figure 4.8) we discuss the motion of an object moving in a circle at a constant speed. This motion, called **uniform circular motion**, is the result of the fact that centripetal force always acts toward the center of the circle. The key question when thinking about uniform circular motion is, How hard does something have to pull to keep the object moving in a circle? Part of the answer to this question is pretty obvious: the more massive an object is, the harder it will be to keep it moving on its circular path. According to Newton's second law of motion, $F = ma$, or in this case, the centripetal force equals the mass multiplied by the centripetal acceleration. The larger the mass, the greater the force required to keep it moving in its circle.

The centripetal force needed to keep an object moving in constant circular motion also depends on two other quantities: the speed of the object and the size of the circle. The faster an object is moving, the more rapidly it has to change direction to stay on a circle of a given size. The second quantity that influences the needed acceleration is the radius of the circle. The smaller the circle, the greater the pull needed to keep it on track. You can understand this by looking at the motion. A small circle requires a continuous "hard" turn, whereas a larger circle requires a more gentle change in direction. It takes more force to keep an object moving faster in a smaller circle than it does to keep the same object moving more slowly in a larger circle. (To get a better feel for how this works, think about the difference between riding in a car that is taking a tight curve at high speed and a car that is moving slowly around a gentle curve.)

To arrive at the circular velocity and other results discussed in Chapter 4, these intuitive ideas about uniform circular motion are turned into a quantitative expression of exactly how much centripetal acceleration is needed to keep an object moving in a circle with radius r at speed v. **Figure A7.1** shows a ball moving around a circle of radius r at a constant speed v at two different times. The centripetal acceleration that is keeping the ball on the circle is a. Remember that the acceleration is always directed toward the center of the circle, whereas the velocity of the ball is always perpendicular to the acceleration. The ball's velocity and its acceleration are always at right angles to each other. As the object moves around the circle, the direction of motion and the direction of the acceleration change together in lockstep.

Figure A7.1 contains two triangles. Triangle 1 shows the velocity (speed and direction) at each of the two times. The arrow labeled "Δv" connecting the heads of the two velocity arrows shows how much the velocity changed between time 1 (t_1) and time 2 (t_2). This change is the effect of the centripetal acceleration. If you imagine that points 1 and 2 are very close together—so close that the direction of the centripetal acceleration does not change by much between the two—then the centripetal acceleration equals the change in the velocity divided by the time between the two, $\Delta t = t_2 - t_1$. So, $\Delta v = a\Delta t$.

Triangle 2 shows something similar. Here the arrow labeled "Δr" indicates the change in the position of the ball between time 1 and time 2. Again, if you imagine that the time between the two points is very short, Δr is equal to the velocity multiplied by the time, or $\Delta r = v\Delta t$.

The line between the center of the circle and the ball is always perpendicular to the velocity of the ball. So if the direction of the ball's velocity changes by an angle θ, then the direction of the line between the ball and the center of the circle must also change by the same angle α. In other words, triangles 1 and 2 are "similar triangles." They have the same *shape*. If the triangles are the same shape, the ratio of two sides of triangle 1 must equal the ratio of the two corresponding sides of triangle 2. Then:

$$\frac{a\Delta t}{v} = \frac{v\Delta t}{r}$$

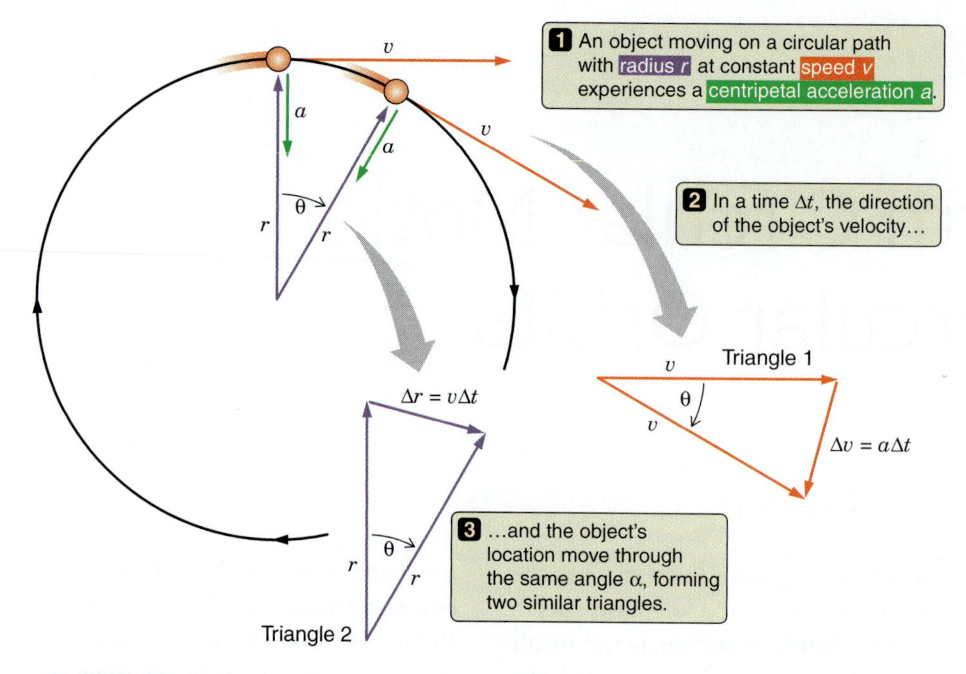

FIGURE A7.1 Similar triangles are used to find the centripetal force that is needed to keep an object moving at a constant speed on a circular path.

If we divide by Δt on both sides of the equation and then cross-multiply, we obtain:

$$ar = v^2$$

which, after dividing both sides of the equation by r, becomes:

$$a_{\text{centripetal}} = \frac{v^2}{r}$$

The subscript "centripetal" is added to a to signify that this is the centripetal acceleration needed to keep the object moving in a circle of radius r at speed v. The centripetal force required to keep an object of mass m moving on such a circle is then:

$$F_{\text{centripetal}} = ma_{\text{centripetal}} = \frac{mv^2}{r}$$

Circular Orbits

In the case of an object moving in a circular orbit, there is no string to hold the ball on its circular path. Instead, this force is provided by **gravity**.

Think about an object with mass m in orbit about a much larger object with mass M. The orbit is circular, and the distance between the two objects is given by r. The force needed to keep the smaller object moving at speed v in a circle with radius r is given by the previous expression for $F_{\text{centripetal}}$. The force actually provided by gravity (see Chapter 4) is:

$$F_{\text{grav}} = G\frac{Mm}{r^2}$$

If gravity is responsible for holding the mass in its circular motion, then it should be true that $F_{\text{grav}} = F_{\text{centripetal}}$. That is, if mass m is moving in a circle under the force of gravity, the force provided by gravity must equal the centripetal force needed to explain that circular motion. Setting the two expressions for $F_{\text{centripetal}}$ and F_{grav} equal to each other gives:

$$\frac{mv^2}{r} = G\frac{Mm}{r^2}$$

All that remains is a bit of algebra. Dividing by m on both sides of the equation and multiplying both sides by r gives:

$$v^2 = G\frac{M}{r}$$

Taking the square root of both sides then yields the desired result:

$$v_{\text{circular}} = \sqrt{\frac{GM}{r}}$$

This is the **circular velocity** we presented in Chapter 4. It is the velocity at which an object in a circular orbit must be moving. If the object were not moving at this velocity, then gravity would not be providing the needed centripetal force, and the object would not move in a circle.

APPENDIX 8*

IAU 2006 Resolutions: "Definition of a Planet in the Solar System" and "Pluto"

August 24, 2006, Prague

Resolutions

Resolution 5 is the principal definition for the IAU usage of "planet" and related terms. Resolution 6 creates, for IAU usage, a new class of objects, for which Pluto is the prototype. The IAU will set up a process to name these objects.

Resolution 5: Definition of a Planet in the Solar System

Contemporary observations are changing our understanding of planetary systems, and it is important that our nomenclature for objects reflect our current understanding. This applies, in particular, to the designation "planets." The word "planet" originally described "wanderers" that were known only as moving lights in the sky. Recent discoveries lead us to create a new definition, which we can make using currently available scientific information.

The IAU therefore resolves that planets and other bodies, except satellites, in our Solar System be defined into three distinct categories in the following way [**Figure A8.1**]:

(1) A "planet"[1] is a celestial body that
 (a) is in orbit around the Sun,
 (b) has sufficient mass for its self-gravity to overcome rigid body forces so that it assumes a hydrostatic equilibrium (nearly round) shape, and
 (c) has cleared the neighborhood around its orbit.

(2) A "dwarf planet" is a celestial body that
 (a) is in orbit around the Sun,
 (b) has sufficient mass for its self-gravity to overcome rigid body forces so that it assumes a hydrostatic equilibrium (nearly round) shape,[2]
 (c) has not cleared the neighborhood around its orbit, and
 (d) is not a satellite.

(3) All other objects,[3] except satellites, orbiting the Sun shall be referred to collectively as "Small Solar-System Bodies."

Resolution 6: Pluto

The IAU further resolves:

Pluto is a "dwarf planet" by the above definition and is recognized as the prototype of a new category of Trans-Neptunian Objects.[4]

*SOURCE: International Astronomical Union.

1 The eight planets are: Mercury, Venus, Earth, Mars, Jupiter, Saturn, Uranus, and Neptune.

2 An IAU process will be established to assign borderline objects to the dwarf planet or to another category.

3 These currently include most of the Solar System asteroids, most Trans-Neptunian Objects (TNOs), comets, and other small bodies.

4 An IAU process will be established to select a name for this category.

FIGURE A8.1 Planets and dwarf planets of the Solar System.

Glossary

A

aberration of starlight The apparent displacement in the position of a star that is due to the finite speed of light and Earth's orbital motion around the Sun.

absolute magnitude A measure of the intrinsic brightness, or luminosity, of a celestial object, generally a star. Specifically, the apparent magnitude an object would have if it were located at a standard distance of 10 parsecs (pc). Compare *apparent magnitude*.

absolute zero The temperature at which thermal motions cease. The lowest possible temperature. Zero on the Kelvin temperature scale.

absorption The process by which an atom captures energy from a passing photon. Compare *emission*.

absorption line A minimum in the intensity of a spectrum that is due to the absorption of electromagnetic radiation at a specific wavelength determined by the energy levels of an atom or molecule. Compare *emission line*.

acceleration The rate at which the speed and/or direction of an object's motion is changing.

accretion disk A flat, rotating disk of gas and dust surrounding an object, such as a young stellar object, a forming planet, a collapsed star in a binary system, or a black hole.

achondrite A stony meteorite that does not contain chondrules. Compare *chondrite*.

active comet A comet nucleus that approaches close enough to the Sun to show signs of activity, such as the production of a coma and tail.

active galactic nucleus (AGN) A highly luminous, compact galactic nucleus whose luminosity may exceed that of the rest of the galaxy.

active region An area of the Sun's chromosphere anchoring bursts of intense magnetic activity.

adaptive optics Electro-optical systems that largely compensate for image distortion caused by Earth's atmosphere.

AGB See *asymptotic giant branch*.

AGN See *active galactic nucleus*.

albedo The fraction of electromagnetic radiation striking a surface that is reflected by that surface.

algebra A branch of mathematics in which numeric variables are represented by letters.

alpha particle A ^{4}He nucleus, consisting of two protons and two neutrons. Alpha particles are given off in the type of radioactive decay referred to as *alpha decay*.

altitude The location of an object above the horizon, measured by the angle formed between an imaginary line from an observer to the object and a second line from the observer to the point on the horizon directly below the object.

Amors A group of asteroids whose orbits cross the orbit of Mars but not the orbit of Earth. Compare *Apollos* and *Atens*.

amplitude In a wave, the maximum deviation from its undisturbed or relaxed position. For example, in a water wave the amplitude is the vertical distance from the wave's crest to the undisturbed water level.

angular momentum A conserved property of a rotating or revolving system whose value depends on the velocity and distribution of the system's mass.

annular solar eclipse The type of solar eclipse that occurs when the apparent diameter of the Moon is less than that of the Sun, leaving a visible ring of light ("annulus") surrounding the dark disk of the Moon. Compare *partial solar eclipse* and *total solar eclipse*.

Antarctic Circle The circle on Earth with latitude 66.5° south, marking the northern limit where at least one day per year is in 24-hour daylight. Compare *Arctic Circle*.

anthropic principle The idea that this universe (or this bubble in the universe) must have physical properties that allow for the development of intelligent life.

anticyclonic motion The rotation of a weather system resulting from the Coriolis effect as air moves outward from a region of high atmospheric pressure. Compare *cyclonic motion*.

antimatter Matter made up of antiparticles.

antiparticle An elementary particle of antimatter identical in mass but opposite in charge and all other properties to its corresponding ordinary matter particle.

aperture The clear diameter of a telescope's objective lens or primary mirror.

aphelion (pl. **aphelia**) The point in a solar orbit that is farthest from the Sun. Compare *perihelion*.

Apollos A group of asteroids whose orbits cross the orbits of both Earth and Mars. Compare *Amors* and *Atens*.

apparent magnitude A measure of the apparent brightness of a celestial object, generally a star. Compare *absolute magnitude*.

arcminute (arcmin) A minute of arc (′), a unit used for measuring angles. An arcminute is 1/60 of a degree of arc.

arcsecond (arcsec) A second of arc (″), a unit used for measuring very small angles. An arcsecond is 1/60 of an arcminute, or 1/3,600 of a degree of arc.

Arctic Circle The circle on Earth with latitude 66.5° north, marking the southern limit where at least one day per year is in 24-hour daylight. Compare *Antarctic Circle*.

asteroid Also called *minor planet*. A primitive rocky or metallic body (planetesimal) that has survived planetary accretion. Asteroids are parent bodies of meteoroids.

asteroid belt The region between the orbits of Mars and Jupiter that contains most of the asteroids in our Solar System.

astrobiology An interdisciplinary science combining astronomy, biology, chemistry, geology, and physics to study life in the cosmos.

astrology The belief that the positions and aspects of stars and planets influence human affairs and characteristics, as well as terrestrial events.

astronomical seeing A measurement of the degree to which Earth's atmosphere degrades the resolution of a telescope's view of astronomical objects.

astronomical unit (AU) The average distance from the Sun to Earth: approximately 150 million kilometers (km).

astronomy The scientific study of planets, stars, galaxies, and the universe as a whole.

astrophysics The application of physical laws to the understanding of planets, stars, galaxies, and the universe as a whole.

asymptotic giant branch (AGB) The path on the H-R diagram that goes from the horizontal branch toward higher luminosities and lower temperatures, asymptotically approaching and then rising above the red giant branch.

Atens A group of asteroids whose orbits cross the orbit of Earth but not the orbit of Mars. Compare *Amors* and *Apollos*.

atmosphere The gravitationally bound, outer gaseous envelope surrounding a planet, moon, or star.

atmospheric greenhouse effect A warming of planetary surfaces produced by atmospheric gases that transmit optical solar radiation but partially trap infrared radiation. Compare *greenhouse effect*.

atmospheric probe An instrumented package designed to provide on-site measurements of the chemical and/or physical properties of a planetary atmosphere.

atmospheric window A region of the electromagnetic spectrum in which radiation is able to penetrate a planet's atmosphere.

atom The smallest unit of a chemical element that retains the properties of that element. Each atom is composed of a nucleus (neutrons and protons) surrounded by a cloud of electrons.

AU See *astronomical unit*.

aurora Emission in the upper atmosphere of a planet from atoms that have been excited by collisions with energetic particles from the planet's magnetosphere.

autumnal equinox 1. One of two points where the Sun crosses the celestial equator. 2. The day on which the Sun appears at this location, marking the first day of autumn (about September 22 in the Northern Hemisphere and March 20 in the Southern Hemisphere). Compare *vernal equinox*. See also *summer solstice* and *winter solstice*.

axion A hypothetical elementary particle first proposed to explain certain properties of the neutron and now considered a candidate for cold dark matter.

B

backlighting Illumination from behind a subject as seen by an observer. Fine material such as human hair and dust in planetary rings stands out best when viewed under backlighting conditions.

bar A unit of pressure. One bar is equivalent to 10^5 newtons per square meter—approximately equal to Earth's atmospheric pressure at sea level.

barred spiral galaxy A spiral galaxy with a bulge having an elongated, barlike shape. Compare *elliptical galaxy*, *irregular galaxy*, *S0 galaxy*, and *spiral galaxy*.

basalt Gray to black volcanic rock, rich in iron and magnesium.

beta decay 1. The decay of a neutron into a proton by emission of an electron (beta ray) and an antineutrino. 2. The decay of a proton into a neutron by emission of a positron and a neutrino.

Big Bang The event that occurred 13.7 billion years ago that marks the beginning of time and the universe.

Big Bang nucleosynthesis The formation of low-mass nuclei (H, He, Li, Be, B) during the first few minutes following the Big Bang.

Big Crunch A hypothetical cosmic future in which the expansion of the universe reverses and the universe collapses onto itself.

Big Rip A hypothetical cosmic future in which all matter in the universe, from stars to subatomic particles, is progressively torn apart by expansion of the universe.

binary star A system in which two stars are in gravitationally bound orbits about their common center of mass.

binding energy The minimum energy required to separate an atomic nucleus into its component protons and neutrons.

biosphere The global sum of all living organisms on Earth (or any planet or moon). Compare *hydrosphere* and *lithosphere*.

bipolar outflow Material streaming away in opposite directions from either side of the accretion disk of a young star.

black hole An object so dense that its escape velocity exceeds the speed of light; a *singularity* in spacetime.

blackbody An object that absorbs and can reemit all electromagnetic energy it receives.

blackbody spectrum See *Planck spectrum*.

blue-green algae See *cyanobacteria*.

blueshift The Doppler shift toward shorter (bluer) wavelengths of light from an approaching object. Compare *redshift*.

Bohr model A model of the atom, proposed by Niels Bohr in 1913, in which a small positively charged nucleus is surrounded by orbiting electrons, similar to a miniature solar system.

bolide A very bright, exploding meteor.

bound orbit An orbit in which an object is gravitationally bound to the body it is orbiting. A bound orbit's velocity is less than the escape velocity. Compare *unbound orbit*.

bow shock The boundary at which the speed of the solar wind abruptly drops from supersonic to subsonic in its approach to a planet's magnetosphere; the boundary between the region dominated by the solar wind and the region dominated by a planet's magnetosphere.

brightness The apparent intensity of light from a luminous object. Brightness depends on both the *luminosity* of a source and its distance. Units at the detector: watts per square meter (W/m^2).

brown dwarf A "failed" star that is not massive enough to cause hydrogen fusion in its core. An object whose mass is intermediate between that of the least massive stars and that of supermassive planets.

bulge The central region of a spiral galaxy that is similar in appearance to a small elliptical galaxy.

C

C See *Celsius*.

C-type asteroid An asteroid made of material that has remained mostly unmodified since the formation of the Solar System; the most primitive type of asteroid. Compare *M-type asteroid* and *S-type asteroid*.

caldera The summit crater of a volcano.

Cambrian explosion The spectacular rise in the number and diversity of biological species that occurred between 540 and 500 million years ago.

carbon-nitrogen-oxygen (CNO) cycle One of the ways in which hydrogen is converted to helium (hydrogen burning) in the interiors of main-sequence stars. Compare *proton-proton chain*.

carbon star A cool red giant or asymptotic giant branch star that has an excess of carbon in its atmosphere.

carbonaceous chondrite A primitive stony meteorite that contains chondrules and is rich in carbon and volatile materials.

Cassini Division The largest gap in Saturn's rings, discovered by Jean-Dominique Cassini in 1675.

catalyst An atomic and molecular structure that permits or encourages chemical and nuclear reactions but does not change its own chemical or nuclear properties.

CCD See *charge-coupled device*.

celestial equator The imaginary great circle that is the projection of Earth's equator onto the celestial sphere.

celestial sphere An imaginary sphere with celestial objects on its inner surface and Earth at its center. The celestial sphere has no physical existence but is a convenient tool for picturing the directions in which celestial objects are seen from the surface of Earth.

Celsius (C) Also called *centigrade scale*. The arbitrary temperature scale, formulated by Anders Celsius (1701–1744), that defines 0°C as the freezing point of water and 100°C as the boiling point of water at sea level. Unit: degrees Celsius (°C). Compare *Fahrenheit* and *Kelvin scale*.

center of mass 1. The weighted average location of all the mass in a system of objects. The point in any isolated system that moves according to Newton's first law of motion. 2. In a binary star system, the point between the two stars that is the focus of both their elliptical orbits.

centigrade scale See *Celsius*.

centripetal force A force directed toward the center of curvature of an object's curved path.

Cepheid variable An evolved high-mass star with an atmosphere that is pulsating, leading to variability in the star's luminosity and color.

Chandrasekhar limit The upper limit on the mass of an object supported by electron degeneracy pressure; approximately 1.4 solar mass ($M_\odot$).

chaotic behavior Behavior in complex, interrelated systems in which tiny differences in the initial configuration of a system result in dramatic differences in the system's later evolution.

charge-coupled device (CCD) A common type of solid-state detector of electromagnetic radiation that transforms the intensity of light directly into electric signals.

chondrite A stony meteorite that contains chondrules. Compare *achondrite*.

chondrule A small, crystallized, spherical inclusion of rapidly cooled molten droplets found inside some meteorites.

chromatic aberration A detrimental property of a lens in which rays of different wavelengths are brought to different focal distances from the lens.

chromosphere The region of the Sun's atmosphere located between the *photosphere* and the *corona*.

circular velocity The orbital velocity needed to keep an object moving in a circular orbit.

circumpolar Describing the part of the sky, near either celestial pole, that can always be seen above the horizon from a specific location on Earth.

circumstellar disk See *protoplanetary disk*.

climate The state of an atmosphere averaged over an extended time. Compare *weather*.

closed universe A finite universe with a curved spatial structure such that the sum of the angles of a triangle always exceeds 180°. Compare *flat universe* and *open universe*.

CMB See *cosmic microwave background radiation*.

CNO cycle See *carbon-nitrogen-oxygen cycle*.

cold dark matter Particles of dark matter that move slowly enough to be gravitationally bound even in the smallest galaxies. Compare *hot dark matter*.

color index The color of a celestial object, generally a star, based on the ratio of its brightness in blue light to its brightness in "visual" (yellow-green) light. The difference between an object's blue (B) magnitude and visual (V) magnitude, $B - V$.

coma (pl. **comae**) The nearly spherical cloud of gas and dust surrounding the nucleus of an active comet.

comet A complex object consisting of a small, solid, icy nucleus; an atmospheric halo; and a tail of gas and dust.

comet nucleus A primitive planetesimal, composed of ices and refractory materials, that has survived planetary accretion. The "heart" of a comet, containing nearly the entire mass of the comet. A "dirty snowball."

comparative planetology The study of planets through comparison of their chemical and physical properties.

composite volcano A large, cone-shaped volcano formed by viscous, pasty lava flows alternating with pyroclastic (explosively generated) rock deposits. Compare *shield volcano*.

compound lens A lens made up of two or more elements of differing refractive index, the purpose of which is to minimize chromatic aberration.

concave mirror A telescope mirror with a surface that curves inward toward the incoming light.

conduction The transfer of energy in which the thermal energy of particles is transferred to adjacent particles by collisions or other interactions. Conduction is the most important way that thermal energy is transported in solid matter. Compare *convection*.

conservation law A physical law stating that the amount of a particular physical quantity (such as energy or angular momentum) of an isolated system does not change over time.

conservation of angular momentum The physical law stating that the amount of angular momentum of an isolated system does not change over time.

conservation of energy The physical law stating that the amount of energy of an isolated, closed system does not change over time.

constant of proportionality The number by which one quantity is multiplied to get another number.

constellation An imaginary image formed by patterns of stars; any of 88 defined areas on the celestial sphere used by astronomers to locate celestial objects.

constructive interference A state in which the amplitudes of two intersecting waves reinforce one another. Compare *destructive interference*.

continental drift The slow motion (centimeters per year) of Earth's continents relative to each other and to Earth's mantle. See also *plate tectonics*.

continuous radiation Electromagnetic radiation with intensity that varies smoothly over a wide range of wavelengths.

convection The transport of thermal energy from the lower (hotter) to the higher (cooler) layers of a fluid by motions within the fluid driven by variations in buoyancy. Compare *conduction*.

convective zone A region within a star where energy is transported outward by convection. Compare *radiative zone*.

co-orbital moons Moons that occupy the same orbit.

core 1. The innermost region of a planetary interior. Compare *crust* and *mantle*. 2. The innermost part of a star.

core accretion–gas capture A process for forming giant planets, in which large amounts of surrounding hydrogen and helium gas are gravitationally captured onto a massive rocky core.

Coriolis effect The apparent displacement of objects in a direction perpendicular to their true motion as viewed from a rotating frame of reference. On a rotating planet, different latitudes rotating at different speeds cause this effect.

corona The hot, outermost part of the Sun's atmosphere. Compare *chromosphere* and *photosphere*.

coronal hole A low-density region in the solar corona containing "open" magnetic-field lines along which coronal material is free to stream into interplanetary space.

coronal mass ejection An eruption on the Sun that ejects hot gas and energetic particles at much higher speeds than are typical in the solar wind.

cosmic microwave background radiation (CMB) Also called simply *cosmic background radiation*. Isotropic microwave radiation from every direction in the sky having a 2.73-kelvin (K) blackbody spectrum. The CMB is residual radiation from the Big Bang.

cosmic ray A very fast-moving particle (usually an atomic nucleus) that originated in outer space; cosmic rays fill the disk of the Milky Way.

cosmological constant A constant, introduced into general relativity by Einstein, that characterizes an extra, repulsive force in the universe due to the vacuum of space itself.

cosmological principle The (testable) assumption that the same physical laws that apply here and now also apply everywhere and at all times, and that there are no special locations or directions in the universe.

cosmological redshift (z) The redshift that results from the expansion of the universe rather than from the motions of galaxies or gravity. Compare *gravitational redshift*.

cosmology The study of the large-scale structure and evolution of the universe as a whole.

crescent Any phase of the Moon, Mercury, or Venus in which the object appears less than half illuminated by the Sun. Compare *gibbous*.

Cretaceous-Tertiary (K-T) boundary The boundary between the Cretaceous and Tertiary periods in Earth's history. This boundary corresponds to the time of the impact of an asteroid or comet and the extinction of the dinosaurs.

critical density The value of mass density of the universe that, ignoring any cosmological constant, is just barely capable of halting expansion of the universe.

crust The relatively thin, outermost, hard layer of a planet, which is chemically distinct from the interior. Compare *core* and *mantle*.

cryovolcanism Low-temperature volcanism in which the magmas are composed of molten ices rather than rocky material.

cyanobacteria Also called *blue-green algae*. Single-celled organisms that were responsible for creating oxygen in Earth's atmosphere by photosynthesizing carbon dioxide and releasing oxygen as a waste product.

cyclonic motion The rotation of a weather system resulting from the Coriolis effect as air moves toward a region of low atmospheric pressure. Compare *anticyclonic motion*.

D

Dark Ages The epoch in the history of the universe during which there was no visible "light" from astronomical objects.

dark energy A form of energy that permeates all of space (including the vacuum), producing a repulsive force that accelerates the expansion of the universe.

dark matter Matter in galaxies that does not emit or absorb electromagnetic radiation. Dark matter is thought to constitute most of the mass in the universe. Compare *luminous matter.*

dark matter halo The centrally condensed, greatly extended dark matter component of a galaxy that accounts for up to 95 percent of the galaxy's mass.

daughter product An element resulting from radioactive decay of a more massive *parent element.*

decay 1. The process of a radioactive nucleus changing into its daughter product. 2. The process of an atom or molecule dropping from a higher energy state to a lower energy state. 3. The process of a satellite's orbit losing energy.

declination A measure, analogous to *latitude,* that tells you the angular distance of a celestial body north or south of the celestial equator (from 0° to ±90°). Compare *right ascension.*

density The measure of an object's mass per unit of volume. Possible units include: kilograms per cubic meter (kg/m^3).

destructive interference A state in which the amplitudes of two intersecting waves cancel one another. Compare *constructive interference.*

differential rotation Rotation of different parts of a system at different rates.

differentiation The process by which materials of higher density sink toward the center of a molten or fluid planetary interior.

diffraction The spreading of a wave after it passes through an opening or beyond the edge of an object.

diffraction limit The limit of a telescope's angular resolution caused by diffraction.

diffuse ring A sparsely populated planetary ring spread out both horizontally and vertically.

dispersion The separation of rays of light into their component wavelengths.

distance ladder A sequence of techniques for measuring cosmic distances; each method is calibrated using the results from other methods that have been applied to closer objects.

Doppler effect The change in wavelength of sound or light that is due to the relative motion of the source toward or away from the observer.

Doppler shift The amount by which the wavelength of light is shifted by the Doppler effect.

Drake equation A prescription for estimating the number of intelligent civilizations existing in the Milky Way Galaxy.

dust devil A small tornado-like column of air containing dust or sand.

dust tail A type of comet tail consisting of dust particles that are pushed away from the comet's head by radiation pressure from the Sun. Compare *ion tail.*

dwarf galaxy A small galaxy with a luminosity ranging from 1 million to 1 billion solar luminosities ($L_\odot$). Compare *giant galaxy.*

dwarf planet A body with characteristics similar to those of a planet except that it has not cleared smaller bodies from the neighboring regions around its orbit. Compare *planet* (definition 2).

dynamic equilibrium A state in which a system is constantly changing but its configuration remains the same because one source of change is exactly balanced by another source of change. Compare *static equilibrium.*

dynamo theory A theory postulating that Earth's magnetic field (and those of other planets) is generated from a rotating and electrically conducting liquid core.

E

Earth Similarity Index (ESI) An index for quantifying the habitability of extrasolar planets in which currently available data on an extrasolar planet are used to estimate how much it is like Earth. Factors include the radius, density, escape velocity, and surface temperature. Compare *Planetary Habitability Index.*

eccentricity (*e*) The ratio of the distance between the two foci of an ellipse to the length of its major axis, which measures how noncircular the ellipse is.

eclipse 1. The total or partial obscuration of one celestial body by another. 2. The total or partial obscuration of light from one celestial body as it passes through the shadow of another celestial body.

eclipse season Any time during the year when the Moon's line of nodes is sufficiently close to the Sun for eclipses to occur.

eclipsing binary A binary system in which the orbital plane is oriented such that the two stars appear to pass in front of one another as seen from Earth. Compare *spectroscopic binary* and *visual binary.*

ecliptic 1. The apparent annual path of the Sun against the background of stars. 2. The projection of Earth's orbital plane onto the celestial sphere.

ecliptic plane The plane of Earth's orbit around the Sun. The ecliptic is the projection of this plane on the celestial sphere.

effective temperature The temperature at which a blackbody, such as a star, appears to radiate.

Einstein ring Light bent by gravitational lensing into a ring.

ejecta 1. Material thrown outward by the impact of an asteroid or comet on a planetary surface, leaving a crater behind. 2. Material thrown outward by a stellar explosion.

electric field A field that is able to exert a force on a charged object, whether at rest or moving. Compare *magnetic field.*

electric force The force exerted on electrically charged particles such as protons and electrons, arising from their electric charges. Compare *magnetic force.* See also *electromagnetic force.*

electromagnetic force The force, including both electric and magnetic forces, that acts on electrically charged particles. One of four fundamental forces of nature, along with the *strong nuclear force, weak nuclear force,* and *gravity* (definition 1). The force is mediated by the exchange of photons.

electromagnetic radiation A traveling disturbance in the electric and magnetic fields caused by accelerating electric charges. In quantum mechanics, a stream of photons. Light.

electromagnetic spectrum The spectrum made up of all possible frequencies or wavelengths of electromagnetic radiation, ranging from gamma rays through radio waves and including the portion our eyes can use.

electromagnetic wave A wave consisting of oscillations in the electric-field strength and the magnetic-field strength.

electron (e^-) A subatomic particle having a negative electric charge of 1.6×10^{-19} coulomb (C), a rest mass of 9.1×10^{-31} kilogram (kg), and rest energy of 8×10^{-14} joule (J). The antiparticle of the *positron.* Compare *proton* and *neutron.*

electron-degenerate Describing matter, compressed to the point at which electron density reaches the limit imposed by the rules of quantum mechanics.

electroweak theory The quantum theory that combines descriptions of both the electromagnetic force and the weak nuclear force.

element One of 92 naturally occurring substances (such as hydrogen, oxygen, and uranium) and more than 20 human-made ones (such as plutonium). Each element is chemically defined by the specific number of protons in the nuclei of its atoms.

elementary particle One of the basic building blocks of nature that is not known to have substructure, such as the *electron* or the *quark.*

ellipse A conic section produced by the intersection of a plane with a cone when the plane is passed through the cone at an angle to the axis other than 0° or 90°. The shape that results when you attach the two ends of a piece of string to a piece of paper, stretch the string tight with the tip of a pencil, and then draw around those two points while keeping the string taut.

elliptical galaxy A galaxy of Hubble type "E" class, with a circular to elliptical outline on the sky containing almost no disk and a population of old stars. Compare *barred spiral galaxy*, *irregular galaxy*, *S0 galaxy*, and *spiral galaxy*.

emission The production of a photon when an atom decays to a lower energy state. Compare *absorption*.

emission line A peak in the intensity of a spectrum that is due to the emission of electromagnetic radiation at a specific wavelength determined by the energy levels of an atom or molecule. Compare *absorption line*.

empirical science Scientific investigation that is based primarily on observations and experimental data. It is descriptive rather than based on theoretical inference.

energy The conserved quantity that gives objects and systems the ability to do work. Possible units include: joules (J).

energy transport The transfer of energy from one location to another. In stars, energy transport is carried out mainly by radiation or convection.

entropy A measure of the disorder of a system related to the number of ways a system can be rearranged without its appearance being affected.

equator The imaginary great circle on the surface of a body midway between its poles that divides the body into northern and southern hemispheres. The equatorial plane passes through the center of the body and is perpendicular to its rotation axis. Compare *meridian*.

equilibrium The state of an object in which physical processes balance each other so that its properties or conditions remain constant.

equinox Literally, "equal night." 1. One of two positions on the ecliptic where it intersects the celestial equator. 2. Either of the two times of year (the *autumnal equinox* and *vernal equinox*) when the Sun is at one of these two positions. At this time, night and day are of the same length everywhere on Earth. Compare *solstice*.

equivalence principle The principle stating that there is no difference between a frame of reference that is freely floating through space and one that is freely falling within a gravitational field.

erosion The degradation of a planet's surface topography by the mechanical action of wind and/or water.

escape velocity The minimum velocity needed for an object to achieve a parabolic trajectory and thus permanently leave the gravitational grasp of another mass.

ESI See *Earth Similarity Index*.

eternal inflation The idea that a universe might inflate forever. In such a universe, quantum effects could randomly cause regions to slow their expansion, eventually stop inflating, and experience an explosion resembling the Big Bang.

event A particular location in spacetime.

event horizon The effective "surface" of a black hole. Nothing inside this surface—not even light—can escape from a black hole.

evolutionary track The path that a star follows across the H-R diagram as it evolves through its lifetime.

excited state Any energy level of a system or part of a system, such as an atom, molecule, or particle, that is higher than its ground state. Compare *ground state*.

exoplanet See *extrasolar planet*.

extrasolar planet Also called *exoplanet*. A planet orbiting a star other than the Sun.

extremophile A life-form that thrives under extreme environmental conditions.

F

F See *Fahrenheit*.

Fahrenheit (F) The arbitrary temperature scale, formulated by Daniel Gabriel Fahrenheit (1686–1736), that defines 32°F as the melting point of water and 212°F as the boiling point of water at sea level. Unit: degrees Fahrenheit (°F). Compare *Celsius* and *Kelvin scale*.

fault A fracture in the crust of a planet or moon along which blocks of material can slide.

filter An instrument element that transmits a limited wavelength range of electromagnetic radiation. For the optical range, such elements are typically made of different kinds of glass and take on the hue of the light they transmit.

first quarter Moon The phase of the Moon in which only the western half of the Moon, as viewed from Earth, is illuminated by the Sun. It occurs about a week after a new Moon. Compare *third quarter Moon*. See also *full Moon* and *new Moon*.

fissure A fracture in the planetary lithosphere from which magma emerges.

flat rotation curve A rotation curve of a spiral galaxy in which rotation rates do not decline in the outer part of the galaxy, but remain relatively constant to the outermost points.

flat universe An infinite universe whose spatial structure obeys Euclidean geometry, such that the sum of the angles of a triangle always equals 180°. Compare *closed universe* and *open universe*.

flatness problem The surprising result that the sum of Ω_{mass} plus Ω_Λ is extremely close to unity in the present-day universe; equivalent to saying that it is surprising the universe is so close to being exactly flat.

flux The total amount of energy passing through each square meter of a surface each second. Unit: watts per square meter (W/m^2).

flux tube A strong magnetic field contained within a tubelike structure. Flux tubes are found in the solar atmosphere and connecting the space between Jupiter and its moon Io.

flyby A spacecraft that first approaches and then continues flying past a planet or moon. Flybys can visit multiple objects, but they remain in the vicinity of their targets only briefly. Compare *orbiter*.

focal length The optical distance between a telescope's objective lens or primary mirror and the plane (called the focal plane) on which the light from a distant object is focused.

focal plane The plane, perpendicular to the optical axis of a lens or mirror, in which an image is formed.

focus (pl. **foci**) 1. One of two points that define an ellipse. 2. A point in the focal plane of a telescope.

force A push or a pull on an object.

frame of reference A coordinate system within which an observer measures positions and motions.

free fall The motion of an object when the only force acting on it is gravity.

frequency The number of times per second that a periodic process occurs. Unit: hertz (Hz), or cycles per second (1/s).

full Moon The phase of the Moon in which the near side of the Moon, as viewed from Earth, is fully illuminated by the Sun. It occurs about two weeks after a *new Moon*. See also *first quarter Moon* and *third quarter Moon*.

G

galaxy A gravitationally bound system that consists of stars and star clusters, gas, dust, and dark matter; typically greater than 1,000 light-years across and recognizable as a discrete, single object.

galaxy cluster A large, gravitationally bound collection of galaxies containing hundreds to thousands of members; typically 3–5 megaparsecs (Mpc) across. Compare *galaxy group* and *supercluster*.

galaxy group A small, gravitationally bound collection of galaxies containing from several to a hundred members; typically 1–2 megaparsecs (Mpc) across. Compare *galaxy cluster* and *supercluster*.

gamma ray Also called *gamma radiation*. Electromagnetic radiation with higher frequency, higher photon energy, and shorter wavelength than all other types of electromagnetic radiation.

gamma-ray burst (GRB) An brief, intense burst of gamma rays from a distant energetic explosion.

gas giant A giant planet formed mostly of hydrogen and helium. In the Solar System, Jupiter and Saturn are the gas giants. Compare *ice giant*.

gauss A basic unit measuring the strength of a magnetic field.

general relativistic time dilation The verified prediction that time passes more slowly in a gravitational field than in the absence of a gravitational field. Compare *time dilation*.

general relativity See *general theory of relativity.*

general theory of relativity Sometimes referred to as simply *general relativity.* Einstein's theory explaining gravity as the distortion of spacetime by massive objects, such that particles travel on the shortest path between two events in spacetime. This theory deals with all types of motion. Compare *special theory of relativity.*

geocentric model A historical cosmological model with Earth at its center, and all of the other objects in the universe in orbit around Earth. Compare *heliocentric.*

geodesic The path an object will follow through spacetime in the absence of external forces.

geometry A branch of mathematics that deals with points, lines, angles, and shapes.

giant galaxy A galaxy with luminosity greater than about 1 billion solar luminosities ($L_{\odot}$). Compare *dwarf galaxy.*

giant molecular cloud An interstellar cloud composed primarily of molecular gas and dust, having hundreds of thousands of solar masses.

giant planet Also called *Jovian planet.* One of the largest planets in the Solar System (Saturn, Jupiter, Uranus, or Neptune), typically 10 times the size and many times the mass of any *terrestrial planet* and lacking a solid surface.

gibbous Any phase of the Moon, Mercury, or Venus in which the object appears more than half illuminated by the Sun. Compare *crescent.*

global circulation The overall, planetwide circulation pattern of a planet's atmosphere.

globular cluster A spherically symmetric, highly condensed group of stars, containing tens of thousands to a million members. Compare *open cluster.*

gluon The particle that carries (or, equivalently, mediates) interactions due to the strong nuclear force.

grand unified theory (GUT) A unified quantum theory that combines the strong nuclear, weak nuclear, and electromagnetic forces but does not include gravity.

granite Rock that is cooled from magma and is relatively rich in silicon and oxygen.

grating An optical surface containing many narrow, closely and equally spaced parallel grooves or slits that spectrally disperse reflected or transmitted light.

gravitational lens A massive object that gravitationally focuses the light of a more distant object to produce multiple brighter, magnified, possibly distorted images.

gravitational lensing The bending of light by gravity.

gravitational potential energy The stored energy in an object that is due solely to its position within a gravitational field.

gravitational redshift The shifting to longer wavelengths of radiation from an object deep within a gravitational well. Compare *cosmological redshift.*

gravitational wave A wave in the fabric of spacetime emitted by accelerating masses.

gravity 1. The mutually attractive force between massive objects. One of four fundamental forces of nature, along with the *electromagnetic force,* the *strong nuclear force,* and the *weak nuclear force.* 2. An effect arising from the bending of spacetime by massive objects.

GRB See *gamma-ray burst.*

great circle Any circle on a sphere that has as its center the center of the sphere. The celestial equator, the meridian, and the ecliptic are all great circles on the sphere of the sky, as is any circle drawn through the zenith.

Great Red Spot The giant, oval, brick red anticyclone seen in Jupiter's southern hemisphere.

greenhouse effect The solar heating of air in an enclosed space, such as a closed building or car, resulting primarily from the inability of the hot air to escape. Compare *atmospheric greenhouse effect.*

greenhouse gas One of a group of atmospheric gases such as carbon dioxide that are transparent to visible radiation but absorb infrared radiation.

greenhouse molecule A molecule such as water vapor or carbon dioxide that transmits visible radiation but absorbs infrared radiation.

Gregorian calendar The modern calendar. A modification of the Julian calendar decreed by Pope Gregory XIII in 1582. By this time the less accurate Julian calendar had developed an error of 10 days over the 13 centuries since its inception.

ground state The lowest possible energy state for a system or part of a system, such as an atom, molecule, or particle. Compare *excited state.*

GUT See *grand unified theory.*

H

H II region A region of interstellar gas that has been ionized by ultraviolet radiation from nearby hot, massive stars.

H-R diagram The Hertzsprung-Russell diagram, a plot of the luminosities versus the surface temperatures of stars. The evolving properties of stars are plotted as tracks across the H-R diagram.

habitable zone The distance from its star at which a planet must be located in order to have a temperature suitable for water to exist in a liquid state.

Hadley circulation A simplified, and therefore uncommon, atmospheric global circulation that carries thermal energy directly from the equator to the polar regions of a planet.

half-life The time it takes for half a sample of a particular radioactive parent element to decay into a daughter product.

halo The spherically symmetric, low-density distribution of stars and dark matter that defines the outermost regions of a galaxy.

harmonic law See *Kepler's third law.*

Hawking radiation Radiation from a black hole.

Hayashi track The path that a protostar follows on the H-R diagram as it contracts toward the main sequence.

head The part of a comet that includes both the nucleus and the inner part of the coma.

heat death The possible eventual fate of an open universe, in which entropy has triumphed and all energy- and structure-producing processes have come to an end.

heavy element Also called *massive element.* Any element more massive than helium.

Heisenberg uncertainty principle The physical limitation that the product of the position and the momentum of a particle cannot be smaller than a well-defined value, Planck's constant (h).

heliocentric model A model of the Solar System, with the Sun at its center, and the planets, including Earth, in orbit around the Sun. Compare *geocentric.*

helioseismology The use of solar oscillations to study the interior of the Sun.

heliosphere A region surrounding the Solar System in which the solar wind blows against the interstellar medium and clears out an area like the inside of a bubble. The heliosphere protects the Solar System from cosmic rays.

helium flash The runaway explosive burning of helium in the degenerate helium core of a red giant star.

Herbig-Haro (HH) object A glowing, rapidly moving knot of gas and dust that is excited by bipolar outflows in very young stars.

heredity The process by which one generation passes on its characteristics to future generations.

hertz (Hz) A unit of frequency equivalent to cycles per second.

Hertzsprung-Russell diagram See *H-R diagram.*

HH object See *Herbig-Haro object.*

hierarchical clustering The "bottom-up" process of forming large-scale structure. Small-scale structure first produces groups of galaxies, which in turn form clusters, which then form superclusters.

high-mass star A star with a main-sequence mass of greater than about 8 solar masses ($M_{\odot}$). Compare *intermediate-mass star* and *low-mass star.*

homogeneous In cosmology, describing a universe in which observers in any location would observe the same properties. Compare *isotropic.*

horizon The boundary that separates the sky from the ground.

horizon problem The puzzling observation that the cosmic background radiation is so uniform in all directions, even though widely separated regions should have been "over the horizon" from each other in the early universe.

horizontal branch A region on the H-R diagram defined by stars burning helium to carbon in a stable core.

hot dark matter Particles of dark matter that move so fast that gravity cannot confine them to the volume occupied by a galaxy's normal luminous matter. Compare *cold dark matter.*

hot Jupiter A large, Jupiter-type extrasolar planet located very close to its parent star.

hot spot A place where hot plumes of mantle material rise near the surface of a planet.

Hubble constant (H_0) The constant of proportionality relating the recession velocities of galaxies to their distances. Compare *Hubble time.*

Hubble time An estimate of the age of the universe from the inverse of the *Hubble constant*, $1/H_0$.

Hubble's law The law stating that the speed at which a galaxy is moving away from Earth is proportional to the distance of that galaxy.

hurricane A large tropical cyclonic system circulating counterclockwise in the Northern Hemisphere and clockwise in the Southern Hemisphere. Hurricanes can extend outward from their center to more than 600 kilometers (km) and generate winds in excess of 300 kilometers per hour (km/h).

hydrogen burning The release of energy from the nuclear fusion of four hydrogen atoms into a single helium atom.

hydrogen shell burning The fusion of hydrogen in a shell surrounding a stellar core that may be either degenerate or fusing more massive elements.

hydrosphere The portion of Earth that is largely liquid water. Compare *biosphere* and *lithosphere.*

hydrostatic equilibrium The condition in which the weight bearing down at a particular point within an object is balanced by the pressure within the object.

hypernova (pl. **hypernovae**) A very energetic supernova from a very high-mass star.

hypothesis A well-considered idea, based on scientific principles and knowledge, that leads to testable predictions. Compare *theory.*

Hz See *hertz.*

I

ice The solid form of a volatile material; sometimes the *volatile material* itself, regardless of its physical form.

ice giant A giant planet formed mostly of the liquid form of volatile substances (ices). In the Solar System, Uranus and Neptune are the ice giants. Compare *gas giant.*

ideal gas law The relationship of pressure (P) to density of particles (n) and temperature (T) expressed as $P = nkT$, where k is Boltzmann's constant.

igneous activity The formation and action of molten rock (magma).

impact crater The scar of the impact left on a solid planetary or moon surface by collision with another object. Compare *secondary crater.*

impact cratering The process in which solid planetary objects collide with each other, leaving distinctive scars.

index of refraction (n) The ratio of the speed of light in a vacuum (c) to the speed of light in an optical medium (v).

inert gas A gaseous element that combines with other elements only under conditions of extreme temperature and pressure. Examples include helium, neon, and argon.

inertia The tendency for objects to retain their state of motion.

inertial frame of reference 1. A frame of reference that is moving in a straight line at constant speed—that is, not accelerating. 2. In general relativity, a frame of reference that is falling freely in a gravitational field.

inflation An extremely brief phase of ultra-rapid expansion of the very early universe. Following inflation, the standard Big Bang models of expansion apply.

infrared (IR) radiation Electromagnetic radiation with frequencies, photon energies, and wavelengths between those of visible light and microwaves.

instability strip A region of the H-R diagram containing stars that pulsate with a periodic variation in luminosity.

integration time The time interval during which photons are collected and added up in a detecting device.

intensity Of light, the amount of radiant energy emitted per second per unit area. Units for electromagnetic radiation: watts per square meter (W/m^2).

intercloud gas A low-density region of the interstellar medium that fills the space between interstellar clouds.

interference The interaction of two sets of waves producing high and low intensity, depending on whether their amplitudes reinforce (*constructive interference*) or cancel (*destructive interference*).

interferometer Linked optical or radio telescopes whose overall separation determines the angular resolution of the system.

interferometric array An interferometer that is made up of several telescopes arranged in an array.

intermediate-mass star A star with a main-sequence mass between 3 and 8 solar masses ($M_\odot$). Compare *low-mass star* and *high-mass star.*

interstellar cloud A discrete, high-density region of the interstellar medium made up mostly of atomic or molecular hydrogen and dust.

interstellar dust Small particles or grains (0.01–10 microns [μm]) of matter, primarily carbon and silicates, distributed throughout interstellar space.

interstellar extinction The dimming of visible and ultraviolet light by interstellar dust.

interstellar gas The tenuous gas, far less dense than air, comprising 99 percent of the matter in the interstellar medium.

interstellar medium The gas and dust that fill the space between the stars within a galaxy.

inverse square law The rule stating that a quantity or effect diminishes with the square of the distance from the source.

ion An atom or molecule that has lost or gained one or more electrons.

ion tail A type of comet tail consisting of ionized gas. Particles in the ion tail are pushed directly away from the comet's head in the antisolar direction at high speeds by the solar wind. Compare *dust tail.*

ionization The process by which electrons are stripped free from an atom or molecule, resulting in free electrons and a positively charged atom or molecule.

ionosphere A layer high in Earth's atmosphere in which most of the atoms are ionized by solar radiation.

IR Infrared. See *infrared radiation.*

iron meteorite A metallic meteorite composed mostly of iron-nickel alloys. Compare *stony-iron meteorite* and *stony meteorite.*

irregular galaxy A galaxy without regular or symmetric appearance. Compare *barred spiral galaxy, elliptical galaxy, S0 galaxy,* and *spiral galaxy.*

irregular moon A moon that has been captured by a planet rather than having formed along with that planet. Some irregular moons revolve in the opposite direction from the rotation of the planet, and many are in distant, unstable orbits. Compare *regular moon.*

isotopes Forms of the same element that have different numbers of neutrons.

isotropic In cosmology, having the same appearance to an observer in all directions. Compare *homogeneous.*

J

J See *joule.*

jansky (Jy) The basic unit of flux density. Unit: watts per square meter per hertz ($W/m^2/Hz$).

jet 1. A stream of gas and dust ejected from a comet nucleus by solar heating. 2. A stream of material that moves away from a protostar or active galactic nucleus at hundreds of kilometers per second.

joule (J) A unit of energy or work. 1 J = 1 newton meter.

Jovian planet See *giant planet.*

Jy See *jansky.*

K

K See *kelvin.*

K-T boundary See *Cretaceous-Tertiary boundary.*

KBO See *Kuiper Belt object.*

kelvin (K) The basic unit of the Kelvin scale of temperature.

Kelvin scale The temperature scale, formulated by William Thomson, better known as Lord Kelvin (1824–1907), that uses Celsius-sized degrees, but defines 0 K as absolute zero instead of as the melting point of water. Unit: kelvins (K). Compare *Celsius* and *Fahrenheit.*

Kepler's first law A rule of planetary motion, inferred by Johannes Kepler, stating that planets move in elliptical orbits with the Sun at one focus.

Kepler's laws The three rules of planetary motion inferred by Johannes Kepler from the data collected by Tycho Brahe.

Kepler's second law Also called *law of equal areas.* A rule of planetary motion, inferred by Johannes Kepler, stating that a line drawn from the Sun to a planet sweeps out equal areas in equal times as the planet orbits the Sun.

Kepler's third law Also called *harmonic law.* A rule of planetary motion, inferred by Johannes Kepler, that describes the relationship between the period of a planet's orbit and its distance from the Sun. The law states that the square of the period of a planet's orbit, measured in years, is equal to the cube of the semimajor axis of the planet's orbit, measured in astronomical units: $(P_{years})^2 = (A_{AU})^3$.

kinetic energy (E_K) The energy of an object due to its motions. $E_K = \frac{1}{2}mv^2$. Possible units include: joules (J).

Kirkwood gap A gap in the main asteroid belt related to orbital resonances with Jupiter.

Kuiper Belt A disk-shaped population of comet nuclei extending from Neptune's orbit to perhaps several thousand astronomical units (AU) from the Sun. The highly populated innermost part of the Kuiper Belt has an outer edge approximately 50 AU from the Sun.

Kuiper Belt object (KBO) Also called *trans-Neptunian object.* An icy planetesimal (comet nucleus) that orbits within the Kuiper Belt beyond the orbit of Neptune.

L

Lagrangian equilibrium point Also called simply *Lagrangian point.* One of five points of equilibrium in a system consisting of two massive objects in nearly circular orbit around a common center of mass. Only two Lagrangian points (L_4 and L_5) represent stable equilibrium. A third, smaller body located at one of the five points will move in lockstep with the center of mass of the larger bodies.

Lambda-CDM The standard model of the Big Bang universe in which most of the energy density of the universe is dark energy (similar to Einstein's cosmological constant) and most of the mass in the universe is cold dark matter.

lander An instrumented spacecraft designed to land on a planet or moon. Compare *rover.*

large-scale structure Observable aggregates on the largest scales in the universe, including galaxy groups, clusters, and superclusters.

latitude The angular distance north (+) or south (−) from the equatorial plane of a nearly spherical body. Compare *longitude.*

lava Molten rock flowing out of a volcano during an eruption; also the rock that solidifies and cools from this liquid.

law of equal areas See *Kepler's second law.*

law of gravitation See *universal law of gravitation.*

leap year A year that contains 366 days. Leap years occur every 4 years when the year is divisible by 4, correcting for the accumulated excess time in a normal year, which is approximately 365¼ days long.

length contraction The relativistic compression of moving objects in the direction of their motion.

Leonids A November meteor shower associated with the dust debris left by comet Tempel-Tuttle.

life A biochemical process in which living organisms can reproduce, evolve, and sustain themselves by drawing energy from their environment. All terrestrial life involves carbon-based chemistry, assisted by the self-replicating molecules ribonucleic acid (RNA) and deoxyribonucleic acid (DNA).

light All electromagnetic radiation, which comprises the entire electromagnetic spectrum.

light-year (ly) The distance that light travels in 1 year—about 9.5 trillion kilometers (km).

limb The outer edge of the visible disk of a planet, moon, or the Sun.

limb darkening The darker appearance caused by increased atmospheric absorption near the limb of a planet or star.

limestone A common sedimentary rock composed of calcium carbonate.

line of nodes 1. A line defined by the intersection of two orbital planes. 2. The line defined by the intersection of Earth's equatorial plane and the plane of the ecliptic.

lithosphere The solid, brittle part of Earth (or any planet or moon), including the crust and the upper part of the mantle. Compare *biosphere* and *hydrosphere.*

lithospheric plate A separate piece of Earth's lithosphere capable of moving independently. See also *continental drift* and *plate tectonics.*

Local Group The group of galaxies that includes the Milky Way and Andromeda galaxies as members.

long-period comet A comet with an orbital period of greater than 200 years. Compare *short-period comet.*

longitude The angular distance east (+) or west (−) from the prime meridian at Greenwich, England. Compare *latitude.*

longitudinal wave A wave that oscillates parallel to the direction of the wave's propagation. Compare *transverse wave.*

look-back time The amount of time that the light from an astronomical object has taken to reach Earth.

low-mass star A star with a main-sequence mass of less than about 3 solar masses ($M_\odot$). Compare *intermediate-mass star* and *high-mass star.*

luminosity The total amount of light emitted by an object. Unit: watts (W). Compare *brightness.*

luminosity class A spectral classification based on stellar size, ranging from supergiants at the large end to white dwarfs at the small end.

luminosity-temperature-radius relationship A relationship among these three properties of stars indicating that if any two are known, the third can be calculated.

luminous matter Also called *normal matter.* Matter in galaxies—including stars, gas, and dust—that emits electromagnetic radiation. Compare *dark matter.*

lunar eclipse An eclipse that occurs when the Moon is partially or entirely in Earth's shadow. Compare *solar eclipse.*

lunar tide A tide on Earth that is due to the differential gravitational pull of the Moon. Compare *solar tide.*

ly See *light-year.*

M

μm See *micron.*

M-type asteroid An asteroid made of material that was once part of the metallic core of a larger, differentiated body that has since broken into pieces; made mostly of iron and nickel. Compare *C-type asteroid* and *S-type asteroid.*

MACHO Short for *massive compact halo object.* MACHOs include brown dwarfs, white dwarfs, and black holes and are candidates for dark matter. Compare *WIMP.*

magma Molten rock, often containing dissolved gases and solid minerals.

magnetic field A field that is able to exert a force on a moving electric charge. Compare *electric field.*

magnetic force The force exerted on electrically charged particles such as protons and electrons, arising from their motion. Compare *electric force.* See also *electromagnetic force.*

magnetometer A device that measures magnetic fields.

magnetosphere The region surrounding a planet that is filled with relatively intense magnetic fields and plasmas.

magnitude A system used by astronomers to describe the brightness or luminosity of stars. The brighter the star, the lower its magnitude.
main asteroid belt See *asteroid belt*.
main sequence The strip on the H-R diagram where most stars are found. Main-sequence stars are fusing hydrogen to helium in their cores.
main-sequence lifetime The amount of time a star spends on the main sequence, fusing hydrogen into helium in its core.
main-sequence turnoff The location on the H-R diagram of a single-aged stellar population (such as a star cluster) where stars have just evolved off the main sequence. This location is determined by the age of the stellar population.
mantle The solid portion of a rocky planet that lies between the *crust* and the *core*.
mare (pl. **maria**) A dark region on the Moon composed of basaltic lava flows.
mass 1. Inertial mass: the property of matter that determines its resistance to changes in motion. Compare *weight*. 2. Gravitational mass: the property of matter defined by its attractive force on other objects. According to general relativity, the two are equivalent.
mass-luminosity relationship An empirical relationship between the luminosity (L) and mass (M) of main-sequence stars—for example, $L \propto M^{3.5}$.
mass transfer The transfer of mass from one member of a binary star system to its companion. Mass transfer occurs when one of the stars evolves to the point that it overfills its Roche lobe, so that its outer layers are pulled toward its binary companion.
massive element Also called *heavy element*. Any element more massive than helium.
matter 1. Objects made of particles that have mass, such as protons, neutrons, and electrons. 2. Anything that occupies space and has mass.
Maunder Minimum The period from 1645 to 1715, when very few sunspots were observed.
megabar A unit of pressure equal to 1 million bars.
megaparsec (**Mpc**) A unit of distance equal to 1 million parsecs, or 3.26 million light-years.
meridian The imaginary arc in the sky running from the horizon at due north through the zenith to the horizon at due south. The meridian divides the observer's sky into eastern and western hemispheres. Compare *equator*.
mesosphere The layer of Earth's atmosphere immediately above the stratosphere, extending from an altitude of 50 kilometers (km) to about 90 km. Compare *troposphere, stratosphere*, and *thermosphere*.
meteor The incandescent trail produced by a small piece of interplanetary debris as it travels through the atmosphere at very high speeds. Compare *meteorite* and *meteoroid*.
meteor shower A larger-than-normal display of meteors, occurring when Earth passes through the orbit of a disintegrating comet, sweeping up its debris. Compare *sporadic meteor*.
meteorite A piece of rock or other fragment of material (a meteoroid) that survives to reach a planet's surface. Compare *meteor* and *meteoroid*.
meteoroid A small cometary or asteroidal fragment, ranging in size from 100 microns (μm) to 100 meters. When entering a planetary atmosphere, the meteoroid creates a *meteor*. Compare *meteor* and *meteorite*; also *planetesimal* and *zodiacal dust*.
micrometer (**μm**) See *micron*.
micron One-millionth (10^{-6}) of a meter; a unit of length used for the wavelength of infrared light.
microwave radiation Electromagnetic radiation with frequencies, photon energies, and wavelengths between those of infrared radiation and radio waves.
Milky Way Galaxy The galaxy in which the Sun and Solar System reside.
minor planet See *asteroid*.
minute of arc See *arcminute*.
modern physics Usually, the physical principles, including relativity and quantum mechanics, developed after 1900.
molecular cloud An interstellar cloud composed primarily of molecular hydrogen.
molecular-cloud core A dense clump within a molecular cloud that forms as the cloud collapses and fragments. Protostars form from molecular-cloud cores.
molecule Generally, the smallest particle of a substance that retains its chemical properties and is composed of two or more atoms.
momentum The product of the mass and velocity of a particle. Possible units include: kilograms times meters per second (kg m/s).
moon A less massive satellite orbiting a more massive object. Moons are found around planets, dwarf planets, asteroids, and Kuiper Belt objects. The term is usually capitalized when referring to Earth's Moon.
Mpc See *megaparsec*.
multiverse A collection of parallel universes that together comprise all that is.
mutation In biology, an imperfect reproduction of self-replicating material.

N

N See *newton*.
nadir The point on the celestial sphere located directly below an observer, opposite the *zenith*.
nanometer (**nm**) One-billionth (10^{-9}) of a meter; a unit of length used for the wavelength of visible light.
natural selection The process by which forms of structure, ranging from molecules to whole organisms, that are best adapted to their environment become more common than less well-adapted forms.
NCP See *north celestial pole*.
neap tide An especially weak tide that occurs around the time of the first or third quarter Moon, when the gravitational forces of the Moon and the Sun on Earth are at right angles to each other. Compare *spring tide*.
near-Earth asteroid An asteroid whose orbit brings it close to the orbit of Earth. See also *near-Earth object*.
near-Earth object (**NEO**) An asteroid, comet, or large meteoroid whose orbit intersects Earth's orbit.
nebula (pl. **nebulae**) A cloud of interstellar gas and dust, either illuminated by stars (bright nebula) or seen in silhouette against a brighter background (dark nebula).
nebular hypothesis The first plausible theory of the formation of the Solar System, proposed by Immanuel Kant in 1734, which stated that the Solar System formed from the collapse of an interstellar cloud of rotating gas.
NEO See *near-Earth object*.
neutrino A very low-mass, electrically neutral particle emitted during beta decay. Neutrinos interact with matter only very feebly and so can penetrate through great quantities of matter.
neutrino cooling The process in which thermal energy is carried out of the center of a star by neutrinos rather than by electromagnetic radiation or convection.
neutron A subatomic particle having no net electric charge, and a rest mass and rest energy nearly equal to that of the proton. Compare *electron* and *proton*.
neutron star The neutron-degenerate remnant left behind by a Type II supernova.
new Moon The phase of the Moon in which the Moon is between Earth and the Sun, and from Earth we see only the side of the Moon not being illuminated by the Sun. Compare *full Moon*. See also *first quarter Moon* and *third quarter Moon*.
newton (**N**) The force required to accelerate a 1-kilogram (kg) mass at a rate of 1 meter per second per second (m/s^2). Unit: kilograms times meters per second squared (kg m/s^2).
Newton's first law of motion The law, formulated by Isaac Newton, stating that an object will remain at rest or will continue moving along a straight line at a constant speed until an unbalanced force acts on it.
Newton's laws The three physical laws of motion formulated by Isaac Newton.
Newton's second law of motion The law, formulated by Isaac Newton, stating that if an unbalanced force acts on a body, the body will have an acceleration proportional to the unbalanced force and inversely proportional to the object's mass: $a = F/m$. The acceleration will be in the direction of the unbalanced force.
Newton's third law of motion The law, formulated by Isaac Newton, stating that for every force there is an equal force in the opposite direction.
nm See *nanometer*.

normal matter See *luminous matter.*
north celestial pole (NCP) The northward projection of Earth's rotation axis onto the celestial sphere. Compare *south celestial pole.*
North Pole The location in the Northern Hemisphere where Earth's rotation axis intersects the surface of Earth. Compare *South Pole.*
nova (pl. **novae**) A stellar explosion that results from runaway nuclear fusion in a layer of material on the surface of a white dwarf in a binary system.
nuclear burning Release of energy by nuclear fusion of low-mass elements.
nuclear fusion The combination of two less massive atomic nuclei into a single more massive atomic nucleus.
nucleosynthesis The formation of more massive atomic nuclei from less massive nuclei, either in the Big Bang (Big Bang nucleosynthesis) or in the interiors of stars (stellar nucleosynthesis).
nucleus (pl. **nuclei**) 1. The dense, central part of an atom. 2. The central core of a galaxy, comet, or other diffuse object.

O

objective lens The primary optical element in a telescope or camera that produces an image of an object.
oblateness The flattening of an otherwise spherical planet or star caused by its rapid rotation.
obliquity The inclination of a celestial body's equator to its orbital plane.
observational uncertainty The fact that real measurements are never perfect; all observations are uncertain by some amount.
Occam's razor The principle that the simplest hypothesis is the most likely, named after William of Occam (circa 1285–1349), the medieval English cleric to whom the idea is attributed.
Oort Cloud A spherical distribution of comet nuclei stretching from beyond the Kuiper Belt to more than 50,000 astronomical units (AU) from the Sun.
opacity A measure of how effectively a material blocks the radiation going through it.
open cluster A loosely bound group of a few dozen to a few thousand stars that formed together in the disk of a spiral galaxy. Compare *globular cluster.*
open universe An infinite universe with a negatively curved spatial structure (much like the surface of a saddle) such that the sum of the angles of a triangle is always less than 180°. Compare *closed universe* and *flat universe.*
orbit The path taken by one object moving around another object under the influence of their mutual gravitational or electric attraction.
orbital resonance A situation in which the orbital periods of two objects are related by a ratio of small integers.
orbiter A spacecraft that is placed in orbit around a planet or moon. Compare *flyby.*
organic Containing the element carbon.

P

P wave See *primary wave.*
pair production The creation of a particle-antiparticle pair from a source of electromagnetic energy.
paleoclimatology The study of changes in Earth's climate throughout its history.
paleomagnetism The record of Earth's magnetic field as preserved in rocks.
palimpsest The flat circular patch of bright terrain that remains after a crater has been deformed over time.
parallax Also called *parallactic angle.* The displacement in the apparent position of a nearby star caused by the changing location of Earth in its orbit.
parent element A radioactive element that decays to form more stable *daughter products.*
parsec (pc) Short for *parallax second.* The distance to a star with a parallax of 1 arcsecond (arcsec) using a base of 1 astronomical unit (AU). One parsec is approximately 3.26 light-years.
partial solar eclipse The type of eclipse that occurs when Earth passes through the penumbra of the Moon's shadow, so that the Moon blocks only a portion of the Sun's disk. Compare *annular solar eclipse* and *total solar eclipse.*
pc See *parsec.*
peculiar velocity The motion of a galaxy relative to the overall expansion of the universe.
penumbra (pl. **penumbrae**) 1. The outer part of a shadow, where the source of light is only partially blocked. Compare *umbra* (definition 1). 2. The region surrounding the umbra of a sunspot. The penumbra is cooler and darker than the surrounding surface of the Sun but not as cool or dark as the umbra. Compare *umbra* (definition 2).
penumbral lunar eclipse A lunar eclipse in which the Moon passes through the penumbra of Earth's shadow. Compare *total lunar eclipse.*
perihelion (pl. **perihelia**) The point in a solar orbit that is closest to the Sun. Compare *aphelion.*
period The time it takes for a regularly repetitive process to complete one cycle.
period-luminosity relationship The relationship between the period of variability of a pulsating variable star, such as a Cepheid or RR Lyrae variable, and the luminosity of the star. Longer-period pulsating variable stars are more luminous than shorter-period ones.
Perseids A prominent August meteor shower associated with the dust debris left by comet Swift-Tuttle.
phase One of the various appearances of the sunlit surface of the Moon or a planet caused by the change in viewing location of Earth relative to both the Sun and the object. Examples include crescent phase and gibbous phase.
PHI See *Planetary Habitability Index.*
photino An elementary particle related to the photon. One of the candidates for cold dark matter.
photochemical Resulting from the action of light on chemical systems.
photodissociation The breaking apart of molecules into smaller fragments or individual atoms by the action of photons. Compare *recombination* (definition 1).
photoelectric effect The emission of electrons from a substance that is illuminated by electromagnetic radiation greater than a certain critical frequency.
photometry The process of measuring the brightness of a source of light, generally over a specific range of wavelength.
photon Also called *quantum of light.* A discrete unit or particle of electromagnetic radiation. The energy of a photon is equal to Planck's constant (h) multiplied by the frequency (f) of its electromagnetic radiation: $E_{\text{photon}} = h \times f$. The photon is the carrier of the electromagnetic force.
photosphere The apparent surface of the Sun as seen in visible light. Compare *chromosphere* and *corona.*
physical law A broad statement that predicts a particular aspect of how the physical universe behaves and that is supported by many empirical tests. See also *theory.*
pixel The smallest picture element in a digital image array.
Planck era The early time, just after the Big Bang, when the universe as a whole must be described with quantum mechanics.
Planck spectrum Also called *blackbody spectrum.* The spectrum of electromagnetic energy emitted by a blackbody per unit area per second, which is determined only by the temperature of the object.
Planck's constant (*h*) The constant of proportionality between the energy and the frequency of a photon. This constant defines how much energy a single photon of a given frequency or wavelength has. Value: 6.63×10^{-34} joule-second.
planet 1. A large body that orbits the Sun or other star that shines only by light reflected from the Sun or star. 2. In the Solar System, a body that orbits the Sun, has sufficient mass for self-gravity to overcome rigid body forces so that it assumes a spherical shape, and has cleared smaller bodies from the neighborhood around its orbit. Compare *dwarf planet.*

planet migration The theory that a planet can move to a location away from where it formed, through gravitational interactions with other bodies or loss of orbital energy from interaction with gas in the protoplanetary disk.

Planetary Habitability Index (PHI) An index for quantifying the habitability of extrasolar planets that aims to be less Earth-centric than the *Earth Similarity Index* and to broaden the options for habitability. It depends on factors not yet measured (or measurable) for most extrasolar planets, including the availability of energy, the presence of some kind of liquid, the type of surface, and the chemical makeup.

planetary nebula The expanding shell of material ejected by a dying asymptotic giant branch star. A planetary nebula glows from fluorescence caused by intense ultraviolet light coming from the hot, stellar remnant at its center.

planetary system A system of planets and other smaller objects in orbit around a star.

planetesimal A primitive body of rock and ice, 100 meters or more in diameter, that combines with others to form a planet. Compare *meteoroid* and *zodiacal dust*.

plasma A gas that is composed largely of charged particles but also may include some neutral atoms.

plate tectonics The geological theory concerning the motions of lithospheric plates, which in turn provides the theoretical basis for *continental drift*.

positron A positively charged subatomic particle; the antiparticle of the *electron*.

power The rate at which work is done or at which energy is delivered. Possible units include: watts (W) and joules per second (J/s).

precession of the equinoxes The slow change in orientation between the ecliptic plane and the celestial equator caused by the wobbling of Earth's axis.

pressure Force per unit area. Possible units include: newtons per square meter (N/m^2) and bars.

primary atmosphere An atmosphere, composed mostly of hydrogen and helium, that forms at the same time as its host planet. Compare *secondary atmosphere*.

primary mirror The principal optical mirror in a reflecting telescope. The primary mirror determines the telescope's light-gathering power and resolution. Compare *secondary mirror*.

primary wave Also called *P wave*. A longitudinal seismic wave, in which the oscillations involve compression and decompression parallel to the direction of travel. Compare *secondary wave*.

principle A general idea or sense about how the universe is that guides us in constructing new scientific theories. Principles can be testable theories.

prograde motion 1. Rotational or orbital motion of a moon that is in the same sense as the planet it orbits. 2. The counterclockwise orbital motion of Solar System objects as seen from above Earth's orbital plane. Compare *retrograde motion*.

prominence An archlike projection above the solar photosphere often associated with a sunspot.

proportionality A relationship between two things whose ratio is a constant.

proton (*p* or p^+) A subatomic particle having a positive electric charge of 1.6×10^{-19} coulomb (C), a rest mass of 1.67×10^{-27} kilogram (kg), and a rest energy of 1.5×10^{-10} joule (J). Compare *electron* and *neutron*.

proton-proton chain One of the ways in which hydrogen burning can take place. This is the most important path for hydrogen burning in low-mass stars such as the Sun. Compare *carbon-nitrogen-oxygen cycle*.

protoplanetary disk The remains of the accretion disk around a young star from which a planetary system may form. Sometimes called *circumstellar disk*.

protostar A young stellar object that derives its luminosity from the conversion of gravitational energy to thermal energy, rather than from nuclear reactions in its core.

pulsar A rapidly rotating neutron star that beams radiation into space in two searchlight-like beams. To a distant observer, the star appears to flash on and off.

pulsating variable star A variable star that undergoes periodic radial pulsations.

Q

QCD See *quantum chromodynamics*.

QED See *quantum electrodynamics*.

quantized Existing as discrete, irreducible units.

quantum chromodynamics (QCD) The quantum theory describing the strong nuclear force and its mediation by gluons. Compare *quantum electrodynamics*.

quantum efficiency The likelihood that a particular photon falling on a detector will actually produce a response in the detector.

quantum electrodynamics (QED) The quantum theory describing the electromagnetic force and its mediation by photons. Compare *quantum chromodynamics*.

quantum mechanics The branch of physics that deals with the quantized and probabilistic behavior of atoms and subatomic particles.

quantum of light See *photon*.

quark The building block of protons and neutrons.

quasar Short for *quasi-stellar radio source*. The most luminous of the active galactic nuclei, seen only at great distances from the Milky Way.

R

radial velocity The component of velocity that is directed toward or away from the observer.

radian The angle at the center of a circle subtended by an arc equal to the length of the circle's radius; 2π radians equals 360°, and 1 radian equals approximately 57.3°.

radiant The direction in the sky from which the meteors in a meteor shower seem to come.

radiation belt A toroidal ring of high-energy particles surrounding a planet.

radiative transfer The transport of energy from one location to another by electromagnetic radiation.

radiative zone A region within a star where energy is transported outward by radiation. Compare *convective zone*.

radio galaxy A type of elliptical galaxy that has an active galactic nucleus at its center and very strong emission (10^{35} to 10^{38} watts [W]) in the radio part of the electromagnetic spectrum. Compare *Seyfert galaxy*.

radio telescope An instrument for detecting and measuring radio frequency emissions from celestial sources.

radio wave Electromagnetic radiation in the extreme long-wavelength region of the spectrum, beyond the region of microwaves.

radioisotope A radioactive element.

radiometric dating Use of the radioactive decay of elements to measure the ages of materials such as minerals.

ratio The relationship in quantity or size between two or more things.

ray 1. A beam of electromagnetic radiation. 2. A bright streak emanating from a young impact crater.

recombination 1. The combining of ions and electrons to form neutral atoms. Compare *photodissociation*. 2. An event early in the evolution of the universe in which hydrogen and helium nuclei combined with electrons to form neutral atoms. The removal of electrons caused the universe to become transparent to electromagnetic radiation.

red giant A low-mass star that has evolved beyond the main sequence and is now fusing hydrogen in a shell surrounding a degenerate helium core.

red giant branch A region on the H-R diagram defined by low-mass stars evolving from the main sequence toward the horizontal branch.

reddening The effect by which stars and other objects, when viewed through interstellar dust, appear redder than they actually are. Reddening is caused by the fact that blue light is more strongly absorbed and scattered than red light.

redshift The Doppler shift toward longer (redder) wavelengths of light from an approaching object. Compare *blueshift*.

reflecting telescope A telescope that uses mirrors for collecting and focusing incoming electromagnetic radiation to form an image in their focal planes. The size of a reflecting telescope is defined by the diameter of the primary mirror. Compare *refracting telescope.*

reflection The redirection of a beam of light that strikes, but does not cross, the surface between two media having different refractive indices. If the surface is flat and smooth, the angle of incidence equals the angle of reflection. Compare *refraction.*

refracting telescope A telescope that uses objective lenses for collecting and focusing incoming electromagnetic radiation to form an image. Compare *reflecting telescope.*

refraction The redirection or bending of a beam of light when it crosses the boundary between two media having different refractive indices. Compare *reflection.*

refractory material Material that remains solid at high temperatures. Compare *volatile material.*

regular moon A moon that formed together with the planet it orbits. Compare *irregular moon.*

reionization A period following the Dark Ages during which objects formed that radiated enough energy to ionize neutral hydrogen, at redshift $6 < z < 20$.

relative humidity The amount of water vapor held by a volume of air at a given temperature compared (stated as a percentage) to the total amount of water that could be held by the same volume of air at the same temperature.

relative motion The difference in motion between two individual frames of reference.

relativistic Describing systems that travel at nearly the speed of light or are located in the vicinity of very strong gravitational fields.

relativistic beaming The effect created when material moving at nearly the speed of light beams the radiation it emits in the direction of its motion.

remote sensing The use of images, spectra, radar, or other techniques to measure the properties of an object from a distance.

resolution The ability of a telescope to separate two point sources of light. Resolution is determined by the telescope's aperture and the wavelength of light it receives.

rest wavelength The wavelength of light that is seen coming from an object at rest with respect to the observer.

retrograde motion 1. Rotation or orbital motion of a moon that is in the opposite sense to the rotation of the planet it orbits. 2. The clockwise orbital motion of Solar System objects as seen from above Earth's orbital plane. Compare *prograde motion.* 3. Apparent retrograde motion is a motion of the planets with respect to the "fixed stars," in which the planets appear to move westward for a period of time before resuming their normal eastward motion.

rift zone A region created by a geological fault, in which mantle material rises up, cools, and slowly spreads out, forming new crust.

right ascension A measure, analogous to *longitude*, that tells you the angular distance of a celestial body eastward along the celestial equator from the vernal equinox. Compare *declination.*

ring An aggregation of small particles orbiting a planet or star. The rings of the four giant planets of the Solar System are composed variously of silicates, organic materials, and ices.

ring arc A discontinuous, higher-density region within an otherwise continuous, narrow ring.

ringlet A narrowly confined concentration of ring particles.

Roche limit The distance at which a planet's tidal forces exceed the self-gravity of a smaller object—such as a moon, asteroid, or comet—causing the object to break apart.

Roche lobe The hourglass- or figure eight–shaped volume of space surrounding two stars, which constrains material that is gravitationally bound by one or the other.

rotation curve A plot showing how the orbital velocity of stars and gas in a galaxy changes with radial distance from the galaxy's center.

rover A remotely controlled instrumented vehicle designed to move and explore the surface of a terrestrial planet or moon. Compare *lander.*

RR Lyrae variable A variable giant star whose regularly timed pulsations are good predictors of its luminosity. RR Lyrae variables are used for distance measurements to globular clusters.

S

S-type asteroid An asteroid made of material that was once part of the outer layer of a larger, differentiated body that has since broken into pieces. Compare *C-type asteroid* and *M-type asteroid.*

S wave See *secondary wave.*

S0 galaxy A galaxy with a bulge and a disk-like spiral, but smooth in appearance like ellipticals. Compare *barred spiral galaxy, elliptical galaxy, irregular galaxy,* and *spiral galaxy.*

satellite An object in orbit about a more massive body—for example, a human-made satellite, or a moon of any planet.

scale factor (R_U) A dimensionless number proportional to the distance between two points in space. The scale factor increases as the universe expands.

scattering The random change in the direction of travel of photons, caused by their interactions with molecules or dust particles.

Schwarzschild radius The distance from the center of a nonrotating, spherical black hole at which the escape velocity equals the speed of light.

scientific method The formal procedure—including hypothesis, prediction, and experiment or observation—used to test (attempt to falsify) the validity of scientific hypotheses and theories.

scientific notation The standard expression of numbers with one digit (which can be zero) to the left of the decimal point and multiplied by 10 to the exponent required to give the number its correct value. Example: $2.99 \times 10^8 = 299{,}000{,}000$.

SCP See *south celestial pole.*

second law of thermodynamics The law stating that the entropy or disorder of an isolated system always increases as the system evolves.

second of arc See *arcsecond.*

secondary atmosphere An atmosphere that forms—as a result of volcanism, comet impacts, or another process—sometime after its host planet has formed. Compare *primary atmosphere.*

secondary crater A crater formed from ejecta thrown from an *impact crater.*

secondary mirror A small mirror placed on the optical axis of a reflecting telescope that returns the beam back through a small hole in the *primary mirror,* thereby shortening the mechanical length of the telescope.

secondary wave Also called *S wave.* A transverse seismic wave, which involves the sideways motion of material. Compare *primary wave.*

sedimentation A process in which material carried by water or wind deposits layers of material and buries what lies below.

seismic wave A vibration due to an earthquake, a large explosion, or an impact on the surface that travels through a planet's interior.

seismometer An instrument that measures the amplitude and frequency of seismic waves.

self-gravity The gravitational attraction among all parts of the same object.

semimajor axis Half of the longer axis of an ellipse.

SETI The Search for Extraterrestrial Intelligence project, which uses advanced technology combined with radio telescopes to search for evidence of intelligent life elsewhere in the universe.

Seyfert galaxy A type of spiral galaxy with an active galactic nucleus at its center; first discovered in 1943 by Carl Seyfert. Compare *radio galaxy.*

shepherd moon A moon that orbits close to rings and gravitationally confines the orbits of the ring particles.

shield volcano A volcano formed by very fluid lava flowing from a single source and spreading out from that source. Compare *composite volcano.*

short-period comet A comet with an orbital period of less than 200 years. Compare *long-period comet.*

sidereal day Earth's period of rotation with respect to the stars—about 23 hours 56 minutes—which is the time it takes for Earth to make one rotation and face the exact same star on the meridian. It differs from the *solar day* because of Earth's motion around the Sun.

sidereal period An object's orbital or rotational period measured with respect to the stars. Compare *synodic period.*

silicate One of the family of minerals composed of silicon and oxygen in combination with other elements.

singularity The point where a mathematical expression or equation becomes meaningless, such as the denominator of a fraction approaching zero. See also *black hole.*

solar abundance The relative amount of an element detected in the atmosphere of the Sun, expressed as the ratio of the number of atoms of that element to the number of hydrogen atoms.

solar day The day in common usage—24 hours, which is Earth's period of rotation that brings the Sun back to the same local meridian where the rotation started. Compare *sidereal day.*

solar eclipse An eclipse that occurs when the Sun is partially or entirely blocked by the Moon. Compare *lunar eclipse.*

solar flare An explosion on the Sun's surface associated with a complex sunspot group and a strong magnetic field.

solar maximum (pl. **maxima**) The time, occurring about every 11 years, when the Sun is at its peak activity, meaning that sunspot activity and related phenomena (such as prominences, flares, and coronal mass ejections) are at their peak.

solar neutrino problem The historical observation that only about a third as many neutrinos as predicted by theory seemed to be coming from the Sun.

Solar System The gravitationally bound system made up of the Sun, planets, dwarf planets, moons, asteroids, comets, and Kuiper Belt objects, along with their associated gas and dust.

solar tide A tide on Earth that is due to the differential gravitational pull of the Sun. Compare *lunar tide.*

solar wind The stream of charged particles emitted by the Sun that flows at high speeds through interplanetary space.

solstice Literally, "sun standing still." 1. One of the two most northerly and southerly points on the ecliptic. 2. Either of the two times of year (the *summer solstice* and *winter solstice*) when the Sun is at one of these two positions. Compare *equinox.*

south celestial pole (**SCP**) The southward projection of Earth's rotation axis onto the celestial sphere. Compare *north celestial pole.*

South Pole The location in the Southern Hemisphere where Earth's rotation axis intersects the surface of Earth. Compare *North Pole.*

spacetime A concept that combines space and time into a four-dimensional continuum with three spatial dimensions plus time.

special relativity See *special theory of relativity.*

special theory of relativity Sometimes referred to as simply *special relativity.* Einstein's theory explaining how the fact that the speed of light is a constant affects nonaccelerating frames of reference. Compare *general theory of relativity.*

spectral type A classification system for stars based on the presence and relative strength of absorption lines in their spectra. Spectral type is related to the surface temperature of a star.

spectrograph Also called *spectrometer.* A device that spreads out the light from an object into its component wavelengths.

spectrometer See *spectrograph.*

spectroscopic binary A binary star system whose existence and properties are revealed to astronomers only by the Doppler shift of its spectral lines. Most spectroscopic binaries are close pairs. Compare *eclipsing binary* and *visual binary.*

spectroscopic parallax Use of the spectroscopically determined luminosity and the observed brightness of a star to determine the star's distance.

spectroscopy The study of an object's electromagnetic radiation in terms of its component wavelengths.

spectrum (pl. **spectra**) Waves sorted by wavelength. See also *electromagnetic spectrum.*

speed The rate of change of an object's position with time, without regard to the direction of movement. Possible units include: meters per second (m/s) and kilometers per hour (km/h). Compare *velocity.*

spherically symmetric Describing an object whose properties depend only on distance from the object's center, so that the object has the same form viewed from any direction.

spin-orbit resonance A relationship between the orbital and rotation periods of an object such that the ratio of their periods can be expressed by simple integers.

spiral density wave A stable, spiral-shaped change in the local gravity of a galactic disk that can be produced by periodic gravitational kicks from neighboring galaxies or from nonspherical bulges and bars in spiral galaxies.

spiral galaxy A galaxy of Hubble type "S" class, with a discernible disk in which large spiral patterns exist. Compare *barred spiral galaxy, elliptical galaxy, irregular galaxy,* and *S0 galaxy.*

spoke One of several narrow radial features seen occasionally in Saturn's B Ring. Spokes appear dark in backscattered light and bright in forward, scattering light, indicating that they are composed of tiny particles. Their origin is not well understood.

sporadic meteor A meteor that is not associated with a specific *meteor shower.*

spreading center A zone from which two tectonic plates diverge.

spring tide An especially strong tide that occurs around the time of a new or full Moon, when lunar tides and solar tides reinforce each other. Compare *neap tide.*

stable equilibrium An equilibrium state in which the system returns to its former condition after a small disturbance. Compare *unstable equilibrium.*

standard candle An object whose luminosity either is known or can be predicted in a distance-independent way, so its brightness can be used to determine its distance via the inverse square law of radiation.

standard model The theory of particle physics that combines electroweak theory with quantum chromodynamics to describe the structure of known forms of matter.

star A luminous ball of gas that is held together by gravity. A normal star is powered by nuclear reactions in its interior.

star cluster A group of stars that all formed at the same time and in the same general location.

static equilibrium A state in which the forces within a system are all in balance so that the system does not change. Compare *dynamic equilibrium.*

Stefan-Boltzmann constant (σ) The constant of proportionality that relates the flux emitted by an object to the fourth power of its absolute temperature. Value: 5.67×10^{-8} W/(m^2 K^4) (W = watts, m = meters, K = kelvins).

Stefan-Boltzmann law The law, formulated by Josef Stefan and Ludwig Boltzmann, stating that the amount of electromagnetic energy emitted from the surface of a body (flux), summed over the energies of all photons of all wavelengths emitted, is proportional to the fourth power of the temperature of the body: $\mathcal{F} = \sigma T^4$.

stellar-mass loss The loss of mass from the outermost parts of a star's atmosphere during the course of its evolution.

stellar occultation An event in which a planet or other Solar System body moves between the observer and a star, eclipsing the light emitted by that star.

stellar population A group of stars with similar ages, chemical compositions, and dynamic properties.

stereoscopic vision The way an animal's brain combines the different information from its two eyes to perceive the distances to objects around it.

stony-iron meteorite A meteorite consisting of a mixture of silicate minerals and iron-nickel alloys. Compare *iron meteorite* and *stony meteorite.*

stony meteorite A meteorite composed primarily of silicate minerals, similar to those found on Earth. Compare *iron meteorite* and *stony-iron meteorite.*

stratosphere The atmospheric layer immediately above the troposphere. On Earth it extends upward to an altitude of 50 kilometers (km). Compare *troposphere, mesosphere,* and *thermosphere.*

string theory See *superstring theory.*

stromatolite A fossilized mass of simple microbes.

strong nuclear force The attractive short-range force between protons and neutrons that holds atomic nuclei together. One of the four fundamental forces of nature, along with the *electromagnetic force,* the *weak nuclear force,* and *gravity* (definition 1). The force is mediated by the exchange of gluons.

subduction zone A region where two tectonic plates converge, with one plate sliding under the other and being drawn downward into the interior.

subgiant A giant star that is smaller and lower in luminosity than normal giant stars of the same spectral type. Subgiants evolve to become giants.

subgiant branch A region of the H-R diagram defined by stars that have left the main sequence but have not yet reached the red giant branch.

sublimation The process by which a solid becomes a gas without first becoming a liquid.

subsonic Moving within a medium at a speed slower than the speed of sound in that medium. Compare *supersonic.*

summer solstice 1. One of two points where the Sun is at its greatest distance from the celestial equator. 2. The day on which the Sun appears at this location, marking the first day of summer (about June 20 in the Northern Hemisphere and December 21 in the Southern Hemisphere). Compare *winter solstice.* See also *autumnal equinox* and *vernal equinox.*

Sun The star at the center of the Solar System.

sungrazer A comet whose perihelion is within a few solar diameters of the surface of the Sun.

sunspot A cooler, transitory region on the solar surface produced when loops of magnetic flux break through the surface of the Sun.

sunspot cycle The approximate 11-year cycle during which sunspot activity increases and then decreases. This is one-half of a full 22-year cycle, in which the magnetic polarity of the Sun first reverses and then returns to its original configuration.

supercluster A large conglomeration of galaxy clusters and galaxy groups; typically, more than 100–300 megaparsecs (Mpc) in size and containing tens of thousands to hundreds of thousands of galaxies. Compare *galaxy cluster* and *galaxy group.*

super-Earth An extrasolar planet with about 2–10 times the mass of Earth.

superluminal motion The appearance (though not the reality) that a jet is moving faster than the speed of light.

supermassive black hole A black hole of 1,000 solar masses ($M_{\odot}$) or more that resides in the center of a galaxy, and whose gravity powers active galactic nuclei.

supernova (pl. **supernovae**) A stellar explosion resulting in the release of tremendous amounts of energy, including the high-speed ejection of matter into the interstellar medium. See also *Type Ia supernova* and *Type II supernova.*

supersonic Moving within a medium at a speed faster than the speed of sound in that medium. Compare *subsonic.*

superstring theory The theory that conceives of particles as strings in 10 dimensions of space and time; the current contender for a theory of everything.

surface brightness The amount of electromagnetic radiation emitted or reflected per unit area.

surface wave A seismic wave that travels on the surface of a planet or moon.

symmetry In theoretical physics, the properties of physical laws that remain constant when certain things change, such as the symmetry between matter and antimatter even though their charges may be different.

synchronous rotation The case that occurs when a body's rotation period equals its orbital period around another body. A special type of spin-orbit resonance.

synchrotron radiation Radiation from electrons moving at close to the speed of light as they spiral in a strong magnetic field; named because this kind of radiation was first identified on Earth in particle accelerators called synchrotrons.

synodic period An object's orbital or rotational period measured with respect to the Sun. Compare *sidereal period.*

T

T Tauri star A young stellar object that has dispersed enough of the material surrounding it to be seen in visible light.

tail A stream of gas and dust swept away from the coma of a comet by the solar wind and by radiation pressure from the Sun.

tectonism Deformation of the lithosphere of a planet.

telescope The basic tool of astronomers. Working over the entire range from gamma rays to radio waves, astronomical telescopes collect and concentrate electromagnetic radiation from celestial objects.

temperature A measure of the average kinetic energy of the atoms or molecules in a gas, solid, or liquid.

terrestrial planet An Earth-like planet, made of rock and metal and having a solid surface. In the Solar System, the terrestrial planets are Mercury, Venus, Earth, and Mars. Compare *giant planet.*

theoretical model A detailed description of the properties of a particular object or system in terms of known physical laws or theories. Often, a computer calculation of predicted properties based on such a description.

theory A well-developed idea or group of ideas that are tied solidly to known physical laws and make testable predictions about the world. A very well-tested theory may be called a *physical law,* or simply a fact. Compare *hypothesis.*

theory of everything (TOE) A theory that unifies all four fundamental forces of nature: strong nuclear, weak nuclear, electromagnetic, and gravitational forces.

thermal conduction See *conduction.*

thermal energy The energy that resides in the random motion of atoms, molecules, and particles, by which we measure their temperature.

thermal equilibrium The state in which the rate of thermal-energy emission by an object is equal to the rate of thermal-energy absorption.

thermal motion The random motion of atoms, molecules, and particles that gives rise to thermal radiation.

thermal radiation Electromagnetic radiation resulting from the random motion of the charged particles in every substance.

thermosphere The layer of Earth's atmosphere at altitudes greater than 90 kilometers (km), above the mesosphere. Near its top, at an altitude of 600 km, the temperature can reach 1000 K. Compare *troposphere, stratosphere,* and *mesosphere.*

third quarter Moon The phase of the Moon in which only the eastern half of the Moon, as viewed from Earth, is illuminated by the Sun. It occurs about one week after the full Moon. Compare *first quarter Moon.* See also *full Moon* and *new Moon.*

tidal bulge Distortion of a body resulting from tidal stresses.

tidal force A force caused by the change in the strength of gravity across an object.

tidal locking Synchronous rotation of an object caused by internal friction as the object rotates through its tidal bulge.

tide On Earth, the rise and fall of the oceans as Earth rotates through a tidal bulge caused by the Moon and the Sun. See also *lunar tide, neap tide, solar tide,* and *spring tide.*

time dilation The relativistic "stretching" of time. Compare *general relativistic time dilation.*

TOE See *theory of everything.*

topographic relief The differences in elevation from point to point on a planetary surface.

tornado A violent rotating column of air, typically 75 meters across with 200- kilometer-per-hour (km/h) winds. Some tornadoes can be more than 3 km across, and winds up to 500 km/h have been observed.

torus (pl. **tori**) A three-dimensional, dough-nut-shaped ring.

total lunar eclipse A lunar eclipse in which the Moon passes through the umbra of Earth's shadow. Compare *penumbral lunar eclipse.*

total solar eclipse The type of eclipse that occurs when Earth passes through the umbra of the Moon's shadow, so that the Moon completely blocks the disk of the Sun.

Compare *annular solar eclipse* and *partial solar eclipse*.

transform fault The actively slipping segment of a fracture zone between lithospheric plates.

trans-Neptunian object See *Kuiper Belt object*.

transverse wave A wave that oscillates perpendicular to the direction of the wave's propagation. Compare *longitudinal wave*.

triple-alpha process The nuclear fusion reaction that combines three helium nuclei (alpha particles) together into a single nucleus of carbon.

Trojans A group of asteroids orbiting in the L_4 and L_5 Lagrangian points of Jupiter's orbit.

tropical year The time between one crossing of the vernal equinox and the next. Because of the precession of the equinoxes, a tropical year is slightly shorter than the time that it takes for Earth to orbit once about the Sun. Compare *year*.

Tropics The region on Earth between latitudes 23.5° south and 23.5° north, where the Sun appears directly overhead twice during the year.

tropopause The top of a planet's troposphere.

troposphere The convection-dominated layer of a planet's atmosphere. On Earth, the atmospheric region closest to the ground within which most weather phenomena take place. Compare *stratosphere*, *mesosphere*, and *thermosphere*.

tuning fork diagram The two-pronged diagram showing Hubble's classification of galaxies into ellipticals, S0s, spirals, barred spirals, and irregular galaxies.

turbulence The random motion of blobs of gas within a larger cloud of gas.

Type Ia supernova A supernova explosion with a calibrated peak luminosity that occurs as a result of runaway carbon burning in a white dwarf star that accretes mass from a companion and approaches the Chandrasekhar mass of 1.4 $M_\odot$.

Type II supernova A supernova explosion in which the degenerate core of an evolved massive star suddenly collapses and rebounds.

U

ultrafaint dwarf galaxy A dim dwarf galaxy with only 1,000–100,000 times the Sun's luminosity. Ultrafaint dwarf galaxies differ from globular clusters in that they are composed of large amounts of dark matter.

ultraviolet (UV) radiation Electromagnetic radiation with frequencies and photon energies greater than those of visible light but less than those of X-rays, and wavelengths shorter than those of visible light but longer than those of X-rays.

umbra (pl. **umbrae**) 1. The darkest part of a shadow, where the source of light is completely blocked. Compare *penumbra* (definition 1). 2. The darkest, innermost part of a sunspot. Compare *penumbra* (definition 2).

unbalanced force The nonzero net force acting on a body.

unbound orbit An orbit in which an object is no longer gravitationally bound to the body it was orbiting. An unbound orbit's velocity is greater than the escape velocity. Compare *bound orbit*.

uncertainty principle See *Heisenberg uncertainty principle*.

unified model of AGN A model in which many different types of activity in the nuclei of galaxies are all explained by accretion of matter around a supermassive black hole.

uniform circular motion Motion in a circular path at a constant speed.

unit A fundamental quantity of measurement. The meter is an example of a metric unit; the foot is an example of an English unit.

universal gravitational constant (*G*) The constant of proportionality in the universal law of gravitation. Value: 6.67×10^{-11} meters cubed per kilogram second squared [$m^3/(kg\ s^2) = N\ m^2/kg^2$].

universal law of gravitation The law, formulated by Isaac Newton, stating that the gravitational force between any two objects is proportional to the product of their masses and inversely proportional to the square of the distance between them: $F_{grav} \propto (m^1 m^2/r^2)$.

universe 1. All of space and everything contained therein. 2. Our own universe in a collection of parallel universes that together comprise all that is.

unstable equilibrium An equilibrium state in which a small disturbance will cause a system to move away from equilibrium. Compare *stable equilibrium*.

UV Ultraviolet. See *ultraviolet radiation*.

V

vacuum A region of space that contains very little matter. In quantum mechanics and general relativity, however, even a perfect vacuum has physical properties.

variable star A star with varying luminosity. Many periodic variables are found within the instability strip on the H-R diagram.

velocity The rate and direction of change of an object's position with time. Possible units include: meters per second (m/s) and kilometers per hour (km/h). Compare *speed*.

vernal equinox 1. One of two points where the Sun crosses the celestial equator. 2. The day on which the Sun appears at this location, marking the first day of spring (about March 20 in the Northern Hemisphere and September 22 in the Southern Hemisphere). Compare *autumnal equinox*. See also *summer solstice* and *winter solstice*.

virtual particle A particle that, according to quantum mechanics, comes into existence only momentarily. According to theory, fundamental forces are mediated by the exchange of virtual particles.

visual binary A binary system in which the two stars can be seen individually from Earth. Compare *eclipsing binary* and *spectroscopic binary*.

void A region in space containing little or no matter. Examples include regions in cosmological space that are largely empty of galaxies.

volatile material Sometimes called *ice*. Material that remains gaseous at moderate temperature. Compare *refractory material*.

volcanism A form of geological activity on a planet or moon in which molten rock (magma) erupts at the surface.

W

W See *watt*.

waning Describing the changing phases of the Moon as it becomes less fully illuminated between full Moon and new Moon as seen from Earth. Compare *waxing*.

water cycle The flow of water on, above, and through Earth's surface.

watt (W) A measure of *power*. Unit: joules per second (J/s).

wave A disturbance moving along a surface or passing through a space or a medium.

wavefront The imaginary surface of an electromagnetic wave, either plane or spherical, oriented perpendicular to the direction of travel.

wavelength The distance on a wave between two adjacent points having identical characteristics. The distance a wave travels in one period. Possible units include: meters (m).

waxing Describing the changing phases of the Moon as it becomes more fully illuminated between new Moon and full Moon as seen from Earth. Compare *waning*.

weak nuclear force The force underlying some forms of radioactivity and certain interactions between subatomic particles. It is responsible for radioactive beta decay and for the initial proton-proton interactions that lead to nuclear fusion in the Sun and other stars. One of the four fundamental forces of nature, along with the *electromagnetic force*, the *strong nuclear force*, and *gravity* (definition 1). The force is mediated by the exchange of *W* and *Z* particles.

weather The state of an atmosphere at any given time and place. Compare *climate*.

weight The gravitational force acting on an object; that is, the force equal to the mass of an object multiplied by the local acceleration due to gravity. In general relativity, the force equal to the mass of an object multiplied by the acceleration of the frame of reference in which the object is observed. Compare *mass*.

white dwarf The stellar remnant left at the end of the evolution of a low-mass star. A typical white dwarf has a mass of 0.6 solar mass ($M_\odot$) and a size about equal to that of

Earth; it is made of nonburning, electron-degenerate carbon.

Wien's law A law, named for Wilhelm Wien, stating that location of the peak wavelength in the electromagnetic spectrum of an object is inversely proportional to the temperature of the object.

WIMP Short for *weakly interacting massive particle*. A hypothetical massive particle that interacts through the weak nuclear force and gravity but not with electromagnetic radiation. WIMPS are a candidate for dark matter. Compare *MACHO*.

winter solstice 1. One of two points where the Sun is at its greatest distance from the celestial equator. 2. The day on which the Sun appears at this location, marking the first day of winter (about December 21 in the Northern Hemisphere and June 20 in the Southern Hemisphere). Compare *summer solstice*. See also *autumnal equinox* and *vernal equinox*.

X

X-ray Electromagnetic radiation having frequencies and photon energies greater than those of ultraviolet light but less than those of gamma rays, and wavelengths shorter than those of UV light but longer than those of gamma rays.

X-ray binary A binary system in which mass from an evolving star spills over onto a collapsed companion, such as a neutron star or black hole. The material falling in is heated to such high temperatures that it glows brightly in X-rays.

Y

year The time it takes Earth to make one revolution around the Sun. A solar year is measured from equinox to equinox. A sidereal year, Earth's true orbital period, is measured relative to the stars. Compare *tropical year*.

Z

zenith The point on the celestial sphere located directly overhead from an observer. Compare *nadir*.

zero-age main sequence The strip on the H-R diagram plotting where stars of all masses in a cluster begin their lives.

zodiac The 12 constellations lying along the plane of the ecliptic.

zodiacal dust Particles of cometary and asteroidal debris less than 100 microns (μm) in size that orbit the inner Solar System close to the plane of the ecliptic. Compare *meteoroid* and *planetesimal*.

zodiacal light A band of light in the night sky caused by sunlight reflected by zodiacal dust.

zonal wind The east–west component of a wind.

Credits

These pages contain credits for all 24 chapters in the complete volume of *21st Century Astronomy*, Fourth Edition. The Solar System Edition does not include Chapters 15–23. The Stars and Galaxies Edition does not include Chapters 8–12.

Chapter 1

PHOTOS: p. 2: Rogelio Bernal Andreo DeepSkyColors.com; p. 6 (from top down): Neil Ryder Hoos; Neil Ryder Hoos; Owen Franken/Corbis; Monkey Business Images /Dreamstime.com; American Museum of Natural History; American Museum of Natural History; NASA/JPL/Caltech; p. 7 (top): NASA and STScI; (bottom): Michael J. Tuttle NASM/NASA; p. 8 (top, both): SOHO ESA/NASA; (bottom): CERN; p. 11 (center): RMN-Grand Palais/Art Resource, NY; (right): Robbie Jack/Corbis; p. 12: Courtesy of the Archives, California Institute of Technology; p. 13: NASA, ESA, S. Beckwith (STScI), and The Hubble Heritage Team (STScI/AURA); p. 14 (top): Matheisl/Getty Images; (a): ESA/ Hubble and NASA; (b): NASA and The Hubble Heritage Team (STScI/AURA); p. 17: Copyright © Tribune Media Services, Inc. All rights reserved. Reprinted with permission.

Chapter 2

PHOTOS: p. 24: Yannis Behrakis/REUTERS/ Newscom; p. 26 (all): British Library; p. 27 (a): Roger Ressmeyer/Corbis; (b): Photo Researchers/Getty Images; (c): Henry Westheim Photography/Alamy; p. 35 (left): Pekka Parviainen/Photo Researchers; (right): David Nunuk/Photo Researchers; p. 41 (both): Larry Landolfi/Photo Researchers; p. 43 (top left): AP Photos; (bottom left): Lindsay Hebberd/Corbis; (right):Lynn Goldsmith/ Corbis; p. 44: Arnulf Husmo/Getty Images; p. 51 (top left): Nick Quinn; (top right): Dennis L. Mammana; (bottom a): Reuters/Corbis; (bottom b): Laura Kay and Lisa Rand; p. 53 (a): Johannes Schedler; (b): Anthony Ayiomamitis (TWAN).

Chapter 3

PHOTOS: p. 62: NASA; p. 64: Photo Researchers; p. 65: Nicolaus Copernicus Museum/Giraudon/The Bridgeman Art Library; p. 66: Bettmann/Corbis; p. 68: Tunc Tezel; p. 69 (a): Granger Collection; (b): De Mundi Aetherei Recentioribus Phaenomenis, 1588. The Tycho Brahe Museum, Sweden. www.tychobrahe.com; p. 70: SSPL/The Image Works; p. 75 (left): Galleria Palatina, Palazzo Pitti, Florence/ The Bridgeman Art Library; (right): SSPL/Jamie Cooper/The Image Works; p. 77: Lebrecht Music and Arts Photo Library/Alamy.

Chapter 4

PHOTOS: p. 88: NASA; p. 90: Courtesy of Alan Bean; p. 107 (both): Christopher Mackay; p. 111 (both): NASA/AURA.

Chapter 5

PHOTOS: p. 116: Yva Momatiuk & John Eastcott/Minden Pictures/National Geographic Stock; p. 143: NASA/NSSDC/ GSFC.

Chapter 6

PHOTOS: p. 150: © Laurie Hatch; p. 152: (left): Junenoire Photography; (right): Gianni Tortoli/Photo Researchers; p. 154 (top): Yerkes Observatory photographs: Richard Dreiser; (bottom): Jim Sugar/Corbis; pp. 156-157 (all): ASU Physics Instructional Resource Team; p. 158: © Laurie Hatch; p. 159 (top): Courtesy TMT Observatory Corporation; (bottom): Data courtesy Marc Imhoff of NASA GSFC and Christopher Elvidge of NOAA NGDC. Image by Craig Mayhew and Robert Simmon, NASA GSFC; p. 162 (a): Gemini Observatory; (b, both): The Palomar Observatory; p. 165 (b): Jean-Charles Cuillandre (CFHT); (c): Junenoire Photography; p. 167 (from left to right): NASA E/PO, Sonoma State University, Aurore Simonnet; NASA/CXC/SAO; NASA/ JPL-Caltech; © Laurie Hatch; NASA/JPL/ Caltech; Joint Astronomy Center in Hilo, Hawaii; NRAO VLA Image Gallery; The National Radio Astronomy Observatory, Green Bank; Dave Finley, Courtesy National Radio Astronomy Observatory and Associated Universities, Inc; p. 168 (a): Roger Ressmeyer/ Corbis; (b): David Parker/Photo Researchers; (bottom): Photo by Dave Finley, courtesy National Radio Astronomy Observatory and Associated Universities, Inc; p. 169 (top): ESO; (bottom left): NASA Photo/ Carla Thomas; (bottom right): NASA/Tom Tschida; p. 172 (top): NASA/JPL; (bottom): NASA/Johns Hopkins University Applied Physics Laboratory/Carnegie Institution of Washington; p. 173 (top): NASA/JPL-Caltech; (bottom): NASA; p. 174: CERN; p. 176 (left): Jim Haugen/NSF; (right): LIGO; p. 177 (top): Volker Steger/Science Source/Photo Researchers; (center): Photo courtesy of the LIGO Laboratory; (bottom): EADS Astrium; p. 178 (all): Patrik Jonsson, Greg Novak & Joel Primack, UC Santa Cruz, 2008.

Chapter 7

PHOTOS: p. 184: Caltech/NASA/JPL; p. 187 (both): STScI/ Karl Stapelfeldt (JPL); p. 188: Photograph by Pelisson, SaharaMet; p. 190: Reuters/Corbis; p. 199: NASA/Johns Hopkins University Applied Physics Laboratory/ Carnegie Institution of Washington; p. 201: NASA, ESA, J. R. Graham and P. Kalas (University of California, Berkeley), and B. Matthews (Hertzberg Institute of Astrophysics); p. 203: NASA Ames/Dan Fabrycky, UC Santa Cruz; p. 205 (a-b): NRC-HIA, Christian Marois, and the W.M. Keck Observatory; (bottom): ESO/A.-M. Lagrange et al.; p. 207 (top): NASA/JPL-Caltech/ ASU; (bottom): NASA/Ames/JPL-Caltech. LINE ART: p. 204: Figure 7.20: Graph: "Radial Velocity/Year" from Exoplantets.org. Reprinted by permission of the Department of Terrestrial Magnetism, Carnegie Institution of Washington.

Chapter 8

PHOTOS: p. 214: NASA/Goddard/MIT/ Brown; p. 216 (all): NASA; p. 217 (top): Stockli, Nelson, Hasler, Goddard Space Flight Center/NASA; (bottom): NASA; p. 218 (a): Jim Wark/Visuals Unlimited/Corbis; (b): Frank Lukasseck/Corbis; (c): USGS Hawaiian Volcano Observatory; (d): Stuart Wilson/ Photo Researchers; p. 219 (b): NASA/JSC;

(c): NASA/Johns Hopkins University Applied Physics Laboratory/Carnegie Institution of Washington; p. 220 (left): Montes De Oca & Associates; (right) Photograph by D.J. Roddy and K.A. Zeller, USGS, Flagstaff, AZ; p. 221: NASA/JPL/Caltech; p. 222: Dennis Flaherty/ Photo Researchers; p. 224: NASA/JPL/ Caltech; p. 228: Joe Tucciarone/Science Photo Library/Photo Researchers; p. 231 (b): Art Directors & TRIP/Alamy; p. 233: Donald Duckson/Visuals Unlimited/ Corbis; p. 237 (left): NASA/JSC; (right): NASA/Johns Hopkins University Applied Physics Laboratory/Carnegie Institution of Washington; p. 238 (a): ESA/DLR/FU Berlin (G. Neukum); (b): NSSDC/NASA; p. 239: NASA/Magellan Image/JPL; p. 241 (left): NASA/JSC; (right): K.C. Pau; p. 242 (left): NASA/Johns Hopkins University Applied Physics Laboratory/Arizona State University/ Carnegie Institution of Washington. Image reproduced courtesy of Science/AAAS; (right): ESA/DLR/FU Berlin (G. Neukum); p. 243 (a): NASA/MOLA Science Team; (b): NASA/JPL/ Malin Space Science Systems; p. 244 (left): NASA/JPL-Caltech/University of Arizona/ Texas A&M University; (right): HiRISE, MRO, LPL (U. Arizona), NASA; p. 245 (top, both): NASA/JPL-Caltech; (bottom): NASA/JPL/ University of Arizona; p. 246 (both): NASA/ JPL/Malin Space Science Systems; p. 247 (top left): ESA/DLR/FU Berlin (G. Neukum); (top right): NASA/JPL-Caltech/University of Arizona/Texas A&M University; (bottom): NASA/JPL-Caltech/UTA/UA/MSSS/ESA/ DLR/JPL Solar System Visualization Project; p. 248: NASA/JPL-Caltech/Univ. of Arizona; p. 249: © Don Davis.

Chapter 9

PHOTOS: p. 256: NASA/JPL-Caltech/UA; p. 258 (from left to right): NASA/NSSDC; NASA/NSSDC; NASA/STScI; p. 272 (a): NASA; (b): © Dennis C. Anderson www.auroradude.com; p. 275: NOAA; p. 276 (left): ESA © 2007 MPS/DLR-PF/IDA; (right): © Ted Stryk 2007; p. 277 (top): NASA/JPL-Caltech/University of Arizona; (bottom left): NASA/JPL/Caltech; (bottom right): NASA/ JPL-Caltech/MSSS; p. 278: NASA/STScI; p. 279: NASA/JPL/University of Arizona. LINE ART: p. 280: Figure 9.21: "Global temperature and CO2 variations over the last 800,000 years of Earth's history," from NBI.com, May 14, 2008. Reprinted with permission from the Centre for Ice and Climate, Niels Bohr Institute, University of Copenhagen, Denmark, www.iceandclimate.dk.

Chapter 10

PHOTOS: p. 290: NASA/JPL/Space Science Institute; p. 293 (top and bottom left): NASA/ JPL/University of Arizona; (right, both): NASA/JPL; p. 296: NASA/JPL/Caltech; p. 299 (from left to right): Credit: M. Wong and I. de Pater (University of California, Berkeley); NASA/JHU/APL; NASA/JPL/ Caltech; p. 300 (top): CICLOPS/NASA/JPL/ University of Arizona; (bottom): NASA/ JPL/Caltech; p. 301 (left, insert): NASA/ JPL/Space Science Institute; (right): NASA/ JPL-Caltech/Space Science Institute; p. 302 (top, both): NASA, L. Sromovsky, and P. Fry (University of Wisconsin-Madison); (bottom left): NASA/JPL; (bottom center): NASA/JPL/ Caltech; (bottom right): NASA; p. 307: NASA, ESA, L. Sromovsky and P. Fry (University of Wisconsin), H. Hammel (Space Science Institute), and K. Rages (SETI Institute); p. 313: http://www.freenaturepictures.com/; p. 315 (top): N.M. Schneider, J.T. Trauger, Catalina Observatory; (bottom left): J. Clarke (University of Michigan), NASA; (bottom right): Courtesy of John Clark, Boston University and NASA/STScI; p. 321: NASA/ JPL.

Chapter 11

PHOTOS: p. 322: NASA/JPL/Space Science Institute; p. 324 (left): NASA/JPL/University of Arizona; (right): ESA/DLR/FU Berlin (G. Neukum); p. 325: NASA/JPL/Caltech; p. 326: NASA/JPL/Space Science Institute; p. 328: NASA/JPL/Caltech; p. 329 (both): NASA/JPL/ Caltech; p. 330 (all): NASA/JPL/Space Science Institute; p. 331 (top left): NASA/JPL/Space Science Institute; (top right); NASA/JPL/ Caltech; (bottom): NASA/JPL/Caltech; p. 333 (top): NASA/JPL/Caltech; (a): B.E. Schmidt/ P.M. Schenk/S.P. Carter/USGS/NASA; (b): B.E. Schmidt & Dead Pixel VFX/UTIG; p. 334 (a-c): NASA/JPL/Space Science Institute; p. 335 (top): NASA/JPL; (bottom left): NASA/ JPL/ESA/University of Arizona; (bottom right): NASA/JPL; p. 336: NASA/JPL/ESA/ University of Arizona; p. 337: NASA; p. 338 (top and bottom left): NASA/JPL/Space Science Institute; (right); NASA/JPL/Space Science Institute; p. 339 (top): NASA/JPL/Caltech; (bottom): NASA/JPL/Ted Stryk; p. 342 (top): NASA/JPL/Caltech; (bottom): NASA/JPL/ Caltech; p. 343 (a): NASA/JPL-Caltech/Univ. of Virginia; (b): NASA/JPL-Caltech; p. 344 (a): Stephanie Swartz/Dreamstime; (b): NASA/ JPL/Space Science Institute; p. 345 (top left): NASA/JPL/Caltech; (top right): NASA, ESA, and M. Showalter (SETI Institute); (bottom right): NASA/JPL/Caltech; p. 348: NASA/ JPL/Space Science Institute; p. 349: (top a, b): NASA/JPL/Space Science Institute; (bottom a-c): NASA/JPL/Space Science Institute; p. 351 (both): NASA/JPL/Space Science Institute; p. 357: NASA/JPL/Caltech.

Chapter 12

PHOTOS: p. 358: T. A. Rector (University of Alaska Anchorage), Z. Levay and L.Frattare (Space Telescope Science Institute) and WIYN/NOAO/AURA/NSF; p. 361 (all): Dr. R. Albrecht, ESA/ESO Space Telescope European Coordinating Facility/NASA; p. 362: M. Brown (Caltech), C. Trujillo (Gemini), D. Rabinowitz (Yale), NSF, NASA; p. 365: M. Brown (Caltech), C. Trujillo (Gemini), D. Rabinowitz (Yale), NSF, NASA; p. 369 (top): NASA/JPL/USGS; (bottom): NASA/JPL/Caltech; p. 370 (bottom right): NASA/JPL-Caltecj/UCLA/MPSDLR/ IDA; p. 370 (top left): NASA/Johns Hopkins University Applied Physics Laboratory (JHU/ APL); p. 371 (bottom): NASA/JPL-Caltech/ UCLA/MPS/DLR/IDA; p. 375 (top left): Courtesy of Terry Acomb; (bottom left): Dr Robert McNaught; p. 377: NASA/NSSDC/ GSFC; p. 378 (top): NASA/JPL/Caltech; (bottom, both): NASA/JPL/Caltech/UMD; p. 379: NASA/JPL-Caltech/UMD; p. 381: Courtesy of the Wolbach Library, Harvard-Smithsonian Center for Astrophysics, Cambridge, MA; p. 383 (bottom): Barrie Rokeach; p. 384: Tony Hallas/Science Faction/ Corbis; p. 385 (top, all): Courtesy of Ron Greeley; (bottom): NASA/JPL/Cornell; p. 386: Courtesy of Joe Orman; p. 387: NASA; p. 393 (all): Larry Marschall and Christy Zuidema, Gettysburg College.

Chapter 13

PHOTOS: p. 394: Tyler Nordgren; p. 414: Bettmann/Corbis; p. 414: Norman Lockyer Observatory, Sidmouth.

Chapter 14

PHOTOS: p. 426: NASA/SDO/AIA; p. 440 (a): Brookhaven National Laboratory; (b): Lawrence Berkeley National Laboratory; (c): © CNRS Photothèque/IN2P3/ MONTANET François; p. 441 (right): NOAO/AURA/NSF; p. 443: Nigel Sharp, NOAO/NSO/Kitt Peak FTS/ AURA/NSF; p. 444: (top a): Hinode JAXA/ NASA; (top b): 2001 by Fred Espenak, courtesy of www.MrEclipse.com; (top c): 2001 by Fred Espenak, courtesy of www.MrEclipse. com; (bottom): NASA/Photo Researchers; p. 446 (both): NASA/SDO/Solar Dynamics Observatory; p. 447: SOHO/ESA/NASA; p. 448 (bottom): NASA; p. 449 (left): SOHO/ESA/ NASA; (right): NASA SDO; p. 450: SOHO/ESA and NASA; p. 451: JPL/NASA.

Chapter 15

PHOTOS: p. 458: ESA and the PACS, SPIRE & HSC consortia, F. Motte (AIM Saclay,CEA/ IRFU - CNRS/INSU - U. ParisDidedrot) for the HOBYS key programme; p. 461 (a): © Axel Mellinger; (b): NASA/JPL-Caltech/ IRAS/2MASS/COBE; p. 462: AP Photos; p. 465: NASA/JPL-Caltech/UCLA; p. 466: Masashi Kimura/MAXI/SSC; p. 467 (top left): Dr. Douglas Finkbeiner; (right a): Courtesy of Stefan Seip; (right b): Courtesy of Emmanuel Mallart www.astrophoto-mallart.com & www.axisinstruments.com; (right c): ESO; p. 468 (left): X-ray: NASA/CXC/PSU/L. Townsley et al.; Infrared: NASA/JPL/PSU/ L.Townsley et al.; (right): Image courtesy of NRAO/AUI; p. 469 (both): ESO; p. 470: ESA/ Planck Collaboration; p. 475: Copyright: Far-infrared: ESA/Herschel/PACS/SPIRE/Hill, Motte, HOBYS Key Programme Consortium; ESA/XMM-Newton/EPIC/XMM-Newton-SOC/Boulanger; optical: MPG/ESO; near-infrared: VLT/ISAAC/McCaughrean & Andersen/AIP/ESO; p. 478: NASA/IPAC/ R. Hurt (SSC); p. 481 (b): NASA/JPL-Caltech/ UIUC; (insert): Karl Stapelfeldt /JPL/ NASA; p. 482 (top a): NASA, ESA, and P. Hartigan (Rice University) (top b): NASA, ESA, and the Hubble Heritage Team (STScI/AURA); (bottom right): Anglo-Australian Observatory/ Royal Observatory, Edinburgh, photograph from UK Schmidt, plates by David Malin; p. 484 (top): X-ray: NASA/CXC/Penn State/ L.Townsley et al.; Optical: Pal Obs. DSS; (bottom): NASA/JPL-Caltech/T. Megeath (University of Toledo, Ohio).

Chapter 16

PHOTOS: p. 490: X-ray: NASA/CXC/SAO/J. Hughes et al., Optical: NASA/ESA/Hubble Heritage Team (STScI/AURA); p. 506 (Ant Nebula): NASA, ESA and The Hubble Heritage Team (STScI/AURA) Acknowledgment: R. Sahai (Jet Propulsion Lab) and B. Balick (University of Washington); (butterfly): NASA, ESA, and the Hubble SM4 ERO Team; (Cat's Eye): NASA, ESA, HEIC, and The Hubble Heritage Team (STScI/AURA); (helix): NASA/JPL-Caltech/Univ. of Arizona; (little ghost): NASA and The Hubble Heritage Team (STScI/AURA); (ring nebula): Minnesota Astronomical Society; p. 512: NASA/GSFC/D. Berry; p. 514 (a): NASA/CXC/Rutgers/K. Eriksen et al.; Optical: DSS; (b): X-ray: NASA/ CXC/SAO/D.Patnaude, Optical: DSS.

Chapter 17

PHOTOS: p. 520: NASA/ESA/JPL-Caltech/ UCLA/CXC/SAO; p. 531 (a): Illustration: NASA/CXC/M. Weiss; Chandra X-ray Observatory Center; (b): X-ray: NASA/CXC/ SAO; Optical: NASA/STScI; Infrared: NASA/ JPL-Caltech/Steward/O. Krause et al.; p. 532 (a-b): Anglo-Australian Observatory, photographs by David Malin; (c): ESA/Hubble & NASA; p. 533 (a): ESO; (b): X-ray: NASA/ CXC/GSFC/M. Corcoran et al.; Optical: NASA/ STScI; Society; p. 534 (a): Joachim Trumper, Max-Planck-Institut für extraterrestrische Physik; (b): NASA/JPL-Caltech; (c): Courtesy Nancy Levenson and colleagues, American Astronomical; (d): Jeff Hester, Arizona State University, and NASA; p. 540 (top, from left to right): X-ray: NASA/CXC/SAO/T. Temim et al. Chandra X-ray Observatory Center; X-ray: NASA/CXC/SAO/T. Temim et al. Infrared: NASA/JPL/Caltech; X-ray: NASA/CXC/ CfA/P. Slane et al.; (bottom): X-ray: NASA/ CXC/SAO/F. Seward; Optical: NASA/ESA/ ASU/J. Hester & A. Loll; Infrared: NASA/ JPL-Caltech/Univ. Minn./R.Gehrz; p. 541 (top left): ESA/Hubble & NASA; (top right): NASA & ESA; (bottom left): European Space Agency & NASA; (bottom right): NASA/CXC/Univ. de Liège/Y. Naze et al.; p. 542 (top right): NOAO/ AURA/NSF/Photo Researchers; (top left): The Electronic Universe Project; (bottom right): Barbara Mochejska Hedberg, J. Kaluzny (Copernicus Astronomical Center).

Chapter 18

PHOTOS: p. 550: ESA/Hubble & NASA; p. 573 (top, a-b): Optical: DSS; Illustration: NASA/ CXC/M.Weiss; (bottom): Illustration: NASA/ CXC/M.Weiss; p. 574 (a): CXC/M.Weiss; Inset: NASA/CXC/Cal; Chandra X-ray Observatory Center; (b): ESO/A. Roquette.

Chapter 19

PHOTOS: p. 580: NASA, ESA, G. Illingworth, D. Magee, and P. Oesch (University of California, Santa Cruz), R. Bouwens (Leiden University), and the HUDF09 Team; p. 582: Time Pix/Pars International; p. 583: NASA, ESA, and the Hubble Heritage Team (STScI/ AURA); p. 588: STScI/Nasa; p. 591 (a): BJ Fulton (LCOGT) / PTF/Byrne Observatory at Sedgwick Reserve; (b, both): NASA/Swift/ Peter Brown, Univ. of Utah; p. 601: Roger Ressmeyer/Corbis; p. 602 (a,b): Ann and Rob Simpson; p. 603 (a-c): NASA COBE Science Team; p. 609: Courtesy of Stacy Palen.

Chapter 20

PHOTOS: p. 610: NASA, ESA, and The Hubble Heritage Team (STScI/AURA); p. 612 (top left): NASA; (top right): Canada-France-Hawaii Telescope/J.-C. Cuillandre/ Coelum; (center, both): NASA; (bottom. both): NASA; p. 613: Gemini Observatory/AURA; p. 616 (left): NOAO/AURA/NSF; p. 617 (left); Anglo-Australian Observatory, David Malin; (right): NASA and The Hubble Heritage Team (STScI/AURA); p. 618 (a-b): European Southern Observatory (ESO); p. 619 (top: a-b): Johannes Schedler/Panther Observatory; (bottom a): NASA/JPL-Caltech; (bottom b): Credit & Copyright: Robert Gendler www.robgendlerastropics.com; p. 620 (a): Todd Boroson/NOAO/AURA/NSF; (b): Richard Rand, University of New Mexico; p. 622: image100/SuperStock; p. 623: Anglo-Australian Observatory, David Malin; p. 624: NASA/CXC/Penn State/G. Garmire; Optical: NASA/ESA/STScI/M. West; p. 627 (top, both): HST; (bottom): X-ray: NASA/CXC/ CfA/R. Kraft et al.; Radio: NSF/VLA/Univ. Hertfordshire/M. Hardcastle; Optical: ESO/ WFI/M.Rejkuba et al.; p. 632 (top): X-ray (NASA/CXC/KIPAC/N. Werner, E. Million et al); Radio NRAO/AUI/NSF/F. Owen; (bottom): Holland Ford, STScI/Johns Hopkins University; Richard Harms, Applied Research Corp.; Zlatan Tsvetanov, Arthur Davidsen, and Gerard Kriss at Johns Hopkins University; Ralph Bohlin and George Hartig at STScI; Linda Dressel and Ajay K. Kochhar at Applied Research Corp. in Landover, MD; and Bruce Margon from the University of Washington, Seattle/NASA; p. 633: F. Owen, NRAO, with J. Biretta, STScI, and J. Eilek, NMIMT; p. 634: NASA/Swift/NOAO/Michael Koss and Richard Mushotzky (Univ. Maryland); p. 639: NASA, ESA, and the Hubble Heritage Team (STScI/AURA).

Chapter 21

PHOTOS: p. 640: Tony Rowell / www. Astrophotostore.com; p. 642 (a): Axel Mellinger; (b): C. Howk and B. Savage (University of Wisconsin); N. Sharp (NOAO)/ WIYN, Inc.; p. 643 (top a): US Patent Office Library; (top b): © Lynette Cook, all rights reserved; (bottom left): NOAO/AURA/ NSF; (bottom right): Hubble Heritage Team (AURA/STScI/NASA); p. 644: NASA/JPL-Caltech; p. 646 (a): NASA, ESA, and A. Feild (STScI); p. 651: ESA; p. 653: NASA/CXC/M. Weiss; NASA/CXC/Ohio State/A. Gupta et al.; p. 655 (top a): NASA/CXC/MIT/F.K. Baganoff et al./E. Slawik; (top b): Hubble: NASA, ESA, and Q.D. Wang (University of Massachusetts, Amherst); Spitzer: NASA, Jet Propulsion Laboratory, and S. Stolovy (Spitzer Science Center/Caltech); (bottom): Keck/UCLA Galactic Center Group; p. 656 (top): NRAO/ AUI; (bottom): Credit: David A. Aguilar (CfA); p. 658 (top, both): Anglo-Australian Observatory, David Malin; (bottom): Amanda Smith, Institute of Astronomy, University of Cambridge; p. 659 (top): NASA, ESA, and A. Feild and R. van der Marel (STScI); (bottom):

NASA, ESA, Z. Levay and R. van der Marel (STScI), T. Hallas, and A. Mellinger.

Chapter 22

PHOTOS: p. 666: Courtesy of Brookhaven National Laboratory; p. 672: STScI/NASA; p. 673: Robert Kirshner, Harvard University; p. 684: From *The Elegant Universe* by Brian Greene. ©1999 by Brian R. Greene. Used by permission of W. W. Norton & Company, Inc.

Chapter 23

PHOTOS: p. 694: Robert Gendler, Jan-Erik Ovaldsen, Allan Hornstrup, IDA Image data: ESO/Danish 1.5m telescope at La Silla, Chile – 2008; p. 696: Credit: X-ray (NASA/CXC/CfA/E. O'Sullivan); Optical (Canada-France-Hawaii-Telescope/Coelum); p. 697 (left): Image courtesy of NRAO/AUI and Chung et al., Columbia University; (right, a): Max Tegmark/SDSS Collaboration; (right, b): M. Blanton and the Sloan Digital Sky Survey; p. 698 (bottom): IfA; p. 699 (a): NASA/CXC/MIT/E.-H Peng et al.; Optical: NASA/STScI; (b): NASA; ESA; L. Bradley (Johns Hopkins University); R. Bouwens (University of California, Santa Cruz); H. Ford (Johns Hopkins University); and G. Illingworth (University of California, Santa Cruz); (c): NASA/CXC/MIT/E.-H Peng et al.; NASA/STScI; p. 700 (b): NASA, A. Fruchter and the ERO team (STScI, ST-ECF) NASA, ESA; p. 705: Simulation: Matthew Turk, Tom Abel, Brian O'Shea Visualization: Matthew Turk, Samuel Skillman; p. 706 (top left): Gemini Observatory /AURA/Andrew Levan (Univ. of Warwick, UK); (top right): NASA /Swift /Stefan Immler; p. 707 (a, b): NASA/JPL-Caltech/GSFC; p. 708 (left): NASA, ESA, G. Illingworth (University of California, Santa Cruz), R. Bouwens (University of California, Santa Cruz, and Leiden University), and the HUDF09 Team; (right, both): Guedes, Javiera; Callegari, Simone; Madau, Piero; Mayer, Lucio *The Astrophysical Journal*, Volume 742, Issue 2, article id. 76 (2011); p. 709 (top): NASA, ESA, and N. Pirzkal (STScI/ESA); (bottom): NASA, ESA, Sloan Digital Sky Survey, R. Delgado-Serrano and F. Hammer (Observatoire de Paris); p. 710 (left): X-ray (NASA/CXC/SAO/P. Green et al.), Optical (Carnegie Obs./Magellan/W.Baade Telescope/J.S.Mulchaey et al.); (right): NASA, ESA, M. Trenti (University of Colorado, Boulder, and Institute of Astronomy, University of Cambridge, UK), L. Bradley (STScI), and the BoRG team; p. 712 (top): NASA, ESA, J. Lotz (STScI), M. Davis (University of California, Berkeley), and A. Koekemoer (STScI); (bottom): NASA, ESA, the Hubble Heritage (STScI/AURA)-ESA/Hubble Collaboration, and A. Evans (University of Virginia, Charlottesville/NRAO/Stony Brook University); p. 713 (top left): NASA/CXC/CfA/M. Markevitch et al.; and NASA/STScI; Magellan/U. Arizona/D. Clowe et al.; Lensing Map: NASA/STScI; ESO WFI; Magellan/U. Arizona/D. Clowe et al.; (top right): Credit: X-ray (NASA/CXC/IfA/C. Ma et al.); Optical (NASA/STScI/IfA/C. Ma et al.); (bottom): Courtesy Joel Primack and George Blumenthal; p. 714 (all): Courtesy Joel Primack and George Blumenthal; p. 715: NASA, ESA, and A. Feild (STScI). LINE ART: Figure 23.12: Figure from *Journal of Annual Reviews of Astronomy and Astrophysics* 49 (373–407), by Bromm V, Yoshida N. 2011. Reprinted with permission via the Copyright Clearance Center.

Chapter 24

PHOTOS: p. 722: Seth Shostak/SETI Institute; p. 725 (a): Dr. Michael Perfit, University of Florida, Robert Embley/NOAA; (b): Woods Hole Oceanographic Institution, Deep Submergence Operations Group, Dan Fornari; (bottom): Gaertner/Alamy; p. 726 (top left): Chris Boydell/Australian Picture Library/Corbis; (top right): National Park Service Photo by Jim Peaco; (bottom): Michael Abbey / Science Source/Photo Researchers; p. 728: Courtesy of National Science Foundation; p. 730 (skeleton): Stephen Frink Collection/Alamy; (dinosaur): Shutterstock; (fish): Shutterstock; (flower): AP Photo; p. 731: Kenneth Eward/Biografx/Photo Researchers; p. 732: Jeff J. Daly/Painet; p. 735 (left): NASA/JPL-Caltech/MSSS; (right): Planetary Habitability Laboratory @ UPR Arecibo (phl.upr.edu)/ NASA; p. 737: (bottom): The Habitable Exoplanets Catalog, Planetary Habitability Laboratory, University of Puerto Rico at Arecibo; p. 738 (right): NASA; p. 741: Twentieth Century Fox Film Corporation / Photofest; p. 743: Seti/NASA. LINE ART: p. 737: Figure 24.16: "Exoplanet Atmospheres" by Sara Seager, http://seagerexoplanets.mit.edu/images/transitschematic.gif Reprinted by permission of Sara Seager; Figure 24.17: "Earth Similarity Index (ESI)" from "The ESI and HZD Exoplanets Habitability Indices," Nov. 25, 2011. Reprinted by permission of The Planetary Habitability Laboratory, University of Puerto Rico at Arecibo.

Index

This index contains page references for all 24 chapters in the complete volume of *21st Century Astronomy*, Fourth Edition. The Solar System Edition does not include Chapters 15–23 (pp. 458–721). The Stars and Galaxies Edition does not include Chapters 8–12 (pp. 214–393).

Page numbers in *italics* refer to illustrations. An italic *n* after a number refers to a footnote.

aberration of starlight, 552–53, *553*
absolute magnitude, 401, *413* (table)
absolute zero, 139
absorption, *130,* 132, 466
absorption lines, *131,* 135, 139, 403–5, 622, 626, *626*
  defined, 132
  interstellar, 466
  as spectral fingerprints of atoms, 132
  in spectrum of Sun, 139, 441–43, *443*
  star classification and, 403–5, *404*
  star composition and, 403–5, 466
acceleration, 408
  defined, *77,* 78
  gravitational, 90, 94, 98, 103, 104, *104, 109,* 496, 504, 558, *558*
  miniature solar system model of atom and, 127
  Newton's second law of motion and, 78–80, *78,* 81
accretion disks:
  in binary system, 510, 536, *537*
  defined, 192
  formation of, *187, 188,* 192–98, *193,* 473, *473,* 510, 652
  around supermassive black holes, 629–30, *630*
  around white dwarfs, 510, 629
  see also protoplanetary disks; protostellar disks
achondrites, 385, *385*
active comets, 371, 372, 375–76, *375*
active galactic nuclei (AGNs), *626,* 627–34, 635
  normal galaxies and, 630–34
  perspective on, *630,* 632–33, *633*
  size of, 628–29, *628*
  unified model of, 629, *630,* 632–33
  see also quasars
Adams, Douglas, 667
Adams, Fred, 715, *716*
Adams, John Couch, 295
Adams Ring, 345, *345,* 349
adaptive optics, 160, *162*
Adrastea, 346
Aeschylus, 25
AGB, *see* asymptotic giant branch (AGB) star
AGNs, *see* active galactic nuclei
air pressure, Coriolis effect and, 273
albedo, 143, 279, 283
algebra, defined, 13, 15
Allen, Paul, 741
Allen Telescope Array (ATA), 169, *722,* 741
Alpha *Canis Majoris, see* Sirius
Alpha Centauri, 399, 592
Alpha Magnetic Spectrometer, 174
alpha particles, 501
Alpher, Ralph, 600–601
altitude:
  atmospheric pressure and, 304
  temperature and, 268, *269,* 272, 301, 304
Alvin, 309
Amalthea, 346
amino acids, 724, 731, 734
ammonia, 197, 334, 724
  giant planets and, 297, 304, 310
  terrestrial planets and, 260, 261, 264
ammonium hydrosulfide, 297, 304
Amors, 366
amplitude, of wave, 123, *123*
Andromeda Galaxy, *583, 619,* 657, 658, 696
Andromeda Nebula, 583, 589
angle of incidence, 156, *157*
angle of reflection, 156, *157*
angular momentum, 190–92, 198, 510, 570
  law of conservation of, 190–92, 195, 470–73, *473,* 536, 711
  of molecular clouds, 470–73, *473*
angular resolution, 167, 168
annular solar eclipse, 50, *50, 51, 53*
Antarctica, 352, 384
Antarctic Circle, 44
ANTARES experiment, 174
anthropic principle, 688–89
anticyclonic motion, 299, *300, 308*
antimatter, 680
  symmetry between matter and, 682, 684
antiparticles, 432, 571, *572,* 679, 680, *680*
  see also positrons
aperture, 153, 162, 167
aphelion, 364, 374
Apollinaire, Guillaume, 641
Apollo missions, 173, *173,* 217, *219,* 222, *224,* 232, *237*
  Apollo 10, 237
  Apollo 15, 7, 241
Apollos, 366, *367*
apparent daily motion:
  latitude and, 32–36, *33, 34, 35, 36*
  of stars, 27–36, *31, 32, 34, 35, 36,* 37–38, *41*
  of sun, 30, *38*
  view from the poles of, *30,* 31–32, *32*
apparent magnitude, 401
Aquarius, 38
Arab culture, 12
archaeans, 727
archaeology, sky patterns and, 26
arcsecond, 160
Arctic Circle, 44
Arecibo radio telescope, 167–68, *167, 168,* 739
argon, *133,* 440, 469
Ariel, *325,* 338
Ariel 1 satellite, 170
A Ring, 340, 342, *342,* 349
Aristotle, 27
Arizona, Meteor Crater in, 220, *220*
Armstrong, Neil, 215
Arrhenius, Svante, 282
Artemis, 46
Artemis mission, *175* (table)
arts, science compared with, 11
Asimov, Isaac, 645
A stars, 403, 541
asteroid belt, 17, 200, 369, 370
  Kirkwood gaps in, 365, *365*
  main, 324, 360, 364, 367, *367*
asteroids, 4, 7, 17, 171, 221, 295, 360, 364–65, 508
  close-up view of, 369–70, *369*
  comet nuclei compared with, 372
  C-type, 368, 369
  defined, 365
  density of, 367, 369, 382
  discovery of, 364
  as fractured rock, 367–69

asteroids (*continued*)
ice on, 368–69
impacts and, 220–21, 379–80, 386
key concepts about, 360
landslides on, 244
mass of, 367
meteorites related to, *368*, 384–86
M-type, 368
near-Earth, 366, *367*
orbital resonances and, 365–66, *365*, *366*
orbit of, 109, 365–66, *365*, *366*, 367, *367*, 380, 384
S-type, 368, 369, 370
Trojan, 366
astrobiology, 18, 734
astrology, 26, 38
astrometric method, 202
astronauts, *90*, 271
Apollo, 217, *219*, 222, *224*, 228, 232, *241*, 248
in free fall, 98, *98*
Newton's laws of motion and, 81, *81*
astronomical seeing, 160
astronomical unit (AU), 73, 118, 198, 199, 362, 371, 396, 445, 508, 553
astronomy:
amateur, *152*, *165*
defined, 4
overview of, 4–7
tools of, 202
see also specific topics
asymptotic giant branch (AGB) star, 503, *507*, 508, 533, 534, 541, 543
mass loss in, 504, *505*, 545
structure of, *503*, 504
Atacama Desert, 352
Atacama Large Millimeter/Submillimeter Array (ALMA), 169
Atens, 366
atmosphere, 737
escape of, 259–64
formation of, 198, 258–63, *259*
of giant planets, 198, 258, 292, 297, 299–308, *300*, *302*, *303*, *305*
of moons, 324, 328, 329–35, *334*
pressure and, 737
primary, 198, 199, 258, 263
of protostars, 479, 496–97
secondary, 198, 199, 259–66, 267
see also specific planets
atmosphere of Earth, 7–8, 217, 244, 258–66, *258*, 266–78, 279, 283, 387
atmospheric greenhouse effect and, 257, 264–66, 309
chemical composition of, 227–28, 261–66, *263* (table), 266–68, *267*
comets and, 371
distortion caused by, 160–61, *162*
electromagnetic radiation blocked by, 161, *167*, 466
impacts and, 220
layers of, 268–70, *269*
meteorites and, 384
primary, 199, 258
role of life in, 266–68, *267*
secondary, 199, 259–66, 267
solar activity and, 450
weather and, 272–73, *273*
wind and, 272–73, *273*
atmospheric distortion, 160–61, *162*
atmospheric escape, 261–62
atmospheric greenhouse effect, 257, 264–66, *264*, 265–66, 307, 332
on Venus, 264, 265, 274–75
atmospheric probes, 172–73
atmospheric windows, 166, 167, *167*
atomic bomb, impacts compared with, 220
atoms, 125–34, 260–61
Bohr model of, 127
decay of, 128–32, *129*, 133–34, *133*
emission and absorption determined by energy levels of, 128–32, *129*, *130*, *131*
empty space in, 496
energy state of, *see* energy states, atomic
ground state of, 128, *129*
in human body, 5, *7*
ideal gas law and, 260–61
interaction of light with, 125–34
isolated vs. molecular, 469
miniature solar system model of, 127, *127*
in neutrino detectors, 437
pressure of gas and, 260–61, *261*
in primary atmosphere, 258
in protoplanetary disk, 195
spectral fingerprints of, 117–45, *133*
thermal motions of, 138, *138*, 498–501
AU Microscopii, *201*
auroras, 272, *272*, 315, *315*, 451
autumn, 43
autumnal equinox, 43
axes, 16
axions, 702

Baade, Walter, 537
backlighting, 344, *344*, *345*, *351*
bacteria, 352, 727
"ball in the car" experiment, 554–55, *554*
Baltic Sea, 106
Barnard 68, *469*
barred spirals (SB), 613, *617*, 621
basalt, 227, 240
Bay of Fundy, tides in, 106, *107*
Bean, Alan, *90*
Bell, Jocelyn, 539
Bell Laboratories, 164, 167, 600, *601*
beryllium, 5, 18, 534, 535, 603, 686
beryllium-8 nucleus, 500, 501
Bessel, F. W., 399
beta decay, 432, 679
Beta Pictoris, *205*
Betelguese, *394*
Big Bang theory, 581, 593–604, *595*, *650*, 666, 683, 684, *685*, 686, *686*, 700, 701, 702, 705, *705*, 706, 707, 708, *708*, *710*, 716
abundance of the least massive elements predicted by, 604, *604*
center of, 598
confirmation of predictions of, 581
expanding universe, 668, *673*, 686
nucleosynthesis and, 603–4, *604*, *685*, 686
problems with, 675–78, *676*, *678*
radiation left over from, 178, 600–603, *601*, *602*, *603*
spacetime and, 671, 674–75, 684
"Big Crunch," 673, 714
Big Dipper (Ursa Major of the Great Bear), *26*, 49
Big Rip, 673, *674*
binary stars, 408–12
eclipsing, 409, 411, *411*, *412*
evolution of, 506, 508–13, *509*, *511*, 536
formation of, 483
mass transfer and, 508–13, *509*, *511*
neutron stars in, 536, *537*, 552, 629
orbits of, 408–9, *409*, *410*, 510, 568
Roche lobes of, 509–10, 536, *537*
total mass of, 411
visual, 409, *410*, 411
white dwarf, 506, 510–13, *511*, 536, *537*, 591
X-ray, 536, *537*, 571, 629
binding energy, 527, *527*, 528
biosphere, 217
bipolar outflow, 481, *481*
blackbody, 139, 178
blackbody spectrum, *see* Planck (blackbody) spectrum
Black Hole Era, 715, *716*
black holes, *537*, 545, 568–69, 622, 625, 626, 631, *634*, 635, 656, 705, 706, 707, 710, 715–16
defined, 553
event horizon of, 570–71, *572*, 599
in Milky Way Galaxy, 641, 654–57, *655*, 656
"seeing," 571–73
as singularity, 570, *570*
supermassive, 570–73, 611, 629–31, *630*, *632*, 654–57
trip into, 570–71, *571*
wind and, 571, 573
"blueberries," 246
blue light, 402
interstellar dust and, 462, *463*
of quasars, 626
temperature and, 139–40
blueshifted light, *134*, 135, *135*, *204*, 409, 602, 618
blue subdwarfs, 508
body, human:
as stardust, 7, *7*
temperature of, 137
thermal equilibrium of, 137
see also brain, human; eyes, human
Bohr, Niels, atomic model of, 127
Bolshoi simulation, 712, *713*
Boltzmann's constant (k), *see* Stefan-Boltzmann constant
bookcase, energy state of atoms as, 128, *129*, 133
BOOMERANG experiment, 179
boron, 5, 18, 534, 544, 716
in early universe, 603, 686
Borrelly, Comet, 377–78, *377*
BO supergiant, *573*
bound orbits, 101, *101*, 102
bowling ball, law of gravity and, 92

bow shock, 271
Bradley, James, 121, 553
Bragg, William Henry, 117
Brahe, Tycho, 65, 69–70, *69*, 70, 371
brain, human:
  color and, 402
  stereoscopic vision and, 396, *396*
brightness:
  of CCD images, 176
  defined, 141, 400
  distance and, *142*, 400, *400*
  luminosity vs., 141
  of stars, 163, 202, *202*, *203*, 396, 400–401, *400*
  surface, 163
B Ring, 340, 342, *342*, 351, *351*
brown dwarfs, 201, *201*, 478, *478*, 715
Bruno, Giordano, 185, 186
B stars, 403, 483, 492, 526–27, 541, 620, 642
bubble, bobbing of, electromagnetic waves compared to, 122, *122*
bubbles, atmospheric, 161, *162*
bulge, 613, 616, *617*, 618
Bullet Cluster of galaxies, 626, *713*

Cabeus Crater, 248
Caesar, Julius, 49–50
calcium, 368
calculus, 100
calendars:
  cultures and, 49–50
  lunar, 49
  lunisolar, 49
California, San Andreas Fault in, 236
Callisto, *325*, 338–39, *339*, 352
Caloris Basin, 241, *242*
Cambrian explosion, 728, 736
Canada, 106, *107*, 231
canali, 245
Cancer, 44
Canis Major, 26
Cannon, Annie Jump, 403, 419
cannonball, in Newton's thought experiment, 97–99, *97*, 103
Capricorn, 44
carbon, 5, 18, 132, 197, 297, 315, 385, *406* (table), 429, 535, 545, 603, 653, 706, 708, 731–34
  in comets, 379
  in interstellar medium, 460–61, 470
  in planetary nebulae, 545
  in stellar core, 500, *500*, 503, *503*, 505, 512
  in white dwarfs, 507, 512
carbonaceous chondrites, 385
carbonates, 264, 267
carbon burning:
  in high-mass stars, 523, *525*, 527, 528, *528* (table), 529, *529*
  in white dwarfs, 512
carbon dioxide, 246, 267, 279, *280*, 282–83, 376, 724, 737
  in atmosphere, 198, 261–66, *263* (table), *264*, 265, *266* (table), 267, 274, 277
  greenhouse effect and, 264–66, *264*, 265–66, 274
carbon monoxide, 197, 260, 374, 470, 620, *620*
carbon-nitrogen-oxygen (CNO) cycle, 522–23, *522*, 533, 535
carbon stars, 545
car speeds:
  acceleration and, 77–78, *77*
  direction of rainfall and, 552–53, *552*
  frame of reference and, 552, 554–55
  relative motions and, 28, 552, 554–55
Carter, Jimmy, 738
Cassegrain telescope, 154
Cassini, Jean-Dominique, 339
Cassini Division, 339, 340, *342*, 343, 347
Cassini–Huygens missions, *175* (table)
Cassini missions, *290*, *293*, 295, 300, *301*, 307, 315, *322*, *326*, 330, *330*, *331*, 334, *334*, 336, *338*, *342*, 735
Cassiopeia A, *531*
Catalina Observatory, *165*
catalysts, 268, 522
Catholics, 50
Cat's Eye Nebula, *506*
cause-and-effect relationships vs. quantum probabilities, 134
celestial equator, *31*, *33*, *34*, 35
  defined, 30
  seasons and, 39–43
  wobbling of, 45
celestial poles, 30–36, *31*, *33*, *34*
celestial sphere, 27
  as useful fiction, 30, *31*, *36*
Centaurus A (radio galaxy), *627*
center of mass, 408, *408*, *409*
centripetal force, 99, *99*
Cepheid variables, 525, *525*, *526*, 533, *583*, 591, 592, 643, 645–46
Ceres, 17, *74* (table), *186*, 200, 360, 362–64, *365*, 370, *370*
CERN (European Organization for Nuclear Research), *174*
Chandrasekhar, Subrahmanyan, 512
Chandrasekhar limit, 512, *512*, 552, 571, 591
Chandra X-Ray Observatory, *167*, 170, *170* (table), *468*
Chandra X-ray telescope, *540*, *710*, *713*
Chandrayaan-1, 248
Chandrayaan mission, *175* (table)
Chang'e orbiters, *175* (table)
chaos, chaotic behavior, *333*
  orbits and, 326
chaotic system, 326
charge-coupled devices (CCD), 164, *165*, 176
Charon, 17, 109, *325*, 326, *326*, 361, *361*
chemical reactions, self-sustaining, 527
chemosynthesis, 352
Chile, 352
China:
  astronomy in, 537, *540*
  culture of, 12
China National Space Administration (CNSA), 171
chlorofluorocarbons (CFCs), 265
CHON (carbon, hydrogen, oxygen, nitrogen), 731
chondrites, 385, *385*
chondrules, 384
chromatic aberration, 154, 156, *157*
chromosphere, 441, *444*, 447
Churyumov/Gerasimenko, *377*
circle, defined, 72
circular orbits, 39, 72, 97–102
  Newton's prediction of, 97–99, *97*, *99*
  of stars, 616
circular velocity, 99, 100
circumpolar portion of sky, 35, *35*, 44
circumstellar disk, *201*
classical mechanics, *see* Newtonian (classical) mechanics
classical planets, 74, *75*, *143*, 171, 216
Clementine mission, 248
climate, 267, 268, 307–8
  defined, 278
  latitudinal changes in, 40
  modification of, 265, 278–79
  seasonal change in, 38–40, 43
climate change, planetary, 278–83
clocks, 118
  cratering, 224
  in Einstein's thought experiment, 557, *557*
  relativity theory and, 557, *557*, 558–59, 568, *569*
  see also time
closed universe, 675, *675*
clouds:
  of giant planets, 297, 299–304, *302*, *303*, 305
  interstellar, *see* interstellar clouds; molecular clouds
  see also intercloud gas
CMB, *see* cosmic microwave background radiation
CNO cycle, *see* carbon-nitrogen-oxygen (CNO) cycle
cobalt, 731
COBE, *see* Cosmic Background Explorer
coins, compared to galaxies, 612, *612*
cold, cooling, 137–39, 272
  defined, 138
  in evolution of planetary interiors, 228–31
  subjective experience of, 138
  of universe, 684–86, *685*, 686
cold dark matter, 702
colliders, high-energy, 174, *174*
color index, 401
Columbus, Christopher, 27
coma, *375*, 376
Coma cluster, 696
Comet Churyumov-Gerasimenko, 370
comets, 4, 7, 249, 293–95, *358*, 369, 371–79, 742
  abode of, 371–73
  active, 371, 372, 375–76, *375*
  atmosphere and, 198, 371, 379
  coma of, *375*, 376
  composition of, 197, 374, 387
  head of, *375*, 376
  impacts and, 249, 260, *380*, *381*
  Jupiter family of, 378
  key concepts about, 360
  Kuiper Belt, 260, 360, *372*, 373–74, 379, 387

comets (*continued*)
long-period, 742
mass of, 377
meteor showers and, 382–84
nuclei of, 200, 260, 360, 371, 375, *375*, 381, 382, 387
Oort cloud, 260, 371–73, *372*, 373, 375, 387
orbits of, 260, 373–75, *373*, 376, 380, 382, 383
planetary rings and, 346
shattered by tides, 109, 352
short-period, 373–74, *373*, 378
superstitions about, 371
tail of, 312, 374, 375, *375*, 376, *376*
see also specific comets
comparative planetology, 217
composite volcano, 240
compound lens, 157, *157*
computers, computer models, 176, *178*
of galaxy–galaxy interactions, 631
in simulating structure of universe, 712–14
concave mirrors, 156, *157*
conduction, 231
cones, 162
connections boxes:
aberration of starlight, 552
Andromeda and the Milky Way, 659
backlighting phenomenon, 344
brown dwarfs, 478
convection, 274–75
Copernicus, 66–67
determining the ages of rocks, 222
direction of light, 156–57
energy, 196
equilibrium, 136–37
forever in a day, 729
gas, 260–61
gravitation, 96, 110, 565
interference and diffraction, 166
Kirkwood gaps, 365–66
life, the universe, and everything, 732
magnitude system, 401
neutrino astronomy, 440
nightfall, 645
parallels, between galaxy and star formation, 711
proton-proton chain, 432–33
relative motions, 28–29
superstring theory, 684
synchrotron radiation, 314
unified model of AGN, 632–33
variations by stellar mass, 533
what happens to the planets, 508
when redshift exceeds 1, 599
conservation laws, 190
for angular momentum, 190–92, 195, 470–73, *473*, 536, 711
for energy, 190, 195, 196
see also motion, Newton's laws of, first
constants of proportionality, 79
constellations, 26, 38, *38*
in human mind, 26, *26*
see also specific constellations
constructive interference, 166
contamination, planetary samples and, 174
continental drift, 233–34
continuous radiation, 139
convection, 231, 238
atmosphere and, 264–65, 268, *270*, 274, 304, *305*, 306, *308*
in energy transport, *434*, 435, 481
plate tectonics and, 234–35, *235*
in stars, *434*, 435, 523, *523*, 528, 545
thunderstorms and, 274, *274*, 306
Convection Rotation and Planetary Transits (COROT) space telescope, *170* (table)
convective zone, *434*, 435, 441, *442*
Copernicus, Nicolaus, 27, 65, *65*, 66–67, 76, 103, 186
Sun-centered theory of, 65, 103, 324
copper, 545, 731
Cordelia, *349*
core, 198
of Earth, 5, *226*, *227*, 230, 232, 239, 309
of Ganymede, 324, *325*
of giant planets, 199, 309–10, *309*
of other terrestrial planets, 232–33, 309
core, stellar, 493, 494–97, *497*, 498–502, *498–500*
collapse of, 523, 529–31, *530*
electron-degenerate, *see* electron-degenerate core
of high-mass stars, 523, 527, 529–31, 535–37
neutron-degenerate, *see* neutron stars
of Sun, *429*, 430–37, *438*, 496
core accretion, 199
Coriolis effect, 29, *29*, *273*
on giant planets, 304–8, *305*, *308*
wind patterns and, 272–73, *273*, 304–8, *305*, *308*
corona, 441, *442*, *443–45*, *450*
coronal holes, 444, *445*
coronal mass ejections, 449, *450*
Cosmic Background Explorer (COBE), 178, *178*, 601–3, *603*, 701
cosmic microwave background radiation (CMB), 178, *178*, 179, 581, 600–603, *601*, *602*, 675, 701
COBE map of, 601–2, *603*
COBE measurement of, 601–2, 603, *603*
excessive smoothness of, 676–77
horizon problem and, 677–78
structure of universe and, 701–2
cosmic rays, 278, 558, *559*, 574, 653
cosmological constant, *669*, *670–75*, 671, 689
identical to dark energy, 673
cosmological principle, 9, 405, 584, 669, 687
Copernicus and, 103
homogeneity and, 584, *585*, 586–87, *589*
isotropy and, 584, *585*, 586–87
Newton and, 97, 102–3
as testable theory, 9, 97
cosmological redshift, 599
cosmology, 584–604
Big Bang, *see* Big Bang theory
defined, 668
Hubble as pioneer of, 584–91
particle physics vs., 683
see also expanding universe; structure of the universe; universe
Crab Nebula, 537–40, *540*, 627
Crane, Stephen, 581
Crater, 602
craters:
on asteroids, 369
impact, *see* impact cratering, impact craters
on Mars, *221*, *222*, 246, *246*, *247*
secondary, 219, *220*
creationism, 732
crescent moon, 47
Cretaceous period, 248
Cretaceous–Tertiary boundary (K–T boundary), 248–49
C Ring (Crepe Ring), 340–41, 342, 347
critical density, 669
critical mass density, 668, 676
crust, 309
of Earth, *226*, *227*, 230, 236, 239
cryovolcanism, 329–30, *330*, 334
C-type asteroids, 368, 369
culture:
calendars and, 49–50
myth and legends, 26
science and, 12–13
winter festivals, 43, *43*
Curiosity Mars rover, 172, *173*
Curtis, H. D., 583
cyanobacteria, 267, 726
cyclonic motion, 273, 299
Cygnus, 571
Cygnus Loop, *534*
Cygnus X-1, 571, 572, *573*
cynobacteria, *726*

Dactyl, 369, *369*
Dante Alighieri, 257
Dark Ages, 705, 716
dark energy, 4, 597, 673, *674*
identical to cosmological constant, 673, *674*
Dark Era, 716, *716*
dark matter, 4, 597, 604, 611, 622–26, *624*, 670, 698–99, 701–14, *705*, 709, 715, 717
classes of, 702–5
collapse of, 702–14
defined, 623
in galaxy formation, 625, 701–14, *702*, *713*
in Milky Way, 641, 646–48, *647*, 658, 702
searching for signs of, *699*
dark matter halo, 623–24, *624*, 702, *714*
of Milky Way, 625, 626, 641, 702
daughter products, 222, *223*
Dawn mission, 364
Dawn orbiter, *175* (table)
Dawn spacecraft, 370, *370*
day, 272
length of, 40, 108, 297–98
Day the Earth Stood Still, The (1951), *741*
Dead Sea, 352
decay, atomic, 128–32, *129*, *133*, 330, 334
spontaneous, 133
Deep Impact/EPOXI, *175* (table)
Deep Impact mission, 378, *378*
Deep Space 1 mission, 377
Defense Department, U.S., 248

Degenerate Era, 715, *716*
Deimos, *325*, *352*
Delta Cephei, 525
deoxyribonucleic acid (DNA), 352, 514724, 726–28, 731, 734
deserts, 265
de Sitter, Willem, 671, *673*
destructive interference, 166
deuterium, 432, 603, 604
Dialogo sopra i due massimi sistomi del mondo (Dialogue Concerning the Two Chief World Systems) (Galileo), 76
Diana, 46
Dicke, Robert, 600
differential rotation, 446
differentiation, 227, 309, 368, 384
  of asteroids, 368, *368*, 384
  in gas giants, 309
  of meteorites, *227*, 384
diffraction, 160, *160*, 161, 166
diffraction grating, 165, 166, *166*
diffraction limit, 160, 161
diffuse nebulae, 582
diffuse rings, 342
dinosaurs, extinction of, 248, *249*, 387, 728
Dione, *325*, 337
direct imaging, 202–5
Discorsi e dimostrazioni mathematiche intorno a due nuove scienze attenenti alla meccanica ("Discourses and Mathematical Demonstrations Relating to Two New Sciences") (Galileo), 76
disk instability, 200
dispersion, 156–57, *157*, 165
distance:
  brightness and, 141–42, *142*, 400–401, *400*
  from Earth to stars, 396–401, *397*, *398*, 415–16, *419* (table)
  expressed as units of time, 118–19
distance ladder, 590–92, *590*
distortion, atmospheric, 160–61
Doppler, Christian, 135
Doppler effect, 134–35, 599, 618, 622
  AGNs and, 627, 630, 632
  relativistic beaming, 633
Doppler shift, *134*, 201, 204, *204*, 405, 468, 537, 571, 630–31, 699
  of binary stars, 409–10, *410*, *412*
  defined, 135
  gravitational redshift compared with, 568
  helioseismology and, 438, *441*
  structure of Milky Way and, *647*
  see also blueshifted light; redshifted light
Draconids, *383* (table)
Drake, Frank, 738, 740
Drake equation, 738–40, *740*
Draper, John W., 163, *163*
D Ring, 341–42
dust:
  in comets, 374, 376, *376*
  in diffuse nebulae, 582
  in spiral galaxies, 611, 616, *617*, 620, *620*, 622
  zodiacal, 386–87, *386*
dust, interstellar, 7, *184*, *188*, 460–62, *461*, *463*, 473
  light blocked by, 460–62, *461*, *463*
dust devils, 275
dust storms, on Mars, 276, 278
dust tail, *375*, *376*, *376*
dwarf galaxies, 618, *619*, 625, 653, *706*, 709
  ultrafaint, 708
  see also Large Magellanic Cloud; Small Magellanic Cloud
dwarf planets, 4, 17, 72, 74, *75*, *186*, 200, 324, *325*, 360–64, *361* (table), *362*, *365*
dynamic equilibrium, 136, 230
dynamo:
  magnetic fields of planets as, 447
  in protostar, 481–82
  in Sun, 447
dynamo theory, 232

Eagle Nebula, 474, *475*
Earth, 4, *5*, 207, *249*, 374, 379–80, 386
  atmosphere of, *see* atmosphere of Earth
  axis of, 279
  carbon dioxide concentrations on, *283*
  as center of universe, 12, 27, 64, 65, *65*
  climate change on, 278, 279–83
  core of, 5, *226*, 227, 230, 232, 239, 309
  cosmic background radiation and, 601–3, *603*
  craters on, 217, 220, *220*, 222
  crust of, *226*, 227, 230, 236
  density of, 224–27
  eclipses and, 50–54, *52*, *54*, 55, *55*
  effects of solar activity on, 428, 444, 445, 450
  erosion on, 217, *218*, 220, 222, 243–44
  evolution of life on, 55–56, 265, 267, 387, 660, 724–31
  extinctions on, 248–49, *249*, 387
  formation of, 198, 199, 200
  fossil record of, 232, 233, 248–49
  giant planets as viewed from, 293, 295, *296*, 299
  giant planets compared with, 292, 299, 305, 306, 308–9, 311–12, 316
  global average temperature of, *283*
  gravity of, *92*, 94–95, *96*, 97–98, 99–102, 103–8, 187, 224, 227–28, 509, 558, *558*, 561, 562
  impacts in shaping of, 217–22, 725
  interior of, 108, 224–32, *226*, *231*, 438
  life on, 483, 514, 660, 742–43
  location of, 265, 266
  magnetic field of, 231–32, *231*, 233–34, 270–71, *271*, 311–12, 451
  magnetosphere of, 270–71, *271*, *272*, 312, 445, 451
  mantle of, 5, *226*, 227, 234–35, *235*, 239, 309
  mass of, 17, 92, 94, 96, *96*, 103, 108, 199, 224–25, 263, 265, 266, 295
  measurement of distance to stars from, 396–401, *397*, *398*, 415–16, *419* (table)
  Moon's orbit around, 45–49, *46*, *47*, *48*, 51, 54, 55, 103–4, 108–9, *108*, *109*, 200, 346
  night sky as seen from, 642, *642*
  obliquity of, 55–56
  oceans of, *see* oceans of Earth
  origins of, 283–84
  ozone layer of, 660
  radiation belts of, 271, 314
  satellite images of, 159
  as seen from space, *217*
  size of, 5, *201*, 224, 230, 292, *299*
  as spherically symmetric, 96
  Sun and, 508
  tectonism on, 217, 222, 233–39, *233*, *234*, *235*, *236*, *237*, 279, 328
  temperature cycles on, 279, 280, *280*
  temperature of, 137, 143, *197*, 228–30, 264, 265, 267
  thermal equilibrium of, 137
  tidal locking of Moon to, 108–9, *108*
  tides and, 103–11, *104–9*, 230
  Venus compared with, 237–38, 265–66, 274–76
  volcanism on, 217, *218*, 222, 232, 239–42, 263, 724
  water on, 283–84
  winds on, 244, 272–73
  wobbling axis of, 45
Earth, orbit of, 15, 27, 28, 37–45, 55, 82, 99, 104, 202, 263, 280, 382, 386, 508, *553*, 565
  aberration of starlight in measurement of, 552–53, *553*
  eccentricity of, 72
  as ellipse, 70–72
  Kepler's laws and, 70–73
  seasons and, 38–44
  speed of light and, 118
  tilt of axis and, 38–40, *39*, *41*
Earth, rotation of, 27, 28, *280*
  change of seasons and, 38–44, 56
  Coriolis effect and, *273*
  latitude and, 32–36, *33*
  magnetic field and, *231*, 232
  tides and, 104–9, *106*
  view from the poles and, 27–36, *30*, *32*
earthquakes, 217, 225–27, 438
  plate tectonics and, 236, *237*
Earth Similarity Index (ESI), 737, *737*
eccentricity, of orbit, 72, *72*, 101, *101*, *409*
eccentric orbits, 364
eclipses, 50–55
  stellar, *see* stellar occultations
  Stonehenge and, 50, *50*
  varieties of, 50–54; *see also* lunar eclipses; solar eclipse
eclipse seasons, 55
eclipsing binary, 409, 411, *411*, *412*
ecliptic, *31*, 38, *38*, 40, 44, 386
ecliptic plane, 39
Eddington, Sir Arthur, 567–68, *567*
82 G. Eridani (star), 560, 561
Einstein, Albert, 3, 121, 430, 536, 551, 555, 561, 588, 672
  gravitation law of, 669
  particle theory of light and, 125, 126
  quantum mechanics and, 12, 134
  scientific revolutions and, 11–12, *12*, 134
  see also general relativity theory; special relativity theory

Einstein ring, *550,* 568
ejecta, 219, 221, *221*
electric field, 119–20, *120,* 679, *680*
 in electromagnetic waves, *120,* 122, *123*
 neon signs and, 130–31
electric force, 119, 555
electricity:
 lightning and, 274
 potential, 274
electromagnetic force, 119–20, 679–80, 682, 684, *685*
electromagnetic radiation, 119, 164, 174, 314–15, 627, 701
 see also light; spectrum
electromagnetic waves, 275
 defined, 120
 energy of photons and, 126, 129–30
 gravity waves compared with, 568
 light as, 119–25, *122,* 125–26, 555, 556
electron-degenerate core, 496, 523
 atomic nuclei in, 498–500, *500*
 carbon, 503, *503*
 of post-AGB stars, *505,* 506
 of red giants, 497, *497,* 498–502, *498, 499, 500, 503,* 510–11, 529
 of white dwarfs, 506, 552, 591
electron microscopes, 161
electron-positron pair annihilation, 680, *680, 682, 685*
electrons, 132, *165,* 270, 307, 312, 315, 432, 433, 467–68, 537, 630, 716
 antiparticle of, *see* positrons
 in atomic structure, 127, *127,* 128, 496, 679
 "clouds" or "waves" of, 127, *127,* 128
 deflected by electric field from proton, *680*
 in early universe, 680, *682,* 686
 magnetized, 467–68, *468*
 of neutron stars, 537
 radiation blocked by, 601, *602*
electroweak theory, 679, 682
elementary particles, 432, 702
elements, chemical, 5–7, 18
 isotopes of, 222, 432, 510–11, 527, 529, 604, *604,* 701
 least massive, Big Bang theory and, 603–4, *604,* 668
 massive, *see* massive (heavy) elements
 origin of, 5–7, *7,* 18, 506
 parent, 222
 periodic table of, *298,* 535
 relative abundance of, 535, *535*
 solar abundance and, 297
 see also specific elements
ellipses:
 defined, 70
 eccentricity of, 70–72
 focus of, 70–72, *72,* 103
 semimajor axis of, *72, 72*
elliptical galaxies, 582–84, 613, *619,* 634
 bulge in, *614* (table), 631
 classification of, 612–19, *613, 614* (table), 615, *616*
 color of, 617
 dark matter in, 623–25
 disk of, 616
 formation of, 711–12
 globular clusters compared with, 643–45
 mass of, 623–25
 radio, *see* radio galaxies
 stellar orbits in, 614, 616, *616,* 648
elliptical nebulae, 582
elliptical orbits, 15, 73, 99–102, *101,* 613
 of binary stars, 408–9, *409*
 as bound orbits, 102
 of comets, 373
 eccentricity of, 72, *72,* 101, *101, 409*
 focus of, 15, 72, *72*
 Kepler's first law and, 70–72, *72*
 Newton's prediction of, 97, 100, 567
 tides and, *109*
El Niño, 283
emission, 128–34, *129,* 130, *130,* 469
 defined, 130
 stimulated, 133
emission lines, 128–34, *130, 133,* 135, 139, 405, 537, 568, 618, 626, *626,* 627
 of AGNs, *626,* 627, 630
 defined, 130
 interstellar, *467,* 469
 molecular, 469
 as spectral fingerprints of atoms, 132, *133*
 star classification and, 403, *404*
 Type Ia supernova and, 512
empirical rules:
 Kepler's discovery of, *see* Kepler's laws
 physical laws vs., 76
empirical science, 69, 103
Enceladus, 315, 328, 330, *330, 331,* 342, 347, 352, 362, 735
Encke, Comet, *383* (table)
Encke Division, *342,* 349, *349*
energy, 12
 equivalence between mass and, 430, 558, 680
 grand unified theory and, 683
 kinetic, *see* kinetic energy
 law of conservation of, 190, 195, 196
 light in transport of, 122, 126
 potential, *see* potential energy
 primordial, 308
 relativity theory and, 558
 thermal, *see* thermal energy
energy states, atomic, *129*
 emission and absorption determined by, 128–32
 excited, 128–32, 133–34, 468
energy transport, 433
Ensisheim, meteorite fall at, 382
entropy, 716, 732
epicycle, *64,* 65
Epimetheus, 347
EPOXI spacecraft, 378, *379*
Epsilon Ring, 344, 348, 349, *349*
equator, 30, 33
 celestial, *see* celestial equator
 of giant planets, *292* (table), 297–98, 305
 seasons near, 44
 wind patterns and, 272–73
equilibrium:
 basic properties of, 136–37
 dynamic, 230
 hydrostatic, *227,* 266, 428, 436, 470
 Lagrangian points and, 110, *110,* 442
 temperature, 143, *143, 145,* 307, *463,* 465
 thermal, 137–38
equinoxes:
 autumnal, 43
 precession of, 45
 vernal, 40, 49, *396*
equivalence principle, 562, *562, 563*
Eratosthenes, 95
Eridanus, *610*
E Ring, 330, *331,* 342, 346, 347
Eris, 17, *74* (table), 200, 324, 360, 362, *362, 372*
 orbit of, 364
Eros, 369, *370*
erosion, *218,* 240
 defined, 217
 in shaping terrestrial planets, 218, 220, 222, 243–47, *244, 245, 246*
 by water, 217, 243–47, 265
 by wind, 217, 243–47, *244, 245*
escape velocity, 102, 258
 black holes and, 552, 570
 defined, 102
Eta Aquariids, *383* (table)
Eta Carinae, *7,* 533, *533,* 574
eternal inflation, 687, 688
ethane, 334, 336
eukaryotes, 727, *727*
Europa, *325,* 332–36, *333,* 735, *735,* 737
European Extremely Large Telescope (E-ELT), *155,* 158
European GALLEX experiment, 440
European Organization for Nuclear Research (CERN), 683
European Southern Observatory (ESO), 158, 169, *169*
European Space Agency (ESA), 171, 202, 238, *242,* 246, 442, 603
event, in relativity theory, 555–56
event horizon, 570–71, *572,* 599
evolution:
 biological, 514
 on Earth, 660
 of life, 265, 267, 352, 387, 452, 492, 574, 688–89
 stellar, *see* high-mass stars, evolution of; low-mass stars, evolution of; stellar evolution
 of universe, *685,* 686, 696; *see also* universe, chemical evolution of
evolutionary track, 480, *480*
excited states of atoms, 128–32, 133–34, 468
expanding universe, 581–605, 684–86, 696, 700–701
 acceleration of, 670–75, *670*
 fate of, 581, 668–75, *669–75,* 714–16
 gravity and, 668, *669,* 671, 700
 Hubble's law and, 584–97, *589, 593,* 594, *594, 595, 596*
 inflation period of, 677–78, *678,* 684, *685,* 700

key concepts about, 581
recombination in, 601, *602, 685*, 686, 701–2
scale factor of, 597–98, 599, 668, *669*, 671, *671, 674*
shape of, 582, 674–78, *675, 676*, 700
see also Big Bang theory
exponential growth, 730
extinction of species, 248–49, *249*, 387, 452
extrasolar planets, 200–201, 316, 737
estimating size of, 206
extremophiles, 352, 726
eyepiece, 152, 153, 158
eyes, human, *153*, 162
color and, 402
limitations of, 161, 166
stereoscopic vision of, 396, *396, 397*, 402

Fahrenheit (F), 138
Fairbanks, Alaska, 36
faults, 233
Fermi, Enrico, 740
Fermi Gamma-Ray Space Telescope, *170* (table), 656, *656*
festivals, winter, 43, *43*
filters, 402
fire truck siren, Doppler effect and, 134
first quarter Moon, 48, 103, 107, *107*
fissures, 239
flatness problem, 676
flat rotation curves, 623, *624*, 646–48, *647*
flat universe, 675, *675, 676*
flux, 140
flux tube, *315*, 316, *316*, 443
prominences as, 448
flyby, 171
focal length, 153, *153*, 156, *157*, 160
focus, of ellipse, 15, 70, *70*, *72*, 103
focus, focal plane, 153, *153*, 156, *157*
fold, 233
forces, 679–80, *679*
centripetal, 99, *99*
electric, 119, 555
electromagnetic, 119–20, 679–80, 682, 684, *685*
electroweak theory and, 679, 682
grand unified theory and, 682–86, *685*
gravitational, *see* gravity
magnetic, 119, 555
in Newton's third law, 80, *80*, 81, *81*
strong nuclear, 430, 501, 522, 531, 679, 680, 682, 684, *685*, 689
unbalanced, *see* unbalanced force
weak nuclear, 679, 682, 684, *685*
fossil fuels, 282
fossil record, 232, 233, 248–49, 514
frame of reference, 28, *28*
accelerated, 558–59, 563, *563*
change in, 558–59
direction of rainfall and, 552–53, *552*
equivalence principle and, *562*, 563, *563*
inertial, *see* inertial frame of reference
motion of object and, 28–29, *41*, 554–55
relativity theory and, 556–59, 561–63, *562, 563*
stationary vs high velocity, 561–63, *562*
free fall, 98, *98*
frequency, of wave, 123, 126, 129, 134
energy of photon and, 126, 129
for sound, 134
Friedmann, Alexander, 584, 669
F Ring, 348–49, *348, 349*
F stars, 541
full Moon, 46, 48, 54–55, *55*
tides and, 103, 107, *107*
future, of expanding universe, 581, 668–75, *669–75*, 714–16

Gaia, *170* (table)
"galactic fountain" model of disk of spiral galaxy, *651*, 652
Galápagos Islands, 352
Galatea, 345
galaxies, 544, *550*
appearance of, 612–18, *612*
centers of, 573, 611, 626–34, *627*, 642, *647*
classification of, 582, 612–18, *612, 613, 614* (table), *618, 619*; *see also* elliptical galaxies; spiral galaxies
clusters of, 103, 657, 696, *697*, 698, 699, *699*, 711
coins compared to, 612, *612*
dark matter in, 698–99
defined, 110, 582
discovery of, 582–84
dwarf, 618, *619*, 625, 653; *see also* Large Magellanic Cloud; Small Magellanic Cloud
elliptical, *see* elliptical galaxies
evolution of, 708–14, 717
expansion of space and, 593–95, *596, 597*
formation of first, 706–8, *707*
giant, 618, *629*
gravity of, 590, 614, 620, 646, 652, 698
habitability in, 634–35
intergalactic space compared with, 463
irregular, *613*, 614, *614* (table), 617–18
key concepts about, 611
Local Group of, 4, 696
luminosity of, 618
mass of, 103, 611, 618, *618*, 622–26, *623, 624*, 670, 699
mergers and interactions of, *111, 178*, 590, 611, 620, 626, 631, 634, 650, *712*
peculiar velocities of, 590
recession velocity of, 585–86, 592, *593*
redshift of, 585, 590, 592, 598–99, *598, 599*, 670, *670*, 696, *697*
S0 (S-zero), 613–14, *613, 614* (table), 618
size of, 618, *618, 619*
spiral, *see* spiral galaxies
starburst, 618
in structure of universe, 696–98, *697*
superclusters of, 4, 657, 696, *697*, 700
supermassive black holes in, 573, 611, 629–31, *630, 632*, 654–57
tidal interactions and, *111*, 631, *712*
ultrafaint dwarf, 708
see also Milky Way Galaxy
Galaxy Evolution Explorer (GALEX), *170* (table)
galaxy formation, 648–49, 657–60, *685*, 696, 703
dark matter in, 625, 701–14, *702*
galaxy groups, 657
Galaxy Redshift Survey, *697*
Galilean moons, *see* Jupiter, moons of
Galileo Galilei, 63, 65, *75*, 151, 152, 293, 324
gravity and, 90, *91*, 92, 94–95
law of inertia of, 76
motion of objects as viewed by, 75–76, 98
speed of light and, 118
telescope of, *75*, 152, *152*
Galileo missions, 328, 329, *339*
Galileo probe, *175* (table), 295, 304, *329*, 332, 369, *369*, 379
moons and, *328*, 329, *329, 333*, 336
planetary rings and, *345*
Galle, Johann, 293
gallium, 440
Galveston, Tex., 275
gamma-ray bursts (GRBs), 573–74, *574*, 660, 705, 706, *706*, 707
gamma rays, 124, *124*, 170, 278, 448, *500*, 529
photons of, 432, 680, *680*
Gamow, George, 600
Ganymede, 324, *325*, 336–37, *337*
gas, interstellar, *184, 188*, 460, 463–73, 507, 520, 649, 706
ionization of, 466–67
jets from protostars and, 481, *482*
in Milky Way, 648–52, *651, 654*
temperatures and densities of, 463–70, *467*
see also intercloud gas; interstellar clouds
gases:
in comet, 374, 376
definition and properties of, 260–61
in elliptical galaxies, 616–17, 624, *624*
in galaxy clusters, 698, *699*
inert, 297
in mass of galaxies, 611, 622, *623*, 624, *624*
photons absorbed by, *131*, 132
photons emitted by, 130–31, *130*
in spiral galaxies, 616–17, 620, *620*, 621–22
see also specific gases
gas giants, 292, *292*
Gaspra, 369, *369*
gauss, 312
Gauss, Carl Friedrich, 312
Gemini, 43, 44
Geminids, *383* (table)
Gemini telescopes, *155* (table)
general relativistic time dilation, 568
general relativity theory, 11–12, 561–69, *562*, 675
confrontation between quantum mechanics and, 683
cosmological constant and, 671
equivalence principle and, 562, *562, 563*
expansion of space and, 595–96
frame of reference and, 561–63, *562, 563*
as geometrical theory, 563–65, *564*
MACHOs and, 625
Newton's law of gravity and, 565, 566, 567
Newton's laws of motion and, 565

general relativity theory (*continued*)
observable consequences of, 565–69
redshift of galaxies and, 598–99
structure of universe and, 568, 569, 683
Geneva, Switzerland, *8*
geocentric model of the universe, *27*, 64–65
see also Earth, as center of universe
geodesic, 562–63, *564*
geology, geologists, 217–48
history of Solar System as viewed by, 187
model of Earth constructed by, 227
see also erosion; impact cratering, impact craters; tectonism; volcanism
geometry, 93
defined, 13, 15
of spacetime, 563–65, *564*
George III, King of England, 295
geothermal vents, *725*
germanium, 440
Giacobini-Zinner, Comet, *377*, *383* (table)
giant dust ring, *343*
giant galaxies, 618, *619*
Giant Magellan Telescope (GMT), *155*, 158
giant molecular clouds, 470
giant planets, 198, 199, 292–316
atmosphere of, 198, 258, 259, *275*, 292, 297, 299–308, *300*, *302*, *303*, *305*
cometary orbits disrupted by, 260
defined, 292
extrasolar, 202–5
formation of, 316
gas vs. ice, 292, *293*, 310
gravity on, 347
interiors of, 309–10, *309*
as magnetic powerhouses, 311–16, *311*, *313*, *314*, *316*
migration of, 316–17, *317*
rings around, *see* rings, planetary; *specific rings and planets*
terrestrial planets compared with, 292, 295–99, 305, 306, 307–8, 309
see also Jupiter; Neptune; Saturn; Uranus
gibbous, *47*, 48
Giotto spacecraft, 377, *377*
global circulation, 273, 308
Global Oscillation Network Group (GONG), 438
Global Positioning System (GPS), 568
global warming, 265
globular clusters, 543, 583, *617*, 625, 641, 643–46, *643*, *652*
age of, 541, 645, 657, 674
distribution of, 646
massive elements in, 544–45, 650
open clusters compared with, 540, *541*
gluons, 680
Gold, Thomas, 539
golf ball analogy, for interstellar medium, 460
gossamer rings, 346
Grand Canyon, 222, *222*, 237
grand unified theory (GUT), 682–86, *685*
granite, 227
Gran Telescopio Canarias, *155* (table)
granules, *447*
graphs, reading, 16
grating, diffraction, 165, 166, *166*
gravitational constant, universal (G), 79, 93, 94, 688, 689
gravitational lensing, 202, *550*, 567–68, *567*
in search for dark matter, 625, *625*, 698, *713*
gravitational potential energy, 196–97, 308
gravitational redshift, 568, *569*, 599, 603
gravitational wave detectors, 174
gravitational waves, 176, 568, 569–70
gravity, 89–111, 679, *685*
acceleration and, 90, 94, 98, 103, 104, *104*, 108, *109*, 496, 504, 558, *558*
atmosphere and, 198, 258–59
in binary system, 509–11, 512
of black hole, 552, 570, 571, *571*
centripetal force and, 99, *99*
as distortion of spacetime, 561–69
Einstein's law on, 669
electron-degenerate core and, 496, 501
equivalence between effect of inertia and, 95, 562, *563*
expansion of the universe and, 668, *669*, 671, 700
Galileo's insight and observation about, 90, 92, 94–95
general relativity and, 565, 683
grand unified theory and, 683
of high-mass stars, 522, 526, 529
of horizontal branch stars, 503
as increasing outward from object's center, *96*
of interstellar cloud, 187, 470–73, *471*, 473, 620
as inverse square law, 93–95, *95*, 436, 471, 476
of molecular cloud, 473, *474*
on neutron stars, 536
orbital resonance and, 110, 365–66, *366*
place-to-place differences in, 96
planetary rings and, 340, 346, 348–49
of planetesimals, 194–95, *194*, 198
of planets, *see specific planets*
in protoplanetary disk, 198
of protostars, 188, 200, 473–78, *476*, 493, 630
protostellar disk and, 188
of Sun, 92–93, 103, 107, 365–66, 372, 428, *428*, 436, 443, 493, 496
symmetries and, 96, 682
tectonism and, 233
tides and, 103, 104–9, *104*, *111*
gravity, Newton's law of, 90–102, *93*, 176, 224, 260, 295, 316, 374, 504, 565, 566, 567, 671, 688
Newton's laws of motion applied to, 90–92, 94–95
playing with, 94–95
reasoning toward, 90–93
testing of, 97
Gravity Recovery and Interior Laboratory (GRAIL), *175* (table)
Great Attractor, 697, *698*
great circle, 38, 40
Great Dark Spot (GDS), 301, *302*, 306
Great Debate, 583–84
Great Red Spot (GRS), 299–300, *299*, *302*, 306–7
anticyclonic motion of, *299*, *300*
Greeks, ancient, 27, 46, 64, 76, 382
Green Bank radio telescope, *167*
greenhouse effect, 264–66, 282, 283, 284, 737
atmospheric, *see* atmospheric greenhouse effect
ozone hole vs., 267–68
warming trend and, 265, 274–75
greenhouse gases, 279, 282–83, *282*
greenhouse molecules, 265
Gregorian calendar, 49
Gregory XIII, Pope, 49–50
G Ring, 342, 345
ground state of atom, 128–31
Gusev crater, 246
GUT, *see* grand unified theory
Guth, Alan, 677

habitable zones, 207, *207*, 419–20, *420*, 508, 660, 735–38, 742
Hadley cells, 272
Hadley circulation, 272, *273*, 276
hail, 274
Hale, George Ellery, 447
Hale-Bopp, Comet, 374, 375, *375*
half-life, 223, *223*
Halley, Comet, 374, 377, *377*, *383* (table)
Halley, Edmund, 374
halo, 645
halogens, 268
Halo Ring, 346
Harmony of the Worlds (Kepler's third law), 74, *74* (table), *75*, 100, 103, 295, 655
Hartley 2, Comet, *377*, 378–79, *379*
Harvard Observatory, 403, 582
Harvard-Smithsonian Center for Astrophysics, *697*
Haumea, 360, 362
Hawaii, 370
Hawaiian Islands, 238
volcanism in, *218*, 240, *240*, 242, *243*
Hawking, Stephen, 571
Hawking radiation, 571, *572*, 716
Hayabusa mission, *175* (table), 370
Hayashi, Chushiro, 479
Hayashi track, 480, *480*, 481, 483, 492, 497, *498*
HD226868 (star), 571
head (comet), defined, 376
heat, heating, 137–40
blue light and, 139–40
defined, 138
in evolution of planetary interiors, 228–31
luminosity and, 139
subjective experience of, 138
tidal, 230
Wien's law and, 140–41
heat death, entropy and, 716
heavy elements, *see* massive (heavy) elements
Heisenberg, Werner, 127, *128*
Heisenberg uncertainty principle, 127, *128*
heliocentric model of the universe, 65, 66
see also sun, as center of universe

helioseismology, 438–41, *441*
heliosphere, 451, *451*
helium, 6, 18, 132, *133*, 537, 545, 634, 653, 705, 731
  atmosphere formation and, 198, 199, 258
  birth of universe and, 534, 603, 604, 686
  CNO cycle and, *522*, 523
  in giant planets, 198, 199, 292, 297, 309–10
  in interstellar medium, 460
  liquid, 308
  in nuclear fusion, 429–32, *433*, 483, 492–96
  stellar build-up of, 496, 498, 523
  in stellar composition, 405, *406* (table), 441, 443, *495*, 496
helium burning:
  in high-mass stars, 523, *525*, 527–28, *528* (table), 529, *529*, 534
  in horizontal branch star, 501–2
  in intermediate-mass stars, 524
  in low-mass stars, 498–506, *500*, *502*
  triple-alpha process and, 500, *500*, 504–5, 527, 535, 603
helium flash, 501–2, *502*, *507*, 511, 533
helium shell burning, 501, *503*, 510, 523, 533
hematite, 246
hemglobin, 545
Herbig–Haro (HH) objects, 481, *482*, 630
heredity, 729
Herman, Robert, 600
Herschel, Caroline, 364, 582
Herschel, Sir William, 17, 293–95, 337, 364, 582, 611, 734
  model of Milky Way proposed by, 642, *643*
Herschel Crater, 337, *338*
Herschel infrared space telescope, 330
Herschel Space Observatory, *170* (table), 379, *458*, 483, *484*, 708
hertz, 123
Hertz, Heinrich, 123
Hertzsprung, Einar, 412, 419
Hertzsprung–Russell (H–R) diagram, 395, 412–15, *413*, *544*
  age of stars and, 540–44, *542*, *543*, *544*
  of globular clusters, 645
  instability strip of, 524, *524*
  main sequence and, 415–19, *415*, *416*, *417*, *417* (table), 478, 480, 510, 541–43, *542*, *543*, 645
  non–main sequence and, 417–18, 497, *498*
  post–main sequence stellar evolution and, 497, *498*, 502–5, *503*, *505*, 506, *507*, 510, 523, 524, *524*, 533, 542–43
  of pre–main sequence stars, 480, *480*
  of star clusters, 540–44, *542*, *543*, *544*
  star formation and, 478–80, *480*
Hewish, Anthony, 539
Hickson Compact Group, *613*
hierarchical clustering, 700
Higgs boson, 683
high-energy colliders, 174, *174*
high-mass stars, evolution of, 421–546
  CNO cycle in, 522–23, *522*, 535
  core collapse in, 523, 529–31, *530*
  evolution of intermediate–mass stars compared with, 533
  evolution of low–mass stars compared with, 522–24, 525, 526–27, 529, 541, 542–43, 544
  final days in life of, 527–29
  H–R diagram and, 523, *524*, 533, 540–43, *542*, *543*
  instability strip and, 524, *524*
  main sequence, 522–23, *522*, *523*, 526–27, 540–43
  mass loss in, 526–27, *533*
  neutrino cooling in, 529
  post–main sequence, 523–31, *524*, *527*, *530*, 533, *533*
  stages of nuclear burning in, 522–24, *525*, 527–31, *528* (table), 731
  see also black holes
high–mass stars, in spiral galaxies, 622
high–velocity stars, 652, 653, 657
Hi'iaka, 362
Hipparchus, 401
Hipparcos mission, 399, 415, *415*
Hobby–Eberly telescope, *155* (table)
Holmes, Comet, 376
Homestake neutrino detector, 440, *440*
homogeneity, of distribution of galaxies, 584, *585*, 586–87
horizon, 30–36
horizon problem, 677, 678
horizontal branch stars, 502, *502*, 504, *507*, 523, 543
  instability strip and, 524, 645
Horsehead Nebula, *467*
hot dark matter, 702
  see also neutrinos
"hot Jupiters," 205
hot spots, 238–40, 242
H–R diagram, *see* Hertzsprung–Russell (H–R) diagram
HST, *see* Hubble Space Telescope
H II regions, 466–67, *468*, *475*, 505, 620, 649
Hubble, Edwin P., 583–84, 588, 599, 657, 670
  Cepheid variables discovered by, 643
  classification of galaxies by, 613, *613*, *614* (table), 615, 618
  on cover of *Time*, *582*
  Great Debate and, 583
Hubble constant, 586, 587, 589–92, 674
Hubble flow, 590
Hubble's law, 581, 584–97, *596*, 698
  discovery of, 585–86, *589*
  structure of the universe and, 593, *594*, *595*, 657
  universe mapped in space and time by, 581, 592–93
Hubble Space Telescope (HST), 160, 161, 164, 170, *170* (table), 171, *187*, 193, *278*, *293*, 340, *342*, 401, 450–51, *475*, 591, 592, *627*, 628, 670, 708, *708*, *710*, 711–12
  AGNs and, 631, 632, *632*
  deep image of "blank" sky made by, *580*, 582
  giant planets and, 295, *299*, *302*, *315*
  HH objects and, 481, *482*
  moons and, 361, *361*, 362
  Shoemaker–Levy 9 and, 379
  stellar evolution and, *506*, *534*, *540*
Hubble time, 674
  origin of, 593–95, *594*
humidity, relative, 269
hurricanes, 275, *275*
  Coriolis effect and, 273, 274, 275, 306
Huygens, Christiaan, 339, 340
Huygens probe, 334–35, *336*
Hyakutake, Comet, 380
hydroelectricity, 196
hydrogen, 5, 18, 132, 198, 545, 634, 689, 705, 706, 731
  in atmosphere of terrestrial planets, 199, 258, 272
  energy states, of, 132, *133*, 467–68
  in giant planets, 198, 199, 292, 297, 309–10, *309*
  in interstellar medium, 460, 466–67, *467*, *468*, 483
  molecular, 468–70
  in protostellar and stellar atmospheres, 479, 496–97
  quasars and, 626, *626*
  radiation and, 601
  spectral lines of, 132, *133*, 622, 626, *626*
  spiral galaxies and, 618, 620, *620*, 646, *647*, *651*, 652
  star classification and, 403
  in stellar composition, 405, *406* (table), 492–97, *495*
hydrogen burning, 501–2, 505, 508, 510–11, *511*, 706
  CNO cycle and, 522, *522*, 535
  in high-mass stars, 522–24, *522*, 523–24, *525*, 527–28, *529*, 533, 535
  in Sun, 430–32, *433*, 483, 492–93
hydrogen cyanide, 352, 378
hydrogen shell burning, 496–98, *497*, 498, *498*, *499*, 503, *503*, 512, 523, 533
hydrogen sulfide, 297
hydrosphere, 217
hydrostatic equilibrium, 227, 266, 428, 436, 470
hydrostatic jump, *622*
hydrothermal vents, 352
hyperbolic orbits, 102
Hyperion, 326, *326*
hypernovae, 573
hypothesis, defined, 9

Iapetus, *337*, *338*
IBEX satellite, 452
ice, *197*, 198, 266, 274, 384
  in comets, 198, 372, 374, 387
  dry, 376
  on dwarf planets, 362, 364
  in giant planets, 197, 292, *293*, 304, 310
  on Moon, 248
  in moons, 324–25, *325*, 327, 328, 330, *336*, 337, 338, 346
  in terrestrial planets, 247, *247*
ice ages, 278, 279, 282
IceCube neutrino observatory, 174–76, *176*, *440*
ice giants, 292, *293*
Iceland, *234*, 239
ice skater, angular momentum of, 190–91, *190*

Ida, 369
ideal gas law, 260–61
igneous activity, 217, 239–42
  see also volcanism
Ikeya–Seki, Comet, 374, 377
impact cratering, impact craters, 217–22, *218*
  comets and, 249, 378, 379–80, *381*
  defined, 217, 219
  on Earth, 217, *218*, 220, *220*, 222, 236, 379–80
  formation of features in, 219–20, *220*
  on Moon, 217, *219*, 220, *221*, 222, 224, 241, 247
  on moons, 328, 330, 332, 336, *337*
  in shaping terrestrial planets, 198–99, *199*, 200, 217–22, *219*, *220*, *221*, *237*, *249*
  stages in formation of, *219*
incidence, angle of, 156, *157*
index of refraction, 156
Indian Space Research Organization (ISRO), 171
inert gases, 297
inertia, 76–80, 98, 558
  defined, 76
  equivalence between effect of gravity and, 95, 562, *563*
  Galileo's law of, 76
  measure of, *see* mass
  Newton's first law of motion and, 77, 562
inertial frame of reference, 77, *77*, 554, 555, 596
  preferred, 602
  special relativity and, 561, *563*
inflation model of universe, 677–78, *678*, 684, *685*, 700
Infrared Astronomical Satellite (IRAS), 462
infrared (IR) radiation, 124, 166, 171, 268, 387
  atmospheric greenhouse effect and, *264*, 274
  interstellar dust and, 461–62, *461*, *463*
  of protostar, 473–74
  star formation and, 473, *475*
infrared telescope, *169*, 368
Inquisition, 76
instability strip, 524, *524*, 645
integration time, 163
intelligent design, 732
intensity of light, 126, *126*
intercloud gas, 463–68, *466*, 469, 470
interference, 166, *166*
interferometer, interferometric array, 168
intergalactic space, galaxies compared with, 463
intermediate-mass stars, 492, 533
International Astronomical Union (IAU), 17, 201, 360, 362
International Space Station (ISS), *88*, 170, 626
interstellar clouds, *184*, 387, 451, 532, *532*, *534*, 706
  collapse of, 187–93, *192*, 195, 470–73, *472*, *473*, 536, 540, 620
  defined, 463
  gravity of, 187, 470–73, *471*, 473, 620
  in origins of solar system, 7, 187–93, *188*
  size of, 192
  in star formation, 187–93, *192*, 195, 459, 470–73, *471–75*, 620
  see also molecular clouds
interstellar extinction, 461, *461*
interstellar gas, 484, 520, 706
interstellar medium, 445, *445*, *451*, 452, 532, *534*, 544, 545
  components of, *465*
  dust in, *see* dust, interstellar
  gas in, *see* gas, interstellar; intercloud gas; interstellar clouds
  of Milky Way, 649, *651*, 652, 653
  star formation and, 187–93, 460–84, 649–50, *651*
inverse square law:
  defined, 93
  distance of star determined with, 643
  gravity as, 93–95, *95*, 436, 471, 476
  radiation as, *142*, 407, 415, *419* (table)
  stellar luminosity and, 407, 415, *419* (table)
Io, 315, *325*
  flux tube and, *315*, 316, *316*, 328
  plasma torus associated with, 315, *315*, 316, *316*, 328
  volcanism on, 328–29, *328*, *329*, 346
ionization, 274, 405, 466
ionosphere, 270, 272
ion tail, *375*, 376, *376*
IR, *see* infrared (IR) radiation
iridium, 249
iron, 197, 309, 368, 370, 378, 460–61, 512, 545, 660, 708, 731, 737
  chemical vs. nuclear burning of, *527*
  on Earth, 5
  in high-mass stars, 527, 529–31, 534, 574
  in Jupiter, 297
  magnetic field and, *231*, 232
  in planetesimals, 368
  in Sun, *406* (table)
  in terrestrial planets, 227, 309, 386
iron meteorites, 368, 384, *385*
irregular galaxies, *613*, 614, *614* (table), 617–18
irregular moons, 327
Irwin, James B., *7*
"island universes," 582
isotopes, 222, 432, 510–11, 527, 529, 701
  in Big Bang, 604, *604*
isotropy:
  of cosmic background radiation, 602
  of distribution of galaxies, 584, *585*, 586–87
Itokawa, 370

James Webb Space Telescope, *170* (table), 171, 451, 700
jansky, 167
Jansky, Karl, 167
Janus, 347
Japan Aerospace Exploration Agency (JAXA), 171
Jefferson, Thomas, 89, 382
jet, 377
jet stream, 300, *300*, 305
joule, 140
Juno, 17
Juno orbiter, *175* (table)
Jupiter, 7, 64, 111, 171, 292–315, *293*, *299*, 316, 386, 508
  atmosphere of, 109, 198, 275, 297, 299–304, *300*, *302*, 379
  auroras of, *315*, 316
  chemical composition of, 292, 297, 304, 309–10
  colorful bands of, 299, 304
  comet crash on, 109, 379, *380*
  density of, *292* (table), 297, 309–10
  differentiation in, 309
  formation of, 199, 297, 310
  gravitational influence of, 279
  gravity of, 198, 199, 366, *366*
  Great Red Spot (GRS) on, 299–300, *300*, *302*, 306–7
  interior of, 309–10, *309*
  magnetic field of, 311–12, *311*, 316, 346
  magnetosphere of, 312–16, *313*
  mass of, 198, 199, 201, *205*, *292* (table), 295, 297, 308, 310, 316
  migration of, 316, 317
  moons of, 111, 118, *119*, 152, 324, *325*, *327*, 328–36, *328–31*, *335*, 338–39, *339*, 346; *see also* Europa; Io
  orbital resonance and, 365–66, *365*, *366*
  orbit of, 73, 201–2, *202*, 292, *292* (table), 316, *317*, 365–66, *366*, 374, 386
  radiation belts of, 314–15
  rings of, 340, *341*, 345, *345*, 346, 348
  Roche limit of, 109
  rotation of, *292* (table), 298
  size of, *201*, *292*, 295, *296*, 310
  temperature of, *197*, 200, *293*, 307
  winds on, 304–5, *305*, 306–7

Kant, Immanuel, 187, 582
Katrina, Hurricane, 275, *275*
Kayuga orbiter, *175* (table), 248
Keck telescopes, *150*, *155* (table), 158, *158*, 160, 670
Keeler Gap, 351
kelvin (K), 139, 195, 197
Kelvin scale, 139
Kepler, Johannes, 65, 70–74, *70*, 92–93, 103
Kepler-22, 207, *207*
Kepler exoplanets, 634, 635
Kepler Mission, *203*, 207
Kepler orbital spacecraft, 202
Kepler's laws, 70–74, 82, 97–103, 179, 192, 204, 206, 260, 316, 324, 361, 364, 622, 623, 624, 648
  first, 70–72, *72*
  mass of binary stars and, 408, *419* (table)
  moons and, 327
  Newton's explanation of, 97–99, 101, 103, 118, 327, 655
  planetary rings and, 348
  second (Law of Equal Areas), 73, *73*, 100–101, 346–47
  third (Harmony of the Worlds), 74, *74* (table), *75*, 100, 103, 295, 655
Kepler space observatory, *203*
Kepler space telescope, 409

Kepler's supernova, 514
Kepler telescope, *170* (table), 207, *207*
Kilmer, Joyce, 427
kinetic energy, 196, 219, 381, 681
  of atmospheric atoms and molecules, 258–59, 260
  of supernovae, 531–32
Kirkwood, Daniel, 365
Kirkwood gaps, 365, *365*
Kitt Peak, *358*
knowledge, 9–11, 684
  as power, 26
  scientific revolution and, 12, 555
  scientific theory and, 9–10
  sky patterns and, 26
Krakatoa, 239
Kuhn, Thomas, 11
Kuiper, Gerard, 371
Kuiper Airborne Observatory (KAO), 169
Kuiper Belt, 260, 360, 372–73, *372*, 373–74, 379, 387
Kuiper Belt objects (KBOs), 197, 360, 362

laboratories, physics, 437
Lagrange, Joseph, 110
Lagrangian equilibrium points, 110, 442
Lamba Ring, 345
Lambda-CDM model, 701, *708*
Landau, Lev, 695
landers, 172–73
landslides, 244
Laplace, Pierre-Simon, 187
Large Binocular Telescope, *155* (table)
Large Hadron Collider, *8*, 174, *174*, 626, 702
Large Magellanic Cloud (LMC), 440, 467, 532, *532*, *653*, *658*
  orbit of, 625, *647*
large-scale structure, *685*, *697*
  defined, 697
Large Synoptic Survey Telescope (LSST), 158
Laser Interferometer Gravitational-Wave Observatory (LIGO), 176, *176*, 177, 569
latitude, 158
  apparent daily motion and, 32–36, *33*, *34*, *35*, *36*
  climate and, 40
Laughlin, Gregory, 715, *716*
Law of Equal Areas (Kepler's second law), 73, *73*, 100–101
lead, 223
leap years, 49
Leavitt, Henrietta, 525, 584, 645
legends, 26
Leibniz, Gottfried, 76
lenses, 152, *157*, *163*
  compound, 157, *157*
  convex, 157
  objective, 160
Leo, *38*
Leonids, 383, *383*, *383* (table), *384*
Le Verrier, Urbain-Jean-Joseph, 295
life, 483–84, 634–35
  chemistry of, 731–34
  defined, 724
  on Earth, 483, 514, 660, 742–43
  Earth's atmosphere reshaped by, 266–68, *267*
  evolution of, 267, 352, 387, 452, 492, 575, 688–89, 724–31, *725*, 732
  extraterrestrial, 278, 324
  fate of, 742–43
  key concepts about, 724
  origins of, 17–18, 110–11
  patterns and, 13, 26
  solar wind and, 451–52
  in universe, 688–89
light, 117–45
  astronomical importance of, 118
  brightness of, 141–42, *142*
  diffraction of, *160*, 166
  dispersion of, 156–57, *157*, 165
  Doppler effect and, 134–35, *134*
  as electromagnetic wave, 119–25, *122*, 125–26, 555, 556
  energy transported by, 118, 122, 126, *126*
  heat and, 139–40
  intensity of, 126, *126*
  interactions of atoms with, 125–34
  interference and, *166*
  particle theory of, 119, 125–35, 142
  of pulsars, 537, *538*
  quantized, 126; *see also* photons
  reflection of, 156, *157*
  refraction of, 156, *156*
  spectrum of, *see* spectrum
  visible, 124, 126, 132, *133*, 165
  wavelengths of, *see* wavelengths of light
  wave/particle nature of, 125, 127, 160
light, speed of, *120*, 156, 437, 552, 555
  AGNs and, 627, 633
  Doppler effect and, 135
  electromagnetic waves and, 120, *123*
  Galileo's failure and, 118
  measurement of, 118–19, *119*, 555
  Newtonian physics and, *554*, 555
  photons and, 126
  relativity theory and, 555–59, *557*, 561
  Rømer's work on, 118, *119*
  size of universe and, 4–5, *6*
  twin paradox and, 559, 560, 561
  as ultimate speed limit, 558, *558*, 741
lightbulbs:
  brightness of, 141–42
  luminosity of, 141–42
  spectrum produced by, 131–32, *131*, 139
lightning, 274, 307
light-years, 4, 398, 416, 440, 470
  defined, 119, 191, 398
limb darkening, 441–42, *442*
limestone, 265–66
Lincoln, Abraham, 491
Linde, Andrei, 687
linearity, 164
line of nodes, 55, *55*
Lippershey, Hans, 152
liquid:
  in core of giant planets, 309–10, *309*
  defined, 260
LISA mission, 177
lithium, 5, 535, 603, 604, 686, 705, 716
lithosphere, 217, 227, 234–40
lithospheric plates, 234–40
Local Group, 4, 657, *657*, 696, 697, 708
local midnight, 31
Loeb, Avi, 715
logarithm, 16
Loihi Volcano, 240, *240*
longitudinal wave, defined, *225*
long-period comets, 373–74, *373*, 374
look-back time, 593, 599
Lorentz, Hendrik, 560
Lorentz factor, 560, *560*
low earth orbit (LEO), 170
Lowell, Percival, 245, 731
Lowell Observatory, 584
low-mass stars, 625, 731
  defined, 492
  in spiral galaxies, 622
low-mass stars, evolution of:
  in binary systems, 507, 508–13, *509*, *511*
  evolution of high-mass stars compared with, 522–24, 525, 526–27, 529, 541, 542–43, 544
  evolution of intermediate-mass stars compared with, 533
  H-R diagram and, 496–98, *498*, 502–5, *503*, *505*, 506, 510, 524
  hydrogen shell burning in, 496–98, *497*, 498, *498*, *499*, 503, *503*, 512
  last evolutionary stages of, 502–8, *503*, *505*, *506*, *507*
  main sequence, *495*, *495* (table), 507, *507*, 512, 540–43, *542*, *543*
  mass loss in, 502–7, *505*, *506*
  post-main sequence, 494–508, *507*, *508*, 525; *see also specific stages*
  see also red giants, red giant branch; white dwarfs
luminosity, 139–40
  of active galactic nuclei, 627
  brightness vs., 141, 401
  defined, 139
  of galaxies, 618, *618*, *619*
  of globular clusters, 645–46
  intrinsic peak, 512
  of protostars, 197, 473, 479
  Stefan's law and, 140
  of Type Ia supernovae, 591, *591*
luminosity, stellar, 708
  of asymptotic giant branch stars, 502–4, 505
  distance determined by, 643
  electron-degenerate core and, 496–98
  after helium flash, 501
  of high-mass stars, 522, 523, 524–25, 544
  H-R diagram and, *413*, 415–19
  of main sequence stars, 415–18, *417*, *417* (table), *418*, 493, *495* (table)
  mass and, 412, 493, 544
  measurement of, 395, 400–401, *400*, *419* (table)
  of red giants, 497, *499*, 501
  size of stars and, 406

luminosity, stellar (*continued*)
of Sun, 400–401, 428, 436, 493, 497, 504, 512
of white dwarf, 506
luminosity class, 418
luminosity–temperature–radius relationship, 406, 407, 415, *419* (table), 524–25
luminous blue variable (LBV) stars, 533, *533*
luminous (normal) matter, 622, 623–25, 648, 651, 670, 709
density of, 603–4, *604*
in galaxy formation, 626
structure of the universe and, *702*
Luna 3 mission, 171
lunar calendars, 49
Lunar Crater Observation and Sensing Satellite (LCROSS), *175* (table), 248
lunar crescent, 49
lunar cycles, 49
lunar eclipses, 54–55, *54*
penumbral, 54
total, 54
Lunar Prospector, 248
Lunar Reconnaissance Orbiter (LRO), *175* (table), *214*, 241, 248
lunar rocks, 248
lunisolar calendars, 49
Lyell, Sir Charles, 323
Lyrids, *383* (table), *640*

M13 (star cluster), 738
M55 (star cluster), 541, *542*, 543
M80 (globular cluster), *643*
M87 (radio galaxy), 632–33, *632*, *633*
M104 (galaxy), *617*
MACHOs (massive compact halo objects), 625, *625*, 698
McNaught, Comet, 374–75, *375*, 376, *377*
Magellan, Ferdinand, *658*
Magellan mission, 238, *239*, 276
Magellan telescopes, *155* (table)
magma, 217, 239–40, *240*
magnesium, 227, *406* (table), 507, 545
in high-mass stars, 523, 527
magnetic bottle, 271, *271*
magnetic field, 119, 679
in electromagnetic waves, *120*, 122
in Milky Way, 653–54, *654*
of molecular clouds, 471, 540
motion of charged particles in, 270–71, *271*
of neutron stars, *538*, 540
of planets, 231–33, *311*, 312–16, 447; *see also specific planets*
of protostar, 481
synchrotron radiation and, 312–14, *314*, 627
magnetic force, 119
magnetism, 119, 555
magnetometers, 232
magnetosphere, 232
of Crab Nebula, 538–39
of Earth, 270–71, *271*, *272*, 312, 445, 451
of giant planets, 312–16, *313*
of neutron stars, 536–37, 538–39
magnification, 153
magnitudes, 400
magnitude system of brightness, 401
main-sequence lifetime, 492, 493, *495* (table)
main sequence of stars, 483, 508
binary stars on, 509–10
defined, 415, 492
evolution of, *495*, *495* (table), 507, *507*, 512, 522–23, *522*, *523*, *525*, 541–44, *542*, *543*
H–R diagram and, 415–19, *415*, *417*, *417* (table), 478, 480, 510, 541–42, *542*, *543*, 645
life of, 492–94
properties of, 415–19, *417*, *417* (table), *418*
see also Sun
main sequence turnoff, 542–44, *543*, *544*, 645
Makemake, 360
mantle, 234–40
of Earth, 5, *226*, *227*, 234–35, *235*, 239, 309
marbles, random motions of, 195, *195*
marching band, AGNs compared with, 628–29, *628*
Mare Ibrium, *241*
maria, 240
Mariana Trench, 236
Mars, 7, 64, *68*, 82, 172, 221, 227, *228*, 249, 352, 508, 734–35, 737, 742
atmosphere of, 199, 217, 244, 258, *258*, 261–65, *263* (table), *269*, 272, 276–78, *277*
axial tilt of, *279*
climate change on, 278, 279
craters on, 221, *221*, *242*, *245*, 246, *246*, *247*
dust on, *256*
dust storms on, 276, *277*, 278, *278*
erosion on, *244*, *245*, 246, *246*
formation of, 199, 200
gravity of, 264
iron meteorite on, *385*
Kepler's second law and, 73
life on, 278
magnetic field of, 233, 451
mass of, 263, 316, 317
meteorites compared with, 386
moons of, 279, 324, *324*, *325*, 352
orbit of, 277–78, 279, 366, 376–77
samples of, 386
surface of, 267
tectonics on, *237*, *238*
temperature of, 143, *197*, *269*, 276–77
Tycho's observations of, 70
volcanism on, 237, 241–42, *242*, *243*, 263
water on, 221, *221*, 245, 246, *246*, 266, *277*, 283–84
wind on, 244, *244*, *245*, 275, 278
Mars Atmosphere and Volatile EvolutioN (MAVEN) mission, *175* (table), 735
Mars Exploration Rover, *175* (table)
Mars Express, *175* (table)
Mars Express mission, *242*, *246*, *247*
Mars Global Surveyor, *175* (table), 233, *243*
Mars Odyssey, *175* (table)
Mars Orbiter laser altimeter, *243*
Mars Reconnaissance Orbiter (MRO), *175* (table), *245*, 246, *247*, 275, *324*
Mars rovers, 172–73, *173*, *244*, 246, 247
Mars Science Laboratory, *175* (table)
Mars Science Laboratory Curiosity, 247, 735, *735*
mass, *78*, 80
atmosphere formation and, 198
of atoms, 126–27
black holes and, 553, 570, 571, 611, 629–31, *630*, *632*, 654–57, 715–16
center of, 408, *409*
defined, 558
equivalence between energy and, 430, 558, 680
escape velocity and, 102
of galaxies, 103, 611, 618, *618*, 622–26, *623*, *624*, 670
of galaxy clusters, 698
geometry of spacetime distorted by, 564–65, *564*, 698
gravity and, *96*, 103–4, 295, 408, 471, 492, 522, 552, 623–24
inertial vs. gravitational, 95, 561–62
momentum and, 127
of neutrinos, 437, *438*
Newton's laws of motion and, 80, 92, 95
Newton's theory as tool for measurement of, 103
particle–antiparticle pair annihilation and, 680
planetary, *see specific planets*
planetary, measurement of, 295–97
of protostar, 474–75, 478
relativity theory and, 12, 558, 562, 563, 564–65, *564*, 698
required for star formation, 200
of universe, 103, 668, *669*, 676, 715–16
weight vs., 92
mass, stellar, 200
of asymptotic giant branch, 502–7, *505*
Chandrasekhar limit and, 512
H–R diagram and, *417*, *417* (table), *418*, 645
lifetime of star and, *495* (table)
measurement of, 395, 408–9, *409*, *410*, *412*, *419* (table), 509, 571, 572, 573
of post-AGB stars, 504–5, *505*
of red giants, 497, 499
star formation and, 480–83, 492
of Sun, 100, 103, 406, 408, 428, 436–37, 480, 482, 492, 493, 494, 596
of white dwarfs, 506, 510, 591
massive (heavy) elements, 544–45, 634, 689, 706, 708, 716, 731
defined, 297, 405
origin of, 534, 537, 545, 574
in stars, 405, *406* (table), 526–27, 534, 625–26, *651*, 652, 657, 660, 731
mass–luminosity relationship, 417, 495
mass transfer:
in binary systems, 509–13, *509*, *511*
as "tidal stripping," 509
mathematics, 13–15
discomfort with, 13
tools of, 15, 74
Mather, John, 178
Mathilde, *369*
math tools boxes:
achieving circular velocity, 100
angular momentum, 191
atmosphere retention, 262

binding energy of atomic nuclei, 528
boxcar experiment, 557
calculating escape velocities, 102
calculating the recessional velocity and distance of galaxies, 587
computing the ages of rocks, 223
critical density, 669
diffraction limit, 161
Drake equation, 739
dust in the infrared, 465
eccentric orbits, 364
escaping surface of evolved star, 504
estimating main-sequence lifetimes, 494
estimating size of Earth, 37
estimating size of extrasolar planet, 206
estimating size of planetary orbits, 204
estimating sizes of stars, 407
expansion and the age of the universe, 594
exponential growth, 730
feeding an AGN, 631
feeding the rings, 347
finding distance from Type Ia supernovae, 592
gravity on a neutron star, 536
heating the giant planets, 307
how planets cool off, 230
how wind speeds on distant planets are measured, 306
impact energy, 381
Kepler's Third Law, 74
making use of the Doppler effect, 136
masses in X-ray binaries, 572
mass of galaxy clusters, 699
mass of Milky Way in Sun's orbit, 648
mass of Milky Way's central black hole, 656
mathematical tools, 15
moons and Kepler's law, 327
observing high-redshift objects, 707
pair production in early universe, 681
parallax and distance, 399
playing with Newton's Laws of Motion and Gravitation, 94–95
proportionality, 79
protostars, 479
reading a graph, 16
sidereal and synodic periods, 68
the source of the sun's energy, 431
sunspots and temperature, 446
supermassive black holes, 629
telescope aperture and magnification, 153
tidal forces, 105
tidal forces on moons, 332
twin paradox, 560
using a bathroom scale to measure the masses of an eclipsing binary pair, 411–12
using Newton's Laws, 81
using radiation laws to calculate equilibrium temperatures of planets, 144
working with electromagnetic radiation, 125
working with the Stefan-Boltzmann and Wien's Laws, 141

matter, 12
atomic theory of, 126–28, *127*, 260–61; *see also* atoms
defined, 126
electron-degenerate, 496
forms of, 260–61
interaction of electromagnetic waves with, 122
interaction of light with, 125–34
luminous, *see* luminous (normal) matter
symmetry between antimatter and, 682, 684

Mauna Kea, 370
Mauna Kea Observatory, *150*, 158, 169, 268, 270
Mauna Loa, *240*, 242, *243*
Maunder minimum, 447, *449*
Maxwell, James Clerk, 340
electromagnetic wave theory of, 119–23, 125, 126, 555, 556, 679

Mediterranean Sea, 106
megabars, 309
megaparsecs (Mpc), 586, 619
Mercury, 221, *242*, 249, 258, 261, 293, *325*, 336
craters of, *199*, 200, 222, 224, *237*, 247
formation of, 198, 200
ice on, 247
interior of, *228*, *232*, 237
lack of atmosphere on, 199, 244, 247, 258, 261–63, 278, 283, 734, 737
lack of moon for, 324
lack of rings for, 352
magnetic field of, 233
orbit of, 73, 92, 109, 374, 567, *567*
spin-orbit resonance of, 109
tectonics on, 237
temperature of, 143, 197
volcanism on, 241, *241*

meridian, 30, 48
mesosphere, *269*, 270
Messenger mission, *175* (table)
Messenger spacecraft, 171, *172*, 248, 278, *175* (table)
Messier, Charles, 582
Meteor Crater, 220, *220*
meteorites, 173, 187, *188*, 328, 384–86
categories of, 384–85
chemical composition of, 220, 225, 227, 249, 346
defined, 382
differentiation and, 227, 384
Earth hit by, 218, 220–21, 249, 360, 382, 384, 386
Earth's interior and, 225
key concepts about, 360

meteoroids, 217, 220, 351, 360, 382
meteors, 217, 382, 384
sporadic, 382

meteor showers, 382–83, *383*, *383* (table), *384*
methane, 197, 260, 279, *280*, 282, 283, 297, *302*, 304, 310, 331–35, 336, 362, 724
Metis, 346
metonic cycle, 49
metric system, 15
Mexico, 249, *249*
micrometeoroids, 244
micrometer (micron), 125
Microsoft, 741
microwave radiation, 125, 170, 178
microwave telescopes, 178–79
Mid-Atlantic Ridge, 239
midnight, 48
local, 31

Milanković, Milutin, 279
Milankovitch cycles, 279, 280, *280*
Milky Way Galaxy, 4, 5, *5*, 12–13, *13*, 152, 159, 167, 176, 186, 206, 611, 630, 635, *640*, 641–60, *643*, *698*, *722*, 737, 742
arms of, 642–43
black hole in, 641, 654–57, *655*, 656
bulge of, 645, 646, *646*, *652*, 657, 658, 660
chemical composition of, 460, 641, 648–51
chemical evolution of, *650*
dark matter in, 641, 646–48, 658, 702
disk of, 641, 642, *642*, 645, 646, *646*, *647*, 648–57, *652*, *654*, 711
dust in, 616, 645, 646, 648
gas in, 648–52, *651*, 653
globular clusters in, 641, 643–46, *643*, 648–52, 657
halo of, 625, 626, 641, 645, 646, *646*, 649, 650, 652–53, *652*, *653*, 657, 705, *708*, 709
Herschel's model of, 642, *643*
interstellar medium in, 460–70, *461*, *468*, 642
location of Sun in, 645, 646, *646*
magnetic fields and cosmic rays in, 653–54, *654*
mass of, 103, 648
novae in, 511
rotation curve of, 647–48, *647*, 655
rotation of, 647–48, *647*
size of, 582–84, 642–46
star formation in, 648–52, *651*
structure of, 642–43, 646–48, 696, 711

Miller, Stanley, 724
Milton, John, 291
Mimas, *325*, 330, 337–38, *338*, 347
minihalos, 706, *706*, 707, 710
minute of arc (arcminute), 398
Miranda, *325*, 338, *339*
mirrors, in reflecting telescopes, 152, 154, *155*, 156, *157*, 160
modern physics, 12, 683
Modified Newtonian Dynamics (MOND), 626
molecular cloud cores, 472, *472*, 473, 480–83, 711
molecular clouds, 460, 469, *469*, 532, *656*, 660, 706, 731
collapse of, 470–73, *472*, *473*, *474*, 480–82, 483, 536, 705
giant, 470
star formation and, 460, 470–73, *471–75*, 480–83, 540, 542, 652

molecules, 125, 138, *138*, 679
in atmosphere, 261–65
emission lines of, 469
forms of matter and, 260
greenhouse, 265
and pressure of gas, 260–61, *261*
in protoplanetary disk, 195–96

Molyneux, Samuel, 553
momentum, 127–28, *128*
Moon, 5, *24*, 152, *163*, 171, *172*, 249, 279, 324, *325*, 326, 401, 734
apparent size of, 51
craters of, *214*, 217, *219*, 220, *221*, 222, 224, 241, 247
dark side of, 46

Moon (*continued*)
eclipses and, 50–55
formation of, 200, 221, 229, 324
gravity of, *90, 92,* 103–4, 108, 509
ice on, 247, 248
interior of, 227, 228, *228,* 232
lack of atmosphere on, 199, 244, 247, 258, 261–63, 278, 283, 734, 737
lack of magnetic field on, 232, 451
orbit of, 45–49, *46, 47, 48,* 51, 54–55, 104, 108, *108, 109,* 200, 346
phases of, 46–49, *47, 48,* 54–55, *55,* 103, 107, *107*
rotation of, 108–9, *108*
same face of, 45–46, *46*
samples of, 220, 222, 224, *224,* 386
sidereal period of, 48, *48*
size of, 232
synchronous rotation of, 45–46, *46,* 108, 109
synodic period of, 48, *48*
tectonics on, 237, *237*
tidal effects of, 103, 104, 105, *106, 107,* 108, *108,* 230
as tidally locked to Earth, 108–9, *108*
volcanism on, 241, *241*
walk on, 7, *7, 90,* 216
waning of, 48
water on, 248
waxing of, 47–48
moons, 324–27, 352, 369
atmosphere of, 324, 328, 331–36, *334*
composition of, 197, 324, 346
co-orbital, 347
dead, 338–39
defined, 200
formation of, 198, 200, 324–27
formerly active, 336–38
geological activity on, 324, 327–33, *328–31,* 336–39, 346, 352
irregular, 327
Kepler's law and, 327
landslides on, 244
oceans on, 324, *325,* 332, 352
as origin of ring material, 346
regular, 325–26
as satellites, 98
shepherd, 348–49, *349*
as small worlds, 328
tidal forces on, 332
tidal locking of, 325–26
volcanism on, 347
see also specific moons and planets
motion:
anticyclonic, 299, *300, 308*
apparent, *see* apparent daily motion
of charged particles in magnetic field, 270–71, *271*
cyclonic, 273, 299
Doppler effect and, 134–35, *134, 135*
frame of reference and, 28–29, *41,* 554, 555
Galileo's views on, 75–76, 98
length of object in, 559
of Moon, 27, 45–49, *46, 47, 48,* 51, 54–55
planetary, *see* elliptical orbits; *specific planets*
prograde, 65–69
relative, *see* relative motions
retrograde, 64
uniform circular, 98
motion, Newton's laws of, *75,* 76–80, 95, 97, 99–103, 260, 295, 361, 408
as approximations to special relativity and quantum mechanics, 565
first, 77, *81, 99,* 190
playing with, 94–95
second (a–F/m), *78,* 80, *81,* 90, 94, 562
speed of light and, *554,* 555
third, 80, *80,* 81, *81*
M stars, 405, 482–83, 541
Mt. Wilson Observatory, 583
M-type asteroids, 368
multiverses, *see* universe, multiple
muons, 558, *559*
mutation, 728
myths, stars in, 26

nadir, 30
Namaka, 362
nanometer (nm), 124
NASA, *170* (table), 171, 202, 207, 228, 246, *247,* 248, 364, 366–67, 377–78, *387,* 442, 451, 462, 706, *706,* 712
natural selection, 730
nature, science in relation to, 11, 12
Naval Observatory, U.S., 17
navigation, stars used for, 35–36
neap tides, *107,* 108
NEAR Shoemaker spacecraft, 369, *370*
nebulae, 583, 584
use of term, 12–13, 187, 582
nebular hypothesis, 187
neon, *133, 297, 406* (table), 507
neon burning, in high-mass stars, 523, 528, *528* (table)
neon signs, emission and, 130–31
Neptune, 17, 292–97, *293*
atmosphere of, 275, 299–304, *302,* 306
chemical composition of, 292, 297, 304, 310
density of, *292* (table), 297, 310
discovery of, 295, 345
formation of, 199, 297, 310
Great Dark Spot (GDS) on, 301, *302,* 306
interior of, 309, *309*
magnetic field of, 311, *311,* 312
magnetosphere of, 312
mass of, *292* (table), 295, 310, 316
moon of, 293, *325,* 327, 330–31
orbit of, 292, *292* (table), 316, 360, 374
rings of, 340, *341,* 345, *345,* 346, 348, 349
rotation of, *292* (table), 298, 312
size of, 292–93, 295, *296*
winds on, 306, 308
neutrino cooling, 529
neutrinos, 174, 439, 702–3
defined, 432
detection of, 174
from high-mass stars, 528, 529, 531
solar, 427, 432, 437–38, *438, 440*
from supernovae, 437, 440, 532
neutrino wave detectors, 174
neutron-degenerate core, *see* neutron stars
neutrons, 604
in atomic structure, 126, *127,* 429, 430, *527,* 679
decay of, 715
free, 534–35
in isotopes, 222
neutron stars, 536–37, *537,* 539, 552–53, 573, 625, 715
in binary systems, 536, *537,* 552, 568, 629
general relativistic time dilation and, 568
gravity on, 536
pulsars, 537–38, *538, 540*
New Horizons mission, *175* (table), *299,* 326, 329, 361, 372
new Moon, 47, 54, 55, *55*
tides and, 103, 107, *107*
New Orleans, La., 275
Newton, Sir Isaac, 11–12, 65, 76–81, *77,* 327, 374, 555, 648
Kepler's laws derived by, 97–99, 101, 103, 118
law of gravity of, *see* gravity, Newton's law of
laws of motion of, *see* motion, Newton's laws of
light as viewed by, 119
telescope of, 154, *154*
Newtonian (classical) mechanics, 11–12, *75,* 553
clockwork universe and, 134
special relativity and, 556, 561
see also motion, Newton's laws of
newtons (N), 92
Newton's laws, 111, 179, 316
NGC 891 (spiral galaxy), *642*
NGC 1132 (elliptical galaxy), *624*
NGC 1532/1531 (galaxy pair), *694*
NGC 2362 (star cluster), 544
NGC 3198 (spiral galaxy), *624*
NGC 6530 (star cluster), 541, *542*
nickel, 5, 197, 227, 368, 512, 545, 660
Nietzsche, Friedrich, 459
night, 272, 297
Nightfall (Asimov), 645
nitrates, 574
nitrogen, 5, 18, *406* (table), 535, 545, 706, 724, 731
in atmosphere, 261–63, *263* (table), 266, *266* (table), 272, 277, 331–35
in giant planets, 297, 315
in planetary nebulae, 545
nitrogen oxides, 574
noon:
local, 31
phases of the moon at, 48
normal matter, *see* luminous (normal) matter
north celestial pole, 30–36, *31, 34,* 44
Northern Hemisphere, 35, 272
phases of the moon in, 50
seasons in, 39–43, *39,* 40–43, *41*
tilt of Earth's axis and, 39–40, *39*
North Pole, 30–36, *30, 47,* 231
latitude of, 32–33

seasons near, 43–44
and tilt of Earth's axis, 39–40, *39*, 43
view of sky from, 30, *30*, *32*
north star (Polaris), 36, 44–45
novae, 511, *511*, 535–36, 629
nuclear force, strong, 689
nuclear fusion, 7, 483, 689
defined, 430
effects of photodisintegration on, 529
in star formation, 460, *477*
stellar evolution and, 493
in Sun, *430*, *433*, 437, 679
see also carbon burning; helium burning; hydrogen burning; neon burning
Nuclear Spectroscopic Telescopic Array (NUSTAR), *170* (table)
nucleons, 527
nucleosynthesis, 689, 705, 706, 707, 716
Big Bang, 603–4, *604*, *685*, 686
stellar, 534, *535*, 537, 544, 545, 650, *651*; *see also* stellar mass loss
nucleus, atomic, 126–27, *127*, 132, 493, 496, 535, 686
binding energy of, 527, *527*, 528
in electron–degenerate core, 498–500, *500*

Oberon, *325*, 338
objective lens, 152
objective truth, 12
oblateness, 297, 362
obliquity, 56, *292* (table), 298
observational uncertainty, 399
Occam's razor, 9, 539
occultations, stellar, 295, *296*, 340
oceans:
of giant planets, 292, 297, 309
on moons, 324, *325*, 332, 352
oceans of Earth, 217, 221, 245, 265, 279, 283, 297, 309, 352, 508
development of life in, 265
formation of, 261, 265
magnetic structure of, 233–34, *234*
rifts in floor of, 233–34
tidal forces and, 104–9, *106*, *107*
waves of, 461, *462*
Olympus Mons, 242, *243*
Oort, Jan, 371
Oort Cloud, 260, 371–73, *372*, 373, 375, 379, 387
opacity, 434
open clusters, 540, *541*
open universe, 675, *675*
Ophelia, *349*
Opportunity Mars rover, 246, *246*, 385
orbital resonance (orbital interaction), 327, 351, 361, 365–66
asteroids and, 365–66, *365*, *366*
Cassini Division and, 347
Lagrangian points and, 110
orbiter, 171
orbits, 97–103
of binary stars, 408–9, *409*, *410*, 510, 568–69
bound, *101*, 102
changing of, 316
circular, *see* circular orbits
of comets, 260, 373–75, *373*, 376, 382, 383, 387
decay of, 450–51
defined, 98
eccentric, 206, 364
elliptical, *see* elliptical orbits
estimating size of, 204
gravity and, 97–103, *97*, *98*, *99*, *101*
hyperbolic, 102
of Jupiter, 201–2, *202*
Kepler's second law and, 73, *73*
of Moon, 45–49, *46*, *47*, *48*, 51, 54–55, 104, 108–9, *108*, *109*, 200, 346
of moons, 325–26, 330
parabolic, 102
period of, 73–74, *74* (table), 100, 108–9, *108*, *365*, *366*, *373*, 376, 411
planetary, *see specific planets*
of planetary rings, 340, 349–51
prograde vs. retrograde, 373–74
of protostellar disk, 188
retrograde, 327
stellar galaxies shaped by, 611, 613, *614* (table), 616, *616*, *617*, 648
unbound, 102
Ordovician mass extinction, 574
organic materials, 197, 198, 334, 346, 360, 470, 483
in comets, 374, 376, 378, 387
prebiotic, 724
in terrestrial life, 734
Orion, *394*, 483, *484*
Orionids, *383* (table)
Orion Nebula, 467, *467*
O stars, 418, 483, 492, 505, 526–27, 541, 620, 642
formation of, 466
life span of, 466
spectra of, *404*, 405
oxygen, 5, 18, 132, 259, 263, 315, 368, *406* (table), 483, 507, 535, 545, 660, 706, 708, 731, 737
in atmosphere, 261–66, *263* (table), 266, *266* (table), 267, 272, 277
oxygen burning, in high-mass stars, 523, 527, 528, *528* (table), 529
ozone, 267–68, 277, 574, 660
ozone hole, 268

Pacini, Franco, 539
pair production, *680*, 681
paleoclimatology, 279–83
paleomagnetism, 232, 233, *234*
palimpsests, 336, *337*
Pallas, 17
Pandora, 348, *348*
parabolic mirrors, 154, 156, *157*
parabolic orbits, 102
parallax, 398–401, *398*, 415–16, *419* (table)
distance ladder and, 590, *590*
spectroscopic, 415–16
parallel universes, 686–88
parent elements, 222, *223*
parsec (pc), 119, 398–99
partial solar eclipse, 50, *50*
particle accelerators, 174
particles, particle physics, 174, 571, *572*, 668, 679–86, *680*
cosmology vs., 683
massless, 125
quantum electrodynamics and, *680*
standard model of, 680
temperature as measure of movement of, 139
wavelike properties of, 127
see also axions; electrons; neutrinos; neutrons; photinos; photons; protons
patterns:
life and science made possible by, 13, *14*
mathematics as language of, 13–14
understanding and, 13
see also sky patterns
peculiar velocity, 590, 697–98
Pelée, Mount, 239
penumbra, 446, *447*
defined, 50
lunar eclipse and, 54
partial solar eclipse and, 50, *50*, 52
Penzias, Arno, 178, 600, 601, *601*
perihelion, 109, 364, 374, 376
period:
of orbits, 73–74, *74* (table), 100, 108–9, *108*, 118, *365*, *366*, *373*, 376, 411
of wave, 123, *225*
period–luminosity relationship, for Cepheid variables, 525, *525*, 584, 592, 646
permafrost, 283
Perseids, 382, *383*, *383* (table)
Perseus, *643*
perspiration, thermal equilibrium maintained by, 137
Phaethon, Asteroid, *383* (table)
philosophy, Greek, 27
Phobos, *325*, 352
Phoebe, 327
Phoenix, Ariz., 36
Phoenix lander, *244*, 246, *247*
Phoenix mission, *277*
phosphorus, 18, 304, 545, 731
photinos, 702
photochemical, 302
photodisintegration, 529
photodissociation, 267, 334
photoelectric effect, 125
photographic plates, 163–64, 165
photography, in astronomy, 163–64
photometers, 164
photons, 121, 125–26, 139, 142, 163, *165*, 437, 679, 702, 705, 716
absorption of, *130*, 132, 466
atomic decay and, 128–32
defined, 125
in early universe, 680, *682*
emission of, 128–34, *130*, 466, 468–69
gamma ray, 432, 680, *680*, 681
in interstellar medium, 466–67, 468
in quantum electrodynamics, 679, *680*
as quantum of light, 126
in radiative transfer, 433–34, *434*

photosphere, 441–43, *442*, 443, 445–46
photosynthesis, 267
physical laws, 9, 227
defined, 9
discovery of, 69
of Newton, *see* gravity, Newton's law of; motion, Newton's laws of
proportionality and, 79
structure and, 679–84, *680, 682, 684*
physics, modern, 12
physics laboratories, 437
phytoplankton, 279, 574
Piazzi, Giuseppe, 362, 364
Pinwheel Galaxy, *591*, 592
Pioneer 11 probe, 295, 315, 340, 348, 738, *738*
Pioneer Venus mission, 275
Pisces, 40
pixels, 164, *165*
Planck, Max, 125, 126, 139–40
Planck era, 683, *685*
Planck observatory, *470*
Planck satellite, 603
Planck's constant, 125, 128, 129
Planck (blackbody) spectrum, 140, *140*, 314, 402–3, *403, 404, 406*, 571, 681
Big Bang and, 178, *178*, 600, *600*, 601, *602*
of Sun, 443
Planck telescope, *170* (table), 179
plane, 15
planet, defined, 17
planetary configurations, superior vs. inferior, 66–67, *66*
Planetary Habitability Index (PHI), 737
planetary motion, 27, 614
Kepler's first law of, 70–72, *72*
Kepler's second law of (Law of Equal Areas), 73, *73*, 100–101
Kepler's third law of (Harmony of the Worlds), 73–74, *74* (table), *75*, 100, 103, 295
Newton's theory of, 295
origins of Solar System and, 187
Sun's influence on, 94, 96, 97
thought experiment and, 97–99, *97*
see also elliptical orbits; *specific planets*
planetary nebula, 504–6, *505, 506, 507*, 508, 533, 634, *650*
Crab Nebula compared with, 537
planetary systems, 200–205, *203*
birth and evolution of, 185–208
planetesimals, 194, *194*, 197, *197*, 198, *199*, 200, 299, 310, 316, 317, *317*, 324, 326, 346, 360, 364, 367, 387, 725
fate of, 367–68, *368*
see also asteroids; comets
planet migration, 198
planets, 70–72, *197*
albedo of, 143
determination of temperature of, 137, 143, *143, 145*
Earth-like, *see* terrestrial planets
exploration of, 172–73, 244, 246
extrasolar, 200–205, *202*, 737
formation of, 186–200, 297, 309–10, 483
inferior, 66
mass of, *205*
migration of, 508
"naked eye," 64
orbit of, 508
post–main sequence star expansion and, 508
as satellites, 98
superior, 66
surface gravity of, 737
thermal equilibrium of, 137
thermal radiation from, 143, *143, 145*
see also giant planets; terrestrial planets; *specific planets*
plant life, atmospheric oxygen and, 267
plasma, 270, 312, 315, 448–49
associated with Io, 315, *315*, 316, *316*, 328
plate tectonics, 234–40, 242, 279, 352
convection and, 234–35, *235*
defined, 234
paleomagnetism as record of, *233, 234*
volcanism and, 236, *236*, 239–40
Pleiades (Seven Sisters), 49, 482, *482*
Pluto, 72, *143, 186*, 200, 299, 324, *325*, 360–64, *361*, 363, 371
chemical composition of, 361
discovery of, 360–61
as dwarf planet, 17, *17*
mass of, 17, 360
moons of, 109, 324, *325*, 326, *326*, 361, *361*
orbit of, 72, 299, 361
size of, 360–61
temperature of, 143, 197
Polaris (north star), 36, 44–45
polar regions:
on Earth, 30–36, *30*, *47*, 231–32
of giant planets, 305, 306
wind patterns and, 272–73
Pompeii, 239
Poseidon, Temple of, *24*
positrons, 432, 537, 680–82, *680*, 716
in early universe, 680, *682*
post-AGB stars, 504–6, *505*
see also white dwarfs
potential energy, 196–97, 274
gravitational, 196–97, 308, 310
power, defined, 139
prebiotic molecules, 660
prebiotic organic molecules, 724, 735
precession of the equinoxes, 45
pressure, 260–61, *261*, 266, 272
atmospheric, 263, *263* (table), 266, 268, 275
hydrostatic equilibrium and, 227, 266
in protostar, 474–75, *476, 477*, 492, 493
size and, 310
structure of Sun and, 428, *428, 429*, 496
ultrahigh, giant planets and, 310
weight and, 227, 239, 266, 428, 522, 523
primary atmosphere, 198, 199, 258, 263
primary mirror, 154, 155, 160
primary waves, 225–26, *226*
primordial energy, 308
Primordial Era, 715, *716*
primordial gas clouds, 706
principles, scientific, defined, 9
prisms, *157*, 165
probes, robotic, *173*
see also rovers
prograde motion, 65–69
prokaryotes, *727, 727*
Prometheus, 348, *349*
prominences, *442*, 448, *449*
proportionality, 80
constants of, 79
defined, 15, 79
inverse square laws and, 93
in universal law of gravitation, 94
proteins, 352, 724, 731
Protestants, 50
protogalaxies, 658–59
proton–proton chain, 432, *433*, 436, 437, 493, 522, 523
protons, 132, 270, 315, 429–32, 466–67, 535, 630, 653
in atomic structure, 126, *127*, 429, 430, *527*, 679
decay of, 682, 715
in early universe, 603, 604
grand unified theory and, 682
in isotopes, 222
magnetized, 467–68, *468*
in motion vs. at rest, 558
protoplanetary disks, 193–200, *199*, 460
aggregation in, 193–94, *194*
composition of, 197, *197*, 309–10, 387, 460
defined, 190
temperature of, 195–97, *195, 197*, 200
protostars, *197*, 200, 459, 460
atmosphere of, 479, 496–97
collapse of, 190–92, 477–78, 480–82, 493, 497, *498*, 500
defined, 188
evolution of, 473–80, *476, 477, 480*, 496–97, 630
formation of accretion disk around, *see* protoplanetary disks
luminosity of, 479
radius of, 479
temperature of, 479
protostellar disks, 188–90, *192, 473, 474*
material flow patterns in, 481–82, *481*
Proxima Centauri, 399, 561
Ptolemy, 26, 44, 64, *64*, 65, 70
pulsars, 537, *538, 540*, 568
in Crab Nebula, 537–38, *540*
pulsating variable stars, 524
Cepheid, 525, *525, 526*, 533, *583*, 584, 591, 592, 643, 645–46
pupil, 162, *163*

Quadrantids, *383* (table)
quadrature, 66
quantum chromodynamics (QCD), 679–80
quantum efficiency, 162, 163–64
quantum electrodynamics (QED), 679, *680*
quantum mechanics, 12, 132, 467, 571
confrontation between relativity theory and, 683

defined, 126
density of stellar core and, 496, 499–500, 503, 529, 535
early universe and, 677–78, 688, 700
Einstein's problems with, 12, 134
Newton's laws of motion and, 565
probabilities and, *127*, 134
thinking outside the box required by, 126
uncertainty principle in, 677–78
Quaoar, 372
quarks, 680
quartz, 220
quasars, 314, 626–27, *626*, *627*, 628, *628*, 631

radar, 553
radial velocity, 204, *204*
defined, 135
spectroscopic, 201, *202*
radian, 398
radiant, 383, *383*
radiation, 8, 529, 634, 635
continuous, 139
cosmic background, *see* cosmic microwave background radiation
electromagnetic 119; electromagnetic radiation; light
Hawking, 571, *572*, 716
interstellar extinction and, 461, *461*
interstellar medium and, 461–70, *461–69*
as inverse square law, *142*, 407, 415, *419* (table)
structure of universe and, 701–2, *702*
thermal, *see* thermal radiation
21-cm, rotation of Milky Way measured with, 646, *647*
radiation belts, 271, 314–16
radiative transfer, 433–34, *434*
radiative zone, 434, *434*
radioactive (radiometric) dating, 222, *224*, 232, 234
RadioAstron, *170* (table)
radio frequency maps, 167
radio galaxies, 627, *627*, 629, 633
M87, 632–33, *633*
radiometric dating, 222, *224*, 232, 234
radio telescopes, 8, 167–69, *167*, *168*
cosmic background radiation discovered with, *601*
radio waves, *124*, 125, 167, 270, 312, 314, 382, 537, 571, 626–27, *627*
of interstellar medium, *468*, 470
photons of, 125
rain, 270, 274, 307
rainbow, spectrum of, *116*, 124, *124*
rainfall, direction of, frame of reference and, 552–53, *552*
ratios, defined, 15
Reber, Grote, 167
reddening effect, 462, *463*
red giants, red giant branch, 460, 508, 523, 529, 533, 742
asymptotic giant branch compared with, *503*, 504
degenerate cores of, 497, *497*, 498–502, *498*, *499*, *500*, *503*, 510–11, 529
evolution of, *497–500*, *503*, *507*, 508, 511, 523
helium burning in, 498–502, *500*
red light, 402
spiral galaxy viewed in, *620*
Red Rectangle, *14*
redshifted light, *204*, 700, 702, 705, 707
age and, *701*
cosmological, 599
defined, 135
discovery of Hubble's law and, 585, 590, 599
Doppler effect and, *134*, 135, *135*, 409, 598, 599, *599*, 602, 618
of galaxies, 585, *586*, 590, 592, 598–99, *599*, 670, *670*, 696, *697*
gravitational, 568, *569*, 599, 603
of Planck spectrum, 600, *600*, 601, *603*
of quasars, 626, *626*
scale factor and, 598–99, *598*
redshifted objects, 707, 708, *708*
redshift surveys, 696, *697*
red supergiants, 406
reflecting telescopes, *152*, 153, 154, *154*, *155*
reflection:
angle of, 156, *157*
of light, 156, *157*
refracting telescopes, 152, *152*, *153*
refraction, 156, *156*
index of, 156, *156*
refractory materials, 197, *197*
regular moon, 325–26
reionization, 705
relative humidity, 269
relative motions, 28, 552
gravity and, 104
inertia and, 77
relativistic beaming, 633
relativity theory, *see* general relativity theory; special relativity theory
remote sensing, 171
Renaissance, 12
reproduction, 728
resolution, 159–60
angular, 167, 168
rest wavelength, 135, 585
retina, 162, 163, *163*
retrograde motion, 64, *64*, *68*
retrograde orbits, 327
Rheasilvia, *371*
ribonucleic acid (RNA), 724, 734
rift zones, 235
Rigel, *394*
Rima Ariadaeus, *237*
ring arcs, 340
ringlets, 342, *342*
Ring Nebula, *506*
rings, planetary, 339–52, *341*
backlighting and, 344, *344*, *345*, *351*
origin of material for, 346
see also specific planets and rings
RNA (ribonucleic acid), 352
robotic probes, *173*
Roche, Edouard A., 509
Roche limit, 109, 110, 346, 352, 508, 509
Roche lobes, of binary stars, 509–10, 536, *537*
rocket ship, speed of light and, 558, 561
rocks, 197, 352
of Earth, 222, 225–27, 231–32, 236, 243
in giant planets, 309, 310
lunar, 220, 222, 224, *224*, 241, *241*, 244
on Mars, *244*, 246
in moons, 324, 327, 330–31, 336, *336*, 338
in planetesimals, 198
radioactive dating of, 222, *224*, 232, 234
rods, 162
Roentgen, W. C., 124
Romans, ancient, 46
Rømer, Ole, 118, *119*, 121
Rosetta mission, 370
Rosette molecular cloud, *458*
rotation:
of asteroids, 367
equal to orbit, 109
of moons, 325–26
of object, tied to orbit by tides, 103, 108–9, *108*
synchronous, *see* synchronous rotation
see also specific planets
rotation curve, 623, *623*
flat, 623, *624*, *647*, 648
of Milky Way, 647–48, *647*, 655
rovers, 172–73, *173*, *244*, 246
RR Lyrae variables, 525, 645–46
Rubin, Vera, 623, 699
Russell, Henry Norris, 412, 419
Russian Federal Space Agency (Roscosmos), 171

S0 (S-zero) galaxies, 613–14, *613*, *614* (table), 618
Sagittarius, 43, 44, 461, 646
Sagittarius Dwarf, 658, *658*
San Andreas Fault, 236
sand dunes, 244, *244*
Sanduleak, 532
satellite observations, 7–8
satellites, 98–102
artificial, 7–8, 98, *450*
Global Positioning System, 568
see also Moon; moons; planets; *specific moons and planets*
saturation, 270
Saturn, 64, 111, 171, 172, 198, *290*, 292–301, *293*, 305–15, 352, 508
atmosphere of, 275, 297, 299, 300, *300*, *301*, 305, 340–41
auroral rings of, *315*
backlighting and, *344*
Cassini Division and, 340, *342*, 343, 347
chemical composition of, 292, 297, 304, 309–10
density of, *292* (table), 297, 310–11
differentiation in, 310
formation of, 199, 297, 310
giant dust ring of, *343*
interior of, *309*, 310–11

Saturn (*continued*)
magnetic field of, 311, *311*, 312, 313, 351
magnetosphere of, 313, 315
mass of, *292* (table), 295, 309, 310
migration of, 316
moons of, 111, 200, 315, 324, *325*, 326, *326*, 327, 330, *331*, *334*, 337, *338*, 340, 346, 347, 348, *348*, *349*, 352
orbit of, *292* (table), 312, *313*, 316, *317*, 351
rings of, 200, 299, *322*, *341*, *342*, 346, 347, 348, *348*, *349*, *351*
rotation of, *292* (table), 298
size of, 292, 295, *296*, 299
storms on, 299, *301*
temperature of, *197*, 200
winds on, *300*, *301*, 305, *305*
scale factor, of universe, 597–98, *597*, 599, 668, *669*, 671, *671*, *674*
scattering, 276, 297
Schiaparelli, Giovanni, 245, 734
Schmidt, Maarten, 626
Schwarzschild radius, 629
science:
arts compared with, 11, *11*
as collaborative, 414
as creative activity, 11
culture and, 12–13
empirical, 69, 103
mathematics as language of, 13–14
patterns and, 13
process of, 42, 71, 91, 121, 177, 189, 229, 281, 294, 350, 363, 414, 439, 464, 513, 539, 588, 644, 672, 704, 733
technology and, 177
as way of viewing the world, 9–13
science fiction, 645
scientific method, 9, *10*, 103, 279
scientific notation, 15
scientific revolutions, 11–12, 103
of Einstein, 11–12, *12*, 555
Kuhn's views on, 11
scientific theory:
cosmological principle as, 97
failure of, 143
as falsifiable, 9, 569, 688
interplay between observation and, 227
as nonarbitrary, 12–13
old vs. new, 565, 566
rise of, 103
testable predictions of, 9, 569
use of term, 9
see also specific theories
Scott, David, 90, *90*, *91*
Scutum–Centaurus, *643*
Search for Extraterrestrial Intelligence (SETI), 740–41
seasons, 82
change of, 38–45, 297–99
Earth's wobbling axis and, 44–45
on Mars, 277–78
near Earth's poles, 43–44
near Equator, 44
tilt of Earth's axis and, 38–40, 56
secondary atmosphere, 198, 260–65, 267
secondary craters, 219, *220*
secondary mirrors, 154, *155*
secondary waves, 225–26, *226*
second of arc (arcsecond), 398
sedimentation, 222, 246
Sedna, 372
seismic waves, 225–27, *226*
seismometers, 226
selection effect, 205
selenium, 731
self-gravity, 96, 109, 187, 346, 376, 512, 703, 711
of molecular clouds, 470–72, *471*
semimajor axis, 72, *72*
SETI@home, 741
SETI Institute, 741
Seyfert, Carl, 627
Seyfert galaxies, 627, 629
Sgr A, *650*, *655*
Shapley, Harlow, 582–83, 643, 644
size of Milky Way determined by, 646
Shapley Supercluster, 697, *698*
shepherd moons, 348–49, *349*
shield volcanoes, 239–40
Shoemaker–Levy 9, 379, 380
destruction of, 379, *380*
short-period comets, 373–74, *373*, 378
Siberia, 379, *381*
sidereal day, 49
sidereal period, 48, *48*, 67, 68
Sikhote-Alin region, 380
silica, 227
silicates, 197, 324, *325*, 332, 346, 368, 737
silicon, 5, 368, *406* (table), 460–61, 545, 660, 734, 737
in high-mass stars, 527, 528, 529
singularity, 570, *570*
Sirius (Alpha *Canis Majoris* or the Dog Star), 26, 49, 401, 506
61 Cygni, 399
sky patterns:
ancient views of, *14*
Earth's orbit and, 37–38, *553*
Earth's rotation and, 30–36, *30–36*
life patterns and, 26, 27
of phases of the Moon, *see* Moon, phases of
"Slice of the universe, A" (redshift survey), *697*
Slipher, Vesto, 584–85
Sloan Digital Sky Survey (SDSS), 697, 713
Sloan Great Wall, 697
slope, 16
Small Magellanic Cloud, 625, *653*, *658*
Smoot, George, 178
snowball rolling downhill, red giant compared with, 498
sodium, *133*, 315, 328, 523, 545
SOFIA, 484
SOHO orbiter, *175* (table)
solar abundance, 297
Solar and Heliospheric Observatory (SOHO), 442, *450*
Solar Dynamics Observatory (SDO), *170* (table), *426*, 442, *446*, *449*
solar eclipse, 50–55, *50*, *51*, 443, *444*
annular, 50, *50*, *53*
gravitational lensing and, 567–68
lunar eclipse compared with, *52*, 54, *54*
partial, 50, *50*
total, 50–54, *51*
solar flares, 448–49, 451
solar luminosity, 742
solar maxima, 447
solar maximum, 450
solar neutrino problem, 437–38, *438*
solar neutrinos, 427, 432, 437–38, *438*, *440*
solar neutrino telescopes, 437–38, *440*
Solar System, 4, *5*, 15, 167, *186*, 316–17, 463
age of, 385
AGNs compared with, 628
debris in, 380–87
exploration of, 7–8, 171–73, 244, 246, 248
formation of, 508
key concepts about, 186
measurement of distances in, 553
missions of, *175*
nonsolar mass of, 295
origins of, 5–7, 186–200, 385, 460, 473
scale of, 66–67
see also specific planets and moons
solar wind, 173, 278, 376, 444, 445, *445*, 451–52
Earth's magnetic field and, *271*, 272
life and, 451–52
magnetosphere and, 312, *313*, 315, 445
solar year, 49
solids:
defined, 260
thermal conduction in, 433
solstice:
summer, *39*, 43, 44, *396*
winter, 43
sound:
Doppler effect and, 134
speed of, 628–29
south celestial pole, 30, *31*, 33, *34*, 35
Southern African Large Telescope, *155* (table)
Southern Hemisphere, 30, 33, 35, 272, *658*
constellations visible from, 26
o-zone depletion in, 268
seasons in, *39*, 40, 43
tilt of Earth's axis and, *39*, 40
South Pole, 231
latitude of, 32–33
seasons near, 43–44
tilt of the Earth's axis and, 40
view of sky from, 32
Soviet-American Gallium Experiment (SAGE), 440
Soviet Union, 222, 276
spacecraft, 171–73, 246
flyby missions of, 171
information about giant planets supplied by, 295, 297
in measurement of mass of planet, 295, 297
orbital missions of, 171, 202
orbit of, 98, *98*, 450–51
reconnaissance, 171

samples returned by, 173–74, *241*
stationary vs. high velocity, 562–63, *562*
see also specific spacecraft
space exploration, 7–8, 171–73, 244, 246, 248
see also specific missions
space probes, 7–8, 187, 216
"spaceship traveling at half the speed of light" experiment, *554*, 555
space shuttle, 556, 561–62
spacetime, 556, 683
Big Bang and, 671, 674–75
gravity as distortion of, 561–69
mass in distortion of, 564–65, *564*
as rubber sheet, 563–65, *564*, 675
see also time
space travel, 559–61
special relativity theory, 11–12, 430, *554*, 555–59, 561, 599, 741
frame of reference and, 556–59
implications of, 556–59
particle-antiparticle pair annihilation and, 679
species, extinction of, 248–49, *249*, 387
spectral fingerprints, 132, *133*
spectral types, 403–5, *404*, 405, *416*, *417* (table)
spectrographs, spectrometers, 164–65, *165*, 166, 202, 352, 696
spectroscope parallax, 415–16
spectroscopic binary stars, 410
spectroscopic parallex, 590, *590*
spectroscopic radial velocity, 201, *202*, 204
spectroscopy, 164, 186
spectrum, 126, 130–35, 164
of asteroids, 370
composition of stars and, 395, 402–5, *404*, *419* (table)
defined, 124
Doppler effect and, 134–35, *134*
electromagnetic, 124–25, 165, 166, 170, 402; *see also specific kinds of radiation*
Planck, *see* Planck (blackbody) spectrum
of Sun, 441–43, *443*
temperature and, 402–5, *404*, *406*, 462, *463*
Wien's law and, 140–41, 402, 405, *406*, 462
see also absorption lines; emission lines
sphere, net gravitational force at surface of, 96, *96*
spherical symmetry, 96
spin-orbit resonance, 109
spiral density waves, 621–22
spiral galaxies, *13*, *14*, 582–84, 698, 709, 710
arms of, 611, 619–20, *619*, *620*, *621*, 635
barred, *610*, 613, *613*, *614* (table), *617*, 621, 642, *643*, 659
bulges in, *614* (table), 616, *617*, 620–21, 631
classification of, 613–19, *613*, *614* (table)
color of, 617
dark matter in, 623–25, *624*, 646–48, 659
disks of, 616–21
edge-on, 642, *642*
formation of, 620–21
light of, 622, *623*
mass of, 622, *623*, *624*
merger of, 711–12
rotation curve of, *623*, *623*, *624*, 647–48, *647*, 655, 705
rotation of, 620–21, *621*, 622
Seyfert, 627, 629, 631
star formation in, 616–22, *620*, 634
unbarred, *613*, *614* (table)
see also Milky Way Galaxy
spiral nebulae, 582–84, 611
Spirit Mars rover, 246
Spitzer Space Telescope, 170, *170* (table), 342, *482*, 508, *540*, *707*, 708
spoke, 351
spreading centers, 233, 236, 239–40
spring, 43
spring balance, 475, *476*
spring equinox, 49–50
spring tides, 107, *107*
Sputnik I, launch of, 7
Square Kilometre Array, 169
stable equilibrium, 137
standard candles, 645, 646
Type Ia supernovae as, 591, *591*, 592, 670, *670*
use of term, 590
see also globular clusters
standard model, 680
star, low-mass, *201*
starburst galaxies, 618
star clusters, 482, *482*
as snapshots of evolution, 540–45, *542*, *543*, *544*
see also globular clusters; open clusters
Stardust mission, *175* (table), 378, *378*
star formation, 187–93, 197, 200, 308, 413, 483–84, 540, 541, 542, 544, 545, *685*, 705, *705*, 710–11, *710*
accretion disks and, *187*, 188–93, *188*
conservation of angular momentum and, 190–92, 195
in galaxies, 611, 616–22, 634
gravitational energy converted to thermal energy in, 474–78, *476*, *477*, 630, 711
gravitational instability and, 711
interstellar medium and, 460–84, 649–50, *651*
key concepts about, 460
in Milky Way, 648–52, *651*
planetary systems as by-product of, 186, 206–7
see also protostars
stars, starlight, 158, 626
aberration of, 552–53, *553*
age of, 540–44, *542*, *543*, *544*, 645, 648–51
apparent movement of, 27–36, *32*, *34*, *35*, *36*, 37–38, *41*
binary, *see* binary stars
brightness of, 163, 202, *202*, *203*, 396, 400–401, *400*
circumpolar, 35
classification of, 395, 403–5, *404*
color of, 395, 402, *403*, 415, *417* (table), 544
composition of, 174, 395, 405, *406* (table), 417, *419* (table), 435–37, 441, 443, 460, 492–97, *495*, 501–2, 507, 545, 634, 648–51, *651*
core of 492, 706; core, stellar
death of, 7, *7*, 413, 460, 529–34
density of, 402
evolution of, 508
failed, *see* brown dwarfs
halo vs. disk, 652–53, 657
high-velocity, 652, 653, 657
interiors of, 402
key concepts about, 460
in legends and myths, 26
life of, 7; *see also* stellar evolution
luminosity of, *see* luminosity, stellar
luminosity-temperature-radius relationship for, 406, 407
main sequence of, *see* main sequence of stars; Sun
mass of, 508
in mass of galaxies, 611
nuclear fusion reactions within, *see* nuclear fusion
parallax of, *see* parallax
patterns of, 26, 396
properties of, 395, 400–401, *413*, 415–19
protostar vs., 478–79, 497
shape of galaxy due to, 611, 614–16, *614* (table), 616, *616*, *617*
size of, 395, 402, 406, *406*, 415–19, *419* (table), 435–36, 497, 731
as source of chemical elements, 7, *7*, 506, 532–35
spectral type of, 402–5, *404*, 405, *416*, *417* (table)
spectroscopic binary, 410
see also Sun; *specific types of stars*
Star Trek, 680
static equilibrium, 136
Stefan, Josef, 140
Stefan-Boltzmann constant, 140
Stefan's law, 141, 144, 307, 473
size of stars and, 402, 406, *406*, 415, *419* (table), 479
stellar evolution, 492, 634, 689, 705
of intermediate-mass stars, 533
mass and, 511–12; *see also* high-mass stars, evolution of; low-mass stars, evolution of
star clusters as snapshots of, 540–45, *542*, *543*, *544*
stellar explosions, 461
see also supernovae
stellar lifetimes, 514
stellar luminosity, *see* luminosity, stellar
stellar mass, *see* mass, stellar
stellar mass loss:
in high-mass stars, 526–27, *533*
in low-mass stars, 502–7, *505*, *506*
stellar occultations, 295, *296*, 340
stellar population, 544
stellar temperature, *see* temperature, stellar
stellar winds, 621, 634, *650*, 652
Stelliferous Era, 715, *716*
Stephen's Quintet, *696*
STEREO orbiters, *175* (table)
stereoscopic vision, 396, *396*, *397*, 402

Stonehenge, 26, *27*, 50
stony–iron meteorites, 368, *385*, *386*
stony meteorites, 384–85, *385*
storms:
  on giant planets, 299, *301*, 306–7
  see also dust storms; hurricanes; thunderstorms
stovetop convection, 234–35
stratosphere, 268, *269*, 270, 272, 301
Stratospheric Observatory for Infrared Astronomy (SOFIA), 169, *169*
stromatolites, 725, *726*
strong force, 684, *685*
strong nuclear force, 430, 501, 522, 531, 679, 682, 684, *685*, 689
  quantum chromodynamics and, 679
Structure of Scientific Revolutions, The (Kuhn), 11
structure of the universe, 696–717
  distribution of galaxies and, 696–98, *697*
  general relativity and, 567, 569, 683
  hierarchical clustering in, 700
  Hubble's law and, 593, *594*, *595*, 696
  physical law and, 679–84, *680*, *682*, *684*
S-type asteroids, 368, 369, 370
Subaru telescope, *155* (table)
subduction zones, 235–36, 239
subgiant stars, subgiant branch, 497
sublimation, 376
sulfur, 18, 246, 297, 304, 315, 328, 379, *406* (table), 527, 545, 731
sulfur dioxide, 328, 374
sulfuric acid, 275
Sumatra, 239
summer, 38–44, *39*, *41*, 44, 272
summer solstice, *39*, 43, 44, *396*
Sun, 4, 5, 17, 109, 399, 401, *426*, 428–52, *451*, 742
  age of, 508, 649
  apparent motion of, 27, 31, *38*, 40–43, *41*
  apparent size of, 51
  atmosphere of, 441–51, *442*
  backlighting and, 344, *344*, *345*
  balance in structure of, 427, *428*, *429*, 493, 494, 524
  as center of universe, *65*, 66, 75, 103
  chemical composition of, 405, *406* (table), 441, 443, 460, 493–97, *495*, 650–51
  circumpolar, 44
  climate change and, 278
  comets and, 376
  core of, 427, *429*, 430–37, *438*, 494–96
  density of, 428, *429*, 430, 435–37, 440, 441, *442*, 492, 496
  as disk star, 652
  Earth and, 508
  Earth's orbit around, *see* Earth, orbit of
  eclipse of, *see* solar eclipse
  ecliptic path of, 38, 39, 40, 44
  eleven-year cycle of, 447, *448*, *449*
  energy conversion in, 427, *430*, *433*, 437, 679
  energy transport in, 433, *434*, 436–37
  evolution of, 742
  as focus, 15, 70–72, *72*
  formation of, 7, *188*, 193, 199, 360, *474*, 483, 493, *495*
  future evolution of, *see* low-mass stars, evolution of
  giant planets and, 292–93, 297, 310
  gravitational influence of, 316
  gravitational redshift on, 568
  gravity of, 92–93, 103, 107, 365–66, 372, 428, *428*, 436, 443, 492, 493, 496
  great circle of, 38, 40
  greenhouse effect and, 264–65
  hypothetical star compared with, 435–37, *436*
  importance of, 4, 428
  interior of, 428–42, *428*, *429*, *434*, *441*, *442*, 447
  Jupiter compared with, 297, 310
  location of, 646, *646*
  luminosity of, 400–401, 428, 429, 436, 493, 497, 504, 508, 512, 708
  magnetic activity on, 443–49, *443–45*, *448*, *449*
  massive elements in, 545
  mass of, 100, 103, 406, 408, 428, 436–37, 480, 482, 492, 493, 496, 596
  Moon's orbit around, 45
  at noon, 31
  orbit of, 648
  other stars compared with, 400–401, 405, 408, 428
  phases of Moon and, 46–49
  photosphere of, 441–43, *442*, 445–46
  planetary motion influenced by, 92–93, 97
  pressure of, 428, *428*, *429*, 492, 493
  as protostar, 188, 199, 460
  red giant branch star compared with, *497*
  rotation of, 446
  seasonal changes and, 39–44, *39*, *41*
  seismic vibrations on, 438–41, *441*
  size of, *201*, 428, 435–36, *497*
  spectrum of, 130, 139, 141
  telescopic images of, *8*
  temperature of, 141, 428, *429*, 430, *434*, 435, 441–43, *442*, 478, 492, 493, 496
  and temperature of planets, 143, *143*, *145*
  testing models of, 437–41, *440*, *441*
  tidal effects of, 103, 107, *107*, 109, 111, 230
  22-year cycle of, 447, *449*
  volume of, 460
sungrazers, 374
sunlight (solar radiation):
  photons emitted by, 130
  received by Earth, 450, *450*, 732
sunspot cycle, 447, *448*
sunspots, *442*, 446, *446*, *447*, *448*, *449*
  butterfly diagram for, 447, *448*
  historical record of, 447, *449*
superclusters, 4, 696, *697*, *698*, 700
super computers, 712–14, *713*
Super-Kamiokande detector, 440, *440*
superluminal motion, 633
supermassive black hole, 629–31, *632*
supernova 1987A, 532, *532*
Supernova 2011fe, *591*
supernovae, 512–13, *512*, 520, 542, 545, 573, 574, *591*, 621, 655, 660, 706, 707, 731
  binary stars and, *511*, 512
  in "galactic fountain" model, *651*, 652
  intercloud gas and, 465, *466*
  neutrinos from, 437–38, 440, 532
  synchrotron radiation and, 314
  Type Ia, *see* Type Ia supernovae
  Type II, *see* Type II supernovae
superstring theory, 684, *684*
surface brightness, 163
surface waves, 225
Swift Gamma-Ray Burst Mission, *170* (table)
Swift Gamma-Ray observatory, *634*
Swift satellite, 706, *706*
Swift–Tuttle, Comet, 383, *383* (table)
symmetry, 682
  gravity and, 96, 682
  types of, 96
synchronous rotation, 45–46, *46*, 108, 109, 326, *326*
  of Pluto–Charon system, 361, *361*
synchrotron radiation, 312–14, *314*, 316
  from AGNs, 627, 630, 655
  from cosmic rays, 653
  of Crab Nebula, 540, *540*, 627
  in Milky Way, 653, *656*
synodic period, 48, *48*, 54, 67, *67*, 68

Tarantula Nebula, 467, *468*
Taurids, *383* (table)
Taurus, 43, 44, 482, *482*
Taurus–Gemini star field, *61*
technology:
  science and, 177
  see also tools, astronomical
tectonism, *218*, 279
  defined, 217
  on moons, 330, 332, 336
  in shaping terrestrial planets, 217, 222, 233–39, *233*, *234*, *235*, *236*, *237*, 328
telescopes, 8, 372, 628
  airborne, 170
  aperture of, 153, 167–68
  Galileo's use of, 75, 293
  ground based, 171
  history of, 152
  infrared, 167, 170–71
  magnification of, 153
  microwave, 178–79
  neutrino, 437, *440*, 532
  observatory locations for, 158–59
  optical, 152–65, 169
  orbital, 160, *170* (table), 171
  and origin of Solar System, 187, *187*
  planets discovered by, 293
  radio, 167–69, *167*, *168*
  reflecting, *152*, 153, 154, *154*
  refracting, 152, *152*, *153*
  resolution of, 161
  X-ray, 466
  see also specific telescopes
Tempel 1 Comet, *377*, *378*

Tempel–Tuttle Comet, 383, *383* (table)
temperature, 118, 132, 137–44
absolute zero, 139
atmosphere and, 160, *162*, 259, 263, 267, 268–69, *269*, 272, 275, 276, 277, 302, 304
atmospheric greenhouse effects and, 264–65
of cosmic background radiation, 601, *603*
defined, 138
determination of, 138, 143, *145*, 462, *463*
of gas, 260–61
of interstellar medium, 462–68, *463*, *467*
Kelvin scale of, 139
luminosity and, 139–40
as measure of how energetically particles are in movement, 138
of Planck spectrum, 600, *600*, 601, *603*
of planets, *see specific planets*
of protostars, 474, 475–80, *477*, *480*
Stefan's law and, 140, *140*
thermal equilibrium and, 138
Wien's law and, 140–41, 402, 405, *406*, *419* (table), 462
see also cold, cooling; heat, heating
temperature, stellar, 310, 402, 493
and classification of stars, 403–5
color and, 402, *403*, *419* (table)
of high-mass stars, 523, *528* (table)
H-R diagram and, 413–15, *413*, 417–18, *417*, *417* (table)
measurement of, 395, 402, *406*, 417, *419* (table)
of post-AGB stars, 505
of red giants, 497, 500, 501, *502*
size of stars and, 406
of Sun, 141, 428, *429*, 430, *434*, 435, 441–43, *442*, 478, 492, 493, 496
temperature gradient, 435
terrestrial planets, 198–99, 216, *216* (table), 221, 230, 249, 309
atmosphere of, 199, 257–84, *258*, *259*, *263* (table), *264*, *266* (table), 297
erosion on, 218, 222, 243–47, *244*, *245*, *246*
extrasolar, 200–201, 202, 205–6
giant planets compared with, 292, 295–99, 305, 306, 307–8, 309
impacts on, *199*, 200, 217–22, *219*, *220*, *221*, 243, *249*, 386
interiors of, 224–33, *228*, 309
magnetic fields of, *see specific planets*
tectonism on, 217, 222, 233–39, *233*, *234*, *235*, *236*, *237*, *238*, 328
volcanism on, 217, *218*, 222, 232, 239–42, *241*, *242*, *243*
see also Earth; Mars; Mercury; Venus
Tertiary period, 248
Tethys, *325*, 337
Thatcher, Comet, *383* (table)
Thebe, 346
Themis, 368
theoretical models, 9
theory, *see* scientific theory
theory of everything (TOE), 683–84, *685*, 688
theory of evolution, 352
thermal conduction, 433
thermal energy, 137–38, 228, 258, 330–36
in accretion disk, 197
convection and, 234–35, 238, 239, 264–65, 268, *270*, 306
in Earth's interior, 108, 228, 238, 239–40
evolution of binary stars and, 510, 512
gravitational energy converted to, 474–78, *476*, 630
greenhouse effect and, 264–65
hurricanes and, 275
instability strip and, 524
from interior of giant planets, 306, 308
magma and, 239–40
of protostars, 473–78, *476*, *477*, 630
in self-sustaining reactions, 527
of Sun, 428, 429–35, *429*, *430*, *433*, *434*, 436–37
thunderstorms and, 274
transport of, 433, *434*, 435, 501
thermal equilibrium, 137–38
thermal motions, 138, *138*, 498–501
thermal noise, 164
thermal radiation, 139–40, 402
defined, 139
equilibrium temperature of planets and, 143, *143*, *145*
Planck spectrum and, 140, *140*, 402–3, *403*, *404*, *406*
temperature and size of stars and, 402–3, *406*, 531
thermodynamics, second law of, 732, *732*
thermonuclear reactions, 501, 510–11, 512, 732
thermophiles, *726*
thermosphere, *269*, 270
third quarter Moon, 48, 103, 107, *107*
30 Doradus, 467, *468*
Thirty Meter Telescope (TMT), *155*, 158, *159*
thought experiments:
"ball in the car," 554–55
of Einstein, 556
equivalence principle and, 562
of Newton, 97–99, *97*, 103
3C, 273, *626*
thrust faults, *237*
thunderstorms, 274, 275
convection and, *274*
tidal bulge, 104–9, *106*, *108*, *109*
tidal disruption, 110, 367
tidal effects, 110
tidal flexing, 330
tidal forces, 104–10
on moons, 332
origins and, 110–11
tidal locking, 108–9, *108*
of moons, 325–26
tidal stresses, 346
tide pools, *725*
tides, 230
destructiveness of, 109, *111*, 346
and differences in gravity from external objects, 103, 104, *104*
heating and, 230
high vs. low, 106, *107*
interactions between galaxies and, *111*, 631, *712*
lunar, 104–8, *107*
object's rotation tied to its orbit by, 108–9, *108*
oceans flow in response to, 104–8, *106*, *107*
solar, 107, *107*, 109, 111
spring vs. neap, 107–8, *107*
Tiktaalic, *728*
time, 5
atomic decay and, 133–34
Hubble, 593–95, *594*, 674
look-back, 593, 599
Newtonian physics and, 556
relativity theory and, 12, 556–59
see also spacetime
Time, *12*, *582*
time dilation, 558–59, *559*, 560, 561
Titan, 7, *325*, 332–36, *334*, *335*, 336, *336*, 352, 735, *735*, 737
atmosphere of, 333–34
gravitational field of, 336
weather on, 336
Titania, *325*, 338
TOE (theory of everything), 683–84, *685*, 688
Tolkien, J. R. R., 521
Tolstoy, Leo, 359
Tombaugh, Clyde, 361
tools, astronomical, 202
topographic relief, 218
tornadoes, 275
torus, 315
total eclipses:
lunar, 54
solar, 50–54, *51*
toys, "glow in the dark", atomic decay and, 133–34
traffic circle without exits, 193, *193*
transform fault, 236
transit method, *203*, 206, *206*
of planet detection, 202, *202*
trans-Neptunian objects (TNOs), 360
transverse waves, *225*
Triangulum Galaxy, 657–58
trigonometric parallax, 590, *590*
triple-alpha process, 500, *500*, 504, 527, 535, 603
Triton, 293, *325*, *327*, 330–31, *331*
cantaloupe terrain of, 331, *331*
Trojan asteroids, 366
tropical year, 49
Tropic of Cancer, 44
Tropic of Capricorn, 44
tropopause, *269*, 270, 272
troposphere, 268, *269*, 270, 272, 304
Truzzi, Marcello, 723
T Tauri star, 482
Tully-Fisher relation, 618
Tunguska River, 379, *381*
tuning fork diagram, 612–13, *613*, 615
Tuttle, Comet, *383* (table)
Twain, Mark, 374
twin paradox, 559, 560, 561
Tycho's supernova, 514

Type Ia supernovae, *511*, 512, *512*, *513*, 514, 531, *532*, 536, 629
nucleosynthesis and, 534
as standard candles, 591, *591*, 592, 670, *670*
Type II supernovae, *530*, 531–40, *532*
energetic and chemical legacy of, 532–35, *532*, 535, *535*, *537*, *538*, *540*
nucleosynthesis and, 534

ultrafaint dwarf galaxies, 708
ultraviolet (UV) radiation, 124, 166, 170, 261, 277, *315*, 334, *449*, 450, *450*, *534*, 574, 630
interstellar medium and, 461–62, *463*, 466
ozone and, 267
photons of, 126, 304
of post-AGB stars, 505
spiral galaxy viewed in, 619–20, *619*, 620
star formation and, 621
Ulysses orbiter, *175* (table)
umbra, 54, 446, *447*
defined, 50
total solar eclipse and, 50, *50*, 52
Umbriel, 338, 339
unbalanced force, 76
motion is changed by (Newton's second law) (a–F/m), *78*, 80, *81*, 90, 562
unbound orbits, 102
uncertainty principle, 127–28, 677–78
understanding:
patterns and, 13
scientific, 9–11, 12
unified model of AGNs, 629, *630*, 632–33
uniform circular motion, 98
units, defined, 15
universal gravitational constant *(G)*, 79, 93, 94, 688, 689
universality, 91
universe, 4–7, 174
age of, 674, *674*
chemical evolution of, 5–7, *7*, 405, 545, 603–4, *604*, *685*
clockwork, 134
collapse of, *674*; *see also* "Big Crunch"
cosmological principle and, 9, 103, 584
dynamic vs. static, 671
early, AGN activity in, 630, 631
Earth-centered, 27
effect of new tools on view of, 7–8
eras in evolution of, *685*, 686, 696, *716*
expansion of, *see* Big Bang theory; expanding universe
homogeneity of, 584, *585*, 586–87, *589*
inflation model of, 677–78, *678*, 684, *685*, 700
isotropy of, 584, *585*, 586–87, 602
life in, 688–89
local, 4
mass of, 103, 668, *669*, 676, 715–16
measuring distances in, 591–93, 643
multiple, 686–88, *686*, *687*, 689
origins of, 17–18
perspective on, 645
remoteness and difference of, 5, *6*
scientific revolutions and, 11–12
size of, 4–5, *6*, 399, 583–84
structure formation in, *715*
structure of, *see* structure of universe
Sun-centered, 75, 103
uniformity of, 696
University of California, 741
unstable equilibrium, 136
uranium, 545, 706
half-life, 223
isotopes of, 223
Uranus, 17, 292–301, *293*, 310–12
atmosphere of, 275, 297, 299, 301, *302*, 304, 344
chemical composition of, 297, 304
density of, *292* (table), 297, 310
discovery of, 293–94, 734
formation of, 199, 200, 297, 310
interior of, *309*, 310
magnetic field of, 311, *311*, 312
magnetosphere of, 312
mass of, *292* (table), 295
moons of, *325*, 338–39, *339*, *349*
obliquity of, *292* (table), 298
orbit of, 198, *292* (table), 295, 298, 316, 340, 361, 378
rings of, *302*, 340, *341*, 343–45, *345*, 346, 348, 349, *349*
rotation of, *292* (table), 298
size of, 292–93, 295, *296*
temperature of, *197*, 305
winds on, 306
Urey, Harold, 724
Urey-Miller experiment, 724, *724*
Ursa Major, *see* Big Dipper
Ursids, *383* (table)

V838 Monocerotis, *490*
vacuum, 673
Valhalla, 339
Valles Marineris, 237, *238*
variable stars, pulsating, *see* pulsating variable stars
VEGA spacecraft, 377
velocity:
change in, *see* acceleration
defined, 78
escape, *see* escape velocity
momentum and, 127
peculiar, 590
radial, 135
thermal, 195
Venera 14 mission, *276*
vents, hydrothermal, 352
Venus, 7, 64, 152, 171, 172, 220, 221, 227, *228*, 249, *276*, 293, 352, 401, 553, 737
atmosphere of, 199, 217, 220, 242, 244, 258, *258*, 260–65, *263* (table), 265, *269*, 272, 274–76, 283, 284
atmospheric greenhouse effect on, 264, 265–66, 274–75
craters of, 222
Earth compared with, 237–38, 265–66, 274–76
erosion on, *245*
formation of, 198
gravity of, 237, 264
lack of magnetic field on, 232, 233
lack of moon for, 324
lack of rings for, 352
lack of tectonism on, 233
mass of, 237, 263, 264
orbit of, 263, 265, 299, 374
phases of, *75*
rotation of, 276
tectonics on, 237–38, *239*
temperature of, 143, *197*, 264, 265–66, *269*, 274, 276, 734
volcanism on, 238, 243, 263, 265
water on, 247, 265, 283–84
wind on, 244, *245*, 276
Venus Express orbiter, *175* (table), 238, 242, 266, 275
vernal equinox, 40, 49, *396*
Very Large Array (VLA) radio telescope, 168, *168*
Very Large Telescope (VLT), *155* (table), 169, *169*
Very Long Baseline Array (VLBA) radio telescope, 168
Vesta, 17, 370, *370*, *371*
Vesuvius, Mount, 239, *240*
Viking Mars missions, *238*, *243*, 734
Viking orbiters, 237
Virgo, 43
cluster, *2*, 696, *697*
supercluster, 4, *5*, 658, 696
virtual particles, 571, *572*
visual binary, 409, *410*, 411
voids, 696
volatile materials, 197, *197*, 198, 199, 200, 227, 329
atmosphere formation and, 260
in comets, 374, 376
in giant planets, 302, *303*, 304, 307
volatiles, 248
volcanic domes, 240
volcanism, 263
atmosphere and, 198, 724
defined, 217
on moons, 324, 328–33, *328*, *329*, 336–38, 346, 347
plate tectonics and, 236, *236*, 239–40, 352
in shaping terrestrial planets, *218*, 222, 232, 239–42, *241*, *242*, *243*, 724
as source of carbon dioxide, 267
volcanoes, climate change and, 279
vortices, of giant planets, 299–300, *300*, 305, 307
Voyager missions, 171, *172*, *175* (table), 328, 329, 352, 361, 738, 743, *743*
to giant planets, *293*, 295, 299–300, *300*, *302*, 305, 307, 312, 314, 331, 336, *337*, *339*, 340, *342*, 345, *345*, 348
Voyager 1, 307, 314, 340, 342, 348, *349*, 445, *445*
Voyager 2, 300, *300*, 305, 312, 331, 340, *342*, 343, *345*

W3 IRS5 star forming complex, *484*
Waller, Edmund, 395

War of the Worlds (Wells), 734
water, 197, *197*, 198, 330, 387, 419–20, 483, *484*, 689
in atmosphere, 260–61, 263, *263* (table), 265, 269–70, 274
in comets, 374, 376
on dwarf planets, 362, 364
erosion by, 217, 243, 246, *246*, 266
geological evidence for, 244–48
on giant planets, 297, 304, 309
on Mars, 221, *221*, 245, 246, *246*, *247*, 266, 277, *277*
on Moon, 248
on moons, *331*, 332–33, *333*, 336, 338–39
thunderstorms and, 274
on Venus, 247
see also oceans; oceans of Earth
water vapor, 170, 724
wave, waves:
amplitude of, 123, *123*
defined, 119
electromagnetic, *see* electromagnetic waves
frequency of, 123, 126, 129, 134
gravity, 568, 569–70
longitudinal, *225*
period of, 123, *225*
sound, 134
transverse, *225*
wave detectors:
gravitational, 174
neutrino, 174
wavefronts, 161, *162*, 166
wavelength:
defined, 123, *123*
of radio waves, 167
relationship between period and, 123
of sound, 134
wavelengths of light, 165, 402–3
Doppler shift of, *134*, 135
energy levels of atoms in determination of emission and absorption of, 128–32
energy of photon and, 126, 129
Planck spectrum and, 140, *140*, 141, 402–3
rest, 135
scattering and, 276
spectral fingerprints and, 132, *133*
spectrum and, 124, 126
waves, radio, 167
weak nuclear force, 679, 682, 684, *685*
weather, 268, 269–70
on giant planets, 304–8, *305*
interstellar medium compared with, 470
wind and, 272–73
weathering, 243
Weber bar, 177
Wegener, Alfred, 233
weight:
defined, 92
pressure and, 227, 239, 266, 428, 522, 523
Welles, Orson, 734
Wells, H. G., 734
white dwarfs, 406, *505*, 506–8, *512*, 533, 539, 625, 715, 742
accretion disks around, 510, 629
in binary system, 507, 510–13, *511*, 536, *537*, 591
degenerate core of, 506, 552, 591
gravitational redshift of, 568
neutron star compared with, 552
Widefield Infrared Survey Explorer (WISE) mission, 366–67, 462, *465*
Wien's law, 140–41, 179, 402, 405, *406*, *419* (table), 462, 505, 600, 601
Wild 2, Comet, *377*, 378, *378*
Wilkinson Microwave Anisotropy Probe (WMAP), 110, 179, 603, *603*, 701, 712
Wilson, Robert, 178, 600–601, *601*
WIMPs (weakly interacting massive particles), 625–26
wind, 272–73, 526, 545
black holes and, 571, 573, *573*
Coriolis effect and, 272–73, *273*, 304–8, *305*, *308*
erosion by, 217, 243–44, *245*
on giant planets, 304–8, *305*, *308*
measurement of speed of, on distant planets, 306
solar 451–52; solar wind
zonal, 273, 304, 305, 307–8, *308*
see also specific planets
wind streaks, 244, *245*
winter, 38–44, *39*, *41*, 272
festivals in, 43, *43*
winter solstice, 43
Wolf-Rayet (WR) stars, 533, *533*

X-ray binaries, 536, *537*
black holes and, 571, 629
calculating masses in, 572
X-ray Multi-Mirror Mission (XMM-Newton), *170* (table)
X-rays, 126, 166, 170, 314, 510, *514*, 630, *632*, 655–56
of clusters of galaxies, 698, *699*
Crab pulsar observed in, *540*
defined, 124
emitted by gas in elliptical galaxies, 616, 624, *624*
as evidence of black holes, 571, *573*
of interstellar medium, 465–66
of Sun, 443
Sun as source of, 443, 448, 450
supernovae remnants and, *534*

years, 37, 49
leap, 49
Yellowstone National Park, 240, 352
Yerkes 1-m telescope, 154, *154*
Yucatan Peninsula, 249, *249*

Zenith, 30, 44
zinc, 545, 731
zodiac, 38, *38*
zodiacal dust, 386–87, *386*
zodiacal light, 386–87
zonal winds, 273
on giant planets, 304, 305, 307–8, *308*
Zwicky, Fritz, 537